Concise Encyclopedia Biochemistry

Concise Encyclopedia Biochemistry

Second Edition

Revised and expanded by
Thomas Scott and Mary Eagleson

 Walter de Gruyter
Berlin · New York 1988

Concise Encyclopedia of Biochemistry
first published in 1983 based on the original, German language edition,
Brockhaus ABC Biochemie
Copyright © 1976; 1981 VEB F.A. Brockhaus Verlag Leipzig

Translated into English, revised and enlarged by Thomas Scott and Mary Brewer

First Printing, March 1983
Second Printing with corrections, October 1983

Second edition 1988 by

Thomas Scott, Ph.D.
Department of Biochemistry
University of Leeds
Leeds, England

Mary Eagleson, Ph.D.
Berkshire Road
Sandy Hook
Connecticut 06482, USA

CIP-Titelaufnahme der Deutschen Bibliothek

Concise encyclopedia biochemistry / rev. and expanded by Thomas Scott
and Mary Eagleson. [Transl. into English, rev. and enl. by Thomas Scott
and Mary Brewer]. - 2. ed. - Berlin ; New York : de Gruyter, 1988
    Einheitssacht.: Brockhaus-ABC Biochemie ⟨engl.⟩
    Früher u.d.T.: Concise encyclopedia of biochemistry
    ISBN 3-11-011625-1
NE: Scott, Thomas [Bearb.]; EST

Library of Congress Cataloging in Publication Data

Brockhaus ABC Biochemie. English.
    Concise encyclopedia biochemistry. - English language ed., 2nd
ed. / revised and expanded by Thomas Scott and Mary Eagleson.
    Translation of: Brockhaus ABC Biochemie. © 1981.
    Rev. ed. of: Concise encyclopedia of biochemistry. 1983.
    ISBN 0-89925-457-8 (U.S.)
    1. Biochemistry-Dictionaries. I. Scott, Thomas, 1935-   .
II. Eagleson, Mary, 1945-   . III. Concise encyclopedia of
biochemistry. IV. Title.
    [DNLM: 1. Biochemistry-dictionaries. QU 13 B864]
QD415.A25B713 1988
574.19'2'0321-dc19
DNLM/DLC
for Library of Congress

Preface to the second edition

In the preparation of the first edition, much academic energy and real time were consumed by the translation exercise. For the second edition we have been able to devote ourselves exclusively to collecting and classifying new material, and to revising old material. It is a measure of the pace of development in biochemistry that genetic engineering and the cloning of DNA, which are well represented in this second edition, received scant attention in the first edition. Entries on proteins listed physical properties, purification methods, structure and function. Now it is always necessary to ask whether a protein has been studied by recombinant DNA technology and, if so, to what purpose. In fact, the primary structures of many recently studied proteins have not been determined by direct analysis, but by prediction from the nucleotide sequence of the cloned gene. Here lies a cautionary tale: the entry on Lectins describes a newly discovered type of posttranslational protein modification, resulting in a primary structure that could not have been predicted from the nucleotide sequence of the gene!

Our policy has been to include all areas of biochemistry. Most of the new material can be classified as metabolism, metabolic regulation, molecular biology, enzymology, nonenzymatic protein function, or natural products; moreover we have attempted to give fair (but obviously not equal) coverage to animal, medical, microbiological and plant biochemistry.

The EC numbers of enzymes are from the Recommendations (1984) of the Nomenclature Committee of the International Union of Biochemistry on the Nomenclature and Classification of Enzyme-Catalysed Reactions ("Enzyme Nomenclature", Academic Press, 1984).

Thanks are due to colleagues who suggested new entries, provided information from their own areas of expertise, and criticized the manuscript. Comments from readers in different countries have been most useful, and we hope that this second edition will generate a similarly large response.

Mrs. Ingrid Ullrich of de Gruyter publishers oversaw the production of all the galley and page proofs, and the graphic reproduction of a very large number of new diagrams. The time table for proof reading and circulation of the manuscript between USA, Berlin and England was devised by Dr. Rudolf Weber of de Gruyter publishers, and we are grateful for his guidance and support.

March 1988

Mary Eagleson
Sandy Hook, Connecticut, USA

Thomas Scott
Leeds, Yorkshire, England

Preface to the first edition

The "Brockhaus ABC Biochemie" was first published in 1976 in Leipzig, the second edition followed in 1981. When we undertook to translate this book, based on the second German edition, it was clear that our work would also involve considerable updating of existing entries and the introduction of new material. Such a task can, of course, never be complete. It is a rare and fortunate author or editor in the life sciences, and particularly in biochemistry, whose material is still completely up to date at the time of publication; progress in this field is so rapid and shows no sign of abating. Therein, however, lies the excitement and challenge of this venture. Already we have started collecting, classifying and revising in preparation for a subsequent edition.

We have departed from the style of the German edition by quoting a few literature references. These have been included with some of the new material, and we hope they will be useful to readers who want more information than can be fitted into a work of this sort. Where possible, we have also given each enzyme its EC (Enzyme Commission) Number, according to the Recommendations (1978) of the Nomenclature Committee of the International Union of Biochemistry (published in "Enzyme Nomenclature" Academic Press, 1979).

We apologize to any biochemist whose pet compound, mechanism or pathway has been overlooked, and we should be grateful to receive suggestions for new entries. It is also recognized that a reference work should reach into the past, defining terms no longer used, but encountered when using the older literature. In this respect, suggestions from our more "senior" readers would be most welcome.

Finally, thanks are due to Dr. Rudolf Weber of de Gruyter Publishers for his guidance and encouragement in the preparation of the manuscript and the production of this book.

January 1983

Mary Brewer
Menlo Park, California, USA

Thomas Scott
Leeds, Yorkshire, England

Using this book

Cross referencing is indicated by the word "see", and the subject of the cross reference starts with a high case letter, e.g. ... in the Posttranslational modification of proteins (see), or ... see Enzyme induction. Numbers, Greek letters and configurational letters at the beginning of names are ignored in the allocation of alphabetical order, e.g. β-Galactosidase is listed under G; L-Histidine under H; *N*-2-Hydroxyethylpiperazine ... under H. The main entry title is printed in bold type, followed by synonyms in bold italics. The remaining text uses only two further types, normal and italics.

Abbreviations: (The standard biochemical abbreviations, e.g. ATP, NAD, etc. are found as entries in the appropriate alphabetical positions).

abb.	abbreviation
$[\alpha]$	specific optical rotation
b.p.	boiling point
c	concentration
°C	degrees Celsius
(d.)	with decomposition
ρ	density
IP	isoelectric point
M	molar
m.p.	melting point
M_r	relative molecular mass
n	refractive index
syn.	synonym

A

A: a nucleotide residue (in a nucleic acid) in which the base is adenine; abb. for absorbance.

Å: Angstrom unit = 10^{-10} m.

Abrin: see Ricin.

Abscisic acid, abb. *ABA, abscisin, dormin:* (S)-(+)-5-(1'-hydroxy-4'-oxo-2',6',6'-trimethyl-2-cyclohexen-1-yl)-3-methyl-*cis,trans*-2,4-pentadienoic acid, a widely occurring, sesquiterpene plant hormone. Its action is mainly inhibitory. ABA appears to be ubiquitous in plants and acts as antagonist to the auxins, gibberellins and cytokinins. It induces dormancy in seeds and promotes the falling of leaves and fruits. It is thus present in relatively large quantities in fruits, dormant seeds, buds and wilting leaves. The β-D-glucose ester of ABA has been found in the yellow lupine *(Lupinus luteus)*, rose *(Rosa)*, beans *(Phaseolus)*, maple *(Acer pseudoplantanus)*. It is assayed both spectroscopically and by biological tests based on its growth-inhibiting properties. The biosynthesis of ABA is still unknown. A direct path from isopentenyl pyrophosphate via geranyl and farnesyl pyrophosphate, or formation from carotenoids by photochemical cleavage of violaxanthin via xanthoxin have both been proposed. It was first isolated in 1963 from cotton bolls by Addicot and Lyon and from maple leaves by Wareing. Its structure was determined in 1965. It exists in two stereoisomeric forms, depending on the *cis* or *trans* orientation of the $\Delta^{2,3}$ double bond. The *cis*-isomer is the predominant form in all plants; small amounts of the *trans* isomer are occasionally found. The *trans* isomer is active only in bioassays performed in the light, presumably because it undergoes light-induced isomerization to the *cis* form. Both stereoisomers can exist in optically active forms (asymmetric C-atom at C1'), but only the (+)-form is found naturally.

(S)-(+)-Abcisic acid

Absolute oils: see Essential oils.

Absorbance, *extinction, optical density:* a measure of the quantity of light absorbed by a solution. It is equal to log I_0/I, where I_0 is the intensity of the incident light, and I is the intensity of the transmitted light.

Absorptivity, *absorbance index, absorption coefficient:* the proportionality constant ε, in Beer's law for light absorption: $A = \varepsilon l c$, where A is absorbance, l is the length of the light path, and c is the concentration. If concentration is expressed on a molar basis,

ε becomes the *molar absorptivity, molar absorption coefficient* or *molar extinction coefficient,* i.e. $\varepsilon = A/lc$, where l is the length of the light path in centimeters, and c is the molar concentration.

Acatalasia: see Inborn errors of metabolism.

Acceptor RNA: see Transfer RNA.

Acetaldehyde, *ethanal:* CH_3-CHO, important intermediate in the degradation of carbohydrates. In its activated form (see Thiamin pyrophosphate), it is involved in a number of reactions (see Alcoholic fermentation). Two molecules of A. can undergo acyloin condensation to form Acetoin (see).

3'-Acetamido-3'-deoxyadenosine: see 3'-Amino-3'-deoxyadenosine.

Acetate kinase, *acetokinase* (EC 2.7.2.1): see Acetylphosphate and Phosphoroclastic pyruvate cleavage.

Acetic acid, *ethanoic acid:* CH_3-COOH, a very common monocarboxylic acid. A.a. occurs in the free form as the end product of fermentation and oxidation reactions in some organisms. Acetate is formed metabolically by dehydrogenation of acetaldehyde, catalysed either by aldehyde oxidase (EC 1.2.3.1) or a $NAD(P)^+$-dependent aldehyde dehydrogenase (EC 1.2.1.3). The activated form of A.a., Acetyl-coenzyme A (see), is a key substance in intermediary metabolism.

Acetogenins: see Polyketides.

Acetoin, *3-hydroxy-2-butanone, acetyl methyl carbinol:* CH_3-CO-CHOH-CH_3, a reduction product of diacetyl which arises under certain conditions as a side product of the pyruvate decarboxylase (EC 4.1.1.1) reaction. A. is also formed by decarboxylation of acetolactate by acetolactate decarboxylase (EC 4.1.1.5). It is oxidized in a reversible reaction to diacetyl by acetoin dehydrogenase (EC 1.1.1.5), and in some microorganisms it is converted to 2,3-butanediol by D(-)-butanediol dehydrogenase (EC 1.1.1.4).

Acetylcarnitine: see Carnitine.

Acetylcholine: a highly active cholinergic Neurotransmitter (see) in nerves and neuromuscular synapses. After release from the nerve terminal at the synapse, A. binds to and triggers a response by receptors in the postsynaptic neuron; it then leaves the receptor and is rapidly degraded by acetylcholinesterase (EC 3.1.1.7). Nerves which employ acetylcholine for their chemical transmission are called cholinergic nerves.

The following are cholinergic nerves: all motor nerves to skeletal muscle; all preganglionic nerves, including the nerve supply to the adrenal medulla; all postganglionic parasympathetic nerves; postganglionic sympathetic nerves to sweat glands; some postganglionic sympathetic nerves to blood vessels in skeletal muscle.

Acetylcholine

Acetyl-CoA: choline O-acetyltransferase (EC 2.3.1.6)
or choline acetylase

$$H_3C-\overset{\overset{\displaystyle CH_3}{|}}{\underset{\underset{\displaystyle CH_3}{|}}{N^+}}-CH_2-CH_2OH \quad \underset{\text{acetate} \quad \quad H_2O}{\overset{\text{Acetyl-CoA} \quad \quad CoA}{\rightleftharpoons}} \quad H_3C-\overset{\overset{\displaystyle CH_3}{|}}{\underset{\underset{\displaystyle CH_3}{|}}{N^+}}-CH_2-CH_2-O-\overset{\overset{\displaystyle O}{||}}{C}-CH_3$$

Choline

Acetylcholinesterase (EC 3.1.1.7.)

Depending on its concentration, A. exerts two different physiological effects. Injection of small amounts of A. produces the same response as the injection of Muscarin (see), i.e. a fall in blood pressure (due to vasodilation), slowing of the heart beat, increased contraction of smooth muscle in many organs, and copious secretion from exocrine glands; this is therefore referred to as the muscarinic effect of A. The muscarinic effect (of muscarin or A.) is abolished by atropine. After administration of atropine, larger amounts of A. cause a rise in blood pressure, similar to that caused by nicotine; this is therefore called the nicotinic effect of A. This term, nicotinic, refers to the action of A. on autonomic ganglion cells and skeletal muscle fibers, and other actions which can be blocked by ganglion- or neuromuscular-blocking agents.

Nicotinic and muscarinic effects are mediated by nicotinic and muscarinic receptors, respectively. Occupancy of the nicotinic receptor (by A.) triggers a rapid response (1-2 milliseconds) by direct activation of Na^+ channels, thereby causing depolarization of the postsynaptic membrane. The muscarinic receptor response is slower; it operates via inhibition of adenylate cyclase, breakdown of phosphoinositides and modulation of K^+ channels by the action of guanine nucleotide-binding regulatory proteins, known as G-proteins [D. Brown *Nature* **319** (1986) 358-359]. Functional subtypes of the muscarinic receptor may exist, reacting selectively with different G-proteins. Nicotinic cholinergic synapses are found in vertebrate neuromuscular junctions, certain ganglia, central synapses, and the electroplax of *Torpedo*. Muscarinic cholinergic synapses operate in smooth muscle, cardiac muscle, ganglia, and many central brain regions. In the brain and central nervous system, muscarinic synapses outnumber nicotinic synapses by 10-100 fold. They can be differentiated with the aid of drugs which specifically block or stimulate only one type of synapse (Table).

The nicotinic cholinergic receptor from the electroplax of *Torpedo* has M_r 250000. It consists of 4 subunits, M_r 40000 (α), 50000 (β), 60000 (γ) and 65000 (δ) in the ratio 2:1:1:1. All the subunits are glycoproteins with close homologies of amino acid sequence. The receptor from mammalian muscle appears to be similar [B. M. Conti-Troconi et al. *Science* **218** (1982) 1227-1229]. DNA for all four subunits has been cloned and sequenced [M. Noda et al. *Nature* **301** (1983) 251-254]. Nicotinic cholinergic receptor is envisaged as an integral membrane protein complex, which acts as a gated ion channel [J. S. Linstrom et al. *Cold Spring Harbor Symp. Quant. Biol.* **48** (1983) 89-99].

The cDNA of the muscarinic cholinergic receptor from porcine cerebrum has been cloned, sequenced and expressed in *Xenopus* oocytes [T. Kubo et al. *Nature* **323** (1986) 411-426]; the primary sequence shows similarities with those of the β_2 adrenoreceptor and rhodopsin, suggesting variations on a common structural theme.

Histochemical localization of acetylcholinesterase serves to identify cholinergic synapses. It is based on the technique of Koelle and Friedenwald [G. B. Koelle *Handb. Exp. Pharmakol.* **15** (1963) 187-298]. Cholinesterase and acetylcholinesterase are distinguished by using specific inhibitors of each (Table). The substrate used is acetyl- or butylthiocholine, and the product, thiocholine, is visualized by precipitation with lead or copper salts. A more specific marker for cholinergic neurons (acetylcholinesterase is also present in dopaminergic cells of the substantia nigra) is choline acetylase (EC 2.3.1.6) (Fig.).

A. is phylogenetically an ancient hormone which also appears in protists. It may be an evolutionary precursor of the neurohormones.

Table. Drugs which affect cholinergic systems.
*: action mainly at peripheral ganglia.
Formulae not shown in this table may be found under separate entries.

Muscarinic agonists: acetylcholine, muscarin, carbachol, methacholine, bethanechol, pilocarpine, arecoline, oxotremorine.

Muscarinic antagonists: atropine, scopolamine, benztropine (also blocks dopamine uptake), quinuclidinylbromide, pirenzipine.

Nicotinic agonists: acetylcholine, nicotine*, carbachol, arecoline, suberyldicholine, tetramethylammonium*, phenyltrimethylammonium*, dimethylphenylpiperazine*.

Nicotinic antagonists: D-tubocurarine, succinylcholine (depolarizing, desensitizing), gallamine, pempidine*, mecamylamine*, hexamethonium*, pentolinium*, pancuronium, α-bungarotoxin.

Inhibitor of acetylcholine synthesis: 4-naphthylvinylpyridine.

Pump inhibitors (prevent entry of choline into nerve cell, leading to failure to synthesize acetylcholine): triethylcholine, hemicholinium.

Cholinesterase inhibitors (used to eliminate cholinesterase in the histochemical detection of acetylcholinesterase): diisopropylphosphofluoridate, neostigmine, physostigmine, edrophonium.

Release inhibitor: Botulinum toxin.

Specific binding agents: α-bungarotoxin, propylbenzilylcholine mustard, quinuclidinyl benzilate.

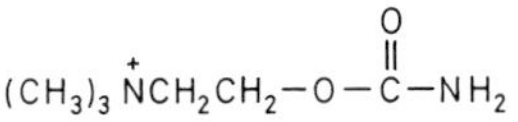

$$(CH_3)_3 \overset{+}{N}CH_2CH_2-O-\overset{O}{\overset{\|}{C}}-NH_2$$

Carbamylcholine (carbachol) (first stimulates skeletal muscle, then blocks neuromuscular transmission)

$$(C_2H_5)_4\overset{+}{N}$$
Tetraethylammonium

$$(CH_3)_3\overset{+}{N}(CH_2)_5\overset{+}{N}(CH_3)_3$$
Pentamethonium

$$(CH_3)_3\overset{+}{N}(CH_2)_6\overset{+}{N}(CH_3)_3$$
Hexamethonium

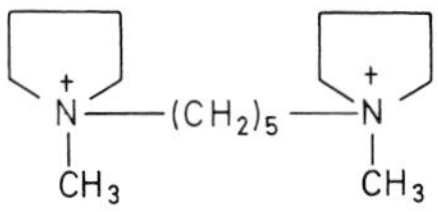

Pentolinium

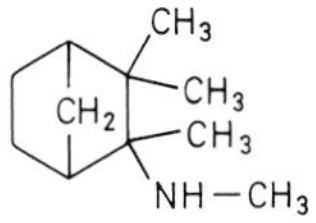

Mecamylamine

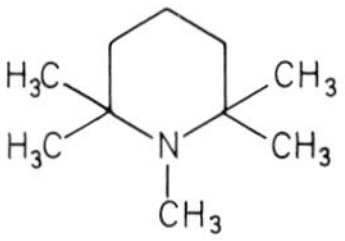

Pempidine

> Ganglion blockers

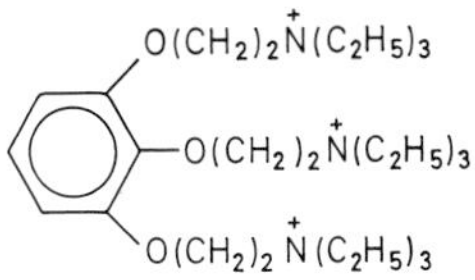

Gallamine (blocks neuromuscular transmission without prior stimulation)

$$(CH_3)_3\overset{+}{N}CH_2CH_2-O-\overset{O}{\overset{\|}{C}}-CH_2$$
$$(CH_3)_3\overset{+}{N}CH_2CH_2-O-\underset{\underset{O}{\|}}{C}-CH_2$$

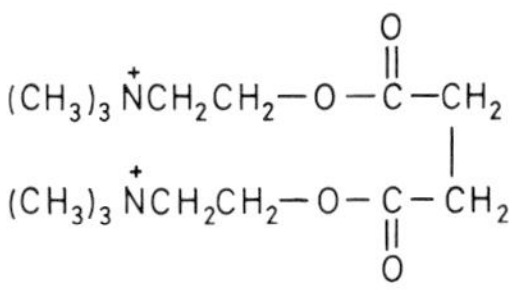

Succinylcholine (suxamethonium) (first stimulates skeletal muscle, then blocks neuromuscular transmission)

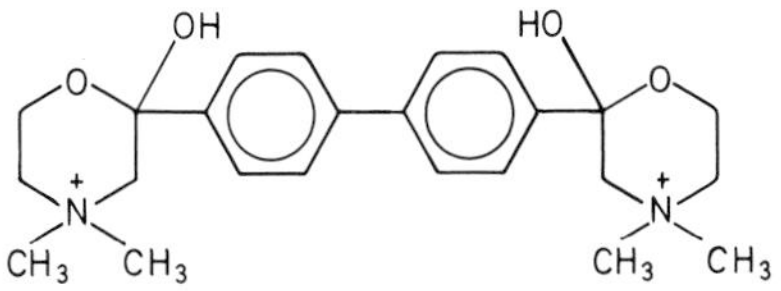

Hemicholinium (prevents entry of choline into nerve cell)

$$(C_2H_5)_3\overset{+}{N}CH_2CH_2OH$$

Triethylcholine (prevents entry of choline into nerve cell)

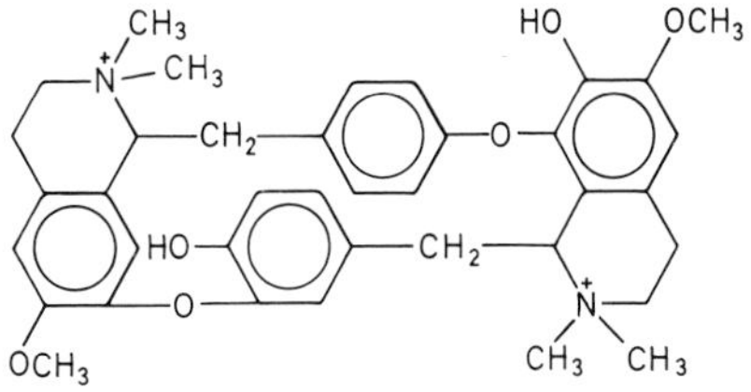

Tubocurarine (blocks neuromuscular transmission without prior stimulation)

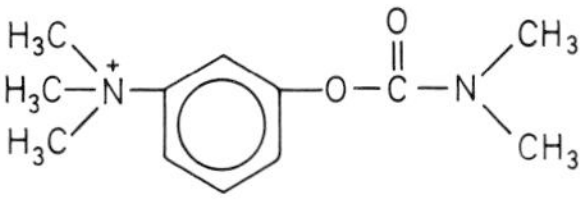

Neostigmine

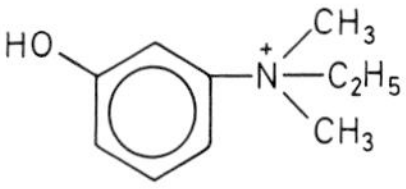

Edrophonium.

Acetylcholine receptor: see Acetylcholine.

Acetylcholinesterase (EC 3.1.1.7): "true cholinesterase", catalyses the hydrolysis of acetylcholine into choline and acetate. Due to the high turnover number of A. (0.5 to 3.0×10^6 molecules substrate per molecule enzyme per min), the acetylcholine released at a synapse is hydrolysed within 0.1 ms. This enzyme is found in the central nervous system, particularly in the postsynaptic membranes of the striated muscles, the parasympathetic ganglia, the erythrocytes and the electric organs of fish. Crystalline A. (M_r 330000) has been isolated from the electric organ of the electric eel *(Electrophorus electricus)*. It consists of 4 identical inactive subunits of M_r 82500; the half-molecules consisting of 2 covalently bound subunits (M_r 165000) are enzymatically active. Proteolytic attack on the subunits produces two fragments of M_r 60000 and 22500.

The active center of A. has two parts, the anionic binding site for the quaternary nitrogen, which is responsible for the alcohol specificity, and the esterase center, where a catalytic serine and histidine lyse the

3

ester bond. The enzyme is inactivated by blockage of either the serine hydroxyl (by organic phosphate esters, such as diisopropylfluorophosphate or diethyl p-nitrophenylphosphate), or the anionic center by trimethylammonium derivatives. If the enzyme has been blocked by organophosphates, it can be reactivated by pralidoxime salts, which are therefore used as antidotes to organophosphate poisoning.

Acetyl-coenzyme A, *acetyl-CoA, active acetate*: $CH_3CO \sim SCoA$, a derivative of acetic acid in which the aetyl residue is bound by a high-energy bond to the free SH-group of coenzyme A. M_r 809.6, $\lambda_{max} = 260$ nm. The very reactive thioester has a high potential for transfer of the acetyl group, and is therefore a universal intermediate which provides the C_2 fragment for numerous syntheses. The free energy of the bond (34.3 kJ/mol = 8.2 kcal/mol), however, has no significance as a form of energy storage. In the transfer reactions mediated by acetyl-CoA, either the carboxyl group (electrophilic reaction) or the methyl group (nucleophilic reaction) can react.

By far the most important pathways for the synthesis of acetyl-CoA (Table) are 1) the oxidative decarboxylation of pyruvate, 2) the degradation of fatty acids and 3) the degradation of certain amino acids. The formation of acetyl Co-A involves either 1) the transfer of an acetyl residue from a suitable donor, such as pyruvate, and simultaneous reduction of NAD^+, or 2) the activation of free acetate in a one or two-step process which requires ATP and free coenzyme A.

Acetyl-CoA is the hub of carbohydrate metabolism and has a central position in overall metabolism. The products of carbohydrate, fat and protein metabolism are channeled via acetyl-CoA into oxidative degradation in the tricarboxylic acid cycle. The acetyl residue is used in the synthesis of esters and amides (e.g. acetylcholine, N-acetylglucosamine, N-acetylglutamate). Acetyl-CoA is also the starting point for isoprenoid synthesis via mevalonic acid and for fatty acid synthesis. The latter path is especially important and was elucidated in 1951 by Lynen and Lipman.

N-Acetylglutamic acid, *N-acetylglutamate*, abb. ***Ac-Glu:*** $HOOC-CH(NHCOCH_3)-CH_2-CH_2-COOH$, the acetylated form of glutamic acid, is the cofactor of carbamoyl phosphate synthetase (ammonia) (EC 6.3.4.16) and allosterically activates this enzyme. See Carbamoyl phosphate.

Acetyl methyl carbinol: see Acetoin.

Acetyl phosphate: $CH_3-COOPO(OH)_2$, an energy-rich acyl phosphate. It is the product of acetate activation in some organisms: Acetate + ATP $\rightleftharpoons$ A.p. + ADP; the reaction is catalysed by acetate kinase (EC 2.7.2.1). The back reaction can be used for ATP synthesis, for example in the phosphoroclastic cleavage of pyruvate.

Acidic α_1-glycoprotein: see Orosomucoid.

Acid lipase deficiency: see Inborn errors of metabolism.

Acidosis: a decrease in the pH of body fluids. It is corrected by excretion of acid via the lungs and kidneys (see Buffer, section on Buffers of body fluids). There are two types of acidosis:
a) *Metabolic acidosis*, which is caused by a decrease in the bicarbonate fraction, with little or no change in the carbonic acid fraction. There are several possible causes. 1) Severe diarrhea results in the loss of gastrointestinal secretions containing high concentrations of bicarbonate (the resulting acidosis is a contributary factor to infant death in the developing world). 2) Acetazolamide (Diamox), a drug used to promote diuresis, inhibits carbonic anhydrase in the brush border of the proximal tubule epithelium. Reabsorption of bicarbonate is therefore retarded, leading to acidosis. 3) Severe renal disease, which may impair the ability of the kidney to remove acids

Table. Reactions in which acetyl-coenzyme A is synthesized

Enzyme	Reaction	Occurrence/Significance
Acetyl-CoA synthetase (EC 6.2.1.1)	$CH_3COO^- + ATP + CoA \rightleftharpoons$ $CH_3CO-CoA + AMP + PP_i$	Yeasts, Animals Higher plants
Acyl-CoA synthetase (GDP-forming) (EC 6.2.1.10)	$CH_3COO^- + GTP + CoA \rightleftharpoons$ $CH_3CO-CoA + GDP + P_i$	Liver
Acetate kinase (EC 2.7.2.1)	$CH_3COO^- + ATP \rightleftharpoons$ $CH_3CO-O-PO_3H_2 + ADP$	Microorganisms
Phosphate acetyltransferase (EC 2.3.1.8)	$CH_3CO-O-PO_3H_2 + CoA \rightleftharpoons$ $CH_3CO-CoA + P_i$	Microorganisms
ATP citrate (*pro-3S*)-lyase (EC 4.1.3.8)	Citrate + ATP + CoA $\rightleftharpoons$ $CH_3CO-CoA$ + oxaloacetate $+ ADP + P_i$	Outside the mitochondria
Pyruvate dehydrogenase complex (EC 1.2.4.1; 2.3.1.12; 1.6.4.3)	Pyruvate + NAD^+ + CoA $\rightleftharpoons$ $CH_3-CO-CoA + CO_2$ $+ NADH + H^+$	Mitochondrial particles Involves TPP, $LipS_2$
Acetyl-CoA transacetylase (EC 2.3.1.9)	Acetoacetyl-CoA + CoA $\rightleftharpoons$ $2CH_3CO-CoA$	Fatty acid degradation

formed in the normal course of metabolism. 4) Vomiting usually leads to the loss of bicarbonate from the upper intestine, as well as the acidic stomach contents. Since the loss of alkali exceeds the loss of acid, the net result is acidosis. 5) Diabetes mellitus results in the excessive formation of acetoacetic acid, which accumulates and causes acidosis in the extracellular fluids; as much as 500-1000 mmoles of acid may be excreted per day.

b) *Respiratory acidosis* is caused by an increase in carbonic acid in relation to bicarbonate. It may occur when alveolar ventilation is impaired, e.g. in pneumonia and asthma, and it can be caused by depression of the respiratory center, e.g. by morphine poisoning.

Acid phosphatase deficiency: see Inborn errors of metabolism.

Acid plants, *ammonium plants:* plants which accumulate organic acids in their leaf cells, which are neutralized by ammonium ions.

Aconitate hydratase, *aconitase,* (EC 4.2.1.3): a hydratase which catalyses one stage of the tricarboxylic acid cycle, the reversible interconversion of citrate and isocitrate. The reaction proceeds via the enzyme-bound intermediate, *cis*-aconitate. At equilibrium, the relative abundances are 90% citrate, 4% *cis*-aconitate and 6% isocitrate. Thus citrate is favored at equilibrium, but in respiring tissues the reaction proceeds from citrate to isocitrate, as isocitrate is oxidized by isocitrate dehydrogenase. The enzyme contains Fe(II) and requires a thiol such as cysteine or reduced glutathione. The Fe(II) ion forms a stable chelate with citric acid. X-ray analysis of Fe(II) complexes of tricarboxylic acids suggested the "ferrous wheel" hypothesis of aconitase action. According to this mechanism, three points on the *cis*-aconitate molecule are bound at separate sites on the enzyme surface; in addition the molecule is also complexed with the Fe(II) atom at the active center. The stereospecific *trans* addition of water to *cis*-aconitate to form either citrate or isocitrate is achieved by rotation of the ferrous wheel, which can add OH to either side of the molecule. Aconitase is inhibited by fluorocitrate. Two isoenzymes are present in animal tissues, one in the cytosol and one in the mitochondria. [Glusker, J.P., in Boyer, P.D. (ed.), *The Enzymes,* 5, 434, Academic Press Inc., 1971.]

Aconitic acid: an unsaturated tricarboxylic acid, usually occurring in the *cis* form, but sometimes in the *trans. cis*-A.a., m.p.130°C, *trans*-A.a., m.p.194 to 195°C. A.a. was discovered in free form in aconite, *Aconitum napellus.* The anionic form of *cis*-A.a. (propene-*cis*-1,2,3-trioic acid) is important as an intermediate in the isomerization of citrate to isocitrate in the Tricarboxylic acid cycle (see).

Aconitine: an Aconitum alkaloid (see Terpene alkaloids) from the roots of aconite *(Aconitum napellus)* and other *Aconitum* and *Delphinium* species. A. is an esterified alkaloid. It is extremely poisonous and can cause death in adults at a dose of 1 to 2 mg by paralysing the heart and respiration. Its hydrolysis products are only slightly toxic. In spite of useful physiological properties, A. is rarely used in medicine, due to its toxicity. It is sometimes used internally as tincture for rheumatism and neuralgias and

externally as a pain-killing salve. In antiquity, aconitine preparations were used as arrow poisons by the Greeks and (East) Indians.

Aconitum alkaloids: a group of terpene alkaloids, some of them very poisonous, from various aconite *(Aconitum)* species. The best-known representative is Aconitine (see).

ACP: abb. for Acyl carrier protein (see).

Acrasin: an attractant secreted by aggregation centers of slime molds (social amebas) which stimulates single-celled amebas to aggregate and form fruiting bodies. The A. of *Dictostelium* is cyclic AMP (see), which attests to the antiquity of the use of this substance as a hormone. The A. of *Polysphondium violaceum* is a dipeptide called "glorin":

ACTH: abb. for adrenocorticotropic hormone. See Corticotropin.

Actinomycins: a large group of peptide lactone antibiotics produced by various strains of *Streptomyces*. These highly toxic red compounds contain a chromophore, 2-amino-4,6-dimethyl-3-ketophenoxazine-1,9-dioic acid (actinocin), which is linked to two 5-membered peptide lactones by the amino groups of two threonine residues. The various A. differ only in the amino acid sequence of the lactone rings. In vivo, A. inhibit DNA-dependent RNA synthesis at the level of transcription by interacting with the DNA. The concentration required for inhibition depends on the base composition of the DNA; more is required for DNA with a low guanine content. A. are pharmacologically very important due to their bacteriostatic and cytostatic effect. Actinomycin D (Fig.) is one of the most widely occurring A. Its spatial structure has been elucidated by NMR studies, and the specificity of its interaction with deoxyguanosine was demonstrated by X-ray analysis. Actinomycin D is used as a cytostatic, e.g. in the treatment of Hodgkin's disease.

Actinomycin D

Actins: Contractile proteins found in many cell types. A. is an essential component of the contractile complex of Muscle proteins (see). Microvilli, microspikes (filopodia), and stereocilia (hair cells in the cochlea of the ear and other related organs) consist of A. associated with other proteins. Microfilaments in the cell cytoplasm consist of F- (polymerized) A. (see Cytoskeleton). Monomeric, or G-A. (M_r 41720) is an irregular mass (Fig.). A consensus model of F-A. shows a helical filament with a diameter of 90-100 Å. The monomers lie with their long axes nearly perpendicular to the filament axis. The positions of the monomers within the filament are flexible, so that binding of proteins (e.g. tropomyosin) to the filament may impose a periodic but non-helical structure; the repeat distance is 7 monomers. [D.J. DeRosier & L.G. Tilney in J.W. Shay, ed. *The Cytoskeleton*, vol 5. of *Cell and Muscle Motility* (Plenum Press, New York, 1985), pp. 139-169; E.H. Egelman, *J. Musc. Res. Cell Motil.* **6** (1985) 129-151]

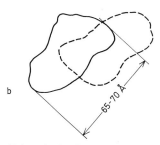

Polymerized actin

a. Idealized model showing arrangement of actin monomers, represented here as two fused spheres. From DeRosier & Tilney; used with permission.
b. Cross sections through two successive monomers. The solid and dashed outlines represent sections in two different planes, about 27 Å apart and rotated through 167 ° around the helix axis. From Egelman; used with permission.

Each A. monomer can bind 1 molecule of ATP; when polymerization occurs, the ATP is hydrolysed and the resulting ADP remains bound to the A. However, the hydrolysis is not coupled to the polymerization itself. Instead, it occurs about 10 seconds after the monomer has been added to the polymer. Growth of a filament produces an "ATP cap", or terminal region in which the monomers still have bound ATP. These monomers dissociate more slowly than ADP-bound monomers, so that the "ATP cap" promotes further growth of the polymer. Conversely, a shrinking polymer, from which monomers dissociate more rapidly than they are being added, has a region of ADP-bound monomers at its ends, and this region increases the rate of dissociation.
Addition and release of monomers can occur at either the "pointed" or the "barbed" end of G-A., but the processes of addition and release are about 10 times faster at the barbed ends. This led to the

earlier inference that F-A. was added only to the barbed ends and released from the pointed ends in a process of "treadmilling". [A. Wegner, *Nature* **313** (1985) 97-98]
The structure of A. has been highly conserved in the course of evolution; this may be due to the large number of proteins with which it interacts specifically.

Activated amino acids: see Aminoacyl adenylate.

Activated fatty acids: fatty acyl coenzyme A thioesters which, as high energy compounds, have a large potential for group transfer. They are formed during fatty acid biosynthesis, or by the activation of free fatty acids. Acyl CoA synthetases catalyse formation of the CoA derivatives according to the reaction: $CH_3(CH_2)_nCOO^- + ATP + HS\text{-}CoA \rightleftharpoons CH_3(CH_2)_nCO \sim SCoA + AMP + PP_i$. The reaction involves acyladenylate as an intermediate, which is cleaved by coenzyme A to form acyl-CoA and AMP. Several such enzymes are known, and they are named according to the length of carbon chain that shows optimal activity, e.g. acetyl-CoA synthetase converts C_2 and C_3 fatty acids, octanoyl-CoA synthetase (C_4 to C_{12}) and dodecanoyl-CoA synthetase (C_{10} to C_{18}). Mitochondria also contain an acyl CoA synthetase that cleaves GTP to GDP and P_i. Acyl CoA derivatives of short chain fatty acids may also be formed in a transfer reaction involving succinyl-CoA, catalysed by thiophorases: Succinyl $\sim$ SCoA + R-COOH $\rightleftharpoons$ succinic acid + R-CO $\sim$ SCoA. Activated fatty acids are in equilibrium with acylcarnitine in the organism. They are the starting point for fatty acid degradation.

Activated succinate: see Tricarboxylic acid cycle, Succinate-glycine cycle, Fatty acid degradation.

Activation hormone: see Insect hormones.

Activator protein: see Calmodulin.

Active acetaldehyde: see Thiamin pyrophosphate.

Active acetate: see Acetyl-coenzyme A.

Active aldehyde: see Thiamin pyrophosphate.

Active carbon dioxide: see Biotin enzymes.

Active center: that part of an enzyme or other protein which binds the specific substrate and converts it to product (enzymes) or otherwise interacts with it. The A.c. thus consists of the actual catalytic center, which is relatively unspecific, and the substrate-binding site, which is responsible for the specificity of the enzyme. Usually only a few amino acids interact directly with the substrate in the A.c.; the rest of the protein molecule serves to hold these few in the proper configuration. The amino acids involved in catalysis may lie at a considerable distance from each other in the absence of substrate; they are brought into play by conformational changes induced by the substrate when it binds (see Cooperativity model, Chymotrypsin and Serine proteases). Information on the amino acids involved in the A.c. is obtained by specific labelling with coenzyme, or by reaction with inhibitors or reagents specific for particular side chains. Some widely used irreversible inhibitors for the catalytic center of the serine proteases are tosyllysine chloromethyl ketone (TLCK), which reacts with histidine, and diisopropylfluorophosphate (DFP) and phenylmethane sulfonyl fluoride (PMSF), which form esters with serine residues.

Active formaldehyde: see Active one-carbon units; Thiamin pyrophosphate.

Active formate: see Active one-carbon units.

Active glucose: see Nucleoside diphosphate sugars.

Active glycolaldehyde: see Thiamin pyrophosphate.

Active methionine: see S-Adenosyl-L-methionine.

Active one-carbon units, abb. C_1 **units:** C_1 fragments which are activated by binding to tetrahydrofolic acid, or less commonly, to thiamin pyrophosphate. The active ethylenediamine group of tetrahydrofolic acid serves as a carrier for the metabolic transfer of a formyl or methyl group. Fig. 1 shows the active forms of tetrahydrofolic acid (THF).

The main source of C_1 units is the hydroxymethyl group of serine, which is transferred to THF by serine hydroxymethyltransferase (EC 2.1.2.1), forming hydroxymethyl-THF (activated formaldehyde). The formation of C_1 units in the course of histidine catabolism or the anaerobic degradation of purines is of particular importance. C_1 units are used in purine biosynthesis and as the donors of the 5-methyl group of thymine.

The various C_1 units can be interconverted while attached to the THF (Fig. 2).

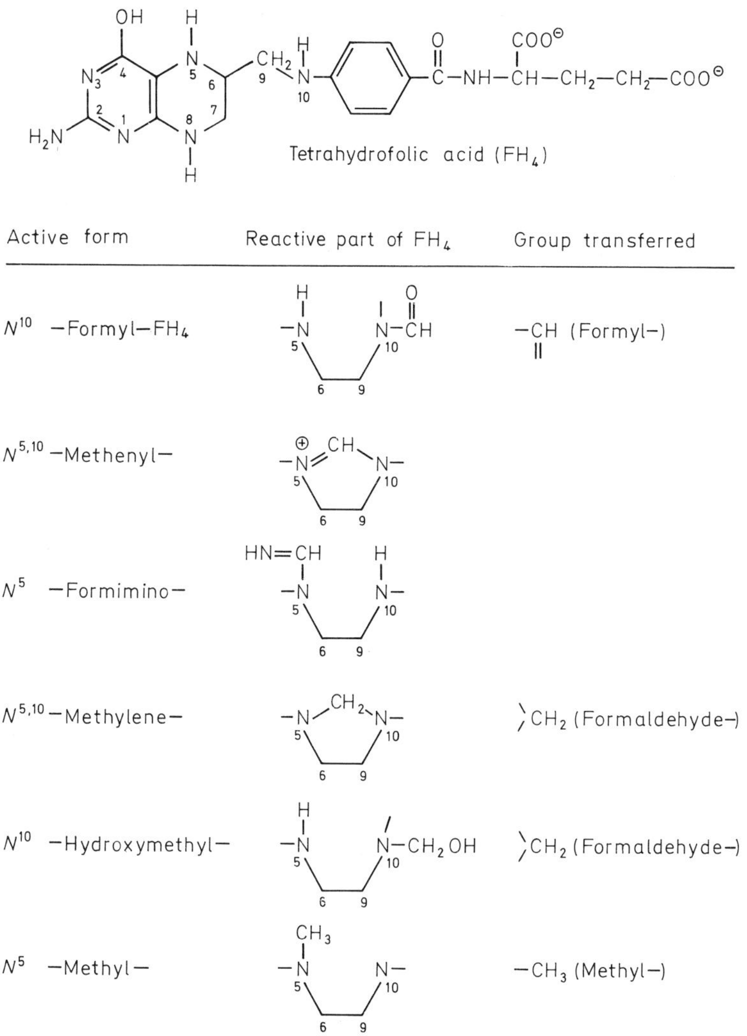

Figure 1. Structure of tetrahydrofolic acid and activated one-carbon units

7

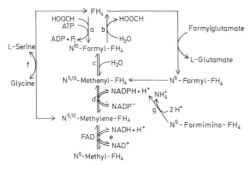

Figure 2. Interconversions of active one-carbon units
a. Formyl-FH₄ synthetase (EC 6.3.4.3)
b. Formyl-FH₄ deformylase (EC 3.5.1.10)
c. Methenyl-FH₄ cyclohydrolase (EC 3.5.4.9)
d. Methylene-FH₄ dehydrogenase (NADP⁺) (EC 1.5.1.5)
e. 5,10-Methylene-FH₄ reductase (FADH₂) (EC 1.7.99.5)
f. Serine hydroxymethyltransferase (EC 2.1.2.1)
g. Formimino-FH₄ cyclodeaminase (EC 4.3.1.4)
Associated systems and reactions:
1. N^5-Formimino-FH₄ is formed from FH₄ and formiminoglycine (from bacterial fermentation of purines) and from FH₄ and formiminoglutamate (see Histidine).
2. In *Clostridium*, reversal of reaction *a* serves to generate ATP.
3. $N^{5,10}$-Methylene-FH₄ acts as a reducing agent as well as a source of one-carbon units in the synthesis of thymidylic acid (see Pyrimidine biosynthesis).
4. The Glycine cleavage system (see) also converts FH₄ to $N^{5,10}$-methylene-FH₄.
5. N^5-Methyl-FH₄ serves as a source of methyl groups for the conversion of L-homocysteine to L-methionine (5-methyltetrahydrofolate-homocysteine methyltransferase, EC 2.1.1.13) (see L-Methionine).
6. See Purine biosynthesis for action of the enzymes EC 2.1.2.2 and 2.1.2.3.
7. N^5-Methyl-FH₄ is a substrate for methane formation in methanogenic bacteria.
8. See One-carbon cycle for other important reactions.
THF and FH₄ are both commonly used abbreviations of tetrahydrofolic acid.

Active pyruvate: see Thiamin pyrophosphate.

Active sulfate: see Phosphoadenosine phosphosulfate.

Active transport: a process in which solute molecules or ions move across a biomembrane from lower to higher concentration, i. e. against the concentration gradient. Since thermodynamic work is involved, A.t. must be coupled to an exergonic reaction. In primary A.t., the coupling is direct. The transport of Na⁺ and K⁺ ions across a cell membrane by the Na⁺,K⁺-ATPase system, for example, requires the simultaneous hydrolysis of ATP. Secondary A.t. utilizes the energy of an electrochemical gradient established for a second solute to transport the first. One form of secondary A.t. is *cotransport*, in which the transport of one solute drives the other. An example is the Na⁺-dependent transport of cer-

tain sugars and amino acids in animal cells: the concentration of Na⁺ in the cell is maintained at a level far below the intercellular concentration by the Na⁺,K⁺ pump. A specific transport protein (carrier) binds both glucose and Na⁺ outside the cell and releases them on the inside. The process is energetically favorable because the Na⁺ is moving from a region of higher concentration to one of lower concentration. In other cases, the membrane potential generated by electron flow along the respiratory chain drives the active transport of sugars or amino acids.
A third form of A.t. is called *group translocation* because the solute is changed in the course of transport. An example is the *phosphotransferase system* in some bacteria, in which sugars are phosphorylated in the course of transport. An interesting feature of this system is that phospho*enol*pyruvate rather than ATP is the phosphate donor.
A.t. processes are highly specific, and they are saturable. This implies that enzyme-like proteins, or *carriers,* mediate the transport. (The term "carriers" also applies to the mechanism of Facilitated diffusion (see)).
The bacterial transport systems called *permeases* have been extensively studied by genetic and other means. The protein product of the lactose permease *(y)* gene has been isolated, as have a number of membrane proteins which bind the substrates of other permeases. These proteins are probably components of the respective permeases.

Actomyosin: see Muscle proteins.

Acylcarnitine: see Carnitine.

Acyl carrier protein, abb. *ACP:* a small, acidic, heat-stable globular protein which is part of the fatty-acid synthesizing complex in *Escherichia coli* and other bacteria, yeast and plants. It is the carrier of the fatty acid chain during its biosynthesis. The primary structure of the ACP from *E. coli* has been determined: it contains 77 amino acids and has M_r 8847. The protein itself contains no sulfur, but it carries a molecule of phosphopantetheine (which possesses -SH) linked via a phosphate ester to the hydroxyl of serine 36. All acyl residues formed during fatty acid biosynthesis are bound as thioesters to the SH-group of this prosthetic group. The M_r of the ACP isolated so far lie between 8600 *(Clostridium butyricum)* and 16000 (yeast).
Synthetic apo-ACP protein, a polypeptide representing amino acids 2 to 74 of the *E. coli* protein, functions as substrate for the holo-acyl-carrier protein synthase (EC 2.7.8.7); the product is biologically as active as natural holo-ACP.

Acylglycerols, *glycerides:* esters of fatty acids with glycerol. Mono- and diacylglycerols usually occur only as metabolic intermediates. Mixtures of triacylglycerols are neutral fats (see Fats). The IUPAC-IUB Commission on Biochemical Nomenclature discourages the use of the terms "mono-, di- and triglycerides" in favor of "acylglycerols".
In the intestine, triacylglycerols are hydrolysed to monoacylglycerols, which are re-esterified to triacylglycerols in the intestinal mucosa (Fig. 1). In other tissues (notably liver and adipose tissue), triacylglycerols are synthesized from glycerol 3-phosphate and fatty acyl-CoA (Fig. 2). In adipose tissue, the

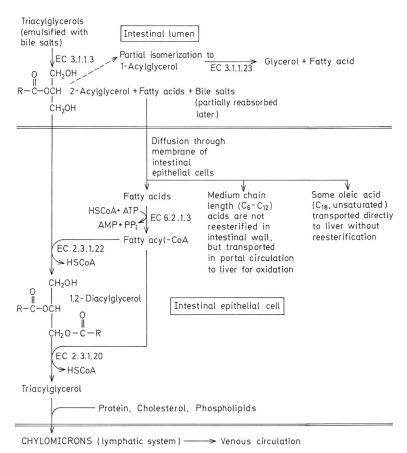

Figure 1. Digestion and resynthesis of triacylglycerols in the duodenum
EC 2.3.1.20: Diacylglycerol acyltransferase
EC 2.3.1.22: Acylglycerol palmitoyltransferase
EC 3.1.1.3: Pancreatic triacylglycerol lipase
EC 3.1.1.23: Acylglycerol lipase
EC 6.2.1.3: Long-chain-fatty-acid-CoA ligase

rates of breakdown and synthesis of triacylglycerols are under hormonal control, resulting in fat storage and/or release of fatty acids, depending on nutritional state, exercise and stress (Fig. 3). Plasma Lipoproteins (see) are responsible for the transport and deposition of triacylglycerols in the body.
Role of triacylglycerol synthesis and degradation in adipose tissue (see figure 3):
Glycerol formed during triacylglycerol degradation cannot be reutilized because adipose tissue does not contain glycerol kinase (EC 2.7.1.30). Synthesis of triacylglycerols therefore depends on a continuous supply of glucose for the production of glycerol 3-phosphate. Conversion of triacylglycerol to diacylglycerol is relatively slow, and is the rate-limiting step of triacylglycerol degradation. The lipase catalysing this reaction is activated by phosphorylation, a process indirectly under hormonal control via a cAMP-dependent protein kinase. In addition, gluco-

corticoids stimulate the lipase independently of cAMP, and this stimulation is prevented by insulin.
In a state of caloric excess, high insulin levels promote glucose uptake, and prevent activation of mobilizing lipase. Glycerol 3-phosphate is therefore abundant, the rate of triacylglycerol synthesis is high, export of free fatty acids is minimal and the quantity of stored triacylglycerols shows a net increase. High insulin levels cause an increase in the activity of lipoprotein lipase; this is a tissue-specific effect, and the lipoprotein lipase activity of other tissues does not increase in response to insulin.
In short-term starvation, low insulin levels lead to a decreased glucose uptake, with a consequent decrease in the supply of glycerol 3-phosphate. Re-esterification is retarded, and fatty acids are exported. Activation of mobilizing lipase is not significant in short-term starvation.
Long-term starvation, exercise or stress leads to in-

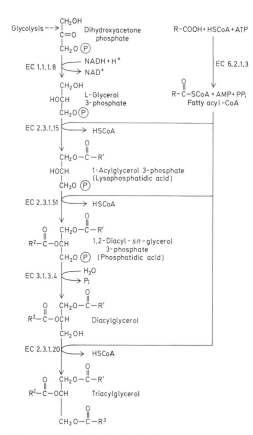

Figure 2. Biosynthesis of triacylglycerols in liver and adipose tissue cells
EC 1.1.1.8: Glycerol-3-phosphate dehydrogenase (NAD⁺)
EC 2.3.1.15: Glycerol-3-phosphate acyltransferase
EC 2.3.1.20: Diacylglycerol acyltransferase
EC 2.3.1.51: 1-Acylglycerol-3-phosphate acyltransferase
EC 3.1.3.4: Phosphatidate phosphatase
EC 6.2.1.3: Long-chain-fatty-acid-CoA ligase

creased activity of mobilizing lipase. In stress and exercise, adrenalin and noradrenalin (catecholamines) are mainly responsible for the observed increase of mobilizing lipase activity by stimulation of adenylate kinase. Insulin reverses this activation by catecholamines. In long-term starvation, lack of insulin and excess growth hormone cause increased cAMP synthesis, leading to stimulation of mobilizing lipase. Re-esterification is also retarded (lack of insulin prevents glucose uptake), and free fatty acids are exported.

Acylmercaptan: see Thioester.
Adair-Koshland-Nemethy-Filmer model: see Cooperativity model.
Adaptive enzymes: obsolete term for inducible enzymes. See Enzyme induction.
Adaptor hypothesis: a suggestion made by Crick to explain the translation of the genetic code.

He proposed that there must be an adaptor between the information-carrying nucleic acid and the protein being synthesized which was able to "recognize" both kinds of molecules. The discovery of tRNA and the corresponding aminoacyl-tRNA synthetases confirmed his hypothesis.

Addison's disease: see Adrenal corticosteroids.
Adenine, abb. *Ade:* 6-aminopurine, one of the common nucleic acid bases. An A. residue is also present in the structure of the adenosine phosphates and other physiologically active substances, including Nicotinamide-adenine-dinucleotide phosphate (see), Flavin adenine dinucleotide (see) and various Nucleoside antibiotics (see). A. is found in free form in various plants, especially in yeasts. It is synthesized de novo via adenosine monophosphate, or is formed by degradation of nucleic acids. Adenine deaminase (EC 3.5.4.2) removes the 6-amino group to yield hypoxanthine.

Tautomeric forms of adenine

Adenine arabinoside: see Arabinosides.
Adenine deaminase, *adenase* (EC 3.5.4.2): see Purine degradation.
Adenine xyloside: see Xylosylnucleosides.
Adenosine, abb. *ado:* 9-β-D-ribofuranosyladenine. Phosphorylated derivatives of Ado are metabolically important. See Adenosine phosphates, Nucleosides.
Adenosine deaminase (EC 3.5.4.4): an enzyme of M_r 217000 (two subunits, M_r 103000 each) which deaminates adenine to inosine. It is present in takadiastase preparations from *Aspergillus oryzae*, and is sometimes confused with Taka amylase (see).
Adenosine 3'-phosphate 5'-phosphosulfate: see Phosphoadenosine-phosphosulfate.
Adenosine phosphates, *adenine ribonucleotides:* important as components of nucleic acids and as the major form in which chemical free energy is stored and transferred. They are also important metabolic regulators, for example in glycolysis and the tricarboxylic acid cycle. The biologically significant derivatives, including cyclic adenosine 3',5'-monophosphate, carry the phosphate ester on the C-5 of the ribose.
1. *Adenosine 5'-monophosphate,* abb. *AMP.* is synthesized de novo from inosinic acid (see Purine biosynthesis) and also arises by cleavage of pyrophosphate from adenosine triphosphate.
2. *Adenosine 5'-diphosphate,* abb. *ADP,* is formed either by adding a second phosphate to AMP (see Adenylate kinase), or by removal of a phosphate from ATP; the latter conversion may be catalysed by one of the adenosine triphosphatases (EC 3.6.1.3) or by an enzyme which transfers the phosphate to another organic molecule (kinase). The energy stored in the

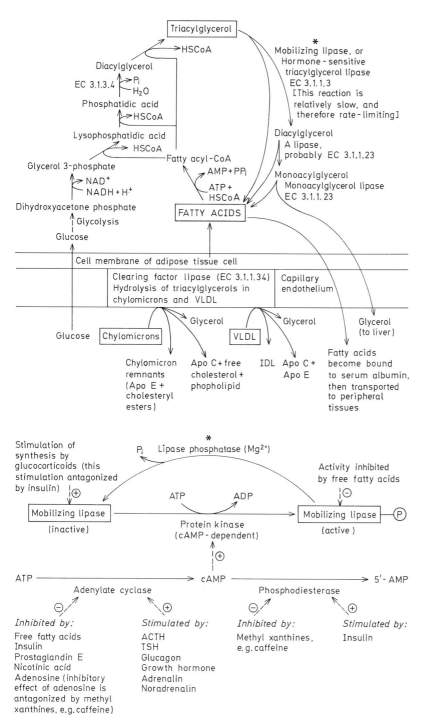

Figure 3. Role of triacylglycerol synthesis and degradation in adipose tissue

anhydride bond of ADP can be made available by the enzyme adenylate kinase, which catalyses the reaction 2 ADP $\rightleftharpoons$ ATP + AMP. ADP is the phosphate acceptor in Substrate-level, Oxidative and Photophosphorylation (see). In all three processes, ADP is converted to ATP.

3. *Adenosine 5'-triphosphate, ATP*, was discovered in 1929 by Lohmann. It is the universal energy "currency" of every living cell (see Energy-rich phosphates.)

Biosynthesis of ATP. ATP is the immediate product of all processes in the cell leading to the chemical storage of energy. It is biosynthesized by phosphorylation of ADP in the course of Substrate phosphorylation (see), Oxidative phosphorylation (see) and non-cyclic Photophosphorylation (see) in plants. Energy in the form of a third phosphate may also be transferred to ADP from other high-energy phosphates, such as creatine phosphate (see Creatine) or other nucleoside triphosphates, or from the adenylate kinase reaction.

Cleavage of ATP. ATP has a high potential for group transfer (Fig. 1, Table 1.):

Table 1. Free energy of hydrolysis of ATP in kJ/mol (kcal/mol)

Removal of orthophosphate:	
ATP → ADP + P$_i$	29.4 (7.0)
Removal of pyrophosphate:	
ATP → AMP + PP$_i$	36.12 (8.6)
PP$_i$ → P$_i$ + P$_i$	28.14 (6.7)

lease of AMP, for example in the synthesis of 5-phosphoribosyl-1-pyrophosphate from ribose 5-phosphate in the course of purine biosynthesis.

c) Transfer of the AMP residue and release of pyrophosphate. The receiving group is given a higher group-transfer potential in this process, which occurs in the activation of fatty acids and amino acids in synthetic pathways. The released pyrophosphate may be hydrolysed by inorganic pyrophosphatase (EC 3.6.1.1), which makes the transfer reaction essentially irreversible.

d) Transfer of the adenosyl residue and release of both orthophosphate and pyrophosphate, for example in the synthesis of *S*-adenosyl-L-methionine.

Uses of ATP. The chemical energy stored in ATP is used in chemical reactions, including the synthesis of macromolecules from the corresponding monomeric compounds and formation of activated compounds. Often an endergonic reaction is driven forward by enzymatic coupling to the hydrolysis of ATP. Many catabolic pathways, including glycolysis, require an "investment" of ATP which is later re-synthesized. Essentially all anabolic pathways require ATP, either directly or indirectly.

ATP provides energy for the contraction of muscles and the motion of cilia and flagella. In some organisms, ATP can provide the energy for Bioluminescence (see), which in turn provides a very sensitive assay for ATP. Electric fish can generate electric current by hydrolysing ATP. Active transport of many substances across membranes depends on a source of ATP. ATP is also known to be released together with acetylcholine at synapses in the periph-

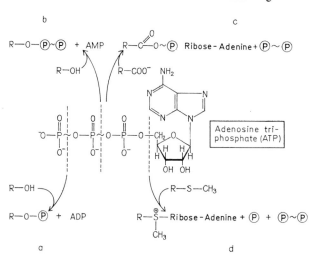

Figure 1. Possible cleavages of adenosine 5'-triphosphate

a) Transfer of orthophosphate to alcoholic hydroxyl groups, acid groups or amide groups and release of ADP. The enzymes which catalyse these reactions are the kinases, which in some cases (e.g. creatine kinase) can also catalyse the synthesis of ATP from ADP.

b) Transfer of the pyrophosphate residue and re-

eral nervous systems of vertebrates. It may be a modulator of nervous transmission, as it inhibits the release of acetylcholine. [E. M. Silinsky & B. L. Ginsborg, *Nature* **305** (1983) 327–328.]

Other nucleoside triphosphates, which are energetically equivalent to ATP, are important in some metabolic reactions: cytidine triphosphate in phos-

Table 2. *Some examples of the occurrence and effect of cyclic adenosine 3′,5′-monophosphate (according to Harde-land)*

Organisms	Effect
1) Protozoa *Paramecium*	Activation of the protein kinase.
2) Bacteria *Escherichia coli*	Release of glucose inhibition of enzyme induction (see Catabolite repression). Initiation of messenger RNA synthesis mediated by a specific cAMP receptor protein (catabolite gene activator protein). Inhibition of the degradation of messenger RNA bound to ribosomes. Stimulation of the synthesis of many enzymes.
Serratia marcescens *Salmonella typhimurium* *Proteus inconstans* *Aerobacter aerogenes* *Brevibacterium liquefaciens*	Release of catabolite repression and stimulation of the synthesis of β-galactosidase. (*Brevibacterium liquefaciens* excretes cAMP into the medium).
Photobacterium fischerei	Release of catabolite repression and production of bioluminescence.
3) Fungi Slime molds *Dictyostelium discoideum*	Extracellular signal transmitter. Cell aggregation as response to chemotactic stimuli.
Polysphondylium pallidium Yeasts	Does not respond to cAMP, has a different acrasin.
Saccharomyces cereviseae	Affects the oscillation and the redox equilibrium in the course of glycolysis. Affects sporulation.
4) Invertebrates e.g. annelids (*Golgingia, Nereis*) starfish	Activation of protein kinases.
Liver fluke (*Fasciola*) Blowfly (*Calliphora*)	Transmission of the effect of serotonin.
5) Vertebrates Frog, toad, turkey, pigeon, rat, mouse, guinea pig, rabbit, human	"Second messenger" in the transmission of hormone stimuli.
6) Higher plants Barley (endosperm) Peas, lettuce Weeds	Enzyme induction during germination. Stimulation of the synthesis of amylase. Cell extension growth, especially in dwarf varieties of *Pisum sativum*. Effects on germination.

phatide biosynthesis, guanosine triphosphate in protein synthesis and oxidative decarboxylation of 2-oxoacids (see Tricarboxylic acid cycle), inosine triphosphate in some carboxylations, uridine triphosphate in polysaccharide synthesis.

In the living organism, the adenosine phosphates are in equilibrium and are regarded collectively as the adenylic acid system. The physiological concentrations for ADP and ATP are around 10^{-3} mol/l. The ratio of the forms is called the *energy charge* and is given by the equation

$$EC = ([ATP] + 0.5[ADP]) \, / \, ([ATP] + [ADP] + [AMP]).$$

The square brackets indicate molar concentrations. If the total complement of A. is in the form of ATP,

the energy charge is 1; otherwise it is smaller than 1.

4. *Cyclic adenosine 3′,5′-monophosphate.* 3′,5′-AMP, cyclo-AMP, cAMP was discovered by Sutherland in 1956 as a heat-stable factor in the liver. It is generated from ATP (Fig. 2) by adenylate cyclase (EC 4.6.1.1), and is degraded to AMP by 3′:5′-cyclic-nucleotide phosphodiesterase (EC 3.1.4.17), which is specific for cyclic nucleotides. The activities of these two enzymes determine the intracellular level of cAMP. Physiological amounts of various substances, for example pyridoxal phosphate in *Escherichia coli*, can reduce the activity of the adenylate cyclase. The phosphodiesterase of mammals is inhibited by nucleoside triphosphates, pyrophosphate, citrate and methylated xanthines (especially theophylline) and stimulated by nicotinic acid. The ad-

13

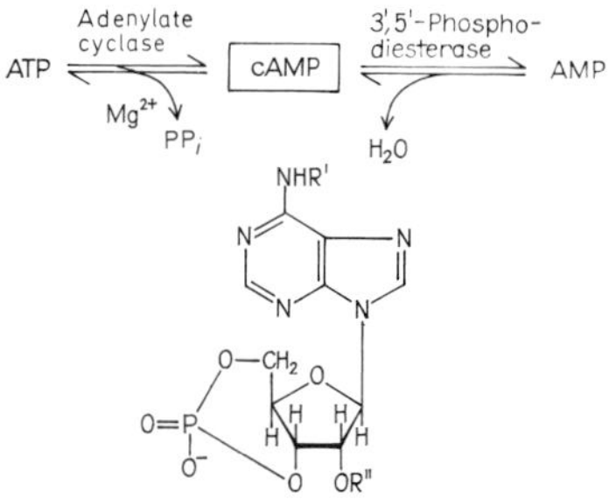

$R'=R''=H$ Cyclic adenosine 3',5'-monophosphate (cAMP)

$R'=R''=CO-(CH_2)_2-CH_3$ $N^6,O^{2'}$-Dibutyryl-cAMP

Figure 2. Synthesis of cyclic adenosine 3',5'-monophosphate; and the structure of cyclic $N^6,O^{2'}$-dibutyryladenosine 3',5'-monophosphate

enylate cyclase in many cells is located just inside the plasma membrane. The receptors for hormones and other chemical signals are located on the outside of the plasma membrane of those cells which are the "targets" of the signal molecules. In many systems, binding of a hormone or other activator to its receptor leads to activation of the adenylate cyclase (by as yet unknown mechanisms) and an increase in the cAMP concentration. The cAMP often serves as an effector molecule which increases the activity of a protein kinase (or other enzyme) that in turn regulates some cellular process by Covalent modification of enzymes (see). In this way cAMP serves as a "second messenger" for a variety of hormones. In addition, it affects the production and release of hormones, e.g. acetylcholine, glucagon, insulin, melanotropin, parathyrin, vasopressin and corticotropin. It also affects the equilibria among various metabolic pathways, e.g. glycogen metabolism. In many cases, the physiological effects of cAMP are only seen in the presence of calcium ions. The exogenous "artificial" control of the intracellular cAMP level is becoming medically important. Substances which raise this level have been successfully used in treatment; for example, psoriasis is treated with the alkaloid papaverine, which inhibits the cyclic nucleotide phosphodiesterase, and the tissue hormone dopamine, which stimulates the formation of cAMP in the epidermis. cAMP also inhibits the growth of certain tumors.

Because it is very polar, cAMP penetrates the cell membrane only in very small quantities. Its synthetic derivatives have better permeability because they have been made more lipophilic by substitution with organic acids. The most commonly used is $N^6,O^{2'}$-dibutyryladenosine 3',5'-monophosphate, abb. DBcAMP. A number of other cyclic 3',5'-mononucleotides with special functions have been found to occur naturally.

Adenosine triphosphatase, *ATPase* (EC 3.6.1.3): enzyme which catalyses cleavage of ATP to $ADP + P_i$. Physiologically, this is generally coupled to another process in the cell so that the free energy released by cleavage of the ATP is used to drive an endergonic process. Two of the best-studied examples are the actomyosin complex (see Muscle proteins), in which cleavage of ATP provides energy for contraction, and the Na^+/K^+ pump in cell membranes, in which the energy required for transport of the ions against their concentration gradients is provided by ATP (see Transport). The mitochondrial ATPase is believed to function, in intact mitochondria, as an ATP synthetase rather than as ATPase; when the organelles are disrupted, however, the enzyme catalyses the reaction in the thermodynamically favorable direction. The mechanisms by which the energy of ATP hydrolysis is converted to mechanical or osmotic energy are not known.

Adenosine triphosphate: see Adenosine phosphates.

S-Adenosyl-L-homocysteine: see *S*-Adenosyl-L-methionine.

S-Adenosyl-L-methionine, *S-(5'-deoxyadenosine-5')-methionine, active methionine, active methyl,* abb. *S-Ado-Met, SAM:* a reactive sulfonium compound which is the most important methylating agent in cellular metabolism (see Transmethylation). M_r of the free cation, 398.4. The natural form is the L-(+)-isomer. Due to the asymmetry of the sulfonium group, there are 4 stereoisomers. SAM is unstable at room temperature, both as the solid and in aqueous solution. It is formed by activation of L-methionine with ATP: $Met + ATP = SAM + PP_i + P_i$. The adenosine residue of the ATP is transferred to the methionine.

Table. Properties of adenylylsulfate reductases from various organisms

Organism	pH optimum	M_r	Comments
Desulfovibrio[1]	7.4	220000	Contains 1 molecule FAD and 6 to 8 atoms nonheme iron
Thiobacillus thioparus[1]	7.4	170000	Contains 1 molecule FAD and 8 to 10 atoms nonheme iron
Thiocapsa roseopersicina[1]	8.0	180000	Contains 1 molecule FAD, 4 atoms non-heme iron and 2 atoms heme iron 60 to 80 fold enrichment to a homogeneous preparation in the ultracentrifuge
Chlorella pyrenoidosa[2]		330000	Partly purified enzyme

1. With $Fe(CN)_6^{3-}$; 2. with a thiol as electron donor; the enzyme from *Chlorella* is active with phosphoadenylylsulfate only in the presence of 3'-nucleotidase.

The transmethylation reaction produces, in addition to the methylated product, *S*-adenosyl-L-homocysteine. It may be reconverted to SAM after cleavage into adenosine and L-homocysteine, which is the substrate of dimethylthetin-homocysteine methyltransferase (EC 2.1.1.3). See Methionine.

S-Adenosyl-L-methionine

Adenylate cyclase (EC 4.6.1.1): see Adenosine phosphates.

Adenylate kinase, *myokinase:* (EC 2.7.4.3): a trimeric enzyme found in the mitochondria of muscles and other tissues. It is resistant to heat and acid. M_r 68 000, subunit M_r 23 000. It catalyses the conversion of two molecules of ADP into ATP+AMP, thus making available the energy of the ADP. At equilibrium, the concentrations of the three adenosine phosphates are nearly equal. In many energy-requiring reactions ATP is converted into pyrophosphate and AMP (see Adenosine phosphates). A.k. is important because it catalyses the first stage (AMP to ADP) of the reconversion of this AMP into ATP.

Adenylic acid: see Adenosine phosphates.

Adenylosuccinate, *N-succinyladenylate,* abb. *sAMP:* 5-aminoimidazole-4-*N*-succinocarboxamide ribonucleotide, an intermediate in purine biosynthesis. M_r 463.3.

Adenylylsulfate reductases: enzymes of sulfur metabolism which reduce either phosphoadenylylsulfate (APS reductase) or adenylylsulfate. Adenylylsulfate reductase (EC 1.8.99.2) is identical with one component of the sulfate reductase in sulfate assimilation, since adenylsulfate is the donor of the sulfate group. The table shows the properties of some of these reductases. The reductase is in every case a complex of three components, an adenylylsulfate transferase (see Sulfate assimilation, Fig. 1), a low-molecular-weight carrier and the actual adenylylsulfate reductase.
Phosphoadenylylsulfate reductase from *Saccharomyces cerevisiae* requires NADPH and has been partly purified and fractionated.

Adermine: vitamin B$_6$. See Vitamins.

ADH: abb. for Antidiuretic hormone. See Vasopressin.

Adiuretin: see Vasopressin.

Adjuvant: a mixture of oils, emulsifiers, killed bacteria and other components which serves to intensify unspecifically the immune response. The A., which is not (supposed to be) itself antigenic, is injected several times, intramuscularly or subcutaneously, into an animal to produce the maximal amount of antibodies. Freund's incomplete A., an emulsion of paraffin oils which protects the antigen from too rapid degradation, and Freund's complete A., which contains in addition killed mycobacteria or tuberculosis bacteria, are most commonly used in experimental immunology. In vaccines, the A. is usually aluminum hydroxide or calcium phosphate gel.

Ado: abb. for Adenosine (see).

ADP: abb. for Adenosine 5'-diphosphate. See Adenosine phosphates.

ADP-ribosylation of proteins: attachment of monomeric or polymeric ADP-ribosyl groups to a protein by transfer from NAD$^+$.

$$\begin{array}{cc} \text{Adenine} & \text{Nicotinamide} \\ | & | \end{array}$$
$$(\text{ribose}\,\textcircled{P}\text{-}\textcircled{P}\text{-ribose})_n + \text{Protein} \longrightarrow$$

$$\begin{array}{c} \text{Adenine} \\ | \end{array}$$
$$\text{Protein-(ribose-}\textcircled{P}\text{-}\textcircled{P}\text{-ribose})_n + \text{Nicotinamide} + \text{H}^+,$$

where n can vary from 1 to 50. Poly ADP-ribosyl groups represent a novel homopolymer of repeating ADP-ribose groups linked 1'-2' between respective ribose moieties:

$$\downarrow 2' \qquad\qquad 1'$$
$$\text{Adenine-ribose-}\textcircled{P}\text{-}\textcircled{P}\text{-ribose}$$
$$\downarrow 2' \qquad\qquad 1'$$
$$\text{Adenine-ribose-}\textcircled{P}\text{-}\textcircled{P}\text{-ribose}$$
$$\downarrow$$

The free energy of hydrolysis of the β-*N*-glycosidic linkage of NAD$^+$ is -34.4 kJ (-8.2 kcal)/mol at pH 7 and 25 °C; it is therefore a high-energy bond, and NAD$^+$ can act as an ADP-ribosyl transferring agent. The transfer of one ADP-ribosyl group ($n=1$ in the above equation) is catalysed by ADP-ribosyltransferase. Formation and concomitant transfer of poly(ADP-ribose) to an acceptor is catalysed by poly(ADP-ribose) synthetase (n is greater than 1 in the above equation).
Diphtheria toxin, produced by strains of *Corynebacterium diphtheriae* that carry β phage, inhibits protein synthesis in eukaryotic cells by catalysing the transfer of an ADP-ribose moiety from NAD$^+$ to elongation factor 2. *Pseudomonas* toxin catalyses a similar reaction. T4 phage catalyses the monomeric ADP-ribosylation of RNA polymerase and other proteins in *Escherichia coli*. Choleragen activates adenylate cyclase by catalysing transfer of ADP-ribose from NAD$^+$ to the enzyme.
Poly ADP-ribose groups are found in eukaryotic chromosomal proteins, mitochondrial proteins, and histones. The biological function of the ADP-ribosylation of proteins in eukaryotic cells is not known, but the occurrence of poly ADP-ribosyl groups in nuclear proteins, particularly in association with chromatin, suggests a regulatory role in nuclear function.
The nature of the linkage to protein is not known, but it appears to involve attachment to basic amino acids. In the choleragen-activated ADP-ribosylation of adenylate cyclase, an arginine residue appears to

ADP-ribosylation of proteins

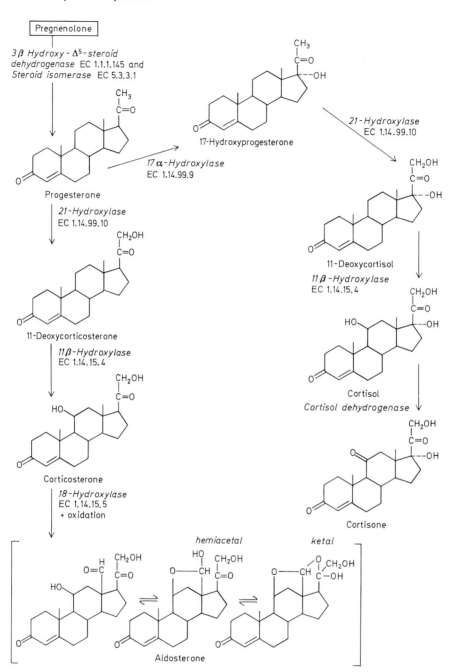

Biosynthesis of adrenal corticosteroids. The primary precursor is acetyl-CoA, and the biosynthesis proceeds via cholesterol (see Terpenes, Steroids). The adrenal cortex also utilizes cholesterol which it receives as cholesterol esters from extra-adrenal sources (see Lipoproteins). The adrenal cortex is differentiated into three concentrically arranged layers. The outermost layer (zona glomerulosa) is primarily responsible for synthesis and secretion of aldosterone; it contains the 18-hydroxylase, and it lacks the 17α-hydroxylase. The intermediate layer (zona fasciculata) and innermost layer (zona reticularis) are responsible for the synthesis of glucocorticoids (mainly cortisol) and adrenal androgens (see Androgens); they possess the 17α-hydroxylase, and lack the 18-hydroxylase.

be the chief receptor for ADP-ribose. O. Hayaishi & K. Ueda *Ann. Rev. Biochem.* 46 (1977) 95-116. M. R. Purnell, P. R. Stone & W. J. D. Whish, *Biochemical Society Transactions* 8 (1980) 215-227]

Adrenal corticosteroids, *adrenocorticoids, corticosteroids, corticoids, cortins:* an important group of steroid hormones, formed in the adrenal cortex in response to adrenocorticotropic hormone (ACTH, Corticotropin (see)). A. c. are structurally related to pregnane (see Steroids); they contain a carboxyl group with a neighboring α,β-unsaturated bond in ring A, a ketol side chain in position 17, and other oxygen functions, particularly in positions 11, 17, 21 and 18.

More than 30 steroids have been found in the adrenal cortex; the following show marked A. c. activity: Cortisol (see), Cortisone (see), Cortexolone (see), 11-Dehydrocorticosterone (see), Corticosterone (see) and Aldosterone (see). Quantitatively the most significant members of this group are cortisol, corticosterone and aldosterone, which are secreted daily into the blood in quantities of 15, 3, and 0.3 mg, respectively.

The production of A. c. is increased in physical and/or psychological stress. Deficiency of A. c., e. g. caused by pathological changes in the adrenal glands, results in Addison's disease (*Morbus Addison*); this condition is characterized by tiredness, emaciation, decrease of blood sugar, and dark pigmentation of those areas of the skin exposed to light. Adrenalectomy of experimental animals leads to rapid death, unless exogenous A. c. are given. A. c. control mineral metabolism by causing the retention of Na^+, Cl^- and water, with a simultaneous K^+ diuresis (mineralocorticoidal or mineralotropic action). They also regulate carbohydrate metabolism, in particular glycogen synthesis in the liver (glucocorticoidal or glucotropic action). Depending on the type of activity that predominates, the A. c. are classified as *mineralocorticoids* (aldosterone, cortexone, cortexolone), or *glucocorticoids* (cortisol, cortisone, corticosterone, 11-dehydro-corticosterone).

For the therapy of adrenal insufficiency, extracts of adrenal cortex are no longer used and have been replaced by pure A. c. High doses reveal other pharmacological properties, especially anti-inflammatory and antiallergic activities. This discovery led to the development of highly active, synthetic derivatives, which are now used widely for the treatment of rheumatism, asthma, allergies, eczema, etc., e. g. prednisone (see Prednisolone), Dexamethasone (see) and Triamcinolone (see).

Adrenal gland, *suprarenal gland, Glandula suprarenalis:* a heavily vasculated, vertebrate endocrine gland, weighing about 15 g in the adult human. There are two A. g., one just above each kidney. The A. g. consists of two developmentally and functionally distinct parts: the mesodermal adrenal cortex (AC) and the ectodermal adrenal medulla (AM). The AC, which contains three histologically distinct zones, produces and exports glucocorticoids (see Cortisol) and mineralocorticoids (see Aldosterone) in response to the action of the pituitary hormones, corticotropin and renin/angiotensin II, respectively. The AC also produces sex steroids (see Androgens). The AM (*Paraganglion suprarenale*) is the largest

(but not the only) ganglion of the sympathetic nervous system. It produces the hormones Adrenalin (see) and Noradrenalin (see). The AM is a model example of the close association of the sympathetic nervous system with an endocrine system. The secretory cells are richly innervated by cholinergic, preganglionic, sympathetic nerve fibers. No nerve supply to the AC has been demonstrated.

Adrenal hyperplasia: see Inborn errors of metabolism.

Adrenalin, *epinephrine:* 4-[1-hydroxy-2-(methylamino)ethyl]-1,2-benzenediol, M_r 183.2, a catecholamine hormone and drug. The L-form is physiologically active, affecting carbohydrate metabolism and the circulatory system.

A. is synthesized in the adrenal cortex from tyrosine (via dopa, dopamine and noradrenalin), stored in the chromaffin granules and released into the blood stream upon nervous stimulation by the nervus splanchnicus. It is also an adrenergic neurotransmitter synthesized in and released from neurons of the sympathetic nervous system. A. activates, via the adenylate cyclase system, the liver and muscle phosphorylases (EC 2.4.1.1) (glycogenolysis) and the lipase of adipose tissue, leading to higher blood concentrations of glucose (hyperglycemia), lactate and free fatty acids. The latter undergo oxidation, resulting in a higher oxygen consumption.

Adrenalin

A. is degraded after *O*-methylation and oxidative deamination by a monoamine oxidase. It is excreted in the urine as 3-methoxy-4-hydroxymandelic acid (vanillinemandelic acid).

Analogs of A. are used to control blood pressure, counteract depression, stimulate the appetite and relieve asthma.

Adrenocorticotropin, *adrenocorticotropic hormone:* see Corticotropin.

Adrenosterone: adrost-4-ene-3,11,17-trione, a steroid derived from androstane. M_r 300.9. A. is synthesized in the adrenal cortex and is considered one of the male gonadal hormones, due to its weak androgenic effect (see Androgens).

Adrenosterone

Adsorption chromatography: see Chromatography.

Affinity chromatography: see Proteins.

Aflatoxins: mycotoxins produced by *Aspergillus flavus*, *A. parasiticus* and *A. oryzae*, as well as some *Penicillium* strains. A. have been found in a variety of foodstuffs, especially in damp, tropical environments, which favor growth of the producer microorganisms. Many liver cancers in the tropics are attributable to ingestion of A., and A. poisoning in malnourished children is the cause of Kwashiorkor (see). Unmodified A. are relatively nontoxic *per se*, but A. become converted into potent toxins and carcinogens by monofunctional oxygenases in the liver (Fig.) (see Cytochrome P_{450}). LD_{50} values for ducklings (µg per 50 g body weight) are A. B_1, 18.2; A. B_2, 84.8; A. G_1, 39.2; A. G_2, 172.5; A. M_1, 16.6; A. M_2, 62. The mechanism of toxicity (Fig.) requires epoxide formation by addition of oxygen at a double bond ($\Delta^{9,10}$ in A. G_1; $\Delta^{8,9}$ in A. B_1 and A. M_1). This double bond is absent from the A_2 compounds, which are probably oxidized to the A_1 compounds in the body, or become bound to DNA by a different mechanism. [R. Langenbach et al. *Nature* **276** (1978) 277-280].

Agar-agar: a polysaccharide plant slime from various red algae. It consists of about 70% agarose and 30% agaropectin. Agarose is a linear polymer of alternating D-galactose and 3,6-anhydrogalactose. Agaropectin consists of D-galactose units, linked β-1,3 glycosidically. In some of them, position 6 is esterified with sulfate. A. is obtained by hot-water extraction of bleached algae, which may contain up to 40% A. It is used as a gelling agent in the pharmaceutical and food industries, and to make solid medium for microorganisms.

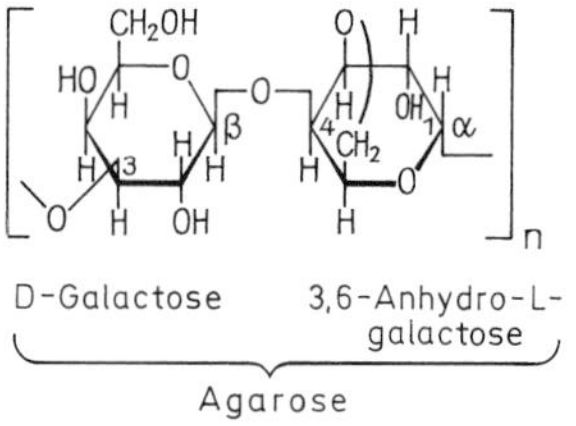

D-Galactose   3,6-Anhydro-L-galactose

Agarose

Agarose

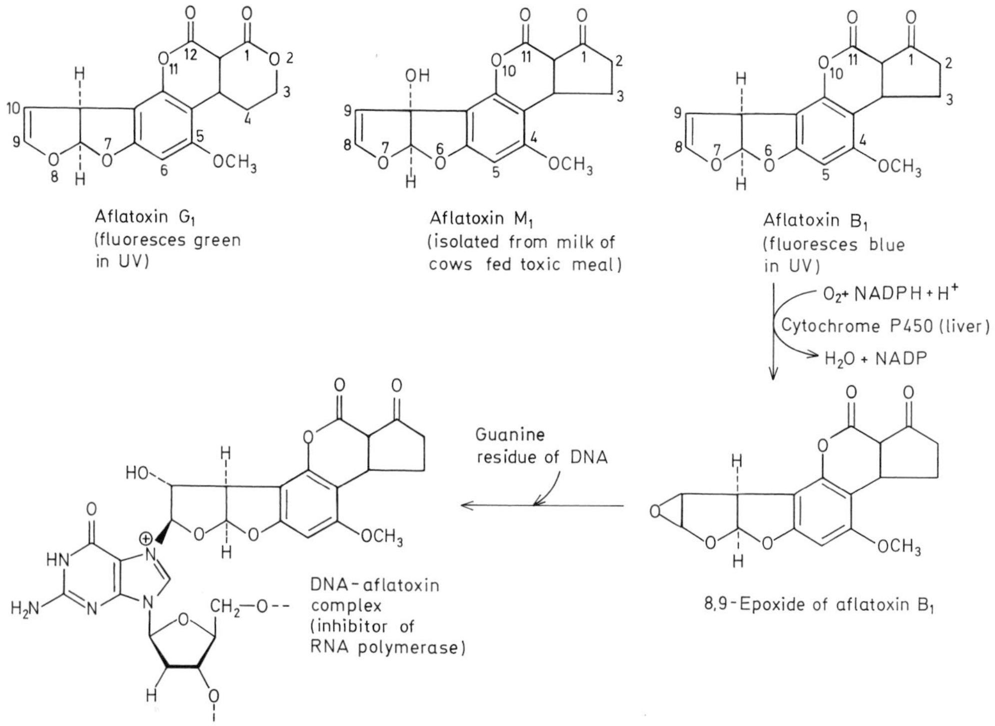

Aflatoxin G_1
(fluoresces green in UV)

Aflatoxin M_1
(isolated from milk of cows fed toxic meal)

Aflatoxin B_1
(fluoresces blue in UV)

O_2 + NADPH + H⁺
Cytochrome P450 (liver)
H_2O + NADP

Guanine residue of DNA

DNA-aflatoxin complex
(inhibitor of RNA polymerase)

8,9-Epoxide of aflatoxin B_1

Aflatoxins and the mechanism of their conversion to carcinogenic and toxic derivatives. Aflatoxins B_2, M_2 and G_2 do not possess a double bond at position 9,8 (B_2, M_2) or 9,10 (G_2).

Afrormosin: 7-hydroxy-6,4'-dimethoxyisoflavone (see Isoflavone).

AGA: abb. for *N*-Acetylglutamate.

Agathisflavone: see Biflavonoids.

Agglutination: the clumping of insoluble antigens bound to particles, such as bacteria, viruses or erythrocytes, by the appropriate antibodies. These must be at least bivalent, in order to bind the antigen-carrying particles together. A. is much more sensitive than precipitation, because the antigen-antibody reaction takes place on the surface of larger particles. The lower limit of recognition for precipitation is about 10 μg/ml serum, but with A. the limit is 0.01 μg/ml.

Passive hemagglutination has a lower limit of detection of 3 to 6 ng antibodies/ml serum. In this technique, soluble antigens are bound to the surface of erythrocytes, which are agglutinated when the antigen-antibody reaction occurs. This method is 1000 times more sensitive than precipitation.

Agglutinins: see Lectins.

Aglycon, *aglycone, genin:* the noncarbohydrate part of a glycoside. Aglycons are released by hydrolysis (e.g. by acid or enzymes) of the C-, N-, or S-glycosidic linkage. See Glycosides, Glucosinolate.

Agnosterol: 5α-lanosta-7,9,(11),24-trien-3β-ol, a tetracyclic triterpene alcohol derived structurally from 5α-lanostane (see Lanosterol). M_r 424.7. A. is a zoosterol (see Sterols) present in the sebaceous oil of sheep's wool.

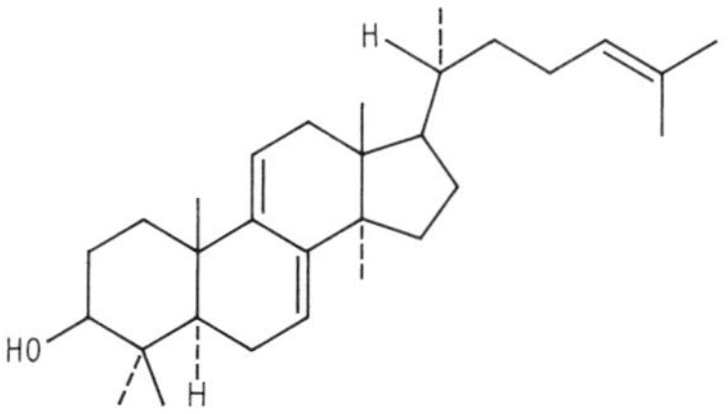

Agnosterol

α₁AGp: see Orosomucoid.

AICAR: abb. for 5(4)-Aminoimidazole-4(5)-carboxamide ribotide. See Purine biosynthesis.

AIR: abb. for 5-Aminoimidazole ribotide. See Purine biosynthesis.

Ajmaline: a Rauwolfia alkaloid. A. is used medicinally to normalize heart rhythm. In high doses it has the tranquilizing effect of Rauwolfia alkaloids.

Alanine, *aminopropionic acid, M_r 89.1.*
1. *L-α-alanine,* abb. *Ala,* CH_3-$CH(NH_2)$-COOH, a proteogenic amino acid. Ala is glucogenic and is closely involved in the metabolism of sugars and organic acids. It is one of the main components of silk fibroin. Free Ala, together with glycine, occurs in relatively large amounts in human blood plasma. It is produced from pyruvate by transamination, and in some microorganisms, e.g. bacilli, by reductive amination, catalysed by alanine dehydrogenase (EC 1.4.1.1). This enzyme has been reported to be a pro-

tomer of the oligomeric glutamate dehydrogenase (EC 1.4.1.2). Ala is degraded to pyruvate and ammonia by alanine dehydrogenase (alanine oxidase: see Flavin enzymes), or it can be converted into pyruvate by Transamination (see).

2. *β-Alanine,* H_2N-CH_2-CH_2-COOH, a nonproteogenic amino acid. β-A. occurs in free form, for example in the human brain, and is a component of the dipeptides carnosine and anserine, and of coenzyme A. It is not usually formed by decarboxylation of L-aspartate, but rather in the course of reductive pyridine degradation. It can be further metabolized to acetate by deamination, decarboxylation and oxidation.

Alar 85: see Succinic acid 2,2-dimethylamide.

Albizziin, *2-amino-3-ureidopropionic acid:* $H_2NCONHCH_2CH(NH_2)$-COOH, a nonproteogenic amino acid, occurring primarily in species of the genus *Albizzia*. It is presumably formed from carbamylphosphate and 2,3-diaminopropionic acid by transcarbamylation. It is an antagonist of glutamine.

Albomycin: an antibiotic synthesized by *Actinomyces subtropicus*. It is a cyclic polypeptide with a pyrimidine base (cytosine) (Fig.) It contains 4.16% iron in the form of a hydroxamate iron(III) complex. A. is one of the sideromycins and interferes with iron metabolism as an antimetabolite of the sideramines. It is similar to or identical to grisein. It is effective against both Gram-positive and Gram-negative bacteria, and inhibits the aerobic metabolism of *Staphylococcus aureus* and *Escherichia coli.*

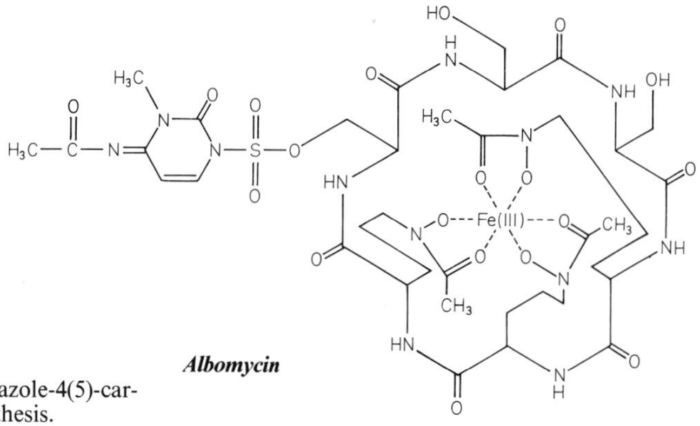

Albomycin

Albumins: a group of simple proteins. They are found in the body fluids and tissues of animals and in some plant seeds. In contrast to the globulins, they are of low molecular weight, water-soluble and easily crystallizable, and they contain an excess of acidic amino acids. A. can only be precipitated by high concentrations of neutral salts. They are rich in glutamate and aspartate (20-25%) and leucine and isoleucine (up to 16%) but contain little glycine (1%). Important representatives of this group are serum albumin, α-lactalbumin (milk proteins) and ovalbumin from animals, and the poisonous ricin (from *Rizinus* seeds), leucosin (from seeds of wheat, rye and barley), and legumelin (from legumes).

Serum albumin (plasma albumin) makes up 55 to 62% of the serum protein, and is one of the few carbohydrate-free proteins in blood plasma, or the serum obtained from it by clotting. Due to its relatively low M_r of 67500 and high net charge (IP 4.9), serum albumin has a good binding capacity for water, Ca^{2+}, Na^+, K^+, fatty acids, bilirubin, hormones and drugs. (Researchers who use serum albumin in defined media should not forget the general "stickiness" of this protein!) Its main function is the regulation of the colloidal osmotic pressure of the blood. Bovine and human serum albumins contain 16% nitrogen and are used as standard proteins for calibration, due to the ease of obtaining them in crystalline, highly purified form.

Human serum albumin consists of a single polypeptide chain of 584 amino acids which is stabilized by 17 disulfide bridges. In contrast, α-lactalbumin and

essential for catalysis) and one coenzyme binding site per subunit (M_r 40000, 374 amino acids, sequence known: Cys 46 is the site of binding and catalysis). In the dehydrogenation process, a ternary complex among ADH, NAD^+ and ethanol is formed in which both the coenzyme and the substrate are bound to the reactive SH group of Cys 46 via a zinc atom. Due to the fact that there are two very similar polypeptide chains E and S, there are three types of liver ADH: the two isoenzymes EE (preferentially dehydrogenates ethanol) and SS (active with sterols), and a hybrid with a M_r of 60000. It also differs from other A. in its subunit structure (8 subunits of M_r 7400).

Alcoholic fermentation: the anaerobic formation of ethanol and carbon dioxide from glucose. The process yields energy: A. f. of 1 mol of glucose yields 2 mol of ATP.

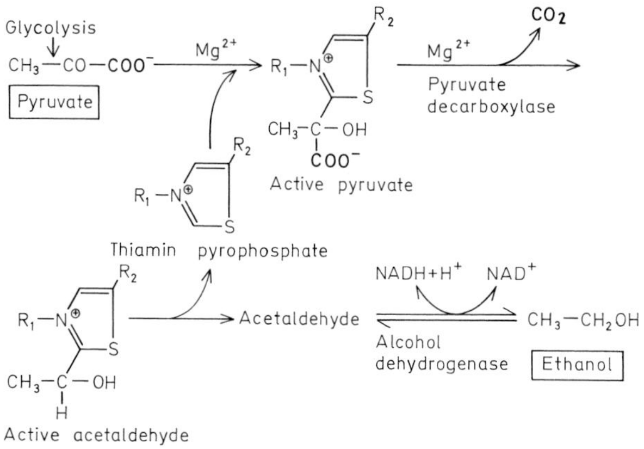

Formation of ethanol from pyruvate

ovalbumin (M_r 44000) contain one oligosaccharide chain each, coupled to the peptide chain via an aspartate residue (3.2% carbohydrate of M_r 1550 in ovalbumin). In ovalbumin, one serine residue is esterified with phosphate.

Alcaptonuria, *alkaptonuria:* see L-Phenylalanine.

Alcohol dehydrogenase, abb. *ADH* (EC 1.1.1.1): a zinc-containing oxidoreductase which, in the presence of NAD^+, reversibly oxidizes primary and secondary alcohols to the corresponding aldehydes and ketones. ADH occurs in bacteria, yeasts, plants and the liver and retina of animals. The ADH from yeast, which is distinguished by its high affinity for ethanol, is of practical significance as the last enzyme in alcoholic fermentation. In the liver, ADH acts in concert with other mechanisms to clear the blood of ethanol. The ADH in the retina, however, serves to convert the vitamin A aldehyde, all-*trans*-retinol, to retinol (see Vitamins). ADH from animal organs and yeast has low substrate specificity, since it dehydrogenates both short (C_2 to C_6) and long-chain alcohols. Yeast ADH (M_r 145000) consists of four catalytically active, zinc-containing subunits (M_r 35000) with four NAD^+ or NADH binding sites per molecule of ADH. The dimeric horse liver enzyme (M_r 80000) contains two zinc atoms (one is

The most important fermenting organisms are yeasts and other microorganisms, but A. f. can also be carried out by the tissues of higher plants, e. g. carrots and maize roots. There is no A. f. in animals, which lack pyruvate decarboxylase (see below). The starting point for A. f. is glucose 6-phosphate, which is converted by the Glycolysis (see) reactions to pyruvate. Pyruvate is decarboxylated by pyruvate decarboxylase (EC 4.1.1.1) to acetaldehyde, which is then reduced to ethanol by alcohol dehydrogenase (EC 1.1.1.1) (Fig.) Balance: $C_6H_{12}O_6 + 2\ P_i + 2\ ADP \rightleftharpoons 2\ CH_3CH_2OH + 2\ CO_2 + 2\ ATP + 2\ H_2O$. A. f. is the longest known and technologically most important form of fermentation. The most important substrates are the monosaccharides D-glucose, D-fructose, D-mannose and sometimes D-galactose. In some cases the disaccharides sucrose and maltose and the polysaccharide starch can serve as substrates. The formation of fusel oils is considered a side reaction of A. f. *Historical.* The simple equation for A. f., 1 glucose → 2 CO_2 + 2 ethanol, was established in 1815 by Gay-Lussac. In 1857, Pasteur proposed that A. f. could only be carried out by living organisms (vitalistic theory of fermentation). This was disproved in 1897 by Buchner, who established that a cell-free filtrate of disrupted yeast cells was capable of A. f.

This discovery was the beginning of modern enzymology. The enzyme system responsible for A.f., which was originally thought to be one enzyme, was called zymase. In 1905 the role of phosphate in A.f. was described by Harden and Young. In 1912, Neuberg proposed the first fermentation scheme, which was revised in 1933 by Embden and Meyerhof.

Alcohols: hydrocarbon derivatives carrying one or more hydroxyl (-OH) groups. The ending "-ol" in a systematic or trivial name of an organic compound indicates that it is an alcohol. A "diol" has two -OH groups, a "triol" has three, and so on. There are primary (RCH_2OH), secondary (R^1R^2CHOH) and tertiary ($R^1R^2R^3COH$) A. In nature, the A. in the form of esters are important components of the essential oils, fats and waxes. A number of lower A., e.g. ethanol, are formed by fermentative processes from carbohydrates (themselves polyalcohols) and proteins.

Aldehyde oxidase (EC 1.2.3.1): see Molybdenum enzymes.

Aldoketomutase: see Lactoyl-glutathione lyase.

Aldolase: see Fructose-bisphosphate aldolase (EC 4.1.2.13).

Aldonic acids: monocarboxylic acids derived from aldoses (sugars containing an aldehyde group) by oxidation of the aldehyde group. The names of these acids are formed by replacing the "-ose" of the parent aldose with "-onic acid". A.a. may form a 1,4-lactone ring (γ-lactones) or the more stable 1,5-lactone (δ-lactones). Some important A.a. are L-arabonic acid, xylonic acid, D-gluconic acid, D-mannonic acid and galactonic acid.

Aldoses: polyhydroxyaldehydes, one of the two main subdivisions of monosaccharides. (The other is the ketoses. See Carbohydrates). A. are characterized by their terminal aldehyde group -CHO, which is always given the number 1 in systematic nomenclature. The A. are formally derived from their simplest representative, glyceraldehyde, by chain extension. They are classified according to the number of carbon atoms in their chain as trioses, tetroses, etc. The pentoses and hexoses are particularly important in biochemistry.

Aldosterone: 11β,21-dihydroxy-3,20-dioxo-pregn-4-en-18-al-18→11 hemiacetal, a highly active mineralocorticoid hormone from the adrenal cortex. In contrast to other adrenal cortex horones, A. has a carbonyl group on C-18, which forms a hemiacetal with the 11β-hydroxyl group. It is the most important mineralocorticoid, regulating NaCl resorption and potassium excretion. It also has a certain degree of glucocorticoid activity. For structure and biosynthesis, see Adrenal corticosteroids.

Alginic acid: a polyuronic acid extracted from seaweeds. It is composed of varying proportions of D-mannuronic and L-guluronic acids, linked β-1,4. The M_r is about 120000. The polymer replaces pectin in the brown algae, from which it is extracted with NaOH. It can absorb up to 300 times its weight of water. Because it is easily digested, A.a. is widely used in the food industry, in surgery as resorbable sutures, and in the pharmaceutical and cosmetic industries.

D-Mannuronic   L-Guluronic
    acid   acid

Alginic acid

Alginic acid

Alizarin: 1,2-dihydroxyanthraquinone, a red dye, m.p. 290 °C. It occurs in the root of the madder plant (*Rubia tinctorum* L.) and other *Rubiaceae* in combination with 2 moles glucose, forming the compound ruberythric acid. A. was an important natural dye. It has been made synthetically since 1871, so the production from madder has died out. A. and several of its derivatives are widely used as alizarin dyes. A. was isolated in 1826 by Colin and Robiquet, and its structure was determined in 1868 by Graebe and Liebermann. The first technical syntheses were developed independently by Caro and W. H. Perkin.

Alizarin

Alkaloids: basic natural products occurring primarily in plants. They contain one or more heterocyclic nitrogen atoms, and are generally found in the form of salts with organic acids. Several thousand A. are known, and the structures of many have been determined. They usually have trivial names based on those of the plants in which they were discovered.

Classification: It is difficult to define A. in such a way as to exclude other nitrogen-containing plant products. If they were classified according to occurrence and function, those of animal and microbial origin would be excluded, and in addition, some A. (e.g. nicotine) are found so widely that a strict classification by botanical origin is not possible. If the A. are classified chemically, on the basis of the structure of their skeletons, the colchicum alkaloids would not be included, because they lack the heterocyclic nitrogen. Recently the A. have been classified on the basis of their biogenesis as protoalkaloids (see Biogenic amines), pseudoalkaloids and the A. in the narrower sense. The protoalkaloids include, for example, the decarboxylation products of amino acids, and the pseudoalkaloids include compounds which are structurally related to other classes of natural products, such as the terpenes. The A. in the narrower sense can be subdivided into derivatives of ornithine, lysine, phenylalanine, tryptophan and anthranilic acid. In the classification given in the table, both structural and biogenetic features are taken into consideration.

Alkaloids

Occurrence. Probably 10 to 20% of all higher plants contain A.; they are particularly frequent in some families of dicotyledons. Closely related families often produce similar A. A plant usually contains a mixture of A. of similar structure (primary and secondary A.) in the form of hydrophilic salts dissolved in the vacuolar sap. A. are found in all parts of the plants, but are particularly abundant in the seeds, bark and roots. Heterocyclic compounds similar to the A. are also found in many microorganisms and a few animals (e.g. salamander alkaloids).

Biosynthesis: A. are the end products of secondary metabolism, and are not subject to significant degradation. They are accumulated because the plant has no excretory organs. Most A. are derivatives of amino acids (Table), which provide the heterocyclic nitrogen. In addition, acetic acid, mevalonic acid and one-carbon units may be involved in the synthesis.

The connection between amino acid metabolism and A. synthesis was discovered around the turn of the century. A new era was opened by experiments attempting to synthesize A. under physiological conditions, i.e. with native reactants, without high pressure or temperature, and at neutral pH. The biosynthesis of A. in vivo was traced by administration of amino acids labelled with ^{13}C, ^{14}C, ^{15}N or ^{3}H. It

Table. The most important classes of alkaloids and their precursors

Class of alkaloid	Structural type (main precursor emphasized)	Biogenetic precursors
Pyrrolidine	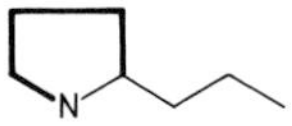 *(structure)*	Ornithine and Acetate
Pyrrolizidine	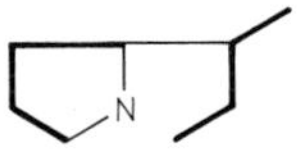 *(structure)*	Ornithine
Tropane	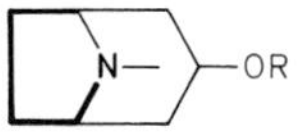 *(structure)* — OR	Ornithine and Acetate
Piperidine Conium	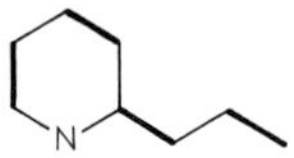 *(structure)*	Acetate
Punica, Sedum and Lobelia	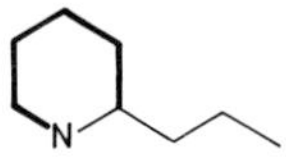 *(structure)*	Lysine, Acetate or Phenylalanine
Quinolizidine	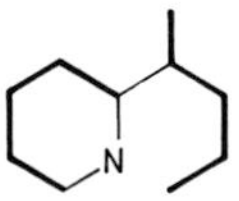 *(structure)*	Lysine
Isoquinoline	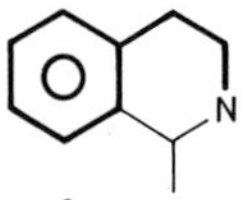 *(structure)*	Phenylalanine or Tyrosine
Indole	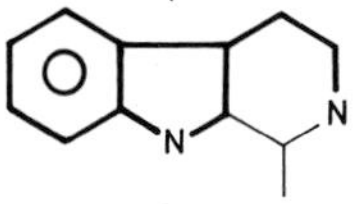 *(structure)*	Tryptophan
Rutaceae	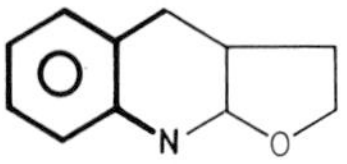 *(structure)*	Anthranilic acid
Terpene	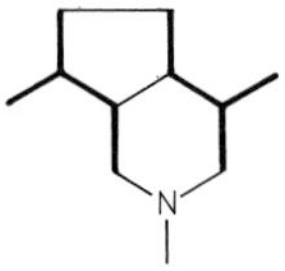 *(structure)*	Mevalonic acid

was found that plants create the very large number of structures from only a few components, using only a few mechanisms of cyclization: *N*-heterocyclic rings are made by Mannich condensation, or by the formation of amides or Schiff's bases. Secondary cyclizations, i. e. ring closures not involving nitrogen, are the result of oxidative coupling (phenol oxidation).

The biosyntheses of many A. have been studied by isotope techniques, but little is known about the enzyme systems. The A. which are built up entirely from acetic or mevalonic acid (conium and terpene alkaloids) occupy a special position: in their case, neither the source of the nitrogen nor the form of ring closure is known.

Synthesis: The first synthesis of an A. was reported in 1886 by Ladenburg, who generated coniine from α-picoline. On the hypothesis that A. are synthesized in the plant as amino acid derivatives, Robinson and Schöpf developed corresponding synthetic pathways and tested them under physiological conditions (see Tropinone). These studies were also fruitful for laboratory syntheses, and some compounds were synthesized for the first time by routes similar to the biosyntheses. Most A. for medical use are prepared from plant sources.

Biological and economic significance: A general explanation of the biological significance of A. cannot yet be given. Their protective function against consumption by animals has been proven in only a few cases, but the fact that most insects are limited to one or a few plant species may be due to their alkaloid contents. Many A. have a strong and very specific effect on certain centers of the nervous system (e.g. opiates; see Endorphins). Therefore A. are widely used therapeutically, as combinations of pure compounds, as extracts of total alkaloids, or as synthetic analogs. However, their use is often accompanied by side effects, primarily toxicity and narcosis.

Historical. Most of the alkaloid plants were known very early in folk medicine for their toxicity or useful pharmacological properties. Morphine was first isolated from poppies as the "sleep-inducing principle" by F. W. Sertuerner in 1806. The term A. was coined in 1819 by C. F. W. Meissner.

Alkalosis: an increase in the pH of body fluids. It is corrected by excretion of bicarbonate and retention of acid (see Buffer, section on Buffers of body fluids.) There are two types of alkalosis:

a) *Metabolic alkalosis*, which is caused by an increase in the bicarbonate fraction, with little or no change in the carbonic acid fraction. There are several possible causes: 1) Ingestion of alkaline drugs, e.g. $NaHCO_3$ for treatment of peptic ulcer may produce A. 2) Excessive vomiting of stomach contents only (and not the alkaline contents of the intestine) results in the loss of HCl secreted by the gastric mucosa. HCO_3^- replaces the lost Cl^-, leading to A., also called hypochloremic alkalosis. Vomiting of stomach contents only is a feature of pyloric obstruction in newborns, caused by hypertrophy of the pyloric sphincter muscle. 3) Excess secretion of aldosterone by the adrenal glands promotes increased reabsorption of Na^+ from the distal tubules of the nephrons. Since this process is coupled with increased secretion of H^+, the result is A.

b) *Respiratory alkalosis* is caused by a decrease in the carbonic acid fraction, with little or no change in the bicarbonate. It can result from hyperventilation, which may occur in central nervous system diseases affecting the nervous control of breathing, and in early stages of salicylate poisoning. It can also be promoted by voluntary hyperventilation.

Alkannin: see Napthoquinones (table).

Alkylating agents: chemical compounds which can donate alkyl groups, usually methyl or ethyl. Monofunctional A. a., like dimethylsulfate or ethylmethanesulfonate, can transfer only a single functional group, while bifunctional A. a., like mustard gas, nitrogen mustard gas or cyclophosphamide, can react with several molecules or parts of a macromolecule, thus cross-linking them. A. a. are frequently carcinogenic and mutagenic, but some of them are nevertheless used in chemotherapy of cancer (see Mitomycin C). The Activated one-carbon units (see) are biological A. a. Chemical A. a. have been widely used in laboratory synthesis of macromolecules, to protect reactive groups, and in the elucidation of Active centers (see) of enzymes and receptor molecules.

Allantoic acid: diureidoacetate, a degradation product of allantoin in aerobic Purine degradation (see) and in anaerobic allantoin degradation. M_r 176.1.

Allantoin: 5-ureidohydantoin, glyoxyldiureide, an intermediate in aerobic purine degradation. M_r 158.1. A. was discovered in 1799 in the allantoic fluid of the cow. It is the end product of purine metabolism in most mammals and some reptiles, and is excreted in their urine. A. is also found widely in plants. In a number of plant families known as ureide plants, A. is abundant in the soluble nitrogen pool. In certain species of bacteria (*Arthrobacter allantoicus* and *Streptococcus allantoicus*), A. can serve as C, N and energy source under anaerobic conditions. It is first converted to allantoic acid by allantoinase (EC 3.5.2.5), and the allantoic acid is degraded by allantoate deiminase (EC 3.5.3.9) to ureidoglycine, NH_3 and CO_2 (anaerobic allantoin degradation).

Allantoin

Allantoinase (EC 3.5.2.5): see Purine degradation (aerobic).

Allelochemicals: see Pheromones.

Allen-Doisy test: see Estrogens.

Allergy: a hypersensitivity of the immune apparatus, a pathological immune reaction induced either by antibodies (immediate hypersensitivity) or by lymphoid cells (delayed type A.). Unlike the delayed type, immediate hypersensitivity can be passively transmitted in the serum. The symptoms begin shortly after contact and decay rapidly, but the delayed type symptoms do not reach their maximum for 24 to 48 hours, and decline slowly over a period of days or weeks. Examples of this type of A. are anaphylaxis, the Arthus reaction and serum sickness. The best known A., anaphylaxia, can occur as a local (cutaneous) reaction (for example a rash with blisters) or as a systemic reaction (anaphylactic shock). Asthma, hay fever and nettle rashes are also examples of local anaphylactic reactions which are induced by reagins (see Immunoglobulins: IgE). Only primates can be sensitized by injection with human reagins. An example of the delayed type A. is the tuberculin reaction, which is based on a cellular immune response.

Allocholane: outdated term for 5α-cholane, see Steroids.

Allodeoxycholic acid: 3α,12α-dihydroxy-5α-cholan-24-oic acid, one of the Bile acids (see). A dihydroxy steroid carboxylic acid, M_r 392.6. Unlike most other bile acids, A. has an A/B-*trans* ring coupling. It was isolated from the bile and feces of rabbits.

Allogibberellic acid: see Gibberellins.

Allomones: see Pheromones.

Allophanate hydrolase: see Urea amidolyase.

Allophanic acid: see Urea amidolyase.

Allopregnane: outdated term for 5α-pregnane, see Steroids.

All-or-nothing model: see Cooperativity model.

Allosteric enzymes: see Cooperative oligomeric enzymes.

Allostery: the phenomenon of changes in conformation of proteins with quaternary structure upon binding to certain low-molecular-weight ligands. A. has an important role in enzyme regulation (see Effectors) and in the uptake of oxygen by hemoglobin.

Alloxan: a compound first discovered by Liebig in mucus secreted during dysentery. It is used to produce experimental diabetes in animals, as it preferentially attacks pancreatic β-cells.

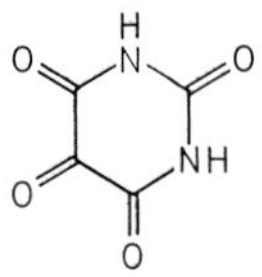

Alloxan

Alnulin: see Taraxerol.

ALS: abb. for Antilymphocyte serum. This has been largely replaced by monoclonal antibodies specific for cell-surface determinants of particular types of lymphocytes, such as T-cells, B-cells or macrophages.

Alternaria alternata toxins: *Alternaria alternata* is a fungus which causes a number of plant diseases. There are several different pathotypes, each infecting a different species and producing different host-specific phytotoxins; several of these have been characterized.

The pathotype attacking apples produces a toxin consisting of at least 6 components, of which two have been identified: the depsipeptide **alternariolide** (Fig. 1) and its demethoxy derivative. In susceptible

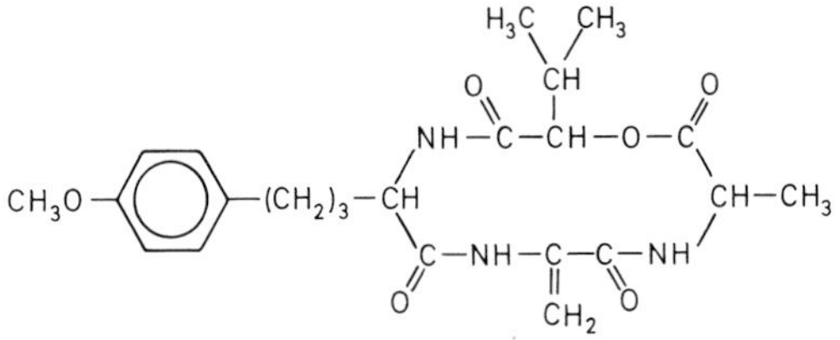

*Figure 1. **Alternariolide** [S. Lee (1976) Tetrahedron Lett. No. 11, 843–846.]*

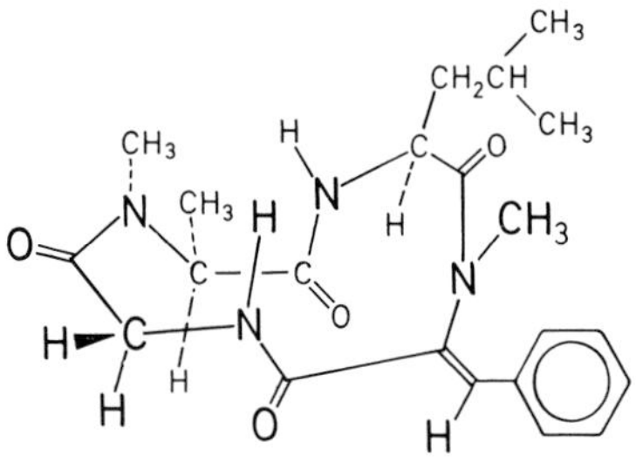

*Figure 2. **Tentoxin** [D. H. Rich & P. K. Bhatnagar (1978) J. Am. Chem. Soc. **100** 2212–2218.]*

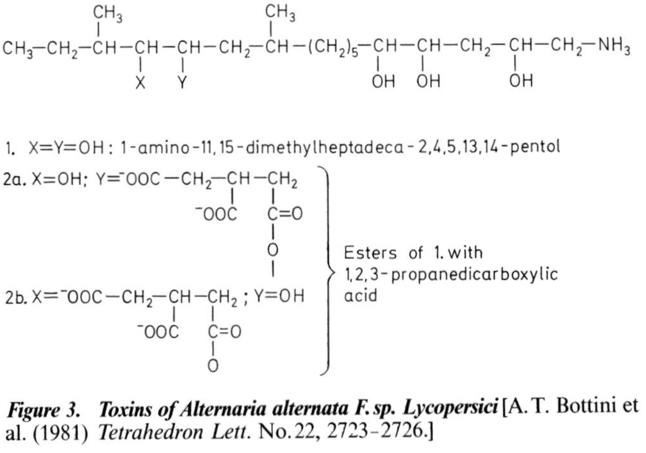

1. X=Y=OH: 1-amino-11,15-dimethylheptadeca-2,4,5,13,14-pentol

2a. X=OH; Y=⁻OOC—CH₂—CH—CH₂
                              | |
                           ⁻OOC   C=O
                                       |
                                       O

2b. X=⁻OOC—CH₂—CH—CH₂ ; Y=OH
                 | |
              ⁻OOC   C=O
                          |
                          O

Esters of 1. with 1,2,3-propanedicarboxylic acid

*Figure 3. **Toxins of Alternaria alternata F. sp. Lycopersici** [A. T. Bottini et al. (1981) Tetrahedron Lett. No. 22, 2723–2726.]*

apple trees, alternariolide causes interveinal chlorosis at nM concentrations, whereas about 1 μM is required in resistant varieties. The toxin causes immediate release of electrolytes, suggesting the plasmalemma as the toxin-sensitive site.

Various pathotypes also produce a cyclic tetrapeptide known as **tentoxin** (Fig. 2), which induces chlorosis in many plants (e.g. lettuce, potato, cucumber, spinach), but not in *Nicotiniana*, tomato, cabbage or radish. It binds to chloroplast coupling factor 1 (one toxin-binding site per αβ subunit complex), inhibiting photophosphorylation and the Ca^{2+}-dependent ATPase. Species specificity is due to different binding affinities for the coupling factor 1 (1.3 to 20 × 10^{-7}M for 50% inhibition in sensitive species, and 20-fold higher for insensitive species).

The pathotype which causes tomato stem canker produces 3 related toxins (Fig. 3). Aspartate and certain products of aspartate metabolism (e.g. orotic acid) protect tomato plants against the toxins, and it is thought that toxicity may be due to the inhibition of aspartate transcarbamylase. Susceptibility to the disease is controlled by a single genetic locus with two alleles.

Amanitin: see Amatoxins.

Amarinthin: a red dye belonging to the betacyanin group. In place of the glucose residue found in betanin, A. has a glycosidically linked disaccharide 5-O-(β-D-glucopyranosyl uronic acid)-5-O-β-D-glucopyranoside. It is found in *Amaranthus* species, for example the foxtail *Celosia argentea*.

Amaryllidaceae alkaloids: a group of complicated alkaloids found only in the plant family *Amaryllidaceae*. They can be classified as phenylisoquinoline alkaloids (see Isoquinoline alkaloids), because their biosynthesis (Fig.) is similar to that of the isoquinoline alkaloids, beginning with phenylethylamine or tyramine and a carbonyl compound. The final structures are generated by secondary ring cleavage and cyclization. The last step of the biosynthesis is catalysed by a phenol oxidase. The main alkaloid galanthamine is isolated from Caucasian snowdrops, *Galanthus woronowii*, and is used therapeutically as an inhibitor of acetylcholinesterase.

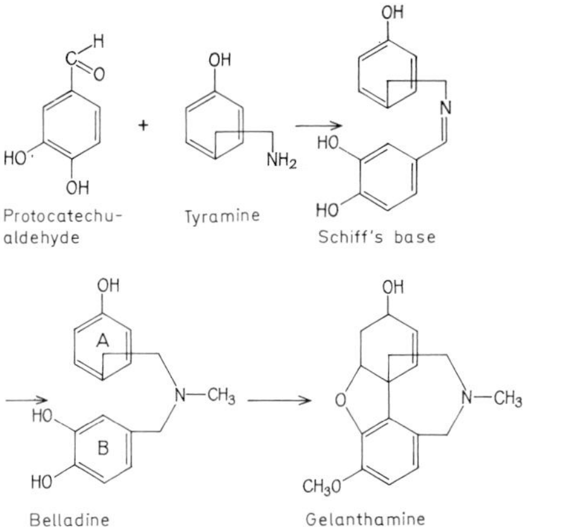

Amatoxins: a group of bicyclic octapeptides which, together with the phallatoxins, are the most important poisons in the death cap fungus, *Amanita phalloides* (Fig.).

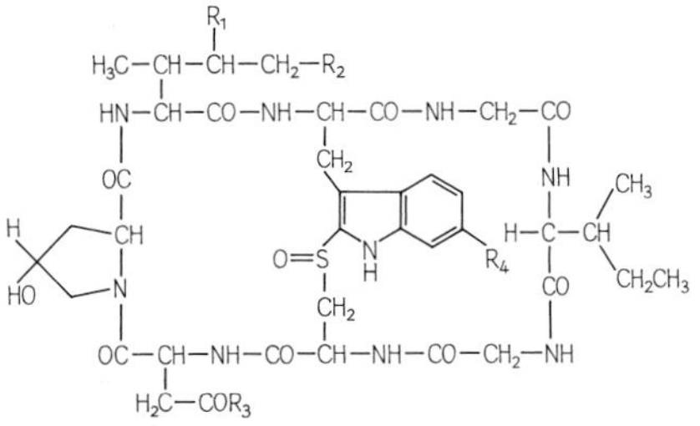

	R_1	R_2	R_3	R_4
α-Amanitin	OH	OH	NH_2	OH
β-Amanitin	OH	OH	OH	OH
γ-Amanitin	OH	H	NH_2	OH
Amanin	OH	OH	OH	H
Amanullin	H	H	NH_2	OH

Amatoxins

These poisons inhibit the nucleoplasmic RNA polymerase II (EC 2.7.7.6) of eukaryotic cells, which leads to necrosis of liver and kidney cells. The poisonous effects of the A. and phallatoxins can be inhibited by a simultaneous application of antamanid. More than 90% of fatal mushroom poisonings are due to consumption of *Amanita phalloides* and related species. The structure of the amanitins was established by Th. Wieland.

Ambanol: the only known naturally occurring isoflavonol (Fig.). A. occurs in the roots of *Neorautanenia amboensis*. [M. E. Oberholzer et al. (1977) *Tetrahedron Lett.* 1165–1168.]

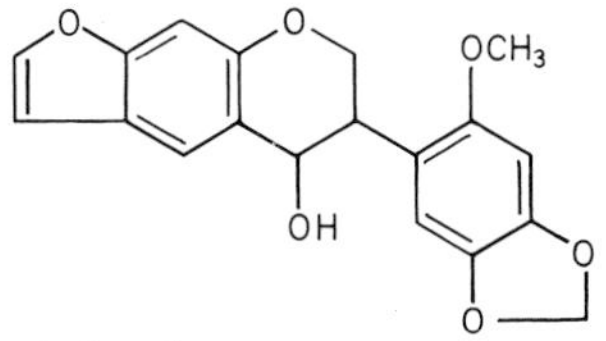

Ambanol

Amber codon, *nonsense codon*: the sequence UAG in a mRNA. It does not code for any of the 20 proteogenic amino acids, and it results in the premature termination of protein synthesis. It may be formed by mutation of a sense codon; potential precursors are the codons UCG (serine), UAU and UAC (tyrosine) and CAG (glutamine).

Amber mutants: mutant bacteria in which the mRNA contains the codon UAG because of a point mutation (see Amber codon). The mutation is not necessarily lethal, because a compensatory suppressor mutation in a tRNA may enable the protein synthesizing system to recognize the amber codon as a sense codon. The term "amber" was arbitrarily chosen by the discoverer of the mutants.

Biosynthesis of belladine and galanthamine alkaloids

Amentoflavone: see Biflavonoids.

Amicetins: pyrimidine antibiotics (see Nucleoside antibiotics) synthesized by various *Streptomyces* species. A. A is formed by *S. fasciculatus* and *S. vinaceusdrappus*. M_r 618.7. It is primarily bacteriostatic, especially against Gram-positive bacteria. 0.5 µg A. A/ml inhibits the growth of *Mycobacterium tuberculosis*.
A. B (plicacetin) was isolated from *Streptomyces plicatus* by Snesi (1957) and Haskell (1958). It is presumed to be the precursor of A. A. M_r 517.6. Its spectrum is the same as that of A. A, but its effect is weaker.
Cytimidine, M_r 331.3, is a degradation product of A. It consists of cytosine, 4-aminobenzoic acid and 2-methyl-D-serine.

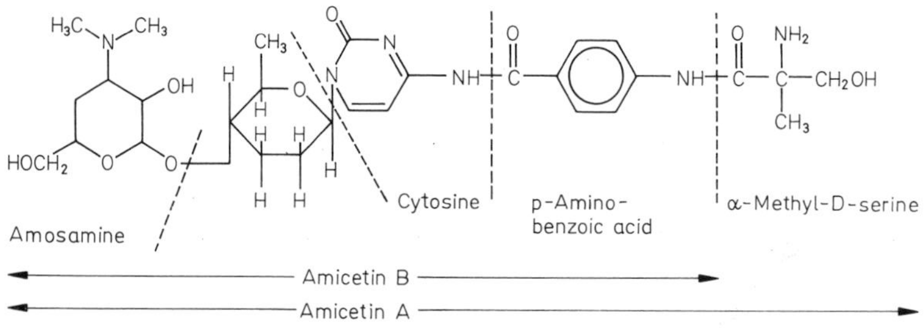

H^+. The latter is inhibited by A., hence the designation "amiloride-sensitive Na^+ pump". [J. B. Smith & E. Rozengurt (1978) *Proc. Natl. Acad. Sci. USA* **75** 5560–5564; E. Rozengurt (1981) *Adv. Enz. Regul.* **19** 61–85]

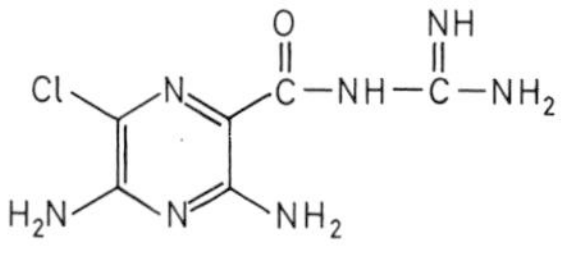

Amination: the introduction of the amino group (-NH_2) into an organic carbon compound. It may be accomplished either by reductive A. or transamination. Reductive A. requires a reduced pyridine nucleotide as reducing agent. The most common reductive A. of 2-oxoglutarate by L-glutamate dehydrogenase (EC 1.4.1.4) requires NADPH.

Amine precursor uptake decarboxylase system, *APUD system:* the nervous system and part of the gastrointestinal tract derived from the neuroectodermal cells of the primitive neural crest. Gut and

Structures of amicetins A and B

Amidinotransferases, *transamidinases,* (EC 2.1.4): a group of enzymes catalysing transamidination. They are involved in the biosynthesis of creatine. An A. from *Streptomyces griseus* and *S. baikiniensis* is involved in the biosynthesis of streptidine. In addition, the presence of A. of unexplored substrate specificity in the biosynthesis of guanidine derivatives has been reported. The transfer of the intact amidine group by A. was demonstrated by double labelling with ^{14}C and ^{15}N of the amidine group of L-arginine. This A. has both transferase and hydrolase activities, and is in this sense a possible arginase.

Amiloride: a drug which inhibits the influx of Na^+ into cells. It was discovered as a natriuretic which increases Na^+ excretion but does not affect K^+ excretion [Baer et al. (1967) *J. Pharmacol. Exp. Ther.* **157** 472]. Animal cells have two systems for Na^+ transport. The first is Na^+/K^+ ATPase, which exports Na^+ and imports K^+, each against its concentration gradient. (The extracellular concentration of Na^+ is much higher than the intracellular concentration, while K^+ is far more concentrated inside the cell than outside.) This Na^+/K^+ pump is inhibited by ouabain. In addition, there is a more recently discovered pump which imports Na^+ and exports

brain produce several identical peptides, e.g. bombesin, cholecystokinin, neurotensin, substance P, enkephalin, gastrin, vasoactive intestinal peptide, somatostatin, etc., which reflects the common embryological origin of the two tissues. These peptides function specifically in both gut and nervous system, but their functions are generally distinct, because the peptides are largely excluded by the blood-brain barrier.
APUD cells are descended from an ancestral neuron; although the axon has been lost, innervation is still possible via synaptic terminals, or nervous control may be only indirect. Six modes of APUD cell expression are recognized: neurocrine (into neurons), neuroendocrine (via axons), endocrine (into the blood stream), paracrine (into the intercellular space), epicrine (into somatic cells), and exocrine (to the externum).
The APUD concept of a diffuse neuroendocrine system has been defined as follows by A. G. E. Pearse [in *Centrally Acting Peptides* (1978), J. Hughes, ed., MacMillan]: "The cells of the APUD series, producing peptides active as hormones or as neurotransmitters, are all derived from neuroendocrine-programmed cells originating from the ectoblast. They constitute a third (endocrine or neuroendocrine) division of the nervous system whose cells act as third line effectors to support, modulate or amplify the actions of neurons in the somatic and autonomic divisions, and possibly as tropins to both neuronal and non-neuronal cells."

Aminoacetic acid: see Glycine.

Amino acid activating enzymes: see Aminoacyl-tRNA synthetases.

Amino acid analyser: see Proteins.

Amino acid oxidases: see Flavin enzymes.

Amiloride

Amino acid reagents: reagents for the colorimetric identification and quantitation of amino acids. One of the most important is the ninhydrin reaction, in which a blue-violet dye, called Ruhemann's purple (absorbance maximum at 570 nm, for proline at 440 nm) is formed by the reaction of 2,2-dihydroxy-1H-indene-1,3(2H)-dione (ninhydrin) with the amino acid.

In the fluorescamine technique, the amino acids are converted by reaction with 4-phenyl[furan-2H (3H)-1'-phthalane]-3,3'-dione (fluorescamine) into strongly fluorescing compounds which can be detected even in nanomole quantities at 336 nm. The reagent itself is not fluorescent, and in contrast to ninhydrin, it is not sensitive to ammonia (Udenfriend, 1972).

Other highly sensitive reagents are 2,4,6-trinitrobenzosulfonic acid, 1,2-naphthoquinone-4-sulfonic acid (Folin's reagent) and 4,4'-tetramethyldiaminodiphenylmethane (TDM). Intensely fluorescing amino acid derivatives are formed by reaction with *o*-phthalaldehyde in the presence of reducing agents; with pyridoxal and zinc^{2+} ions; and with dansyl chloride (5-dimethylaminonaphthalene sulfonyl chloride).

Amino acids: aminocarboxylic acids, organic acids carrying amino groups, usually not more than two. A.a. are classified as α, β, γ, ..., depending on the position of the carbon bearing the -NH$_2$ group with respect to the -COOH (Fig. 1).

The α-A.a., as the components of proteins and peptides, but also in their free form, are one of the most important classes of organic substances in the cell.

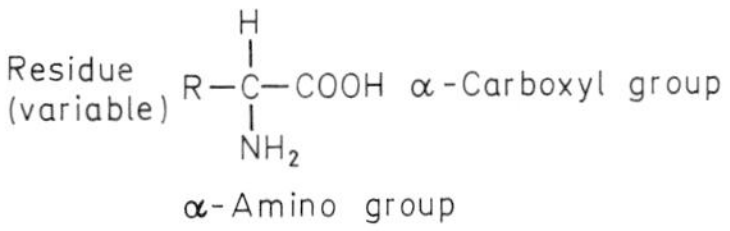

Figure 1. Structure of an α-amino acid

The amino acids are classified as acidic or basic, depending on their isoelectric points, or, depending on the nature of their side chains, they are divided into four groups, indicated in Table 1 by Roman numerals I-IV:

I. A.a. with neutral, hydrophobic (non-polar) side chains

II. A.a. with neutral and hydrophilic (polar) side chains

III. A.a. with acid and hydrophilic (polar) side chains

IV. A.a. with basic and hydrophilic (polar) side chains.

In addition to this chemical classification, A.a. can be divided according to their degradation products into glucogenic and ketogenic A.a. Glucogenic A.a. can be degraded to C$_4$-dicarboxylic acids or pyruvate, which are intermediates in the tricarboxylic acid cycle. This cycle can also provide oxaloacetate for Gluconeogenesis (see), so the carbon chains from this group of A.a. can be incorporated into glucose. Ketogenic A.a., on the other hand, are degraded to ketones which are not part of the tricarboxylic acid cycle, in particular acetoacetic acid. Finally, the A.a. can be classified as essential (indispensible) and

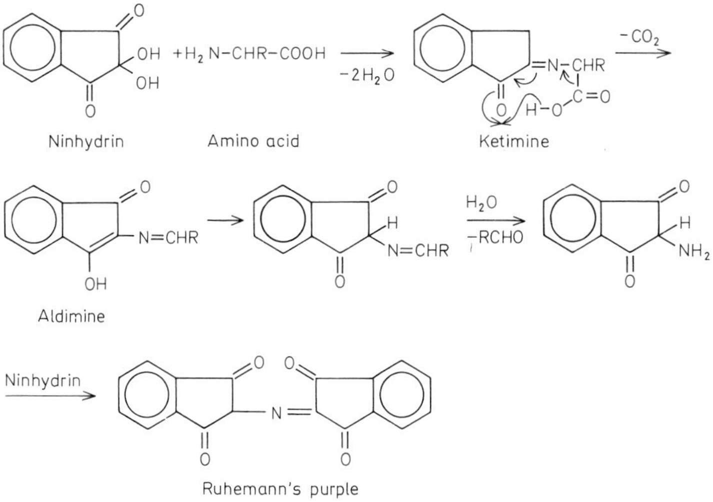

The ninhydrin reaction

Twenty A.a. are encoded in messenger RNA (see Protein biosynthesis, Genetic code); these are commonly found in proteins and are known as proteogenic or proteinogenic A.a. (Table 1). The occurrence of nonproteogenic A.a. residues in proteins is due to Post-translational modification of proteins (see). For further information on individual A.a., the separate entries should be consulted.

nonessential (dispensible), depending on whether the organism in question is able to synthesize them in amounts adequate for its needs. Essential amino acids must be supplied by the diet, and an inadequate supply leads to negative Nitrogen balance (see). Half-essential A.a. are those which can be synthesized, but in insufficient quantity for all physiological requirements, e.g. growth. An inadequate

Table 1. *Proteogenic amino acids.* The three-letter abbreviations are universally accepted and are routinely used for depicting protein and peptide sequences. (See Peptides.) One-letter notations (recommended by the IUPAC–IUB Commission on Biochemical Nomenclature [*Eur. J. Biochem.* **5** (1968) 151–153] should not be used for the publication of sequences, but are intended to facilitate storage of sequence information and sequence comparisons by computer.

Amino acid	Abb. 3-letter	1-letter	Class
L-Alanine	Ala	A	I
L-Arginine	Arg	R	IV
L-Asparagine	Asn*	N*	II
L-Aspartic acid	Asp*	D*	III
L-Cysteine	Cys	C	II
L-Glutamic acid	Glu	E	III
L-Glutamine	Gln	Q	II
Glycine	Gly	G	I
L-Histidine	His	H	IV
L-Isoleucine	Ile	I	I
L-Leucine	Leu	L	I
L-Lysine	Lys	K	IV
L-Methionine	Met	M	I
L-Phenylalanine	Phe	F	I
L-Proline	Pro	P	I
L-Serine	Ser	S	II
L-Threonine	The	T	II
L-Tryptophan	Trp	W	I
L-Tyrosine	Tyr	Y	II
L-Valine	Val	V	I
Unknown or other	–	X	

* When it is not known whether the a.a. in the original protein was Asn or Asp, the abb. Asx or B is used; Glx or Z indicates Glu or Gln. The ambiguities arise because chemical hydrolysis of the peptide bonds between amino acids also converts the amide of the Asn or Gln side chain to the corresponding acid.

human requirements for essential amino acids are listed in Table 2.

Except for glycine, all A.a. contain at least one asymmetric C atom, and therefore occur in nature as optically active compounds. With a few exceptions, natural A.a. have the L configuration. D-A.a. occur in the cell walls, capsules and culture media of some microorganisms, and in many antibiotics. A.a lacking the amino group in the α-position (β-, γ-, δ-A.a., e.g. β-alanine) occur as the free acids or as components of natural products, but not in proteins.

Properties. A.a. are amphoteric, since they carry both NH_2 and COOH groups, and their solutions are ampholytes. In the solid state and in strongly polar solvents, they are zwitterions, H_3^+N-CHR-COO$^-$. With a few exceptions, they are highly soluble in water, ammonia and other polar solvents, but barely so in nonpolar and less polar solvents such as ethanol, methanol and acetone. A.a. with hydrophilic side chains are more soluble in water. The water solubility of an A.a. is lowest at its isoelectric point, because the dominating zwitterionic form reduces the hydrophilic property of the amino and carboxyl groups.

The dissociation of the A.a. is strongly dependent on the pH value of the solution, and the zwitterionic form is only present between pH 4 and 9. In the more acid range, the A.a. are present as cations, H_3^+N-CHR-COOH, and in more alkaline solutions, as anions H_2N-CHR-COO$^-$. The titration curves of the A.a. therefore show two buffering zones, and are also affected by the dissociation behavior of the side chains, especially the acidic and basic ones (Fig. 2).

The acid-base behavior of the A.a. is a model for that of peptides and proteins, and also the basis for separation by electrophoresis and ion-exchange chromatography. The UV absorption of A.a. with chromophoric side chain functions (e.g. tryptophan, tyrosine and phenylalanine) makes it possible to de-

Amino acids formed by post-translational modification, and present only in special proteins

Amino Acid	Occurrence
δ-Hydroxy-L-lysine	Fish collagen
L-3,5-Dibromotyrosine	Skeleton of *Primnoa lepadifera* (coral)
L-3,5-Diiodotyrosine	Skeleton of *Gorgonia cavolinii* (coral)
L-3,5,3′-Triiodothyronine	Thyroglobulin (tissue protein in the thyroid gland)
L-Thyroxin	Thyroglobulin
Hydroxy-L-proline	Collagen, gelatins

dietary supply of any one of the essential amino acids inhibits protein synthesis, because the ribosome-mRNA-nascent polypeptide complex must suspend its operation at the point where the missing amino acid should be incorporated. The other amino acids then accumulate and are shunted into degradative metabolic pathways; hence the loss of nitrogen. The

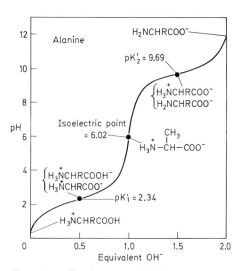

Figure. 2a. *Titration curve of alanine*

Table 2. Minimal requirements of human beings for essential amino acids in mg per kg and day

	Ile	Leu	Lys	Phe(Tyr)	Met	Cys	Thr	Trp	Val
Child	90		90	90[1]	85		60	30	85
Man	10.4	9.9	8.8	4.3[2]	1.5	11.6	6.5	2.9	8.8
				13.3[3]	13.2	0			
Woman	5.2	7.1	3.3	3.1[4]	4.7	0.5	3.5	2.1	9.2
WHO Norms	3.0	3.4	3.0	2.0[5]	1.6	1.4	2.0	3.0	

[1]in the presence of L-tyrosine; [2]Tyr 15.9 mg/kg: Tyr/Trp = 5.5; [3]in the absence of L-tyrosine; [4]Tyr 15.6 mg/kg: Tyr/Trp = 7.4; [5]Tyr 5.0 mg/kg: Tyr/Trp = 2.0. WHO is the World Health Organization.

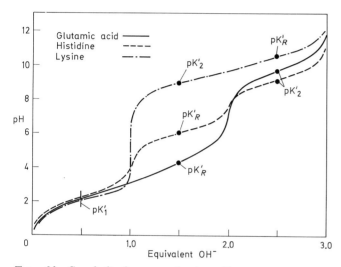

Figure. 2b. Sample titration curves of amino acids

termine them quantitatively, both as the free A.a. and in proteins and peptides. The proteogenic A.a. in the cell make up the amino acid pool (or pools in the case of compartmentation), into which the A.a. from nutritional sources, proteolysis and de novo synthesis mix. This pool also includes nitrogen-containing precursors and intermediates of the proteogenic A.a. and the nonproteogenic A.a.

The proteogenic A.a. can be grouped into families according to the biosynthetic sources of their carbon skeletons:

1. The serine family includes the A.a. derived from triose phosphate: serine, glycine, cysteine and cystine;
2. The ketoglutarate family is made up of those A.a. whose skeletons are derived from oxoglutarate supplied by the tricarboxylic acid cycle: glutamate, glutamine, ornithine, citrulline, arginine (see Urea cycle), proline and hydroxyproline;
3. The pyruvate family is derived from pyruvate and oxaloacetate (Fig. 3);
4. The pentose family includes histidine and the three aromatic A.a. (see Aromatic biosynthesis), phenylalanine, tyrosine and tryptophan. The biosynthetic pathways of the individual A.a. include various nonproteogenic A.a. as intermediates.

The microbial synthesis of A.a. is technically exploited. The microbes (usually bacteria) are grown on synthetic medium. Because of mutations, the strains used have defects in the control of their metabolic pathways which cause massive overproduction (from the organism's point of view) of the A.a. of interest. The A.a. is excreted into the medium and can be harvested from it. [Herman J. Phaff, "Industrial Microorganisms", in *Scientific American* **245** (1981) 76–89.]

The A.a. in the cell's amino acid pool are used for the synthesis of new proteins, or they are degraded, or they may serve as the precursors of special metabolites like hormones. The major metabolic pathways are 1. transamination to 2-oxoacids, 2. decarboxylation, 3. transformation of the side chain, and 4. oxidative deamination to 2-oxoacids (Table 3.)

Nonproteogenic A.a. (nonprotein A.a.) are usually not incorporated into proteins. They include A.a. which are intermediates in the biosynthesis of proteogenic A.a., for example δ-aminoadipic acid, diaminopimelic acid and cystathionine. At present about 200 nonprotein A.a. are known, most of them occurring in plants and limited in each case to certain taxonomic groups. The majority can be grouped according to their biosyntheses into the four groups

29

Amino acids

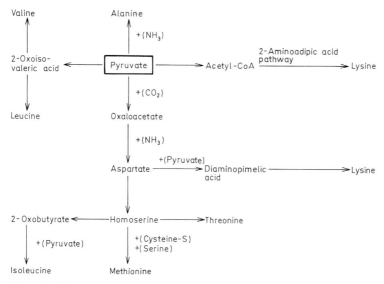

Figure 3. *The pyruvate family of amino acids*

Table 3. *Metabolic reactions of amino acids*

Type of reaction	Equation
Transamination	$\underset{NH_2}{RCHCOOH} + \underset{O}{R'CCOOH} \rightleftharpoons \underset{O}{RCCOOH} + \underset{NH_2}{R'CHCOOH}$
Decarboxylation	$\underset{NH_2}{RCHCOOH} \longrightarrow RCH_2NH_2 + CO_2$
Amination	$\underset{O}{RCCOOH} + NH_3 + NAD(P)H_2 \longrightarrow \underset{NH_2}{RCHCOOH} + NAD(P)$
Deamination	$\underset{NH_2}{RCHCOOH} \xrightarrow{-2[H]} \underset{NH}{RCCOOH} \xrightarrow{+H_2O} \underset{O}{RCCOOH} + NH_3$
Modification of side chain Hydroxyl group α-Amino group	$R-OH \xrightarrow{ATP} R-O \sim PO_2H_2$ (phosphorylation) $R-NH_2 \longrightarrow R-NH-COCH_3$ (acetylation)
α-Carboxyl group	$R-COOH \xrightarrow[NH_3]{ATP} R-CONH_3$ (amide synthesis)
Peptide formation	$\underset{NH_2}{RCHCOOH} + \underset{NH_2}{R'CHCOOH} \xrightarrow{-H_2O} \underset{NH_2}{RCHCO}-\underset{R}{NHCHCOOH}$
Amino acid activation (Protein biosynthesis 1)	$\underset{NH_2}{RCHC}\overset{O}{-}OH + AMP \sim P \sim P \xrightarrow{Enz.} Enz. AMP \sim \underset{NH_2}{CCHR} + PP_i$
Amino acid transfer (Protein biosynthesis 2) (Synthesis of aminoacyl-tRNA)	$Enz. AMP \sim \underset{NH_2}{\overset{O}{CCHR}} + tRNS-OH \longrightarrow AMP + Enz. + tRNS-O \sim \underset{NH_2}{\overset{O}{CCHR}}$
Gramicidin S synthesis	$Enz. AMP \sim \underset{NH_2}{\overset{O}{CCHR}} + E^{-SH} \longrightarrow AMP + Enz. + E^{-S} \sim \underset{NH_2}{COCHR}$

Enz. = Enzyme; E = Protein II of gramicidin S-synthetase

30

of biogenetically related A.a. In exceptional cases, A.a. which are not normally considered to be proteogenic are incorporated during translation, for example L-citrulline in the protein of hair follicles and δ-aminoadipic acid in maize protein. The occurrence of the rare natural A.a. can be used in chemitaxonomy. The terms "rare" or "unusual" refer only to their sporadic occurrence and their structural differences from proteogenic A.a. However, they are structurally related to the proteogenic A.a. More than 20 nonprotein A.a. are known which differ from alanine by substitution of one hydrogen of the methyl group. Many other rare, naturally occurring A.a. are structurally related to proteogenic A.a. Nonprotein A.a. may also act as A.a. antagonists, e.g. azetidine-2-carboxylic acid, a toxic constituent of lily of the valley, is a structural analog of proline, in which the ring is contracted by one C-atom. In lily of the valley, uncontrolled incorporation of azetidine-2-carboxylic acid into the plant's own protein is avoided by the highly specific synthesis of prolyl-tRNA, but in other organisms azetidine-2-carboxylic acid becomes incorporated in place of proline, leading to marked alterations in the tertiary structure and biological activity of proteins. Nonprotein A.a. are particularly common in certain plant families, e.g. the *Mimosaceae* contain 2-diaminopropionic acid and its derivatives, thioether derivatives of L-cysteine, and derivatives of lysine and glutamic acid. Some nonprotein A.a. are biologically active, e.g. Lathyrogenic A.a. (see) and indospicine (L-2-amino-6-amidinocaproic acid) from *Indigofera spicata*, which acts as a liver toxin and causes growth deformities.

New A.a. are usually discovered by chromatography, on the basis of their unusual R_f values or unusual color reactions, e.g. with ninhydrin.

Aminoacyladenylate, *activated amino acid:* the product of the first enzymatic reaction of protein biosynthesis. It consists of an amino acid linked by an acid anhydride bond to the phosphate of AMP. In the cell, these compounds are always associated with aminoacyl-tRNA synthetases, which also catalyse a further reaction, the transfer of the amino acyl residue of the A. to a specific tRNA. AMP is released in this second step. See Aminoacyl-tRNA synthetases.

Aminoacyl-tRNA: a transfer RNA charged with a specific amino acid; the transport form in which the amino acid is brought to the specific acceptor site on the ribosome. The carboxyl group of the amino acid is esterified to either the 2′ or the 3′ OH group of the ribose of the terminal adenosine of the tRNA. The free energy of hydrolysis of this ester bond is 29.0 kJ/mol. See High-energy bonds; Aminoacyl-tRNA synthetases.

Aminoacyl-tRNA synthetases, *amino-acid activating enzymes* (EC 6.1.1): a group of enzymes which activate amino acids and transfer them to specific tRNA molecules as the first step in protein biosynthesis. The process consists of two steps, illustrated here with leucine:
1) Leu + ATP + leucyl-tRNA synthetase →[Leu-AMP-enzyme] + PP_i.
2) [Leu-AMP-enzyme] + tRNA^{Leu} → leucyl-tRNA^{Leu} + AMP + enzyme.

The A. are highly specific with respect to the amino acid they activate, and they also recognize the tRNA with great precision. The mechanism by which the enzyme recognizes the appropriate tRNA is still unclear. Their accuracy in loading the correct amino acid depends on an editing process. The active site which forms the ester bond between the amino acid and the tRNA is unable to distinguish completely between the correct amino acid and a smaller homolog, i.e. valyl-tRNA synthetase may attach Ala or Thr to the tRNA^{Val}. This is compensated by the existence of a second site on the synthetase which has esterase activity and is much more active with the wrong amino acid than the right one.

A. may consist either of one polypeptide chain or of two or four homologous or heterologous subunits. Eukaryotic cells contain more than 20 different A., because the mitochondria and plastids have their own amino-acid specific A., which differ in their specificity toward homologous tRNA from those of the cytoplasm. Some A. are able to load several amino-acid-specific tRNAs, e.g. the leucyl-tRNA synthetase of *Escherichia coli*, which can load 5 tRNA$_{E.\ coli}$^{Leu}.

2-Aminoadipic acid, abb *Aad:* HOOC-CH₂-CH₂-CH₂-CH(NH₂)-COOH, an amino acid which is proteogenic only in maize. M_r161.1. Aad is an intermediate in the biosynthesis of L-lysine by the Aad pathway. The free acid cyclizes in boiling water to piperidone carboxylic acid.

2-Aminoadipic acid pathway: see L-Lysine.

4-Aminobutyrate pathway, *γ-aminobutyrate pathway:* see 4-Aminobutyric acid.

4-Aminobutyric acid, abb. *4-Abu, γ-aminobutyric acid,* abb, *GABA:* H₂N-CH₂-CH₂-CH₂-COOH, a nonproteogenic amino acid. M_r 103.1. Formation of GABA from L-glutamic acid, by the action of glutamate decarboxylase (EC 4.1.1.15), has been demonstrated in brain, various microorganisms (e.g. *Clostridium welchii, Escherichia coli*), higher plants (e.g. spinach, barley) and other animal tissues (liver and muscle). It can also be formed from 4-guanidobutyric acid (see Glutamine derivatives) in higher fungi (*Basidiomycetes*) and *Streptomycetes* by removal of urea. Degradation of GABA proceeds by transamination to succinic semialdehyde and subsequent oxidation to succinic acid, which is oxidized in the tricarboxylic acid cycle. Synthesis of GABA is particularly important in the brain, where it functions as an inhibitory neurotransmitter. On account of its neural activity, GABA is used for the treatment of epilepsy, cerebral hemorrhage, etc. The 4-aminobutyrate pathway (Fig.) represents a bypass of the oxidative decarboxylation of 2-oxoglutarate in the tricarboxylic acid cycle. Only some brain cells make GABA, and only approximately 25% of 2-oxoglutarate produced in these cells is converted to GABA. The 4-aminobutyrate pathway accounts for less than 10% of the total oxidative metabolism of the brain.

Aminocarboxylic acids: see Amino acids.

Aminocitric acid: CH(NH₂)-C(OH)-CH₂
      | | |
      COOH COOH COOH
an amino acid identified in acid hydrolysates of ribonucleoproteins from calf thymus, bovine and human spleen, *Escherichia coli* and *Salmonella typhi*.

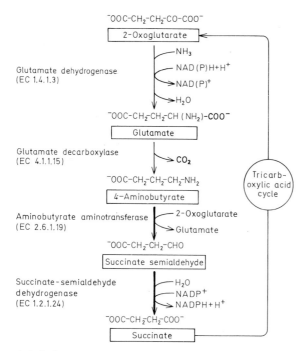

$$^-OOC-CH_2-CH_2-CO-COO^-$$

2-Oxoglutarate

Glutamate dehydrogenase
(EC 1.4.1.3)

—NH₃ ... NH_3

— NAD(P)H+H⁺

↘NAD(P)⁺

↘H₂O

$$^-OOC-CH_2-CH_2-CH(NH_2)-COO^-$$

Glutamate

Glutamate decarboxylase
(EC 4.1.1.15)

↘CO₂

$$^-OOC-CH_2-CH_2-CH_2-NH_2$$

4-Aminobutyrate

Aminobutyrate aminotransferase
(EC 2.6.1.19)

— 2-Oxoglutarate

↘Glutamate

$$^-OOC-CH_2-CH_2-CHO$$

Succinate semialdehyde

Succinate-semialdehyde
dehydrogenase
(EC 1.2.1.24)

—H₂O

—NADP⁺

↘NADPH+H⁺

$$^-OOC-CH_2-CH_2-COO^-$$

Succinate

Tricarb-
oxylic acid
cycle

4-Aminobutyrate pathway

It is an acidic amino acid, elutes before cysteic acid from the amino acid analyser, and gives a characteristic yellow color with ninhydrin. [G. Wilhelm & K. D. Kupka (1981) *FEBS Letters* **123** 141–144.]

3′-Amino-3′-deoxyadenosine: a purine antibiotic synthesized by *Cordyceps militaris* and *Helminthosporium* species (see Nucleoside antibiotics). A. has antitumor activity. The acetylated derivative 3′-acetamido-3′-deoxyadenosine has also been isolated from *Helminthosporium* species.

2-Amino-2-deoxy-D-galactose: see D-Galactosamine.

Aminoethanol: see Ethanolamine.

Aminoethanol phosphoglycerides: see Phospholipids.

L-α-Aminoglutaric acid: see L-Glutamic acid.

5(4)-Aminoimidazole-4(5)-carboxamide ribonucleotide, abb. *AICAR:* 5′-phosphoribosyl-5-amino-4-imidazolecarboxamide; see Purine biosynthesis, Fig. 1.

5(4)-Aminoimidazole-4(5)-carboxyribonucleotide: 5′-phosphoribosyl-5-amino-4-imidazolecarboxylate; see Purine biosynthesis, Fig. 1.

5-Aminoimidazole ribonucleotide, abb. *AIR:* 5′-phosphoribosyl-5-aminoimidazole, an intermediate in Purine biosynthesis (see) and in the formation of thiamin. M_r 295.2. In certain microbial mutants lacking purine synthesis, AIR can polymerize to a red pigment.

5-Aminoimidazole-4-N-succinocarboxamide ribonucleotide: 5′-phosphoribosyl-4-(N-succinocarboxamide)-5-aminoimidazole. See Purine biosynthesis, Fig. 1.

Aminoisobutyric acid: in the β-form (2-methyl-β-alanine), $H_2N-CH_2-CH(CH_3)-COOH$, a product of the reductive degradation of thymine (see Pyrimidine degradation). The α-form, 2-methylalanine, $H_2N-C(CH_3)_2-COOH$, does not occur in nature. Neither A. a. is incorporated into proteins. Since α-A.a. is not metabolized (or only to a negligible extent), it is used as a model substance to study transport and cytokinin effects.

5-Aminolevulinic acid, δ-aminolevulinic acid: an intermediate in the biosynthesis of the porphyrins and part of the Shemin cycle (see Succinate-glycine cycle).

Aminopeptidases: exopeptidases (EC 3.4.11), usually containing metal ions, which shorten proteins and peptides from the N-terminal end of the chain, removing one amino acid residue per step. The best-known A. is leucine aminopeptidase, which can be obtained from the intestinal mucosa, kidneys and lens of the eye in highly purified or crystalline form. The best synthetic substrates for this enzyme are leucinamide, leucine p-nitroanilide, and leucine hydrazide. Its effectors are bivalent metal ions. A characteristic of all A. is their inability to hydrolyse prolylpeptide bonds. A. are large proteins (M_r 230000 to 330000). The liver A. consists of two subunits, the kidney enzyme of four, and the lens A. of six. A. are used in peptide sequencing.

Aminopropionic acid: see Alanine.

Aminopterin: 4-amino-4-deoxyfolic acid (see Vitamins, section on folic acid), a substance used as a cytostatic agent and in the management of some cancers. It inhibits the enzyme dihydrofolate reductase, which reduces the folate coenzymes required for Purine biosynthesis (see) and thymine production (see Pyrimidine biosynthesis), and thus prevents the synthesis of DNA. However, it is toxic to nondividing cells as well and cannot be tolerated indefinitely. *Methotrexate* (amethopterin) has a similar activity.

R = H , Aminopterin
R = CH$_3$,Methotrexate

Amino sugars: monosaccharides in which an hydroxyl group has been replaced by an amino group. The amino group is often acetylated. The 2-amino-2-deoxyaldoses are particularly important as components of bacterial cell walls, of some antibiotics, blood-group substances, milk oligosaccharides and high-molecular weight natural products, such as chitin. Examples are D-galactosamine, D-glucosamine, D-mannosamine, neuraminic acid and muramic acid.

In the synthesis of amino sugars, the amino group is supplied by transamination from glutamine. Fructose 6-phosphate is aminated to D-glucosamine 6-phosphate by a hexose-phosphate transaminase. The glucosamine phosphate is converted by a transacetylase into the *N*-acetyl derivative. This is isomerized to the 1-phosphate, then activated by coupling to UTP, to form UDP-*N*-acetylglucosamine. The latter can be isomerized to UDP-*N*-acetylgalactosamine. Neuraminic acid is often found as cytidine monophosphate-*N*-acetylneuraminate or as *N*-glycoloylneuraminate. It is synthesized in the form of the *N*-acetyl derivative from mannosamine and phospho*enol*pyruvate. Muramic acid is synthesized from UDP-*N*-acetylglucosamine, which is condensed with phospho*enol*pyruvate. This compound is then reduced to UDP-*N*-acetylglucosamine lactate (UDP-muramate).

Ammonia, *NH$_3$:* a colorless gas with a sharp smell. At normal pressure, the condensation temperature is about -40 °C. NH$_3$ is very soluble in cold water, but it is completely driven off by boiling. The aqueous solution is weakly basic, due to the ability of NH$_3$ to take up protons, forming ammonium: NH$_3$ + H$_2$O $\rightleftharpoons$ NH$_4^+$ + OH$^-$. The reaction equilibrium lies far to the left, so that NH$_3$ can be displaced from ammonium compounds by bases. The toxicity of NH$_3$ is related to the high permeation rate of the nonprotonated form and its tendency to become protonated.

Occurrence: NH$_3$ is the end product of the degradation of nitrogen-containing organic matter. It is therefore to be found in the form of ammonium salts in the soil. The concentration of ammonium ions in the body fluids of animals or in any cells is relatively low, since NH$_3$ is eliminated by detoxification reactions (see Ammonia detoxification). In plants and bacteria it is assimilated for the synthesis of nitrogenous compounds.

Metabolism. NH$_3$ is the product of nitrate reduction,

biological nitrogen fixation, and deamination of amino acids and various catabolic pathways, for example oxidative purine degradation and reductive pyrimidine degradation. In this sense, NH$_3$ is the nitrogen-containing, inorganic end product of the degradation of proteins and nucleic acids. It is taken up into the pool of organic nitrogen by various reactions of primary nitrogen assimilation and is further distributed by reactions transferring nitrogen-containing groups (see Group transfer). Through these reactions, NH$_3$ is used for the synthesis of *N*-containing body substances. Plants in particular have a well developed metabolism of inorganic nitrogen. Many microorganisms can grow with ammonium salts as their only source of nitrogen. Higher green plants have only a limited ability to take up ammonium ions from the soil; they depend largely on nitrate, which is formed in the soil by microbial oxidation of ammonium (nitrification). After entering the plant cell cytoplasm, nitrate is reduced to ammonium (see Nitrate reduction), which is then assimilated. Carbamyl phosphate and glutamine can be regarded as "activated NH$_3$", since the biosynthesis of many nitrogen-containing compounds depends on them.

Ammonia assimilation: the utilization of ammonia in the net synthesis of the nitrogen-containing groups of nitrogenous cell constituents, e.g. amino acids, amides, carbamyl and guanido compounds. Incorporation of ammonia into the amide group of glutamine, catalysed by glutamine synthetase (EC 6.3.1.2), is of central importance: L-glutamate + NH$_3$ + ATP $\rightleftharpoons$ L-glutamine + ADP + P$_i$. The amide nitrogen of L-glutamine is then useful in various syntheses:

1. L-glutamine + α-ketoglutarate + 2H$^+$ + 2e$^-$ $\rightleftharpoons$ 2 L-glutamate (glutamate synthase). The glutamate takes part in the synthesis of other amino acids by transamination: thus a series of coupled reactions result in net assimilation of ammonia into the amino groups of amino acids (Fig.)

Reducing power for bacterial glutamate synthase (EC 1.4.1.13) is provided by NADPH, whereas chloroplast glutamate synthase (EC 1.4.7.1) utilizes reduced ferredoxin. Glutamine synthetase and glutamate synthase occur in plant chloroplasts, where ATP and reduced ferredoxin are supplied directly by the light reaction of photosynthesis. Animals lack glutamate synthase, and they cannot achieve the net synthesis of amino groups from ammonia.

2. L-Glutamine + HCO$_3^-$ + 2 ATP + H$_2$O $\rightleftharpoons$ carbamoyl phosphate + L-glutamate + 2 ADP + P$_i$ (see Carbamoyl phosphate). *N*-Acetyl-glutamate is an essential positive allosteric effector for this enzyme (carbamoyl-phosphate synthetase, EC 6.3.5.5). In eukaryotes, the enzyme is located in the cytoplasm. This carbamoyl phosphate provides C-2 and N-3 in the synthesis of pyrimidines and it contributes to the synthesis of the guanido group of arginine in plants and bacteria.

3. The amide group of glutamine is used in purine synthesis, where it provides N-3 and N-9 of the purine ring, and the 2-NH$_2$ group of guanine.

33

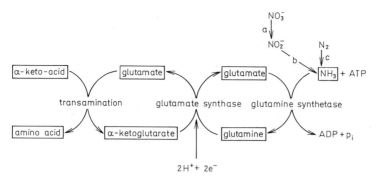

Figure. *Assimilation of ammonia into amino groups of amino acids.* a, nitrate reductase, b, nitrite reductase, c, nitrogenase

4. In several syntheses, nitrogen is derived directly from the amide group of L-glutamine, e.g. histidine synthesis; conversion of chorismate into anthranilate (see Tryptophan synthesis); the synthesis of amino sugars; amination of UTP to CTP.

5. In some organisms, the amide group of glutamine is transferred to aspartate by the action of asparagine synthetase (glutamine hydrolysing) (EC 6.3.5.4): L-glutamine + L-aspartate + ATP $\rightleftharpoons$ L-glutamate + L-asparagine + AMP + PP_i.

In mammalian liver mitochondria, ammonia is converted directly into carbamoyl phosphate: $NH_3 + HCO_3^- + 2\,ATP \rightleftharpoons$ carbamoyl phosphate + 2 $ADP + P_i$. N-Acetyl-glutamate is an essential positive allosteric effector of the enzyme, carbamoyl-phosphate synthetase (ammonia) (EC 6.3.4.16). This reaction introduces ammonia into the urea cycle. It may or may not result in the net synthesis of arginine, depending on the ability of the animal to synthesize ornithine. Much of this ammonia nitrogen may therefore be excreted as urea, without contributing to net synthesis of guanido groups.

In plants, molds and bacteria, assimilation of ammonia into the amino group of glutamate is also possible by NADPH-dependent glutamate dehydrogenase (EC 1.4.1.3); the enzyme is most effective when ammonium salts are available directly from the environment in fairly high concentration. In many organisms, its activity is lower than that of glutamine synthetase, which also has the lower K_m for ammonia. Thus the glutamine synthetase-glutamate synthase system is the more effective, and the more important, especially when the ammonia is derived from nitrate reduction, or nitrogen fixation.

In some microorganisms, ammonia assimilation may also occur by alanine synthesis, catalysed by alanine dehydrogenase (EC 1.4.1.1).

Asparagine is usually synthesized by the action of asparagine synthetase (ADP-forming) (EC 6.3.1.4): L-aspartate + NH_3 + ATP $\rightleftharpoons$ L-asparagine + ADP + P_i.

Asparagine, like its homolog glutamine, is important in the storage and transport of amide nitrogen. It is also a constituent of proteins. Metabolically, it is less versatile than glutamine, and its amide nitrogen does not appear to be transferred directly in synthetic reactions.

Ammonia detoxification: the detoxification of the toxic, non-dissociated compound by formation of ammonium salts and nitrogen-excretion compounds. The ammonia produced by catabolism in animals is either excreted directly, or it is converted into other nitrogenous compounds for excretion; it is not reassimilated. The excretion of ammonia, ammonotelism, is limited to a few aquatic organisms. Most animals rid themselves of catabolic ammonia by synthesizing nitrogen excretion products, such as urea, uric acid, guanine (spiders) or allantoin *(Diptera)*. Those animals which excrete urea are called ureoteles, and those excreting uric acid, uricoteles. The form of nitrogen excretion can change in the course of ontogenesis: the frog larva (tadpole) is ammonotelic, but the adult frog is ureotelic. The ureotelism develops in the course of metamorphosis. The enzymes of the urea cycle are induced, having been repressed during the larval stage. The form of nitrogen excretion also depends on the ecological conditions, but there is still a characteristic pattern of distribution over the animal phyla, or in some cases it is relatively taxon-specific (Table 1).

Ureoteles and uricoteles make use of pre-existing, very ancient biochemical mechanisms for nitrogen excretion: urea synthesis by the urea cycle, which was first developed for the synthesis and degradation of arginine; and uric acid formation, which represents oxidative purine degradation, adapted for the purpose of nitrogen excretion, i.e. the existing reaction sequences of purine synthesis and oxidative purine degradation have been combined to form an ammonia detoxication system. The particular form of nitrogen excretion in organisms using the purine pathway depends on the last step of purine degradation possessed by the organism. In uricoteles, it is uric acid; in other animals, oxidative purine degradation comes to an end at various levels, because in the course of evolution, various enzymes have been lost, resulting in a shortening of the reaction chain (Table 2).

There is a striking correlation between the form of nitrogen excretion and the pathway by which an animal deaminates amino acids produced by proteolysis: ureoteles form ammonia by transamination to glutamate, followed by the action of glutamate de-

Table 1. *Forms of ammonia detoxification and nitrogen excretion*

Nitrogen excretion	Excretion type	Synthesis pathway	Occurrence
Ammonia	Ammonotelia	Deamination of amino acids	Octopi, marine mussels, crustaceans, etc.
Trimethylamine oxide			Bony sea fish
Urea	Ureotelia	Urea cycle	Amphibians, mammals, cartilaginous marine fish
		Purine cycle	Lungfish
Uric acid	Uricotelia	Purine synthesis and degradation	Terrestrial reptiles, birds, insects, except for Diptera
Guanine		Purine synthesis and degradation	Spiders

Table 2. *End products of the degradation of nucleic acid purine in animals*

End product	Enzyme presumed to have been lost	Animal groups
Urea		Fish, amphibians, mussels
Allantoin	Allantoinase, etc.	Mammals (except for primates), snails, dipterans
Uric acid	Uricase, etc.	Primates, birds, reptiles, insects (except for dipterans)
Guanine	Guanase, etc.	Spiders
Adenine	Adenase, etc.	Flat worms, annelids

Table 3. *Parallels between nitrogen excretion (animals) and storage (plants)*

N-excretion compound	N-storage compound	Occurrence in plants
Urea	Urea	Bovists and puff-balls *(Gasteromycetales)*
Urea	L-Arginine	*Malaceae* (apples), *Saxifragaceae,* etc.
Urea	L-Citrulline	Birches *(Betulaceae),* Walnuts *(Juglandaceae)*
Uric acid	Allantoin	Borage *(Boraginaceae)*
Uric acid	Allantoic acid	Maples *(Aceraceae)*
	L-Glutamine	Many plants
Urea	L-Canavanine	Legumes

hydrogenase, while uricoteles make use of oxidative deamination by amino acid oxidases.

Unlike animals, plants generally have no excretory mechanisms, and as nitrogen is usually a growth-limiting factor for them, they transform ammonia into nitrogen compounds which can be reused. These are deposited in the plant's storage organs. When needed, the ammonia can be released from them and completely reassimilated. Nitrogen-storage compounds can also serve as nitrogen-transport forms in the plant. According to the theory of A. d., nitrogen excretion forms and stable nitrogen storage compounds have analogous functions (Table 3.) These substances are formed by the same reaction sequences in animals and plants: allantoin in the plant corresponds to animal uric acid, citrulline and arginine to urea, etc.

Ammonia fixation: see Ammonia assimilation.
Ammonium plants: see Acid plants.
Ammonoteles: see Ammonia detoxification.
AMO 1618: (2-isopropyl-5-methyl-4-trimethyl-ammonium chloride) phenyl-1-piperidinocarboxylate, a synthetic growth retardant of plants. Because it inhibits growth of the shoot, it is used on nursery plants to make them bushy. It is an antagonist of gibberellin and inhibits the biosynthesis of gibberellin A_3.
AMP: abb. for Adenosine 5'-monophosphate. See Adenosine phosphates.
3',5'-AMP: abb. for Cyclic adenosine-3',5'-monophosphate. See Adenosine phosphates).
Amphetamine: see Antidepressants.
Amphibian toxins: a group of chemically very heterogeneous toxins (biogenic amines, peptides, al-

AMO 1618

kaloids, steroids) produced by toads, salamanders, frogs (see Batrachotoxin) and newts (see Tarichatoxin). Pharmacologically, they include heart, muscle and nerve toxins, sympathicomimetica, cholinomimetica, local anaesthetics and hallucinogens. The A.t. serve as a protection against both natural enemies and bacteria and fungi which might attack the animal's skin.

Amphibolic pathways: metabolic pathways serving both breakdown and synthesis. See Metabolic cycle.

Amygdalin: see Cyanogenic glycosides (table).

Amylases, *diastases* (EC 3.2.1.1, 3.2.1.2 and 3.2.1.3): a widely occurring group of hydrolases which cleave the α-1,4-glycosidic bonds in oligosaccharides (from trisaccharides upwards) and polysaccharides like starch, glycogen and dextrins. α-A. is an endoamylase, while β-A. (saccharogenic A.) and γ-A. (glucoamylase) are exoamylases. The activity of α-A. produces first dextrins, which are secondarily cleaved to maltose (87%), glucose (10%) and branched oligosaccharides (3%). The β-A. and γ-A. attack the substrate from the non-reducing end of the chain, the β-A. producing maltose units (in their β configuration after inversion) and the γ-A. producing glucose units (even from α-1,6 glycosidic bonds, if they are adjacent to a 1,4 bond). The A. differ in occurrence, structure and mechanism of action. While α-A. and γ-A. occur both in animals (salivary glands of omnivores, and pancreas for α-A., liver for γ-A.) and plants, β-A. are found only in plant seeds. Mammalian α-A. are dependent on chloride, and all animal and plant α-A. contain calcium. The α-A. of *Bacillus subtilis* contains zinc. Plant β-A. are characterized by their formation as insoluble zymogens during the process of seed maturation.

α-Amylase test: see Gibberellins.

Amylo-1,6-glucosidase, *debranching enzyme* (EC 3.2.1.33): an endoglucosidase which lyses the 1→6 glycosidic bonds at the branching points in glycogen and amylopectin. In mammals and yeast, it is associated with a glycosyl transferase which removes all the glucose residues up to the 1→6 bond. The M_r of the complex is 273000 in muscle and 210000 in yeast. The muscle complex consists of two subunits of M_r 130000, while the yeast complex has three subunits of M_r 120000, 85000 and 70000.

Amyloid protein: a pathological, fibrillar, low-molecular-weight protein. In amyloidosis, it is deposited together with glycoproteins and proteoglycans, primarily in the spleen, liver and kidneys. There are two types of A.p., differing in amino acid composition, one occurring in chronic amyloidosis, the other

in paramyloidosis: 1) amyloid protein A, a single chain composed of 77 amino acids, and containing no disulfide bridges. It has no structural relationship to the immunoglobulins. The primary structure of ape amyloid protein A (M_r 8500) has been determined. 2) Amyloid protein IV, an amyloid-like protein containing 40 to 45 amino acids (M_r 6000) which is both structurally and immunologically very similar to the kappa and lambda chains of the immunoglobulins.

There are morphological similarities between A.p. and infectious protein in scrapie and Creutzfeld-Jakob disease, and protein deposits in Alzheimer's disease, all of which are associated with progressive and irreversible degeneration of nervous function leading to death. [P. A. Merz, et al. *Nature* **306** (1983) 474–476. H. Diringer et al., *ibid.* 476–478.]

Amylopectin: a component of starch (the other is amylose). A. is a branched, water-insoluble polysaccharide consisting of a main chain of D-glucose units linked α-1,4 with side chains attached to every 8th or 9th glucose by an α-1,6 bond. The side chains consist of 15 to 25 D-glucose units, M_r 500000 to 1000000. Iodine forms violet to red-violet inclusion compounds with A. A. swells in water; upon heating, it forms a paste.

β-Amylase (EC 3.2.1.2) degrades A. up to the limit dextrins, while α-amylase (EC 3.2.1.1) degrades it to about 70% maltose, 10% isomaltose and 20% glucose. Hydrolysis with dilute acid yields D-glucose.

Amylopectinosis: see Glycogen storage disease.

Amylose: a component of starch (the other is amylopectin). A. is an unbranched, water-soluble polysaccharide consisting of 100 to 300 D-glucopyranoside residues. M_r 10000 to 50000. The basic disaccharide of A. is maltose. The polysaccharide chain is held in a spiral arrangement by hydrogen bonds, with 6 monosaccharide units per turn of the helix. This arrangement differs from that of cellulose. A. forms blue inclusion compounds with iodine. Degradation by α-amylase (EC 3.2.1.1) produces about 90% maltose and 10% D-glucose. Gentle degradation yields dextrins.

Amylose

β-Amyrenol: see Amyrin.

Amyrin: a pentacyclic triterpene alcohol with one double bond. α-A. is found free, esterified and as the aglycon of triterpene saponins in many plants and has been isolated from many latexes, e. g. that of the dandelion, *Taraxacum officinale*. The hydrocarbon skeleton is called 5-α-ursane. β-A. (β-amyrenol, α-viscol) is found in mistletoe leaves, grape seed oil, latex of *Taraxacum officinale*, etc. It is found free, esterified and as the aglycon of triterpene saponins in many other plants, especially in species which yield caoutchouc and gutta percha. The skeleton hydrocarbon is called 5-α-oleane.

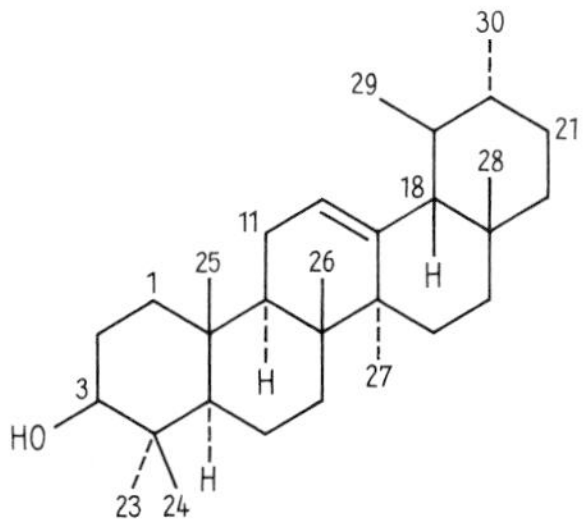

α-Amyrin

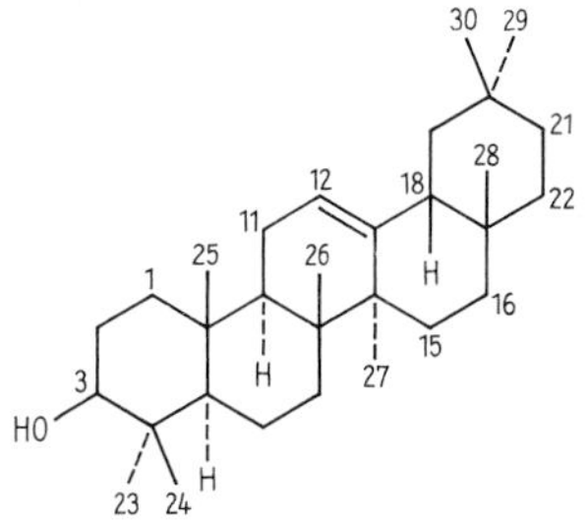

β-Amyrin

Anabasine: a Nicotiana alkaloid. It is isolated in the L-form, primarily from *Nicotiniana glauca* and the asiatic *Anabasis aphylla*, in which it is the main alkaloid. In other tobacco species it is only a secondary alkaloid. M_r 162.2. Its physiological effects are similar to those of nicotine, and like nicotine, it is used as an insecticide. Biosynthesis, see Nicotiana alkaloids.

Anabolic steroids: a group of synthetic steroids which stimulate the production of body protein. Male gonadal hormones have this effect, as was first demonstrated in 1935 for testosterone. It is possible, by structural modification of natural androgens, to separate the anabolic from the androgenic effects, leading to therapeutically very valuable A.s., for example methenolone, 4-chlorotestosterone and androstanozole. A.s. are used in conditions in which there is increased protein turnover or degradation, for example after extensive operations, in osteoporosis, burns, nutritional and growth impairments and rickets. The activity of A.s. is measured in the Hershberg test, in which castrated infant male rats are treated with the substance being tested; after a certain time the ratio of the increase in mass of the musculus levator ani (anabolic effect) to that of the seminal vesicles (androgenic activity) is measured.

Anabolism: the sum of synthetic metabolic reactions. See Metabolic cycle; Metabolism.

Anaplerotic sequence: see Metabolic cycle.

Anchorin: see Ankyrin.

Anderson's disease: see Glycogen storage disease.

Androcymbin: see Colchicum alkaloids.

Androgens: a group of male gonadal hormones, including testosterone, androsterone and androstenolone, which are formed in the intermediary cells of the testes tissue, and a number of less active A. produced in the adrenal cortex, e.g. androstenedione and andrenosterone (Fig. p. 38). Since A. are the precursors of Estrogens (see), they are also synthesized in the ovaries and the fetoplacental unit. At present more than 30 naturally occurring A. have been discovered, all of them structurally related to the skeleton hydrocarbon androstane (see Steroids).

A. are found in the sperm, blood or urine; in the latter case a part of them is bound to glucuronic acid, sulfuric acid or protein. The biological function of the A. is to induce the development of secondary male sex characteristics; they are also required for the maturation of the sperm and the activity of the accessory glands of the genital tract. Aside from these sex-specific effects, A. stimulate anabolic processes, such as protein buildup, and increase the nitrogen retention (see Anabolic steroids).

Castration causes the secondary sex characteristics to disappear, which can be counteracted by administration of A. This is the basis of the capon comb test, which is used as a biological assay for A. Administration of A. to a capon causes its degenerated comb to grow. A capon unit is that amount of androgen which produces an increase in the comb surface of 20% (e.g. 15 mg testosterone). For clinical testing, Immunoassays (see) are now available.

Highly effective oral A. can be obtained by structural modification, e.g. methyltestosterone and mesterolone. A. are used therapeutically, especially to correct for deficiency symptoms following castration, hypogenitalism, impotency due to lack of hormones, climacterium, mammary carcinoma and peripheral circulatory impairment. Some synthetic testosterone analogs with a 1,2-cyclopropane ring have antiandrogenic effects. Improper administration of anabolic steroids (e.g. to athletes) can lead to impotence or sterility.

Androisoxazole: see Androstanazole.

Androstanazole: a synthetic and highly active anabolic steroid. It is used, for example, in the therapy of inflammation and tumors, and in convalescence. The similarly used androisoxazole differs from A. in having an oxygen atom in place of the NH group.

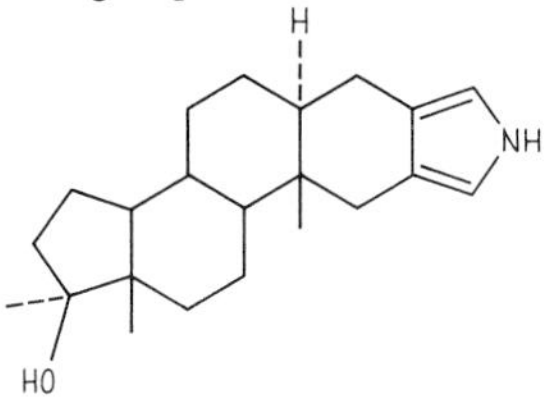

Androstanazole

Androstane: see Steroids.

Androstenedione: androst-4-ene-3,17-dione, a weak androgen, and an intermediate in testosterone biosynthesis (see Androgens).

Androstenolone, *dehydroepiandrosterone:* 3β-hydroxyandrost-5-en-17-dione, an androgen less potent than testosterone, but with similar physiological effects. It is an intermediate in the biosynthesis of testosterone in the adrenal cortex. For structure and biosynthesis, see Androgens.

Androstenolone

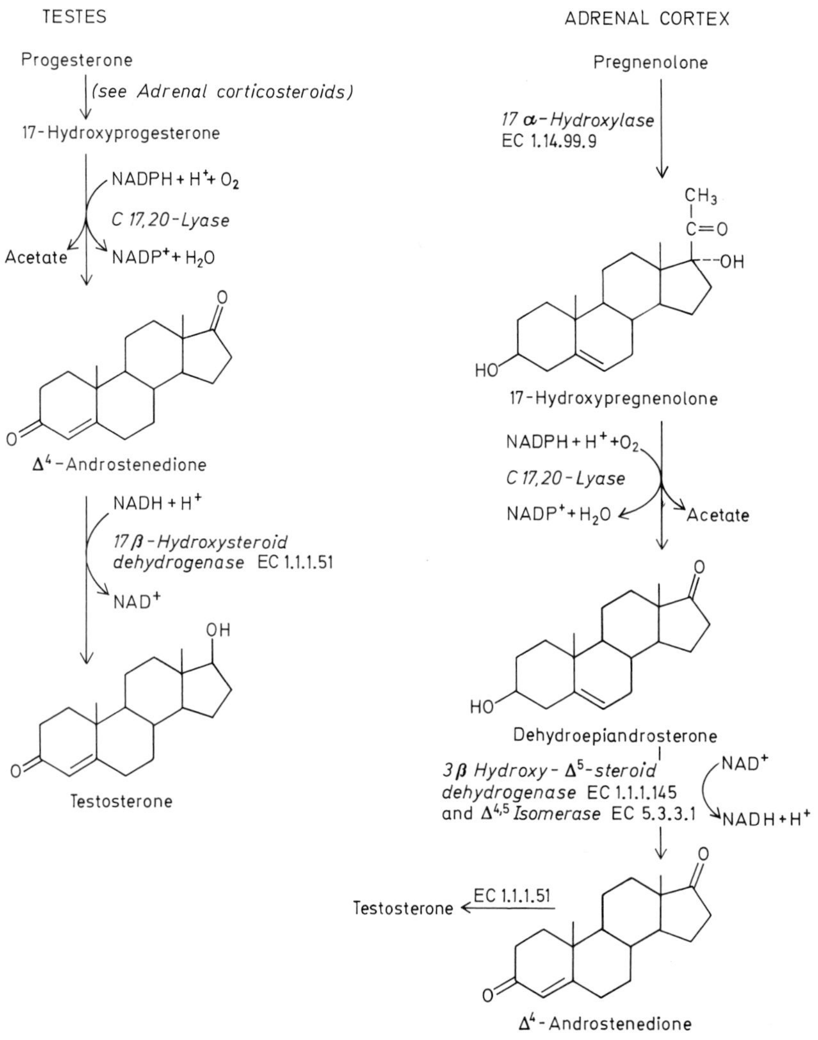
TESTES
Progesterone
(see Adrenal corticosteroids)
17-Hydroxyprogesterone
NADPH + H⁺+ O2
C 17,20-Lyase
Acetate
NADP⁺+ H2O
O
O
Δ4-Androstenedione
NADH + H⁺
17β-Hydroxysteroid
dehydrogenase EC 1.1.1.51
NAD⁺
OH
O
Testosterone
ADRENAL CORTEX
Pregnenolone
17 α-Hydroxylase
EC 1.14.99.9
CH3
C=O
OH
HO
17-Hydroxypregnenolone
NADPH + H⁺+O2
C 17,20-Lyase
NADP⁺+ H2O
Acetate
O
HO
Dehydroepiandrosterone
3β Hydroxy-Δ5-steroid
dehydrogenase EC 1.1.1.145
and Δ4,5 Isomerase EC 5.3.3.1
NAD⁺
NADH + H⁺
O
Testosterone
EC 1.1.1.51
O
Δ4-Androstenedione

Biosynthesis of testosterone. The primary precursor is acetyl-CoA, and the biosynthesis proceeds via cholesterol (see Terpenes, Steroids). The testes and the adrenal cortex also utilize cholesterol which they receive as cholesterol esters from plasma lipoproteins (see Lipoproteins). A second androgen, more potent than testosterone, is 5α-dihydrotestosterone; this is not formed, however, by the Leydig cells of the testes, but at peripheral sites of action from circulating testosterone.

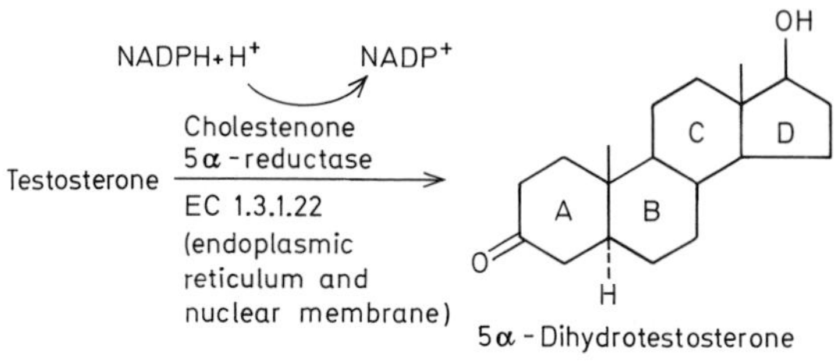
NADPH+H⁺
NADP⁺
Cholestenone
5α-reductase
Testosterone
EC 1.3.1.22
(endoplasmic
reticulum and
nuclear membrane)
OH
C
D
A
B
O
H
5α-Dihydrotestosterone

Ring A of 5α-dihydrotestosterone cannot be aromatized, so that this hormone cannot serve as an estrogen precursor.

Androsterone: 3α-hydroxy-5α-androstan-17-one, an androgen. M_r 290.5. A. is formed in the interstitial cells of the testes. It is similar to testosterone, but its androgenic activity is 7 times weaker. It is biosynthesized from progesterone via 17α-hydroxyprogesterone and androstenedione. It is excreted in human urine, from which it was first isolated in 1931 by Butenandt. He obtained 15 mg A. from 15 000 l urine.

Androsterone

Aneurin: see Vitamins (Vitamin B_1)

Aneurin pyrophosphate: see Thiamin pyrophosphate.

Angiokeratoma corporis diffusum: see Inborn errors of metabolism.

Angiotensin, *angiotonin, hypertensin:* a tissue peptide hormone affecting the blood pressure. The kidney protease renin (EC 3.4.99.19) releases the decapeptide A. I. from a plasma protein of the α_2-globulin fraction. The sequence of A. I is Asp-Arg-Val-Tyr-Ile-His-Pro-Phe-His-Leu. The active A. II is then produced by enzymatic cleavage of the His-Leu bond. A. II strongly increases the blood pressure, much more effectively than noradrenalin, and also stimulates the production of aldosterone in the adrenal cortex. A. is inactivated by an angiotensinase (EC 3.4.99.3) in the blood.

Angiotonin: see Angiotensin.

Angolamycin: see Macrolide antibiotics.

Angstrom unit, Å: unit of length widely used for light wavelengths and atomic dimensions. $1 Å = 10^{-8}$ cm.

Angustmycins: purine antibiotics (see Nucleoside antibiotics) synthesized by various species of *Streptomyces*. Angustmycin A (decoyinin; 9-β-D-5,6-didehydropsicofuranosyl adenine) was isolated by Hsu in 1954 from *Streptomyces hygroscopius* var. *angustmyceticus*. M_r 279.2. It is specific for mycobacteria. Gram positive and Gram negative bacteria and fungi are not sensitive to A.A. It acts by inhibiting the formation of 5-phosphoribosyl-1-pyrophosphate in purine biosynthesis.

Streptomyces hygroscopicus also synthesizes the structurally related angustmycin C (9-β-D-psicofuranosyladenine), M_r297.3. A.C is identical with psicofuranine, which was isolated in 1959 from *Streptomyces hygroscopicus* var. *decoyicus* by Eble. It specifically inhibits the XMP aminase in purine biosynthesis and has antibacterial and antitumoral activity.

Anhaline: see Hordenine.

Anhalonidine: see Anhalonium alkaloids.

Anhalonium alkaloids: cactus alkaloids, isoquinoline alkaloids found primarily in cacti. They all have a tetrahydroisoquinoline skeleton carrying

several phenolic hydroxyl groups on Ring B. These may be etherified.

Both the biosynthesis and the chemical synthesis are based on a Mannich condensation of a β-phenylethylamine derivative with a carbonyl component (Fig.). Depending on the latter, the substituent at the C-1 can be H-, CH_3- or an isoprenoid residue. Several molecules of A.a. can be linked to oligomers by phenol oxidation as in pilocereine. As secondary alkaloids, the cacti also produce derivatives of β-phenylethylamine, e.g. hordenine and mescaline.

The A.a. are weak narcotics and paralysing agents. Pellotine has a cramp-relaxing effect similar to that of acetylcholine. The most toxic A.a. is lophophorin. The heterocyclic A.a. are not directly responsible for the intoxication following ingestion of cactus preparations.

Tyramine   Acetaldehyde   Anhalonidine

Biosynthesis of the anhalonium base, anhalonidine, by a Mannich condensation

Animal protein factor: see Vitamins (Vitamin B_{12}).

Ankyrin, *anchorin, syndein:* a membrane protein, found in erythrocytes and brain, which binds Spectrin (see) to the membrane. The spectrin-binding activity was first isolated as a proteolytic fragment of M_r 72000; the M_r of the entire protein is 215,000. There are about 10^5 copies of A. per erythrocyte ghost, i.e. the number required to bind all the spectrin dimers in 1:1 complex. In the erythrocyte, A. links spectrin to the cytoplasmic part of Band 3 (a transmembrane protein), to the anion-transport protein, and to microtubules. A. will also bind unpolymerized tubulin. [V. Bennett (1985) *Ann. Rev. Biochem.* **54** 273-304]

Annealing of nucleic acids: see Hybridization; Melting point.

Anomers: see Carbohydrates.

Anserine: see Peptides.

Antagonists: see Inhibitors.

Antamanide: a cyclic decapeptide (all the amino acid residues have the L-configuration) isolated from the death cap fungus, *Amanita phalloides*. In experimental animals it acts as an antidote to phallatoxins and amatoxins, providing it is administered no later than the toxins. Like valinomycin, it is able to bind alkali metal ions, but in contrast to valinomycin it shows a strong preference for sodium over potassium. It readily forms stable complexes with Na^+ or Ca^{2+} ions. On account of its sodium selectivity and its highly lipophilic nature, it also has the properties of a sodium ionophore. A. was discovered in 1968 by Wieland et al., and the formula (Fig.) was determined in 1969 by Prox et al. The conformation of the free peptide and its Na^+ complex have been determined by direct X-ray diffraction. Replacement of any Pro residue or of Phe residues 9 or 10 abolishes the antidotal activity. Replacement of other amino acid residues has less effect. Since the imino nitrogen and carboxyl of proline are part of a relatively rigid structure (Pro does not permit α-helix formation in protein and is known as a "helix breaker"), replacement of Pro by any other amino acid will alter the conformation of the molecule. It is suggested that Phe 9 and Phe 10 in correct alignment occupy a target receptor on the liver cell membrane, and thereby prevent entry of phalloidin into the cell. See Silybin. [I. L. Karle et al. *Proc. Nat. Acad. Sci. U.S.A.* **76** (1979) 1532–1536; H. L. Lotter, *Zeitschrift für Naturforsch.* **39c** (1984) 535–542]

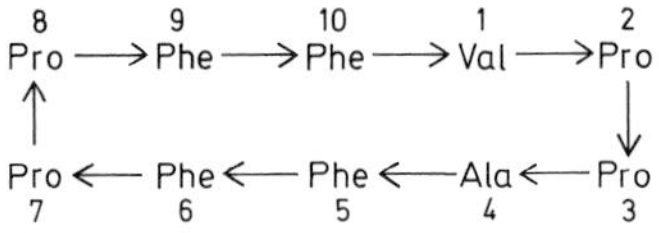

$$
\begin{array}{ccccc}
8 & 9 & 10 & 1 & 2 \\
\text{Pro} \longrightarrow & \text{Phe} \longrightarrow & \text{Phe} \longrightarrow & \text{Val} \longrightarrow & \text{Pro} \\
\uparrow & & & & \downarrow \\
\text{Pro} \longleftarrow & \text{Phe} \longleftarrow & \text{Phe} \longleftarrow & \text{Ala} \longleftarrow & \text{Pro} \\
7 & 6 & 5 & 4 & 3
\end{array}
$$

Antamanide

Antheraxanthin: see Zeaxanthin.

Antheridiogen: a phytohormone which is chemically a diterpenoid. M_r 330. It is derived structurally and biogenetically from the gibberellins. It was isolated from the fern *Aneimia phyllitidis* and stimulates the formation of antheridia in this plant, even at a dilution of 10 µg/l. Other A. have been found in other ferns of the families *Schizaeaceae* and *Polypodiaceae*.

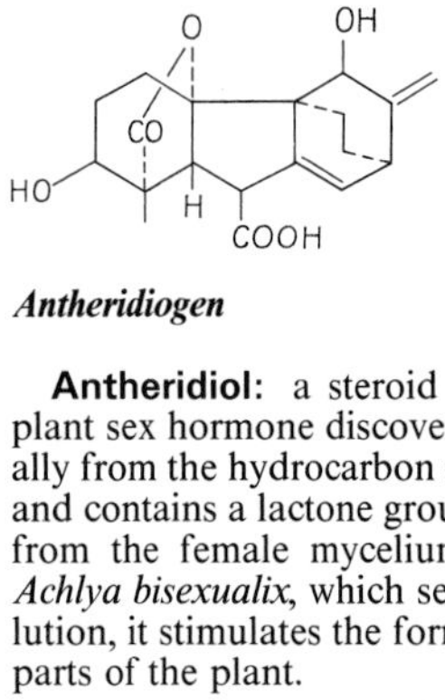

Antheridiogen

Antheridiol: a steroid plant hormone, the first plant sex hormone discovered. It is derived structurally from the hydrocarbon stigmastane (see Steroids) and contains a lactone group. It was isolated in 1971 from the female mycelium of the aquatic fungus *Achlya bisexualix*, which secretes it. Even at high dilution, it stimulates the formation of hyphae in male parts of the plant.

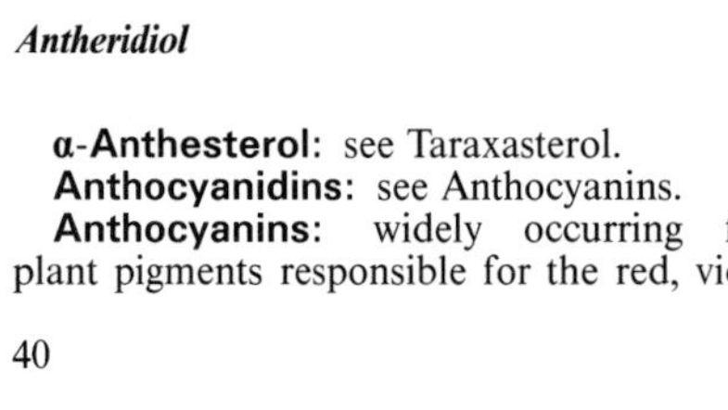

Antheridiol

α-Anthesterol: see Taraxasterol.
Anthocyanidins: see Anthocyanins.
Anthocyanins: widely occurring flavonoid plant pigments responsible for the red, violet, blue

or black colors of blossoms, leaves and fruits of higher plants. The variety of colors and patterns is due to the occurrence of the A. either alone or in combination with other pigments, usually other flavonoids (copigmentation). A. are the water-soluble glycosides of hydroxylated 2-phenylbenzopyrylium salts; they may be cleaved into the carbohydrate and the water-insoluble and unstable aglycon (anthocyanidin) by treatment with acid or enzymes.

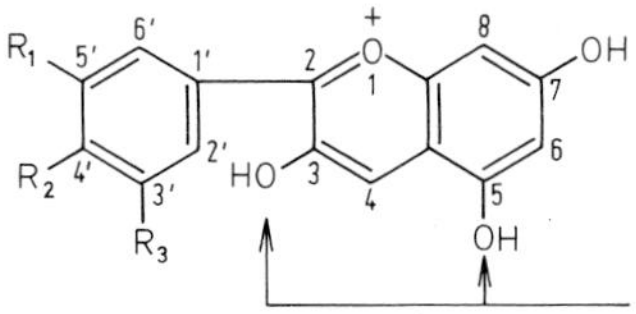

Glycosylation at one or both these sites with glucose, galactose, rhamnose or arabinose, or various oligosaccharides gives the water-soluble anthocyanins.

Anthocyanidins
$R_1 = H$, $R_2 = OH$, $R_3 = OH$: Cyanidin (blue) (cornflowers)
$R_1 = H$, $R_2 = OH$, $R_3 = H$: Pelargonidin (red) (geraniums)
$R_1 = OH$, $R_2 = OH$, $R_3 = OCH_3$: Peonidin (red) (peonies)
$R_1 = OH$, $R_2 = OH$, $R_3 = OCH_3$: Petunidin (red) (petunias)
$R_1 = OCH_3$, $R_2 = OH$, $R_3 = OCH_3$: Malvidin (blue) (mallow)
$R_1 = OH$, $R_2 = OH$, $R_3 = OH$: Delphinidin (blue) (delphiniums)

The basic structure of the A. is the C_{15} flavylium cation. Ring B and the C-atoms 2, 3 and 4 of this structure are biologically synthesized, as other flavonoids are, from a C_6-C_3 unit. Ring A is derived from a C_6 unit formed from 3 molecules of acetate (see Stilbenes). The anthocyanidins differ in the number and positions of hydroxyl groups, which may be replaced by methoxy groups or esterified, for example with *p*-cumaric or ferulic acid. The great majority of the anthocyanidins is hydroxylated at positions 3, 5 and 7. Those lacking the 3-hydroxyl group, e. g. 3-deoxypelargonidin, 3-deoxycyanidin or 3-deoxydelphinidin are less common. The three basic types, pelargonidin (4'-hydroxy), cyanidin (3',4'-dihydroxy) and delphinidin (3',4',5'-trihydroxy) are formed by the addition of more hydroxyl groups to the B ring; with respect to quantity and distribution, the glycosides of these compounds are the most important representatives of this class of pigments.

The carbohydrate moiety of the A. is usually linked β-glycosidically to the C-3, more rarely to the C-5, or, in the diglycosides, to both. The most common carbohydrate residues are the monosaccharides glucose, galactose and rhamnose. Xylose and arabinose are less common, and di- and trisaccharides are rare. The latter are called biosides or triosides. There are more than 20 known types of A., differing from each other in the number, kind and position of their carbohydrate residues. Well over 100 natural A. have been isolated and structurally characterized.

The A. are amphoteric. Their salts with acids are red, at neutral pH they form colorless pseudo-bases, and in the alkaline range, they form unstable, violet anhydrobases. A. are dissolved in the cytoplasm and excreted into the vacuole. The absorption maxima of the A. lie in the range from 475 to 550 nm and around 275 nm.

Intensive research on the A. has been done by R. Willstätter, R. Robinson and P. Karrer.

Anthraquinones: yellow, orange, red, red-brown or violet derivatives of anthraquinone (9,10-anthracenedione), the largest group of naturally occurring quinones. With a few exceptions, e.g. 2-methylanthraquinone, all natural A. are hydroxylated. Other common substituents are methyl, hydroxymethyl, methoxy, formyl, carboxyl, long-chain alkyl and benzyl groups. A number of A. are dimeric.

Of the more than 170 known natural A., which occur either in the free form or as glycosides, more than half are found in lower fungi, particularly in the *Penicillium* and *Aspergillus* species, and in lichens. The others are found in higher plants, and, in isolated instances, in insects. The *Rubiaceae, Rhamnaceae, Leguminosae, Polygonaceae, Bignoniaceae, Verbeniaceae* and *Liliaceae* are particularly rich in A. The best known insect A. are carminic acid, kermesic acid and laccaic acid. The most important vegetable A. are emodin, alizarin, rhein, purpurin and morindon. Helminthosporin, skyrin and julichrom are found in fungi. The roots of madder (*Rubia tinctorum*), the bark of *Coposma australis* and the mycelium of *Helminthosporin gramineum* have particularly large contents of A.

Due to the laxative qualities of A., a number of drugs, such as *Radix Rhei, Cortex Frangulae, Fructus* and *Folia Sennae* are pharmaceutically significant. A few A., such as carminic acid and alizarin, are still used to some extent as dyes.

Anthraquinone

Antiandrogens: a group of chemical compounds which reversibly inhibit the effect of male sex hormones by competing with them for their receptors. Several testosterone and gestagen analogs with 1α,2α-methylene groups, for example cyproterone acetate, are especially effective.

Antianemia factor, *Vitamin B₁₂:* see Vitamins.

Antiauxins: see Auxin antagonists.

Anti-beri-beri-factor, *Vitamin B₁:* see Vitamins.

Antibiotics: substances produced by microbes which kill or inhibit the growth of other microorganisms. In contrast to general cell poisons, the A. are selective. The term A. is derived from "antibios" (Greek, "against life"), which was coined in 1889 by Vuillemin to describe processes in which one organism destroys the life of another. A. are synthesized both by bacteria and by fungi. The genus *Bacillus*

yields a number of therapeutically important A. (bacitracin, gramicidin, polymyxin, tyrocidine), as do *Streptomyces* and *Actinomyces* species (streptomycin, tetracycline, actinomycin, chloramphenicol, macrolides, neomycin) and molds of the genera *Penicillium* and *Aspergillus* (penicillin, griseofulvin, xanthocillin, helvolinic acid). Some of these are peptides (see Peptide antibiotics).

Chemically speaking, the A. are heterogeneous. Their most important components are amino acids, often of the non-proteogenic D-series (gramicidin), acetate/malonate units (griseofulvin, tetracycline), sugars and sugar derivatives (streptomycin), tetracyclic triterpenes (fusidinic acid, helvolinic acid, cephalosporin P₁). A large group of A. is derived from nucleosides (see Nucleoside antibiotics). The mechanisms by which A. act vary. Many interfere with the process of protein biosynthesis, but others, e.g. penicillin, inhibit the production of cell walls in bacteria. The A. are produced industrially by chemical synthesis, and more often, by microbial techniques. These techniques are equivalent to those described in general for fermentation (see Industrial microbiology). It is particularly important to have a highly productive inoculum so the microbes are maintained in a culture collection. The activity of the parent culture can then be tested in the laboratory before it is used to generate the inoculum.

The composition of the nutrient broth varies for the individual A. and for the stage of production. Various sugars, starch, soybean millings and often the water in which corn has been soaked are used as raw materials. The yield of penicillin G can be increased by adding phenylacetic acid to the medium as a precursor. As is typical for the products of secondary metabolism, the production of A. becomes significant only toward the end of the logarithmic growth phase. Therefore the medium is designed so that cell division is limited by lack of some nutrient before the sugar has been consumed. Under these conditions, the production of the A. begins and can continue until the energy supply is exhausted. The productivity of the microbial strain, the management of fermentation and the recovery of the products are of equal importance in the economical production of A.

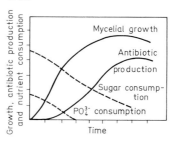

Kinetics of growth and antibiotic production

Antibodies: see Immunoglobulins.

Anticodon: a sequence of three nucleotides in one loop of a transfer RNA molecule which recognize the nucleotides of the codon in messenger RNA by forming H-bonded base pairs with them. This anticodon loop thus insures that the correct amino acid is built into the polypeptide sequence. See Wobble hypothesis. See Transfer RNA.

Anticytokinins: antagonists of the cytokinins which can partially negate the physiological effects of the hormones.
A synthetically available kinetin inhibitor is 6-methylpurine.

Antidepressants: drugs with stimulatory and antidepressant action. They decrease fatigue, reduce appetite and decrease sleeping time. These effects are the result of reduced autonomic activity, especially of cholinergic systems, and of central sympathetic activation. A. are inhibitors of monoamine oxidase (E.C.1.4.3.4) and they are all amines, presumably acting as analogs of natural monoamine oxidase substrates; inhibition of monoamine oxidase delays or prevents destruction of the natural catecholamines (adrenalin, dopamine, noradrenalin) which therefore persist at elevated concentrations and cause stimulation. Examples of A. are amphetamine (β-phenylisopropylamine), ephedrine, harmine, etc. The first A. was iproniazid (*N*-isonicotinyl-*N'*-isopropylhydrazide); during the original tests of this compound as an antitubercular drug, it was observed that patients became elated. A. are used for the treatment of mental depression; they have also been used for doping race horses and athletes. A. can be addictive, especially amphetamine.

Antidermatitis factor: see Vitamins.

Antidiuretic hormone: see Vasopressin.

Antidiuretin: see Vasopressin.

Antienzymes: polypeptides or proteins which act as enzyme inhibitors, including antibodies formed in response to antigenic enzyme proteins or coenzymes in the blood. The inhibitors include the many animal and plant protease inhibitors (see Toxic proteins), such as soybean trypsin inhibitor and serum antitrypsin. These form complexes with the corresponding enzymes which do not easily dissociate.
The induction of antibody A. by injection of enzyme-containing solution has become very important in the purification and characterization of enzymes. Injection of the whole enzyme induces antibodies against the specificity-determining apoenzyme, but injection of an antigenic coenzyme raises antibodies against the coenzyme which inactivate all enzymes with that coenzyme.

Antigen-antibody reaction, abb. *AAR:* together with phagocytosis, the most important protective mechanism of the animal organism against invading foreign substances. The AAR is the specific formation of an insoluble antigen-antibody complex. Soluble antigens are precipitated, and cells carrying antigen on their surfaces are agglutinated. Since the specificity and sensitivity (in the picogram range) of the AAR is very great, even in vitro, it is used in diagnosis. The bonds between antigen and antibody are almost exclusively non-covalent, i.e. ionic, hydrophobic or hydrogen bonding between the amino, carboxyl, hydroxyl and aliphatic groups is involved. See Immunoglobulins.

Antigens: substances which induce an immune response. They are normally foreign to the body, and may be natural or synthetic macromolecules, particularly proteins and polysaccharides ($M_r >$ 2000), or surface structures on foreign particles which can be phagocytosed. An A. consists of a high-molecular-weight carrier and, usually, several low-molecular-weight groups which are responsible for the specificity of the immune response and the reaction of the A. with the corresponding immunoglobulins. These small groups, called antigenic determinants, or haptens, lie on the surface of the molecule and determine the valence of the A. Nearly all A. are polyvalent, and therefore they induce more than one kind of antibody. The largest determinant groups occur in protein A., and may include up to 30 amino acids. Simple polysaccharide determinants involve from 2 to 6 or 7 sugar residues. Much structural work on mouse immunoglobulins has been done on antibodies to the artificial haptens phosphorylcholine (PC) and *p*-azophenylarsonate (Ars) coupled to various proteins. The determinants on protein A. can be either conformational or sequential; the former are destroyed when the protein is denatured.

Antigibberellins: see Gibberellin antagonists.

Anti-gray-hair factor: a member of the B_2 complex, see Vitamins.

Antihemophilic factor, *factor VIII:* an oligomeric $β_2$-glycoprotein (6% carbohydrate) which activates factor X in the process of blood clotting, and is thereby completely consumed. A.f. is stabilized by calcium ions, but is inactivated when blood is stored. M_r 1.12×10^6 (human) or 1.2×10^6 (bovine). The factor is composed of covalently bound subunits. DNA clones encoding the complete 2351 amino acid sequence of human A.f. subunits have been isolated. The recombinant protein (M_r 267039) decreases the clotting time of hemophiliac plasma, and it is biochemically and immunologically similar to serum-derived A.f. It has structural similarities with factor V and ceruloplasmin. [J. Gitscher et al., *Nature* 312 (1984) 326–330; W.I. Wood et al. *Nature* 312 (1984) 330–337; G.A. Vehar et al. *Nature* 312 (1984) 337–342; J.J. Toole et al. *Nature* 312 (1984) 342–347]

Antihemorrhagic vitamin: see Vitamins.

Antilymphocyte globulin: see Antilymphocyte serum.

Antilymphocyte serum, abb. *ALS:* an immune serum with an opsonizing (see Opsonization) and cytotoxic effect on lymphocytes. It was used for immune suppression before Monoclonal antibodies (see) to specific classes of lymphocytes became widely available. The antibodies were raised in other species of animals by injection of human lymphocytes from blood, thymus, spleen or the thoracic duct. To prevent undesired side effects, the A.s. was purified and administered primarily as antilymphocyte globulin, abb. ALG. Treatment with ALG affects primarily cellular immunity, particularly that provided by long-lived circulating lymphocytes.

Antimetabolite: a compound so similar in structure to a metabolite that it can occupy the enzyme binding sites specific for the latter. This results either in the inhibition of the enzyme or the substitution of the A. for the metabolite in the reaction, possibly leading to the incorporation of the A. in cell components. Both effects lead to metabolic disturbances or inhibition of cell division. Many A. are used therapeutically; for example, analogs of purines and pyrimidines and folic acid antagonists are used as carcinostatic agents.

Antineuritic vitamin, *vitamin* B_1**:** see Vitamins.

Antipain: see Inhibitor peptides.

Antiparallel arrangement: see Strand polarity.

Antipurines: see Purine analogs.

Antipyrimidines: see Pyrimidine analogs.

Antipyrine: 1,2-dihydro-1,5-dimethyl-2-phenyl-3H-pyrazol-3-one, a weak base, pK_a 1.4. After oral administration A. is rapidly absorbed and becomes distributed throughout the total body water within 2 hours. It was first synthesized in 1884, used as an antipyretic and later as an analgesic, then went out of favor in the 1930's as new analgesics became available.

Antirachitis vitamin, *vitamin D:* see Vitamins.

Antiscorbutic vitamin, *vitamin C:* see Vitamins.

Antiserum, *immune serum:* the serum of an animal (or human being) which has been immunized against an antigen. An A. can be monovalent or polyvalent, i.e. contain antibodies specific for one or several antigenic determinant(s), depending on whether the animal was immunized with a purified antigen or a mixture of antigens. In addition to the antibodies, the A. contains all other serum proteins. These can be removed by ion-exchange or immune adsorption chromatography.

Antisterility factor, *vitamin E:* see Vitamins.

Antitumor enzymes: enzymes which stimulate either the irreversible degradation of amino acids which cannot be synthesized by tumor cells, or the inhibition of tumor-specific DNA, leading to a stopping of tumor growth. These enzymes include asparaginase (EC 3.5.1.1) and glutaminase (3.5.1.2) (both effective against leukemia), leucine dehydrogenase (EC 1.4.1.9), isoenzyme A of glutaminase, arginase (EC 3.5.3.1), phenylammonium lyase, urease (EC 3.5.1.5), glutamyl hydrolase (EC 3.4.22.12) (splits folic acid conjugates) and ascorbate oxidase (EC 1.10.3.3) (inhibits growth and synthesis of tumor cell DNA). With the exception of urease, only A.e. of microbial origin, which are available in large quantities, are used. A.e. have been successfully used to treat certain forms of leukemia.

Antitumor proteins: proteinaceous antibiotics isolated from culture filtrates of various *Streptomyces* strains which inhibit tumor growth. The most thoroughly examined A.p., neocarcinostatin, is an acidic, single-chain protein (M_r 10700, 109 amino acids, not including histidine or methionine; primary structure determined 1972) with typical antibiotic activity against Gram-positive bacteria, such as *Sarcina lutea* and *Staphylococcus aureus*. However, it is also highly effective against experimental tumors and seems particularly well suited for treatment of tumors of the rectum, stomach, gall bladder and penis. The main effect of the A.p. is inhibition of mitosis through inhibition of DNA synthesis and accelerated degradation of existing DNA.

Knowledge of the primary structure of the A.p. makes possible the chemical synthesis of A.p. which are highly effective against tumors. For example, reaction of neocarcinostatin with fluorescein iso-

thiocyanate considerably reduces its toxicity without affecting its high antitumor activity. The phytotoxins abrin and ricin must also be classified as A.p., because they inhibit the growth of certain tumor cells due to their inhibitory effect on protein synthesis.

Antivitamins: antimetabolites of the vitamins which inhibit the growth of vitamin-dependent microorganisms and cause symptoms of vitamin deficiency in animals.

Apigenin: see Flavones (Table).

Apiin: see Flavones (Table).

D-Apiose: a monosaccharide with a branched carbon chain. M_r 150.13, $[\alpha]_D^{19}$ +9 ° (pure syrup). A. is found in various glycosides and as a component of various polysaccharides. The biosynthesis starts from D-glucuronic acid.

Apoenzyme: see Coenzyme.

Apoprotein: the protein component of a conjugated protein.

Aporepressor: see Enzyme repression.

APP: abb. for Aneurine pyrophosphate; see Thiamin pyrophosphate.

APS: abb. for Adenosine 5′-phosphosulfate (see Sulfate assimilation).

APS reductase: see Adenylsulfate reductase.

APUD system: see Amine precursor uptake decarboxylase system.

Apurine acids: polynucleotides which have been subjected to short treatment with mild acid, which removes the purines and leaves the phosphate, pentose and pyrimidines.

Apyrimidine acids: nucleic acids from which the pyrimidines have been removed by chemical treatment, such as exposure to hydrazine.

Arabans: high-molecular-weight, branched polysaccharides composed of L-arabinose linked 1,5 and 1,3 in furanose form (see Carbohydrates). A. are found widely as components of hemicelluloses in plants.

1β-D-Arabinofuranosyladenine: see Arabinosides.

1-β-D-Arabinofuranosylcytosine: see Arabinosides.

1-β-D-Arabinofuranosylthymine: see Arabinosides.

1-β-D-Arabinofuranosyluracil: see Arabinosides.

Arabinose: a pentose occurring naturally in both the D- and the L-forms. M_r 150.13. L-Arabinose, β-form, m.p. 160 °C, $[\alpha]_D^{20}$ +190 ° → +105 ° (water) is a component of hemicelluloses, for example the araban of cherry gum, and is found in plant mucilages, glycosides and saponins.

D-Arabinose, β-form m.p. 160 °C, $[\alpha]_D^{20}$ −175 ° → −105.5 ° (water) has been isolated from some bacteria and is a component of glycosides.

```
       CHO              CHO
HO—C—H           H—C—OH
H—C—OH           HO—C—H
H—C—OH           HO—C—H
     CH2OH            CH2OH
 D-Arabinose      L-Arabinose
```

Arabinosides, *arabinonucleosides:* structural analogs of the ribonucleotides in which the sugar is arabinofuranose rather than ribofuranose. 1-β-D-Arabinofuranosylcytosine (cytosine arabinoside) inhibits the reduction of cytidine diphosphate to the corresponding deoxynucleoside in the course of DNA synthesis. It is rapidly deaminated to the corresponding uracil derivative. 1-β-D-Arabinofuranosyluracil (spongouridine) and 1-β-D-arabinofuranosylthymine (spongothymidine) have been found as natural pyrimidine analogs in various sponges. A. are therefore often called spongonucleosides.

Purine bases can also be components of A., for example 1-β-D- and 9-β-D-arabinofuranosyladenine (Ara-A) and 9-[(2-hydroxyethoxy)methyl]guanine (Acyclovir). All of these are phosphorylated by a viral thymidine kinase to triphosphate esters which inhibit the viral DNA polymerase. They are thus interesting as antiviral agents, but unfortunately they are rapidly deaminated by adenine deaminase. Esterification of the sugar hydroxyls makes the compounds penetrate the cell membrane more easily, which enhances their effectiveness. Other synthetic A. are being developed as antiviral agents; the compound 2′-fluoro-5-iodoarabinosylcytosine is potent against herpes virus in mice and is being tested in humans. [T. H. Maugh II, "ACS Highlights" *Science* **220** (1983) 292–293]

Structures of known pyrimidine arabinosides

Arabinoside	R_5	R_6
1-β-D-Arabinofuranosylcytosine	NH_2	H
1-β-D-Arabinofuranosyluracil	H	OH
1-β-D-Arabinofuranosylthymine	OH	CH_3

Arachidic acid: icosanoic acid, CH_3-$(CH_2)_{18}$-COOH, a fatty acid. M_r 312.5, m. p. 75.3 °C, b. p. 205 °C. A. a. occurs widely as a component of glycerides, but is usually present only in low concentrations. In sunflower oil, soybean oil, milk fat and peanut oil, A. a. may represent up to 3% of the fatty acids.

Arachidonic acid: all-*cis*-5,8,11,14-eicosatetraenoic acid, the precursor of several groups of regulatory substances: the Prostaglandins (see), Thromboxanes (see), Leukotrienes (see), and oxidized eicosatrienoic and eicosatetraenoic acids. A. is derived either from the diet or from linoleic acid, by desaturation and chain elongation. It is incorporated into membrane phospholipids, especially the phosphatidylinositols (see Inositol phosphates); when released from these by phospholipase A_2, it is available for oxidation by several enzymes. Since the phosphatidylinositols are cleaved in response to the occupation of hormone receptors by their hormones (see Second messengers), A. and its metabolites are able to amplify hormonal signals and to "rebroadcast" them into the surrounding tissue.

There are two types of enzyme for which A. is a substrate, the lipoxygenases and the cyclo-oxygenase. The lipoxygenases specific for positions 15 or 12 produce 15- or 12-hydroperoxyeicosatetraenoic acids (HPETE), which may be further metabolized to hydroxyeicosatetraenoic acids (HETE). These substances are pharmacologically active, and may inhibit production of the leukotrienes. The introduction of a hydroperoxy group at the 5 position by a third lipoxygenase leads to 5-HPETE, which is the precursor of the Leukotrienes (see). The cyclo-oxygenase forms endoperoxides, which, after addition of oxygen at position 15, become Prostaglandins (see). Finally, the endoperoxide PGH_2 is the precursor of the Thromboxanes (see). [S. Moncada & J. R. Vane, *Pharmacol. Rev.* **30** (1979) 293–331]

Arachin: peanut protein composed of 6 subunits, M_r 34000. Each subunit is composed of 2 equal-sized, covalently bound polypeptide chains. A. is very similar to edestin.

Arcain: 1,4-diguanidobutane, H_2N-C($=NH$)-NH-$(CH_2)_4$-NH-C($=NH$)-NH_2, a strongly basic guanidine derivative first isolated from a mussel *(Arca noae)*, but also found in higher fungi, for example *Panus tigrinus*. A. is biosynthesized from putrescine via agmatine. The process consists of two transamidations from L-arginine. A. reduces the blood sugar level.

Areca alkaloids: pyridine alkaloids in which the pyridine ring is partially hydrated (Fig.). The A. a. are obtained from betel nuts, the seeds of the betel palm *(Areca catechu)*. In the plant, the A. a. are bound to tannins. The main alkaloid is arecoline. The A. a. are probably derived from nicotinic acid or its precursors, but the exact biosynthetic pathway is unknown.

The significance of the A. a. in human and veterinary medicine has declined greatly, but as a mild stimulant, betel is still widely used. The betel nuts are chewed together with lime (to release the alkaloids) and the leaves of the betel pepper *(Piper betle)*. The practice, at least 2000 years old, is common in East africa, India and Oceania. The number of users of betel nuts, who can be recognized by the red stain on their teeth, is estimated to be about 200 million.

	$R_1 = H$	$R_1 = CH_3$
$R_2 = H$	Guvacine	Arecaidine
$R_2 = CH_3$	Guavacoline	Arecoline

Areca alkaloids

Arecaidine: see Arecoline.

Arecoline: 1,2,5,6-tetrahydro-1-methyl-3-pyridinecarboxylic acid methyl ester, the most important of the areca alkaloids. M_r 155.19. b. p. 209 °C. The compound is the methyl ester of the alkaloid arecaidine (M_r 141.19 m. p. 232 °C). A. is responsible for the physiological effect of betel nuts. It acts as a parasympatheticomimetic, but due to its high toxicity, it is used only in veterinary medicine.

Arginase (EC 3.5.3.1): a highly active and specific liver enzyme which catalyses the last reaction in the urea cycle (L-arginine + H_2O → L-ornithine + urea) in ureotelic animals. In land ureoteles such as mammals, frogs, and swamp turtles, A. is found practically only in the liver, with traces in the pancreas, mammary glands, testes and kidneys. In the ureotelic cartilagenous fish (sharks and rays) the enzyme is not limited to the liver. In these fish it serves to generate a high urea concentration in the blood (2 to 2.5%) which is needed to maintain its osmotic pressure.
Liver arginase is present in larger amounts when the diet is rich in protein, and in subnormal amounts in patients with liver carcinoma. It is a tetrameric molecule (M_r 118000) binding four Mn(II) ions. Removal of the metal ions, for example by EDTA, causes the enzyme to dissociate into its 4 inactive subunits (M_r per subunit 30000). The process can be reversed by addition of manganese. A. is stable for months at pH 7.0 and +4 °C. Below pH 6 it is increasingly inactivated as it reversibly dissociates, into dimers at pH 4 and monomers at pH 2. The enzyme is highly specific, hydrolysing only canavanine and L-arginine, but not D-arginine or other guanido compounds. Its pH optimum is dependent on metal ions, lying around pH 10 in the presence of Mn^{2+} ions. It is inhibited by L-ornithine and L-lysine. The uricotelic vertebrates (birds, reptiles) also have an L-arginine-specific A., but its activity is very low and it differs from the ureotelic liver A. in K_m (50 to 100 times that of the rat liver A.), in M_r (276000), in susceptibility to inhibition by ornithine and in its immunological behavior.

L-Arginine, abb. *Arg:* 2-amino-5-guanidovaleric acid, the most strongly basic amino acid. M_r 174.2. m. p. 238 °C (d.) $[\alpha]_D^{20} = +26.9$ ° ($c = 1.65$ in 6N HCl). Arg is unstable in hot alkaline solutions and forms nearly insoluble nitrates, picrates and picrolonates, and a particularly insoluble salt with flavianic acid. The Sakaguchi reaction is used for the detection and quantitative determination of Arg. It is found in particularly large amounts in protamines and histones. High concentrations of free Arg are found in many plants, including red algae, *Curcurbitaceae* and conifers, where it serves as a nitrogen storage and transport form. For this reason it is found in particularly high concentrations in the seedlings and reserve organs.
Arg is glucogenic and half-essential for humans, i.e. it is not required by adults. Various experimental animals also show limited growth on an Arg-free diet. Young rats and chicks require the amino acid because they cannot synthesize it in adequate amounts. Arg is an important member of the urea cycle. Various enzymes attack it, depending on the species: 1. arginase (EC 3.5.3.1) hydrolyses proteolytically released Arg; it is an important step in ammonia detoxification via the urea cycle; 2. Transamidinases (EC 2.1.4) transfer the amidine group to various

amino acids and amines, forming guanidine derivatives and are involved in the formation of phosphagens; 3. Arginine deiminase (EC 3.5.3.6) hydrolyses Arg to L-citrulline and NH_3; 4. L-Amino-acid oxidase (EC 1.4.3.2) deaminates Arg to 2-oxo-5-guanidovaleric acid; 5. Arginine decarboxylase (EC 4.1.1.19) decarboxylates Arg to agmatine; 6. Arginine 2-monooxygenase (EC 1.13.12.1) decarboxylates and oxidizes Arg to 4-guanidobutyramide.
Arg is synthesized by the reactions of the urea cycle from carbamoylphosphate, L-ornithine and the α-amino-group of L-aspartate. This same pathway is generally responsible for the biosynthesis of the guanido group, because all guanidine derivates are generated from Arg. The last step in the synthesis of Arg is catalysed by argininosuccinate lyase (EC 4.3.2.1).
Arg increases spermatogenesis. Various natural products contain the amino acid, for example the phosphagen arginine phosphate, and octopine.
Arg was first isolated from lupin seedlings in 1886 by Schulze and Steiger.
Argininemia: see Inborn errors of metabolism.
Arginine phosphate: see Phosphagens.
Arginine-urea cycle: see Urea cycle.
Argininosuccinic aciduria: see Inborn errors of metabolism.
Aristolochic acids: a group of related aromatic nitro compounds from *Aristolochia spp.* The most important is aristolochic acid I, m. p. 173 °C.
Biosynthesis. The A. a. are formed from isoquinoline alkaloids of the norlaudanosine type by oxidation of the nitrogen-containing ring (Fig.).
Aristolochia drugs are among the oldest known, but due to their toxicity, have become less important.

Norlaudanosoline   Aristolochic acid I

Biosynthesis of Aristolochic acid I

Arogenic acid: see Pretyrosine.
Aromatic biosynthesis, *aromatization:* The most important mechanisms are 1. the shikimic acid/chorismic acid pathway, in which the aromatic amino acids L-phenylalanine, L-tyrosine and L-tryptophan, 4-hydroxybenzoic acid (precursor of ubiquinone), 4-aminobenzoic acid (precursor of folic acid), and the phenylpropanes C_6-C_3, including the components of lignin, cinnamic acid derivatives and flavonoids are synthesized, and 2. the polyketide pathway in which acetate molecules are condensed and aromatic compounds (e. g. 6-methysalicylic acid) are synthesized via poly-β-keto acids. The biosynthesis of flavonoids (e. g. the anthocyanidins) can occur by either pathway.
Shikimic acid and chorismic acid are common precursors of all aromatics synthesized by the shikimic acid/chorismic acid pathway (Fig. 2). Figure 3

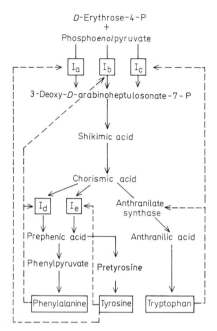

Figure 1. Simplified scheme of the regulation of aromatic biosynthesis. I_a-I_c = isoenzymes of phospho-2-keto-3-deoxyheptonate aldolase (EC 4.1.2.15) I_d, I_e = isoenzymes of chorismate mutase (EC 5.4.99.5).

shows the sequence of reactions up to chorismic acid, which is the branching point in the biosynthesis of the various aromatics.

The regulation of aromatic biosynthesis is achieved by feed-back mechanisms. The first step in the synthesis of the three aromatic amino acids is catalysed by phospho-2-keto-3-deoxy-heptonate aldolase (EC 4.1.2.15). For each amino acid there is a separate isoenzyme which is subject both to endproduct inhibition and to repression of its synthesis by the corresponding amino acid (Fig. 3). Regulatory isoenzymes are also involved in the transformation of

chorismic acid into prephenic acid. The mechanisms of feed-back control and repression vary from species to species.

Arrow poisons: natural toxins used for coating arrows, spears or blow pipe darts. In ancient Greece, extracts of aconitin were used. Ouabain (see) and similar cardiac glycosides were (are) used in Africa. The jungle Indians of South America use various kinds of Curare alkaloids (see), and tribes in Columbia use Batrachotoxin (see) from the Columbian arrow poison frog.

Arylamidases: an ubiquitous group of aminopeptidases which preferentially cleave amino acid arylamides, e.g. alanine β-naphthylamide. The majority of A. are bound to particles or membranes, and are therefore thought to play an important role in resorptive and secretory protein transport. It is also generally assumed that the A. are involved in the degradation of peptide hormones and in the final phase of intracellular protein degradation. The A. can be classified on the basis of their specificity: A.A. hydrolyse Asp or Glu arylamides, A.B, Lys or Arg arylamides and A.N, Ala or Leu arylamides. High activities of the A. are found in the brush borders of the intestinal mucosa, the kidney tubules and liver plasma membranes. The M_r values of the particulate A. isolated to date lie between 125000 and 280000. Determination of A. in the serum and urine is considered to have a certain diagnostic value in some liver diseases.

2-Arylbenzofurans: natural compounds with the ring system shown (Fig.). Biosynthetically, they form an heterogeneous group. Thus, egonol (5-(3-hydroxypropyl)-7-methoxy-3':4'-methylenedioxy-2-arylbenzofuran), a constituent of the unsaponifiable fraction of the seed oil of *Styrax japonicum*, may be derived by loss of a carbon atom from a bisarylproponoid, i.e., of lignan or neolignan origin. 6,3',5'-trihydroxy-2-arylbenzofuran occurs with the stilbene oxyresveratrol, and is probably derived from it by oxidative cyclization. Other 2-arylbenzofurans are found in members of the *Leguminoseae*, and several are phytoalexins. They are accompanied by isoflavonoids with similar substitution patterns, and therefore appear to be derived from isoflavonoids by the loss of a carbon atom, e.g. vignafuran

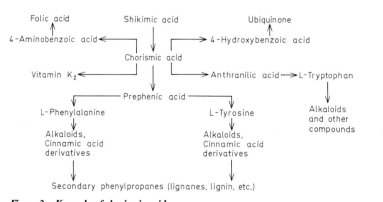

Figure 2. Key role of chorismic acid

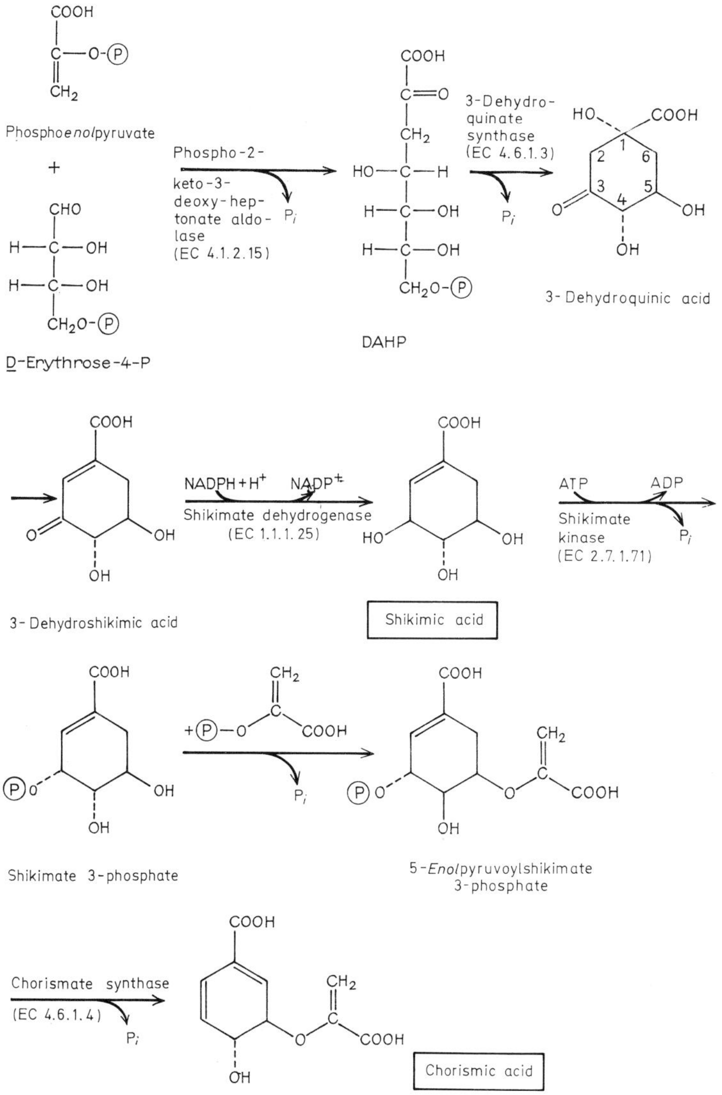

Figure 3. Synthesis of chorismic acid. DAHP = 3-deoxy-D-arabinoheptulosonic acid-7-phosphate.

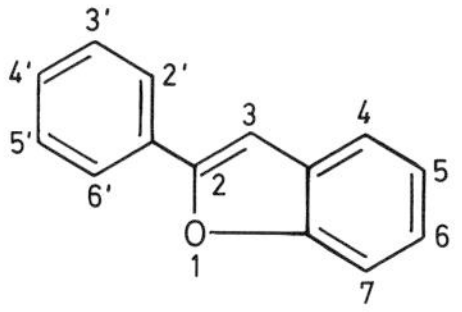

2-Arylbenzofuran ring system

(6,2'-dimethoxy-4'-hydroxy-2-arylbenzofuran), from *Lablab niger* and *Vigna unguiculata*. [P. M. Dewick in J.B. Harborne & T.J. Mabry (eds.), *The Flavonoids: Advances in Research* (Chapman and Hall, 1982) 602–605]

3-Aryl-4-hydroxycoumarins: a closely related group of naturally occurring Isoflavonoids (see), e.g. *scandenin* (Fig.) from *Derris scandens.* [C.P. Falshaw et al. *J. Chem. Soc.* (C) (1969) 374–382]

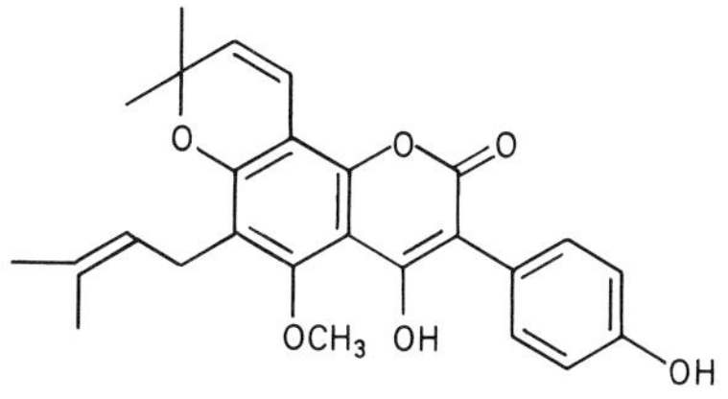

Scandenin

Arylsulfatases: the best-studied group of sulfatases. A. hydrolyse aromatic sulfate esters at the O-S bond in the reaction $R\text{-}O\text{-}SO_3^- + H_2O \rightarrow R\text{-}OH + H^+ + SO_4^{2-}$, where R-OH is a phenol. Typical substrates of A. are 2-hydroxy-5-nitrophenylsulfate (nitrocatechol sulfate) and 2-hydroxy-4-nitrophenyl sulfate. The A. can be divided into two groups, those which are inhibited by sulfate (Type II) and those which are not (Type I). Most microbial A. are of Type I; the only other member of this group is the microsomal A. of vertebrates (A.C), which is usually contaminated by other sulfatases. The A. of the mold *Aspergillus aerogenes* (M_r 40700; high substrate specificity) and *Aspergillus oryzae* (M_r 90000; low substrate specificity) have been thoroughly studied. Type II A. are found primarily in animal lysosomes. The A. from beef liver has been obtained in highly purified form. There are two forms with different M_r and isoelectric points, the abundant A.A. (M_r 107000; 4 subunits of 27000 each, IP=3.4) and the minor form A.B. (M_r 45000, single-chained (?); IP=8.3).

Ascorbate oxidase (EC 1.10.3.3): see Vitamins.

Ascorbate shuttle: a shuttle system in which cytosolic ascorbate provides reducing equivalents for regeneration of ascorbate in the matrix of adrenal chromaffin granules. Ascorbate in chromaffin granules is oxidized during noradrenalin biosynthesis by intragranular Dopamine β-hydroxylase (see). Electrons are transported across the granule membrane by membrane-bound cytochrome b_{561}. The process is driven by the proton motive force generated by a membrane ATP-ase, which transports protons from the cytosol to the matrix of the granule. Cytosolic ascorbate is regenerated by mitochondrial semidehydroascorbate reductase. Since the interconversion of ascorbate and its semidehydro free radical involves a proton, the pH gradient across the granule membrane ($\Delta pH = 1.5$ units) cooperates with the transmembrane potential ($\Delta\psi = +0.06$ V) to shift the E_h of the ascorbate/semidehydroascorbate couple inside the granule ($E_{h(inside)} = E_{h(outside)} + \Delta p$)

from $+0.07$ V (cytosol) to $+0.22$ V, thereby favoring flow of electrons into the granule (Fig.) [M.F. Beers et al., *J. Biol. Chem.* **261** (1986) 2529-2535; L.M. Wakefield et al., *J. Biol. Chem.* **261** (1986) 9739-9745 and 9746-9752]

Ascorbic acid, *vitamin C:* see Vitamins.

Asn, *Asp-NH₂:* abb. for L-asparagine.

L-Asparaginase (EC 3.5.1.1): a widely occurring enzyme which hydrolyses L-asparagine to L-aspartate and ammonia. The enzyme is used as an antitumor agent, particularly against lymphosarcomas and lymphatic leukemia. The cells of these cancers, unlike normal cells, cannot compensate for the lack of asparagine. The tumor cells still have an asparagine synthetase, but they have lost the ability to synthesize glycine from serine, so they cannot synthesize asparagine from glycine via glyoxylic acid and oxaloacetate.

To date, only microbial L-asparaginases have been highly purified (from *Escherichia coli, Actinetobacter, Erwinia carotovora*). They have M_r in the range of 135000 to 138000 and are composed of four identical subunits (M_r 33000).

L-Asparagine, abb. *Asn* or *Asp-NH₂:* $H_2N\text{-}OC\text{-}CH_2\text{-}CH(NH_2)\text{-}COOH$, the β-half-amide of L-aspartic acid. M_r 132.1, m.p. (hydrate) 236 °C (d.). Asn and aspartic acid occur ubiquitously, both in free form and as protein components. Asn plays a role in the metabolic control of cell functions in nerve and brain tissue. Asn and L-glutamine are used by many plants as soluble nitrogen reserve substances. Asn is synthesized from L-aspartic acid and ammonia by an asparagine synthetase (EC 6.3.1.1 or 6.3.1.4); cleavage of the amide bond by asparaginase (EC 3.5.1.1) is usually the first step in its degradation.

Aspartate ammonia-lyase, *aspartase* (EC 4.3.1.1): an enzyme found in bacteria, higher plants and a few lower animals which catalyses the reversible interconversion of aspartate and fumarate plus ammonia:

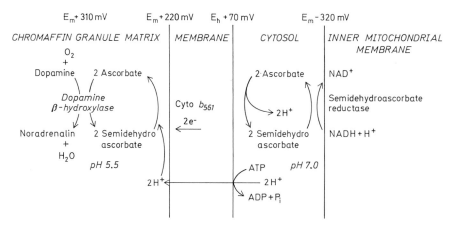

Ascorbate shuttle. E_m, midpoint potential; E_h, standard half-cell potential. Chromaffin granules contain 40 nmol ascorbate/mg protein, representing a concentration of 20 mM if all is in solution.

$$^-OOC-CH_2-CH-COO^- \rightleftharpoons {}^-OOC-CH=CH-COO^- + NH_4{}^+.$$
$$\underset{NH_3{}^+}{\mid}$$

At high temperatures, the deamination reaction occurs spontaneously. This pathway, which exists only for aspartate and not for the other amino acids found in proteins, is lacking in vertebrates, where it would contribute to the formation of ammonia, which is toxic to them. The better studied bacterial A. (M_r 180000) is composed of four equal-sized subunits (M_r 45000).

L-Aspartic acid, abb. *Asp:* 2-aminosuccinic acid, $HOOC-CH_2-CH(NH_2)-COOH$, a proteogenic, acidic amino acid which is not essential for mammals. M_r 133.1, m.p. 269 to 271 °C, $[\alpha]_D^{25}$ +5.05 ($c=2$, water). Asp and oxaloacetic acid are interconvertible through transamination. The enzyme aspartate ammonia-lyase (EC 4.3.1.1) is more important for the deamination of Asp than for its synthesis from fumaric acid and ammonia. Asp plays an important role in the urea cycle and in purine and pyrimidine biosynthesis, the latter by way of orotate. In the synthesis of urea and purines, Asp donates its α-amino group in a transamination which is dependent on ATP, but does not require pyridoxal phosphate. The N-1 atom of the purine ring system and the 6-amino group of adenine are derived from the amino group of Asp.

Aspartylglycosamine: see Inborn errors of metabolism.

Aspartyl glycosaminuria: see Inborn errors of metabolism.

Aspergillic acid: an antibiotic, M_r 224.3, produced by *Aspergillus flavus*. See Hydroxamic acid.

Aspiculamycin: see Gougerotin.

Assimilate: in the wider sense, a product of assimilation. In the narrower sense, a stabilized end-product of Photosynthesis (see).

Assimilation: the incorporation of nutrients into an organism. See Carbon dioxide assimilation.

Assimilatory power: see Photosynthesis.

Assimilatory sulfate reduction: see Sulfate assimilation.

Astaxanthine: 3,3′-dihydroxy-β,β-carotene-4,4′-dione, M_r 596.82, m.p. 216 °C. The compound occurs widely as a red animal pigment, especially in crustaceans, echinoderms and tunicates. It is also found in the feathers and the skin of flamingos and other birds, but is not found in many plants. In native form the compound can occur free, as a red pigment, as an ester, such as the dipalmitate, or as a blue, green or brown chromoprotein. The dark, blue-black pigment in the shell of the lobster *Astacus*

gammarus is a complex of A. with protein which, upon denaturation (for example cooking) releases red A. and colorless protein. R. Kuhn isolated A. and determined its structure.

Asteromycin: see Gougerotin.

Asterosaponin A: a steroid saponin (see Saponins). The aglycon is 3β-,6α-dihydroxypregn-9(11)-en-20-one, which is derived from pregnane (see Steroids), and it is linked to two molecules each of 6-deoxy-D-glucose and 6-deoxy-D-galactose and one molecule of sulfuric acid. A. was first isolated from the starfish *Asterias amurensis*.

AT-content: see GC-content.

Atmungsferment: cytochrome oxidase.

Atomic mass unit: see Dalton.

ATP: abb. for Adenosine 5′-triphosphate (see Adenosine phosphates).

ATPase: see Adenosine triphosphatase.

ATP citrate (pro-3S)-lyase, *citrate cleavage enzyme* (EC 4.1.3.8): a cytosolic enzyme converting citrate to acetyl-CoA and oxaloacetate and simultaneously cleaving one ATP to ADP: Citrate + ATP+CoA→Acetyl-CoA+oxaloacetate+ADP+P_i.

ATP-imidazole cycle: see L-Histidine.

ATP: urea amidolyase: see Urea amidolyase.

Atractyloside: a glucoside from the Mediterranean thistle *Atractylis gummitera*. It is a competitive inhibitor of adenine nucleotide binding and transport across the inner mitochondrial membrane. The closely related carboxyatractylate binds with higher affinity (K_d 10^{-8} M) and is not displaced by adenine nucleotides.

Atrial natriuretic factor, *ANF:* a polypeptide hormone produced by atrial heart muscle. ANF is a potent diuretic (natriuretic) and hypotensive agent, and inhibits renin and aldosterone secretion. Immunocytochemical studies show that ANF is stored within specific granules of atrial cardiocytes. Immunocytochemistry and radioimmunoassay indicate the existence of ANF in the central nervous system and the kidney, but in mammals the quantity of ANF in the atria is orders of magnitude higher than in extracardiac tissues. Atria also contain the greatest amount of ANF precursor mRNA. In nonmammals, ANF synthesis may also occur in ventricular muscle, and its production in tissues other than the heart may be quantitatively different from that in mammals.

Cloning and sequence analysis of cDNA for ANF precursor show that ANF is synthesized in a prepro

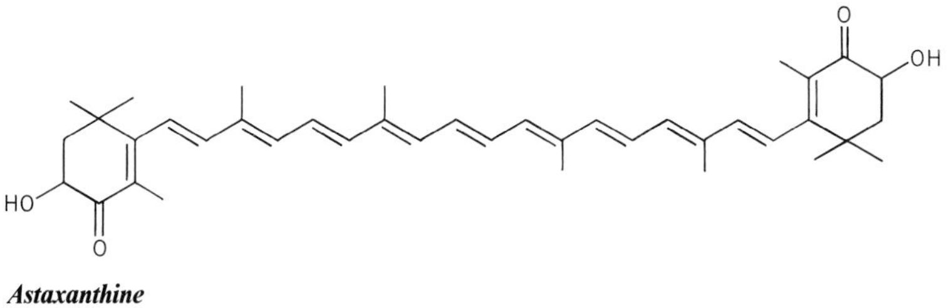

Astaxanthine

form containing 152 amino acids (rat) or 151 amino acids (human). A disulfide-looped sequence of 17 amino acids with various C- and N-terminal extensions is necessary for activity. In rat and human atrial granules and homogenate, the main form of ANF is a 126-residue peptide called cardionatrin IV or γ-atrial natriuretic peptide (γ-ANP); this is derived from prepro-ANF by processing. Circulating ANF appears to be a 28-residue peptide, called cardionatrin I or α-atrial natriuretic peptide (α-ANP). A previously described β-ANP is probably an artifact formed by proteolysis during isolation. Studies with cultured cardiocytes suggest that, in the intact animal, pro-ANF is secreted from the atria, then cleaved to α-ANP by a blood protease.

Specific receptors for ANF have been demonstrated in kidney, blood vessels and adrenal cortex. In vitro studies show that ANF release is stimulated by adrenalin, arginine vasopressin, acetylcholine and atrial distension, but the true nature of the physiological control of ANF release is not clear. Circulating plasma levels vary between 25 and 100 pg/ml in humans, and 100 and 1000 pg/ml in rats. ANF probably functions in the short- and long-term control of water and electrolyte balance. [A.J. deBold, *Science* **230** (1985) 767–770 (rat ANF); K. Kangawa et al. *Nature* **313** (1985) (human ANF); K.D. Bloch

et al. *Science* **230** (1985) 1168–1171; T.G. Flynn & P.L. Davies, *Biochem. J.* **232** (1985) 313–321]

Atrial peptide: see Atrial natriuretic factor.

Atromentin: see Benzoquinones.

Atropine, DL-hyoscyamine: a racemate formed during alkaline treatment of L-hyoscyamine. M_r 289.48. A. is the ester of DL-tropic acid with tropine, m.p. 114–116 °C. It specifically inhibits those cholinergic neurons which are activated by muscarine. L-Hyoscyamine is the tropine ester of L-tropic acid, m.p. 108–111 °C, $[\alpha]_D$ −21° (ethanol). The pupil-dilating effect is due only to the (−)-form, but only racemic atropine preparations are used medicinally. It is also used to inhibit salivation and sweating and to relax cramps in the gastrointestinal tract and bronchi (in asthma). A. is highly toxic. For formula, occurrence and biosynthesis, see Tropane alkaloids.

Atroscine: see Scopolamine.

Attenuation: a regulatory mechanism employed by the bacterial cell. Whereas Enzyme repression (see) allows the cell to respond to extreme concentrations of metabolites, A. probably represents a means of "fine tuning" to relatively mild fluctuations in the concentrations of metabolites. A. sites are probably present in all amino acid synthesis operons in bacteria; so far, A. has been demon-

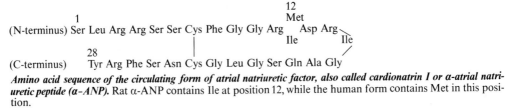

(N-terminus) Ser Leu Arg Arg Ser Ser Cys Phe Gly Gly Arg ... Asp Arg ...
1 ... 12 Met ... Ile ... Ile
28 (C-terminus) Tyr Arg Phe Ser Asn Cys Gly Leu Gly Ser Gln Ala Gly

Amino acid sequence of the circulating form of atrial natriuretic factor, also called cardionatrin I or α-atrial natriuretic peptide (α-ANP). Rat α-ANP contains Ile at position 12, while the human form contains Met in this position.

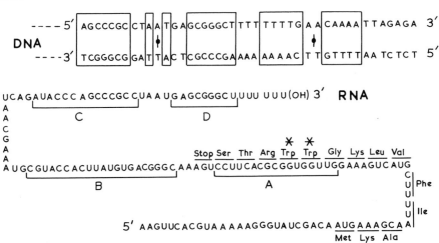

Figure 1. Attenuator site in the leader of the tryptophan synthesis operon of Escherichia coli, and the complete sequence of the terminated leader RNA. Two regions in the *trp* attenuator DNA have a two-fold axis of symmetry. A, B, C, D in the RNA correspond to similarly labelled regions in Fig.2. Base pairing with resulting stem and loop formation is possible between A and B (free energy of formation −46.9 kJ/mol), B and C (−49 kJ/mol), C and D (−83.7 kJ/mol). Codons for the leader peptide are also shown and the two strategic Trp codons are indicated by asterisks.

strated and studied in the tryptophan and phenylalanine operons of *Escherichia coli,* and the histidine, leucine and tryptophan operons of *Salmonella typhimurium.*

The first structural gene of the operon is separated from the promoter-operater by a length of DNA called the leader. Transcription of the operon proceeds via this leader and into the structural genes. For example, in the case of the *trp* operon of *E. coli,* a single continuous 7000 nucleotide *trp* mRNA transcript is formed, which includes leader RNA (162 nucleotides) at the 5' end, followed by mRNA

sequences for all the enzymes in the biosynthetic pathway. This only occurs, however, when tryptophan is relatively scarce. When the tryptophan level is high, transcription proceeds only part of the way through the leader and is then terminated, producing a small transcript consisting of 140 nucleotides. The controlling factor is not tryptophan itself, but tryptophan-aminoacyl-tRNA, the level of which reflects the availability of tryptophan. Part of the leader RNA is translated into a peptide of 14 amino acid residues, two of which are tryptophan residues. Codons for these two Trp residues occupy strategic

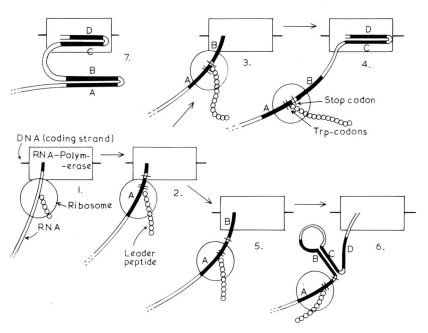

Figure 2. Diagrammatic representation of attenuation during coordinate transcription and translation, based on attenuation of the tryptophan synthesis operon in Escherichia coli.

1. RNA-polymerase has left the operater-promoter site, and has started to transcribe the leader region as it proceeds in the direction of the first structural gene of the operon (component I of anthranilate synthetase). The start codon for the leader peptide has appeared, a ribosome has become attached, and translation has started.

2. As more RNA is synthesized, translation keeps pace with transcription. Region A has been synthesized and B is just beginning. The ribosome is about to translate the Trp codons of the leader peptide message.

3. Tryptophan-aminoacyl-tRNA is plentiful, so translation continues and keeps pace with transcription. The ribosome has just reached the stop codon, and partly covers region B which is now fully synthesized.

4. The ribosome can proceed no further, but transcription continues. When region C first appears, it cannot base pair with B, due to the presence of the ribosome. As D appears, it base-pairs with C. The resulting secondary structure (CD loop and stem) acts as a signal for the termination of transcription. The RNA-polymerase and its transcript leave the DNA.

5. Tryptophan-aminoacyl-tRNA is scarce, so the ribosome pauses at the Trp codons. As B is synthesized it is not partly covered by the ribosome. Translation is not keeping pace with transcription.

6. As region C emerges from the RNA-polymerase, it is able to base-pair with B, forming a stem-loop structure, known as a preemptor (it preempts the formation of CD). As D is formed, C is not available for base-pairing, so the transcription termination signal (CD loop and stem) cannot be formed. Transcription continues into the first structural gene, and the total operon is transcribed.

7. If other aminoacyl-tRNA species are scarce, the ribosome will not even reach A by the time D has appeared. The thermodynamically favored structure then consists of the respective stem and loop structures AB and CD. Transcription is terminated by CD. Thus, a deficiency of other amino acids can attenuate the synthesis of tryptophan.

positions on the leader RNA (see Figs.1 and 2). Analysis of leader RNA reveals more than one possible secondary structure. Which secondary structure actually occurs is determined by the progress of ribosomes along the RNA. If tryptophan-aminoacyl-tRNA is plentiful (it is assumed that all the other aminoacyl-tRNA species needed are also plentiful), translation proceeds smoothly and the ribosome reaches the stop codon. At this point, the ribosome physically overlaps two regions of the RNA, that are potentially capable of taking part in secondary structure formation (regions A and B in Figs.1 and 2). The only possible secondary structure is therefore the stem and loop formed by regions C and D, which acts as a transcription termination signal. If tryptophan-aminoacyl-tRNA is scarce, the ribosome halts earlier at the trp codons, only region A is covered, and a noninhibitory secondary RNA structure is formed; transcription then proceeds into the structural genes. The mechanism is better appreciated by considering the coordinate progress of transcription and translation as shown in Fig.2. It can be seen that A. represents a graded response to slight changes in the level of the control amino acid. This is not an "all or nothing" or threshold type of control, unlike repression or feedback inhibition which regulate by an on-off switch mechanism. It is unlikely that the ribosome actually stops at the *trp* codon; it may slow down or pause, depending on the relative supply of amino acyl-tRNA of the control amino acid. In view of the speed of transcription and coordinate translation (45 nucleotide residues incorporated per second; about 4 min for transcription of the total *trp* operon; about 3 min for complete degradation of the *trp* operon transcript) a mere slowing of the ribosome will be sufficient to favor the noninhibitory structure of the leader RNA. Even the two secondary RNA structures must be considered as two thermodynamic extremes that are, to a certain extent, in equilibrium.

In all the other A. systems studied so far, the mechanism of A. is analogous, codons for the control amino acid being located at strategic positions in the leader RNA. Four (e.g. *leu* A.) or seven (e.g. *phe* and *his* A.) such strategic codons may be present. [Oxender, D.L., Zurawski, G. and Yanofsky, C. (1979) *Proc. Natl. Acad. Sci. USA, 76,* 5524–5528. Gemmill, R.M., Wessler, S.R., Keller, E.B. and Calvo, J.M. (1979) *Proc. Natl. Acad. Sci. USA, 76,* 4941–4945. Johnston, H.M., Barnes, W.M., Chumley, F.G., Bossi, L. and Roth, J.R. (1980) *Proc. Natl. Acad. Sci. USA, 77,* 508–512].

Aucubin: see Iridoids.

Auranofin , *(2,3,4,6-tetra-O-acetyl-1-thio-β-D-glucopyranosato-S)(triethylphosphine)gold(I):* an orally active antiarthritic agent, which modulates the functional activities of macrophages in vitro and in vivo. Studies on ^{195}Au-labelled A. have provided evidence for a sulfhydryl shuttle, in which gold is removed from the drug and becomes associated with cellular protein.

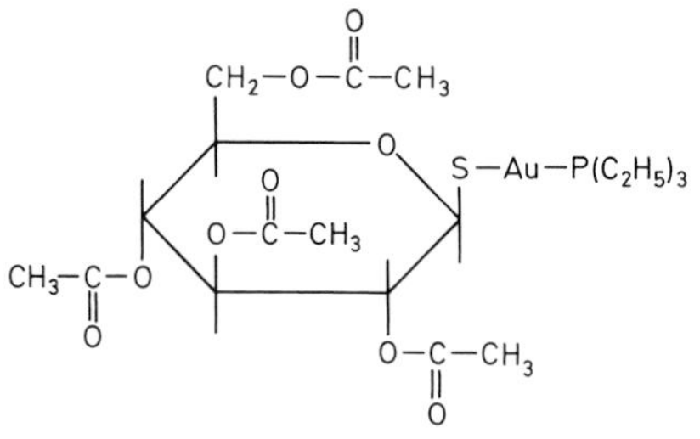

Auranofin

Association of Au with the cell membrane

Auranofin + Mem-SH → Tetraacetylglucose + Mem-S-Au-Et$_3$P
   +
R-SH
   ↓

Tetraacetylglucose + R-S-Au-Et$_3$P + Mem-SH ⇌ R-SH + Mem-S-Au-Et$_3$P

Internalization

Mem-S-Au-Et$_3$P + Cyt-SH ⇌ Mem-SH + Cyt-S-Au-Et$_3$P $\xrightarrow{\text{ox.}}$ Cyt-S-Au$^-$ + OEt$_3$P

Cyt-S-Au$^-$ + Cyt-SH → Cyt-S-Au-S-Cyt

Efflux

Cyt-S-Au-S-Cyt + Mem-SH ⇌ Cyt-S-Au-S-Mem and/or Mem-S-Au-S-Mem
   ↓   +
Cell lysis   R-SH
   ↓   ⇅
Cyt-S-Au-S-Cyt   R-S-Au-S-R + Mem-SH

Proposed sulfhydryl shuttle for the cellular internalization and efflux of the gold atom of auranofin. Mem-SH, membrane sulfhydryl; R-SH, extracellular sulfhydryl; Cyt-SH, cytosolic sulfhydryl; Et$_3$P, triethylphosphine. [R.M. Snyder et al. *Biochem. Pharmacol.* **35** (1986) 923–932]

Auriculoside: 7,5'-dihydroxy-4'-methoxy-3'-O-β-D-glucosylflavan, see Flavan.

Aurones, *2-benzylidene-3-cumarones:* yellow flavonoid plant pigments. A. are very common in *Compositae, Leguminosae, Hepaticae* and *Anacardiaceae,* and have been found sporadically in a few other plant families. Individual A. differ with respect to the substituents in rings A and B; common substituent groups are hydroxy, methoxy, methylenedioxy and furano. Examples are sulfuretin (6,3',4'-trihydroxyaurone, a petal pigment in many members of the *Compositae*), and furano (2″, 3″, 6,7):3,4-methylene-dioxyaurone from *Derris obtusa.* A. frequently occur as glycosides.

A checklist of 19 known, naturally occurring A. is given by B. A. Bohm in *The Flavonoids: Advances in Research* (J. B. Harborne & T. J. Mabry, eds) (Chapman and Hall, 1982) pp. 336–337.

Aurone ring system

Autoimmune diseases: conditions resulting from a lack of immune tolerance for the organism's own components. They may arise by 1. cross reactions between body substances and foreign materials, 2. by lack of immune tolerance for body substances which are normally not exposed to the immune apparatus, or 3. by structural changes in body substances or through reactivation of certain cells which lead to a breakdown of tolerance. A. d. include neurological conditions like multiple sclerosis, bacterial eye inflammation (ophthalmia sympathica), myasthenia gravis, erythematodes (a skin disease), chronic rheumatism, chronic liver inflammation and certain forms of chronic kidney inflammation.

Autolysis: the self digestion of dying cells which occurs when the cathepsins (see Proteolysis) are released from lysosomes and attack cytoplasmic proteins. In the living cell these proteases are sequestered inside the lysosome membranes, but at death the membranes break down. (See Intracellular digestion).

Autophagy: see Intracellular digestion, Proteolysis.

Autotrophy: the ability to synthesize all organic components from simple inorganic compounds like carbon dioxide, ammonia, nitrate and sulfate. The term originally applied only to *carbon autotrophy,* the ability of green plants to assimilate CO_2 from the air (see Carbon dioxide assimilation). *Nitrogen autotrophy* refers to nitrate assimilation (see Nitrate reduction) or to Nitrogen fixation (see), processes by which ammonia can be formed and assimilated (see Ammonia assimilation). *Sulfur autotrophy* depends on the ability of plants and microbes reductively to assimilate sulfate (see Sulfate assimilation). The opposite of autotrophy is heterotrophy.

Auxin antagonists: inhibitors of auxins whose effects can be at least partly reversed by auxins. The term is used independently of the mechanism of inhibition; competitive inhibitors are called antiauxins. The A. a. include many structurally different kinds of compounds, including some synthetic auxins, e. g. phenylacetic acid and phenylbutyric acid.

Auxin conjugates: see Auxins.

Auxins: a group of plant hormones which regulate growth. They stimulate extension growth and cell division in the cambium and root; they influence certain enzyme activities. Natural A. are indole derivatives biosynthesized from tryptophan. The most important A. is indole-3-acetic acid (I. A. A., heteroauxin) (Fig.), which is used as the standard for comparison of the activity of other growth stimulants. I. A. A. was isolated in 1934 from human urine by Kögl and was recognized as a natural plant hormone in 1950. The existence of auxin effects in plants had been established in 1926/1928 by Went. I. A. A. is synthesized in all higher plants and has also been found in many lower plants and even in bacteria. Pineapple plants, for example, contain 6 µg/kg fresh weight. A. is determined quantitatively by the Oat coleoptile test (see). Other natural A. include

2-(3-indolyl)ethanol (tryptophol)
2-(3-indolyl)acetaldehyde
2-(3-indolyl)acetonitrile
2-(3-indolyl)acetamide
3-(3-indolyl)propionic acid
4-(3-indolyl)butyric acid
3-(3-indolyl)succinic acid
2-(3-indolyl)glycolic acid
2-(3-indolyl)acetyl-β-D-glucose
[2-(3-indolyl)acetyl]aspartic acid
2-(3-indolyl)acetyl-mesoinositol
2-(3-indolyl)acetyl-mesoinositol galactoside
2-(3-indolyl)acetyl-mesoinositol arabinoside

The A. at the end of the list, containing sugars or amino acids, are called auxin conjugates and are probably transport forms.

2-(3-Indolyl)acetic acid

Auxochromes: see Pigment colors.

Auxotrophic mutants, *deficiency mutants:* mutants with nutritional requirements not present in the parental or wild-type form. The substances which cannot be synthesized by the mutant must be supplied as vitamins or substrates in the growth medium. A. m. have an important role in the elucidation of biosynthetic pathways (see Mutant technique). They may be monoauxotrophic, requiring only one substance as a growth supplement, or polyauxotrophic. In the latter case, the defects may be polygenic, or the loss of one gene may affect the synthesis of several products. This can happen (a) when the block occurs before the branching point of a branched synthesis chain, or (b) an enzyme required in two or more parallel synthetic paths may be defective. An example of the first case (a) are methionine-isoleucine double auxotrophs, which re-

53

quire supplementation of a minimal medium with methionine + isoleucine (or threonine, since threonine can be converted to isoleucine). The metabolic block occurs before the formation of homoserine, the branching point in this synthetic pathway:

Precursors → L-homoserine → L-threonine
↓ ↓
L-methionine L-isoleucine.

An example of the second type (b) is the mutation leading to loss of the ketol-acid reductoisomerase (EC 1.1.1.86) in the parallel pathways leading to valine and isoleucine. This enzyme catalyses an isomerization coupled to a reduction, producing 2,3-dihydroxyisovaleric acid in the valine pathway, or 2,3-dihydroxy-3-methylvaleric acid in the isoleucine pathway. A single mutation of the gene for this enzyme produces val^-/ile^- doubly auxotrophic organisms.

$$\begin{array}{ccccccccc} & (1) & & (2) & (2) & & (3) & & (4) \\ \text{Pyruvate} & \to & I_1 & \to & I_2 & \to & I_3 & \to & I_4 & \to & \text{L-valine} \end{array}$$

Reductoisomerase

$$\begin{array}{ccccccccc} \text{Ketobu-} & \to & I_a & \to & I_b & \to & I_c & \to & I_d & \to & \text{L-isoleucine} \\ \text{tyrate} & & (1) & & (2) & & (2) & & (3) & & (4) \end{array}$$

Here (2) is the reductoisomerase, I_3 is 2,3-dihydroxyisovaleric acid and I_c is 2,3-dihydroxy-3-methylvaleric acid.

Polyauxotrophic nutritional requirements cannot be induced by the mutation of a single gene and loss of a single protein if the metabolic step in question is catalysed by isoenzymes encoded in different genes.

Auxotrophy: the condition of requiring nutritional supplements for growth (see Growth factors). A. can arise through mutation (see Auxotrophic mutants). Opposite: Prototrophy (see).

Avena coleoptile test: see Oat coleoptile test.

Avenasterol, *28-isofucosterol:* an isomer of Fucosterol (see) found in green marine algae (see Sterols).

Avermectins: a class of macrocyclic lactones produced by the actinomycete *Streptomyces avermitilis.* The A. are toxic at low dosage to nematode and arthropod parasites of cattle, horses, sheep, dogs and swine. They may eventually be used against human parasites as well. The A. appear to interfere with the 4-aminobutyric acid (GABA) (see) receptors in those synapses for which GABA is the neurotransmitter. In mammals, such neurons occur only in the central nervous system, where they are protected by the blood-brain barrier. [*Science* 221 (1983) 823–827]

Avidin: a basic glycoprotein found in the egg whites of many birds and amphibia. The primary structure of chicken A. is known: M_r 66000, IP 10, 10.5% threonine. There are 4 identical subunits, 128 amino acids each, M_r 14332 (without carbohydrate). A. forms a stoichiometric but non-covalent complex with 4 molecules of the vitamin biotin; the complex is not attacked by proteolytic enzymes and thus not resorbed. Feeding of A. or raw egg white can therefore result in an experimental biotin deficiency (see Vitamins, Biotin). A. is denatured and thus inactivated by heating. Like the unrelated lysozyme and conalbumin (see Siderophilins), A. protects the egg white against bacterial invasion.

Avitaminosis: see Vitamins.

Axerophthol: see Vitamins, Retinol, etc.

5-Azacytidine, *1-β-D-ribofuranosyl-5-azacytosine:* a pyrimidine antibiotic (see Nucleoside antibiotics) synthesized by *Streptoverticillius lakadamus* var. lakadamus. 5-A. is effective against Gram-negative bacteria.

5-Azacytidine

8-Azaguanine, *pathocidin:* a purine antagonist first synthesized in 1945 (Roblin et al. [1945], *J. Am. chem. Soc.* 67, 290–294). Systematic name, using the system of Patterson and Capell ("The Ring Index" Reinhold Publishing Corp. [1940], System 702) is 5-amino-7-hydroxy-1-*v*-triazolo [d] pyrimidine. M_r 152.04; it decomposes without melting above 300 °C. Later, it was shown to be identical with pathocidin, an antibiotic from *Streptomyces spectabalis.* 8-Azaguanine affects many different enzymes of purine metabolism and synthesis. In particular it inhibits translation and causes errors of translation, due to its incorporation into mRNA. It is converted by the cell into the corresponding nucleoside 5′-triphosphate, which feedback inhibits the first enzyme of purine biosynthesis, 5-phosphoribosylpyrophosphate amidotransferase (EC 2.4.2.14). 8-Azaguanine is inhibitory and toxic to a wide variety of living systems, including bacteria, protozoa and higher animals. It inhibits the growth of several murine adenocarcinomas, but has found only limited application in the treatment of human cancers.

A-Azahomosteroids: see Salamander alkaloids.

L-Azaserine, *O-diazoacetyl-L-serine:* $\bar{N} = N = CH-CO-OCH_2-CH(NH_2)-COOH$, a glutamine analog. It inhibits the transfer of the amide group (see Transamidination) from L-glutamine to formylglycinamide ribotide. It is mutagenic and has antitumorigenic activity. L-A. is synthesized by *Streptomyces* species and is effective against *Clostridia, Mycobacterium tuberculosis* and *Rickettsias*.

Azofer: see Nitrogenase.

Azofermo: see Nitrogenase.

Azomycin, *2-nitroimidazole:* an antibiotic synthesized by *Nocardia mesenterica* and *Streptomyces eurocidicus.* m.p. 281 to 283 °C. A. was isolated in 1953 by Maeda and in 1957 by Taguchi and Nakano. Both Gram-negative and Gram-positive bacteria are highly sensitive to the compound. 0.5 μg/ml inhibits the growth of *Trichomonas vaginalis,* 3 μg/ml inhibits the growth of *Salmonella paratyphi*.

A-Z solution: see Nutrient medium, Table 3.

Azulene, *cyclopentacycloheptene:* the parent compound of a group of blue to violet, nonbenzoid aro-

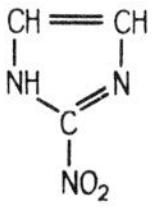

Azomycin

matics. The fused five and seven-membered rings are stabilized by a π-electron sextet.

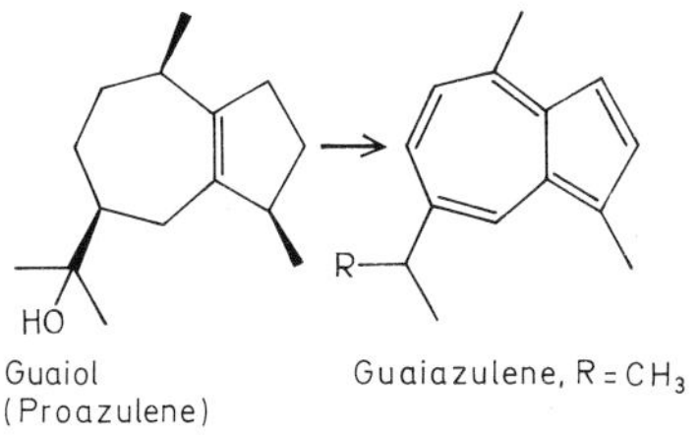

Guaiol   Guaiazulene, R = CH₃
(Proazulene)

Formation of azulenes

Originally, the term was used for the blue, high-boiling fraction of camomile oil. These compounds are artefacts, however, produced from colorless sesquiterpenes, the proazulenes (Fig.). The compounds have antiinflammatory properties. The most impor-

tant are chamazulene (R = H) and guaiazulene (R = CH₃).

Azulene-forming compounds: see Proazulenes.

Azulenogens: see Proazulenes.

Azurin: a family of blue, copper-containing proteins from *Pseudomonas, Alicaligenes* and *Bordetella* species. A. from *Pseudomonas fluorescens* contains 128 amino acid residues of known sequence, and one intrachain disulfide bond. A. from all sources has M_r 14000–16000, and homologous primary and tertiary structure, but different species of A. have different redox potentials.

A. contains one atom of Cu^{2+} per molecule, bound in a trigonal pyrimidal array, involving a distorted N_2SS donor set of Cys (S), Met (S) and 2 His (N). With respect to size and the nature of the Cu binding, A. is similar to plastocyanin (green plants) and stellacyanin (latex of Chinese laquer tree). Together with cytochrome c_{551}, A. is believed to function in a terminal respiratory network, using either O_2 or nitrate as the electron acceptor. On account of its blue color (λ_{max} 625–630 nm; ε about 7000) and its IP (pH 5.65), A. from *Pseudomonas aeruginosa* is a useful standardization marker in isoelectric focussing. [P. Rosen et al. *Eur. J. Biochem.* **120** (1981) 339–344; P. Frank et al. *J. Biol. Chem.* **260** (1985) 5518–5525]

B

B 9: see Succinic acid 2,2-dimethylamide.

B 995: see Succinic acid 2,2-dimethylhydrazide.

Bacimethrin: 4-amino-2-methoxy-5-pyrimidine-methanol, an antibiotic synthesized by *Bacillus megatherium*. M_r 155.16, m.p. 174 °C. B. was first isolated by Tanaka in 1961. It is antagonized by thiamin and pyridoxine, and is active against some yeasts and bacteria.

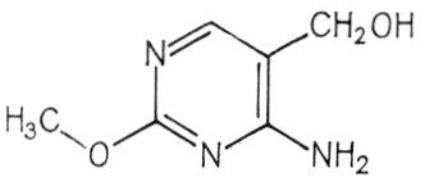

Bacimethrin

Bacitracins: branched, cyclic peptides produced by various strains of *Bacillus licheniformis*. They are effective against Gram-positive bacteria, in which they interfere with Murein (see) synthesis. The most important of the group is bacitracin A, which contains an unusual thiazoline structure synthesized from the *N*-terminal isoleucine and the neighboring cysteine. Bacitracin F is a rearrangement product of bacitracin A in which the amino group of the heteroproduct is oxidatively removed and the thiazoline ring system is dehydrated. It has little antimicrobial activity.

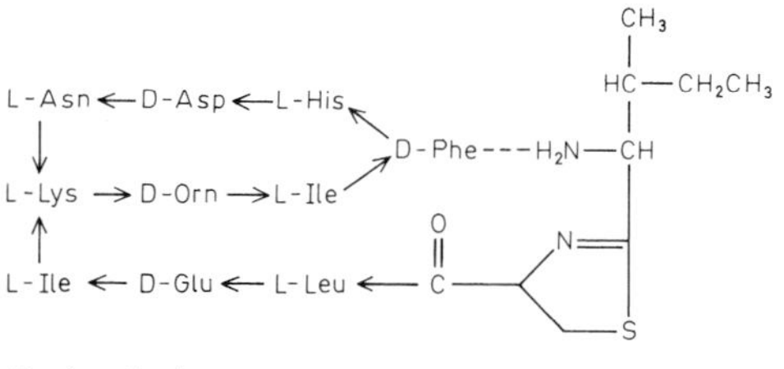

Bacitracin A

Bacterial chlorophylls: see Photosynthetic bacteria; Chlorophyll.

Bacterial photosynthesis: a primitive form of photosynthesis using H_2S, thiosulfate, fatty acids or other organic reducing agents rather than water as the source of hydrogen for carbon fixation. Unlike Photosynthesis in green plants (see), B.p. involves only one photosystem. It can thus carry out only cyclic photophosphorylation, not the noncyclic photophosphorylation seen in green plants. The formation of reducing equivalents (NADH instead of NADPH) is facultative and independent of light. B.p. is also characterized by Reversed electron transport (see).

The photosynthetic pigments involved differ from those in green plants. Chlorophyll *a* is replaced by bacteriochlorophyll *a*, which has absorption maxima between 800 and 1000 nm. In *Rhodospirillum rubrum*, cytochrome P_{700}, a spectral modification of chlorophyll *a*, is replaced by a modification of bacteriochlorophyll *a*, cytochrome P_{890}. It presumably plays the same role in bacteria as P_{700} in plants.

Bacteriochlorin: 7,8,17,18-tetrahydroporphyrin. See also Chlorin.

Bacteriocide: see Growth.

Bacteriophage: see Phages.

Bacteriorhodopsin: a retinaldehyde-containing, purple membrane protein first discovered in the halophilic bacterium *Halobacterium halobium*. M_r 26000, 248 amino acids. The primary structure of B. is not homologous to that of vertebrate rhodopsin (M_r 41000), but the tertiary structures of the two proteins are similar. B. consists of 7 α-helical regions which lie in the membrane. They are presumably connected by non-helical regions which protrude into the cytoplasm and extracellular space. The membrane portions of the molecule are elongated ovoids which have their long axes roughly perpendicular to the plane of the membrane. The retinaldehyde is linked to the protein via a Schiff's base with lysine 216. When activated by light, the protein acts as a proton pump. The physiological function of B. is unknown. Mutants which lack it have been isolated; these are able to extrude Na^+ and to generate ATP in the light by means of another retinaldehyde-containing protein, *haloopsin*. [M. A. Keniry, H. S. Gutowsky & E. Oldfield, *Nature* **307** (1984) 383–386]

Bacteriostatic agents: see Growth.

Bacteroids: symbiotic, nitrogen-fixing, intracellular forms of *Rhizobium* spp. in the root nodules of leguminous plants (see Rhizobia). A root nodule of the soybean contains several thousand cells. Each cell is tightly packed with membrane vesicles (syn. membrane envelope; the membrane of the vesicle consists of a double layer of phospholipid, and is derived from the plasmalemma of the host cell), and each vesicle contains about five bacteroids bathed in a solution of Leghemoglobin (see). Bacteroids will not redifferentiate into free-living *Rhizobia*; cultures of *Rhizobia* obtained from nodules are derived from unchanged bacterial cells remaining in the infection threads of the host tissue.

L-Baikiain: see Pipecolic acid.

Balata: a rubber-like polyterpene of low molecular weight obtained from the latex of tropical trees, especially from *Mimusops balata*. The double bonds in the compound are in the trans-configuration (see Polyterpenes, Fig.), and it is thus not very elastic, in contrast to rubber. It softens when heated.

Balbiani rings: see Giant chromosomes.

Baldrianal: see Valtratum.

Balsams, *oleoresins*: solutions of resins in volatile oils. B. are produced by plants, either normally or pathologically. The most important is turpentine, which is produced by conifers in response to bark injury in amounts up to 1 or 2 kg per tree and year. Steam distillation of the crude balsam yields turpentine oils; the residue is colophony (rosin). Other balsams are usually named after the country of origin (e.g. Peru balsam, Canada balsam) and are used in perfumes and to some extent in pharmaceuticals.

Bamicetin: a pyrimidine antibiotic (see Nucleoside antibiotics) synthesized by *Streptomyces plicatus*. M_r 604.65, m. p. 240 to 241 °C $[\alpha]_D^{26}$ +123 ° ($c = 0.5$ in 0.1 N HCl). B. has one fewer CH_2 group in its glycoside moiety than the related Amicetins (see). It is effective in low concentrations against *Mycobacterium tuberculosis*.

Base pairing: the specific hydrogen bonding between adenine and uracil (RNA) or thymine (DNA), and cytosine and guanine in a double-stranded molecule of RNA or DNA (Fig.). B.p. is the basis for the formation of a double-helical structure of DNA from two complementary single strands (see Deoxyribonucleic acid).

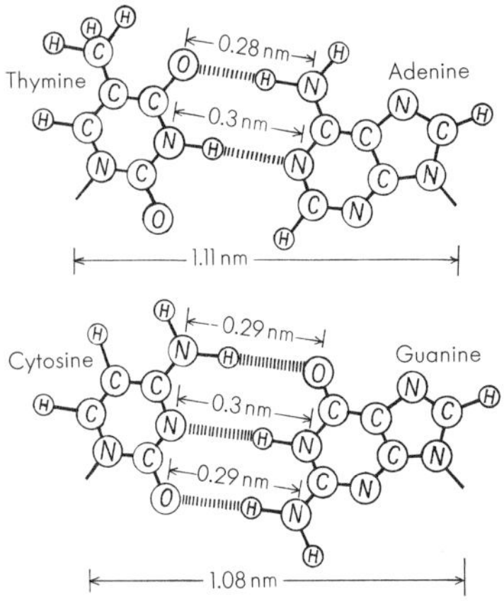

Hydrogen bonding between complementary bases in DNA

Basic proteins: a group of small proteins rich in arginine and lysine found in the cell nucleus (Histones; see) and in sperm (Protamines, see). They form complexes with the nucleic acids. The exact function of these complexes is unknown, but is probably structural (see Proteins).

Batatasins: see Orchinol.

Batch culture: see Fermentation technology.

Bathochrome: see Pigment colors.

Batrachotoxin: a neurotoxin from the skin of the Columbian arrow poison frog, genus *Phyllobates*. It causes a selective and irreversible increase in the permeability of nerve membranes to sodium ions. LD$_{50}$ in mice 2 μg/kg i. v. In order to be effective, B. must be injected or gain entry via damaged tissue. It is nontoxic when ingested, providing there are no scratches or ulcers in the gastrointestinal tract. Batrachotoxin A from the same source is less toxic, LD$_{50}$ in mice 1 mg/kg i. v.

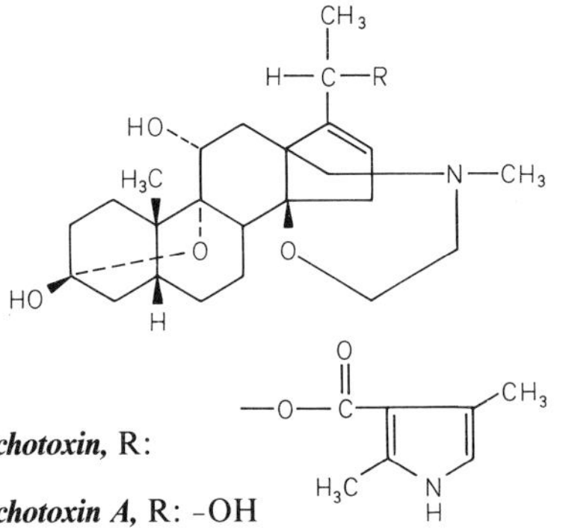

***Batrachotoxin,* R:**

***Batrachotoxin A,* R: –OH**

Bavachin: 7,4′-dihydroxy-6-prenylflavanone, see Flavanone.

Bay-region theory of carcinogenesis: an hypothesis that polycyclic hydrocarbons possessing a bay region are more potent procarcinogens than those lacking a bay region, e.g. phenanthrene is more carcinogenic than anthracene, and benz[a]pyrene is more potent than benz[c,d]pyrene. The carcinogens are epoxides formed from polycyclic hydrocarbons by the action of Cytochrome P450 (see), and in the most powerfully carcinogenic of these, the epoxide group is situated on the rim of a bay region. Thus, the ultimate carcinogen formed metabolically from benz[a]pyrene is the 7,8-diol-9,10-epoxide. Carcinogenesis is due to alkylation of DNA by a carbonium ion derived from the epoxide. Perturbational molecular orbital calculations suggest that formation of benzylic carbonium ions is facilitated if the carbonium ion is part of a bay region of a polycyclic hydrocarbon. In further support of the theory, synthetically prepared 1,2,3,4-tetrahydrophenanthrene-3,4-epoxide (epoxide group on the rim of the bay) is more carcinogenic than 1,2,3,4-tetrahydrophenanthrene-1,2-epoxide (epoxide group distal to the bay); similarly, 7,8,9,10-tetrahydrobenz[a]pyrene-9,10-epoxide is more carcinogenic than 7,8,9,10-tetrahydrobenz[a]pyrene-7,8-epoxide. Exceptions include chrysene and benz[e]pyrene, which possess bays, but are only weakly carcinogenic; these compounds are readily attacked in regions distal to the bays (regions of relatively high electron density and double bond character, known as "K" regions) forming diols which are rapidly conjugated with sulfate or glucuronate and excreted. See Fig. See Cytochrome P450, Aflatoxins. [M.C. MacLeod et al. *Cancer Res.* **39** (1979) 3463–3470]

BC: see Biotin enzymes.

B-cell growth factor, *BCGF*: see Lymphokines.

B$_{12}$-coenzyme: see 5′-Deoxyadenosylcobalamine.

Bdellins: a group of protease inhibitors from the leech. Especially high activities of B. are found in the region of the outer sex organs and in the salivary glands of the leech, *Hirudo medicinalis*. B. inhibit trypsin and plasmin, and they show a strong inhib-

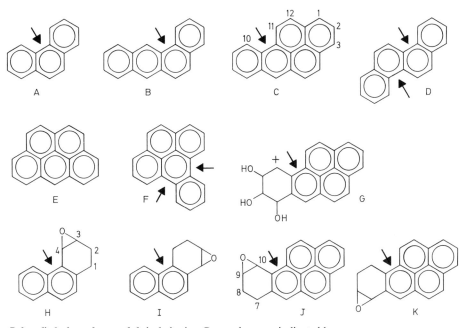

Polycyclic hydrocarbons and their derivatives. Bay regions are indicated by arrows.
A, Phenanthrene. *B*, Benzanthracene. *C*, Benz[a]pyrene. *D*, Chrysene. *E*, Benz[c,d]pyrene. *F*, Benz[e]pyrene.
G, Carbonium ion derived from benz[a]pyrene-7,8-diol-9,10-epoxide. *H*, 1,2,3,4-Tetrahydrophenanthrene-3,4-epoxide. *I*, 1,2,3,4-Tetrahydrophenanthrene-1,2-epoxide. *J*, 7,8,9,10-Tetrahydrobenz[a]pyrene-9,10-epoxide. *K*, 7,8,9,10-Tetrahydrobenz[a]pyrene-7,8-epoxide.

itory activity towards the trypsin-like protease, acrosin, present in the acrosomes of spermatozoa.

Bee toxin: a defense secretion produced in an abdominal gland of queen and worker bees (*Apis mellifera* L.) and delivered by the sting. It contains three types of active principle: (1) biogenic amines, including histamine, which cause pain and dilate the blood vessels, allowing wider penetration; (2) biologically active peptides such as Mellitin (see) and apamin, and (3) enzymes like hyaluronidase and phospholipase A. B.t. is applied intracutaneously or percutaneously for treatment of neuralgias, rheumatism and allergies.

Belladonna alkaloids: see Tropane alkaloids.

Bence-Jones proteins: proteins excreted in the urine of patients with multiple myeloma, a malignancy of antibody-producing cells. The proteins were first noticed because they precipitate on heating to 50 °C, but redissolve on further heating to 100 °C. They are now known to consist of dimers of immunoglobin L chains. The proteins are synthesized by clones of identical cells, so each patient produces identical polypeptides in large enough quantities for sequence determination. The proteins have been valuable in studies of the structure of Immunoglobins (see). Each light (L) chain consists of 214 amino acids and has M_r about 22500. They do not contain methionine and they have no regions of α-helix. The dimers are held together by disulfide bridges.

Benzoquinones: compounds derived from *p*-

benzoquinone. *p*-Benzoquinone, and its mono-, di- and trimethyl, ethyl, methoxy and 2-methoxy-3-methyl derivatives are found in the defense secretions of certain arthropods. More than 90 different B. have been isolated from higher plants, fungi and molds. Among them are the yellow skin irritant primin (m. p. 62–63 °C) from *Primula obconica,* the yellow perezone (m. p. 102–103 °C) from various Mexican *Perezia* species, the orange embelin (m. p. 142–143 °C), with antihelminthic and antibiotic ef-

p-Benzoquinone

fects, and the orange rapanon (m. p. 142–143 °C). Embelin and rapanon are found in many members of the *Myrsinaceae.* The fungal B. include the maroon fumigatin (m. p. 116 °C) which is excreted in the culture medium of *Aspergillus fumigatus* and *Penicillium spinulosum,* the bronze-colored polyporinic acid (m. p. 310–312 °C), which is synthesized by the parasitic fungus *Polyporus nidulans,* the bronze-colored atromentin from *Paxillus atromentosus,* and the red spinulosin (m. p. 203 °C) from *Penicllium* spp. and *Aspergillus fumigatus.*

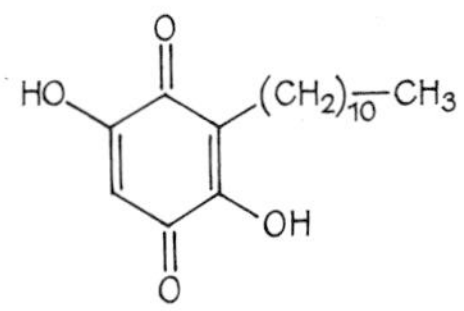

Embelin

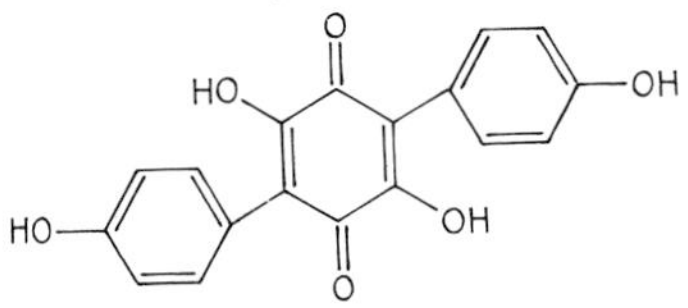

Atromentin

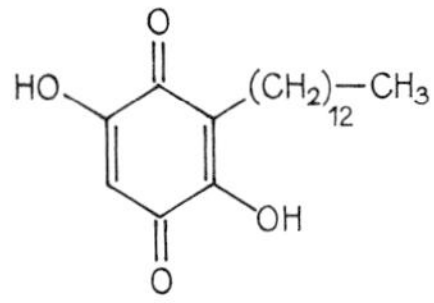

Rapanone

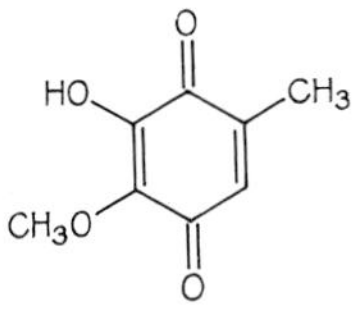

Fumigatin

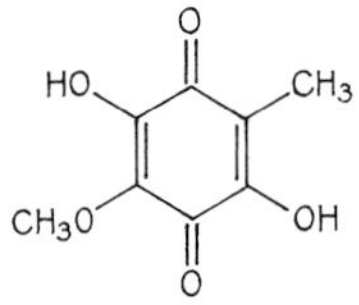

Spinulosin

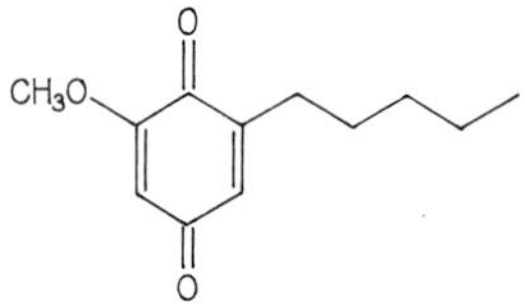

Primin

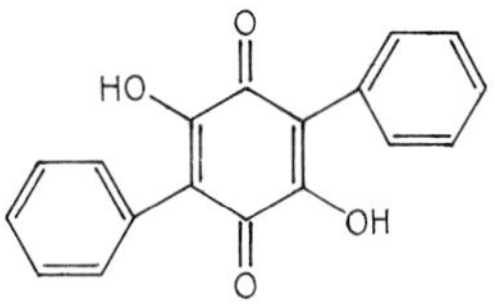

Polyporic acid

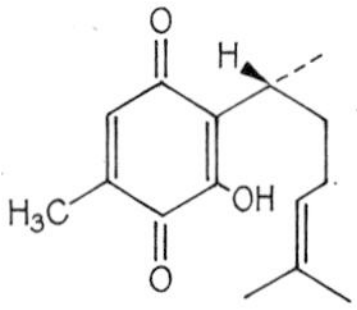

Perezone

The benzoquinone ring is also an essential feature of the quinones involved in electron transport in the respiratory chain and in the light reactions of photosynthesis (ubiquinone and plastoquinone).

The essential quinones of the electron transport chains of respiration and photosynthesis, ubiquinone and plastoquinone, are characterized by isoprenoid side chains and a functionally necessary benzoquinone ring.

N-Benzyladenine: see 6-Benzylaminopurine.

6-Benzylaminopurine, *N-benzyladenine:* a frequently used synthetic cytokinin. In B. the furfuryl residue of natural kinetin is replaced by a benzyl group. In the plant, B. is metabolized to 6-benzylamino-7-glucofuranosylpurine, which lacks cytokinin activity. 6-(2-Hydroxybenzylamino)purine riboside has been isolated as a natural cytokinin from the poplar.

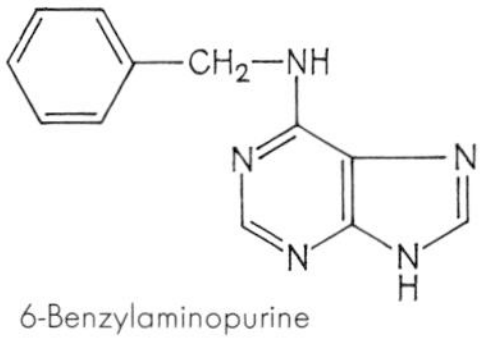

Benzylisoquinoline alkaloids: a group of alkaloids found mainly in poppy plants *(Papaveraceae).* The benzyl substituent on the C1 atom of the isoquinoline radical can enter various secondary cyclizations through phenol oxidation. The ring structures of the therapeutically important Papaveraceae alkaloids (see), Erythrina alkaloids (see) and Curare alkaloids (see) arise in this way. (The curare alkaloids are bisbenzylisoquinolines). *Biosynthesis.* The precursors of papaverine and the alkaloids derived

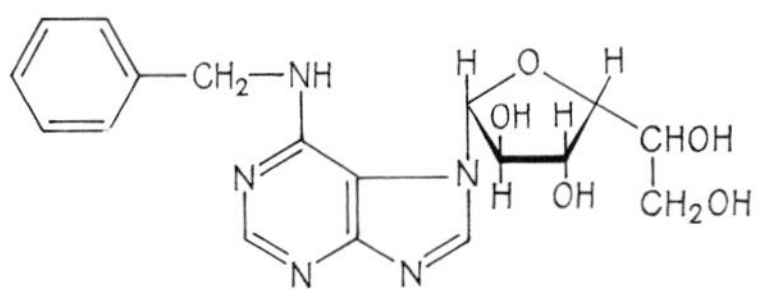

6-Benzylaminopurine

6-Benzylamino-7-glucofuranosylpurine

from it are two molecules of dopa, one of which is converted to dopamine, the other to 3,4-dihydroxyphenylacetaldehyde. Mannich condensation of these two leads first to norlaudanosine, which is then dehydrated to papaverine (Fig. 1).

The tetrahydroisoquinoline bases are also precursors of those alkaloids containing the morphine skeleton (e. g. thebaine, codeine and morphine). The precursor, reticuline, is formed via a biradical generated by phenol oxidases. It is transformed via the tetracyclic alkaloid I into thebaine, from which co-

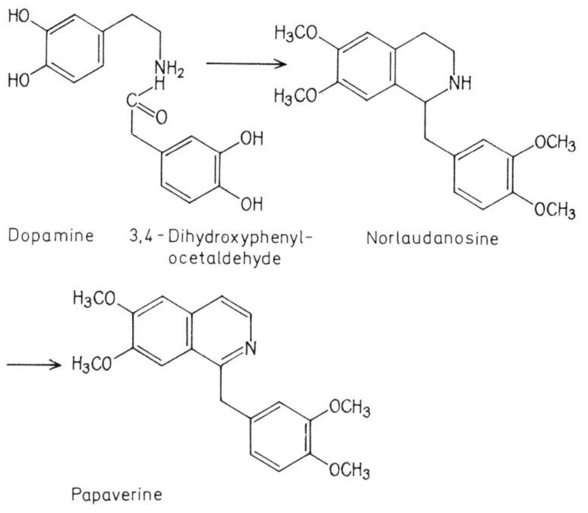

Figure 1. Biosynthesis of papaverine

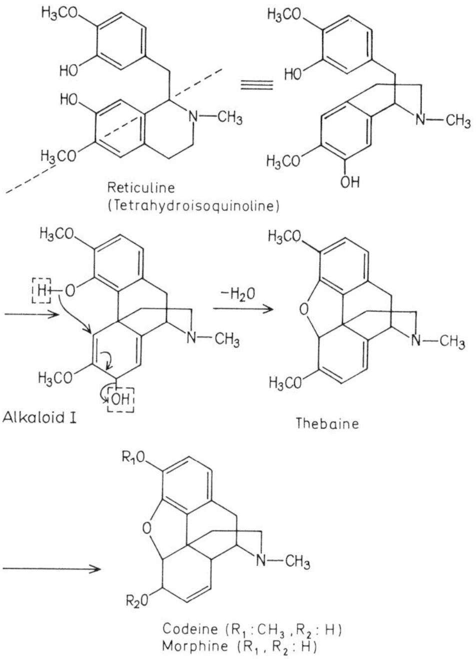

Figure 2. Biosynthesis of the morphine skeleton

deine and morphine are derived by hydrolysis of the methoxy group(s) (Fig. 2).

The bisbenzylisoquinoline derivatives, including the curare alkaloids (see), are formed by oxidative coupling of two benzylisoquinoline molecules.

Beri-beri: deficiency disease resulting from lack of vitamin B_1. See Vitamins.

Betacyanins: members of the betalain group of plant pigments with absorption maxima between 534 and 552 nm. They are derived from betanidin or isobetanidin and differ from one another with respect to their glycosylation. Betanin and amaranthin are important representatives of the group.

Betaines: widely occurring biogenic amines. The simplest is betaine (glycine betaine, trimethylglycocoll betaine), $(CH_3)_3N^+-CH_2COO^-$. M_r 117.2, m. p. of the hydrochloride 227–228 °C (d). B. are synthesized in the ethanolamine cycle and are metabolically related to mono- and dimethylglycine. They can serve as methyl donors in methylation. The betaine structure, which characterizes all betaines, is the peralkylated zwitterionic form $R_3N^+-CHR'-COO^-$.

Betalains: a group of nitrogen-containing plant pigments found almost exclusively in the family *Centrospermae*. They do not occur in the same plant with anthocyanins, and are therefore taxonomically useful. The red pigment in the cap of the fly agaric mushroom is also a betalain derivative. All B. are based on the Betalaminic acid (see) skeleton, which is joined to various imino and amino acids or amines to form the pigments. The most important B. are Betanin, Indicaxanthin (see) and Muscaaurin (see) from fly agaric. The red pigments of this type are called Betacyanins (see), and the yellow ones, Betaxanthins (see).

Betalaminic acid: the skeleton of the betalains. It is biosynthesized from tyrosine via dopa. Bonding to various amino acids, such as cyclodopa or proline, leads to the Betacyanins (see) and Betaxanthins (see).

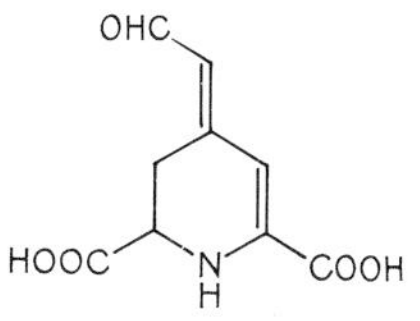

Betalaminic acid

Betamethasone: 9-fluoro-11β,17,21-trihydroxy-16β-methylpregna-1,4-diene-3,20-dione, a synthetic pregnane derivative (see Steroids) with high anti-inflammatory activity. It differs from Dexamethasone (see) only in the configuration of the 16 methyl group. It is used as a drug, e.g. for arthritis.

Betanidin: see Betanin.

Betanin: the red pigment in beets *(Beta vulgaris)*. It is a member of the Betalain (see) group, and is highly water-soluble. The aglycon of B. is betanidin. B. is a zwitterion; it exists as a violet cation below pH 2 and the inflection point to the red form is at pH 4. B. is the best known of the betacyanins.

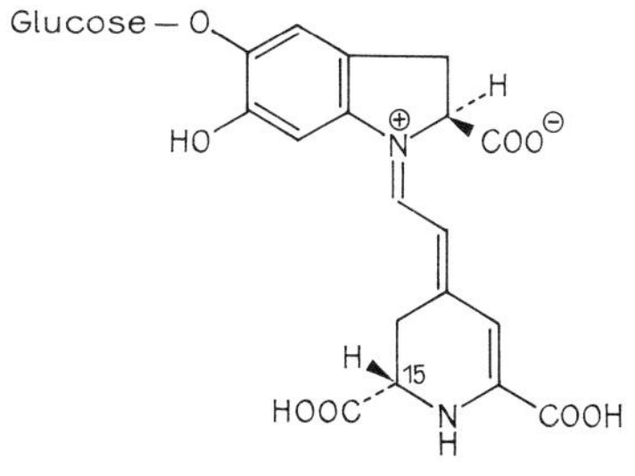

Betanin

61

Beta structure: see Proteins.

Betaxanthins: yellow Betalain (see) plant pigments with absorption maxima between 474 and 486 nm. The best known B. are Indicaxanthin (see), the Miraxanthins (see) and the Vulgaxanthins (see).

Betel: see Areca alkaloids.

Betonicine: see Pyrrolidine alkaloids.

Betulaprenols: see Polyprenols.

Betulenols: isomeric bicyclic sesquiterpene alcohols from birch bud oil. The B. are derived from caryophyllene and are optically active. Their structures are species-specific for *Betula alba* and *Betula lenta*.

Betulin: a pentacyclic triterpene diol. M_r 442.73, m. p. 261 °C, $[\alpha]_D$ +15 ° (chloroform). B. differs from Lupeol (see) in having a second hydroxyl group at position 28. It is found in birch and hazelnut bark, in rose hips and in the cactus *Lemaireocereus griseus.*

Betulinic acid: a pentacyclic triterpene carboxylic acid found in many plants, e. g. *Gratiola officinalis, Melaleuca* spp., various cacti, the bark of *Platanus* spp. and the bark of the pomegranate tree, *Punica granatum.*

Biapigenin: see Biflavonoids.

Bicuculline: an alkaloid from *Dicentra cucullaria, Adlumia fungosa* and several spp. of *Corydalis.* m. p. (from chloroform/methanol) 215 °C; solidifies and remelts at 193–195 °C. B. is a neurotoxin and a convulsant, which acts as a specific antagonist of the neurotransmitter 4-Aminobutyric acid (see).

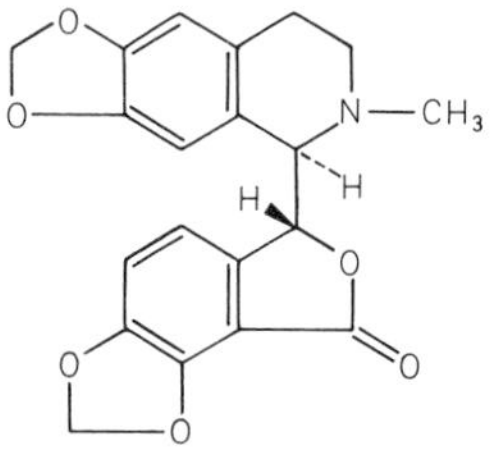

the electrophilic attack of one radical on the phloroglucinol nucleus of a chalcone or flavone [B. Hackson et al. *J. Chem. Soc. C* (1971) 3791–3804]. Both of these suggested biosynthetic routes seem very plausible, but experimental support is lacking.

Most B. are dimers of apigenin, and the precursor chalcone is generally thought to be the similarly substituted naringenin chalcone. Different ring substitution patterns are, however, found (e. g. xanthochymusside, Fig.). Most B. are biflavones, i. e. each flavonoid moiety possesses a C-2,3 double bond, but flavanone-flavones (e. g. C-C-linked volkensiflavone, a dimer of naringenin(3) and apigenin(8″)) and biflavanones (e. g. xanthochymusside, Fig., and succedaneaflavone) are also found. Glycosides of B. are rare, e. g. xanthochymusside (Fig.)

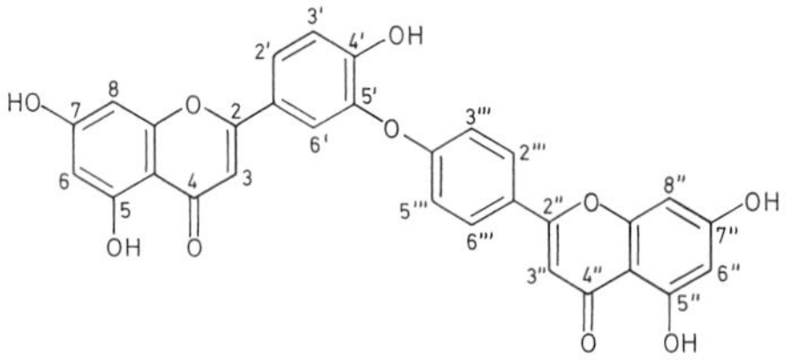

Ochnaflavone (a biflavonoid from *Ochna squarrosa*)

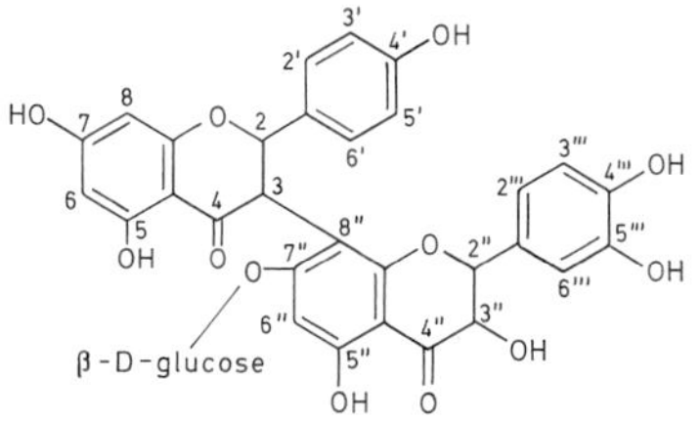

Xanthochymusside (a biflavonoid from *Garcinia*)

Biflavonoids, *biflavenyls*: dimers of flavonoid units. The interflavonoid linkage may be a carbon-carbon bond or an ether bond. Ten types of B. have been reported, the classification depending on the sites and type of linkage:

5′,4‴(ether), e. g. ochnaflavone (Fig.) (apigenin dimer)

4′,6″(ether), e. g. hinekiflavone (apigenin dimer)

5′,6″(C-C), e. g. robustaflavone (apigenin dimer)

5′,8″(C-C), e. g. amentoflavone (apigenin dimer)

8,6″(C-C), e. g. agathisflavone (apigenin dimer)

8,8″(C-C), e. g. cupressuflavone (apigenin dimer)

3,3″(C-C), e. g. biapigenin (apigenin dimer)

3,3‴(C-C), e. g. taiwaniaflavone (apigenin dimer)

6,6″(C-), e. g. succedaneaflavanone (naringenin dimer)

3,8″(C-C), e. g. xanthochymusside (Fig.) Other sites of linkage in naturally occurring B. probably remain to be discovered.

Formation of B. can be explained by oxidative coupling (pairing of radicals) of two chalcone units, with subsequent cyclization of the C_3 chains, or by

The ability to form B. appears to be an early evolutionary development in vascular plants, which has been lost in most angiosperms and some gymnosperms. B. have been isolated from 4 of the 5 orders of gymnosperms, from every family of the *Coniferales*, with the exception of the *Pinaceae*, and from 11 angiosperm families. B. are absent from the *Gnetales*, which supports the exclusion of this order from the gymnosperms and its classification as a separate group, the chlamydosperms. [Geiger & Quinn in *The Flavonoids: Advances in Research* (ed. J. B. Harborne & T. J. Mabry) (Chapman and Hall, 1982) pp. 525–534]

Bile acids: the components of bile which serve as emulsifying agents for fats. They are steroid carboxylic acids bound in peptide linkage to taurine or glycine. Depending on the amino acid to which they are linked they are classified as *glycocholic* or *taurocholic* acids; the former are predominant in human and bovine bile, the latter in canine bile. B. a. are typical of mammals. Often the pattern of B. a. is specific for the given species. In lower vertebrates, the Bile alcohols (see) have the same function.

The salts of B.a. reduce the surface tension and emulsify fats, so that they can be absorbed in the intestine. The lipases are also activated by B.a. In humans, the daily production is 20 to 30g. 90% of this is resorbed in the intestine and returned to the liver in the enterohepatic circulation. 1 liter bile contains 30 g B.a.

B.a. are biosynthesized from cholesterol by 7α hydroxylation, reduction of the double bond at position 5, and epimerization at position 3. The C_{27} side

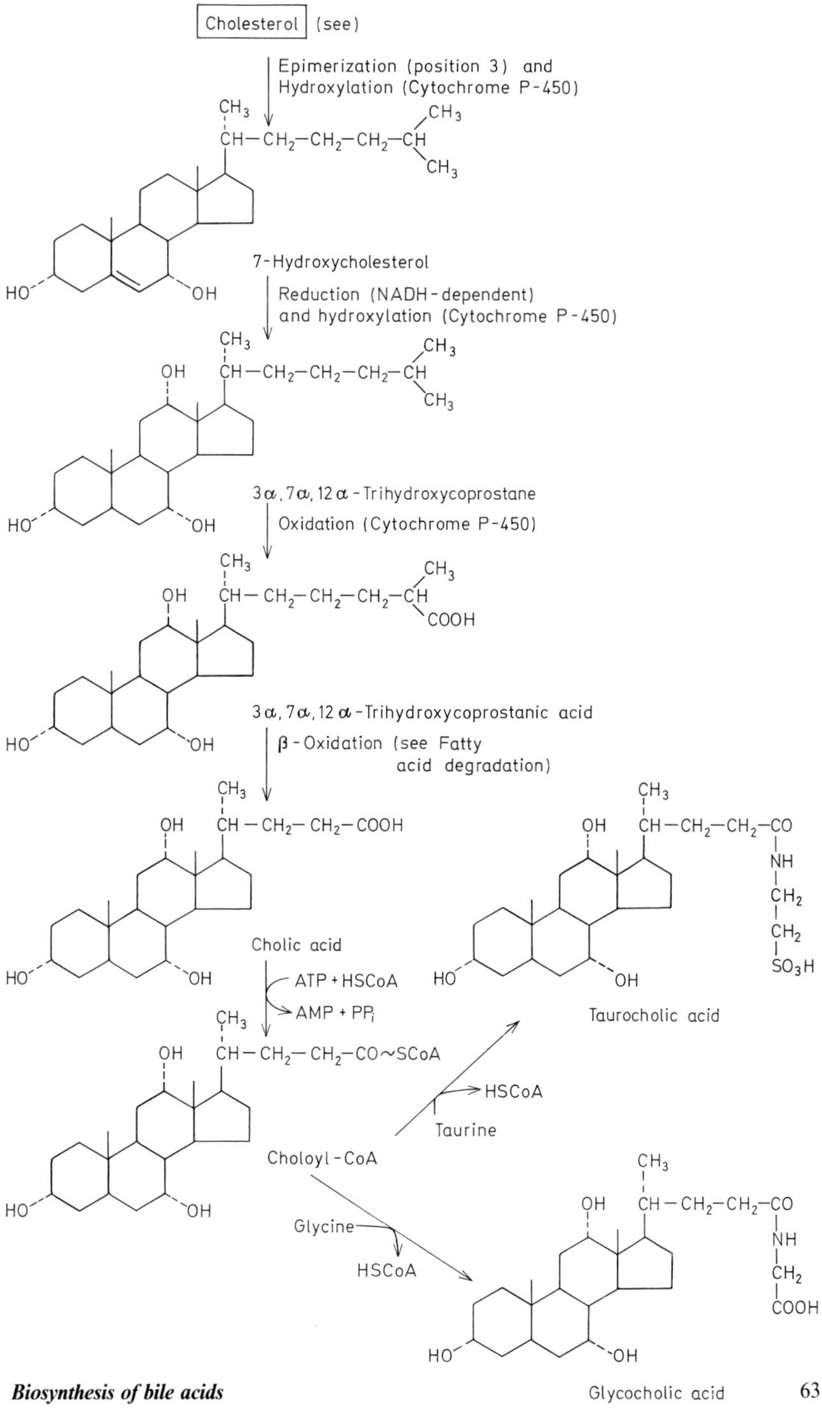

Biosynthesis of bile acids

Glycocholic acid

chain is shortened by β-oxidation (see Fatty acid degradation). Free B.a. are obtained by alkaline hydrolysis of animal bile. They are important as starting materials for partial synthesis of therapeutically important steroid hormones (see Cholic acid, Deoxycholic acid and Lithocholic acid).

Bile alcohols: a group of polyhydroxylated steroids derived structurally from cholestane. B.a. occur as sulfuric acid esters in the bile of lower vertebrates. A typical example is *scymnol,* 3α, 7α,12α,24,25,27-hexahydroxy-5β-cholestane, from shark bile.

Bile pigments: degradation products of porphyrins, especially heme. The α-methine bridge between rings A and B is oxidatively cleaved to form CO and *biliverdin IX* by microsomal hydroxylases. (In *meso-biliverdin,* the two vinyl substituents are reduced to ethyl.) Biliverdin is reduced to *bilirubin* (Fig. 1) and transported to the liver as a complex with serum albumin. Bilirubin is produced mainly from the hemoglobin of aged erythrocytes in the reticuloendothelial system (spleen), bone marrow and liver. In pernicious anemia, some of the bilirubin is also formed by degradation of myoglobin and cytochromes.

Bilirubin is transported as a complex with serum albumin; in free form it is highly toxic, especially for newborn infants, in whom it can cross the blood-brain barrier more easily than in adults. It acts as an uncoupler of respiration.

In the liver, bilirubin is released from serum albumin and concentrated in the liver cells, where it is conjugated to two glucuronic acid residues (Fig. 2) by the glucuronosyltransferases, EC 2.4.1.76 and 2.4.1.77. These are located mainly in the smooth endoplasmic reticulum. Conjugated bilirubin is excreted in bile, where it is probably bound to lecithin or bile proteins. In the large intestine, most of it is hydrolysed back to free bilirubin, then reduced by the intestinal bacteria to urobilinogen, stercobilinogen and *meso*-bilirubinogen. These compounds are colorless, but are oxidized by oxygen to stercobilin and urobilin, which give feces their brown color.

Tetrapyrrole compounds can form metal complexes in which the metal ion is bound to all four nitrogen atoms in a nearly planar ring structure. This is possible because the pyrrole ring can exist both in the lactim and the lactam forms. The ring formed by the metal complex is presumably stabilized by hydrogen bonds.

Biliproteins: see Phycobilisome.
Bilirubin: see Bile pigments.
Biliverdin: see Bile pigments.

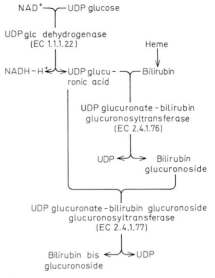

Figure 2.

Biochanin A: see Isoflavone.
Biochemical oxygen demand, BOD: the rate at which the oxygen dissolved in water is consumed for the oxidation of organic compounds in the water by microorganisms.

Biochromes: see Natural pigments.
Biocytin: see Vitamins (Biotin).
Bioelectronics: the incorporation of biological materials into electronic devices, e.g. Enzfets (see Field effect transistors). See Biosensor. The term is also used in a more general sense to mean the application of electronics to the investigation of biological processes.

Bioelements: those chemical elements required by organisms. The elemental composition of organisms is considerably different from that of the earth's crust (Table 1); only about 40 of the 90-odd crust elements are represented in living matter, and the six elements C, O, H, N, S and P together account for about 90% of it. The six main elements are present both as constituents of biomolecules and in inorganic matrix substances and in water, the medium of organic processes. Minerals (see) are rarely present in large amounts. 99.9% of the biomass is accounted for by the above six B. plus Ca, K, Na, Cl, Mg and Fe. The remaining elements occur chief-

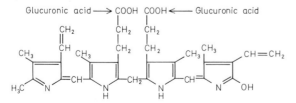

Figure 1. Bilirubin

ly as Trace elements (see), because they are needed only in catalytic quantities. While the light metals are usually present in organisms as mobile cations (see Minerals), the heavy metals are generally fixed as stable components of biocomplexes. Table 2 shows the elemental composition of the human body.

Carbon (C) forms the skeleton of all organic molecules. All C in the biomass is ultimately derived from CO_2 fixed by Photosynthesis (see), which means that all biomolecules are derived biosynthetically from carbohydrates.

Oxygen (O) is a component of nearly all biomolecules, providing a reactive "handle" for metabolic transformations (hydrocarbons are not generally biodegradable!) of acids, aldehydes and ketones, alcohols and ethers; it is also part of the hydroxylapatite in bones and, very importantly, of water. The oxygen in the atmosphere is the result of the photolytic activity of plants, which has transformed the originally reducing atmosphere of the earth to an oxidizing one.

Hydrogen (H) is present in all biomolecules, attached to C, N, O, and S. Removal of H is equivalent to oxidizing a substrate; when the H is combined with O_2 via the electron transport chain of respiring cells, ATP is generated. In most biological reactions, H participates as the ion $H^+ + e^-$; the coenzymes NADH and NADPH are carriers of $H^+ + 2e^-$. Reactions involving H_2 are rare (see Hydrogen metabolism).

Nitrogen (N) is a component of many biomolecules, especially of proteins and nucleic acids. Molecular N is reduced by certain free-living and symbiotic microorganisms (see Nitrogen fixation) to ammonia, which is the first and final product of nitrogen metabolism.

Sulfur (S) is present in two amino acids, cysteine and methionine, and in certain coenzymes. S is assimilated by plants in the form of sulfate (see Sulfate assimilation); animals must obtain it from plants.

Phosphorus (P) is present in both inorganic and organic compounds as phosphate. The nucleotides in nucleic acids are linked by phosphate bonds, and energy is transferred from one molecule to another in the form of a high-energy phosphate bond, very often in ATP. Many coenzymes also contain phosphate.

Biogenic amines: a biologically and pharmacologically important class of compounds characterized by the presence of an amine group. They occur widely in plants and animals. The group can be subdivided into derivatives of *ethanolamine,* e.g. choline, acetylcholine and muscarine; *polymethylene diamines,* e.g. putrescine and cadaverine; *polyamines,* e.g. spermine; *imidazolylalkylamines,* e.g. histamine; *phenylalkylamines,* e.g. mescaline, tyramine and hordenine; *catecholamines,* e.g. adrenalin, noradrenalin and dopamine; *indolylalkylamines,* e.g. tryptamine and serotonin; and *betaines,* e.g. carnitine.

The B.a. can be precursors of alkaloids (which is why they are also called protoalkaloids) and hormones. In addition, some are Neurotransmitters (see) or components of phospholipids, vitamins, ribosomes and bacteria.

The biosynthesis and metabolism of the B.a. are generally similar. With the exception of the betaines, they are synthesized by decarboxylation and hydroxylation of amino acids. The hormones tyramine and the catecholamines are synthesized from tyrosine; tryptamine, serotonin and melatonin from tryptophan; histamines from histidine, and 4-aminobutyric acid from glutamic acid. Other B.a. are precursors for vitamins. Propanolamine, a precursor of vitamin B_{12}, is synthesized from threonine, while the coenzyme A precursors cysteamine and β-alanine are made from cysteine and aspartic acids, respectively. Cadaverine (from lysine) and putrescine (from ornithine) are found in ribosomes and bacteria. Spermidine (or spermine) is synthesized from methionine and occurs in sperm. Ethanolamine, which is a component of phospholipids, is made from serine.

Some B.a. are hallucinogens, e.g. mescaline and psilocybin. Many are toxic, e.g. cadaverine and putrescine.

Bioluminescence: the emission of visible light by an organism. The light is emitted as a result of a redox reaction catalysed by the enzyme luciferase (see Luciferin). The energy obtained from this reac-

Table 1. Relative abundance of the chemical elements in the earth's crust and in the human body (adapted from Rapoport)

Element	Earth's crust %	Human body %	Concentration (-fold)
Oxygen	50	63	–
Silicon	28	0	–
Aluminum	9	0	–
Iron	5	0.004	–
Calcium	3.6	1.5	–
Potassium	2.6	0.25	–
Magnesium	2.1	0.04	–
Hydrogen	0.9	10	10
Carbon	0.09	20	200
Phosphorus	0.08	1	10
Sulfur	0.05	0.2	4
Nitrogen	0.03	3	100

Table 2. Elemental composition of the human body, based on dry weight

Element	Percent
Carbon	50
Oxygen	20
Hydrogen	10
Nitrogen	8.5
Sulfur	0.8
Phosphorus	2.5
Calcium	4.0
Potassium	1.0
Sodium	0.4
Chlorine	0.4
Magnesium	0.1
Iron	0.01
Manganese	0.001
Iodine	0.00005

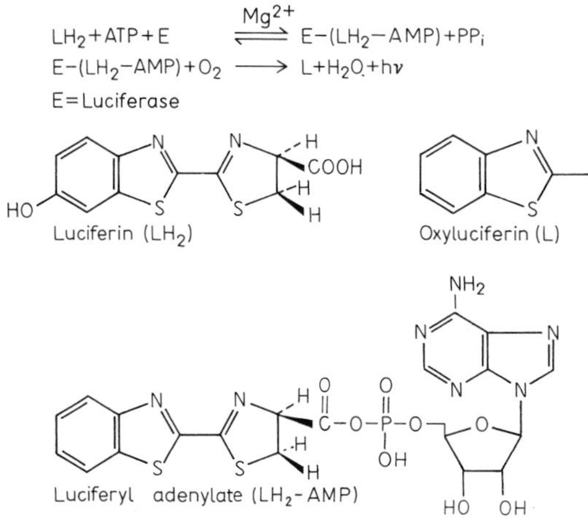

$$LH_2 + ATP + E \underset{Mg^{2+}}{\rightleftharpoons} E-(LH_2-AMP) + PP_i$$
$$E-(LH_2-AMP) + O_2 \longrightarrow L + H_2O + h\nu$$
$$E = Luciferase$$

Luciferin (LH_2)

Oxyluciferin (L)

Luciferyl adenylate (LH_2-AMP)

Mechanism of light emission by Photinus luciferin

tion is used to excite the electrons of an oxidation product of luciferin to a higher electronic state. Light is emitted as the oxyluciferin returns to the ground state (Fig.).

In warm regions B. is fairly common, especially among marine animals. Of the vertebrates, only a few fish are luminescent, as are a few fungi and bacteria. B. can be generated either extracellularly, in which case the luciferin and luciferase are secreted by glands, or intracellularly. In the latter case, the components react in special cells. The light can also be generated by symbiotic bacteria, which are usually harbored by marine animals in small pockets on the body surface.

The luciferins of different animals vary widely in chemical structure. Luciferases are species-specific, and may be peroxidases, mono or di-oxygenases. Various types of reaction mechanism and cofactor requirements have been found (Table). The system of the firefly *Photinus pyralis* (Fig.) requires O_2, ATP and Mg^{2+} in addition to luciferin and luciferase. The light is generated by an intermediate of the pro-

cess (probably a peroxide) with a quantum yield of 1 (i.e. 1 photon per molecule of luciferin). The bacterial luciferase catalyses a 2-step reaction:
$E-FMNH_2 + O_2 \rightarrow E-FMNH-OOH$
$E-FMNH-OOH + RCHO \rightarrow E + FMN + RCOOH + h\nu$. (Here E is the enzyme, RCHO is a long-chain aldehyde, and $E-FMNH-OOH$ is a 4-hydroperoxy-flavin.) [M. Kurfürst et al. *Eur. J. Biochem* (1982) **123** 355-361]

The biological significance of B. has not been completely clarified. In some cases it serves as protection from enemies or as bait for prey. Many species use B. for communication. The information is encoded in the spectrum, frequency or rhythm of flashing or the arrangement of the light organs on the body. For example, each species of *Photinus* uses a different flashing frequency to attract mates; and some predatory species mimic the flashing pattern of other species to attract them as prey.

The B. of *Photinus* is used as the basis of an extremely sensitive assay for ATP in biological samples. On being mixed with a sample containing ATP, a luciferin-luciferase system (or powdered firefly abdomens) will generate an amount of light proportional to the amount of ATP, and this can be measured in a spectrophotometer. The assay is sensitive in the nanogram range, and thus can even be used to count bacteria on the basis of their ATP content. Similarly, preparations from luminescent bacteria are used for NADH assays and the photoprotein from the jellyfish *Aequorea* for calcium or strontium determination.

Horseradish peroxidase can act as a luciferase in the presence of H_2O_2, a cyclic hydrazide such as luminol and synthetic firefly luciferin (which is now commercially available). This enzyme can be coupled to many proteins, including antibodies, without loss of activity. Thus B. can be used in a wide variety of immunoassays, many of which are clinically important. [T. P. Whitehead et al. Nature 305 (1983) 158-159.]

Biomacromolecule: see Biopolymer.

Biomass: the amount of organic substance in living organisms on a given area. The term is used in ecology.

Table. Bioluminescent systems

Reaction characteristics	Organism	λ max (nm)
Requires NADH and FMNH	*Photobacterium*	470...505
Requires ATP	*Photinus* (Firefly)	552...582
	Renilla reniformis (Sea pansy, a coelenterate)	509
No cofactors	*Cypridina* (An ostracod crustacean)	460
	Latia (Limpet)	535
Photoprotein	*Aequora* (Jellyfish)	469

Biomembrane: a structure containing lipids, glycolipids, proteins and glycoproteins, bounding the cell (cell membrane) or subdividing it into compartments. It is a sheet-like structure about 60 to 100 Å thick. The lipids have hydrophilic "head" groups (indicated by spheres in the Fig.) and hydrophobic "tail" regions. In bulk aqueous solution, they can spontaneously form bilayers, in which the molecules are lined up side by side and tail to tail, with the heads pointing towards the water phase on each side of the bilayer, but excluding water from the tail regions. This structure, when stained for electron microscopy with osmium tetroxide or uranyl acetate, is seen as two dark lines separated by an unstained gap. (This image was previously referred to as the *unit membrane,* but the term is now obsolete.) Mitochondria and plastids are surrounded by two membranes, and nuclei by a membrane that effectively doubles back on itself. The cytoplasm of eukaryotic cells is characterized by extensive membranous structures, for example the Endoplasmic reticulum (see), the Golgi apparatus (see) and assorted vacuoles. Prokaryotes, in contrast, have no internal membranes, though in some the cell membrane is extensively invaginated.

The major classes of lipids are Phospholipids (see), Glycolipids (see) and Cholesterol (see) and its esters; but there are many minor components, and the exact composition depends on the species and type of cell. The amount and kinds of proteins depend on the function of the membrane. Myelin membranes, for example, which are thought to serve as electrical insulators, have very little protein (18%), whereas the inner mitochondrial membrane, which is the site of Oxidative phosphorylation (see), contains about 75% protein. The membrane proteins have a wide variety of functions. They act, for example, as mediators of both the active and passive transport of lipid-insoluble substances across the membrane, as receptors for hormones and other informational molecules, and as enzymes. In certain cases they may also have a structural role.

The currently accepted model of the structure of B. is based on the *fluid mosaic* concept. There is evidence from several kinds of experiment that the lipid molecules and the membrane proteins are free to

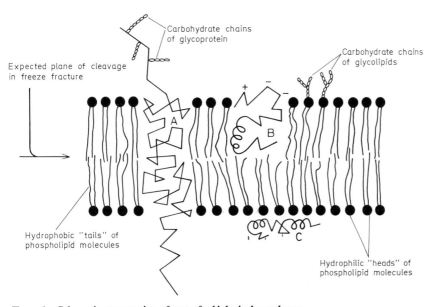

Figure 1. Schematic cross section of part of a biological membrane
A represents an intrinsic (integral) protein that completely spans the membrane; in the example shown, the protein chain extends beyond the membrane surface on both sides; it is a glycoprotein, and carbohydrate residues are present only on that segment of the protein protruding from the outer membrane surface. The model illustrated is roughly equivalent to glycophorin, an erythrocyte membrane protein; the protein chain protruding from the inside surface of the erythrocyte membrane is thought to be associated with an extrinsic protein called spectrin.
B represents an intrinsic (integral) membrane protein partly buried and partly exposed at the membrane surface.
C is representative of many extrinsic (peripheral) proteins that are more or less firmly associated with membranes, but appear not to be integrated into the phospholipid layer, e.g. spectrin of the erythrocyte membrane. Extrinsic proteins are usually associated with intrinsic proteins, rather than simply adhering to the hydrophilic heads of the phospholipid molecules as shown here.
Cytochrome *c* may represent an association intermediate between **B** and **C**; it is an important functional constituent of the inner mitochondrial membrane, but extremely easily removed.

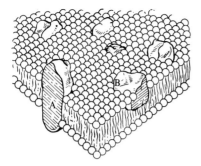

Figure 2. *Three dimensional artistic impression of a phospholipid bilayer with intrinsic proteins* A and B as in Figure 1.

diffuse laterally and to spin within the bilayer in which they are located. However, a flip-flop motion from the inner to the outer surface, or vice versa, is energetically unfavorable, because it would require the movement of hydrophilic substituents through the hydrophobic phase. Hence this type of motion is almost never seen in the case of proteins, and occurs much less readily than translational motion in the case of lipids. The fact that there is little movement of material from the inner to the outer half (or vice versa) of the bilayer means that the two faces of the B. can have different compositions. For the membrane proteins the asymmetry is absolute and, at least in the case of the plasma membrane, different proportions of lipid classes exist in the two monolayers. The attached carbohydrate residues appear to be located only on the noncytosolic surface. It is known that the carbohydrate groups extending from the B. participate in cell recognition, cell adhesion, possibly in intercellular communication, and that they often contribute to the distinct immunological character of a cell.

Physically, the surface of a B. is an interface between a polar solvent (water) and a nonpolar solvent (the hydrophobic tails of the lipid molecules). Hydrophobic molecules may cross B. because they can dissolve in the hydrophobic milieu of the membrane interior. Ionic and hydrophilic molecules, however, are solvated in aqueous solution. In order to penetrate the membrane, they must shed their solvation shells, and this is energetically unfavorable. Pure phospholipid membranes (see Liposomes) are therefore effectively impermeable to many biomolecules. There are a number of active and passive mechanisms by which a B. can be selectively permeable to those hydrophilic substances that it needs to transport. In these processes, proteins are thought to act as carriers of pores responsible for Facilitated diffusion (see), Active transport (see) or ionic pumping (see Ion pumps).

Integral or *intrinsic* proteins of the B. cannot be removed except by dissolution of the B. with organic solvents or detergents; these proteins are responsible for the bumps observed in freeze-fractured B. preparations under the electron microscope. *Peripheral* or *extrinsic* proteins can be stripped from the B. by altered ionic conditions or extremes of pH.

Biopolymer, *biomacromolecule:* a biological molecule which can be regarded as a chain (polymer) of identical or similar subunits (monomers). *Periodic* B. are formed from identical subunits; an example is cellulose, which is built of glucose monomers. *Aperiodic* B. are composed of similar but not identical monomers. The following classes of B. are listed under separate entries: Proteins, Nucleic acids, Carbohydrates, Lignins, Polyprenols and Polyterpenes. The monomer sequences of nucleic acids and proteins are determined by nucleic acid templates, and organisms have elaborate systems for synthesizing these B. with very low rates of error. The overall composition of polysaccharides and lignins is determined by enzymes catalysing their synthesis, but their size and exact monomer sequence may be determined by factors such as the juxtaposition of other structural molecules and rate of growth (e.g. cell wall polysaccharides and lignin) or the rate of degradation and availability of precursor monomers (e.g. storage polysaccharides). Blood group antigens (see), and presumably other cell-surface recognition oligo- and polysaccharides, however, have precise sizes and monomer sequences. There are classes of mixed B.: Glycoproteins (see) have carbohydrate monomers linked to amino-acid side chains; Glycolipids (see) have carbohydrates linked to some part of a lipid.

Biopterin: see Pteridines.

Biosensor: a biological component (e.g. enzyme, antibody) in combination with a transducing system. Interaction of an analyte with the biological component generates an electrical signal, which is amplified then recorded, or displayed. In most B., the biological component is immobilized on the surface of the transducer. The signal from the biological material to the transducer may be a change in charge, potential, heat, light, mass or other parameter. An amperometric B. functions by producing an electric current on application of a potential difference between two electrodes, whereas a potentiometric B. produces a potential gradient. In both cases the electrical effect is due to a biological reaction. A typical amperometric device is a glucose sensor, using glucose oxidase, whereas urease is used in a typical potentiometric sensor for urea. The term B. is sometimes used in the broader sense of sensors which detect biological changes or materials, but which do not incorporate biological material in their sensing mechanism, e.g. oxygen electrodes. See Field effect transistor, Immunosensor, Oxygen electrode, Piezoelectric sensors.

Bios II: see Vitamins (Biotin).

Biosynthesis, *biogenesis:* the biological synthesis of a compound, i.e. an enzymatic synthesis. B. is to be distinguished from the abiotic formation of chemical compounds in the course of geological processes, in the course of prebiotic Evolution (see) or in the laboratory under simulated primitive earth conditions and from chemical syntheses carried out in the laboratory or on a technical scale.

B. can occur either in vivo, that is, in a living organism or in a cultivated part of it (such as a perfused organ), or in vitro, in cell homogenates, extracts, enzyme preparations and reconstituted systems. Although the mechanism of a B. may be the same in vivo and in vitro (insofar as there are no artifacts), the conditions are usually quite different (see Methods of biochemistry).

The sum of all B. in an organism is referred to as anabolism.

Biotechnology: The application of living organisms or systems derived from them in manufacturing industry. The subject has ancient origins, embracing agriculture, cheese production, fermentation of alcoholic beverages, leather tanning, etc. However, B. is a modern term, referring to particular types of biological exploitation, and the scope of the subject has been greatly expanded by developments in molecular biology. Important areas of B. are:

1. Genetic engineering (see) or Recombinant DNA technology,
2. Process engineering (e.g. production of antibiotics by fermentation; microbiological conversion of simple chemical source materials like methanol and ammonia into animal feed; production of fuel alcohol by fermentation),
3. Concentration of minerals, or "microbiological mining",
4. Biosensors (see),
5. Biocatalysts (see Immobilized enzymes),
6. Waste treatment and biodegradation,
7. Production of monoclonal antibodies for medical research,
8. Study of the relationship between protein structure and function, and the nature of molecular interactions, leading to designed (nonbiological) synthesis of model catalysts, and of new drugs.

Biotest, *biological assay:* a procedure in which the biological activity of a compound is tested by observing its effect on an appropriate organism. B. are frequently used for hormones, inhibitors and antibiotics. The test objects are usually whole organisms, or isolated organs or tissues, sometimes in tissue culture.

Biotin: see Vitamins.

Biotin carboxylase: see Biotin enzymes.

Biotin enzymes: enzymes which catalyse carboxylation reactions, using biotin as a cofactor. The prosthetic group (biotin) is bound via an amide bond to the ε-amino group of a specific lysine residue in the enzyme protein. The latter is called the biotinyllysyl group of *biocytin.* The carboxyl group being transferred is linked to N atom 1′ of the biotin ("active CO_2"). The loading of biotin with CO_2 is endergonic and only occurs with the simultaneous hydrolysis of ATP:

$ATP + HCO_3^- +$ biotinyl-enzyme (I) →
$ADP + P_i +$ carboxybiotinyl-enzyme (II).

Carboxylation reactions occur in the biosynthesis of the fatty acids, in the degradation of leucine and isoleucine and in the degradation of fatty acids with uneven numbers of C atoms. *Acetyl-CoA carboxylase* (EC 6.4.1.2) catalyses the following reaction in the biosynthesis of fatty acids:

$Acetyl-CoA + HCO_3^- + H^+ + ATP →$
$Malonyl-CoA + ADP + P_i$

The monomer of this enzyme is composed of 4 different subunits. One of these, *biotin carboxylase* (III) catalyses the carboxylation of the biotin residue. This biotin residue is covalently bound to the second subunit and is called the *biotin-carboxyl carrier protein* (biotin-CCP).

$Biotin-CCP + HCO_3^- + H^+ + ATP \xrightarrow{III}$
$Carboxy-biotin-CCP + ADP + P_i$

In the second step of the reaction, the third subunit, the *carboxyl transferase* (IV) catalyses the transfer of the carboxyl group to acetyl-CoA:

$Carboxy-biotin-CCP + acetyl-CoA \xrightarrow{IV}$
$Biotin-CCP + malonyl-CoA$

In the biotin-carboxyl-carrier protein, the biotin residue acts as a swinging arm which transfers the hydrogen carbonate from the biotin carboxylase to the acetyl-CoA which is bound to the active center of the carboxyl transferase.

Another example of a carboxylation reaction is the formation of oxaloacetate from pyruvate. *Pyruvate carboxylase* (EC 6.4.1.1) consists of 4 subunits, each of which is covalently bound to one molecule of biotin and also contains one Mg^{2+} ion.

$Biotinyl-enzyme + ATP + CO_2 + H_2O →$
$Carboxybiotinyl-enzyme + ADP + P_i$
$Carboxybiotinyl-enyzme + pyruvate →$
$Biotinyl-enzyme + oxaloacetate$

In the degradation of fatty acids with an odd number of C atoms, the carboxylation of propionyl-CoA to methylmalonyl-CoA is also catalysed by B.:

$$Carboxybiotinyl-enzyme + CH_3-CH_2-C\overset{\displaystyle O}{\overset{\|}{}}\sim SCoA$$
$$→$$
$$Biotinyl-enzyme + CH_3-\underset{\underset{\displaystyle COOH}{|}}{CH}-C\overset{\displaystyle O}{\overset{\|}{}}\sim SCoA$$

The same reaction is one of the steps in the degradation of aliphatic amino acids (isoleucine, leucine and valine).

Bisabolane: see Sesquiterpenes (Fig.).

Bisbenzylisoquinoline alkaloids: a group of isoquinoline alkaloids with a skeleton formed by fusion of two isoquinoline units by a phenol oxidation. The curare alkaloids, for example tubocurarine, belong to this class.

2,3-Bisphosphoglycerate, *glycerate 2,3-bisphosphate, 2,3-diphosphoglycerate:* see Glycolysis, Hemoglobin, Rapoport-Luerbing shuttle.

Bitter peptides: bitter tasting peptides that may spoil the palatability of certain foodstuffs, e.g. B.p. sometimes arise during the ripening of certain types of cheese, and they have also been found in fermented soybean products. B.p. have also been isolated from controlled enzymatic digests of pure proteins, e.g. chymotryptic hydrolysis of casein. It is claimed that bitterness is related to the average hydrophobicity of the peptide, so that peptides with a high content of Val, Leu, Phe and Tyr are likely to be bitter. B.p. isolated from peptic hydrolysis of soya protein are Leu-Phe, Leu-Lys, Phe-Ile-Leu-Glu-Gly-Val, Arg-Leu-Leu and Arg-Leu. On the other hand various markedly hydrophilic peptides from spinach, e.g. Glu-Gly, Asp(Glu,Gly,Ser$_2$), Ala(Glu$_2$,Gly,Ser), are not bitter.

Bitter principles: bitter-tasting substances, found especially in *Compositae, Gentianaceae* and *Labiatae,* which produce a reflexive increase in the secretion of saliva and digestive juice. Extracts from such plants are used as bitter spices to increase the appetite and promote digestion, and they are used in the preparation of stomachic bitters.

Chemically, the B.p. are very diverse. Many are terpenes with lactone groups, like the gentiopicroside from gentian root. Bitter-tasting principles which have other physiological effects, such as quinine alkaloids and curcurbitacines, are not included in the B.p.

Bixin: the monomethyl ester of a C$_{24}$ dicarboxylic acid called norbixin. M_r of bixin, 394, m.p. 198 °C. Norbixin has 4 methyl side branches, 2 terminal carboxyl groups and 9 conjugated double bonds. The C$_{20}$ chain corresponds to the middle part of β-carotene. In naturally occurring B. the Δ^{16} double bond is *cis,* but it easily isomerizes to form the more stable all-*trans.* B., m.p. 217 °C. B. is a diapocarotenoid formed by oxidation of a C$_{40}$ carotenoid. It is a yellow to red-orange pigment found in the seeds of *Bixa orellana.* The extract of these seeds is used as a food color.

Bleomycins: a group of glycopeptide antibiotics produced by *Streptomyces verticillus,* and used widely for treatment of squamous cell carcinomas, lymphomas and testicular cancer. About 200 different B. are known, differing mostly in the nature of the C-terminal substituent, which can be varied by adding different amines to the bacterial culture medium. The clinically used preparation is a mixture of 11 B., known by the trade name "Blenoxane"; the chief constituents being B.A$_2$ (60-70%) and B.B$_2$ (20-25%) (Fig.).

B. cause strand breakage in double-stranded DNA in solution, and in DNA in intact cells. All types of DNA (mammalian, viral, bacterial and synthetic polymers) are attacked. B. also bind to RNA, but do not cause degradation. In RNA-DNA hybrids, B. degrade the DNA and not the RNA. This specificity is not related to the presence of thymine or uracil, and is determined by the presence of 2-deoxyribose. DNA degradation by B. requires oxygen, and it is prevented by metal-chelating agents, e.g. EDTA and the iron-specific reagent, deferoxamine. It is generally accepted that the biologically active form of B. is a complex with Fe(II). Stopped flow spectroscopy shows two distinct kinetic events occurring after the Fe(II).complex is exposed to O$_2$. The first event forms an unstable intermediate complex, Fe(II)-B.-O$_2$. This decomposes to an ESR-active species, known as "activated bleomycin", formulated as Fe(III)-B.-(O$_2$H) or Fe(III)-B.-(O$_2^{2-}$), which attacks DNA.

Phleomycins do not possess the double bond at C-4' of the bithiazole moiety, and are otherwise identical to B., but they are too toxic for use as antibiotics. Other related antibiotics are zorbamycin, zorbonomycin, victomycin and platomycin, which have not

Bixin

Blasticidines: pyrimidine antibiotics synthesized by *Streptomyces griseochromogenes* (see Nucleoside antibiotics). B. were first isolated by Fukunaga in 1955. An important representative of the group is *blasticidine S.,* m.p. 235 to 236 °C, [α]$_D^{11}$ +108.4 ° (*c*=1 in water). B. inhibit the growth of fungi, for example the rice fungus *Piricularia oryzae,* and a few bacteria. The antibiotic effect is due to the suppression of the elongation of polypeptide chains during protein biosynthesis.

Blasticidin S

been extensively studied. [J.C. Dabrowiak in *Advances In Inorganic Biochemistry* vol. 4 (Elsevier Biomedical, 1982) pp.69-113].

Blitz blot: see Southern blot.

Blood-brain barrier: the barrier that must be crossed when water and solutes are exchanged between blood and extracellular fluid of the central nervous system. Many substances when injected intravenously become distributed "throughout" in the various tissues and organs of the body, "except for" the central nervous system, e.g. acidic dyes (Ehrlich, 1885), bile acids (Biedl and Kraus, 1898), tetanus toxin (Roux and Borell, 1898) sodium ferricyanide (Lewandowsky, 1900), trypan blue (Goldmann, 1908).

The permeability properties of cerebral capillaries are quite different from those of capillaries in other organs and tissues. Water-filled channels, which can be demonstrated in the walls of most non-nervous capillaries, are absent from cerebral capillaries. A hydrostatically promoted transcapillary flow of fluid into tissue on the arterial side, with the reverse pro-β-Aminoalanine amide

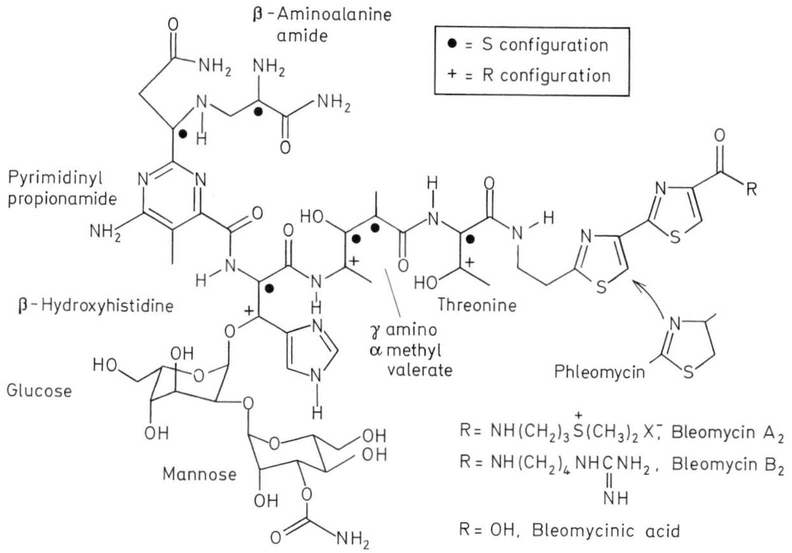

• = S configuration
+ = R configuration

Pyrimidinyl propionamide

β-Hydroxyhistidine

Glucose

Mannose

Threonine

γ amino α methyl valerate

Phleomycin

R = NH(CH₂)₃S⁺(CH₃)₂ X⁻, Bleomycin A₂

R = NH(CH₂)₄NHCNH₂, Bleomycin B₂
                    ‖
                    NH

R = OH, Bleomycinic acid

Bleomycins

cess on the venous side (i.e. Starling-type transcapillary flow) has never been demonstrated in brain. The central nervous system does not possess lymphatics. Cerebral capillary endothelial cells have no pinocytotic activity. Furthermore, the central nervous system and various compartments of cerebrospinal fluid are also excluded from the extracellular fluid of the rest of the body by the tight choroid epithelium and the tight layer of arachnoid.

The endothelial cells possess mitochondria which probably provide the necessary energy for facilitated diffusion of sugars and other transported metabolites, e.g. amino acids. In the absence of a specific transport mechanism, ions and low M_r polar nonelectrolytes are also unable to cross the endothelial cells of cerebral capillaries.

The lipid membranes of cerebral capillary endothelial cells do not present a barrier to lipid-soluble materials, including oxygen and carbon dioxide. The ability of drugs and other xenobiotics to cross the B.b.b. is strongly correlated with their lipid solubility.

All vertebrates and insects possess a B.b.b. It is present in the embryonic brain, and it can even be demonstrated a few hours after brain death. The B.b.b. clearly serves a protective function, preventing exposure of the central nervous system to a wide variety of electrolytes and macromolecules which circulate in the blood in health and disease. However, it may be that its most important function is the maintenance of a constant environment in the central nervous system; homeostasis of potassium, calcium, magnesium and hydrogen ions in cerebrospinal fluid and cerebral interstitial fluid is probably necessary for the correct function of the central nervous system. Other solutes, e.g. vitamins, may also be subject to homeostasis in the brain and cerebrospinal fluid. In addition, the B.b.b. prevents loss of cer-

tain active compounds, e.g. neurotransmitters, from the brain.

For the history of the B.b.b. concept, see *The Blood Brain Barrier*, by L.Bakay, (Charles C.Thomas, 1956); and *The Concept of a Blood-Brain Barrier* by M.Bradbury (John Wiley, 1979). The latter publication also contains a complete discussion of the supporting experimental evidence for the modern B.b.b. concept.

Blood coagulation, *blood clotting:* a process in which blood forms a gel (the clot) sufficiently dense to prevent bleeding from a wound. In most invertebrates the process only involves clumping of agglutinized blood cells, but in vertebrates and a few crustaceans the clot is formed from Fibrin (see) as well as trapped blood cells. Fibrin is an insoluble polymer derived from Fibrinogen (see), a large but soluble protein present in the plasma. Fibrinogen is converted to fibrin by the action of Thrombin (see), a serine protease with a high degree of specificity, and Factor XIII_a. The latter is an enzyme which catalyses transamidation reactions between glutamine and lysine residues and thus forms links between adjacent polypeptide chains in the precipitated fibrin.

The release of thrombin from its inactive zymogen, Prothrombin (see), is the penultimate step in a series of reactions, each of which releases an active serine protease from an inactive precursor in the blood. It is an example of cascade regulation, in which each activated protease activates the next precursor down the line, with increasing amounts of material being involved at each step. The clotting factors are listed in the table. The Roman numerals assigned to them are historical, and the activated form is indicated by the subscript a (i.e. XII_a).

Traditionally, two systems of B.c. have been recognized, the intrinsic and extrinsic systems. Blood

71

plasma will clot slowly (in several minutes) in the presence of a foreign surface like kaolin or glass. (For this reason blood for clinical use is withdrawn into vessels coated with Heparin (see) or a calcium-binding agent like citrate or EDTA.) The factors involved are all intrinsic to the plasma; hence the name. The physiological significance of the activation of the intrinsic system by foreign surfaces is not known. The proteins involved are F.XII, XI, HMW kininogen and prekallikrein. The mechanism of activation of the latter to kallikrein is not known; however, kallikrein is known to activate F.XII, which activates F.XI, and so on (Fig.).

The extrinsic system, which is activated by a specific proteinaceous tissue factor and phospholipids, produces a clot in seconds rather than minutes. The only plasma factor which belongs exclusively to the extrinsic system is F.VII (Fig.). However, it may be activated by F.XII$_a$ or kallikrein, both of which belong to the intrinsic pathway. In addition, it has recently been discovered that purified platelets, in the presence of purified F.V$_a$ and F.X, can be stimulated by low concentrations of prostacyclin to activate F.X. [A.K. Dutta-Roy et al. *Science* **231** (1986) 385–388] Thus the distinction between intrinsic and extrinsic pathways may well be artificial.

Under physiological conditions, the conversion of prothrombin to thrombin by F.X appears to require binding to a cell membrane. F.V is the platelet cell-membrane receptor for F.X, and increases the proteolytic activity of the latter.

The extent of B.c. is limited in vivo by the activation of vitamin-K-dependent protein C. This is another serine protease made up of 2 polypeptide chains, M_r 21 000 and 35 000, linked by disulfide bond(s). It has a high degree of sequence homology to the Gla-containing clotting factors, and like them, exists in the plasma as a zymogen. It is activated by thrombin

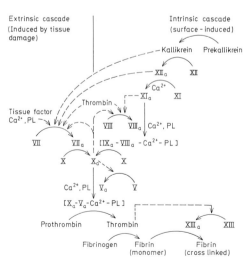

The coagulation cascade. PL=phospholipid or cell membrane. Solid arrows indicate reactions; dotted arrows, catalysis. Adapted from L. Lorand, *Methods Enzymol.* **45B** (1976) 31–37.

No.	Name	Properties and functions
I	Fibrinogen	M_r 340000, composed of 6 chains: $(A\alpha)_2(B\beta)_2\gamma_2$. Converted to fibrin by removal of 2 A and 2 B peptides. Fibrin is dissolved by plasmin, a protease related to coagulation factors and released from plasminogen by its own activation cascade.
II	Prothrombin (see) II$_a$ is Thrombin (see)	M_r 72500, 582 amino acids with 12 disulfide bridges and 10 Gla* residues. Glycoprotein. Converted to thrombin by F.X; thrombin is inhibited by antithrombin III, α_2-macroglobulin, α_1-antitrypsin and hirudin.
III	Tissue factor	A specific protein + phospholipids
IV	Calcium ions	Mediate binding of Factors IX, X and VII, and prothrombin, to acidic phospholipids of cell membranes, where activation occurs. Stabilize F.V, fibrinogen and possibly other proteins involved in activation and subunit dissociation of F.XIII.
V	Proaccelerin V$_a$ is accelerin	Labile glycoprotein, M_r 350000. F.V$_a$ mediates binding of F.X and prothrombin to platelets, where F.X is activated and prothrombin is converted to thrombin.
VII	Proconvertin	M_r 45–54000; contains Gla*; part of extrinsic cascade; released by tissue trauma.
VIII	Antihemophilic factor (see)	Several forms, M_r 10^5 to 2×10^6. Controversy over which is active in coagulation. F.VIII is an accessory in activation of F.X by IX$_a$. Its absence causes hemophilia A.
IX	Christmas factor	M_r 55400 (bovine), 57000 (human); single chain. IX$_a$ contains heavy chain homologous to serine proteases and light chain containing Gla*. Its absence causes hemophilia B.
X	Stuart factor	M_r 54500 (bovine), 59000 (human); 12 Gla* residues. Activated by F.IX$_a$ + F.VIII$_a$ + Ca^{2+} or

No.	Name	Properties and functions
		VII_a + tissue factor + Ca^{2+} or platelet membrane protease.
XI	Plasma thromboplastin antecedent	M_r 124000 (bovine), 160000 (human), glycoprotein composed of 2 similar or identical polypeptides joined by disulfide bond(s). Activates F. IX. XI_a is inhibited by antithrombin III, trypsin inhibitors, α_1-trypsin inhibitor and C1 inhibitor.
XII	Hageman factor	M_r 74000 (bovine), 76000 (human), glycoprotein, single chain. Activated by plasmin, kallikrein and $F. XII_a$. Inhibited by antithrombin III, (inhibition accelerated by heparin), C1 esterase inhibitor and lima bean trypsin inhibitor. Activation of F. XII initiated by contact with abnormal surfaces. F. XII is the first factor in the intrinsic pathway.
XIII	Fibrin-stabilizing factor (Laki-Lorand factor)	M_r 320000 (tetrameric plasma form; 160000 (dimeric platelet form). Both forms contain 2 a subunits, which are cleaved on activation. Plasma form also contains 2 b subunits. XIII_a is the transpeptidase responsible for cross-linking fibrin fibers.
	Prekallikrein	Activated to kallikrein, a serine protease which activates F. XII.
	HMW (high molecular weight) kininogen, contact activation cofactor, Fitzgerald factor, Williams factor, Flaujeac factor.	Activated to a kinin involved in activation of F. XII, at least in vitro.

* Gla = γ-carboxyglutamic acid

bound to an endothelial cell-surface receptor. Activated protein C inactivates F. V_a and VIII_a by limited proteolysis; these two factors are required for efficient conversion of F. X to X_a. Thus protein C, which is identical to autoprothrombin IIA, serves as a negative feedback control element for B. c. [T. Drakenberg et al. *Proc. Nat. Acad. Sci. USA* **80**

(1983) 1802–1806; L. M. Jackson & Y. Nemerson, *Ann. Rev. Biochem.* **49** (1980) 765–811; L. Lorand, *Methods Enzymol.* **45 B** (1976) 31–37]
The B. c. factors were discovered by analysis of hemophilias, the most common of which is caused by lack of F. VIII. The other factors were discovered by complementation in mixtures of non-clotting plasmas. Prekallikrein, prothrombin and factors XII, XI, X, IX and VII all give rise to Serine proteases (see) when activated by cleavage of specific peptide bonds. Prothrombin and the factors VII, IX, X and XI have homologous N-terminal regions containing multiple Gla residues. These are synthesized, post-translationally, by a carboxylase which uses vitamin K as a cofactor. The Gla residues are essential for the Ca^{2+}-mediated binding of these proteins to phospholipids (e. g. platelet membranes) where activation occurs. Hemorrhage is the major symptom of vitamin K deficiency.

Blood group antigens: specific oligosaccharide structures attached to glycoproteins in the membranes of blood cells which are recognized as antigens by the immune systems of other individuals or organisms. The antigens are attached to the protein glycophorin in erythrocytes and to both proteins and lipids in other parts of the body. In humans, five systems of antigens have been identified, the ABO system, the MN, P, rhesus and Lutheran systems. Only the ABO and rhesus systems affect blood transfusions between humans; the other systems have been identified using animal antibodies against human blood.
The structural differences between the ABO oligosaccharides are shown in Fig. 1. The genetic basis for these groups is the existence of three alleles of a gene coding for the synthesis of a glycosyltransferase. In A type individuals, the enzyme transfers *N*-acetylgalactosamine onto the terminal positions of the oligosaccharide chains, while in B type individuals, the enzyme is specific for galactose. The O gene appears to produce an inactive enzyme. Another gene, the H gene, codes for a fucosyltransferase which places an L-fucose on the oligosaccharide. When the H gene is inactive, the individual has the rare type I blood, or, if he has an active Le gene, which codes for an enzyme adding fucose to the *N*-acetylglucosamine, he has type Le^a. Individuals with both active H and Le genes have type Le^b. (Le is for Lewis factor.)

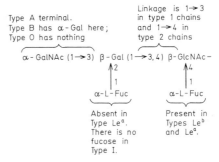

Figure 1. Ends of oligosaccharide chains in individuals with different blood groups

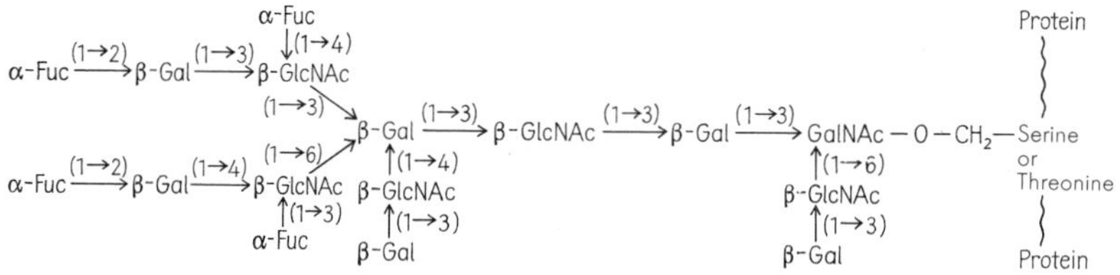

$$\begin{array}{c}
\alpha\text{-Fuc} \\
\downarrow(1\to4) \\
\alpha\text{-Fuc} \xrightarrow{(1\to2)} \beta\text{-Gal} \xrightarrow{(1\to3)} \beta\text{-GlcNAc}
\end{array}$$

Figure 2. Structure of the carbohydrate component of the Le^b glycoprotein of human blood
GlcNAc = N-acetyl-D-glucosaminyl; GalNAc = N-acetyl-D-galactosaminyl; Gal = D-galactosyl; Fuc = L-fucosyl.

About 80% of the population has an active Se (secretion) gene, so that they secrete glycoproteins bearing the blood group substance into saliva and other body fluids. The structure of the carbohydrate portion of such a glycoprotein is shown in Fig. 2.

Blood sugar: see D-Glucose.

Blue-green bacteria, *cyanobacteria, blue-green algae, Cyanophyta, Cyanophyceae*: a group of photosynthetic prokaryotic organisms using H_2O as hydrogen donor (see Photosynthesis). Many are also able to fix atmospheric nitrogen (see Nitrogen fixation). Since they are prokaryotes, they have no nucleus, mitochondria or chloroplasts (see Cell, 1.). They were originally classified as plants on the basis of their capacity for photosynthesis, but they are phylogenetically unrelated to the green algae, which are eukaryotes. There are many similarities between B.g.b. and chloroplasts, and the endosymbiotic theory of evolution proposes that chloroplasts are descended from symbiotic B.g.b. In the B.g.b. photosynthesis takes place on the thylakoids or thylakoid stacks, which are derived from the cell membrane. The photosynthetic pigments are chlorophyll *a*, carotenoids and biliproteins (phycocyanin and phycoerythrin). The cytoplasm contains prokaryotic-type ribosomes. Typical of most B.g.b. are cyanophycin granules; these are colorless and spherical or polyhedral, and visible under the light microscope. Cyanophycin granules contain storage material consisting of a copolymer of arginine and aspartic acid. The storage carbohydrate of B.g.b. is a polymer of glucose with a degree of branching intermediate between that of glycogen and amylopectin. Cytochemically it resembles glycogen: it stains brown with iodine and occurs as discrete granules of diameter 25-30 nm located between the thylakoids. Polyphosphate granules (syn. metachromatin, volutin, metachromatic granules) are also present in the cytoplasm of B.g.b.; these are spherical and vary in size from sub-light microscopic to several microns in diameter; they are concretions of the potassium salts of high M_r linear polyphosphates.

The inner cell wall, like that of bacteria, consists of murein anchored in the cell membrane. It is attacked by lysozyme. Outside the murein layer, there is a plasmatic layer, and beyond that there may be a slime capsule.

There are two major classes of B.g.b.: the *chroococcals,* in which the cells are solitary (e.g. *Anacystis*) or colonial, held together by mucoid hulls; and the *hormogonals,* which grow in filaments (trichomas), often enclosed in a sheath. The cells of hormogonal B.g.b. communicate with each other and form a physiological unit. In the trichomas there is a certain degree of specialization: the heterocysts, characterized by thick, highly refractory cell walls, are the site of nitrogen fixation.

BOD: see Biochemical oxygen demand.

Bohr effect: see Hemoglobin.

Bombesin: a peptide isolated from frog skin by Anastasi et al. The amino acid sequence is Pyr-Glu-Arg-Leu-Gly-Asn-Glu-Trp-Ala-Val-Gly-His-Leu-Met-NH$_2$. B. stimulates the release of gastric, pancreatic and adenohypophyseal hormones in mammals, and causes contraction of smooth muscles in the gastric and urinary tracts, and in the uterus. When injected peripherally, B. inhibits the secretion of growth hormone. The explanation of this action is the homology of B. with a number of peptides found in mammalian tissues, including the hypothalamus. For example, 9 of the 10 carboxy-terminal amino acids of B. are identical to the 10 carboxy-terminal amino acids of porcine gastrin-releasing hormone. [W. A. Murphy et al., *Endocrinology* **117** (1985) 1179-1183]

Bombykol, *10-trans-12-cis-hexadecadienol-(1)*: a pheromone exuded by female silk moths *(Bombyx mori)* to attract males. B. is an oil, n_D^{20} 1.4835. The first determination of structure was made on 15 mg B. isolated from the abdominal glands of 500 000 female moths. The configuration was established by Butenandt by comparison of the biological activities of synthetic compounds with the natural product.

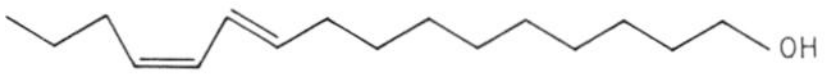

Bombykol

Bongkrekic acid: 3-carboxymethyl-17-methoxy-6,18,21-trimethyldocosa-2,4,8,12,14,18,20-heptaenedioic acid, M_r 486.61. One of two toxic antibiotics produced by *Pseudomonas cocovenenans* in spoiled bongkrek (a coconut product consumed in Indonesia). It is an inhibitor of adenine nucleoside translocation and affects carbohydrate metabolism.

Bornane: see Monoterpenes, Fig.

Boron: an element essential for growth of higher plants. It is absorbed by the roots in the form of borate. Lack of boron causes heart rot of sugar beet and other roots.

Botulin: see Toxic proteins.

Bowman-Birk inhibitor: see Soybean trypsin inhibitor.

Bradykinin, *kallidin I, kinin 9*: Arg-Pro-Pro-Gly-Phe-Ser-Pro-Phe-Arg, one of a group of plasma hormones called kinins. Like the other kinins, it is produced from a plasma precursor by the action of Kallikrein (see), Trypsin (see) or Plasmin (see). It causes dilation of blood vessels, and thus a reduction of blood pressure, causes the smooth muscles of the bronchia, intestines and uterus to contract, and is a potent pain-producing agent. Lysylbradykinin has similar activity.

Brassicasterol, *ergosta-5,22-dien-3β-ol*: a plant sterol (see Sterols). M_r 398.69, m.p. 148 °C, $[\alpha]_D$ − 64 °. B. was first isolated from rapeseed *(Brassica campestris)* oil.

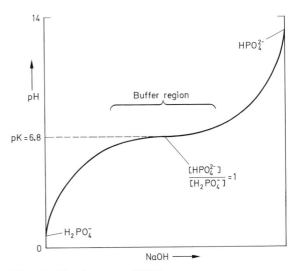

Brassicasterol

Brevifolin carboxylic acid: see Tannins.
Brimacombe fragments: see Ribosomal proteins.
Bromelain: a thiol enzyme (EC 3.4.22.4) from the stems and fruits of the pineapple plant. The stem enzyme is a basic glycoprotein (M_r 33 000; IP 9.55) structurally and catalytically similar to papain. B. is activated by mercaptoethanol and other SH compounds. It is irreversibly inhibited by agents which block SH groups. It is an endopeptidase and is used in protein chemistry to hydrolyse polypeptide chains into large fragments.
Broussin: 7-hydroxy-4′-methoxyflavan; see Flavan.

Brucine: 2,3-dimethoxystrychnine. M_r 394.47, m.p. 105 °C (tetrahydrate), m. p. 178 °C (anhydrous). It is used in preparative chemistry to separate racemic acids into optical isomers. B. is highly poisonous, but 10 times less so than strychnine, from which it is derived. The main physiological effects is a paralysis of the smooth musculature. For formula and biosynthesis, see Strychnos alkaloids.
Bryokinin: see N^6-(γ,γ-dimethylallyl)Adenosine.
Bufadienolides: see Toad poisons, Cardiac glycosides.
Buffer: a system that resists and minimizes changes of pH caused by addition or loss of acid and base. Physiologically, buffers stabilize the pH of cells and body fluids during metabolic under- or overproduction of acids and bases, or during environmental changes of hydrogen ion concentration. For laboratory purposes, buffers are used to maintain constant and favorable pH values for enzymatic reactions, to prevent protein denaturation, and to provide appropriate pH conditions for culture of microorganisms and tissues. Buffered solutions are also important as chromatography eluants and electrophoresis media, and for the performance of some chemical reactions, such as the Edman degradation of proteins.

Buffer action is illustrated (Fig. 1) by the titration of a weak acid (NaH_2PO_4) with a strong base (NaOH): $OH^- + H_2PO_4^- \rightleftharpoons HPO_4^{2-} + H_2O$. As NaOH is added, the ratio $[HPO_4^{2-}]/[H_2PO_4^-]$ (square brackets indicate concentrations) increases, starting at almost zero and becoming progressively larger until it is approximately infinity at the end point of the titration. The resulting titration curve is a graphic representation of the Henderson-Hasselbach equation:

$$pH = pK + \log \frac{[salt]}{[acid]}, \text{ or}$$

$$pH = pK + \log \frac{[conjugate\ base]}{[acid]}$$

Figure 1. Titration curve of $H_2PO_4^-$

Therefore, in the present case, $pH = pK + \log$ $[HPO_4^{2-}]/[H_2PO_4^-]$. When $[HPO_4^{2-}]/[H_2PO_4^-]$ is unity, i.e. one half of the $H_2PO_4^-$ has been titrated, then $pH = pK$. The intermediate plateau of the titration curve, where pH is relatively unaffected by addition of acid or base, is known as the buffer region, and is defined by the pK value (See also titration curves in the entry Amino acids). As a rule, the useful buffer region is the pK' value $\pm$ 1 pH unit. (pK is an absolute dissociation constant corrected to zero ionic strength, whereas pK' is the apparent or concentration dissociation constant determined at known ionic strength and, where appropriate, in the presence of other solutes.)

The degree of dilution and the presence of other ions may affect pH, e.g. if NaCl is added to phosphate buffer, the pH falls because Na^+ tends to associate with the HPO_4^{2-}, and Cl^- with H^+, thereby increasing the degree of dissociation of $H_2PO_4^{2-}$. In very dilute solution, this effect may be unrecognizable because the ions are too far apart for interaction of their electric fields. Thus the pK of $H_2PO_4^-$ in infinitely dilute solution is 7.2, whereas in blood the value is 6.8. This phosphate system is one of the most popular and widely used biochemical buffers, but it suffers from the disadvantages that it has poor buffering capacity above pH 7.5, it tends to precipitate polyvalent cations, and phosphate is a metabolite in some systems. The buffer may be conveniently prepared by titrating solutions of KH_2PO_4 or NaH_2PO_4 with NaOH or KOH to the required pH. Usually it is prepared by mixing standard solutions of NaH_2PO_4 and Na_2HPO_4 as shown in Table 1.

An ideal buffer should have a pK between 6 and 8, which is the pH range required for most biological reactions. It should also be very soluble in water, and should not cross biological membranes. If the buffer is used in a spectrophotometric assay, it should not absorb at the measurement wavelength. Minimal effects of temperature, concentration and ionic composition of the medium are also desirable. No buffer system satisfies all these requirements, and more than one buffer should be tested in the investigation of biochemical reactions and other biological systems. Buffers such as borate, imidazole, veronal (5,5-diethylbarbiturate), maleate and others are suitable for specific biological systems, but all have potential disadvantages. For example, borate forms complexes with cis-diols, which includes many sugars and respiratory metabolites. On the other hand, this property has been exploited in the chromatographic separation and analysis of sugars. Buffer systems more suitable for biochemical purposes are listed in Table 2.

It must be remembered that all buffers show a temperature effect, so that the pH recorded at room temperature may be significantly different from that at the working temperature. For example, TRIS (see Table 2) has a high temperature/pH gradient, so that 0.05 M TRIS, pH 7.05 (adjusted with HCl) at 37 °C has pH 7.20 at 23 °C. Since the logarithmic pH scale covers a very wide range of hydrogen ion concentration, apparently small changes of pH should not be dismissed as unimportant; e.g. at pH 7.05 and pH 7.2, the respective hydrogen ion concentrations are $[H^+] = 8.912 \times 10^{-8}$M and 6.309×10^{-8}M, a difference of 41%.

Buffer capacity. Whereas pH depends on pK' and the ratio [proton acceptor]/[proton donor], buffer capacity depends on the quantity of buffer components present. Buffer capacity is defined as the minimal quantity of acid or base that must be added or removed in order to cause a significant pH change. For example, two HEPES solutions of pH 7.05 and of equal volumes may contain different quantities of zwitterion and base.

$pH (7.05) = pK(7.55) + \log [base]/[zwitterion]$.

Solution 1 (0.05 M): [base]/[zwitterion] = 0.038/0.012

Solution 2 (0.10 M): [base]/[zwitterion] = 0.076/0.024

Solution 2 has the same pH but twice the buffering capacity of solution 1, i.e. solution 2 is able to absorb twice as many protons as solution 1 before a significant change of pH occurs. In medical physiology, buffer capacity is represented by the amount of acid or base that body fluids can accommodate before the pH becomes dangerously low or high.

Buffers of body fluids. Acids and bases, produced by metabolism, pass into body fluids and are ultimately excreted. The body fluids are temporarily protected (more commonly against acidity, because normal metabolism results in a net production of acid) by their buffer systems. The most important body fluid buffers are plasma (HCO_3^-/CO_2, protein$^{(n-)}$/protein(nH^+), $HPO_4^{2-}/H_2PO_4^-$), erythrocytes (HCO_3^-/CO_2, deoxyhemoglobin/deoxyhemoglobin·H^+, oxyhemoglobin/oxyhemoglobin·H^+, $HPO_4^{2-}/H_2PO_4^-$), interstitial fluid (similar to plasma, but with lower protein concentration) and intracellular fluid (excluding erythrocytes), which contains much higher concentrations of phosphate and protein than plasma or interstitial fluid.

In the blood, the main buffer system is bicarbonate at a concentration of $[HCO_3^-] = 0.02$–0.03 M (20–30 mEq/l). Hemoglobin provides a further 10 mEq/l buffer capacity, and phosphate makes a small contribution of 1.5 mEq/l. The 5 liters of blood in an average adult human are thus able to absorb about 0.15 mole H^+ before the pH becomes dangerously low. The major buffers of the body are, however, present in other tissues. The total musculature of the body, for example, can neutralize about 5 times as much acid as the blood, and the blood HCO_3^-/CO_2

Zwitterion of HEPES　　　　　　　　　　　　　*base of HEPES*

system represents only about a tenth of the total buffer capacity of the body. But all the buffer systems of the body are able to interact with one another, so that all changes in the acid/base balance of the body are reflected in the blood. In other words, the buffer systems of the body buffer each other by shifting H^+ from one to the other, a principle known as the isohydric principle.

Bicarbonate acts not only as a blood buffer system, but also represents the main form in which CO_2 is transported from respiring tissues to the lungs for expiration. Some CO_2 is transported as carbamino groups of proteins:

Protein-NH_2 + CO_2 ⇌ Protein-NH-COOH ⇌ Protein-NH-$\overset{-}{C}OO^-$ + H^+

and about 80% of CO_2 is transported as bicarbonate. The different forms of blood CO_2 are in equilibrium:

Gaseous CO_2
⇅
Carbamino groups ⇌ dissolved CO_2 + H_2O ⇌ H_2CO_3 ⇌ H^+ + HCO_3^-
⇅
CO_2 from respiration

In order to achieve high concentrations of HCO_3^- from dissolved CO_2, the resulting protons must be removed, i.e. accommodated by a buffer system. The principal buffers serving this function are plasma proteins (accounting for about 10% of the protons), erythrocyte phosphate (about 20%), and erythrocyte hemoglobin (60-70%). For the role of hemoglobin, see Bohr effect; Hemoglobin.

At any time, the extracellular fluids contain about 1.2 mmol/l CO_2 (in its various forms) in transit between respiring tissues and lungs.

Proteins as buffers. Many of the acidic and basic amino acid side chains on the surface of proteins possess pK' values that permit them to contribute to physiological buffer systems (Fig. 2).

Control of pH by excretion

a) *via the lungs*. The respiratory center in the medulla oblongata responds to increased $[H^+]$ by increasing alveolar ventilation, and vice versa. An increased rate of alveolar ventilation removes CO_2 more rapidly from the blood, thereby increasing blood pH. This regulatory feedback system for the control of $[H^+]$ is 50-75% efficient, with a feedback gain of 1-3, i.e. if the blood pH is suddenly decreased from

7.4 to 7.0, the resulting increase in alveolar ventilation adjusts the pH to 7.2-7.3 in 3-12 minutes. As a rough approximation, a decrease in the pH of extracellular fluids of about 0.23 causes a twofold increase in the rate of alveolar ventilation. The alveolar ventilation rate can be varied from about zero to fifteen times normal.

b) *via the kidneys*. The inorganic composition of glomerular filtrate is almost the same as that of plasma, but it has only a very low protein concentration (0.03%). In transit through the nephron, water and various solutes are reabsorbed and returned to the blood. The mechanism for the reabsorption of bicarbonate also plays a role in the loss or retention of H^+. The epithelial cells of most of the nephron (Fig. 3) (proximal tubule, thick segment of the ascending loop of Henle, distal tubule, collecting tubule and collecting duct) all secrete H^+ into the tubular fluid (Fig. 4) in exchange for the uptake of Na^+.

The bicarbonate ions in the tubular fluid (from the plasma via glomerular filtrate) are therefore in the presence of an increasing concentration of H^+, which favors the formation of H_2CO_3. A carbonic anhydrase in the brush border of the epithelial cells in the proximal, but not the distal, tubules accelerates formation of CO_2 and H_2O from the H_2CO_3 in the tubular fluid. The CO_2 (which diffuses readily across all biological membranes) diffuses into the tubular cells, and there forms H_2CO_3. (This reaction is catalysed by intracellular carbonic anhydrase.) The H_2CO_3 dissociates to H^+ and HCO_3^-.

The Na^+ taken up from the tubule is removed from the epithelial cell into the extracellular fluid by active transport; charge neutrality is maintained because the export of Na^+ is accompanied by export of HCO_3^-. The net effect is that for every H^+ formed and secreted into the tubule, bicarbonate ion diffuses into the extracellular fluid in combination with Na^+ absorbed from the tubule. Thus in effect, bicarbonate is reabsorbed from the glomerular filtrate.

In the collecting ducts, the pH of the tubular fluid can fall as low as 4.5, i.e. $[H^+]$ is 900 times greater in

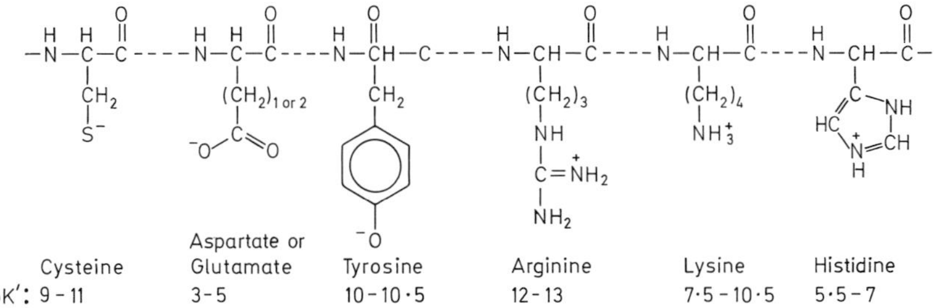

Figure 2. Ionizable amino acid side chains of proteins and their pK' values

Buffer

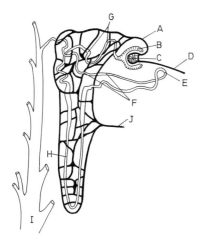

Figure 3. Diagrammatic representation of a single nephron.
A, efferent arteriole; B, Bowman's capsule; C, glomerulus, D, afferent arteriole; E, juxtaglomerular apparatus; F, distal tubule; G, promiximal tubule; H, descending or thin loop of Henle; I, collecting duct; J, vein.

the tubular fluid than in the extracellular fluid. The proximal tubule secretes about 84% of all H^+, but it can only maintain a 3-4 fold gradient, i.e. a tubular fluid pH of 6.9 compared with pH 7.4 for the extracellular fluid. Distal tubules occupy an intermediate position, being able to decrease the pH of tubular fluid to 6.0-6.5. In Acidosis (see), the ratio

Table 1. Preparation of 0.1 M phosphate buffer. See Table 2 (Orthophosphate).
Solution A: 0.2 M NaH_2PO_4 (27.8 g/1000 ml)
Solution B: 0.2 M Na_2HPO_4 (53.65 g $Na_2HPO_4 \cdot$ $7H_2O$ or 71.7 g $Na_2HPO_4 \cdot 12H_2O$/1000 ml)

Mix the volumes of A and B shown below, then dilute to 200 ml to give 0.1 M. The values of pH shown are accurate to $\pm$ 0.05 pH at 23 °C.

A (ml)	B (ml)	pH
93.5	6.5	5.7
92.0	8.0	5.8
90.0	10.0	5.9
87.7	12.3	6.0
85.0	15.0	6.1
81.5	18.5	6.2
77.5	22.5	6.3
73.5	26.5	6.4
68.5	31.5	6.5
62.5	37.5	6.6
56.5	43.5	6.7
51.0	49.0	6.8
45.0	55.0	6.9
39.0	61.0	7.0
33.0	67.0	7.1
28.0	72.0	7.2
23.0	77.0	7.3
19.0	81.0	7.4
16.0	84.0	7.5
13.0	87.0	7.6
10.0	90.0	7.7
8.5	91.5	7.8
7.0	93.0	7.9
5.3	94.7	8.0

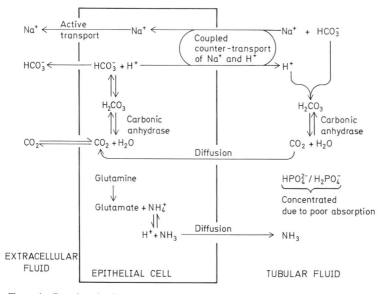

Figure 4. Reactions leading to reabsorption of bicarbonate from the kidney tubule. Excess H^+ in the tubular fluid associates with HPO_4^{2-} and NH_3 and is excreted in the urine as $H_2PO_4^-$ and NH_4^+. Excess HCO_3^- in the tubular fluid passes into the urine.

Table 2. Laboratory buffers used in biochemistry. The nonprotonated forms of amine buffers (e.g. TRIS) and zwitterionic buffers (e.g. MES, MOPS, HEPES, etc.) may be inhibitory in some biochemical systems, so that a working pH above the pK_a is often advisable.

Name	Accepted abbreviation	Formula	pK_a	$\Delta pK_a/{}^\circ C$	Useful range	Comments
Pyrophosphate	–	$H^+ + {}^-O-\overset{\overset{O}{\|\|}}{\underset{\underset{O^-}{\|}}{P}}-O-\overset{\overset{O}{\|\|}}{\underset{\underset{O^-}{\|}}{P}}-OH$	5.8	−0.01	4.8–6.8	Binds or precipitates many polyvalent cations. Product and/or substrate in many biochemical reactions.
2-(N-morpholino) ethanesulfonic acid	MES	$O\langle\text{ring}\rangle\overset{+}{N}HCH_2CH_2SO_3^-$	6.15	−0.011	5.8–6.5	Metal binding constants: $Mg^{2+}(0.8)$, $Ca^{2+}(0.7)$, $Mn^{2+}(0.7)$, Cu^{2+} (negligible). Suitable for isolation of photosynthetically active, intact chloroplasts. Does not interfere in Folin protein assay. Decomposes when autoclaved in presence of glucose.
Bicarbonate	–	$H^+ + HCO_3^- \rightleftharpoons H_2CO_3 \rightleftharpoons CO_2 + H_2O$	6.3	?	5.3–7.3	Excellent, but usually impractical, because it must be used in a closed system.
Orthophosphite	–	$H^+ + {}^-O-\overset{\overset{O}{\|\|}}{\underset{\underset{H}{\|}}{P}}-O^-$	6.5	+0.003	5.5–7.5	Forms complexes with polyvalent cations, but not as strongly as does orthophosphate. Inhibits oxygen uptake and nitrogen fixation in Azotobacter. May be useful, but not widely tested.
Orthophosphate	–	$H^+ + {}^-O-\overset{\overset{O}{\|\|}}{\underset{\underset{OH}{\|}}{P}}-O^-$	6.8	−0.005	5.7–8.0	May precipitate Ca^{2+}, Mg^{2+}, Fe^{3+}, unless care taken. It is an important metabolite. Despite disadvantages, it is generally very useful and very widely used.
Bis(2-hydroxyethyl)amino-tris(hydroxymethyl)methane	BIS-TRIS	$(HOCH_2CH_2)_2\overset{+}{\underset{H}{N}}(CH_2OH)_3$	6.5	?	5.8–7.2	Intended as an addition to the "TRIS" range to extend pH coverage. Has usual disadvantages of amine buffers, and should not be used for study of electron transport or ATP synthesis in organelles or preparations thereof.
N-(2-acetamido)-iminodiacetic acid	ADA	$H_2NCOCH_2\overset{+}{\underset{H}{N}}\overset{\diagup CH_2COO^-}{\diagdown CH_2COOH}$	6.6	−0.011	6.0–7.2	Metal binding constants: $Mg^{2+}(2.5)$, $Ca^{++}(4.0)$, $Mn^{2+}(4.9)$, $Cu^{2+}(9.7)$. Like other buffers based on glycine, may be used in place of TRIS.
Piperazine-N, N-bis-(2-ethane-sulfonic acid)	PIPES	$HO_3SCH_2CH_2N\langle\text{ring}\rangle\overset{+}{N}HCH_2CH_2SO_3^-$	6.8	−0.0085	6.4–7.2	Metal binding negligible. Fairly widely used and generally satisfactory for most purposes.
N-(2-acetamido)-2-aminoethane-sulfonic acid	ACES	$H_2NCOCH_2\overset{+}{N}H_2CH_2CH_2SO_3^-$	6.9	−0.020	6.4–7.4	Metal binding constants: $Mg^{2+}(0.4)$, $Ca^{2+}(0.4)$, Mn^{2+}(negligible), $Cu^{2+}(4.6)$. Not widely tested. Satisfactory for uncoupled electron transport in chloroplasts.

Buffer

Name	Accepted abbreviation	Formula	pK_a	$\Delta pK_a/°C$	Useful range	Comments
3-(N-morpholino)-2-hydroxypropane-sulfonic acid	MOPSO	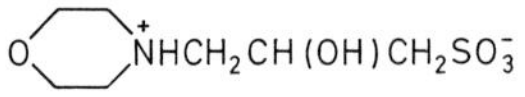 $O\!\!\overbrace{}\!\!\overset{+}{N}HCH_2CH(OH)CH_2SO_3^-$	6.95	-0.015	6.2–7.6	Metal binding negligible. Generally satisfactory for most purposes. Does not interfere in Folin or biuret protein assays. Unstable when autoclaved with glucose. Same comments apply as for DIPSO
Imidazole	–	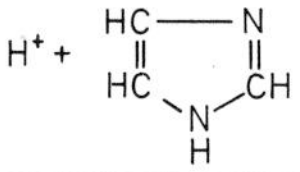 $H^+ + \begin{array}{c}HC\!\!-\!\!N\\ \parallel \quad \parallel \\ HC \quad CH \\ \diagdown N \diagup \\ H\end{array}$	7.05	?	6.1–8.0	Very reactive and unstable. Uncoupler of oxidative phosphorylation. Best avoided.
Cholamine chloride	–	$(CH_3)_3\overset{+}{N}CH_2CH_2NH_2\,Cl^-$	7.1	-0.027	6.2–7.6	Metal binding negligible. Uncoupler and inhibitor of Hill reaction in chloroplasts.
N,N-bis(2-hydroxyethyl)-2-amino-ethanesulfonic acid	BES	$(HOCH_2CH_2)_2\overset{+}{N}HCH_2CH_2SO_3^-$	7.15	-0.016	6.6–7.6	Metal binding negligible for Mg^{2+}, Ca^{2+}, Mn^{2+}; binding constant for $Cu^{2+}=3.5$. Not widely tested. Satisfactory for electron transport studies in chloroplast lamellae.
3-(N-morpholino)-propanesulfonic acid	MOPS	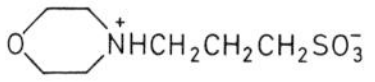 $O\!\!\overbrace{}\!\!\overset{+}{N}HCH_2CH_2CH_2SO_3^-$	7.2	-0.013	6.5–7.9	Metal binding negligible. Inexpensive. Generally satisfactory for most purposes. Does not interfere in Folin or biuret protein assays. Unstable when autoclaved with glucose.
N-[tris(hydroxymethyl)methyl]-2-amino-ethanesulfonic acid	TES	$(HOCH_2)_3C\overset{+}{N}H_2CH_2CH_2SO_3^-$	7.5	-0.020	6.8–8.2	Negligible binding of Mg^{2+}, Ca^{2+}, Mn^{2+}. Binding constant $Cu^{2+}=3.2$. Excellent general purpose buffer. Preserves electron transport and phosphorylation of plant mitochondria. Does not interfere in Folin protein assay.
4-(2-hydroxyethyl)-piperazine-1-ethanesulfonic acid	HEPES	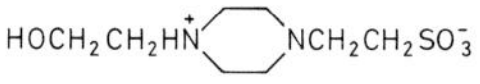 $HOCH_2CH_2H\overset{+}{N}\!\!\overbrace{}\!\!NCH_2CH_2SO_3^-$	7.55	-0.014	7.0–8.0	Metal binding negligible. One of the best buffers for general biochemical purposes. Beneficial for preserving electron transport and phosphorylation of plant mitochondria. Interferes in Folin protein assay, but not in the biuret assay.
3-[N-bis(hydroxyethyl)amino]-2-hydroxypropane-sulfonic acid	DIPSO	$(HOCH_2CH_2)_2\overset{+}{N}H_2CH_2CH(OH)CH_2SO_3^-$	7.6	-0.015	6.9–8.1	One of a series of buffers synthesized by reaction of appropriate amine with inocuous sodium 3-chloro-2-hydroxy-propanesulfonate, designed to replace related propane sulfonate buffers synthesized from carcinogenic propane sulfone. Tested in mammalian tissue culture, bacterial growth media, electron transport and ATP synthesis in chloroplast lamellae, and experimental virus infection of plant protoplasts, and found to be equivalent to or better than other buffers.

Name	Accepted abbreviation	Formula	pK_a	$\Delta pK_a/°C$	Useful range	Comments
3-[tris(hydroxy-methyl)methyl-amino]-2-hydroxy-propane sulfonic acid	TAPSO	$(HOCH_2)_3 C\overset{+}{N}H_2 CH_2 CH(OH)CH_2 SO_3^-$	7.7	−0.018	7.0–8.2	As for DIPSO.
Acetamidoglycine	–	$H_2NCOCH_2\overset{+}{N}H_2CH_2COO^-$	7.7	?	7.2–8.2	Like other buffers based on glycine, it may be used in place of TRIS. Not advised for work with chloroplast and mitochondrial systems.
Piperazine-N, N-bis(2-hydroxy-propanesulfonic acid	POPSO	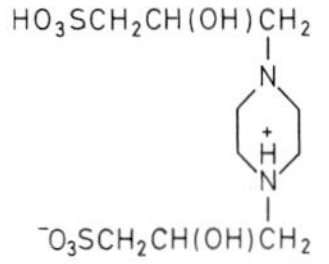 $HO_3SCH_2CH(OH)CH_2$ [piperazine ring] $^-O_3SCH_2CH(OH)CH_2$	7.85	−0.013	7.2–8.5	As for DIPSO.
N-hydroxyethyl-piperazine-N′--2-hydroxypropane-sulfonic acid	HEPPSO	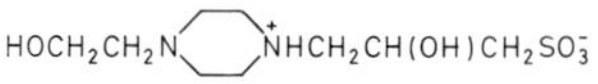 $HOCH_2CH_2N$ [piperazine ring] $\overset{+}{N}HCH_2CH(OH)CH_2SO_3^-$	7.9	−0.01	7.1–8.5	As for DIPSO.
4-(2-hydroxyethyl)-piperazine-1--propanesulfonic acid	HEPPS or EPPS	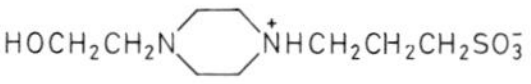 $HOCH_2CH_2N$ [piperazine ring] $\overset{+}{N}HCH_2CH_2CH_2SO_3^-$	8.0	−0.015	7.3–8.7	Like MOPS, a generally satisfactory propanesulfo-nate buffer.
N-[tris(hydroxy-methyl)methyl] glycine	Tricine	$(HOCH_2)_3 C\overset{+}{N}H_2 CH_2 COO^-$	8.15	−0.021	7.4–8.8	Metal binding constants: $Mg^{2+}(1.2)$, $Ca^{2+}(2.4)$, $Mn^{2+}(2.7)$, $Cu^{2+}(7.3)$. May act protectively by com-plexing heavy metal ions. Subject to photo-oxidation in presence of flavins and flavoproteins. Becoming standard buffer for chloro-plast lamellae.
Pyrophosphate	–	$H^+ + {}^-O-\underset{O^-}{\overset{O}{P}}-O-\underset{O^-}{\overset{O}{P}}-O^-$	8.2	−0.006	7.2–9.2	See comments for pK_a 5.8.
Glycinamide	–	$H_2NCOCH_2\overset{-}{N}H_2$	8.2	−0.029	7.4–8.8	Potent uncoupler and pow-erful inhibitor of Hill reac-tion in chloroplasts.
Tris(hydroxyme-thyl)aminomethane	TRIS	$(HOCH_2)_3CNH_2$	8.3	−0.031	7.0–9.0	Metal binding negligible. Widely used, inexpensive. Penetrates membranes. In-terferes in photosynthesis. Weak uncoupler of oxida-tive phosphorylation.
N,N-bis(2-hydroxy-ethyl)glycine	Bicine	$(HOCH_2CH_2)_2 \overset{+}{N}HCH_2COO^-$	8.35	−0.018	7.8–8.8	Metal binding constants: $Mg^{2+}(1.5)$, $Ca^{2+}(2.8)$, $Mn^{2+}(3.1)$, $Cu^{2+}(8.1)$. Sat-isfactory for Hill reaction. More easily oxidized by hexacyanoferrate than tri-cine.
Glycylglycine	–	$H_3\overset{+}{N}CH_2CONHCH_2COO^-$	8.4	−0.028	7.5–8.9	Metal binding constants: $Mg^{2+}(0.8)$, $Ca^{2+}(0.8)$, $Mn^{2+}(1.7)$, $Cu^{2+}(5.8)$. Ex-cellent for all purposes in this pH range, but expen-sive. No advantages over Tricine or TAPS.
3-[tris(hydroxy-methyl)methyl-amino]-1-propane-sulfonic acid	TAPS	$(HOCH_2)_3 C\overset{+}{N}H_2 (CH_2)_3 SO_3^-$	8.55	−0.027	7.7–9.1	Does not bind Mg^{2+}, Ca^{2+}, Mn^{2+} or Cu^{2+}. Satisfactory for study of phosphoryla-tion and electron transport in chloroplast preparations.

Buffer

Name	Accepted abbreviation	Formula	pK_a	$\Delta pK_a/°C$	Useful range	Comments
3-[dimethyl(hydroxymethyl)methylamino]-2-hydroxypropanesulfonic acid	AMPSO	$HOCH_2C(CH_3)_2\overset{+}{N}H_2CH_2CH(OH)CH_2SO_3^-$	9.0	?	8.3–9.7	Like all the 2-hydroxy-1-propanesulfonate buffers, it is generally satisfactory for the appropriate pH range.
2-[N-cyclohexylamino]ethanesulfonic acid	CHES	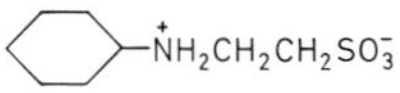 $\overset{+}{N}H_2CH_2CH_2SO_3^-$	9.3	?	8.6–10.0	Like HEPES, a generally satisfactory ethanesulfonate buffer.
3-[cyclohexylamino]-2-hydroxy--1-propanesulfonic acid	CAPSO	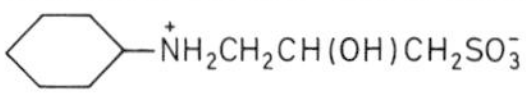 $\overset{+}{N}H_2CH_2CH(OH)CH_2SO_3^-$	9.6	?	8.9–10.3	Like all the 2-hydroxy-1-propanesulfonate buffers, it is generally satisfactory for the appropriate pH range.
Glycine	–	$H_3\overset{+}{N}CH_2COO^-$	9.9	?	9.2–10.6	A good buffer for this high pH range. Satisfactory for study of electron transport and phosphorylation in chloroplasts; such high pH, however, rarely used.
Carbonate	–	$H^+ + CO_3^{2-} \rightleftharpoons HCO_3^-$	10.25	?	9.3–11.2	Useful for study of electron transport and photophosphorylation at high pH where amine buffers, e.g. Tris, are inhibitory. Acts also as CO_2 buffer, producing constant low concentrations of CO_2, which can be calculated. Algal cells readily perform photosynthesis in carbonate-bicarbonate-buffered systems.
3-[cyclohexylamino]-1-propanesulfonic acid	CAPS	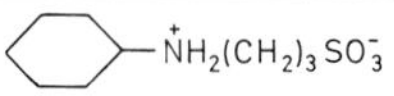 $\overset{+}{N}H_2(CH_2)_3SO_3^-$	10.4	?	9.7–11.1	Like MOPS, a generally satisfactory propanesulfonate buffer.
1,3-Bis[tris(hydroxymethyl)methylamino]propane	BIS-TRIS-propane	$CH_2[CH_2NHC(CH_2OH)_3]_2$	6.8 and 9.0	?	6.3–9.5	As for BIS-TRIS.
2-Hydroxyethylimino-tris(hydroxymethyl) methane	Mono-TRIS	$HOCH_2CH_2NHC(CH_2OH)_3$	7.93	?	7.1–8.5	As for BIS-TRIS.

$[HCO_3^-]/[CO_2]$ decreases in the extracellular fluid and in the glomerular filtrate. The rate of H^+ secretion therefore exceeds the rate of glomerular filtration of HCO_3^-. The results are twofold:

a) There is a net increase in the quantity of HCO_3^- and Na^+ in the extracellular fluid, which corrects the acidosis by increasing pH, and

b) Excess H^+ associates with other buffer anions (mainly HPO_4^{2-}) and with ammonia, and is excreted in the urine. HPO_4^{2-} and $H_2PO_4^-$ are poorly reabsorbed by the tubule, so that they become concentrated in the tubular fluid, forming a strong buffer system, pK 6.8. H^+ reacting with anions other than HCO_3^- in the tubule contributes to urinary titratable acidity, which is equivalent to the amount of alkali required to return the urine pH to 7.4 (i.e. the pH of glomerular filtrate).

All tubule epithelial cells and those of the thin segment of the loop of Henle continually produce ammonia, 60% arising from the action of glutaminase on glutamine, and 40% from other amino acids (e.g. by action of glutamate dehydrogenase on L-glutamate). The ammonia reacts with H^+, forming NH_4^+, which is excreted with Cl^- and other anions. Local acidosis of tubular cells promotes the production of ammonia by inducing the production of glutaminase.

In Alkalosis (see), the ratio $[HCO_3^-]/[CO_2]$ increases in the extracellular fluid, so that the glomerular filtrate contains an excess of HCO_3^-, compared with the H^+ that is secreted from the tubule epithelium. The excess HCO_3^- passes into the urine together with Na^+ and other cations. Thus part of the bicarbonate complement of the blood is removed, and the pH decreases, correcting the alkalosis. [G. Gomori, *Methods in Enzymology* **1** (1955) 138–146; N. E. Good et al., *Biochemistry* **5** (1966) 467–477; N. E. Good & S. Izawa, *Methods in Enzymology* **24B** (1972) 53–68; W. J. Fergusun et al. *Anal. Biochem.* **104** (1980) 300–310.]

Bufogenins: see Cardiac glycosides.

Bufotenine, *5-hydroxy-N-dimethyltryptamine*: a Toad poison (see). It is also found in toadstools from which it was first isolated by Wieland.

Bufotoxin: the main toxin in the venom of the European toad *Bufo vulgaris*. M_r 757. It is a steroid derivative. The minimum lethal dose (in cats) is 390 µg/kg.

Buoyant density: see Density gradient centrifugation.

2,3-Butanediol: see Fermentation products.

2,3-Butylene glycol: see Fermentation products.

n-Butyric acid, *butanoic acid*: $CH_3-(CH_2)_2-COOH$, a fatty acid. M_r 88.1, m.p. -5 °C. B.a. accounts for 3 to 5% of the fatty acids esterified to glycerol in butterfat. When butter becomes rancid, it is the free B.a. produced by hydrolysis which is responsible for the unpleasant odor. B.a. is found in free form in many plants and fungi, and in traces in sweat. Esters of B.a. are found in many essential oils.

Buxus alkaloids: a group of steroid alkaloids which are characteristic of plants in the boxwood

Cyclobuxamine H

genus *(Buxus)*. The B.a. are derived structurally from pregnane (see Steroids), having additional methyl groups on positions 4 and 14, amino or methylated amino functions at positions 3 and 20, and usually a 16α-hydroxyl group and a 9,10-cyclopropane ring. A typical example is *cyclobuxamine H* (Figure).

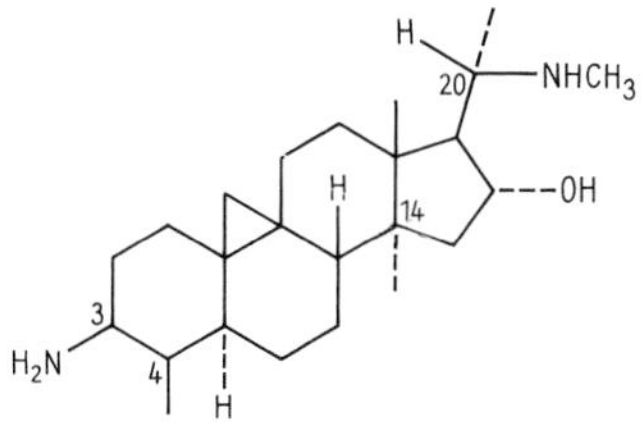

The B.a. are biosynthesized via cycloartenol or a similar triterpenoid precursor.

C

C: see Cytosine.

Cachectin: see Tumor necrosis factors.

C₄-acid cycle: see Hatch-Slack-Kortschack cycle.

Cactus alkaloids: see Anhalonium alkaloids.

Cadaverine, *1,5-diaminopentane:* a biogenic amine produced enzymatically by decarboxylation of lysine. It is a precursor of a few alkaloids. C. is one of the compounds responsible for the odor of decaying meat and fecal matter, and it is poisonous. It is a preferred substrate of the amine oxidase (EC 1.4.3.6).

Cadinane: see Sesquiterpenes, Fig.

β-Cadinene: an optically active sesquiterpene found in the essential oils of junipers and cedars. Together with its isomers and their hydroxyl derivatives (cadinols), C. is representative of the cadinolenes, which are the best known and most widespread sesquiterpenes. They are called *cadaline precursors*, because they can be dehydrated to this aromatic compound. For formula and biosynthesis, see Sesquiterpenes.

Cadinols: see β-Cadinene.

Caffeine, *1,3,7-trimethylxanthine:* a purine derivative (Fig., see Methylated xanthines) found in coffee beans and leaves, tea leaves, cacao beans and cola nuts. It is usually produced from tea leaves (1.5 to 3.5% caffeine content) and as a byproduct from the production of decaffeinated coffee. Due to its stimulatory effects on the central nervous system, C. and C.-containing beverages are used to stimulate the heart and circulation. Its effects are chiefly due to the inhibition of the phosphodiesterase which degrades cyclic AMP to AMP in adrenalin-producing cells. This prolongs adrenalin action.

Calciferol: same as vitamin D. See Vitamins.

Calcitonin, *thyreocalcitonin:* a polypeptide hormone containing 32 amino acids. M_r (human) 3420. It is formed in the parafollicular cells of the thyroid in mammals, and in the ultimobranchial gland of nonmammalian species. (In both cases, the gland is derived from the 5th gill pocket of the embryonic gut). C. causes a rapid but short-lived drop in the level of calcium and phosphate in the blood by promoting the incorporation of these ions in the bones. C. is released in response to a rising Ca^{2+} level and is antagonized by Parathormone (see). It is detected by radioimmunological methods. The total synthesis of C. from various species has been accomplished.

Calcium: an alkaline earth element which, as its divalent cation, is ubiquitous in nature. Ca^{2+} has a number of vital functions in organisms. As $CaPO_4$ and $CaCO_3$, it provides rigidity and hardness to shells and bones. Because of its chelating ability, Ca^{2+} provides structural stability to proteins and lipids in cell membranes, cytoplasm and organelles, and in chromosomes. It is required as a cofactor for a number of extracellular enzymes, including prokaryotic and eukaryotic digestive enzymes, factors II, VII, IX and X in Blood coagulation (see), and for complement activation by antigen-antibody complexes. Ca^{2+} binds to tubulin with high affinity, and is required for the entry of a cell into the S phase of the Cell cycle (see). Contractility, secretion, chemotaxis and aggregation of cells are regulated by Ca^{2+} and arachidonic acid metabolites. Ca^{2+} has an essential part in the propagation of nerve impulses and in muscle contraction.

Ca^{2+} serves as second messenger (see Hormones) in animal cells in a somewhat complex fashion. When the calcium-system receptor is activated by the arrival of a molecular stimulus, 4,5-diphospho-phosphatidyl inositol (4,5-PIn) is hydrolysed to inositol 4,5-bisphosphate and diacylglycerol. Diacylglycerol is also a second messenger, and promotes phosphorylation of several proteins, by activating Protein kinase C (see). At the same time, Ca^{2+} is either allowed to enter the cell from the outside or it is mobilized from reserves within the cell. The intracellular Ca^{2+} concentration rises, briefly, to about 1.0 μmolar. At this concentration, Ca^{2+} activates some proteins directly, and others indirectly, when it has bound to Calmodulin (see). Some of the calmodulin-activated proteins are kinases which phosphorylate a different group of proteins from those phosphorylated by protein kinase C. Protein kinase C is itself activated by high concentrations of Ca^{2+}. If it has first been sensitized by diacylglycerol, however, it may be activated by Ca^{2+} concentrations which are only slightly higher than those in the unstimulated cell. Thus once a cell has been activated by a brief surge of calcium ions, the activation may persist at much lower calcium concentrations because protein kinase C has been activated.

In many cases, a given stimulus will activate both cAMP and Ca^{2+}/diacylglycerol as second messengers. H. Rasmussen has proposed the term "synarchic" to describe the interactions which then occur. In some cases, artificial stimulation of just one of the two systems (e.g. calcium influx through an ionophore or injection of cAMP) leads to phosphorylation of the same proteins. Often the influx of calcium leads to a brief stimulation, while the injection of cAMP produces a slow but long-term stimulation; normal stimulation of both systems simultaneously produces a response which is both rapid and long-term. The presence of two mutually regulatory second messenger systems allows for great plasticity in the response of the cell to the primary stimulus.

There is considerable evidence that Ca^{2+} serves as a second messenger in plants, which contain calmodulin. (cAMP is not a second messenger in plants.) Both gravitropic and phototropic responses in plants appear to depend on Ca^{2+}.

Excess calcium is toxic to cells, which therefore have efficient mechanisms for removing it. One means is a calcium pump, which efficiently removes it to the outside. This pump is activated by high intracellular calcium concentrations. Under normal conditions, Ca^{2+} may be exchanged for Na^+ leaking into the cell. Another means of handling large amounts of Ca^{2+} is sequestration within the mitochondria. The normal intracellular concentration of Ca^{2+} is 0.1 μmolar; the concentration of free Ca^{2+} in the blood is about 1.5 μmolar. [A.K. Campbell, *Intracellular Calcium, Its Universal Role as Regulator* (Wiley & Sons Ltd, 1983); D. Marmé, ed. *Calcium and Cell Physiology* (Springer, 1985); W.Y. Cheung, ed. *Calcium and Cell Function, Vol. IV* (Academic Press, 1983)]

Calcium-dependent regulator protein: see Calmodulin.

C-alkaloids: see Curare alkaloids.

***Table.** Reactions regenerating ribulose 1,5-bisphosphate in the Calvin cycle*

Reaction	Enzyme	Reaction	Enzyme
Glyceraldehyde-P $\rightleftharpoons$ Dihydroxyacetone-P	Triosephosphate isomerase (EC 5.3.1.1)	Sedoheptulose-1,7-P_2 $\rightleftharpoons$ Sedoheptulose-7-P + P_i	Sedoheptulose-bisphosphatase (EC 3.1.3.37)
Glyceraldehyde-P + Dihydroxyacetone-P $\rightleftharpoons$ Fructose-1,6-P_2	Fructose-bisphosphate aldolase (EC 4.1.2.13)	Sedoheptulose-7-P + Glyceraldehyde-P $\rightleftharpoons$ Ribose-5-P + Xylulose-5-P	Transketolase (EC 2.2.1.1)
Fructose-1,6-P_2 $\rightleftharpoons$ Fructose-6-P + P_i	Fructose-bisphosphatase (EC 3.1.3.11)	Ribose-5-P $\rightleftharpoons$ Ribulose-5-P	Ribosephosphate isomerase (EC 5.3.1.6)
Fructose-6-P + Glyceraldehyde-P $\rightleftharpoons$ Xylulose-5-P + Erythrose-4-P	Transketolase (EC 2.2.1.1)	Xylulose-5-P $\rightleftharpoons$ Ribulose-5-P	Ribulosephosphate 3-epimerase (EC 5.1.3.1)
Erythrose-4-P + Dihydroxyacetone-P $\rightleftharpoons$ Sedoheptulose-1,7-P_2	Aldolase (EC 4.1.2.13)	Ribulose-5-P + ATP $\rightleftharpoons$ Ribulose-1,5-P_2 + ADP	Phosphoribulokinase (EC 2.7.1.19)

Sum: 5 Triose-P $\longrightarrow$ 3 Pentose-P

Reaction 1: reaction of ribulosebisphosphate carboxylase (EC 4.1.1.3) (carboxydismutase); the intermediate shown in brackets is enzyme-bound

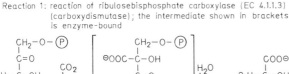

Reaction 2:

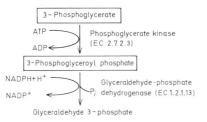

Calvin cycle reactions

Callistephin: see Pelargonidin.

Calmodulin: a calcium-binding protein which mediates the functions of Ca^{2+} in eukaryotes. It consists of 148 amino acids in a sequence which has been highly conserved throughout eukaryote evolution, and which is about 70% homologous with troponin C, the Ca^{2+}-binding unit in the muscle contractile apparatus. C. has 4 binding sites for Ca^{2+}; on binding the ion, the protein undergoes a change in conformation which exposes a hydrophobic site. This site interacts with the protein which is regulated by C.

Although C. is apparently ubiquitous, it mediates different cellular responses to Ca^{2+} in different tissues. This specificity of C. effects is due to the presence of different C.-activated proteins in different tissues. Some of the proteins activated by interaction with the C.-Ca^{2+} complex are cyclic nucleotide phosphodiesterase, adenylate cyclase and guanylate cyclase. These enzymes are active in the metabolism of cAMP and cGMP, which, like Ca^{2+}, serve as second messengers for hormones (see Calcium, Adenosine phosphates). Their regulation by C. allows for interaction of the two messenger systems. Some other proteins activated by C.-Ca^{2+} are muscle phosphorylase kinase and myosin light chain kinase, glycogen synthase kinase, and a variety of other protein kinases which provide for regulation of different metabolic pathways in different tissues. C. also binds to non-enzyme proteins, such as histones, basic protein, spectrin, tubulin and dynein ATPase. In mitotic cells, most of the C. is localized on the mitotic spindle (which consists largely of tubulin). (Since C. can substitute for other Ca^{2+}-binding proteins in vitro, it is possible that some of the proteins reported to be activated by C. in vitro are activated by other Ca^{2+}-binding proteins in vivo.) [L. J. Van Eldik & D. M. Watterson in Marmé, ed., *Calcium and Cell Physiology* (Springer, 1985) pp. 105-126; R. C. Brady et al. *ibid.* pp. 140-147; D. Marmé & P. Dieter, in W. Y. Cheung, ed. *Calcium and Cell Function*, Vol. IV, (Academic Press, 1983) pp. 246-311]

Calvin cycle, *photosynthesis cycle, reductive pentose phosphate cycle:* a series of at least 15 enzymatic reactions which, taken together, generate 1 molecule of hexose phosphate from 6 molecules of CO_2. In the process, which does not require light (dark reactions), 12 molecules NADPH and 18 molecules of ATP are consumed per molecule of hexose phosphate. The C. can be divided into 4 phases (Figure): 1) In the *carboxylation phase,* 1 molecule of ribulose 1,5-bisphosphate and 1 molecule of CO_2 produce 2 molecules of 3-phosphoglyceric acid in a reaction catalysed by ribulose-bisphosphate carboxylase (see Photosynthetic carboxylation).

2) In the *reduction phase,* the carboxyl group of 3-phosphoglyceric acid is reduced to the aldehyde group of glyceraldehyde 3-phosphate. In this reaction, the carboxyl group must be activated by ATP before it can be reduced by NADPH. In principle, the reaction is the reverse of the oxidation of glyceraldehyde 3-phosphate by triose-phosphate dehydrogenase in the course of glycolysis. The reduction consumes the products of the light reaction, ATP and NADPH, and is thus the point at which the light and dark reactions are coupled. 3) In the *regeneration phase,* the acceptor of CO_2, ribulose 1,5-bisphosphate, is regenerated by a series of steps (Table). Two molecules of triose phosphate are condensed to fructose 1,6-bisphosphate, which after removal of one phosphate, enters the 4) *synthetic phase* as fructose 6-phosphate. In this phase, sucrose or starch is generated. The C. can be summarized as follows: $6 CO_2 + 12$ NADPH $+ 18$ ATP $+ 6 H_2O \rightarrow$ hexose phosphate $+ 18$ ADP $+ 17$ $P_i + 12$ NADP$^+$.

Calvin plants: see C_3 Plants.

Calycosin: see Pterocarpans.

CAM: abb. for Cell adhesion molecule (see) or Crassulacean acid metabolism (see).

cAMP: abb. for cyclic adenosine 3′,5′-monophosphate. See Adenosine phosphates.

Campesterol *(24R)-ergost-5-en-3β-ol:* a plant sterol (see Sterols). M_r 400.68, m. p. 158 °C, $[\alpha]_D^{23}$ -33 ° (22.5 mg in 5 ml chloroform). C. is found in the oils of rapeseed *(Brassica campestris),* soybean and wheat germ and in some molluscs.

Campesterol

Camphor: a bicyclic monoterpene ketone found widely in plants. Both optical isomers occur naturally: $(+) - $C. (Japan camphor), m. p. 180 °C, b. p.

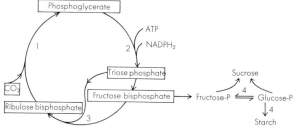

Calvin cycle reactions

Phosphoglycerate

ATP

NADPH$_2$

Triose phosphate

CO_2

Fructose bisphosphate → Fructose-P ⇌ Glucose-P

Sucrose

Ribulose bisphosphate

Starch

204 °C, $[\alpha]_D^{20} + 43.8$ ° ($c = 7.5$ in abs. ethanol), and (−)−C. (Matricaria camphor), m.p. 178.6 °C, b.p. 204 °C, $[\alpha]_D^{20} - 44.2$ ° (ethanol). C. is obtained commercially from camphor trees *(Cinnamomum camphora)* native to the coastal areas of Eastern Asia. The partial synthesis from pinene is also important, although the product is a racemic mixture used mostly in plastics. Natural C. is used pharmaceutically in salves. For formula and biosynthesis, see Monoterpenes.

cAMP receptor protein: see Adenosine phosphates, Table 2.

Camptothecin: the main alkaloid from the wood and bark of the Chinese tree *Camptotheca acuminata*. M_r 348, m.p. 264–267 °C (d.), $[\alpha]_D^{25}$ +31.3 ° (in chloroform/methanol, 8:2). Its total synthesis from an indole compound has been reported. C. is one of the most active natural substances against leukemia and tumors.

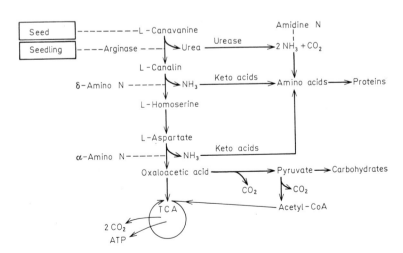

Camptothecin

L-Canalin: 2-amino-4-aminooxy-butyric acid, M_r 131.4, m.p. 214 °C (d.). It is a hydrolysis product of L-Canavanine (see), and is found in a few canavanine-containing legumes.

L-Canavanine, *2-amino-4-guanidinohydroxybutyric acid:* a structural analog of Arginine (see) found only in certain legumes. The presence or absence of C. is a useful trait in the Chemical taxonomy (see) of the legumes. C. is a non-proteogenic amino acid, and in fact is poisonous to most animals and to C.-free plants grafted onto C.-containing plants. The toxicity of C. is due in part to the fact that it is incorporated into protein in place of arginine. However, C. is much less basic than A., and under physiologi-

cal conditions it has less positive charge. Thus it can be expected to disrupt the secondary and tertiary structures of proteins, and in those enzymes in which the arginine side chain is part of the active site, C. probably produces a less active or an inactive enzyme.

In those plants which produce it, C. may be a major nitrogen storage compound in the seeds - it accounts for 55% of the nitrogen in seeds of *Dioclea megacarpa*, a tropical legume. The production of poisonous C. is apparently an adaptive advantage for *D. megacarpa*, as it has only one insect predator, the beetle *Caryedes brasiliensis*, which is totally dependent on it. The legumes and *C. brasiliensis* avoid being poisoned by C. by having arginyl-tRNA synthetases which distinguish between arginine and C., and thus do not incorporate C. into protein. The degradation of C. is similar in germinating seeds and the beetle, which both utilize the nitrogen in it (Fig.) [G. A. Rosenthal, *Scientific American* **249** (1983) 164–171]

Cancer: a *cancer* is a malignant growth. A *tumor*, strictly speaking, is any lump or swelling, while a *neoplasm* results from uncontrolled growth of cells. Some authors consider a neoplasm benign if it does not *metastasize*, or shed cells capable of colonizing other parts of the body, and reserve the term "malignant" for a metastasizing growth. There are three major types of malignant neoplasm: a) carcinomas, which arise from epithelial tissue, b) sarcomas, which arise from connective tissue, muscle, cartilage, fat or bone, and c) leukemias and lymphomas, which arise from hematopoietic structures. The most common type is the carcinoma.

C. research includes the following areas: etiology, including the epidemiology and prevention of C.; mechanisms of development; transformation of cells in culture; genetics and the phenotypes of cancer cells; carcinogens, metabolic activation of carcinogens and tumor promotion; and treatment of C. The study of animal C. viruses and their transformation of cells in tissue culture has led to the discovery of Oncogenes (see).

It has long been recognized, from both epidemiolo-

gy and laboratory experiments, that C. can be caused by viruses, certain chemical substances called *carcinogens*, and by ionizing radiation. The effective viruses belong to the group of *retroviruses* in which the genetic material is RNA; this is transcribed into DNA which may be inserted into the genome of the host cell. Such insertions may disrupt the control of gene expression, in some cases leading to uncontrolled growth. In other cases, the virus may insert a cancer-causing gene (Oncogene, see) acquired from a previous host. Similarly, ionizing radiation is thought to induce C. by causing mutations which release growth-promoting genes from the repressed state. The action of carcinogens is often subtle. Some of these compounds are Mutagens (see), which act at random on the genome, occasionally altering genes which are responsible for maintenance of growth arrest. Other compounds, called *tumor promoters* or *cocarcinogens*, evidently do not interact directly with the chromosomes, but make them more susceptible to the effects of carcinogens. Such compounds, e.g. Phorbol esters (see) may interact with membrane systems whose physiological function is to trigger growth and division in the presence of appropriate signals, such as hormones.
The transformation of a cell or clone into a C. is rarely, if ever, a single-step event. Cell growth and division is evidently under the control of several to many genes, no one of which is able, by itself, to induce uncontrolled cell division. Correspondingly, there may be clones of cells in a normal body which are "precancerous"; if they are exposed to further injury or mutation these may become malignant. There is some evidence that precancerous (benign) tumors are polyclonal, and that the most aggressively dividing of these clones multiply at the expense of the others, as a result of natural selection.
The literature on C. is vast. A useful starting point is E.C. Friedberg, ed., *Cancer Biology: Readings from Scientific American* (W.H. Freeman, New York, 1986).

Candicine, *N,N,N-trimethyltyramine:* a biogenic amine found especially in grasses and cacti.

Cane sugar: see Sucrose.

Cantharidin: toxic agent produced by the beetle *Cantharis vesicatoria* (Spanish fly, blistering beetle), which is native to Southern and Central Europe M_r 196.21, m.p. 218 °C. It is biosynthesized from mevalonic acid. Its use as an aphrodisiac has caused fatalities.

700000. X-ray and IR data have shown that its double bonds are *cis* orientied, while in the caoutchouc-like polyterpenes, gutta and balata, they are *trans* (see polyterpenes, Fig.). As an unsaturated hydrocarbon, C. reacts easily with oxidizing agents. Complete ozonolysis yields ups to 90% levulinic acid derivatives. C. is biosynthesized from mevalonic acid and this process can be carried out with isolated enzymes.
Occurrence. There are hundreds of species of plants which contain C. in their latex, but it can only be obtained on a large scale from a few representatives of the spurge *(Euphorbiaceae)*, dogbane *(Apocyznaceae)* and milkweed *(Asclepiadaceae)* families. The following types of C. are utilized: C. from the rubber tree *Hevea brasiliensis,* which is native to the Amazon region but is also grown in India and Indonesia, and which is most important with regard to the amounts produced. Guayule C. is similar to that from *Hevea* and differs from it only in its lower molecular weight. It is obtained from *Parthenium argentatum* in Mexico and California. Kok-saghys is a C. from the roots of the dandelion species *Taraxacum kok-saghys,* which has been cultivated since 1935 in the USSR.
C. is harvested by making cuts in the bark of the trees, without injuring the cambium. Latex flows from the cut latex channels, in amounts of 40 to 80 ml/per tree and harvest. On the average, latex contains 25 to 35% dry weight of C., 60 to 75% water, 2% protein, 2% resin, 1.5% sugar and 1% ash. The C. is in the form of fine droplets which are prevented from coagulating by protein. It is precipitated with dilute acid or sodium fluorosilicate, pressed out in sheets to remove the water, and sometimes smoked.
2) Synthetic C. is similar in structure and properties to natural C. It is produced by technical polymerization of unsaturated monomers. The most important starting material is butadiene.

Capnine: 2-amino-3-hydroxy-15-methylhexadecane-1-sulfonic acid, or 1-deoxy-15-methylhexadecasphinganine-1-sulfonic acid, the main sulfonolipid from the bacterial genera, *Cytophaga, Capnocytophaga, Sporocytophaga* and *Flexibacter* (all noted for their gliding motility). Other sulfonolipids in these genera are probably *N*-acylated derivatives of C. Sulfonolipids constitute up to 20% of the cellular lipids of these microorganisms, and are major components of the cell envelope. C. is a structural analog

Cantharidin

Capnine

Caoutchouc: elastic, high-polymer hydrocarbons (elastomers) which become rubber on vulcanization. 1) Natural caoutchouc, $(C_5H_8)_n$, is the most important representative of the polyterpenes. It is a mixture of polyisoprenes with varying molecular weights, which usually range from 300000 to

of sphinganine; similarly, *N*-acylcapnines are analogs of ceramides. The only other known source of a similar sulfonolipid is the diatom *Nitzschia alba* (a eukaryote), which contains *N*-acylated 2-amino-3-hydroxy-4-*trans*-octadecene-1-sulfonic acid, i.e. the nonacylated moiety is an analog of sphingosine.

[W. Godchaux III & E. R. Leadbetter, (1984) *J. Biol. Chem.* **259**, 2982-2990].

Capon-comb unit: see Androgens.

Capon test: see Androgens.

n-Caproic acid, *decanoic acid:* a fatty acid, $CH_3-(CH_2)_8-COOH$. M_r 172.3, m.p. 31.3 °C, b.p. 268 °C. It occurs in milk fat (2%), coconut oil (< 1%) and various other seed and essential oils.

n-Capronic acid, *hexanoic acid:* $CH_3-(CH_2)_4-COOH$, a fatty acid. M_r 116.16, m.p. -1.5 °C, b.p. 207 °C. Found in milk fats (2%) and in small amounts in coconut oil and other palm oils. It also occurs in essential oils from plants and plant fats.

n-Caprylic acid, *octanoic acid:* $CH_3-(CH_2)_6-COOH$, a fatty acid. M_r 144.2, m.p. 16.5 °C, b.p. 237 °C. It is found in various glycerides, e.g. 1 to 2% in milk, fat, and as 6 to 8% of the coconut oil fats. It is also found in other plant fats.

Capsaicin: a pungent principle in the fruits of some peppers *(Capsicum)*. M_r 305.42, m.p. 64 to 65 °C. The aromatic part is biosynthesized from phenylalanine. It is occasionally used as a counter-irritant.

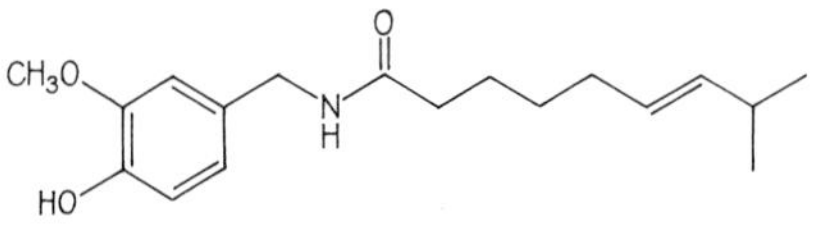

Capsaicin

Capsanthin: a carotenoid pigment isolated from paprika *(Capsicum annuum)*. M_r 584.85, m.p. 176 °C, $[\alpha]_{Cd} = +36°$ (chloroform). It is characterized by a terminal five-membered ring. The secondary hydroxyl groups have the R configuration on C 3 and the S-configuration on C 3'. C. is the main red pigment in paprika, where it is accompanied by Cap-

sorubin (see), cryptocapsin and capsanthin-5,6-epoxide. Ripe fruits contain 9.6 mg C. per 100 g fresh weight. C. was isolated and its structure was elucidated by Zechmeister and von Cholnoky.

Capsid: see Viral coat protein.

Capsidiol: see Phytoalexins.

Capsorubin: a carotenoid pigment found in paprika *(Capsicum annuum)*. M_r 600.85, m.p. 218 °C. C. contains two identical cyclopentanol rings. The hydroxyl groups have the S-configuration on C-atom 3 and the R-configuration on C-atom 3'. C. is usually present in paprika fruits in the esterified form. Ripe fruits contain about 1.5 mg per 100 g fresh weight. C. was isolated in 1943 by Zechmeister.

Ca^{2+} pump: see Membrane transport.

Caran: see Monoterpenes, Fig.

Carbamate: see Carbamoyl phosphate.

Carbamic acid: see Carbamoyl phosphate.

Carbamide: see Urea.

Carbamino group: see Buffer, section on Buffers of body fluids.

6-Carbamoylglycyl purine nucleoside: see 6-Carbamoylthreonyl purine nucleoside.

Carbamoyl phosphate: $H_2N-COO \sim PO_3H_2$, an energy-rich phosphorylated carbamate which is an important metabolic intermediate. Carbamic acid, NH_2COOH, is unstable in free form. Carbamate removed hydrolytically from carbamyl compounds, for example ureidopropionic acid (see Pyrimidine degradation), decomposes immediately into CO_2 and NH_3. C. p. is also a relatively labile compound. It is the starting point for the biosynthesis of arginine and urea (see Urea cycle), and of pyrimidines via the orotic acid pathway (see Pyrimidine biosynthesis).

Metabolism: C. p. is synthesized de novo or generated by phosphorolysis of ureido compounds. De novo synthesis involves three different enzymes: 1) carbamoyl-phosphate synthetase (ammonia) (EC

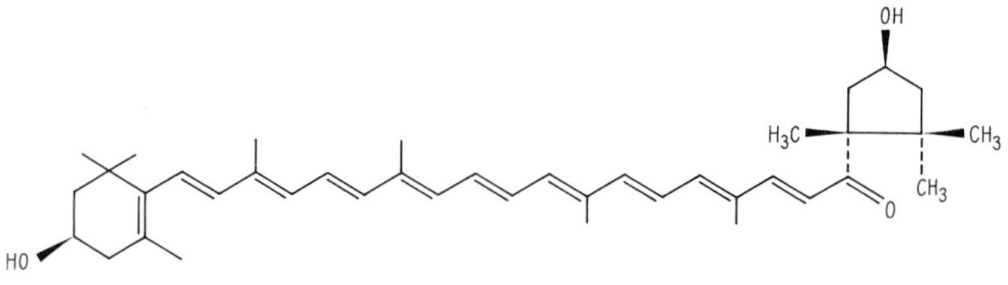

Capsanthin

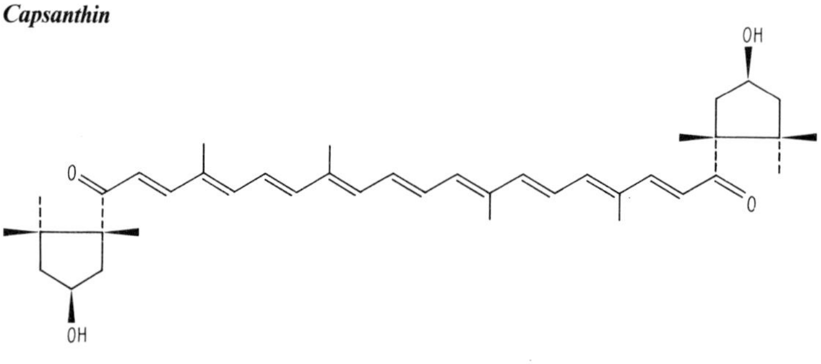

Capsorubin

6.3.4.16), which is found in the vertebrate liver, catalyses the formation of C.p. from ammonium hydrogen carbonate at the expense of 2 ATP. It requires N-acetyl-L-glutamate (AGA) as cofactor:

$$NH_4HCO_3 + 2 \ ATP \xrightarrow[\text{AGA}]{Mg^{2+}} H_2N\text{-}COO\text{-}PO_3H_2 + 2$$

ADP + P_i. The reaction is irreversible. AGA is required for the active enzyme conformation, not for the activation of the carbon dioxide. The enzyme is localized in the mitochondria and is used for the synthesis of arginine and urea.
2) Carbamoyl-phosphate synthetase (glutamine-hydrolysing) (EC 6.3.5.5) requires the amide group of L-glutamine as an N donor: $HCO_3- + L\text{-gluta-}$ mine $+ ATP + H_2O \xrightarrow{Mg^{2+}} C.p. + L\text{-glutamate} + ADP$.

The overall reaction is irreversible because it involves an hydrolytic step. This enzyme is located in the cytoplasm and is used for pyrimidine synthesis. Free ammonia ions can replace the glutamine, but only at higher than physiological concentrations. The enzyme was first discovered in the cultivated mushroom, but it occurs widely.
3) Carbamate kinase (carbamyl phosphokinase) (EC 2.7.2.2) catalyses the phosphorylation of carbamate by ATP: $NH_2\text{-}COO^- + ATP \rightleftharpoons NH_2\text{-}COO\text{-}PO_3H_2 + ADP$. It is found in various microorganisms (*Streptococcus, Neurospora,* etc.). Thermodynamically, the formation of ATP is favored, so the enzyme is thought to be used for the generation of ATP rather than the synthesis of C.p. There are various microorganisms which catalyse the phosphorolysis of allantoin and citrulline, forming C.p. which can then be utilized for ATP formation. Phosphorolysis of citrulline produces ornithine and C.p. In the course of allantoin fermentation by *Streptococcus allantoicus* and *Arthrobacter allantoicus,* carbamyloxamic acid (oxaluric acid) is formed and then phosphorolysed:
$NH_2\text{-}CO\text{-}NH\text{-}CO\text{-}COOH + P_i \rightarrow$
(Oxaluric acid)
$NH_2CO\text{-}COOH + NH_2\text{-}COO\text{-}PO_3H_2$
(Oxamic acid)
The formation of ATP from C.p. obtained from citrulline or allantoin is an example of Substrate phosphorylation (see).
All transcarbamylation reactions require C.p. as donor of the carbamyl group, i.e. transcarbamylation directly from a ureido compound is not possible. C.p. is probably also the carbamyl donor in the biosynthesis of *O*- and *N*-carbamyl derivatives, such as Albizziin (see). C.p. is a metabolically active form of ammonia used as the starting material for the synthesis of other nitrogen compounds.

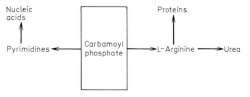

Carbamoyl-phosphate synthetase: see Carbamoyl phosphate.

6-Carbamoylthreonyl purine nucleoside, *N-(nebularin-6-ylcarbamoyl)-threonine:* one of the rare nucleic acid bases. It has so far been detected in six specific transfer RNAs. The analogous compound 6-carbomoylglycyl purine nucleoside has been isolated from yeast tRNA. See Figure.

$R = -NH-CH_2-COOH$	6-Carbamoylglycyl purine nucleoside
$R = -NH-\underset{\underset{COOH}{\mid}}{\overset{\overset{CH_3}{\mid}}{\underset{HCOH}{\mid}}}-$ 	6-Carbamoylthreonyl purine nucleoside

Carbamylcholine, *carbachol:* see Acetylcholine.
Carbamyl phosphokinase: see carbamoyl phosphate.
Carbohydrases: see Glycosidases.
Carbohydrate metabolism: the constant formation, transformation and degradation of the carbohydrates in the organism. The most important reactions in C.m. are 1. Interconversions of the polymeric storage forms (glycogen and starch) and the monomeric transport and substrate form (glucose), 2. reactions of carbohydrate degradation and interconversion, and 3. reactions for the synthesis of glucose from noncarbohydrate substances (glucogenic amino acids, fats). Glucose 6-phosphate has a central position in the entire C.m. Aside from the minor pathways (see Glucuronate pathway; Entner-Doudoroff pathway; Phosphoketolase pathway) there are four main pathways for glucose 6-phosphate: 1. Glycolysis (see), 2. Glycogen synthesis (see Glycogen metabolism), 3. Pentose phosphate cycle (see), and 4. enzymatic hydrolysis to free glucose. The effectiveness of these pathways depends, in the animal organism, on the function of the tissue in question. A change in the activity of the tissue, as in illness, has a large effect on its C.m. (Figure 1).
Under certain conditions, carbohydrates may be resynthesized from the degradation products of C.m. The starting materials for this Gluconeogenesis (see) are lactate and glucogenic amino acids.
There are three different phases of C.m. 1. mobilization, in which poly-, oligo- and disaccharides are cleaved and phosphorylated to hexose phosphates, particularly to glucose 6-phosphate. In digestion, the cleavage is achieved by hydrolysis.
2. Interconversions: the mutual transformations of monosaccharides involve the following types of reactions: a) epimerization, the reversal of the steric arrangement on a C-atom by epimerases (e.g. in ga-

Carbohydrate metabolism

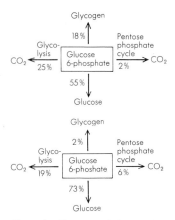

Figure 1. Turnover of glucose 6-phosphate under normal (above) and disease (diabetes mellitus) conditions (below), as percent of total

lactose metabolism); b) isomerization, the reversible transformation by isomerases of aldoses into ketoses (e.g. glyceraldehyde 3-phosphate into dihydroxyacetone phosphate); c) transfer of C_3 (see Transaldolation) and C_2 (see Transketolation) fragments in the form of a dihydroxyacetone phosphate residue or "active glycolaldehyde"; d) oxidation of an aldose to an acid and its subsequent decarboxylation. In this second phase of C.m., the intermediary products of the first phase are incompletely degraded. The main products are triose phosphates. In these processes, about a third of the total free energy potential is released and partly used for the synthesis of the energy-rich adenosine triphosphate.

3. Amphibolic reaction chains. The degradation products of C.m. flow into the general metabolism as pyruvate and acetyl-CoA (Figure 2).

Biosyntheses in C.m. Plants are the main producers of carbohydrates in nature. In photosynthesis, a series of enzymatic reactions produces phosphorylated monosaccharide derivatives, which can be hydrolysed to free sugars or converted to Nucleoside diphosphate sugars (see).

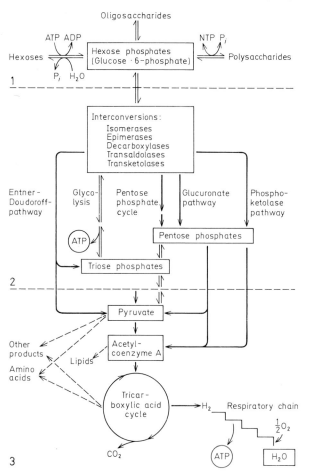

Figure 2. The three phases of carbohydrate metabolism. NTP is nucleotide triphosphate

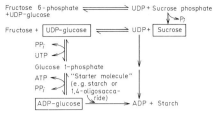

Figure 3. Reactions of starch and sucrose synthesis

In the biosynthesis of polysaccharides like starch, glycogen and cellulose, the monosaccharide units are activated by the formation of nucleotide derivatives, and they are transferred in this form by the appropriate enzymes to the non-activated, growing end of the polysaccharide chain. The nucleotide is split off in the coupling reaction. The biosynthesis depends on the presence of a highly polymerized starter molecule. Uridine diphosphate glucose (UDPG) and adenosine diphosphate glucose (ADPG) play the most important role in the biosynthesis of the oligosaccharide sucrose and the polysaccharide starch (Figure 3). Guanosine diphosphate glucose can also serve as a precursor for the synthesis of the 1,4-glucosyl chain of cellulose.

The oligo- and polysaccharides are degraded in organisms by specific enzymatic hydrolysis (hydrolases) or phosphorolysis (phosphorylases).

Regulation of C.m. is characterized by tight interactions between the individual metabolic pathways mediated by the metabolic products. Glycolysis is controlled by allosteric regulation of the enzymes phosphofructokinase and pyruvate kinase. The controlling factor is the ATP/ADP ratio. Increased production of ATP in the respiratory chain leads to inhibition of phosphofructokinase, while an increased consumption of ATP stimulates the glycolytic turnover of glucose 6-phosphate via the increased level of ADP, which stimulates the phosphofructokinase. This in turn causes increased ATP formation by substrate chain phosphorylation. Another regulatory parameter of this system is the amount of available oxygen (Pasteur effect). Glucose 6-phosphate activates glycogen synthetase. Therefore, a high concentration of glucose 6-phosphate leads to increased production of the storage carbohydrate glycogen. Intermediary products of the pentose phosphate cycle inhibit the first enzyme of glycolysis, phosphoglucose isomerase, which reduces the amount of glycolytic degradation. The NADPH produced in larger amounts by an active pentose phosphate cycle can be used in fatty acid synthesis. Excessive fat synthesis is prevented by the inhibitory action of long-chain acyl-coenzyme A compounds on the key enzyme of the pentose phosphate cycle, glucose-6-phosphate dehydrogenase.

Carbohydrates: a large class of natural substances, structurally the polyhydroxycarbonyl compounds and their derivatives. In general they correspond to the composition $(C)_n(H_2O)_n$. They were originally characterized as hydrated forms of carbon and were named C. in 1844 by K. Schmidt. The

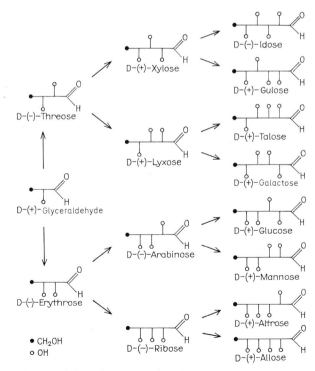

Figure 1. Schematic representation of the D-aldoses

name has been retained, although it is inaccurate from a chemical point of view and now compounds are included in the C. which have other elemental compositions, e.g. the aldonic acids, uronic acids and deoxysugars, or which contain additional elements, e.g. amino sugars, mucopolysaccharides. Mono- and oligosaccharides are also called sugars. Individual C. have trivial names or systematic names derived from them which end in -ose, e.g. glucose, fructose. The IUPAC Commission on the Nomenclature of Organic Chemistry and the IUPAC-IUB Commission on Biochemical Nomenclature published tentative rules for carbohydrate nomenclature in 1969.

C. are present in every plant or animal cell and make up the largest portion, in terms of mass, of organic compounds present on earth. They are formed in plants in the course of assimilation processes. Together with fats and proteins, they are the organic nutrients for humans and animals. The C. are subdivided on the basis of their molecular size.

1. *Monosaccharides* (simple C.) cannot be further hydrolysed into simpler types of C. They can be regarded as the primary oxidation products of aliphatic polyalcohols, usually with unbranched carbon chains. If the oxidation occurs at the terminal primary alcohol group, the resulting polyhydroxy-aldehyde is called an aldose (Figure 1). Oxidation of a secondary hydroxyl, usually the C-atom 2, produces a polyhydroxyketone, or ketose.

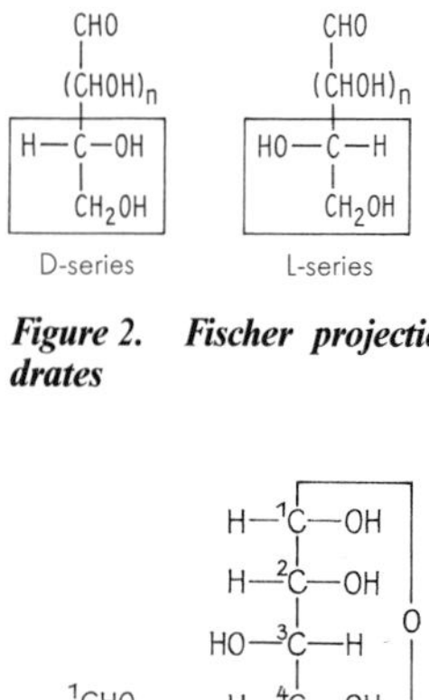

Figure 2. *Fischer projections of D- and L-carbohydrates*

a) *Structure.* Nearly all naturally occurring monosaccharides have unbranched carbohydrate chains; exceptions are hamamelose, apiose, streptose, etc.

The configuration is indicated by the prefix D or L (Fig. 2); these have nothing to do with the optical rotation, which is indicated by $(+)$ or $(-)$, according to Wohl and Freudenberg, e.g. $(+)$-D-glucose. According to Rosanow, Wohl and Freudenberg, a D-sugar is derived from the reference substance D-glyceraldehyde, which was chosen arbitrarily. In this system, the asymmetric carbon atom farthest from the carbonyl group has its hydroxyl group on the right in the Fischer projection; in an L-sugar, the corresponding hydroxyl group is on the left. In general, each carbon atom of the C. carries a hydroxyl function or a derived functional group. Replacement of the hydroxyl group by a hydrogen yields a deoxysugar; replacement by an amino group, an amino sugar. C. are usually characterized by a number of asymmetric centers, and the number of stereoisomeric forms of a monosaccharide s given by the formula 2^n, where n is the number of asymmetric carbon atoms.

In the Fischer projections of C. (Fig. 3a), the formulas are written vertically as chains, with the aldehyde group at the top and the hydroxymethyl group at the bottom. This representation is easy to comprehend, but it does not indicate the spatial structure of the monosaccharide. Furthermore, the open-chain form does not correspond to all the properties of a monosaccharide. For example, sodium hydrogen sulfite and ammonia do not react with the aldehyde group of an aldose.

According to Tollens, this is due to the fact that the monosaccharides do not exist in the open-chained form, or only a small fraction of the molecules is in this form. Instead, the carbonyl group forms a half-acetal bond with one of the hydroxyl groups, so that the molecule is an oxygen-containing ring. If the half-acetal bond is from C-atom 1 to C-atom 4, the resulting five-membered ring is a furanose; a six-

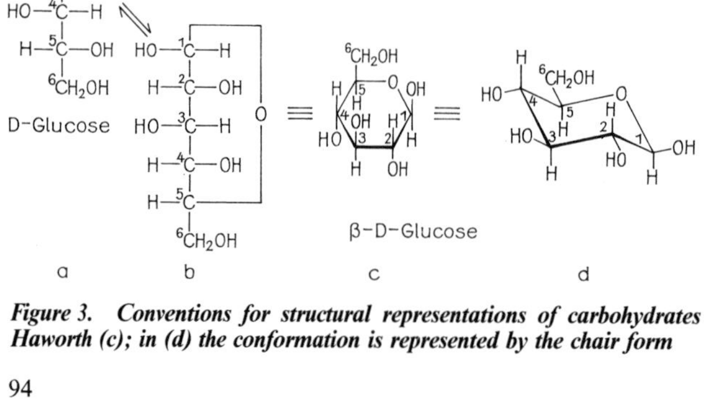

Figure 3. *Conventions for structural representations of carbohydrates according to Fischer (a), Tollens (b) and Haworth (c); in (d) the conformation is represented by the chair form*

membered ring (from C-atom 1 to C-atom 5) is called a pyranose. Most monosaccharides are pyranoses, but the furanose form is present in some oligosaccharides, e.g. sucrose, and in a few polysaccharides, e.g. arabans. The cyclic half-acetal form may be indicated according to Tollens (Figure 3b), or, better and more comprehensibly, according to Haworth (Figure 3c). The ring carbon atoms may be left out in the perspective representation, in which the heavy line between the ring atoms indicates which side is in the foreground. The substituents are drawn perpendicular to the plane of the ring.

The furanose ring is nearly planar, but the pyranose ring is puckered, because the C–O–C angle in the pyranose ring, 111 °, is very close to the tetrahedral angle of 109 ° 28′.

Thus the spatial relationships are similar to those in cyclohexane. The asymmetry caused by the heteroatom makes possible two chair and six boat forms. However, the pyranoses spend the most time in the energetically favorable chair form (Figure 4).

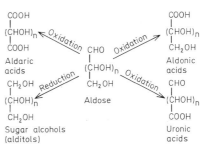

Figure 4. *Chair form of the pyranoses.* (a axial, e equatorial positions of OH-groups)

Examples are D-glucose, D-galactose and D-mannose. Of the 10 substituents on the 5 ring atoms, 5 are axial and 5 equatorial; in β-D-glucose, for example, all hydroxyl groups and the hydroxymethyl group are equatorial. The conformation formulas given in Figure 3d come closest to reality, because they best express the spatial arrangements of the substituents and make the chemical and biochemical reactions and the physical properties of the molecule more understandable.

The formation of a ring creates a new center of asymmetry on the atom originally carrying the carbonyl group (C-atom 1 in the aldoses, C-atom 2 in the ketoses), and thus two new series of isomers, which are called the α and β forms.

They differ with respect to solubility, melting point, optical rotation, etc. According to Hudson, the diastereoisomer in the D-series with the more highly positive optical rotation is the α-form, and the more levorotatory form is the β-form; the reverse holds for the L-series. In the Tollens conformational formulas, the hydroxyl group attached to C-atom 1 in the D-series is on the right side in the α-form and on the left in the β-form. In the Haworth projection, the corresponding hydroxyl group points down in the α-form and up in the β-form. In the L-series, the hydroxyl groups have the opposite arrangement. The same holds for the conformational formulas. One can see that in the α-form the hydroxyl groups on C-atoms 1 and 2 are *cis* to each other, and in the β-form, they are *trans*.

In solution, the α and β isomers are in equilibrium via the open-chain form (oxo-cyclo tautomerism). The phenomenon, known as mutarotation, is based on the attainment of equilibrium between the α and β forms; it is observed that the optical rotation of a freshly prepared aqueous solution changes until it reaches a constant end value.

If two diasteromers differ only in the configuration about C-atom 1, they are called anomers, e.g. α- and β-glucose. Epimers are diastereomeric monosaccharides which have the opposite configuration of the hydroxyl group at only one position, e.g. D-glucose and D-mannose at C-atom 2, or D-galactose and D-glucose at C-atom 4.

b) *Reactions.* The chemical properties of the monosaccharides are due to the presence of reactive keto or aldehyde groups and the alcoholic hydroxyl groups. Aldoses and ketoses give osazones and oximes, respectively, with an excess of phenylhydrazine or hydroxylamine. These are well suited to characterization of the compounds. Mild oxidation leads to the aldonic acids, and stronger oxidation to the aldaric acids. Suitable oxidation of glycosides in which the sensitive carbonyl function is protected yields uronic acids. Reduction of a glycoside with uptake of two atoms of hydrogen produces an alditol (Figure 5). The glycosidic hydroxyl group of the half-acetal is especially reactive and reacts with OH, NH or SH groups to form glycosides. The alcoholic hydroxyl groups can form esters or ethers. Sugar anhydrides are formed by intramolecular exclusion of water.

COOH
|
(CHOH)$_n$ ←Oxidation
|
COOH

Aldaric acids

CHO Oxidation→

|
(CHOH)$_n$
|
CH$_2$OH

COOH
|
(CHOH)$_n$
|
CH$_2$OH

Aldonic acids

CH$_2$OH Reduction
|
(CHOH)$_n$
|
CH$_2$OH

Sugar alcohols (alditols)

CHO
|
(CHOH)$_n$
|
CH$_2$OH

Aldose

Oxidation→

CHO
|
(CHOH)$_n$
|
COOH

Uronic acids

Figure 5. *Types of carbohydrate*

For the chemical degradation of C. the methods of Wohl, Ruff and Weerman are usually used, while chemical synthesis of sugars is usually achieved by the cyanohydrin method discovered by Kiliani. For biochemically important metabolic processes, see Carbohydrate metabolism.

c) *Detection and determination.* Monosaccharides are normally isolated and identified by suitable chromatographic techniques, including paper, thin-layer, column and gas chromatography, and by electrophoresis. The methods of detection are mostly based on color reactions with phenols, such as naphthol (violet) or resorcin (green), or on the reduction of metal salt solutions in the presence of acid or alkali. With heating, pentoses lose water to form furfural, while hexoses form 5-hydroxymethyl-furfural. The latter decomposes, taking up water in the process, into levulinic acid and formic acid. Furfural and levulinic acid condense easily with phenols to form dyes, which can be used for identification or quantitative determination. One important biochemical reaction is fermentation by yeast. Since pentoses are generally not fermentable, they can be differentiated from the hexoses.

d) *Occurrence.* Monosaccharides occur naturally in the free form, especially D-glucose and D-fructose, and bound as the basic units of many oligo- and polysaccharides. At present more than 100 naturally occurring monosaccharides are known. Some of them play a decisive role in many metabolic processes and are, in the form of their phosphoric acid esters, important intermediate products. The pentoses and hexoses are especially important.

2. *Oligosaccharides* are made up of 2 to 10 monosaccharides α- or β-glycosidically linked. They are classified as di-, tri-, tetrasaccharides, etc. depending on the number of subunits. They can be hydrolysed by acids or enzymes into their subunits, which they resemble in physical and chemical properties. Oligosaccharides are widespread in the plant and animal kingdoms, and occur in free and bound forms. They are synthesized via the nucleoside diphosphate sugars. The disaccharides, which are composed of two glycosidically linked monosaccharide units of the same or different types, are especially important. These are divided according to the nature of the glycosidic bond into the trehalose and maltose types. In the trehalose (see Trehalose) type, the hydroxyl groups on the two C-1 carbon atoms of two monosaccharides are linked with the exclusion of a water molecule. Due to this 1,1 linkage, both sugars are full acetals and do not enter the typical sugar reactions, such as reduction, hydrazone and oxime formation, or mutarotation. Sucrose and trehalose are typical representatives of this group. In the maltose type (see Maltose), however, the glycosidic hydroxyl group of one monosaccharide is linked to an alcoholic hydroxyl group of a second monosaccharide, usually the one on C-atom 4 or 6. C. with this type of structure have a free reducing group and therefore show reductive and mutarotatory properties; they are able to form hydrazones and oximes. Maltose, cellobiose, gentiobiose and melibiose are representatives of this structural type.

In principle, higher oligosaccharides or polysaccharides can also be constructed either as trehalose or as maltose types. Disaccharides are sometimes present in free form, but in general they are bound as glycosides or as subunits of oligo or polysaccharides. Other oligosaccharides are the trisaccharides raffinose, gentianose and melecitose, etc., the tetrasaccharide stachyose and the pentasaccharide verbascose.

3. *Polysaccharides.* In this quantitatively very large group of C., 10 or more monosaccharide units are bound, according to the same structural principles as in oligosaccharides, α- or β-glycosidically to branched or unbranched chains. The chains may be arranged linearly, spirally or spherically. The most common components are the hexoses D-glucose, D-fructose, D-galactose and D-mannose, the pentoses D-arabinose and D-xylose, and the amino sugar D-glucosamine.

Polysaccharides built of only one type of component are called homoglycans; those composed of different types of C. units, heteroglycans. The polyuronides are composed of uronic acids.

Polysaccharides generally contain hundreds or thousands of monosaccharide units and have very high molecular weights. The various types thus differ not only in the types of units of which they are composed, but also above all in their degrees of polymerization and in the nature of their bonds. Their chemical and physical properties are different from those of their mono or oligosaccharide components. Water solubility, reductive capacity and sweetness diminish with increasing molecular size. The skeletal C. cellulose and chitin, for example, are water-insoluble and resist enzymatic degradation. In contrast, reserve C. like starch, lichenin and glycogen are colloidally soluble in water and are more easily hydrolysed by enzymes. Acid hydrolysis of polysaccharides produces first oligosaccharides, then the monosaccharide subunits. Polysaccharides are not hydrolysed by yeast. They have optical activity in colloidal solution and generally have a microcrystalline structure. Polysaccharides are widespread in the plant and animal kingdoms as reserve substances or skeletal material. The structures of many polysaccharides have not been elucidated in detail.

Carboline: see Indole alkaloids (Table).

Carbomycin: see Macrolide antibiotics.

Carbon dioxide assimilation; *carbon fixation:* the incorporation of CO_2 into larger organic compounds. C. is sometimes used as a synonym for photosynthesis, but this is, strictly speaking, incorrect. Photosynthesis is a series of reactions which generate ATP and NAD(P)H required to drive the reactions of C. By far the greatest amount of C. occurs in green plants and blue-green bacteria. Most of this takes place via the Calvin cycle (see), and a lesser amount by the Hatch-Slack-Kortschack cycle (see). C. can also be driven by chemical energy in auxotrophic or chemoautotrophic organisms, especially bacteria. Some of these use a cycle similar to the Calvin cycle. The methanogenic bacteria *Methanobacterium thermoautotrophicum*, however, fixes CO_2 by a series

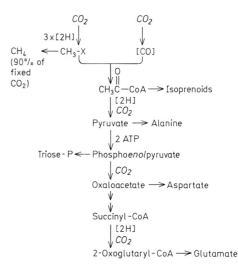

Carbon dioxide assimilation in Methanobacterium thermoautotrophicum. [adapted from M. Rühlmann et al. *Arch. Microbiol.* **141** (1985) 399–406]

of 5 major reactions which do not include the Calvin cycle (Fig.) There is also anaplerotic CO_2 fixation (see Carboxylation; Biotin enzymes).

Carbonic acid anhydrase, *carboanhydrase, carbonate dehydratase* (EC 4.2.1.1): a widely occurring, zinc-containing enzyme, usually monomeric. C. catalyses the reversible hydration of carbon dioxide ($CO_2 + H_2O \rightleftharpoons H^+ + HCO_3^-$) with one of the largest known turnover rates. C. also has a weak hydrating effect on aldehydes and an esterase effect. As an important regulator of the acid-base balance, C. plays an important role in respiration, CO_2 transport and other physiological processes where the rapid conversion of carbon dioxide to hydrogen carbonate is vitally necessary. High concentrations of C. are found not only in erythrocytes and the mucous membrane of the stomach, but in the kidneys and lenses of the eyes, and in the gills, digestive glands, and swim bladders of fish. Furthermore, C. activity has been found in all classes of nonvertebrates, higher and lower plants and bacteria. The best studied erythrocyte C. is that of human origin. In addition to the two isoenzymes C.B and C.C, it has a third (similar to the B form) called C.A.; C.B and C.C differ markedly in the sequence of their 256 (M_r 28000) and 259 (M_r 28500) amino acids, their chain conformation and catalytic activity. With the exception of the C. from parsley (M_r 180000, 6 subunits with M_r 29000 and one zinc atom each), the C. consist of single polypeptide chains (M_r 28000 to 30000) and contain one zinc atom essential for the catalytic activity. The zinc is located in a hydrophobic pocket where three of its four ligand sites are occupied by histidine residues. Other characteristics common to C.B and C.C are the lack of disulfide bridges and a relatively large amount of β-structures, which amount in C.C to 37%, compared to 20% α-helix in the conformation. Of the many inhibitors for C., the monovalent sulfide, cyanide and cyanate ions are the strongest (e.g. K_i for sulfide ions is 210^{-6}M). Of the sulfonamides, acetazolamide is the inhibitor most frequently used for therapeutic purposes (for treatement of glaucoma).

Carbon monoxide: CO, a colorless and practically odorless, combustible gas. C. is extremely poisonous, because its affinity for hemoglobin (abb. Hb) is 300 times that of oxygen. Otherwise, CO is completely inert. The reaction between Hb and CO is reversible: $HbO_2 + CO \rightleftharpoons HbCO + O_2$. The CO displaces O_2 from hemoglobin, so that the red blood cells cannot perform their normal function, oxygen transport to the tissue. O_2 can only displace CO

from Hb when it is present in large excess. Death by CO poisoning occurs after the following series of events: 1. the transport capacity of blood for oxygen is reduced by formation of HbCO; 2. intoxication of oxygen-sensitive tissues, especially in the brain, occurs (symptom, headache); 3. the respiratory center in the brain is incapacitated (symptom, unconsciousness); 4. the heart stops beating for lack of adequate oxygen supply. CO can only enter the body through the alveoli of the lungs. Concentrations $> 0.01\%$ are considered toxic. The maximal allowable concentration at a place of work is 55 mg CO/m^3 air.

Carbonyl **cyanide-*p*-trifluoromethoxyphenyl-hydrazone,** *FCCP:* an uncoupler of oxidative phosphorylation. For structure and mode of action, see Ionophore.

Carboxydismutase: see Ribulose-1,5-bisphosphate carboxylase.

4-Carboxyglutamic acid, *γ-carboxyglutamic acid, Gla:* an amino acid residue found in certain proteins, and formed by posttranslational carboxylation of glutamate residues (see 4-Glutamyl carboxylase). It is present in the blood clotting factors, prothrombin, factor VII, factor X and factor IX; in low M_r protein isolated from the bones of several vertebrates (bovine bone protein M_r 6800 contains 3 residues of Gla); in the calcium-binding protein from chick embryonic chorioallantoic membrane; in a new plasma protein (protein C) not concerned in the blood coagulation cascade; and in protein of the kidney cortex. The doubly charged side chain of Gla acts as a chelator of Ca^{2+}; many of the proteins known to contain Gla bind and/or transport Ca^{2+}.

Carboxylase: 1. see Carboxylation. 2. Old term for pyruvate decarboxylase.

Carboxylation: the transfer of carbon dioxide, frequently in activated form. The C. of pyruvate to dicarboxylic acids (Table) was discovered by Wood and Werkman as a balanced reaction ($C_3 + C_1 = C_4$), and serves to renew the pool of oxaloacetate (anaplerotic CO_2 fixation). Oxaloacetate is used in various syntheses (see Tricarboxylic acid cycle). The C. of pyruvate occurs in two ways: 1. direct addition of "activated CO_2" (Wood-Werkman reaction) and 2. reduction of C. by means of the "malic" enzymes (EC 1.1.1.38, 1.1.1.39 and 1.1.1.40) (Ochoa). This enzyme reduces pyruvate with NAD(P)H and fixes CO_2 in a single step. The product is malate, from which oxaloacetate can be made.

The most important form of C. is photosynthetic fixation of CO_2 (see Calvin cycle), but C. is also re-

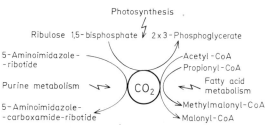

Metabolic carboxylations

Table. *Carboxylation reactions synthesizing oxaloacetate or malate*

Enzyme	Reaction
Phospho*enol*pyruvate carboxylase (EC 4.1.1.31)	Phospho*enol*pyruvate $+ CO_2$ $\rightarrow$ Oxaloacetae $+ P_i$
Pyruvate carboxylase (EC 6.4.1.1)	Pyruvate $+ CO_2 + ATP$ $+ H_2O \xrightarrow{Mn^{2+}}$ Oxaloacetate $+ ADP + P_i$
Malate dehydrogenase (decarboxylating) (NADP) (EC 1.1.1.39) (Malic enzyme)	Pyruvate $+ CO_2$ $+ NAD(P)H + H^+ \rightleftarrows$ Malate $+ NAD(P)^+$

quired for fatty acid and purine metabolism (Fig.). The C. in purine metabolism does not require preliminary activation of the CO_2.

Carboxyl carrier protein: see Biotin enzymes.

Carboxylic acid esterases: see Esterases.

Carboxylic acids: organic compounds containing the carboxyl group, COOH. Alkane and alkene monocarboxylic acids are also called fatty acids. In aqueous solution, the C.a. dissociate into hydroxonium (H_3O^+) and carboxylate ions ($R\text{-}COO^-$). C.a. and their derivatives, the esters and amides, are metabolically very important. Esters are formed by condensation of an alcohol and an acid. Although the equilibrium under physiological conditions favors the free components, esters such as acetyl-CoA and fats are present in large amounts in cells. Amides are formed by replacing the OH group of the C.a. by an NH_2 group. They are synthesized from activated acid derivatives, such as anhydrides or thioesters (for example, acetyl-coenzyme A). Other C.a. derivatives are the Hydroxyacids (see) and the Ketoacids (see). In cells, at physiological pH values, C.a. are largely present as anions.

Carboxylic esterases: 1. a sub-group of esterases acting on carboxylic esters; 2. carboxylesterase (EC 3.1.1.1), a class of enzymes with wide specificity, usually for a short-chain acid and an alcohol with only one hydroxyl group. The Lipases (see) belong to the group in sense (1). C.e. in sense (2) differ from the lipases in that they are not activated by Ca^{2+} or taurocholate. These enzymes are widely distributed in vertebrate tissues, blood serum, digestive juices (especially in the arthropods and molluscs) and in plant seeds, citrus fruits, mycobacteria and fungi. In mammals, the highest activities of C.e. are found in the liver, kidneys, duodenum and, in males, in the testes and epididymis. The best-studied liver and kidney carboxylesterases are localized in the microsomes. In addition, the lysosomes of these organs contain an arylesterase with an acid pH optimum. The physiological significance of these enzymes is the inactivation of pharmacological ester or amide compounds (for example atropine or phenacetin) and in their effects as aminoacyl group transferases.

Carboxyl transferases: see Biotin enzymes.

Carboxypeptidases: single-chain exopeptidases which contain zinc. They remove successive

amino acids from the *C*-terminal ends of proteins. The animal C. play a role in the digestion of protein in the small intestine, into which they are secreted as inactive precursors (zymogens) from the pancreas. They are converted to the active form by trypsin. They are classified according to their substrate specificity into two groups, C.A and C.B. C.A preferentially hydrolyse amino acids with aromatic or branched chain aliphatic side chains, while C.B attack peptide bonds in which the NH group belongs to a lysine or arginine. With respect to structure and mechanism, bovine C.A is the best examined metalloprotease. The zinc atom is located in a pocket-like active center and participates in the catalytic process. The molecule contains 8 hydrophobic segments with β-structure, which include 20% of the amino acid residues. Another 35% compose surface α-helices, and 20% are part of random coil areas.

When C.A is released from pro-C.A (M_r 87000), there are two secondary products in addition to C.A (M_r 34409, 307 amino acids, primary and tertiary structure known). These are the active form C.Aβ (305 amino acids), C.Aγ and C.Aδ (300 amino acids each), which are formed by further enzymatic shortening of the amino terminal end. C.B (300 amino acids, M_r 34000, M_r of the zymogen, 57400) has, as does C.A, a high esterase activity in addition to its peptidase activity.

Other C. with less specificity have been isolated from citrus fruits and leaves (C.C), yeast (C.Y), *Pseudomonas* (M_r 92000, two polypeptide chains), *Aspergillus* species, germinating barley and cotton seeds. Unlike C. and C. b, these less specific C. are inhibited by the serine protease inhibitor diisopropylfluorophosphate.

Carcinogen: see Cancer.

Cardenolides: see Cardiac glycosides.

Cardiac glycosides: a group of poisonous vegetable glycosides with specific cardiac effects, and related compounds, which are also found in animals, e.g. Toad poisons (see). The aglycones are steroids characterized by a 17β unsaturated lactone ring, *C*/*D-cis* ring coupling and a 14β-hydroxyl group. In addition to a 3β-hydroxyl, which is always present, there may be other oxygen substituents in positions 1, 11, 12, 16 and 19 (see Steroids, Fig. 2). C.g. with a 5-membered lactone ring are classified as *cardenolides*, e.g. Strophanthins (see) and *Digitalis* glycosides (see), while those with a 6-membered lactone ring are *bufodienolides* (bufogenins), e.g. Scillarenin A (see). The sugar component is bound to the 3β-hydroxyl group. In addition to D-glucose, 2-deoxy sugars, for example D-digitoxose, are often found. To date, C.g. have been found in 12 plant families, including the dogbanes (*Apocyanaceae*), figworts (*Scrophulariaceae*), lilies (*Liliaceae*), mulberries (*Moraceae*) and buttercups (*Ranunculaceae*). Some toad poisons, e.g. bufotoxin, are also bufadienolides. C.g. have also been found in insects: the grasshopper *Poekilocerus bufonius* and the butterfly *Danaus plexippus* take them up with their vegetable diet. Pregnenolone is a precursor of C.g. synthesis.

C.g. in therapeutic doses strengthen the contraction of heart muscle and have long been used in therapy. In higher doses, they are toxic. The mechanism of action is unknown, but they are known to bind initially to specific membrane receptors, which are thought to be part of the $(Na^+ + K^+)ATPase$ complex. The existence of receptors suggests that of endogenous compounds which bind to them, and such compounds, called "ouabain-like compounds" or OLC, have been purified from several sources. [Y. Shimoni et al., "Endogenous ouabain-like compound increases heart muscle contractility", *Nature* **307** (1984) 369–371.]

Cardiotoxins: in the widest sense, substances which, in toxic doses, cause heart damage and may lead to heart stoppage. They may interfere with the generation or conduction of stimuli or with the heart's own blood supply, or they may directly attack the heart muscle. C. in the narrower sense include the cardiac glycosides and their aglycones from plants, and a group of toad poisons.

Carminic acid: a red glucoside pigment from the scale insect *Coccus cacti* L. which lives on Central American cacti of the genus *Opuntia*. m.p. 130 °C (d.) $[\alpha]^{15}_{654} + 51.6$ ° (water). C. is the principle component of cochineal, which was formerly one of the most prized dyes for wool and silk.

Carmine is an insoluble complex of C. with alkaline earth or heavy metals, such as zinc. Carmine dyes are used in cosmetics, artists' colors, inks and food colors.

Carminic acid

Carnitine, *vitamin B_T:* serves as a carrier of acetyl and acyl groups through the mitochondrial membrane (see Fatty acid degradation). C. is a characteristic component of muscle tissue and has been isolated in crystalline form from yeast. Mammals are able to synthesize C., but insects must obtain it from their food; otherwise they are unable to complete metamorphosis. C. is biosynthesized from lysine and γ-butyrobetaine (Fig.) and degraded via glycine betaine and 2-methylcholine.

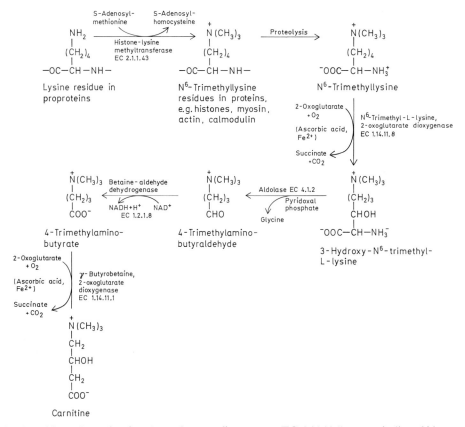

Biosynthesis of carnitine. γ-Butyrobetaine, 2-oxoglutarate dioxygenase (EC 1.14.11.1) occurs in liver, kidney and brain, whereas the other enzymes occur in most tissues. The aldolase cleaving 3-hydroxy-N^6-trimethyl-L-lysine may be identical with glycine hydroxymethyltransferase (EC 2.1.2.1). [W. A. Dunn et al. *J. Biol. Chem.* **259** (1984) 10764–10770]

Carnosine: see Peptides.

Carotenes: a group of isomeric unsaturated hydrocarbons with nine conjugated *trans* double bonds, four methyl branches and a β-ionone ring at one end. The isomers differ in the arrangement of the other end of the chain (Fig.).

(+)-α-C., M_r 536.85, m.p. 188 °C, $[α]_{643}^{18} + 385$ ° ($c = 0.08$ in benzene) has an α-ionone ring as the second terminus. It has the R configuration at C′6. α-C. makes up 15% of the carotene in carrots. It is as widespread as β-C., but is present in smaller amounts. Its vitamin A activity is only half that of β-C. β-C., M_r 536, m.p. 183 °C, has two terminal β-ionone rings and is optically inactive. The intense yellow color is due to the eleven conjugated *trans* double bonds. β-C. is the most common C. in plants and is also found in bacteria, fungi and animals (milk, fat, blood serum, etc.). The C. mixture of the carrot contains about 85% β-C. β-C. can be enzymatically hydrolysed to two molecules of vitamin A, and is therefore important as a provitamin. It is now produced synthetically on a large scale and is used as a dye and in the food industry. It is also used in pharmaceuticals and cosmetics.

γ-C., M_r 536, m.p. 178 °C, is optically inactive. It has a β-ionone ring at one end and an open chain at the other. It has been found in bacteria, fungi and various higher plants, including the carrot (about 0.1%) but it appears to be less widespread than the α and β-isomers. For the history of its discovery, see Carotenoids.

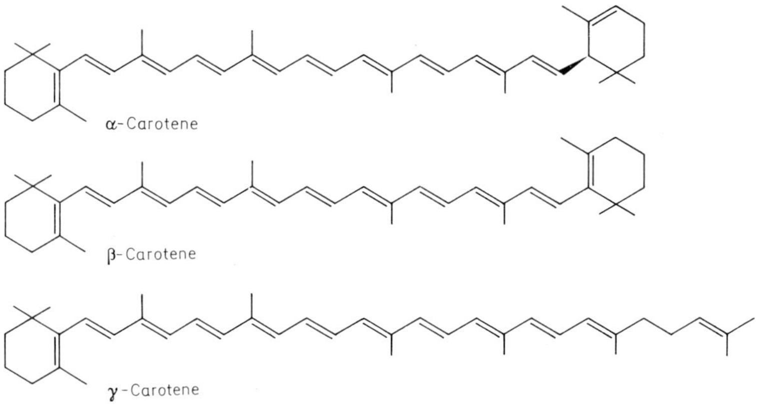

gen-containing, yellow xanthophylls, such as violaxanthin, zeaxanthin, fucoxanthin, lutein, neoxanthin, cryptoxanthin, astaxanthin, capsanthin, capsorubin, rubixanthin and rhodoxanthin. In the native state, the xanthophylls often occur as carotenoproteins (i.e. the carotenoid is bound to a protein component to form a soluble chromoprotein which is stabilized against the effects of air and light), or esterified to fatty acids or glucose. C. contain 9 to 15 (usually 9 or 11) conjugated double bonds, usually *all* trans, and are therefore planar polyenes. Their usually intense yellow to red color is due to this chromophoric system.

Most C. are based on an unsaturated C_{22} chain with methyl branches and with a 9-membered unit on each end which may be cyclic, as in carotene, or acyclic, as in lycopene. When present, the terminal rings are usually the 6-membered α- and β-ionone systems. The median 22 C atoms are generally constant, as the oxidations leading to the xanthophyll structures usually occur on the two C_9 end sections. Most of the 300 known C. are xanthophylls with one, two or many hydroxyl groups, and ether, aldehyde ketone or acid substituents. Stereochemical differences center on *cis-trans* isomerisms at the $C=C$ double bonds and on any chiral centers present. The prototype C. is Lycopene (see), an acyclic C_{40} compound, from which the other members of this group of compounds can be formally derived (Fig.).

C. are one of the most important groups of natural

α-Carotene

β-Carotene

γ-Carotene

Carotenoids: a large class of yellow and red pigments which are highly unsaturated aliphatic and alicyclic hydrocarbons and their oxidation products. The C. arise biogenetically from isoprene units (C_5H_8), and therefore have the methyl branches typical of isoprenoid compounds. The majority of C. have 40 C atoms and are thus members of the tetraterpene family, which are built of 8 isoprene units. In addition, there are a few C. with 45 and 50 C atoms, especially in non-photosynthetic bacteria. C. with lower numbers of C atoms are referred to as nor-, seco- or apocarotenoids.

The C. are subdivided into the carotenes, i.e. the pure hydrocarbons such as lycopene, carotene, neurosporene, phytofluene and phytoene, and the oxy-

pigments, occurring very widely in plants and animals. They occur in animals mainly in surface tissues such as skin, shells, scales, feathers and beaks, but also in birds' egg yolks and as visual pigments. Those C. found in animals are of plant origin, as animals are incapable of de novo synthesis of C. However, they do transform plant C. into their own forms. For the biosynthesis of C., see Tetraterpenes.

The concentration of C. in natural materials is usually on the order of 0.02 to 0.1% of the dry weight. The C. content of the eye ring of the pheasant *Narcissus majalis* is extremely high, being 16%, or about 10^4 times the amount in the carrot. It has been estimated that about 10^8 metric tons of C. are produced per year by the earth's organisms, the

Lycopene

most abundant being fucoxanthin, lutein, violaxanthin and neoxanthin, followed by β-carotene, zeaxanthin, lycopene, capsanthin and bixin. More than 90% of the C. in a plant are found in the leaves, usually as a mixture of 20 to 40% carotenes, of which more than 70% are β-carotene, and 60 to 80% xanthophylls like lutein, violaxanthin, cryptoxanthin and zeaxanthin.

The significance of the C. lies in their contribution to energy transfer in photosynthesis and in their protective properties against light. They are also extremely important as precursors of vitamin A, and thus of visual pigments.

C. are isolated from plant material by extraction and chromatography. Partial and total syntheses of β-carotene are carried out on an industrial scale. C. are used as food dyes and in the pharmaceutical and cosmetic industries.

Historical. The first carotenoid to be isolated was the ruby-red pigment of the carrot, which was crystallized in 1831 by H.F.Wackenroder. From 1931 to 1933, R.Kuhn succeeded in separating chromatographically the three isomers of carotene (α, β, γ), which had previously been thought to be a single substance, and in clarifying their structures. The intensive study of the carotenoid area is linked especially with the names of R.Willstätter, P.Karrer, R.Kuhn, L.Zechmeister, S.Liaaen-Jensen, B.C. L.Weedon and T.W.Goodwin.

Carrageenan, *carrageen*: a mixture of polysaccharides, similar to agar-agar, obtained from red algae by hot-water extraction. C. contains about 45% carragenin, a polysaccharide containing galactose and galactose sulfate. λ-Carragenin is a branched-chain compound of 3,6-anhydro-α-D-galactose and D-galactose-4-sulfate in which 1,3 and 1,6 links alternate. λ-Carragenin is composed of α-1,3-glycosidically linked D-galactose-4-sulfate residues. C. is used as a gelling agent in the food, pharmaceutical and cosmetic industries.

Carrier: see Biomembranes; see Metabolic cycle.

Caryophyllane: see Sesquiterpenes, Fig.

Caryophyllenes: isomeric cyclic sesquiterpene hydrocarbons found in many essential oils. M_r 204.36. α-C. is the same as Humulene (see). β-C., b.p.$_{-10}$ 123 to 125 °C, $[α]_D$ −9 ° p$_4$ 0.9074, n_D^{17} 1.4988. C. has a *trans*-linked cyclobutane-cyclononane ring system, tends to acid-catalysed cyclizations and forms a crystalline epoxide.

γ-C., isocaryophyllene, b.p.$_{14}$ 125 °C, $[α]_D$ −26.2 °, occurs together with the β-isomer and has a *cis* endocyclic double bond. For the biosynthesis of the C., see Sesquiterpenes.

Caryoplasm: see cell Nucleus.

Caseins: see Milk proteins.

Cassaine: see Erythrophleum alkaloids.

Castaprenols: see Polyprenols.

Castoramine: see Nuphara alkaloids.

Catabolism, *dissimilation*: the sum of all degradative metabolic processes. Amino acids, purines, pyrimidines, etc. formed in the turnover of cell components can be degraded, in some cases to inorganic compounds (water, carbon dioxide, ammonia, etc.). Catabolic processes are linked with the formation of ATP by substrate phosphorylation and respiratory chain phosphorylation. C. is therefore, in a certain sense, identical to energy metabolism. Hydrolysis and oxidation play the major roles in the degradation of biopolymers. The C. of carbohydrate chains occurs by oxidation followed by removal of a carboxyl group (decarboxylation), which shortens the chain. The oxidation usually involves the removal of hydrogen (dehydrogenation) from the substrates of fermentation or respiration.

Catabolite repression: the inhibition of enzyme synthesis by increased concentrations of certain metabolic products. The enzymes subject to C.r. are formed in response to metabolic events (utilization of new nutrients by catabolic enzymes, synthesis of secondary products in certain developmental phases of microorganisms). C.r. is probably present in all organisms, although the molecular mechanisms are diverse. Examples of C.r. are glucose repression of catabolic enzymes in *Escherichia coli* and the enzymes of secondary metabolism in microorganisms, e.g. those for the synthesis of penicillin, actinomycin and riboflavin.

The mechanism which prevents induction of β-galactosidase in *Escherichia coli* when glucose as well as the inducer is present in the medium has been elucidated. The regulatory key intermediate in this case is cAMP. When β-galactosidase is induced, the initiation of transcription at the lac promotor is effected by cAMP interacting with a regulatory protein. When glucose is present in excess (as well as the inducer lactose), the cAMP level is very low, so that the initiation complex at the promoter is less likely to be formed or cannot be formed. This regulatory protein is called catabolite-gene activator protein (CAP). The structure of CAP from *Escherichia coli* has been determined by Steitz and McKay. The cAMP-CAP complex binds to DNA near the genes for enzymes that convert other sugars to glucose. X-ray data indicate that cAMP-CAP can bind tightly to a left handed helix, rather than the normal right handed helix. Steitz and McKay propose that by forcing the DNA of the binding site into a left handed configuration, the cAMP-CAP complex causes the nearby region to unwind, thus promoting transcription. The interaction between glucose and the cAMP concentration is still unexplained.

Catalase: a tetrameric heme enzyme (EC 1.11.1.6, M_r 245 000) which catalyses the removal of the highly poisonous hydrogen peroxide from the cell: $H_2O_2 \rightarrow H_2O + \frac{1}{2}O_2$. Each subunit of C. ($M_r$ 60000) contains a heme group in the form of ferri-protoporphyrin IX. C. is found in all animal organs, particularly the liver (here in special cell organelles, the peroxisomes) and erythrocytes, all plant organs and in nearly all aerobic microorganisms. The turnover number of C., $5 \cdot 10^6$ molecules H_2O_2 per min and C. molecule, is one of the highest known for an enzyme. At low concentrations of H_2O_2, C. acts as a peroxidase. It is inhibited by hydrogen sulfide, hydrogen cyanide and N_3 compounds, e.g. NaN_3.

Catalytic center: see Active center.

Catalytic constants: see Michaelis-Menten equation.

Catechins: flavan-3-ols. (+)-Catechin and (-)-epicatechin are widespread in the plant kingdom. Other C. have a more limited distribution (Fig.). C. have been implicated in the biosynthesis of condensed tannins (see Tannins).

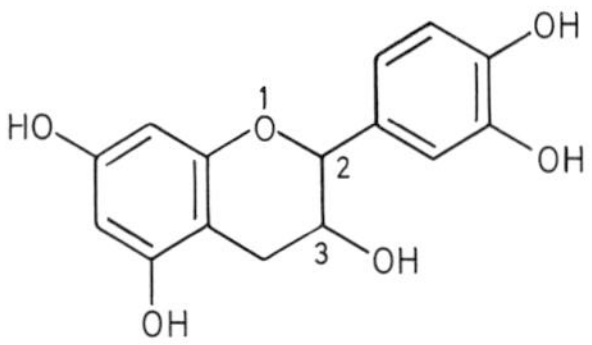

(+)-*Catechin* (2R, 3S)
(-)-*Epicatechin* (2R, 3R)

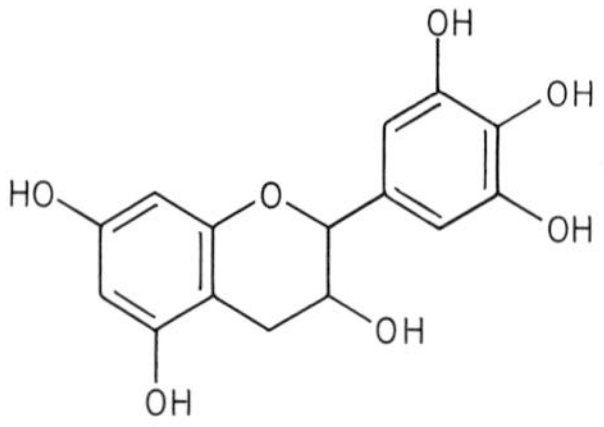

(+)-*Gallocatechin* (2R, 3S)
(*Camellia sinensis*)

(-)-*Epigallocatechin* (2R, 3R)
(*Camellia sinensis*)

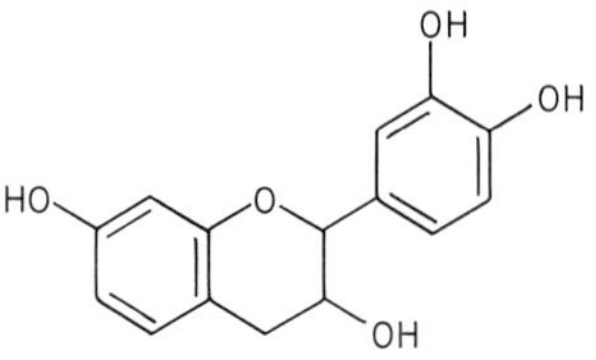

(+)-*Fisetinidol* (2R, 3S)
(*Schinopsis quebracho-colorado*)

(-)-*Fisetinidol* (2R, 3R)
(heartwood and bark of *Acacia mearnsii*)

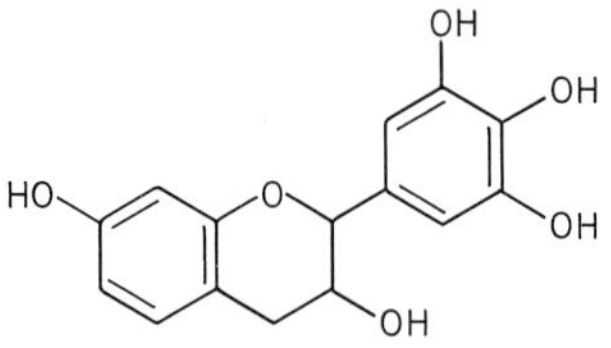

(-)-*Robinetinidol* (2R, 3R)
(*Robinia pseudoacacia, Acacia mearnsii*)

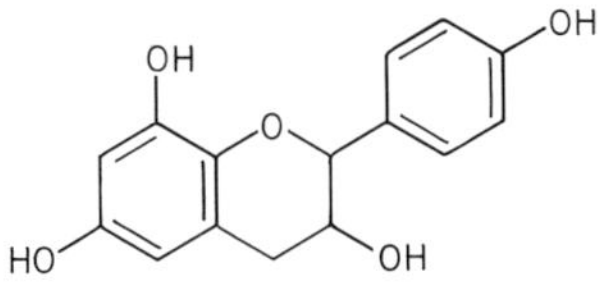

(+)-*Afzelechin* (2R, 3S)
(*Cochlospermum gillivraei, Desmoncus polycanthos, Eucalyptus calophylla*)

(+)-*Epiafzelechin* (2S, 3S)
(*Livinstoma chinensis*)

(-)-*Epiafzelechin* (2R, 3R)
(Afzelia wood, *Larix sibirica, Actinidia chinensis, Juniperis communis, Cassia javanica*)

Catecholamines: alkylamino derivatives of pyrocatechol (*o*-dihydroxybenzene), a group of substances including the hormones Adrenalin (see), Noradrenalin (see) and Dopamine (see). The C. are derived from tyrosine. They affect the blood vessels, the intermediary metabolism and nerve transmission. Their degradation products are excreted in the urine and can be detected there. Derivatives of the C. are used as drugs.

Catechol estrogens: 2-hydroxylated derivatives of estrogens. Estrogens are hydroxylated by the action of the microsomal cytochrome P450 system of the liver. The 2-hydroxyl group then becomes methylated by the action of a methyl transferase and S-adenosyl-L-methionine in the liver cytosol. The corresponding methoxy derivatives are excreted in the urine. C.e. and their methylated derivatives can account for up to 50% of excreted estrogen metabolites; they lack estrogenic activity, and were discovered much later than other known estrogen metabolites (e.g. sulfates and glucuronides of estradiol and estrone), because they were not detectable by biological tests. Synthetic estrogens used in the antibaby pill (e.g. ethinylestradiol, see Ovulation inhibitors) are also subject to 2-hydroxylation, often to a greater extent than the natural estrogens. The resulting C.e. may delay inactivation of catecholamines by competing with them for methylation by methyl transferase. This may explain the hypertension that sometimes accompanies the use of oral contraceptives.

Since C.e. are nonestrogenic (i.e. nonuterotropic), they are generally regarded as inactivation products of the estrogens. There is, however, evidence that the C.e. (in particular 2-hydroxyestrone) are hormones in their own right, and active in the suppression of prolactin secretion [Fishman, J. and Tulchinsky, D. *Science* (1980) *210*, 73-74].

Naturally occurring catechol estrogens and methylated derivatives.

$R_1 = H$, $R_2 = \alpha H\beta OH$: 2-Hydroxyestradiol-17β
$R_1 = CH_3$, $R_2 = \alpha H\beta OH$: 2-Methoxyestradiol-17β
$R_1 = H$, $R_2 = O$: 2-Hydroxyestrone
$R_1 = CH_3$, $R_2 = O$: 2-Methoxyestrone

Catenane, *catenated DNA:* two or more molecules of circular DNA interlocked like the links of a chain. The frequency of occurrence of catenated DNA is increased by inhibitors of protein synthesis. It has been found in cells infected with SV40 or ΦX174, in mouse mitochondria, and in human leukocyte mitochondria. Catenation and decatenation (unlinking) of circular DNA is catalysed in vitro by Topoisomerases (see). The physiological function of catenated DNA is unknown, and it may sometimes occur as a result of errors in DNA replication. The highly catenated networks of DNA found in the single kinetoplast of trypanosomes appear, however, to represent the normal state of the DNA. This kinetoplast DNA (kDNA) contains thousands of catenated DNA circles. The majority of these are so-called minicircles (0.8-2.5 kilobases, depending on the species of trypanosome) with a few maxicircles (20-40 kilobases). For example, kDNA from *Trypanosoma brucei* consists of 6000 catenated minicircles interlaced with 25-50 maxicircles. The diameter of the total network varies from about 5 fm in African trypanosomes (agents of human sleeping sickness) to about 15 fm in *Crithidia* (a trypanosome that parasitizes insects). [P. T. England et al., *Ann. Rev. Biochem.* **51** (1982) 695-726.]

Catharanthus alkaloids: see Vinca alkaloids.

Cathepsins: see Proteolysis.

C_1-bodies: 1) an abb. for Activated one-carbon units (see); 2) in the wider sense, activated carbon dioxide (see Biotin enzymes) or the methyl group (see Transmethylation, S-Adenosyl-L-methionine, L-methionine).

CCC: see Chlorocholine chloride.

CCP: see Biotin enzymes.

cDNA: see Complementary DNA; Hybridization.

CDP: abb. for cytidine 5'-diphosphate. See Cytidine phosphates.

CDP-choline: abb. for cytidine diphosphocholine. See Phosphatide biosynthesis.

CDP-glyceride: abb. for cytidine diphosphoglyceride. See Phosphatide biosynthesis.

CDP-ribitol: see CDP-sugars.

CDP-sugars: a metabolically activated form of sugars and sugar derivatives (see Nucleoside diphosphate sugars). The sugar alcohol CDP-ribitol plays a role in the synthesis of bacterial cell walls.

Cell: the smallest living unit of an organism. Living organisms are classified as prokaryotic (prokaryotes) or eukaryotic (eukaryotes), depending on their cell type.

Prokaryotic organisms are unicellular, i.e. bacteria and blue-green bacteria (erroneously called blue-green "algae"). Prokaryotic cells characteristically contain circular DNA not associated with histones. It is associated with other proteins and is often attached to the cell membrane. The DNA is not enclosed by a nuclear membrane, but it can be demonstrated by cytochemical and radiometric methods, and visualized by electron microscopy. Prokaryotic genes are rarely (if ever) interrupted by non-coding sequences of nucleotides (Introns, see), in contrast to eukaryotic genes. Prokaryotic cells also contain characteristic ribosomes and have prokaryotic-type RNA and protein synthesis (see Protein biosynthesis). Cell division is much simpler than in eukaryotes, as there is no nuclear membrane and no mitotic spindle; the daughter chromosomes are evidently carried into the daughter cells by means of their attachment to the plasma membrane. Prokaryotes have a cell wall containing mucopolysaccharides. Compartmentation into organelles is absent, although there is functional compartmentation e. g. the Mesosome (see), and the photosynthetic membranes of autotropic forms. Respiratory enzymes are located in the cytoplasmic membrane. Photosynthetic prokaryotic cells contain thylakoids and chromatophores. When flagella are present, they consist of a single fibril. The cell wall has certain specific characteristics (see Blue-green bacteria). The bacterial cell wall is based on a giant sac-shaped molecule of Murein (see), which typically contains muramic acid and the amino acids diaminopimelate and D-alanine.

Prokaryotes are morphologically primitive, but biochemically highly adaptable.

Eukaryotic organisms may be unicellular, e.g. flagellates and green algae, but this group is especially represented by all higher plants and animals. These organisms show a high degree of morphological development and differentiation and have a large content of genetic information. Their genes are frequently interrupted by Introns (see), so that their messenger RNA must be processed before transcription (see Processing of messenger RNA).

Eukaryotic cells characteristically contain a Nucleus (see) with Chromosomes (see), and they divide by mitosis and meiosis. They contain organelles, such as Mitochondria (see), Ribosomes (see), Lysosomes (see), Golgi apparatus (see), Endoplasmic reticulum (see) and (in plants) Plastids (see) and Vacuoles (see). In contrast to the prokaryotic type, eukaryotic flagella are multifibrillar; and eukaryotic mucopeptides contain sialic acid, not muramic acid. Eukaryotic cells are bounded by a Cell membrane (see), and plant cells have in addition a Cell wall (see).

The total cell material, including the nucleus, is known as the *protoplasm*. The material outside the nucleus is called (after Strasburger) the *cytoplasm*. The cytoplasm contains the various inclusions and organelles that can be demonstrated by light or electron microscopy. The *hyaloplasm* is (after Höfler) the cytoplasm containing those organelles that can only be demonstrated by electron microscopy or histochemically (ribosomes, mitochondria, Golgi apparatus, endoplasmic reticulum, lysosomes), i.e. it is homogeneous by light microscopy.

By convention, mitochondria and chloroplasts are called Plastids (see).

Cell adhesion molecules, abb. *CAM:* cell surface glycoproteins which promote adhesion between cells of a tissue. Neural (N-CAM), neural-glial (Ng-CAM), and liver (L-CAM) CAM have been isolated. N-CAM exists in at least two forms, distinguished by the difference in the amount of sialic acid residues bound to them. The embryonic form has 30 g sialic acid/100 g protein, while the adult form has 10 g sialic acid/100 g protein. Adhesion between neurons is specifically inhibited by antibodies to N-CAM, but adhesion between neurons and glial cells was not inhibited by purified anti-N-CAM. In chick embryos, both N- and L-CAM appear very early, before the formation of germ layers. Later, the distribution of the two types of CAM correlates with the fate map of blastodermal cells: N-CAM labels the future neural plate, notochord, somites and parts of the lateral plate mesoderm; while L-CAM labels non-neural ectoderm and endoderm. It is likely that embryological development is guided by the expression (sometimes transitory) of one or more types of CAM on the various cell types. [*Science* **222** (1983) 60-62.; G.M. Edelman (1985) *Ann. Rev. Biochem.* **54** 135-169]

Cell count determination: see Growth.

Cell cycle, *mitotic cycle:* the sequence of phases during nuclear and cell division in a mitotically dividing eukaryotic cell. The following phases are recognized: the G_1 (G stands for "gap") or postmitotic phase, without DNA synthesis; the S phase, when DNA is synthesized; the G_2 or premitotic phase, without DNA synthesis; the M or mitosis phase, culminating in cell division. The duration of each phase is variable, depending on the organism, e.g. in the root meristem of the broad bean, $G_1 = 12$ h; $S = 6$ h; $G_2 = 8$ h; $M = 4$ h. Sometimes G_1 or G_2 may be too brief to measure.

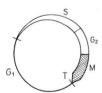

Schematic representation of the cell cycle

Cell homogenate: see Proteins.

Cell mass determination: see Growth.

Cell membrane, *plasmalemma:* a Biomembrane (see) which serves as the outer boundary of the cell. In addition to a C.m., plant and bacterial cells also have Cell walls (see).

Cellobiase: see Disaccharidases.

Cellobiose: a reducing disaccharide consisting of two molecules of D-glucose linked β-1,4. C. differs from maltose in having a β- rather than an α-glycosidic bond. It is not fermented by yeast or hydrolysed by maltase. C. is the disaccharide unit on which cellulose and lichenin are built, but it is not found free in nature. It is occasionally found as a component of a glycoside.

β-Cellobiose

Cell plate: see Cell wall.

Cell sap: see Vacuole.

Cellular enzymes: see Enzymes.

Cellular metabolism: see Primary metabolism.

Cellulases: enzymes found in plants, microbes and fungi which hydrolyse cellulose to cellobiose. The best studied C. comes from *Penicillium notatum* and consists of 324 amino acids (M_r 35000) with a disulfide bridge and no free SH groups. C. are used to produce digestion tablets, to remove undesired cellulose in foods, and to convert cellulose to sugar.

Cellulose: a plant polysaccharide built of glucose units linked β-1,4. The chains are not branched and have M_r between 300000 and 500000, corresponding to 3000 to 5000 glucose units. C. is enzymatically hydrolysed to the disaccharide cellobiose. It can be hydrolysed to D-glucose by treatment with concentrated acids, such as 40% HCl or 60 to 70% H_2SO_4 at high temperature. This process, called saccharification of wood, is used to produce fermentable sugar from wood.

C. is the main component of the plant cell wall. Certain plant fibers, such as cotton, hemp, flax and jute, are almost pure C. Wood, in contrast, is only 40 to 60% C. In the cell wall, the C. is arranged in microfibrils, which are 100 to 300 Å wide and about ½ as thick. The C. chains are parallel, stabilized by interchain hydrogen bonding. The microfibrils are embedded in a matrix of other polysaccharides, including pectins, hemicelluloses and lignin, and small amounts of a protein, extensin.

The microfibrils are arranged transverse to the long axis of elongating cells, with a fair amount of angular dispersion. In thicker walls, they are arranged in helical lamellae, in which the pitch of the C. spirals changes from one lamella to the next.

C. accounts for more of the biomass than any other compound. The total amount is equivalent to about 50% of the carbon dioxide in the atmosphere; about 100 billion metric tons are produced each year. It is biodegraded by organisms which possess Cellulases (see): lower plants, wood-destroying fungi and some bacteria. Termites, ruminants and some rodents harbor symbiont bacteria in their digestive tracts which enable them to utilize cellulose. Humans and carnivores do not digest C., so that it is a ballast substance for them. C. is a very important industrial product. It is obtained primarily by acid (sulfite pro-

Cellulose

cess) or alkaline (sulfate process) hydrolysis. It is used in the manufacture of paper, textiles, plastics, explosives, feeds and fermentation products. Some of the most important derivatives of C. are ethers, esters and xanthogenates.

Cellulose ion-exchangers: see Ion-exchangers; Column chromatography.

Cell wall: a rigid structure external to the cell membrane, and synthesized by the protoplasm. Animal cells do not have C. w.; the cells of prokaryotes, green plants, fungi and some protists do.
Bacterial C. w. The classification of bacteria as Gram positive or negative on the basis of their reaction to the Gram stain corresponds to a fundamental difference in the structures of their cell walls. Gram positive bacteria have relatively simple walls, usually consisting of two layers. The outer layer is often a Teichoic acid (see), though in some species it may be a neutral polysaccharide or an acidic one called teichuronic acid. The inner layer is Murein (see). Gram negative C. w. are more complicated. Under the electron microscope they appear to consist of at least five layers, comprising lipoproteins, lipopolysaccharides, proteins and murein; the murein again forms the innermost layer, or it may be separated from the cell membrane by an extra layer of protein.
Plant C. w. are exceedingly complex structures which have largely defied detailed analysis. The precursor of the C. w. is the cell plate, a non-cellulose structure formed between the two daughter nuclei during mitotic division of the cell. After the new cell membranes have formed on either side, the cell plate (a strongly hydrated structure) matures to a "middle lamella". As growth proceeds, the area of the new wall is increased by intussusception, i.e. incorporation of new material within the existing matrix. The thickness increases by apposition as new layers of wall material are added.
The primary C. w. consists of an extremely complicated array of carbohydrates. The hemisubstances are hemicelluloses and polyuronides, plant gums, mucilages (e.g. fucoidin, laminarin, alginates, agar, carrageenans) and reserve carbohydrates (e.g. arabans, xylans, mannans and galactoarabans). [M. McNeil, A. G. Darvill, S. C. Fry and P. Albersheim, *Ann. Rev. Plant Biochem.* **53** (1984) 625–663] These carbohydrates form the thick cell walls of certain plants, such as date palms and vegetable ivory. Fragments derived from them (see Oligosaccharins) may be potent effectors of cell function. The composition of the primary C. w. appears not to be random, and to require the expression of numerous carbohydrate-transferring enzymes to achieve the correct structures.
Secondary and tertiary C. w. contain structural materials. In most plant cells, the chief structural material is Cellulose (see), laid down as a network of submicroscopic microfibrils. In *Basidiomycetes* and *Phycomycetes*, this material is chitin. The interstices of the structural network contain incrustation materials, such as lignin, silicic acid (*Equisetum* and diatoms), calcium carbonate (brittlestars) or calcium oxalate (cypress).
Accrustation materials of the C. w. of boundary tissues (epidermis, periderm) are cutin and suberin. Cutin forms a cuticle of varying thickness and

strength. Suberin is the material of cork, e.g. in the cork oak, *Quercus suber*. Some plant C. w. secrete wax, e.g. the Andean wax palm, *Copernica cerifera*. The walls of pollen grains and cryptogam (e.g. fern) spores consist of Sporopollenin (see).

Cembranes, *cembranoids:* monocyclic diterpenes isolated from several plants, especially from the gum resins of pines. C. have also been found in marine coelenterates and insects.
The cembrane skeleton (1-isopropyl-4,8,12-trimethylcyclotetradecane, or octahydrocembrene) consists of a 14-membered carbocyclic ring with an isopropyl group at position 1, and 3 methyl groups at positions 4,8, and 12, but cembrane itself does not occur naturally. Cembrene (Fig.) was isolated from pine oleoresins, and eunicin (Fig.) from the Caribbean gorgonian *Eunicea mammosa*. Cembrene A (or neocembrene) ($- C[CH_3] = CH_2$ at position 1) was isolated from the gum resin of *Commiphora mukul,* which was used in the ancient Ayurvedic system of Indian medicine. The same source also yielded mukulol (3,7,11-cembatriene-2-ol). Isocembrene (no 4,5-double bond, and $= CH_2$ at position 4) is present in *Pinus sibirica*. 2,7,11-Cembatriene-4-ol has been isolated from several pines.
The biosynthesis of crassin acetate (Fig.) of the Caribbean gorgonian *Pseudoplexaura porosa* appears to be carried out by its algal symbionts (zooxanthellae). Cell extracts of the zooxanthellae incorporate mevalonic acid and geranyl pyrophosphate into crassin acetate.
The first studies on C. were reported in 1962. Since then, many naturally occurring representatives of the group have been discovered. The literature up to 1977 has been reviewed by Weinheimer, A.J., Chang, C.W.J. and Matson, J.A.: "Naturally Occurring Cembranes" in *Progress in the Chemistry of Organic Natural Products* (1979) 36, 285–387. Springer.

Structures of some naturally occurring cembrane derivatives

Central dogma of molecular biology: the fact that genetic information can be transferred from DNA to protein, but not in the reverse direction (Fig.). The discovery that RNA can code for the synthesis of DNA (see RNA-dependent DNA-synthetase), does not alter the validity of the dogma.

Schematic representation of the central dogma of molecular biology

Centromere: see Chromosomes.

Cephalins: see Phospholipids.

Cephalosporin P₁: a tetracyclic triterpene antibiotic. M_r 574.73, m.p. 147 °C, $[\alpha]_D^{20}+28$ ° ($c=2.7$ in chloroform). C. differs from the related Fusidic acid (see) in that it lacks an 11 α-hydroxyl group and has an additional 6α-acetoxy function. It is obtained from culture filtrates of *Cephalosporium* sp. and is effective against Gram-positive organisms.

Ceramide: see Glycolipids.

Ceramide lactoside lipidosis: see Inborn errors of metabolism.

Ceramide 1-phosphorylcholine: see Phospholipids.

Cerebrocuprein: see Superoxide dismutase.

Cerebronic acid, 2-hydroxytetracosanoic acid, α-hydroxylignoceric acid: $CH_3-(CH_2)_{21}-CHOH-COOH$, an hydroxy fatty acid. M_r 384.63, m.p. 101 °C. C. is a component of various glycolipids.

Cerebroside: see Glycolipids.

Cerotinic acid, n-hexacosanoic acid: $CH_3-(CH_2)_{24}-COOH$, a fatty acid. M_r 396.7, m.p. 87.7 °C. C. is a component of beeswax, wool grease, carnuba wax and montan wax, and occurs in traces in plant fats.

Ceruloplasmin: a blue, copper-containing glycoprotein found in mammalian blood plasma. It was long held to be a transport protein for copper, but has now been discovered to be an oxidase as well. Its substrates are unsaturated compounds, including such indole derivatives as amphetamine, adrenalin and dopamine. C. has a central role in the metabolism of copper; when the organism has no C., as in Wilson's disease, copper is deposited in the tissues, which causes death.
The first studies of C. suggested an octameric structure, then a tetrameric α₂β₂ structure. However, gel filtration of reduced and alkylated C. in 6 M guanidine hydrochloride showed that the protein from several species is a single chain of M_r 120000. The earlier results are probably artifacts of partial proteolysis. The amino acid sequence consists of a 340-residue unit repeated twice (i.e. three copies in all). [L. Ryden in R. Lontie (ed.) *Copper Proteins and Copper Enzymes* (CRC Press, Boca Raton, 1984)] C. is phylogenetically related to plant laccases and to ascorbate oxidase, both of which are blue. Recently it was discovered that the 340-residue unit is 30% homologous to the A subunit of the blood-clotting protein Factor VIII. [R. M. Lawn, *Cell* **42** (1985) 405-406].

Ceveratrum alkaloids: see Veratrum alkaloids.

Cevine: see Germine.

cGMP: abb. for cyclic guanosine 3',5'-monophosphate. (see Guanosine phosphates).

Chain conformation: see Proteins.

Chaksine: see Guanidine derivatives.

Chalcone isomerase (EC 5.5.1.6): a plant enzyme purified from several sources, catalysing the stereospecific isomerization of chalcones to the corresponding (-) (2S)flavanones (Fig.), an important early stage in the biosynthesis of Flavonoid (see) compounds.

Proposed mechanism for action of chalcone isomerase [M. J. Boland & E. Wong, *Bioorg. Chem.* **8** (1979) 1-8]. Nucleophilic addition of an imidazole group at the active site of the enzyme to the double bond is followed by nucleophilic attack by the 2'-phenolate ion. A-H may be an acidic side chain or simply a water molecule.

Chalcones: flavonoids with the ring system shown (Fig.). C. are widely distributed in the plant kingdom, particularly in the *Compositae* and *Leguminoseae*, and they contribute to the flower color (yellow to orange) in certain members of the *Compositae, Oxalidaceae, Scrophulariaceae, Generiaceae, Acanthaceae* and *Liliaceae*. C. are the first detectable C_{15} precursors in flavonoid biosynthesis (see Flavonoids; Chalcone synthase). A. B. Bohm (in *The Flavonoids: Advances in Research*, J. B. Harborne & T. J. Mabry, eds. (Chapman and Hall, 1982) pp. 410-412) gives a checklist of 11 naturally occurring chalcone aglycons and 28 chalcone glycosides. Examples are *butein* (2',4',3,4-tetrahydroxychalcone, from e.g. *Acacia*), *coreopsin* (4'-glucoside of butein, from e.g. *Cereopsis*), *pedicin* (2',5'-dihydroxy-3',4',6'-trimethoxychalcone, from *Didymocarpus*), *ovalichalcone* (2'-hydroxy-3'-prenyl-4',6'-dimethoxychalcone, from *Milletia ovalifolia*), *ψ-isocordein* (2',4'-dihydroxy-3'-(α,α-dimethylallyl)-chalcone, from *Lonchocarpus* spp.).

Chalcone ring system

Chalcone synthase, CHS: a plant enzyme (formerly called flavanone synthase) catalysing the synthesis of chalcones from one molecule of the CoA ester of a substituted cinnamic acid and 3 molecules of malonyl CoA, a key reaction in flavonoid biosynthesis (see Stilbenes, Fig., and Flavonoid, Fig.). Enzyme specificity varies depending on the source, e.g. from *Tulipa* stamens and *Cosmos* petals, CHS forms naringenin, eriodictyol or homoeriodictyol from 4-coumaroyl-CoA, carreoyl-CoA or feruloyl-CoA, respectively, whereas CHS from *Petroselinum hortense* uses only 4-coumaroyl-CoA at pH 8.0, but also attacks caffeoyl-CoA at pH 6.0. (The products named here are flavanones, which are formed from the chalcone product by the action of chalcone isomerase, and to some extent spontaneously. Rigorous removal of chalcone isomerase during purification of CHS is necessary for the clear demonstration of chalcone formation. Formulae of these compounds can be found under Flavonoid, Flavanones, Lignin. M_r of CHS: 80000 (*Phaseolus vulgaris*), 55000 (*Tulipa, Cosmos*), 77000 (*Petroselinum hortense, Brassica oleraea, Haplopappus gracilis*). Most active CHS enzymes appear to consist of two identical subunits. CHS strongly resembles 3-oxoacyl-(acyl carrier protein) synthase (syn. β-ketoacyl-ACP synthase) from type II (non-aggregated) fatty acid synthase, and it is suggested that CHS arose by gene duplication [F. Kreuzaler et al. *Eur. J. Biochem.* **99** (1979) 89–96]. CHS from all sources investigated so far catalyse formation of chalcones with phloroglucinol-type A-ring hydroxylation patterns, the three OH groups originating from the CoA-esterified carboxyl groups of the malonyl CoA. It is therefore necessary to postulate a separate 6'-deoxychalcone synthase to account for most of the isoflavonoid phytoalexins. It is suggested that synthesis of 5-deoxyisoflavonoids is catalysed by an enzyme system similar to 6-methylsalicylic acid synthase from *Penicillium patulum*, i.e. a multienzyme complex using acetyl CoA primer and 3 molecules of malonyl CoA, similar to type I (aggregated) fatty acid synthase from yeast and *P. patulum*. This suggestion is supported by the pattern of incorporation of [^{13}C]acetate into ring A of pisatin in *Pisum sativum* and phaseolin in *Phaseolus vulgaris* [M. Steele et al. *Z. Naturforsch* **37c** (1982) 363–368].

Chalinasterol, *ostreasterol, 24-methylenyl cholesterol, ergosta-5,24(28)-dien-3β-ol:* an animal sterol (see Sterols). M_r 398.66, m.p. 192 °C, $[\alpha]_D$ −35 ° (chloroform). C. is a characteristic sterol in pollen, and has also been found in sponges, oysters and mussels, and in honeybees.

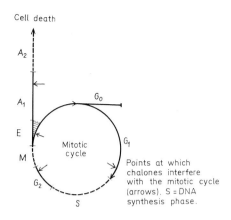

Chalinasterol

Chalones: antitemplate substances, tissue-specific, endogenous mitosis inhibitors. C. are proteins produced by mature or differentiated cells which inhibit cell division in primordial cells by a kind of negative feedback (P. Weiss, 1952). The effect of C. is reversible and not species-specific. They have been isolated from epidermis, lymphocytes, granulocytes, fibroblasts, erythrocytes and liver. The C. from lymphocytes, granulocytes and fibroblasts are glycoproteins with a M_r of 30000 to 50000, while those from erythrocytes and the liver are low-molecular-weight polypeptides with M_r of about 2000. As can be seen from the Fig., C. are active in the late presynthetic phase G_1 and in the premitotic phase G_2 of the cell division cycle. The C. also inhibit the decision phase E and the first postmitotic maturation phase A_1, which precedes the aging phase A_2. Thus the C. not only inhibit cell division, they also retard the process of aging and thus increase the life expectancy of the cells. G_0 is the mitotically inactive phase.

The C. are thought to play a central role in homeostasis (maintenance of the metabolic balance in the body), regeneration and healing of wounds. In general, they are involved in all normal growth, but also in cancerous growth. No C. has yet been purified.

Chanoclavine: see Ergot alkaloids.

CHAPS: 3-[(3-cholamidopropyl)-dimethylammonio]-1-propanesulfonate, a nondenaturing, zwitterionic detergent, combining the properties of sulfobetaine-type detergents and bile salt anions. C. is more effective than sodium cholate or Triton X-100 at breaking protein-protein interactions. It is useful for the solubilization of membrane proteins without denaturation, e.g. solubilization of brain opiate receptors with retention of reversible opiate binding was first achieved with C., which is still (1986) the only successful detergent for this purpose, and the detergent of choice for the isolation of many other membrane proteins, especially receptors. [L. M. Hjelmeland et al. *Anal. Biochem.* **130** (1983) 72–82; W. F. Simonds et al. *Proc. Natl. Acad. Sci. U. S. A.* **77** (1980) 4623–4627]

CHAPSO: 3[(3-cholamidopropyl)-dimethylammonio]-2-hydroxy-1-propanesulfonate, a nondenaturing, zwitterionic detergent. It is chemically similar to CHAPS (see), and is equally useful as a biochemical detergent in the isolation of membrane proteins. [L. M. Hjelmeland et al. *Anal. Biochem.* **130** (1983) 72–82]

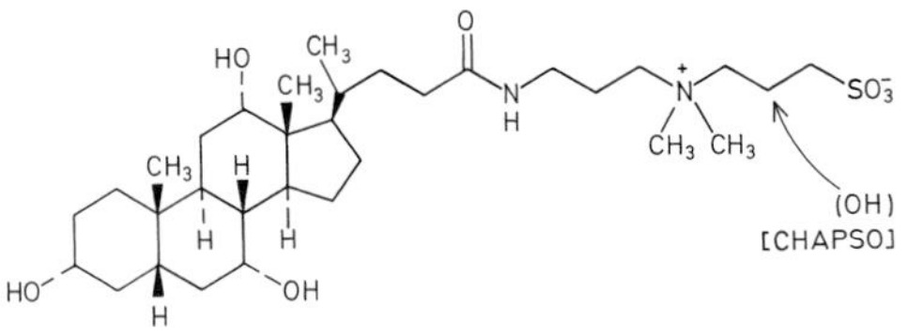

CHAPS and *CHAPSO*

Charge-relay system: a network of hydrogen bridges in the catalytic center of chymotrypsin and other serine proteases which is responsible for the high degree of nucleophilicity of the Ser_{195} hydroxyl group. In addition to Ser_{195}, the system includes the Asp_{102} residue in the hydrophobic interior of the molecule and the imidazole ring of the His_{57}, which lies between the other two. It makes possible an electron flow from the negatively charged carboxyl group of Asp_{102} over the polarized imidazole ring of the His_{57} to the oxygen of the Ser_{195}, which lies on the surface of the molecule. This makes the oxygen highly nucleophilic, but only in an alkaline medium (Fig.). In a weakly acid medium, the Asp and His are ionically bonded rather than hydrogen-bonded. The charge relay system is a series of hydrogen bonds between the three amino acids which are catalytically active in serine proteases at pH 8.0.

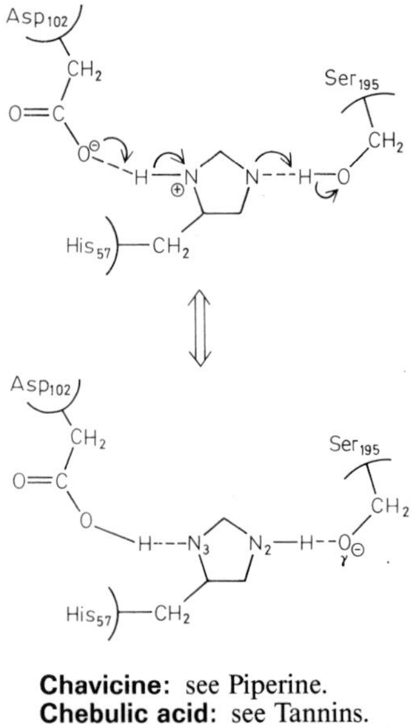

Chavicine: see Piperine.
Chebulic acid: see Tannins.
Chemical taxonomy: the deduction of taxonomic relationships from the distribution of certain natural products. In the absence of chemical convergence, the presence of the same natural products indicates the presence of the same metabolic pathways, enzyme complements and genes, which can only have arisen through a common evolutionary history. Unfortunately, there are many cases of chemical convergence, so that the simple presence of the same natural product does not prove a taxo-

nomic relationship. On the other hand, it is a nearly impossible task to compare the enzyme patterns and reaction sequences in large numbers of organisms. Evolutionary relationships can also be determined by serological methods, by comparison of the amino acid sequences of homologous proteins, etc. (see Evolution). However, these methods are too time-consuming to be of great value for C.
Compared to the number of species of plants (400 000 recent plants and 100 000 lower plants), the number of taxa studied by C. is small. Nevertheless, the amount of data which has been gathered is enormous (Hegenauer). The results of C. have partly confirmed the results of morphological comparison (especially of flowers), and in some cases they have led to corrections in taxa where the relationships were uncertain. However, they have often brought more confusion than clarification.
Chemolithoautotrophic metabolism: see Nutritional physiology of microorganisms.
Chemolithotrophy: see Chemosynthesis.
Chemostat: see Fermentation techniques.
Chemosynthesis: 1) chemical synthesis. 2) The utilization of inorganic compounds or ions (ammonia, nitrite, hydrogen sulfide, thiosulfate, sulfite, iron(II) or manganese(II)ions) and of hydrogen or elemental sulfur to obtain reducing equivalents and ATP. Most organisms capable of C. (water and soil bacteria) fix CO_2 autotrophically. Substrates are oxidized through aerobic or anaerobic respiration. This autotrophic way of life was also called inorganic oxidation.
The term C. should be replaced by "chemolithotrophy".
Chemotherapy: C. was defined by its founder, Paul Ehrlich, as "the use of a drug to combat an invading parasite without damaging the host". The term is also used for treatment of cancer with chemicals which are more damaging to dividing cells than to differentiated ones. In either case, it is always possible that a resistant clone of bacteria or cancer cells will arise, and, since it has no competition from the wild type, multiply rapidly. This has made it necessary to develop a large arsenal of chemotherapeutic agents. The first were the sulfonamides (sulfa drugs). They were later followed by penicillin and other antibiotics.
Chenodeoxycholic acid: 3α, 7α-dihydroxy-5β-cholan-24-oic acid, M_r 392.56, m.p. 119 °C, $[\alpha]_D$ +11 ° (ethanol). C. is the main component of the bile of hens, geese and other fowl, and occurs in small amounts in the bile of ox, guinea pig, bear, pig and human.
Chicken pancreas hormone: a polypeptide hormone (see Peptide hormones) which occurs in the amide form. It consists of 36 amino acid residues. Its primary structure is known, but not its function. M_r 4237.
Chicle: see Gutta.
Chinese gallotannin: see Tannins.
Chirality: the necessary and sufficient condition for optical activity (rotation of the plane of rotation of polarized light). C. means "handedness" (from Greek, κείρ = hand). Chiral molecules have no second-order symmetry element (center, plane or axis of symmetry) and exist in two mirror-image forms

(enantiomers) which cannot be rotated in such a way as to coincide. Most chiral compounds contain an asymmetrically substitued carbon atom, that is, a tetrahedral (sp^3) carbon with four different substituents.

Chitin: a nitrogen-containing polysaccharide which is a major component of the exoskeletons of arthropods. It consists of straight chains of N-acetyl-D-glucosamine residues linked β-1,4. C. is found in the cell walls of diatoms, fungi and higher plants. It usually occurs together with other polysaccharides, proteins or inorganic salts (calcium deposits). C.-hydrolysing enzymes, such as chitinase (EC 3.2.1.14) and chitobiase (β-N-acetyl-D-glycosaminidase, EC 3.2.1.30) are widespread among microorganisms, animals and plants. The biosynthesis of C. from UDP-N-acetylglucosamine is catalysed by chitin synthase (EC 2.4.1.16).

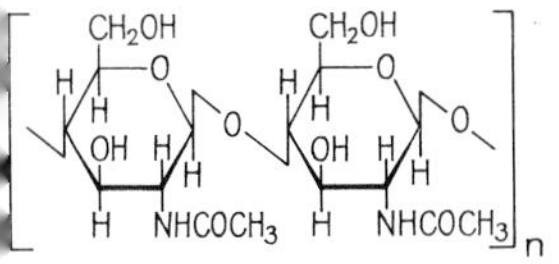

Chitin

Chitosamine: see D-Glucosamine.

Chloramphenicol: an antibiotic isolated from *Streptomyces venezuelae*. M_r 323. There are four different stereoisomers, of which only the D(-)-*threo*-C. shown here is antibiotic. C. inhibits protein synthesis on the 70S ribosomes of prokaryotes, and also on the mitochondrial ribosomes of eukaryotic cells. Protein synthesis on the 80S ribosomes of eukaryotes is not affected. C. inhibits the formation of peptide bonds and translocalization on the 50S subunit of the ribosomes, possibly through specific binding to one of the ribosomal proteins involved in these reactions. The protein in question is probably localized in the acceptor-donor region of the ribosome. C. is used as a broad-spectrum antibiotic in the treatment of typhoid fever, paratyphus, spotted fever, infectious hepatitis, dysentery, malaria, diphtheria, small pox and "viral influenza". Because it inhibits protein synthesis in mitochondrial ribosomes, C. is relatively toxic. It is now produced entirely synthetically.

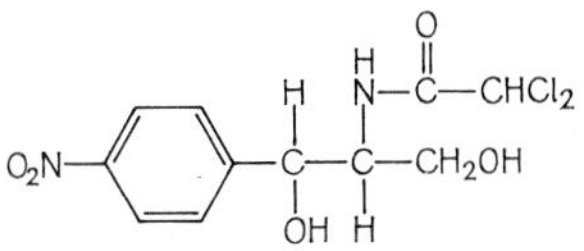

Chloramphenicol

Chlorin: one of the basic ring structures of the porphyrins. Chlorin is 17,18-dihydroporphyrin, using the numbering system of the Commission of Nomenclature of Biological Chemistry, 1960 (see Porphyrins). With the introduction of the term porphyrin, terms in the original Fischer nomenclature, such as chlorin, porphin, phorbin, bacteriochlorin, etc., are now superfluous.

Chlormadinone acetate, *6-chloro-17α-hydroxypregna-4,6-diene-3,20-dione acetate:* a synthetic gestagen. Administered orally, C. has high progesterone activity and is used in oral contraceptives. It is also used in animal breeding to bring on heat and synchronize the ovulation cycle.

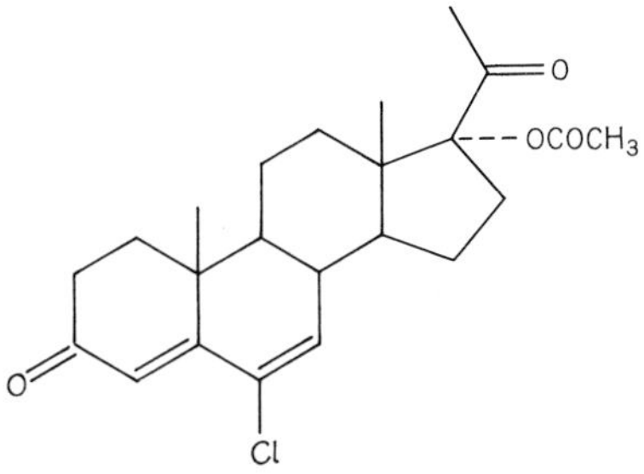

Chlormadione acetate

Chlorobium chlorophylls: see Photosynthetic bacteria.

Chlorocruorin: a green respiratory protein containing heme(II) iron found in the hemolymphs of some marine annelids. That of *Spirographis* (M_r 3 × 10^6, M_r of the 12 subunits, 250000, IP 4.3) has an heme component more closely related to that of cytochrome oxidase than to the heme of hemoglobin.

Chlorocruoroheme: the heme prosthetic group of the chlorocruorins, green respiratory pigments of certain invertebrates, such as the annelid *Spirographis*.

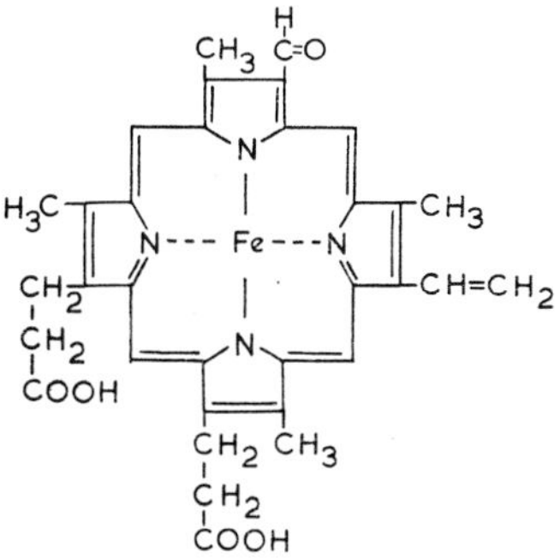

Chlorocruoroheme

Chloroethyl choline chloride, abb. *CCC,* *2-chloroethyl trimethyl ammonium chloride:* [Cl-CH$_2$-CH$_2$-N$^+$(CH$_3$)$_3$]Cl$^-$, a synthetic growth retardant used on grain to produce shorter, thicker straw and to prevent lodging. In many biotests, C. acts as an inhibitor. As a gibberellin antagonist, it blocks the synthesis of gibberellins.

2-Chloroethylphosphonic acid, *Ethrel:* Cl-CH$_2$-CH$_2$-PO(OH)$_2$, a synthetic growth regulator. It is an *ethylene generator;* ethylene is a ripening hormone for fruit. C. is used on *Prunus* fruits to loosen the fruits, and on pineapple to stimulate blossoming.

Chloroflurenol: see Morphactins.

Chlorofucin: see Chlorophyll.

Chlorohemin crystals: see Teichmann's crystals.

Chlorophyll: green photosynthetic pigments found in the Chloroplasts (see) of all higher plants. C. are magnesium complexes of tetrapyrroles and can be considered derivatives of protoporphyrins (see Porphyrins). C. differ from other porphyrins in that 1) they have a saturated rather than a double bond between C atoms 7 and 8; 2) they have a pentanone ring carrying a methylated carboxyl group fused to ring III of the pyrrole; 3) atom 7 carries an esterified propionic acid residue; in bacteriochlorophyll *a* and C. *a*, this is esterified to phytol (Fig. 1), and in other bacteriochlorophylls, it is esterified to farnesol (see Photosynthetic bacteria). The long-chain alcohol is responsible for the waxy quality of C. and makes it difficult to crystallize. It also provides a lipophilic anchor by which the molecule is held in place in the thylakoid membrane. The tetrapyrrole ring is hydrophilic.

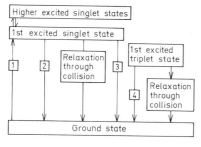

Figure 1. Chlorophyll a

Removal of the Mg from C. generates *pheophytin*; hydrolysis of the phytol or farnesol ester creates a water-soluble *chlorophyllide*. This reaction is catalysed by the enzyme chlorophyllase (EC 3.1.1.4). The metal-free chlorophyllide is called a *pheophorbide*. All these compounds are intensely colored and fluoresce strongly in strictly anhydrous solution in

Higher excited singlet states

1st excited singlet state

Relaxation through collision

1st excited triplet state

Relaxation through collision

Ground state

1 Absorption
2 Fluorescence
3 Photochemistry
4 Phosphorescence

Figure 2

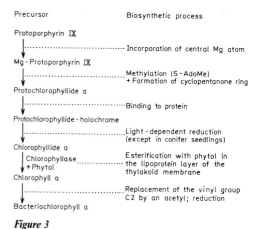

Precursor	Biosynthetic process
Protoporphyrin IX	
	Incorporation of central Mg atom
Mg - Protoporphyrin IX	
	Methylation (S – AdoMe) + Formation of cyclopentanone ring
Protochlorophyllide a	
	Binding to protein
Protochlorophyllide - holochrome	
	Light - dependent reduction (except in conifer seedlings)
Chlorophyllide a	
Chlorophyllase + Phytol	Esterification with phytol in the lipoprotein layer of the thylakoid membrane
Chlorophyll a	
	Replacement of the vinyl group C2 by an acetyl; reduction
Bacteriochlorophyll a	

Figure 3

Table.   Occurrence of the chlorophylls

Species
Occurrence
Differences
from chlorophyll *a*

Chlorophyll *a*
All photosynthetic organisms except photosynthetic bacteria
Fig. 1

Chlorophyll *b*
All higher plants except the orchid *Neottia nidus avis*, green algae *(Euglenophyta, Chlorophyta)* and the *Charophyta*.
–CHO instead of
–CH$_3$ at position 3

Chlorophyll *c*
(Bacillariophyta, Dinophyta), brown algae *(Phaeophyta)*, and a few red algae *(Rhodophyta)*

Chlorophyll *d*
Red algae
–CHO at position C-2 instead of a vinyl group.

Bacteriochlorophyll *a*
In purple sulfur bacteria *(Thiorhodaceae)*, non-sulfur purple bacteria *(Athiorhodaceae)* and in green sulfur bacteria *(Chlorobacteriaceae)*.
2-acetyl-2-devinyl-dihydrochlorophyll a

Bacteriochlorophyll *b*
Thiococcus, one strain of *Rhodopseudomonas (Athiorhodaceae)*
Structure unknown

Bacteriochlorophyll *c* + *d* (*Chlorobium* chlorophylls)
Green sulfur bacteria
Phytol ester replaced by farnesol ester

Protochlorophyll
Biosynthetic precursor of chlorophyll *a*.
A double bond between C atoms 7 and 8; a porphyrin

Chlorophyllide
No phytol residue, water-soluble

an organic solvent. The characteristic absorption spectra are used to identify and quantify the C. and their derivatives. In vivo, the absorption maxima are strongly affected by binding to proteins. In the thylakoid membrane, the C. form protein complexes, some of which can be isolated and identified.

Only a very small part of the plant's C. is directly active in Photosynthesis (see), and these molecules are of C.a or bacteriochlorophyll a. Active centers consisting of C.a_I or C.a_{II} complexed with Plastoquinone (see) are absolutely necessary for the primary process of photosynthesis. More than 99% of the C. serve as accessory pigments, as do the thylakoid Carotenoids (see), which trap and funnel light to the active centers of Photosystems I and II (see). Fig. 2 shows the possibilities for conversion of the electronic energy resulting from the trapping of a photon by a molecule of C. Some work has been done on the spatial arrangement of C. in the membrane proteins. [W. Kuehlbrandt, "Three-dimensional structure of the light-harvesting chlorophyll a/b-protein complex" *Nature* **307** (1984) 478–480]

The properties of the various C. are shown in the table. C.c (chlorofucin) is probably a mixture of the monomethyl esters of magnesium hexadehydropheoporphyrin a_5 and magnesium tetrahydropheoporphyrin a_5. C.d differs from C.a only in having a formyl group instead of a vinyl group on C atom 2. C.b has a formyl group in place of a methyl group at C atom 3.

The biosynthesis of C. is the same as that of the Porphyrins (see) up to the level of protoporphyrin IX. There are a number of alternative pathways from there to C.a (Fig. 3). The derivation of the other C. from C.a is not clear. The last steps in the synthesis appear to take place in situ in the thylakoid membrane.

Chlorophyllase (EC 3.1.1.4): a plant carboxyesterase which catalyses the reversible transformation of chlorophyllides to chlorophyll in the last step of chlorophyll biosynthesis. Chlorophyllide a + phytol $\leftrightarrows$ chlorophyll + H_2O. The pH optimum for the reaction is 6.2. The enzyme is found in all plants, in both the green and nongreen parts, such as roots. It is localized in the lipoprotein layer of the thylakoid membrane, which contains all the enzymes and pigments of the photosynthetic apparatus.

Chlorophyllide: see Chlorophyll.

Chloroplasts: the photosynthesis organelles of higher plants. The whole C. is lens-shaped and contains stacks of membranes (thylakoids) embedded in an aqueous phase (stroma). The chlorophyll is embedded in the thylakoids, and here the transformation of light energy to ATP and reductive equivalents (NADPH) occurs (both are required for reduction of CO_2). Hydrogen (as protons + electrons) is removed from water, setting free O_2. This is called the light reaction (see Photosynthesis). In the dark reactions, which take place in the stroma, CO_2 is reduced to carbohydrate. The stroma also contains enzymes for the synthesis of starch, fatty acids and amino acids, as well as the genetic system of the C., including the components of transcription and translation.

The C. of higher plants and green flagellates contain 10^{-15} to 10^{-14} g of DNA, which is organized in 20 or more redundant copies of a circular molecule. The individual molecules have M_r of 90×10^6 and a length of 40 μm (peas, lettuce, *Euglena*). The DNA and some of the 70S ribosomes are associated with the thylakoid membrane.

4-Chlorotestosterone: 4-chloro-17β-hydroxyandrost-4-en-3-one, an anabolic steroid obtained by partial synthesis from testosterone. 4-C. is used therapeutically as an ester, for example in convalescence, inflammation and tumors.

7'-Chlorotetracycline: see Tetracyclines.

Chlorotriazine dyes: dyes containing chlorotriazinyl groups. They react readily with polysaccharide matrices under alkaline conditions, forming stable dyed products. A large number of such dyes is available, covering a wide range of colors. Two have found application in biochemistry: *Cibacron Blue F3GA* (Giba-Geigy) and *Procion Red HE3B* (ICI). Cibacron Blue is the chromophore of *Blue Dextran*, a high M_r polysaccharide used to measure the void volumes of gel filtration colums. Blue Dextran is made from a dextran fraction of average M_r 2 million. It is completely excluded from all types of gel filtration columns. Some proteins (e.g. pyruvate kinase, phosphofructokinase and lactate dehydrogenase) bind to Blue Dextran, and therefore emerge earlier than expected from gel filtration columns. This observation led to the discovery that the dye moiety (i.e. Cibacron Blue) of Blue Dextran interacts specifically with the nucleotide binding domain (see Dinucleotide fold) in lactate, malate and glyceraldehyde 3-phosphate dehydrogenases, and with the ATP-binding site in phosphoglycerate kinase. Columns of Blue Sepharose (Sepharose with bound Cibacron Blue) can therefore be used for the affinity chromatography (see Proteins) of NAD^+- or ATP-dependent enzymes; the coenzyme is used as the competing ligand to elute the protein from the column. Procion Red has similar affinity properties to Cibacron Blue; immobilized Procion Red shows a higher specificity for binding $NADP^+-$ dependent enzymes, and it has been used to advantage in the affinity chromatographic purification of glucose 6-phosphate dehydrogense.

Cholane: see Steroids.

Cholanoic acid: 5β-cholan-24-oic acid; see Bile acids.

Cholecalciferol: Vitamin D_3; see Vitamins.

Cholecystokinin, *pancreozymin*: a tissue hormone consisting of 33 amino acids (porcine) with a M_r 3838. It is formed in the mucosa of the upper intestine in response to the presence of chymus (acid mixture of partially digested food) or to a nervous stimulus, and promotes contraction of the gall bladder and secretion of pancreatic juice.

Cholera toxins: see Toxic proteins.

Cholestane: see Steroids.

Cholestanol: 5α-cholestan-3β-ol, a zoosterol (see sterols). C. is a 5,6-dihydro derivative of cholesterol and occurs in small amounts with cholesterol in animal cells. In some sponges, C. is the main sterol. It differs from the stereoisomeric Coprostanol (see) in the configuration at the C atom 5 and the *trans* configuration of rings A and B.

Cholesterol: the most important sterol in higher animals (see Sterols). C. is found free or esterified to

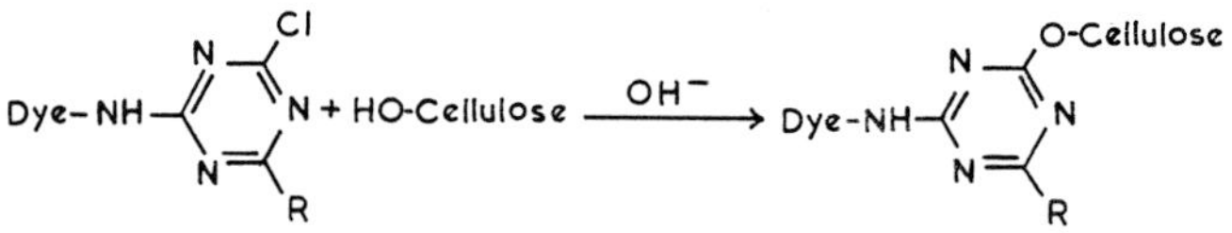

Cholestyramine

Coupling reaction of a chlorotriazine dye with cellulose

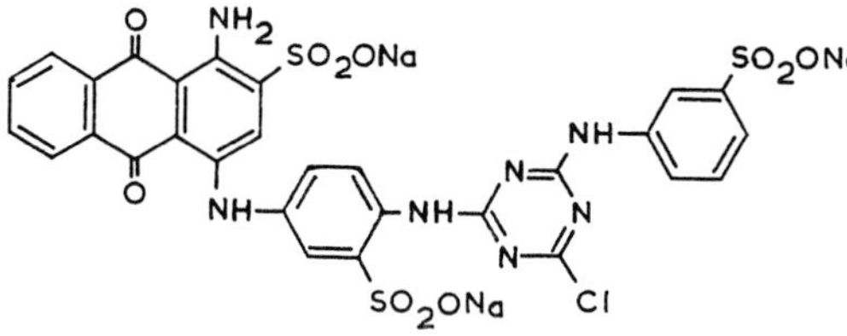

Cibacron Blue F3GA

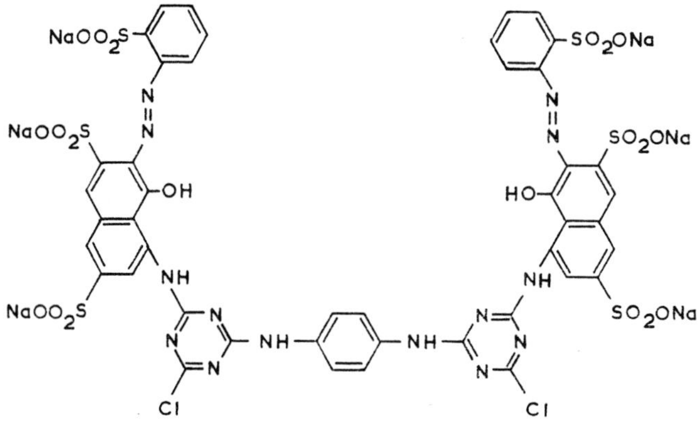

Procion Red HE3B

fatty acids in all mammalian tissues, often together with phospholipids. It is especially abundant in brain, (about 10% of dry matter) adrenals, egg yolk and wool grease. Blood contains about 2 mg/ml, bound to Lipoproteins (see). C. is a component of biomembranes and of the myelin sheaths around nerve axons. It has a detoxifying effect on hemolytically active Saponins (see), with which it forms insoluble complexes, thus protecting the red blood cells from lysis. The human skin excretes up to 300 mg C. daily, as a protective agent. The deposition of C. on the interior of artery walls (arteriosclerosis) and in gall stones is pathological. (See Lipopro-

teins.) C. was first isolated from the latter by Green, in 1788.
C. has also been isolated in small amounts from plants, e.g. from potato plants, from many pollens, isolated chloroplasts and from bacteria. C. is a vitamin for many insects, which require it as a precursor of ecdysone and related molting hormones. Commercially, C. is obtained from the spinal cords of cattle, or from wool grease. It was first synthesized by Robinson and Woodward in 1951. For biosynthesis, see Terpenes.
C. is synthesized ultimately from acetyl-coenzyme A, via the triterpene lanosterol (see Steroids, Fig. 4) and zymosterol. In turn, it is a key intermediate in the biosynthesis of many other steroids, including steroid hormones, steroid sapogenins and steroid alkaloids.

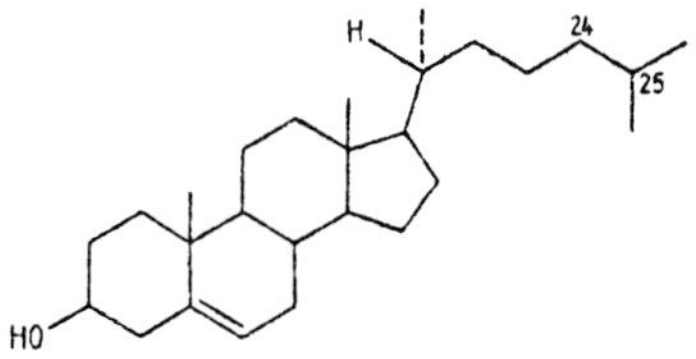

Cholestyramine: a quaternary ammonium anion exchange resin, in which the basic groups are attached to a styrene-divinyl benzene copolymer by C-C bonds. It is a lyophilic solid, equivalent weight about 230. C. binds bile acids in the intestine, thereby preventing their reabsorption and return to the liver in the enterohepatic circulation. This results in an increase in the conversion of cholesterol to bile acids in the liver, and an increased uptake of choles-

Cholesterol

terol-containing Phospholipids (see) by the liver. Dietary administration of C. decreases plasma cholesterol in cockerels, dogs and humans, and decreases aortic plaque formation in cholesterol-fed cockerels. It has been used to decrease plasma cholesterol levels in human heterozygotes for familial hypercholesterolemia. [J. W. Huff et al. *Proc. Soc. Exp. Biol. Med.* **114** (1963) 352–355; M. S. Brown & J. L. Goldstein *Science* **232** (1986) 34–47]

Cholic acid: $3\alpha,7\alpha,12\alpha$-trihydroxy-5β-cholan-24-oic acid, a bile acid found in conjugation with lysine or taurine in the bile of most vertebrates. M_r 408.56, m.p. 196 to 198 °C (anhydrous), $[\alpha]_D^{20}$ +37 ° ($c=0.6$ in alcohol). It forms salts with fatty acids and other lipids, such as cholesterol and carotene. It is used as starting material for the partial synthesis of therapeutically important steroid hormones.

Cholic acid

Choline: $[(CH_3)_3N^+-CH_2-CH_2-OH]\ OH^-$, m.p. 180 °C (d.). As the natural hydrolysis product of lecithin, it is found in many plants and animals, in the brain, egg yolk, hops, *Belladonna* and *Strophanthus*. It is a methylating agent in metabolic processes. It reduces the deposition of body fat, lowers the blood pressure and causes the uterus to contract. It is required for the resynthesis of methionine from homocysteine.

Choline acetylase (EC 2.3.1.6): see Acetylcholine.

Cholinergic receptor: see Acetylcholine.

Cholinesterase, *pseudocholinesterase,* (EC 3.1.1.8): an unspecific acylcholinesterase which hydrolyses butyroyl and propionoyl choline much faster than acetylcholine. C. is found primarily in the serum (M_r of horse serum C., 315000, 4 subunits of M_r 78000; human serum C., M_r 348000, 4 subunits of M_r 86000), the liver and pancreas. It is also found in cobra venom. Because it is secreted by the liver, the serum concentration of C. is measurably reduced in cases of liver parenchyma damage. The enzyme is inhibited by the carbamate esters physostigmine (eserine) and prostigmine, which have no effect at 10^{-4}M on the type A and B caroxylesterases of the serum and organs. Because C. has a catalytically important serine* in the active center (Gly-Gly-Asp-Ser*-Gly), it is stoichiometrically and irreversibly inhibited by organic phosphate esters (DFP, E 600, E 605, Wofatox and other nerve and tissue poisons).

Chondroitin sulfate: a water-soluble mucopolysaccharide found in animals. M_r around 250000. Chondroitin A and C consist of equimolar amounts of D-glucuronic acid and *N*-acetyl-D-galactosamine linked in alternating β-1,3 and β-1,4 bonds; they dif-

fer in the position of the sulfate ester. In C. A it is on the C4 hydroxyl, while in C. C, it is on the C6 hydroxyl. C. B (dermatan sulfate) contains L-iduronic acid instead of D-glucuronic acid. The C. are the main component of cartilage tissue, where they make up 40% of the dry weight. They are also found in skin, tendons, umbilical cord, heart valves and other connective tissue. The C. are bound to proteins in vivo.

Chondroitin sulfate A: R=H, R'=SO$_3$H
C: R=SO$_3$H, R'=H

Chondroitin sulfate B

Chondrome: the genetic information contained in the mitochondria of a cell. Since the number of mitochondria per cell varies as widely as the amount of DNA in each mitochondrion, the genetic capacity of the individual mitochondrion is highly variable.

Chondrosamine: see D-Galactosamine.

Choriogonadotropin, *placental gonadotropin, human chorionic gonadotropin,* abb. *hCG:* the most important hormone formed in the placenta during pregnancy. C. is a glycoprotein (M_r 30000) containing about 30% carbohydrate and two polypeptide chains of 92 (α, M_r 10205) and 139 (β, M_r 14902) amino acids. The α subunit is common to all glycoprotein hormones (follicle-stimulating hormone, luteinizing hormone and thyreotropin) and is encoded by a single gene. The β chain has 82% sequence homology with the β chain of luteinizing hormone and probably evolved from it. The function of C. is to stimulate the ovaries to produce the steroid hormones necessary to maintain pregnancy. The presence of hCG in maternal urine or plasma can be detected as early as one week after implantation of the fetus, using monoclonal antibodies (see Immunoassays, enzyme-linked), and it is therefore used as a pregnancy test.

Choriomammotropin, *placentalactogen,* abb. *PL, human lactogen:* a single chain polypeptide hormone of known primary structure (191 amino acid residues, M_r 22308). It is synthesized in increasing amounts by the placenta during pregnancy, and secreted into the maternal circulation. Its action is similar to that of Somatotropin (see).

Chorismic acid: see Aromatic biosynthesis.

Christmas factor: see Blood coagulation.

Chromatid: see Chromosomes.

Chromatin: the stainable material of the interphase nucleus, consisting of DNA, RNA and several specialized proteins, which is dispersed randomly in the nucleus. The chromosomal DNA is intact in this phase; it is merely uncoiled or "relaxed". Immediately prior to cell division, C. condenses into dense bodies (chromosomes) which can be intensely stained. Euchromatin is tightly coiled chromosomal material which is not being transcribed, while heterochromatin has a looser structure and is the site of transcription. See Nucleosomes.

Chromatography: a type of method used for analytical or preparative separation of mixtures of compounds. There are many kinds of C., and except for Gas chromatography (see), high performance liquid chromatography, and the automated ion exchange systems for amino acid analysis, they require only a modest amount of apparatus. They are suitable for separations on the micro and ultramicro scales; the limiting factor is the means of detection of the separated material after the C. is complete. This group of techniques has revolutionized organic and biological chemistry by making possible separations of closely related compounds which could not be resolved by any other method.

For any chromatographic technique, there must be a stationary and a mobile phase. The mixture to be separated is carried through or past the stationary phase by the mobile phase; separation occurs because the components of the mixture are retarded to differing degrees by the stationary phase. There are a number of different kinds of interactions between the components and the stationary phase which may be utilized. In *partition C.*, the substances are distributed between two immiscible phases on the basis of their relative solubilities in the two phases. This form of chromatography is analogous to extraction by countercurrent distribution, but in C. the number of steps approaches infinity. *Adsorption C.* depends on the difference in degree of adsorption of the components to the solid stationary phase. The most elegant application of this principle is *affinity C.*, in which one component of a mixture is separated by its specific affinity to the column material. This stationary phase is prepared by linking a specific substrate (or antigen, or biological binding partner) to an inert solid such as Sepharose. The desired enzyme (or antibody, hormone, repressor protein, etc.) is then selectively bound to its substrate, while all others pass through. Other forms of interaction with the stationary phase are found in Ion-exchange C. (see) and Gel-filtration C. Electrophoresis (see) is a related technique in which the components to be separated are moved through the stationary phase by electromotive force rather than by a mobile phase.

C. is also named according to the type of stationary phase: 1) Paper chromatography (see), 2) Thin-layer chromatography (see), 3) Column chromatography (see) and 4) Gas chromatography (see).

C. can be refined further by the creation of more specific absorbing agents for the stationary phase, by further reduction in scale to the atomic level, and by improving the sensitivity and selectivity of the detection techniques for the separated components.

Chromatophores: 1) in botany, plastids: Chloroplasts (see), Chromoplasts (see) and Leucoplasts (see).
2) The photosynthetic organelle of the Photosynthetic bacteria (see). The bacterial C. are intraplasmatic membranes originating from the cell membrane. They may exist as closed vesicles or as flattened stacks. Their membranes carry the photosynthetic pigments and the components of photosynthetic electron transport and of photophosphorylation.

Chromium, Cr.: an essential dietary constituent for animals. Very small amounts are required (at least 100 parts per billion in the diet of the rat), but the human dietary requirement has not been determined. Although relatively large amounts of Cr are found associated with isolated fractions of RNA, the relationship of Cr to RNA structure or function is not known. Cr is known to be a constituent of glucose tolerance factor (GTF), a water-soluble, relatively stable organic complex of Cr, with M_r about 500, which is essential in animals and humans for normal glucose tolerance. The earliest symptom of Cr deficiency is impaired glucose tolerance. More severe deficiency leads to glycosuria, fasting hyperglycemia, impaired growth, shortening of life span; cases of diabetes refractory to insulin may be the result of Cr deficiency. In such cases children respond more readily than adults to infused Cr salts, suggesting that the ability to convert Cr into GTF decreases with age, although the ability of humans to convert Cr into GTF is not unequivocally proven. Many natural foods contain GTF, especially brewer's yeast, black pepper, liver, cheese, bread and

Definition of terms used in chromatography

Term	Definition
Equilibration	Saturation of the paper to be used with the vapor of the elution solvent.
Detection	Visualization of the separated material: staining of spots or bands, UV or light absorption, radioactivity, etc.
Elution	Washing out of the components
Eluant	The material which has been eluted
Eluate	A solution emerging from a chromatography column
Developer, Eluent, or Elutant	The mobile (usually liquid) phase
Development	The process of chromatography, or, treatment with the reagent(s) used for detection
Solvent	Either a pure liquid or a mixture of solvents
Front	The position of the leading edge of the solvent
Running time	The time during which C. occurs
Standard substance	A pure, authentic substance used for identification

beef. Foods highest in GTF are not necessarily highest in Cr. Only Cr^{3+} is physiologically active.

Chromogranins: proteins which specifically bind catecholamines (e.g. adrenalin, noradrenalin). C. are associated with the catecholamines in storage vesicles in those cells which produce them.

Chromomeres: see Chromosomes.

Chromophores: see Pigments.

Chromoplast: a chromatophore filled with carotenoids, and therefore red-orange to yellow in color. The pigments may have crystallized out of solution within the C., as in the carrot root. In the petals of the forsythia, the C. are developed from green plastids (see Chloroplasts). The pigment vesicles of the red algae, which are colored red to violet by Biliproteins (see), are called Rhodoplasts (see).

Chromoproteins: proteins which contain a colored prosthetic group bound either covalently or noncovalently. The group includes heme proteins and iron porphyrin enzymes, flavoproteins, chlorophyll-protein complexes and the non-porphyrin iron and copper protein in the blood of vertebrates (for example, transferrin and ceruloplasmin) and invertebrates (for example hemerythrin and hemocyanin).

Chromosomal RNA: see Chromosomes.

Chromosome: a cellular structure for storage and transmission to the next generation of genetic information. The term was originally applied to the stainable material in eukaryotic nuclei, but it has been expanded to include the gene carrier(s) of any cell. The genes in any C. are sequences of nucleotide base pairs in the DNA molecule which is the core of the C. (see Deoxyribonucleic acid). Genetic mapping shows that the genes of a C. form a single, linear array, which indicates that the C. contains only one very long molecule of DNA. Prokaryotes contain only a single, circular C. It is complexed with regulator proteins, but does not contain structural proteins (histones) and therefore is not stainable. It is attached to the cell membrane (see Replicon).

Eukaryotes have more than one C. per cell, the number and shapes being species-specific. Eukaryotic C. are complexes of DNA (10-30%), RNA (3-15%) and protein (40-75%). The amount of DNA per C. is constant and species-specific, but the amount of RNA varies with the transcription activity in the cell (see Ribonucleic acid). In addition to the RNA involved in protein synthesis, there is also an organ-specific fraction called *chromosomal RNA*, which is probably part of the structure of the C. and may have a regulatory function in transcription. The chromosomal proteins can be divided into two classes, the basic histones, and the more acidic non-histone proteins. There are six classes of Histones (see), but the non-histone proteins, when subjected to gel electrophoresis, yield very complicated patterns with hundreds of bands. The histones combine with the DNA to form a fibrillar DNA-histone complex, or supercoil. In this form, the DNA cannot be transcribed. The non-histone proteins are probably involved in the regulation of transcription (they include the polymerases), but owing to their complexity, they are not well understood.

The eukaryotic C. form compact structures prior to cell division (mitosis or meiosis). At this time they consist of two identical longitudinal halves, the *chromatids*, which are joined at one point, the *centromere*. The centromere is also the point of attachment for the spindle fibers which pull apart the daughter chromatids during cell division. The nucleotide sequences of yeast centromere DNA have been analysed and even inserted into Plasmids (see), which confers on the latter mitotic stability which is nearly as great as that of C. (Segments of C. which have no centromeres are not evenly distributed between daughter cells and are rapidly lost from a dividing population.)

Other specialized DNA sequences, the *telomeres*, are found at the ends of linear C. Insertion of telomere sequences into a circular plasmid converts it to a linear structure. Another type of sequence, ARS (autonomously replicating sequence), confers on plasmids the ability to replicate independently of the C., and may well be the sites at which replication of the C. is initiated. [A. W. Murray & J. W. Szostak *Nature* **305**, (1983) 189-193]

In stages when the C. are not tightly coiled (prophase), bead-like *chromomeres* can be seen along the length of the C. In some specialized cells where protein synthesis occurs at very high rates, Giant chromosomes (see) or Lamp-brush chromosomes (see) are found.

Chrysanthemin: see Cyanidin.

Chrysin: see Flavones (Table).

Chylomicrons: see Lipoproteins.

Chymopapain (EC 3.4.22.6): an enzyme from the latex of the papaya tree. It is made up of several equal sized segments (C. A and B, M_r 35000). C. resembles papain in the structure of its active center and its action, but it differs in its acid stability and its very high isoelectric point (C. A pH 10, C. B pH 10.4; papain, pH 8.75).

Chymosin (EC 3.4.23.4): see Rennin.

Chymostatin: see Inhibitor peptide.

Chymotropic pigment: a pigment dissolved in the vacuole of a plant cell.

Chymotrypsin: a family of structurally and catalytically homologous serine proteases (see Proteases) formed and stored in the pancreas in the form of precursors. Chymotrypsinogen A (IP 9.1, 245 amino acids, M_r 25670) is cationic at pH 8, and chymotrypsinogen B (IP 5.2, 248 amino acids, M_r 25760) is anionic at pH 8. The latter is lacking in swine pancreas, from which chymotrypsinogen C (281 amino acids, Trp-rich, M_r 31800) has been isolated. There are differences in the substrate specificities of the activated forms. All C. hydrolyse preferentially phenylalanyl, tyrosyl and tryptophanyl peptide and ester bonds with a pH optimum at pH 8 to 8.5. C. B also attacks other bonds, for example in glucagon, and C. C attacks leucyl and glutaminyl bonds.

Unlike the activation of trypsinogen, the activation of chymotrypsinogen A is a very complicated, multistep process leading to several active forms (Fig. 1). Chymotrypsinogen A consists of a polypeptide chain with 5 intrachain disulfide bridges (1-122, 42-58, 136-201, 168-182, 191-221).

Trypsin activates chymotrypsinogen A by hydrolysing the peptide bond between Arg^{15}-Ile^{16}, producing active π-C. This enzyme removes the dipeptides

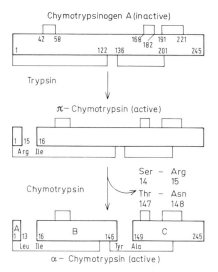

Figure 1. Activation of chymotrypsinogen

Ser[14]-Arg[15] and Thr[147]-Asn[148] from other molecules of π-C. to produce the stable, active δ-C. In it, the A, B and C chains are held together by S-S bridges. The catalytically active amino acids His[57], Asp[102] and Ser[195] are located in the B and C chains.

The tertiary structure of α-C. has been extensively studied. It consists mostly of extended chain segments held together by hydrogen bonds and 5 disulfide bridges. Except for the 11 amino acids at the carboxyl end of the C chain, there are no α-helices in the molecule. As in trypsin, the active site is located in a loosely structured region of the molecule and has a pocket-like cleft for the side chain of the amino acid of the substrate. (Fig. 2). This pocket, which determines the substrate specificity of the enzyme, has the Ser[189] at the bottom. The catalytically active amino acids Ser[195] and His[57] are identical in C., elastase and trypsin.

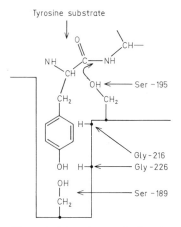

Figure 2. The substrate binding site of chymotrypsin

In the course of activation, the chain segment 187 to 194 is rotated by 180°, bringing the catalytic amino acids to within 0.3 nm of the surface of the molecule. Hydrolysis of the substrate involves formation of an acyl-enzyme intermediate between the acid group of the peptide substrate and the hydroxyl of the Ser[195]. The strongly nucleophilic character of this hydroxyl is due to the neighboring proton donor (or acceptor) His[57]. The effect is amplified by Asp[102] (see Charge-relay system).

Chymotrypsinogen B is activated to C. B_π by a similar trypsin cleavage, but it is not further transformed to a δ-form. A dipeptide is slowly removed from C. B_π by autocatalysis, forming the major form of C. B (243 amino acids, M_r 25 400). Little is known about the activation of chymotrypsinogen C, except that the M_r of the product C. C is about 25% lower than that of chymotrypsinogen C.

The nucleophilic attack of the active serine hydroxyl is indicated by the curved arrow.

Cinchona alkaloids: a group of about 30 alkaloids from the bark of tropical trees, especially *Cinchona succiruba*. They are included among the indole alkaloids on account of their precursors, although the main representatives have a quinoline structure. Cinchona, the dried bark, contains 7 to 10% alkaloids. The main alkaloid is Quinine (see), which makes up 5 to 7%; secondary alkaloids include quinidine, epiquinine and epiquinidine. Other important C.a. are cinchonine, M_r 294.40, m.p. 264°C, $[\alpha]_D$ +224° (alcohol), and its stereoisomers,

Biosynthesis of the Cinchona alkaloids

and cinchonamine, M_r 296.40, m.p. 186 °C, $[\alpha]_D^{20}$ ($c = 0.66$ in methanol). In contrast to the above C.a., cinchonamine has an indole rather than a quinoline structure. Extracts of cinchona as well as pure quinine and quinidine have many medical applications, in particular as antimalarial agents. *Biosynthesis:* The quinoline ring of the C.a. is synthesized from tryptophan via tryptamine, and the quinuclidine nucleus from iridoid compounds. Cinchonamine is synthesized from a carboline type compound by cleavage of the C ring and reaction of the N atom with the para side chain of the D ring. This is followed by oxidation of the primary alcohol group, hydroxylation and opening of the indole ring (cpd. I, Fig.).

Cinchonamine: see Cinchona alkaloids.

Cinchonine: see Cinchona alkaloids.

1,8-Cineole, *eucalyptol:* the main constituent oil of oil of eucalyptus. M_r 154.25, m.p. +1.3 °C, b.p. 176–177 °C, ρ_{20} 0.9267, n_D^{20} 1.455–1.460. It is used in cough syrup.

1,8-Cineole

Cinnamic acid 4-hydroxylase (EC 1.14.13.11): a mixed function mono-oxygenase, present in plants, which catalyses the insertion of an atom of oxygen into cinnamic acid to form 4-hydroxycinnamic acid (4-coumaric acid) with concomitant oxidation of one molecule of NADPH; this is an early reaction in Flavonoid (see) biosynthesis. The enzyme is a cytochrome P450 system associated with the microsomal fraction, and is specific for the *trans* isomer of cinnamic acid. During the hydroxylation, hydrogen at position 4 (experimentally, tritium at position 4) is retained, i.e. there is an NIH shift (see). In vitro, a thiol, e.g. 2-mercaptoethanol, is required for activity. [P.R. Rich & C.J. Lamb *Eur. J. Biochem.* **72** (1977) 353–360]

Cisplatin *diamminedichloroplatinum, cis-DDP:* a platinum-containing antitumor drug, first licensed in 1979 and now (1986) one of the most extensively used of all chemotherapeutic agents against cancer. It is especially effective in the control of testicular tumors, and is a common component of drug combinations against other tumors, notably ovarian tumors. Serious side effects include nephrotoxicity, and peripheral neuropathy has caused permanent disability in some patients. The cellular target of C. is thought to be DNA or chromatin. [D.C.H. McBrien and T.F. Slater eds., *Biochemical Mechanisms of Platinum Antitumor Drugs* (IRL Press, Oxford, 1986)].

Cisplatin

Cistron: a section of DNA which codes for the amino acid sequence of a polypeptide chain. It thus corresponds to a gene.

Citral: a doubly unsaturated monoterpene aldehyde. A mixture of the *cis* and *trans* isomers is a component of many essential oils. *C. A, trans-citral, geranial:* M_r 152.24, b.p.$_{12}$ 110–112 °C, ρ_{20} 0.8898, n_D^{17} 1.4894; *C. B, cis-citral, neral:* b.p.$_{12}$ 102–104 °C, ρ_{20} 0.8888, n_D^{20} 1.4891. C. is a component of complex insect pheromone mixtures. When heated, C. is converted to isocitral, and in the light, it cyclizes to *photocitral A*. The conversion of C. to pseudoionone with acetone is important as the first step in the industrial synthesis of vitamin A. In the perfume and food industries, C. is the most important of the aliphatic monoterpenes.

trans-Citral   cis-Citral

Citrate cleavage enzyme: see ATP citrate (pro-3S)-lyase.

Citrate condensing enzyme: see Citrate (si)-synthase.

Citrate cycle: see Tricarboxylic acid cycle.

Citrate lyase: see Citric acid.

Citrate (si)-synthase, *citrate condensing enzyme, citrogenase* (EC 4.1.3.7): the tricarboxylic-acid-cycle enzyme which catalyses the synthesis of citrate from oxaloacetate and acetyl-coenzyme A in an aldol condensation. C. from *Escherichia coli* (M_r 248000) consists of 4 subunits (M_r 62000). The C. from swine or rat heart (M_r 98000) consists of only 2 subunits (M_r 49000).

Citric acid: a key metabolic intermediate. m.p. 153–155 °C. Citrate is the starting point of the tricarboxylic acid cycle. Its concentration also coordinates several other metabolic pathways. At sufficiently high levels of C.a., acetyl-coenzyme A car-

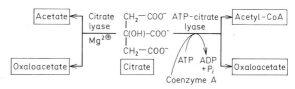

Cleavage of citrate

boxylase (EC 6.4.1.2), the key enzyme in fatty acid biosynthesis, is allosterically activated. C.a. is also a negative allosteric effector for 6-phosphofructokinase (EC 2.7.1.11), the key enzyme of glycolysis. C.a. can form complexes with various cations, particularly with iron and calcium. In animals, C.a. improves the utilization of nutritional calcium.

In bacteria, C.a. can be hydrolysed by ATP citrate lyase (EC 4.1.3.8) and citrate pro-(3S)-lyase (EC 4.1.3.6) (Fig.).

C.a. was first isolated in 1784 from lemon juice by Scheele. It is found in many different plants, especially in fruits, but also in leaves and roots. Various microorganisms, for example *Aspergillus niger*, are used to produce it on an industrial scale, usually from molasses, with yields as high as 60% of the sugar used. C.a. is separated from the medium by precipitation as the calcium salt.

Citrinin: see Mycotoxin.

Citrogenase: see Citrate (*si*)-synthase.

Citronellal, an unsaturated monoterpene aldehyde found in both optical isomers and the isopropylidene form. M_r 154.25. The more important form is (+)-C., b.p. 205 to 206 °C, $[\alpha]_D^{25}$ +11.5 °. C. is the chief component of citronella oil and the essential oils of various eucalyptus species. It is an alarm pheromone of ants of the genus *Lasius*. Its tendency to cyclize is utilized in the synthesis of monocyclic monoterpenes, such as menthol.

(+)-Citronellal

Citrostadienol

Citrostadienol: 4α-methyl-5α-stigmasta-7, 24(28)diene-3β-ol, a phytosterol (see Sterols). M_r 426.7, m.p. 162 °C, $[\alpha]_D$ +24 ° (chloroform). C. is found in the oils of grapefruit and orange peels. It is also considered one of the tetracyclic triterpenes and is an intermediate in the synthesis of sterols.

L-Citrulline: N^5-(aminocarbonyl)-L-ornithine, α-amino-δ-ureidovaleric acid, $H_2N-CO-NH-(CH_2)_3-CH(NH_2)-COOH$, a nonprotein amino acid. M_r 175.2, m.p. 220 °C (d.), $[\alpha]_D^{25}$ +4.0 (c=2 in water). It occurs in the free form in plants and animals, and in large amounts in the sap of birches, alder, and walnut trees. It is synthesized in the liver

from carbamoyl phosphate and L-ornithine by ornithine carbamoyltransferase (EC 2.1.3.3). The citrulline phosphorylase complex consists of ornithine carbamoyltransferase and carbamate kinase (EC 2.7.2.2) (see Carbamoyl phosphate) and catalyses the reaction L-C. + P_i + ADP ⇌ ATP + ornithine + HCO_3^- + NH_4^+. Arginine is synthesized from L-C. (see Urea cycle) via the intermediate argininosuccinate, which is synthesized from L-C. and L-aspartic acid in an ATP-requiring reaction. L-C. was first isolated by Wada in 1930 from the juice of the watermelon, *Citrullus vulgaris.*

Citrullinemia: see Inborn errors of metabolism.

Clathrin: a protein associated with "coated pits" on the surface of the plasma membrane, and with transport vesicles associated with the Golgi apparatus (see). C. from bovine brain is a hexamer of 3 heavy (M_r 180,000) and 3 light chains; there are two forms of the latter, LC–A (M_r 36,000) and LC–B (M_r 33,000). The overall ratio of A : B chains is 1 : 2, but they are apparently randomly distributed among the population of C. molecules. The molecule is Y-shaped. Purified C. can reassemble into "cages" in vitro; these "cages" surround the coated pits or vesicles in vivo.

C.-coated pits on the cell surface take up hormones, serum proteins, lysosomal hydrolases and presumably many other molecules for which the cell has specific receptors. The receptors are concentrated in the pits, either before or after association with their substrates. The pits pinch off to form vesicles, and within minutes, the contents are acidified by an ATP-dependent proton pump. In addition, the vesicles shed their C. They do not immediately fuse with lysosomes, unlike the vesicles formed by micropinocytosis; thus there is a chance for the contents to be transported to other targets within the cell. A transglutaminase which cross-links the contents of the pits or vesicles is required for endocytosis; inhibition of this enzyme prevents ingestion of the proteins. [T. Kirchhausen et al. *Proc. Natl. Acad. Sci. USA* **80** (1983) 2481-2485; B.M.F. Pearse & M. Bretscher *Ann. Rev. Biochem.* **50** (1981) 85-101]

Clauberg test: see Progesterone.

Claviceps alkaloids: see Ergot alkaloids.

Clavine alkaloids: see Ergot alkaloids.

Cleland's short notation: a nomenclature for representing reaction mechanisms among several substrates. The symbols A, B, C . . . are used for substrates; P, Q, R, . . . for products; I. J . . . for inhibitors, E, F, G . . . for stable enzyme forms; EA, EAB, FB . . . for enzyme-substrate complexes; (EAB), (EPQ), . . . for short-lived intermediate complexes. The molecularity (uni-, bi-, ter- . . . molecular) of the reaction is determined by the number of reactants which are kinetically significant. The enzyme forms are written from left to right below solid horizontal lines, while reactants and products are indicated by vertical arrows (Fig.). In *sequential mechanisms*, all the substrates are associated with the enzyme before the first product is released. If the substrates must bind in a particular order, it is an *ordered* mechanism; otherwise, it is a random mechanism. In *ping-pong mechanisms*, one or more products dissociate from the enzyme before all the substrates have been bound.

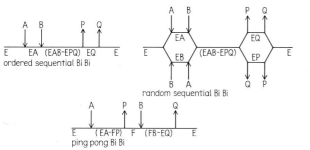

Cleland's notation

Clinical chemistry: clinical chemical laboratory diagnostics, part of clinical and medical biochemistry. The results of C. c. are important for the diagnosis, therapy and prophylaxis of diseases.

Clionasterol: (24S)-stigmast-5-ene-3 β-ol, a marine zoosterol (see Sterols). M_r 414.7, m. p. 138 °C, $[\alpha]_D$ − 42 ° (chloroform). C. differs from β-sitosterol (see Sitosterols) in its stereochemistry at C atom 24. It is found in sponges, for example *Cliona celata* and *Spongilla lacustris.*

Clonal deletion theory: see Immunoglobulins.

Clonal selection theory: see Immunoglobulins.

Clone bank: see Recombinant DNA technology.

Clostripain (EC 3.4.22.8): SH-dependent, trypsin-like protease (M_r 50000) with endopeptidase and amidase-esterase activity isolated from culture filtrates of *Clostridium histolyticum.* The endopeptidase activity hydrolyses proteins, while the amidase-esterase activity cleaves synthetic amino acid amides and amino acid esters. Because C. attacks only arginyl and lysyl residues, it can be used to isolate large peptide fragments without previous chemical modification of the substrate.

Cloverleaf model: see Transfer RNA.

Clupeine: see Protamines.

CMP: abb. for cytidine 5′-monophosphate.

CoA, *CoA-SH:* abb. for coenzyme A.

Coagulation factors: see Blood coagulation.

Coagulation vitamin: obsolete term for vitamin K. See Vitamins.

Cobalamine: vitamin B_{12}. See Vitamins.

Cobalt, *Co:* an important bioelement which is present in traces in plants, animals and microorganisms. The ligands in cobalt complexes in living cells are often the corrin ring system of vitamin B_{12}, benzimidazole or sugar components. As a component of certain coenzymes, Co is also required for the symbiotic fixation of atmospheric nitrogen. Traces of Co are required for microbial growth. Co can also serve as cofactor of several enzymes, e. g. in pyrophosphatases, peptidases and arginase.

Cobamide coenzyme: see 5′-Deoxyadenosyl-cobalamine.

Cobramine: see Snake venoms.

Cobra toxin: see Snake venoms.

Coca alkaloids: see Tropane alkaloids.

Cocaine: a tropane alkaloid, the main alkaloid of the coca plant, *Erythroxylon coca,* and related forms growing in the tropics. (−) C. is a bitter white powder, m. p. 98 °C, b. p.-0.1 187 °C, $[\alpha]_D^{20}$ − 16 ° ($c=4$ in chloroform). Both C. and its secondary alkaloids are based on the alkamine ecgonine. C. is obtained by extraction of the leaves of the coca plant. The extract is hydrolysed, and the ecgonine so obtained is easily converted to C. by esterification with methanol and benzoic acid. C. is sometimes used as a local anaesthetic. Due to its euphoric or hallucinogenic (at higher doses) effects, the drug is a popular (though illegal) intoxicant. In South America the leaves of the coca plant are chewed with lime to release the alkaloids. The number of addicts has been estimated to be 5 million.

Cocarboxylase: 1) see Thiamin pyrophosphate; 2) obsolete term for the prosthetic group of the decarboxylating yeast enzyme pyruvate decarboxylase. The name is confusing, however, because C. does not participate in a carboxylation, but in a decarboxylation dependent on thiamin pyrophosphate.

Cochineal: see Carminic acid.

Cochliobolin B: see Sesterterpenes.

Cock's comb test: see Androgens.

CO_2-compensation point: the concentration of CO_2 at which the rate of photosynthesis (CO_2 incorporation) and the rate of respiration (CO_2 production) are balanced. The value varies with illumination and must be quoted for a given light intensity. C3-plants have a CO_2-compensation point of 40–60 ppm CO_2 at 25 °C. For C4-plants the value is often less than 10 ppm. The CO_2-compensation point increases with temperature, thus resulting in a loss of efficiency of photosynthesis in C3-plants as the day temperature increases. C4-plants, however, are not affected. Also, with increasing light intensity, the CO_2 concentration of the air decreases around growing plants, making a further contribution to the loss of efficiency. See Photorespiration; Light compensation point.

Code: see Genetic code.

Codehydrogenase I: see Nicotinamide adenine dinucleotide.

Codeine: an opium alkaloid. Opium consists of about 4% C., but it is found in other poppy species as well. M_r 299.37, m. p. 154 to 156 °C, $[\alpha]_D^{20}$ −137.7 ° (alcohol). C. is morphine 3-methyl ether, and is converted to morphine as the poppy ripens. About 80% of the world production of morphine is methylated to the therapeutically more important C. In contrast to morphine, C. is only slightly analgesic,

but it can increase the effects of other analgesics, It strongly inhibits coughing, and the danger of habituation or addiction is slight.
For formula and biosynthesis, see Benzylisoquinoline alkaloids.

Code triplet: see Codon.

Codogenic strand: the strand of a DNA double helix from which the genetic information is transcribed onto RNA. The other, complementary, strand serves only to preserve the information in semiconservative replication. In some viruses and bacteria, alternating sections of both strands may be codogenic.

Codon, *code triplet:* linear sequence of three adjacent nucleotides in RNA which specify a particular amino acid. In the course of translation, the C. on a messenger RNA is paired with the anticodon on a tRNA which carries a specific amino acid. (see Genetic code).

Coenzyme: in the narrow sense, the dissociable, low-molecular-weight active group of an enzyme which transfers chemical groups (see Group transfer) or hydrogen or electrons. C. in this sense couple two otherwise independent reactions, and can thus be regarded as transport metabolites.
In a wider sense, a C. can be thought of as any catalytically active, low-molecular-weight component of

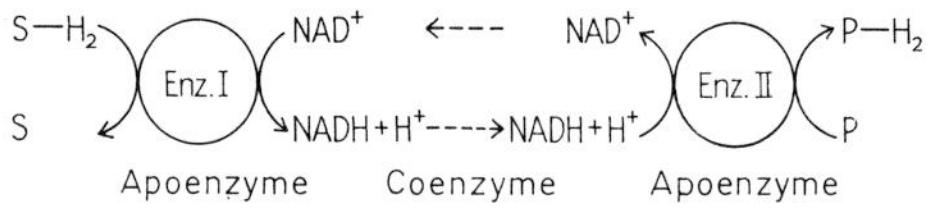

Enz. I = triosephosphate dehydrogenase,
Enz. II = alcohol dehydrogenase;
S and S-H_2 = oxidized and reduced substrate, respectively; P and P-H_2 = oxidized and reduced products.

an enzyme. This definition includes C. which are covalently bound to the enzyme as a prosthetic group. In this case, the C. and the *apoenzyme (enzyme protein)* together make up the active *holoenzyme.*
C. in the narrow sense enters the reaction stoichiometrically, in that it reacts sequentially with two enzyme proteins, and thus catalyses substrate turnover. An example is nicotinamide adenine dinucleotide (NAD), the working group of dehydrogenases and reductases. It first forms an active complex with a dehydrogenase (see, e.g. Glycolysis) and accepts the hydrogen removed from the substrate. Reduced NADH then dissociates from this enzyme I and associates with a reductase (enzyme II) (see, e.g. Alcoholic fermentation), donating the hydrogen to the substrate of this enzyme (Fig.). Since the C. acts as a

Table. Classification, metabolic function and source of the coenzymes

Coenzyme	Function	Vitamin source
1) *Oxidoreduction coenzymes*		
NAD	Hydrogen and electron transport	Nicotinic acid
NADP	Hydrogen and electron transport	Nicotinic acid
FMN	Hydrogen and electron transport	Riboflavin
FAD	Hydrogen and electron transport	Riboflavin
Ubiquinone (Coenzyme Q)	Hydrogen and electron transport	–
Lipoic acid	Hydrogen and acyl transfer	–
Heme coenzymes (Cytochromes)	Electron transport	–
Ferredoxins	Electron transport Hydrogen activation	–
Thioredoxins	Hydrogen transport	
2) *Group transfer coenzymes*		
Nucleoside diphosphates	Transfer of phosphoryl-choline (CDP) and sugars (UDP, GDP, TDP, CDP)	–
Pyridoxal phosphate	Transamination, Decarboxylation, etc.	Vitamin B_6
Phosphoadenosine phosphosulfate	Sulfate transfer	
Adenosine triphosphate	Phosphorylation, Pyrophosphorylation, Transfer of adenosyl and adenyl groups	
S-Adenosyl-L-methionine	Transmethylation	(Methionine)
Tetrahydrofolic acid and conjugates	Transfer of formyl, hydroxymethyl and methyl groups	Folic acid
Biotin or CO_2-biotin enzymes	Carboxylation, transcarboxylation, decarboxylation	Biotin
Coenzyme A	Transacylation, etc.	Pantothenic acid
Thiamin pyrophosphate	C_2-group transfer, rarely, C_1-group transfer	Thiamin (Aneurin)
3) *Isomerization coenzymes*		
Coenzyme form of vitamin B_{12}	Carboxyl shifts	Vitamin B_{12}
Uridine diphosphate	Sugar isomerization	

second substrate, it can be called a *cosubstrate*. It must be able to react reversibly with the apoenzymes of two different enzymes.

Flavin, heme and pyridoxal phosphate are examples of C. in the wider sense. Metals are considered inorganic complements of enzyme reactions and are not termed C., but rather *cofactors*.

Many C. in the wider sense are synthesized from Vitamins (see). The relationships of some C. to vitamins and metabolic function are listed in the table. Strictly speaking, ATP, which commands a special position in metabolism, does not fit the definition of a C. The C. of C_1-unit transfer are S-Adenosylmethionine (see), Tetrahydrofolic acid (see) and Biotin (see). The C. of C_2 transfers are Coenzyme A (see) and Thiamin pyrophosphate (see). Vitamin B_{12} is involved in various metabolic reactions in free form, as methyl-vitamin B_{12} and as 5'-Deoxyadenosylcobalamine (see).

Nearly all C. contain a phosphate group. They often bind non-ionized (uncharged) molecules or groups. In the Nucleotide coenzymes (see), the phosphate is part of the nucleotide structure.

Coenzyme I: see Nicotinamide adenine dinucleotide.

Coenzyme II: see Nicotinamide adenine dinucleotide phosphate.

Coenzyme A, abb. *CoA*, also *CoA-SH*: the coenzyme of acylation. M_r 767.6 λ_{max} 257 nm at pH 2.5 to 11.0 Solutions of CoA between pH2 and pH6 are relatively stable. CoA consists of adenosine 3',5'-diphosphate linked via the 5'-phosphate to the phosphate of pantotheine 4'-phosphate. (Fig. 1). The Thiol group (see) of the cysteamine is responsible for the biological activity of CoA. Practically all the pantothenic acid in a cell is bound as CoA. The metabolically active form of CoA is acyl-CoA, which serves as a donor of the acyl group.

The metabolic significance of CoA rests on its ability to form high-energy thioester bonds. The methylene group next to the activated thioester tends to dissociate into a carbanion and a proton (Fig. 2). The carbanion is subject to electrophilic attack. Therefore the formation of the thioester activates both the carboxyl group (electrophilic form, nucleophilic reactions) and the neighboring position (nucleophilic form, electrophilic reactions). The following reactions of the carboxyl group are biochemically important: 1) reduction to the aldehyde, 2) transacylation in the formation of acetylcholine, hippuric acid, acetylated amino sugars, S-acetylhydrolipoic acid (see Pyruvate dehydrogenase) and acetylphosphate on the enzyme phosphotransacetylase (Fig. 2a, see Phosphoroclastic pyruvate cleavage), 3) exchange of the sulfhydryl components in the thiophorase reaction, in which, for example, the CoA is transferred from succinyl-CoA to acetyl-CoA.

There are numerous condensation reactions typical of the α-methyl group in acetyl-CoA, especially 1) the carboxylation of acetyl-CoA to malonyl-CoA by the biotin-dependent acetyl-CoA carboxylase (EC 6.4.1.2), which plays a role in the synthesis of fatty acids, 2) aldol condensations like those in citrate synthesis (Fig. 2b) in the tricarboxylic acid cy-

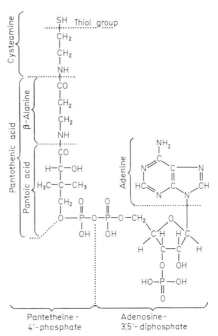

Figure 1. Coenzyme A

2a

2b

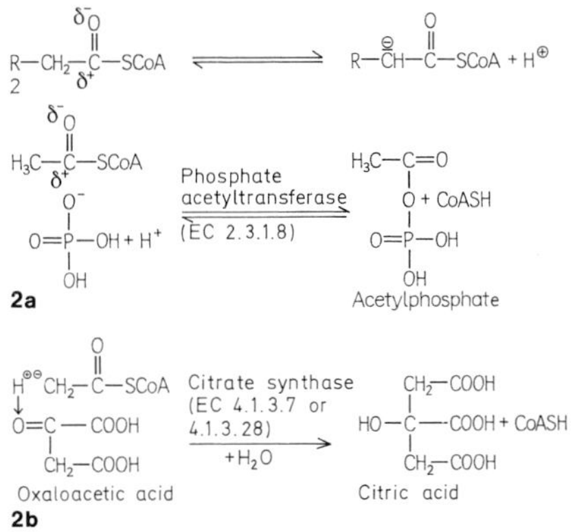

Figure 2. Active forms of acyl-CoA (thioesters)
2a Synthesis of acetylphosphate (activated carboxyl group)
2b Citrate synthesis (activated α-methylene group)

cle. When acetoacetyl-CoA is synthesized from two molecules of acetyl-CoA (ester condensation), one molecule enters the reaction in the electrophilic form, and the other in the nucleophilic form.

The intermediates in the β-oxidation of fatty acids (see Fatty acid degradation) are derivatives of CoA, as are some intermediates in the synthesis of some alkaloids.

Coenzyme F: see Tetrahydrofolic acid.

Coenzyme form of vitamin B_{12}: see 5'-Deoxyadenosylcobalamine.

121

Coenzyme Q: see Ubiquinone.
Coenzyme R: see Vitamins (vitamin H).
Colamine: see Ethanolamine.
Colamine cephalins: see Phospholipids.
Colcemid, *demecolcine:* see Colchicum alkaloids.
Colchicine: an alkaloid extracted from *Colchicum autumnale* L. Because it binds specifically to tubulin, C.prevents "treadmilling" of microtubules (see Cytoskeleton), including those of the mitotic or meiotic spindle. In plants, C.induces polyploidy by preventing the separation of chromosomes during cell division. In small doses, C.relieves pain and suppresses inflammation, but it is highly toxic, 20 mg being a lethal dose. It has been used against neoplastic growth.

Colchicum alkaloids: a group of Isoquinoline alkaloids (see) in which the nitrogen is present as a substituted amino group on a tricyclic skeleton. The latter consists of an aromatic ring, a tropolone ring and another 7-membered ring. C.a. are synthesized only by a few genera of the lily family. The main comes from *Colchicum autumnale* L., the meadow saffron. The main representatives of the group are Colchicine (see) and *demecolcine*; the former often

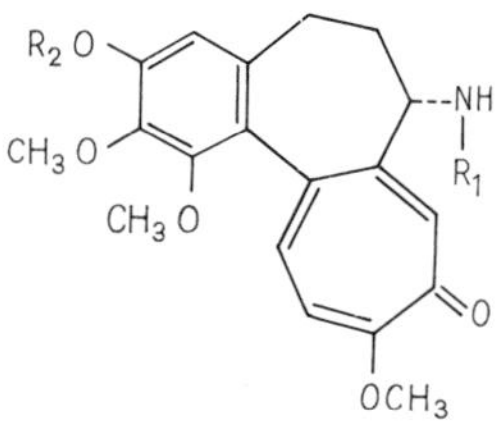

	R₁	R₂
Demecolcine	CH₃	CH₃
Colchicine	COCH₃	CH₃
Colchicoside	COCH₃	Glucose

Figure 1. The most common colchicum alkaloids

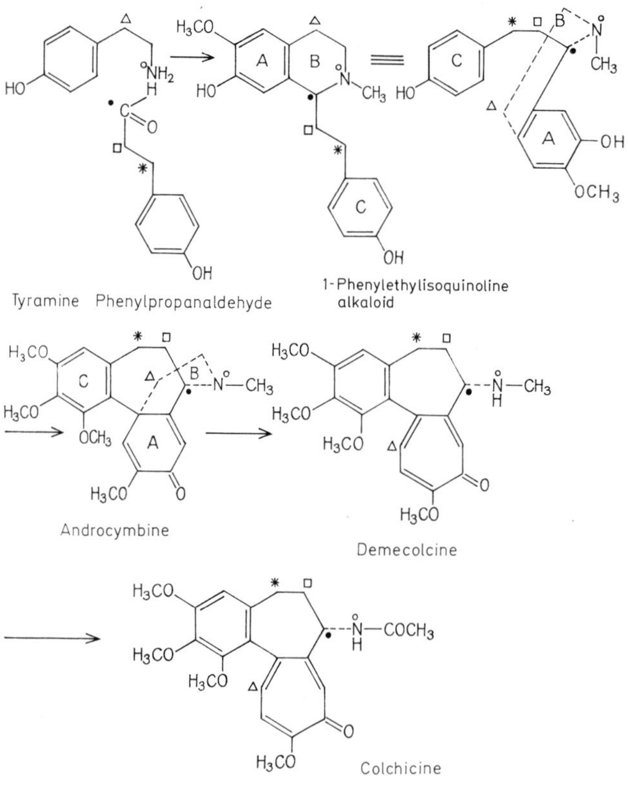

Tyramine Phenylpropanaldehyde

1-Phenylethylisoquinoline alkaloid

Androcymbine

Demecolcine

Colchicine

△,✳,□, O, • corresponding atoms

Biosynthesis of colchicine

occurs as the glucoside (colchicoside). In the light, the tropolone ring of the C. a. may rearrange to a C_4 and a C_5 ring (*lumicolchicines*).

The C. a. are biosynthesized via a 1-phenylethyl-isoquinoline alkaloid from which androcymbin is formed by hydroxylation, methoxylation, attack of phenol oxidases and oxidative couplings. Androcymbin is then converted to demecolcine and further to colchicine (Fig.).

Cold-sensitive enzymes: a group of oligomeric (consisting of several polypeptide chains) enzymes, of which about 25 are presently known, which lose their stability and thus enzymatic activity as the temperature is decreased. The cause of this effect, which is often reversible, is the dissociation of the enzymes into their inactive subunits due to the weakening of hydrophobic and/or electrostatic and ionic interactions. Examples of this class are the mitochondrial adenosine triphosphatase and pyruvate carboxylase, and glyceraldehyde-phosphate dehydrogenase, yeast pyruvate kinase, fructose bisphosphatase and carbamylphosphate synthetase from muscle.

Colicins: see Toxic proteins.

Collagen: an extracellular protein which is responsible for the strength and flexibility of connective tissue. It accounts for 25 to 30% of the protein in an animal. C. in its mature form is insoluble under physiological conditions, although it can be readily denatured by heat, mild acid or alkaline treatment.

Structure: C. is arranged in fibrils visible under a light microscope, which are seen under an electron microscope to be composed of microfibrils. These have a characteristic striation with a repeat distance of about 670 Å, due to the end-to-end alignment of the basic molecular unit, tropocollagen (Fig.1). The tropocollagen molecule is a right-handed triple helix composed of two identical polypeptide chains (α_1) and one slightly different (α_2) chain. Each α-chain is itself a left-handed helix with a pitch of 9.5 Å, while the superhelix has a pitch of 104 Å. The helical and superhelical structures are stabilized by hydrogen bonds between the HN group of glycine in one chain and the O=C group of a proline or other amino acid in an adjacent chain. The tropocollagen molecules are also crosslinked (see below). The unusual structural properties of collagen are due to its amino acid sequence. The α-chains consist of about 1000 amino acids each, and thus have a molecular weight of about 100000. The sequence can be summarized as $(Gly-X-Y)_n$, where the X is frequently Pro, and the Y is often Hyp (hydroxyproline).

Biosynthesis: Collagen is synthesized in fibroblasts as a precursor, procollagen, which also consists of three chains. Each of these has a M_r of about 140000. The polypeptides synthesized on the ribosomes do not contain hydroxyproline or hydroxylysine; these are generated as post-translational modifications before the procollagen is extruded. Some of the hydroxylysines are further modified intracellularly by the addition of galactose or glucosylgalactose to the hydroxyl group. The chains are then extruded into the intracellular space. There N-terminal peptides of M_r 20000 and C-terminal peptides of 35000 are removed by two procollagen peptidases.

The remaining tropocollagen then spontaneously forms microfibrils. Specific residues of lysine and hydroxylysine are oxidized by lysyl oxidase, which converts their side-chain $-NH_2$ to an aldehyde. These aldehydes spontaneously undergo Schiff base and aldol condensations with neighboring side chains, forming a variety of crosslinks that contribute to the strength of C. (Fig.2).

Collagen disorders: There are a number of hereditary and environmental disorders due to impairment of collagen synthesis. Hereditary enzyme deficiencies are responsible for the Ehler-Danlos syndrome, in which the skin is hyperextensible; the Marfan syndrome, characterized by a tendency of the aorta to rupture; osteogenesis imperfecta, in which the bones are very brittle; and dermatosparaxis, a cattle disease in which the skin is very brittle. Since ascorbic acid is required for the formation of hydroxyproline, a dietary deficiency (scurvy) of the vitamin blocks collagen formation. Lathyrism, a condition resulting from ingestion by young animals of sweet peas (*Lathyrus odoratus*), of nitriles, or from copper deficiency, is due to inhibition of lysyl oxidase and the consequent lack of crosslinks in the collagen. Degradation of normal collagen fibers is involved in rheumatoid arthritis, osteo-arthrosis, scleroderma and alkaptonuria. The increasing brittleness of skin and bones with age appears to be due to progressive formation of crosslinks, but the nature of these is not certain.

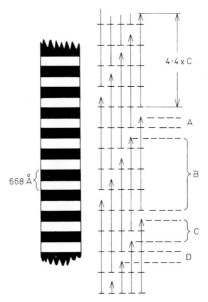

Figure 1. Schematic structure of the collagen microfibril. A: a region of short overlap. B: a long overlap region. C: an overlap region corresponding to one hole zone and one region of short overlap, and giving rise to the distance of 668 Å on the banded structure of the microfibril. D: a hole zone. Each single arrow (length 4.4 × C) represents a tropocollagen molecule.

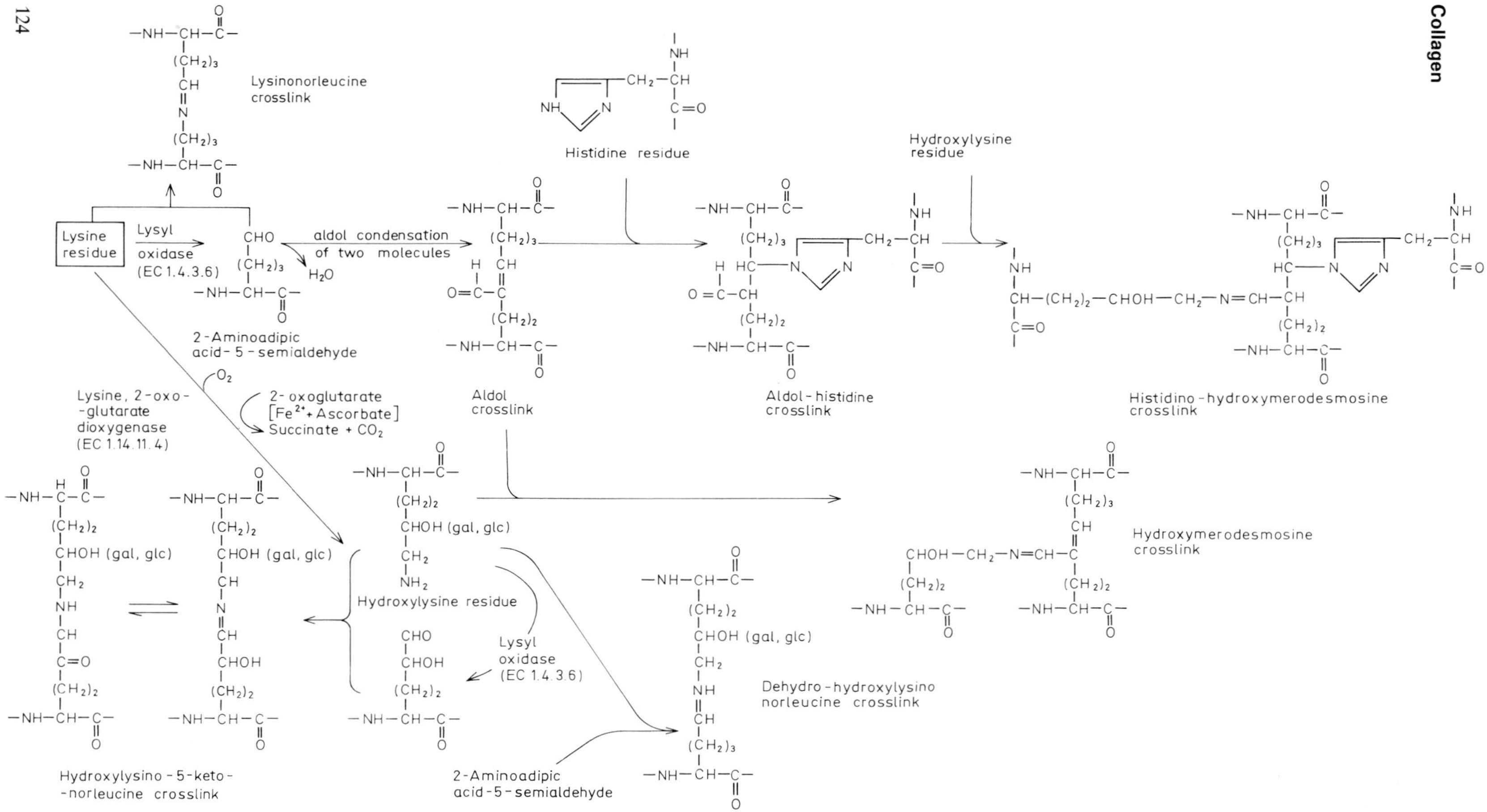

Figure 2. Formation and structure of collagen crosslinks. The lysinonorleucine crosslink is common to both collagen and elastin. Aldol crosslinks occur only at the N-termini of collagen chains. Hydroxylysino-5-keto-norleucine crosslinks (with or without carbohydrate) are most abundant in mineralized collagens. Most collagens contain large amounts of dehydro-hydroxylysinonorleucine crosslinks. The aldol-histidine is only found in appreciable quantities in cow skin colla-

Collagenase: a proteolytic enzyme, the only enzyme capable of degrading native collagen to soluble, low-molecular-weight peptides. Of the more than 20 enzymes from bacteria, fungi, arthropods, amphibia and mammals which have been described as C., only the extracellular C. of the bacterium *Clostridium histolyticum* and the C. from the skin and tail of the tadpole have been examined in detail. The main point of attack of the C. from *Clostridium histolyticum* on the collagen chain is the peptide bond in front of a Gly-Pro sequence. The bonds in front of Gly-Leu and Gly-Ala are hydrolysed to a lesser extent, e. g. in the α_1-chain.

The M_r of the two C. in *Clostrigium histolyticum* are 105 000 for C. A., and 57 000 for C. B. The inactive subunit has a M_r of 25 000.

↓Gly(16) Pro Ser↓Gly Pro Arg↓Gly Leu Hyp↓Gly Pro Hyp↓GlyAla Hyp↓Gly Pro Gln(33)
Points of attack of the collagenase from Clostridium histolyticum on the collagen chain.

Colony stimulating factors, *CSFs:* a group of glycoproteins active in stimulating proliferation and differentiation of hematopoietic progenitor cells, at concentrations of 10^{-11} to 10^{-13} M. Functional subclasses of CSF are M-CSF (macrophage-stimulating), G-CSF (granulocyte-stimulating), GM-CSF (granulocyte- and macrophage-stimulating) and multi-CSF (also called interleukin 3, burst promoting activity, P-cell stimulating factor and hem(at)opoietic cell growth factor), which also stimulates proliferation of eosinophils, megakaryocytes, erythroid and mast cells, as well as neutrophilic granulocytes and macrophages. Impure material is also called colony stimulating activity (CSA) or macrophage-granulocyte inducer (MGI). CSFs are probably produced by all animal tissues and by most normal cell types. Pluripoietin (also called human pluripotent hematopoietic colony stimulating factor, or pluripotent CSF) is produced constitutively by human bladder carcinoma cell line 5637; it supports growth of human mixed colonies, granulocyte/macrophage colonies, and early erythroid colonies, and induces differentiation of human promyelocytic leukemic cell line HL-60 and the murine myelomonocytic leukemic cell line WEHI-35 (D+).

Primary structures of several CSFs have been reported, based on partial protein sequencing and on sequence prediction from cloned cDNA: murine GM-CSF (M_r 23 000) [N. M. Gough et al. *Nature* **309** (1984) 763-767]; murine multi-CSF (M_r 23-28 000) [M. C. Fung et al. *Nature* **307** (1984) 233-237; T. Yokata et al., *Proc. Natl. Acad. Sci. U. S. A.* **81** (1984) 1070-1074]; human GM-CSF (M_r 22 000) [G. G. Wong et al. *Science* **228** (1985) 810-815]; human M-CSF (M_r 45 000 of the homodimer; also called CSF-1) [E. S. Kawasaki *Science* **230** (1985) 291-296]. The CSF amino acid sequences determined so far show no homologies, and they display considerable differences in predicted secondary and tertiary structures, which is surprising in view of their extensive functional overlap and the similar response of progenitor cells to different CSFs.

[125]I-labelled CSFs have been used to demonstrate the existence and to determine the number of specific receptor sites on the cell surface. Separate and specific receptors for each murine CSF are coex-

pressed on granulocytes and monocytes, and they show no cross competition for their respective ligands. However, a hierarchical down modulation has been observed, e. g. occupancy of multi-CSF receptor decreases the binding by all other receptors of their specific ligands; occupancy of GM-CSF receptors causes a similar down modulation of G-CSF and M-CSF receptors.

M-CSF receptor is structurally related and probably identical to the c-*fms* proto-oncogen product. It has been suggested that autocrine production of CSFs might be involved in the genesis of myeloid leukemia, but evidence for this is very conflicting.

The receptor for murine M-CSF has been isolated by cross-linking ^{125}I-M-CSF to its receptor, using disuccinimidyl suberate, followed by release from the membrane with detergent. [C. J. Morgan & E. R. Stanley, *Biochem. Biophys. Res. Commun.* **119** (1984) 35-41] Subsequent sequencing and cloning studies on all CSF receptors are to be expected. [A. W. Burgess & D. Metcalf *Blood* **56** (1980) 947-958; D. Metcalf, *Cell* **43** (1985) 5-6]

Colophony: the residue from turpentine distillation, see Balsams.

Column chromatography: a chromatographic separation method, in which the carrier material is packed in the form of a column inside a tube (usually of glass, and known as a chromatography column). The term C. c. therefore indicates the mode of use of the chromatographic material, and does not indicate the type of chromatography, which may be partition, adsorption, ion exchange, gel filtration, affinity, etc., depending on the nature of the column packing. The column packing is usually equilibrated with the running solvent or *elutant*. Sample solution is placed on the top of the column packing and washed into the column with running solvent, followed by a continuous supply of the same solvent. Discrete samples of the effluent or *eluate* are collected from the bottom of the column (usually automatically with the aid of a *fraction collector),* then analysed for separated substances.

For the column materials used in ion exchange C. c., see Ion exchangers. To increase the sharpness and speed of separations, *gradient elution* is often employed in ion exchange C. c.; a gradient mixer, supplied by two or more different buffer solutions, produces a continuous change in the pH or ionic strength (or both) of the running solvent entering the top of the column. Such a gradient may be linear, convex or concave; concentration gradients are usually of increasing concentration; pH gradients may fall (become more acidic) or rise (more alkaline).

Commensalism: the close spatial coexistence of two organisms of different species which is neither beneficial for the partners (commensals), as in symbiosis, nor detrimental to either partner, as in parasitism. C. can at most be described as a sort of spatial parasitism.

Compactin, *ML-236B,* *6-demethylmevinolin:* 1,2,6,7,8,8a-hexahydro-β,δ-dihydroxy-2-methyl-8(2-methyl-1-oxobutoxy)-1-naphthalene-heptanoic acid δ-lactone (for structural formula, see Mevinolin), a fungal metabolite from the culture media of *Penicillium citrinum* and *P. brevicompactum.* The parent hydroxyacid of C. is a potent competitive inhibitor (K_i 1.4 nM) of 3-hydroxy-3-methylglutaryl-CoA reductase (EC 1.1.1.34). C. acts like Mevinolin (see) in reducing plasma LDL cholesterol levels, but it is rather less potent than the latter. [A. Endo et al. *FEBS Letters* **72** (1976) 323–326; A. W. Alberts et al. *Proc. Nat. Acad. Sci.* **77** (1980) 3957–3961]

Compartment: a geometrically bounded portion of the cell which is structurally or biochemically separate from the rest of the cell space. Compartmentation is the division of the cell into areas with different enzymatic equipment. The formation of C. makes possible the simultaneous and temporally organized metabolic processes which, in their complicated interactions, are the basis of life.
Biomembranes (see) are particularly important for compartmentation. They are both barriers to free diffusion of metabolites and a means of communication between separate C., in that they make possible directed transport of material. Thus, the presence of biomembranes results in vectorial biochemical reactions.
There are plasmatic and nonplasmatic C. Both are enclosed by biomembranes like the vacuole and plasma membranes (tonoplast or plasmalemma). Plasmatic C. are surrounded by a double membrane, contain active nucleic acid, are the site of ATP formation and protein synthesis, and form α-glucans (glycogens or starch). The plasmatic C. are the nucleoplasm, the cytoplasm and the organelle plasma in the inner C. of mitochondria and plastids. The above processes do not occur in the nonplasmatic C. If a cell is capable of forming β-glucans (cellulose, callose), these are formed in the nonplasmatic C. The nonplasmatic C. are the vacuoles, the dictyosomes, the various cytosomes like lysosomes, uricosomes, peroxisomes, glyoxysomes, etc. and the outer C. of mitochondria and plastids. The formation of C. can only be partially demonstrated by light or electron microscopy. For example, one can only localize the lysosomal system of the cell histochemically, using detection methods for the lysosomal enzymes. C. and the cell structure or ultrastructure are only partly identical.
In a wider sense, compartmentation is possible through purely chemical means. The embedding of enzyme proteins in lipid layers and aggregation of enzymes to form multienzyme complexes are intermediate between the formation of C. by cell structures and metabolic compartmentation.

Compartmentation: see Compartment; Biomembranes.

Competitive inhibition: see Effectors.

Complementary DNA, cDNA: DNA complementary to a mRNA. It is prepared in the laboratory as a probe for hybridization studies by incubating the mRNA with dATP, dGTP, dTTP and dCTP in the presence of a reverse transcriptase (see RNA-dependent DNA-polymerase).

Complementary structures: two structures which define one another, for example, the two polynucleotide chains in the double helix of DNA. The base pairs adenine and thymine (or uracil, in RNA) and guanine and cytosine are complementary, which results in the fact that the nucleotide sequence in one polynucleotide chain defines a unique sequence in the complementary strand through the formation of base pairs.

Complement binding reaction: the binding of the C1 component of complement to the Fc fragments (see Immunoglobulins) of antibodies which are bound to the surface antigens of erythrocytes or bacteria. The Complement system (see) is activated by this step.

Complement system: A heat-labile (100% inactivation after 30 min at 56 °C) cascade system in the sera of all vertebrates, and composed in mammals of at least 20 glycoproteins, 7 of which control localization of the effect. Each activated component (or complex) is a highly specific protease acting only on the next component of the cascade. Two pathways are involved, the *classical* and the *alternative*. The classical pathway is activated by binding of C1 to immune complexes containing IgG or IgM (C1 binding sites in IgG and IgM reside in the respective Fc regions – see Immunoglobulins – and binding is dependent on Ca^{2+}). C1 consists of 3 subunits: C1q, C1r and C1s. The binding sites for IgG and IgM reside on C1q. Occupation of these binding sites confers proteolytic activity, which cleaves a single peptide bond of C1r. The resulting "activated" C1r in turn hydrolyses a peptide bond in C1s, thus yielding the fully active C1 complex, which initiates the sequential assembly of circulating components into a surface-bound protein complex (Fig.). The alternative pathway is initiated by a repertoire of activators, including antibodies (IgA and IgE) which do not activate the classical pathway when complexed with antigens, high M_r polysaccharides of bacteria and yeasts, fragments of plant cell walls and protozoa. This pathway is thought to provide the initial response to bacterial invasion, since it can be activated in the absence of antibodies. Factor D of the alternative pathway is already proteolytically active in non-activated serum, i. e. its formation from an inactive precursor is not part of the amplification system of the cascade.
Early components of each pathway are primarily concerned with the formation of 2 protein complexes, which function respectively as C3 and C5 convertases. Proteolytic activation of C5 is the final enzymatic event which triggers the spontaneous association of the late components (C6-C9) to form a nonenzymatic lytic complex capable of puncturing cell membranes (Figs. 1 & 2). Thus, the main function of the C. s. is lysis of foreign invasive cells. Also, through the anaphylatoxic and chemotactic properties of some components (notably C3a, C4a and C5a), foreign cells are rendered susceptible to phagocytosis. Particulate antigens bound to C3b and IgG are engulfed by phagocytosis. The C. s. is also involved in solubilization of immune complexes, and in development of the cellular immune response. Deficiencies of C. s. components are associated with repeated bacterial infections, and are also implicated in certain autoimmune diseases.
The amino acid sequences of most C. s. components have been determined by a combination of protein and cDNA sequencing techniques.
For experimental purposes, either fresh guinea pig serum or serum from patients genetically deficient in one component is used as a source of C. s. (hemolyt-

ic system). Sensitized sheep erythrocytes serve as immune aggregates for study of the classical pathway, and rabbit erythrocytes for the alternative pathway. [R.R. Porter, P.J. Lachmann & K.B.M. Reid (eds.) *Biochemistry and Genetics of Complement* (Cambridge University Press, 1986)]

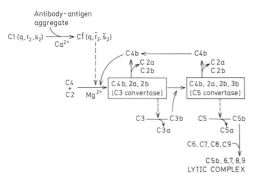

Figure 1. Classical pathway of complement activation. A bar over the number or letter of a factor (e.g. C1̄) indicates that the factor is activated (i.e. proteolytically active). Activation by proteolysis occurs near to the *N*-terminus, producing a small (a) and a large (b) cleavage product, e.g. C3 → C3a + C3b.

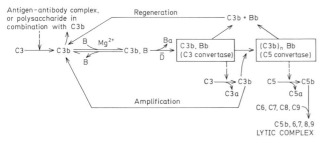

Figure 2. Alternative pathway of complement activation. See legend to Fig. 1.

death in humans at a dose of 0.5 to 1 g. The poison hemlock was used in ancient Athens to put Socrates to death.
The synthesis of C. from α-picoline and paraldehyde by Ladenburg in 1886 was the first synthesis of any alkaloid.

Conium alkaloids: a group of simple piperidine alkaloids found only in poison hemlock, *Conium maculatum*. The main alkaloids are Coniine (see) and γ-coniceine (M_r 125.22, b.p. 168 °C); the secondary alkaloids are *N*-methyl and hydroxy derivatives of coniine. In contrast to other piperidine alkaloids, the ring system of the C. is synthesized from acetate rather than from lysine (Fig.).

Conotoxins: a series of peptide neurotoxins isolated from marine, fish-hunting snails of the genus *Conus*. They contain 13 to 29 amino acids, are strongly basic, and highly cross-linked by disulfide bonds. The ω-C. inhibit voltage-activated entry of calcium into the presynaptic membrane and thus the release of acetylcholine; the α-C. inhibit the postsynaptic acetylcholine receptor, and the μ-C. prevent muscle action potentials. Their small size is probably an advantage in promoting rapid diffusion through the tissues of the prey; fish stung by the snails are paralysed within seconds. C. are of interest for the elucidation of nerve mechanisms. [B.M. Olivera et al., *Science* **230** (1986) 1338–1343]

Constitutive enzymes: enzymes which, in contrast to the inducible enzymes, are constantly produced by the cell, irrespective of the growth conditions.

Contraceptives: see Ovulation inhibitors.
Contractile proteins: see Muscle proteins.
Convallatoxin: see Strophanthins.
Cooperative oligomeric enzymes, *allosteric enzymes:* enzymes composed of several subunits and displaying cooperativity.

Cooperativity: this is displayed by oligomeric or monomeric enzymes which possess more than one binding site for a certain ligand. The cooperative binding may be negative or positive, and it may

Conalbumin: see Siderophilins.
Concanavalin A: see Lectins.
Concentration variables, *fundamental variables, primary variables:* those substances in an enzymatic system whose concentrations can be directly controlled by the experimenter, for example the substrates, products and effectors. They are thus to be distinguished from the enzyme species, the concentrations of which can be calculated from the kinetic equations at steady state for the given values of the C.v. Usually, in kinetic experiments, one C.v. is varied and the others are held constant.

Conchiolin: see Paleoproteins.
Concretion oils: see Essential oils.
Conessine: see Holarrhena alkaloids.
γ-Coniceine: see Conium alkaloids.
Coniine, *2-propylpiperidine:* the most important of the Conium alkaloids (see). M_r 127.22, m.p. −2.5 °C, b.p. 166 °C, $[\alpha]_D^{17} \pm 16$ °C is the toxic-principle of the poison hemlock, *Conium maculatum*. The largest quantities of C. are found in the unripe seeds. It is extremely poisonous and can cause

occur for the same ligand (homotropic cooperativity) or for a different ligand (heterotropic cooperativity). Purely phenomenologically, this means that the dissociation constant for each successive ligand bond is lower (positive cooperativity) or higher (negative C.) than the preceding one. C. is also involved if the binding of one substrate or effector molecule changes the configuration and thereby the reactivity, or catalytic constant (see Michaelis-Menten equation) for other substrate molecules. The degree of C. is usually determined from a Hill plot (see).

The binding curves (saturation curves) are sigmoidal (S-shaped) in the case of positive C. Various models have been developed to describe C. (see Cooperativity model). It is assumed that C. is caused by changes in the three-dimensional structure of the enzyme protein and that each subunit of an oligomeric enzyme can exist in at least two configurations which react differently with effector molecules. Further, the change in configuration of one subunit is thought to induce changes in the configurations of the other subunits in the same molecule. The most general sigmoidal rate equation is: $v^{-1} = a + bS^{-1} + cS^{-2} + \ldots$ Such an equation, and thus sigmoidicity of the binding curves, can also be obtained from a number of mechanisms which are not included in the C. model.

Cooperativity model: a functional and structural model of cooperative oligomeric enzymes which is intended to describe and explain the Cooperativity (see) in the turnover of substrates or binding of effectors. The C.m. can be used to derive binding potentials and equations which can be compared with the experimental data. A number of hypotheses must be made in order to arrive at special C.m. for particular enzymes: 1) the number of subunits, 2) the geometric hypothesis, or arrangement of the subunits (tetrahedral, square planar, etc.), 3) the number of configurations of the subunits and the nature of their interactions with effectors (configurational hypothesis), and 4) the way in which the change in the configuration of one subunit affects the other subunits (interaction hypothesis).

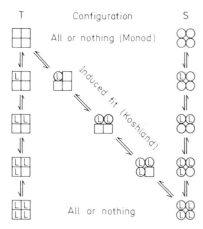

2-Configuration cooperativity model of a tetrameric enzyme. L, ligand (substrate or effector).

The C.m. is characterized by these hypotheses. The C.m. for a tetrameric enzyme, for which the Monod model and the Koshland models are limiting cases (Fig.), is often called the general C.m. This assumes only two subunit configurations (S and T), but even this is somewhat specialized. The configurational hypothesis of the Monod model (Monod-Wyman-Changeux model, MWC model, all-or-nothing-model) presumes 2 configurations (the S and T states of the subunits), in which the effector binding of the second state can be exclusive or nonexclusive. The configurational hypothesis of the Koshland model (Adair-Koshland-Nemethy-Filmer model, abb. AKNF model, induced-fit model) assumes induced fit, i.e. the ligand-binding configuration is present only in negligible quantities in the absence of ligands and is induced by the presence of the latter. The interaction hypothesis postulated by the Monod model is an all-or-nothing transition of the subunit configuration. This is also called a concerted transition. The Koshland model ascribes to each contact between subunits a free energy of interaction which is characteristic for the 2 configurations of the subunits.

Copolymer: a polymeric molecule containing more than one kind of monomer unit. In biochemistry it often refers to polynucleotides made by incubating two or more nucleoside di- or triphosphates with the appropriate polymerase. Using RNA polymerase, Khorana was able to synthesize C. according to the following scheme (the choice of starter molecule was arbitrary and could be varied according to need):

$$AUG + [ATP + UTP + GTP] \xrightarrow{\text{RNA polymerase}}$$
(starter)

$$AUGAUGAUGAUG\ldots$$

The synthesis of such C. with defined triplets played a major role in the cracking of the genetic code.

Copper, Cu: an important bioelement, frequently involved in electron transport processes in membranes and mitochondria. In plants, Cu is necessary for the synthesis of chlorophyll and is a component of a number of enzymes (see Copper proteins). Large amounts of Cu are toxic, although tolerant strains of plants and microorganisms arise near copper mines and dumps. The threshold limits of exposure for workers in foundries and smelters in the United States are 0.1 mg/m^3 fumes and 1.0 mg/m^3 dust and mists. The recommended daily intake for humans is 2 mg, and the amount in the body is 100 to 150 mg. The highest concentrations are found in the liver and the bones. Mammalian blood contains a number of copper proteins. The synthesis of hemoglobin is dependent on Cu, although it does not contain the metal. The oxygen-transport pigments in mollusc and crustacean blood contain Cu rather than Fe. [C. A. Owen, Jr. *Copper Deficiency and Toxicity* (Noyes Publications, Park Ridge, N.J., 1981)].

Copper proteins: metalloproteins, often blue in color, which usually contain a mixture of mono- and mostly divalent copper in their molecules. Exceptions are the plastocyanins of chloroplasts (M_r 11 000, 2 Cu^{2+}) and Hemocyanins (see), the oxygen-transport proteins of arthropod and molluscan

blood, which contain only Cu^{2+} or Cu^+. With the exception of the copper thiolate proteins (see Metallothioneins), the known C.p. are either enzymes for which oxygen is a substrate, or they are oxygen transport proteins. Even Ceruloplasmin (see), which was long thought to be merely a transport protein, has oxidase activity with unsaturated compounds, including indoles. Some of these oxidase reactions incorporate the entire O_2 molecule into hydrogen peroxide, e.g. the Cu-containing Amine oxidases (see) catalyse the reaction $RCH_2NH_2 + O_2 + H_2 \rightarrow RCHO + H_2O_2 + NH_3$, and galactose oxidase catalyses $O_2 + galactose \rightarrow H_2O_2 + galactohexosedialdose$. A more common type of Cu oxidase incorporates only one atom of the O_2 into product, while the other is reduced to water. Some important examples are: 1) dopamine β-monooxygenase, EC 1.14.17.1, which hydroxylates dopamine to noradrenalin (ascorbate donates the hydrogens for reduction of the second O atom);
2) the monophenol monooxygenases, EC 1.14.18.1, which oxidize tyrosine and other phenols;
3) the laccases (EC 1.10.3.2) found in higher plants and fungi, especially white rot fungi, are involved in the metabolism of Lignin (see); and
4) classical tyrosinase (catechol oxygenase, E.C.1.10.3.1) catalyses the first step in the synthesis of Melanin (see) from tyrosine. Cytochrome *c* oxidase (EC 1.9.3.1) and ascorbate oxidase (1.10.3.3) reduce O_2 to 2 molecules of H_2O. Superoxide dismutase (EC 1.15.1.1; see) has an unusual substrate, the superoxide radical ion. See Azurin. [R.Lontie (ed.) *Copper Proteins and Copper Enzymes* (CRC Press, Boca Raton, 1984); T.G. Spiro (ed.) *Copper Proteins* (Wiley, New York, 1981)]

Coprogen: a Siderochrome (see) synthesized by *Penicillium* and *Neurospora* spp.

Coprostane: obsolete term for 5β-cholestane, see Steroids.

Coprostanol: 5β-cholestan 3β-ol, a sterol alcohol. M_r 388.64, m.p. 101 °C, $[\alpha]_D + 28$ ° (chloroform). C. is the main sterol in the feces, where it arises by reduction of cholesterol by intestinal bacteria. C. differs from its stereoisomer Cholestanol (see) in the configuration at C-atom 5 and has A/B *cis* bonding.

Cordycepin, 3'-deoxyadenosine, adenine 9-cordyceposide: a purine antibiotic synthesized by *Cordyceps militaris* and *Aspergillus nidulans* (see Nucleoside antibiotics). M_r 251.24, m.p. 225-226 °C, $[\alpha]_D^{20} = -47$ ° (water). C. was first isolated in 1951 by Cunningham. As an antimetabolite of adenosine, it inhibits purine biosynthesis, but is not very toxic. It can be phosphorylated to the monophosphate.

Core particles: particles released from Chromatin (see) by partial enzymatic digestion. They contain about 140 base pairs of DNA and 2 molecules each of the "inner histones" H2A, H2B, H3 and H4. C.p. are separated by about 40 base pairs of DNA associated with the "outer histones" H1 and H5. See also Nu bodies.

Corepressor: see Enzyme repression.

Coriandrol: see d-Linalool.

Cori ester: see Glucose 1-phosphate.

Corilagin: see Tannins.

Coronatine: a chlorosis-inducing toxin produced by several members of the *Pseudomonas* ge-

nus of bacteria. These infect a variety of grasses, notably Italian ryegrass, and soybeans. The bioassay of C. is based on its ability to cause hypertrophic growth of potato tuber tissue. Structural and conformational studies of C. have all been performed on material from liquid cultures of bacteria. Evidence for the isolation of C. from infected leaves of Italian ryegrass is fairly conclusive (chromatography, activity in biological test), but confirmatory mass spectroscopy is lacking. Later work on culture filtrates of *Pseudomonas syringae* [R. E. Mitchell, *Phytochemistry* **23** (1984) 791-793] revealed that C.is accompanied by approximately equal quantities of a structural analog, *N*-coronafacoylvaline, which is also a chlorosis-inducing toxin.

Coronatine [A. Ichihara et al., *Tetrahedron Lett.* No.4 (1979) 365-368]

N-Coronafacoylvaline

Corpus luteum hormone: see Progesterone.

Corrinoids: chemical compounds based on the corrin ring, which is similar to the porphyrin ring. It consists of four pyrrole rings linked in a large ring. Three of the links between pyrroles are methenyl groups; the fourth is a direct bond between the two pyrroles. The pyrroles carry acetate, propionate and methyl substituents. The central atom is a covalently bound cobalt. Vitamin B_{12} is a C. (see Vitamins).

Cortexolone, *Reichstein's substance S, 11-deoxycortisol:* 17α-hydroxy-11-deoxycorticosterone, 17α,21-dihydroxypregn-4-ene-4,20-dione, a mineralocorticoid from the adrenal cortex. For structure and biosynthesis, see Adrenal corticosteroids.

Cortexone, *Reichstein's substance Q, 11-deoxycorticosterone,* abb. **DOC:** 21-hydroxypregn-4-ene-3,20-dione, a mineralocorticoid from the adrenal cortex. The acetate or glucoside is used in the treatment of Addison's disease, and of shock. For structure and biosynthesis, see Adrenal corticosteroids.

Corticoids: see Adrenal corticosteroids.

Corticosteroids: see Adrenal corticosteroids.

Corticosterone, *Reichstein's substance H, Kendall's substance B:* 11β,21-dihydroxypregn-4-ene-3,20-dione, an adrenal cortex hormone and a glucocorticoid. C. is biosynthesized from progesterone, which is hydroxylated first to cortexone and then in the 11 position to C. For structure and biosynthesis, see Adrenal corticosteroids.

Corticotropin, *adrenocorticotropin, adrenocorticotropic hormone,* abb. *ACTH:* a polypeptide hormone secreted by the pituitary. The primary structure of the human hormone is Ser-Tyr-Ser-Met-Glu-His-Phe-Arg-Trp-Gly-Lys-Pro-Val-Gly-Lys-Lys-Arg-Arg-Pro-Val-Lys-Val-Tyr-Pro-Asn-Gly-Ala-Glu-Asp-Glu-Leu-Glu-Phe, M_r 4541. Only the sequence from 31 to 33 is species-specific; the biological activity is determined by the first (fixed) 20 amino acids. The sequence of the first 13 amino acids is identical to that of α-melanotropin (see Melanotropin).

C. is the smallest of the hormones produced by the anterior lobe of the pituitary. It is made in the γ-cells when they are stimulated by corticotropin releasing hormone (see Releasing hormones). The target organ of C. is the adrenal cortex, which it stimulates to growth and increased production of glucocorticoids. This action is mediated by the adenylate cyclase system.

The concentration of C. in the blood can be determined by radioimmunological techniques. It lies in the range of ng/ml, and fluctuations are observed with daily and seasonal periodicities. The blood concentration of C. is controlled via the hypothalamus, in response to the blood level of glucocorticoids.

The total synthesis of C. was reported in 1963 by Schwyzer.

Corticotropin-like peptide, *CLIP:* a peptide from the pars intermedia of rat and pig pituitaries. CLIP is composed of the amino acid sequence 18–39 of ACTH. No definite function of CLIP is known. It may be a cleavage product of ACTH unavoidably formed during production of α-MSH. See Peptides, Fig. 3.

Corticotropin releasing hormone: see Releasing hormones.

Cortine: see Adrenal corticosteroids.

Cortisol, *Reichstein's substance M, Kendall's substance F:* 11β,17α,21-trihydroxypregn-4-ene-3,20-dione, a glucocorticoid hormone from the adrenal cortex. C. is found in the blood. For structure and biosynthesis, see Adrenal corticosteroids.

Cortisone, *Reichstein's substance F, Kendall's substance E:* 11-dehydro-17α-hydroxycorticosterone, 11α,21-dihydroxypregn-4-ene-3,11,20-trione, a glucocorticoid hormone from the adrenal cortex. C. differs from Cortisol (see) in having a keto group instead of a hydroxy in the 11 position. Like cortisol, it stimulates the formation of carbohydrate from proteins, promotes glycogen storage in the liver and raises the blood sugar level. It is biosynthesized from cortisol by enzymatic dehydrogenation of the 11β-hydroxyl group. C. can be obtained by partial synthesis from other pregnane compounds, bile acids and steroid sapogenins. The first total synthesis was reported in 1951 by Woodward. C. was first isolated from the adrenal cortex in 1935 by Reichstein (substance F), Kendall (substance E) and Wintersteiner (substance F) simultaneously. It is also found in the blood and urine. *Cortisone acetate* is used as a drug for rheumatic arthritis (Hench and Kendall, 1948/49) and allergic skin reactions, but it is surpassed in these properties by synthetic derivatives such as prednisone (see Prednisolone) and triamcinolone. For structure and biosynthesis, see Adrenal corticosteroids.

Corynantheine: see Yohimbine.

Cosmid: a DNA molecule made by fusing DNA from a phage and a bacterial plasmid and used as a cloning vector (see Recombinant DNA technology). It contains the plasmid gene(s) for antibiotic resistance (for selection purposes), the plasmid replication origin, and the *cos* site of the phage. The *cos* site is required for packaging of the DNA into the phage protein coat. Many phage genes are lacking. After transfection, Cs. become packaged in multiple, infective, phage-like particles within the host bacterial cell, but cell lysis does not occur. Cs. are relatively small, consisting of about 8 kb, and they can carry more than 50 kb of foreign DNA by insertion.

Cosubstrate: a Coenzyme (see), which enters an enzymatic reaction as a second substrate.

Cotransport: see Membrane transport.

Cotylenins: leaf growth-promoting substances from *Cladosporium*. All C. are glycosides of cotylenol. The absolute stereochemistry of cotylenol is the same as that of the Fusicoccins (see), with which they probably share an essentially identical biosynthetic pathway. The formulas of known C. are shown (Fig.). Previously isolated C. B, D and G are now known to be artifacts. [T. Sassa et al. *Agric. Biol. Chem.* **39** (1975) 1729–1734 (A, C, E); T. Sassa & A. Takahama *ibid.* (1975) 2213–2215 (C, F); A. Takahama et al. *ibid.* (1979) 647–650 (H, I); A. Bottalico et al. *Phytopathol. Mediterr.* **17** (1978) 127–134 (structure-activity relationships)]

Cotylenins structures: cotylenol (R = H), cotylenin A, cotylenin C, cotylenin E, cotylenin H, cotylenin I, cotylenin F

Couepic acid: see Licanic acid.

4-Coumarate:CoA ligase, *hydroxycinnamoyl-CoA ligase* (EC 6.2.1.12): a plant enzyme catalysing a two-step process:

$$\text{Enzyme} + \text{R-CH}=\text{CH-COOH} + \text{ATP} \xrightarrow{\text{Mg}^{2+}} \text{Enzyme [R-CH}=\text{CH-CO·AMP]} + \text{PPi}$$

Enzyme[R-CH=CH-CO·AMP]+CoA-SH ⇌ R-CH=CH-COS-CoA+AMP+Enzyme where R-CH=CH-COOH represents *trans*-4-hydroxycinnamic acid. From most sources, the enzyme has M_r in the range 55000-67000. Isoenzymes have been detected and separated by ion exhange chromatography. The enzyme from any source is inhibited competitively by AMP. An increase in activity is associated with phytoalexin production in response to biotic elicitor from *Phytophthora megasperma* in *Glycine max*, and in cell suspension cultures of *Phaseolus vulgaris*. The catalysed reaction is an early step in Flavonoid (see) biosynthesis. [R. A. Dixon et al. *Advances in Enzymology* 55 (1983) 1-136]

Coumarins: lactones of *cis*-o-hydroxycinnamic acid derivatives, which are widely distributed in plants (Table), especially in the *Umbelliferae* and *Rutaceae*. Their lactone structure is opened by treatment with alkali to give *cis*-o-hydroxycinnamic acids, which spontaneously recyclize in acid. Most C. are formal derivatives of umbelliferone, i.e. they are hydroxylated at C-7. Some also possess a second

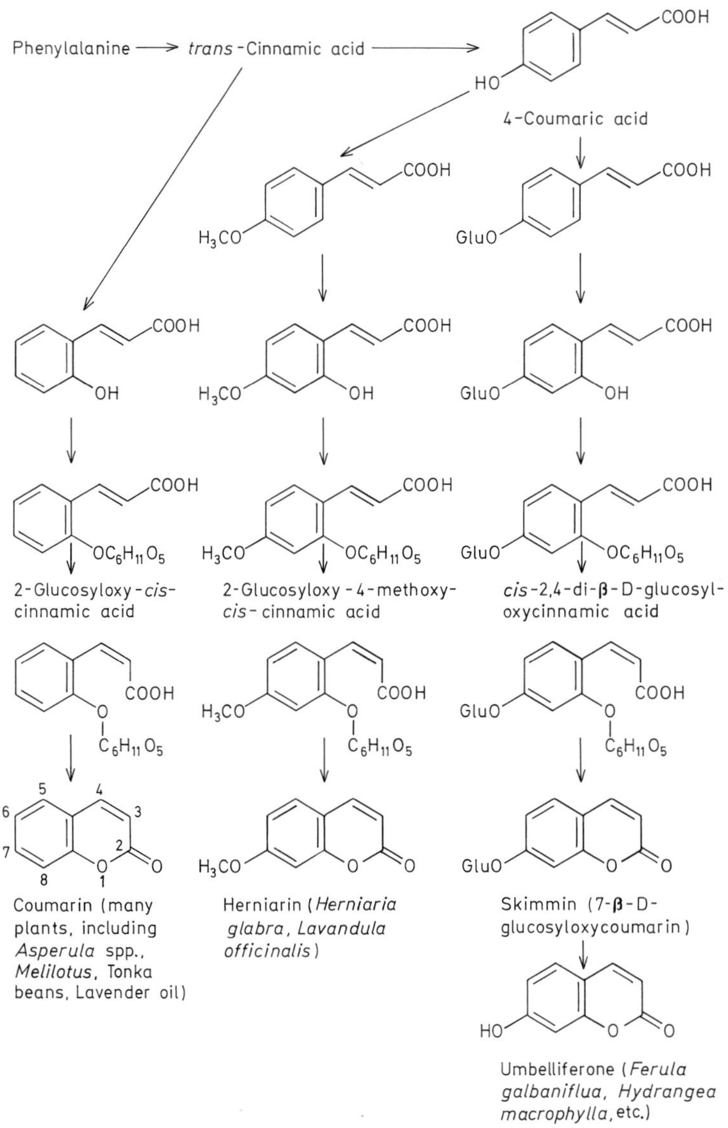

Figure 1. Biosynthesis of coumarins. *Trans*-Cinnamic acid is converted to 4-coumaric acid by Cinnamate 4-hydroxylase (see). Cyclization to the coumarin ring system occurs spontaneously after removal of the 2-glucosyl group by a specific glucosidase when the plant is damaged.

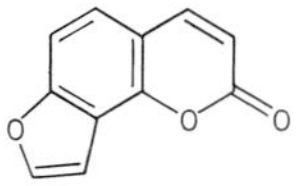

OH-group (or alkoxyl group) at C-5 or, more rarely, at C-4.

Coumarin itself (2H-1-benzopyran-2-one; 1,2-benzopyrone; *cis*-o-coumarinic acid lactone; o-hydroxycinnamic acid lactone) is a pleasant smelling compound.

Biosynthesis. C. are products of the shikimic acid pathway of Aromatic biosynthesis (see). The key intermediate is *trans*-cinnamic acid, which may be converted either to coumarin itself, or become hydroxylated in the *para* position as a prelude to the synthesis of other C. (Fig. 1). The appropriately substituted *trans*-cinnamic acid precursor is then hydroxylated in the *ortho* position. Glycosylation of this hydroxyl group may be important for the subsequent *trans-cis* isomerization of the side chain, because strong intramolecular hydrogen bonding between the carboxyl group and 2-hydroxyl group of the glucose residue is only possible in the *cis*-form of the 2-glucosyloxycinnamic acid. There is some evidence for the existence of an isomerase enzyme, but it seems probable that the isomerization is largely photo-catalysed. Cinnamic acid 4-hydroxylase (see), present in the microsomal fraction of disrupted cells, converts *trans*-cinnamic acid to p-coumaric acid, whereas the *ortho* hydroxylation is catalysed by an enzyme present in chloroplasts and stimulated by NADPH and 2-amino-4-hydroxy-6,7-dimethyl-5,6,7,8-tetrahydrobiopteridine. [H. Kindl, *Z. physiol. Chem.* **352** (1971) 78–84] It is noteworthy that the joint occurrence of coumarin itself and 7-hydroxylated C. is rare, i.e. synthesis normally proceeds to 7-hydroxylated C., unless the membrane system catalysing the *para* hydroxylation of *trans*-cinnamic acid is absent or of low activity.

Free C. do not occur in significant amounts in healthy plant tissue. When plant cells are damaged or the plant wilts, specific glucosidases remove glucose from the 2-glucosyloxy-*cis*-cinnamic acid precursors, which then spontaneously cyclize to the corresponding C. Some C. may also be stored as glucosides, if an hydroxyl group is available for glucose

Angelicin   **Angustifolin**

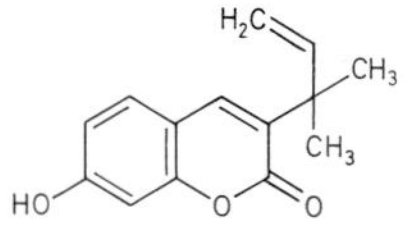

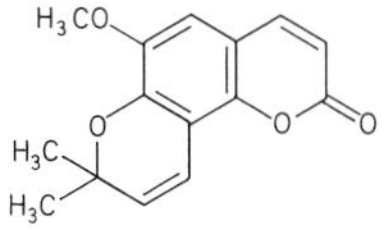

Braylin   **Ceylantin**

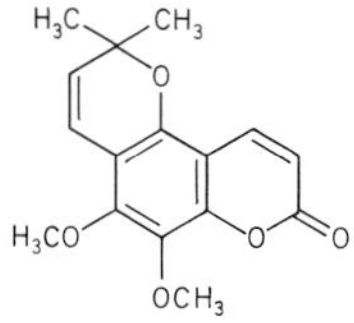

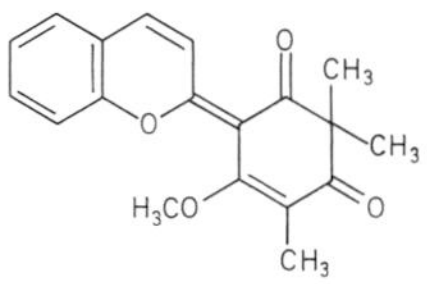

Dalrubone   **Pereflorin B**

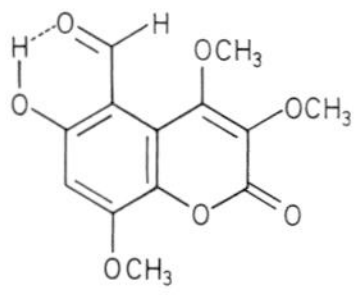

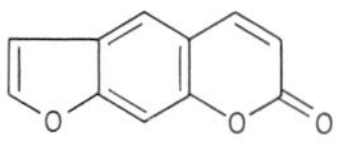

Psoralen   **Suberosin**

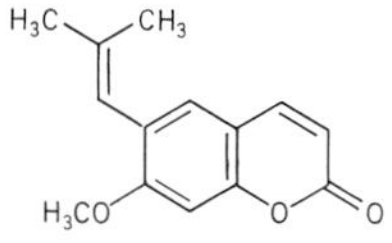

Structures of some coumarins
(see Table for further description).

attachment; thus, both skimmin and *cis*-2,4-di-β-D-glucosyloxycinnamic acid represent stored precursors, which are converted into umbelliferone when leaves of *Hydrangea macrophylla* are damaged. [D. J. Austin & M. B. Meyer, *Phytochemistry* **4** (1965) 255–262]

Chloroplasts of *Saxifraga stolonifera* convert *cis*-caffeic acid into esculetin. [M. Sato, *Phytochemistry* **6** (1967) 1363–1373] The enzyme responsible is a phenolase (ortho-diphenol:O$_2$ reductase), which can be prepared in soluble form by detergent treatment of chloroplasts. It is proposed that the *ortho*-quinone from *cis*-caffeic acid converts spontaneously to esculetin, following spontaneous hydroxylation by water in the *ortho* position (Fig. 2). During synthesis in the plant, the 3,4,6-trihydroxy-*cis*-coumaric acid is presumably trapped as its 2-glucosyloxy derivative. A different biosynthetic route has been proposed for dalrubone, and it has been suggested that some C. may arise from the A ring of anthocyanins. [D. L. Dreyer et al., *Tetrahedron* **31** (1975) 287–293]

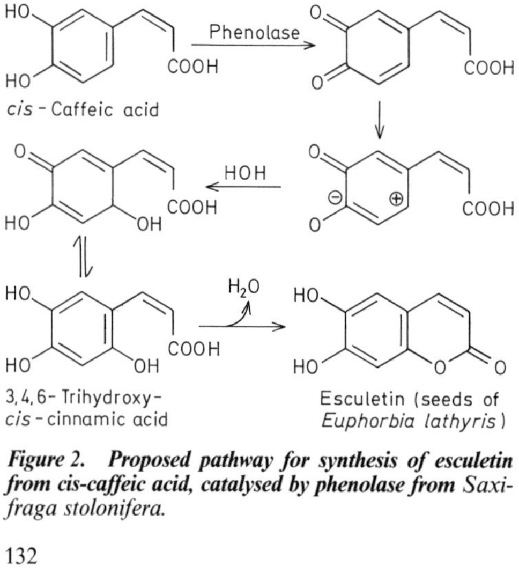

cis - Caffeic acid

3,4,6- Trihydroxy-
cis - cinnamic acid

Esculetin (seeds of
Euphorbia lathyris)

Figure 2.   Proposed pathway for synthesis of esculetin from cis-caffeic acid, catalysed by phenolase from Saxifraga stolonifera.

Table. A selection of naturally occurring coumarins

Trivial name	Substituents on coumarin ring system (see Fig. 1 for numbering)	Source
Angelicin	furano (2':3'-8:7)	*Angelica archangelica* (roots)
Angustifolin	3-(1',1'-dimethylallyl)-7-hydroxy	*Ruta angustifolia* (leaves) [*Phytochemistry*, 1984, **23,** 2095–2096]
Ayapin	6,7-methylenedioxy	*Eupatorium ayapana* (leaves)
Bergapten	5-methoxy-furano (2':3'-6:7)	*Citrus bergamia* and many other plants
Braylin	6-methoxy-7,8-pyrano	*Flindersia brayleyana* (bark)
Calycanthoside	6,8-dimethoxy-7-glucosido	*Calycanthus occidentalis* (twigs)
Ceylantin	7,8-dimethoxy-5,6-pyrano	*Atalantia ceylanica* (heartwood)
Chicoriin	6-hydroxy-7-glucosido	*Chicorium intybus* (flowers)
Coumarin	see Fig. 1	–
Dalrubone	see formula	*Dalea* spp.
Dalbergin	3-phenyl-6-hydroxy-7-methoxy	*Dalbergia sissoo* (heartwood)
Daphnetin	7,8-dihydroxy	*Euphorbia lathyris* (seeds)
Daphnin	7-glucosido-8-hydroxy	*Daphne alpina* (bark)
Esculetin	see Fig. 2	–
Esculin	6-glucosido-7-hydroxy	*Aesculus hippocastanum* (bark)
Fraxetin	6-methoxy-7,8-dihydroxy	*Fraxinus intermedia* (bark)
Fraxin	6-methoxy-7-hydroxy-8-glucosido	*Fraxinus intermedia* (bark)
Fraxinol	5,7-dimethoxy-6-hydroxy	*Fraxinus excelsior* (bark)
Isobergapten	5-methoxy-furano(2':3'-6:7)	*Heracleum lanatium* (roots)
Isofraxidin	6,8-dimethoxy-7-hydroxy	*Fraxinus intermedia* (bark)
Limettin	5,7-dimethoxy	*Citrus limetta* (fruit)
Osthenol	7-hydroxy-8-*iso*pent-2'-enyl, or 7-hydroxy-8-dimethylallyl	*Angelica archangelica* (roots)
Osthole	7-methoxy-8-*iso*pent-2'-enyl, or 7-methoxy-8-dimethylallyl	*Imperatorium ostruthum* (rhizomes)
Pereflorin	5-methyl-4-methoxy	*Perezia multiflora* (roots)
Psoralen	furano (2':3'-6:7)	*Psoralea corylifolia* (seeds) *Xanthoxylum flavum* (wood)
Skimmin	see Fig. 1	–
Suberosin	7-methoxy-6-*iso*pent-2'-enyl, or 7-methoxy-6-dimethylallyl	*Xanthoxylum suberosum* (bark), *Xanthoxylum flavum* (wood)
Trimethoxy-coumarin	6,7,8-trimethoxy	*Fagara macrophylla* (heartwood)
Umbelliferone	see Fig. 2	–
Vellein	7-glucosido-8-*iso*pent-2'-enyl, or 7-glucosido-8-dimethylallyl	*Vellia discophora*
Xanthotoxin (used by indigenous medical centers in India for treatment of leucoderma)	8-methoxy-furano (2':3'-6:7)	*Fagara xanthoxyloides* (fruit)

The coumarin ring system in Novobiocin (see) from *Streptomyces spheroides* is derived from tyrosine, i.e. the hydroxyl group at C-7 is introduced at an earlier stage in the bacterial than in the plant synthesis. Other details of the bacterial synthesis of the coumarin ring system are lacking. [K. B. G. Torsell, *Natural Products Chemistry* (John Wiley, 1983); G. Bilek (ed.), *Biosynthesis of Aromatic Compounds* (Academic Press, 1969)]

Coumestans, *coumaranocoumarins:* compounds with the ring system shown (Fig.), which represents the highest possible oxidation level of the Isoflavonoid (see) skeleton. Like other isoflavonoids, C. are largely restricted to the *Leguminoseae*, and they are accompanied by the corresponding 6a,11a-dehydropterocarpans (see Pterocarpans). Examples: *lucernol* (2,3,9-trihydroxycoumestan, *Medicago sativa*), *sativol* (4,9-dihydroxy-3-methoxycoumestan, *Medicago sativa*), *psoralidin* (3-hydroxy-9-methoxy-2-γ,γ-dimethylallylcoumestan, *Psoralea corylifolia*), 2-hydroxy-1,3-dimethoxy-8,9-methylenedioxycoumestan (*Swartzia leiocalycina*). [J. B. Harborne, T. J. Mabry & H. Mabry, eds., *The Flavonoids* (Chapman and Hall, 1975)].

Coumestan ring system

Coupling factors: see Respiratory chain.

Covalent modification of enzymes, *modulation of enzymes, interconversion of enzymes, enzyme modulation, enzyme interconversion:* Oligomeric (i.e. multichain) enzymes may exist in two or more forms, which are interconvertible by enzyme-catalysed covalent modifications. These various forms of the enyzme differ in their catalytic properties, e.g. activity, substrate affinity and dependence on effectors. Usually the difference in activity is such that one form is active and the other inactive. The activities of the conversion enzymes are in turn regulated by other enzymes, metabolites and/or effectors. Covalent modifications are therefore important in physiological regulation, in addition to Allostery (see). Whereas allostery can provide fine adjustment of metabolic rates, C. can provide on/off switching of cellular functions which is very sensitive to environmental fluctuation.

The most common type of C. appears to be a phosphorylation/dephosphorylation cycle, although other types are also observed. It has been found that many receptor molecules in the cell membrane are kinases, as are the products of some oncogenes.

The potential for such systems to respond radically to small changes in the concentrations of effector molecules is a result of the fact that their kinetics can be described by the Michaelis-Menton equation (see). At steady state, the fraction of a protein P that is in active form (P^*/P_{tot}) depends on the rate constants of the activation and inactivation reactions, P

$\overset{k_c}{\underset{k_b}{\rightleftharpoons}}$ P*. If the concentration of substrate (the regulated protein) is less than the K_m values of the regulating enzymes, the kinetics are first order. For the forward reaction, $v_f = (V_{mf}/K_{mf})P$, and for the back reaction, $v_b = (V_{mb}/K_{mb})P^*$. If the two reactions are plotted as in Fig. 1 a, the steady-state fraction of activated protein is found at the intersection of the plots for the two enzymes. A minor change in the activity (V_m) of one of the two enzymes does not greatly affect the fraction of the substrate protein which is active.

By contrast, if the concentration of the regulated protein is high enough to saturate the two modifying enzymes, giving zero-order kinetics, a relatively small change in the V_m value of one of them causes a large change in the fraction of the regulated protein which is in the active form (Fig. 1 b). This makes the system very sensitive to changes in the concen-

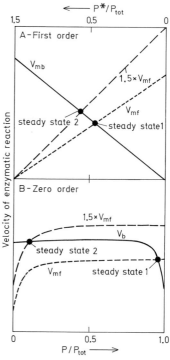

Figure 1. Steady state fractions of activated (P*) and inactive (P) protein. A) If the K_m values for the enzymes catalysing the forward and back reactions are greater than P_{tot}, the reaction kinetics are first-order. Increasing the rate of the forward reaction by 50% does not shift the fraction of P in the active form (P*/P) very much. B) By contrast, if P_{tot} is much larger than K_m, so that the reaction kinetics are zero-order, a 50% increase in the rate of the forward reaction shifts the fraction of active protein from about 0.05 to 0.9 (in this example). [Adapted from D. C. LaPorte & D. E. Koshland, Jr., *Nature* **305** (1983) 286–290]

tration of an allosteric regulator for one of the modifying enzymes. An example of a system in which a zero-order regulation occurs is the isocitrate dehydrogenase in *Escherichia coli*. [D.C. LaPorte & D.E. Koshland, Jr., *Nature* **305** (1983) 286–290] This enzyme determines the distribution of acetyl-CoA between the Tricarboxylic acid cycle (see) and the Glyoxylate cycle (see). Isocitrate dehydrogenase, which is part of the TCA cycle, is inactivated by phosphorylation. The phosphatase which removes the phosphate groups and thus activates the dehydrogenase is in turn activated by 3-phosphoglycerate, a metabolic precursor of acetyl-CoA. When 3-phosphoglycerate is abundant, the active isocitrate dehydrogenase allows the acetyl-CoA to enter the TCA cycle and to be oxidized to CO_2. However, when the bacteria are growing on acetate, the relative lack of 3-phosphoglycerate causes the phosphatase, and thus the dehydrogenase as well, to be inactive. Now the TCA cycle is shut down and the glyoxylate shunt provides carbon for the synthesis of cell substance. The relationship between the K_m of the phosphatase and the concentration of isocitrate dehydrogenase is such that a small change in the concentration of 3-phosphoglycerate can shift the equilibrium of isocitrate dehydrogenase from nearly all active to nearly all inactive, and vice versa.

Phosphorylase, which catalyses the breakdown of glycogen, is an example of C. in response to the concentration of the hormone "second messenger", cyclic AMP (see Hormones). Phosphorylase in its inactive (*b*) form is activated by phosphorylation of a serine residue. In muscle phosphorylase, this causes the dimeric *b* form of the enzyme to aggregate to a tetramer, the active *a* form. (In liver, both *a* and *b* forms have the same M_r.) The enzyme which catalyses activation, phosphorylase *b* kinase, must itself first be activated by a specific kinase.

Some other enzymes subject to regulation by C. by phosphorylation and dephosphorylation are pyruvate dehydrogenase, glycogen synthetase, phosphofructokinase and glutamate dehydrogenase. C. accompanied by a change of M_r, i.e. association and dissociation, affects glutamine phosphoribosylpyrophosphate amidotransferase, pancreatic lipase (F- and S-lipase), human glucose 6-phosphate dehydrogenase and rat kidney pyruvate kinase. See also ADP-ribosylation of proteins. [O.M. Rosen and E.G. Krebs, eds., *Protein Phosphorylation* (Cold Spring Harbor Conferences on Cell Proliferation, Cold Spring Harbor Laboratory, Cold Spring Harbor, N.Y. (1982) vol.8 A and B)]

There is evidence that most of the "second messenger" effects of cyclic AMP and cyclic GMP on neuronal function, as well as those of calcium (see Calmodulin) are achieved by activation of specific protein kinases. [E.J. Nestler & P.Greengard, *Nature* **305** (1983) 583–588] Another well studied example of C. is found in glutamine synthetase (EC 6.3.1.2), which catalyses the synthesis of L-glutamine: $ATP + L\text{-glutamate} + NH_3 = ADP + P_i + L\text{-glutamine}$. Control of this enzyme by C. has been studied chiefly in *Escherichia coli* and *Klebsiella* spp. (Fig.2).

The enzyme consists of 12 subunits (each of M_r about 50000) arranged in the form of two hexagons attached face to face. The subunits are identical and each contains a tyrosyl residue that is capable of adenylylation by the glutamine-synthetase adenylyltransferase (EC 2.7.7.42). A phosphoester linkage is formed between the phosphate of the AMP residue and the phenolic OH group of the tyrosyl residue. When all twelve subunits are adenylylated (E_{12}), the enzyme is totally inactive, while full activity is expressed when no subunits are adenylylated (E_0). Between E_{12} and E_0, there are theoretically 382 possible forms of adenylylated glutamine synthetase, but the true order or arrangement of adenyl groups is not known for the intermediate levels of adenylylation. The plot of the number of AMP residues per molecule of enzyme (as extracted from cells grown under different conditions of nitrogen supply) against specific enzymatic activity is not a straight line; a plateau at about one-half maximal activity is found between E_3 and E_7. This suggests threshhold values in a control system. Glutamine-synthetase ad-

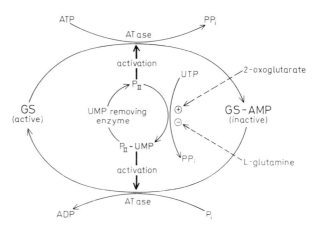

Figure 2. **Control of the activity of glutamine synthetase by adenylylation and deadenylylation.** GS = glutamine synthetase; GS–AMP = adenylylated glutamine synthetase; ATase = adenylyltransferase.

enylyltransferase has been purified from *E. coli*. It appears to contain two active centers, one adenylylating and the other deadenylylating the glutamine synthetase. The transferase therefore activates or inactivates the synthetase, depending on the relative activities of its two active sites. These, in turn, depend on a regulatory protein P_{II}, which consists of four identical subunits. Each subunit contains a tyrosyl residue that is subject to uridylylation by the action of a uridylyltransferase. The uridylylated form activates the deadenylylating activity, while the non-uridylylated form stimulates the adenylylating activity of the adenylyltransferase. The uridylyltransferase is activated by 2-oxoglutarate and inhibited by L-glutamine, so that the ratio of 2-oxoglutarate/L-glutamine ultimately controls the intracellular activity of glutamine synthetase.

Cozymase: see Nicotinamide adenine dinucleotide.

C_3 plants, *Calvin plants*: green plants which produce C_3 compounds as the first product of CO_2 fixation. The reaction is catalysed by the ribulose 1,5-bisphosphate carboxylase (EC 4.1.1.39): ribulose 1,5-bisphosphate + $CO_2 \rightarrow$ 2(3-phosphoglycerate). The products of the light reaction, ATP and NADPH, are used to reduce the 3-phosphoglycerate to 3-phosphoglyceraldehyde (see Carbon dioxide assimilation), which enters the Calvin cycle (see). The vast majority of green plants are C_3 plants.

C_4 plants: green plants which fix CO_2 into 4-carbon compounds, such as oxaloacetate, aspartate and malate. The carboxylation enzyme is phospho*enol*pyruvate carboxylase (EC 4.1.1.31) which carboxylates phospho*enol*pyruvate to oxaloacetic acid (see Hatch-Slack-Kortschak cycle). These plants also have the Calvin cycle (see); the C_4 compounds are decarboxylated to supply CO_2 for the Calvin cycle in the cells of the vascular bundles. The C_4 reactions are found in many tropical plants and in unrelated dicotyledons which are adapted to a hot, dry climate. These plants must close their stomata during the hot part of the day in order to avoid excessive water loss. However, this limits the availability of CO_2 to the mesophyll cells. The affinity of phospho*enol*pyruvate carboxylase for CO_2 is much higher than that of ribulosebisphosphate carboxylase, the first enzyme of the Calvin cycle. Therefore the C_4 plants are able to make use of the scarce CO_2 within their leaves during the part of the day when light, and thus ATP, is abundant.

Crabtree effect: a reduction in the rate of respiration after addition of glucose. The cause is the increased intracellular glucose 6-phosphate level, which inhibits hexokinase, the first enzyme in glycolysis. Glucose also inhibits the enzymes of the tricarboxylic acid cycle and the respiratory chain.

Crassulacean acid metabolism; *CAM plants:* Some plants fix large amounts of CO_2 at night and store it as malic acid by means of the Hatch-Slack-Kortschak cycle (see). This causes their tissues to become acidic at night; hence the effect was named "acid metabolism". The term "crassulacean" was added because it was first observed in members of the *Crassulaceae*, but it is by no means limited to this family. Some plants have obligate CAM, while in others it is facultative. It is common in plants adapted to dry areas, which open their stomata only at night, but is also found in some submerged plants for which CO_2 is growth-limiting. [G. Edwards, D. A. Walker *C_3, C_4 Mechanisms and Cellular and Environmental Regulation of Photosynthesis* (Blackwell, London, 1983); I. P. Ting, *Ann. Rev. Plant Physiol.* **36** (1985) 595–622]

Creatine: β-methylguanidoacetic acid, methylglycocyamine, *N*-amidosarcosine,

H$_2$N-C-N(CH$_3$)-CH$_2$-COOH
‖
NH

a biochemically important amino acid derivative. m.p. 291 °C (d.). C. is very easily converted to creatinine. Over 90% of the C. in an adult human is localized in muscle. The concentration is particularly high when a large amount of chemical energy is being converted to mechanical energy. The creatinine concentration in the blood and body fluids is normally very low, and reaches higher values only if the kidneys are damaged. Creatinuria occurs in muscular dystrophy and other diseases; creatinine, on the other hand, is one of the main nitrogen compounds in normal urine.

C. is phosphorylated by creatine kinase and ATP to creatine phosphate, a phosphagen, and is regenerated in the reverse reaction, which serves to produce ATP (Lohmann reaction). C. is metabolically synthesized from arginine, glycine and methionine. In the kidney, the first step is Transamidination (see) of glycine from arginine, catalysed by the arginine: glycine transamidinase to form the guanidine derivative glycocyamine (guanidoacetic acid), which is converted to creatine by transmethylation (Fig.).

Creatine kinase, *creatine phosphokinase:* a phosphotransferase, EC 2.7.3.2., found in brain and muscle. C. k. is a dimeric enzyme of M_r 82000. There are four known isoenzymes, comprised of three different subunits: M (muscle), B (brain) and Mi (mitochondria). The BB dimer is found in brain, smooth muscle and embryonic skeletal muscle. As the latter develops, there is a gradual change from BB through MB to MM dimers. Most C. k. is soluble, but about 5% of the MM-C. k. in striated muscle is located in the M line (see Muscle proteins). MiMi-C. k. is found on the outer side of inner mitochondrial membranes, where it is coupled to an ADP/ATP translocase.

C.k. catalyses the formation of ATP from ADP and creatine phosphate in a reversible reaction dependent on Mg(II) or Mn(II). Muscle contraction is dependent on ATP, but during prolonged work, the level of ATP remains high due to the C.k. reaction. In effect, the muscle cell "burns" creatine phosphate, which is stored in it in large amounts. This prevents the accumulation of protons (which are released in ATP hydrolysis) in the sarcoplasm. The immobilized C.k. in the M lines of muscle is sufficient to maintain the ATP level even during vigorous contraction. It is thought that the remaining 95% of the C.k., which is soluble, serves to rebuild the creatine phosphate supply rapidly after a period of depletion. Similarly, the MiMi-C.k. of the mitochondria makes possible efficient conversion of mi-

duced by addition of bifunctional alkylating agents to molecules, such as DNA or soluble proteins, which are not naturally cross-linked. They are formed enzymatically in insoluble structural proteins (see Collagen, for example) and in the cell walls of bacteria.

Crotactin: see Snake venoms.
Crotamine: see Snake venoms.
Crotoxin: see Snake venoms.
Crustecdysone: see Ecdysterone.
Cryptosterol: see Lanosterol.
Cryptoxanthin: *(3R)-β*, β-caroten-3-ol, a carotenoid belonging to the group of xanthophylls. M_r 552.85, m.p. 169 °C. C. is a common pigment in plants and is found especially in fruits and berries, e.g. bell peppers, oranges, tangerines and papayas, as well as in maize and egg yolk. As a provitamin A, C. is half as effective as β-carotene. Its structure was established in 1933 by R. Kuhn.

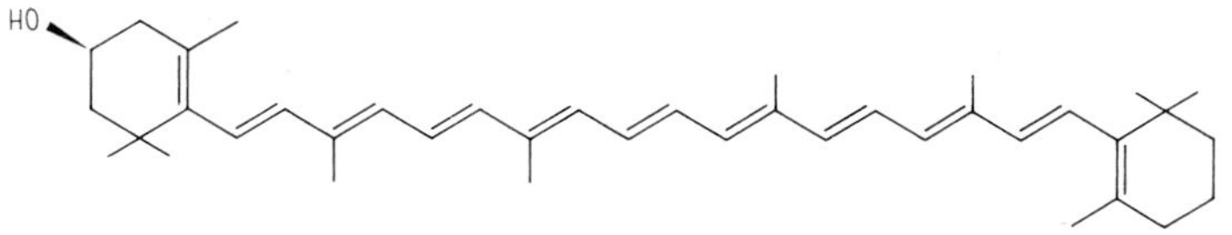

Cryptoxanthin

tochondrial ATP to creatine phosphate. [T. Wallimann & H. M. Eppenberger in *Cell and Muscle Motility* 6 (J. Shay, ed.) (Plenum Press, New York & London, 1985) pp. 239-285]

When there is damage to the skeletal or heart musculature, the level of C. k. in the serum is increased. It is therefore used in the early diagnosis of heart infarction, the detection of muscle wastage and to distinguish heart infarction from a lung embolism, in which there is no increase in the serum C. k.

In the muscles of invertebrates, creatine phosphate is replaced by other energy-rich phosphates; see Phosphagens. There the generation of ATP is catalysed by phosphotransferases homologous to C. k.

Creatine phosphate: see Phosphagens.

CRH: abb. for corticotropin releasing hormone (see Releasing hormones).

Crigler-Najjar syndrome: see Inborn errors of metabolism.

Crocetin: a brick-red C_{20} dicarboxylic acid with seven conjugated *trans* double bonds, 4 methyl branches and two terminal carboxyl groups. M_r 328.39, m.p. 285 °C. C. is an apocarotenoid, an oxidation product of carotenoids. Its structure was dis-

Crystallins: soluble proteins which account for almost 90% of the total vertebrate lens protein. Lens cells are not replaced, so they must survive the lifespan of the animal. Moreover, C. cannot be replaced, because lens cells lose their nuclei and other organelles, presumably because these would cause discontinuities in refractive index. C. must therefore be exceptionally stable, withstanding all abuses which could lead to denaturation or aggregation. C. must also have short range order in solution to ensure a smoothly changing refractive index.

C. are classified into three major groups, α, β and γ, originally on the basis of their precipitability, but now according to their M_r range as determined by gel filtration. α-C. comprise two subgroups, one with average M_r 800000, the other much higher. All α-C. are oligomeric and differ only in the relative quantities of four subunits: αA_1, αA_2, αB_1 and αB_2. There are two primary gene products: αA_2 and αB_2, while αA_1 and αB_1 arise by post-translational modification. Lens mRNA from calves has been translated in vitro into α-C. and transcribed into cDNA [A. J. M. Berns & H. Bloemendal, *Methods in Enzymology* 30 (1974) 675-694].

Crocin

covered in 1932 by P. Karrer et al. The digentiobiose ester of C., *crocin* (Fig.), is found as a yellow pigment in crocusses and a few other higher plants. M_r 976, m.p. 186 °C.

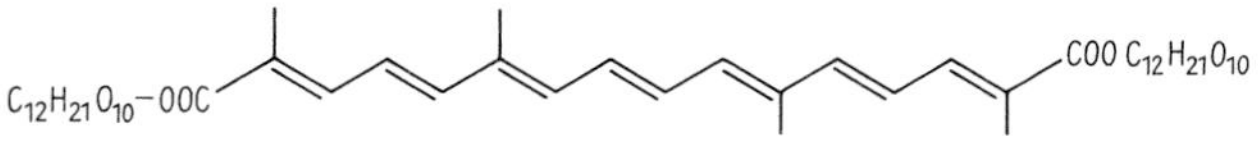

Cross-link: a covalent bond formed between two polymers. Cross-links may be artificaly intro-

β-C. comprise at least seven gene products. All β-C. are oligomers, ranging from dimers to aggregates of up to 8 chains, and all share one major subunit, βBp, but they may differ considerably with respect to other subunits. γ-C. are all monomeric, M_r less than 28000, representing a group of 4 or 5 gene products.

In addition to α, β and γ, reptile and bird lens contain tetrameric δ-C., which may be related to the softness of the lens and its ability to accommodate in these species. The proportions of different C. vary greatly between species; similarly, the size and charge of C., the stoichiometry of the component monomers and their isoelectric points show species differences. Species differences in amino acid sequence are, however, relatively small, and there is considerable immunological cross reactivity between C. from different species. C. are therefore highly conserved and, compared with many other proteins, have evolved relatively slowly since the vertebrate lens came into existence about 500 million years ago. [W. W. DeJong et al. *J. Mol. Evol.* **10** (1977) 123-135] All β- and γ-polypeptides sequenced so far appear to be homologous, most probably derived from an ancestral protein by point mutation, insertion, deletion and, in the β-C., by extension at the N- and C-termini. γ-C. contain an unusually large proportion of cysteine (7 Cys per 174 amino acid residues) and it is suggested that the free SH groups serve as a source of reducing power, preventing oxidative cross-linking in other lens proteins.

The principle subunit of β-C. (dimeric βBp), tetrameric avian δ-C. and several γ-C. have been crystallized. X-ray crystallography of calf γ-II-C. to a refinement of 1.6 Å reveals a very symmetrical molecule with two lobes, constructed from four similar polypeptide domains, each consisting of about 40 amino acid residues. Many Met, Cys, Trp, Tyr and Phe residues are very close to one another (less than 6.5 Å), which suggests an interaction of aromatic side chains, Arg residues, the polarizable sulfur of Met residues and the SH of Cys residues. This may contribute to protein stability by the delocalization of π electrons, and may also serve as an electron-transfer system. [H. Bloemendal & W. W. DeJong *Trends in Biochemical Sciences* **4** (1979) 137-141; L. Summers et al. *Peptide and Protein Reviews* **3** (1984) 147-168]

CSF: abb. for Colony stimulating factor (see).

C-terminal amino acids: see Peptides.

C-toxiferin I: calabash toxiferin 1, see Curare alkaloids.

CTP: abb. for cytidine 5'-triphosphate.

Cucumber hypocotyl test: see Gibberellins.

Cucurbitacins: a group of tetracyclic triterpenes found as glycosides in *Cucurbitaceae* and *Cruciferae*. These toxic, bitter compounds are structurally related to the parent hydrocarbon cucurbitane, 19(10-9β)abeo-5β-lanostane, which differs from la-

Cururbitacin B

nostane (see Lanosterol) in the formal shift of the 10 methyl group to the 9β-position. Cucurbitacin E was isolated in 1831 in crystalline form under the name *elaterin*.

C. are known as laxatives; some serve as insect attractants, and a few have antineoplastic and antigibberelin activities.

5α-Cucurbitane: see Cucurbitacins.

Cultivation of microorganisms, *culture techniques*: deliberate propagation of an organism. Cultivation of an organism may be discontinuous (static) or continuous. The technical culture of microorganisms is discussed under Fermentation techniques (see). In discontinuous culture, microorganisms are grown in a closed system until one factor in the nutrient medium becomes limiting. The culture conditions therefore change constantly with time. The growth can be compared with the genetically limited growth of a multicellular organism, whose life is divided into the phases of youth, maturity, ageing and death. The corresponding growth phases in discontinuous culture are the lag, exponential growth, stationary and post-stationary (decline) phases. These are represented graphically in a growth curve by plotting the logarithms of the total cell number, or of the number of viable cells (see Growth) against time. Semi-logarithmic representation is useful because the phase of exponential growth is represented by a straight line, whose slope corresponds to the rate of cell division. Because of the linear relationship between the logarithm of cell numbers and time, exponential growth is also referred to as logarithmic growth. The lag phase is the period between inoculation and the attainment of the maximal rate of cell division. The exponential phase merges slowly into the stationary phase, in which the cells no longer grow, for any of a number of reasons. There may still be significant metabolic activity in the form of turnover during the stationary phase. It may last a long time or be artificially maintained (resting cells). The transition to the post-stationary phase is also gradual.

The important growth parameters are the lag time, the growth rate (doubling time), and the yield. The ratio of the mass produced to the substrate consumed is called the coefficient of yield, or economic coefficient. It can also be defined as the energy yield coefficient, if one compares the actual yield of ATP to the theoretically possible yield. Because the composition of the medium changes continuously during discontinuous culture, the cells may already begin to change during the exponential growth phase, as the substrate concentration decreases, the cell density increases and metabolic products accumulate or are excreted.

It has been observed that there is a more or less clear relationship between the growth phase or growth rate and the synthesis of secondary metabolic products. The latter begins, in many cases, during the transition from the exponential to the stationary growth phase. One can therefore distinguish the tropho or nutritional from the idioproduction phase (Bu'Lock).

In continuous culture, a growing population of microorganisms is continuously supplied with new nutrient solution at the same rate that microorganism suspension is removed from the culture vessel. The system is open and tends toward a steady state. The cells can be allowed to grow exponentially for an indefinite period (exclusion of the time factor) under constant environmental conditions. The system is stabilized by limiting one growth substrate.

Culture techniques: see Cultivation of microorganisms.

Cupressuflavone: see Biflavonoids.

Curare alkaloids: the toxic principles of arrow and bait poisons used by South American Indians. At the end of the last century they were classified by Boehm as pot, tube or bamboo, and gourd or calabash curares, according to the vessels in which the Indians stored them. The terms have been retained because the poisons come from different sources and differ in their chemical and pharmacological properties. The poisons are extracted from the plants with water, which is then boiled off to concentrate them.

Pot and *tube* curare are rather similar, and not highly poisonous. They are obtained from plants of the *Chondodendron* genus and are stored in clay pots or bamboo tubes. The alkaloids are isoquinolines, the most important being the quaternary bisbenzyl-isoquinoline alkaloid *tubocurarine*.

Calabash curare is obtained from plants of the *Strychnos* genus and is packed in hollow gourds. It contains a large number of alkaloids and is extremely poisonous. The alkaloids are derived from indole, and can be divided into types, the yohimbine type (e.g. mavacurine), the strychnine type (e.g. Wieland-Gumlich aldehyde) and the bisindole type (e.g. calabash toxiferin-I).

In current scientific usage, curare refers only to those C. a. with muscle-relaxing (paralysing) effects, namely the dimeric indole alkaloids with a strychnine skeleton and two quarternary nitrogen atoms. These compounds are extremely poisonous; calabash alkaloids E and G are among the most toxic compounds known. They cause death by respiratory paralysis.

Pharmacological effects: The C.a. interrupt nervous impulses at the end plates of motor nerves by displacing acetylcholine, which leads to paralysis of the striated muscles. Because the C.a. are absorbed very slowly from the intestines, animals which have been killed by the arrow poisons can be eaten with impunity. The curarizing effects set in sooner and last longer when the compounds are injected subcutaneously or intramuscularly. Because the composition of the natural preparations varies, and because of the unpleasant side effects, they have been replaced in medicine by the pure alkaloids or synthetic or semisynthetic analogs (e.g. Alloferin). They are used as muscle relaxants in operations, severe tetanus and nervous muscle cramps.

Curcumin, *turmeric yellow, diferuloylmethane:* a yellow pigment from the roots and pods of *Curcuma longa* L., which is cultivated in Southeast Asia. m.p. 183 °C. The dried root is used as medicine for liver and bile ailments, and is a component of curry powder. C. is used as a food color, as a dye for textiles and as an indicator (curcumin-boric acid paper). It has been shown that phenylalanine and acetate/malonate are specific biosynthetic precursors.

Curcumin

Cuscohygrine: see Pyrrolidine alkaloids.

Cyanides: salts of hydrogen cyanide (HCN) with the general formula Me^1CN. The soluble C. are hydrolysed in water to the metal hydroxide and hydrogen cyanide, and are thus highly poisonous. The C. of heavy metals are generally insoluble and form complex ions with excess CN^- ions. HCN and alkali cyanides are often components of insecticides and other biocides. HCN occurs in nature as an addition product of an aldehyde or ketone in the cyanide glycosides. For example, in amygdalin, it takes the form of an α-hydroxynitrile (cyanhydrin). Nitriles $(R-C \equiv N)$ are fairly rare in organisms, although the nitrile group does occur in the alkaloid ricinine and the rare amino acid β-cyanoalanine. Cyanide can be assimilated by plants in the form of β-cyanoalanine, which is enzymatically hydrolysed to L-Asparagine (see) (Fig.).

C. reversibly inhibit respiratory chain enzymes. In cases of cyanide poisoning, the venous blood is bright red, because the oxygen has not been used. Hemoglobin does not react with C. The average lethal dose of C. for humans, taken orally, is 60 to 90 mg HCN (200 mg KCN). An antidote is methemoglobin (MHgb), which has a higher affinity for C. than cytochrome oxidase. It forms cyanomethemoglobin, which is slowly converted to normal hemoglobin as the CN^- is converted to thiocyanate by the enzyme rhodanese (EC 2.8.1.1.): $CN^- \rightarrow SCN^-$. Another detoxification reaction (which is quantitatively insignificant, however) is the formation of cyanocobalamine (vitamin B_{12}, see Vitamins).

Cyanide assimilation in plants

Cyanidin: 3,5,7,3',4'-pentahydroxyflavylium cation, the aglycon of many Anthocyanins (see). m.p. 200 °C (d.). Glycosides of C. and a few acylated derivatives are found in many plants; the oxonium salts are responsible for the deep red color of many

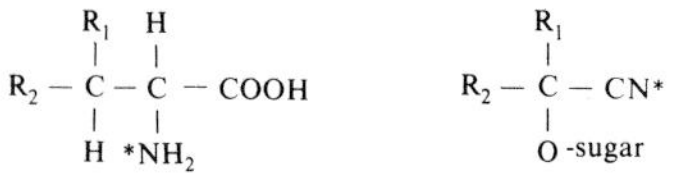

$$R_2 - \overset{\overset{\displaystyle R_1}{|}}{\underset{\underset{\displaystyle H}{|}}{C}} - \overset{\overset{\displaystyle H}{|}}{\underset{\underset{\displaystyle *NH_2}{|}}{C}} - COOH \qquad R_2 - \overset{\overset{\displaystyle R_1}{|}}{\underset{\underset{\displaystyle O \text{ -sugar}}{|}}{C}} - CN*$$

Amino acid	Cyanogenic glycoside	R_1	R_2	Sugar
L-Valine	Linamarin	$-CH_3$	$-CH_3$	Glucose
L-Isoleucine	Lotaustralin	$-CH_2CH_3$	$-CH_3$	Glucose
L-Phenylalanine	Prunasin	$-C_6H_5$	$-H$	Glucose
L-Phenylalanine	Amygdalin	$-C_6H_5$	$-H$	Gentiobiose
L-Tyrosine	Dhurrin	$-C_6H_4OH$	$-H$	Glucose

flowers and fruits, such as red roses, geraniums, tulips, poppies and zinnias. Chelates with iron(III) or aluminium ions are deep blue in color. When bound to a polysaccharide carrier, they form chromosaccharides, such as *protocyanin*, the blue pigment of cornflowers.
The structures of more than 20 different natural glycosides of C. are known, including *chrysanthemin* (3-β-glucoside) chloride, m.p. 210 °C (d.), from the red autumn leaves of some maples, wild strawberries and blackberries; *idaein* (3-β-galactoside) chloride, m.p. 210 °C (d.), from apples and cranberries; *mekocyanin* (3-gentiobioside) from hibiscus blossoms and sour cherries; *keracyanin* (3-rhamnoglucoside) from snapdragons, canna lilies, tulips and cherries; *cyanin* (3.5-di-β-glucoside), chloride m.p. 204 °C from blue cornflowers, violets, dahlias, red roses, etc.

Cyanin: see Cyanidin.

Cyanocobalamin: one of the B_{12} vitamins (see Vitamins).

Cyanogenic glycosides: a group of O-glycosides formed from decarboxylated amino acids (Table). The cyano group arises from the α-C atom and the amino group. Hydrocyanic acid is generated from C.g. by β-glycosidases (such as emulsin) and oxinitrilases.

Cyanophycin granules: see Blue-green bacteria.

Cyasterone: a phytoecdyson (see Ecdysone), M_r 520.67, m.p. 164 to 154 °C, $[\alpha]_D + 64.5$ ° (pyridine). It has been isolated from various plants, including *Cyathula capitata,* family *Amaranthaceae,* and *Ajuga decumbens,* family *Labiatae.* Unlike other ecdysones, C. has a stigmastane skeleton (see Steroids) and a γ-lactone group in the side chain.

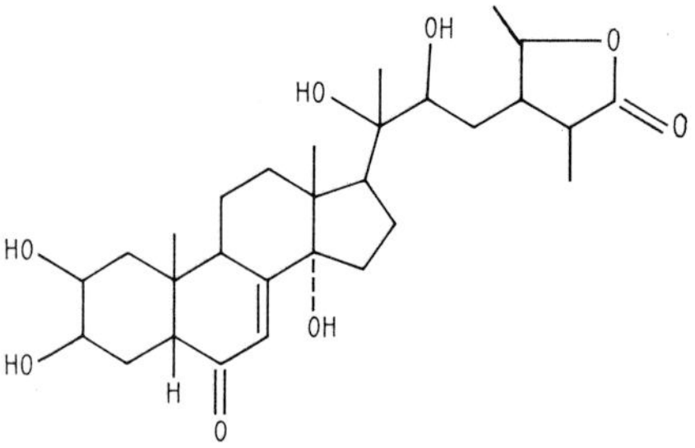

Cyasterone

Cyclic adenosine 3′,5′-monophosphate: see Adenosine phosphates.

Cyclic N^6, $O^{2'}$-dibutyryladenosine 3′,5′-monophosphate, DBCAMP: a synthetic derivative of cAMP (see Adenosine phosphate).

Cyclic guanosine 3′,5′-monophosphate: see Guanosine phosphates.

Cyclic inosine 3′,5′-monophosphate: see Inosine phosphates.

Cyclic nucleotides: see Nucleotides.

Cyclic uridine 3′,5′-monophosphate: see Uridine phosphates.

Cyclitols: compounds derived from 1,2,3,4,5,6-hexahydroxycyclohexane. They have the same chemical formula, $C_6H_{12}O_6$, as the hexoses, and are biosynthetically related to them, but they have an isocyclic rather than a heterocyclic ring. The hexane ring has the chair configuration in the C. The hydroxyl groups can lie either in the equatorial or the axial position, which makes possible 8 *cis trans* isomers, of which 7, as *meso*forms, are optically inactive. Another isomer pair results from the D and L forms of the optically active 8th *cis, trans* isomer. All 9 isomers are found in nature, and Myoinositol (see) is very common. In plants, *inositol (M_r 180.16, m.p. 227 °C, $[\alpha]_D + 65$ ° in water), scyllitol (m.p. 348 °C) and two methyl ethers of D-inositol, pinitol (M_r 194.2, m.p. 186 °C) and quebrachitol (M_r 194.2, m.p. 192 °C, $[\alpha]_D^{25} + 81$ ° in water) are most common.

Cycloalkanes: saturated hydrocarbon rings with the general formula C_nH_{2n}. The tendency to form and the stability of the rings depends on the number of carbon atoms, with 5 and 6-membered rings being most stable and most easily formed. C. are components of many natural products, especially the cyclic terpenes, which also include derivatives of cyclobutane and cyclopropane.

Cyclo-AMP: abb. for cyclic adenosine 3′,5′-monophosphate (see Adenosine phosphates).

Cycloartenol: 9,19-cyclo-5α,9β-lanost-24-en-3β-ol, a tetracyclic triterpene alcohol. M_r 426.73, m.p. 115 °C, $[\alpha]_D + 54$ °. C. is found in many plants, including the latex of *Euphorbiaceae,* the potato *Solanum tuberosum)* and nux vomica *(Strychnos nux vomica).* It is biosynthesized from squalene via 2,3-epoxysqualene (see Steroids, Fig. 4).

Cyclobuxamine H: a *Buxus* alkaloid.

Cyclo-GMP: abb. for cyclic guanosine 3′5′-monophosphate (see Guanosine phosphates).

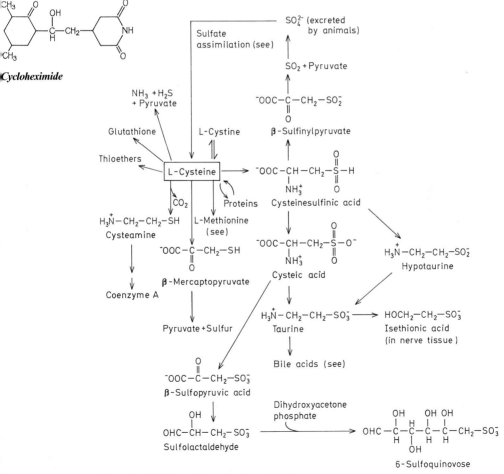

Cycloartenol

Cycloheximide, *actidione:* an antibiotic isolated from *Streptomyces griseus.* M_r 281. C. is chemically a derivative of glutarimide and is water soluble. It inhibits protein biosynthesis on 80S ribosomes in eukaryotes by preventing the initiation and elongation reactions. The latter increases the number of monosomes among the polyribosomes. The exact mechanism is unknown. C. is used as a fungicide to control cherry leaf spot, turf diseases and rose powdery mildew.

Cyclo-IMP: abb. for cyclic inosine 3′-5′-monophosphate (see Inosine phosphates).

Cyclopenase: see Viridicatine.

Cyclopentanoperhydrophenanthrene: see Steroids.

Cyclophosphamide, *Cytoxan, Endoxan:* the most important of the alkylating agents used as immune suppressives. C. inhibits cellular and humoral immune reactions by alkylating the SH and NH_2 groups of proteins and the N^7 of guanine in nucleic acids. It leads to a long-lasting suppression of antibody synthesis, provided that the C. is administered for 1 day before and 15 days after the antigen is introduced. It has also been proposed as an antineoplastic agent.

Cyclophosphamide

Figure 1.   Position of L-cysteine in metabolism

Cyclo-UMP: abb. for cyclic uridine 3',5'-monophosphate (see Uridine phosphates).

Cyd: abb. for Cytidine.

Cyproterone acetate: 17α-acetoxy-6-chloro-1α,2α-methylenepregna-4,6-diene-3,20-dione, an antiandrogen steroid derived from pregnane. C.a. counteracts the effects of the male sex hormone testosterone, and is used in cases of hypersexuality ("hormonal castration").

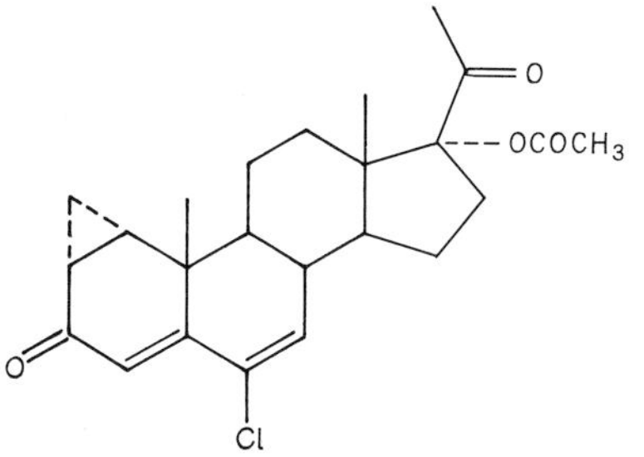

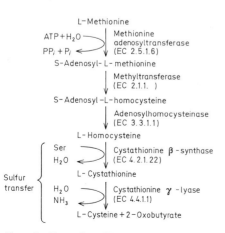

Cyproterone acetate

Cys: abb. for L-Cysteine.

Cystathioninuria: see Inborn errors of metabolism.

L-Cysteine, abb. *Cys:* L-2-amino-3-mercaptopropionic acid, HS-CH$_2$-CH(NH$_2$)-COOH, a sulfur-containing, proteogenic amino acid which is the central compound in Sulfur metabolism (see). M_r 121.2, m.p. 240 °C (d.), $[\alpha]_D^{25} +6.5$ ($c=2$ in 5 N HCl). At neutral or alkaline pH, solutions of Cys are oxidized in the air to L-cystine. Cys also plays an important role in redox reactions in the organism. In proteins, the thiol group-SH or the disulfide bond –S–S– of cystine can be very important for the tertiary structure and/or the enzymatic activity. Cys is synthesized in the course of Sulfate assimilation (see) or from Methionine (see) by Transsulfuration (see) (Fig. 2). It is presumably the precursor of the nonproteogenic sulfur-containing amino acids. It is degraded either reductively or oxidatively (see

Fig. 1). The end products of Cys degradation are usually oxidized sulfur compounds like taurine, NH$_2$-CH$_2$-CH$_2$-SO$_3$H.

Cysteine sulfinic acid: see Cysteine.

Cysteine synthase: see Sulfate assimilation.

L-Cystine, *dicysteine:* 3,3'-dithiobis(2-aminopropionic acid), the dimer of cysteine formed by oxidation of the -SH group to a disulfide –S–S–. M_r 240.3, m.p. 258 to 261 °C (d.), $[\alpha]_D^{25} -232$ ° ($c=1$ in 5 N HCl). In proteins, Disulfide bridges (see) are formed by C. However, it is always incorporated into the polypeptide chain as cysteine.

Cystine bridges: see Disulfide bridges.

Cyt: abb. for cytosine.

Cytidine, abb. *Cyd, cytosine riboside, 3-D-ribofuranosyl-cytosine:* a β-glycosidic Nucleoside (see) consisting of D-ribose and the pyrimidine base cytosine. M_r 243.22, m.p. 220-230 °C (d.), $[\alpha]_D^{25} +34.2$ ° ($c=2.0$ in water). The Cytidine phosphates (see) have a very important role in the metabolism of all organisms.

Cytidine diphosphocholine, abb. *CDP-choline, active choline:* the activated form of choline formed from phosphocholine and CTP (see Phosphatide biosynthesis).

Cytidine diphosphoglyceride, abb. *CDP-glyceride:* see Phosphatide biosynthesis.

Cytidine phosphates, *phosphoric esters of cytidine: Cytidine 5'-monophosphate,* abb. *CMP, cytidylic acid,* M_r 323.2, m.p. 233 °C (d.), *cytidine 5' diphosphate,* abb. *CDP,* M_r 403.19, and *cytidine 5'-triphosphate,* abb. *CTP,* M_r 483.16 are important in Phosphatide biosynthesis (see). CDP has been called the coenzyme of this synthesis. Activated choline is CDP-choline, and the sugar alcohol ribitol is also activated by bonding to CDP.
The reduction of ribose to deoxyribose in the synthesis of deoxyribonucleotides also occurs most often at the level of CDP (see Nucleotides).

Cytidylic acid: see Cytidine phosphates.

Cytimidine: see Amicetins.

Cytisine: see Lupine alkaloids.

Cytochalasins: a group of mold metabolites with cytostatic activity. Six chemically similar C. have been isolated from *Helminthosporium dematioideum, Metarrhizium anisopliae* and *Rosellinia necatrix.*

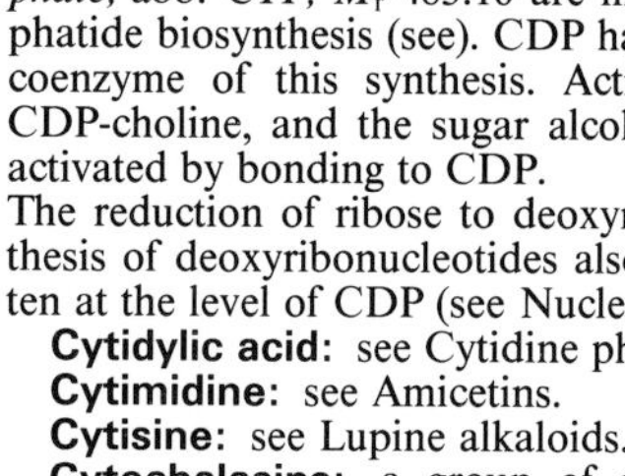

Structure of phomin, a typical cytochalasin.

Cytochrome oxidase: the last member of the electron transport chain. C.o. normally reacts with oxygen, but can be inhibited by reaction with CN$^-$ or CO. Its prosthetic group is hemin a, or cytohemin, which has a lipophilic C$_{12}$ side chain, an aldehyde and a vinyl group on the porphyrin ring. The reaction with O$_2$ involves both the heme iron and copper: $2 \text{ Fe}^{2+} \xrightarrow{\frac{1}{2}O_2} 2 \text{ Fe}^{3+}$ and $2 \text{ Cu}^{2+} \xrightarrow{2e^{-2}} \text{Cu}^+$.

Methylation pathway:

L-Methionine

ATP+H$_2$O ⟶ ⟍ Methionine adenosyltransferase
PP$_i$ + P$_i$ ⟸ ⟋ (EC 2.5.1.6)

S-Adenosyl-L-methionine

⟍ Methyltransferase
⟋ (EC 2.1.1.)

S-Adenosyl–L-homocysteine

⟍ Adenosylhomocysteinase
⟋ (EC 3.3.1.1)

L-Homocysteine

Ser ⟶ ⟍ Cystathionine β-synthase
H$_2$O ⟸ ⟋ (EC 4.2.1.22)

L-Cystathionine

H$_2$O ⟶ ⟍ Cystathionine γ-lyase
NH$_3$ ⟸ ⟋ (EC 4.4.1.1)

L-Cysteine + 2-Oxobutyrate

(Sulfur transfer)

Figure 2. Formation of L-cysteine from L-methionine

Electron microscopic studies have shown that the protein part of the molecule undergoes conformational changes during the redox process; the oxidized form is crystalline, but the reduced form is amorphous.

In the native state, C.o. is a component of a phospholipid-containing supermolecule (M_r 3×10^6) bound to the mitochondrial membrane. This complex can only be dissociated by treatment with detergent (2% deoxycholate). Once isolated, C.o. (IP 4 to 5) forms a stable, active, noncovalent complex with cytochrome c (IP 10.1), from which it can be separated by treatment with detergent (0.1% Emasol) and gel filtration.

The cytochrome aa_3 appears to be a single heme protein which exists in two functional states, rather than a complex of two proteins, as was thought earlier. The a state is not autoxidizable, and does not react with O_2, CO or CN^-. The a_3 form does react with these.

Purified heart muscle C.o. is a tetrameric heme lipoprotein, which contains 4 heme and 4 copper chromophores per lipid-containing (M_r 440000) or lipid-free (M_r 350000) molecule. It is converted in the presence of low concentrations of guanidine (1 M) or dodecyl sulfate (0.5%) to a dimeric form (M_r 190000) which is 2 to 3 times as active as the tetramer. The monomer can be obtained by raising the pH, but, like the tetramer, it is not very active. Higher concentrations of dodecyl sulfate (1 to 5%) cause the C.o. monomer (M_r 90000 to 100000) to dissociate into 4 to 6 nonidentical polypeptide chains (main species M_r 11500, 14000, 20000, 39000).

In contrast to the heart muscle C.o., the C.o. monomer from bacteria, e.g. *Pseudomonas*, the autoxidizable and copper-free cytochrome a_2 (M_r 120000) contains only two subunits (M_r 58000). It is not yet known which of the chains carries the prosthetic groups.

Cytochrome P450: a cytochrome whose reduced complex with carbon monoxide absorbs maximally at 450 nm. C.P450s function as terminal oxidases in C.P450-dependent monooxygenase systems (Figs. 1 & 2). The system comprises C.P450 reductase (a flavoprotein), C.P450, phospholipid and possibly nonheme iron. It is uncertain whether the phospholipid is directly involved in electron transfer, or whether it has only a structural role. In some systems electrons may be also transferred from cytochome b_5. The ultimate source of reducing power for the system operating via C.P450 reductase is NADPH. Prokaryotic C.P450s are soluble, whereas eukaryotic C.P450s are membrane-bound within the endoplasmic reticulum (e.g. in liver) or in the inner mitochondrial membrane. C.P450s in the mitochondria of the mammalian adrenal cortex are responsible for some of the hydroxylations in corticosteroid biosynthesis [S. Takemori and & S. Kominami *Trends Biochem. Sci.* **9** (1984) 393–396]. In steroidogenic tissues, like adrenals, placenta and brain, the C.P450 responsible for cholesterol side chain cleavage (EC 1.14.15.6) (see Steroids) is present in the inner mitochondrial membrane. [B. Walther et al. *Arch. Biochem. Biophys.* **254** (1987) 592–596] Mammalian liver contains a superfamily of C.P450s with overlap-

ping substrate specificities, which is responsible for the oxidative conversion of various endogenous metabolites (e.g. bile acid biosynthesis) and xenobiotics into more polar compounds. The latter are more easily excreted, either alone or conjugated with sulfate or glucuronate. Reactions catalysed by C.P450s are shown in the Table. C.P450s are also involved in fatty acid metabolism, prostaglandin biosynthesis, leukotriene biosynthesis, steroid biosynthesis, 1,25-dihydroxyvitamin D_3 biosynthesis and many other endogenous processes.

C.P450-mediated oxidation of some xenobiotics may render them toxic or carcinogenic, e.g. Aflatoxins (see), Estragole (see) and Safrole (see) are converted into carcinogens. Polycyclic hydrocarbons

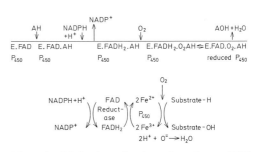

Figure 1. *Mechanism of action of cytochrome P450 systems.* above: Cleland notation, showing order of binding and removal of substrates and products. below: Electron transfer to oxygen and substrate via C.P450 reductase and C.P450.

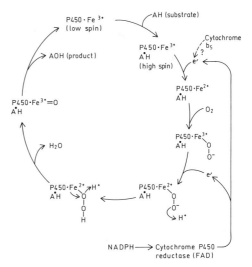

Figure 2. *Mechanism of action of cytochrome P450.* AH, substrate. AOH, hydroxylated product. Binding of oxygen to the Fe(II) form of the hemoprotein, followed by delocalization of an electron, is an essential stage in all Fe-dependent oxygenase reactions, and in the binding of oxygen to carrier hemoproteins like hemoglobin.

Cytochrome P450

Table. Reactions catalysed by Cytochrome P450.

1. *Hydroxylation of aromatic compounds,* e.g. salicylic acid, phenobarbital, acetanilide, synthetic and natural estrogens, diphenyl.

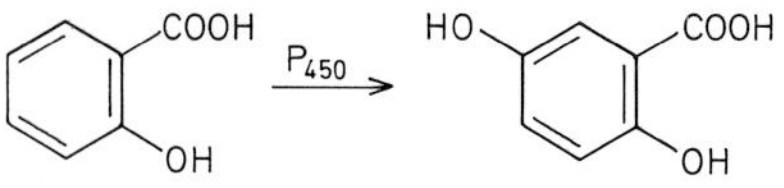

Salicylic acid

2. *Hydroxylation of aliphatic compounds,* e.g. pentobarbital, antipyrine, tolbutamide, imipramine.

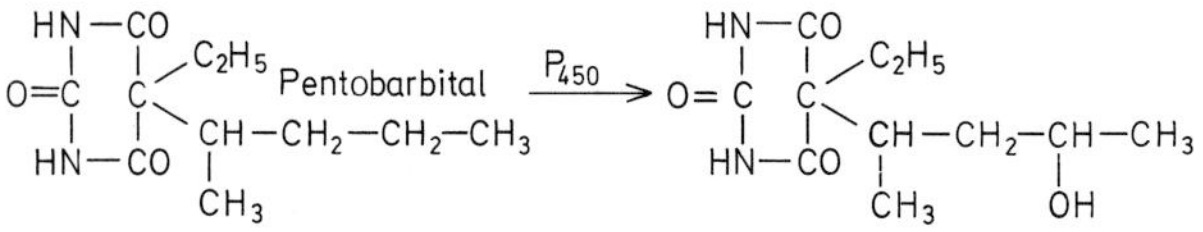

3. *O-Dealkylation,* e.g. phenacetin, griseofulvin, codeine.

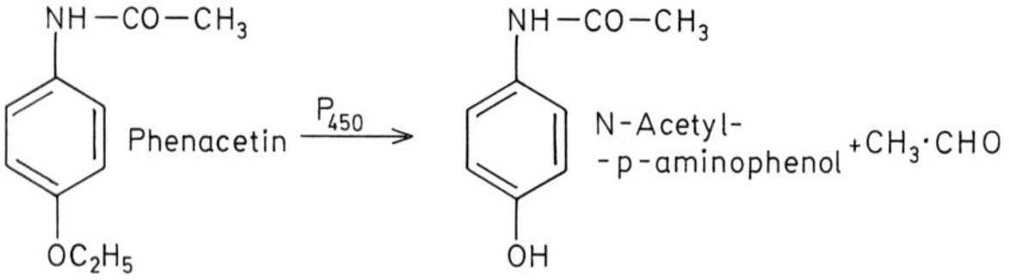

4. *N-Dealkylation,* e.g. aminopyrine, chlorpromazine, ephedrine, morphine.

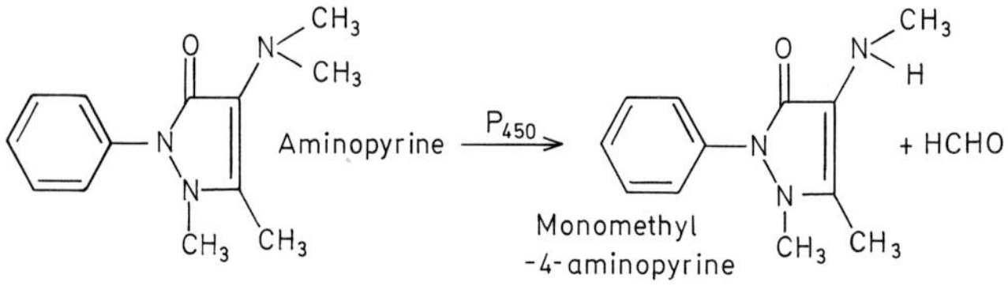

5. *S-Dealkylation,* e.g. 6-methylmercaptopurine, methylthiobenzylthiazole, dimethylsulfide.

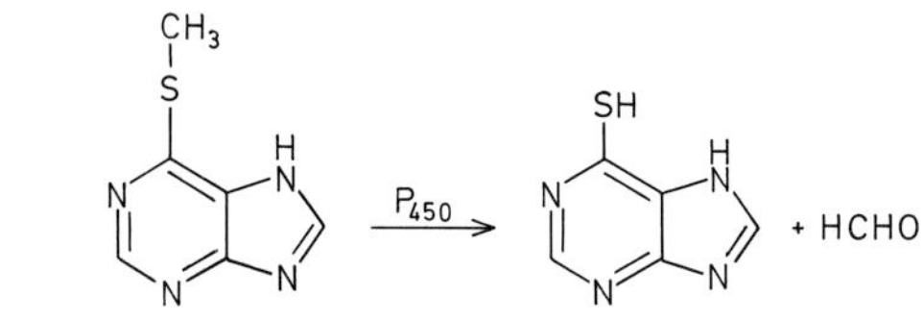

6-Methylmercaptopurine   6-Mercaptopurine

6. *N-Oxidation,* e.g. 2-acetylaminofluorene, nicotinamide, trimethylamine, guanethidine, chlorpromazine, imipramine.

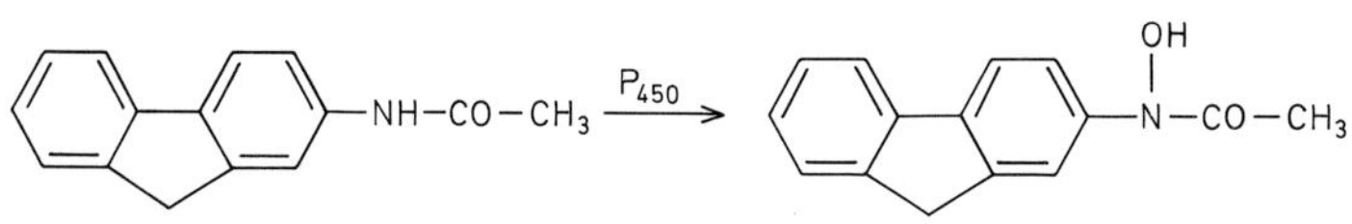

2-Acetylaminofluorene   N-Hydroxy-2-acetylaminofluorene

144

7. *Dehalogenation*, e.g. halothane, carbon tetrachloride, DDT, triiodothyronine.

$$CF_3-CHBrCl \xrightarrow{P_{450}} CF_3-COOH \ + \ Br^- + Cl^-$$

Halothane   Trifluoroacetic acid

8. *Sulfoxidation*, e.g. chlorpromazine.

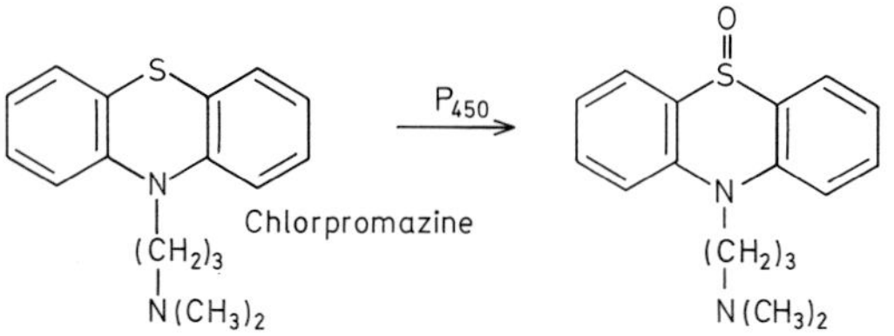

Chlorpromazine

$(CH_2)_3$

$N(CH_3)_2$

P_{450}

$(CH_2)_3$

$N(CH_3)_2$

9. *Phosphothionate oxidation*, e.g. parathion.

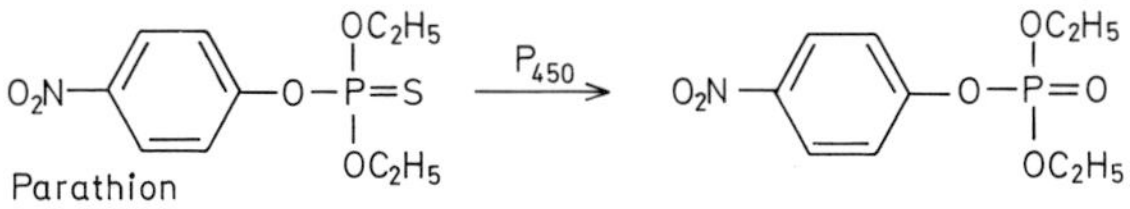

Parathion

10. *Deamination*, e.g. amphetamine, ephedrine.

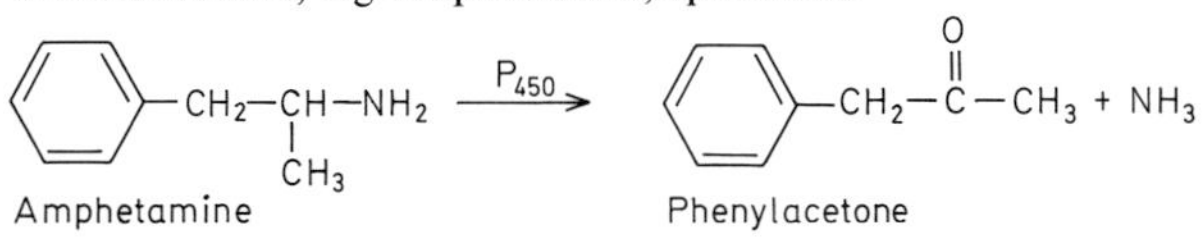

Amphetamine   Phenylacetone

11. *Epoxidation*, e.g. benzpyrene, Aflatoxin (see).

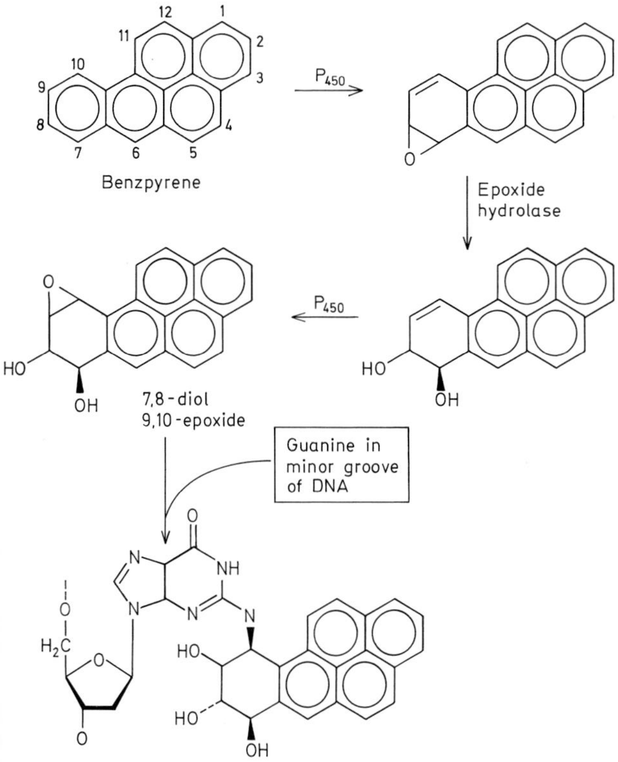

Benzpyrene

Epoxide hydrolase

P_{450}

7,8-diol
9,10-epoxide

Guanine in minor groove of DNA

and halogenated polycyclic hydrocarbons are also rendered carcinogenic by the action of C. P450s (see Table, see Bay-region theory of carcinogenesis).

Administration of a xenobiotic to an animal promotes the hepatic synthesis of an appropriate C. P450. Thus phenobarbital feeding causes an increase in the hepatic C. P450 content in rats, and the increase is largely accounted for by C. P450b, which has a high specificity for the hydroxylation of phenobarbital. On the other hand, a different species (C. P450c), specific for the epoxidation of polycyclic hydrocarbons, is promoted by administration of benzpyrene or methylcholanthrene. Isosafrole promotes the synthesis of C. P450d. In these stimulation studies, several species of C. P450 are affected, but usually one species increases above the others. In contrast, Arochlor (a commercial mixture of halogenated polycyclic hydrocarbons, which is strongly carcinogenic) causes up to 40-fold increases in hepatic C. P450b, c, d and e, with little or no effect on C. P450a and other minor species.

Stimulation of synthesis of C. P450s occurs at the transcriptional level, and different cytosolic receptors have been identified for benzpyrene, dioxin, isosafrole and 3-methylcholanthrene. The analogy with the mode of action of steroid hormones is striking. It must be remembered, however, that the existence of cytosolic steroid receptors in vivo is doubted [J. Gorski et al. *Molec. Cell. Endocrinol.* **36** (1984) 11–15]. Cytosolic receptors are probably artifacts of cell disruption, and steroid receptors are thought to reside only in the nucleus. This may also apply to cytosolic receptors of C. P450 substrates. [S. K. Yang et al. *Science* **196** (1977) 1199–1201; D. Pfeil & J. Friedrich *Pharmazie* **40** (1985) 217–221; J. Friedrich & D. Pfeil *Pharmazie* **40** (1985) 228–232; H. V. Gelboin & P. O. P. Ts'o (eds.) *Polycyclic Hydrocarbons and Cancer* Vol. 1, *Environmental Chemistry and Metabolism* (Academic Press, 1978); S. Arnold et al. *Biochem. J.* **242** (1987) 375–381]

Cytochromes: a group of heme proteins which serve as particle-bound redox catalysts in respiration, energy conservation, photosynthesis and some processes in anaerobic bacteria. They act as electron donors and acceptors by reversible valence changes in the iron atom at the center of their porphyrin complex:

$$Fe^{3+} \underset{-e^-}{\overset{+e^-}{\rightleftarrows}} Fe^{2+}.$$

The importance of the cytochromes can be seen from the fact that they are very old proteins, dating back more than 2 billion years, and their structures have been modified only insignificantly by point mutations. They are found in all organisms. On the basis of structures and their spectra, especially the α, β and γ bands, the C. fall into three main groups, *a*, *b* and *c*. About 30 different C. are known, and they are distinguished by subscripts, e. g. C. b_1. All three types are found in the mitochondria of higher plants and animals, where they are essential components of the Respiratory chain (see).

The *cytochrome a/a₃* complex is identical with Cytochrome oxidase (see). It is assumed by some, on the basis of absorption spectra and behavior in the

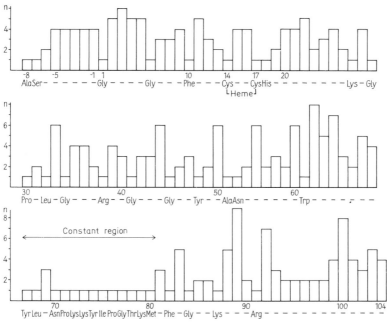

Figure 1. The primary structures of cytochrome c in 40 species (13 mammals, 5 birds, 2 reptiles, 1 frog, 5 fish, 1 snail, 4 insects, 5 higher plants, 4 fungi). The names of the 35 or 37 invariable residues (in cytochromes containing 104 and 112 residues, respectively) are given. The number of different amino acids found in each position is given by the height of the column.

presence of inhibitors, that there are two components, a and a_3, but they appear to be part of the same protein molecule. Others feel there is one C. which oscillates between two very different configurations.

The prosthetic group of *cytochrome b* is the same as that of hemoglobin, iron(II)-protoporphyrin IX. Of the respiratory chain C., it has the lowest redox potential and therefore lies between ubiquinone and C.c. C.b is very firmly bound to the mitochondrial membrane, and can only be extracted with detergents. It is a dimeric protein (M_r 60000) with 1 heme group per monomer. Its central iron atom, like that of C.c, is not autooxidizable and does not react with CO or cyanide. It has been reported that C.b exists in two forms, C.b_K and b_T, which have different redox potentials. C.b_T is believed to have a function in the transfer of energy in the course of electron transport.

Another C.b is the C.b_5 from the microsomal fraction of bird and mammal livers. This C. is thought to deliver electrons to the fatty acid desaturase system of the endoplasmic reticulum. The heme group is noncovalently bound to the protein via histidine and does not react with O_2. It also protects the C.b_5 molecule from denaturation and proteolytic attack. C.b_5 solubilized by detergent treatment is an oligomer (M_r 120000) of several monomers (M_r 16000, 126 amino acids), while C.b_5 solubilized by protease or lipase treatment has only 82 to 98 amino acids, depending on the species. The sequences of these fragments are known. They have 50% α-helix and 25% β-structure.

Other C.b_5 are the soluble C.b_{562} from *Escherichia coli* (110 amino acids, M_r 12000, sequence and heme binding similar to myoglobin), which has the highest potential of the known C.b; and C.b_6 and C.b_{559} from the chloroplast grana of higher plants. These can only be solubilized by combined treatment with the detergent Triton X-100 (0.1%) and 4M urea, and are obtained in disaggregated form. They are both part of the photosynthetic electron transport chain. C.b_1 and b_2, from *Escherichia* coli and yeast, respectively, are both tetramers with dehydrogenase activity.

C. P-450 (see), a mixed-function mitochondrial oxidase, also belongs to the C.b group. The term P-450 refers to the unusual absorption maximum at 450 nm of the CO-complex. This C. is involved in a number of steroid hydroxylation reactions in the adrenal cortex and is a prosthetic group in the monooxygenase which removes the side chain from cholesterol. The structure of the active oxygen on C. P-450 is unknown. It has an unusually high partial specific volume of 0.765, a M_r 850000, is composed of 16 identical subunits (M_r 53000 each), and contains 8 heme units.

Cytochrome c is the most widespread and best studied C. It is a central component of the Respiratory chain (see) in mitochondria. Because it is easy to extract and relatively small (C.c. of vertebrates, 104 amino acids, M_r 12400; of higher plants, 111 amino acids, M_r 13100) C. c. has been used for phylogenetic studies (Fig. 1). The heme group (Fig. 2) is joined to the apoprotein by two thioether bonds to cysteine residues in the interior of the molecule. The iron

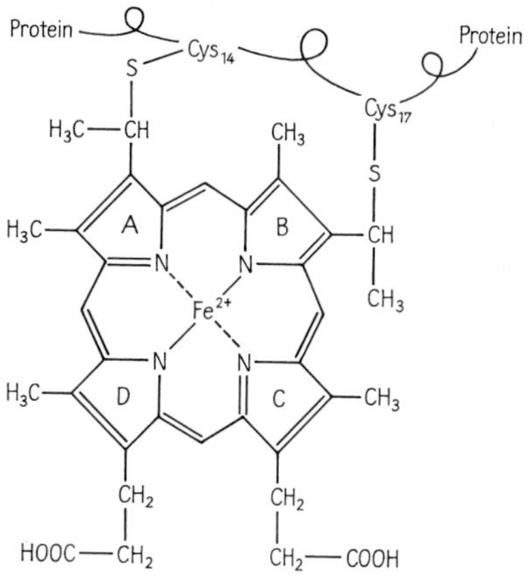

Figure 2. Prosthetic group of the cytochromes c

atom is complexed with two other interior residues, methionine 80 and histidine 18, which prevent native C.c from reacting with O_2 or other heme-complexing agents, such as CO. However, aggregates of C.c are biologically inactive and autooxidizble. They are deaggregated and simultaneously reactivated by addition of urea or guanidine HCl. All C.c from higher organisms whose sequences are known have an N-acetylalanine or N-acetylglycine at the N-terminus. In invertebrates, there is a chain of at most 7 amino acids in place of the acetyl residue. Lysines 72 and 73 play a role in the interaction of C.c with cytochrome oxidase. The basicity of C.c (IP=10) is due to lysines 27, 79, 87 and 100, and arginines 38 and 91.

The primary sequences of more than 50 C.c from phylogenetically distant species have been compared (Fig. 1). The molecule has been extremely conserved; 35 of the 104 to 112 residues are invariant, with most of these invariant residues being located between positions 17 and 32, and 67 and 80. The configurations of the chains of four species studied, horse muscle, bonito, tuna and *Rhodospirillum*, were also found to be very similar. The molecule is unusual in undergoing a major structural change from the compact iron(III) form to the expanded iron(II) form. Some of the structural characteristics of C.c, such as the large helix regions found only at the N- and C-termini, are shared by the prokaryotic C.c_2 and C.c_{550} (137 amino acids) from *Rhodospirillum* and *Micrococcus* (Fig. 3).

In addition to the classical C.c, eukaryotes have a second, C.c_1, which is larger (M_r 37000), insoluble, and has a different amino acid composition. C.c_1 is part of cytochrome c reductase.

Bacterial C.-c are similar to eukaryotic C.c, but they do not react with the mammalian cytochrome c oxidase. The following bacterial C.c have been studied: C.c_2 (112 amino acids, M_r 13500, IP 6.3, primary and tertiary structure known) from *Rhodospirillum rubrum*; C.c_3 (102, 107 or 111 amino acids, with varying primary structure, 2 heme groups) from *Desulfovibrio* and other sulfate-reducing bacteria; C.c_4 (M_r 24000, a dimer with 2 heme groups) and C.c_5 (M_r 24400, a dimer) from *Azotobacter*. C.c_5 from

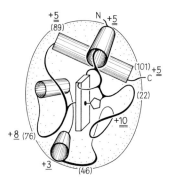

Figure 3. Folding of the polypeptide chains in vertebrate cytochrome c (104 residues) and two bacterial cytochromes, c_2 (112 residues) and c_{550} (137 residues). The segments of α-helix are indicated by cylinders, and the heme by the rectangular plate. The numbers in parentheses show the numbering of the amino acid sequence in horse cytochrome c.

Pseudomonas (87 amino acids, M_r 10100, sequence known) is an acid protein; C.*cc'* (monomer obtained from guanidine treatment of the dimer, 125 amino acids, M_r 14000, 2 heme groups) from *Chromatium* and C.*c'* (127 amino acids, primary structure similar to that of C.*cc'*) from *Alcaligenes*, both photosynthetic bacteria; C.*c551* (82 amino acids, M_r 9600, sequence known for 5 species, acidic protein), an electron carrier for the nitrite reductases from *Pseudomonas*. C.*c553* (82 amino acids, M_r 9600, basic, sequence known) and C.*c555*, also called C.*c6* or C.*f*, are components of the sulfide reductases of green sulfur bacteria. C.*f* is also found in the chloroplast membranes of euglena *(Euglena gracilis)* and in higher plants, where it is part of the photosynthetic electron transport chain, along with C.*b559*. C.*f* from spinach (M_r 270000) consists of 8 subunits (M_r 34000), of which only 4 carry a heme group. C.*f* from *Euglena gracilis* (IP 5.5, M_r 13500) is functionally similar to C.*c*.

Historical: The C. were discovered by Mac Munn and rediscovered in 1925 by Keilin. The discovery that cellular respiration is based on an iron catalysis was made by Otto Warburg.

Cytokinins, phytokinins, kinins: a group of plant hormones which promote cell division and generally stimulate plant metabolism, in particular RNA and protein synthesis. C. are generally *N*-substituted derivatives of adenine (Fig.1). Together with other plant hormones (gibberellins and auxins), they mediate the plant's response to environmental factors, such as light.

The C. are synthesized mainly in the roots of higher plants, and they are not subject to much transloca-

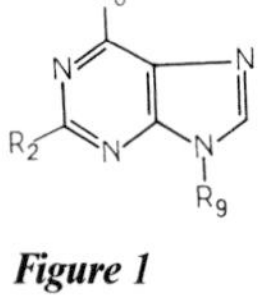

R6, Zeatin and N-Benzyladenine structures

N-Benzyladenine

SD 8339

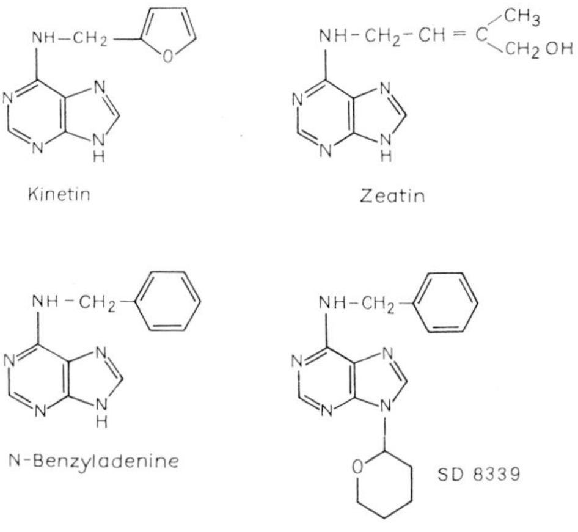

Figure 2

tion. It is interesting that they occur in certain transfer RNA molecules: 0.05 to 0.1% of the purine bases in tRNA have cytokinin activity. In the intestinal bacterium *Escherichia coli,* the tRNA species for phenylalanine, leucine, serine, tyrosine and tryptophan contain C. N^6-(γ,γ-dimethylallyl)-Adenosine (see) also occurs in the free form.

The most important C. are Kinetin (see), Zeatin (see) and Dihydrozeatin (see). The synthetic C., *N*-benzyladenine (see 6-Benzylaminopurine) and SD 8339 (Fig.2) also have high cytokinin activity.

Biotests for the detection and quantitative measurement of C. activity measure one of the following: 1. stimulation of cell division in tissue cultures, for example in tobacco or soybean callus tissue; 2. enlargement of cells in the leaf disc test, often with bean or radish leaves; 3. inhibition of the loss of chlorophyll from detached leaves; 4. promotion of the germination of seeds in the dark.

The first C. to be discovered was kinetin, which was isolated from hydrolysed herring sperm DNA by Skoog and Miller in 1955. The first C. obtained from a plant was zeatin, which was isolated in 1964 from maize kernels.

Cytolipin K: see Inborn errors of metabolism.

Cytolysosomes: see Intracellular digestion.

Cytoplasm: see Cell, 2.

Cytoplasmic inheritance: transfer of genetic information in eukaryotic sexual reproduction, which is not carried by the chromosomes of the nucleus. C.i. is due to extrachromosomal genetic carriers, e.g. mitochrondrial and plastid DNA. C.i. does not obey Mendelian rules, and it permits mixing of cytoplasmic genetic factors during mitosis. Certain petite mutations of yeast, the killer property of certain strains of *Paramecium* and leaf pigmentation in *Antirrhinum majus* are examples of properties transmitted by C.i.

Cytoplasmon: the total extranuclear genetic information of a eukaryotic cell, excluding that in the mitochondria and plastids.

Cytosine, abb. *C* or *Cyt, 6-amino-2-hydroxypyrimidine:* one of the pyrimidine bases of DNA and RNA. M_r 111.1, m.p. 320 to 325 °C (d.). C. is synthesized as the triphosphate from UTP by an

R6, R2, R9 purine structure

Figure 1

ATP-dependent amination. The donor of the amino group is glutamine in animal tissues, and ammonia in bacteria. C. is also a component of various nucleoside antibiotics.

Cytosine arabinoside: see Arabinosides.

Cytoskeleton: a three-dimensional network of fibrous proteins in the cytoplasm which provide structural support, motility and a scaffolding along which intracellular bodies can be moved. There are three distinct components of the C., the microtubules (25 nm diameter), the microfilaments (6-8 nm diameter) and the intermediate filaments (10 nm diameter).

Microtubules are polymers of the protein *tubulin* which, in vivo, are probably always associated with other proteins, the *Microtubule Associated Proteins* (MAPs). The functions of microtubules are diverse. They make up the mitotic and meiotic spindles, cilia and flagella, and in axons and dendrites, they serve as cables along which organelles are moved by kinesin and the still unnamed retrograde translocator [R. D. Vale et al. *Cell* (1986) **43** 623-632]. It may be assumed that different MAPs are associated with microtubules which are performing different functions. In addition, the tubulins are a family of related proteins which are the products of several genes. Although the sequences of amino acids have been highly conserved throughout metazoan evolution, there has been a divergence among different tubulin genes within a species, and these genes are differentially expressed during development. Tubulins have been amenable to isolation and purification because they bind specifically to the drug colchicine. Native tubulin is a dimer of α and β subunits, each of M_r 50000. The C-terminal regions of both subunits are very acidic; the last 40 amino acid residues of the α-subunit include 16 Glu and 3 Asp. The tubulin dimer contains 2 bound guanosine nucleotides. One of these, in the E site, exchanges readily with exogenous GTP, but the other, in the N site, does not. After polymerization, neither site exchanges with the medium. Blocking or oxidation of the SH groups in tubulin prevents assembly of microtubules in vitro.

At relatively high concentrations, purified tubulin dimers will polymerize in vitro, and they are capable of hydrolysing GTP. Polymerization requires GTP, Mg^{2+} and EGTA to chelate Ca^{2+}, and occurs only at 35 °C. The microtubules depolymerize when cooled. In the presence of MAPs, polymerization occurs at lower tubulin concentration, and the resulting microtubules are more stable.

Although tubulin has GTPase activity, and although GTP hydrolysis accompanies microtubule assembly, the addition of a tubulin dimer to the polymer does not require GTP hydrolysis. Instead, the GTP appears to be hydrolysed after incorporation of the dimer into the polymer. A region of GTP-tubulin dimers forms near the plus end of the microtubule; somewhat further back the dimers bear GDP in the E site. The dimers released from the minus end have one GTP and one GDP bound to them. It seems likely that the hydrolysis of GTP stabilizes the polymer.

Assembly of the microtubules is asymmetric. At one end, the plus end, tubulin dimers associate with the tubule faster than they dissociate, while at the other end, the minus end, dissociation occurs more rapidly than association. At equilibrium, the rate of addition of subunits at one end of the microtubule is equal to the rate of loss at the other end; the process is referred to as "treadmilling", because a given tubulin dimer will pass slowly from the plus to the minus end of the microtubule. In many cells, microtubules are stabilized by being anchored, presumably at the minus end, to microtubule organizing centers (MTOC). This is true of the neuronal axon, polar and mitotic microtubules, but kinetochore microtubules appear to be attached to the kinetochore by their plus ends. (This may be determined by decoration of the microtubules with Dynein (see) or C-shaped protofilament sheets under appropriate conditions.)

MAPs may be defined as proteins with specific binding sites for tubulins. It is not known whether dynein, kinesin and the retrograde translocator (see Kinesin) are MAPs by this definition, because their affinity to microtubules may be mediated by other MAPs. The classic MAPs are listed below.

Protein	M_r	Properties
MAP$_1$	350000	Projection on microtubule surface
MAP$_2$	270000	Projection on microtubule surface
	70000	Protein associated with MAP$_2$
Tau	55000-62000	Asymmetric protein
Type II cAMP-dependent protein kinase	54000-39000	Enzyme associated with MAP$_2$
LMW MAPs	28000-30000	Light chains of MAP$_1$

The biological functions of MAPs are not known. Although they promote microtubule assembly in vitro, they have not been observed to do so in vivo. MAP$_2$ will bind to actin filaments, secretory granules, coated vesicles and neurofilaments. However, it is not certain that this binding is specific. It has been demonstrated by immunofluorescence that MAP$_2$ binds to intermediate filaments (vimentin). [T. W. McKeithan & J. L. Rosenbaum in J. W. Shay, ed. *The Cytoskeleton*, vol. 5 of *Cell and Muscle Motility* (Plenum Press, New York, 1985) pp. 255-288; R. B. Vallee, *ibid.* pp. 289-311; D. W. Cleveland & K. F. Sullivan, *Ann. Rev. Biochem.* **54** (1985) 331-365]

Microfilaments consist of Actin (see) and actin-binding proteins. *Stress fibers*, which were first observed by bright-field microscopy, were eventually shown to be microfilaments. Stress fibers are bundles containing microfilaments with opposite polarities; they contain actin, α-actinin and myosin or myosin-like protein. Tropomyosin, filamen and vinculin are located in some domains of stress fibers. Stress fibers can contract in vitro, but in vivo they probably do not. Instead, they probably act as tensile elements which resist motion. Their absence from rapidly moving cells suggests this interpretation. Stress fibers probably anchor the cell to its substrate. *Vinculin*, M_r 130000, is associated not only with the ends of stress fibers, but with substrate adhesion plaques and cell-cell contact regions. Fibronectin (see), which also mediates cell adhesion, is another organizer of microfilaments. Finally, the patches formed by cross-linking of cell-surface proteins by antibodies or lectins are associated with stress fibers. In addition to stress fibers, microfilaments make up polygonal nets within the cell; these have been dramatically demonstrated by immunofluorescence photographs of cells treated with anti-actin antibodies. Filamen (M_r 250000) has been shown to cross-link microfilaments in vitro. Fascin and fimbrin are proteins which promote formation of actin bundles, while actin gelation protein and actin binding protein have been found to promote formation of an irregular mesh. [H. R. Byers et al. in J. W. Shay, ed. *The Cytoskeleton*, vol. 5 of *Cell and Muscle Motility* (Plenum Press, New York, 1985) pp. 83–137; D. J. DeRosier & L. G. Tilney, *ibid*. pp. 139–169]

Intermediate filaments (IF) are polymers of several different proteins which have very similar secondary structures and form fibers about 10 nm in diameter. They may be isolated because they are specifically bound by colcemid (demecolcine; see Colchicum alkaloids). The expression of these proteins is linked to differentiation of cell lineages, each of which has a characteristic complement of IF proteins when fully differentiated. The following five proteins have been recognized as components of IF for some time: *cyto- (or pre-)keratins*, (from tonofilaments) subunit M_r 40–68000, *neurofilament triplet proteins*, M_r 68000, 160000 and 200000, *glial fibrillary acidic protein*, (from glial filaments) M_r 51000, *desmin*, M_r 53000, and *vimentin*, M_r 53500. (The subunit M_r values given for the first three were determined by SDS–PAGE, while the last two are derived from the known amino acid sequences.) In addition to the above, there are several proteins which may be IF proteins, because they bind to colcemid, but which have not been polymerized in vitro: synemin (M_r 230000, from avian muscle), paranemin (M_r 280000, from embryonic skeletal muscle), a 66 kdalton protein (rat brain, squid axoplasm, possibly identical to the 66 kd heat-shock protein), a 68 kd protein (associated with brain and spinal cord microtubules and skeletal muscle myofibrils, identical to the 68 kd heat-shock protein), a 95 kd protein (myofibrils, IF from various cultured cells), 50 kd proteins (cold-labile neurofilaments), and 60–70 kd proteins associated with IF in cultured cell lines.

IF usually form a network of insoluble protein throughout the cytoplasm. It may be revealed by dissolving the membranes in a detergent like Triton X-100, which allows soluble cytoplasmic components to diffuse away from the cytoskeleton. IF bind the MAPs of microtubules, and form end-to-side linkages with microfilaments (the pointed ends of the IF bind to the actin). Actin and the IF are also linked by 3-nm filaments which may be made up of spectrin (fodrin). In pathological conditions like Alzheimer's disease, tangles of neurofilaments are found in degenerate neurons.

On the basis of their distribution in the cell and their insolubility, IF have long been thought to have a structural function. Recently, P. Traub has proposed that in addition to this function, or instead of it, the IF may have a nuclear function. His hypothesis is based on the observation, firstly, that microinjection of antibodies against vimentin or cytokeratins or both causes the IF system to collapse into a dense ring around the nucleus. However, the treated cells survive for several days and may even divide once or twice. This indicates that the structural function of the IF is not essential to cell survival. Secondly, IF are involved in receptor-mediated endocytosis, vectorial transport of the endosomes, and membrane capping, which occurs when cell-surface antigens are cross-linked by antibodies or lectins. (Capping precedes internalization of the antigen-antibody complexes and is a necessary step in activation of lymphocytes.) Thirdly, vimentin, desmin, glial fibrillary acidic protein and the neurofibrillary triplet proteins bind both single-stranded DNA and ribosomal RNA. In the case of vimentin, the avidity of binding to DNA depends on the GC content of the latter. At neutral pH and physiological ionic strength, vimentin will bind to purified core histones, but not to H1 histone. A mixture of all 5 histones was resistant to hydrolysis by Ca^{2+} activated proteinase, but in the presence of vimentin, they were hydrolysed. This suggests that vimentin could serve to loosen the chromatin structure and make DNA available for transcription.

Traub hypothesizes that the Ca^{2+}-activated proteinase associated with IF may, when activated, convert the proteins to a form which does not readily polymerize, but which binds more readily to histones. The normal intracellular concentration of Ca^{2+} is far too low to activate the proteinase, but it is known that endocytosed vesicles and other membrane structures may release Ca^{2+}. Thus the local concentration could be high enough to activate the proteinase and to cause conversion of the IF protein from the fibrous to the "signal" form. [P. Traub, *Intermediate Filaments* (Springer, Heidelberg, 1985)]

Cytoxan: see Cyclophosphamide.

D

Daidzein: 4′,7-dihydroxyisoflavone, see Isoflavone.

Daidzin: daidzein 7-glucoside, see Isoflavone.

Dalton (symbol Da), **_atomic mass unit_** (symbol u): the symbol, u, for this unit is possibly confusing, in that it also stands for unit; also multiples and submultiples, e. g. mu, could be misleading. The term, atomic mass unit, is also unwieldy. In 1981, the NC-IUB and JCBN therefore asked the IUB and IUPAC to apply to the International Committee of Weights and Measures (Comité International des Poids et Mesures, CIPM) to approve the name "dalton" (symbol Da) as an alternative to the name "atomic mass unit". The dalton is one twelfth of the mass of the nuclide ^{12}C. It is equal to $1.6605655 \times 10^{-27}$ kg within about 6 parts per million.

Dalton complex: see Dictyosomes.

Dark reaction: see Photosynthesis.

Datura alkaloids: see Tropane alkaloids.

DBC coenzyme: see 5′-Deoxyadenosylcobalamine.

DCMU: see Dichlorophenyl dimethyl urea.

ddATP: see Dideoxyribonucleotide triphosphates.

ddCTP: see Dideoxyribonucleotide triphosphates.

ddGTP: see Dideoxyribonucleotide triphosphates.

ddTTP: see Dideoxyribonucleotide triphosphates.

Deamination: removal of the amino group, $-NH_2$ from a chemical compound. a) _Oxidative_ D. is catalysed by Flavin enzymes (see) and pyridine nucleotide enzymes. Amino acids may be oxidatively deaminated to ketoacids (see Amino acids, table 3). b) Transamination involves the transfer of an amino group from an amino to a keto compound, and is important in the synthesis of amino acids from tricarboxylic acid cycle intermediates. The reverse reactions feed excess amino acids into the cycle for oxidation. Ammonia may also be removed from a compound, leaving a double bond, as in the D. of L-aspartate to fumarate.

Debranching enzyme: see Amylo-1,6-glucosidase.

Decarboxylases: enzymes which catalyse the cleavage of CO_2 (see Decarboxylation) from α-ketoacids. One of the most important is Pyruvate decarboxylase (see), which decarboxylates pyruvate to acetaldehyde. It requires thiamin pyrophosphate as a coenzyme.

Decarboxylation: the cleavage of the carboxyl group from a carboxylic acid to produce CO_2. The process occurs several times in the course of the Tricarboxylic acid cycle (see). The D. of β-ketoacids often occurs spontaneously. In biological systems, the oxidative D. of α-ketoacids requires coenzymes such as thiamin pyrophosphate, lipoic acid, coenzyme A, flavin adenine dinucleotide or nicotinamide adenine dinucleotide. The oxidative D. of Pyruvate (see) to acetyl coenzyme A and of α-ketoglutarate to succinyl coenzyme A are nodes at which many metabolic pathways cross. The D. of amino acids is catalysed by pyridoxal phosphate enzymes and produces biogenic amines.

Decoyinin: see Angustmycin.

Defective organism: a term sometimes applied to an Auxotrophic mutant (see).

Deficiency mutants: see Auxotrophic mutants.

Dehydrobufotenin: a Toad poison (see) found in _Bufo marinus_. M_r 202. Minimal lethal dose in mice 6 mg/kg.

7-Dehydrocholesterol, _provitamin D_$_3$_, cholesta-5,7-dien-3β-ol:_ a zoosterol (see Sterols). M_r 384.6, m. p. 150 °C, $[\alpha]_D^{20}$ -114 ° ($c=1$ in chloroform). D. occurs in relatively high concentrations in animal and human skin, where it can be converted to vitamin D_3 by ultraviolet radiation. The prevention and cure of rickets by UV irradiation is due to this conversion.

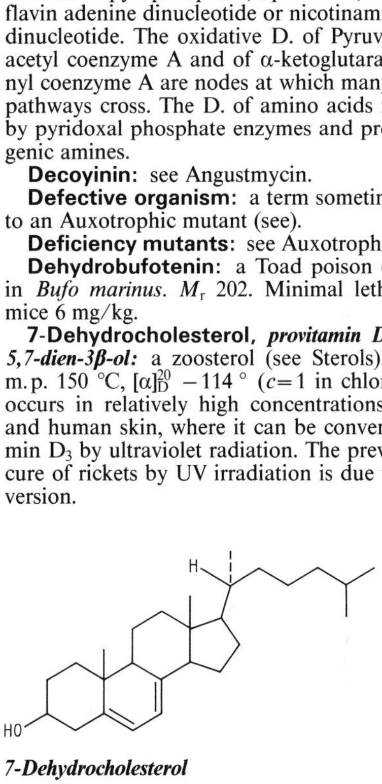

7-Dehydrocholesterol

24-Dehydrocholesterol: see Desmosterol.

11-Dehydrocorticosterone, _Kendall's substance A, 21-hydroxypregn-4-ene-3,11,20-trione:_ a glucocorticoid hormone from the adrenal cortex. M_r 344.43, m. p. 178 to 180 °C, $[\alpha]_D^{25}$ $+258$ ° (ethanol).

Dehydrogenases: redox enzymes which extract hydrogen ($2 H^+ + 2 e^-$) from a substrate (hydrogen donor) and transfer it to a second substrate (hydrogen acceptor). There are two main groups of D., those requiring pyridine nucleotides and those requiring flavin coenzymes (see Flavin enzymes). Those requiring pyridine nucleotides transfer the hydrogen to NAD^+ or $NADP^+$ as the primary acceptor.

Dehydrogenation: removal of hydrogen from a reduced substrate, which thereby becomes oxidized. Enzymatic D. is catalysed by dehydrogenases or oxidases. In general, 2 H atoms (and 2 electrons) are removed at once. The opposite reaction is called *hydrogenation*. The following metabolic D. are common:

Saturated compounds	→ Unsaturated compounds
Alcohols	→ Carbonyls
Aldehyde hydrates	→ Carboxylic acids
Dihydropyridine nucleotides	→ Pyridine nucleotides
Reduced flavin nucleotides	→ Oxidized flavin nucleotides
Hydroquinones	→ Quinones
Amines	→ Imines

D. of amino acids and amines produces unstable imino compounds which spontaneously react with water, producing ammonia and the corresponding carbonyl compounds. The amino acids are converted in this way to the analogous oxoacids.

Deletion: loss of one or more nucleotides of DNA, or of an entire segment of a chromosome; a form of mutation.

Delphinidin: 3,3′,4′,5,5′,7-hexahydroxyflavylium cation, the aglycon of many Anthocyanins (see). D. glycosides are widespread among plants and are responsible for the mauve and blue colors of many flowers and fruits. Some of the more important are tulipanin (3-rhamnoglucoside), the pigment of various tulips, violanin from pansies *(Viola tricolor)* and delphin (3,5-di-β-glucoside) from salvia and delphiniums.

Demecolcine: see Colchicum alkaloids.

Demissine: one of the Solanum alkaloids, a steroid found in wild potatoes *(Solanum demissum)*. It is a glycoalkaloid, consisting of the steroid base demissidine, 5α-solanidan-3β-ol, M_r 399.67, m.p. 221 to 222 °C, $[\alpha]_D$ +30 (chloroform) and the tetrasaccharide β-lycotetrose (D-galactose, D-xylose and 2 D-glucose residues). The aglycon demissidine differs structurally from solanidine (see α-Solanine) in that it has no double bond at C5, and has the 5 α-configuration. D. is repellant to the larvae of potato beetles, so it protects the wild potato from them.

Denaturation: structural change in biopolymers which destroys the native, active configuration. It is brought about by heat, pH changes or chemical agents. See Nucleic acids; Proteins.

Dendrobium alkaloids: a group of terpene alkaloids from various species in the orchid family *Dendrobium*. The D. are sesquiterpenes with one heterocyclic nitrogen atom. They are interesting because alkaloids are seldom encountered in monocotyledonous plants.

Denitrification: see Nitrate reduction.

De novo purine synthesis: see Purine biosynthesis.

De novo pyrimidine synthesis: see Pyrimidine biosynthesis.

Density gradient centrifugation: a method of separating macromolecules on the basis of their density.

In a very high-speed centrifuge, a solution of CsCl will form a stable density gradient after a sufficient time (24–48 h). Macromolecules will come to rest in a layer, or *isopycnic zone,* which corresponds to their buoyant density. This parameter, expressed in g/cm^3, can be measured exactly after the centrifuga-

tion. The densities of the CsCl solution range from 1.3 to 1.8 g/cm^3, which covers the range of most biomolecules, e.g. DNA, 1.7; RNA, 1.6; proteins, 1.35–1.4. Cell organelles, which have lower densities, can be separated in sucrose density gradients. In contrast to the above equilibrium method, sucrose gradient centrifugation can be used as a dynamic method in which the buoyant density is estimated from the rate at which macromolecules sediment in the sucrose gradient. In this case, the gradient is established by carefully layering sucrose solutions of linearly or exponentially decreasing density into the centrifuge tube. The macromolecules are layered onto the top of the solution, and the run is timed so that they do not have time to move all the way to the bottom.

G. B. Cline and R. B. Ryel, in *Methods in Enzymology* (W. B. Jokoby, ed.) Vol. 22 p. 38–50. (1971).

T. J. Bowen, *An Introduction to Ultracentrifugation* Wiley (Interscience) New York, 1970.

5′-Deoxyadenosine: a β-glycosidic deoxynucleoside (see Nucleosides) which is important as a component of vitamin B_{12} (see 5′-Deoxyadenosylcobalamine).

***S*-(5′-deoxyadenosine-5′)-methionine:** see *S*-adenosyl-L-methionine.

5′-Deoxyadenosylcobalamin, B_{12} *coenzyme,* **DBC coenzyme** (DBC = dimethylbenzimidazole cobamide): one of the coenzyme forms of vitamin B_{12} (see Vitamins). In this compound, the 6th coordination position of the cobalt atom in the center of the corrinoid ring is covalently bound to the 5′C atom of the deoxyadenosine. Other cobamide coenzymes contain an *N*-heterocyclic base other than dimethylbenzimidazole. D. is the coenzyme of certain isomerization reactions (Fig. 1). In the isomerization of L-glutamic acid to β-methylaspartic acid (I), there is a reversible transfer of the glycine portion of L-glutamate from the C2 to C3 of the propionic acid moiety, and at the same time, an H atom is shifted in the opposite direction. In the conversion of methylmalonyl-CoA to succinyl-CoA (II), the thioester group is shifted from the 2nd to the 3rd C of the propionic acid part of methylmalonyl-CoA. This reaction plays a part in the biological degradation of branched-chain amino acids and in the propionic acid metabolism of *Propionibacterium.*

Cobamide coenzymes are also involved in the degradation of L-lysine to fatty acids and ammonia in

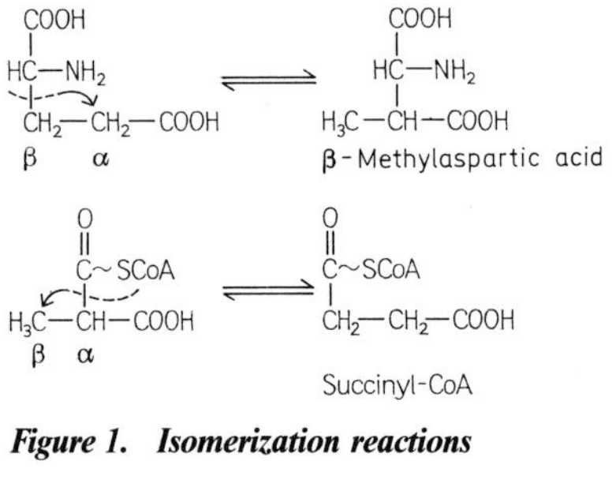

Figure 1. Isomerization reactions

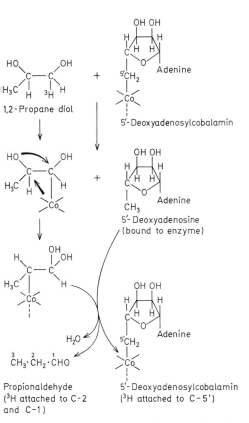

1,2-Propane diol

5'-Deoxyadenosylcobalamin

5'-Deoxyadenosine
(bound to enzyme)

H_2O

$\overset{3}{C}H_3 \cdot \overset{2}{C}H_2 \cdot \overset{1}{C}HO$

Propionaldehyde
(^{3}H attached to C-2
and C-1)

5'-Deoxyadenosylcobalamin
(^{3}H attached to C-5')

Figure 2. Mechanism of action of 5'-deoxyadenosylcobalamin as the cofactor of propanediol dehydratase (EC 4.2.1.28). The mechanism is supported by the migration of ^{3}H from C-1 of the substrate to C-2 of the substrate and C-5' of the cofactor. The same enzyme also dehydrates ethylene glycol to acetaldehyde.

Clostridium and the conversion of 1,2-diols to aldehydes in various microorganisms (Fig. 2).

Deoxycholic acid, *3α,12α-dihydroxy-5β-cholan-24-oic acid:* a bile acid. M_r 392.56, m.p. 176 to 177 °C, $[\alpha]_D^{20}$ +55 ° (ethanol). D. is found in the bile of most mammals, including man, dog, ox, sheep, and rabbit. It can be used as starting material for the partial synthesis of therapeutically important steroid hormones.

3-Deoxygibberellin C: see Gibberellins.

Deoxyhemoglobin: see Hemoglobin.

6-Deoxyhexoses: see 5-Methylpentoses; Deoxy sugars.

Deoxynucleoside: see Nucleosides.

Deoxynucleoside phosphorylases: see Nucleosides; Salvage pathway.

Deoxynucleotide: see Nucleotides.

Deoxynucleotide biosynthesis: see Nucleotides.

Deoxyribonuclease I (EC 3.1.21.1): an enzyme which preferentially attacks double-stranded DNA

and produces 5'-phosphodi- and -oligonucleotides by endonucleolytic cleavage. It is inhibited by a protein (M_r 49000).

Deoxyribonuclease II (EC 3.1.22.1): a pancreatic enzyme (M_r 31000, single chain, pH optimum at 7) which requires calcium for activity and stability.

Deoxyribonucleic acid, abb. *DNA:* previously also called thymus DNA; a polymer of deoxyribonucleotides found in all living cells and some viruses. It is the carrier of genetic information which is passed on from generation to generation by an exact replication of the DNA molecule.

Structure: Each mononucleotide unit of the DNA polymer consists of phosphorylated 2-deoxyribose which is glycosidically linked to one of four bases: adenine, guanine, cytosine or thymine (abb. A, G, C or T, respectively). In the DNA of higher organisms, some of the cytosine is methylated in the 5 position, and in bacteriophage DNA, some is replaced by hydroxymethylcytosine. The structures of the bases are shown under Nucleosides.

The mononucleotides are linked by 3',5'-diester bridges (see Nucleic acids; Phosphodiesters) in an unbranched polynucleotide chain. The base contents of DNA from different organisms vary widely (from 18% A in tuberculosis bacteria to 30% A in calf thymus), but the amount of A is always equal to the amount of T, and the amount of G is always equal to the amount of C (see GC content). This fact and the results of X-ray diffraction studies by Wilkins and Franklin led Watson and Crick to propose the double-helix model of the DNA structure [*Nature* 171 (1953) 737.] The life sciences were revolutionized by this concept of DNA structure.

According to the Watson-Crick model, the DNA molecule consists of two complementary, but not identical strands which spiral around an imaginary common axis. The two spiral bands are made up of sugar phosphate chains from which the bases protrude at regular intervals into the interior of the helix. The two strands are held together by hydrogen bonds between bases on opposite strands. In order for the strands to fit together in the helix, a purine on one strand must always be opposed by a pyrimidine on the other. Hydrogen bonds can only form (within the constraints of the double helix) between adenine and thymine (A-T) or guanine and cytosine (G-C), so that the sequence of bases along one strand determines the sequence of the other (see Base pairing). The genetic information is encoded in the sequence of bases in the DNA (See Genetic code). The two strands are antiparallel, meaning that the phosphate diesters between the deoxyribose units read 3' to 5' on one chain and 5' to 3' on the other. In most organisms, only one of the two strands is transcribed; thus it is "codogenic", while the other is "noncodogenic".

The double helix is not symmetrical; it has a broad and a narrow groove between the chains. These provide steric orientation for the processes of replication and transcription (Fig. 1).

This right-handed double helix with 10 base pairs per helical turn is called B-DNA. It probably approximates closely the structure of relaxed (i.e. unstrained) DNA. It is generally accepted, however,

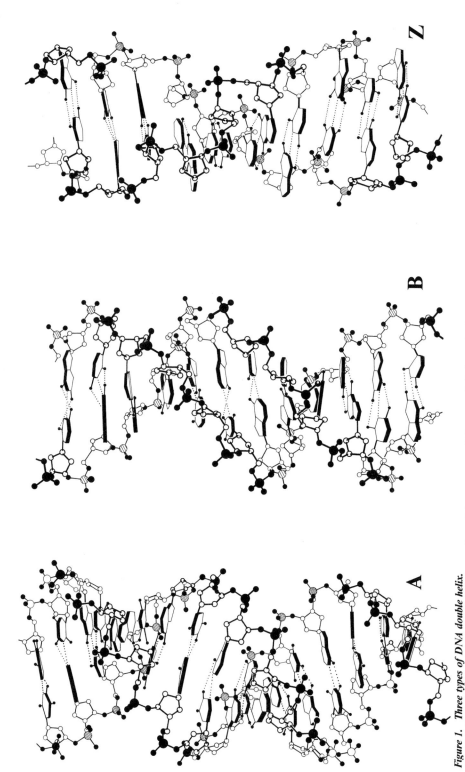

Figure 1. ***Three types of DNA double helix.***
A-DNA (left) structure was generated by extension of the structure of the central six bases in the octamer *GGTATACC*, the crystal structure of which has been determined. B-DNA (center) was generated by repeating the central 10 base pairs of the dodecamer CGCGAATTCGCG. Z-DNA is a left-handed helix of alternating guanines and cytosines; the structure here was generated by extension of the central 4 base pairs of CGCGCG. Hydrogen atoms are not shown. Bases are represented as laminae with black sides. All oxygen atoms bonded to phosphorus are black. Phosphorus atoms are black (front of helix) or hatched (rear of helix). The carbon atoms and ring oxygen of deoxyribose are all represented as open circles. For dimensions see Table 1.

hat DNA is a dynamic molecule, with different conformations in equilibrium with one another. This equilibrium is affected by nucleotide sequence, ionic strength of the environment, presence of proteins e.g. Histones - see - and other DNA binding proteins, see) and the extent to which the molecule is under topological strain.

It has been shown that the DNA fragment d(CpGpCpGpCpG) crystallizes as a left-handed double helix A.Wang, et al., *Nature* **282** (1979) 680-686]. This structure is known as Z-DNA, because an imaginary line joining the phosphate groups around the outer surface of the molecule describes a zig-zag course. (In B-DNA, the phosphate groups follow a smooth spiral) (Fig.1). The two strands of Z-DNA are antiparallel, and complementary bases are hydrogen bonded as in B-DNA; however, the orientation of the bases to the backbone of the molecule is different from that in right-handed B-DNA. In Z-DNA the flat planes of the base ring systems are still more or less at 90 ° to the long axis of the molecule, and parallel with one another, but they are effectively rotated through 180 ° in comparison with bases in B-DNA. In the case of guanine this occurs by rotation about the glycosidic bond, so that guanine residues have a *syn* configuration; in the case of cytosine both the base and the deoxyribose are rotated, so that the cytosine residues retain the *anti* conformation. The result is an alternating orientation of adjacent sugars: the O1' of the dG points down, and the O1' of the dC points up. Because of this, the repeating unit of Z-DNA is a dinucleotide (in B-DNA it is a mononucleotide). (See Table 1 for further comparisons.)

A third type of helix is observed in X-ray studies of DNA made under conditions of low (75%) humidity. A-DNA is right-handed, like B-DNA, but it has nearly 11 base pairs per helix twist, and they are inclined by 13 ° with respect to the plane perpendicular to the helix axis. A-DNA has a very deep major groove, in contrast to the B form. Since the 2'-OH of the ribose moieties of RNA prevent RNA from adopting a B conformation, it has been suggested that DNA-RNA hybrids must have the A conformation.

The mean helix parameters of the three forms of DNA are given in the table.

The "propeller twist" of a base pair is the angle between the planes of the two bases. (One can visualize the "rung" of the DNA ladder as being shaped like a two-bladed propeller rather than like a flat plank). The "base roll" and the "base inclination" refer to the average plane of the base pair. The base inclination is the angle between this average plane and the plane perpendicular to the helix axis, while the base roll is the angle between two successive base pairs. As can be seen from the table, the standard deviations for propeller twist, base roll and base inclination are large. This is probably due to steric interactions between the bases; some combinations can be more tightly packed together than others. Van der Waals attraction between the bases insure that each pair will fit as closely as possible. The variations in the helix parameters are thus due to the base sequence, and can presumably be recognized by the proteins whose function demands recognition of specific sequences.

[Richard E. Dickerson, The DNA Helix and How It is Read, *Scientific American* **249** (1983) 94-112.]

The axis of B-DNA passes through base pairs, whereas the axis of Z-DNA is practically empty, and a single deep, narrow groove extends into the center of the Z-DNA molecule. Part of the imidazole ring of guanine is exposed on the outer convex surface of the Z-DNA molecule. Thus, the double helix of Z-DNA resembles a ribbon of material with a serrated edge of phosphate groups wrapped around an imaginary central axis. In solution, poly(dG-dC)(dG-dC) exists in the B- or Z-conformation. Interconversion of the two forms can be measured from the inversion of the circular dichroism spectrum of the molecule. [F.M. Pohl & T.M. Jovin, *J. Mol. Biol.* **67** (1972) 375-396.] Z-DNA and B-DNA are immunologically distinct. Thus, by using

Table 1. Comparison of B-, A- and Z-DNA.

	B-DNA	A-DNA	Z-DNA
Handedness of helix	right	right	left
Residues per turn	10.0	10.9	12.0
Diameter of helix	20Å		18Å
Distance between adjacent base pairs along the axis (= rise per residue)	3.4Å	2.9Å	G-C, 3.5Å C-G, 4.1Å
Length of one helix turn (= helix pitch)	34Å	32Å	45Å
Rotation of adjacent base pairs relative to each other (= helical twist)			
mean	36 °	33 °	G-C – 51 ° C-G – 8.5 °
observed range	16–44 °	28–42 °	
Propeller twist	11.7 ± 4.8	15.4 ± 6.2	4.4 ± 2.8
Base roll	– 1.0 ± 5.5 °	5.9 ± 4.7 °	3.4 ± 2.1 °
Base inclination	– 2.0 ± 4.6 °	13.0 ± 1.9	8.8 ± 0.7

specific antibodies, it has been shown that the negatively supercoiled plasmid pBR322 contains a section of left-handed Z-DNA sequence, d(CpAp-CpGpGpGpTpGpCpGpCpApTpG) [A. Nordheim et al. *Cell* **31** (1982) 309–318]; the Z-conformation is therefore not restricted to poly(dG-dC) sequences. In this case, formation of the Z-structure is favored because it releases the strain caused by supercoiling. Both B- and Z-DNA probably exist as families of closely related structures, the members of which differ from one another by slight modifications of conformation. Two such conformations of Z-DNA (Z_I and Z_{II}) have been described.

Some viral DNA forms a single-stranded coil rather than a double helix. Both single- and double-stranded DNA can form ring-shaped molecules (bacteria, mitochondria). The rings can be more tightly coiled (superhelices, hypertwisted configurations), the number of "supercoils" depending on the size of the molecule (33 to 44 for mitochondrial DNA). See Superhelix.

Because the double helix must unwind in order for the strands to separate when the molecule is replicated, some scientists have suspected that the double helix observed in crystalline DNA by X-ray diffraction might not exist in solution. In 1976, for example, Rodley et al. proposed the "side-by-side" model, in which the two strands are not twisted around each other. However, in 1983 Iwamoto and Hsu showed that the two chains of a 39-base-pair segment of double-stranded DNA in solution must have been twisted around each other 3 to 4 times in a right-handed direction. This is consistent with the double helix, but not the side-by-side model. [S. Iwamoto & M.-T. Hsu, Nature 305 (1983) 70–72. F.H.C. Crick, *Proc. Natl. Acad. Sci. U.S.A.* 73 (1976) 2639–2643]

On the basis of conformation studies on double-stranded DNA, Cyriax and Gäth concluded in 1978 that transitions are possible between double-helical and non-coiled structures. The non-coiled transition conformation was called the *cis-ladder conformation* because the sugar-phosphate chains have a *cis*-like position with respect to the base pairs, which are arranged like the rungs of a ladder. The DNA strands can go from the *cis*-ladder conformation into other conformations, including helices, without generating strains in the molecule, so this structure may be regarded as a transitional form between the double helix and the single-stranded state, and also between the double helix and other highly ordered structures.

The molecular weight of DNA is difficult to measure, because the molecules break very easily in the course of extraction. The highest measured M_r was 10^9 (about 2×10^6 base pairs), but the estimated M_r of *Escherichia coli* DNA is 2.8×10^9 (about 3 to 4×10^6 base pairs). (The ring-shaped molecule of *E. coli* DNA has been photographed in the electron microscope.) The M_r measured for mammalian DNA is much smaller (10^8), but this may reflect only the difficulty of separating it from chromosomal proteins. There is genetic evidence that the DNA of a single chromosome is one gigantic chain, which would then have to be 10 to 20 times as long as *E. coli* DNA. The *E. coli* DNA molecule is about 1000

μm long, while that of the mitochondria is about 50 μm long.

Prokaryotic DNA is bound to the membrane as giant rings; sometimes the cells also contain smaller fragments, the Plasmids (see) or episomes. About 95% of the DNA of an eukaryotic cell is located in the nucleus, where it is complexed with proteins in the form of Chromosomes (see). The mitochondria and chloroplasts also contain DNA (*extrachromosomal DNA*); mitochondrial DNA accounts for 1 to 2% of the total, while chloroplast DNA may account for up to 5%. The amount of DNA per cell varies widely among different species, but within a given organism it is constant (except for the germ cells of diploid organisms, which contain only half as much as the somatic cells).

The DNA of a cell can be fractionated according to its base composition (see GC content) by density gradient centrifugation in CsCl. The buoyant density, given as g CsCl/cm³, is used to characterize the fractions. Mouse liver DNA, for example, has a main fraction (1.702) and Satellite DNA (see) (1.691). Nuclear DNA from *Euglena* has a buoyant density of 1.707, the mitochondrial DNA, 1.691, and chloroplast DNA, 1.686.

Replication of DNA: Each single strand of the parent duplex acts as a template for the synthesis of a new partner strand, i.e. replication is *semiconservative*. This was shown experimentally by the classical Meselson and Stahl experiment (see). Replication must therefore involve rotation of the entire parent molecule, or a mechanism which breaks and rejoins single strands, or both. Circular double strands cannot replicate without at least one breaking and rejoining event.

When the strands are separated, each base binds a complementary base from the pool of nucleotides (base pairing), so that the sequence of nucleotides in the new strand is determined by that of the old strand. This mechanism allows information (the sequence of bases) to be reproduced and passed on to subsequent generations. The DNA of eukaryotic chromosomes is divided into replicative subregions (*replicons*; Fig. 2), but bacterial DNA is replicated as a single unit.

The circular DNA remains attached to the cell membrane during the entire process, so that when the cell divides, each daughter cell receives a DNA molecule. However, the binding to the membrane could have a regulatory effect on initiation, especially since the attachment point lies near the replicator region. On the other hand, the DNA polymerase and the initiator proteins are associated with the membrane.

The replication process must occur at extremely high speed, since *E. coli* can divide every 20 minutes under optimal conditions. This requires the replication of at least 133 000 base pairs (133 kbases) or 50 μm DNA per min. DNA replication in eukaryotes occurs much more slowly. The process is catalysed by DNA polymerase (EC 2.7.7.7) which joins the four 5'-nucleoside triphosphates into the linear DNA molecule, using a single-stranded DNA as a template. Each nucleotide loses a pyrophosphate group (PP_i), and the energy stored in these bonds drives the reaction.

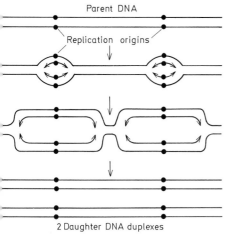

Parent DNA

Replication origins

2 Daughter DNA duplexes

Figure 2. *Replication of a eukaryotic chromosome.* Replication occurs simultaneously in different subregions, which eventually join with one another, producing two separate daughter duplexes.

After *initiation,* which probably involves breaking one of the strands, the ring is progressively unwound and replicated in the clockwise direction (Fig. 3).

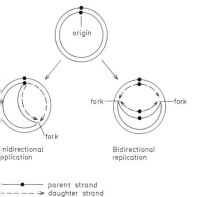

origin

fork — — fork

fork

Unidirectional
replication

Bidirectional
replication

———●——— parent strand
- - - - -> daughter strand

Figure 3. *Unidirectional and bidirectional replication of prokaryotic and viral DNA.* The bidirectional mechanism is found in *Escherichia coli* and other bacteria and many DNA viruses. The unidirectional mechanism has been observed in certain plasmids.
Autoradiographic studies indicate that the ring structure is retained during the entire process. Some viral DNAs are replicated by a unidirectional rolling circle mechanism (Fig. 4).

The enzymology of DNA synthesis has not been completely clarified. DNA synthesis always occurs in the 5'→3' direction (since association between strands is always antiparallel, the template strand is read in the 3'→5' direction). Replication depends on the action of 20 or more enzymes and proteins. The several sequential steps of replication include recog-

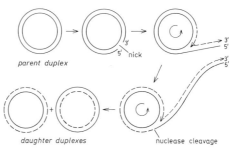

parent duplex

nick

daughter duplexes

nuclease cleavage

Figure 4. *Rolling circle mechanism of replication of some viral DNAs.* The circular parent DNA is first cleaved ("nicked") enzymatically. New nucleotide units are added to the 3'-terminus of the broken strand, and the continuous growth of the new strand displaces the 5' tail of the broken strand from the rolling circular template. Thus, the 5' tail becomes a linear template for the synthesis of a new complementary strand. The duplex originating from the 5' tail is then cleaved from the other daughter duplex by a nuclease.

According to Jacob and Brenner, the synthesis of DNA is repressed during part of the cell cycle, apparently by specific repressors. Replication is initiated by the synthesis of an initiator protein encoded by the regulator gene J. This protein activates the replicator, the starting point of synthesis (Fig. 5).

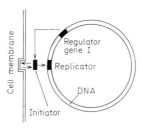

Cell membrane

Regulator gene I

Replicator

DNA

Initiator

Figure 5. *Regulon model of the regulation of replication in circular, bacterial DNA.*

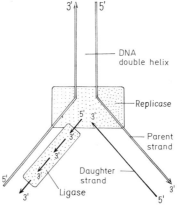

DNA double helix

Replicase

Parent strand

Daughter strand

Ligase

Figure 6. *Replication fork of DNA.* The arrows indicate the direction of synthesis.

nition of the origin or starting point of the process, unwinding of the parent duplex (by helicases, see Topoisomerases), maintenance of strand separation in the replicating region, initiation of daughter strand synthesis, elongation of daughter strands, rewinding of the double helix (see Topoisomerases), and termination of the process. The complex of factors involved in replication is called the *DNA replicase system*, or *replisome* (Table 2; Fig. 6).

It is clear from autoradiographic studies that synthesis occurs simultaneously from the 3' end of one parent strand and the 5' end of the other, but all three of the known DNA polymerases work only in the 5'→3' direction. The 5'→3' parent strand is replicated in short pieces (about 2000 nucleotides in bacteria, less than 200 nucleotides in animal cells). These short pieces, known as *Okazaki fragments*, are later linked by a DNA ligase. Each Okazaki fragment is synthesized as an extension of a short RNA primer (about 10 nucleotides). The RNA primer is synthesized in the 5'→3' direction of the template of the replicating DNA strand by the action of a specialized RNA polymerase, called a *primase*. The pri-

mase is associated with other proteins in a *primo some complex*, which may also be considered a par of the replisome complex (Table 2). DNA synthesis proceeds from the 3' end of the primer. The RNA primer is then removed, one nucleotide at a time, by the 5'→3' exonuclease activity of DNA polymerase I. As each ribonucleotide is removed, it is replaced by a corresponding deoxyribonucleotide by the polymerase activity of the same enzyme. Final splic ing of the extended Okazaki fragments is due to *DNA ligase*, which catalyses phosphodiester bond synthesis between the 3'-phosphate group of the elongating DNA and the 5'-hydroxyl group of the newly synthesized Okazaki fragment. Ligation i coupled to pyrophosphate bond cleavage (in bacte ria, NAD is cleaved to nicotinamide mononucleo tide and AMP; in animal cells, ATP is cleaved to AMP and pyrophosphate) (Fig. 7).

Theoretically, discontinuous synthesis is not neces sary for the progress of replication along the 3'→5 replicating parent strand, but there is evidence tha this may also be discontinuous. In addition to nor mal DNA-dependent DNA synthesis, RNA-depen

Table 2. Some constituents of the replisome of Escherichia coli

Protein	$M_r \times 1000$	No. of subunits	Function	No. of molecules per cell
Single strand DNA binding protein (SSB)	74	4	Binding to single strand after opening of helix.	300
Protein i	66	3		50
Protein n	28	2		80
Protein n'	76	1	Primosome assembly.	70
Protein n''	17	1	Initiation of primosome	–
dnaC	29	1	action. Formation of	100
dnaB	300	6	primer RNA.	20
Primase	60	1		50
DNA polymerase III holoenzyme, which consists of:	760	2		20
α	140	1		–
ε	25	1		–
θ	10	1		–
β	37	1	Chain elongation.	300
γ	52	1		20
δ	32	1		–
τ	83	1		–
DNA polymerase I	102	1	Excision of primer and replacement with DNA.	300
Ligase	74	1	Ligation of extended Okazaki fragments.	300
Gyrase, comprising:	400	4	Supercoiling.	–
GyrA	210	2		250
GyrB	190	2		25
Helicase I (rep protein)	65	1	Helicase.	50
Helicase II	75	1	Helicase.	5000
dnaA	48	–	Binds at the origin of replication.	

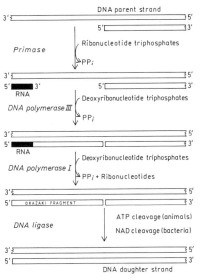

Figure 7. Role of RNA primer and Okazaki fragments in DNA replication.

dent synthesis can occur in some cases (see RNA-dependent DNA polymerase). [A. Kornberg *DNA Replication* (W. H. Freeman & Co., 1980); A. Kornberg *1982 Supplement to DNA Replication* (W. H. Freeman & Co., 1982)]

DNA repair: The cell has a complement of enzymes whose function is to repair damage to DNA caused by chemicals, radiation or replication errors. The damaged spot is recognized and excised by a nuclease, the correct sequence is synthesized by a DNA polymerase, and the ends of the new section are joined to the old by a ligase. UV-damaged DNA can also be repaired by Photoreactivation (see). *Excision repair,* which occurs in bacteria, is similar to the above, except that the DNA polymerase is also an endonuclease, which removes the defective oligonucleotide base by base as it synthesizes the replacement for it.

DNA degradation: In vital cells, DNA is normally not degraded, but there are various enzymes in tissues which degrade the DNA of dead or broken cells (see Nucleases, Deoxyribonuclease, Phosphodiesterase). When bacteriophages enter bacterial cells, they produce nucleases which degrade the host's DNA. (Some bacteria have a significant number of bases which have been methylated, e.g. 5-methylcytosine and 6-methyladenine, which appears to protect their DNA from viral degradation. The T-even bacteriophages, which attack *Escherichia coli,* contain hydroxymethylcytosine in place of cytosine, which protects their DNA from their own deoxyribonucleases.) In the laboratory, DNA can be degraded by acid hydrolysis. It is completely hydrolysed to phosphate, bases and deoxyribose by strong acid, but under milder conditions it can be hydrolysed to nucleosides or nucleotides.

Biological significance: The information for the synthesis of all the cell's proteins is contained in the base sequence of its DNA (see Genetic code). This was proved directly by experiments on Transformation (see) and Transduction (see), and by demonstrating the role of phage DNA (see Phage development). More recent work in genetics is elucidating the role of DNA sequences which do not themselves encode proteins, but which regulate their expression (see Operon, Intron, DNA-binding proteins.)

Deoxyribonucleotides: see Nucleotides.

2-Deoxy-D-ribose: a pentose lacking one hydroxyl group. M_r 134.13, α-form m. p. 82 °C, $[\alpha]_D$ -58 ° (water); β-form, m. p. 98 °C $[\alpha]_D^{20}$ -91 ° $\rightarrow$ -58 ° (water). D. is the carbohydrate part of deoxyribonucleic acids.

β-2-Deoxy-D-ribose

Deoxyribose phosphates: phosphorylated derivatives of the deoxypentose 2-deoxy-D-ribose. They are synthesized biologically by reduction of ribose phosphates in the course of nucleotide synthesis. There are two enzymes in *Escherichia coli* which reduce cytidine diphosphate to deoxycytidine diphosphate. Deoxyribose 5- and 1-phosphate are equilibrated by the enzyme phosphopentomutase (EC 2.7.5.6).

Deoxyribosyltransferases: enzymes which catalyse the transfer of deoxyribose from purine and pyrimidine deoxyribosides to free bases in the synthesis of deoxynucleosides.

Deoxyribotide: see Nucleotide.

Deoxy sugars: monosaccharides in which one or more hydroxyl groups have been replaced by hydrogen. There are two types, those with a methyl group in the terminal position, such as the 6-deoxyhexoses L-fucose and L-rhamnose, and those with the hydroxyl missing from the middle of the molecule, such as the DNA component 2-deoxy-D-ribose. D. are often components of glycosides; D-digitoxose, for example, is the sugar component of many digitalis glycosides.

Deoxythymidine: see Thymidine.

Deoxythymidylic acid: see Thymidine phosphate.

Deposiston: see Ovulation inhibitors.

Depsipeptide: polypeptides which contain ester bonds as well as peptides. The naturally occurring D. are usually cyclic peptides, also called *peptolides,* which generally have α- or β-hydroxyacids as heterocomponents. In the wider sense, this class includes O-peptides and peptide lactones. The most important peptide lactones are the Actinomycins (see), Etamycin (see) and Echinomycin (see); the peptolides include the Enniatins (see), Valinomycin (see), Sporidesmolides (see), Serratamolide (see), Esperin (see), etc. D. are metabolic products of microorganisms which often have very high antibiotic activity. Many D. can be chemically synthesized by methods which are very similar to those used in the chemical synthesis of peptides.

Derepression: the release of an operon from repression of transcription. In prokaryotic cells it occurs by inactivation of a repressor. This may be done either by removal of a corepressor (see Enzyme repression) or by binding of an inducer (see Enzyme induction). The mechanism of D. in eukaryotic cells is unknown, but must involve regulatory proteins and effectors, such as hormones (see Gene activation).

Dermatan sulfate: formerly called β-heparin, and chondroitin sulfate B. D.s. is a mucopolysaccharide containing L-iduronic acid linked α1,3 to N-acetyl-D-galactosamine 4-sulfate; the latter is linked α1,4 to the next iduronic acid residue. See Chondroitin sulfate.

Desmin: a fibrous protein in muscle cells; a component of intermediate filaments (see Cytoskeleton). The subunit M_r is 52000; the protein is structurally related to the α-keratins (see Keratins).

Desmoplakins: fibrous proteins found on the cytoplasmic side of desmosomes (cell-cell contact regions of the plasma membranes of epithelial cells). D. mediate the attachment of tonofilaments (see Cytoskeleton) to the plasma membrane.

Desmosine: see Elastin.

Desmosterol, *24-dehydrocholesterol, 5α-cholesta-5,24-diene-3β-ol:* a zoosterol (see Sterols). D. differs from Cholesterol (see) in having an extra double bond between C-24 and C-25. It is an intermediate in the biosynthesis of cholesterol (see Steroids). D. has been isolated from the barnacle *Balanus glandula,* chicken embryos and rat skin.

Desulfurase: see Desulfurication.

Desulfuricants: anaerobic bacteria of the genera *Desulfovibrio* and *Desulfotomaculum,* whose Sulfate respiration (see) contributes to the process of Desulfurication (see). The most important of these bacteria is *Desulfovibrio desulfuricans. Desulfovibrio* are thought to be responsible for the hydrogen sulfide and other sulfides in the Black Sea.

Desulfurication: the anaerobic degradation of sulfur-containing organic compounds to inorganic sulfur. In these processes, the sulfhydryl groups are removed from proteins by *desulfurases* to yield hydrogen sulfide. D. is also the formation of hydrogen sulfide by desulfuricants.

Detergent degradation: the processes by which microorganisms digest synthetic detergents, thus removing them from the environment. Unbranched hydrocarbon chains can be degraded, while branched-chain compounds resist degradation. Thus it is possible to design biodegradable detergents.

Determinant group: see Antigens.

DETPP: see Thiamin pyrophosphate.

Dexamethasone, *9α-fluoro-16α-methylprednisolone, 9α-fluoro-11β,17,21-trihydroxy-16α-methylpregna-1,4-diene-3,20-dione:* a synthetic pregnane derivative (see Steroids) which is highly antiinflammatory but has little mineralocorticoid activity. It is used for arthritis. D. is synthesized from cortisole.

Dextranase: see Enzymes table 2.

Dextrans: high-molecular-weight polysaccharides synthesized by certain microorganisms. They consist of D-glucose linked α-glycosidically, primarily in 1,6 bonds, but with some 1,3 and 1,4. The M_r

Dexamethasone

of the microbial product is several million. The colloidal osmotic pressure of D. with a M_r of about 75000 corresponds to that of blood, so they are used as a substitute for blood plasma. These smaller D. are produced by controlled microbial synthesis or by partial hydrolysis of larger molecules. The organisms used are the lactic acid bacteria *Leuconostoc mesenteroides* and *Leuconostoc dextranicum,* which are grown anaerobically in sucrose-containing media. D. are the starting material for the dextran gels used for molecular sieves. The polysaccharides are cross-linked to a three-dimensional network, which contains a large number of hydroxyl groups. The degree of cross-linkage determines the size of the pores and the capacity of the gel for water uptake. (More cross-linking means smaller pores and smaller water capacity). These cross-linked D. are sold under the trade name Sephadex.

Dextrin 6-α-D-glucanohydrolase, *oligo-1,6-glucosidase* (EC 3.2.1.10): a glycosidase present in the digestive juices of the small intestine; formerly known as limit dextrinase, or isomaltase. It is specific for the hydrolysis of the α-1,6 bonds in isomaltose and in the oligosaccharides produced by the action of α-amylase on starch and glycogen. Products of the hydrolysis are glucose, maltose and unbranched oligosaccharides, which can then be degraded further by α-amylase.

Dextrins: water-soluble degradation products of starch. They are classified according to the color of their reaction product with iodine as *amylodextrins* (blue), *erythrodextrins* (red), and low-molecular-weight *achroodextrins,* which have no colored reaction product with iodine. Heating starch with 3% HCl or HNO_3 produces *acid dextrins. Schardinger dextrins* are the result of the action of *Bacillus macerans* on starch; these are rings of 6 to 8 glucose units, linked α-1,4. Depending on the size of the ring, they are called α-, β- or γ-D. D. (in particular British gum, starch gum or gommelin; produced by dry heating starch at 160 °–200 °C) are used for dry extracts and pills, emulsifiers, thickeners for fabric dyes, sizing in paper and textiles, printing inks, glues, matches, fireworks and explosives.

Dextrose: see D-Glucose.

DHF: abb. for dihydrofolic acid.

Diabetes mellitus: a disease caused by partial or complete lack of Insulin (see), or by decreased numbers or sensitivity of cellular insulin receptors. D.m. is the most common human endocrine disorder (at least 1% of the population of Europe and North America) except for obesity.

Insulin regulates the uptake of blood glucose, ions and other substances into the cells. In D. very high blood glucose levels (8-60 mM) occur, with the result that glucose is lost through the kidneys. D. is characterized by copius, sweet urine. In uncontrolled D., the muscles and liver, which are unable to absorb glucose from the hyperglycemic blood, burn proteins and fats to supply their metabolism. The result is wasting similar to that in starvation. Two types of D. are known, juvenile (type I) and adult (type II) onset. A propensity to type I is inherited; the gene or genes are located in the major histocompatibility complex. In type I D., the islets of Langerhans are destroyed; there is massive infiltration by lymphocytes, and the disease is thought to be autoimmune. Alloxan D. produced experimentally mimics this condition because alloxan preferentially destroys the pancreatic β-cells. Type II D. is strongly correlated with obesity, and may develop through loss of fully active insulin receptors in cell membranes.

Mild D. can be controlled by diet, but in severe

directly with acetyl-coenzyme A to produce D. Little is known about the further metabolism of D. It is used as an aroma carrier in the food industry.

Diacylglycerol: see Acylglycerols.

Diaminopimelic acid pathway: see L-Lysine.

6-Diazo-5-oxo-L-norleucine, abb. *DON:* $\overset{\ominus}{N} = \overset{\oplus}{N} = CH-CO-CH_2-CH(NH_2)-COOH$ an antagonist of glutamine. It inhibits de novo purine synthesis in bacteria and mammals. It prevents the growth of experimental tumors, but is toxic for animals.

Dicarboxylic acid cycle: a cyclic pathway for the utilization of glyoxylic acid or one of its precursors (e.g. glycolic acid) as a carbohydrate source for the growth of microorganisms. The D. a. c. includes some of the reactions of the tricarboxylic acid cycle, and malate synthase (EC 4.1.3.2), of the glyoxylate cycle, acts here as a respiratory enzyme (Fig.). The D. a. c. also provides starting materials for the synthesis of cell components. If the concentration of D. a. c. intermediates is too severely reduced by diversion to synthetic pathways, they can be replenished from the Glycerate pathway (see).

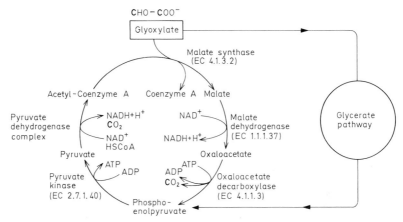

Oxidation of glyoxylate in the dicarboxylic acid cycle.

cases, insulin must be injected. Until the discovery of insulin early in this century, severe D. was invariably fatal. Milder D. is accompanied by a number of side effects, including damage to retinal capillaries, cataracts, dwindling and demyelination of neuronal axons, which lead to motor, sensory and autonomic dysfunction, arteriosclerosis and renal disease. The severity of these side effects correlates directly with the degree to which blood glucose exceeds normal levels. [M. Bliss, *The Discovery of Insulin* (University of Chicago Press, 1982); M. Brownlee & A. Cerami, *Ann. Rev. Biochem.* **50** (1981) 385-432; M. Hattori et al., *Science* **231** (1986) 733-735]

Diacetyl, *butane-2,3-dione:* $CH_3-CO-CO-CH_3$, a diketone produced as a byproduct of carbohydrate degradation. m. p. −3 °C, b. p. 88.8 °C. D. is a component of the butter aroma, and has been found in many biological materials. It is produced by dehydrogenation from acetoin, the decarboxylation product of pyruvate. In microorganisms there is another pathway in which active acetaldehyde reacts

Dichlorophenyl dimethyl urea, *DCMU, diuron:* an herbicide which blocks electron transport from photosystem II to photosystem I (see Photosynthesis). m. p. 158-159 °C.

Dichrostachinoic acid: a sulfur amino acid which contains both reduced and oxidized sulfur.

Dictyosomes: components of the Golgi apparatus, especially in plants. The terminology varies. D. have also been called *lipochondria* and osmiophilic material. Sitte has suggested that the total of the dictyosomes in a plant cell should be called the Golgi apparatus (see) and the *Dalton complex.* Sometimes D. and the Golgi apparatus are considered to be the same thing.

Dicumarol, *dicoumarol, dicoumarin:* 3,3'-methylenebis(4-hydroxy-2H-1-benzopyran-2-one); 3,3'-methylenebis(4-hydroxycoumarin): a vitamin K antagonist formed by the action of microorganisms on coumarin and/or coumarin precursors (see Coumarins) in improperly cured (spoiled) sweet clover (*Melilotus*) hay. Cattle eating this hay are subject to hemorrhage (sweet clover disease). D.is used clinically to prevent thromboses.

Dicumarol

Various organic syntheses have been reported. Of particular interest are those designed for synthesis of [^{14}C]D. for metabolic investigations, e.g. [2-^{14}C]4-hydroxycoumarin (from reaction of *o*-hydroxyacetophenone with [^{14}C]diethylcarbonate in presence of sodium ethoxide) is converted to [2-^{14}C]D. by treatment with formaldehyde [H. R. Eisenhauer et al. *Can. J. Chem.* **30** (1952) 245–250 (this is a useful reference source for other synthetic methods). Isolation and structural elucidation, H. A. Campbell & K. P. Link *J. Biol. Chem.* **138** (1941) 21–33; M. A. Stahmann et al. *J. Biol. Chem.* **138** (1941) 513–527]

Dicysteine: see L-Cystine.

Dideoxyadenosine triphosphate: see Dideoxyribonucleotide triphosphates.

Dideoxycytidine triphosphate: see Dideoxyribonucleotide triphosphates.

Dideoxyguanosine triphosphate: see Dideoxyribonucleotide triphosphates.

Dideoxyribonucleotide triphosphates, *terminating triphosphates:* synthetic substrates of DNA polymerase I, which catalyses their incorporation into growing oligonucleotide chains in place of the normal deoxyribonucleotide triphosphate substrates. Incorporation of a terminating triphosphate results in chain termination, since the 3'-hydroxyl group is lacking. They are used in the Sanger method for DNA sequencing. In the reaction mixture of template DNA, DNA polymerase and the four deoxyribonucleotide triphosphate substrates, one terminating triphosphate is included at about 1% of the concentration of its normal counterpart. This results in a family of variously extended DNA sequences, which can be sized by gel electrophoresis. By performing separate incubations for each terminating triphosphate, four patterns of bands are obtained from which the DNA sequence can be read off.

Dideoxyribonucleotide triphosphates
R = Adenine: Dideoxyadenosine triphosphate (ddATP)
R = Guanine: Dideoxyguanosine triphosphate (ddGTP)
R = Cytosine: Dideoxycytidine triphosphate (ddCTP)
R = Thymine: Dideoxythymidine triphosphate (ddTTP)

[F. Sanger et al., *Proc. Natl. Acad. Sci.* **74** (1977) 5463–5467].

Dideoxythymidine triphosphate: see Dideoxyribonucleotide triphosphates.

Didymocarpin: see Humulenes.

Differential gene activation: see Gene activation.

Differential gene expression: see Metabolic regulation.

Diffutin 7-hydroxy-3',4'-dimethoxy-5'-*O*-β-D-glucosylflavan, see Flavan.

Digestion: the totality of mechanical and chemical processes occurring in the digestive tract, that result in the degradation of foodstuffs to low M_r, absorbable, nonantigenic substances. The digestive tract, especially in mammals, shows considerable structural and biochemical adaptation to the nutritional physiology of the organism, e.g. in carnivores, herbivores and omnivores. Generally a distinction is made between *buccal D., gastric D.* and *duodenal D.* The chemical processes of buccal D. have little significance since salivary amylase occurs only in man, apes, pig and some rodents, and food is present for too short a time in the buccal cavity. The first chief site of D. is the stomach. Despite the many different designs of this organ between individual species, it always serves as the site for the degradation of dietary proteins to peptones, and of starch (in animals with salivary amylase) to water-soluble dextrins. A special situation is found in ruminants and other herbivores, where the rumen, reticulum, omasum and other compartments serve as bacterial fermentation chambers. Rumen bacteria perform the anaerobic degradation of cellulose to absorbable end products; these are not glucose, but short chain fatty acids, like acetic, propionic, butyric and valeric acids. In adult mammals the stomach is not absolutely essential for life, but in young suckling animals, including man, HCl and rennin in the gastric juice perform the important process of milk coagulation. The most important stage of D. occurs in the small intestine, which contains all the Digestive enzymes (see) for the continuation and completion of D. The digestive enzymes are secreted by the walls of the small intestine (succus entericus) and by the pancreas (pancreatic juice). In herbivores, a small proportion of the digestive enzymes is derived from the food (dietary enzymes). The bile (secreted by the gall bladder situated in the liver) provides activator substances, in particular bile acid salts, which together with the $NaHCO_3$ of the pancreatic juice provide an optimal environment for the digestive processes.
The absorbed endproducts of D. are L-amino acids, monosaccharides (glucose, fructose, galactose, mannose and pentoses), sodium salts of fatty acids, glycerol, monoglycerides and nucleosides. The remaining undigested and nonabsorbed material passes to the large intestine (colon) where it is concentrated by absorption of water. In the colon it is also subjected to various bacterial fermentation and putrefaction processes, which result in the production of lactic acid, acetic acid, various gases, poisonous amines and phenols. If passage through the intestinal tract takes too long, toxins may be absorbed in the latter half of the colon.

Table. Digestive enzymes of vertebrales.

Enzyme	M_r	Site of attack ($\downarrow$)	pH-optimum
(I) Proteases			
1. Proteinases			
a) Gastric			
Pepsin A (alkali-labile)	34500	$-$Gly$\downarrow$Try$-$Phe, $-$Glu$\downarrow$Phe	1.8
Pepsin B (Gelatinase)	36000	hydrolyses only gelatin	
Pepsin C (Gastricsin)	31500	$-$Tyr$\downarrow$Ser$-$, $-$Phe$\downarrow$Ser	3.0
Rennin	30700	$-$Phe$\downarrow$Met (in κ-casein)	4.8
b) Pancreatic (in the duodenum)			
Trypsin	23400	$-$Arg$\downarrow$R, $-$Lys$\downarrow$R	8.0
Chymotrypsin A (α and γ)	25170	$-$Try$\downarrow$R, $\downarrow$Phe$-$R, $\downarrow$Try$-$R, $-$Met$\downarrow$R	8.0
δ-Chymotrypsin	25400	as chymotrypsin A	
Chymotrypsin B	25400	as chymotrypsin A	
Chymotrypsin C	23900	as chymotrypsin A; also $-$Leu$\downarrow$R, $-$Glu(Asp)$\downarrow$R	8.0
Elastase	25700	R$-$neutral amino acid$\downarrow$R	8.0
Collagenase	?	hydrolyses only collagen	5.5
c) Duodenal secretion			8.0
Enterokinase	196000	H_2N-Val$-$(Asp)$_{2-5}$Lys$_6\downarrow$Ile$_7-$Tryp-sin$-$COOH	
2. Peptidases			
a) Pancreatic			
Carboxypeptidase A	34400	Peptidyl$\downarrow$Phe, $\downarrow$Try, $\downarrow$Trp, $\downarrow$Leu	8.0
Carboxypeptidase B	34000	Peptidyl$\downarrow$Lys, $\downarrow$Arg	8.0
b) Duodenal secretion			
Leucine aminopeptidase	300000	H_2N-Leu$\downarrow$peptide, or $\downarrow$polypeptide	8.9
Aminotripeptidase	300000	Ala$\downarrow$dipeptide	8.0
Dipeptidases	100000	Gly$\downarrow$Gly, Gly$\downarrow$Leu, Cys$\downarrow$Gly	7.8
Prolidase	?	Gly$\downarrow$Pro	7.8
Prolinase	?	Pro$\downarrow$Gly	7.8
(II) Glycosidases			
a) Salivary and pancreatic, acting in the buccal cavity, stomach and duodenum			
α-Amylase	50000	α-glycosidic $1 \rightarrow 4$ bonds	6.5
b) Duodenal secretion			
α-Glycosidases			
5 Specific maltases	~ 200000	α-glycosidic $1 \rightarrow 4$ bonds	7.0
A specific sucrase	~ 200000	α-glycosidic $1 \rightarrow 2$ bonds	7.0
A trehalase	~ 200000	α-glycosidic $1 \rightarrow 1$ bonds	7.0
α-1,3-Glycosidase	~ 200000	α-glycosidic $1 \rightarrow 3$ bonds	7.0
β-Galactosidase (Lactase)	~ 200000	β-glycosidic $1 \rightarrow 4$ bonds	6.0
Oligo-α($1 \rightarrow 6$)-glucosidase	~200000	α-glycosidic $1 \rightarrow 6$ bonds in starch and glycogen	7.0
(III) Esterases			
a) Gastric			
Gastric Lipase	35000	Ester bonds in triglycerides, especially milk fat	5.0
b) Pancreatic			
Pancreatic lipase	35000	Ester bonds in triglycerides	7.5
Phospholipase A + B	14000	Ester bonds in phospholipids	7.5
Cholesterol esterase	400000	Cholesterol fatty acid esters	7.5
c) Duodenal secretion			
Monoacylglycerol lipase	?	Ester bonds of monoglycerides	7.5
Carboxylic acid esterase	160000	Esters of aliphatic fatty acid	7.8
Alkaline phosphatase	140000	Phosphate ester bonds	9.0
(IV) Nucleases from the pancreas			
Ribonuclease	13700	3′-Phosphate ester bonds	7.3
Deoxyribonuclease	31000	3′-Phosphate ester bonds	7.0

Degradation of the cell's own biopolymers is Intracellular D. (see), and is a process distinct from D.

Digestive enzymes: hydrolases present in the digestive tract of all animals, which catalyse hydrolysis of mechanically disrupted foodstuffs (proteins, carbohydrates, fats, nucleic acids) to their absorbable components. These low M_r components are absorbed as rapidly as they are formed, so that the equilibrium of digestive processes is continually displaced in favor of hydrolysis. With the exception of disaccharidases and certain peptidases, D.e. are Secretory enzymes (see), which are synthesized in high concentrations in accessory glands, like the pancreas or salivary glands, or in the gastric or intestinal mucosa. D.e. include some of the most thoroughly investigated enzymes. They are classified as 1. Proteases (see) and Peptidases (see); 2. Glycosidases (see), which cleave carbohydrates; 3. Esterases (see), especially lipases; and 4. Nucleases, which cleave nucleic acids. See Table.

Digestive vacuole: see Intracellular digestion.

Digifolein: see Diginin.

Digifologenin: see Diginin.

Diginigenin: see Diginin.

Diginin: a digitanol composed of the pregnane derivative (see Steroids) diginigenin, M_r 344.45 m.p. 115 °C, $[\alpha]_D$ −126 °, and the deoxysugar D-diginose. D. occurs together with cardiac glycosides in Digitalis, e.g. Digitalis purpurea, from which it was isolated in 1936 by Karrer.

Digifolein, which also occurs in Digitalis species, has the aglycon digifologenin, M_r 360.45, m.p. 176 °C, $[\alpha]_D$ −269 ° which differs from diginigenin in having an extra 2β-hydroxyl group.

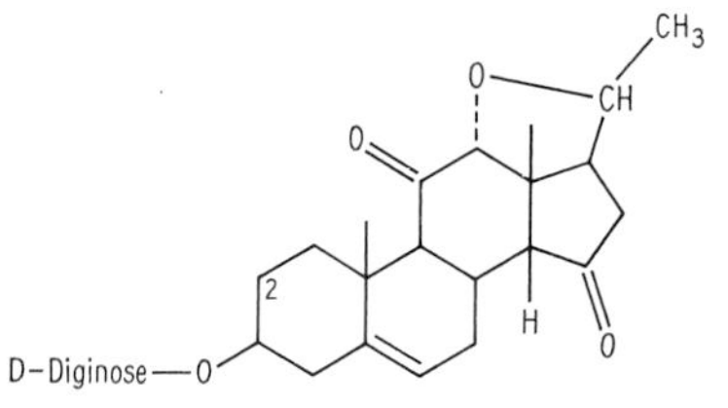

Diginin

Digitalis glycosides: cardiac glycosides of the cardenolide group found in the leaves of foxgloves, Digitalis purpurea and Digitalis lanata. The three most important D. are digitoxin, M_r 764.92, m.p. 256–257 °C (anhydrous), $[\alpha]_D^{20} + 4.8$ ° (c = 1.2 in dioxan); digoxin, M_r 780.92, m.p. 265 °C (d.),

$[\alpha]_D + 11$ °; and gitoxin, M_r 780.92, m.p. 285 °C (d.), $[\alpha]_D + 22$ °. The D.g. are secondary glycosides formed during preparation of the Digitalis leaves.

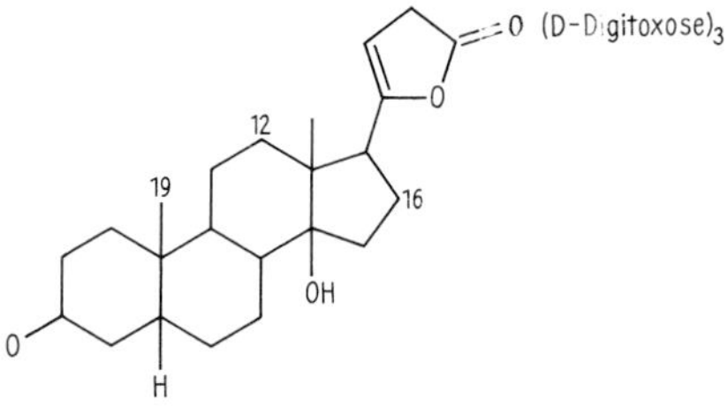

Digitoxin

The primary glycosides found naturally in the plants, the lanatosides, carry a D-glucose group and an acetic ester. The aglyca of digitoxin, digoxin and gitoxin are digitoxigenin, M_r 374.50 m.p. 253 °C, $[\alpha]_D^{17} + 19.1$° (c = 1.36 in methanol), digoxigenin, M_r 390.53, m.p. 222 °C (anhydrous), $[\alpha]_{546}^{20} + 27.0$ ° (c = 1.77 in methanol), and gitoxigenin, M_r 390.50, m.p. 234 °C, $[\alpha]_{545}^{20} + 38.5$ ° (c = 0.68 in methanol), respectively. The latter two differ from digitoxigenin only in having an extra hydroxyl group at C-12 or C-16. The sugar component is always three molecules of D-digitoxose.

The D.g. are obtained by gentle extraction of the fresh plant matter with ethyl acetate or chloroform. To remove tannic acids, the alcoholic solution is precipitated with lead salts. The D.g. are released enzymatically from the lanatosides and separated chromatographically.

Some of the color reactions of the D.g. are: red with sodium nitroprusside in sodium hydroxide solution, orange with alkaline picric acid solution, and blue-violet with alkaline m-dinitrobenzene solution.

The D.g. are indispensible cardiotonic agents, used for long-term treatment of chronic heart weakness and defective heart valves. The pure glycosides are now used instead of leaf powders or extracts.

Digitanols, digitanol glycosides: a group of plant glycosides with pregnane type Steroids (see) as aglyca, for example, Diginin (see) and digifolein. D. occur together with cardiac glycosides, but have no cardiotonic activity themselves. They are biosynthesized from pregnenolone.

Digitogenin: see Digitonin.

Digitonin: a mixture of four different steroid saponins from the seeds of the purple foxglove, Digitalis purpurea (see Saponins). The main compo-

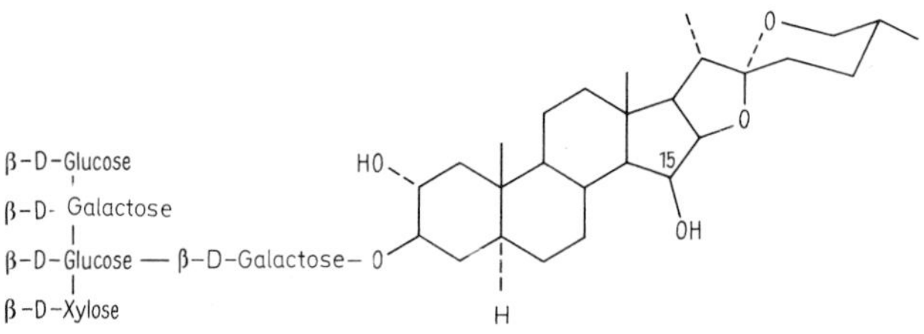

Digitonin

nent, also called D., makes up 70 to 80% of the mixture. M_r 1 229.30, m.p. 235 °C, $[\alpha]_D^{20} - 54$ ° ($c =$ 0.45 g in 15.8 ml methanol). The aglycon is digitogenin, (25R)-5α-spirostan-2α, 3β, 15β -triol, M_r 448.62, m.p. 296 ° (d.) $[\alpha]_D^{10} - 81$ ° ($c = 1.4$ in chloroform). D. is a strong hemolytic poison, due to its affinity to blood cholesterol. It is used as a precipitating agent for cholesterol and other sterols, in their isolation and quantitative determination.

Digitoxigenin: see Digitalis glycosides.

Digitoxin: see Digitalis glycosides.

Diglyceride: see Acylglycerols.

Digoxigenin: see Digitalis glycosides.

Digoxin: see Digitalis glycosides.

Dihydrofolic acid: see Tetrahydrofolic acid.

Dihydroorotate: an intermediate in Pyrimidine biosynthesis (see).

Dihydrouracil: an intermediate in Pyrimidine degradation (see). M_r 114.10, m.p. 274 °C. 5,6-D. is found as a rare base in some nucleic acids.

Dihydroxyacetone phosphate: see Triose phosphates.

20,22-Dihydroxycholesterol: see Cholesterol.

20,26-Dihydroxyecdysone: a steroid which acts as a molting hormone. It has been isolated together with Ecdysone (see) and ecdysterone from pupae of the tobacco hornworm, *Manduca sexta*.

Dihydrozeatin, *6-(4-hydroxy-3-methyl-butylamino)purine:* a cytokinin from corn *(zea mays)*. It is a derivative of zeatin and has also been isolated as a riboside and a ribotide.

Dimers: double molecules consisting of two identical subunits, or monomers. The *pyrimidine dimers* formed by UV irradiation of nucleic acids are particularly important because of their genetic consequences. The most frequently found after irradiation of DNA solutions is the thymine dimer (TT), which can only be formed between adjacent bases in the DNA chain. Their formation depends on the wavelength; 265 nm promotes dimerization, but at 235 nm, previously formed dimers revert to the monomers. In living cells, enzymatic repair is also possible (see Photoreactivation; see Excision repair, under Deoxyribonucleic acid).

Dimethazide: see Succinic acid mono-N-dimethylhydrazide.

N⁶(-γ, γ-dimethylallyl)Adenosine, N⁶-*isopentenyladenosine: one of the Rare bases (see) in nucleic acids found in certain transfer RNAs, for example, serine tRNA. It also acts as a Cytokinin (see), and is found in free form in the culture media of *Corynebacterium* and *Agrobacterium*.

Bryokinin, a cytokinin found in the callus cells of moss sporophytes, is identical with the free base N^6-γ, γ-dimethylallyladenine.

Dimethylallylpyrophosphate: an intermediate in Terpene biosynthesis (see).

3,7-Dimethyloctane type: see Monoterpenes, Fig.

Dinucleotide fold: a characteristic folded protein structure constituting part or all of the structure of four NAD-dependent dehydrogenases, and certain other enzymes, some of which do not bind nucleotides. The D.f. was first identified in the tertiary structures of liver alcohol dehydrogenase (EC 1.1.1.1), glyceraldehyde 3-phosphate dehydrogenase

(EC 1.2.1.12), lactate dehydrogenase (EC 1.1.1.28) and malate dehydrogenase (EC 1.1.1.37). All four of these dehydrogenases contain between 327 and 374 amino acid residues, which are folded into two distinct domains. One domain binds the NAD cofactor, while the other domain carries the binding and catalytic sites for the substrate. In each case, the NAD-binding domain has a fold consisting of a core of β-pleated sheet structure containing six parallel strands (strand order CBADEF), with the α-helical intrastrand loops above or below the sheet. A similar structure exists in phosphoglycerate kinase (EC 2.7.2.3), where the D.f. is responsible for binding ATP. Other enzymes with tertiary structures resembling the D.f. are phosphoglycerate mutase (EC 2.7.5.3), adenylate kinase (EC 2.7.4.3), phosphorylase a (EC 2.4.1.1) and pyruvate kinase (EC 2.7.1.40). Not all of these enzymes are known to bind nucleotides, and it is possible that the D.f. was present in an ancestral protein and was later exploited for nucleotide binding where this was advantageous. On the other hand, the D.f. may be an especially stable structure, which has arisen in more than one enzyme family. It is noteworthy that enzymes so far shown to possess a D.f. bind either dinucleotides or 2-oxotrioses. There may in fact be an "oxotriose fold" sharing a common ancestry with the D.f. [Blake, C.C.F. *Nature* (1972) *267*, 482–483].

Dioscin: a steroid saponin (see Saponins) M_r 869.08, m.p. 275 to 277 °C (d.), $[\alpha]_D^{13} - 115$ ° ($c=0.373$ in ethanol). The aglycon of D. is *diosgenin*, (25R)-spirost-5-en-3β-ol, M_r 414,61, m.p. 204 to 207 °C. It is found in yams *(Dioscorea)* and trilliums. Diosgenin is an important starting material for partial synthesis of steroid hormones.

α-L - Rhamnose
|
β-D- Glucose-O
|
α-L - Rhamnose

Dioscin

Diosgenin: see Dioscin.

Dioxygenases: see Oxygen metabolism; Oxygenases.

Dipentene: see p-Menthadienes.

Diphosphatidylglyceride: see Phospholipids.

2,3-Diphosphoglycerate, *glycerate 2,3-bisphosphate, 2,3-bisphosphoglycerate:* see Glycolysis, Hemoglobin, Rapoport-Luerbing shuttle.

Diphosphopyridine nucleotide: see Nicotinamide adenine dinucleotide.

Diptheria toxin: see Toxic proteins.

Disaccharidases: a group of enzymes which hydrolyse disaccharides. They are most abundant in ripe fruits, microorganisms (yeasts) and the intestinal mucosa. Some of the best known are β-D-fructofuranosidase (invertase or saccharase) (EC 3.2.1.26) from yeast, α-1,4-glucosidases (maltase) (EC 3.2.1.20), which hydrolyse α-D-glucosides like maltose, sucrose and turanose; α-1,4-glucosidases (gentiobiase, cellobiase) (EC 3.2.1.21), which hydrolyse cellobiose and gentiobiose; and β-galactosidase (lactase) (EC 3.2.1.23), which hydrolyses lactose.

Disaccharides: see Carbohydrates.

Disc electrophoresis: see Proteins.

Discontinuous process: see Fermentation techniques.

Dissimilation: see Catabolism.

Dissimilatory sulfate reduction: see Sulfate respiration.

Disulfide bridges, *cystine bridges:* a term referring to disulfide bonds, -S-S-, in proteins and peptides formed by oxidation of two sulfhydryl groups: 2-SH → -S-S-. D.b. are the major factors responsible for formation and maintenance of secondary structure in proteins. Proteins which contain a large number of D.b. are very resistant to denaturation by heat, acid or alkali, detergents, etc. and to hydrolysis by proteolytic enzymes. D.b. can be cleaved either reductively, for example by 2-mercaptoethanol, or oxidatively, for example by performic acid. In organisms, the formation and cleavage of D.b. is catalysed by enzymes, the best known being protein-disulfide reductase (glutathione) (EC 1.8.4.2).

Diterpene alkaloids: a group of Terpene alkaloids (see).

Diterpenes: terpenes built from four isoprene units ($C_{20}H_{32}$). Phytol, an aliphatic D., is important as the ester component of chlorophyll and as a part of vitamins K and E. Aside from a few hydrocarbons and alcohols, most of the cyclic D. are acids, and have a variety of biological properties (Table).

Biosynthesis: The starting compound is geranyl pyrophosphate (see Terpenes). The acyclic D. are formed by hydolysis of the pyrophosphate residue (phytol, for example). Geranylgeranylpyrophosphate is probably converted easily to geranyl linalool, which is then converted to bi- and tricyclic compounds (Fig.). In a few of the cyclic D., for example abietic acid, there is a migration of the substi-

Table. *Diterpenes and their significance*

Class	Representatives
Hormones	Gibberellins, trisporic acids, antheridiogens
Vitamins	Vitamin A
Chromophore of visual purple	Retinol
Resenic acids	Abietic acid
Alkaloids	Cassaine, aconitine
Sweet substance	Stevioside

tuents. The gibberellins are derived from the labdadiene type of D.

Dityrosine: a dimer of L-tyrosine found in acid hydrolysates of several biological materials: tussa silk, fibroin, insect cuticle resilin, spore coat of *Bacillus subtilis*, and the fertilization membrane of sea urchin egg. In *Saccharomyces cerevisiae*, D. is sporulation-specific, being found only in spores and not in vegetative cells or non-sporulating cells under sporulating conditions. It is now thought that dityrosyl residues exist in vivo, acting as cross links in structural proteins. D. can be synthesized in vitro by the action of horse radish peroxidase on L-tyrosine. NMR analysis shows that the earlier assigned structure (phenolic OH-groups *ortho* to the inter-ring bond) is wrong; in the correct structure, the phenolic OH-groups are *meta* to the inter-ring bond (Fig.). [P. Briza *J. Biol. Chem.* **261** (1986) 4288–4294]

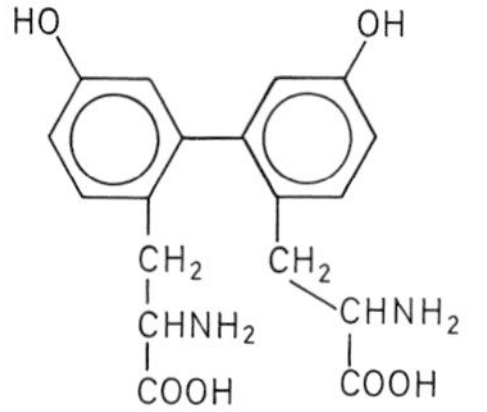

Dityrosine

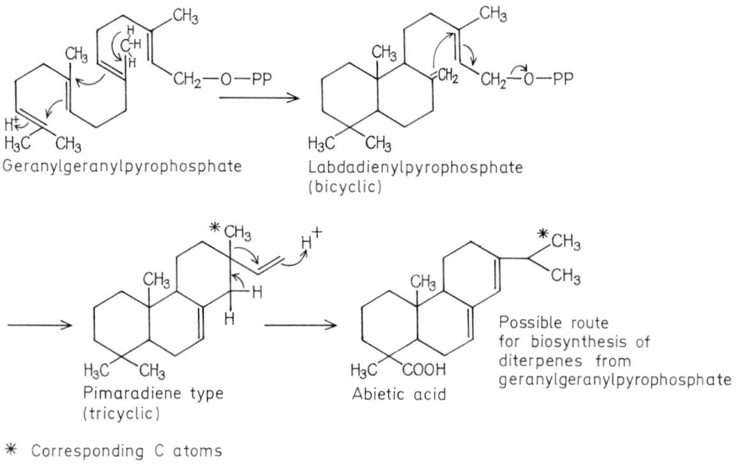

Geranylgeranylpyrophosphate

Labdadienylpyrophosphate (bicyclic)

Pimaradiene type (tricyclic)

Abietic acid

Possible route for biosynthesis of diterpenes from geranylgeranylpyrophosphate

* Corresponding C atoms

Possible route for biosynthesis of diterpenes from geranylgeranylpyrophosphe

Diurnal acid rhythm: see Crassulacean acid metabolism.

DNA: abb. for Deoxyribonucleic acid (see).

DNA-binding proteins: Proteins bound to DNA may have either structural or regulative functions, or both. Proteins which repress or induce the transcription of particular genes (see Operon) must be able to recognize specific nucleotide sequences, while histones (see Chromosomes) may recognize more general features, such as areas of higher or lower GC content (see). Variations in the propeller twist and base roll (see Deoxyribonucleic acid) of the base pairs are highly correlated with the base sequences of synthetic DNA oligomers. It is likely that these variations affect the binding of hydrogen-bonding proteins, and thus are the basis of sequence recognition by the proteins. The Z structure should also be easily recognizable by proteins. It is usually a higher-energy configuration than A or B DNA, but it is stabilized by alternating purine and pyrimidine sequences, or by proteins. It is of great interest that in the genome of the SV40 virus and others, alternating purine-pyrimidine sequences are clustered in the control regions.

The structures of three proteins which regulate gene expression have been determined from their crystals: Cro, which is a repressor of the repressor maintenance promoter P-RM in the bacteriophage lambda; the lambda repressor, which can act as a repressor but can also stimulate expression of its own gene; and CAP, the "catabolite gene activator protein", which promotes transcription of several genes in the presence of cyclic AMP, but in other circumstances it can also act as a repressor. Each of the three includes two sections of α-helical structure (see Proteins) which are spaced in such a way that they can bind two sequential turns of the major groove of the DNA helix. These helical regions have similar primary structures, and considerable sequence homology exists with other regulator proteins which have been sequenced. A model of the Cro protein and its tightest known binding site shows multiple hydrogen bonding between the nucleotide bases and amino acid side chains; hydrophobic interactions are also important. This type of specificity is presumed to be responsible for the recognition of a particular nucleotide sequence by the protein. [Y. Takeda et al. *Science* **221** (1983) 1020–1026; F.A. Jurnak & A. McPherson, eds. *Biological Macromolecules and Assemblies, vol. 2* (Wiley, New York, 1985)]

DNA gyrase: a type II topoisomerase. See Topoisomerases.

DNA-ligase: see Polynucleotide ligase.

DNA nucleotidyltransferase: see DNA polymerase.

DNA polymerase, *DNA nucleotidyltransferase* (EC 2.7.7.7): an enzyme which catalyses the synthesis of DNA polynucleotide chains on a pre-existing DNA matrix (DNA replication). The precursors are the four 3'deoxyribonucleotide triphosphates. In vitro, the enzyme can also synthesize homo- and copolymers from triphosphates. The D.p. from different organisms have different specificities, some for single and others for double strands of DNA as primers.

Three D.p. have been isolated from *Escherichia coli.* The best studied is *polymerase I* (Kornberg enzyme), which can join deoxyribonucleoside triphosphates to high-molecular-weight polynucleotides, simultaneously splitting out pyrophosphate. The chain grows from the 5' phosphate to the 3' end. In vitro, the reaction requires an oligonucleotide primer; the nucleotides are added to its 3'-hydroxyl end. A matrix DNA is also required to give the correct sequence. In 1967, Kornberg reported the total synthesis of the DNA of the single-stranded phage ΦX 174 using this enzyme. However, it is probably not the enzyme responsible for replication of DNA in vivo; it is more likely to be a repair enzyme (see Deoxyribonucleic acid). The function of D.p. II in the cell is still unclear. DNA replication is probably catalysed by D.p. III.

DNA-relaxing enzyme: a type I eukaryotic topoisomerase isolated from mammalian tissue culture cells. [W. Keller, *Proc. Natl. Acad. Sci. U.S.A.* **72** (1975) 2550–2554.] The term may also be loosely applied to all type I and type II Topoisomerases (see).

DNA-RNA hybrids: double-stranded molecules, of which one strand is DNA, and the other the complementary RNA. They are presumably the intermediate form in RNA transcription (see Ribonucleic acids) and in the multiplication of oncogenic RNA viruses (see RNA-dependent DNA polymerase), but they can also be produced in vitro by Hybridization (see). They are resistant to ribonuclease.

DNA-swivelase: a type I topoisomerase. See Topoisomerases.

DNA synthesis: synthesis of oligodeoxyribonucleotides of specified base sequence, using chemical methods. In the original phosphodiester method, the 5'-phosphate of one nucleotide (with other functional groups protected) is condensed with the 3'-hydroxyl of another protected nucleotide (Fig. 1). Reaction times are long, and the yield decreases rapidly with the length of the synthesized chain. This method is historically important, since it was used to perform the first total synthesis of a gene, a biologically functional suppressor transfer RNA gene [H.G. Khorana *Science* **203** (1979) 614–625].

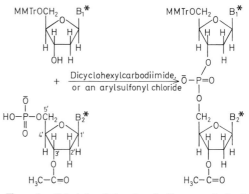

Figure 1. Principle of the phosphodiester method of DNA synthesis.
B* = protected base. For other abbreviations and formulae, see Fig. 5.

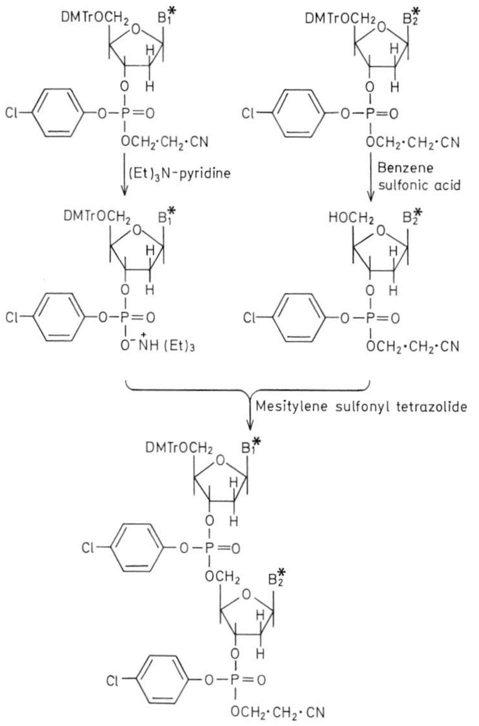

DNA synthesis

Figure 2.   The phosphotriester method of DNA synthesis.
B* = protected base. For other abbreviations and formulae, see Fig. 5.

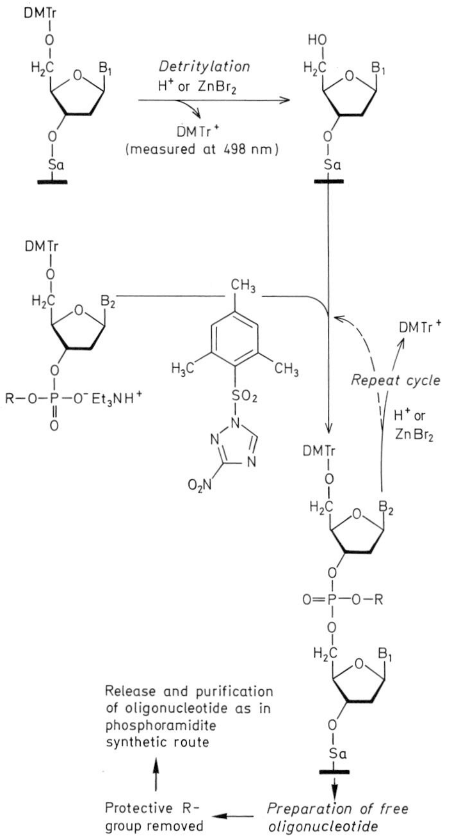

Figure 3.   Phosphotriester method of DNA synthesis, using a solid phase system. The brilliant orange color (λ_{max} 498 nm) of the DMTr cation can be used to monitor coupling efficiency between synthetic cycles. Sa = spacer arm. B_1, B_2 = protected bases. Condensation is activated by mesitylene-2-sulfonyl-3-nitro-1,2,3-triazole (formula shown). R is a phosphoryl protecting group (see Fig. 2).

The phosphotriester method overcomes some of the disadvantages of the phosphodiester method by blocking each intermediate phosphodiester function during the synthesis of the required sequence. This method was used to synthesize 67 different oligonucleotides of chain length 10–20, which were then spliced to generate a 517-base-pair α-interferon gene [M. D. Edge et al. *Nature* **292** (1981) 756–762]. The reactions may be performed in solution, as shown in Fig. 2, but solid phase synthesis is more efficient (Fig. 3).

In solid phase synthesis, the oligodeoxyribonucleotide is synthesized while covalently attached to a solid support, and excess soluble protected nucleotides and coupling reagents are used to drive each stage of the synthesis to completion. Solid phase synthesis is easily automated, and instruments for this purpose are technically advanced liquid-dispensing devices, in which the delivery of reagents to the solid support is metered and controlled by computer.

The modern method of choice (manual or automated) is a solid phase system, using phosphoramidite chemistry (Fig. 4). The most advanced automated systems employ β-cyanoethylamidites, rather than methylamidites. Cyanoethyl protection avoids the potential hazard of thymine methylation by internucleotide methyl phosphate, which is present when methylamidites are used; the cyanoethyl group is al-

so more easily removed than methyl, so that deprotection of the phosphate, deprotection of the bases, and release from the solid support can be performed in a single stage.

The solid support may be controlled pore glass or silica (phosphoramidite method), or polystyrene-divinylbenzene (phosphotriester method). Attachment is via a succinyl residue to the amino terminus of a spacer arm (Fig. 5).

The product of chemical DNA synthesis is a single-stranded oligonucleotide. Short oligonucleotides can be converted to the double-stranded form in vitro by suitable enzymatic methods (see Recombinant DNA technology). For the synthesis of very long double-stranded DNA (e.g. a whole gene), overlapping oligonucleotides of both strands may be synthesized, then formed into the total polynucleotide by annealing and ligase action (Fig. 6). [R. L. Letsinger & W. B. Lunsford *J. Amer. Chem. Soc.* **98** (1976) 3655–3661;

168

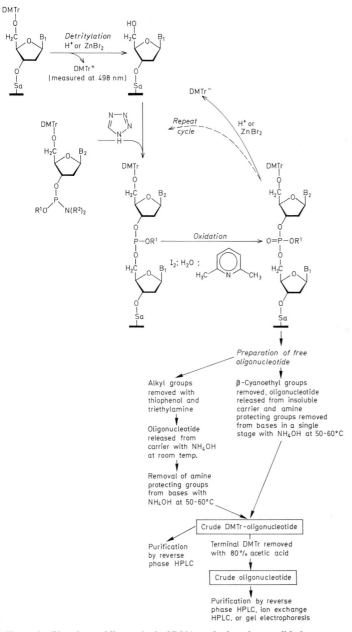

Figure 4. Phosphoramidite method of DNA synthesis, using a solid phase system. Coupling efficiency is measured by monitoring the DMTr cation, as in Fig. 3. R^1 is a methyl or β-cyanoethyl group. R^2 is methyl, ethyl or isopropyl. Sa = spacer arm. The reactive phosphite is oxidized to a stable triester by aqueous iodine in the presence of lutidine. At the beginning of each new cycle, any unreacted, carrier-bound 5′-hydroxyl groups are acetylated with acetic anhydride, a process known as capping.

DNA synthesis

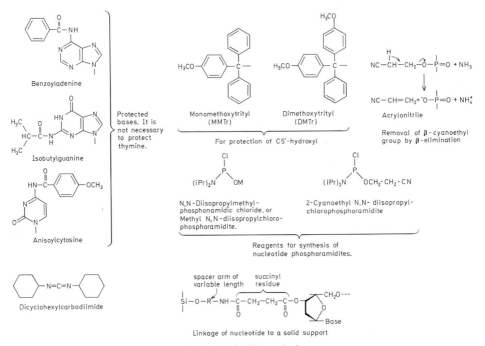

Figure 5. Some structures and reagents in chemical DNA synthesis.

DNA - Synthesis

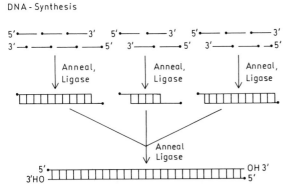

Figure 6. Synthesis of double-stranded DNA from overlapping oligonucleotides of both strands. The 5'-end of each oligonucleotide is first phosphorylated by the action of polynucleotide kinase in the presence of ATP. —• = 5'-phosphate.

N. D. Sinha et al. *Nucleic Acids Res.* **12** (1984) 4539-4557; R. Newton, *Internat. Biotech. Lab.* **5** (1987) 46-53]

DNA technology: see Recombinant DNA

DOC: see Cortexone.

Dolichol phosphate: a coenzyme involved in the glycosylation of proteins and lipids. D. p. is a membrane-bound polyprenol phosphate which accepts glycosyl units from soluble donors (uridine- or guanosine-diphosphate glycosides) and donates

$$H-\left[H_2C-\underset{\underset{CH_3}{|}}{C}=CH-CH_2-\right]_n CH_2-\underset{\underset{CH_3}{|}}{CH}-CH_2-CH_2O-PO_3H_2$$

n = 13 - 20 (mainly) in mammals

Dolichol phosphate

them in turn to membrane proteins or lipids. The highest concentrations of D. p. are found in nuclear, Golgi and rough endoplasmic reticulum membranes. The D. p. contain 14 to 24 isoprene units.

DOM: see Mescaline.

DON: abb. for for 6-diazo-5-oxo-L-norleucine.

Donor position: the binding site for the peptidyl-tRNA on the ribosome during Protein synthesis (see).

Dopa: abb. for 3,4-dihydroxyphenylalanine. See Dopamine.

Dopamine, *hydroxytyramine, 3,4-dihydroxyphenethylamine:* one of the catecholamines, M_r 153.2. D. is formed by decarboxylation of 3,4-dihydroxyphenylalanine (dopa), which in turn is formed by hydroxylation of tyrosine. D. is the precursor of the hormones Noradrenalin and Adrenalin (see). In the liver, lungs and intestines, it is the end product of tyrosine metabolism. D. is a neurotransmitter in the central nervous system. The highest concentration of D. neurons occurs in the nigrostriatal system, which degenerates in Parkinson's disease. D. cannot cross the blood-brain barrier, but the precursor dopa can, and it is used to treat Parkinson's disease. [H. N. Wagner et al., *Science* **221** (1983) 1264-1266; E. S. Garnett, G. Firnau & C. Nahmias *Nature* **305** (1983) 137-138.] Agonists of D., such as amphetamines (see Antidepressants), can elicit psychotic symptoms similar to those of schizophrenia, while antagonists of D. (neuroleptics) are used in treatment of schizophrenia. It has been shown that post-synaptic D. receptors are functionally linked to an adenylate cyclase via an intrinsic membrane protein, the G/F protein. When the D. receptors are occupied, the G/F protein releases a molecule of GDP and binds a molecule of GTP. This is the rate-limiting step in

activation of adenylate cyclase. It has been suspected for many years that hypersensitivity to D. is a cause of schizophrenia, but comparison of brain tissue from schizophrenics and control persons failed to reveal a difference in the activation of the adenylate cyclase by D. More recently, two types of D. recognition sites have been described: the D_1 type mediates stimulation of the adenylate cyclase, and the D_2 type mediates inhibition. Both types are coupled to the cyclase via the G/F protein. It has now been observed that 2,3,4,5-tetrahydro-7,8-dihydroxy-1-phenyl-1H-3-benzazepine, a selective D_1 agonist, stimulates more AMP formation in homogenates of the nucleus caudatus from schizophrenic brains than in homogenates from normal brains. This difference was observed only in the nucleus caudatus. [M. Memo, J. Kleinman & I. Hanbauer, *Science* **221** (1983) 1304-1306.] See also Ascorbate shuttle.

Dopamine β-hydroxylase, *3,4-dihydroxyphenylethylamine, ascorbate: oxygen oxidoreductase (β-hydroxylating)* (EC 1.14.17.1): a mono-oxygenase catalysing the hydroxylation of dopamine to noradrenalin, and tyramine to octopamine. DβH is a copper protein, stimulated by fumarate and specifically inhibited by disulfiram. It has been purified and its physical properties studied [S. Friedman & S. Kaufman, *J. Biol. Chem.* **240** (1965) 4763-4773]. Ascorbate, which is required as an electron donor, is converted into the free radical, Semidehydroascorbate (see) [T. Skotland & T. Ljones, *Biochim. Biophys. Acta* **630** (1980) 30-35]. In vitro, two free radicals then dismute rapidly to form one molecule of ascorbate and one of dehydroascorbate. Under physiological conditions, e. g. during noradrenalin synthesis in the adrenal medulla, the semidehydroascorbate free radical is re-reduced to ascorbate. See Ascorbate shuttle.

Doping: the administration of foreign substances to increase the performance of athletes. Even in antiquity, a number of means were known to stimulate physical and psychological strength. About the turn of the century, the practice began with race horses and greyhounds, and in the 1930's, with human athletes. The agents used usually have detrimental effects in the long run, and the practice is strictly forbidden. The International Association of Athletic Federations (IAAF) has prepared a comprehensive list of forbidden drugs. The list includes 1) central stimulating agents, such as psychostimulants, hallucinogens, etc., heart and circulatory stimulants, and hormones; and 2) central depressive agents, such as tranquillizers, opiates, hypnotics, sedatives, etc. Glucose, vitamins and salt, and caffeine in the form of coffee, tea or cola beverages are not forbidden. However, caffeine and cocaine may not be injected.

Dormin: see Abscisic acid.

Dose: the amount of a pharmaceutical agent administered at a time. In animal experiments, the dose is given as the amount having a specified effect on a stated fraction of the individuals tested (Table and Fig.). The D. should be given in mg per kg body weight; and the form in which it is administered should be stated explicitly, for example, LD_{50} 20 mg/kg mouse, subcutaneously. The safety of an agent is indicated by the *therapeutic range* (therapeutic index), LD_{50}/ED_{50}.

$$H_2N-CH_2$$
$$|$$
$$CH_2$$

HO

OH

Dopamine

Table. *Dose-effect relationships*

Term	Abb.	Definition
Median effective dose	ED_{50}	D. at which 50% of the experimental subjects show an effect.
Effective, therapeutic or curative D.	CD_{50}	Same as above; used with therapeutically applied compounds.
Minimal lethal D	LD_{05}	D. causing death of 5% of the subjects
Median lethal D.	LD_{50}	D. which causes death in 50% of the experimental animals.
Lethal D. Absolute lethal D.	LD_{100}	D. which kills 100% of the experimental animals.

Double helix: see Deoxyribonucleic acid.
Double membrane: see Biomembrane.
Double strand break: a break in a double-stranded DNA molecule in which both strands are broken without separating from each other. A D.s.b. can be made by mechanical forces or by radiation, chemicals or enzymes.

DPN: abb. for diphosphopyridine nucleotide; see Nicotinamide adenine dinucleotide.

DPX 1840: 3,3a-dihydro-2-(p-methoxyphenyl)-8H-pyrazolo-[5,1-a]-isoindol-8-one, a synthetic growth regulator affecting auxin transport, the formation of fruit-ripening hormone and root growth, etc. in cotton and soybeans.

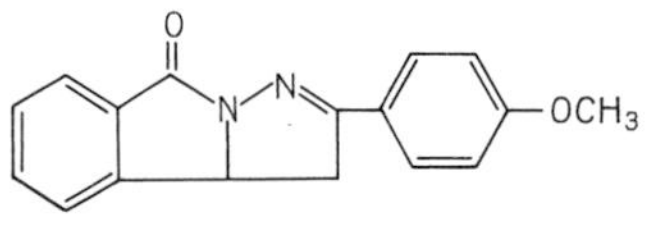

DPX 1840

D-RNA: see Messenger RNA.
Droplet countercurrent chromatography, *DCC chromatography, DCCC:* a separation method based on partition chromatography. The stationary and mobile phases are liquid and immiscible, and of different densities. Droplets of the mobile phase pass through a battery (normally about 300) of glass columns (2.0-3.0 mm internal diam., length 40 cm) containing the stationary phase. DCCC is very protective, and gives 100% recovery of microgram or gram quantities of separated material. It has been successfully used for practically every type of natural product, including alkaloids, cardiac glycosides, saponins, anthraquinones, terpenoids, sugars, amino acid derivatives, peptides, etc.

Dry weight: the weight of material, in g or mg, remaining after drying tissues, organisms, etc. at temperatures somewhat above 100 °C.

dTDP: see Thymidine phosphates.

dTDP-sugars: sugars or sugar derivatives activated by bonding to deoxythymidine diphosphate sugars.

dThd: abb. for deoxythymidine: see Thymidine.

dTMP: see Thymidine phosphates.

dTTP: see Thymidine phosphates.

(-)-Duartin: (3 S)-7,3'-dihydroxy-8,4',2'-trimethoxyisoflavan, see Isoflavan.

Duicitol, *galactitol:* an optically inactive sugar alcohol derived from galactose. It has 6 C atoms, M_r 182.17, 189 °C. It is found in algae, fungi and the sap and bark of various higher plants. It is produced synthetically by reduction of galactose or by isolation from dulcite or Madagascar mannu (*Melampyrum nemorosum* L.).

Dwarf maize test: see Gibberellins.

Dwarf pea test: see Gibberellins.

Dynein: an ATPase responsible for motility in cilia and flagella. It causes the outer doublet microtubules in these organelles to slide with respect to one another. In cilia, the D. is firmly attached to the A subfiber of the outer doublet, and interacts transiently with the B subfiber of the adjacent fiber in such a way as to generate force. ATP is hydrolysed in the process. From the clearest electron micrographs available, the D. from cilia consists of 3 globular "heads", each of which is attached by a filament to a common, root-like base. The particle has M_r approx. 2×10^6. The three "heads" may not be identical. In cilia, the base is attached to the A subfiber of the microtubule, and the "heads" interact with the B subfiber. Micrographs of bovine brain microtubules have been obtained which show 7 D. molecules surrounding 14-protofilament microtubules, with the "heads" pointing toward the microtubule. [K. A. Johnson & J.S. Wall *J. Cell Biol* (1983) **96** 669-678]

E

Eadie-Hofstee plot: see Kinetic data analysis.
EAG: abb. for electroantennogram. See Pheromones.
Early proteins: see Phage development.
Early RNA: see Phage development.
Ecdysone, *α-ecdysone, molting hormone,*
22R)-2β,3β, 14, 22, 25-pentahydroxy-5β-cholest-7-en-
-one: a steroid hormone which stimulates molting
f caterpillars, pupa formation and emergence from
he pupa. Its first recognizable effect is the activa-
on of certain genes (see Gene activation). E. was
he first insect hormone to be isolated in crystalline
orm. This was accomplished in 1954 by Butenandt
nd Karlson, who isolated it from silk-worm pupae,
Bombyx mori (25 mg/550 kg). It has also been de-
ected in other insects. The structure was clarified in
963 using X-ray crystallography. E. was the first of
number of related insect and crustacean molting
ormones to be discovered. Similar compounds, the
hytoecdysones, are synthesized by some plants, for
xample *Lemmaphyllum microphyllum* Presl. and
olypodium vulgare L. (see Ecdysterone). E. and the
elated hormones are synthesized from cholesterol
r phytosterols (see Sterols), which are required by
he insects as vitamins.

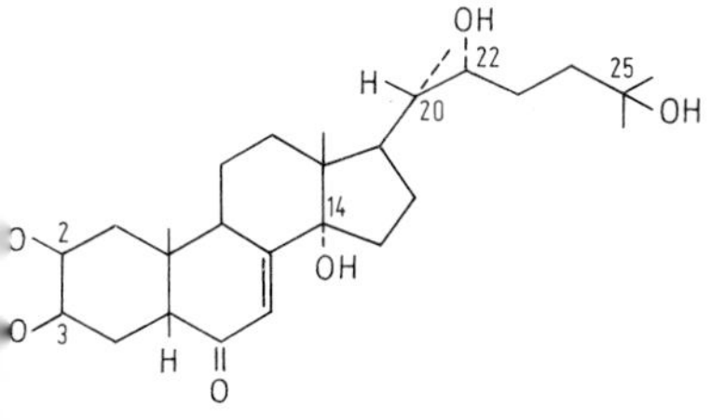

cdysone

β-Ecdysone: see Ecdysterone.
Ecdysterone, *β-ecdysone, crustecdysone, 20-hy-*
droxyecdysone: a molting hormone, M_r 480.65, m.p.
38 °C, found together with Ecdysone (see) in pu-
ae of silk-worms *(Bombyx mori)* in amounts of
.5 mg/500 kg. It has also been isolated from other
nsects and crustaceans and, more recently, from
nany plants, including *Lemmaphyllum microphyllum*
resl., Podocarpus elatus and *Trillium smalli.* E. dif-
ers from ecdysone in having an extra hydroxyl
roup at position 20.
Ecgonine: the principal part of the cocaine
molecule and the basis of many Coca alkaloids. M_r
85.22, (–) form, m.p. 205 °C (d.), $[\alpha]_D^{15} - 45$ ° $(c=5)$.
. has four chirality centers, and therefore exists in a
umber of naturally occurring stereoisomers. It is
roduced commercially by hydrolysis of the raw al-
aloids of coca leaves, and converted to cocaine by

esterification with methanol and benzoic acid. Formula and biosynthesis, see Tropane alkaloids.
Echinochromes: see Spinochromes.
Echinoderm saponins: see Echinoderm toxins.
Echinoderm toxins, *echinoderm saponins:* low
M_r steroid toxins, produced in the glands of echinoderms. Sea cucumbers *(Holothuria)* secrete highly
toxic sulfated steroid glycosides, called holothurins.
Starfish *(Asteroidea)* produce asteriotoxins or
asteriosaponins, in which the main aglycon is preg-
nene diolone. See Asterosaponin A.
Echinomycin, *quinomycin A:* a depsipeptide antibiotic isolated from *Streptomyces echinatus* and effective against Gram-positive bacteria. It contains
two quinoxalinoic acid residues as heterocomponents.
EC nomenclature: see Enzymes.
Ecological chemistry: 1. the investigation and
optimization of the effects of man-made chemicals
(e.g. pesticides, fertilizers) on the environment. E.c.
also includes the development of chemicals and
chemical processes that are not harmful to the environment, and the identification of harmful pollutants and their origins.
2. the study of the interactions of living organisms
with each other, and with their environment, at a
chemical level. This is therefore a wide subject, including such topics as the chemistry and biochemistry of pheromones, insect defense secretions, plant
antifeeding substances, attraction and warning pigmentation in plants and animals, etc., and biochemical adaptation to drought, salinity water-logging,
high temperature, etc. Several names are used for
this relatively new area of biochemistry: *Ecological*
Chemistry, Chemical Ecology, Ecological Biochem-
istry, Biochemical Ecology.
Economic coefficient: see Cultivation of microorganisms.
Ectocarpene, *all-cis-(1-cyclohepta-2′,5′-dienyl)-*
but-1-ene: a sexual attractant excreted by the female
gametes of the brown algae *Ectocarpus stiliculosus.*
b.p. 80 °C, $[\alpha]_D^{22} + 72$ $(c=0.03$, chloroform).
Ectotoxins: see Toxic proteins.
Edestin: a hexameric, globular protein from
hemp seeds, *Cannabis sativa.* M_r 300000. Each of
the 6 subunits consists of two nonidentical polypeptide chains (M_r 27000 and 23000) joined by disulfide bridges. E. is very similar to arachin and excel-
sin from brazil nuts.
Edrophonium: see Acetylcholine.
EDTA: see Ethylenediaminetetraacetic acid.
Effectors: chemical compounds, such as metabolites, hormones or cyclic AMP, which regulate the
activity of a gene or enzyme, usually by allosteric interaction with a regulator protein or the enzyme protein. They may increase the rate of an enzyme reac-

tion (activators) or decrease it (inhibitors). An inhibitor may occupy the active center of an enzyme and thus prevent the substrate from occupying it (competitive inhibition), or it may bind either the free enzyme or the enzyme-substrate complex (noncompetitive inhibition), or it may bind only to the enzyme-substrate complex (uncompetitive inhibition). E. which bind to the enzyme at a site other than the catalytic center are called allosteric E.

EGF: abb. for Epidermal growth factor (see).

Ehlers-Danlos syndrome: see Inborn errors of metabolism.

EIDA: abb. for Enzyme immunodetection assay. See Recombinant DNA technology.

EL-531, *a-cyclopropyl-a-(4-methoxyphenyl)-5-pyrimidine methanol:* a synthetic growth retardant. It is a gibberellin antagonist which delays the growth of lettuce hypocotyls.

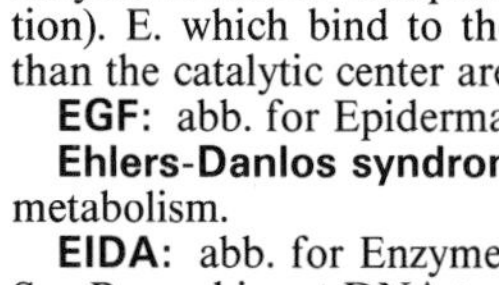

EL-531

Elaidic acid: see Oleic acid.

Elastase (EC 3.4.21.11): an endopeptidase specific for the Elastin (see) in animal elastic fibers. Its inactive precursor, *proelastase,* is formed in the vertebrate pancreas and transformed in the duodenum to elastase by the action of trypsin. The natural substrate of E. is elastin, an insoluble protein rich in valine, leucine and isoleucine. E. attacks the peptide bond adjacent to a non-aromatic, hydrophobic amino acid. The best synthetic substrates are therefore acetyl-Ala-Ala-Ala-OCH_3 and benzoylalanine methyl ester. The trypsin substrate benzoylarginine ester and the chymotrypsin substrate acetyltyrosine ester are not attacked by E.

The primary and tertiary structures of E. are very similar to those of the other pancreas proteinases. Of the 240 amino acids in E. (M_r 25700), 52% are identical to those in trypsin and chymotrypsin A and B. These include the catalytically important amino acids His_{57}, Asp_{102} and Ser_{195}, the ion pair Val_{16} Asp_{197}, which is important for the conformation, and the four disulfide bridges. The sequence of the *N*-terminal peptide removed from proelastase (M_r 27000, 251 amino acids) is still unknown. In contrast to the acid trypsin activation peptide, it must be extremely basic, because the IP of the precursor (10.7) is higher than that of E. (9.5). As might be expected, the spatial structure of E. is very similar to that of the other pancreatic serine proteases (see Chymotrypsin, Trypsin).

A second elastolytic enzyme (M_r 21900) has been isolated from the porcine pancreas. It hydrolyses the chymotrypsin substrate acetyltyrosine ester even better than chymotrypsin does. Another E.-like enzyme, α-lytic proteinase (M_r 19900, 198 amino acids) has been isolated from the soil bacterium *Myxobacter 495.* This enzyme is remarkably similar to the pancreatic E. both in structure (41% homology, sequence in the active center Gly-Asp-Ser-Gly, 3 ho-

mologous disulfide bridges) and substrate specificity. Since an E. (M_r 22300) has been isolated from yet another microorganism, *Pseudomonas aeruginosa,* the enzyme appears to be important for other organisms than animals.

Elastin: a Structural protein (see) which is the main component of the elastic fibers of the tendons, ligaments, bronchi and arterial walls. It owes its elasticity to a high content of glycine, alanine and proline, and of valine (17%), leucine and isoleucine (12% together). The sequences -Gly-Val-Pro-Gly- and -Gly-Gly-Val-Pro- are frequent in the protein. In addition, it is cross-linked by two unusual, blue fluorescing amino acids, *desmosine* and *isodesmosine.* The elasticity, yellow color, insolubility in water and sodium hydroxide solution, and resistance to denaturation and to proteases (except elastase) are all due the three-dimensional network created by these cross-links. E. can be hydrolysed to the water soluble *α-elastin* (M_r 70000) by treatment with hot oxalic acid. The precursor of E. is soluble tropoelastin, which has no desmosine or isodesmosine cross links.

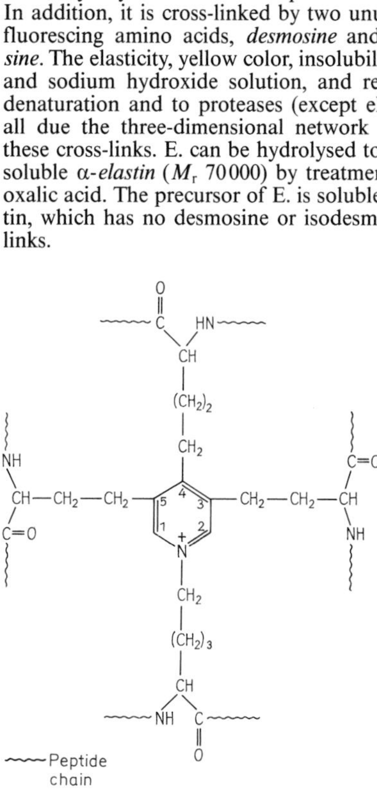

Desmosine
In isodesmosine, there is a chain attached to the C2 of the pyridine ring instead of the C4.

Elaterin: see Cucurbitacins.
Electroantennogram: see Pheromones.
Electroblot: see Southern blot.
Electrofocusing: see Proteins.
Electron transfer flavins, abb. *ETF:* flavoproteins (see Flavin enzymes) which mediate electron transport from reduced FADH to the cytochrome system. Flavoproteins can oxidize those substrates of the Respiratory chain (see) whose oxidation does not involve pyridine nucleotides. However, the substrate must have a more positive redox potential than the $NADH + H^+/NAD^+$ system. ETF from pork liver has 6 molecules flavin per atom of iron and copper is also present.

Electron transport chain: see Respiratory chain.

Electron transport particles, abb. *ETP:* fragments of mitochondria, obtained by ultrasonic or detergent treatment, which are capable of transporting electrons. ETP contain the complete electron transport system of the Respiratory chain (see). In the electron microscope, ETP appear as membrane-enclosed vesicles, which are thought by some authors to be a giant molecule of defined composition. *Heavy ETP* (abb. ETP$_H$) also include succinate dehydrogenase. Different preparations of ETP contain different enzymes and cofactors and structural lipids. When the lipids are extracted, the activity of the ETP is lost, but it can be regenerated by addition of appropriate lipids. Further degradation of the ETP produces the complexes of the respiratory chain.

Electrophoresis: a method of separating charged particles or macromolecules by allowing them to migrate in an electric field. The method is used most frequently in biochemistry to separate delicate macromolecules, usually proteins or nucleic acids. In *free E.* (Tiselius, 1937), the substances to be separated are placed in a solution in a U-shaped tube. In *carrier E.,* the separation is done on a carrier such as paper, cellulose powder, glass powder, starch agar or polyacrylamide gels, or other materials. These stabilize the separated bands against diffusion or convection, but they also interact with the particles, so that there may be molecular sieving or electrostatic effects as well as pure electrophoresis. E. can be automatized and reduced in scale to the µg range. The separation of Plasma proteins (see) is an example of the sensitivity of the method.

Special variants of E. are *immunoelectrophoresis, disc electrophoresis* and *isolectric focussing,* which are discussed under Proteins (see).

Eledoisin: an undecapeptide, Pyr-Pro-Ser-Lys-Asp-Ala-Phe-Ile-Gly-Leu-Met-NH$_2$, from the salivary glands of a cephalopod (*Eledone moschata* and *E. Aldrovandi*). Its biological activities in vitro are similar to those of Substance P (see), with which it also shares certain sequence similarities. [V. Erspamer & A. Anastasi, *Experenia* **18** (1962) 58–59; E. Sandrin & R. A. Boissonas, *ibid.* 59–61.]

(9R,11S)-Eleutherin: see Naphthoquinones (table).

Elicitor: any factor, biotic or abiotic, which induces formation of Phytoalexins (see) in plant tissue, e. g. heavy metal salts like HgCl$_2$ and CuCl$_2$ are abiotic E. The term E. is often used specifically for a glucan fraction, released by heat treatment from the cell wall of a phytopathogenic fungus. Glucan E. is used experimentally to induce the synthesis of phytoalexins. E. activity does not appear to be species or variety specific, and is probably a general defensive response caused by any fungal wall. Specific Oligosaccharins (see) released from the cell wall of either the plant or the invading fungus have been identified as E. [A. R. Ayers et al. *Plant Physiol* **57** (1976) 751–759, 760–765, 766–774, 775–779; U. Zähringer et al. *Z. Naturforsch.* **360** (1981) 234–241; A. G. Darvill & P. A. Albersheim, *Ann. Rev. Plant Physiol.* **35** (1984) 243–275]

ELISA: acronym for Enzyme-Linked Immunosorbent Assay. See Immunoassays.

Ellagic acid: see Tannins.

Elongation: the phase of Protein biosynthesis (see) in which the amino acid chain is extended by addition of new residues.

Elongation factors, *transfer factors:* proteins catalysing the elongation of peptide chains in Protein biosynthesis (see). Three have been isolated from bacteria: EFT, which is a mixed dimer of two proteins, Ts (M_r 42000) and Tu (M_r 44000), and EFG. The bacterial elongation factors do not interact with the 80S ribosomes of eukaryotes. They can be obtained by saline treatment of the 70S ribosomes. (See Protein biosynthesis).

Embden ester: a mixture of D-glucose-6-phosphate and D-fructose 6-phosphate, both of which are intermediates in glycolysis.

Embden-Meyerhof-Parnas pathway: see Glycolysis.

Embelin: see Benzoquinones.

Embryonal inducers: compounds which induce the differentiation of organs in the course of embryonic development. A low-molecular-weight protein has been isolated from chick embryos which, when injected into the ectoderm of an amphibian gastrula, can induce the formation of kidney and muscle primordia (mesodermal factor) and of notochord tissue (neural factor).

Emerson effect, *enhancement effect:* an increase in the photosynthetic quantum yield from long-wave red light (700 nm) obtained by simultaneous irradiation with shorter wavelengths (< 670 nm). There is a sharp drop-off in photosynthetic efficiency around 700 nm, but light of this wavelength can be utilized synergistically with light of shorter wavelengths. The effect indicates that there are two photosystems participating in the generation of oxygen, and that these have different light-collecting pigments. The conclusion is supported by action spectra of the E. e.

Emetine: a dimeric isoquinoline alkaloid which is the principal alkaloid of ipecac, the ground roots of *Uragoga ipecacuanha.* M_r 480.63, m. p. 74 °C $[\alpha]_D^{20} - 50$ ° ($c=2$ in chloroform). E. is very poisonous, exerting a strong stimulus on mucous membranes. In large doses, it leads to vomiting. It is used in the treatment of amebic dysentery and its complications. The alkaloids of the E. group are biosynthesized by a Mannich condensation of two molecules of phenylethylamine with an iridoid C$_9$ body (Fig.).

Emetine

Emodin: an orange anthraquinone pigment, m. p. 225 °C. It often occurs as a glycoside, or as the dimer, skyrin. The 5,5′ dimer, (+)-*skyrin,* m. p. 380 °C, is found in *Penicillium* species. E. is found in many higher plants including rhubarb root, alder buckthorn (*Rhamnus frangula* L.), *Cascara sagrada,* etc. It is used as a cathartic.

Emodin

Encephalitogenic protein, *myelin protein A 1:* the most important myelin protein of the mammalian central nervous system. On injection into guinea pigs, rabbits or rats, it induces allergic autoimmune encephalomyelitis, abb. EAE, an inflammation of the brain and spinal column. The structure of E.p. from the myelin sheath of humans, cattle, rabbits and guinea pigs has been determined. M_r 18000, 170 amino acids, including 11% Arg, 8% Lys and 6% His, but no Cys. EAE can be induced by a relatively small active region of E.p., the location of which varies in the E.p. of different species.

Endocrine hormones: hormones produced by specialized cells in endocrine glands, and released into the blood stream. They stand in contrast to Tissue hormones (see), which are produced by individual cells in tissues specialized for other functions.

Endocrinology: the study of hormones. It includes the chemistry, metabolism, effects on cells, organs, individuals and populations of hormones. *Ecoendocrinology* is concerned with the interactions of endocrine systems with the environment.

Endogenous minimum: see Minimum protein requirement.

Endolysin: see Lysozyme.

Endomembrane system: see Endoplasmic reticulum.

Endonucleases: see Nucleases.

Endopeptidases: see Proteases.

Endoplasmic reticulum, abb. *ER:* a net-like system of double membranes located in the cytoplasm of eukaryotic cells. The membranes form tubes or channels with cross sections of 50 to 500 nm. The ER appears to be continuous with the outer nuclear membrane and with the secretory system of the cell (the Golgi apparatus, see). If the ER is associated with ribosomes, it is called rough ER; otherwise it is called smooth ER. Rough ER is engaged in the synthesis of membrane proteins and proteins for export. The growing polypeptides pass through the membrane. When the cell is homogenized, the membranes of the ER form vesicles which can be isolated as the *microsome* fraction, which is characterized by the presence of glucose-6-phosphatase and the mixed-function oxidases (marker enzymes). Other enzymes of the ER are responsible for the biosynthesis of triacylglycerols, glycerophospholipids, mucopolysaccharides and glucuronides. Various steps of steroid biosynthesis, including formation of mevalonic acid from hydroxymethylglutarate, synthesis of squalene from farnesyl pyrophosphate and cholesterol from lanosterol occur on or in the ER. (Other steps in cholesterol biosynthesis occur in the cytoplasm or mitochondria). Steroid conversions and degradation occur in the ER, and it is also involved in sulfur metabolism.

Endorphins: endogenous peptides with morphine-like effects (*endo*genous mo*rphin*e); the natural ligands for the opiate receptors. See Opioid peptides.

Endotoxins: see Toxic proteins.

Endoxan: see Cyclophosphamide.

End oxidation, *terminal oxidation:* the last step in catabolism. In aerobically respiring cells, E.o. is carried out via the tricarboxylic acid cycle.

End product: the last compound in a metabolic pathway, which is irreversible as written. The E.p. may either be the starting material for another metabolic pathway, or it may be accumulated or excreted. An E.p. can control its own rate of synthesis, either as an allosteric Effector (see) of one of the enzymes at the beginning of the pathway, or as a repressor of the operon coding for the enzymes.

End product inhibition: see Metabolic regulation.

End product repression: inhibition of the synthesis of the enzymes of a reaction sequence (see Enzyme repression), or inhibition of the activity of the first enzyme in the sequence (see Allostery).

Energy charge: see Adenosine phosphates.

Energy metabolism: those reactions which serve to release energy by degradation of carbohydrates and fats. The most important link between the energy-producing and energy-consuming reactions is ATP. It is produced in the largest amounts by Respiration (see), although some is produced by Glycolysis (see). In photosynthetic organisms, ATP is also produced in the light reactions (see Photosynthesis). Since the intermediate products of glycolysis and the tricarboxylic acid cycle are also starting materials for many syntheses there can be no sharp division between energy metabolism and synthetic metabolism.

Energy-rich bonds: see High energy bonds.

Energy-rich phosphates, *high energy phosphates:* phosphorylated compounds with high energies of hydrolysis of the phosphate ester. Chemical energy in biological systems is stored in energy-rich phosphates. These may be acid anhydrides of phosphoric acid (e.g. ATP), *enol* phosphates (e.g. see Phosphoenolpyruvate) or amidine phosphates (e.g. see Phosphagens). A high energy phosphate bond is represented as R ~ ℗, instead of the usual hyphen between groups. See High energy bonds.

The terms high energy phosphate and phosphate bond energy, and the so-called "squiggle" bond (~) are contrary to the purity and austerity of classical thermodynamics, but have proved useful in conveying the importance of phosphate groups in the storage and transfer of chemical energy in biochemical systems. These concepts were introduced in 1941 by Lipmann ("Metabolic Generation and Utilization of Phosphate Bond Energy", Lipmann, F. *Advances in Enzymology, 1,* (1941) 99–162).

Enhancement effect: see Emerson effect.

Enkephalins: Pentapeptides with affinity for the opiate receptors in the brain (see Opioid peptides). Met-E. has the structure Tyr-Gly-Gly-Phe-Met; Leu-E. has the sequence Tyr-Gly-Gly-Phe-Leu. These two sequences are also found at the *N*-termini of a number of opioid peptides. The physiological source of the E., however, is preproenkephalin A, a protein of 267 amino acid residues. Its sequence has been deduced from cloned DNA. [M. Noda et al., *Nature* **295** (1982) 202–206; M. Comb et al., *ibid.* 663–666; V. Gubler et al., *ibid.* 206–208]

Enniatins: ring-shaped depsipeptide antibiotics produced by the fungus *Fusarium orthoceras* var. enniatum and other *Fusaria*. Enniatin A, cyclo-(D-Hyv-MeIle-)₃ and enniatin B, cyclo-(-D-Hyv-Me-Val-)₃ are found together. (Hyv = hydroxyisovaleric acid, MeIle = methylisoleucine, MeVal = methylvaline.) These compounds act as artificial pores in membranes, permitting potassium ions to penetrate them. See Ionophore.

Enolase (EC 4.2.1.11): an enzyme of Glycolysis (see) which catalyses the reversible dehydration of 2-phosphoglycerate to phospho*enol*pyruvate. It is a dimeric metalloenzyme, and requires 2 molecules magnesium to maintain its structure, or 4 molecules Mg for activity. Zinc(II) and manganese(II) ions also activate E., while F⁻ inhibits it. The M_r of several E. are (M_r of the subunit in parentheses): 82000 (41000) for rabbit liver and muscle; 100000 (48000) for salmon; 88000 (44000) for yeast and 90000 (46000) for *Escherichia coli*. Like most of the glycolysis enzymes E. exists as several Isoenzymes (see) which can be separated by electrophoresis.

Enteroamine: see Serotonin.

Enterobacteriaceae: a family of bacteria capable of formic acid fermentation. They are Gram-negative rods which do not form spores. Motility is provided by peritrichal flagella. E. are facultatively anaerobic and can grow on simple synthetic media containing mineral salts, sugars (carbon and energy sources) and ammonium salts for nitrogen. Some of the E. including that favorite research organism, *Escherichia coli* (see), and *Aerobacter aerogenes* and *Proteus vulgaris,* are intestinal bacteria. The family also includes the plant pathogens of the genus *Erwinia,* which cause white rots, the organisms of food poisoning, *Salmonella typhimurium,* typhus, *Salmonella typhi,* and dysentery, *Shigella dysenteriae.*

Enterodiol: 2,3-bis(3-hydroxybenzyl)butane-1,4-diol, a lignan which occurs as its glucuronide in primate urine. It is accompanied by the glucuronide of the related lignan, Enterolactone (see). [S. R. Stitch, *Nature* **287** (1980) 738–740; D. R. Setchell et al., *Biochem. J.* **197** (1981) 447–458]

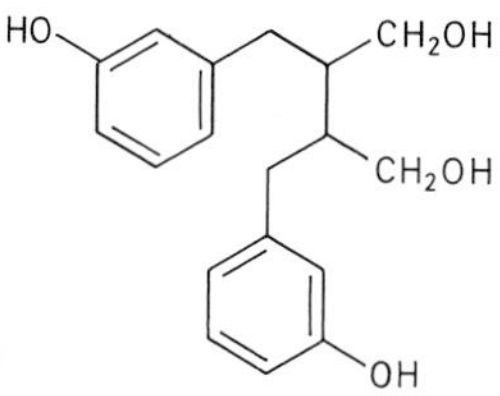

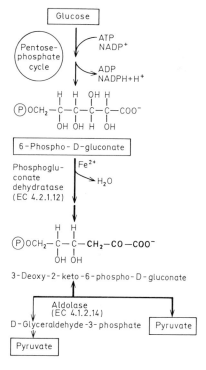

Enterodiol

Enterogastrone: see Gastrin-inhibiting hormone.

Enterohepatic circulation: see Bile acids.

Enterokinase: see Enteropeptidase.

Enterolactone: *trans*-(±)-3,4-bis[(3-hydroxyphenyl)methyl]dihydro-2-(3 H)furanone; *trans*-2,3-bis(3-hydroxybenzyl)-γ-butyrolactone, a lignan which occurs as its glucuronide in primate urine. It is accompanied by the glucuronide of the related lignan, Enterodiol (see). [S. R. Stitch, *Nature* **287** (1980) 738–740; D. R. Setchell et al., *Biochem. J.* **197** (1981) 447–458]

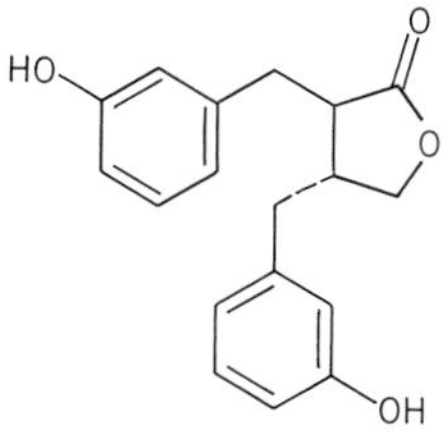

Enterolactone

Enteropeptidase, *enterokinase* (EC 3.4.21.9): a highly specific duodenal protease which acts only on trypsinogen. M_r 196000, 1100 amino acids. E. consits of two covalently bound glycopeptides (M_r 134000 and 62000). It removes the *N*-terminal peptide from trypsinogen (Val-[Asp]₄-Lys) to activate it to trypsin. E. contains the most carbohydrate (37%) of any digestive enzyme.

Enterotoxins see Toxic proteins.

Entner-Douderoff pathway: a degradation

pathway for carbohydrates in microorganisms, especially *Pseudomonas* species, which lack the enzymes hexokinase, phosphofructokinase and glyceraldehyde-3-phosphate dehydrogenase (Fig.)
Balance: $Glucose(C_6H_{12}O_6) + ADP + P_i + 2NAD(P)^+$
$\rightarrow 2$ pyruvate $(CH_3COCOO^-) + ATP + 2NAD(P)H + 2H^+ + H_2O$.

Enzyme analogs. see Synzymes.

Enzyme graph, *enzyme network:* a way of representing the stoichiometry of an enzyme reaction as a network. The nodes of the network are the Enzyme

well understood. Hormones or substrates can cause a large increase in the synthesis of a particular enzyme. For example, the hormone ecdysone induces dopa decarboxylase in insects, and nitrate (as substrate) induces nitrate reductase in higher plants. (See Gene activation).

Enzyme interconversion: see Covalent modification of enzymes.

Enzyme isomerization: reversible changes in enzyme conformation in the course of a catalytic cycle. See Enzyme kinetic parameters.

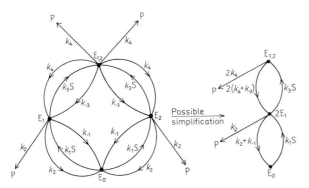

Enzyme graph for an enzyme with two active centers. E_i are enzyme species; P is product; S is substrate.

species (see), which are connected by arrows showing the direction of the reaction. These are labelled with the corresponding rate constants. An E.g. simplifies the writing of Rate equations (see) for the reactions. The Fig. shows an example of an E.g. for an enzyme with two active centers. E denotes enzyme species with zero, one, or two (E_0, E_1, E_2) substrate molecules bound in a Michaelis-Menten complex.

Enzyme immunodetection assay: see Recombinant DNA technology.

Enzyme induction: stimulation of the synthesis of enzymes by inducers. In bacteria, many catabolic enzymes are only synthesized when the appropriate substrate is available in the medium. The classical example is the induction of the enzymes of the lac operon in *Escherichia coli* by their substrate, lactose (Fig.) (the true inducer is allolactose, derived in small quantities by rearrangement of lactose). The regulator gene R codes for a specific repressor protein which, in the absence of an inducer, binds to the corresponding operator O and blocks the transcription of the structural genes z, y and a (negative control). Since the mRNA for these genes is not produced, neither are the proteins. If lactose (or an artificial inducer) is added to the medium, the inducer binds to the repressor and inhibits its binding to the operator. Now the transcription of the structural genes is possible, and the enzymes are synthesized. The concentration of β-galactosidase is about 1000 times higher in induced cells than in those growing on other media. The presence of glucose prevents induction of β-galactosidase by lactose; this is due to Catabolite repression (see). E.i. also occurs in higher organisms, although the mechanism is not

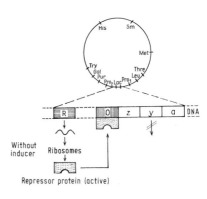

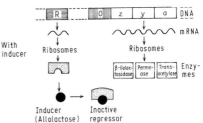

Induction of the lac operon. The circle at the top represents the *Escherichia coli* chromosome map showing the lactose and other regions.

Enzyme kinetic parameters, *enzyme parameters:* the parameters of the rate equations which remain constant, so long as temperature, pressure, pH value and buffer composition are constant. They are derived from the rate constants of the rate equations, and are frequently used to characterize the enzyme functionally. Some can be interpreted in physical terms: K_i as the dissociation constant for the enzyme-inhibitor complex, for example, V_m as the maximal velocity at a saturating substrate concentration K_s (substrate constant) as the concentration of substrate which half saturates the enzyme, K_m (Michaelis constant) as the substrate concentration for half-maximal velocity.

Enzyme kinetics: the mathematical treatment of enzyme-catalysed reactions. A great deal of information about reaction mechanisms can be obtained from kinetic experiments and evaluation of the data so obtained (see Kinetic data evaluation). A kinetic experiment consists of measuring the rate of disappearance of substrate, or of the appearance of product under controlled conditions of temperature, pH, substrate and enzyme concentration, buffers, etc. The simplest graphical representation of the data is a *progress curve,* a plot of $\Delta[S]$ vs. time (where $[S]$ is the concentration of substrate). One is usually interested in the *instantaneous velocity* $v = d[S]/dt$, which may not be easy to determine from the progress curve. In this case an *integrated rate expression,* or the time course of product formation may be useful. The *activity* of an enzyme is expressed as the amount of substrate consumed (or product formed) in a given time. In 1961, the Nomenclature Committee of the International Union of Biochemistry recommended the use of the *Unit (U),* which is the amount of enzyme required to turn over 1 micromole substrate per minute under standard conditions. In accordance with the shift to mks system, in 1972 they recommended the *katal,* the amount of enzyme turning over 1 mole substrate per second. This is a very large unit, so that in practice one uses the micro-, nano- and picokatal. *Specific activity* is given in U/mg or kat/kg.

Single-substrate enzymes (see) have the simplest kinetics, namely first-order. The *rate equation* for such a *unimolecular* or *pseudounimolecular* reaction is $v = -d[S]/dt = k[S]$. The reaction is characterized by a *half-life* $t_{\frac{1}{2}} = \ln 2/k = 0.693/k$, where k is the first-order rate constant. The *relaxation time,* or the time required for $[S]$ to fall to $(1/e)$ times its initial value is τ. $\tau = 1/k = t_{\frac{1}{2}}/\ln 2$.

When there is more than one substrate in a reaction (see Multi-substrate enzymes), the kinetics may be *second order* (or pseudo-second-order. See Cleland's short notation). The equation for a second order reaction is $A + B \xrightarrow{k_2} P$, where k_2 is the bimolecular rate constant, and $v = k_2[A][B]$. All chemical reactions are reversible and eventually reach an equilibrium in which the rates of the forward and reverse reactions are equal. At this point, $A + B \underset{k_2}{\overset{k_1}{\rightleftharpoons}} P$, where k_1 is the rate of the forward reaction and k_2 is that of the reverse. The equilibrium constant is given by

$$K = \frac{[P]}{[A][B]} = \frac{k_1}{k_2}.$$

The Michaelis-Menten treatment of E. assumes that the enzyme and substrate bind temporarily to form an *enzyme-substrate complex,* which may break down either to substrate or product.
$E + S \rightleftharpoons ES \rightleftharpoons E + P$. For a short period of time, it can be assumed that the rate of change in $[ES]$ is small compared to the change in $[S]$ (the *steady state approximation*) because the rate at which it is formed is equal to the rate at which it breaks down. It follows that

$$[E][S] = \frac{(k_2 + k_3)}{k_1}[ES] = K_m[ES],$$ where K_m is the *Michaelis constant.* The *Michaelis-Menten equation* for the initial rate of reaction (when $P = O$) is

$$v = \frac{V_{max}}{1 + K_m/[S]} = \frac{V_{max}[S]}{K_m + [S]}.$$

The quantities K_m and V_{max} are called the *kinetic parameters* of an enzyme (see Enzyme kinetic parameters).

A plot of initial rate against substrate concentration is not linear (Fig.), so it is difficult to estimate the kinetic parameters from it (except by computer analysis).

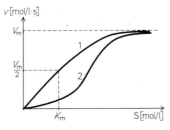

Plot of initial rate vs. substrate concentration, sometimes called a characteristic curve. 1 is a Michaelis-Menten hyperbola; 2 is a sigmoidal plot characteristic of a cooperative enzyme. For linear trasnformations, see Kinetic data evaluation.

In the treatments discussed so far, it has been assumed that the back reaction could be neglected. The reactions catalysed by many enzymes are essentially irreversible or the products are immediately subject to further reaction, so that the assumption of irreversibility is valid. However, if the reaction is reversible, the Michaelis equation must be modified. Haldane has suggested a notation in which V_f and V_r are the maximum velocities in the forward and reverse directions, and K_{mS} and K_{mP} are the Michaelis constants for the substrate and product. The *Haldane relationship* for a system with a single substrate and single product is then $K_{eq} = V_f K_{mP}/V_r K_{mS}$.

Multi-substrate enzymes (see) catalyse reactions of two or more substrates. Such enzymes can form a number of different complexes with one or both substrates and/or products which are called *enzyme species.* The order in which these species are formed may be *random* or *ordered.* Cleland's short notation (see) is a convenient way to represent the possibilities. The kinetics of such reactions become extremely complicated; *enzyme networks* (see Enzyme graphs) provide a means of summarizing them. To evaluate the kinetic data for such systems, one must

179

resort to a computer. Furthermore, the information gained from steady-state experiments may not be sufficient. A number of methods of very rapid measurement have been used to investigate the *pre-steady-state* condition of reactions, including stopped flow, temperature jump and flash methods.

Enzyme modulation: see Covalent modification of enzymes.

Enzyme network: see Enzyme graph.

Enzyme parameter: see Enzyme kinetic parameter.

Enzyme protein: see Coenzymes.

Enzyme repression: the blockage of the synthesis of anabolic enzymes by the end product of the biosynthetic pathway to which they contribute. This type of regulation is found in prokaryotes for operons which synthesize various amino acids which may be present in the medium. If the amino acid is present, the synthesis of all the enzymes in the operon is turned off, but if it is in short supply, the operon is derepressed (see Derepression). See also Attenuation.

The mechanism of E. is shown in the Fig. The regulator gene R codes for an *aporepressor* which cannot bind to the operator gene O unless it forms an active

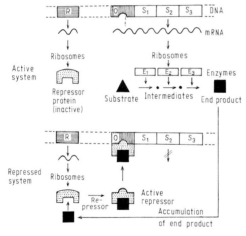

Repression of enzyme synthesis

complex with the *corepressor* (the end product of the synthetic chain). In the absence of the corepressor, the structural genes are transcribed and the enzymes synthesized. See Gene activation.

Enzymes: protein catalysts. About 90% of cellular proteins are E., and some structural proteins (e.g. actin and myosin) are also E. A protein is classified as an E. if it is known to catalyse a reaction, but it is always possible that a given protein catalyses an unrecognized reaction. The Nomenclature Committee of the IUB lists well over 2000 E. which have been well characterized; however, these are classified according to the reaction catalysed, not the identity of the proteins. The same reaction may be catalysed by non-homologous proteins in different organisms, and thus the number of E. actually characterized is much higher than 2000. The Nomenclature Commit-

tee publishes an enzyme catalog (abb. EC) in which each E. receives a four-part number. The first part indicates the general category of reaction (Table), the second and third, the subgroup and sub-subgroup, and the fourth part of the number is assigned arbitrarily, in consecutive order.

Table 1. The main enzyme groups in the enzyme catalog.

Group	Nature
1	Oxidoreductases
2	Transferases
3	Hydrolases
4	Lyases
5	Isomerases
6	Ligases (synthetases)

The committee also establishes a recommended trivial name for each E., to be used in text, and a systematic name as the basis for classification. The latter should be mentioned at least once in each publication. Since the properties of E. from different sources vary, often substantially, it is also necessary to indicate the source in a publication.

The most striking difference between E. and chemical catalysts is the specificity of E, with regard both to their substrates and to the reactions catalysed. E. distinguish stereoisomers absolutely, due to the three-dimensional nature of the binding between substrate and enzyme. The *lock and key* model suggests that the E., like a complicated lock, can only be fit by a substrate (key) of precisely the correct shape. The *induced fit model* hypothesizes that when the substrate binds to the E., the conformation of the latter changes in such a way as to make the fit more complete. In either case, the difference between D- and L-forms of a substrate is analogous to the difference between right and left hands; the E. binding site corresponds to a glove which will fit only one of the two hands.

The substrate adheres to (and often reacts with) the E. because the arrangement of charged or hydrophobic groups in the enzyme precisely complements the arrangement of charges or hydrophobic sections of the substrate, or because hydrogen bonds between the two can form easily.

The mechanisms by which E. catalyse reactions are usually fairly simple. In general, molecules do not react until they have been activated to a higher-energy electronic or vibrational state called the transition state. In spontaneous reactions, the necessary free energy is supplied by molecular collisions or light. However, when the molecule is activated in this way, it can usually rearrange itself, break down or react with another molecule in a number of different ways, so that there are usually a number of reaction products. Any catalyst increases the rate of a reaction by reducing its free energy of activation. A chemical catalyst may also promote the "right" reaction; E. invariably do. This may be accomplished by an E. in several ways: by bringing the reactants together in the appropriate orientations; by providing acidic and/or basic groups at the right positions to promote acid or base catalysis of the "desired" reaction; by providing electrophilic or nucleophilic

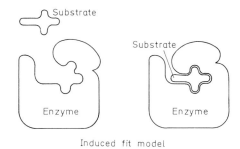

Lock and key
model

Induced fit model

Models of enzyme-substrate binding

groups which form a temporary covalent bond with the substrate; and/or by providing for concerted mechanisms in which two catalytic groups on the E. attack the substrate molecule. Any of these forms of catalysis may be accompanied by conformational changes in the E. molecule which strain the substrate molecule and facilitate the breaking of bonds within it.

The amino acid residues which take part in catalysis and their immediate neighbors constitute the *active center* of the E. The *substrate binding site* is close to or includes the active center, but if the substrate is large, the binding site may extend beyond the active center. Allosteric E. (see Allostery) also have *effector binding sites* which are distinct from the substrate binding site.

Most E. are strictly intracellular, but some are excreted into the body fluids or the medium of unicellular organisms. Proteolytic E. are secreted in the form of inactive precursors which are only activated after they are safely out of the cell; those which are not secreted are sequestered into a special organelle,

the lysosome. Some E. are specific for particular organs or tissues, and their presence in the blood can be used as a diagnostic test for damage to the tissue of origin. Within the eukaryotic cell there is a fair amount of compartmentation of the E. (and reactions) within organelles. When the cell is fractionated, these E. can be used as *markers* for the various fractions (mitochondria, cytosol, liposomes, etc.). In some cases, a given catalysis may be performed by two or more *isoenzymes* in different tissues or cell fractions.

In addition to their use as symptoms of tissue damage, E. are applied as reagents for the specific detection of other metabolites, such as glucose, ATP, lactate, etc., which are diagnostically significant. The proteases, in particular, may be applied to relieve poor digestion, problems of blood coagulation and fibrinolysis. Pathologically increased levels of proteases, on the other hand, may be treated with protease inhibitors from bovine organs, which do not provoke an immune response in humans.

With the development of appropriate methods for

Table 2. A selection of technical and medical applications of enzymes.

Enzyme	Reaction	Use
Glucose oxidase from *Aspergillus niger* or *Penicillium*	Glucose $\xrightarrow{O_2}$ Gluconolactone $+ H_2O_2$	Preserving of foods and drinks by removal of O_2; Prevention of brown discoloration (e.g. in dried eggs) by removal of glucose.
Catalase from microorganisms	$2 H_2O_2 \rightarrow$ $2 H_2O + O_2$	a) Together with glucose oxidase, for food preservation. b) Removal of excess H_2O_2 in milk preservation by treatment with H_2O_2.
Glucose isomerase from yeast and other microorganisms	Glucose $\rightleftharpoons$ Fructose	Fructose production to increase the sweetness of drinks without adding carbohydrate, and fructose to add to paper to increase its plasticity.
Invertase (β-fructosidase) from yeast and *Aspergillus*	Sucrose (S.) $\rightarrow$ Glucose + Fructose (Invert sugar, I.)	I. is sweeter and more easily digested than S. and is used in artificial honey, ice cream, chocolate creams, etc.
Lactase (β-galactosidase)	Lactose $\rightarrow$ Galactose + Glucose	Production of milk products for adults who do not possess the enzyme.
"Naringinase"	Naringin $\rightarrow$ Naringenin + carbohydrate	Removal of the bitter taste of grapefruit juice. Naringenin is less bitter than naringin.
Lipase from pancreas or *Rhizopus nigricans*	Triglyceride $\rightarrow$ Glycerol + Fatty acids	Isolation of labile fatty acids, improvement of cheese aroma, cocoa processing, digestive aid

Table 2. *A selection of technical and medical applications of enzymes.*

Enzyme	Reaction	Use
L-Amino acid acylase from kidneys	DL-Acylamino acid → L-amino acid + D-acylamino acid	Production of essential amino acids for human and animal nutrition
Penicillin-amidase from microorganisms	Penicillin G → 6-Aminopenicillanic adid (6AP)	6AP is the starting material for semisynthetic penicillins
L-Asparaginase from E. coli	L-Asparagine → L-Aspartate	Used in therapy of leukemia and lymphosarcomas
α-Amylase (Endoamylase from pancreas, bacteria, Aspergillus)	Hydrolysis of α-1,4-glucans, e.g. starch, amylose, amylopectin, to dextrins and maltose	Digestion and baking aid, cleavage of starch in beverages, production of starch pastes and non-sweetening syrups, removal of starch in the textile industry.
α-Amylase + Glucoamylase	Removal of glucose from the non-reducing end of amylase products	Production of glucose from starch in high yield and purity
Cellulase from Aspergillus niger and Stachybotrysatra	Hydrolysis of cellulose to cellobiose	Aid to extraction of pharmaceutically active principles, removal of cellulose from food for special diets, softening of cotton
Pectin esterase from Aspergillus niger und polygalacturonidase	Pectin methyl ester → pectinic acid + CH₃OH Hydrolysis of α-1,4-glycosidic bonds	Removal of pectin sheaths from plant fibers, removal of cloudiness in fruit juices and beer, production of fruit juices and purees.
Pectin lyase	Elimination of α-4,5-unsaturated galacturonate residues from pectin	
Dextranase	Hydrolysis of α-1,6-glycosidic bonds	Prevention of caries
Lysozyme from microorganisms	Hydrolysis of the murein of bacterial cell walls	Removal of bacteria from cow's milk for infant formulas and prevention of caries
Proteases (Trypsin, pepsin, elastase, papain, microbial enzymes, etc.)	Hydrolysis of peptide bonds,	Aid to digestion, removal of necrotic tissue, acne treatment, meat tenderizers, preparation of special diets and peptone media for microorganisms, removal of proteins from carbohydrates and fats, prevention of cloudiness in beer, laundry additive, tanning agent for leather
Rennet from calf stomach and similar enzymes from microorganisms	Hydrolysis of the Phe-Met bond in κ-casein	Cheese production
Streptokinase from streptococci	Plasminogen → Plasmin	Removal of clots (fibrinolysis).

large-scale cultivation of microorganisms and isolation of E.in very large quantities, it has become economically feasible to use them in a number of industries, including food processing, textiles, paper, laundry products and pharmaceuticals. With the advent of genetic engineering, there is interest in developing artificial E. to catalyse various reactions in organic and pharmaceutical chemistry.

Enzyme species, *intermediates:* all covalent and noncovalent complexes between an enzyme and a substrate and/or product or effector, and all the various enzyme isomers (see Enzyme isomerization). The concentrations of the E.s. cannot be measured directly by means of steady-state Enzyme kinetics (see), but their steady-state concentrations can be calculated from the kinetic equations for definite values of the concentration variables. In principle, the concentrations of the E.s. could be determined by the methods of presteady-state kinetics.

Enzyme units: see Enzyme kinetics.

Ephedrin: see Antidepressants.

Epidermal growth factor, *EGF*: a mitogenic polypeptide, M_r 6045 of known structure (Fig.) with growth stimulating activity for a wide variety of epidermal and epithelial cells in vitro and in vivo. Injected into newborn mice, EGF causes precocious opening of the eyelids and precocious tooth eruption. EGF accounts for about 0.5% of the protein content of adult male mouse submaxillary gland. An homologous polypeptide, human EGF, has been isolated from human urine. It seems highly probable that Urogastrone (see) is identical with human EGF. EGF is isolated from mouse submaxillary gland as a high M_r complex, containing two molecules of EGF and 2 molecules of EGF binding protein (M_r 29300). The binding protein has Arginine esterase activity, and is thought to represent the processing enzyme that converts proEGF into EGF. The concentration of EGF in submaxillary gland is androgen-dependent, and the concentration is much lower in the glands of female mice. As measured by radi-

oimmunoassay, mouse serum contains 1 ng/ml EGF. This may rise to 150 ng/ml when adrenergic receptors are stimulated by phenylephrine. Thus EGF seems to serve an as yet unidentified physiological function, and its secretion is controlled. Radioimmunoassay of human EGF shows a normal 24 h excretion of 63.0 ± 3.0 (2S.D.) (males) and 52.0 ± 3.5 (females) with no diurnal or postprandial variation. Human saliva contains 6–17 ng/ml, human milk contains 80 ng/l, and it is undetectable in amniotic fluid. Circulating plasma levels are 2–4 ng/ml. Elevated levels are excreted by females taking oral contraceptives.

Analysis of a cDNA clone for mouse EGF suggests that a large protein precursor (1 168 amino acids) is first synthesized, then processed to EGF (Fig.). It seems unlikely that such a large polypeptide serves only as the precursor of EGF. There are several potential proteolytic cleavage sites within the large precursor, but if processing produces other physiologically active peptides, they have not yet been recognized. [J.A. Downie et al. *Ann. Rev. Biochem.* **48** (1979) 103–131; A. Gray et al. *Nature* **303** (1983) 722–725]

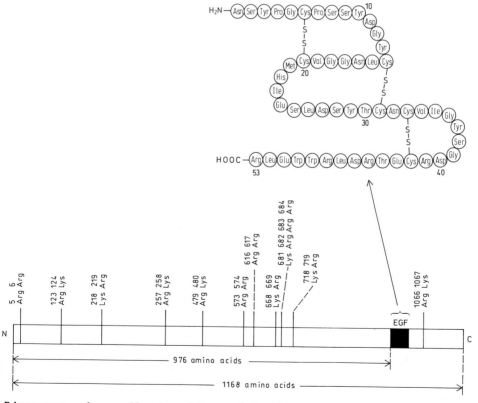

Primary structure of mouse epidermal growth factor, and schematic representation of its large protein precursor (pre-pro-EGF). The signal sequence is uncertain, but the amino terminus of pro-EGF is probably residue 23 (Ile), 26 (Val) or 29 (Trp). The indicated sequences of basic residues represent potential sites of proteolytic cleavage.

Epimers: see Carbohydrates.

Epinephrine: see Adrenalin.

Epiphysis: see Pineal gland.

Episome: see Plasmid.

2,3-Epoxysqualene: see Squalene.

Equilenin: 3-hydroxyestra-1,3,5(10),6,8-pentaen-17-one, an estrogen. M_r 266.32, m.p. 259 °C, $[\alpha]_D^{16} + 87$ ° (12.8 mg made up to 1.8 ml in dioxane). E. is found together with Equilin (see) in the urine of pregnant mares. It has ⅟₂₅ of the biological activity of estrone. E. differs from equilin in having an aromatic ring B. It was the first natural steroid to be totally synthesized, in 1939.

Equilin: 3-hydroxyestra-1,3,5(10),7-tetraen-17-one, an estrogen. M_r 268.34, m.p. 240 °C, $[\alpha]_D^{25} + 308$ ° ($c = 2$ in dioxane). E is isolated from the urine of pregnent mares and has ⅟₂₀ the biological acitivity of estrone.

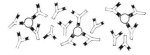

Equilin

Equivalence point, *equivalence zone:* the region of the precipitation curve in which all the antibody binding sites are saturated with the antigenic determinants, and the antibodies have been quantitatively precipitated. Neither antigen nor antibody remains in the supernatant. At this point the antigen-antibody complex forms a complicated network structure (Fig.). Otherwise only smaller structures or binary complexes, which are soluble, are formed.

Equivalence zone: see Equivalence point.

Equol: see Isoflavone.

ER: see Endoplasmic reticulum.

Erepsin: an outdated term for the amino- and dipeptidases secreted by the mucous membranes of the small intestine. Tripeptidases are not included in the erepsin complex, but are present in the mucous membrane itself.

Ergastoplasm: see Endoplasmic reticulum (rough).

Ergocalciferol: see Vitamins (vitamin D₂).

Ergochromes, *secalonic acids:* a group of weakly acid, bright yellow natural pigments based on a dimeric 5-hydroxychromanone skeleton. They have been isolated from a number of molds and lichens and are poisonous. They are synthesized from acetate via the anthraquinone emodin. The most common is secalonic acid A; others are secalonic acid B, C, and D, ergoflavin, ergochrysins A and B and the ergochromes AD, BD, CD and DD. The ergot mold contains two anthraquinone carboxylic acids, the red-orange endocrocin and the red clavorubin, which has an additional hydroxyl group on C5.

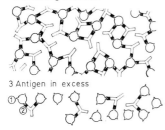

Structural system of ergochromes

Ergoline alkaloids: see Ergot alkaloids.

Ergometrine: see Ergot alkaloids.

Ergosomes: see Polyribosomes.

Ergostane: see Steroids.

Ergosterol, *provitamin D₂, ergosta-5,7,22-trien-3β-ol:* the most abundant mycosterol (fungal sterol). In most fungi, E. is an important component of cell or mycelium membranes. E. is also synthesized by some protozoa (notably Trypanosomatidae and soil amebas), *Chlorella,* and the primitive tracheophyte *Lycopodium complanatum.* It is not synthesized by all fungi, and is, for example, absent from *Pythium* and *Phytophthera,* which do not synthesize any sterols. In some fungi, E. is synthesized but is not the most abundant sterol, e.g. the predominant sterols in rusts (Uredinales) possess a C₂ substituent at C-24. E. is converted by UV irradiation into vitamin D₂ (see Vitamins, cholecalciferol). The main commercial source of E. is yeast. For biosynthesis, see Steroids. [E. I. Mercer *Pestic. Sci.* **15** (1984) 133–155]

Ergot alkaloids, *Claviceps alkaloids, ergoline alkaloids:* a group of over 30 indole alkaloids, possessing the ergoline ring system. They are produced by various species of the fungal genus *Claviceps* (family *Ascomycetes*), which parasitize rye and wild grasses. Following infection by spores of *Claviceps purpurea,* the ears of rye develop 1–3 cm long dark violet (almost black), highly poisonous sclerotia, containing

The basic types of antigen-antibody complex section:

1 Antibodies in excess

2 Antibodies and antigens mutually saturated (equivalence point); interlocking structure

3 Antigen in excess

① Trivalent antigen
② Bivalent antibody

Basic types of antigen-antibody complex, demonstrated for the case of a trivalent antigen and bivalent antibody. The variable part with the antigen-binding site is shown in black.

up to 1% E.a. More recently, E.a. have been found in higher plants.

There are two subgroups of E.a., depending on the oxidation state of the substituent at D-8 of the tetracyclic ergoline ring system (Fig. 1), namely lysergic acid derivatives and clavine alkaloids. The first

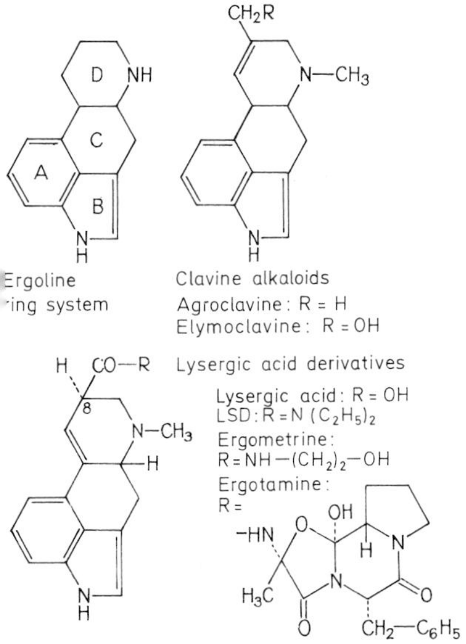

Figure 1. *Ergot alkaloids.*

Ergoline ring system

Clavine alkaloids
Agroclavine: R = H
Elymoclavine: R = OH

Lysergic acid derivatives
Lysergic acid: R = OH
LSD: R = N (C₂H₅)₂
Ergometrine:
R = NH—(CH₂)₂—OH
Ergotamine:
R =

group includes simple amide derivatives (e.g. ergometrin) and cyclic tripeptides (e.g. ergotamine). The tripeptide is formed from three of the following: D-proline, L-leucine, L-valine, L-phenylalanine and L-alanine, and it is linked to the carboxyl group of lysergic acid via alanine (ergotamine type), or valine (ergotoxin type).

All E.a. derived from lysergic acid are levorotatory. The dextrorotatory isomers are derived from isolysergic acid, and their trivial names carry the ending -inine (e.g. ergotaminine). In aqueous solution, the derivatives of lysergic acid slowly isomerize to isolysergic acid derivatives by a change in configuration at C-8. E.a. are detected by fluorescence or by indole reagents.

Biosynthesis: In *Claviceps purpurea,* and presumably also in higher plants, E.a. are biosynthesized from tryptophan and isopentenylpyrophosphate (see Terpenes). The synthesis proceeds via 4-dimethylallyl-tryptophan, which is converted into the alkaloid chanoclavine (by hydroxylation, methylation, decarboxylation and formation of a new C-C bond). All the other clavine alkaloids and the lysergic acid derivatives are derived from chanoclavine. The peptide moieties of the ergotamine and ergotoxin alkaloids are formed by a multienzyme complex (Fig. 2).

E.a. are prepared from the sclerotia of rye, previously inoculated with *Claviceps;* and they are also produced by culture of the fungus on artificial growth media.

Isolysergic acid and clavine alkaloids are physiologically inactive, whereas the lysergic acid derivatives (especially the peptide alkaloids) exhibit a variety of useful pharmacological properties. Extracts of ergot

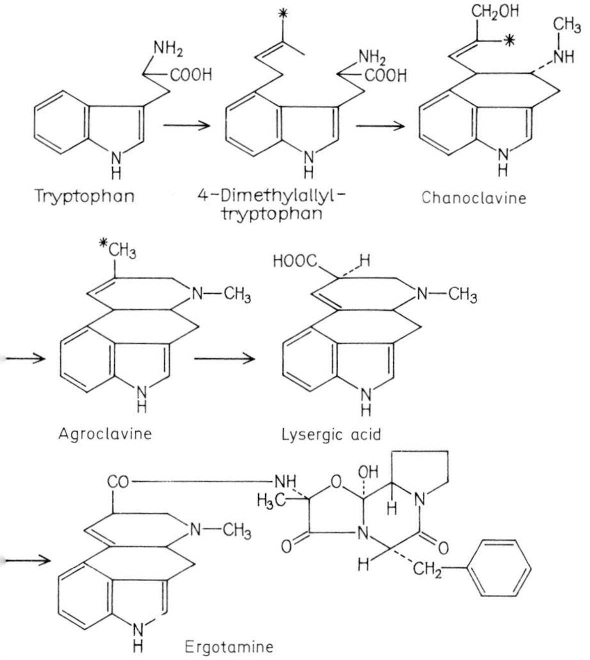

Tryptophan

4–Dimethylallyl–tryptophan

Chanoclavine

Agroclavine

Lysergic acid

Ergotamine

Figure 2. *Biosynthesis of clavine and lysergic acid alkaloids. The asterisk labels carbon atoms of the same origin.*

were used earlier, but owing to their variable alkaloid content, they have now been totally replaced by pure alkaloids (especially ergotamine), or semisynthetic analogues. E.a. stimulate contraction of smooth muscle, especially of uterus and of arterioles in peripheral parts of the body; they are used in gynecology for the control of hemorrhage after childbirth. Owing to their wide spectrum of pharmacological activity E.a. are also used in combination preparations, e.g. with tropane alkaloids for the suppression of the autonomic nervous system. Whereas the natural E.a. have vasoconstrictor activity, the semisynthetic compounds produced by hydrogenation of the Δ^9 – double bond are vasodilatory. *Historical:* The first description of the drug, ergot, with directions for its use dates from 1582. In the nineteenth century, the drug was introduced into many pharmacopeas. In 1816, Vauquelin started work on the identification of the active principle. 100 years later Stoll first succeeded in isolating crystalline ergotamine. For many centuries it was not realized that ergot was poisonous, and in Europe contamination of rye flour with ergot led to periodic widespread outbreaks of poisoning, known as "St. Anthony's fire", now called ergotism. Toxic symptoms are convulsions, permanent mental damage, limb gangrene, often followed by death.

Ergotamine: see Ergot alkaloids.

Ergotoxin: see Ergot alkaloids.

Eriodictyol: 5,7,3',4'-tetrahydroxyflavanone, see Flavanone.

Erucic acid, *Z-13-docosenoic acid*, Δ^{13-14}docosenoic acid: $CH_3-(CH_2)_7-CH=CH-(CH_2)_{11}-COOH$, a fatty acid. M_r 338.56, *cis*-form, m.p. 34 °C, b.p.$_{10}$ 254.5 °C. E.a. is a component of the glycerides in many seed oils of *Cruciferae* and *Tropaeolaceae*. It constitutes 40 to 50% of the total fatty acids of rapeseed, mustard and wallflower seeds, and 80% of fatty acids of nasturtium seeds.

Erythrina alkaloids: a group of isoquinoline alkaloids, usually tetracyclic, found exclusively in the legume genus *Erythrina*.

D-Erythritol: $CH_2OH-CHOH-CHOH-CH_2OH$, an optically inactive sugar alcohol derived from D-erythrose. M_r 122.12, m.p. 122 °C. It is found in some algae, fungi, lichens and grasses.

Erythrocruorin: a hemoglobin-like protein found in many invertebrates. In some snails and worms (e.g. *Cirraformis*, M_r 310^6, 162 heme-bearing polypeptide chains of M_r 18500) it is a high-molecular-weight extracellular respiratory pigment. In sea cucumbers, mussels, polychaete worms and some primitive vertebrates, like the river lamprey, it occurs as an intramuscular, low-molecular-weight protein (M_r 16700 to 56500).

Erythrocuprein: see Superoxide dismutase

Erythromycin: see Macrolide antibiotics.

Erythrophilic γ-globulin: predominantly IgG and a minor amount of IgM (see Immunoglobulins) which coats the surface of erythrocyte membrane, and is necessary for the integrity and normal survival of the erythrocyte. Normal plasma contains about 3000 mg E.per liter, and about 250 mg of bound E.is present in 1 liter of packed erythrocytes. After splenectomy, production of E.is markedly decreased, and the half-life of the erythrocyte is de-

creased by up to 50%. During the 4–8 month period after splenectomy, E.levels gradually return to normal, as does the half-life of the erythrocyte.

Erythrophleum alkaloids: a group of terpene alkaloids from *Erythrophleum guineense* and *Erythrophleum ivorense*. They are esters or amides of the diterpene cassainic acid with substituted ethanolamines. The most important is *cassaine* (Fig.), which is a cardiotonic equal in strength to the digitalis glycosides.

Cassaine

Erythropoietin: a glycoprotein hormone which stimulates production and release of erythrocytes in response to a lack of oxygen. E.is produced in the kidney and liver of adults, and in the liver of fetuses and neonates. E.increases the rate of mRNA formation in bone marrow, resulting in increased heme synthesis and erythroblast production, and ultimately in larger numbers of circulating reticulocytes and erythrocytes. The hormone is quickly degraded and excreted by the kidneys. Human E.has 166 amino acid residues (M_r 18398 without carbohydrate), including 4 Cys and 3 potential sites of N-glycosylation (based on the consensus site -Asn-X-Ser(Thr)-, see Glycoproteins). It is not known if O-linked glycans are present. The glycosylated protein has M_r 34–39000, so that about one half of the mass of the native hormone is due to carbohydrate. The human E.gene has been cloned, and the cDNA has been expressed in COS cells, giving a secreted product with biological activity. (The cell lines COS-1, 3 and 7 are derived from CV-1, an established line of simian cells, by transformation with an origin-defective mutant of SV-40 virus which codes for wild-type T antigen. [Y. Gluzman, *Cell* 23 (1981) 175–182]) The amino acid sequence of E.and the nucleotide sequence of its cDNA show no significant homology with any published sequences. [K.Jacobs et al. *Nature* 313 (1985) 806–810]

D-Erythrose: $CH_2OH-CHOH-CHOH-CHO$, an aldotetrose, M_r 120. Erythrose 4-phosphate plays a role in the intermediary metabolism of carbohydrates.

Escherichia coli: an intestinal bacterium of the family *Enterobacteriaceae*. It is the most common experimental organism in molecular biology. A single cell has dimensions of about $1 \times 1 \times 3$ μm, a volume of about 2.25 μm^3, and a wet weight of about 10^{-12} g, which corresponds to a dry weight of about 2.510^{-13} g. Under optimal conditions, the organism doubles about every 20 minutes. The wild type can grow on a completely synthetic medium containing salts, glucose and ammonia. The chemical composition, biosynthetic capacity and energy expenditures are listed in the table. The majority (70%) of the dry mass is protein, but the number of (smaller) lipid molecules is much larger. When the cell is growing logarithmically, it must synthesize 12500 lipid moleules and 1400 protein molecules per second. As-suming an average of 400 peptide bonds per protein, this amounts to 560000 peptide bonds per second, or which the cell expends 88% of its biosynthetic nergy. (These estimates are based on certain simplifying assumptions. The energy for synthesis of monomeric precursors was ignored, and it was assumed that there is no turnover of macromolecules.)

Estradiol, *oestradiol:* 17β-estradiol, estra-1,3,5(10)-triene-3,17β-diol, the most potent natural estrogen. E. occurs in high concentration in pregnancy urine, Graafian follicles and the placenta. In the organism, E. and Estrone (see) are interconvertible. The stereoisomeric 17α-E. has only 0.29% of the activity of E. The E. used therapeutically is derived totally from chemical synthesis (see Estrone). For

Table. *Chemical composition and biosynthetic capacity of Escherichia coli in the logarithmic phase of growth (modified from Lehninger).* DM dry mass, M_r av average molecular weight.

Class of substance	M_r av	%DM	Molecules per cell	Molecules synthesized per second	Molecules ATP used per second	% biosynthetic energy
DNA	2×10^9	5	4	0.033	6.0×10^4	2.5
RNA	1×10^6	10	15000	12.5	7.5×10^4	3.1
Protein	6×10^4	70	1700000	1400	2.1×10^7	88.0
Lipid	1×10^3	10	15000000	12500	8.8×10^4	3.7
Polysaccharide	2×10^5	5	39000	32.5	6.5×10^4	2.7

The chemical composition figures given in the table are approximations. Generalizations of this sort apply only to logarithmically growing bacteria, because in the stationary state, they accumulate reserve substances (lipids, polysaccharides, polyphosphates, sulfur, etc.), depending on the species), and slime and capsule substances increase the proportion of polysaccharides.

The nuclear material of *E. coli* is a single, ring-shaped DNA molecule, about 1 mm long. It is called the bacterial chromosome (genome, lineome, genophore), and is transferred during bacterial conjugation. The number of DNA molecules per cell is given as 4, in accordance with Lehninger, although the number and molecular weights depend on how actively the cells are growing and dividing.

Eserine: see Physostigmine.

Esperin: a cyclic antibiotic depsipeptide from *Bacillus mesentericus.* There is a higher α-hydroxy fatty acid on the *N*-terminus of the sequence Glu-Leu-Leu-Val-Asp-Leu-Leu which simultaneously forms a lactone structure with the γ-carboxyl group of the Asp residue.

Essential fatty acids: see Vitamins (vitamin F).

Essential oils: an extremely heterogeneous mixture of volatile, lipophilic plant products with characteristic odors. The International Standard Organization (ISO) has defined the E. o. in the strict sense as the steam distillates of plants or oils obtained by pressing out the rinds of certain citrus fruits. However, in practice the products of extraction with organic solvents, enfleurage or maceration of blossoms *(flower oils)* and the *resinoids* which can be extracted from other plant parts, resins and balsams are included by the term E. o.

Ester alkaloids: see Steroid alkaloids.

Esterases: a large group of hydrolases acting on ester bonds: $R^1COOR^2 + H_2O \rightarrow R^1COOH + R^2OH$ (R^1 is the acid residue and R^2 the alcohol residue). The acid may be a carboxylic, a phosphoric or a sulfuric acid, and the ester may be an alcoholic or a thiol ester.

Esters: see Carboxylic acids.

structure, biosynthesis and therapeutic uses, see Estrogens.

Estrane: see Steroids.

Estriol, *oestriol:* estra-1,3,5(10)-trien-3,16α, 17β-triol; an estrogen. M_r 288.39, m.p. 280 °C, $[\alpha]_D + 61°$ (chloroform). E. has been isolated from human female urine (especially during pregnancy) and placenta, from mare's urine, and from willow catkins. It is formed in the organism from estrone and estradiol. E. differs from Estradiol (see) by the presence of an additional hydroxyl group (α-conformation) at position 16.

Estrogens, *oestrogens:* a group of female sex hormones. The ring system of E. is estrane (see Steroids). They have an aromatic A ring with a phenolic 3-hydroxy group, and an oxygen function on C-17. The chief E. are Estrone (see), Estradiol (see) and Estriol (see); see also Equilenin and Equilin.

E. are produced in the Graafian follicles of the ovary, and in the corpus luteum; during pregnancy, they are also formed by the placenta. Small quantities of E. occur in male gonads. In the female, they control the course of the menstrual cycle (in humans and apes), or the estrus or mating cycle (other animals). E., acting together with progesterone (see) and the Gonadotropins (see), are responsible for the proliferation of the uterine mucosa, growth of the mammary glands, and the appearance of secondary sexual characteristics. E. are metabolized in the liver and kidney. They are excreted in the urine in free form and as their sulfate and glucuronide esters. Total E. excretion: 11-47 μg/day in nonpregnant women; 20-40 mg/day in pregnant women; a maximum of 10 μg/day in men.

The Allen-Doisy test was a biological test for E., in which the response of ovarectomized mice to urinary or plasma E. was observed. It was replaced in clinical practice by a colorimetric test, the Korber method, in which the E. are treated with H_2SO_4 and hydroquinone to produce intensely fluorescing derivatives. However, the development of Immunoassays (see) for the steroid hormones has made the Korber method obsolete for routine clinical assays.

Estrogens

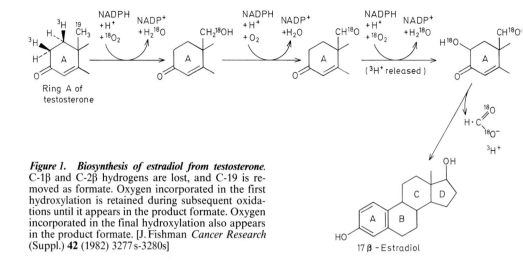

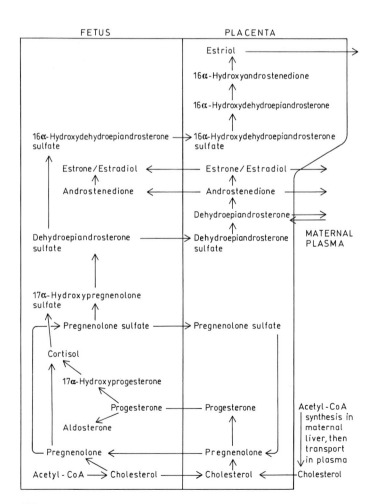

Figure 1. *Biosynthesis of estradiol from testosterone.*
C-1β and C-2β hydrogens are lost, and C-19 is removed as formate. Oxygen incorporated in the first
hydroxylation is retained during subsequent oxidations until it appears in the product formate. Oxygen
incorporated in the final hydroxylation also appears
in the product formate. [J. Fishman *Cancer Research*
(Suppl.) **42** (1982) 3277s-3280s]

17 β -Estradiol

Figure 2. *Steroid metabolism in the fetoplacental unit.*
Plasma estriol levels are determined by fetal metabolism, so that maternal urinary or plasma estriol levels provide an index of fetal health.

E. are biosynthesized from androgens by aromatization of ring A and loss of C-19. Three sequential hydroxylations are involved. The first two occur at the C-19 methyl group. The final and rate-limiting hydroxylation occurs at position 2β, followed by rapid nonenzymatic rearrangement to estradiol (Fig.1). The aromatase system is in the endoplasmic reticulum, and has not yet been solubilized in active form. In the synthesis of E. during pregnancy there is an intricate interplay of fetus and placenta (Fig.2). In particular, desulfatation of dehydroepiandrosterone 3-sulfate produced by the fetus can only be performed by the placenta. Estriol is not formed by direct hydroxylation of estradiol, but by hydroxylation of more oxidized precursors, followed by reduction, e.g. estrone → 16-hydroxyestrone → estriol; 16α-hydroxyandrostenedione → estriol.
E. have also been isolated from plants, e.g. estrone from pomegranate seeds and date stones, and estriol from willow catkins. The roots of a Burmese tree, *Pueraria mirifica*, contain Mirestrol (see) which has the same E. potency as 17β-estradiol.
E. are used therapeutically for the treatment of menstrual disorders and menopausal problems, and as Ovulation inhibitors (see). Some synthetic steroids (see Ethinylestradiol; Mestranol) and nonsteroids (see Stilbestrol; Hexestrol) show E. activity.

Estrone, *oestrone:* 3-hydroxyestra-1,3,5(10)-trien-17-one, an estrogen. M_r 270.38, m.p. 259 °C, $[\alpha]_D + 160$ ° (chloroform). E. occurs in human urine, ovaries and placenta, and it has also been found in pomegranate seeds (17 mg/kg) and palm oil. It was isolated independently by Doisy and by Butenandt in 1929 from pregnancy urine. E. differs from Estradiol (see) by the presence of an oxo-group at position 17. E. and estradiol are interconvertible in the organism. The total synthesis of E. was achieved in 1948 by Anner and Miescher. Nowadays some E. is obtained from horse urine, but most E. is prepared on a large scale by pyrolysis of androsta-1,4-diene-3,17-dione (obtained from cholesterol or sitosterol by microbial oxidation), or from diosgenin by the Marker synthesis (see). E. is the starting material for the chemical synthesis of other pharmacologically important steroids, e.g. estradiol and Ethynylestradiol (see).

Etamycin, *viridogrisein:* a cyclic peptide lactone antibiotic from *Streptomyces griseus* and related species. It is effective against Gram-positive bacteria and *Mycobacterium tuberculosis*.

Etephon: see Chloroethylphosphonic acid.

Ethanal: see Acetaldehyde.

Ethanaloic acid: see Glyoxylic acid.

Ethane dicarboxylic acid: see Succinic acid.

Ethanoic acid: see Acetic acid.

Ethanol, *ethyl alcohol:* CH_3-CH_2-OH, the end product of Alcoholic fermentation of carbohydrates (see). E. is produced by decarboxylation of pyruvate, and is found in small amounts in many organisms. In humans the normal level is 0.002 to 0.005% in the blood. It is degraded in the liver to acetaldehyde and acetate. In certain other tissues which lack the capacity for oxidative degradation of E., it is converted to fatty acid ethyl esters. E. causes short-term intoxication; when it is consumed in large amounts, the liver, and often the pancreas, heart and brain are damaged. The fatty acid ethyl esters are presumably responsible in some way for the damage to the pancreas, heart and brain, which do not form acetaldehyde. [E. A. Laposata & L. G. Lange, *Science* **231** (1986) 497-501]
Alcoholic fermentation is started by adding yeast (*Saccharomyces cerevisiae*) to any substrate containing starch or sugar, such as fruit, grain, potatoes, sugar cane or beets, or even wood hydrolysates. Fermentation stops at an alcohol content of 8 to 10%, so the contents of the fermenter are distilled several times to obtain pure E. (see Fermentation techniques).

Ethanolamine, *aminoethanol:* H_2N-CH_2-CH_2-OH, a biogenic amine produced by decarboxylation of L-serine. It is a common constituent of phospholipids in which it is esterified to an acylglycerol phosphate moiety to form phosphatidyl ethanolamine (see One-carbon cycle).

Ethrel: see Chloroethylphosphonic acid.

Ethyl alcohol: see Ethanol.

Ethylene, *ethene, fruit-ripening hormone:* a gaseous plant hormone found widely in plant tissues,

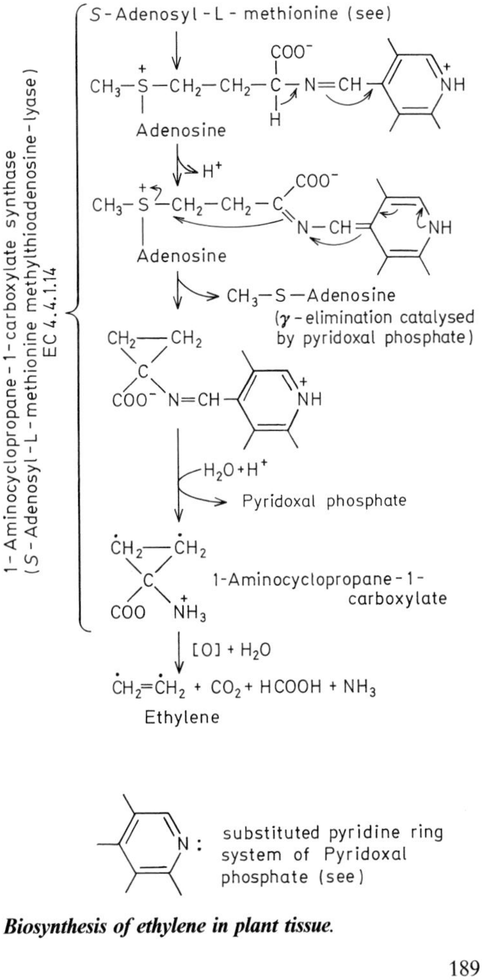

Biosynthesis of ethylene in plant tissue.

which accelerates fruit ripening, leaf and fruit drop, and plant ageing. Endogenous E. formation is promoted by stress and exogenous auxins. Exogenous E. accelerates the ripening and coloring of fruit, and induces blossoming and seed germination. E. is synthesized in plants from L-methionine (Fig.). Conversion of S-adenosyl-L-methionine to 1-aminocyclopropane 1-carboxylate is catalysed by a well characterized enzyme, 1-aminocyclopropane 1-carboxylate synthase (EC 4.4.1.14), M_r 50000 (from tomato pericarp), which is dependent on pyridoxal phosphate. Conversion of 1-aminocyclopropane 1-carboxylate to ethylene, CO_2, formate and ammonia requires molecular oxygen, but the nature of the oxidant is not established (H_2O_2 ?). The rate-limiting step of ethylene biosynthesis is the synthesis of 1-aminocyclopropane 1-carboxylate. Factors promoting the production of E. (stress, auxins) cause an increase in 1-aminocyclopropane 1-carboxylate synthase. [A. B. Bleecker et al. *Proc. Nat. Acad. Sci.* **83** (1986) 7755-7759]

Ethylene generator: see Chloroethylphosphonic acid.

Ethylenediaminetetraacetic acid, abb. *EDTA, ethylenedinitrolotetraacetic acid:* a chelating agent, used in biochemical systems in vitro for the chelation of divalent metal ions. It is commonly used as the disodium salt (Na_2EDTA). Oral administration of EDTA may be used for the chelation of lead in cases of lead poisoning. In the older literature, EDTA was known by its trade name, versene.

Ethynylestradiol, *19-nor-17α-pregna-1,3,5(10)-trien-20-yne-3,17β-diol:* a synthetic estrogen. M_r 296.41, m.p. 146 °C, $[α]_D + 1°$ (dioxan). Subcutaneously, E. has the same biological activity as the natural hormone estradiol, but when administered orally, it is much more active and is therefore used in oral contraceptives. It is synthesized by addition of ethyne (acetylene) to estrone in the presence of sodium in liquid ammonia. Small scale synthesis of E. (e.g. for synthesis of radio-labelled E.) is achieved by addition of lithium acetylide to estrone in dimethyl sulfoxide.

Ethynylestradiol

Etiocholane: see Steroids.

ETP: abb. for Electron transport particle (see). See also Mitochondria.

Eucalyptol: see 1,8-Cineol.

Eukaryote: an organism with eukaryotic cells (see Cell, 2).

Eumelanins: brown or black melanin pigments found in animals. E. are heterogeneous cross-linked polymers built of 5,6-dihydroxyindole units, and usually linked to proteins. They are very widespread

in skin, hair, feathers, insect cuticula, etc. They are contained in melanocytes and are synthesized from tyrosine by oxidation and polymerization. See Melanins.

Evolution: the process by which organisms have come into existence. Charles Darwin recognized the fundamental mechanisms of E.: organisms reproduce themselves, in such numbers that not all of the offspring can survive; and the traits of the parents are generally present in the offspring. The process of reproduction is highly accurate, but not absolutely so, so that Mutants (see) occasionally arise. *Natural selection* (survival of the fittest) occurs when the survival and reproduction of certain individuals, rather than others, is not entirely random but due to some trait or traits which their offspring inherit.

Chemical evolution was the process by which life arose from inorganic matter, apparently soon after the earth had cooled sufficiently to accumulate liquid water. There is much evidence that the atmosphere at that time contained little or no free oxygen, and thus no ozone, so that its components, CH_4, NH_3, H_2 and H_2O, were exposed to the ultraviolet radiation of the sun. Experiments by Miller and others have shown that organic molecules like sugars, amino acids, purines, pyrimidines, nucleotides and even polypeptides can form *abiotically* under *simulated primitive earth* conditions ("chemical evolution"). It is not known exactly how polynucleotides came to be self-replicating, but Kuhn suggested that the alternation between single- and double-stranded forms must have been driven at first by periodic changes in the environment. Selection would have begun to operate as soon as there were systems capable of moderately accurate self-replication, and the traits presumably selected for at first were efficiency and accuracy of replication and stability of the nucleic acid molecules between replication cycles. The E. of nucleic-acid-directed protein synthesis (see Genetic code) also remains a riddle. Proteins could not be selected for their efficiency as enzymes or structural elements until their structures had been encoded in the genetic material of "protoorganisms". Furthermore, the entire system must by this time have been enclosed in some kind of membrane, because otherwise the proteins would have diffused away from their nucleic acid "parents" and their stabilizing qualities would have no effect on the survival of their genes *("organic E.").*

After the stage of nucleic-acid-directed protein synthesis had been reached, the organisms must have multiplied rapidly. Selection would now favor the development of enzymes, first to synthesize nucleic acids and proteins from abiotically generated precursors, and later, as the *"primal soup"* was exhausted, to synthesize the immediate precursors of proteins and nucleic acids from available pre-precursors. The universal metabolic pathways would thus have arisen step by step, going backwards from the complex to the simple, as the growing populations of cells exhausted their immediate abiotic resources. The culmination of this process, called *"biochemical E."* was the E. of photosynthesis by blue-green bacteria, which finally produced an oxidizing atmosphere and put an end to abiotic formation of organic compounds. The oldest fossil bacteria and bluegreens are about 3 billion years old; the atmosphere became oxidizing roughly 2 billion years ago. *Mechanisms of molecular evolution:* An organism with precisely the amount of DNA needed to encode a minimum complement of enzymes required for life would be very inflexible. The mechanisms which increase the complement of DNA per cell are thus essential for variation, adaptability and further E. The most important of these is *gene duplication,* which can be caused by an uneven crossing over or by a single locus duplication of a gene. At first the two duplicate genes are identical, but since mutation in one cannot be lethal if the other is functional, one at least begins to accumulate point mutations, deletions or additions (see Mutants). Selection begins to occur if the new protein assumes some function in the organism which better adapts it to its environment. There are a number of protein families which have arisen in this way: the cytochromes *c,* the series of myoglobin, α-, β-, γ-, and δ-hemoglobin chains (Fig. 1), the fibrinopeptides, some protein hormones, the pancreatic serine proteases and probably many others.

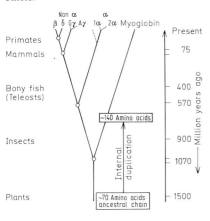

o = Gene duplication

Figure 1. Evolution of human hemoglobin chains (α, β, γ, δ). The separation of the horse and human α-chains is indicated by the dotted line.

Intragenal or *daughter-gene doubling* is a similar process, in which the duplication involves only a part of a gene. Proteins which have arisen in this way have two or more regions of similar sequence along their length.

The most easily tolerated mutations are *point mutations,* in which a single nucleic acid base is exchanged. Since the genetic code is degenerate, it can happen that the new codon codes for the same amino acid as the old; in this case, nothing happens. If the new codon is specific for another amino acid, the substitution may be either conservative, meaning that the new amino acid is chemically similar to the old (Lys for Arg, Glu for Asp, etc.) or radical (Lys for Asp). Conservative changes are of course more easily tolerated, and comparison of enzyme primary structures from different species reveals many such substitutions. Some amino acids cannot be changed without destroying the enzyme's activity, e.g. Ser$_{195}$ in the serine proteases. Such a mutation, when it occurs, is generally lethal. Different families of proteins have accumulated substitutions at different rates. It is possible to calculate an average rate of mutation, given the number of substitutions between the primary sequences of enzymes from two species and the approximate time which has elapsed since their phylogenetic lines diverged. The table gives a few examples, including the extremely rapidly changing fibrinopeptides (90 point mutations per 100 amino acids and 100 million years) and the very conservative histones (2 mutations per 1.5 billion years). It is not that mutations are more frequent in the DNA coding for fibrinopeptides than for histones; rather, the structure of the histones is so rigidly determined by their function that no other will do, and mutations are almost invariably lethal.

The other forms of mutation which are observed in the laboratory, *deletions* and *additions,* can be recognized in the sequences of existing proteins, such as the variable regions of the immunoglobulins, the α- and β-chains of hemoglobins, and the snake venom neurotoxins. Such radical changes seem not to have been tolerated in myoglobin, calcitonin and cytochrome c_{551}. *Hybridization* occurs when there is crossing over between related but non-allelic genes, such as those for β- and δ-chains of hemoglobins. *Fusion* of two neighboring genes can produce bifunctional enzymes, such as tryptophan synthase. In *Escherichia coli,* the α- and β-subunits of this enzyme are separate, but in *Neurospora* they are a single chain. In *Salmonella,* two enzymes of histidine biosynthesis have been fused to a single dimeric protein.

The processes of E. on a molecular level are fairly well understood. The processes by which the cells of a multicellular organism communicate with one another are not well understood, and neither are the processes of embryonic development and tissue differentiation. We thus do not know in detail how a particular mutation effects a particular change in

Table. Mutation rates of different protein families (abbreviated from Dayhoff: Atlas of Protein Sequence, 1972)

n = successful point mutations per 100 amino acid residues in 100 million years.

Protein or peptide	n
Fibrinopeptides	90
Growth hormone	37
Pancreatic RNAase	33
Immunoglobulins	32
Lactalbumin	25
Hemoglobin chains	14
Myoglobin	13
Pancreatic trypsin inhibitor	11
Animal lysozyme	10
Trypsinogen	4
Cytochrome *c*	3
Histone IV	0.06

morphology and behavior, although this is the level at which selection operates. (Selection also insures that all individuals are biochemically functional – a mutant lacking basic functions will generally die as an embryo.) We can assume, however, that the difference between man and monkey is the sum of a number of subtle differences in the actions of a relatively small number of development-regulating genes, enzymes and structural proteins. At both the molecular and the organism level, one can observe *convergent* and *divergent E*. Convergent E. occurs when similar environmental pressures favor the development of similar traits in unrelated organisms: streamlining in fish and sea mammals, for example, or enzymes with quite different sequences which catalyse the same reaction. Divergent E. is the process by which one species of organism or protein gives rise to two or more. The mechanisms discussed above (gene duplication, accumulation of muta-

tions) account for molecular divergence. The divergence of organisms must begin with a change in one or several enzymes which, acting together with all the others in the organism, change it sufficiently so that it cannot interbreed with the members of the parent species.

Excision repair: see Deoxyribonucleic acid.
Exon: see Intron.
Exonucleases: see Nucleases.
Exopeptidases: see Proteases.
Exotoxins: see Toxic proteins.
Extinction: see Absorbance.
Extinction coefficient: see Absorptivity.
Extrachromosomal genes: DNA molecules located outside the nucleus. The most important are in the Plastids (see) and Mitochondria (see). The corresponding E.g. in bacteria are the Plastids (see).
Extrinsic factor: see Vitamins (vitamin B_{12}).

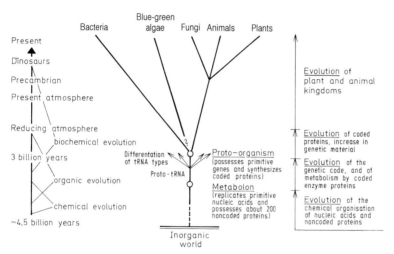

Figure 2. Phases of molecular evolution

F

Fab fragment: see Immunoglobulins.

Fabry's disease: see Inborn errors of metabolism.

Facilitated diffusion: passive transport, the movement of specific compounds across a biomembrane from higher to lower concentration, but at a rate greater than simple diffusion. F.d. is *saturable,* meaning that above a certain concentration, the rate is not dependent on the substrate concentration. Furthermore, it is stereospecific and susceptible to competitive inhibition. Together, these properties indicate that the process is mediated by a "carrier" or pore protein in the membrane. F.d. differs from Active transport (see) in not requiring energy. A class of substances called Ionophores (see) mimic the carriers of F.d. by making membranes permeable to certain ions. Antibiotics that act in this way are called *transport antibiotics.*

Factor II: prothrombin, see Blood coagulation.

Factor VIII: antihemophilic factor, see Blood coagulation.

Factor XIII, *fibrin-stabilizing factor:* the last clotting factor to act. It is an α_2-plasma globulin of M_r 350000 and contains 2α and 2β-chains of M_r 100000 and 77000, respectively. It is activated by thrombin in the presence of calcium ions to factor XIIIa, which catalyses the formation of γ-glutamyl-ϵ-lysine peptide bonds in a calcium-dependent transamidation reaction. These bonds serve to crosslink the fibrin chains into a three-dimensional network, the clot.

FAD: abb. for Flavin adenine dinucleotide (see).

Farber's disease: see Inborn errors of metabolism.

Farnesol: 3,7,11-trimethyl-2,6,10-dodecatrien-1-ol, an acyclic sesquiterpene alcohol. *Cis-trans* isomerism about two of the double bonds produces four isomeric forms, of which the most important is *trans-trans*-F. (M_r 222.36, an oil smelling of lilies of the valley, b.p.$_{10}$ 160 °C, ρ_4 0.895, n_D^{25} 1.4872). F. is

trans-trans-Farnesol

unstable. It is easily oxidized and tends to cyclize. It occurs widely in nature in essential oils, and also takes the place of phytol in the bacterial chlorophyll of the genus *Chlorobium*. It is a pheromone for bumblebees. *Farnesyl pyrophosphate* is an important intermediate in the synthesis of many Terpenes (see).

Farnesyl pyrophosphate: see Farnesol.

Farnoquinone: see Vitamins (Vitamin K_2).

Fascin: see Cytoskeleton (microfilaments).

Fat biosynthesis: see Acylglycerols.

Fat degradation: hydrolysis of the ester bonds in neutral fats to form fatty acids and diacylglycerol, which is further degraded to fatty acids and glycerol. The enzymes responsible are lipases, which are present in large amounts in the pancreas, intestinal wall and liver. The pancreatic lipase hydrolyses only the 1- and 3-acyl esters, but the intestinal enzyme also attacks the 2 position. Glycerol and fatty acids are further degraded in the cells.

Fats, *triacylglycerols:* esters of glycerol with saturated or unsaturated fatty acids. Mono- and diacylglycerols exist, but triacylglycerols are by far the most common in nature. Natural F.are neutral compounds, usually consisting of a mixture of esters of different fatty acids. They are accompanied by Phospholipids, Sterols, Triterpenes, Carotenoids, Tocopherols, Fatty acids (see individual entries in the above; tocopherols are found in the entry on Vitamins), aliphatic alcohols, hydrocarbons and other hydrophobic compounds.

Almost without exception, the fatty acids of natural F.are unbranched and have an even number of carbon atoms, usually between 4 and 26. In plant F., the primary hydroxyl groups in the 1 and 3 positions of the glycerol are usually occupied by saturated fatty acids, while the 2 position carries an unsaturated acid. In animal F., the reverse situation is often found. More than 50 fatty acids have been identified as components of natural fats. The most common are the saturated palmitic (C_{16}) and stearic (C_{18}) acids and the unsaturated oleic, linoleic and linolenic (all C_{18}) acids. The double bonds in naturally occurring fatty acids are always in the *cis* configuration. Branched or hydroxylated fatty acids, or those with an odd number of carbon atoms are found occasionally, such as ricinolic, cerebronic and hydroxynervonic acids.

Hydrolysis of F., either by alkali (saponification) or by lipases produces glycerol and fatty acids (or soaps, their alkali salts) as end products.

The melting point of a fat depends on the nature of the fatty acids in it. Shorter hydrocarbon chains and unsaturation give the fats lower melting points. Those which are liquid at room temperature are called fatty oils. These are further subdivided according to their tendency to undergo autocatalytic oxidation in the presence of oxygen into drying oils, semi-drying oils and non-drying oils. Drying depends on polymerization and cross-linking by oxygen, peroxy- and hydrocarbon bridges between the fatty acids. Only polyunsaturated acids can undergo these reactions. Linseed and poppy seed oils are examples of drying oils; peanut and rapeseed oils are half-drying, and olive oil is a non-drying oil.

Plant and animal fats can be distinguished analytically because they are associated with different sterols – phytosterol in plant fat and cholesterol, for example, in animal fat. They are of equal nutritional value, however, provided that the same vitamins and essential fatty acids are present. Plant fats are most abundant in seeds (40 to 45% in rapeseed, poppy seed and linseed). Olives contain up to 25% F. The most important fruit oils are palm and olive oils; the most important solid seed fats are coconut and palm seed fats and cocoa butter. The seed oils of cotton, corn, sunflowers, peanuts, soy, almonds, sesame, flax (linseed), poppies, rape, mustard and ricinus are economically important.

Animal fats are stored in subcutaneous and omentum tissue, in the peritoneum and the region around the kidneys. The body fats of pigs, cattle, sheep and geese, and of sea animals (whale, seal, fish liver) and the milk F. of cattle, goats and sheep are economically significant.

Liquid oils intended as food (e.g. margarine) are converted to solids by hydrogenation of their double bonds, transesterification and fractionation, and by removal of the lower-melting fraction. F. spoil rather easily. The fatty acids may be released by hydrolysis and if unsaturated, they are easily oxidized to aldehydes and ketones (they become rancid). Antioxidants (vitamin E, butylated hydroxyanisole [BHA], butylated hydroxytoluene [BHT], propyl gallate) or oxygen traps can slow this process.

F. provide more metabolic energy than proteins and carbohydrates: 37.7 kJ/g (9.0 kcal/g). They are also important as body insulation and cushioning materials for organs, and as components of cell membranes. F. are used not only for food, but in the production of fatty acids, glycerol, soaps, salves, candles, fuel and lubricants. The drying oils are used in paints, varnishes and textile dyes. For biosynthesis and degradation of F., see Acylglycerols.

Fatty acid biosynthesis: a process catalysed by fatty acid synthase, in which the fatty acid carbon chain is formed stepwise from 2-carbon units (derived from malonyl groups, with subsequent decarboxylation). The intermediates of F. a. b. are thioesters of Acyl carrier protein (see) (ACP) and not of coenzyme A as in fatty acid degradation.

Malonyl-CoA is synthesized in a biotin-dependent carboxylation of acetyl-CoA (see Biotin, under Vitamins):

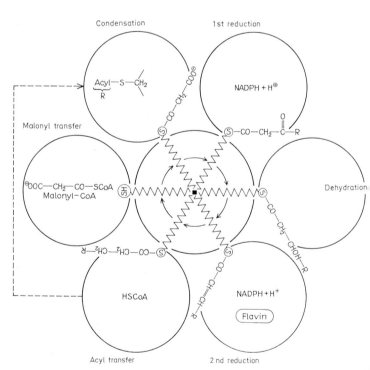

Figure 1. Mechanism of fatty acid biosynthesis. The zig-zags represent the movable pantetheine arm of the acyl carrier protein. The central circle represents acyl carrier protein, and the other six circles represent enzymes of fatty acid biosynthesis (see Table 1). This purely diagrammatic representation does not necessarily show the true juxtaposition of the proteins, which may be part of a multienzyme complex, or located on a single multifunctional protein (see Table 2).

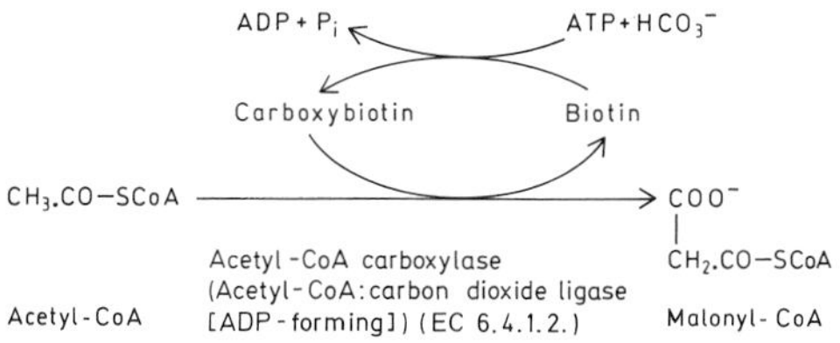

The malonyl group of malonyl-CoA is transferred to ACP, where it forms a thioester linkage with the SH of covalently bound 4-phosphopantetheine. This phosphopantetheine arm serves as a carrier for substrates and intermediates, which are acted upon in turn by the other catalytic activities of fatty acid synthase shown in Table 1 and Fig. 1. Each newly formed saturated acyl group is transferred from the phosphopantetheine arm to the SH group of a cys-

Table 1. Reactions of fatty acid synthesis. The enzyme possesses a peripheral SH-group (which belongs to a cysteine residue), and a central SH-group (which belongs to pantetheine). With the exception of acetyl-CoA carboxylase, the catalytic activities listed below (together with acyl carrier protein) may belong to discrete proteins in a multienzyme complex, or they may all be present in a single multifunctional protein (see Table 2).

Enzyme	Function	Reaction
1) Acetyl-CoA-Carboxylase	Malonyl-CoA-synthesis	$CH_3-CO{\sim}SCoA + HCO_3^- + ATP \xrightarrow[Biotin]{H_2O,\ Mg^{2+}} {}^-OOC-CH_2-CO{\sim}SCoA + ADP - P_i$ Acetyl-CoA → Malonyl-CoA
2) Acetyl-trans-acylase	Acyl transfer (initiating reaction)	$CH_3-CO{\sim}SCoA$ + (HS)/HS–Enzyme ⇌ (HS)/$CH_3-CO{\sim}S$–Enzyme + HSCoA Acetyl-β-ketoacyl-ACP synthetase
3) Malonyl-trans-acylase (Chain elongation)	Malonyl-transfer	${}^-OOC-CH_2-CO{\sim}SCoA$ + (HS)/$CH_3-CO{\sim}S$–Enzyme ⇌ ${}^-OOC-CH_2-CO{\sim}$(S)/$CH_3-CO{\sim}S$–Enzyme + HSCoA Acetyl-β-ketoacyl-ACP synthetase and Malonyl-ACP
4) β-Ketoacyl-ACP-Synthetase (Chain elongation)	Condensation	${}^-OOC-CH_2-CO{\sim}$(S)/$CH_3-CO{\sim}S$–Enzyme ⟶ $CH_3-CO-CH_2-CO{\sim}$(S)/HS–Enzyme + CO_2 Acetoacetyl-ACP
5) β-Ketoacyl-ACP-Reductase (Chain elongation)	1st Reduction	$CH_3-CO-CH_2-CO{\sim}$(S)/HS–Enzyme ⇌ (NADPH + H^+ → $NADP^+$) $CH_3-CHOH-CH_2-CO{\sim}$(S)/HS–Enzyme D-β-Hydroxybutyryl-ACP
6) Enoyl-ACP-Hydratase (Crotonase) (Chain elongation)	Dehydration	$CH_3-CHOH-CH_2-CO{\sim}$(S)/HS–Enzyme ⇌ $CH_3-CH=CH-CO{\sim}$(S)/HS–Enzyme + H_2O Crotonyl-ACP
7) Enoyl-ACP-Reductase (Chain elongation)	2nd Reduction	$CH_3-CH=CH-CO{\sim}$(S)/HS–Enzyme ⇌ (NADPH + H^+ → $NADP^+$; Flavin) $CH_3-CH_2-CH_2-CO{\sim}$(S)/HS–Enzyme
8) Transacylase	Acyl transfer (followed by repeated malonyl transfer, i.e. reaction 3)	$CH_3-CH_2-CH_2-CO{\sim}$(S)/HS–Enzyme ⇌ (HS)/$CH_3-CH_2-CH_2-CO{\sim}S$–Enzyme
	Repeated cycles of chain elongation	C_4 6 8 10 12 14 16
9) Transacylase	Palmityltransfer (terminating reaction)	$CH_3-(CH_2)_{14}-CO{\sim}$(S)/HS–Enzyme + HSCoA ⇌ (HS)/HS–Enzyme + $CH_3(CH_2)_{14}-CO{\sim}SCoA$ Palmityl-CoA (Alternative termination reactions)
9) Thioesterase	Hydrolysis (terminating reaction)	$CH_3-(CH_2)_{14}-CO{\sim}$(S)/HS–Enzyme + H_2O ⇌ (HS)/HS–Enzyme + $CH_3(CH_2)_{14}COOH$ Palmitate (Alternative termination reactions)

HS-peripheral SH-group (cysteinyl)
(HS) central SH-group (4-phosphopantetheine)

Table 2. Fatty acid synthases.

Source	Description
Rat liver, rat adipose tissue, rat and rabbit lactating mammary gland, chicken liver, goose uropygial gland, *Ceratitis capitata* (an insect), *Crypthecodinium* (a marine dinoflagellate).	Type I fatty acid synthase. Enzyme activities are arranged as a series of globular domains in a single multifunctional protein, M_r 4–5 $\times 10^5$, consisting of two identical subunits, each of M_r 1.8–2.5 $\times 10^5$, i.e. α_2 structure. β-Ketoacyl synthase activity is only present in the dimer, since this activity requires juxtaposition of two thiols, one from each subunit. The enoyl reductase does not use a flavin coenzyme (NADPH is used directly for the reduction). Termination is by hydrolysis (thioesterase), and the products are chiefly palmitate with some stearate.

Source	Description
Saccharomyces cerevisiae (The fatty acid synthases from *Neurospora crassa, Penicillium patulum:* and *Pythium debaryanum* – all filamentous fungi – have M_r 2.2–4.0 $\times 10^6$, and appear to be similar to the yeast enzyme)	Seems to be an evolutionary intermediate between types I and II. M_r 2.4 $\times 10^6$, consisting of equal numbers of nonidentical subunits, M_r 213000 (α) and M_r 203000 (β), i.e. $\alpha_6\beta_6$ structure. The enoyl reductase in the β-subunit uses FMN. Products are palmitoyl-CoA or stearoyl-CoA.

α	β-Ketoacyl synthase		Acyl carrier		β-Ketoacyl reductase
β	Enoyl reductase	Dehydratase	Acetyl transacylase	Malonyl and palmitoyl transacylase	

Distribution of catalytic activities in the α and β-subunits of yeast fatty acid synthase

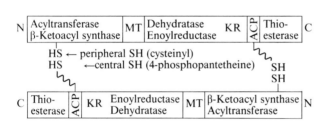

Proposed structure of the dimer of type I fatty acid synthase (based on studies of the chicken liver enzyme). ACP = acyl carrier protein, KR = β-ketoacyl reductase, MT = malonyl transacylase.

Source	Description
Euglena gracilis	Cytoplasm contains type I enzyme (α_2, M_r 200000), producing primarily palmitoyl-CoA. Chloroplasts contain type II system (distinct enzymes), producing primarily stearoyl-ACP.

Source	Description
Mycobacterium smegmatis, Corynebacterium diphtheriae, Streptomyces coelicolor, Brevibacterium ammoniagenes.	Resembles type I enzyme, i.e. subunits are identical, and each is a multifunctional protein, M_r 2 $\times 10^6$, consisting of six identical subunits, each of M_r 290000, i.e. α_6 structure. The enoyl reductase uses FMN. F.a.b. is dependent on both NADH (enoyl-acyl reduction) and NADPH (β-ketoacyl reduction). Palmitate and tetracosanate are major products of *M. smegmatis* enzyme.

Source	Description
Most bacteria, including *Escherichia coli, Clostridium butyricum, Bacillus subtilis, Pseudomonas aeruginosa, Phormium lunidum.* Plants, e.g. avocado, barley, spinach, safflower, parsley. *Chlamydomonas* (alga)	Type II fatty acid synthase. 6–7 discrete enzymes and an acyl carrier protein, associated noncovalently. In plants the enzymes are only in the plastids, and fatty acid synthesis does not occur in the cytoplasm. Similarity of plant and bacterial type II systems supports endosymbiont theory of origin of chloroplasts. Enoyl reductase uses FMN. Primary product is palmitate.

teinyl residue ("peripheral" SH group) of β-ketoacyl synthetase. A further malonyl group is then attached to the freed central thiol, and another cycle of reactions (condensation, reduction, dehydration, reduction) extends the acyl group by two more carbon atoms. These cycles are repeated until a palmitoyl (C_{16}) residue is formed, which is then released as the free acid or as the fatty acyl-CoA, depending on whether the system possesses a thioesterase or CoA-transacylase (see Tables 1 and 2).

For palmitate synthesis, the process can be represented stoichiometrically as:

Acetyl-CoA + 7malonyl-CoA + 14NADPH + $14H^+$ → Palmitate + $7CO_2$ + $14NADP^+$ + 8HSCoA + $6H_2O$.

In most bacteria and in chloroplasts, the acyl carrier protein and the enzymes of F.a.b. are discrete proteins in a noncovalently associated multienzyme complex (type II fatty acid synthase), whereas the fatty acid synthase of animals (type I) is a dimer of a single multifunctional protein. Yeast fatty acid synthase is an intermediate type I/type II enzyme (see Table 2).

In animals, F.a.b. occurs in the cytoplasm, but the starting material, acetyl-CoA, is produced in the mitochondria by pyruvate dehydrogenase (see Multienzyme complex). The main route for provision of cytosolic acetyl-CoA is shown in Fig. 2.

For every acetyl-CoA transferred from the mitochondria to the cytosol, one NADPH is generated. Thus, in the conversion of 8 molecules of acetyl-CoA into palmitate, 8 of the required 14 NADPH molecules are provided by malate dehydrogenase (EC 1.1.1.40, Fig. 2). The additional 6 NADPH are provided by the Pentose phosphate cycle (see). The rate-limiting step in the biosynthesis of fatty acids from acetyl-CoA is the synthesis of malonyl-CoA, catalysed by acetyl-CoA carboxylase (see above). Active acetyl-CoA carboxylase from animal tissues is a polymer (M_r $4-8 \times 10^6$), which can be dissociated into inactive monomers or dimers of the M_r 230000 subunit. Each subunit is a multifunctional protein containing the catalytic activities of biotin

carboxylase, biotin-carboxy-carrier protein and transcarboxylase, as well as the regulatory allosteric site (see Biotin). Citrate and isocitrate activate the enzyme by promoting aggregation of the subunits. This effect of citrate is antagonized by palmitoyl-CoA, which also inhibits the carrier that transports citrate across the mitochondrial inner membrane (Fig. 2).

Elongation of the fatty acid carbon chain. The specificity of β-ketoacyl-ACP synthase is such that the enzyme normally binds all fatty acyl groups up to a chain length of 14 carbons (tetradecanoyl). The hexadecanoyl (C_{16}, palmitoyl) cannot be bound, so that palmitate or palmitoyl-CoA is released as the end product of F.a.b. The chain length can be extended by elongation reactions which occur in the mitochondria and in the endoplasmic reticulum in animals. In mitochondria, elongation occurs by the addition of acetyl-CoA units (not malonyl-ACP). This pathway closely resembles a reversal of the β-oxidation (see Fatty acid degradation), except that the unsaturated bond at C-2 is reduced by NADPH and not by $FADH_2$. In the endoplasmic reticulum, C_2 units for elongation are derived from malonyl-CoA and all intermediates are in the form of their CoA esters (ACP esters are not involved) (Fig. 3). The microsomal elongation system is probably physiologically more important than the mitochondrial system. The mitochondrial system is only likely to operate when the [NADH]/[NAD$^+$] ratio is high in the mitochondria, i.e. under anaerobic conditions,

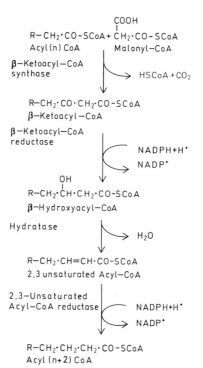

Figure 3. Endoplasmic reticulum system for the elongation of fatty acyl chains.

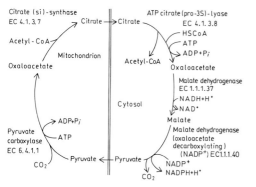

Figure 2. Production of cytosolic acetyl-CoA at the expense of mitochondrial acetyl-CoA by the formation, transport and breakdown of citrate.

or in the liver when excessive quantities of ethanol are oxidized.

In bacteria and plants, elongation occurs by continuation of the reactions of F.a.b. beyond C_{16}, employing synthases of different specificity. Two β-ketoacyl synthases have been isolated from plants, the first producing primarily palmitate, while product distribution shifts primarily to stearate in the presence of the second synthase. Unsaturated acids are elongated in the same way as saturated acids.

Unsaturated fatty acid biosynthesis. Two apparently mutually exclusive pathways exist for the formation of unsaturated fatty acids, and together they account for essentially all naturally occurring unsaturated fatty acids. The *anaerobic pathway* operates only in bacteria. It can be considered as part of the pathway

of saturated F.a.b., which branches at the level of a C_{10} intermediate to produce either palmitate or stearate on the one hand, or palmitoleate or vaccenate on the other. The branch point lies at D-β-hydroxydecanoyl-ACP, which is a normal intermediate in the synthesis of palmitate. On the pathway of palmitate synthesis, this intermediate is dehydrated 2,3 (α,β) to produce a *trans* double bond, and the reactions of F.a.b. continue as described in Table 1. In the biosynthesis of unsaturated fatty acids, this intermediate is dehydrated to produce a *cis* double bond between C-3 and C-4 (β and γ), by the action of bacterial β-hydroxydecanoyl thioester dehydratase (Fig.4). The *cis*-3 double bond is then carried through the subsequent chain lengthening steps of F.a.b. to yield palmitoleate (3 cycles) or vaccenate

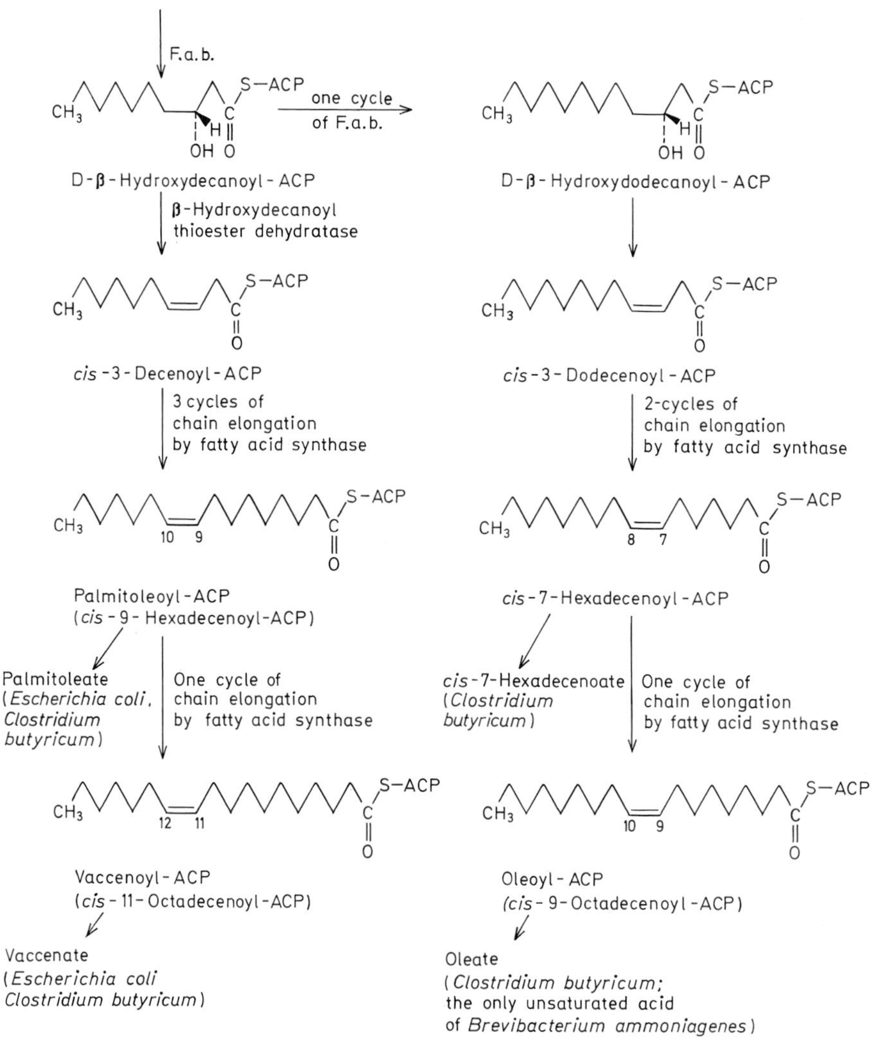

Figure 4. Synthesis of unsaturated fatty acids by the anaerobic pathway in bacteria.

(4 cycles). *Escherichia coli* possesses two distinct β-ketoacyl-ACP synthases (I and II). Enzyme I functions mainly in the elongation of *cis*-3-decenoate to palmitoleate, whereas enzyme II catalyses the condensations leading to palmitate and the chain extension of palmitoleate to vaccenate. *Brevibacterium ammoniagenes* produces oleate as the exclusive unsaturated fatty acid of the anaerobic pathway; this organism is also exceptional in using a type I fatty acid synthase complex for synthesis by the anaerobic pathway. In addition to palmitoleate and vaccenate, *Clostridium butyricum* also produces 7-hexadecenoate and oleate by chain extension of 3-dodecenoyl-ACP (Fig.4).

The *oxygen-dependent pathway of desaturation* operating in plants and animals produces a wide variety of unsaturated fatty acids differing in chain length, branching and position and number of double bonds. This is in marked contrast to the limited number and types of unsaturated fatty acids produced by bacteria in the anaerobic pathway. In animals, the oxygen-dependent desaturase is found in the endoplasmic reticulum; it is a mixed function oxygenase system containing cytochrome b_5 reductase, cytochrome b_5 and the desaturase (Fig.5), and it is specific for fatty acyl-CoA substrates.

Animals are unable to synthesize linoleate or linolenate, which are therefore "essential" and must be supplied by the diet. Animals can desaturate stearoyl-CoA to oleoyl-CoA. Starting from linoleoyl-CoA, animals are able to produce a variety of polyunsaturated fatty acids by desaturation and chain elongation, including the prostaglandin precursor, arachidonate (Fig.6).

Plant desaturase systems appear to be soluble, may contain ferredoxin in place of cytochrome b_5, and act upon fatty acyl-ACP or upon acyl groups already incorporated into membrane lipids (e.g.

Table 3. Different starter molecules for fatty acid biosynthesis. The acyl group of the CoA derivative is transferred to the peripheral SH group (SH group of β-ketoacyl-ACP synthase) (Table 1, reaction 2).

Starting molecule	Fatty acid product
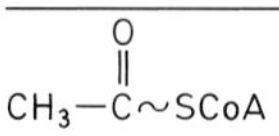 $CH_3-\overset{O}{\underset{\Vert}{C}}\sim SCoA$ Acetyl – CoA	Even numbered, straight carbon chain
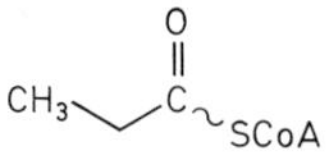 Propionyl – CoA	Odd numbered, straight carbon chain
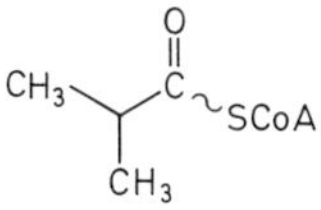 Isobutyryl – CoA (from valine)	Even numbered, branched chain (Iso series)
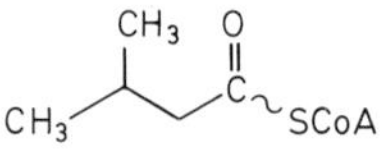 Isovaleryl – CoA (from leucine)	Odd numbered, branched chain (Iso series)
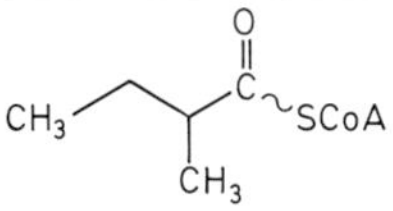 2 – Methylbutyryl – CoA (from isoleucine)	Odd numbered, branched chain (Anteiso series)

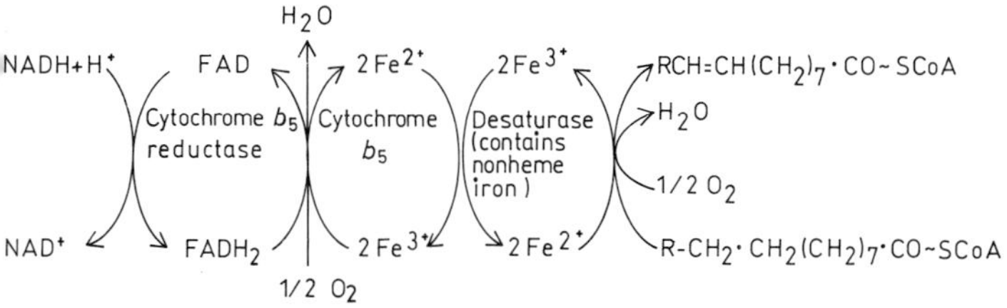

Figure 5. Desaturase system of rat liver endoplasmic reticulum.

[1-oleoyl]-diacylgalactosylglycerol desaturated to [1-linoleoyl]-diacylgalactosylglycerol and then to [1-linolenoyl]-diacylgalactosylglycerol by spinach chloroplast desaturase).

Only a few bacteria operate the oxygen-dependent desaturase system, e.g. *Mycobacterium smegmatis* and *phlei*, *Alcaligenes faecalis*, *Bacillus megaterium*. The system from *M. smegmatis* is particulate, converts palmitoyl-CoA or stearoyl-CoA to the corresponding unsaturated Δ^9-derivatives, and requires NADPH. The substrate for desaturation by the *Bacillus megaterium* system is the acyl group of phosphatidylglycerol.

Synthesis of fatty acids with odd numbers of carbon atoms, and fatty acids with branched chains. Incorporation of various primers or starter molecules in

F.a.b. in place of acetyl-CoA results in odd-numbered carbon chains or in branching (Table 3).

Elongation of the branched starter molecules in Table 3 results in a branched fatty acid with branching distal to the carboxyl group. For the synthesis of branched fatty acids with branching adjacent to the carboxyl group, the starter molecule must be unbranched and the final stages of elongation must employ a branched substrate, e.g. methylmalonyl-CoA. The mycocerosic acids are a family of multimethyl-branched acids produced by mycobacteria, such as *Mycobacterium tuberculosis* var. *bovis* Bacillus Calmette-Guérin. They mostly possess a straight chain of 18–20 carbons with a multimethyl-branched region at the carboxy end. Three are well characterized:

199

Fatty acid biosynthesis

$$\underset{\text{2,4,6,8-Tetramethyloctacosanic acid}}{CH_3(CH_2)_{19}-\overset{\overset{\displaystyle CH_3}{|}}{CH}\cdot CH_2\cdot \overset{\overset{\displaystyle CH_3}{|}}{CH}\cdot CH_2\cdot \overset{\overset{\displaystyle CH_3}{|}}{CH}\cdot CH_2\cdot \overset{\overset{\displaystyle CH_3}{|}}{CH}\cdot COOH}$$

$$\underset{\text{2,4,6-Trimethylhexacosanic acid}}{CH_3(CH_2)_{19}-\overset{\overset{\displaystyle CH_3}{|}}{CH}\cdot CH_2\cdot \overset{\overset{\displaystyle CH_3}{|}}{CH}\cdot CH_2\cdot \overset{\overset{\displaystyle CH_3}{|}}{CH}\cdot COOH}$$

$$\underset{\text{2,4,6-Trimethyltetracosanic acid}}{CH_3(CH_2)_{17}-\overset{\overset{\displaystyle CH_3}{|}}{CH}\cdot CH_2\cdot \overset{\overset{\displaystyle CH_3}{|}}{CH}\cdot CH_2\cdot \overset{\overset{\displaystyle CH_3}{|}}{CH}\cdot COOH}$$

$\left.\rule{0pt}{60pt}\right\}$ Mycocerosic acids

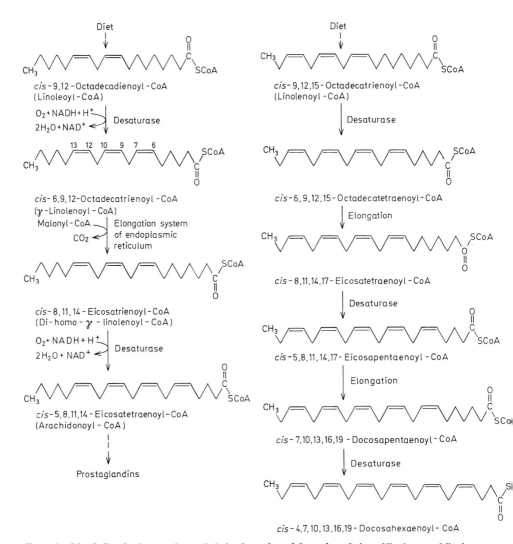

Figure 6. Metabolism by desaturation and chain elongation of the carbon chains of linoleate and linolenate.

A novel fatty acid synthase, mycocerosic synthase, has been isolated from *Mycobacterium tuberculosis* var. *bovis* BCG, which elongates *n*-fatty acyl-CoA with methylmalonyl-CoA [D.L. Rainwater & P.E. Kolattukudy, *J. Biol. Chem.* **260** (1985) 616–223]. It consists of two identical subunits (M_r subunit 238000, M_r dimer 490000). It elongates straight fatty acyl-CoA esters from *n*-C_6 to *n*-C_{20}, to generate the corresponding tetramethyl-branched mycocerosic acids, and it does not utilize malonyl-CoA.

Branch methyl groups located near the center of the fatty acid chain are derived by carbon transfer from *S*-adenosyl-L-methionine to an unsaturated fatty acyl derivative in a phospholipid, followed by a reduction, e.g. synthesis of 10-methylstearic acid (tuberculostearic acid) in mycobacteria, *Nocardia*, *Streptomyces* and *Brevibacterium*:

$$CH_3(CH_2)_7CH=CH(CH_2)_7-COO-X + S\text{-Adenosyl-L-methionine}$$

$$\downarrow$$

$$CH_3(CH_2)_7\underset{\underset{CH_2}{\|}}{C}-CH_2(CH_2)_7-COO-X + S\text{-Adenosyl-L-homocysteine}$$

$$\left\{\begin{array}{l} \curvearrowright NADPH + H^+ \\ \curvearrowright NADP^+ \end{array}\right.$$

$$CH_3(CH_2)_7\underset{\underset{CH_3}{|}}{CH}-CH_2(CH_2)_7-COO-X$$

Tuberculostearoyl group in a phospholipid

Cyclopropane fatty acids (Gram-negative bacteria, some green plants, zooflagellates) are also formed by carbon transfer from *S*-adenosyl-L-methionine to a *cis* monounsaturated fatty acid derivative:

$$CH_3(CH_2)_nCH=CH(CH_2)_xCOO-X + S\text{-Adenosyl-L-methionine}$$

$$\underset{CH_2}{\diagdown} \quad \downarrow$$

$$CH_3(CH_2)_n\underset{CH_2}{CH}-CH(CH_2)_xCOO-X + S\text{-Adenosyl-L-homocysteine}$$

Hydroxy fatty acids may be intermediates of F.a.b. which have been diverted (3-hydroxy fatty acids with D configuration). They may also arise by oxygenation catalysed by cytochrome P-450 systems (especially ω-oxidation in eukaryotes; see Fatty acid degradation). Hydroxy fatty acids can also be biosynthesized by hydration of double bonds in unsaturated fatty acids; specific fatty acid hydratase systems catalysing this reaction have been characterized in bacteria, especially pseudomonads. [J.B. Ohlrogge *Trends Biochem. Sci.* **7** (1982) 386–387; S.J. Wakil et al. *Ann. Rev. Biochem.* **52** (1983) 537–579; A.J. Fulco *Prog. Lipid Res.* **22** (1983) 133–160; A.D. McCarthy & D.G. Hardie *Trends Biochem. Sci.* **9** (1984) 60–63]

Fatty acid degradation: catabolic pathways for fatty acids. The most common is β-oxidation; minor pathways are α- and γ-oxidation.

β-Oxidation is the stepwise degradation of fatty acid from the carboxyl end. In each turn of the cycle, two carbon atoms are removed as acetyl-CoA, and the β-carbon atom is oxidized:

$$\begin{array}{c} HS\text{-}CoA \\ R\text{-}CH_2CH_2\text{-}CO\text{-}SCoA \dashrightarrow R\text{-}CO\text{-}CoA \\ + CH_3\text{-}CO\text{-}CoA. \end{array}$$

Fig. 1 shows the "fatty acid spiral". The initiating reaction is the activation of the fatty acid to acyl-CoA by an acyl-CoA synthetase (EC 6.2.1.3), a process which requires ATP: $R\text{-}CH_2\text{-}CH_2\text{-}COOH + ATP + HS\text{-}CoA \rightarrow R\text{-}CH_2\text{-}CH_2\text{-}SCOA + AMP + PP_i$. The acyl-CoA is reduced by acyl-CoA dehydrogenase (EC 1.3.99.3), an FAD-containing enzyme, to 2,3-dehydroacyl-CoA. This compound is then hydrated, oxidized to the 3-keto-acyl-CoA, and thiolysed to acetyl-CoA and an acyl-CoA which is two carbons shorter than the original one. The new acyl-CoA is immediately subject to further β-oxidation. Each turn of the "fatty acid spiral" thus produces one molecule of acetyl-CoA, which is further oxidized in the tricarboxylic acid cycle. Complete oxidation of 1 molecule of stearic acid yields 148 molecules of ATP (18 C atoms yield 9 molecules of acetyl-CoA; 1 acetyl-CoA yields 12 molecules of ATP in the tricarboxylic acid cycle; in addition, 5 molecules ATP are formed in each of the 8 β-oxidation steps). The energy stored in the terminal phosphate bonds of 148 molecules of ATP is equivalent to 50% of the heat of combustion of the fatty acid.

Various pathways are employed for the degradation of branched-chain fatty acids. Some of these arise from the metabolism of the branched-chain amino acids (see Leucine). Short-chain fatty acids are converted to their fatty acyl derivatives within the mitochondria, but long-chain fatty acids can be activated only by the endoplasmic reticulum and outer mitochondrial membrane. Long-chain acyl-CoA cannot penetrate the inner mitochondrial membrane, and must be transported into the mitochondria as acyl-carnitine (Fig. 3).

α-Oxidation of fatty acids occurs in germinating plant seeds. A fatty acid peroxidase (EC 1.11.1.3) catalyses the decarboxylation and simultaneous for-

Fatty acid degradation

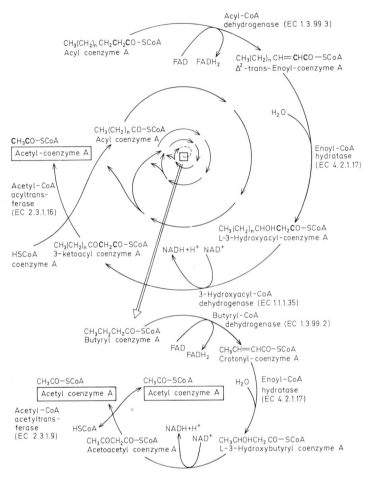

Acyl-CoA
dehydrogenase (EC 1.3.99.3)

$CH_3(CH_2)_n CH_2CH_2CO-SCoA$
Acyl coenzyme A

FAD FADH$_2$

$CH_3(CH_2)_n CH=CHCO-SCoA$
Δ^2-trans-Enoyl-coenzyme A

H$_2$O

Enoyl-CoA
hydratase
(EC 4.2.1.17)

$CH_3(CH_2)_n CO-SCoA$
Acyl coenzyme A

$CH_3CO-SCoA$
Acetyl-coenzyme A

Acetyl-CoA
acyltrans-
ferase
(EC 2.3.1.16)

$CH_3(CH_2)_n CHOHCH_2CO-SCoA$
L-3-Hydroxyacyl-coenzyme A

$CH_3(CH_2)_n COCH_2CO-SCoA$
3-ketoacyl coenzyme A   NADH+H$^+$ NAD$^+$

HSCoA
coenzyme A

3-Hydroxyacyl-CoA
dehydrogenase (EC 1.1.1.35)

Butyryl-CoA
dehydrogenase (EC 1.3.99.2)

$CH_3CH_2CH_2CO-SCoA$
Butyryl coenzyme A

FAD
FADH$_2$

$CH_3CH=CHCO-SCoA$
Crotonyl-coenzyme A

$CH_3CO-SCoA$
Acetyl coenzyme A

$CH_3CO-SCoA$
Acetyl coenzyme A

H$_2$O Enoyl-CoA
hydratase
(EC 4.2.1.17)

Acetyl-CoA
acetyltrans-
ferase
(EC 2.3.1.9)

HSCoA

NADH+H$^+$

$CH_3COCH_2CO-SCoA$
Acetoacetyl coenzyme A

NAD$^+$ $CH_3CHOHCH_2 CO-SCoA$
L-3-Hydroxybutyryl coenzyme A

Figure 1. Fatty acid spiral
In the degradation of odd-numbered fatty acids, the last step produces acetyl-CoA and propionyl-CoA. There are three different pathways for the further metabolism of propionyl-CoA (Fig.2). a) In the methylmalonyl pathway, propionyl-CoA is converted to methylmalonyl-CoA by an ATP-dependent carboxylation. S-Methylmalonyl-CoA mutase (EC 5.4.99.2), a cobamide-requiring enzyme, then converts methylmalonyl-CoA to succinyl-CoA. b) In the lactate pathway, the propionyl-CoA is dehydrogenated to acrylyl-CoA. α-Hydration gives L-lactoyl-CoA, which is hydrolysed to lactate. c) In the third pathway, which is found in plant mitochondria, the acrylyl-CoA is hydrated to 3-hydroxypropionyl-CoA, which is then deacylated and oxidized to malonic acid semialdehyde. This compound is converted directly to acetyl-CoA by oxidative decarboxylation or indirectly via malonyl-CoA.

mation of an aldehyde, in which H$_2$O$_2$ acts as hydrogen acceptor. The aldehyde can either be oxidized to a fatty acid or reduced to a fatty alcohol.
$R-CHOH-COO^- \rightarrow R-CO-COO^- \rightarrow R-COO^- + CO_2$.
ω-Oxidation is oxidation of the terminal methyl group of a fatty acid by enzymes localized in the microsomal fraction of animal and microbial cells. The substrate is usually a C$_8$ to C$_{12}$ fatty acid, which is converted to the dicarboxylic acid in two steps. The first is hydroxylation to an ω-hydroxy fatty acid, which requires oxygen and NADPH. The second step is catalysed by a soluble, non-microsomal enzyme, which is usually NAD$^+$-dependent.

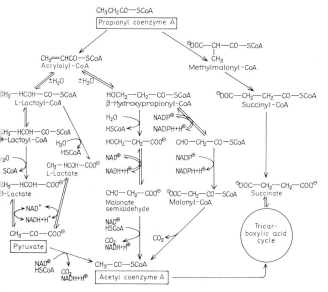

Figure 2.   Degradation of propionyl coenzyme A.

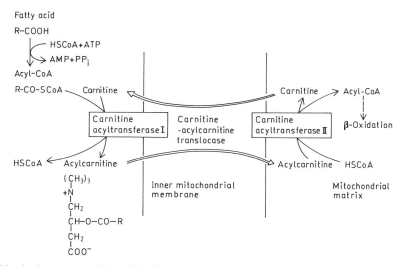

*Figure 3.   **Role of carnitine in the transport of long-chain fatty acids through the inner mitochondrial membrane.** Carnitine-acylcarnitine translocase is an integral membrane exchange transport system. Carnitine acyltransferases I and II are located on the outer and inner surfaces, respectively, of the inner mitochondrial membrane. [S. V. Pande & R. Parvin in R. A. Frenkel & J. D. McGarry, eds., Carnitine Biosynthesis, Metabolism and Functions (Academic Press, New York, 1980)]*

Fatty acids: the carboxylic acids found in fats and oils with the general formula $C_nH_{2n+1}COOH$ for saturated F.a. The unsaturated F.a. have double bonds between two or more carbons, and correspondingly fewer H atoms. Some of the common saturated acids are butyric, palmitic, lauric and stearic acids. Oleic, linoleic and linolenic acids are un-saturated or olefinic acids. *Essential F.a.:* see Vitamins (vitamin F).

Fatty acid synthetase complex: see Multienzyme complexes.

Fatty alcohols: unbranched, aliphatic mono-alcohols with 10 to 20 C atoms. They are found in nature as components of waxes. They are produced by

203

reduction of fatty acids, and are used in the production of surface-active agents and emulsifiers.

Fatty liver, *fatty degeneration of the liver:* see Lipotropic substances.

Favism: see Glucose 6-phosphate dehydrogenase.

FCCP: abb. for carbonyl cyanide-*p*-trifluoromethoxyphenylhydrazone. See Ionophore.

Fc fragment: see Immunoglobulins.

Fd: abb. for ferredoxin.

Febrifugine: a quinolazine alkaloid isolated from the shrub *Dichroa febrifuga,* which has been used in Chinese folk medicine since ancient times, and the common hydrangea. m. p. 139 to 140 °C, $[\alpha]_D^{25} + 6$ ° ($c = 0.5$ in chloroform). F. is a strong antimalarial and fever-sinking agent, but is too toxic for human use.

Feedback mechanisms: see Metabolic regulation.

Feedforward mechanisms: see Metabolic regulation.

Fermentation: a form of metabolism in which the end products could be further oxidized. Per unit of substrate, F. yields far less energy than respiration. For example, a yeast cell obtains 2 molecules ATP per molecule of glucose when it ferments it to ethanol, while complete respiration would yield 38 molecules of ATP. Pasteur's definition of F. as "life without air" is not strictly observed because, in addition to anaerobic processes, some aerobic ones, such as acetic acid fermentation, are called F.

Fermentation products: end products of anaerobic microbial metabolism. The products listed in the table are (or were) industrially significant, but in view of the finite amounts of fossil fuels, it seems likely that F. will become more important in the future.

Fermentation techniques, *microbial production techniques:* techniques for large-scale production of microbial products. F.t. must both provide an optimum environment for the microbial synthesis of the desired product and be economically feasible on a large scale (see Cultivation of microorganisms). They can be divided into surface (emersion) and submersion techniques. The latter may be run in batches or continuously.

In the surface techniques, the microorganisms are cultivated on the surface of a liquid or solid substrate. These techniques are very complicated and are rarely used in industry. They are used for the production of enzymes on solid substrates and, in a few factories, for citric acid. The microbes are grown in flat dishes which are stacked in large containers. A special form of surface technique is the production of acetic acid or sewage treatment in vertical containers of carrier material. The organisms grow on the carrier bed, and the medium or sewage flows past them.

In submersion processes, the microorganisms grow in a liquid medium. Except in traditional beer and wine fermentation, the medium is held in Fermenters (see) and stirred to obtain a homogeneous distribution of cells and medium. Most processes are aerobic, and for these the medium must be vigorously aerated. All important industrial processes (production of biomass and protein, antibiotics and en-

Table. Fermentation products.

Product	Organism	Use
Ethanol	*Saccharomyces cerevisiae*	Industrial solvent beverages
Glycerol	*Saccharomyces cerevisiae*	Production of explosives
Lactic acid	*Lactobacillus delbrueckii* *Lactobacillus bulgaricus*	Food and pharmaceuticals
Acetone and Butanol	*Clostridium acetobutylicum* *Clostridium butylicum*	Solvents
2,3-Butylene glycol (2,3-butane diol)	*Bacillus polymyxa* *Bacillus subtilis* *Aerobacter aerogenes*	After conversion butadiene, production of synthetic rubber; Antifreez

zymes and sewage treatment) are carried out by su mersion processes.

In batch techniques, the fermenter is filled with m dium and innoculated; the cells multiply and sy thesize the product, simultaneously consuming t nutrients in the medium. At the appropriate tim the culture is drained off and worked up. Discont uous processes are used, for example, in the prod tion of antibiotics and amino acids which are p duced only at particular stages of growth of t culture. They are not economical as industrial p cesses because of the discontinuity of production.

In continuous processes, the microorganisms continuously supplied with new medium and t suspension of old medium and microbes is remov simultaneously by overflow. The overflow can worked up continuously. These techniques are us especially for the production of biomass, prot and baker's yeast. A continuous process for brewi beer has been developed, and sewage treatment also continuous. These processes have in comm that the desired product is either the cells themsel (biomass, proteins) or a metabolic product which continuously produced (fermentation product However, continuous processes can also be adapt to products which are synthesized after cell grow has stopped, for example by setting up a series fermenters to produce the desired physiologi state in the cells. The main advantage of continuo processes is that they can be automated. In *chem stats,* the ratio of cell growth to synthesis of secon ary products can be controlled by the rate at whi new medium is added to the culture. *Turbidost* have so far been used more in the laboratory maintain a culture at a defined density of cells. light is shone through the fermenter, and when intensity falling on a photocell on the other side has decreased below a preset threshhold, the culture is diluted with fresh medium. On the industrial scale, continuous processes are difficult to keep sterile over long periods, so they are generally used only for non-sterile production.

Fermenters: large-scale vessels in which microorganisms are cultured to obtain some product of their metabolism (see Industrial fermentation). The main vessel is usually cylindrical; the requirements of the particular microorganism and the desired product dictate the accessories required for stirring, cooling, aeration, input of medium and harvesting. The simplest F. is a large tank like the vats originally used for wine and beer fermentation. A large vat must be cooled by a water jacket or heat exchanger so that the organisms' metabolic heat does not raise the temperature beyond their level of tolerance. Stirring is required in large F. to keep the organisms from settling and to provide a uniform supply of nutrients, including oxygen in aerobic fermentations. There are many types of stirrers, turbines, baffles and recirculating pumps. In an *air-lift* or *bubble F.,*

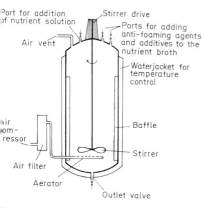

Port for addition of nutrient solution — Stirrer drive — Ports for adding anti-foaming agents and additives to the nutrient broth — Air vent — Waterjacket for temperature control — Air compressor — Baffle — Stirrer — Air filter — Aerator — Outlet valve

Cross section of a fermentor

stirring and aeration are provided by a stream of gas entering the bottom of the tank. In a *cyclone column F.,* a pump continuously circulates the culture from the bottom to the top of the vessel. The medium is ejected from a cyclone head tangentially to the wall of the F., so that it spirals down the wall as a thin film. Air or gas is injected at the bottom of the vessel and rises in a counter current to the culture flow. *Tower F.* are used for beer and other cultures in which the desired products are not produced continuously over the growth cycle. F. range in size from a few liters for laboratory use to more than 500 m^3 for industrial applications. The working volume of medium is usually about ⅔ of the total capacity of the F.

Ferments: see Enzymes.

Ferredoxins, abb. *Fd.:* low-molecular-weight iron-sulfur proteins which transfer electrons from one enzyme system to another, without possessing any enzyme activity themselves. The name was coined by Mortenson for iron-containing proteins from *Clostridium pasteuranum.* The *8-Fe-Fd* take part in many electron transport processes in organisms like clostridia and photosynthetic bacteria (Table). The primary structures of many of these proteins have been determined. They consist of about 55 amino acids, including 8 cysteines, which occupy the same positions in the chains of this family of proteins. The three-dimensional structure has also been determined; the molecule contains two identical 4Fe-4S clusters, each one forming a cube and covalently bonded to 4 cysteine residues in the peptide chain. Each 4Fe-4S center can transfer one electron.

4Fe-Fd have a single 4Fe-4S cluster as an active center. As in the clostridial Fd, the iron atoms are bound to the only 4 cysteine residues in the protein. These Fd have been isolated primarily from bacteria; the primary structures of the *Desulfovibrio gigas* and *Bacillus stearothermophilus* Fd are known.

High potential iron-sulfur protein (HiPIP) is a special type of Fd which so far has been isolated from only four photosynthetic bacteria. However, it has been detected by ESR spectroscopy in other bacteria as

Table. Ferredoxin-dependent metabolic reactions.

Reaction	Ferredoxin type
Nitrogen fixation $N_2 + 6ATP + 6e^- \rightarrow 2NH_3 + 6ADP + 6P_i$	Nitrogenase complex: Molybdoferredoxin, contains 18 to 28 FeS and 1 Mo + Azoferredoxin (4 Fe-4S center) + 8 Fe-Fd. Found in bacteria, blue-green bacteria
Hydrogen metabolism $2H^+ + 2e^- \rightleftharpoons H_2$	Hydrogenase (4 Fe-S center) + 4 Fe-Fd., 8 Fe-Fd, or 2 Fe-Fd. Found in bacteria, blue-green bacteria
Phosphoroclastic reaction $CH_3\text{-}CO\text{-}COO^- + P_i \xrightarrow{CoA}$ $CH_3\text{-}CO\text{-}O\text{-}PO_3H_2 + CO_2$	8 Fe-Fd. + complex Fe-S enzyme Found in Clostridia
Synthesis of α-ketoacids (pyruvate, α-ketoglutarate, α-ketobutyrate), reductive CO_2 fixation $CH_3\text{-}CO\text{-}CoA + CO_2 \rightarrow CH_3\text{-}CO\text{-}COO^- + CoA$	8-Fe-Fd. (photosynthetic and fermenting bacteria, e.g. *Chromatium, Clostridium*)
Photosynthetic nicotinamide-nucleotide reduction $NADP^+ + H_2O \xrightarrow{light} NADPH + H^+ + \frac{1}{2}O_2$	2-Fe-Fd. Found in higher plants, algae.

well. HiPIP also contains a single 4Fe-4S cluster, but it differs from the other Fd in having a positive standard potential of about $+350$ mV. (Most Fd have standard potentials in the range of the hydrogen electrode, about -420 mV). Furthermore, the HiPIP from *Chromatium* is paramagnetic in the oxidized state.

2Fe-Fd from blue-green bacteria, green algae and higher plants are highly homologous (96 to 98 amino acid residues with 4 to 6 cysteines). Four of the six cysteines are found in all plant Fd in positions 39, 44, 47 and 77, and are covalently bound to the Fe-S center, X-ray crystallographic studies of these proteins are in progress. A proposed model of 2Fe-Fd is shown in the Fig. to Iron-sulfur proteins (see). 2Fe-Fd serve as electron-transfer catalysts in both cyclic (1) and noncyclic photophosphorylation (2):

$$NADP^+ + H_2O + ADP + P_i \xrightarrow{Fd/light} NADPH + H^+ + ATP + \tfrac{1}{2}O_2 \ (1)$$
$$ADP + P_i \xrightarrow{Fd/light} ATP \ (2)$$

The main metabolic reactions involving Fd are listed in the table.

A number of chemical structures for Fd have been suggested. The chelate structure (Fig.1) is based on chemical analysis, ESR and Mossbauer spectroscopy (see also Iron-sulfur proteins). Model experiments led to the construction of metal-atom island structures, or clusters. In this model, the "labile" sulfur atoms cause charge delocalization, and electrons are transferred from the sulfur to the iron which is bound between neighboring cysteine residues. The α-helical structure of the polypeptide prevents the formation of a disulfide bond until the electron is transferred to another substrate (Fig.2).

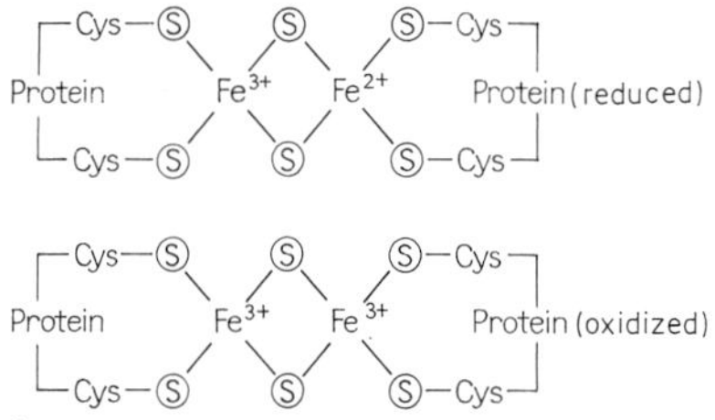

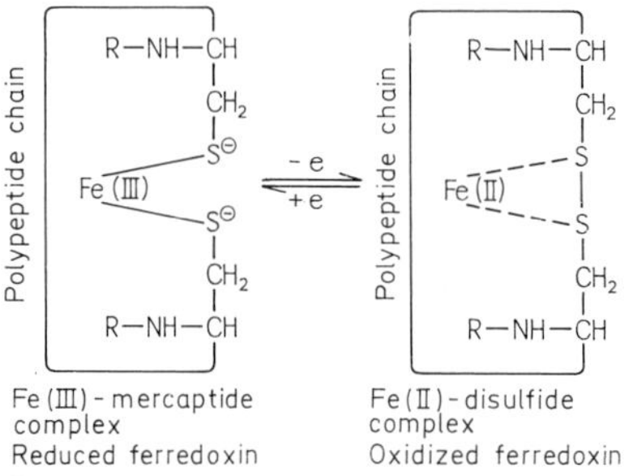

Proposed structures for ferredoxins:
1 Chelate model (only 2 Fe are shown); 2 M atom island structures

Ferreirin: 5,7,6'-trihydroxy-4'-methoxyisoflavanone, see Isoflavanone.

Ferrichrome: see Siderochromes.

Ferrimycin: see Siderochromes.

Ferrioxamines: see Siderochromes.

Ferritin: the most important iron-storage protein in mammals. Together with the related Memosiderin (see), it contains 25% of the iron in the body. F. is found in the spleen, liver, bone marrow and reticulocytes, where excess iron is stored intracellularly and can be mobilized at need. F. has also been found in molluscs, plants and fungi.

F. consists of a shell-like protein, apoferritin, with a diameter of 12 nm. The interior is a chamber with a diameter of 7 nm in which the iron is deposited as iron-hydroxide-oxide micelles. One molecule of F. can hold up to 4300 iron(II) atoms (Fig. 1). The mi

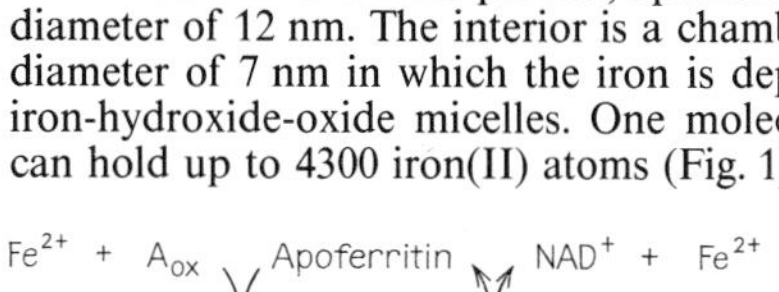

Figure 1. Storage of iron in apoferritin in the presence of an oxidizing agent A and mobilization of iron from ferritin, which requires NADH.

cellar iron has a composition of $8 \ FeO(OH) \cdot FeO(PO_3H_3)$. However, the phosphate is not always present. Iron-free apoferritin can be isolated by chemical reduction of the iron or by centrifugation. It consists of 24 equal-sized subunits (M_r 18500) which are noncovalently bonded to one another in an oligomer of M_r 445000. Apoferritin can be dissociated into subunits by treatment with 1% sodium dodecylsulfate or 5 M guanidine hydrochloride which disrupt the strong hydrophobic bonding between the subunits.

F. is a major storage depot for iron. Both the resorption of Fe^{2+} from the small intestine and the transfer of the iron (Fe^{3+}) from transferrin to apoferritin involve reduction, chelate formation, a transfer step and oxidation (Fig.2). In the presence of a suitable

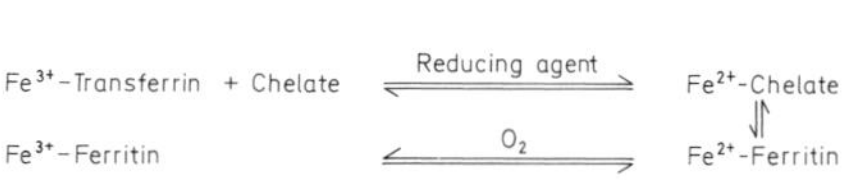

Figure 2. Model for iron transfer from transferrin to ferritin (according to Crichton).

oxidizing agent, such as molecular oxygen, apoferritin catalyses the oxidation of Fe^{2+} to Fe^{3+}. The reverse reaction, the reductive mobilization of the ferritin iron, is catalysed by an NADH-dependent oxidoreductase which may be either membrane bound or soluble.

Fervenulin: 6,8-dimethylpyrimido-5,4-e-1,2,4-triazine-5,7-(6H,8H)-dione, M_r 193.17, m.p. 178 to 179 °C, a pyrimidine antibiotic synthesized by *Streptomyces fervens*. Its structure is analogous to that of toxoflavin, but F. is less toxic. It has a broad spectrum of action, especially against cocci, Gram negative and phytopathogenic bacteria and trichomonads.

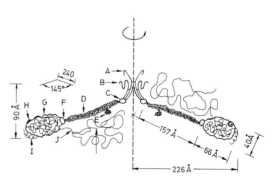

Fervenulin

Fe-S-protein: see Iron-sulfur proteins.

α-Fetoprotein: The first α-globulin formed in the serum of mammalian embryos; M_r 70000. It is synthesized first by the yolk sac, and later by the fetal liver. It is replaced at birth by serum albumin. The presence of α-F. in adult serum is a symptom of liver carcinoma; it is detected by radioimmunoassay. The escape of α-F. into the amniotic fluid is a symptom of spinal bifida (failure of the neural fissure to close) and/or anencephaly. The condition can be detected by an immunoassay of amniotic fluid obtained by amniocentesis. [F. Jacob, "Expression of Embryonic Characters by Malignant Cells," 1983 Ciba Foundation Symposium 96: Fetal Antigens and Cancer (Pitman, London) pp. 4-27.]

FGAR: see N-Formylglycinamide ribotide.

FH$_4$: see Tetrahydrofolic acid.

Fibrin: the protein endproduct of blood coagulation. It is generated from the plasma protein Fibrinogen (see) by thrombin in the presence of Ca(II) ions. Thrombin removes two fibrinopeptide pairs A_2 and B_2 from fibrinogen. After the A peptide has been removed, the protein polymerizes in a pH-dependent, linear fashion. Thereafter, the B peptide is removed. At this point the fibrin bundle is still soluble in urea and has a cross-banding pattern visible in the electron microscope. It is stabilized by factor XIII, which must first be activated by thrombin and Ca^{2+}. Factor XIIIa removes the carbohydrate and generates two intermolecular ε-(γ-glutamyl)lysine bonds (see Isopeptide bonds) per molecule be-

tween the γ and α chains of the fibrin bundle. A cross-linked fibrin clot is insoluble both in water and in 8 M urea, and provides a stable wound closure (see Blood coagulation). It can be degraded to soluble cleavage products by Plasmin (see).

Fibrinogen: the precursor of Fibrin (see). It is the only coagulable protein in the blood plasma of vertebrates and some arthropods. Its concentration in human plasma is 200 to 300 mg/100 ml. On electrophoresis of the plasma proteins, it migrates between the β and γ fractions; its M_r is 340000, of which 15% is carbohydrate. A model for the structure of F. which is now widely accepted was proposed by Doolittle et al. [*Horizons Biochem. Biophys.* **3** (1977) 164-191] (Fig.). The molecule consists of 6 polypeptide chains, two each of Aα, Bβ and γ. These are mutually cross-linked by disulfide bonds in their N-terminal segments. A coiled coil made up of one of each kind of polypeptide emerges on each side of the central knot. In this region, the polypeptide chains are parallel and in register with one another. The coiled coil is separated by a second region of disulfide bonds ("disulfide swivel") from the terminal, globular region of the protein. The α chains extend as "random coils" from this globular region and serve as cross-linkers in the fibrin clot. Thrombin first removes the A peptides from the Aα chains by cleavage at Arg 16; this forms fibrin I monomers. The second cleavage occurs at Arg 14 in the Bβ chain after the fibrin I monomers have polymerized end-to-end to form protofibrils.

F. is also present in platelet α-granules and is secreted from stimulated platelets. It is a major regulator of platelet aggregation, binding to specific receptor sites on the platelet surface. In this respect, F. has features in common with other "adhesive" proteins: Fibronectins (see), von Willebrand factor (see), Thrombospondin (see). [*Molecular Biology of Fibrinogen and Fibrin. Annals of the New York Academy of Sciences* **408** (1983)]

Detailed model of fibrinogen drawn to scale. A, fibrinopeptide A at amino terminus of α-chain; B, fibrinopeptide B at amino terminus of β-chain; C, first disulfide swivel holding all three chains together and in register; D, coiled-coil connection segment holding domains I and III together; E, carbohydrate attached to residue 52 of γ-chain; F, second disulfide swivel holding all three chains together; G, domain III, which corresponds approximately to fragment D; H, carbohydrate on β-chain; I, carboxy-terminal segment of γ-chains which becomes involved in intermolecular cross-linking; J, plasmin attack point and α-chain cross-linking region; K, carboxy terminus of α-chain. [From Doolittle et al.; used with permission.]

Fibrinolysin: see Plasmin.

Fibrinolysis: see Plasmin.

Fibrinopeptides: the two pairs of peptides (A and B) cleaved off the amino ends of the 2α and 2β chains of fibrinogen by thrombin. F. arise by cleavage of Arg-Gly bonds, so that Arg is the C-terminal end of the F. and Gly is the N-terminal end of the α and β chains of fibrin. Human F.A. is Ala-Asp-Ser-Gly-Glu-Gly-Asp-Phe-Leu-Ala-Glu-(Gly)$_3$-Val-Arg and human F.B. is Pyr-Glu-Gly-Val-Asn-Asp-Asn-(Glu)$_2$-Gly-(Phe)$_2$-Ser-Ala-Arg. The size of F.A. ranges from 14 amino acids (horse, lizard) to 19 (cattle), and F.B. has from 9 (rhesus monkeys) to 21 (cattle, elk and kangaroo). The sequences of the F. were used to establish a detailed phylogenetic tree for the mammals which is very similar to the classical one. The F. have a vasoconstrictive effect which serves to keep the coagulation principles from being removed too quickly from an injury.

Fibronectins: Adhesive glycoproteins, M_r 200000-250000 found both as cell surface proteins and in plasma. They have high binding affinity to fibrous proteins such as collagen and fibrin. In tissue cultures, cellular F. is found in fibrillar arrays on cell surfaces, between cells and as part of the intracellular matrix. In vivo, it is found as a component of connective tissue matrices, i.e. in basement membranes and as thin fibers in loose connective tissue. Plasma fibronectin is incorporated into fibrin clots, and is cross-linked to the fibrin, probably via a Gln-Lys transamidation reaction. [M. Okada et al., *J. Biol. Chem.* **260** (1985) 1811-1820] Cellular F. is also known as *Cell Surface Protein, Galactoprotein A, Large External Transformation Sensitive Protein, Surface Fibroblast Antigen,* and *Zeta* or *Z Protein.* It is involved in cell adhesion and cell spreading of fibroblasts. Transformed cells (which do not adhere to substrate) lack F., and addition of F. to some lines of transformed fibroblasts induces them to adhere to the dish and spread out. [E. Pearlstein, L. I. Gold & A. Garcia-Pardo, *Molec. Cell. Biochem.* **29** (1980) 103-128.] The adhesion-promoting activity can be mimicked by the tetrapeptide Arg-Gly-Asp-Ser, a sequence which is found in the cell attachment domain of F., or by Arg-Gly-Asp-Cys. The same tetrapeptides block adhesion of cells to a F.-coated surface, which implies that the cell has a receptor specific for this sequence. [M. D. Pierschbacher & E. Ruoslahti, *Nature* **309** (1984) 30-33.]

F. is a component of a fibrillar matrix underlying the blastocoel roof in the blastula stage of amphibian embryos, and has been shown to be necessary for mesodermal cell migration in gastrulation. [J. C. Boucaut et al., *Nature* **307** (1984) 364-366.]

Fragments of F. have been isolated from normal human plasma and other sources. Enhanced levels of these fragments are characteristic of diseases with abnormally high rates of proteolysis, including cancer. This has led to the suggestion that proteolysis of F. releases "activation peptides" which regulate cell division; evidence for this is the observation that proteolysis of F. by cathepsin D releases fragments which induce DNA synthesis in cultured normal fibroblasts. [M. J. Humphries & S. R. Ayad, *Nature* **305** (1983) 811-813.] There is evidence that cell-surface F. is a "receptor" for binding of the infectious

stage of trypanosomiasis parasites, and for some bacterial pathogens. [M. A. Ouaissi et al., *Nature* **308** (1984) 380-382.]

Fibrous proteins: see Structural proteins.

Ficaprenols: see Polyprenols.

Ficifolinol: 3,9-dihydroxy-2,8-di-γ,γ-dimethylallylpterocarpan, see Pterocarpans.

Field effect transistor, *FET*: an electronic device in which the conductivity of the semiconductor material is controlled by the electrical field at a particular part of its surface, known as the "gate". This electrical field may be varied by attachment (and subsequent reaction) of enzymes (ENZFET) or antibodies (IMMUNOFET) to the gate. The gate may also be sensitized to specific ions (ISFET) or other chemicals (CHEMFET).

Filaggrin: a basic protein, M_r 26500 to 49000, depending on the species of animal, from the stratum corneum of the epidermis. It is probably identical to the interfilamentous keratin matrix protein; its function is to organize prekeratin into filaments as the epidermal cells move outward from the basal layer and become keratinized. F. is synthesized as a highly phosphorylated precursor which cannot organize keratin. Later the phosphate-bearing region is removed by proteolysis; at this stage the protein has a large number of Arg residues. These presumably interact with the acidic regions of the keratins. Still later in its development, the Arg residues are converted to citrulline, and as this happens, F. loses the ability to aggregate pre-keratins. [P. Traub, *Intermediate Filaments* (Springer, Heidelberg, 1985)]

Filamen: see Cytoskeleton (microfilaments).

Fimbrin: see Cytoskeleton (microfilaments).

Finger print technique: see Proteins.

First messenger: see Hormones.

Fisetin: see Flavones (Table).

Flagellin: the main protein component of bacterial flagella. The filaments of F. have an α-keratin structure, which is reversibly converted to the β-keratin structure on stretching of the aggregate. The F. monomer (M_r 33000 to 40000) is dissociated by mild acid treatment (pH 3 to 4). It has 304 amino acid residues, but no cysteine or tryptophan and only traces of proline and histidine.

Flavan: a naturally occurring Flavonoid (see) with the ring system shown (Fig. 1)

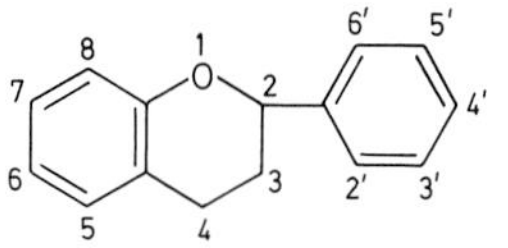

Figure 1. Flavan ring system

F. unsubstituted at positions 3 or 4 are unstable in solution, forming polymeric products, which presumably accounts for the relatively few reports of naturally occurring F. In contrast, the more stable 3- or 4-hydroxylated F. (e.g. flavan-3-ols) or catechins, and flavan-3,4-diols or leukoanthocyanidins) are frequently encountered as plant products. By October, 1984, only 18 free F. and 4 F. glycosides had been reported. It is to be expected that more F. will be found; reports of newly discovered plant constituents often appear in the journal *Phytochemistry* (Pergamon Press). The 4 glycosides are **diffutin** (7-hydroxy-3',4'-dimethoxy-5'-O-β-D-glucosylflavan, *Canscora diffusa*), **auriculoside** (7,5'-dihydroxy-4'-methoxy-3'-O-β-D-glucosylflavan, *Acacia auriculiformis*), **koaburarin** ((2 R)-5-hydroxy-7-O-β-D-glucosylflavan, *Enkianthus nudipes*) and 7,4'-dihydroxy-5-O-β-D-xylosylflavan *Buckleya lanceolata*). F. have so far been found in 7 families: *Ericaceae, Gentianaceae, Leguminoseae, Liliaceae, Myristicaceae, Santalaceae* and *Amaryllidaceae*. Some F. have antimicrobial activity and some are also phytoallexins, e. g. **broussin** (7-hydroxy-4'-methoxyflavan) is formed by wounded xylem tissue of *Broussonetia papyrifera* (paper mulberry), and is absent from healthy tissue; it inhibits growth of the bacterium *Bipolaris leersiae* at $10^{-4} - 10^{-5}$ M. 7-Hydroxyflavan, 7,4'-dihydroxyflavan and 7,4'-dihydroxy-8-methylflavan inhibit the growth of *Botrytis cinerea*; all three are produced by *Narcissus pseudonarcissus* L. in response to infection by the fungus. They are also active against fungal spore germination in liquid culture.

Diffutin (see above) has pronounced adaptogenic (anti-stress and anti-anxiety) activity in addition to mild CNS depressant (barbiturate potentiation) action. It also has a marked inotropic effect in perfused frog heart, and shows no arrhythmogenic properties. In addition, it potentiates the contractile response of guinea pig vas deferens to catecholamines, without inhibition of the uptake of adrenalin. At 500 mg/kg, diffutin is nontoxic to dogs. The use of *Canscora* in Indian medicine as an herbal remedy for certain mental disorders is supported by these observations.

Three dimers containing the F. ring system have also been isolated (Fig. 2).

[S. K. Saini & S. Ghosal (1984) *Phytochemistry* **23** 2415-2421]

Flavan-3,4-diol: see Leucoanthocyanidins.

Flavanone: a flavonoid with the ring system shown (Fig.). F. hydroxylated at position 3 may be called flavanonols, but are more commonly known as dihydroflavonols. F. are amongst the earliest products of flavonoid biosynthesis (see Flavonoid, Fig.) Many F. have fungistatic or fungitoxic properties. F. occur free or as glycosides in many angiosperm families and in many gymnosperms, but insufficient data are available for a useful taxonomic

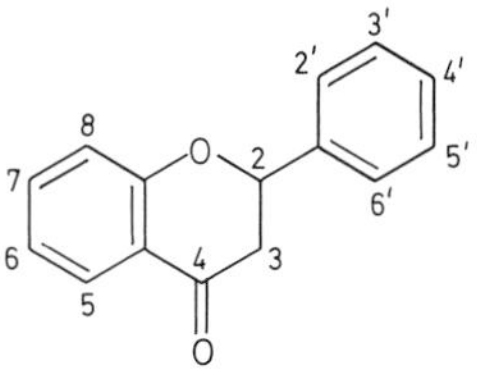

Flavanone ring system

analysis. Examples are **naringenin** (5,7,4'-trihydroxyflavanone), found in several plant families and a growth inhibitor in dormant peach flowers; **naringin** (naringenin 7-neohesperidoside), a bitter constituent of grapefruit and bitter oranges; **prunin** (naringenin 7-glucoside), from *Prunus* and others; **pinostrobin** (5-hydroxy-7-methoxyflavanone), from *Prunus* and others; **pinocembrin** (5,7-dihydroxyflavanone), from *Pinus* and others; **bavachin** (7,4'-dihydroxy-6-prenylflavanone), from *Psoralea*; **eriodictyol** (5,7,3',4'-tetrahydroxyflavanone), from *Eriodictyon*; **hesperetin** (5,7,3'-trihydroxy-4'-methoxyflavanone), from *Citrus*; **hesperidin** (hesperetin 7-rutinoside), from *Citrus*. For lists of known F., see B. A. Bohm in *The Flavonoids: Advances in Research*, J. B. Harborne & T. J. Mabry, eds., (Chapman and Hall, 1982) pp. 350-416.

Flavin: a compound containing the 6,7-dimethylisoalloxazine ring system (Fig.). The biologically important F. are Flavin adenine dinucleotide (see), Flavin mononucleotide (see) and Riboflavin (see, under Vitamins). See Flavin enzymes.

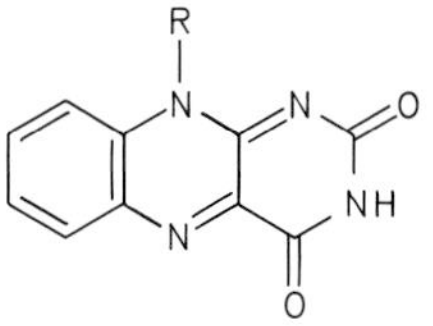

Flavin ring

Flavanone synthase: see Chalcone synthase.

Flavin adenine dinucleotide, abb. *FAD, riboflavin adenosine diphosphate:* a flavin nucleotide which is the active group of many Flavin enzymes (see). M_r 785.6, $E_o' = -0.219$V (pH 7.0, 30 °C), fluorescence maximum at 530 nm. ε_{450} of FAD$^+$ =

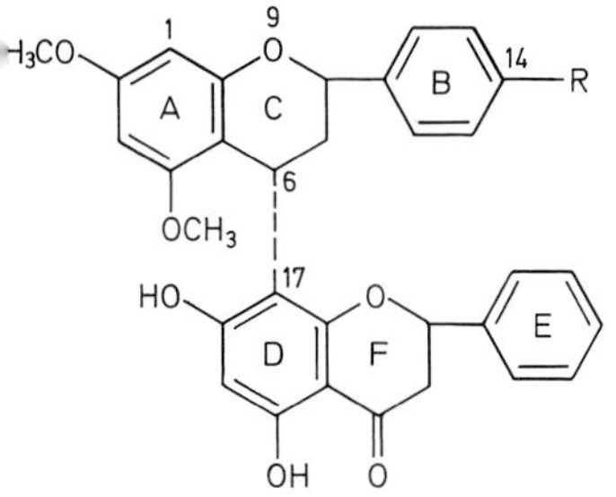

Figure 2. R = H: *Xanthorrhone*
R = OH: *14-Hydroxyxanthorrhone*,
both from *Xanthorrhoea* (*Liliaceae*)

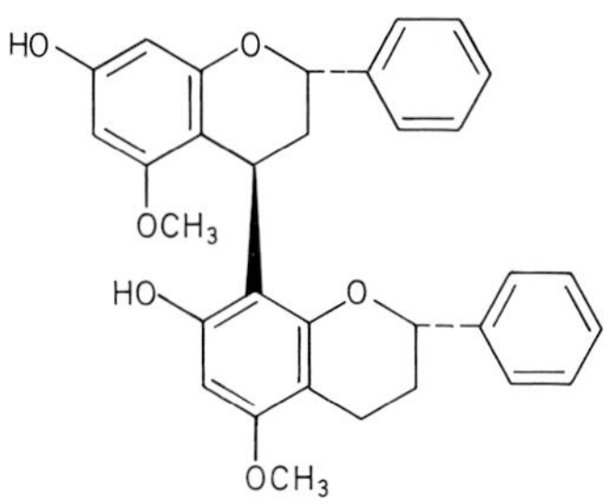

Biflavan 12 from "dragon's blood",
the resin of *Daemonorops draco* (*Palmae*)

11 300, ε_{450} of FADH = 980. When a dilute solution is heated, FAD is partly hydrolysed to Flavin mononucleotide (see). In alkaline solution it is quickly converted to cyclic riboflavin 4′,5′-phosphate. FAD is less photolabile than flavin mononucleotide or riboflavin.

The isoalloxazine ring of the flavin acts as a reversible redox system (see Flavin mononucleotide). The colorless dihydro compound is formed by addition of hydrogen to the N atoms 1 and 10; oxidized FAD is yellow.

In the strict sense, FAD is not a dinucleotide, because it contains the sugar alcohol ribitol instead of glycosidically linked ribose.

Flavin catalysis may involve three oxidation levels of the ring, and each of these may exist as a cation, anion or neutral compound, so there are 9 forms altogether. The oxidation levels are flavoquinone (oxidized) ⇌ flavosemiquinone (half-reduced) ⇌ flavohydroquinone (reduced). There are some catalytic mechanisms involving only one electron in which the ring goes from the flavoquinone to the semiquinone, or from the latter to the hydroquinone. Others involve two electrons, and the compound goes back and forth between the quinone and the hydroquinone states. FAD is synthesized from flavin mononucleotide by the FAD pyrophosphatase (EC 3.6.1.8): FMN + ATP ⇌ FAD + PP$_i$ (Fig.).

Riboflavin — (P) — (P) — Ribose–Adenine
└─── FMN ───┘ └─── AMP ───┘

Structure of FAD

Flavin enzymes, *flavoproteins, yellow enzymes:* a diverse group of more than 70 oxidoreductases found in animals, plants and microorganisms which have Flavin adenine dinucleotide (see) or Flavin mononucleotide (see) as prosthetic groups. These coenzymes can be reversibly reduced by hydrogen atoms, which they can accept either directly from a substrate (for example, succinate dehydrogenase) or from NAD(P)H. The group name refers to the yellow color of the oxidized riboflavin portion. The properties of the flavin adenine dinucleotide (FAD) or flavin mononucleotide (FMN) are strongly dependent on the protein molecule around them. This group of proteins is structurally and functionally very heterogeneous, and there is no single generalized type of flavin enzyme. The *metalloflavin enzymes (metalloflavoproteins)* also contain metals, such as Fe, Mg, Cu or Mo, which are involved in the redox reaction and in the binding of the F.e. to the mitochondrion (see Molybdenum enzymes). F.e. can be divided into the following groups:

1. *Oxidases* can use oxygen as electron acceptor and transfer two or four electrons. Those which transfer two electrons form H_2O_2. They include Glucose oxidase (see), the iron- and molybdenum-containing Xanthine oxidases (see), and the D-amino acid oxidases (containing FMN or FAD). The latter catalyse the irreversible formation of the corresponding α-ketoacid. Oxidases which transfer four electrons contain copper; they oxidize the substrate and form water. Examples are laccase, ascorbate oxidase and *p*-diphenol oxidase.

2. *Reductases* react primarily with cytochromes. They include cytochrome b_5 and c reductases and glutathione reductase (all contain FAD), GMP reductase and the molybdenum-containing nitrate reductase.

3. *Dehydrogenases*; the natural hydrogen acceptor of some of these is unknown. A well-known example of this group is Succinate dehydrogenase (see) of the tricarboxylic acid cycle, and the respiratory chain NADH and NADPH dehydrogenases (FMN) and acyl-CoA dehydrogenase are further examples.

In addition, there are more complex F.e., such as the hemoflavin enzymes (e.g. formic acid dehydrogenase from *Escherichia coli*) which contain metals, sulfhydryl-disulfide systems and heme groups in addition to the flavin component.

Flavin mononucleotide, abb. *FMN, riboflavin 5′-phosphate:* the prosthetic group of various flavin enzymes. M_r 456.4, $E_o' = -0.219$ V (pH 7.0, 30 °C), fluorescence maximum at 530 nm. FMN is composed of 6,7-dimethylisoalloxazine (flavin) and a ribitol residue linked to N 9 (Fig.). FMN occurs as the free acid or the sodium salt and usually contains 2 to 3 molecules H_2O. The phosphate ester bond is hydrolysed in acid solution, and the isoalloxazine-ribitol bond is unstable in alkaline solution. The compound is photolabile over the entire pH range, but is particularly so in alkaline solutions.

FMN is formed from riboflavin and ATP by a flavokinase. It is hydrolysed by acid and alkaline phosphatases.

├─────── Riboflavin ───────┤
Flavin mononucleotide

Flavin mononucleotide

Flavin nucleotides, *flavocoenzymes:* the coenzymes of the Flavin enzymes (see). Strictly speaking, they are Prosthetic groups (see), but some flavin enzymes can be easily separated from them. The two F. are Flavin mononucleotide (see) and Flavin adenine dinucleotide (see).

Flavocoenzymes: see Flavin nucleotides.

Flavones, *flavone pigments:* a group of plant pigments containing the flavone ring system. This consists of two substituted phenyl rings (A and B) and the pyrone ring C, fused to ring A, which is responsible for the typical reactions of the F. The structures of about 300 natural F. are known; except for *flavone* (Fig.), they all carry hydroxyl groups. Some also have methyl and methoxy substituents. The F. may have 1 to 7 OH groups (Table). The favored positions are 3,5,7,3′ and 4′. Those which are hydroxylated in position 3 are often called *flavonols* and are considered a subgroup of the F.

F. occur in the plant either in free form or as glyco-

Pigment	OH at positions	m.p. °C	Occurrence	Derivatives
Chrysin	5,7	285	Poplar buds, heartwood of many pines and putaceous trees	Chrysin 5- and 7-glucosides, 7-methyl ether, 6-methylchrysin
Primetin	5,8	231	In primroses	
Galangin	3,5,7	221	In the wood of various pines and the roots of *Alpina officinarum*	
Apigenin	5,7,4'	345–350	In white and yellow blossoms	Apiin (the 7-glucoside) in parsley
Kaempferol	3,5,7,4'	276–278	In many plants, including 50% of all angiosperms	6 naturally occurring methyl ethers, 30 glycosides, e.g. *robinin* (3-rhamnosylgalactoside-7-rhamnoside)
Fisetin	3,7,3',4'	330 (d.)	In wood and blossoms of many higher plants	
Luteolin	5,7,3',4'	328–330 (d.)	In flowering plants, e.g. mignonette, dahlia, broom	Glycosides
Morin	3,5,7,2',4'	285–290 (d.)	In various members of the mulberry family, e.g. old fustic	
Quercetin	3,5,7,3',4'	314 (d.)	Occurs widely, in 56% of all angiosperms	Mono and dimethyl ethers, 35 glycosides, e.g. *rutin* (3-rutinoside) in garden rue, yellow pansies and forsythia; *quercitrin* (3-rhamnoside) in quercitron bark
Myricetin	3,5,7,3',4',5'	357	In higher plants, e.g. in the bark of *Myrica nagi*	Glycosides, methyl ethers.

sides or esters. The most common positions for a glycosidic bond are 3 and 7. F. and flavonols are vacuolar pigments with a large absorption from 240 to 270 nm and 320 to 380 nm. Consequently they are yellow, and often occur together with anthocyanins, leading to combinations of red and yellow in the blossom. They used to be used in dying and printing, especially those from quercitron bark, old fustic (*Chlorophora tinctoria* L.), buckthorn berries and camomile.

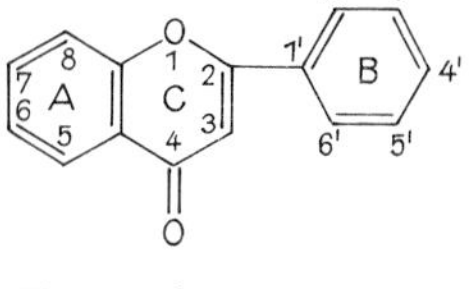

Flavone ring system

The F. are biosynthesized from a phenylpropane unit such as cinnamic acid, which becomes the aromatic ring B and the C atoms 2,3 and 4. The remaining C atoms are added by head-to-tail condensation of acetate and malonate (see Stilbenes).

Flavonoid: member of a large group of natural products with a $C_6C_3C_6$ skeleton. In most F., this skeleton takes the form of a phenylchroman ring system, in which the phenyl group may be attached at position 2 (normal flavonoids), 3 (isoflavonoids) or 4 (neoflavonoids) of the pyran ring. Further classification is based on the degree of oxidation of the pyran ring; see Anthocyanidins, Flavan, Flavanone, Flavones, Isoflavonoid, Isoflavan, Isoflavanone, Isoflavone, Neoflavonoid, Leucoanthocyanidins, Catechins. The oxygen-containing ring may be contracted (see Aurones), or absent (see Chalcones; α-Methyldeoxybenzoins).

Biosynthesis. It is well established by radiotracing that all classes of F. are biosynthetically closely related (Fig.). A chalcone is the latest intermediate in the biosynthetic sequence which is common to all F. The first stages of biosynthesis from phenylalanine also lead to the biosynthesis of other phenylpropyl compounds, e.g. coumarins, lignans, lignin, benzoic acid derivatives, aromatic esters, etc. Chalcones are also the precursors of stilbenes. The various patterns of hydroxylation and methoxylation in the F. may be partly established at an early stage by 3 and/or 5 hydroxylation as well as O-methylation of 4-coumaric acid, e.g. *S*-adenosyl-L-methionine:caffeate 3-O-methyltransferase (EC 2.1.1.?) has been characterized from plants. Hydroxylation, O-methylation and glycosylation also occur at various stages after formation of the F. ring system. Some typical conversions are shown in the Fig., which also shows the bi-

osynthetic relationships between the main classes of
F. [Ebel, J. & Hahlbrock, K. in *The Flavonoids: Advances in Research* (ed. J. B. Harborne & T. J. Mabry)
(Chapman and Hall, 1982) pp. 641–675]

Flavonol: a flavone hydroxylated at position
3. F. and flavones are the two most abundant classes
of flavonoids. For examples, see Flavones.

Flavoproteins: see Flavin enzymes.

Florigen: see Flowering hormone.

Flowering hormone, *anthesin, florigen, vernaline:* a principle which has been demonstrated physiologically but never isolated or chemically characterized. The original name "florigen" has been
dropped in favor of "anthesin". The hormone is synthesized in the leaves under conditions which induce
flowering and transported to the shoot tips, where it
transforms the apical meristem to flower primordia.
The hormone interacts with the gibberellins. It may
be a specific mixture of other plant hormones and
other still unidentified endogenous growth regulators.

Fluid mosaic model: see Biomembrane.

Fluorescamine: see Amino acid reagents.

Fluoride, *F⁻*: an anion found in bone and tooth
apatite. Small quantities are beneficial in lowering
the incidence of caries, and this cariostatic effect of
F⁻ has been clearly demonstrated in humans. Fluoridation of water is now a public health measure
(optimal level in drinking water 1–2 ppm F⁻). The
role of F⁻ in the inhibition of osteoporosis is less
certain. High levels of F⁻ are toxic (fluorosis). F⁻
affects several enzymes. Excess F⁻ decreases fatty
acid oxidase in rat kidney, and partially inhibits intestinal lipase. Fatty acid utilization is generally impaired in fluorosis. Carbohydrate metabolism is also
affected, probably due to inhibition of enolase and a
shift of the NAD/NADH ratio in favor of NADH.
Cow's milk contains 1–2 μg F⁻ per g. dry weight.
Cereals contain 1–3 μg F⁻/g. Tea (100 μg F⁻/g.) is
especially high in F⁻. Sea foods contain 5–10 μg
F⁻/g.

Fluoroacetic acid: CH₂F-COOH, a very poisonous carboxylic acid which is converted in the tricarboxylic acid cycle to *fluorocitrate,* m. p. 35 °C,
b. p. 165 °C. This is a strong inhibitor of aconitase,
thereby stopping the Tricarboxylic acid cycle (see),
with lethal results. F. a. occurs in free form in the
leaves of *Dichapetalum cymosum,* a poisonous African plant.

Fluorocitric acid: see Fluoroacetic acid.

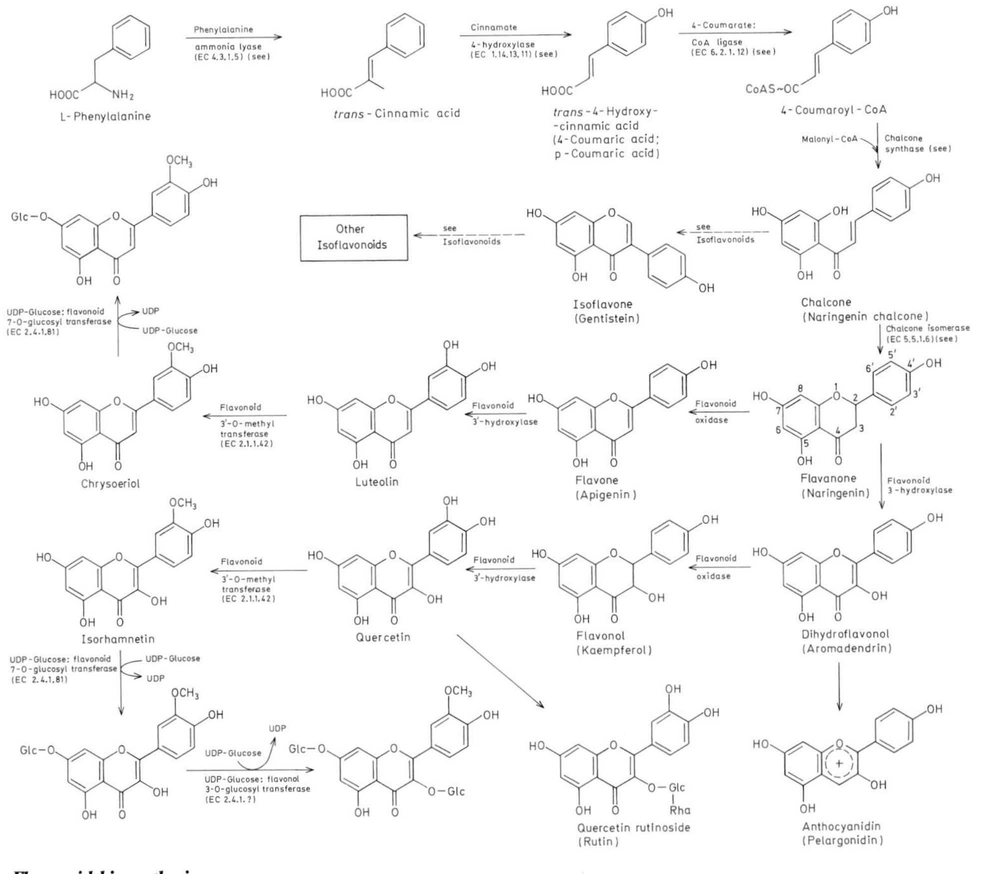

Flavonoid biosynthesis

Flurenol: see Morphactins.

Fly agaric toxins: the toxins of *Amanita muscaria*. Poisonings by this mushroom are rarely fatal. The F.a.t. include Muscarin (see) and other quaternary ammonium bases, such as muscaridin; indole compounds; Ibotenic acid (see) and its easily formed derivatives, Muscimol (see) and Muscazone (see) (Fig.). Muscimol and ibotenic acid inhibit mo-

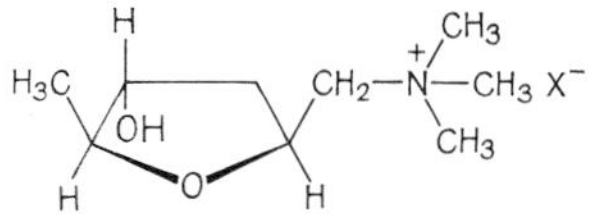

(+)-Muscarine

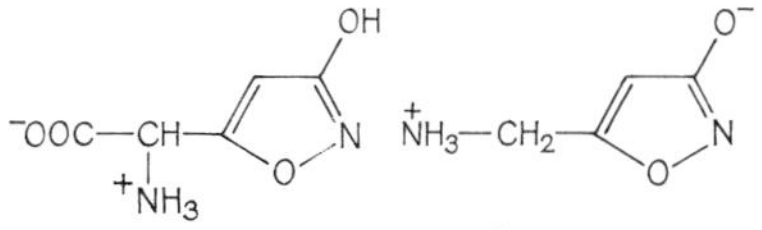

Ibotenic acid   Muscimol

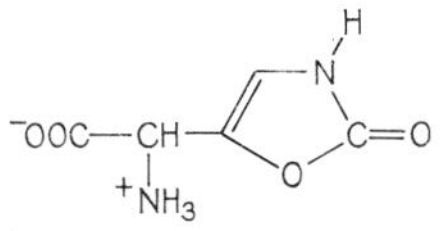

Muscazone

tor functions, and muscimol is psychotropic. This explains the use of the fly agaric as a psychedelic agent in some regions. The fly-killing power long ascribed to this fungus (hence its name) is due to the weak insecticidal action of ibotenic acid and muscimol, which, however, are only effective when consumed by the fly.

FMN: see Flavin mononucleotide.

Folate-H$_2$: see Tetrahydrofolic acid.

Folate-H$_4$: see Tetrahydrofolic acid.

Folic acid: see Vitamins (Vitamin B$_2$ complex).

Follicle-stimulating hormone, *follitropin:* an acidic glycoprotein containing many glutamate, threonine and cysteine residues. It contains 27% carbohydrate which consists chiefly of sialic acid, galactose, mannose and glucosamine, with smaller amounts of galactosamine and fucose. All of the sialic acid is acetylated. The carbohydrate is necessary for biological activity, as removal by enzymes inactivates the hormone. The complete amino acid sequence of follitropin is known. There are two peptide chains. The α chain has 92 amino acids, with carbohydrate attached to asparagine at positions 52 and 78. This chain is almost identical to the α chain of Human chorionic gonadotropin (see), but it differs from the α chain of human luteinizing hormone. The β chain of F. shows microheterogeneity at the *N* and *C* terminals. It is 108 to 115 residues long, and is unique to this hormone.

Follicle-stimulating-hormone releasing hormone: see Releasing hormones.

Follitropin: see Follicle-stimulating hormone.

Forbes' disease: see Glycogen storage disease.

Formaldehyde dehydrogenase, *formaldehyde:NAD$^+$ oxidoreductase (glutathione formylating)* (EC 1.2.1.1): an enzyme catalysing the NAD-dependent

formation of *S*-formylglutathione from glutathione and formaldehyde. This is the first stage in the conversion of formaldehyde to formate (Fig.), which has been demonstrated in beef, chicken, human, monkey and rat liver, human and animal retinas, and in yeast. The second stage is catalysed by *S*-formylglutathione hydrolase. Both enzymes have been purified from rat liver [L. Uotila & M. Koivusalo *J. Biol. Chem.* **249** (1974) 7653–7663, 7664–7672]. Glutathione is also a substrate for Glyoxalase (see), but in the glyoxalase system the substrate is modified by an intramolecular hydride shift, whereas F.d. involves an NAD-linked oxidation (NADP also serves as a cofactor, but NAD is more efficient).

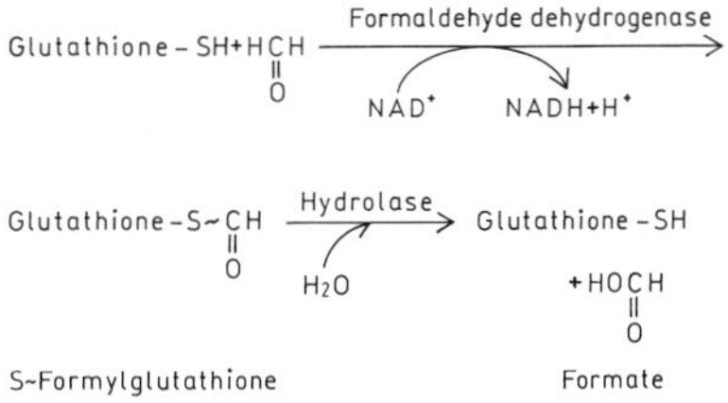

Oxidation of formaldehyde to formate by the coupled action of formaldehyde dehydrogenase and S-formylglutathione hydrolase.

5-Formamidoimidazole-4-carboxamide ribotide: see Purine biosynthesis.

Formaminotransferase deficiency: see Inborn errors of metabolism.

Formononetin: 7-hydroxy-4'-methoxyisoflavone, see Isoflavone.

Formycins: a group of pyrimidine antibiotics (see Nucleoside antibiotics) synthesized by *Nocardia interforma*. Formycin, 3-β-D-ribofuranosyl-7-amino-pyrazolo-(4,3-d)pyrimidine, was first isolated in 1964 by Hori. m.p. 141 to 142 °C. $[\alpha]_D^{20} - 35.5$ ° ($c = 1$ in 0.1 M HCl). The biogenetic precursor is adenosine. F. is converted to formycin 5'-triphosphate, which acts as an ATP analog. *Mycobacterium* and *Xanthomonas oryzae* are particularly sensitive to formycin.

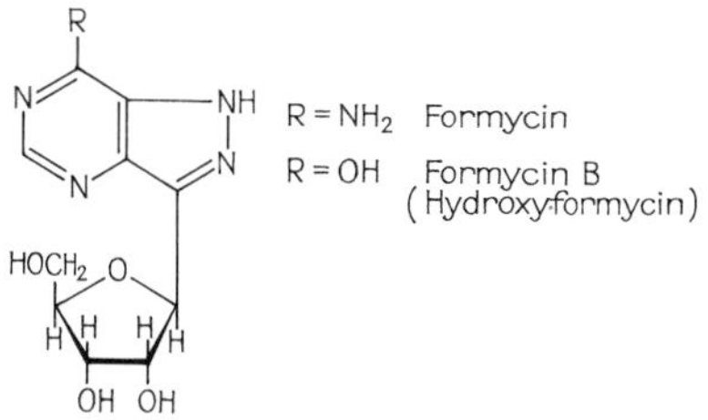

R = NH$_2$ Formycin
R = OH Formycin B (Hydroxyformycin)

Formycin B, hydroxyformycin, is also synthesized by *Streptomyces lavendulae* and *St. roseochromogenes* var. *cyaensis*. m.p. 247 °C (d), $[\alpha]_D^{20} - 51.5$ °C ($c = 1$, water). Formycin B is the deamination product of formycin and is less toxic than the latter. Its effect is limited mostly to *Mycobacterium* and a few viruses.

N-Formylglycinamide ribotide, abb. *FGAR:* an intermediate in Purine biosynthesis (see). M_r 314.2.

N-Formylglycinamidine ribotide: an intermediate in Purine biosynthesis (see).

Formylmethionyl-tRNA: see Initiation tRNA.

N^{10}-Formyltetrahydrofolic acid: see Active one-carbon units.

Fragarin: see Pelargonidin.

Fragment reaction: a reaction used to assay the activity of peptidyl transferase. In a cell-free system containing 70 or 80 S ribosomes, the growing peptide chain is transferred to Puromycin (see) and released as peptidyl-puromycin.

FRH: abb. for follicle-stimulating hormone releasing hormone. See Releasing hormones.

5 α-Friedelane: see Friedelin.

Friedelin: a pentacyclic triterpene ketone. M_r 426.70, m.p. 263 °C. $[α]_D -27.8$ (chloroform). F. is abundant in the bark of the cork oak (1%), in grapefruit rinds and some lichens. The hydrocarbon skeleton of F. is 5α-friedelane.

Fructans: high-molecular-weight polysaccharides of D-fructose linked 1,2- or 1,6-glycosidically. They are common in plants. Examples are inulin and phlein, and the branched triticin, hordecin and graminin.

β-Fructofuranosidase: see Disaccharidases.

D-Fructose, *fruit sugar, levulose:* a ketohexose, M_r 180.16, m.p. 103 to 105 °C (d.), $[α]_D^{20} -135 ° \rightarrow -92 °$ ($c=2$, water). F. tastes sweeter than any other carbohydrate and is fermentable by yeast. It crystallizes as β-pyranose, but forms compounds as furanose (see Carbohydrates). Chemical reduction yields a 1:1 ratio of D-sorbitol and D-mannitol. Its metabolically important derivatives are fructose 1,6-bisphosphate and the 1-phosphate.

F. is found together with glucose and sucrose in many sweet fruits and in honey. It is a component of many oligosaccharides, including sucrose, raffinose, stachyose and gentianose, and of various polysaccharides, such as inulin and levan.

β-D-Fructose   Fructofuranose   Fructopyranose

F. is used as a sweetener by diabetics, because it does not raise the blood sugar level, even in large amounts. *Metabolism:* F. is phosphorylated to fructose 1-phosphate by ketohexokinase (EC 2.7.1.3). Only a small amount is phosphorylated in the 6 position. In the liver, F. 1-phosphate is split into dihydroxyacetone phosphate, which enters Glycolysis (see) directly, and glyceraldehyde, which is phosphorylated either to 2-phosphoglyceric acid (requires ATP and NAD^+) or to glyceraldehyde 3-phosphate. Either of these can enter general carbohydrate metabolism. Because the reactions for F. degradation are different from those for glucose, the two processes can be independently regulated. The liver can also convert F. to glucose, via the sugar alcohol sorbitol.

Fructose 1,6-bisphosphate, *fructose 1,6-diphosphate, Harden-Young ester:* a derivative of fructose in which the OH groups of C atoms 1 and 6 are esterified to phosphoric acid. It is an important metabolic intermediate (see Glycolysis).

Fructose 2,6-bisphosphate: the low M_r stimulator of phosphofructokinase. It has been purified from rat liver.

[Van Schaftingen, E., Hue, L. and Hers, H-G. (1980) *Biochem. J. 192,* 897–901].

Fructose-bisphosphate aldolase, *aldolase* (EC 4.1.2.13): a tetrameric lyase which reversibly cleaves fructose 1,6-bisphosphate into the two triose phosphates dihydroxyacetone phosphate and D-glyceraldehyde phosphate. The reaction is analogous to the aldol condensation $(CH_3CHO + CH_3CHO \rightarrow CH_3-CHOH-CH_2-CHO)$, hence the name of the enzyme. The equilibrium concentrations lie at 89% fructose bisphosphate and 11% triose phosphate. The enzyme catalyses the condensation of a number of aldehydes with dihydroxyacetone phosphate, and can also cleave fructose 1-phosphate. Liver aldolase (aldolase B, M_r 156000, 4 subunits of M_r 39000) cleaves fructose 1,6-bisphosphate and fructose 1-phosphate at nearly the same rate. Muscle aldolase (aldolase) A, M_r 160000, 4 subunits of 41000, IP 6,1), however, is more active with the bisphosphate. The aldolase from yeast is inhibited by cysteine. Fe^{2+}, Zn^{2+} and Co^{2+} ions lead to reactivation. The aldolase from spinach leaves has a M_r of only 120000 (M_r of the subunits, 30000).

Of the animal and human organs, skeletal muscle has the highest aldolase activity, 5 times that in brain, liver and heart muscle. For this reason the determination of aldolase in the serum is of diagnostic value in muscle diseases such as myoglobinuria, or progressive muscular dystrophy.

Fructose intolerance: see Inborn errors of metabolism.

Fructose 6-phosphate, *Neuberg ester:* a phosphoric acid ester of fructose. F. is an intermediate in Glycolysis (see) produced by isomerization of glucose 6-phosphate. It can also be produced by transketolation from erythrose 4-phosphate.

β-h-Fructosidase: see Invertase.

Fructosuria: see Inborn errors of metabolism.

Fruit ripening hormone: see Ethylene.

Fruit sugar: see D-Fructose.

FSH: see Follicle-stimulating hormone.

L-Fucose: 6-deoxy-L-galactose, M_r 164, m.p. (α-form) 140 °C, $[α]_D^{20} -153 ° \rightarrow -76 °$ ($c=9$). L-F. is a component of the blood-group substances A, B and O and of various oligosaccharides in human milk, sea weed, plant mucilages. It is also found in assorted glycosides and antibiotics. Some of the latter also contain D-fucose. L-F. is synthesized as the activated derivative guanosine diphosphate fucose

L-Fucose

rom GDP-D-mannose by dehydrogenation, isomerization and reduction.

Fucosidosis: see Inborn errors of metabolism.

Fucosterol: (24E)-stigmasta-5,24(28)-dien-3β-ol, a phytosterol (see Sterols). M_r 412.67, m.p. 124 °C, $[\alpha]_D^{20} -38.4$ ° (chloroform). F. is characteristic of marine brown algae and has also been isolated from resh-water algae.

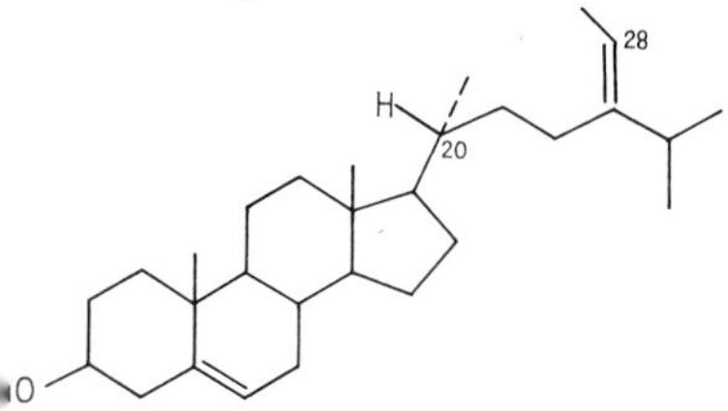

ucosterol

Fucoxanthin: a carotenoid pigment with an allene, an epoxy and a carbonyl group, and three hydroxyl groups, one of them acetylated. M_r 658.88, m.p. 160 °C, $[\alpha]_D^{18} +72.5$ ° ± 9 ° (chloroform). The configurations of the chiral centers are 3 S, 5 R, 6 S, 3′ S, 5′ R, 6′ R. F. is found in many marine algae, especially brown algae *(Phaeophyta)*, and it is the most abundant of the naturally occurring carotenoids.

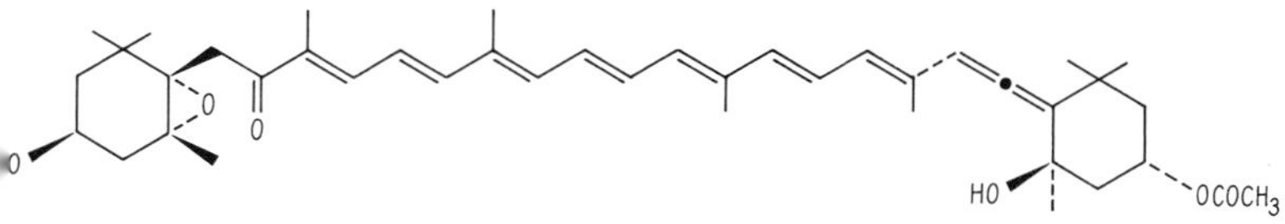

ucoxanthin

Fugu poison: see Tetrodotoxin.

Fumarase, *fumarate hydratase* (EC 4.2.1.2): the tricarboxylic acid cycle enzyme which converts fumarate to malate by adding water to the double bond. The reaction is reversible. In contrast to other hydrolyases, which require either pyridoxal phosphate or metal ions as cofactors, F. has no cofactor requirement. It exists as a number of isoenzymes. F. (M_r 194000, 1784 amino acids) is a tetramer of identical subunits (M_r 48500). It is noteworthy that the molecule contains no disulfide bridges.

Fumaric acid: *trans*-ethylene dicarboxylic acid. m.p. 286–287 °C in closed tube; sublimes at 200 °C F.a. was isolated from mushrooms in 1810 and from *Fumaria officinalis* in 1833. It occurs widely in the free form in plants. F.a. is an intermediate in the Tricarboxylic acid cycle (see), and is the form in which the carbon skeletons of aspartate (see Urea cycle), phenylalanine and tyrosine (via fumarylacetoacetate) are fed into the cycle. The *cis*-isomer is maleic acid.

umaric acid

Fumigatin: see Benzoquinones.

Fundamental variable: see Concentration variable.

Funtumia alkaloids: a group of steroid alkaloids characteristic of the genus *Funtumia* of the dogbane *(Apocynaceae)* family. The F.a. are derived from pregnane (see Steroids) and have an amino or methylamino group on carbon 3 and/or 20. The most important representatives are *Funtumine* (3α-amino-5α-pregnan-20-one) and *funtumidine* (3α-amino-5α-pegnen-20α-ol) from *Funtumia latifolia*. The F.a. are biosynthesized from cholesterol and pregnenolone.

Funtumidine: see Funtumia alkaloids.

Funtumine: see Funtumia alkaloids.

Furanoses: see Carbohydrates.

Fusel oil: unpleasant-tasting side product of alcoholic fermentation. F.o. consists mainly of amyl, isoamyl, isobutyl and propyl alcohols. The compounds are formed from amino acids, especially leucine, isoleucine and tyrosine, by deamination and decarboxylation. Tyrosol, which is formed from tyrosine, is a component of beer.

Fusicoccin, *fusicoccin A:* the major toxin isolated from culture filtrates of *Fusicoccum amygdali*, a pathogenic fungus responsible for a wilt disease of almond and peach. It is thought that F. specifically activates a single central transport system (possibly by interacting with the plasmalemma ATPase, thereby stimulating the conversion of phosphate bond energy into proton gradient energy) and that all other effects of the toxin are consequences of this fundamental process. F. has no effect on fungi, bacteria or animals. In higher plants, it generally causes cell enlargement, proton efflux, K^+ efflux and stomatal opening. It also promotes seed germination in antagonism with abscisic acid.

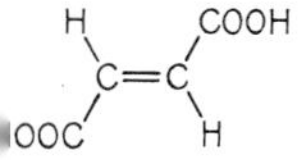

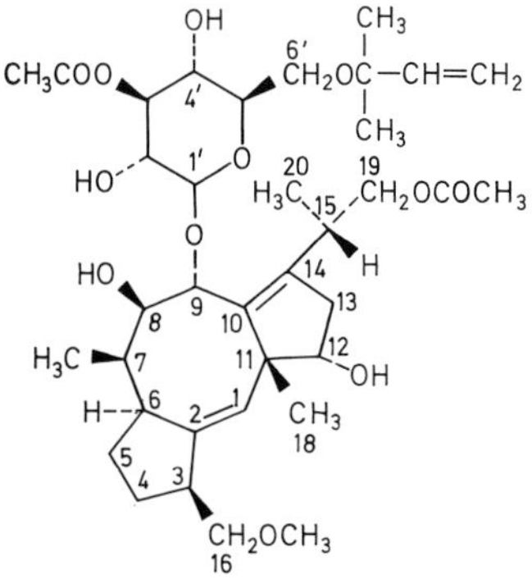

Fusicoccin [K.D. Barrow et al. *Chem. Commun.* No 19 (1968) 1198–1200]

There are two series of related fusicoccins, all of which are present as co-metabolites in culture filtrates of *Fusicoccum amygdali:* 1.fusicoccin A and 10 cometabolites differing only in the number and position of acetyl groups (probably resulting from nonenzymic, in vitro migration of acetyl groups), and 2.19-deoxyfusicoccins.

Labeling studies show that the fusicoccins are true diterpenes (note that the structurally similar ophiobolanes are sesterterpenes). Three of four possible 4-*pro-R* hydrogens of mevalonic acid are incorporated into the aglycon, including one at C-6 and one at C-15, but none at C-3. One of the H-atoms at position 2 of mevalonic acid is incorporated at C-8 (cf. ophiobolin, where this hydrogen migrates to C-15). Six out of eight possible hydrogens at position 5 of mevalonic acid are incorporated into the aglycon, two of them at C-9 and C-13. Together with labeling patterns from [13]C- and ^{14}C-labeled mevalonic acid, these results are consistent only with synthesis via all-*trans*-geranylgeranylpyrophosphate. Final ring closure by two consecutive 1,2 hydride shifts was elegantly demonstrated by the incorporation of [3-^{13}C, 4-[2]H$_2$]-(3RS)mevalonolactone; NMR analysis of the resulting F.showed that the signals due to [13]C-7 and [13]C-15 were strongly depressed, owing to the presence of ^{2}H on each of them.

The fusicoccins and Cotylenins (see) constitute a class of compounds with no other known, naturally occurring representatives. The glycosidic sugar residue is also uncommon in fungal metabolites. [E. Marrè (1979) *Ann. Rev. Plant Physiol.* **30**, 273–278 (mode of action and physiology); A. Banerji et al. (1978) *J. Chem. Soc. Chem. Comm.,* 843–845 (biosynthesis)]

Fusidic acid: a tetracyclic triterpene antibiotic. M_r 516.69, m.p. 192 °C, $[\alpha]_D^{20}$ −9 ° (chloroform). F.a. is isolated from culture filtrates of *Fusidium coccineum* and, like the structurally related antibiotics Cephalosporin P$_1$ (see) and helvolic acid, it is effective against Gram-positive organisms. F.a. is biosynthesized from squalene via 2,3-epoxysqualene. It inhibits protein synthesis by preventing the reaction of the elongation facor EFG with the small ribosomal subunit.

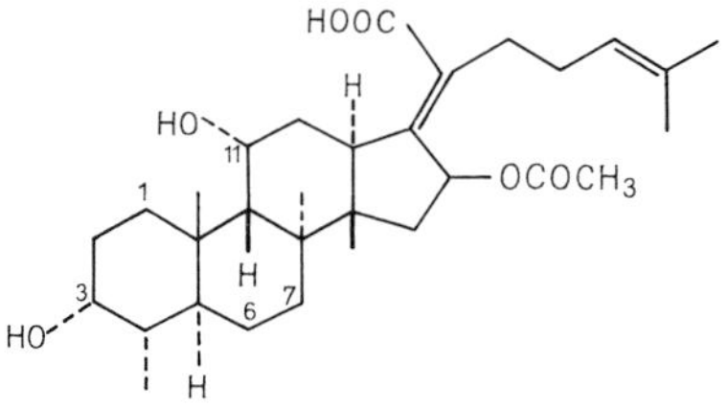

Fusidic acid

Futile cycle, *substrate cycle:* a sequence of metabolic reactions which, in sum, do nothing but degrade ATP or another energy-providing molecule. An example is provided by 6-phosphofructokinase (EC 2.7.1.11), which phosphorylates fructose 6-phosphate to fructose 1,6-bisphosphate (consuming ATP); and fructose-bisphosphatase (EC 3.1.3.11), which removes the 1-phosphate group from fructose 1,6-bisphosphate. The enzymes of Glycolysis (see) and Gluconeogenesis (see) provide another example. There are many others. It is believed that F.c. do not ordinarily consume a large amount of cellular energy, because the enzymes of the opposing reactions are under tight Metabolic control (see), so that the overall reaction runs in only one direction at any given time. However, a measurable rate of F.c. such as the fructose-phosphate and the pyruvate → oxaloacetate → PEP → pyruvate cycle is observed in vivo in liver. F.c. also provide heat in thermogenic tissues such as brown fat and insect thoracic muscle.

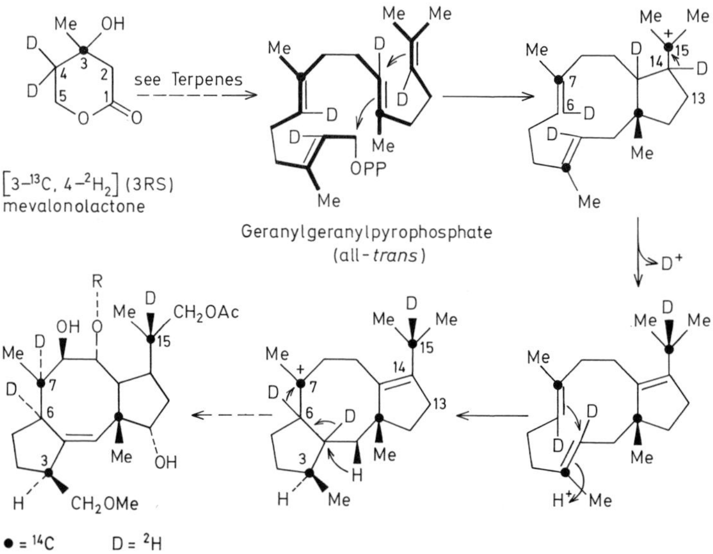

$[3-^{13}C, 4-^2H_2]$ (3RS)
mevalonolactone

Geranylgeranylpyrophosphate
(all-*trans*)

$\bullet = {}^{14}C \qquad D = {}^2H$

Biosynthesis of the aglycon ring system of the fusicoccins

G

G: see Guanine.

GA$_3$: see Gibberellins.

GABA: see 4-Aminobutvric acid.

Gadoleinic acid, ***Δ^9-eicosenoic acid:*** CH_3-$(CH_2)_9$-$CH=CH$-$(CH_2)_7$-COOH, a fatty acid. M_r 310.5, m.p. 39 °C. G.a. is found as a component of acylglycerols in plant and fish oils, and of phosphatides.

Galactans: polysaccharides of D-galactose found in plants. They are usually unbranched and have high molecular weights. Examples are agar-agar and carrageenan.

Galactitol: see Dulcitol.

Galactokinase deficiency: see Inborn errors of metabolism.

D-Galactosamine, ***chondrosamine, 2-amino-2-deoxy-D-galactose:*** an amino sugar. M_r 179.17, m.p. of the hydrochloride, 185 °C $[\alpha]_D^{20}+125° \rightarrow +98°$ (water). G. is derived from D-galactose by replacement of the OH group on C 2 by an amino group. It is usually found in nature as the *N*-acetyl derivative and is a component of a few mucopolysaccharides such as chondroitin sulfate, blood group substance A, etc. It is also found in mucoproteins.

Galactose epimerase deficiency: see Inborn errors of metabolism.

Galactose: an aldohexose occurring naturally in D and L-forms. M_r 180.16, D-galactose, m.p. 167 °C, α-form, $[\alpha]_D^{20}+151° \rightarrow +80°$ (water); β-form, $[\alpha]_D^{20}+53° \rightarrow +80°$ (water). L-Galactose, m.p. 165 °C. G. is especially widespread in animals and is a component of oligosaccharides, such as lactose, and of the cerebrosides and gangliosides of nervous tissues. In plants, it is a component of melibiose, raffinose and stachyose, and of the Galactans (see); it is also the sugar component of some glycosides.

Metabolism: G. is synthesized as uridine diphosphate (UDP)-G. from UDP-glucose. The epimerization on C 4 is catalysed by UDPglucose 4-epimerase (EC 5.1.3.2). This reaction is reversible and the degradation of UDPgalactose occurs via UDPglucose. G. is fed into general glucose metabolism by the pathway shown in the figure. Galactose 1-phosphate can also react directly with UTP to form UDPgalactose.

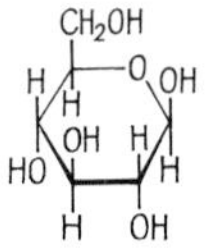

α-D-Galactose

Galactosemia: see Inborn errors of metabolism.

β-**Galactosidase** (EC 3.2.1.23): a disaccharidase, which hydrolyses lactose to galactose and glucose (Fig. 1). The lactose operon (*lac* operon) of *Escherichia coli* contains structural genes for β-G., galactoside permease and thiogalactoside transacetylase. Induction of transcription of the *lac* operon (manifested as induction of the synthesis of β-G. and the other two enzymes) permits the bacterium to use lactose as its sole source of carbon. The true inducer is 1,6-allolactose, which is formed from lactose by transglycosylation (Fig. 1). Nonmetabolizable or gratuitous inducers of the *lac* operon are also known, e.g. isopropylthiogalactoside (Fig. 2). Classical studies on the induction of β-G. led Jacob and Monod to propose the operon model (see Operon) for the regulation of protein synthesis. β-G. activity is conveniently measured with the colorless substrate, *o*-nitrophenyl-β-D-galactoside, which is hydrolysed to galactose and the colored product *o*-nitrophenol. The latter can be determined spectrophotometrically. See Enzyme induction; Enzyme repression. [J. H.

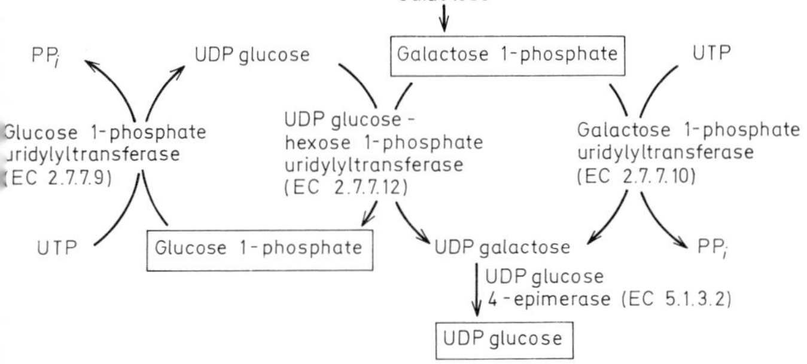

Relationship between galactose and glucose metabolism

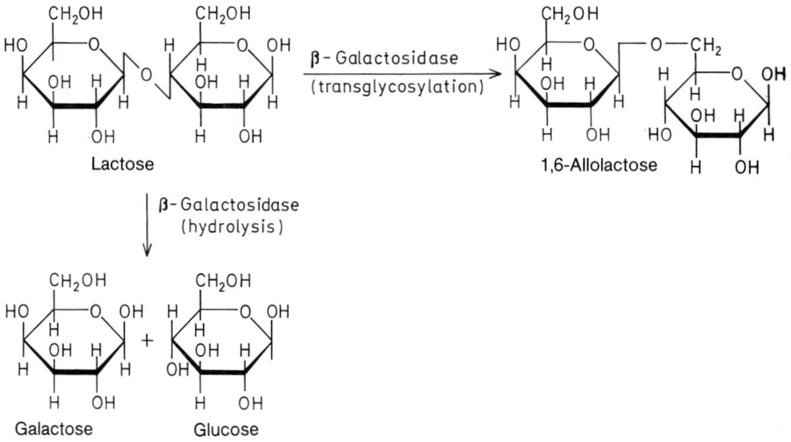

Figure 1. Hydrolysis of lactose to galactose and glucose, and transglycosylation of lactose to 1,6-allolactose. Both reactions are catalysed by β-galactosidase.

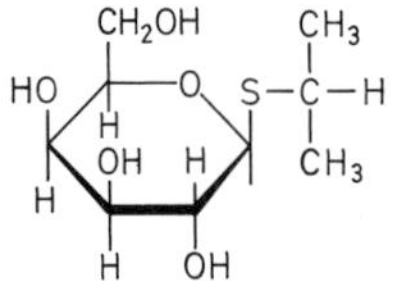

Figure 2. Isopropylthiogalactoside, a gratuitous inducer of β-galactosidase.

Miller, *Experiments in Molecular Genetics* (Cold Spring Harbor, New York, 1972)]

Galactosuria: see Inborn errors of metabolism.

Galactosylceramide lipidosis: see Inborn errors of metabolism.

D-Galacturonic acid: a uronic acid derived from D-galactose. M_r 194.14, m.p. 159 °C, α-form $[\alpha]_D^{20} +98°\rightarrow 50.8°$ (water), β-form $[\alpha]_D^{20} +27°\rightarrow 50.8°$ (water). Pectins contain 40 to 60% G.a.; the compound is a component of a few other plant polysaccharides as well.

Galangin: see Flavones (table).

Galanthamine: see Amaryllidaceae alkaloids.

Galegine, *(3-methyl-2-butenyl)guanidine:* a guanidine derivative found together with 4-hydroxygalegine in the seeds of goat's rue, *Galega officinalis.* G. is synthesized in the shoot and accumulated in the seeds. Its isoprenoid hydrocarbon chain does not derive from the mevalonic acid-isopentenyl pyrophosphate sequence of terpenoid biosynthesis. The guanidino group is added by a Transamidination (see).

Gallamine: see Acetylcholine.

Gallic acid: see Tannins.

Gamete attractants: see Sexual attractants.

Gamones: plant Sexual attractants (see).

Ganglioside: see Glycolipids; Inborn errors of metabolism (Sphingolipidoses).

Gangliosidosis: see Inborn errors of metabolism.

GAR: see Glycinamide ribotide.

Gas chromatography: a separation technique based on the distribution of gaseous compounds between a mobile gaseous phase and a stationary ad-sorbent phase. The method was developed by Martin and James in 1952 [A.J.P. Martin in R.Porter, ed. *Gas Chromatography in Biology and Medicine. A CIBA Foundation Symposium* (J.&A. Churchill, Ltd., London, 1969)].

The GC system consists of five components: 1) the carrier gas supply, 2) the sample injection system, 3) the column and oven, 4) the detector and 5) signal processing and control electronics.

The carrier gas must be chemically inert with respect both to the sample and the stationary phase, and must not interfere with the detector function. H_2, He, N_2, O_2, Ar and CO_2 are commonly used. The sample must be vaporized before it enters the column. Thus for solid or liquid samples, the inlet is kept at a high temperature in order to vaporize the entire sample instantly. Liquid and solid samples are commonly introduced in solution; in this case the solvent must be carefully chosen so that it does not react chemically with either the sample or the stationary phase of the column at high temperature. It must also leave the column at a different time from the sample and not overload the column. If the sample is not volatile, it may be converted to a volatile chemical derivative, e.g. by reaction with trimethylchlorosilane or hexamethyldisilazane.

The column may be made of glass or metal, and it is either filled with a porous packing or it may be a capillary. Packed columns have inner diameters 2 to 6 mm, and lengths of 1-3 m. Capillaries are 0.2-0.5 mm in inner diameter, and 10-100 m long. Materials to be used as stationary phase must be chemically inert, thermally stable and non-volatile. In gas-liquid chromatography (GLC), the liquid stationary phase is either adsorbed to the packing or to the capillary walls. Some comon liquids used as stationary phase are Apiezon, SE 30 (methyl silicone), Carbowax 20 M, and mixtures of phenyl methyl silicone or trifluoropropyl silicone. In gas-solid chromatography (GSC), the packing is an active solid material, such as alumina, silica gel, activated carbon or zeolites. These packing materials can greatly increase retention times, and thus make possible the separation of very volatile substances which move too quickly through a GLC column.

The separation process involves an equilibrium partition of the solute between the stationary and mobile phases. The partition coefficient depends on the solute vapor pressure and is a function of temperature: $\ln(p) = \Delta H/RT + C$, where p is the solute vapor pressure, ΔH is the molar heat of solution and C is a constant. The higher the temperature, the less the solute is retained by the stationary phase and thus, the more quickly it leaves the column. This fact is utilized to drive compounds with longer retention times out of the column by gradually increasing the temperature.

Detection is based on a change in a physical property of the gas emerging from the column. The signal must be proportional to the amount of material eluted. For routine analyses, a flame ionization detector is used. It consists of a small H_2-air flame burning at a metal jet. The eluant gas mixes with the H_2 before the latter emerges from the jet. The combustion of H_2 produces free radicals, but no ions; however, the combustion of carbon produces positive ions in the flame as well as free radicals. The ions travel to the collector electrode, where they are measured by the current they induce.

GC is of enormous value for the separation of small to medium sized molecules and is routinely used in clinical chemistry, e.g. to detect drugs or anaesthetics in serum. [A. Braithwaite & F.J. Smith *Chromatographic Methods* (4th ed.) (Chapman & Hall, London, 1985)]

Gastric inhibitory peptide, *GIP:* a polypeptide hormone (for structure, see Secretin) purified from crude preparations of Cholecystokinin (see). GIP has potent enterogastrone activity, i.e. it inhibits secretion of acid and pepsin by the stomach, and inhibits gastric motility. It possesses no significant secretin or cholecystokinin activity. [J.C. Brown & J.R. Dryburgh, *Canad. J. Biochem.* 49, (1971) 867–872]

Gastrin: a hexadecapeptide hormone from the gastric antrum. Human G.I is Pyr-Gly-Pro-Trp-Leu-Glu]₅-Ala-Gly-Tyr-Trp-Met-Asp-Phe-NH₂, M_r 2116. Human G.II has an additional sulfate group on Tyr 12. Leu 5 is replaced by Met in porcine gastrin and by Val in the sheep and bovine hormones. In bovine gastrin, there is also an Ala instead of Glu in position 10. G.is homologous to the C-terminal end of cholecystokinin. The protected C-terminal peptide, *tetragastrin* (Trp-Met-Asp-Phe-NH₂), is used to test secretion of digestive juice. A biologically active *big G.* with 34 amino acids can be detected by radioimmunology.

G.is formed in the antrum mucosa in response to alkaline pH, mechanical stimulation and vagus (acetylcholine) stimulation. Acidic stomach juice inhibits its secretion. The presence of G.in the blood stimulates the stomach mucosa to produce and secrete hydrochloric acid and the pancreas to secrete its digestive enzymes. Cholecystokinin is a competitive inhibitor of G. G.can be detected by radioimmunoassay.

Gaucher's disease: see Inborn errors of metabolism.

Gay-Lussac equation: see Alcoholic fermentation.

Gazaniaxanthin: see Rubixanthin.

GC content: the amount of guanine + cytosine in nucleic acids, expressed in mol % of the total bases. The GC content plus the AT (adenine + thymine) content of a nucleic acid is thus equal to 100 mole %, provided the molecule is double-stranded. The GC contents for higher organisms lie between 28 and 58%, while those for prokaryotes range from 22 to 74%. The density of a double-stranded molecular species depends on its GC content, as does its melting temperature, which increases almost linearly with increasing GC content. The flexibility of the DNA helix is also a function of GC content: poly(dG)·poly(dC) is much stiffer than random-sequence DNA, which in turn is stiffer than poly(dA)·poly(dT). This applies both to torsional and bending flexibility, and these differences are probably important in the supercoiling of the helices in Chromosomes (see). [M. Hogan, J. LeGrange & B. Austin, *Nature* **304** (1983) 752–754.]

Methylation of DNA (see) is a feature of many vertebrate genomes. There is some evidence that the sequence mCpG is susceptible to deamination of the methylcytosine to thymidine, so that the bulk DNA of vertebrates is has 4–5-fold fewer CpG than would be expected if the distribution were random; it is correspondingly enriched in TpG. This mechanism must contribute to the species differences in overall GC content. [A. Bird et al., *Cell* **40** (1985) 91–99].

GDP: abb. for Guanosine 5'-diphosphate. See Guanosine phosphates.

Gelatins: proteins obtained by extraction from tissues rich in Collagen (see).

Gel filtration: see Proteins.

Gene: a section of DNA coding for a single polypeptide chain (structural gene), a particular species of transfer or ribosomal RNA, or a sequence which is recognized by and interacts with regulator proteins (regulatory gene). A structural G. is the same as a *cistron*. In prokaryotes, several cistrons often comprise a single regulatory unit, the Operon (see). In eukaryotes, nucleotide sequences which encode amino acid sequences (exons) are often interrupted by sequences which do not code for amino acid sequence (introns) and which are excised during Post-transcriptional modification of RNA (see). Such G.are sometimes called "split G."

Gene activation, *gene expression:* No organism continuously synthesizes all the proteins encoded in its genome. Even in prokaryotes, a cell (clone) may undergo many division cycles without producing the enzymes for catabolic or anabolic pathways it does not need, and other enzymes are made only at specific times during the cell cycle. In multicellular eukaryotes, the situation is further complicated by the existence of genes which are expressed only in specific tissues or at specific stages of development. In both types of cell, the rate of transcription of a gene may vary widely in response to environmental conditions. G.a. in prokaryotes is achieved largely by the binding of specific proteins (repressers or activaters) to the DNA adjacent to the site of initiation of transcription (the promoter). See Operon.

In eukaryotes, chromosomal DNA is associated with a large number of proteins. The most abundant are the histones, very basic proteins of which there

Gene amplification

are five types: H1 - H5. The DNA double helix is wound around a core structure formed by histones H2 - H5; about 200 base pairs wrapped around such a core comprise a *nucleosome*. The H1 histones are associated with the DNA strand between nucleosomes. There is evidence that the nucleosomes play a role in G.a., in that RNA polymerase evidently cannot bind to DNA within the nucleosome. Thus if the initiation site of a gene is buried in a nucleosome, the gene cannot be transcribed.

Although operons have not been demonstrated in eukaryotic genomes, the expression of individual genes may be subject to a similar kind of regulation. Eukaryotic genes, like prokaryotic genes have promoters, which are untranslated stretches of DNA adjacent to the structural gene. These are involved in the binding of RNA polymerase to the DNA, and probably in causing it to locate the exact site at which transcription begins: the initiation site. In bacteria, specific proteins bound to a promoter can cause the associated gene (or operon) either to be transcribed or not transcribed; it seems very likely that such activator and represser proteins also exist for eukaryotic genes. In fact the sites at which hormone-receptor complexes bind to DNA have been shown, for the prolactin gene family, to be located just "upstream" (on the 5'-side) of the structural gene.

In addition to their promoters, the genes of some viruses which infect eukaryotic cells are associated with *enhancers*. These can operate over long distances, enhancing transcription from a promoter up to 3000 base pairs away, either upstream or downstream from the enhancer. They are effective when inserted into the genome in either the $5' \rightarrow 3'$ or the $3' \rightarrow 5'$ orientation relative to the direction of transcription. Enhancers have tissue and species specificity, but there is a conserved core sequence, GGTGTGG$\frac{A}{T}\frac{A}{T}\frac{A}{T}$G. Homologous structures have been found in the introns of immunoglobulin genes, and may exist in other cellular genes as well.

Another form of regulation of G.a. was discovered in yeast, which normally contain genes for both of the two mating types α and *a*. In vegetatively reproducing cells, neither mating type is expressed, but under certain circumstances, either the Matα or the Mat*a* genes (there are two α and two *a* genes) are moved to a different location near the center of the chromosome which carries both sets of genes, and in this position, they are expressed. The nontranslocated genes remain unexpressed. The flanking sequences of the genes are the same in both positions, out to 800 bp downstream (on the 3' end) from the structural genes. It was found that repression of the α genes in their non-active position requires activity of another gene, SIR, whose product presumably binds to the DNA at least 800 bp downstream from the gene. The mechanism by which a protein bound to the DNA at such a distance from the structural gene can prevent transcription is a mystery; one suggestion is that it causes the arrangement of the nucleosomes to shift in such a way that the initiation site of the gene is not available for RNA polymerase. [K. A. Nasmyth et al. *Nature* **289** (1981) 245–250]

The product of the Matα 2 gene of yeast, the Homeobox (see) sequence in *Drosophila* and homologous sequences in other organisms appear to function as control elements, similar to bacterial activators, for scattered genes which are induced simultaneously in the course of development. It is presumed that the genes induced by the products of the homeotic genes have sequences analogous to bacterial promoters or possibly to viral enhancers where these regulatory proteins can bind and stimulate transcription.

Mechanisms for transposing genes on a chromosome exist in many types of eukaryotic cell. For immunoglobulin and T-cell antigen receptor molecules, as for yeast mating type genes, rearrangement of the genes is a necessary step in activation of the respective lymphoid clones. How widespread this mechanism of G.a. may be remains to be discovered.

Finally, the methylation of DNA appears to be related to G.a. in eukaryotes. It is found that stretches of DNA in which a large proportion of the thymidine residues are methylated are not transcribed; genes which are methylated (and inactive) in one tissue can be non-methylated and expressed in another tisssue.

A brief introduction to this field is given by J.D. Hawkins in *Gene Structure and Expression* (Cambridge University Press, Cambridge, 1985). The classic book by Benjamin Lewin, *Gene Expression* (Wiley, New York) is now available in a second edition (1980). See also A. Kumar, ed. *Eukaryotic Gene Expression* (Plenum Press, New York, 1984).

Gene amplification: production of extrachromosomal copies of the genes for ribosomal RNA. In frog's egg cells, G.a. leads to the formation of numerous extrachromosomal nucleoli. G.a. is thus a special regulatory mechanism for RNA. It is not the same as Redundancy (see), which is the presence of multiple copies of the same gene on the chromosome.

Gene bank: see Recombinant DNA technology.

Gene cloning: see Recombinant DNA technology.

Geneserine: see Physostigmine.

Gene synthesis, *cell-free gene synthesis:* The term includes cell-free replication of a bacteriophage genome, which was first achieved in 1967 by Goulian, Kornberg and Sinsheimer with ΦX-174. Khorana and coworkers later synthesized the gene for alanine tRNA by a combination of chemical and enzymatic methods. They first synthesized by chemical means a series of polynucleotides of 8 to 12 bases. The free hydroxyl groups at the 5' ends were then phosphorylated with polynucleotide kinase. Taken together, the sequences of these segments made up both strands of the DNA for the gene. Furthermore, they were designed so that each segment of one strand overlapped the ends of two segments on the other strand, each overlap extending for at least five bases. The double-stranded DNA was first organized, segment by segment, by hybridization; then the breaks in the strands were fused by a polynucleotide ligase. Using a similar approach, H. Köster et al. reported the total synthesis of the structural gene for the octapeptide angiotensin II (M_r 21 800) in 1978. The gene for the peptide ormone Somatostatin (see) was synthesized in 1977 by chemical means. It was joined to the *E. coli* β-galactosidase gene in the plasmid pBR322 and introduced into the bacteria. The product was a chimeric polypeptide containing the amino-acid sequence of omatostatin. When subjected to cyanogen bromide cleavage in vitro, the polypeptide yielded biological-y active somatostatin. Later the hormone Insulin see) was synthesized analogously.

Genetic code: the rules for translation of base equences in nucleic acids into amino acid sequences in polypeptides (see Protein synthesis). The nucleic acid bases are read off as triplets. There are four bases, and thus 64 (4^3) different permutations "words"). As there are only 20 amino acids specified by the code, many triplets can be, and are, redundant. The table lists the amino acids by triplet, or *codon*.

	U	C	A	G	
U	Phe	Ser	Tyr	Cys	U
	Phe	Ser	Tyr	Cys	C
	Leu	Ser	term	term	A
	Leu	Ser	term	Trp	G
C	Leu	Pro	His	Arg	U
	Leu	Pro	His	Arg	C
	Leu	Pro	Gln	Arg	A
	Leu	Pro	Gln	Arg	G
A	Ile	Thr	Asn	Ser	U
	Ile	Thr	Asn	Ser	C
	Ile	Thr	Lys	Arg	A
	Met	Thr	Lys	Arg	G
G	Val	Ala	Asp	Gly	U
	Val	Ala	Asp	Gly	C
	Val	Ala	Glu	Gly	A
	Val	Ala	Glu	Gly	G

Three of the codons ("term" in the table) cause translation to stop (see Termination codon). Initiation of the chain is indicated by longer sequences which are recognized by the initiation factors (see Protein biosynthesis).

The code has no commas. The "reading frame", the way in which a sequence is divided into triplets, is determined by the precise point at which translation is initiated. For example, the sequence CATCATCAT could be read CAT CAT CAT, C ATC ATC AT, or CA TCA TCA T in the three possible reading frames. A mutation which changes the reading frame of a gene, e.g. the deletion of one or two nucleotides, is called a *frameshift* mutation. The code is normally not overlapping, but in the bacteriophage ΦX 174, two genes have been found which are entirely contained within other genes. These are translated in different reading frames, so the amino acid sequences of the proteins encoded by them are different.

Historical: The G.c. was cracked between 1961 and 1963, primarily by Khorana, Matthaei, Nirenberg and Ochoa, by the analysis of the translation products of synthetic copolymers of pyrimidine and purine nucleotides. Brenner and Crick showed that defined chemical changes in bacteriophage DNA produced certain types of mutations, which were in accordance with the proposed code. Comparison of the primary sequences of viral coat proteins (Wittmann) and bacterial lysozymes (Streisinger) from wild types and mutants confirmed the code. In 1954, Gamov tried to explain the G.c. by steric fitting of nucleotides and amino acids. In 1957, Crick suggested his Adaptor hypothesis (see), which was later confirmed experimentally.

Genetic code in mitochondrial mRNA: In mitochondrial mRNA, some of the codons are different from those listed in the table, which was established for cytoplasmic and bacterial mRNA. For example, in human mitochondrial mRNA, AGA and AGG are used as termination codons. Other differences are as follows:

Some codons in mitochondrial mRNA

mRNA	5′ CAA	AUA	CUA	UGA	3′
Yeast mitochondria	Gly	Ile	Thr	Trp	
Human mitochondria	Gly	Met	Leu	Trp	
Neurospora mitochondria	Gly	Ile	Leu	Trp	

In addition, mitochondrial systems also use fewer species of tRNA (maximum 24) than nonmitochondrial systems (estimated 32–40). This would appear to be the result of a simpler decoding strategy in mitochondria, where tRNA with U in the wobble position pairs with any of the four bases (see Wobble hypothesis). Thus the eight amino acids which are encoded by at least four codons each can be served by only eight tRNAs.

Genetic code in ciliates: The G.C. in ciliates is similar to that described in other eukaryotes, except that UAA and UAG code for glutamine, and do not act as stop signals. Comparison of gene structure and protein sequence shows that TAA and TAG triplets in DNA (noncoding strand) correspond to glutamine in the surface antigen protein of *Paramecium*, α-tubulin of *Stylonchia*, and a histone of *Tetrahymena*. In ciliate mRNA, the only termination codon appears to be UGA. [F. Caron & E. Meyer, *Nature* **314** (1985) 185–188; J. R. Preer et al., *Nature* **314** (1985) 188–190]

Genetic engineering: see Recombinant DNA technology.

Genetic material: the carrier of hereditary information. In higher organisms, bacteria and some virus, it is double-stranded DNA; in other virus it is single-stranded DNA, and in still other virus, it is RNA. See Deoxyribonucleic acid; Chromosomes.

Genin: see Aglycon.

Genistein: 5,7,4′-trihydroxyisoflavone, see Isoflavone.

Genistin: genistein 7-glucoside, see Isoflavone.

Genome: the sum of all chromosomal genes in a haploid cell (including prokaryotes) or the haploid set of chromosomes in an eukaryotic cell.

Genomic library: see Recombinant DNA technology.

Genotype: the sum of an individual's genes, both dominant and recessive.

Gentamycin: see Streptomycin.

Gentiana alkaloids: terpene alkaloids with a pyridine skeleton (they therefore are also pyridine alkaloids) found primarily in gentian *(Gentiana)* species. The biogenetic precursors of G. a. are probably bicyclic monoterpenes. This assumption is strengthened by the fact that Gentianine (see) can easily be obtained from gentiopicroside in the presence of ammonium ions (Fig.).

Gentianic acid: see Gentianine, 2.

Gentianine: 1. a terpene and pyridine alkaloid, M_r 175.18, m.p. 82 to 83 °C. G. occurs in many plants of the gentian family *(Gentianaceae).* Its existence was long disputed, because it is easily formed in the presence of ammonium ions from *gentiopicroside,* a glucoside found in the roots of many of these plants, M_r 356.32, m.p. (anhydrous) 191 °C or (hemihydrate) 121 °C, $[\alpha]_D^{20} - 199$ ° (ethanol).
2. *Gentisin, gentianic acid, 1,7-dihydroxy-3-methoxy-9H-xanthene-9-one,* a yellow pigment from the yellow gentian *(Gentiana lutea).* M_r 258.22, m.p. 266 to 267 °C.
3. An anthocyan pigment with a delphinidin structure from *Gentiana acaulis.*

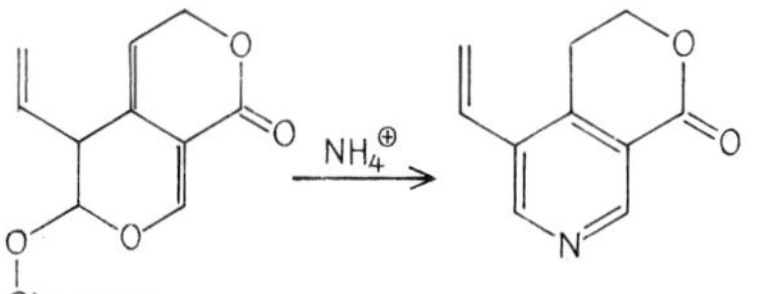

Gentiopicroside   *Gentianine*

Gentianose: a nonreducing trisaccharide composed of two D-glucose and one D-fructose units. m.p. 211 °C, $[\alpha]_D^{20} + 33.4$ °. One glucose and the fructose are joined as in sucrose; the second glucose is linked β-1,6 to the first. G. is a storage compound found in the roots of gentians.

Gentiobiose: a reducing disaccharide consisting of two molecules of D-glucopyranose linked β-1,6. M_r 342.20, α-form, m.p. 86 °C, $[\alpha]_D^{22} + 16$ ° (3 min) → +8.3 ° (3.5 h, $c=4$); β-form, m.p. 190 to 195 °C, $[\alpha]_D^{22} - 5.9$ ° (6 min) → +9.6 ° (6 h, $c=3$). G. differs from isomaltose only in having a β-glycosidic bond instead of an α-link. It is found naturally only in bound form, e.g. in glycosides such as amygdalin and gentiopicrin, and as the ester component of crocin.

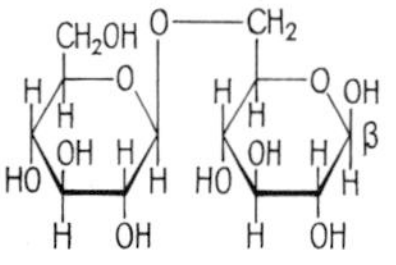

Gentiobiose

Gentiopicroside: see Gentianine 1).
Gentisin: see Gentianine 2).
Geranial: the *trans* isomer of Citral (see).
Geraniol, *2,6-dimethylocta-2,6-dien-8-ol:* the most important of the doubly unsaturated monoterpene

alcohols. M_r 154.24, m.p. -15°C, b.p.$_{757}$ 299 to 230 °C, $p^{20}0.8894$, n_D^{20} 1.4766. The Δ^2 double bond has the *trans* configuration. *Nerol* is the double bond isomer with the *cis* configuration at position 2. The structural isomer linaool has its OH group on C3 instead of C1. G. is a component of many essential oils, and makes up to 60% of oil of roses. It has the fragrance of roses. It is used, primarily as the acetate (b.p. 242 °C), in the perfume industry.

The pyrophosphate ester of G. is an important intermediate in the synthesis of Terpenes (see). Head-to-tail coupling of two molecules produces geranylgeranylpyrophosphate, the precursor of the tetraterpenes.

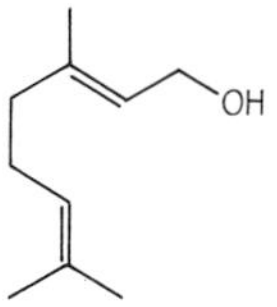

Geraniol 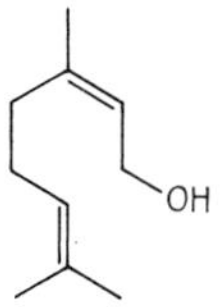   *Nerol* 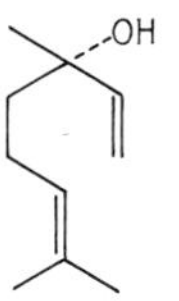   *(+)—Linalool*

Geranyl pyrophosphate: see Geraniol.
Germacrane: see Sesquiterpenes, Fig.
Germine: a Veratrum alkaloid with a C-nor-D-homo ceveratrum type structure. M_r 509.62, m.p. 221.5 to 223 °C, $[\alpha]_D^{25} + 4.5$ ° (95% ethanol), $[\alpha]_D^{16} + 23.1$ ° ($c=1.13$ in 10% acetic acid). G. has been found in *Veratrum album, Veratrum viride* and *Veratrum nigrum* and *Zygadenus venosus* in the form of ester alkaloids. The most common acid components are acetate, angelate and tiglate. These ester alkaloids reduce blood pressure. The isomeric Veratrum base *veracevine* (m.p. 183 °C $[\alpha]_D - 33$ °), which occurs as an ester alkaloid in *Veratrum sabadilla,* differs from G. in having a 12α hydroxyl group instead of the 15 hydroxyl. Alkaline isomerization of veracevine yields *cevine,* with a 3α-hydroxyl group.

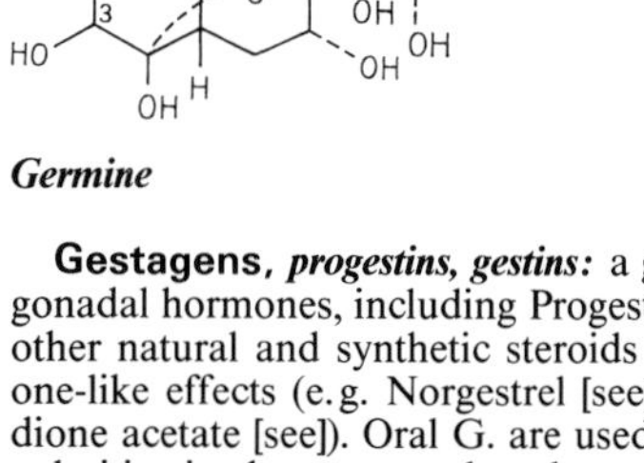

Germine

Gestagens, *progestins, gestins:* a group of female gonadal hormones, including Progesterone (see) and other natural and synthetic steroids with progesterone-like effects (e.g. Norgestrel [see] and Chlormadione acetate [see]). Oral G. are used to correct irregularities in the menstrual cycle, repeated abortion and as components of Ovulation inhibitors (see). They are also being used increasingly to regulate animal reproduction.

GH: abb. for growth hormone. See Somatotropin.

Giant chromosome, *polytene chromosome:* an especially large type of chromosome present in the *Diptera.* G.c. in the salivary glands of *Drosophila* have been extensively studied. G.c. have a cable-like appearance, due to multiple endomitotic duplication of the chromosomes, without separation of the chromatids. They may consist of many thousand single chromosomes, and may reach a length of 0.5 mm and a thickness of 25 µm. The supercoiled DNA molecules are much more extended than in ordinary chromosomes, and the duplicated molecules remain in register as they lie side by side. This gives rise to a banded structure, consisting of DNA-rich bands with DNA-poor regions in between.

G.c. have facilitated the study of chromosome morphology, and made possible the direct demonstration of the involvement of chromosomes in RNA synthesis. Individual bands (about 3000 along the length of the G.c.) undergo periodic changes of shape: they become more loosely packed, and produce a swelling which encircles the axis of the G.c. This swollen structure is called a puff or Balbiani ring (Fig.). Formation of puffs (puffing) is accompanied by a greatly increased RNA synthesis. DNA in the normal bands is tightly coiled and repressed by

Gibberellin antagonists: inhibitors of the effects of gibberellins on plants. Their effects can be at least partially overcome by gibberellins. The term is used independently of the mechanism of inhibition. Competitive inhibitors of gibberellins, compounds which bind to the same active sites as the hormones, are called *antigibberellins.* The G.a. include the phytohormone Abscisic acid (see) and a number of growth retardants, including Chlorocholine chloride (see), Morphactins (see), AMO 1618 (see), EL 531 (see) and Succinic mono-N-methylhydrazide (see). A number of natural products, for example the tannins, also act as G.a.

Gibberellin glucosides: see Gibberellins.

Gibberellinic acid: see Gibberellins.

Gibberellins: a class of widely occurring plant hormones which stimulate extension growth. The first G. to be discovered was isolated from *Gibberella fujikuroi (Fusarium moniliforme),* the pathogen of the rice disease Bakanae, in 1938. The first pure G. was G.A.₃ (gibberellic acid), crystallized in 1954. Further research revealed that G. are ubiquitous natural plant growth regulators. Since many fungal metabolites function as plant growth promoters, claims for G. production by other fungi, based only on biological activity, are unacceptable. G.A₄ has, however, been unequivocally identified (mass spectrometry)

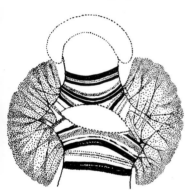

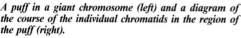

A puff in a giant chromosome (left) and a diagram of the course of the individual chromatids in the region of the puff (right).

specific proteins, whereas DNA in the puffs is relaxed and derepressed, and it can function in the transcription of RNA. Specific incorporation of RNA precursors in the puffs can be demonstrated by autoradiography. Puff formation is reversible. The number, time of occurrence, duration and shape of puffs (i.e. the puff pattern) depend on the cell, the organ and the developmental stage of the insect. Studies of G.c. have made an essential contribution to the understanding of differential Gene activation (see).

Giant messenger-like RNA: see Messenger RNA.

Gibbane: see Gibberellins.

Gibberellane: see Gibberellins.

Gibberellic acid: see Gibberellins.

as a toxin of pathogenic strains of *Sphaceloma manihoticola* [R.S. Ziegler et al. *Phytopathology* **70** (1980) 589–593.] All of the known 52 G. contain the tricyclic *gibbane* ring system. The IUPAC now recommends that nomenclature and numbering be based on the *ent-gibberellane* system (Fig.1).

The G. are referred to, in order of their discovery, as G.A₁ through A₅₂. There are two main groups: G. with 20 C atoms (*ent-gibberellanes*) and those with 19 C atoms (*ent-20-nor-gibberellanes*). Other differences between G. lie in the presence or absence of hydroxyl groups at positions 1, 2, 3, 11, 12 and 13. The substance present in a given sample of plant material can often be characterized only by combined gas chromatography and mass spectroscopy of the methyl esters or trimethylsilyl ethers. The most important of the G., G.A₃, is prepared commercially from culture filtrates of *G. fujikuroi.* Some of the known degradation products of G.A₃ are gibberellenic acid, allogibberic acid, gibberic acid, gibberene and gibberellin C. In addition to these free G., water-soluble G. glucosides and G. glucose esters have been isolated from plants. These are probably transport and storage forms of the biologically active free G.

The activities of the G. in various biotests differ widely. Some appear to be physiologically unimportant, while others may be intermediates in the synthesis or degradation of active hormones. The most important effect of the G. is the stimulation of extension growth and cell division. Dwarf mutants of peas and corn with genetically blocked G. synthesis are used in assays for G. which have a sensitivity of 1 ng (dwarf maize or pea test). The lettuce and cucumber hypocotyl tests are also based on the stimulation of shoot growth by G. In addition, G. inhibit root growth and influence seed dormancy and germination. Parthenocarpy can be induced in toma-

toes, grapes and cucumbers by G. treatment. The acceleration of germination by G. is the basis of the *seed germination* test, and is used practically to obtain uniform and rapid germination in malting barley. G. also stimulates blossom formation and fruit growth, and both properties are utilized in commercial agriculture.

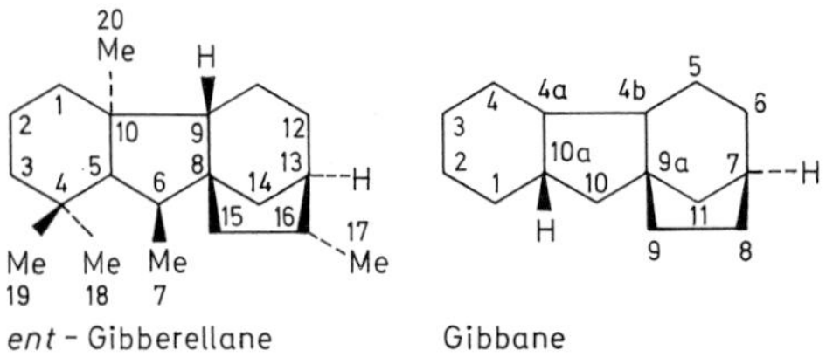

G. A₃ has been shown to activate genes in barley from which the embryo had been removed, suggesting that the embryo releases G. into the aleurone layer, where it activates the genes for hydrolytic enzymes, particularly α-amylase. Another test for G., the α-amylase test, is based on this property. A few G. produced by partial synthesis, for example 3-deoxygibberellin C and pseudogibberellin A₁, act as antigibberellins (see Gibberellin antagonists).

Biosynthesis. G. are diterpenes, synthesized from geranylgeranylpyrophosphate (Fig. 2). The pivotal, common intermediate in the biosynthesis of all G. is G. A₁₂ aldehyde. Thereafter, the biosynthetic pathways diverge considerably. [A. Lang, *Ann. Rev. Plant Physiol.* **21** (1970) 537-570: comprehensive review and reference source for all work up to 1970. J. E. Graebe & H. J. Ropers in *Phytohormones and Related Compounds - A Comprehensive Treatise* Vol. 1 (D. S. Letham, P. B. Goodwin & T. J. V. Higgins, eds.) (North Holland, 1978) pp. 107-204: all aspects. P. Hedden et al. *Ann. Rev. Plant Physiol* **29** (1978) 149-192: biosynthesis.]

> **Gibberene:** see Gibberellins.
> **Gibberic acid:** see Gibberellins.
> **Gibbs effect:** After the brief photosynthetic assimilation of ¹⁴CO₂ by *Chlorella,* the resulting fructose bisphosphate is labelled symmetrically, as expected from the operation of the Calvin cycle (see).

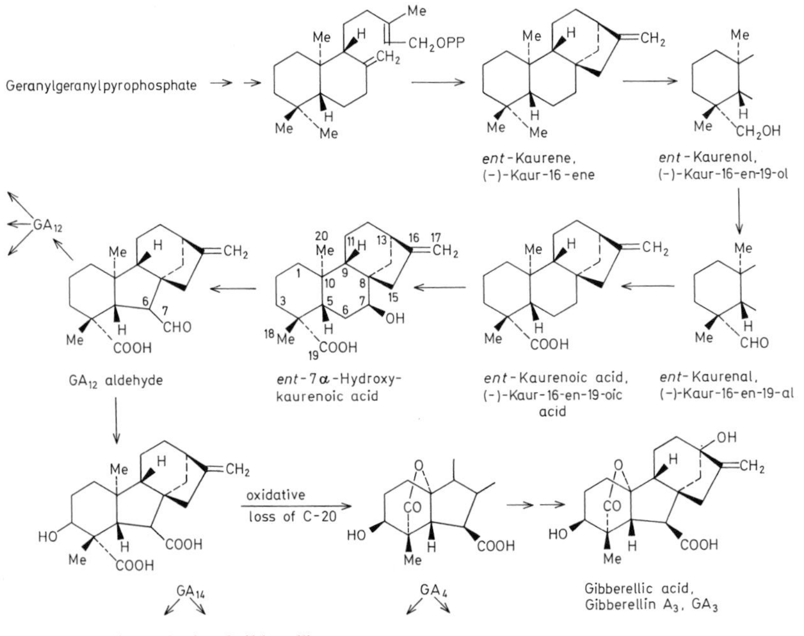

Figure 1.

Figure 2. Biosynthesis of gibberellins

In contrast, glucose phosphates and the glucose moiety of starch are asymmetrically labelled, C-4 containing significantly more ^{14}C than C-4, and C-1 and 2 containing significantly more than C-5 and 6. This apparently anomalous labelling of glucose is known as the Gibbs effect [Kandler, O., & Gibbs, M. (1956) Plant Physiol., 31, 411–412]. It indicates that the two halves of the glucose molecule are not derived from the same pool of triose phosphates as fructose bisphosphate; it also suggests that fructose bisphosphate is not the precursor of glucose phosphates. Reactions in the revised version of the Pentose phosphate cycle (see) may explain the Gibbs effect.

Gitogenin: see Gitonin.

Gitonin: a steroid saponin (see Saponins). M_r 1051.21, m.p. 272 °C, $[\alpha]_D$ − 51 °C (pyridine). The aglycon is *gitogenin,* (25 R)-spirostan-2α,3β-diol, M_r 432.62, m.p. 272 to 275 °C (d.), $[\alpha]_D^{20}$ − 70 ° (c = 1.02 in CHCl₃); the sugar chain consists of 2 galactose, 1 glucose and 1 xylose units. Gitogenin differs from digitogenin (see Digitonin) in lacking the 15β-hydroxyl group. G. has been isolated from *Digitalis purpurea* and *Digitalis germanicum.* Free gitogenin has also been isolated from agave and yucca species.

Gitoxigenin: see Digitalis glycosides.

Gitoxin: see Digitalis glycosides.

Gla: abb. for L-4-carboxyglutamic acid.

Glabrene: see Isoflav-3-ene.

Gliadin: see Prolamines.

Gln: abb. for L-glutamine.

Globoid cell leukodystrophy: see Inborn errors of metabolism.

Globoside: see Inborn errors of metabolism (Fig. Catabolism of glycososphingolipids and sphingomyelin).

Globulins: a group of simple proteins which are insoluble in pure water but soluble in dilute salt solutions (salting in effect). They are found in all animal and plant cells and body fluids, including serum and milk. The group includes many enzymes and most glycoproteins. G. can be precipitated by ammonium sulfate: fibrinogen at 20 to 25%, the euglobulins at 33% and the pseudoglobulins at 50%. The most familiar are the serum G., which are separated by electrophoresis into the α_1, α_2, β_1, β_2 and γ-G. (see Plasma proteins).

Glomerine: a quinazoline alkaloid in the defense secretion of the insect *Glomeris marginata.* m.p. 204 °C. G. is one of the few animal alkaloids. The secretion is exuded from 8 pores arranged in a row on the 1.5 cm animal. It contains about 50 µg G. and another alkaloid, *homoglomerine* (m.p. 149 °C).

Glomerine: R = C₂H₅
Homoglomerine: R = CH₃

< **Glu:** recommended abb. for Pyroglutamic acid (see).

⌐**Glu:** recommended abb. for Pyroglutamic acid (see).

Glu: abb. for L-glutamic acid.

Glucagon: a pancreatic polypeptide hormone consisting of a single chain of 29 amino acids. The primary structure is given in the Fig. M_r 3485. G. is produced in the A cells of the islets of Langerhans in the pancreas in response to a drop in the blood sugar concentration. It promotes the hydrolysis of glycogen and lipids, and raises the blood sugar level. G. is degraded in the liver. It can be detected in blood serum by radioimmunological techniques with a sensitivity in the pg/ml range.

His—Ser—Gln—Gly—Thr—Phe—Thr—Ser—

Asp—Tyr—Ser—Lys—Tyr—Leu—Asp—Ser—

Arg—Arg—Ala—Gln—Asp—Phe—Val—Gln—

Trp—Leu—Met—Asn—Thr

Glucagon
The boxed amino acids are identical with the secretin sequence.

Glucans: polysaccharides composed of D-glucose. They may be either straight or branched chains. The glycosidic linkages may be α-1,4, as in amylose and bacterial dextran, β-1,4, as in cellulose, β-1,3, as in leucosin and callose, or 1,6, as in pustulan. The branched glucans include amylopectin (α-1,4 and α-1,6 bonds), dextran, laminarin and lichenin.

Glucocorticoids: see Adrenal corticosteroids.

Glucokinase: see Kinases.

Gluconeogenesis: the synthesis of glucose from pyruvate or Amino acids (see). G. cannot occur by a simple reversal of Glycolysis (see), because the equilibria of those reactions are too unfavorable under physiological conditions. Instead, the pyruvate is carboxylated to oxaloacetate (either directly or indirectly, via malate) (see Carboxylation). The oxaloacetate is then decarboxylated and phosphorylated simultaneously by the phospho *enol*pyruvate carboxykinase (ATP) (EC 4.1.1.49) to form phospho *enol*pyruvate. (In *Escherichia coli* a direct phosphorylation of pyruvate by ATP is possible.) Reversal of the glycolytic reactions then yields fructose 1,6-bisphosphate from the phospho *enol*pyruvate. The phosphofructokinase reaction is not reversible; instead fructose-bisphosphatase (EC 3.1.3.11) removes one phosphate group to form fructose 6-phosphate. This is converted readily to glucose 6-phosphate. If the blood sugar level is low, glucose 6-phosphate is hydrolysed to glucose by glucose-6-phosphatase (EC 3.1.3.9). Otherwise, the glucose 6-phosphate is used directly for glycogen synthesis. Overall reaction: 2 pyruvate $(2\ CH_3COCOO^-) + 2$ NADH $+ 4\ H^+ + 6$ ATP→ glucose $(C_6H_{12}O_6) + 2$ NAD⁺ $+ 6$ ADP $+ 6\ P_i$. The energy required for G. can be obtained by oxidizing 20 to 30% of the lactate to CO_2 and H_2O.
G. can proceed from the carbon skeleton of any amino acid which can be converted to a C_4 carboxylic acid (glucoplastic amino acids). These intermediates of the Tricarboxylic acid cycle (see) can be converted to oxaloacetate (see above).

225

G. in liver is promoted by Glucagon (see) and Adrenalin (see), the effects being mediated by cAMP. When the organism fasts, glucocorticoids (e.g. Cortisol, see) are released from the adrenal glands. These hormones induce the enzymes of G. in the liver, and also appear to make the cells more sensitive to cAMP, and thus to glucagon. The result is an increased G. from amino acids in a fasting animal.

Glucoplastic amino acids: amino acids whose degradation products can contribute to Gluconeogenesis (see). See Amino acids.

Glucosamine, *2-amino-2-deoxyglucose, chitosamine:* a widely occurring aminosugar. M_r 179.17, α-form m.p. 88 °C, $[\alpha]_D^{20}+100\,° \to +47.5\,°$ after 30 min (water); β-form m.p. 110 °C (d.), $[\alpha]_D^{20}+28\,° \to +47.5\,°$ after 30 min (water). G. is a component of chitin, mucopolysaccharides like heparin, chondroitin and mucoitin sulfate, and of blood group substances and other complex polysac-

β-Glucosamine

to various forms of anaerobic and aerobic fermentation to alcohol, lactate, acetate, or citrate. G. is the most important animal monosaccharide, and is also the most abundant natural organic compound. In free form it is found in many sweet fruits, honey and nectar, and blood (up to 0.1%). G. is also a component of many oligo- and polysaccharides (sucrose, lactose, maltose, starch, glycogen, cellulose, etc.) and glycosides. The phosphoric acid esters of G. are extremely important metabolic intermediates. The activated form, ADPglucose, is used in starch synthesis in plants; UDPglucose is the G. donor in the synthesis of many saccharides (see Nucleoside diphosphate sugars).

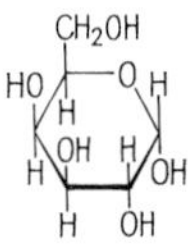

G. is produced commercially by acid or enzymatic hydrolysis of potato or corn starch or cellulose. It is used as a sweetener in beverages and as a nutrient in medicine.

Glucose 1,6-bisphosphate: a glucose derivative which is an important intermediate in glycolysis. It is synthesized in yeast, plants and muscles by the reaction glucose 1-phosphate + ATP $\xrightarrow{Mg^{2+}}$ glucose 1,6-bisphosphate + ADP; and in *Escherichia coli* and muscles by the reaction 2 glucose 1-phosphate ⇌ glucose 1,6-bisphosphate + glucose. It is the cosubstrate of phosphoglucomutase (EC 2.7.5.1), which catalyses the interconversion of glucose 1- and 6-phosphates (Fig.) (see Glycolysis).

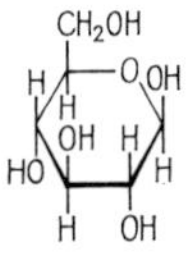

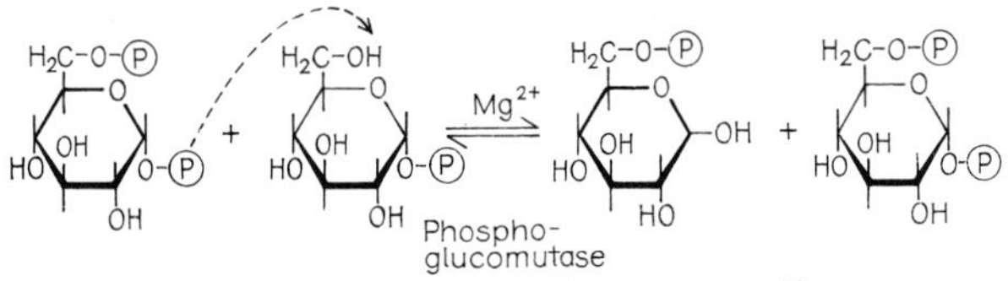

Phosphoglucomutase reaction

charides. It is usually present as *N*-acetylglucosamine.

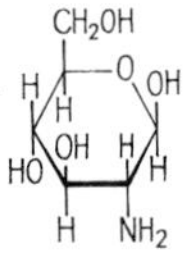

Glucosaminoglycans: see Proteoglycans.

D-Glucose, *dextrose, grape sugar, blood sugar:* a hexose. M_r 180.16, α-form m.p. 146 °C, $[\alpha]_D + 112.1 \to +52.7\,°$ ($c = 10$ in water); β-form, m.p. 148 to 155 °C, $[\alpha]_D + 18.7 \to +52.7\,°$ ($c = 10$ in

α-D-Glucose   *β-D-Glucose*

water). The most stable configuration for the pyranose form is the chair, in which all the hydroxyl groups of the β-form are equatorial (see Carbohydrates, Fig.4). In the α-form, the two hydroxyl groups at positions 1 and 2 are *cis*. G. is susceptible

Glucose isomerase: see Enzymes, Table 2.

D(+)-Glucose oxidase (EC 1.1.3.4): a plant and microbial flavin enzyme which oxidizes β-D-glucose in the presence of O_2 to gluconic acid and H_2O_2. G.o. is used for a specific enzymatic assay of glucose, for example in blood. G.o. from *Aspergillus* is a dimeric flavoglycoprotein (16% carbohydrate, M_r 160000, 2 FAD per molecule, 2 subunits of M_r 80000 each) which is inhibited by *p*-mercuribenzoate.

Glucose 1-phosphate, *Cori ester:* the product of phosphorolysis of Glycogen (see) and Starch (see). It is converted to glucose 6-phosphate by phosphoglucomutase (EC 2.7.5.1).

Glucose 6-phosphate, *Robinson ester:* the key intermediate in Carbohydrate metabolism (see).

Glucose-6-phosphate **dehydrogenase,** *GPDH* (EC 1.1.1.49): the key enzyme of the Pentose phosphate cycle (see). It occurs widely in plants and animals and has been shown to be formed from inactive precursor subunits. GPDH is a tetrameric enzyme with M_r ranging from 206000 in *Neurospora* to

240000 in erythrocytes. Its dimers are held together by NADP. In humans, 50 hereditary variants of the erythrocyte GPDH are known. It is lacking in some individuals who, after consuming certain legumes (e.g. fava beans), breathing bean pollen or taking certain medicines, suffer a severe hemolytic anemia (favism). The condition is especially widespread in the Mediterranean region, Asia and America.

Glucose tolerance factor: see Chromium.

Glucosinolate, *mustard oil thioglucoside, mustard oil glucoside:* a natural plant product with the structure shown in Fig. 1. G. are particularly abundant in members of the *Cruciferae, Capparidaceae* and *Resedaceae.* The two longest known G. are sinalbin (crystallized from white mustard in 1831) and sinigrin (isolated from black mustard in 1840). More than 70 different G. are now known. The carbohydrate residue is always a single glucose unit; all G. are 1-β-D-thioglucopyranosides. The term "glucosinolate" for the anion of the compound has been in use since 1961. The cation is usually potassium, but in sinalbin it is the basic organic molecule, sinapine. An alternative and earlier nomenclature uses the prefix "gluco-" attached to an appropriate part of the Latin binomial of the plant of origin, e.g. glucotropaeolin from *Tropaeolum majus,* etc.

Damage to the plant results in hydrolysis of G. and release of volatile, pungent, lacrimatory isothiocyanates, also known as mustard oils. This ability to form mustard oils has led to the adoption of several members of the *Cruciferae* (e.g. horseradish, mustard) as condiments. Mustard oils are normally absent from the plant, and are first formed when tissue damage permits interaction of G. and thioglucoside glucohydrolase (E.C. 3.2.3.1); activation of this enzyme by its cofactor, ascorbic acid, may also be promoted by tissue damage. Enzymatic hydrolysis of the *S*-glucoside bond is followed by molecular rearrangement of the aglycon with concomitant production of sulfate and isothiocyanate (Fig. 1). Small quantities of the corresponding thiocyanate and nitrile may also be produced (Fig. 1).

Like the cyanogenic glycosides, G. are biosynthesized from amino acids. The pathway suggested in Fig. 2 is strongly supported by isotopic tracing studies with $^{14}C,^{15}N$-labelled amino acids or the corresponding aldoximes. The following G. are derived from proteogenic amino acids: methyl-G. (from alanine; not yet found in *Cruciferae,* but widely distributed and dominant in the *Capparaceae*), isopropyl-G. and sec-butyl-G. (from valine and isoleucine, respectively; both compounds found especially in the genera *Cardamine, Cochlearia, Lunaria* and *Sisymbrium*), isobutyl-G. (from leucine; found in *Conringia orientalis, Cochlearia officinalis* and 2 spp. of *Thelypodium*), 4-hydroxybenzyl-G. (from tyrosine; the salt with sinapine is sinalbin; especially rich in, but rather restricted to the genus *Sinapis*), benzyl-G. (from tryptophan; in several plant families, but restricted to seedlings and young vegetative tissue.) The G. that would be derived from the remaining 13 proteogenic amino acids have not been encountered; owing to its cyclic structure, proline is not even a potential G. precursor.

G. are also known that arise from proteogenic amino acids after homologization (Fig. 3), which consists of a series of type reactions already well known in the Tricarboxylic acid cycle (see) and leucine biosynthesis. For example, ethyl-G. (reported only in one species of *Lepidum*) is derived from alanine after one cycle of homologization. Similarly, 2(S)-methylbutyl-G. is derived from 2-amino-4-methylhexanoic acid, which in turn is derived from isoleucine by homologization. The *Cruciferae* have so far yielded 9 homologous ω-methylthioalkyl-G. (side chain $CH_3-S-(CH_2)_n-$, where n varies from 3 to 11); these are derived from methionine, following 1-9 cycles of homologization. The corresponding sulfoxides and sulfones are also known, e.g. 3-(methylsulfonyl)-propyl-G. from *Cheiranthus.*

The biosynthetic origin of sinigrin is less obvious, but isotopic tracing studies have shown that the carbon atoms of the side chain are derived from methionine after one cycle of homologization (Fig. 4).

Allylisothiocyanate is highly toxic to the pathogen *Peronospora parasitica* (downy mildew). Varieties of cabbage bred for milder flavor (and hence lower sinigrin content) lack resistance to the pathogen. On

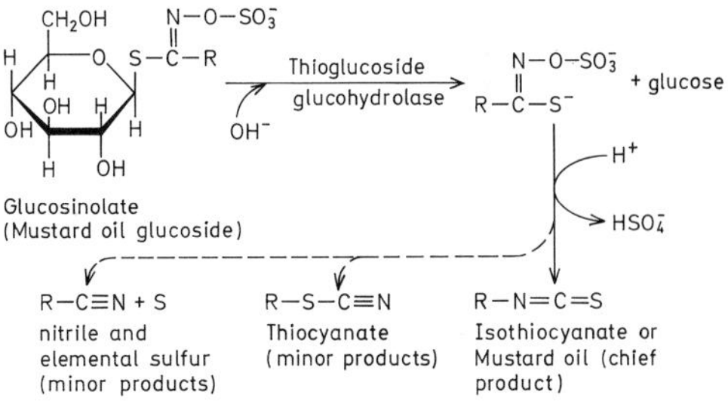

Figure 1. Conversion of a glucosinolate to the corresponding isothiocyanate by the action of thioglucoside glucohydrolase.

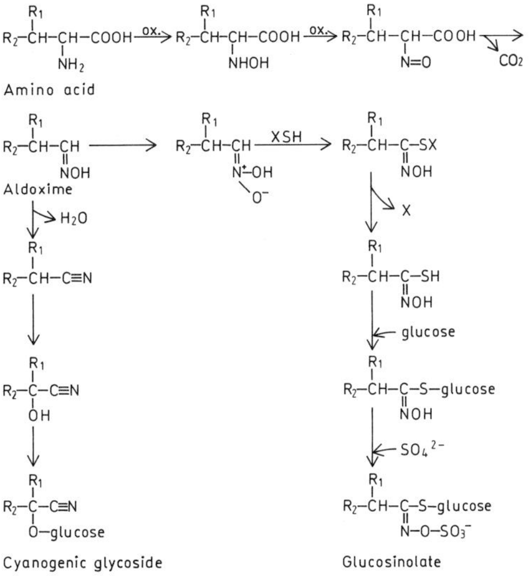

Figure 2. Proposed routes for the biosynthesis of glucosinolates and cyanogenic glycosides from amino acids.

227

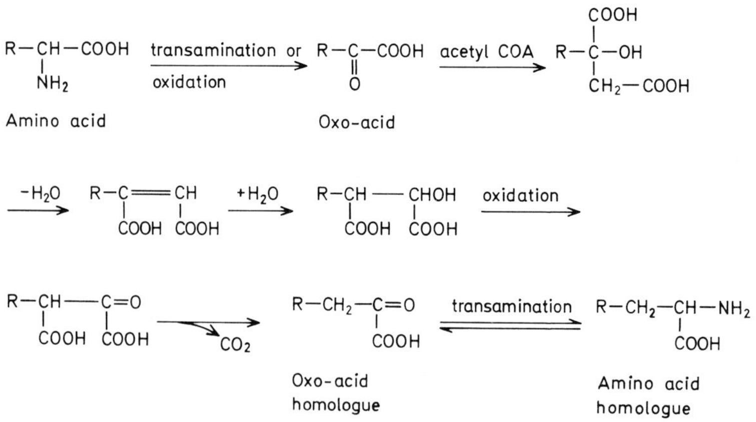

Glucosinolate

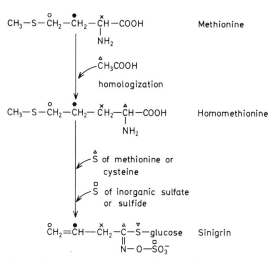

$$R-\overset{\underset{\displaystyle NH_2}{|}}{C}H-COOH \xrightarrow[\text{oxidation}]{\text{transamination or}} R-\overset{\underset{\displaystyle O}{\|}}{C}-COOH \xrightarrow{\text{acetyl COA}} R-\overset{\underset{\displaystyle CH_2-COOH}{|}}{\underset{|}{C}}-OH$$

Amino acid   Oxo-acid

$$\xrightarrow{-H_2O} R-\overset{\underset{\displaystyle COOH}{|}}{C}=\overset{\underset{\displaystyle COOH}{|}}{C}H \xrightarrow{+H_2O} R-\overset{\underset{\displaystyle COOH}{|}}{C}H-\overset{\underset{\displaystyle COOH}{|}}{C}HOH \xrightarrow{\text{oxidation}}$$

$$R-\overset{\underset{\displaystyle COOH}{|}}{C}H-\overset{\underset{\displaystyle COOH}{|}}{C}=O \xrightarrow{CO_2} R-CH_2-\overset{\underset{\displaystyle COOH}{|}}{C}=O \underset{\longleftarrow}{\overset{\text{transamination}}{\longrightarrow}} R-CH_2-\overset{\underset{\displaystyle COOH}{|}}{C}H-NH_2$$

Oxo-acid   Amino acid
homologue   homologue

Figure 3. *Homologization.* One cycle of this process extends the side chain of the precursor amino acid by one CH_2 group; it is analogous to the reactions of the tricarboxylic acid cycle and of leucine biosynthesis.

$$CH_3-S-\overset{o}{C}H_2-CH_2-\overset{x}{C}H-COOH \qquad \text{Methionine}$$
$$| \atop NH_2$$

$$\swarrow \overset{\vartriangle}{C}H_3COOH$$

homologization

$$CH_3-S-\overset{o}{C}H_2-CH_2-\overset{x}{C}H_2-\overset{\vartriangle}{C}H-COOH \qquad \text{Homomethionine}$$
$$| \atop NH_2$$

$$\swarrow \overset{\vartriangle}{S} \text{ of methionine or}$$
cysteine

$$\swarrow \overset{\square}{S} \text{ of inorganic sulfate}$$
or sulfide

$$\overset{o}{C}H_2=\overset{\bullet}{C}H-\overset{x}{C}H_2-\overset{\vartriangle}{C}-\overset{\triangledown}{S}-\text{glucose} \qquad \text{Sinigrin}$$
$$\overset{\|}{N}-O-\overset{\square}{S}O_3^-$$

Figure 4. *Summary of isotope (^{14}C and ^{35}S) tracing studies on the biosynthesis of sinigrin.*

the other hand, sinigrin is a feeding attractant for the cabbage butterfly larva (*Pieris brassicae*) and an oviposition stimulant for the adult female. It is also a feeding attractant to the cabbage aphid, which prefers mature leaves of medium sinigrin content, and avoids feeding on young leaves that are very rich in this G.

Work on G. up to 1960 has been comprehensively reviewed by A. Kjaer in *Progress in the Chemistry of Organic Natural Products* XVIII (L. Zechmeister, ed., Springer (1960)), pp. 122–176. Later work is discussed by various contributors to *The Biology and Chemistry of the Cruciferae,* (J. G. Vaughan et al., ed., Academic Press (1976)).

Table. *Some examples of naturally occurring glucosinolates*

Glucosinolate	Source	Formula
Sinalbin, Sinapine glucosinalbate, Sinapine 4-hydroxy-benzylglucosinolate	*Sinapis alba* (White or yellow mustard)	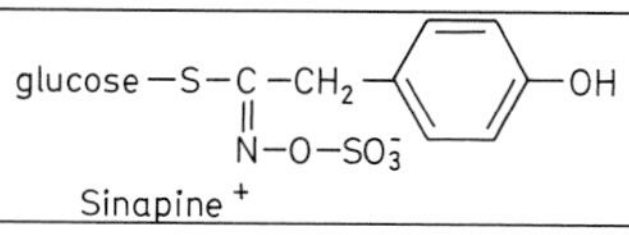 glucose—S—C—CH$_2$—⟨ring⟩—OH, ‖ N—O—SO$_3^-$, Sinapine$^+$
Sinigrin, Allyl-glucosinolate, Sinigroside, Potassium myronate	*Brassica nigra* (Black mustard), *Amoracia rusticana* (Horse radish)	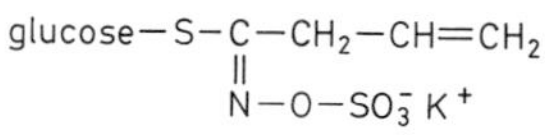 glucose—S—C—CH$_2$—CH=CH$_2$, ‖ N—O—SO$_3^-$ K$^+$
Benzylglucosinolate Glucotropaeolin	*Tropaeolum majus* (Nasturtium)	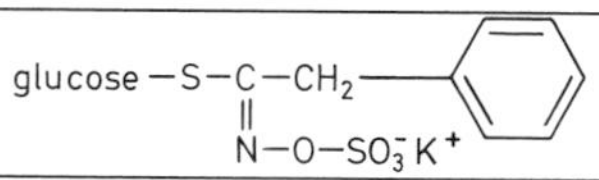 glucose—S—C—CH$_2$—⟨ring⟩, ‖ N—O—SO$_3^-$ K$^+$
Phenylethyl glucosinolate, Gluconasturtiin.	*Nasturtium officinale* (Watercress)	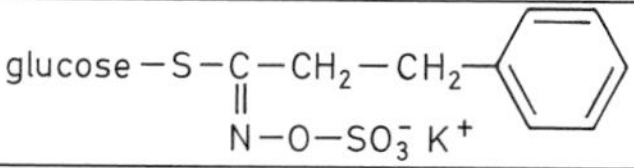 glucose—S—C—CH$_2$—CH$_2$—⟨ring⟩, ‖ N—O—SO$_3^-$ K$^+$
3-(Methylsulfonyl)-propylglucosinolate Glucocheirolin	*Cheiranthus cheiri* (Wallflower)	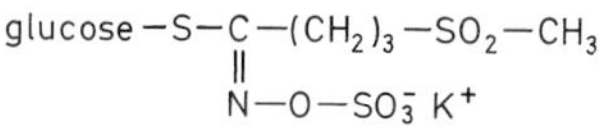 glucose—S—C—(CH$_2$)$_3$—SO$_2$—CH$_3$, ‖ N—O—SO$_3^-$ K$^+$
*2-Hydroxy-3-butenyl-glucosinolate, Progoitrin, Glucorapiferin	Various spp. of Brassica (esp. Yellow turnip)	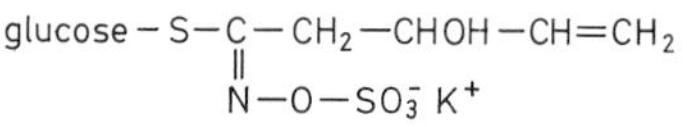 glucose—S—C—CH$_2$—CHOH—CH=CH$_2$, ‖ N—O—SO$_3^-$ K$^+$

* 2-Hydroxy-3-butenylisothiocyanate (S=C=N-CH$_2$-CHOH-CH=CH$_2$), the mustard oil from 2-hydroxy-3-butenylglucosinolate, cyclizes spontaneously to 5-vinyl-2-thiooxazolidone (goitrin), which is responsible for the goitrogenic action (see Goitrogens) of yellow turnips and rapeseed oil meal.

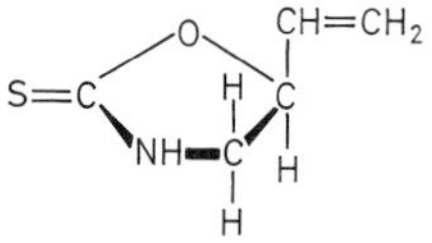

(S)-5-Ethenyl-2-oxazolidinethione (5-vinyl-2-thiooxazolidone, goitrin)

Glucuronate pathway, *glucuronate-xylose cycle,* **D-glucuronate-L-gulonate pathway:** a pathway in carbohydrate metabolism by which *myo*-inositol and ascorbate are synthesized and degraded (Fig. 1). Glucose is oxidized at position 6 to D-glucuronate, probably via UDPglucose (see Nucleoside diphosphate sugars). Glucuronate, which is also the product of *myo*-inositol oxygenase, is the starting material for the synthesis of glucuronides. It is degraded by reduction to L-gulonate. Since the C-6 of glucuronate becomes the C-1 of gulonate, the latter belongs to the L series of carbohydrates. L-Gulonate is diverted into the L-ascorbate pathway (see Vitamins, Fig. 10), or it is oxidized to 3-keto-L-gulonate, which is decarboxylated to form L-xylulose. Xylulose is reduced to the sugar alcohol xylitol, which is reoxidized, to D-xylulose. The change in configuration is again accomplished by an end-for-end shift in which the C-5 of xylulose is formed from the C-1 of xylitol. D-Xylulose is phosphorylated to xylulose 5-phosphate, which is a member of the Pentose phosphate cycle (see). Glucose 6-phosphate, the precursor of UDPglucose, is regenerated from xylulose 5-phosphate in the pentose phosphate cycle. Bacteria possess an alternative pathway for the degradation of glucuronate to glyceraldehyde 3-phosphate and pyruvate (Fig. 2).

D-Glucuronic acid: a biosynthetic derivative of glucose found in mucopolysaccharides like hyaluronic acid and chondroitin sulfate. In animals it is an important conjugation partner for foreign and endogenous substances, especially phenols, which are excreted in "paired" form as G.a. conjugates (*glucuronides*). In most animals, G.a. is the starting material for ascorbic acid (see Vitamin C) synthesis.

L-Glutamic acid, abb. *Glu, L-a-aminoglutaric acid:* a proteogenic amino acid with two carboxyl groups. Since only the α-carboxyl group forms peptide bonds in proteins, the remaining free carboxyl group gives the polypeptide an acid character. M_r 147.13, m.p. 247 to 249 °C (d.), $[\alpha]_D^{25} + 12.0$ ° ($c=2$ in water) or $+31.8$ ° ($c=2$ in 5 NHCl). Glu is present in nearly all proteins, especially seed proteins. It is easily cyclized to L-pyrrolidone carboxylic acid

229

L-Glutamic acid

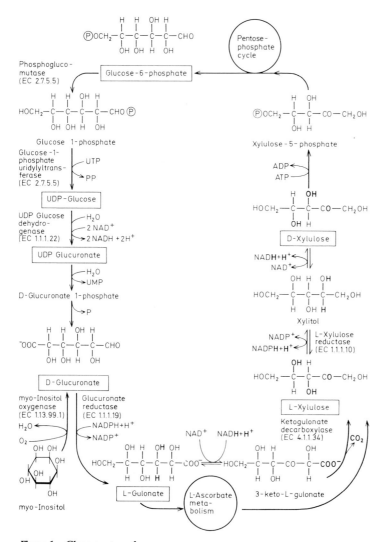

Figure 1. *Glucuronate pathway.*

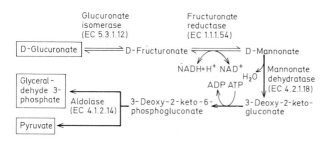

Figure 2. *Alternative catabolic pathway for glucuronate in bacteria.*

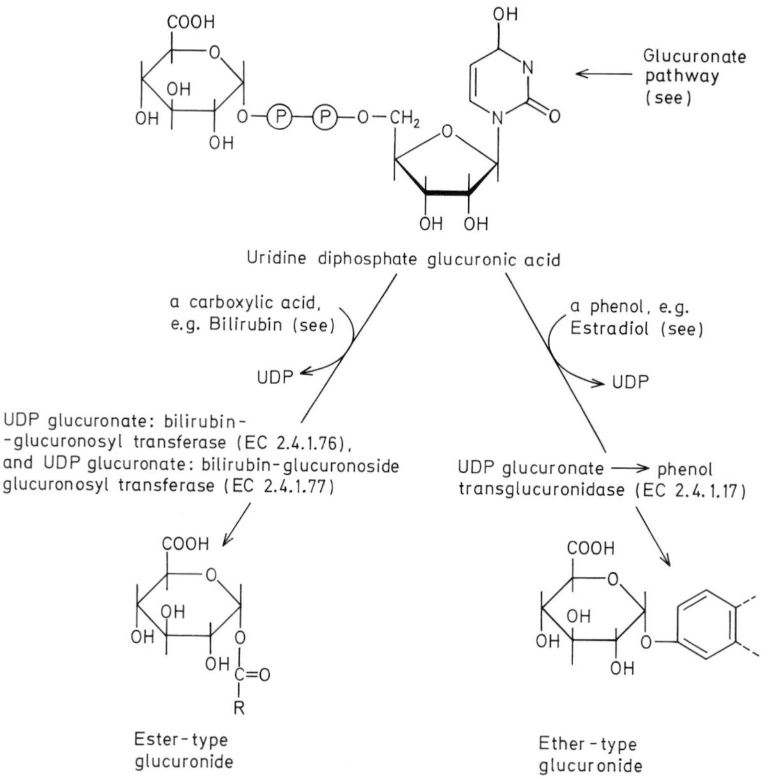

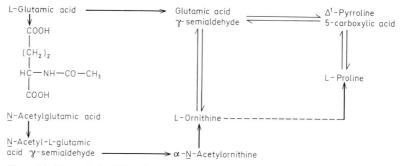

Formation of glucuronic acid conjugates (glucuronides or glucuronosides). UDP glucuronate:phenol transglucu-
ronidase (EC 2.4.1.17) catalyses glucuronidation of a wide range of phenols, alcohols and fatty acids. Glucuronidation of bilirubin is an important special case (see Crigler-Najjar syndrome in Inborn errors of metabolism).

(Fig. 1) on heating; this compound is also formed as a post-translational modification of Glu residues in proteins (see Pyroglutamic acid). The biosynthesis of amino acids belonging to the 2-oxoglutarate (α-ke-toglutarate) family begins with Glu (Fig. 2). Glu is involved in the transport of potassium ions in the brain, and also detoxifies ammonia in the brain by forming glutamine, which can cross the blood-brain barrier. In the central nervous system, Glu is decar-

boxylated to 4-aminobutyric acid by glutamate decarboxylase (EC 4.1.1.15).
Glu is a component of γ-*glutamyl peptides*. A few of

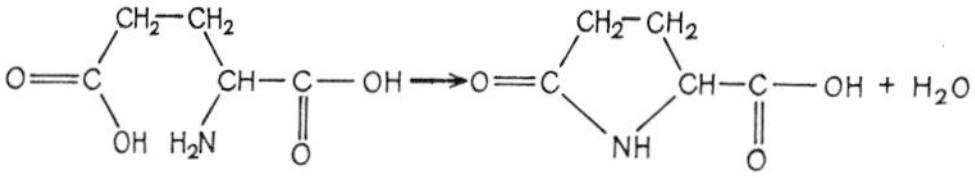

Figure 1. Cyclization of L-glutamic acid

Figure 2. Biosynthesis of L-ornithine and L-proline from L-glutamic acid.

these are found in the central nervous system, but they occur in quantity in plants, where they are storage compounds, and often include rare amino acids. Glu is also a component of Glutathione (see), and its amino group can be enzymatically acetylated.

Other derivatives of Glu are not formed directly from it, for example 4-methylenylglutamic and its amide, 4-methylenyl-glutamine. Glu is degraded via 2-oxoglutaric acid or decarboxylated to 4-aminobutyric acid, then converted to succinic acid which enters the tricarboxylic acid cycle.

L-Glutamine, abb. *Gln* or *Glu-NH₂:* a proteogenic amino acid, the 5-amide of L-glutamic acid. M_r 146.15, m.p. 185 to 186 °C (d.), $[\alpha]_D^{25} + 6.3$ ° ($c = 2$, water) or $+31.8$ ° ($c = 2$, 5 N HCl). In boiling, neutral aqueous solutions or in weak acid, Gln rapidly cyclizes to the ammonium salt of pyrrolidone carboxylic acid (see Glutamic acid).

Gln is synthesized from glutamic acid and ammonium in an endergonic reaction: $Glu + NH_4 + ATP \xrightarrow{Mg^{2+}} Gln + ADP + P_i$ catalysed by glutamine synthetase (EC 6.3.1.2). The synthesis is made thermodynamically favorable by coupling with hydrolysis of ATP. There appear to be no free intermediates. The synthesis plays a role in Ammonia detoxification (see) and the formation of nitrogen excretion products, because Gln is both the N donor in the synthesis of carbamoylphosphate and purines, and a nitrogen storage compound in plants.

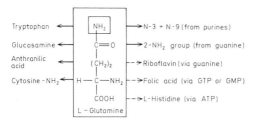

Gln is a central intermediate in nitrogen metabolism (Fig.). The amide nitrogen is transferred to other compounds by transamidation. Gln is also hydrolysed in the kidney to provide free ammonia, which aids the resorption of potassium and sodium ions. Gln is an important nutrient for the brain (see Glutamic acid). Because its amide group is involved in the biosynthesis of hexosamines, Gln promotes the regeneration of mucoproteins and the intestinal epithelium.

Glutamine antagonists, *glutamine analogs:* structural analogs of L-glutamine which competitively inhibit glutamine-dependent enzyme reactions (see L-Glutamine). 6-Diazo-5-oxo-L-norleucine (see) and Albizziin (see) are members of this group of substances.

Glutamine synthetase: see Ammonia assimilation; Covalent modification of enzymes.

Glutamino-carbamoylphosphate synthetase: see Carbamoyl phosphate.

4-Glutamyl carboxylase, *γ-glutamyl carboxylase, vitamin K-dependent γ-glutamyl carboxylase:* the enzyme responsible for the posttranslational carboxylation of glutamate to 4-carboxyglutamate residues

in certain proteins. The reaction requires vitamin K which explains the vitamin K requirement for the synthesis of blood clotting proteins, e.g. prothrombin, which contain residues of 4-carboxyglutamate. It is a microsomal enzyme, which can be solubilized with various detergents. There is an apparent ATP requirement for the microsomal system, but the solubilized enzyme does not require ATP. Biotin is not involved in this carboxylation.

Glutathione, *L-γ-glutamyl-L-cysteinyl-L-glycine, GSH:* a tripeptide found in animals, most plants and bacteria. It serves as a biological redox agent, as a coenzyme and cofactor, and as a substrate in certain coupling reactions catalysed by Glutathione S-transferase (see). The concentration of GSH in cells is 0.1 to 10 mM; it is also exported to the extracellular fluids in the course of the γ-glutamyl cycle (see below). As a redox agent, GSH scavenges free radicals and reduces peroxides ($2\ GSH + ROOH \rightarrow GSSG + ROH + H_2O$), and thus protects membrane lipids from these reactive substances. It is especially important in the lens and in certain parasites which have no catalase to remove H_2O_2.

Metabolism. Within the cell, GSH is formed in two steps: $Glu + Cys \rightarrow γ$-Glu-Cys, and $γ$-Glu-Cys + Gly $\rightarrow$ GSH. These are catalysed, respectively, by γ-glutamylcysteine synthetase (EC 6.3.2.2) and GSH synthetase (EC 6.3.2.3). Some of the GSH is transported out of the cell; in the extracellular space it is converted by membrane-bound γ-glutamyl transferase (EC 2.3.2.2) to Cys-Gly + γ-glutamyl amino acids. The products are transported back into the cell and are cleaved, the Cys-Gly to Gly and Cys, and the γ-glutamyl amino acid to 5-oxoproline and amino acid. 5-Oxoproline is converted to glutamic acid, so that the *γ-glutamyl cycle* can repeat. [A. Meister. *Science* **220** (1983) 472-477]

Glutathione S-transferase, *ligandin,* (EC 2.5.1.18): a group of enzymes of the liver cytosol, which catalyse reaction of glutathione (acting as a nucleophile) with a wide range of electrophilic substrates: $RX + GSH \rightarrow HX + RSG$, where GSH is reduced glutathione, R may be an aliphatic, aromatic or heterocyclic radical, and X may be a sulfate, nitrite, halide, epoxy, or ethene group, or the cyanide group of a thiocyanate. The enzymes are thought to play a physiological role in the detoxication of potential alkylating agents, some pharmacologically active compounds and certain xenobiotics. The electrophilic sites of such compounds are neutralized by reaction with the SH-group of glutathione. The glutathione conjugates are then metabolized further by cleavage of the glutamate and glycine residues, followed by acetylation of the free amino group of the cysteinyl residue, to produce mercapturic acids, which are excreted. G.t. also bind strongly and noncovalently a wide variety of ligands, without catalysing further reaction, e.g. estrogens, steroid conjugates, bilirubin, probenecid, heme, penicillin, chloramphenicol. This latter property gave rise to the name, *ligandin,* before it was discovered that G.t. and ligandin are indentical. A third property of G.t. is its ability to bind covalently several electrophilic substrates of the transfer reaction in the absence of glutathione. Yet a further property of G.t. enzymes is their ability to catalyse certain isomerization reactions for which glutathione is a coenzyme, e.g. maleylacetone to fumarylacetone, and the positional isomerization of Δ^5- to Δ^4-unsaturated 3-ketosteroids.

All G.t. enzymes isolated from rat liver are basic proteins, M_r about 45000. They are designated in the reverse order of their elution from a CM-cellulose column: AA, A, B, C, D and E.
[Jakoby, W.B. and Keen, J.H. (1977) *Trends in Biochemical Sciences 2,* 229–231.]

Glutelins: a group of simple proteins from grain. They are generally insoluble in water, salt solutions and dilute ethanol, but at extreme pH values they do become soluble. They may contain up to 45% glutamic acid. The best known are glutenin in wheat, orycenin in rice and hordenin in barley.

Gluten: a mixture of about equal parts of the simple proteins glutelins and prolamines. G. makes flour capable of rising when made into bread. The presence of G. in wheat and rye flours is most important. Since prolamines are not present in oat and rice grains, their flour is not suitable for baking.

Glutenin: see Glutelins.

Gly: abb. for glycine.

Glyceollins: four phytoallexins produced by *Glycine max* L. in response to infection by *Phytophthora megasperma* var. *sojae,* or in response to treatment of soybean cotyledons or cell cultures with Elicitor (see). A particulate fraction from elicitor-treated soybean cotyledons or cell cultures contains a dimethylallyl transferase activity, which catalyses the synthesis of 4- and 2-dimethylallyl-3,6a,9-trihydroxpterocarpan from (6aS,11aS)-3,6a,9-trihydroxypterocarpan and dimethylallyl pyrophosphate. A biogenetic relationship between these pterocarpans (which also occur naturally in *G.max*) and the pterocarpan G. is strongly implied by these observations (Fig.). Microsomes from elicitor-challenged *Glycine* cell suspensions catalyse 6a-hydroxylation of 3,9-dihydroxypterocarpan to 3,6a,9-trihydroxypterocarpan. The reaction is dependent on NADPH and O_2.
[U. Zähringer et al. *Z. Naturforsch.* **36C** (1981) 234–241; M.-L. Hagmann et al. *Eur. J. Biochem.* **142** (1984) 127–131]

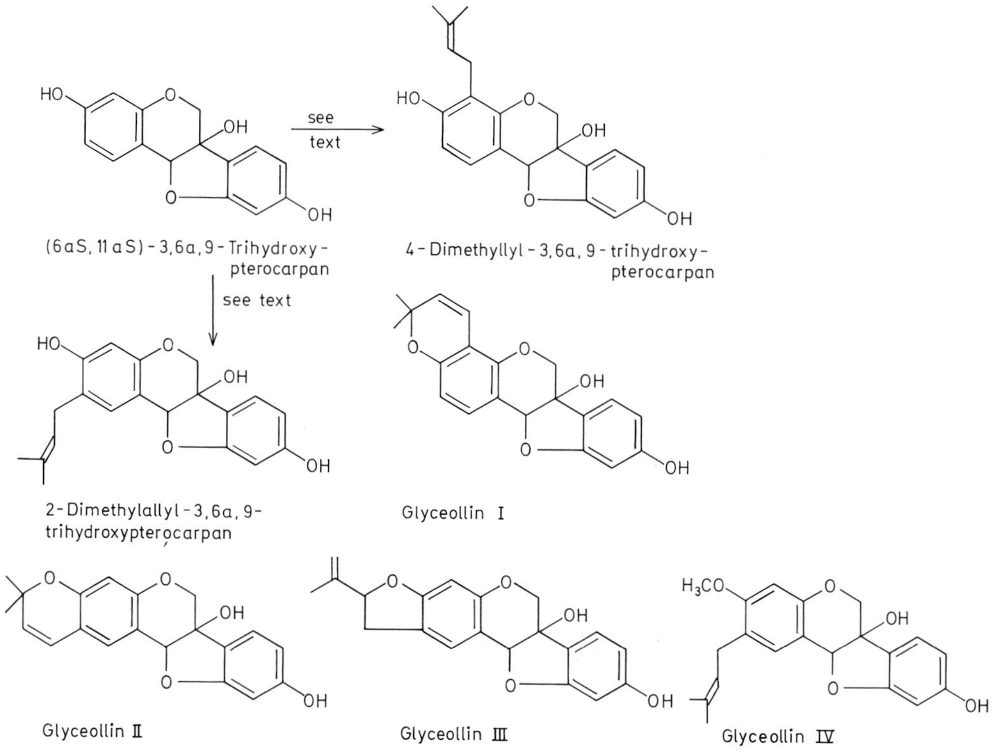

(6aS,11aS)-3,6a,9-Trihydroxy-
pterocarpan

see text →

4-Dimethyllyl-3,6a,9-trihydroxy-
pterocarpan

see text ↓

2-Dimethylallyl-3,6a,9-
trihydroxypterocarpan

Glyceollin I

Glyceollin II

Glyceollin III

Glyceollin IV

Pterocarpans from Glycine max L.

Glyceraldehyde 3-phosphate: see Triose phosphates.

Glycerate 2,3-bisphosphate, *2,3-diphosphoglycerate, 2,3-bisphosphoglycerate:* see Glycolysis, Hemoglobin, Rapoport-Luerbing shuttle.

Glycerate pathway: an anaplerotic pathway for utilization of glyoxylate in plants and microorganisms. Two molecules of glyoxylate are converted to tartronate semialdehyde by tartronate semialdehyde synthase (EC 4.1.1.47). The semialdehyde is then reduced to D-glycerate and phosphorylated by glycerate kinase (EC 2.7.1.31) to 3-phosphoglycerate. This is converted by phosphoglyceromutase (EC 2.7.5.3) and enolase (EC 4.2.1.11) to phospho-*enol*pyruvate, which enters the general metabolism. Balance: 2 glyoxylate $+ ATP + NAD(P)H + H^+ \rightarrow$ phospho *enol*pyruvate $+ ADP + CO_2 + NAD(P)$.

duced technically by addition of sodium hydrogen sulfite (as a natural hydrogen acceptor) to an alcoholic fermentation; this promotes the reduction of dihydroxyacetone phosphate (2nd form of Neuberg fermentation).

Glycerol fermentation: see Neuberg's fermentation.

Glycinamide ribonucleotide, abb. *Gar:* an intermediate of Purine biosynthesis (see).

Glycine, abb. *Gly, aminoacetic acid:* H_2N-CH_2-COOH, the simplest proteogenic amino acid. M_r 75.07, m.p. 233° to 290 °C (d.). Gly is not essential in the diet. Its amino nitrogen can easily be exchanged, so it is added to amino acid diets in large amounts to provide nitrogen for the synthesis of other amino acids. Gly is converted by transamination or oxidative deamination to glyoxylic acid, which is

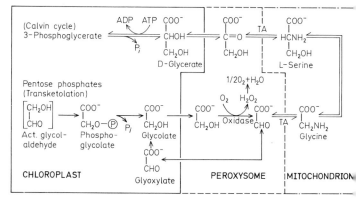

Figure 1. Glycolic acid cycle. TA transamination.

Glyceric acid: $HOCH_2-CHOH-COOH$, an hydroxyacid which is metabolically important, especially in its phosphorylated form. Glycerate is formed from glyoxylate via the Glycerate pathway (see) or from serine via hydroxypyruvate. The G.a. 1-, 2- and 3-phosphates and 1,3-diphosphoglycerate are important intermediates in alcoholic fermentation, glycolysis and photosynthesis. Formation of 3-phospho-D-G.a. from glyceraldehyde 3-phosphate is linked to formation of ATP.

Glycerides, *acylglycerols:* esters of fatty acids with glycerol. Mono- and diacylglycerides usually occur only as metabolic intermediates. Mixtures of triglycerides are Fats (see). It should be noted that the IUPAC-IUB Commission on Biochemical Nomenclature discourages the use of the terms "mono"-, "di-" and "triglyceride" for mono-, di- and triacylglycerols. A glyceride, strictly speaking, is any ester of glycerol; the phosphoric acid esters are particularly important as metabolic intermediates.

Glycerol, *propan-1,2,3-triol:* $CH_2OH-CHOH-CH_2OH$, a syrupy, sweet-tasting fluid. b.p. 290 °C (d.). G. occurs most often in nature as its esters (see Glycerides) in fats, fatty oils and Phospholipids (see). About 3% of G. is formed as a byproduct of alcoholic fermentation by reduction of dihydroxyacetone phosphate or glyceraldehyde 3-phosphate and hydrolysis of the phosphate group. It is pro-

further metabolized to formic acid. It is synthesized from glyoxylate, by transamination, or from serine by removal of the hydroxymethyl group. Glyoxylate and serine are relatively early products of photosynthesis (Fig. 1). The reactions leading to glycine from glyoxylate are part of the *glycolic acid cycle*. The α-C atom of Gly can be used to synthesize Active one-carbon units (see), either directly or via glyoxylic acid. The direct cleavage is dependent on tetrahydrofolic acid (THF) and yields active formaldehyde: $Gly + THF \rightarrow N^{5,10}$-methylene-THF $+ CO_2 + NH_3$. The pathway starting with glyoxylic acid leads to active formic acid. The reaction 2 $Gly \rightarrow$ L-serine $+ CO_2 + NH_3$ requires that one of the Gly be converted to an active one-carbon unit, while the other serves as acceptor in the glycine-serine interconversion. Gly can be completely degraded, either via the Succinate-glycine cycle (see) or by the aminoethanol cycle (Fig. 2). The former cycle is the route by which the α-C atom and the amino nitrogen of Gly are incorporated into Porphyrins (see), including heme. The intact Gly molecule is built into positions 4,5 and 7 of the purine ring, and also provides the C atoms for positions 2 and 8 as one-carbon units (see Purine biosynthesis). Sarcosine and glycine betaine are formed by methylation of Gly. Glycocyamine, a precursor of creatine and creatinine, is synthesized by Transamidination (see) with L-arginine. Gly is ala component of glutathione. It was first isolated, from gelatin, in 1819 by Braconnot.

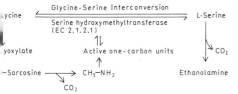

Figure 2. Aminoethanol (ethanolamine) cycle

Glycine allantoin cycle, *purine cycle:* a series of reactions leading to the synthesis of urea (Fig.) in the lungfish *(Dipnoi)* and certain urea-accumulating

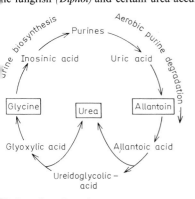

Glycine-allantoin cycle

plants. The glyoxylate and urea produced by purine degradation are reassimilated at different rates. Glyoxylate is converted to glycine, which can be reused for purine synthesis, thus closing the cycle. This cycle is particularly important, according to Krebs and Henseleit, for urea formation in lungfish (e.g. *Protopterus dolloi* and *Protopterus aethiopicus*), which may accumulate to 0.5 to 1% of their dry weight in urea during estivation.

Glycine cleavage system: an enzyme system accounting for a major proportion of glycine metabolism in vertebrate liver. It is composed of 4 proteins: 1. P-protein, which contains pyridoxal phosphate, 2. H-protein, which acts as a carrier and contains lipoic acid, 3. L-protein (lipoamide dehydrogenase) and 4. T-protein, a transferase dependent on tetrahydrofolate. The following reactions occur:

Glycine-succinate cycle: see Succinate-glycine cycle.

Glycinin: chief protein component of soybeans. It is stored in subcellular particles, the "protein bodies". The dimer (M_r 350000) consists of 12 subunits (M_r 28500 each). 6 of these chains are acidic (IP 3.0 to 3.4) and 6 are basic (IP 8.0 to 8.5). G is structurally related to the protein Arachin (see).

Glycoalkaloids: see Steroid alkaloids: Saponins.

Glycocholic acids: see Bile acids.

Glycogen: an animal polysaccharide which, like amylopectin, consists of D-glucose units. Most of the glycosidic bounds are α-1,4, but at branching points there are also α-1,6 links. The side chains consist of 6 to 12 glucose units; G. has about twice as much branching as amylopectin. The M_r ranges from 1 to 16 million, which means a maximum of 10^5 glucose units. G. reacts with iodine to give a brown-violet color. It is most abundant in the liver (up to 10%) and the muscles (up to 1%) and serves as a short-term storage substance. As such, it is subject to continuous synthesis and degradation.

Glycogen metabolism: the formation and

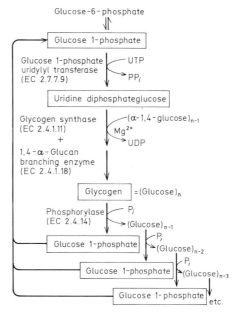

Figure 1. Synthesis and degradation of glycogen.

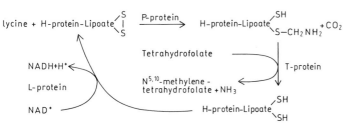

Sum: Glycine + THF + NAD$^+$ = N^{5,10}-methylene-THF + NH$_3$ + CO$_2$ + NADH + H$^+$. [K. Fujiwara et al. *J. Biol. Chem.* **259** (1984) 10664–10668]

breakdown of glycogen. The process has been well studied in mammalian muscle. Glycogen is synthesized from uridine diphosphate glucose (UDPG) and a starter molecule of the structure (α-1,4-glucosyl)$_n$ by glycogen synthase (EC 2.4.1.11). The branching of the molecule is accomplished by 1,4-α-glucan branching enzyme (EC 2.4.1.18), which transfers a segment of chain from a 4-OH to a 6-OH. (The enzyme is sometimes called Q enzyme) (Fig. 1). Glycogen degradation *(glycogenolysis)* is started by the enzyme phosphorylase (EC 2.4.1.1), which transfers a glucose residue from the non-reducing end of the chain to inorganic phosphate. The product is glucose 1-phosphate, which is isomerized to glucose 6-phosphate and then enters Glycolysis (see). The 1,6 bonds are split by a hydrolase rather than by phosphorylase. Therefore, complete degradation of glycogen yields about 10% free glucose and 90% glucose 1-phosphate.

G.m. is controlled by interconversion of the enzymes involved (see Covalent modification of enzymes) (Fig. 2). Glycogen synthase is normally present in muscles in the active form, while phosphorylase is in its inactive form *(b)*. However, phosphorylase *b* can be allosterically activated by a burst of AMP from muscle activity, so that phosphorylation can begin, although hormones and nervous stimulation are more important regulators of G.m. Adrenalin and glucagon activate adenylate cyclase (EC 4.6.1.1), and its product, cAMP, activates a number of protein kinases, including one which activates *phosphorylase kinase* (EC 2.7.1.38); and simultaneously inactivates glycogen synthase. (The inactive form of phosphorylase may also be allosterically activated by Ca$^{2\oplus}$ ions released as a result of muscle contraction.) The activated phosphorylase kinase in turn converts inactive phosphorylase *b* to active phosphorylase *a*. Phosphorylase *a* is subject to allosteric inhibition by glucose 6-phosphate, so that glycogen breakdown is slowed by a buildup of this

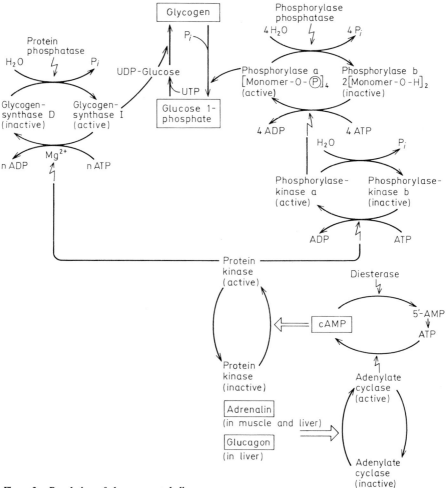

Figure 2. Regulation of glycogen metabolism.
cAMP = cyclic 3′,5′-adenosine monophosphate, 5′-AMP = adenosine 5′-monophosphate.

product. In addition, glucose 6-phosphate allosterically activates glycogen synthase.

The system relaxes to the resting state as a number of phosphatases remove the phosphate groups added by the kinases. Less is known about these enzymes, but they must also be regulated. The total system is thus regulated both allosterically by substrates and products and by posttranslational modification of the enzymes in response to hormones and nerve impulses.

Glycogenolysis: see Glycogen metabolism.

Glycogen storage disease: a group of conditions characterized by excessive storage of glycogen in the liver, muscles or other organs, due to hereditary lack of one of the enzymes of glycogen degradation (Table).

Glycogen storage diseases:

Type	Deficient enzyme and clinical condition
Cori type I G.s.d. (von Gierke's disease)	Glucose-6-phosphatase (EC 3.1.3.9). Enlarged liver and kidneys due to glycogen accumulation. Glycogen structure normal. No accumulation of glycogen in muscle. Hypoglycemia and lactic acidosis. Growth retarded. Prognosis favorable.
Cori type II G.s.d. (Pompe's disease)	Amylo-1,4-α-glucosidase (EC 3.2.1.20). Glycogen structure normal. Glycogen accumulates in many tissues, including heart, which is enlarged. Most of the glycogen is segregated in large vacuoles (distended lysosomes) not seen in other forms of glycogenesis. The enzyme is lysosomal and glycogen enters lysosomes for this stage of degradation. Carbohydrate metabolism is not otherwise markedly affected. Death in first year from heart failure.
Cori type III G.s.d. (Forbes' disease)	Amylo-1,6-α-glucosidase, glycogen debranching enzyme (EC 3.2.1.33). Glycogen has short outer chains and is dextrin-like. Liver enlarged in infancy but may decrease in adolescence. Moderate hypoglycemia. Lactic acidosis mild or absent. Prognosis favorable.
Cori type IV G.s.d. (Andersen's disease, Amylopectinosis)	Amylo-(1,4 → 1,6)-transglucosidase, glycogen branching enzyme (EC 2.4.1.18). Glycogen has long inner and outer chains with very few branch points, and it is readily precipitated in the tissues. Liver cirrhosis. Fatal in early childhood.

Glycogen storage diseases:

Type	Deficient enzyme and clinical condition
Cori type V G.s.d. (McArdle's disease)	Muscle phosphorylase (EC 2.4.1.1). Glycogen structure normal. Glycogen accumulates only in muscle. Exercise causes muscle cramps; myoglobin from damaged muscle may appear in urine. Patients symptomless if they refrain from strenuous exercise. Prognosis favorable.
Cori type VI G.s.d. (Hers' disease)	Liver phosphorylase (EC 2.4.1.1). Glycogen structure normal. Enlarged liver. Moderate hypoglycemia and lactic acidosis. Prognosis favorable.
VII (not listed by Cori)	6-Phosphofructokinase (muscle isoenzyme) (EC 2.7.1.11). Glycogen structure normal. Glycogen accumulation in muscle. Abnormal muscular weakness and stiffness on vigorous or prolonged exercise. Fructose-6-phosphate and glucose-6-phosphate also accumulate. Half of enzyme activity in erythrocytes is lost (remaining 50% is different isoenzyme). Mild hemolysis. Prognosis favorable.
VIII (not listed by Cori)	Phosphorylase kinase (EC 2.7.1.38). Glycogen structure normal. Liver enlarged. Prognosis favorable.

Glycolic acid, *hydroxyacetic acid:* $HOCH_2-COOH$, an hydroxycarboxylic acid found in young plant tissue and green fruits, such as gooseberries grapes and apples. M_r 76.05, m.p. 80 °C. Glycolate is an intermediate in photosynthesis. Its precursor is active glycolaldehyde, which is formed in Transketolation (see) reactions. G.a. is oxidized to glyoxylic acid by glycolate oxidase (EC 1.1.3.1). Glyoxylic acid may be converted to Glycine (see) or reduced back to G.a.

Glycolic acid cycle: see Glycine.

Glycolipids: compounds in which one or more monosaccharide residues are glycosidically linked to a lipid part, either a mono- or diacylglycerol, a long-

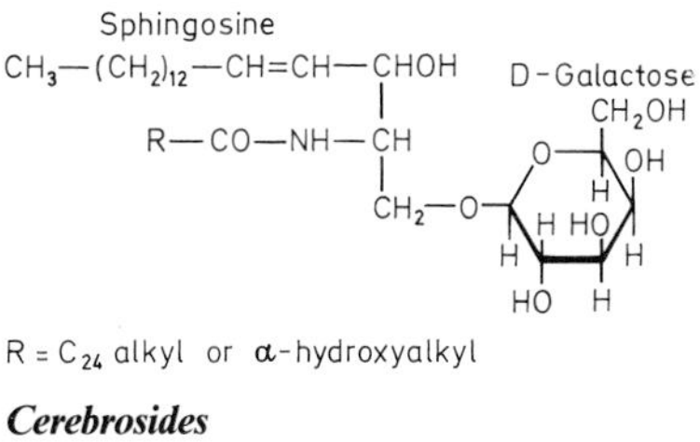

$R = C_{24}$ alkyl or α-hydroxyalkyl

Cerebrosides

Glycolipids

chain base (sphingoid) like sphingosine, or a ceramide. (A *ceramide* consists of a fatty acid linked by an amide bond to the amino group of sphingosine.) In common usage, the term G. refers to sphingosine derivatives which do not contain phosphorus. The *cerebrosides* (Fig.) consist of sphingosine carrying a fatty amide or hydroxy fatty amide and a single monosaccharide unit on its alcohol group, either glucose or galactose. Some examples which have been obtained in pure form are kerasin, containing lignoceric acid; phrenosine, containing cerebronic acid (2-hydroxylignoceric acid); nervone, containing nervonic acid; and hydroxynervone, containing 2-hydroxynervonic acid. Cerebrosides account for 11% of the dry matter in the brain, and are found in smaller amounts in liver, thymus, kidney, adrenals, lungs and egg yolk.

Sulfatides are cerebrosides in which the C-6 of the sugar is esterified to a sulfate residue. They are found in the brain. *Glycosphingolipids* are more complex derivatives of ceramide which contain more than one saccharide unit on the primary alcohol group of the sphingosine. The oligosaccharide unit may be a branched chain. It is usually composed of some combination of galactose, glucose, fucose, N-acetylglucosamine and N-acetylgalactosamine. If the chain contains sialic acid (N-acetyl or N-glycoloyl neuraminic acid), the lipid is a *ganglioside* (sialoglycosphingolipid) (see Sphingolipidoses under Inborn errors of metabolism).

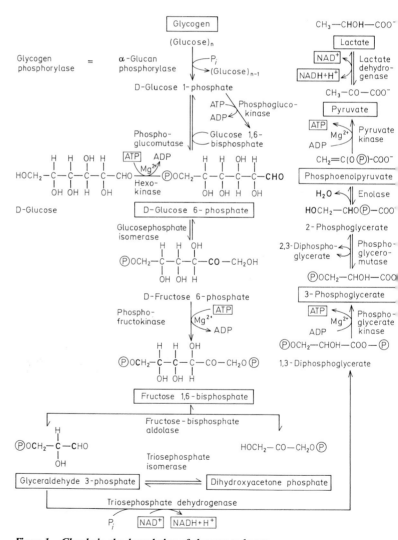

Figure 1. Glycolysis, the degradation of glycogen to lactate.

Gangliosides are especially abundant in the gray matter of the brain and in the thymus, but also are found in erythrocytes, leucocytes, serum, kidneys, adrenals and other organs. They are the characteristic lipid components of some neuronal membranes of the central nervous system and are probably involved in the transmission of impulses along neurons.

Glycolysis, *Embden-Meyerhof-Parnas pathway:* the main pathway for anaerobic degradation of carbohydrates found in all groups of organisms (Fig. 1, Table). For each mole of glucose consumed, 150.7 kJ (36 kcal) energy is released. The organism obtains a net yield of 2 moles ATP per mole glucose. The starting material is glycogen or starch, which is hydrolysed to glucose 1-phosphate or glucose monomers. G. can be divided into four phases: 1. the formation of two molecules of triose phosphate (glyceraldehyde 3-phosphate and dihydroxyacetone phosphate) from one molecule of hexose. Two molecules of ATP are consumed in this step. 2. Dehydrogenation of the triose phosphates to 2-phosphoglycerate; NAD$^+$ is reduced in the process to NADH. One molecule ATP is generated per triose phosphate, which makes up for the ATP invested in step 1.

(Fig. 2). 3. Conversion of 2-phosphoglycerate to Pyruvate (see) via phospho *enol*pyruvate. Another ATP is generated here for each molecule of pyruvate. 4. The reduction of pyruvate to regenerate NAD$^+$. In muscle, the pyruvate is converted to lactate, and in yeast, it is reductively decarboxylated to ethanol (see Alcoholic fermentation). In an aerobically respiring cell, the NADH is oxidized ultimately by the respiratory chain (see Hydrogen metabolism), and the pyruvate is further oxidized in the Tricarboxylic acid cycle (see). Under anaerobic conditions, G. can only be maintained continuously by the redox reactions shown in Fig. 3.

Balance: Glucose $(C_6H_{12}O_6) + 2P_i + 2$ ADP $\rightarrow$ 2 lactate $(C_3H_6O_3) + 2$ ATP.

If the starting material is glycogen, which is degraded to glucose 1-phosphate, the yield is 3 ATP per glucose 1-phosphate consumed.

The key enzyme of G. is 6-phosphofructokinase (EC 2.7.1.11), which is inhibited by high concentrations of ATP and is activated by ADP or AMP. Its product, fructose bisphosphate, activates pyruvate kinase. The Pasteur effect (see) is another form of regulation of G.

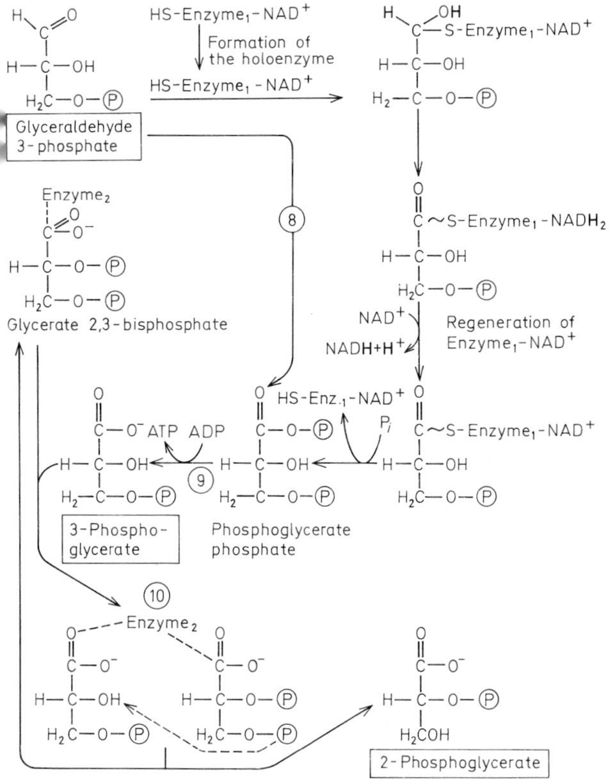

Enzyme$_1$ = triosephosphate dehydrogenase
Enzyme$_2$ = Phosphoglyceromutase

Figure 2. Mechanism of reactions 8 to 10 in glycolysis (see table).

Glycophorins

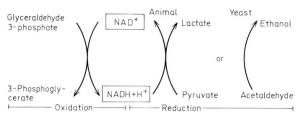

Figure 3. Oxidoreduction cycle of glycolysis.

Table. Reactions of glycolysis.

No.	Chemical equation	Enzyme	Inhibitors	$G^{o'}$ (Free energy) kJ/mol (kcal/mol)
1	Starch or glycogen $+ nP_i \rightarrow n$ glucose-1-P	Phosphorylase (EC 2.4.1.1)		$+3.06$ $(+0.73)$
2	Glucose-1-P $\xrightarrow{\text{glucose 1,6-P}_2}$ Glucose 6-P	Phosphoglucomutase (EC 2.7.5.1)	Fluoride, Organic phosphates	-7.29 (-1.74)
3	Glucose $+$ ATP$\xrightarrow{\text{Mg2}+}$ Glucose-6-P $+$ ADP	Hexokinase (EC 2.7.1.1)		-16.77 (-4.0)
4	Glucose-6-P $\rightleftharpoons$ Fructose-6-P	Glucosephosphate isomerase (EC 5.3.1.9)	2-Deoxyglucose-6-P	$+1.68$ $(+0.4)$
5	Fructose-6-P $+$ ATP $\xrightarrow[\text{(AMP, ADP)}]{\text{Mg}^{2+}, \text{K}^+}$ Fructose-1,6-P$_2$	6-Phosphofructokinase (EC 2.7.1.11)	ATP, Citrate	-14.2 (-3.40)
6	Fructose-1,6-P$_2 \rightleftharpoons$ Dihydroxyacetone-P $+$ Glyceraldehyde-3-P	Fructose-bisphosphate aldolase (EC 4.1.2.13)	Chelating agents (only with microbial enzymes)	$+24.0$ $(+5.73)$
7	Dihydroxyacetone P $\rightleftharpoons$ Glyceraldehyde-3-P	Triosephosphate isomerase (EC 5.3.1.1)		$+7.66$ $(+1.83)$
8	Glyceraldehyde-3-P $+ P_i +$ NAD$^+ \rightleftharpoons$ 1,3-P$_2$-glycerate $+$ NADH $+$ H$^+$	Glyceraldehyde-phosphate dehydrogenase (EC 1.2.1.12)	Threose-2,4-P$_2$	$+6.28$ $(+1.50)$
9	1,3-P$_2$-Glycerate $+$ ADP $+$ H$^+$ $\xrightarrow{\text{Mg}^{2+}}$ 3-P-glycerate $+$ ATP	Phosphoglycerate kinase (EC 2.7.2.3)		-18.86 (-4.5)
10	3-P-glycerate $\xrightarrow{\text{2,3-P-glycerate}}$ 2-P-glycerate	Phosphoglyceromutase (EC 2.7.5.3)		$+4.44$ $(+1.06)$
11	2-P-Glycerate $\xrightarrow{\text{Mg}^2 (\text{Mn}^{2+})}$ Phospho *enol*pyruvate	Enolase (EC 4.2.1.11)	Ca^{2+}, F$^- + P_i$	$+1.84$ $(+0.44)$
12	Phospho *enol*pyruvate $+$ ADP $+$ H$^+$ $\xrightarrow{\text{Mg}^{2+}, \text{K}}$ Pyruvate $+$ ATP	Pyruvate kinase (EC 2.7.1.40)	Ca$^{2+} \rightleftharpoons$ Mg^{2+} Na$^+ \rightleftharpoons$ K$^+$	-31.44 (-7.5)
13	Pyruvate $+$ NADH $+$ H$^+ \rightleftharpoons$ Lactate $+$ NAD$^+$	Lactate dehydrogenase (EC 1.1.1.27)	Oxamate	-25.1 (-6.00)

Note: For each mole of glucose, reactions 8 through 13 must be doubled, because two moles of triose phosphate are produced from one mole of glucose. $-$P$=$phosphate; $P_i =$inorganic phosphate.

Glycophorins: a group of sialoglycopeptides found in human erythrocytes. The major G. is G. A., which comprises about 75% of the total. G. A. is a single polypeptide of 131 amino acids, and in addition carries carbohydrates which make up about 60% of its M_r. The *N*-terminal 70 amino acids protrude from the lipid bilayer (see Biological membranes) outside the cell, and 16 of these are linked to carbohydrate. There are slight differences in the 8 *N*-terminal amino acids; individuals with blood group M have an amino-terminal serine and a glycine at position 5, while group N individuals have leucine and glutamic acid, respectively, at these positions. The patterns of glycosylation of G. A. are responsible for other variations of the MN blood group, and account for the Ss system.

Glycoprotein hormones: a term applied to the two pituitary gonadotropins (luteinizing and follicle-stimulating hormones), choriogonadotropin and thyrotropin, all of which share the following type of structure: 2 peptide subunits (α and β), each of which is glycosylated and internally cross-linked by disulfide bonds. They are thus regarded as a family, and in the narrow sense other glycosylated protein hormones, e.g. erythropoietin, are excluded. It now seems appropriate to include Inhibin (see) in the family of G.h. [J.G. Pierce & T.F. Parsons *Ann. Rev. Biochem.* **50** (1981) 465–495]

Glycoproteins: proteins with covalently linked oligosaccharide. The protein of G. is translated from mRNA, and the addition of oligosaccharide represents a post-translational modification. This is in contrast to the peptidoglycans (see Murein), where the carbohydrate backbone is structurally dominant and cross-linked by relatively short polypeptide sequences synthesized independently of mRNA and ribosomes. G. are typically found in plants and animals, and not in bacteria (a notable exception is a G. from *Halobacterium*, see below). Most G. are either secreted into body fluids or are membrane proteins. They include many enzymes, most protein

hormones, plasma proteins, all antibodies, complement factors, blood group and mucus components and many membrane proteins. The polypeptide chains of G. generally carry a number of short heterosaccharide chains, and these, in turn, almost always include *N*-acetylhexosamines and hexoses (usually galactose and/or mannose, less often, glucose). The last member of the chain is very often sialic acid or L-fucose. The oligosaccharide chains are often branched, and rarely contain more than 15 monomers (2 to 10 monomers, corresponding to a M_r of 540 to 3200 is the usual range). The number of saccharide chains on a polypeptide varies widely. Ovalbumin, at the one extreme, contains only 3% saccharide, while sheep submaxillary G., at the other extreme, is 50% carbohydrate. The sequences of sugar monomers are determined by the specificities

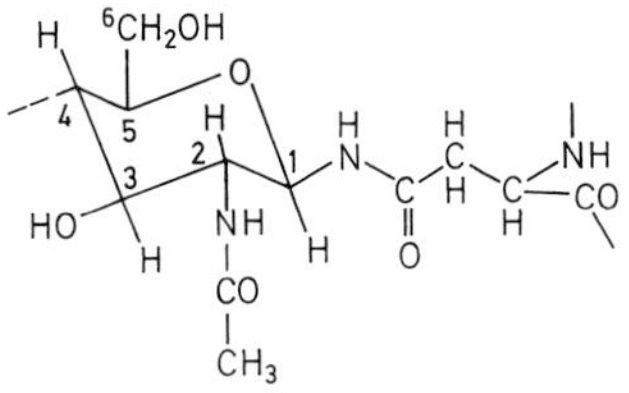

Figure 1. Linkage of N-acetylglucosamine and an L-asparagine residue, representing the site of attachment of N-glycans in glycoproteins.

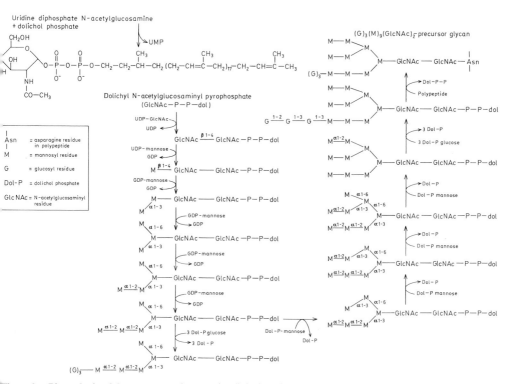

Figure 2. Biosynthesis of the precursor of asparagine-linked N-glycans.

241

of glycosyl transferases, and probably by the concentrations of available activated sugars as well. Consequently, G. display microheterogeneity in their saccharide moieties.

Properties and functions: Due to their high viscosities, the G. serve as lubricants and protective agents, for example against proteolytic enzymes, bacteria and viruses. They play a role in cellular adhesion and contact inhibition of cell growth in tissue culture. They are also responsible for cellular recognition of foreign tissue.

G. are the essential components of the receptors for virus and plant agglutinins, and of blood group substances. The siderophilins and ceruloplasmin are G. Some G. appear to be membrane carrier proteins. The carbohydrate portion of gonadotropic hormones is essential to their biological activity; in many cases, selective removal of the terminal sialic acid residues inactivates the hormone. It is thought that the carbohydrate serves as a label for recognition by a receptor.

G. appear to be synthesized on the rough endoplasmic reticulum, the growing polypeptide chain being extended through the membrane from the cytoplasmic side. Sugar units are added to the chain in the cisternae of the endoplasmic reticulum (ER), preventing it from recrossing the membrane. In the case of N-glycans (see below), a standard complement of sugars is added in the ER, consisting of 2 N-acetylglucosamine, 9 mannose and 3 glucose units. Vesicles which bud off the ER carry the G. to the Golgi apparatus (see), where some of the sugar units are removed and others are added, to produce the structure appropriate to the protein. The processed G. may then be excreted, carried to one of the cell organelles, or incorporated into one of the cell membranes.

The covalent protein-carbohydrate linkage may be N-glycosidic or O-glycosidic. N-Glycosidically linked oligosaccharides are called N-glycans; O-glycosidically linked oligosaccharides are called O-glycans.

N-Glycosidic linkage. Only one type of N-glycosidic linkage is known in animal and plant G. This occurs between the amide nitrogen of an asparaginyl residue and C-1 of an N-acetylglucosamine residue (GlcNAc) (Fig. 1). (A sulfated G., present in the cell wall of *Halobacterium*, contains two types of carbohydrate: 1. a sulfated, repetitive, high-M_r saccharide, resembling an animal glycosaminoglycan, and 2. a sulfated, low M_r saccharide with the composition 1 glucose:3 glucuronic acid:3 sulfate. In this compound, the glucose is linked N-glycosidically to the amide nitrogen of an asparaginyl residue. This is the first report of an asparaginylglucose linkage in any G. [F. Wieland et al. *Proc. Nat. Acad. Sci. USA* **80** (1983) 5470-5474].)

Most N-glycans contain a common core (Fig. 3). Each of the terminal mannoses of the core structure is regarded as an antenna for extension. If each terminal mannose carries at least one GlcNAc residue, the structure is known as complex antennary. If only one core mannose is substituted by GlcNAc, and the other is extended by mannose residues, the structure is known as biantennary hybrid. Attachment of more than one GlcNAc to a core terminal mannose

results in triantennary, tetra-antennary, etc. structures. A bisected antennary structure possesses a GlcNAc residue linked β1→4 to the central mannose of the common core (Fig. 4).

Biosynthesis of N-glycans. A large precursor of the oligosaccharide is first synthesized on the lipid carrier, Dolichol phosphate (see). Glycan synthesis is initiated by the interaction of dolichol phosphate with uridine diphosphate N-acetylglucosamine (UDPGlcNAc) to form dolichyl N-acetylglucosaminylpyrophosphate (Fig. 2). As illustrated in Fig. 2, the subsequent attachment of α-mannosyl and finally glucose units is an ordered process probably requiring separate, specific mannosyl and glucosyl transferases for each stage. The entire final product is transferred to a specific asparaginyl residue of a polypeptide acceptor (with release of dolichol pyrophosphate). Small peptides containing asparagine, as well as native proteins, act as acceptors, and the enzyme is specific for the structure -Asn-X-Ser(or Thr)-, where X may be any amino acid except proline and possibly aspartic acid, and the amino group of Asn and the carboxyl group of Ser (or Thr) must be in a peptide linkage or otherwise blocked.

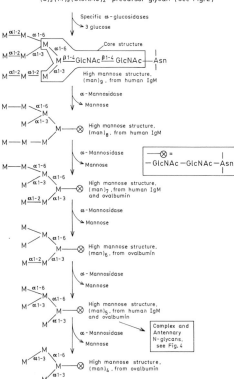

Figure 3. Processing of the asparagine-linked precursor carbohydrate. Removal of the three glucose residues produces a high-mannose structure, designated (man)₉. Subsequent removal of mannose produces high-mannose structures (man)₈ to (man)₄.

Processing (in the Golgi apparatus) then commences with removal of the three glucose residues by specific α-glucosidases, followed by ordered removal of α-mannose residues by specific α-mannosidases, thereby accounting for all high-mannose structures encountered in G.

Complex and antennary N glycans are derived from the (man)₅ high-mannose structure in a series of reactions initiated by transferase I (UDP-GlcNAc:α-D-mannoside(GlcNAc to Manα1-3) β1-2-GlcNAc transferase I) (Fig.4). After attachment of the first GlcNAc to the α1 → 3-linked core mannose of the (man)₅ structure, mannose is removed by α-mannosidases specific for α1 → 3 and α1 → 6 linkages (different enzymes from those involved in earlier processing). Further GlcNAc residues are introduced by specific GlcNAc transferases II, III, and IV, and fucose residues are attached to core GlcNAc residues by fucose transferase (GDP-Fuc:β-N-acetylglucosaminide(Fuc to Asn-linked GlcNAc)α1 → 6-fucosyl transferase). Galactose units are transferred from UDP-galactose to C-4 of GlcNAc residues, producing the N-acetyllactosamine sequence present in many N-glycans. Sialic acid residues (which occur in C-3 or C-6 linkage to penultimate galactose residues) are transferred from CMP-sialic acid by the action of sialyl transferase. After incorporation into the glycan structure, sialic acid may be modified by hydroxylation of the N-acetyl group to N-glycolyl, or by acetylation.

O-Glycans. In the O-glycans, C-1 of a sugar residue is linked glycosidically to the hydroxyl group of the side chain of serine, threonine, hydroxylysine or hydroxyproline. Arabinose linked glycosidically to hydroxyproline is found very commonly in plant G., but has not been identified in G. from any other type of organism. D-Galactose linked β-glycosidically to hydroxylysine has been found only in animal collagen. Collagen also contains 2-O-α-D-glucopyranosyl D-galactose linked to hydroxylysine. Galactose linked glycosidically to serine is found in the plant G., extensin, and in the collagen of the primitive worms, *Lumbricus* and *Nereis.* In the G. of yeasts and fungi, carbohydrate chains are attached by a glycosidic linkage between mannose and serine or threonine, and this type of linkage also makes a mi-

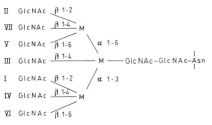

Figure 4 a. Numbering system for antennas and GlcNAc transferases

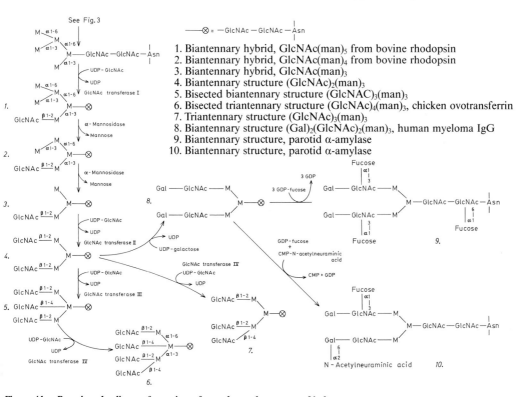

1. Biantennary hybrid, GlcNAc(man)₅ from bovine rhodopsin
2. Biantennary hybrid, GlcNAc(man)₄ from bovine rhodopsin
3. Biantennary hybrid, GlcNAc(man)₃
4. Biantennary structure (GlcNAc)₂(man)₃
5. Bisected biantennary structure (GlcNAC)₃(man)₃
6. Bisected triantennary structure (GlcNAc)₄(man)₃, chicken ovotransferrin
7. Triantennary structure (GlcNAc)₃(man)₃
8. Biantennary structure (Gal)₂(GlcNAc)₂(man)₃, human myeloma IgG
9. Biantennary structure, parotid α-amylase
10. Biantennary structure, parotid α-amylase

Figure 4 b. Reactions leading to formation of complex and antennary N-glycans.

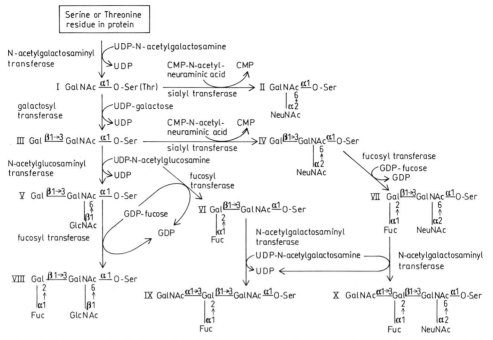

Figure 5. Known reactions in the synthesis of submaxillary gland glycoproteins. The specificity of *N*-acetylgalacto-saminyl transferase responsible for synthesis of I has been studied using basic myelin protein (an unusual substrate, which is not normally glycosylated; partial structure: Val(94)-Thr-Pro-Arg-Thr-Pro-Pro-Pro(101)-). Glycosylation is specific for Thr 98 (Thr 95 is not glycosylated); a proline-rich environment appears to be necessary around the glycosylated hydroxyamino acid residue.

nor contribution to mammalian G. structure. Some established reactions of O-glycan synthesis in submaxillary gland G. are shown in Fig. 5. [R. C. Hughes, *Glycoproteins (Outline Studies in Biology)*, W. J. Brammer & M. Edidin, eds. (Chapman & Hall, 1983); H. Schachter et al. *Can. J. Biochem. Cell Biol.* **61** (1983) 1049–1066]

The disaccharide O-glycan, II, accumulates in sheep salivary gland, which contains only low activity of specific galactosyl transferase. II is not a substrate for other transferases.

Pig submaxillary gland contains high activities of various transferases, and longer glycans are present, e. g. III and IV are prominent.

Formation of the GlcNAc β1-6 linkage in V possibly initiates a series of reactions leading to synthesis of human blood group active products of secretory cells (see Blood group antigens). The GlcNAc transferase responsible for synthesis of V is highly specific for the presence of β 1-3-linked unsubstituted Gal; the enzyme shows only very low activity towards VI. Glycan VIII is human blood group H-active. IX and X show human blood group A antigenic activity.

Glycosaminoglycan: see Mucopolysaccharide.
Glycosaminoglycan degradation: see Inborn errors of metabolism.

Glycosidases: a group of hydrolases which attack glycosidic bonds in carbohydrates, glycoproteins and glycolipids. The G. are not highly specific.

Usually they distinguish only the type of bond, e. g. *O-* or *N*-glycosidic, and its configuration (α or β). See Amylases, Taka amylase Neuraminidase, Oligo-1,6-glucosidase, Amylo-1,6-glucosidase, Disaccharidases, Cellulases, Lysozyme, and Invertase.

Glycosides: a group of compounds in which mono- or oligosaccharide units are linked by acetal bonds with hydroxyl groups of alcohols or phenols (*O*-glycosides) or with amino groups (*N*-glycosides). Acid hydrolysis cleaves these compounds into a sugar and a noncarbohydrate portion, the *aglycon* or *genin*. The Saponins (see), cardiac glycosides and cyanogenic glycosides are examples of *O*-glycosides; the Nucleic acids (see) and Nucleosides (see) are *N*-glycosides.

Glycyrrhetic acid, *glycyrrhetin:* a pentacyclic triterpene with carboxylic acid and ketone functions. M_r 470.7, m. p. 300 °C, $[\alpha]_D +98$ ° (chloroform). G. a. is the aglycon of glycyrrhizic acid, an extremely sweet principle from the root of licorice, *Glycyrrhiza glabra* L. G. a. is the aglycon of Saponins (see) from other plants, e. g. the bark of *Pradosia latescens* and rhizomes of the fern *Polypodium vulgare*. G. a. differs structurally from β-amyrin in having an 11-keto and a 30-carboxyl group (see Amyrin, Fig.).

Glyoxalase, *aldoketomutase:* a system of enzymes found in many organisms. It consists of lactoyl-glutathione lyase (G. I) (EC 4.4.1.5), which condenses methylglyoxal and glutathione to S-lactoyl-glutathione, and hydroxyacylglutathione hydrolase

(G. II) (EC 3.1.2.6), which hydrolyses the condensation product to lactate and glutathione. The significance of the system is unknown.

Glyoxylate carboligase: see Tartronic semialdehyde synthase.

Glyoxylate cycle, *Krebs-Kornberg cycle:* an alternative to the tricarboxylic acid cycle found in microorganisms and plants. In the G. c., oxaloacetate is generated from acetyl-coenzyme A (Fig.). The key enzymes are isocitrate lyase (isocitratase) (EC 4.1.3.1) and malate synthase (EC 4.1.3.2). Balance: 2 acetyl-CoA $+ NAD^+ + 2H_2O \rightarrow$ succinate $+ 2CoA + NADH + H^+$. The intermediates of the G. c. serve as starting materials for various synthetic pathways. Succinate is especially important as the precursor for gluconeogenesis. The G. c. is used by plant seedlings to utilize their fat reserves, and by microorganisms to grow on fatty acids or acetate as their sole carbon sources. The mammalian organism lacks isocitrate lyase and malate synthase, and consequently the G. c.

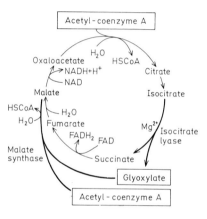

Glyoxylic acid, *oxoacetic acid, glyoxalic acid:* CHO-COOH, a carboxylic acid found in green fruits, seedlings and young leaves. In some plant families (maple, borage, horse chestnut) G. a. is present in the form of allantoin and allantoic acid, which arise in the course of purine catabolism. G. a. is synthesized by transamination or oxidative deamination of glycine, or from sarcosine. Decarboxylation of G. a. yields active formate. G. a. is the starting point of the Glyoxylate cycle (see).

GMP: abb. for guanosine 5'-monophosphate.

GMP reductase (EC 1.6.6.8): a flavin enzyme which catalyses the conversion of guanosine 5'-monophosphate to inosinic acid in a single step. It is part of the system for converting guanine to adenine compounds.

Goitrogens, *antithyroid compounds:* substances which inhibit iodine peroxidase (conversion of iodide to "active iodine": $H_2O_2 + 2I^- + 2 H^+ \rightarrow 2"I" + 2H_2O$), the iodination of tyrosine residues, and the coupling of monoiodotyrosine and diiodotyrosine residues to form T_3 and T_4 (see Thyroxin). The resulting low plasma levels of T_3 and T_4 stimulate the release of thyrotropin from the anterior pituitary. This, in turn, results in a compensatory hypertrophy of the thyroid gland. Such an enlargement, without inflammation or malignancy, is called a goiter. Some G. are 2-thiouracil, thiourea, sulfaguanidine, propylthiourea, 2 mercaptoimidazole, 5-vinyl-2-thiooxazolidone (see Glucosinolate) and allylthiourea (in mustard).

Golgi apparatus: a complex of membranous structures in eukaryotic cells in which proteins are glycosylated. In 1898, Golgi observed a net-like structure which was made visible by staining with OsO_4 or $AgNO_3$, but this is not the same as the G. a. seen in the electron microscope.

The G. a. in differentiated cells consists of three types of structure: cisternae (shaped like flattened balloons), vesicles which form on the edges of cisternae, and groups of vacuoles, which are also called "dictyosomes". The cisternae are stacked; the minimum number in a stack is 4, while in plant cells as many as 20 cisternae may comprise a stack. Mammalian cells typically have 5 or 6 cisternae per stack. There are at least three functionally different compartments in each stack. The cisterna(e) on the *cis* or protein-receiving side of the stack are characterized by the presence of mannosidase I; those in the middle, by mannosidase II and *N*-acetylglucosamine transferases; and those on the *trans* or protein-releasing side, by galactosyl and sialyl transferases (see Glycoproteins for explanation of these enzymes). It has been shown that glycoproteins of the N-glycan type receive an oligosaccharide chain in the lumen of the endoplasmic reticulum. The same chain is added to all the proteins; it consists of 2 *N*-acetylglucosaminyl, 9 mannosyl and 3 glucosyl units. These proteins are carried in vesicles which bud off from the endoplasmic reticulum to the *cis* face of the G. a. Here the proteins destined for different cellular compartments, e.g. lysosomes, plasmalemma, or secretion, are apparently recognized and processed differently, depending on their destinations. Lysosomal proteins are phosphorylated on the carbohydrate chains, but otherwise not greatly modified. Secretory and plasmalemma proteins typically lose all their mannose units and receive other carbohydrate units in a multistep process.

The proteins are transported from one cisterna of the G. a. to the next by the vesicles which form at the edges of the cisternae. Experiments with fused cells have indicated that proteins do not necessarily go through all the steps of glycosylation in the same stack. Proteins radioactively labelled in mutant cells which are unable to complete processing may be correctly completed if the mutant cells are fused with normal ones containing no radioactivity. This implies transit from the mutant to the normal G. a. stacks.

At the end of processing, the glycoproteins may be stored in the dictyosomes, secreted, transported to the appropriate cell organelles, or incorporated into the plasmalemma. The mechanisms by which this sorting and transport are achieved are unknown, but it may be presumed that the nature of the carbohydrate units added in the G. a. serve to direct each type of protein to its correct destination. [J. E. Rothman, *Scientific American* **253** (1985) 74–89; W. G. Dunphy, R. Brands & J. E. Rothman, *Cell* **40** (1985) 463–472]

Gonadal hormones, *sex hormones:* a biologically important group of steroid hormones which are produced in the male (testes) and female (ovaries) gonads. The G.h. determine the male or female character of an organism, in that they effect the normal development and function of the sex organs and the expression of the secondary sexual traits. They are classified as *male G.h.* (see Androgens) and *female G.h.* The latter fall into two groups with different physiological effects, the Estrogens (see) and the Gestagens (see).

Gonadotropin releasing hormone: see Releasing hormones.

Gonadotropins: a group of glycoprotein hormones from the anterior lobe of the pituitary and the placenta. They stimulate the gonads to growth and production of sex-specific hormones (female, see Estrogens and Gestagens; male, see Androgens). The group of G. includes Follitropin, or Follicle-stimulating hormone (FSH) (see), Lutropin, or Luteinizing hormone (LH) (see), Human menopausal gonadotropin (Urogonadotropin, hMG) (see), Prolactin (see), and Chorionic gonadotropin (Choriogonadotropin, hCG) (see). Follitropin, choriogonadotropin and urogonadotropin consist of α and β chains. The primary sequences of the α chains are nearly identical, while the β chains are hormone-specific, which suggests that the three hormones have evolved from a common ancestor. Lutropin and follitropin are produced in the anterior lobe of the pituitary, though not continuously. Choriogonadotropin is synthesized in the syncytiotrophoblasts of the placenta; the maximum occurs in the second month of pregnancy.

The synthesis of lutropin and follitropin is controlled by gonadotropin releasing hormone from the hypothalamus, which in turn is influenced by the sexual center in the brain. The sexual center, hypothalamus, anterior lobe of the pituitary and gonads or placenta form a feed-back loop in which in the sex hormones from the gonads exert negative feedback on the brain. In females, a cyclic pattern of variation in the hormone levels (menstrual cacle) is established. In males, the variations are acyclic. In addition to these long-term variations, there are diurnal fluctuations in the hormone levels.

The G. interact with specific receptors in the theca and follicle cells of the ovary and corpus luteum and with the intestitial cells of Leydig in the testes. cAMP is released as a second messenger to stimulate the production of steroid sex hormones.

G. can be detected by radioimmunoassay, the radioligand-hormone receptor method, or by biological assays (see Hormones).

Gonads, *sex glands:* organs in which the gametes and sex hormones are formed. The male gonads are the *testes,* the female organs, the *ovaries.* The formation and secretion of the sex hormones are controlled by the pituitary gonadotropins.

Gossypol: an aromatic triterpene from cotton seed *(Gossypum hirsutum). M_r* 518.54, m.p. 184 °C from ether, 199 °C from chloroform, 214 °C from ligroin. G. is synthesized from mevalonic acid via neryl pyrophosphate and *cis,cis-*farnesyl pyrophosphate. It is somewhat poisonous, and has been suggested as an insecticide.

Gossypol

GOT: see Transaminases.

Gougerotin, *aspiculamycin, asteromycin:* 1-cytosinyl-4-sarcosyl-D-serylamino-1,4-dideoxy-β-D-glucopyranuramide, an important pyrimidine antibiotic (see Nucleoside antibiotics) synthesized by *Streptomyces gougeroti.* It inhibits protein biosynthesis on both eukaryotic and prokaryotic ribosomes.

Gougerotin

GPT: see Transaminases.

Gradient elution: see Column chromatography.

Gramicidins: cyclic or linear peptide antibiotics produced by *Bacilis brevis.* Gramicidin S, cyclo(-D-Phe-L-Pro-L-Val-L-Orn-L-Leu-)₂, was first isolated in 1942. The primary structure was determined by Synge, and the total synthesis, the first of any peptide antibiotic, was reported in 1965 by R. Schwyzer et al. In the bacteria, the synthesis is accomplished by a small, soluble system of enzymes. The first activates L-Phe. The second activates the other four amino acids by transferring them to the thiol groups of pantotheine molecules covalently bound to the enzyme. The activated phenylalanine, which at some point is converted to D-Phe, attacks the activated L-Pro to initiate synthesis. The Pro attacks the Val, and so on, to form the pentapeptide. Then two pentapeptides are joined and cyclized. G.S is effective against Gram-positive but not Gram-negative bacteria. Its structure is thought to be an antiparallel pleated sheet.

Gramicidin D is a complex of four components, G. A, B, C and D. The structure of gramicidin A is a linear peptide of alternating D and L amino acids: HCO-L-Val-Gly-L-Ala-D-Leu-L-Ala-D-Val-L-Val-D-Val-(L-Trp-D-Leu)₃-L-Trp-NHCH₂CH₂OH. [R. Sarges and B. Witkop, J. Am. Chem. Soc. 86 (1964) 1862.] A second series (the isoleucine G.) has an isoleucine instead of valine in position 1. G. B has a Phe in position 11, while G.C has Tyr at this position. The linear G. act as ionophores. They form

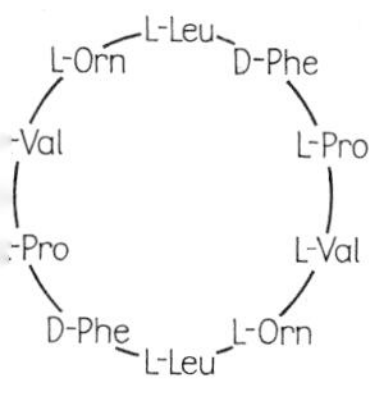

Gramicidin S

channels through biological membranes which allow the passage of monovalent cations, thus allowing equilibration of intra- and extracellular Na^+ and K^+. The G.A channel has been found by physical methods to be a single-stranded helix comprising two G.A molecules . The two CHO residues are adjacent in the middle of the membrane. [D.W. Urry, T.L. Trapane & K.U. Prasad, *Science* **221** (1983) 1064–1067.]

Granatan: see Pseudopelletierine.

Granulocyte-macrophage colony stimulating factors: see Colony stimulating factors.

Granulomatous disease: see Inborn errors of metabolism.

GRH: see Releasing hormones.

Grisein: an iron-containing antibiotic synthesized by *Streptomyces griseus*. It is a cyclic polypeptide containing cytosine. The iron ions are strongly bound as hydroxamate-iron(III) complexes. G. has the same functional groups as Albomycin (see), and like it, G. is a sideromycin. It was first isolated in 1947 by Reynolds and Waksman. It is especially effective against Gram-negative bacteria, but is not effective against fungi.

Griseofulvin: an antifungal agent synthesized by *Penicillin griseofulvi*. It is a polyketide synthesized from one molecule of acetyl-CoA and 6 molecules of malonyl-CoA (Fig.).

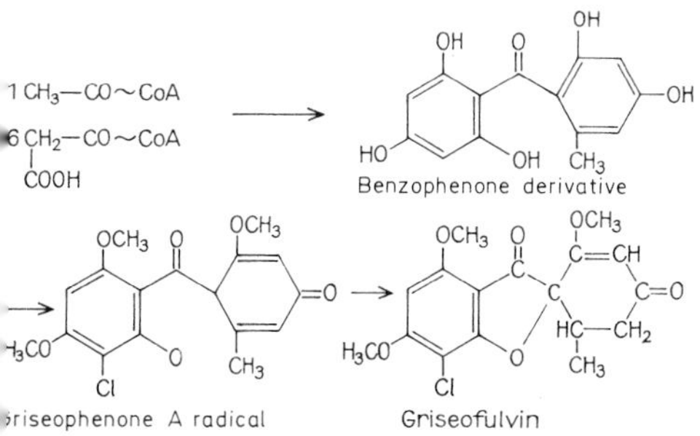

Biosynthesis of griseofulvin

Group transfer: the enzymatic transfer of a functional group from one molecule to another. In both anabolism and catabolism, small groups of atoms are handled as units (Table). Coenzymes (see) often serve as carriers of the groups, hence the term "transport metabolites". Hydrolysis and phosphorolysis can be regarded as special cases of G.t. in which the group is transferred to water or a phosphate ion.

Growth: an irreversible increase in the mass of living material, usually accompanied by an increase

Table. Group transfer reactions

Group transferred		EC number
CH_3-	Methyl	2.1.1
$HOCH_2-$	Hydroxymethyl, formyl, etc.	2.1.2
$-\overset{O}{\overset{\|}{C}}-OH$	Carboxyl and carbamoyl	2.1.3
$R-\overset{O}{\overset{\|}{C}}-$	Aldehyde or ketone residues	2.2.1
$R-\overset{O}{\overset{\|}{C}}-$	Acyltransferases	
	Acyl	2.3.1
	Aminoacyl	2.3.2
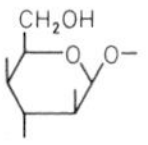	Glycosyltransferases	
	Hexosyl	2.4.1
	Pentosyl	2.4.2
	Other glycosyl groups	2.4.3
$R-CH_2-$	Alkyl or aryl groups other than methyl	2.5.1
	Nitrogenous groups	
NH_2-	Amino-	2.6.1
$HO-N=$	Oximino-	2.6.3
H_2PO_3-	Phosphorous-containing groups	2.7
$S-$	Sulfur-containing groups	2.8
SO_2-		

in the size of a cell or organism, as well as an increase in the number of cells.

Bacterial G. is measured as the increase of total cell material, or the increase of cell numbers, or both (see Cultivation of microorganisms); a strict relationship between cell numbers and cell mass does not necessarily exist (see Fermentation techniques).

In plant physiology, it is important to distinguish between G. by cell division and G. by cell stretching. In cell stretching, water is taken up powerfully by the cell, and the cell wall is physically stretched along the length of the cell. This phenomenon is particularly apparent in the sudden appearance of fruiting bodies of fungi, in the opening of flowers and in the thrusting of shoots.

Total cell numbers in cell suspensions can be determined microscopically in a counting chamber, or with the aid of a membrane filter. Cell mass may be determined directly by weighing the dry weight or fresh weight (dry material + cell water). Other suitable criteria for total cell material are the quantity of protein and the total nitrogen content. The concentration of cells in suspension can be measured indirectly from the absorption of light, i.e. turbidity (500–600 nm in a conventional spectrophotometer), or by the light scattering, i.e. by nephelometry. Such methods must be calibrated against one of the absolute methods of cell counting or mass determination. There may be a difference between *total cell count* and *viable cell count,* i.e. some cells may be dead. The viable cell count of bacterial cultures is determined by plating out suitable dilutions of culture on solid medium and counting the number of colonies formed after incubation. In addition to a correct supply of nutrients, bacterial G. depends on the pH, temperature and osmotic pressure of the culture medium, and on the degree of aeration (aerobes) or the completeness of anaerobiosis (anaerobes).

Certain chemical substances cause a concentration-dependent inhibition of G. In bacteriology, these are classified as *bacteriostatic* and *bacteriocidal* substances. Bacteriostatic substances inhibit growth, but growth resumes after their removal. A bacteriocidal substance (or bacteriocide) kills the bacterium. The underlying mechanisms of G. inhibition are many and varied, depending on the inhibitor.

Growth curve: see Cultivation of microorganisms.

Growth factor: a growth-stimulating substance. The term is used in two distinct ways: a) nutrients, such as vitamins, essential amino acids, essential lipids, inorganic ions, etc.; or b) a variety of non-nutrient, mitogenic substances, such as nerve growth factor, epidermal growth factor, insulin, certain steroids, etc.

Growth hormone: see Somatotropin.

Growth medium: see Nutrient medium.

Growth parameters: see Cultivation of microorganisms.

Growth phases: see Culture methods.

Growth regulators: organic compounds, which in small quantities inhibit, accelerate or in some way influence physiological processes in plants. G.r. include natural (endogenous) substances, e.g. Phytohormones (see) and native inhibitors; and many synthetic compounds, especially herbicides and Growth retardants (see). The Auxins (see) include natural and synthetic compounds.

Growth retardants: synthetic plant growth inhibitors, often used in agriculture for the control of plant growth. In particular, they cause stalk shortening in grasses, and some G.r. are used for controlling stalk length in cereal crops. The following G.r are listed separately: Chlorocholine chloride AMO 1618, Succinic acid mono-*N*-dimethylamide Succinic acid mono-*N*-dimethylhydrazide, EL-531 Maleic hydrazide, Phosphon D.

Growth substances: see Auxins.

Growth vitamin: vitamin A (see Vitamins).

GTF: abb. for glucose tolerance factor. See Chromium.

GTP: abb. for guanosine 5'-triphosphate, see Guanosine phosphates.

Gua: see Guanine.

Guanidine: see Guanidine derivatives.

Guanidine derivatives: compounds containing the strongly basic guanidine group (Fig.1). Guanidine itself, $H_2N-C(=NH)-NH_2$, is found in free form only in a few plants. The guanidino group is synthesized de novo in the course of arginine biosynthesis (see Urea cycle); other G. are formed by Transamidination (see) from arginine (Fig.2).

L-Arginine and γ-guanidobutyric acid are probably present in all living organisms. Some G.d. are limited to plants, e.g. L-Canavanine (see) and Galegine (see); other G.d., such as Phosphagens (see) only occur in animals. The G.d., streptidin, is a component of the antibiotic, streptomycin.

G.d. are degraded by various enzymes: 1. Transamidinase can act as a catabolic enzyme, if the G.d is further degraded by other enzymes and the resulting amino compound is catabolized (Fig.3); 2. Arginase and heteroarginase catalyse the hydrolysis of G.d. to produce urea. Arginase (L-arginine-urea hydrolase) cleaves L-arginine, L-canavanine and γ hydroxyarginine, and it probably occurs in isoenzyme forms that possess heteroarginase activity. Heteroarginases differ markedly from classical argi

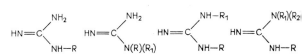

Figure 1. Types of guanidine compounds

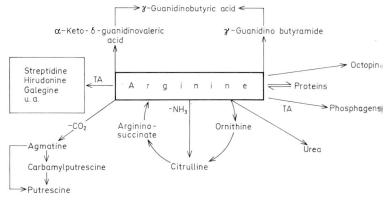

Figure 2. Metabolic fates of arginine. TA = transamidination.

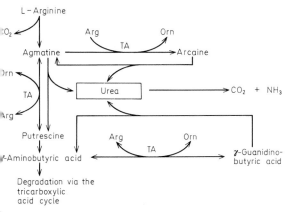

L – Arginine

CO_2

Arg — Orn
TA

Agmatine ————————→ Arcaine

Orn

TA

Urea ————————→ CO_2 + NH_3

Arg

Putrescine

Arg — Orn
TA

γ-Aminobutyric acid ←————————→ γ-Guanidino-butyric acid

Degradation via the tricarboxylic acid cycle

Figure 3. Arginine degradation in a mushroom. TA = Transamination, Arg = L-arginine, Orn = L-ornithine

nase with respect to substrate specificity; they cleave G. d. with chain lengths less than 6 C-atoms, e.g. γ-guanidobutyric acid, and they show wider specificity in that they cleave certain special G. d., e.g. γ-guanidobutyric acid (also cleaved by γ-guanidobutyrase from *Streptomyces griseus*), arcain, agmatine and streptomycin.

Arginine decarboxy-oxidase, hitherto described only from *Streptomyces griseus* and the pond snail *(Limnaea stagnalis)*, degrades arginine to γ-guanidobutyramide, canavanine to β-guanido-oxypropionamide, and homoarginine to δ-guanidovaleramide. Other reactions for the degradation of L-canavanine are found in bacteria.

Guanine, *2-amino-6-hydroxypurine,* abb. **G.** or **Gua:** one of the four nucleic acid bases. M_r 151.13, m.p. 360 °C (d.). G. is also a component of nucleotide coenzymes and the starting material for the biosynthesis of many natural products, including pterins and the vitamins folic acid and riboflavin. Free G. is rare in nature. It was first discovered in 1844 in Peruvian guano. It is the nitrogen excretion compound of spiders. G. is deaminated by guanine deaminase (EC 3.5.4.3) to xanthine; this is the first step in purine degradation.

OH

H₂N

Guanine

Guanosine, abb. *Guo:* 9-β-D-ribofuranosylguanine, a β-glycosidic nucleoside containing D-ribose and guanine. M_r 283.2, m.p. 237–240 °C, $[\alpha]_D^{25}-72$ ° $(c=1.4;\ 0.1\text{M NaOH})$. The Guanosine phosphates (see) play an important role in the metabolism of all organisms.

Guanosine diphosphate sugars, abb. *GDP-sugar:* guanosine-diphosphate-activated forms of various sugars. Their synthesis is analogous to that of other Nucleoside diphosphate sugars (see), i.e. the condensation of a sugar phosphorylated at C-1

and guanosine triphosphate, with the release of pyrophosphate. Guanosine diphosphate mannose is of particular importance; glucose, fucose and rhamnose also occur as GDP derivatives.

Guanosine phosphates: phosphoric acid esters of guanosine. They are nucleotides and are of great importance in metabolism. The biologically important derivatives are those esterified on C-5' of the ribose. According to the phosphoric acid component, the compounds are classified as guanosine mono-, di- and triphosphates. 1. *Guanosine 5'-monophosphate,* abb. *GMP, guanylic acid,* M_r 363.2, m.p. 190–200 °C (d.) is synthesized in the purine pathway from xanthosine monophosphate and is the starting material for the synthesis of the other guanosine phosphates. GMP is used as a flavoring and aroma substance, and is extracted for this purpose from yeast nucleic acids or is produced on a large scale by mutants of certain microorganisms, such as *Corynebacterium glutamicum.*

2. *Guanosine 5' diphosphate,* abb. *GDP,* M_r 443.2, is produced by phosphorylation of GMP by a kinase or by dephosphorylation of guanosine triphosphate. Certain sugars, for example mannose, are activated by binding to GDP (see Nucleoside diphosphate sugars).

3. *Guanosine 5'-triphosphate,* abb. *GTP,* M_r 523.2, like adenosine triphosphate, can provide energy for biochemical reactions. The energy released by the dehydrogenation of α-ketoglutarate (2-oxoglutarate) in the Tricarboxylic cycle (see) flows into the GDP/GTP system, from which it can be transferred to the ADP/ATP system. GTP can also provide the phosphate group for the synthesis of phospho*enol*pyruvate from oxaloacetate in the course of gluconeogenesis. GTP is an important energy source in Protein biosynthesis (see).

4. *Cyclic guanosine 3'-monophosphate,* abb. *cyclo-GMP, cGMP,* M_r 345.2, a structural analog of cyclic adenosine 3',5'-monophosphate (see Adenosine phosphates) which occurs in similar concentrations in many tissues. cGMP is synthesized by a guanylate cyclase which is highly specific for GTP. The biological significance of the cGMP-guanylate cyclase system lies in its mediation of the action of certain hor-

mones and neurohumoral transmitters, such as acetylcholine, prostaglandins and histamine.

Guanylic acid: see Guanosine phosphates.

(+)-Guibourtacacidin: 7,4'-dihydroxyflavan-3,4-diol, see Leucoanthocyanidins.

Guo: abb. for Guanosine

Gutta: a rubber-like polyterpene of about 100 isoprene units in which the double bonds are in the *trans* configuration (see Polyterpenes, Fig.). G. is produced on the Malayan peninsula and the Indonesian islands from the latex of *Palaquium gutta*.

Gutta is less elastic than rubber, but it is more resistant to chemicals and environmental influences (insulating material). Depending on its source, G. occurs in mixtures with other terpenes. The mixture with resins is called guttapercha, and that with triterpene alcohols is chicle (starting material for manufacture of chewing gum).

Guvacine: see Areca alkaloids.

Guvacoline: see Areca alkaloids.

Gyrase: a type II topoisomerase. See Topoisomerases.

H

Haem: same as heme. See Hemes.

Haginin A: 7,4'-dihydroxy-2',3'-dimethoxyiso-flav-3-ene, see Isoflav-3-ene.

Haginin B: 7,4'-dihydroxy-2'-methoxyisoflav-3-ene, see Isoflav-3-ene.

Hallucinogens: a group of drugs which cause changes in mood, perception, thinking and behavior. The group does not include addictive drugs, but some users become dependent on H. Those H. with the strongest action are called *psychedelics*. The H. have been divided into four groups: derivatives of indole alkaloids (tryptamine, harmine, ibogaine, LSD and psyilocybin), derivatives of piperidine (belladonna, atropine, scopolamine, hyoscyamine and phencyclidine), phenylethylamines (mescaline, amphetamine and adrenochrome), and cannabinols from Cannabis plants.
Most of the H. are sympathomimetics and cause a rise of blood pressure, higher pulse rate. dilation of the pupils, sweating, palpitation and increased tendon reflexes. Marijuana (the dried leaves of the Cannabis plant) is the weakest, and remains in the body for 2 to 3 hours if smoked, longer if eaten. LSD is the strongest of this group; 25 µg is an effective dose for an adult. It remains in the body for 6 to 8 hours, with the greatest psychological effects at 3 to 4 hours, by which time most of it has disappeared from the central nervous system.
There are legitimate medical and psychiatric uses for the H., and the nonmedical use of several of them is widespread. For a review see Marihuana Research Findings: 1976, (Robert C. Peterson, ed.) National Institute on Drug Abuse Research Monograph #14 (1977) from the US Department of Health, Education and Welfare.

Haloopsin: see Bacteriorhodopsin.

D-Hamamelose: a monosaccharide with a branched carbon chain. m.p. 111 °C, $[\alpha]_D^{21}$ −7.1 ° (water). D-H. occurs in higher plants, for example hamamelis (witch hazel) bark. Its biosynthesis, which is similar to that of apiose, involves an intramolecular rearrangement of an unbranched hexose.

Hanes-Wilkinson plot: see Kinetic data evaluation.

Haptens: partial or incomplete antigens. H. are either chemically defined molecules, e.g. dinitrophenol, or part of an antigen. They can bind specifically to the corresponding antibodies (see Immunoglobulins), but they do not act by themselves as antigens. Only after coupling to a carrier protein do the H. become full antigens which are capable of eliciting an immune response. After parenteral application of H., the body produces two specific antibodies, one against the protein and one against the bound H. (hapten-specific antibodies). Half-haptens elicit an antigen-antibody reaction, but no precipitation.
One of the most effective H. is the Forsman hapten, a glycolipid substance. The Forsman H. from mammalian (e.g. horse) organs is chemically a pentahexosylceramide with the following sequence in its sugar chain: N-acetyl-α-galactosaminyl-N-acetyl-β-galactosaminyl-(1-3)-[galactosyl-(1-4)]$_2$-glucosyl ceramide. Enzymatic removal of the terminal GalNAc group destroys the Forsman antigenicity. The Forsman antigen, which is composed of the Forsman H. and a specific protein, induces hemolysin formation.

Haptoglobin, abb. *Hp:* an acid α_2-plasma glycoprotein, which binds specifically to free plasma oxyhemoglobin to form a high molecular weight complex, M_r 310000, which cannot be filtered by the kidneys. The associated conformational change in the hemoglobin allows the heme α-methenyl oxygenase of the liver to remove the heme porphyrin ring. Thereafter the globin is degraded by the trypsin-like protease action of the β-chain of the Hp. Hp. is a tetramer consisting of two nonequivalent chain pairs, 2α and 2β held together by disulfide bridges. Human Hp. exists in three genetic variants Hp 1-1, 2-2 and 2-1 which differ in their electrophoretic patterns. While Hp 1-1 moves as a single band, Hp 2-2 and 2-1 display several (up to 14) discrete bands, which represent stable oligomers of the monomer (one β, one α_1 and one α_2 chain in Hp 2-1, and one β and one α_2 chain in Hp 2-2). The M_r of the Hp 2-2 monomer is 57300. This heterogeneity is due to the existence of two different α chain types, α_1 and α_2. While α_1 contains 82 amino acids (M_r 9000), α_2 is nearly twice as large, with a M_r of 17300 and 142 amino acids. Sequence studies of the two chains showed that the α_2 chain, which is only found in human Hp 2-2 and 2-1, is a union of 71 amino acids each from the N and the C terminal ends of the α_1 chain. It is the product of a second autosomal allele, i.e. a homologous gene pair which arose from the α_1 structural gene through an unequal crossing over. Unlike the α-chain, the carbohydrate-carrying and much larger β chain (M_r 40000, without sugar residues, 35000) is the same size in all 3 Hp types. M_r of the Hp 1-1 $(2\alpha_1 + 2\beta)$ 98000; 20% carbohydrate; IP at pH 4, since 19% of the amino acids are aspartate or glutamate. M_r of the main component of Hp 2-2 $(\alpha_2 + \beta)_2$ 114000. The M_r of the Hp 2-2 polymers are even and odd multiples of the monomer $(\alpha_2 + \beta)_n$ ($n = 2$ to maximally 14).
On the basis of sequence homologies between the Hp α chain and the immunoglobulins, Hp appears to be a natural, preformed antibody against hemoglobin.

Har: abb. for Homoarginine.

251

Harden-Young ester: see Fructose 1,6-bisphosphate

Harmaline: see Harman alkaloids.

Harman: see Harman alkaloids.

Harman alkaloids: a group of indole alkaloids with a β-carboline skeleton. The H.a. are formed biosynthetically from tryptophan and a carbonyl component. The same reaction can be performed in vitro using conditions designed to simulate those in the cell. The ubiquitous occurrence of the precursors and the simple synthesis are one reason that H.a. are found in many different plant groups. The main alkaloids are harmine, M_r 212.25, m.p. 262 °C, and harman, m.p. 237-238 °C (Fig.); 3,4-dihydroharmine (harmaline), m.p. 229-231 °C, and 1, 2, 3, 4-tetrahydroharmine. In medicine, the H.a. are occasionally used for encephalitis and Parkinson's disease. Some of the H.a. lead to hallucinations and intoxication.

Harman: R=H
Harmine: R=OCH₃

Harmine: see Harman alkaloids.

Hashish: the dried resin from the glandular hairs of the female hemp plant *(Cannabis sativa L.)*. Due to the psychoactive Δ^3-tetrahydrocannabinol (2 to 8%) which it contains, it is one of the most common Narcotics (see). Marihuana (often used synonymously with hashish) is the dried and chopped tips of the shoots of the female hemp with a content of 0.5 to 2% Δ^3-tetrahydrocannabinol. Both hemp drugs have been used for millenia in folk medicine, due to their intoxicating action. Today they are, along with alcohol, the most widely used drugs; the number of consumers is estimated as 300 million. The drugs are usually smoked alone or mixed with tabacco in cigarettes ("joints") or pipes.

The amount of Δ^1-tetrahydrocannabinol and of the structurally related, non-psychotropic compounds (e.g. cannabinol [m.p. 76 to 77 °C], cannabidiol and cannabidiolic acid) varies considerably in H. The H. produced from European hemp has, in contrast to the tropical culture forms, little Δ^1-tetrahydrocannabinol and much cannabidiol and cannabidiolic acid.

Hatch-Slack-Kortschak cycle, abb. *HSK cycle, C₄ acid cycle:* the carbon-trapping cycle in C_4 plants (see) and Crassulacean acid metabolism (see) (Fig.). The first enzyme in the cycle is phospho*enol*pyruvate carboxylase (1), which carboxylates PEP to oxaloacetate (Fig.). This is reduced to L-malate by NADP-dependent malate dehydrogenase (2). C_4 plants have high concentrations of (1) and (2) in their mesophyll cells, which export the L-malate to the vascular bundle cells. There the L-malate is oxidatively decarboxylated (3) to pyruvate. The released CO_2 is efficiently utilized in these cells for the reactions of the Calvin cycle. The pyruvate is returned to the mesophyll cells, where it is converted back to PEP by pyruvate, orthophosphate dikinase (4): pyruvate + ATP + P_i → PEP + AMP + PP_i. [C.C. Black *Ann. Rev. Plant Physiol.* **24** (1973) 253–286]

Mesophyll cells

Vascular bundle cells

Calvin cycle

Hb: abb. for Hemoglobin.

HbS: abb. for Sickle-cell hemoglobin.

hCG: abb. for Human chorionic gonadotropin, same as Chorionic gonadotropin.

Hcy: abb. for Homocysteine.

HDL: see Lipoproteins.

Heat-shock proteins: a group of proteins synthesized by cells which have been subjected to hyperthermia or other noxious conditions, such as ethanol or sodium arsenite. Their function is unknown, but they may be required for recovery of gene expression. They are the first major proteins transcribed from the embryonic genome in the mouse embryo. [O. Bensaude et al., *Nature* **305** (1983) 331–332.]

Heat-stable enzymes: see Thermostable enzymes.

Heavy metals: all metals with a density greater than 5. In living organisms, they are usually present in stable organic complexes. Biological functions of H.m. are listed in Table 1. Chemical ligands of H.m. are listed in Table 2. Iron (see) is present in a particularly large number of biomolecules. Vanadium is essential for tunicates (a group of strictly marine filter-feeding animals related to the vertebrates) and for lower plants. It is accumulated from sea water by some species of tunicate, notably the *Ascidiidae* and *Perophoridae* (order *Phlebobranchia*), and is found chiefly in the vacuoles of certain types of blood cell, called vanadocytes or morula cells. Vanadocytes of *Ascidia ceratodes* contain approximately 1.2 M V(III), representing a concentration factor over sea water greater than 10^7. The mechanism of this highly selective accumulation process is unknown.

Other H.m., e.g. lead, mercury, cadmium, copper, are toxic (see Lead). Toxicity is usually attributed to the ability of these metals to react irreversibly with free SH-groups of proteins.

Heavy water: see Water.

Heidelberger curve: see Precipitation curve.

Heinz bodies: irregular, refractile inclusions in erythrocytes, attached to the inner surface of the membrane. They may be up to 3 μm in diameter and are stained by vital dyes. Unstained, they are detectable by their green autofluorescence. H.b. are insoluble aggregates of degraded hemoglobin ad-

Table 1. Heavy metals in biomolecules.

Heavy metal	Type of biomolecule	Biological function
Iron	Hemoglobins; cytochromes	O_2-Transport; electron transport
	Flavin enzymes (metalloflavo-enzymes), e.g. xanthine oxidase	Oxidation, dehydrogenation, and/or reduction
	Iron-sulfur protein e.g. ferredoxin, complexes of the respiratory chain	Electron transfer
	Nitrogenase	Reduction of N_2 to ammonia
	Ferretin; conalbumin, transferrin (siderophilin)	Fe-storage; Fe-transport
	Siderochromes, mycobactin; enterobactin, etc.	Fe-transport in microorganisms
Cobalt	Vitamin B_{12} and its coenzyme forms	Reduction and methyl group transfer (methionine synthesis); isomerization
Copper	Laccase, cytochrome oxidase	Oxidation
	Cupreine, ceruloplasmin	Storage and transport of copper
Manganese	Arginase	Arginine hydrolysis, urea cycle
	Decarboxylases and other enzymes	Release of CO_2, etc.
Molybdenum	Nitrogenase	Binding and activation of molecular nitrogen (subsequent reduction also requires iron)
	Nitrate reductase	Nitrate reduction
	Xanthine oxidase	Purine oxidation, etc.
Zinc	Carbonic anhydrase, peptidases, phosphatases, pyridine nucleotide enzymes and other proteins	Zn functions in substrate binding (e.g. for hydride transfer in ternary complexes with pyridine nucleotide enzymes), and in protein structure (e.g. alcohol dehydrogenase)
	Insulin	Aggregation of the polypeptide

Table 2. Ligands of heavy metals in biomolecules.

Heavy metal	Ligand	Example
Iron	Porphyrin, imidazole	Myoglobin
	Sulfur	Ferredoxin
	Phenolate	Transferrin
Cobalt	Corrin, benzimidazole	B_{12} and derivatives
Copper	> N bases	Cupreine
Manganese	Carboxylate, phosphate, imidazole	Glycolysis and proteolysis enzymes
Vanadium	No ligand yet identified	Vanadium proteins of tunicates
Zinc	Imidazole (His), carboxyl groups (Glu)	Carboxypeptidases
	R-S⁻	Dehydrogenases
	Imidazole (His)	Carbonic anhydrase

mixed with fragments of lipid and other protein. Some authors maintain that H.b. formation requires initial formation of methemoglobin, but this is contradicted by the observation that some H.b. producing drugs do not produce a measurable increase in methemoglobin. H.b. formation is promoted by drugs and environmental pollutants which also cause hemolytic anemia, e.g. phenylhydrazine, O-methyl-, O,N,-dimethyl and trimethyl hydroxyl-amine. [H. Martin et al. *Klin. Wochenschr.* **42** (1964) 725–731; G. Rentsch *Biochem. Pharmacol.* **17** (1968) 423–427; E. Beutler *Pharmacol. Rev.* **21** (1969) 73–103; C.C. Winterbourn & R.W. Carrell *Brit. J. Haematol.* **25** (1973) 585–592]

Heliangin: a sesquilactone and plant growth inhibitor first isolated from the leaves of the Jerusalem artichoke *(Helianthus tuberosus).* It is a gibberellin antagonist and inhibits the growth of oat coleoptiles. However, it stimulates root growth in beans (*Phaseolus* spp.).

Helix: the spiral arrangement of a biopolymeric compound, e.g. starch, some proteins and DNA (see Deoxyribonucleic acid, Fig.).

Helix-random-coil transition: see Denaturation.

Helminthosporal: a natural product from the phytopathogenic fungus *Helminthosporium sativum* (syn. *Bipolaris sarokiniana*, Shoemaker), which has a physiological effect on plants similar to that of gibberellins.

Helvolic acid: one of the tetracyclic triterpene antibiotics. M_r 568.7, m.p. 215 °C. It differs structurally from Fusidic acid (see) in having a Δ^1-3,6-diketo and a 7α-acetoxy function, and in lacking the 11α-hydroxyl group. H.a. is produced by *Aspergillus fumigatus* and related fungal species. Like the structurally related triterpene antibiotics, fusidic acid and cephalosporin P_1, it is effective against Gram-positive organisms.

Hemagglutination: see Agglutination.

Hemagglutinins: see Immunoglobulins.

Heme: 1. a metalloporphyrin, which acts as the prosthetic group of a hemoprotein, e.g. cytochrome, hemoglobin, nitrite reductase, etc. The iron atom is coordinately bound to the four pyrrole nitrogen atoms of the porphyrin ring. Biosynthesis of iron porphyrin starts from protoporphyrin IX (see Porphyrins). Iron is added by the action of ferrochelatase (E.C.4.99.1.1) (Fig.).

See Protoheme, Chlorocruoroheme, Siroheme.

2. A complex of iron in which the organic component is not a porphyrin, but a related tetrapyrrole structure, e.g. verdoheme, biliverdin heme.

3. An iron complex of a Chlorin (see).

Heme a: the heme prosthetic group found in cytochromes a/a_3.

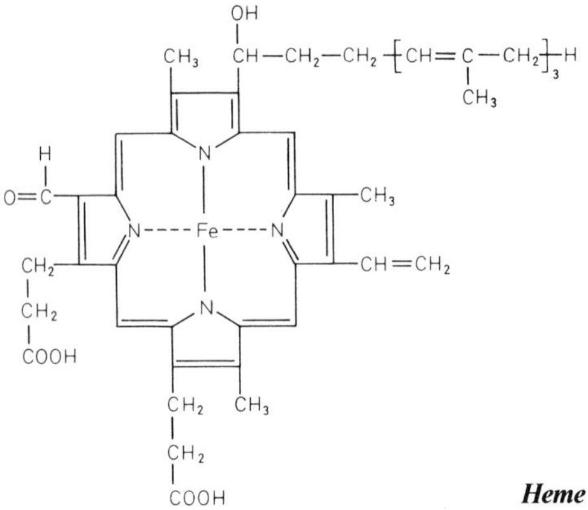

Heme a

Heme c: the heme prosthetic group of cytochromes c, b_4 and f.

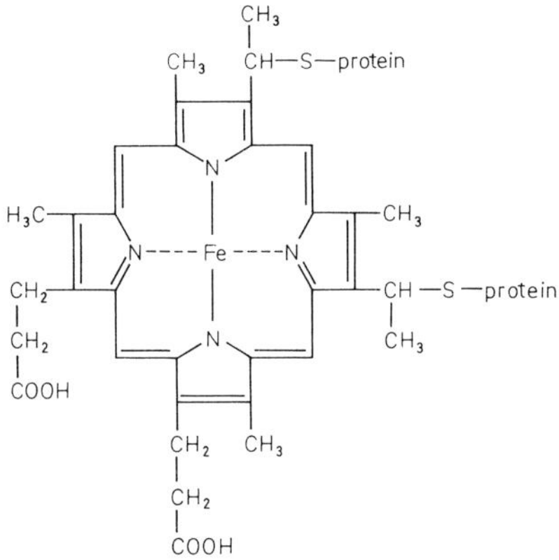

Heme c

Heme d, formerly *heme a₂:* a heme prosthetic group found in the terminal oxidase system of many bacteria.

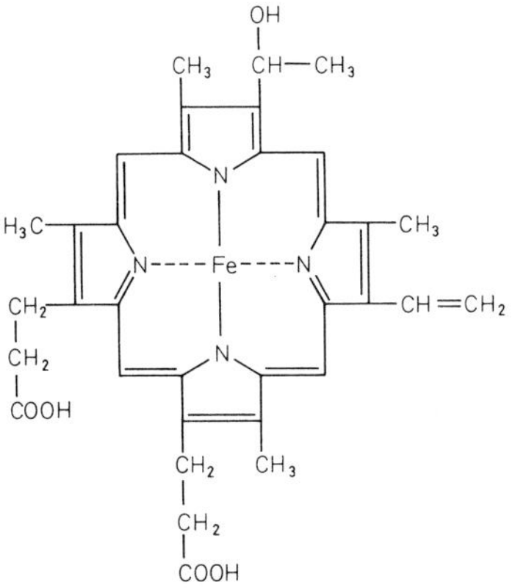

Heme d

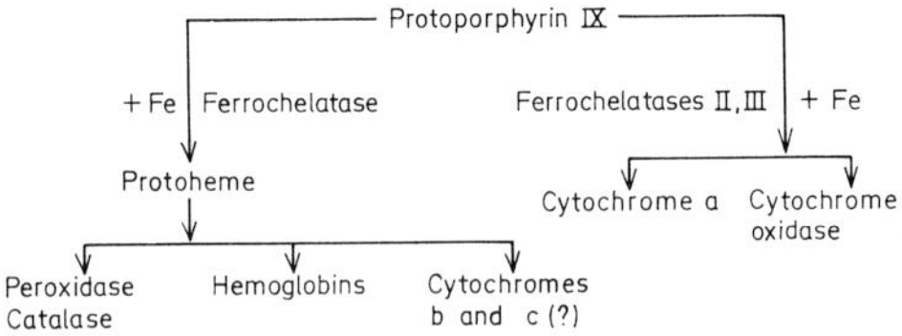

Heme iron: iron which is coordinately bound in porphyrins. It is present, for example, in the Hemoproteins (see).

Hemerythrin: a red-brown chromoprotein containing iron but not porphyrin. It transports oxygen and occurs as respiratory pigment in the blood cells of certain marine invertebrates, such as sipunculoid worms, polychete worms and lamp shells *(Brachiopoda).* The most thoroughly examimed H. is that of the sipunculoid worms. It is an octamer of M_r 108000; the primary structure of its subunit (M_r 13500, 113 amino acids) is known. The iron (2 Fe^{2+}/chain or 16 Fe^{2+}/H.), like the oxygen (1 molecule O_2/chain) is bound directly to the protein.

Hemicelluloses: high molecular weight polysaccharide complexes composed of aldoses which occur in woody parts of plants together with cellulose. H. consist of hexose and pentose residues linked β-1,4, and they often contain uronic acid as well. They are insoluble in water, but soluble in dilute alkali. H.serve as structural compounds and sometimes as reserve substances. Humans and animals cannot digest them. Important H.are arabans, xylans, glucans, galactans, fructans and mannans.

Hemicholinium: see Acetylcholine.

Hemisubstances: see Cell wall.

Hemiterpenes: terpenes composed of a single isoprene unit (C_5H_8). There are few H., the most im-

Isopentenylpyrophosphate   Isoprene

The formation of isoprene from isopentenylpyrophosphate.

254

portant being *isoprene*, which is formed by removal of pyrophosphate from isopentenylpyrophosphate "active" isoprene (Fig.).

Hemocyanins: copper proteins responsible for oxygen transport in arthropod and molluscan hemolymph. They do not contain porphyrin and are not particle-bound. Oxidized H. are blue. The concentration of H. in snail blood is 2 to 4%; the copper content of H. is 0.24%. Two copper (I) ions are required to bind one molecule of oxygen. Molluscan H. are cylindrical oligomers with subunits of M_r 400000. Each subunit contains 8 oxygen-binding sites, i.e. 16 copper atoms. The H. from arthropods consist of hexamers, with subunits of M_r 75000; these may aggregate to form multi-hexamers. Each subunit carries one pair of copper atoms. The oxygen-binding centers of the two classes of H. are believed to be similar, and may resemble the active sites of other copper proteins, such as oxytyrosinase, ascorbate oxidase, laccase and ceruloplasmin. The structure of spiny lobster *(Panulirus interruptus)* H. has been determined to 3.2 Å resolution by Gaykema et al. [W.P.J. Gaykema et al., *Nature* **309** (1984) 23–29.]

Hemoglobin, abb. *Hb:* the most important respiratory protein of the vertebrates. Hb, the colored component of blood, is present as a 34% solution in the erythrocytes (red blood cells) and carries oxygen from the lungs to the other tissues. It is a tetramer composed of two pairs of polypeptide chains and 4 heme groups; M_r 64500. With a Fe^{2+} content of 0.334%, the total of 950 g Hb in the human body represents 3.5 g or 80% of the total body iron. Adult human Hb consists of 96.5 to 98.5% HbA_1 $(\alpha_2\beta_2)$ and 1.5 to 3.5% HbA_2 $(\alpha_2\delta_2)$. At least 7 different structural genes specify globin; these are clustered on chromosomes 11 and 16 (Fig.1). There are two α-

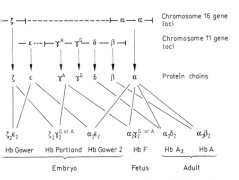

Figure 1. Arrangement of subunits of different hemoglobins and their genetic origin. Hb = hemoglobin.

genes per haploid genome, i.e. a total of 4 α-genes. Three forms of Hb are found very early in embryonic development: $\zeta_2\varepsilon_2$ (Hb Gower), $\zeta_2\gamma_2$ (Hb Portland) and $\alpha_2\varepsilon_2$ (Hb Gower2). After 3 months gestation, ζ and ε polypeptides are no longer synthesized (Fig.2), and all 3 embryonic Hbs (HbEs) disappear to be replaced by fetal Hb (HbF, $\alpha_2\gamma_2$). The ζ-chain can be considered as an embryonic α-chain; it differs from the α-chain by about 60 residues. Most of these differences are conservative, but in particular

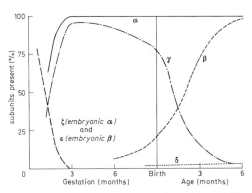

Figure 2. Changes in the concentrations of hemoglobin subunits during human gestation and early development. Since all subunits are part of a hemoglobin tetramer, these changes in turn reflect the developmental changes in the types of hemoglobin present (see Fig.1).

the ζ-chain contains 4 more Arg residues and 3 fewer His residues than the α-chain, the *N*-terminus is acetylated (affecting the Bohr shift), and His(H5) is present. Functionally, the ε-chain is an embryonic β-like chain; it differs from the β-chain by about 32 residues and from the γ-chain by about 25 residues; most of these differences are conservative, but in particular ε77 (E20) is Asn, ε116(G18) is Thr, and there is an extra Lys at ε87(F3) (cf. amino acid sequences in Fig.3). HbE and HbF differ from adult Hb in their higher O_2 affinity. This makes possible the exchange of gas between fetal and maternal blood. The Hb of all human races and even of chimpanzees is identical. Anomalies are pathological and arise through point mutation, which leads to the substitution of amino acids, or, more rarely, to their absence. Of the presently known 153 abnormal Hb, 87 are variations of the β-chain. The most frequent and best known is the Sickle cell Hb (see). Although only a few of the variants lead to disease, such as hemolytic anemia, the absence of entire chains leads to severe damage to the carrier. In β-thalassemia, which occurs in Mediterranean countries, no β-chains are formed (HbA_1 is missing). In its place the blood of these patients contains HbF (85 to 95%) and HbA_2 (5 to 15%). Human α-chains contain 141 amino acid residues, and β-chains, 146 residues; the sequences of human Hb and of many other vertebrate Hb have been determined. In addition to the primary structure of Hb, the tertiary structures of its chains and the quaternary structure of the entire tetrameric molecule were determined by Perutz, using X-ray analysis. Aside from slight differences due to differences in their primary structures, the Hb chains are folded in a manner very similar to that of the myoglobin molecule. The attachment of the heme group by bonding of its iron(II) atom to two histidine residues and the hydrophobic interactions in the heme pocket of Hb also correspond to myoglobin. However, the primarily hydrophobic interactions between the individual Hb chains, which are not present in myoglobin, are much more complicated. There are large regions of hydrophobic contact,

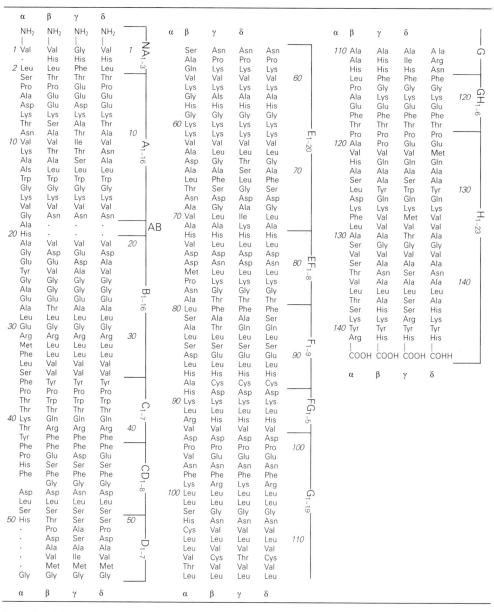

Figure 3. Amino acid sequences of human hemoglobin subunits α, β, γ and δ. Numbers on the left of each column refer to residues in the α-subunit only. Numbers on the right refer to the residues in subunits β, γ and δ. Single letters, A, B, C, etc. refer to helices according to Kendrew's nomenclature (see Figs. 4 and 5 below, and Figs. 9 and 10 in the entry on Proteins). Paired letters AB, CD, etc. refer to nonhelical, interhelical stretches. The number of residues in each region is given as a subscript to the designating letter.

especially between the α- and β-chains, and these are the basis for the interactions of the four spatially separated heme groups in the reversible, cooperative binding of four oxygen molecules to each molecule of Hb. The hydrophobic interactions also make possible the gliding motions of the two αβ dimers with respect to one another during O_2 loading and release. This allosteric effect is responsible for the sigmoidal O_2 binding curve of Hb. The lack of covalent bonds between the Hb chains can be shown by the reversible dissociation of Hb. At pH 4, the dissociation is asymmetric, forming $\alpha_2 + \beta_2$; at pH 11, in 1 M

NaCl, it is symmetric, forming 2 ($\alpha\beta$) units, and in presence of p-chloromercuribenzoate, dissociation into monomers occurs. It is medically significant that the affinity of Hb to the poisonous CO is 325 times greater than its affinity to O_2.

The single-chain, Hb-like respiratory pigments in invertebrates (M_r about 16 000) are similar in primary and tertiary structure to myoglobin and to the α- and β-chains of Hb; they are thus regarded as ancestral to the tetrameric Hb. The single-chain hemoglobin in yeast is unique, because it contains FAD in addition to heme. Its M_r is 50 000.

Conformational states of hemoglobin: In deoxyHb, the four subunits are relatively tightly associated, and the chains are linked by several noncovalent (ionic) bonds, as well as hydrophobic interactions. Deoxy-Hb is therefore known as the T-form (taught, tight or tense). The more important ionic bonds (salt bridges) of the T-form are: free terminal NH_2 of Val-1 (α_2) with the free (terminal) COOH of Arg-141 (α_1), similarly Val-1 (α_1) with Arg-141 (α_2); guanidinium of Arg-141 (α_1) with side chain COOH of Asp-126 (α_2), similarly Arg-141 (α_2) with Asp-126 (α_1); free (terminal) COOH of His-146 (β_1) with ε-NH_2 of Lys-40 (α_2), similarly His-146 (β_2) with Lys-40 (α_1); imidazole of His-146 (β_1) with side chain COOH of Asp-94 (β_1), similarly His-146 (β_2) with Asp-94 (β_2) (Fig. 4).

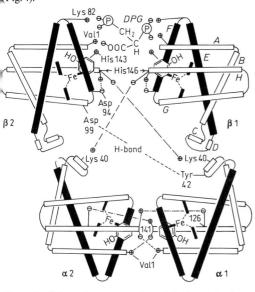

Figure 4. Schematic representation of the four subunits of deoxyhemoglobin. Helical regions are labelled A, B, C, etc., the same designation as in Fig. 3. For the true perspective of the tetramer, see Fig. 6 below, and Figs. 9 and 10 in the entry on Proteins. DPG = 2,3-diphosphoglycerate.

In deoxyHb the penultimate Tyr residues of all 4 chains lie in hydrophobic pockets between helices F and H. The Fe atom of heme is displaced about 0.08 nm from the plane of the porphyrin ring system; the hemes lie in V-shaped pockets formed by helices F and E. In the α subunits, the heme groups

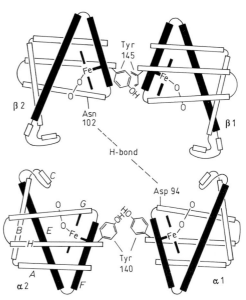

Figure 5. Schematic representation of the four subunits of oxyhemoglobin. Helical regions are labelled A, B, C, etc., the same designation as in Fig. 3. 2,3-Diphosphoglycerate is no longer present and the two β-subunits have moved closer together (cf. Fig. 4). For the true perspective of the tetramer, see Fig. 6 below, and Figs. 9 and 10 in the entry on Proteins.

are in open pockets, permitting access of O_2, whereas the heme pockets of the β subunits are more compressed, preventing entry of O_2. 2,3-Diphosphoglycerate is present and forms salt bridges with positively charged groups of the two β chains (Fig. 4).

In oxyHb the subunits are less tightly associated, so that this form is known as the R-form (relaxed). In it, the dissociation constants of ionizable groups are changed, and salt bridges present in deoxyHb are broken; the most important of these seem to be associated with His-146 (β) and Val-1 (α). Also in oxy-Hb, the β subunits move closer together; the heme pockets of the β-subunits become wider and admit O_2; as the iron of each heme group binds oxygen, it moves 0.08 nm into the plane of the porphyrin ring, pulling His-92 with it. 2,3-Diphosphoglycerate is not present (Fig. 5).

In the transition from the T to the R-form, one pair of subunits ($\alpha_1\beta_1$) rotates through 15 ° relative to the other pair ($\alpha_2\beta_2$). The axis of rotation is eccentric, so that the $\alpha_1\beta_1$ pair also moves slightly towards the axis (Fig. 6).

Bohr effect: a reversible shift in the O_2-binding curve of Hb, which permits O_2 binding and CO_2 release in the alveolar capillaries of the lungs, and the reverse process in respiring tissues (Fig. 7). Carbonic anhydrase in erythrocytes promotes the rapid formation of carbonic acid from dissolved CO_2 in respiring tissue, and each molecule of H_2CO_3 spontaneously dissociates into bicarbonate and a proton. The protons are absorbed by deoxyHb, which thereby acts as a buffer. OxyHb does not bind protons, but for

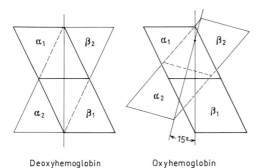

Deoxyhemoglobin   Oxyhemoglobin

Figure 6. Relative positions of the subunits in deoxyhemoglobin (T-form) and oxyhemoglobin (R-form).

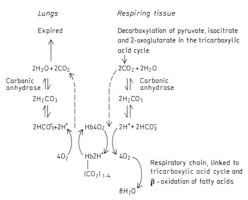

Figure 7. The Bohr effect. The binding of protons and CO_2 in respiring tissue causes a shift in the oxygen binding curve of hemoglobin (Hb) so that oxygen is readily lost. In the capillaries of the lung alveoli, the reverse process occurs, and a high oxygen tension promotes oxygen binding with concomitant loss of protons and CO_2.

every 4 O_2 molecules lost, the resulting deoxyHb binds 2 protons. Also, about 15% of CO_2 carried in the blood is bound to Hb as carbamino groups. The process is reversed in the lungs, where O_2 binds to deoxyHb with release of protons, i.e. binding the O_2 promotes the exhalation of CO_2 from the lungs, whereas proton binding promotes release of O_2 in respiring tissues. This shift in the O_2-binding properties of Hb is caused by the acidity (Fig. 8) and high CO_2 tension (CO_2 alone is effective at constant pH) in respiring tissue, and the low acidity and low CO_2 tension in the lungs. The carbamino groups, formed by reaction of CO_2 with the non-ionized terminal α-amino groups of each of the 4 chains, form salt bridges which help to stabilize the deoxyHb:

$$R-NH_2 + CO_2 \rightleftharpoons R-NH \cdot COO^- + H^+.$$

In the transition from oxyHb to deoxyHb, three pairs of negatively charged (i.e. proton-binding) groups are moved to more negative environments: His-146 (C-terminal) on each β chain; terminal NH_2

groups of the α chains, and His-122 on each α chain. The resulting increases in pK of these groups renders them available for binding protons. The β-chain C-terminal His-146 in oxyHb can rotate freely, but in deoxyHb the β His-146 becomes involved in several interactions, and in particular, it comes into the close proximity of negatively charged Asp-94 on the same β-chain. In deoxyHb the terminal amino group of one α-chain interacts with the carboxyl terminal of the other α-chain, thereby raising the pK of the amino group and increasing its affinity for H^+.

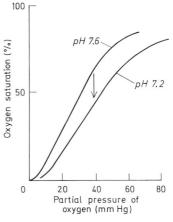

Figure 8. Effect of pH on the oxygen saturation curve of hemoglobin. As acidity increases, i.e. a decrease in pH from 7.6 to 7.2, oxygen release is favored (arrow) because percentage saturation is less at any given oxygen partial pressure.

2,3-Diphosphoglycerate (DPG), glycerate 2,3-bisphosphate: DPG binds noncovalently to deoxyHb, but not to oxyHb. In deoxyHb, one mol DPG is associated with the charged α-amino groups of the N-terminal valine residues of the two β-chains. Other β-chain groups possibly contributing to DPG binding in deoxyHb are His-2, Lys-82 and His-143 (Fig. 3). DPG pulls the equilibrium between oxyHb and deoxyHb + O_2 to the right. The molar concentration of DPG in erythrocytes is about the same as that of Hb; this is sufficient to shift the dissociation curve to the right at all times (Fig. 9):

$$Hb(O_2)_4 + DPG \rightleftharpoons Hb \cdot DPG + 4\,O_2.$$

The erythrocyte DPG concentration may change in response to defective oxygen delivery to the tissues. For example, if airflow in the bronchioles is restricted as in obstructive pulmonary emphysema, the O_2 pressure of the arterial blood is decreased. This is compensated by a shift in the O_2 dissociation curve, due to an increase in the erythrocyte concentration of DPG from 4.5 to as high as 8.0 mM (Fig. 9). DPG may also play a part in high altitude adaptation: transfer from sea level to 4500 meters results, after two days, in an increase of the erythrocyte DPG concentration to 7.0 mM. The O_2 affinity of blood

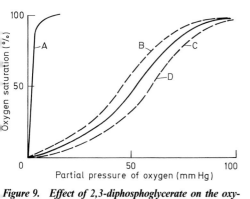

Figure 9. Effect of 2,3-diphosphoglycerate on the oxygen saturation curve of hemoglobin.
A, 2,3-diphosphoglycerate absent.
B, 2,3-diphosphoglycerate concentration low.
C, 2,3-diphosphoglycerate concentration normal.
D, 2,3-diphosphoglycerate concentration high.

increases during storage, due to loss of DPG (DPG decreases to 0.5 mM in 10 days when blood is stored in acid-citrate-dextrose). Although transfused blood can regain DPG (half normal levels achieved after 24 h, following total depletion), this may not be rapid enough in critical cases. Addition of DPG to the blood has no effect, because it cannot cross the erythrocyte membrane. Addition of Inosine (see), however, maintains the DPG level of stored red cells, because inosine crosses the erythrocyte membrane and is converted into DPG, a conversion involving reactions of the Pentose phosphate cycle (see). See Rapoport - Luerbing shuttle. [R. E. Dickerson & I. Geis, *The Structure and Action of Proteins* (W. A. Benjamin, 1969); M. F. Perutz, *Ann. Rev. Biochem.* **48** (1979) 327–386, *Nature* **228** (1970) 726–739, *Sci. Amer.* **239** (6) (1978) 92–125; J. V. Kilmartin, *Brit. Med. Bull.* **32** (1976) 209–222; A. Arnone, *Nature* **237** (1972) 146–149]

Hemoglobinopathy: an inherited abnormality of hemoglobin structure, usually the substitution of a single amino acid in the α or β chains as a result of a point mutation in the α or β gene. Hs. due to amino acid deletions are also known. Hs. are Inborn errors of metabolism (see). Thalassemias (see) may also be classified as Hs. Many structurally abnormal hemoglobins are clinically unimportant. Others may show decreased or increased affinity for oxygen, or display instability leading to Heinz body (see) formation, sickling, decreased erythrocyte survival and hemolytic anemia. Other functional defects include lack of subunit cooperativity, decreased Bohr effect and decreased effect of 2,3-diphosphoglycerate (see Hemoglobin). The main initial screening techniques for abnormal hemoglobins are starch gel and paper electrophoresis (Fig.). Electrophoresis cannot differentiate all mutant hemoglobins. Evidence for the presence of an abnormal hemoglobin is often physiological or clinical, e.g. abnormal oxygen affinity, instability, etc. Ultimately, the α and β chains of the suspect hemoglobin must be purified and analysed by conventional techniques of protein sequence analysis to identify the site of amino acid substitu-

tion. Lettering of abnormal hemoglobins (see Fig. and Table, section G) was at first an initial system (F = fetal, S = sickle, M = methemoglobin), after which letters were assigned in order of discovery. The table does not list every known H., and new examples are continually reported in the literature. About 350 different Hs. are known; about 150 of these are discussed by E. R. Huehns in *Blood and Its Disorders* (R. M. Hardisty & D. J. Weatherall, eds.; 2nd edition, Blackwell Scientific Publications, Oxford, 1982). The *International Hemoglobin Information Center* publishes updated lists of abnormal hemoglobins in the journal *Hemoglobin*. See also R. G. Schneider "Methods for Detection of Hemoglobin Variants and Hemoglobinopathies in the Routine Clinical Laboratory *CRC Critical Review in Clinical Laboratory Sciences* (November 1978).

Abnormal hemoglobins. The names of abnormal hemoglobins usually refer to the geographical location where the hemoglobin was first discovered. The term, *unstable*, refers to those hemoglobins associated with accelerated erythrocyte destruction in vivo, and precipitation of hemoglobin in vitro when warmed to 50 °C. For residue numbering system and comparison with normal hemoglobin A, see Hemoglobin.

A. Amino acid replacements affecting contact with heme
A-1. *Hemoglobins M (M for Methemoglobin).* The iron of heme is irreversibly oxidized, i.e. permanently in the Fe(III) state. The resulting methemoglobin cannot be reduced by methemoglobin reductase. Five types are known. Only heterozygotes survive.

Name	Substitution	Remarks
Boston	α58(E7)His → Tyr	Permanently in deoxy form. Low O_2 affinity. Low cooperativity. Bohr effect decreased.
Saskatoon (= Emory)	β63(E7)His → Tyr	Unstable. High O_2 affinity. Decreased cooperativity. Bohr effect normal.
Milwaukee	β67(E11)Val → Glu	As for Saskatoon.
Iwate	α87(F8)His → Tyr	Permanently in deoxy form. Low O_2 affinity. Cooperativity absent. Bohr effect absent.
Hyde Park	β92(F8)His → Tyr	High O_2 affinity. Decreased cooperativity. Bohr effect normal.

$$\text{His F8: } N{-}Fe^{2+}{-}O_2 \longrightarrow \text{Tyr F8: } \bar{O}{-}Fe^{3+}{-}O{\big\langle}^H_H$$

Replacement of heme-binding His F8 by Tyr destroys the ability of heme to bind oxygen.

A-2. *Other substitutions in the heme binding region.*

Fort de France	α45(CD3)His → Arg	High O$_2$ affinity.
Hirosaki	α43(CD1)Phe → Leu	Unstable
Torino	α43(CD1)Phe → Val	Low O$_2$ affinity.
J Buda	α61(E10)Lys → Asn	Low O$_2$ affinity. Slightly unstable.
Moabit	α86(F7)Leu → Arg	Low O$_2$ affinity. Unstable.
Bibba	α136(H19)Leu → Pro	Unstable.
Hammersmith	β42(CD1)Phe → Ser	Low O$_2$ affinity. Unstable. Slightly decreased cooperativity. Polar Ser probably permits water to enter heme pocket.
Zürich	β63(E7)His → Arg	High O$_2$ affinity. Unstable. Bohr effect normal. Decreased cooperativity.
Shepherd's Bush	β74(E18)Gly → Asp	High O$_2$ affinity. Unstable. Bohr effect normal. Decreased cooperativity. Effect of 2,3-diphosphoglycerate decreased.
Sabine	β91(F7)Leu → Pro	Unstable. Heme group absent.
Casper	β106(G8)Leu → Pro	Unstable. Heme group absent.
Bryn Mawr	β85(F1)Phe → Ser	High O$_2$ affinity. Unstable.

B. *Amino acid replacements affecting contact between α$_1$ and β$_1$ subunits.*

Khartoum	β124(H2)Pro → Arg	Unstable.
Fannin Lubbock	β119(GH2)Gly → Asp	Unstable.
Madrid	β115(G17)Ala → Pro	Unstable.
San Diego	β109(G11)Val → Met	High O$_2$ affinity.
Philly	β35(C1)Tyr → Phe	High O$_2$ affinity. Unstable.

Hemoglobin E	β26(B8)Glu → Lys	Purified HbE has normal O$_2$ affinity. In erythrocytes homozygous HbE has low O$_2$ affinity due to increased 2,3-diphosphoglycerate in cells. Cells containing HbA and HbE (heterozygotes) have normal O$_2$ affinity. See sect. G.
Heathrow	β103(G5)Phe → Leu	High O$_2$ affinity.
Prato	α31(B12)Arg → Ser	No abnormal properties.
Chiapas	α114(GH2)Pro → Arg	No abnormal properties.

C. *Amino acid replacements affecting contact between α$_1$ and β$_2$ subunits.*

Hiroshima	β146(C-term His) → Asp	Bohr proton cannot be donated. High O$_2$ affinity. Bohr effect decreased.
Richmond	β102(G4)Asn → Lys	No abnormal properties.
Kempsey	β99(G1)Asp → Asn	H-bond between Asp(βG1) and Tyr(αC7) absent. High O$_2$ affinity. DeoxyHb less stable.
Setif	α94(G1)Asp → Tyr	Unstable.

D. *Amino acid replacements in cavity between like chains.*

Manitoba	α102(G9)Ser → Arg	Slightly unstable.
Jackson	α127(H10)Lys → Asn	Abnormal properties not reported.
Surenes	α141(HC3)Arg → His	High O$_2$ affinity.
Helsinki	β82(EF6)Lys → Met	Low O$_2$ affinity. Bohr effect decreased. Lys82 is diphosphoglycerate binding site.
Altdorf	β135(H13)Ala → Pro	High O$_2$ affinity. Unstable.
Syracuse	β143(H21)His → Arg	High O$_2$ affinity. Bohr effect decreased. His143 is diphosphoglycerate binding site.

E. Amino acid replacements in the interior of subunits.
Replacement of a nonpolar by a polar residue in the hydrophobic interior, or a small residue by a large one can cause instability. Insertion of Pro into a helix causes distortion (Pro is a "helix breaker") and instability.

Port Phillip	α91(FG3)Leu → Pro	Unstable.
Perth	β32(B14)Leu → Pro	High O₂ affinity. Unstable.
Riverdale-Bronx	β25(B6)Gly → Arg	Unstable. Helices B and E closely packed in this region. Arg too large.

F. Amino acid deletions. All these Hb are unstable.

Freiburg	β23(B5)Gly deleted.	High O₂ affinity.
St. Antoine	β74-75(E18-E19) Gly-Leu deleted	O₂ affinity normal.
Gun Hill	β91-95(F7-FG2) Leu-His-Cys-Asp-Lys deleted	Essential contacts with heme absent. β-chains contain no heme. Hemolytic anemia.

G. Amino acid replacements on the outer surface of the molecule.

Hemo-globin C	β6(A3)Glu → Lys	Relatively common in West Africans and people of West African extraction. Heterozygotes have 30–40% HbC (+ about 60% HbA) and are healthy. Homozygotes may have mild anemia and their lifespan is normal.
Hemo-globin D	β121(GH4)Glu → Gln	Relatively prevalent among Negroes (0.4%), Algerians (2.0%) and Sikhs of north and central India, with sporadic occurrence in other groups. Homozygotes show very minor symptoms (anemia?) and the term "hemoglobin D disease" is probably too strong.
Hemo-globin E	β26(B8)Glu → Lys	See also B. The second most common abnormal Hb in the world, in people of southeast Asian origin. Homozygotes show mild anemia.

Hemo-globin J	α115(GH3)Ala → Asp	Found in Melanesia (New Hebrides) New Guinea. Seems to be of no pathological significance.
Hemo-globin O Arab	β121(GH4)Glu → Lys	Enhances sickling when heterozygous with Hb S. Low O₂ affinity of erythrocytes in homozygotes, which show a sickling condition. Probably originates from non-Arab peoples of presemitic Egypt. Reservoir is possibly Sudanese, spreading through Ottoman Empire. Also found in Jamaica, Roumania, Bulgaria, Hungary.
Korle Bu	β73(E17)Asp → Asn	Described in only one West African family. Homozygotes normal with no pathology

Hemoglobin S (sickle cell hemoglobin). The most common pathologically abnormal hemoglobin. See separate entry.

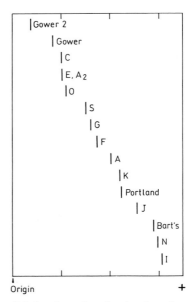

Relative electrophoretic migration of normal and variant human hemoglobins in starch gel electrophoresis in Tris-EDTA-borate buffer, pH 8.6. Hemoglobin I is the same as H or β₄. Hemoglobin Bart's is hemoglobin γ₄.

Hemopexin: a single-chain, heme-binding β_1-plasma glycoprotein, M_r 57000. H. contains 22% carbohydrate. In contrast to Haptoglobin (see), H. binds neither hemoglobin nor cytochrome c, but only their prosthetic group, heme. 100 ml human plasma contains 80 to 100 mg H.

Hemopoietic cell growth factor: see Colony stimulating factor.

Hemoproteins: ubiquitous chromoproteins, which, as respiratory pigments, are involved in oxygen transport (see Hemoglobin, and Hemerythrin) and in oxygen storage (see Myoglobin). As catalase and peroxidase, they effect the reduction of peroxides, and as cytochromes, they are involved in electron transport between dehydrogenases and terminal acceptors. Their prosthetic group, iron porphyrin IX or heme, is tightly bound to the protein component. While 4 of the iron ligands are occupied by the porphyrin ring, the other 2 are used for binding to the protein (via histidine) and to O_2 (in the respiratory H.) or also to the protein (via Cys, Met, Trp, Lys or Tyr). In cytochrome c the porphyrin ring is additionally bound to the protein by covalent bonds between two SH groups and the two vinyl groups of the heme. One of the characteristic properties of the H. in the reduced state (FeII) is a spectrum with three intense bands in the visible range: 1) the α-band, λ_{max} 550 to 565, 2) the β-band, λ_{max} 520 to 535, and 3) the γ or Soret band, λ_{max} 400 to 415 nm.

Hemosiderin: an iron storage protein of the mammalian organism, functionally related to Ferritin (see). H. is deposited in the liver and spleen (hemosiderosis), particularly in diseases associated with increased blood destruction, such as pernicious anemia, or with increased iron resorption (hemochromatosis), or even in hemorrhages. Most of the deposits are located in the liver, which may contain up to 50 g H., compared to the normal content of 120 to 300 mg H. H. from horse spleen consists of 26 to 34% iron(III), and up to 35% protein (aposiderin). The rest is made up of octasubstituted porphyrin, mucopolysaccharides and fatty acid esters.

Heparin: an acid mucopolysaccharide from animal tissues which prevents blood clotting. M_r about 16000. H. consists of equal amounts of D-glucosamine and D-glucuronic acid, α-1,4-glycosidically linked, and contains O- and N-sulfate residues. It is dextrorotatory. Due to its acid character it can form salts; the effective protein component in blood is called heparin complement. H. prevents the clotting of blood by preventing the conversion of prothrombin to thrombin and of fibrinogen to fibrin. Clinically, it is applied parenterally for the treatment of thrombosis, phlebitis and embolism.

Heparin

Heparitin sulfate, *heparan sulfate:* a monosulfate ester of an acetylated (*N*-acetyl) heparin. As isolated from animal tissues (liver), H.s. is probably a mixture of mucopolysaccharides with varying degrees of sulfatation or amino group acetylation.

HEPES, *Hepes:* abb. for *N*-2-Hydroxypiperazine-*N*-2-ethanesulfonic acid, a buffering compound used in the range pH 6.8–8.2. See Buffers.

Heptoses: monosaccharides containing 7 C atoms. The 7-phosphates of D-mannoheptulose and D-sedoheptulose are important in carbohydrate metabolism.

Heroin, *diacetylmorphine, diamorphine:* one of the most dangerous narcotics. m.p. 173 °C, $[\alpha]_D^{25} - 166$ ° (methanol). H. is not very stable and decomposes in boiling water. It is synthesized by acetylation of the two hydroxyl groups of morphine with acetyl chloride. This increases the analgesic effect by a factor of six. Due to the extreme danger of addiction, the therapeutic use of H. is forbidden in most countries. In New York alone, the number of heroin addicts is estimated to be 300000.

Heroin

Hers' disease: see Glycogen storage disease.

Hershberg test: see Auxins.

Hesperetin: 5,7,3'-trihydroxy-4'-methoxyflavanone, see Flavanone.

Hesperidin: hesperetin 7-rutinoside, a Flavanone (see) glycoside which makes up 8% of the dry weight of orange peel (see Vitamin P).

Heteroauxin: see Auxins.

Heterogeneity: in proteins, a term for differences in the structure of a species of protein, which cause no change in the biological activity. H. can be either genetically controlled or can arise through partial chemical or enzymatic modification. Chemical changes include, for example, phosphorylation of serine and threonine residues in casein, or the covalent attachment of carbohydrate chains in the glycoproteins. The most common cause of enzymatic modifications is limited proteolysis, which can produce artifacts. The pseudoisoenzymes β- (single chain) and α- (double chain) trypsin and the erroneously high number of subunits, e.g. of yeast hexokinase (4 instead of 2), are examples of H. The latter are caused during the isolation procedure by proteinase contaminants which are co-purified with the desired protein.

Microheterogeneities are slight differences in the primary structure, which are limited to a few unimportant residues in the peptide chain, or, in glycoproteins, slight differences in the number, length and composition of the carbohydrate chains.

Heteroglycans: polysaccharides composed of two or more different carbohydrate residues, for example, pectins, plant mucilages, plant gums and mucopolysaccharides.

Heterophagy: see Intracellular digestion.

Heteropolar bond: see Noncovalent bonds.

Heteropolypeptides: see Proteinoids.

Heteroside: a compound of one or several carbohydrate residues and a component belonging to a different class of substance, the aglycon or genin. E.g. glycosides are H.

Heterotrophy, *heterotrophic nutrition:* a nutritional dependence on organic compounds. In carbon heterotrophy, organic carbon compounds serve as sources of carbon and energy for the synthesis of body substituents and ATP. The degree of H. varies widely among the various heterotrophic organisms (all animals, including humans, and most microorganisms), and may include a dependence on externally supplied essential amino acids, fatty acids and vitamins. Auxotrophic mutants (see) have special nutritional requirements. Parasitism (see), Saprophytism (see) and Symbiosis (see) are special forms of heterotrophic feeding. The terms autotrophy and heterotrophy do not suffice to describe the widely varying forms of microbial nutrition, since here the nature of the carbon source and the energy source as well as the chemical nature of the reducing agent used for reductive syntheses must be taken into account. The converse of H. is Autotrophy (see).

HETPP: see Active acetaldehyde.

Hexamethonium: see Acetylcholine.

Hexestrol: meso-hexestrol, a synthetic compound with estrogenic activity. It is not a steroid, but is used therapeutically in the same way as natural estrogens.

Hexestrol

Hexitols: sugar alcohols with 6 C-atoms. Of the 10 possible isomers, D-sorbitol, dulcitol, D-mannitol, iditol and allitol are found in nature.

Hexokinase (EC 2.7.1.1): an enzyme catalysing the transfer of a phosphoryl group from ATP to the C-6 oxygen of glucose or another hexose, such as mannose or D-glucosamine. (The physiological substrate is MgATP.) The reaction with glucose is the first step in Glycolysis (see). H. binds specifically to porin, a protein of the outer mitochondrial membrane which permits ADP and saccharides to penetrate this membrane. Although H. is a soluble enzyme, it may therefore be associated with the mitochondrion under physiological conditions. This notion is supported by observations that mitochondrion-associated H. has a higher K_m for MgATP (0.25 mM vs. 0.12 mM for non-associated H.), and is more susceptible to product inhibition by glucose 6-phosphate.

X-ray studies of H. crystals grown in the presence and absence of glucose indicate that when glucose is bound, one lobe of the molecule rotates through 12° to close the substrate binding cleft (Fig.). In this conformation, water is excluded from the cleft. Later studies indicated that the conformational change also occurs in H. in aqueous solution. The conformational change is required for activity; analogs of glucose which are too bulky to allow it are also not substrates, although they may be inhibitors. Although water is able to bind to the substrate pocket in the same position as the 6-OH group of the sugar, it does not induce the conformational change, nor is it readily phosphorylated (this would amount to hydrolysis of ATP; H. is not very active as an ATPase). Interestingly, the enthalpy and heat capacity of the enzyme are nearly unchanged by the binding. [B. I. Kurganov in G. R. Welch, ed., *Organized Multienzyme Systems* (Academic Press, Orlando, 1985) pp. 241–268; K. Takahashi, J. L. Casey & J. M. Sturtevant *Biochem.* **20** (1981) 4693–4697; W. S. Bennett & T. A. Steitz, *Proc. Natl. Acad. Sci. USA* **75** (1978) 4848–4852.]

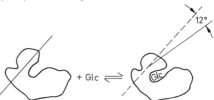

Conformational change in hexokinase induced by glucose. (Adapted from Bennett & Steitz.)

Hexosans: high-molecular-weight plant polysaccharides, belonging to the group of homoglycans, and composed of hexoses. Examples are glucans, fructans, mannans and galactans.

Hexose monophosphate pathway: see Pentose phosphate cycle.

Hexoses: aldoses containing 6 C-atoms; one of the important groups of monosaccharides (see Carbohydrates). All possible stereoisomeric aldohexoses (there are four asymmetric C atoms) have been isolated or synthesized. D-Glucose, D-mannose, D-galactose and L- and D-talose are widespread in nature, both as free sugars and in bound form. Some of the phosphorylated H. are particularly important. The two 6-deoxy-sugars L-rhamnose and L-fucose are also H.

The ketohexoses corresponding to the aldohexoses are called *hexuloses.* They include the naturally occurring monosaccharides D-fructose and L-sorbose.

Hexuloses: see Hexoses.

Hibbert's ketones: see Lignin.

High energy bonds, *energy-rich bonds:* chemical bonds which release more than 25 kJ/mol on hydrolysis. They are usually esters (enol, thiol and phosphate esters), acid anhydrides, or amidine phosphates.

In biological systems, the energy released is used to transfer the hydrolysed residue to other metabolic compounds (group transfer). The H.e.b. are symbolized by ~ instead of the usual hyphen between groups (e.g. $CH_3CO \sim SCoA$). The free energy of hydrolysis of H.e.b. is at the same time a measure of the potential for group transfer. The table summarizes the standard values of the free energy of hydrolysis for some important compounds. See Energy-rich phosphates.

Table ΔG_o of hydrolysis.

Compound	ΔG_o (kJ/mol)	(kcal/mol)
Creatine phosphate	42.7	10.2
Phospho*enol*pyruvate	53.2	12.7
Acetylcoenzyme A	34.3	8.2
Aminoacyl-tRNA	29.0	7.0
ATP → ADP + P_i	30.5	7.3
ATP → AMP + P ~ P	36.0	8.6
P ~ P → 2P_i	28.0	6.7

High-yielding strains: see Production strains.

Hill plot: a graphic method for the determination of the degree of cooperativity of an enzyme (see Cooperativity). The plot of log $Y_s/(1 - Y_s)$ versus log α is a curve with a slope of 1 for large or small α and a finite energy of interaction between the substrate-binding sites. Y_s is the saturation function, that is, the fraction of the enzyme in the enzyme-substrate complex, and $\alpha = S/K_m$. For the values of α usually obtained experimentally, an approximate straight line is obtained with the maximal slope h (Fig.), the *Hill coefficient*. This serves as a measure of the cooperativity, and is not usually identical to the number n of subunits of an enzyme, but is only a minimal estimate of n. If the energy of interaction between the substrate-binding sites is infinite, the Hill plot degenerates into a straight line with a slope equal to the number of subunits. The segment of the ordinate AB, constructed as shown in the Fig., can be used to determine the total energy of interaction ΔG_w between substrate or effector-binding sites: $\Delta D_w = 2.303 \, RT \cdot AB$. Here R is the general gas constant and T is the absolute temperature. The saturation functions for effectors can also be determined. Y_E is the fraction of the enzyme in the enzyme-effector complex. For enzymes which are in equilibrium with their enzyme-substrate complexes, the saturation function Y_s can be replaced by the kinetic saturation v/V_m, so that log $v/(V_m - v)$ can be plotted against log α. Correspondingly, for an effector, log $(v - v_0)/(V - v)$ must be plotted against log α; v is the measured rate of reaction at constant S, v_0 is the rate

in the absence of effector at the same S, and V is the rate at a saturating concentration of effector for the chosen S; S is the substrate concentration.

Hill reaction: light-dependent production of oxygen by the photosynthetic system in the presence of an artifical oxidizing agent (electron acceptor). R. Hill first observed this reaction in illuminated isolated chloroplasts, in the absence of CO_2, and using iron(III) oxalate as oxidizing agent. Iron(III) oxalate (Fe^{3+} is reduced to Fe^{2+} in the reaction) can be replaced by potassium ferricyanide, quinone and other compounds (Hill reagents). Spinach chloroplasts catalyse the following H.: $4 K_3Fe(CN)_6 + 2 H_2O + 4 K^+ \rightarrow 4 K_4 Fe(CN)_6 + 4 H^+ + O_2$. The "natural" Hill reaction is the photolysis of water, and the „natural" Hill reagent is oxidized NADP.

Hinekiflavone: see Biflavonoids.

Hippuric acid: $C_6H_5\text{-}CO\text{-}NH\text{-}CH_2\text{-}COOH$, the *N*-benzoyl derivate of glycine. Mammalian herbivores detoxify benzoic acid by converting it to H. a.

Hircinol: see Orchinol.

His: abb. for L-Histidine.

Histamine: β-imidazol-4(5)ethylamine, a biogenic amine. M_r 114.14. H. is formed by enzymatic decarboxylation of L-histidine. It stimulates the glands in the fundus of the stomach to secrete digestive juices, dilates the blood capillaries (important for increasing blood flow and decreasing blood pressure), increases the permeability (urtication and reddening after local application of histamine), and causes contraction of the smooth muscles of the digestive tract, the uterus and the bronchia (in bronchial asthma). H. is catabolized by diamine oxidases and aldehyde oxidases to imidazolylacetic acid. H. is widely distributed in the plant and animal kingdoms, occurring for example in stinging nettles, ergot, bee venom and the salivary secretions of biting insects. As a tissue hormone, H. is present in the liver, lungs, spleen, striated muscles, mucus membranes of stomach and intestine, and it is stored with heparin in mast cells. The amounts found in tissue are on the order of µg/g fresh weight.

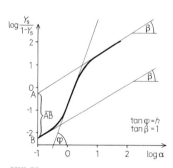

Histamine

L-Histidine, abb. *His:* imidazolylalanine, a proteogenic amino acid, M_r 155.2. It is a component of the catalytic centers of many enzymes, and is also a component of carnosine and anserine. His is an important buffer in the physiological pH range.

His is weakly glucoplastic. In the absence of dietary His, adult animals can maintain their nitrogen balance for a short time, but the amino acid is essential for growing animals. The imidazole ring cannot be synthesized by mammals. In bacteria, His is formed via imidazole glycerophosphate in the last part (Fig. 1) of the ATP-imidazole cycle (Fig. 2). The starting compound in the ATP-imidazole cycle is 5-aminoimidazole-4-carboxamidoribotide, abb. AICAR (see Purine biosynthesis). The intermediate products are inosinic acid, AMP, ATP, phosphorib-

Hill-Plot

osyl-ATP, phosphoribosyl-AMP, and phosphoribosyl-formimino-aminoimidazole carboxamidoribotide. The latter compound is synthesized by an Amadori rearrangement, which is relatively rare in cellular metabolism. It is significant that ATP is the substrate. Only the C-2 atom and the N-1 atom of the purine ring (Fig. 3) are incorporated into the His molecule. His is catabolized by histidase to urocanic acid (imidazoleacrylic acid) and then, via imidazolonepropionic acid and formimino glutamic acid, to glutamic acid. The formimino group is used for the synthesis of Active one-carbon units (see). His is used in the treatment of allergies and anemias. [R. G. Martin et al. *Methods in Enzymology* **XVII B** (1971) 3–44; M. Brenner & B. N. Ames "The Histidine Operon and Its Regulation" in Greenberg, ed., *Metabolic Pathways* (3rd edn.) **5** 349–387]

Histidinemia: see Inborn errors of metabolism.

Histocompatibility antigens: see Major histocompatibility complex.

Histones: a group of basic proteins which form reversible complexes with DNA, called nucleohistones. The primary structures of the H. are for the most part known. They are divided into the following main classes: H1 (or I or f1) is lysine-rich; H2a and H2b (or IIb1 and IIb2 or f2a2 and f2b), are moderately lysine-rich; and H3 (or III or f3) and H4 (or IV or f2a1) are arginine-rich. Octomers containing two molecules each of H2a, H2b, H3 and H4 have been isolated in soluble form, and radioactive labelling experiments indicate that these octomers segregate conservatively during chromatin replication. [I. M. Leffak, *Nature* **307** (1984) 82–85.] The DNA is wound around the octomer cores, giving the chromatin the appearance of a chain of beads. H1 is present in smaller quantities than the other H., and

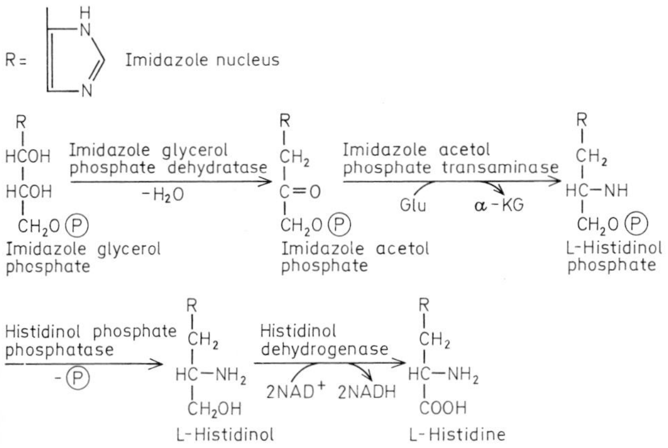

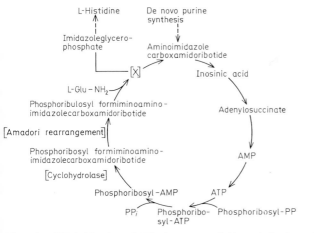

Figure 1. Terminal reactions in histidine biosynthesis.

Figure 2. ATP-imidazole cycle. The latter part of this cycle is shown in greater detail in Fig. 3.

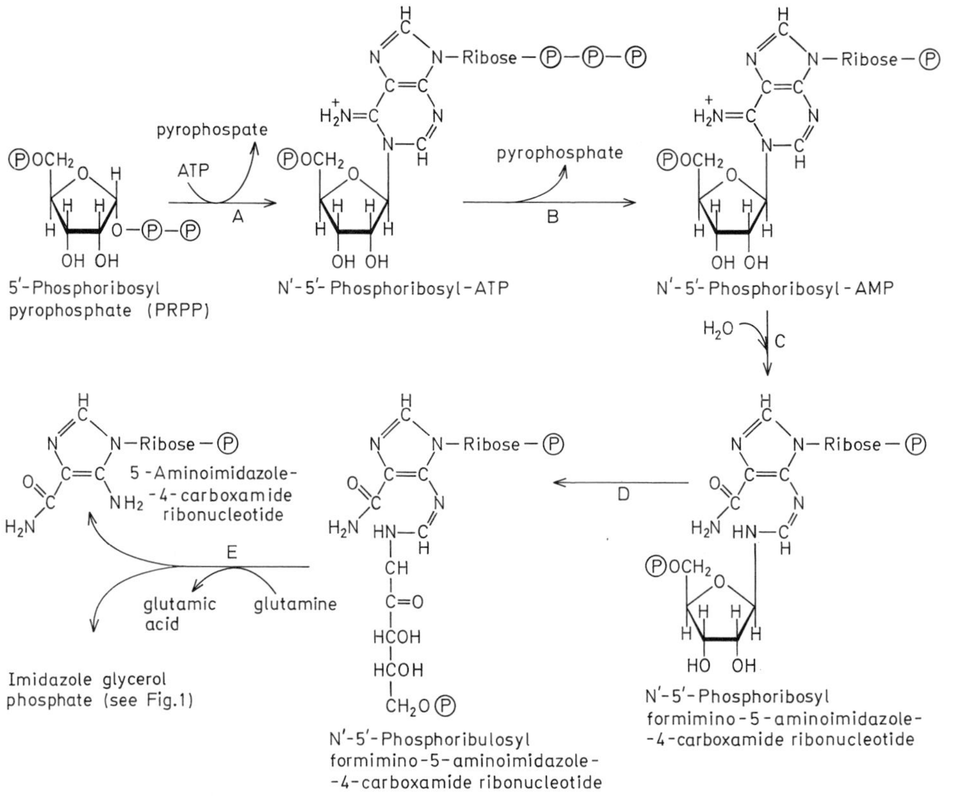

Figure 3. First stages of histidine biosynthesis.
A. N-1-(5'-phosphoribosyl)adenosine triphosphate: pyrophosphate phosphoribosyl transferase, or ATP phosphoribosyltransferase (EC 2.4.2.17).
B. Phosphoribosyl ATP pyrophosphohydrolase (EC 2.4.2.17). In Neurospora, this enzyme is trifunctional, also catalysing reaction A (above) and the conversion of histidinol to histidine (Fig. 1).
C. 1-N-(5'-phospho-D-ribosyl)-AMP 1,6-hydrolase, or phosphoribosyl-AMP cyclohydrolase (EC 3.5.4.19).
E. Amidotransferase and cyclase. Presumably an intermediate is produced by transfer of nitrogen from glutamine, followed by cyclization to form the imidazole ring of imidazole glycerol phosphate. The intermediate is not known, and amidotransferase and cyclase activities have not been separated.

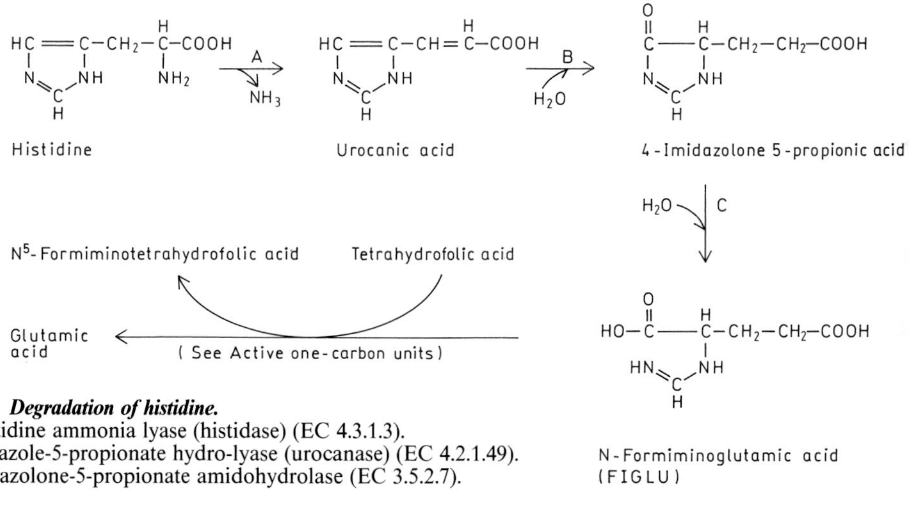

Figure 4. Degradation of histidine.
A. L-Histidine ammonia lyase (histidase) (EC 4.3.1.3).
B. 4-Imidazole-5-propionate hydro-lyase (urocanase) (EC 4.2.1.49).
C. 4-Imidazolone-5-propionate amidohydrolase (EC 3.5.2.7).

is thought to act as a link between "beads" (nu bodies) on the chromatin chain.

Histopine: see D-Octopine.

hMG: abb. for human menopausal gonadotropin.

HMTPP: see Thiamin pyrophosphate.

Holarrhena alkaloids, *kurchi alkaloids*: a group of steroid alkaloids which are the characteristic active substances in plants of the dogbane *(Apocynaceae)* genus *Holarrhena*. The representatives so far isolated (about 50) are formal derivatives of the hydrocarbon pregnane (see Steroids) which is substituted in the 3 and 20 positions with amino or methylamino groups, e.g. holarrhimine and conessine (Fig.). In the most widespread compound, conessine, and its 12β-hydroxy derivative holarrhenine, the 20 amino group is bound to C-atom 18 to form a pyrrolidine ring. Conessine is important as a starting material for the synthesis of aldosterone. H.a. have blood-pressure-reducing, curare-like, diuretic and narcotic properties. The alkaloid-rich bark of the shrub *Holarrhena antidyserenterica* is used for treating dysentery in India. The H.a. are biosynthesized via cholesterol and pregnenolone.

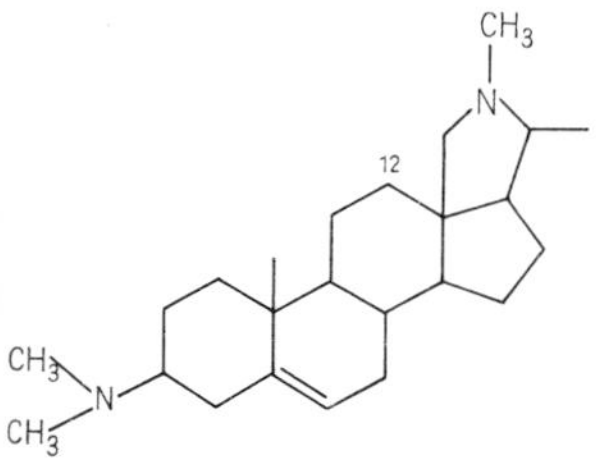

Conessine

Holarrhenine: see Holarrhena alkaloids.

Holarrhimine: see Holarrhena alkaloids.

Holoenzyme: see Coenzyme.

Holosides: compounds consisting of glycosidically linked sugar residues, e.g. oligo- and polysaccharides.

Holothurines: a group of highly toxic compounds from sea cucumbers *(Holothurioidea)*. H. are triterpene saponins (see Saponins), which contain holothurinogenins as aglycons and D-glucose, D-xylose, 3(O)-methylglucose and D-quinose as sugars. An important representative is holothurine A from the sea cucumber *Actinopyga agassizi*, which also contains sulfate, and which has 22,25-oxidoholothurinogenin as its aglycon.

Holothurinogenins: see Holothurines.

Homeo box: A sequence of base pairs in DNA, about 180 kilobases in length. The sequence is highly conserved, and has been identified in insects, other arthropods, annelid worms and chordates, including *Homo sapiens*. The term is derived from "homeosis", the replacement of one body structure by a homologous one, e.g. an antenna by a leg in certain mutant flies. In insects, a number of genes which control the development of body segments have been shown to contain the H.b.; and no gene which contains the box is known to be unrelated to embryonic development. The functions of non-insect genes which contain H.b. are not known. Conceptual translation of the H.b. (i.e. determination of the polypeptides it would encode if it were translated in vivo) suggests that if translated, it would code for a large number of basic amino acids, and thus might encode a DNA-binding domain of those proteins of which it would be a part. [W.J. Gehring *Cell* **40** (1985) 3–5.]

Homoarginine, abb. *Har:* a higher homologue of arginine with an additional methylene group in the side chain.

Homocysteine, abb. *Hcy:* a higher homologue of cysteine with an additional methylene group in the side chain.

Homocysteinemia: see Inborn errors of metabolism.

Homocystinuria: see Inborn errors of metabolism.

Homoferreirin: 5,7-dihydroxy-4′,6′dimethoxyisoflavanone, see Isoflavanone.

Homoglycans: straight or branched chain polysaccharides containing only one kind of monosaccharide residue. H. are widespread in the vegetable kingdom. They include arabans, xylans, glucans, fructans, mannans, galactans, the starch components amylose and amylopectin, cellulose and glycogen.

Table. *Partial sequences of homologous proteolytic (A) and esterolytic (B) enzymes from the region about the active serine residue*.*

Enzyme	Sequence									
A Trypsin (beef, sheep, pig, dogfish, shark, shrimp)	Asp	Ser	Cys	Glu	Gly	Asp	Ser*	Gly	Gly	Pro
Chymotrypsin A and B (beef)	Ser	Ser	Cys	Met	Gly	Asp	Ser*	Gly	Gly	Leu
Elastase (pig)	Ser	Gly	Cys	Glu	Gly	Asp	Ser*	Gly	Gly	Pro
Thrombin (beef)	Asp	Ala	Cys	Glu	Gly	Asp	Ser*	Gly	Gly	Pro
"Trypsin" *(Streptomyces griseus)*	Asp	Thr	Cys	Glu	Gly	Asp	Ser*	Gly	Gly	Pro
B Acetylcholinesterase				Phe	Gly	Glu		Ser*	Ser	Glu	Gly
Pseudocholinesterase (horse)					Gly	Glu		Ser*	Ala	Gly	Gly
Liver esterase (pig, horse, sheep, chicken)					Gly	Glu		Ser*	Ala	Gly	Gly
Pancreatic lipase (pig)					Leu			Ser*	Gly	His	
Alkaline phosphatase *(Escherichia coli)*	Asp	Tyr	Val	Thr	Asp		Ser*	Ala	Ala	Ser

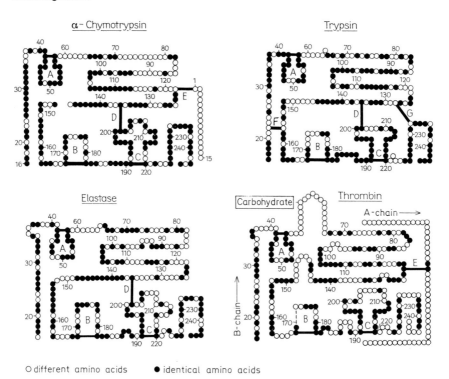

Figure 1. *Comparison of the primary structures and the positions of the disulfide bridges of four serine proteases.* A to G indicate the homologous disulfide bridges.

Homologization: see Glucosinolate.

Homologous proteins: proteins which have arisen through divergent evolution from a common ancestor. They usually have very similar primary and tertiary structures. Examples are the cytochromes, hemoglobin and myoglobin, the ferredoxins (non-heme iron proteins), fibrin peptides, immunoglobulins, peptide hormones (e.g. insulin and hypophyseal hormones), snake venom toxins and enzymes like the serine proteases of the pancreas (trypsin, chymotrypsin, elastase) or the blood-clotting enzymes (e.g. plasmin, thrombin) and lactate dehydrogenase. As examples of homologies in primary structure, the partial sequences around the catalytically important amino acid residues of the serine proteases are given in the table. A comparison of the primary structures and the location of the disulfide bridges of trypsin, chymotrypsin, and elastase is shown in Fig. 1.

Fig. 2 shows that the structural homologies are also reflected in the conformation of the polypeptide chains, taking chymotrypsin and elastase as an example. Although only about 40% of the amino acids of the two proteins are homologous, that is, identical, their spatial folding is similar. NH₂ and COOH indicate the beginning and end, respectively, of the chain. Positions 57, 102 and 195 are occupied by the important amino acids histidine, aspartic acid and serine.

Homopolymer: a polymer built up of identical

monomeric units, for example, amylose and polyphenylalanine. In a narrower sense, H. are synthetic polynucleotides in which all the nucleotides contain the same base, for example polyadenylic acid, polyuridylic acid, polydeoxyadenylic acid. H. in the narrower sense are synthesized in vitro from nucleoside di- or triphosphates using the appropriate polymerases without a matrix. An oligonucleotide is needed as a primer. The diphosphates can be polymerized by polynucleotide phosphorylase. H. are usually single-stranded, but are occasionally double-stranded. H., especially poly(A) sequences, occur naturally in some of the RNA of eukaryotes (see Messenger RNA).

Homopterocarpin: 3,9-dimethoxpterocarpan, see Pterocarpans.

Homosteroids: see Steroids.

Hordein: see Prolamin.

Hordenine, anhaline: *N,N*-dimethyltyramine, one of the biogenic amines, widely distributed in nature. m.p. 117–118 °C, b.p. 173–174 °C. As a derivative of phenylethylamine, H. is one of the amines which increase blood pressure, but it has a low physiological activity.

Hormones: organic compounds produced and interpreted as intercellular signals by nearly all organisms. Very low concentrations of H. are usually encountered – some 10^{-12} to 10^{-15} mole per mg tissue protein in the case of hypothalamic releasing H., pituitary-like H. and gastrointestinal H. in animals.

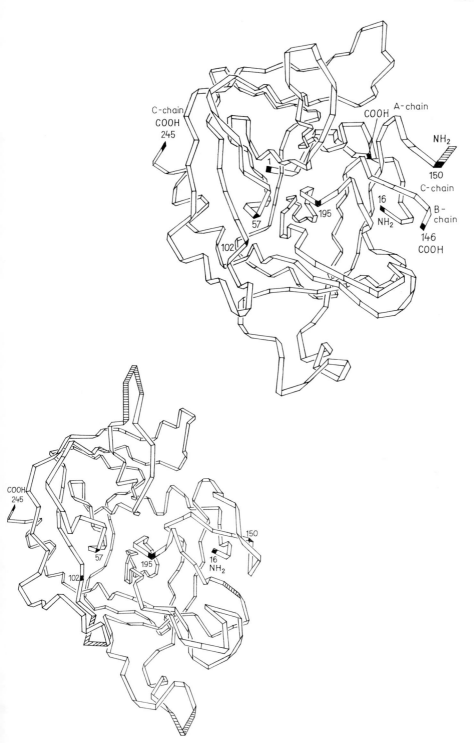

Figure 2.   Conformation of the polypeptide chains of the two homologous pancreas enzymes, α-chymotrypsin (above) and elastase (below).

Plants, unicellular organisms and animals have H.; as nervous systems developed, some H. came to serve as neurotransmitters, e.g. acetylcholine and some peptide H. The existence of a separate endocrine system is phylogenetically more recent, occurring only in the vertebrates. There are close parallels between neurons which secrete neurotransmitters and neurohormones and the endocrine gland cells which secrete the specialized glandular H.

H. have a variety of chemical structures: there are Steroid H. (see), amino acid derivatives (e.g. see Adrenalin, Thyroxin, Auxins), Peptide H. (see), Protein H. (see) and fatty acid derivatives (see Prostaglandins). In plants, the Cytokinins (see) are mostly derivatives of adenine. Insect hormones (see) are discussed in a separate entry.

The common element in the action of all H. is a receptor for the particular H. on the membrane of the "target" cell (e.g. insulin, adrenalin) or within the cell (e.g. steroid H.). What happens next depends on the H. and the cell. In many cases, the H. serves as a "first messenger", which activates "second messenger" (often cyclic AMP, see Adenosine phosphates) within the cell. The cAMP may activate a protein kinase (or the receptor may itself be a protein kinase activated by its H.) which in turn activates some other key enzyme and thus alters the metabolism of the cell (see Post-translational modification of proteins), or the permeability of its membrane to ions or other molecules. Steroid H. bind to their intracellular receptors, forming a complex which interacts with nuclear chromatin. Presumably this is the cause of gene activation by the H. Peptide H. may also affect gene expression, in some cases apparently through the action of histone kinases activated by the H. The H. itself is usually subject to rapid inactivation and/or degradation and excretion.

Biosynthesis, storage and secretion: In animals there are two types of endocrine cells: those synthesizing protein and peptide H., and those synthesizing steroids. The former are stimulated to produce their H. by neurotransmitters (including those released by the neurons synapsing onto the cells), other H., metabolic products or dietary substances. The H. or their precursors (for example, proinsulin) synthesized on the ribosomes enter the Golgi apparatus of the cell and are packed there into small vesicles and stored in the cytosol. On demand, which is signaled by various stimuli (e.g. a high blood glucose level), the vesicles move to the cell membrane and release the H. into the blood stream.

The enzymes required for steroid hormone biosynthesis must be synthesized or activated when the cell is stimulated; for example, corticotropin causes cholesterol to be converted into glucocorticoids in the adrenals. Steroid H. are not stored; the appropriate enzyme system can be rapidly activated and deactivated. Cells which produce steroid H. have large Golgi apparatuses, much smooth endoplasmic reticulum, lipid droplets and lysosomes.

H. are transported in free form or bound to specific or unspecific proteins. Oxytocin for example, is bound to neurophysin, transported within the axon of a nerve from the hypothalamus to the posterior lobe of the hypophysis, and stored there. This binding is loose, non-covalent and easily dissociated.

The same is true of H. transport in the blood stream. For some hormones, the dissociation constants of the H.-transporter complex can be determined. The albumins carry somatotropin; steroid binding globulin possesses a relatively high affinity for steroid H., and some steroid H. have specific transport proteins, e.g. cortisol is transported by transcortin. However, serum albumin, which possesses a relatively low affinity for steroid H. is present in such high concentration that it binds the major fraction of circulating steroid H. The H. is partly inactivated by this binding, but is also stabilized and protected against enzymatic attack. Any method for H. determination in the blood must take into account the fact that the H. is present in both bound and free forms.

Inactivation of the H.: The action of a H. can be immediately stopped if 1) its second messenger is a cyclic nucleotide and is enzymatically hydrolysed to a mononucleotide (e.g. cAMP to 5'-AMP); or 2) if the H. is enzymatically degraded. Peptide and protein H. are inactivated by proteolytic enzymes, catecholamines by monoamine oxidases; steroid H. may be partly inactivated by oxidation or reduction (e.g. about 50% of estrogens in humans are oxidized to nonestrogenic catechols). Both active steroid H. and their inactive metabolites are converted to glucuronides or sulfates, which are excreted via the urine and bile.

Regulation of H. effects: In every H. effect, two points must be kept in mind: 1) there is no such thing as an isolated H. effect. H. have a definite, but not the dominant role in the regulation of intracellular metabolism and of total metabolism. 2) H. effects are usually recognizable in the form of a change in enzyme activity. Every H. effect is subject to fine tuning. The H. themselves, the metabolic products dependent on them, and the nerve system interact in this feedback mechanism. In most cases the synthesis and secretion of the first H. in a system is inhibited, in negative feedback, by the H. whose production it stimulates. A classic example of the coordination of various regulatory circuits is the system composed of hypothalamus (releasing hormones), anterior lobe of the hypophysis (hypophyseal H.) and target organ, in which there is feedback at every level.

Pathobiochemistry: Due to the close coordination of the H. and the nervous system with metabolism, any disturbance in the synthesis, secretion or transport of H., or the lack of receptors, or a disturbance of H. catabolism is reflected by a pathology of the entire metabolism. The lack of mineralocorticoids, for example, leads to a disruption of the mineral and water balance; too much or too little somatotropin, the growth hormone, produces gigantism or dwarfism; imbalances in the thyroid H. system (see Thyroxin) upset the energy metabolism and are reflected in a hyper- or hypothyreosis; sexual functions and the normal course of pregnancy can be severely affected when Gonadotropins (see) and Gonadal hormones (see) are under- or overactive; and there is some evidence that certain psychiatric illnesses are due to imbalances in peptide H. or neurotransmitters.

Methods of determination: H. can be assayed by biological, chemical or immunological methods. The traditional assays are biological, e.g. the pregnancy test in which gonadotropins in the urine cause ripening of the follicles in mouse ovaries. Assays of biological activity are also required to show that a laboratory synthesis has duplicated the natural H. structure. Structurally simple H. can be detected by chemical methods, such as gas chromatography for the detection of prostaglandins. Most work is now done with immunological or radioimmunoassays, because the very small amounts of H. in tissue require the sensitivity of these methods. An indirect method of demonstrating H. synthesis in cells is to use a cloned, radioactive DNA probe for the gene in question. If mRNA for the (peptide or protein) H. is present, it will hybridize with the DNA and bind it to the preparation, which is then subjected to radioautography. [K. Talmadge, N. C. Vamvakopoulos & J. C. Fiddes, *Nature* **307** (1984) 37–40] See Choriogonadotropin.

Hp: abb. for Haptoglobin.

HSK cycle: abb. for Hatch-Slack-Kortschak cycle.

Human chorionic gonadotropin: see Choriogonadotropin.

Human lactogen: see Placenta lactogen.

Human menopausal gonadotropin, *HMG,* *castration gonadotropin:* a glycoprotein of the anterior lobe of the pituitary. M_r 31000. Carbohydrate content 30%. HMG has a similar action to follicle stimulating hormone. Its primary structure is not yet known. Increased quantities of HMG are formed in the pituitary of women in the menopause, or women who have been ovarectomized. The increased synthesis (accompanied by an increased excretion) of HMG is explained by the absence of the negative feedback by sex hormones from the ovary on the hypothalamus and the pituitary.

Humulane type: see Sesquiterpenes (Fig.).

Humulene: isomeric monocyclic sesquiterpene hydrocarbons found in various aromatic oils. M_r 204.36. α-Humulene, α-caryophyllene, b. p.$_{·10}$ 123 °C, p$_4$ 0.8905, n_D 1.5508. β-Humulene, p$_4$ 0.8907, n_D 1.5012, has the same eleven-membered ring system as the α-isomer, but one of the three double bonds is exocyclic. The H. are found primarily in hop oil, together with oxygen-containing derivatives and β-caryophyllene. For the formulas and biosynthesis, see Sesquiterpenes.

Hunter's syndrome: see Inborn errors of metabolism.

Hurler's syndrome: see Inborn errors of metabolism.

Hyaloplasm: see Cell 2).

Hyaluronic acid: an unbranched mucopolysaccharide. M_r 200000 to 400000. The basic subunit is a disaccharide, N-acetyl-D-glucosamine glycosidically linked β-1,4 to D-glucuronic acid. The latter is β-1,3 linked to the next disaccharide unit. H. a. occurs in various animal tissues and joint fluids. Aqueous solutions of it have high viscosity, which explains its biological function as a lubricant. It is synthesized from D-glucose in the fibroblasts.

Hybridization: formation of a hybrid nucleic acid duplex by association of single strands of DNA and RNA (DNA:RNA hybrid) or single strands of DNA not previously associated with each other in a natural duplex (DNA:DNA hybrid). RNA:RNA hybrids are also possible. H. is used to detect and isolate specific nucleotide sequences, and to measure the extent of homology between nucleic acids.
In principle, the two nucleic acids are denatured by heating above the T_m (see Melting point of DNA), then allowed to hybridize at about 25 °C below the T_m. (Single strands must also be denatured by heating to remove intrastrand base pairing.) Usually one component (RNA or cDNA) of the H. mixture is present in relatively low concentration and is radioactive (of known specific radioactivity; labelled with ^{32}P or ^{3}H), while the other component (cellular DNA or fragments thereof) is unlabelled and present in excess. H. is determined by separating single from double-stranded nucleic acids and measuring the specific activity of the latter.
The filter or gel technique is used widely for the measurement of H. Here, DNA is thermally denatured and rapidly cooled so that dimers cannot form (the DNA can also be denatured by adjusting its pH to 12 with NaOH, followed by HCl to pH 7.) The single-stranded DNA is then embedded in agar or polyacrylamide gel, or adsorbed on a disc of cellulose nitrate. The DNA or RNA to be tested for complementarity is radioactively labelled, then applied to the gel or cellulose nitrate filter and allowed to incubate at about 25 °C below the T_m of the native material. Unbound radioactive material is removed by digestion with ribonuclease (DNA:RNA hybrids are resistant to the enzyme) or a deoxyribonuclease which preferentially attacks the single-stranded DNA. The amount of radioactivity remaining on the filter or gel after washing is a measure of the extent of complementarity between the sequences of the two samples. Cellulose nitrate discs are more convenient than other supporting substances; the original method of Gillespie and Spiegelman [*J. Mol. Biol.* **12** (1965) 829–841] is still used with little modification. H. may also be performed in solution (see DNA:DNA hybrids, below), in which case chromatography on hydroxyapatite is used to separate single and double-stranded material. The three methods, filter, gel and solution, have been critically compared by D. Kennel and A. Kotoulas [*J. Mol. Biol.* **34** (1968) 71–84.]
DNA:RNA hybrids. H. of rRNA with DNA has been used extensively to study gene multiplicity. As increasing quantities of radioactive RNA are incubated with cellular DNA, the formation of the DNA:RNA hybrid shows saturation kinetics. The limiting value for the number of RNA binding sites on the DNA is equivalent to the number of RNA cistrons per genome.
DNA:RNA H. is also used for gene isolation. In this technique, the DNA is sheared into shorter lengths (mechanical breakage or treatment with restriction endonucleases), then hybridized with RNA under

Hyaluronic acid

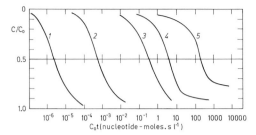

Hybridization plots or "Cot curves" for different DNA samples.
1. Poly U + Poly A.
2. Mouse satellite DNA.
3. Phage T4 DNA.
4. *Rhizobium* DNA.
5. Excess soybean DNA with radioactive root nodule leghemoglobin cDNA (single and double stranded DNA separated on hydroxyapatite, and radioactivity of double stranded DNA monitored as an index of the fraction (C/C_0) of cDNA annealed).

conditions that permit only about 10% renaturation of the DNA. Double- and single-stranded molecules are separated by hydroxyapatite chromatography, and the double-stranded fraction is repeatedly denatured and renatured until single-stranded DNA is effectively removed. Many genes have been partially purified in this way, e.g. rRNA cistrons of *Salmonella typhimurium* [A. Udvardy & P. Venetianer, *Eur. J. Biochem.* **20** (1971) 513–517].
Another technique is cytological H., in which radioactive RNA is hybridized with chromosomal DNA in a histological preparation. This indicates the chromosomal location of the gene(s) for the RNA in question. Conversely, cloned DNA probes can be used to show the histological distribution of mRNA corresponding to the cloned gene. Similarly, ^{32}P-labelled RNA or cDNA is used to locate complementary DNA sequences in Southern blots (see) of electrophoretically separated restriction fragments.
DNA:DNA hybrids. If heat-denatured DNA (see Melting point of DNA) is allowed to cool slowly, double-stranded molecules may be reformed, i.e. the single strands reassociate or become annealed. When two different DNAs are denatured in the presence of each other, the annealed mixture will include hybrid molecules, providing the two DNAs have some base sequences in common. Reassociation and H. (both can be called annealing) are therefore kinetically and mechanistically identical.
The following conditions are required for efficient reassociation or hybridization of denatured DNA:
1) An adequate concentration of cations. (Reassociation is virtually nonexistent below 0.01 M Na$^+$, very dependent on [Na$^+$] below 0.4 M NaCl, and almost independent of [Na$^+$] above 0.4 M NaCl.)
2) An optimal temperature about 25 °C below T_m. (The reaction rate increases as the temperature decreases below T_m, reaching a broad, flat maximum between $(T_m - 15)$ °C and $(T_m - 30)$ °C, then decreases with further decrease in temperature.)
3) Incubation time and DNA concentration must be

sufficient to permit an adequate number of collisions.
4) The statistical probability that the very long complementary strands of denatured eukaryotic DNA will become correctly aligned by random collision is extremely low. The rate of reassociation is therefore conveniently controlled by shearing the DNA and working with relatively small fragments (200–500 base pairs) of known size. On the other hand, certain bacterial genomes can be denatured and reassociated to double-stranded, biologically active DNA; for example, *Escherichia coli* DNA (4.5 million base pairs) is often used to standardize measurements of the kinetics of reassociation.
In many studies of H., it is advantageous to use cDNA (see Complementary DNA) in place of mRNA; using the mRNA as a template, cDNA of high specific radioactivity can be synthesized in relatively large quantities with efficient use of the radioactive nucleotide precursors; moreover, mRNA is constantly at risk from attack by RNAses.
The literature describes many varied applications of DNA:cDNA H., e.g. determination of the number of copies of globin genes in mouse DNA [P. R. Harrison et al., *J. Mol. Biol.* **84** (1974) 539–554]; and the demonstration that the globin of root nodule leghemoglobin is encoded by the plant genome and not by that of the bacterial symbiont [R. Sidloi-Lumbroso et al. *Nature* **273** (1978) 558–560]. In this type of experiment, sheared cellular DNA and cDNA are fractionated to obtain preparations of the same average nucleotide length (200–500 nucleotides or base pairs), and the cellular DNA is present in large excess (about 10^7-fold) over the radioactive cDNA.
As a taxonomic tool, DNA:DNA H. in solution, followed by determination of the T_m of the resulting hybrid, promises to transcend other forms of biochemical taxonomy, e.g. comparison of amino acid sequences of homologous proteins. The difference in T_m (ΔT_m) between conspecific (same species) and heterospecific (different species) DNA hybrids is a measure of the evolutionary closeness of the two organisms. A 1% difference in nucleotide sequence results in a ΔT_m of about 1 °C. Species of the same genus show ΔT_m up to 4 °C. For different genera of the same family, ΔT_m is in the range of 4–11 °C; for different families within the same order, 11–20 °C; and for members of the same class but distantly related orders, about 25 °C. [J. M. Diamond, *Nature* **305** (1983) 17–18].
Kinetics of H. Association of the two strands follows second-order kinetics. Consider the equilibrium:

$$A + B \underset{K_2}{\overset{K_1}{\rightleftharpoons}} AB,$$ where AB represents the hybrid or

reassociated molecule. Let *a* and *b* represent the initial concentrations of the two strands (A and B), respectively, and *x* the concentration of AB at time *t*.

The rate of association, $\dfrac{dx}{dt} = K_1(a-x)(b-x) - K_2 x.$

Since K_2 is much smaller than K_1, this can be rewritten:

$$\frac{dx}{dt} = K_1 (a-x)(b-x)$$

Although one H. partner is often present in relatively low concentration, (i.e. $a >> b$), there are circumstances in which A and B are present in equal concentrations (reassociation of melted DNA; H. of qual quantities of DNA from two different species). The above equation then becomes $\frac{dx}{dt} =$ $K_1 (a-x)^2$, and integration gives $\frac{x}{a} = \frac{aK_1t}{(1+K_1at)}$.

This equation represents the mathematical basis of the Cot method, and is usually expressed as $\frac{C}{C_0} = \frac{1}{1+KCot}$, where C = concentration of DNA (mol/l) remaining at time t (seconds); C_0 = initial concentration of denatured DNA (mol/l). H. or association is expressed graphically by plotting Cot against C/C_0 (Fig.). Since the observed rates range over at least 8 orders of magnitude, it is necessary to plot Cot on a logarithmic scale; the resulting second-order curve is conveniently symmetrical. The units of Cot are moles of nucleotide × seconds per liter (mol·s·l^{-1}). Cot$_{1/2}$ (the half period of reassociation, i.e. the half period for 50% annealing) can be predicted [J. G. Wetmur & N. Davison, *J. Mol. Biol.* **31** (1968) 349-370]:

Let N = complexity of the DNA, i.e. the number of base pairs of nonrepeating sequences; and let $\lrcorner$ = average number of nucleotides per single strand of denatured DNA. The second-order rate constants for all DNAs are then about $3 \times 10^5 L^{0.5}/N$ l·mol^{-1}·sec^{-1}. The reaction rate also increases slightly with GC content and is affected by the viscosity of the matrix.

Reassociation of eukaryotic DNA (especially when fragmented) is often faster than predicted [R. J. Britten & D. R. Kohne, *Science* **161** (1968) 529-540] (this is also a valuable reference for the principles and experimental basis of DNA reassociation). This observation gave rise to the hypothesis that certain sequences are repeated, sometimes hundreds or even thousands of times. In fact it is now known that virtually all eukaryotic DNAs contain families of repetitive sequences ranging in length from 130 to 300 base pairs. (See, e.g. Alu repetitive sequence).

Hybridoma: an immortal cell line prepared by fusing lymphocytes with an appropriate line of transformed cells. H. are the usual source of Monoclonal antibodies (see Immunoglobulins).

Hydroazulene: see Proazulene.

Hydrocarbon degradation, *microbial hydrocarbon degradation:* Some microorganisms are able to degrade hydrocarbons and use them as their sole source of carbon and energy. This ability is important for the microbial production of protein and for the elimination of environmental contamination by mineral oils. H. d. depends greatly on the structure; the unbranched chains (alkanes) of 10 to 18 carbons are most readily degraded. The most important pathway is the oxidation of one end of the chain to CH_2OH by Cytochrome P450 (see), followed by oxidation to -CHO then -COOH by NAD-linked dehydrogenases. Further degradation of the fatty acids is achieved by β-oxidation. Aromatic hydrocarbons are less readily degraded than aliphatic structures. Ring cleavage is always preceded by the formation of a phenol by an oxygenase. Further degradation proceeds via pyrocatechol, *cis,cis*-muconic acid and α-ketoadipic acid to the components of the tricarboxylic acid cycle, acetic acid and succinic acid.

Hydrocyanic acid, *hydrogen cyanide, HCN:* a highly poisonous compound found widely in nature in the form of Cyanogenic glycosides (see). It is released from these by β-glucosidases (such as emulsin) and oxinitrilases. A number of plants, especially those which contain cyanogenic glycosides, can metabolize HCN, usually by binding it to serine or cysteine to form cyanoalanine. Addition of water converts the latter to asparagine.

Hydrogen: see Bioelements, Hydrogen metabolism.

Hydrogenase: see Hydrogen metabolism.

Hydrogenation: see Reduction.

Hydrogen bonding: see Noncovalent bonds.

Hydrogen metabolism: 1. metabolic redox reactions, involving pyridine nucleotide and flavin coenzymes; 2. all metabolic reactions involving hydrogen, i.e. hydrogenation, dehydrogenation, transhydrogenation, activation and formation of molecular hydrogen.

In anaerobic and aerobic respiration, substrates are oxidized by removal of hydrogen (dehydrogenation), a process that does not involve transfer of oxygen. Dehydrogenations are catalysed by dehydrogenases (NAD$^+$ and NADP$^+$ act as coenzymes) and by oxidases. FAD and, less frequently, FMN act as redox prosthetic groups or cofactors of the flavoenzymes. Hydrogen atoms removed from a substrate ($H = H^+ + e^-$) are transferred to the active group of the dehydrogenase or oxidase. Hydrogen is transferred to NAD$^+$ or NADP$^+$ as a hydride ion, i.e. one proton (H^+) and an electron pair ($2e^-$) are transferred from the substrate hydrogen 2[H], leaving one proton free (see Nicotinamide-adenine-dinucleotide). In anaerobic carbohydrate degradation, the reduced coenzyme (NADH) is reoxidized by coupling its oxidation to the reduction of an endproduct of glycolysis; this is an internal oxidoreduction system in which molecular oxygen plays no part. In the complete, aerobic oxidation of glucose via glycolysis, the tricarboxylic acid (TCA) cycle and the respiratory chain, NADH is reoxidized by the respiratory chain with the uptake of molecular oxygen.

NAD and NADP serve different metabolic functions. In aerobic metabolism, most of the NADH produced by the cell is reoxidized by the respiratory chain for the purposes of energy production (see Oxidative phosphorylation). Because NAD$^+$ is the acceptor for most of the hydrogen produced in catabolic reactions, the term "catabolic reduction charge" (CRC) has been proposed for the ratio [NADH]/([NADH]+[NAD$^+$]). NADPH, on the other hand, does not transfer hydrogen to the respiratory chain; it functions as a reducing agent in reductive biosyntheses. Thus the ratio [NADPH]/([NADPH]+[NADP$^+$]) is the "anabolic reduction charge" (ARC). The CRC and the ARC are important parameters in the regulation of cellular metabolism. [K. B. Anderson & K. von Meyenburg, *J. Biol. Chem.* 252 (1977) 4151.]

In the cytoplasm, the CRC is normally maintained at a value much less than 1, i.e. there is much more NAD^+ than NADH. In the mitochondria, however, there is much more of the reduced form, in spite of the fact that this is the site of oxidation of NADH to NAD^+. This suggests an efficient transfer of NADH produced in the cytoplasm to the mitochondria, but the inner mitochondrial membrane is impermeable to NAD^+ and NADH in both directions. What are transferred, then, are reducing equivalents from the cytoplasm to the mitochondrion. This is accomplished by the following *shuttle systems*: 1. *β-Hydroxybutyrate/acetoacetate shuttle*. Acetoacetate in the cytoplasm is reduced to β-hydroxybutyrate by an NADH-dependent reductase (EC 1.1.1.30). β-Hydroxybutyrate enters the mitochondrion, where it is oxidized to acetoacetate by an NAD-dependent dehydrogenase. The resulting NADH is oxidized by the respiratory chain, and the acetoacetate leaves the mitochondrion to be reduced by more cytoplasmic NADH. The reality of this system is in some doubt. Certainly it could not operate in mitochondria that lack β-hydroxybutyrate dehydrogenase, i.e. hepatic mitochondria of ruminants. Vertebrate red muscle contains high levels of mitochondrial β-hydroxybutyrate dehydrogenase, so the shuttle might operate in this tissue.

2. *Dihydroxyacetone phosphate/α-glycerolphosphate shuttle*. Dihydroxyacetone phosphate is reduced in the cytoplasm by glycerol-3-phosphate dehydrogenase (EC 1.1.1.8) and NADH. The resulting α-glycerolphosphate is oxidized in the mitochondria by an efficient glycerol 3-phosphate dehydrogenase (an FAD-flavoprotein, EC 1.1.99.5), and the dihydroxyacetone phosphate returns to the cytoplasm. Whereas the oxidation of NADH by the respiratory chain gives rise to 3 ATP, this shuttle system produces 2 ATP per NADH oxidized, because the mitochondrial dehydrogenase is FAD-dependent. This is an active shuttle system in blowfly flight muscle. The component enzymes are also present in mammalian muscle in amounts compatible with the operation of the shuttle, but it seems doubtful that this particular shuttle is operative in liver.

3. *Malate/oxaloacetate shuttle*. Oxaloacetate is reduced at the expense of NADH by the action of a cytoplasmic malate dehydrogenase (EC 1.1.1.37). Malate enters the mitochondria, where it is dehydrogenated by mitochondrial malate dehydrogenase. Oxaloacetate does not leave the mitochondria, but it is transaminated with glutamate to form aspartate and 2-oxoglutarate. Aspartate crosses the mitochondrial membrane and is transaminated to oxaloacetate in the cytoplasm. This shuttle does not cause an imbalance of charge across the mitochondrial membrane, because the entry of malate is coupled with the exit of 2-oxoglutarate (both dicarboxylic acids) and the entry of glutamate is coupled with the exit of aspartate (both acidic amino acids).

Many other shuttles can be contrived theoretically, but they lack experimental support. Liver cells are probably served by several shuttles, rather than one major system. In contrast to the CRC, the ARC in the cytoplasm is low. The high concentration of NADPH relative to that of $NADP^+$ helps to drive biosyntheses forward by mass action; among them are the reduction of glyceric acid 3-phosphate to glyceraldehyde 3-phosphate in the photosynthetic assimilation of CO_2 (see Calvin cycle), reductive synthesis of glutamate by glutamate synthase (see Ammonia assimilation), and Fatty acid biosynthesis (see).

In heterotrophic organisms, NADPH is formed in the oxidative phase of the Pentose phosphate cycle (see), and in photosynthetic organisms (except photosynthetic bacteria) in the light reaction of Photosynthesis (see). Another important source is the cytoplasmic oxidative decarboxylation of malate by $NADP^+$-linked malate dehydrogenase (EC 1.1.1.40). Pyruvate enters the mitochondria and undergoes ATP-dependent carboxylation to oxaloacetate, which is hydrogenated by NADH-dependent malate dehydrogenase (EC 1.1.1.37). The resulting malate leaves the mitochondrion and is decarboxylated to pyruvate in the cytoplasm by $NADP^+$-linked malate dehydrogenase (EC 1.1.1.40). The operation of this cycle results in the export of reducing equivalents from the mitochondrion to the cytoplasm.

In contrast to the oxidation of NADH, there is no mechanism for the direct oxidation of NADPH by the respiratory chain. In fact the physiological problem faced by animal cells in particular is how to maintain a sufficiently plentiful supply of NADPH for reductive biosynthesis, and how to supplement this by exploiting the reducing equivalents of NADH, e.g. by the malate-pyruvate cycle described above. Nevertheless, some of the hydrogen of NADPH probably does become oxidized by the respiratory chain. For example, NADH and NADPH might become equilibrated by any enzyme that can use both, and such an equilibration has been shown to occur. Enzymes that can use both cofactors with more or less equal facility are glycerol dehydrogenase (pig and rat liver, *Escherichia coli*), glutamate dehydrogenase (muscle, liver, yeast) and 3β-hydroxysteroid dehydrogenase (liver). Liver and heart mitochondria perform a transhydrogenation in which hydrogen is transferred from NADH to $NADP^+$. The process is energy-dependent (ATP) and is catalysed by an enzyme in the inner mitochondrial membrane. The equilibrium is so strongly in favor of the formation of NADPH that the reverse process is not easily measured. Clearly this reaction will not lead to the net production of NADH from NADPH; rather it serves to supply NADPH required for intramitochondrial reductive biosynthesis. Both the donor (NADH) and the acceptor ($NADP^+$) interact with transhydrogenase on the M-side of the mitochondrial membrane. During transhydrogenation, there is no exchange of hydrogen with protons of water. The hydrogen atom is transferred from the A side of NADH to the B side of $NADP^+$, indicating that the planes of the nicotinamide rings of the two cofactors must become closely associated on the enzyme surface. Cytoplasmic transhydrogenases have also been reported, which catalyse transhydrogenation with an equilibrium constant of about unity.

In the majority of living organisms, molecular hydrogen has no significant metabolic role. However, certain enzymes called hydrogenases have been found in a wide range of organisms, including bacteria, plants and animals. It has been suggested (Krebs) that hydrogenase serves to release H_2 from excess NADH, when biosyntheses have a prevalence of oxidative steps and therefore upset the normal redox balance of the cell, e.g. in microorganisms growing anaerobically, excess reducing power may be used to synthesize reduced products that are excreted, or it may simply be released as molecular hydrogen by the agency of hydrogenase.

Some hydrogenases are membrane-bound, and they are often linked to formate dehydrogenase ($HCOOH + NAD^+ \rightarrow CO_2 + NADH + H^+$; $NADH + H^+ \rightarrow NAD^+ + H_2$). In strictly anaerobic bacteria, hydrogenases are linked to ferredoxin, and they catalyse an oxidoreduction reaction between hydrogen and ferredoxin: $H_2 + 2Fd_{ox} \rightarrow 2H^+ + 2Fd_{red}$, where Fd_{ox} and Fd_{red} represent oxidized and reduced ferredoxin, respectively. Certain bacteria, e.g. *Hydrogenomonas, Pseudomonas* and *Alicaligenes*, can oxidize H_2 with O_2; a normal electron transport chain operates and 3 molecules of ATP are generated.

Ferredoxin-dependent hydrogen production catalysed by hydrogenase, e.g. in *Clostridium*, is inhibited by carbon monoxide, and is independent of ATP. It therefore differs from the ATP-dependent, CO-sensitive hydrogen production by Nitrogenase (see). Evolution of hydrogen gas during nitrogen fixation is due to competition between nitrogen and protons for electrons at the reducing center of the enzyme.

In addition to ferredoxin, the Redoxins (see) are important electron transferring agents in hydrogen metabolism.

Hydrogen transfer: see Hydrogen metabolism; Pyridine nucleotide coenzymes.

Hydrolases: see Enzymes, table 1.

Hydrophobic bonds: see Noncovalent bonds.

Hydroxamic acids: derivatives of carbonic acid containing the tautomeric group

$$R-\underset{\underset{O}{\|}}{C}-NHOH \rightleftharpoons R-\underset{\underset{OH}{|}}{C}=N-OH$$

H.a. can form stable, five-membered rings with metal ions. They are especially important in the iron metabolism of many organisms. Well-known examples of H.a. are aspergillic acid (Fig.), synthesized from leucine and isoleucine by *Aspergillus flavus*, and the Siderochromes (see).

Aspergillic acid

Hydroxyacetic acid: See Glycolic acid.

Hydroxyacid: a carboxylic acid, in which one or more hydrogen atoms of the alkyl moiety is replaced by an hydroxyl group. The position of the OH

group in the alkyl chain is indicated by $\alpha, \beta, \gamma, \delta$ etc., or by 2,3,4,5, where the C-atom of the COOH is No.1; thus lactic acid is 2-hydroxypropionic, or α-hydroxypropionic acid. Some important H. are glyceric, malic, lactic and citric acids.

2'-Hydroxy-3-arylcoumarins: a group of naturally occurring Isoflavonoids (see), e.g. *pachyrrhizin* from *Pachyrrhizus erusus* and *Neorautanenia* spp. [L. Crowbie & D. A. Whiting (1963) *J. Chem. Soc.* 1569–1579] (Fig.).

Pachyrrhizin

3-Hydroxy-2-butanone: see Acetoin.

o-Hydroxycinnamic acid lactone: see Coumarin.

Hydroxycinnamoyl-CoA ligase: see 4-Coumarate:CoA ligase.

9-Hydroxy-*trans*-2-decenoic acid: see Queen substance.

N-2-Hydroxyethylpiperazine-N-2-ethanesulfonic acid, Hepes: M_r 238.3, a compound used for the preparation of Buffers (see) in the pH range 6.8–8.2.

Hydroxyisovaleric aciduria: see Inborn errors of metabolism.

Hydroxylases: see Oxygenases.

Hydroxylation: see Oxygenases.

Hydroxylubimin: see Phytoalexins.

5-Hydroxymethylcytosine: a pyrimidine compound, one of the rare nucleic acid bases. M_r 141.1, d. above 200 °C without melting. H. is not synthesized as a modification of cytosine already incorporated in the nucleic acid but is formed de novo in the course of pyrimidine biosynthesis as 5-hydroxymethyldeoxycytidylic acid. H. was isolated in 1952 from DNA. It is found in the DNA of bacteriophages of the T2, T4, T6 series in place of cytosine.

Hydroxymethyl glutarate cycle: see Ketogenesis.

Hydroxynervon: see Glycolipids.

Hydroxynervonic acid: Δ^{15}-2-hydroxytetracosanoic acid, $CH_3-(CH_2)_7-CH=CH(CH_2)_{12}-CHOH-COOH$, an hydroxylated, unsaturated fatty acid. M_r 382.5 m.p. 65 °C. An important component of cerebrosides.

2-Hydroxy-3-oxoadipate synthase, *2-hydroxy-3-oxoadipate glyoxylate-lyase (carboxylating)* (EC 4.1.3.15): An enzyme in bacteria and mammalian liver, which catalyses the decarboxylative condensation of 2-oxoglutarate with glyoxylate. The product, 2-hydroxy-3-oxoadipate, is important in mammals for diverting glyoxylate metabolism from oxalate synthesis (see Oxalic acid; Oxalosis under Inborn errors of metabolism). In vitro, the enzyme catalyses a variety of related decarboxylation and condensation reactions (Fig.), which are analogous to Acetoin (see) formation from pyruvate and acetaldehyde. [M. A. Schlossberg et al. *Biochemistry* **9** (1970) 1148–1153].

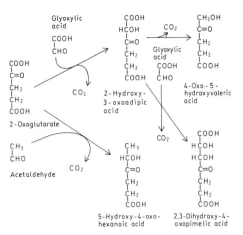

Some reactions catalysed by 2-hydroxy-3-oxoadipate synthase.

3α-Hydroxy-5α-pregnan-20-one: a catabolite of progesterone. Like its stereoisomers, 3β-hydroxy-5α-pregnan-20-one and 3α-hydroxy-5β-pregnan-20-one, 3α-H. appears in the urine of pregnant women.

Hydroxyproline, *Hyp* or *Pro(OH):* Proline residues in collagen are frequently hydroxylated, after incorporation into the polypeptide (see Post-translational modification of proteins), at the 4 or the 3 position. 4Hyp is more common in collagen. After enzymatic degradation of collagen, 4Hyp is degraded by reductive ring cleavage to 4-hydroxy-2-oxoglutarate, and then to pyruvate and glyoxylate (see L-Proline). 3Hyp is also a constituent of *Amanita* toxins (see Amatoxins).

5-Hydroxytryptophan: An intermediate in the synthesis of Melatonin (see) and Serotonin (see). It is used clinically in the treatment of depression and myoclonus. [*Science* **221** (1983) 659–660.]

Hydroxyxanthorrhone: see Flavan.

Hygrine: see Pyrrolidine alkaloids.

Hyocholic acid: 3α, 6α, 7α-trihydroxy-5β-cholan-24-oic acid, one of the bile acids, a trihydroxylated steroid carboxylic acid. M_r 408.58, m.p. 197 °C, $[\alpha]_D$ +5.5 ° (alcohol). H. is a component of pig and rat bile and, like the main component of pig bile, hyodeoxycholic acid (3α, 6α-dihydroxy-5β-cholan-24-oic acid), M_r 392.58, m.p. 197 °C, $[\alpha]_D$ +8 °C (methanol), it is important as a starting material for the chemical synthesis of steroid hormones.

Hyodeoxycholic acid: see Hyocholic acid.

Hyoscine: see Scopolamine.

Hyoscyamine: see Atropine.

Hyp: abb. for Hypoxanthine (see) or Hydroxyproline (see).

Hyperammonemia: see Inborn errors of metabolism.

Hyperbilirubinemia: see Inborn errors of metabolism.

Hyperchromic effect: increase in the extinction of a solution at a particular wavelength due to structural changes in the solute molecules. H.e. is important in the denaturation of DNA when the double

helix held together by hydrogen bonds is transformed into a disordered random coil.

Hyperlysinemia: see Inborn errors of metabolism.

Hyperornithinemia: see Inborn errors of metabolism.

Hyperphosphatasia: see Inborn errors of metabolism.

Hyperprolinemia: see Inborn errors of metabolism.

Hypersarcosinemia: see Inborn errors of metabolism.

Hypersarcosinuria: see Inborn errors of metabolism.

Hypertensin: see Angiotensin.

Hypervitaminosis: see Vitamins.

Hypochloremic alkalosis: see Alkalosis.

Hypochromes: see Pigment colors.

Hypochromic effect: an optical phenomenon in molecules with several chromophores, in which the sum of the optical densities of the individual components is greater than the optical density of the whole molecule. The extinction of the nucleic acids at 260 nm, for example, is less than the calculated sum of the extinctions of the component bases. The H.e. depends on the content of adenine and thymine and is therefore greater in DNA than in RNA. Double-stranded polynucleotides have a greater H.e. than single-stranded, because the effect is intensified by hydrogen bonds.

Hypophysis, *pituitary gland:* a vertebrate organ for hormone production. In humans, it weighs 0.7 g and lies at the base of the brain. It is connected to the midbrain by the hypophyseal stalk. The H. is composed of two parts with different ontogenies. The adenohypophysis, which includes the anterior and middle parts, arises from the roof of the embryonic mouth. The neurohypophysis, or posterior lobe, is formed from an outgrowth of the base of the midbrain. In the anterior lobe, releasing hormones from the hypothalamus stimulate the formation of somatotropin and prolactin in the acidophilic (α-)cells, follicle-stimulating hormone, luteinizing hormone and thyreotropin in the basophilic (β-)cells; and corticotropin in the chromophobic (γ-)cells. Melanotropin is synthesized in the middle part of the H. The neurohormones, oxytocin and vasopressin, are stored in the posterior lobe and released on demand into the blood stream. (See separate entries for each of the above-mentioned hormones.)

Hypotaurine: see Cysteine.

Hypothalamus: the lowest part of the midbrain, which also includes the thalamus and epithalamus. The hypophyseal stalk with the neurohypophysis arises from the underside of the H. As part of the limbic system, the H. is a "gateway to consciousness". Afferent (incoming) stimuli from the breast and abdominal areas and from the circulatory system are processed by the thalamus and passed on to the cerebrum. Conversely, efferent stimuli from the cerebrum flow to the thalamus and H. A series of nuclear regions controlling important vegetative functions, such as blood pressure, temperature, sweat formation, water balance, motions of the digestive tract organs, and sexual function are located in the H. The entire metabolism can be greatly influenced by the appetite and eating center of the H. In addition, in certain nuclear areas nervous excitation is transformed into hormonal signals (releasing hormones and neurohypophyseal hormones), and conversely, hormones from the hypophyseal hormone glandular system exert negative feedback reactions on the formation of hormones in the nerve cells.

Hypovitaminosis: see Vitamins.

Hypoxanthine, abb. *Hyp:* 6-hydroxypurine, a purine derivative (formula, see Inosine). M_r 136.11, m.p. 150 °C (d.). Hyp occurs in the course of aerobic Purine catabolism (see) through deamination of adenine compounds or through hydrolysis of inosine compounds. It is found as a rare base in certain transfer RNAs. It is widely distributed in the plant and animal kingdoms.

Hypoxanthinosine: see Inosine.

Hypusine, *N^ε-(4-amino-2-hydroxybutyl)lysine:* an amino acid present in eukaryotic initiation factor 4D (eIF-4D) and formed by post-translational modification of lysine residues (Fig.).

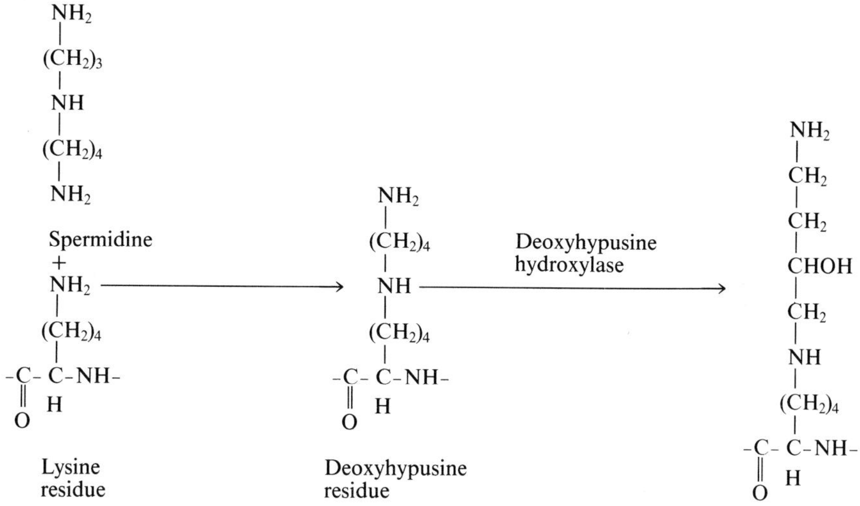

Biosynthesis of an hypusine residue in eukaryotic initiation factor 4D.

Radiotracing confirms that the terminal four carbons of hypusine are derived from the butyl carbons of spermidine, but the mechanism is not known. Deoxyhypusine hydroxylase requires only sulfhydryl reagents for activity. Absence of requirements for Fe^{2+} (which inhibits), 2-oxoglutarate and ascorbic acid indicates that the enzyme is not a 2-oxoacid dioxygenase. [A. Abbruzzese et al., *J. Biol. Chem.* **261** (1986) 3085–3089.]

IAA: see Auxins.

Ibotenic acid: α-amino-3-hydroxy-5-isoxazole-acetic acid. M_r 158, m.p. 145 °C (d.) I.a. is a psychotropic, weakly insecticidal substance which, like its decarboxylation product, Muscimol (see), belongs to the group of amanita poisons. It is found only in a few species of *Amanita* (fly agaric), at concentrations averaging 0.05% of the fresh weight. I.a. is pharmacologically very active, but less so than muscimol in most tests. Another derivative of I.a. is the erythro-dihydroibotenic acid, Tricholomic acid (see).

I-cell disease: see Inborn errors of metabolism.

ICSH: see Luteinizing hormone.

Idaeine: see Cyanidin.

IDP: abb. for Inosine 5′-diphosphate.

I.E.P.: abb. for Isoelectric point.

Ig: abb. for Immunoglobulins.

Ile: abb. for L-Isoleucine.

Imidazole alkaloids: alkaloids of sporadic occurrence, possessing the imidazole ring system. Their most important representative is Pilocarpine (see). The biosynthesis of the I.a. is coupled to histidine metabolism.

Immobilized enzymes: soluble enzymes bound to an insoluble organic or inorganic (e.g. porous glass) matrix, or encapsulated within a membrane in order to increase their stability and make possible their repeated or continual use. The five most commonly used methods of enzyme immobilization are: 1. Adsorption onto an inert or electrically charged carrier (cross-linked dextrans); 2. Covalent binding to a carrier polysaccharide, e.g. Sepharose; 3. Entrapment in a three-dimensional cross-linked carrier, e.g. polyacrylamide gel; 4. Cross linkage, i.e. condensation of several enzyme molecules with bi- or polyfunctional agents, like glutaraldehyde and epichlorhydrin; 5. Microencapsulation. Covalent binding to a carrier is widely used. Since the enzyme is chemically fixed, it is not washed out and can be used repeatedly. Of the organic carriers, Sepharose and large-pore cross-linked dextrans come close to meeting all the requirements of an enzyme carrier, i.e. minimal solubility, high mechanical strength, suitable particle size and shape, high binding capacity for the enzyme, but no adsorption of the substrates and products, and resistance to attack by microorganisms. Being organic, however, they are subject to microbial attack, and they swell and shrink depending on pH and other environmental conditions. Since 1969, silanized porous glass, which has neither of these disadvantages, has been used increasingly for both adsorptive and covalent attachment of enzymes. For covalent attachment, functional groups must be present on the carrier. A frequently used activation procedure, especially with dextran gels, consists of reaction with cyanogen bromide (CNBr), which produces reactive iminocarbonate groups (Fig. 1).

At pH 7–9, these react with the free amino groups of the enzyme protein, with the formation of substituted imidocarbonates ($=C=N$-protein). Many other coupling techniques have been reported for the covalent attachment of enzymes to agar, agarose and Sephadex supports, and to the silanized surface of porous glass. Details of these techniques and other aspects of immobilized enzymes are comprehensive-

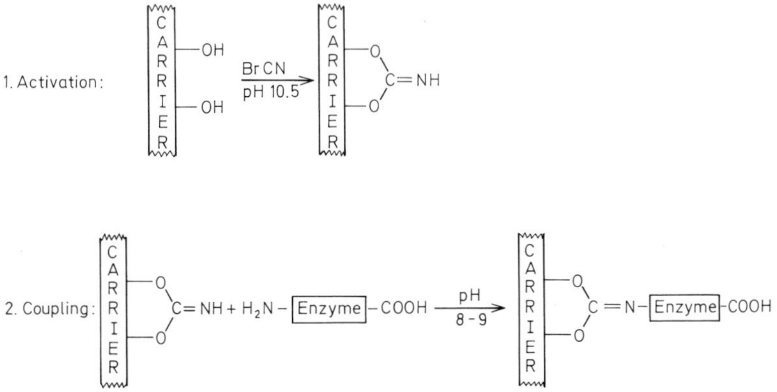

Figure 1. Covalent binding of an enzyme to an unsubstituted polysaccharide (e.g. Sepharose), by the cyanogen bromide method

279

ly treated in *Methods in Enzymology* vol. XLIV, 1976, Klaus Mosbach, ed. Academic Press.

The quantities of covalently bound enzymes are generally low (1–5%); in exceptional cases, especially when carrier and protein are oppositely charged, 10% or more may be bound.

Properties: In general, the free and immobilized enzyme catalyse the same reaction, but depending on the supporting material and nature of the binding, there may be changes of pH and temperature optima (the latter is usually increased), K_m value and specific and maximal activities (the latter is usually decreased) (see Enzyme kinetics). Chief reasons for these alterations are the decreased flexibility and

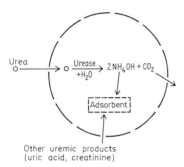

Other uremic products
(uric acid, creatinine)

Figure 2. Schematic representation of an artificial cell, containing urease and albumin-coated active charcoal as an adsorbent for uric acid, ammonia and creatinine. A 10 ml suspension of these 20 μm diameter urease capsules corresponds to a surface area of 20 000 cm², which is larger than that of the conventional artificial kidney.

mobility of the coupled enzyme, and steric factors which interfere with access of the substrate to, and diffusion of product from, the active center. These changes are usually more than compensated by increased stability of the enzyme. They can be avoided or reduced by attaching the enzyme to the support by a side chain, or spacer, which allows greater mobility and unhindered contact with substrates.

Enzymes may also be immobilized by microencapsulation. In this technique, which has medical applications, enzymes are enclosed by various forms of semipermeable membranes, e.g. polyamide, polyurethane, polyphenyl esters and phospholipids. Microcapsules of phospholipids are also called *liposomes.* The microencapsulated enzymes and proteins inside the microcapsule cannot pass the membrane envelope, but low M_r substrates can pass into it, and products can leave. Such encapsulated proteins do not elicit an antigenic response, and they are not attacked by proteases outside the microcapsule. They are therefore suitable for the delivery of enzymes for therapeutic purposes. This area of application is still at an early stage of development, but positive results have been reported from animal experiments and clinical studies, e.g. treatment of inherited catalase deficiency with encapsulated catalase, and treatment of asparagine-dependent lymphosarcoma with encapsulated asparaginase. There are various methods of administration: intramuscu-

lar, subcutaneous or intraperitoneal injection. However, their major area of application is outside the body. For example, microencapsulated urease can be employed as an artificial kidney in hemodiffusion (Fig. 2).

As of 1983, proteases accounted for 59% of the I.e. used commercially; 42% of the proteases were in detergents, and the remainder were used in making cheese (rennins) and other food products. 28% of the I.e. were carbohydrases, used for hydrolysing starch, converting glucose to fructose, etc.; 3% were lipases, also used in food processing, and 10% were "others" used in analysis, production of pharmaceuticals, and development. [*Industrial Enzymology*, Nature Press, New York, 1983]

Immune response: a specific protective or defense reaction against foreign substances, called Antigens (see). In the *humoral* I.r., Immunoglobulins (see), which are also called antibodies, are secreted into the blood, lymph or mucosal secretions. Immunoglobulins (Ig.) are produced by lymphocytes derived from bone marrow in mammals or the Bursa of Fabricius in birds (B cells). B cells display on their surfaces membrane-bound Ig. called surface (s-)Ig. When they are incubated with a fluorescent-labelled antigen (see Immunofluorescence), those B cells with sIg. complementary to the antigen can be seen to undergo "capping". The antigen-sIg. complexes gather together in one area of the cell membrane, and are eventually ingested. This appears to be a necessary step in the activation of B cells to multiply and secrete Ig. Some antigens, including bacterial lipopolysaccharides, can stimulate purified cultures of B cells to produce antibodies, but most antigens require the presence of macrophages and T (thymus) lymphocytes. Particulate antigens, such as bacteria, are engulfed by phagocytes, which then "present" the antigen to the T and B cells. B-cells, when stimulated by "helper" T-cells, multiply and secrete antibodies into the blood or lymph. The multiplication of the stimulated B-cell clones is also regulated by "suppressor" T-cells. In many cases, expression of antibody by B cells requires both "presentation" by a macrophage and activation by helper T-cells. The chemical mediators of the interaction are called Lymphokines (see).

As an infection progresses, many B cells switch from the production of IgM antibodies (see Immunoglobulins) to IgG ("gamma globulins"). The latter usually have higher affinity for the antigen. After the infection has subsided, the serum levels of IgG drop, but do not fall to zero. Some of the B cells are transformed to "memory cells". In the event of a secondary challenge with the same antigen, they are capable of dividing and producing IgG much more quickly than was possible in the primary response.

In *cellular immunity*, some T cells, rather than regulating B cells, directly attack foreign cells or tissue (as in graft rejection). These are called cytotoxic T cells; there are populations of such cells in apparently naive animals which are called "natural killer" cells. The mechanism of their attack is not understood.

Immune serum: see Antiserum.

Immune suppression: the unspecific suppression of the Immune response (see) by corticosteroids, antimetabolites (purine analogs, folic acid antagonists), alkylating substances like Cyclophosphamide (see), ionizing radiation or antilymphocyte serum (see). Although the latter is directed against the cells of the immune system, the other agents have an antiproliferative effect, i.e. they hinder the multiplication of the plasma cells. At the molecular level, these agents directly or indirectly inhibit the synthesis of DNA and RNA. I.s. is an important form of therapy for autoimmune diseases and for preventing the rejection of transplanted organs. However, the suppression of antibody formation results in higher susceptibility to infection and to tumor formation. The pathological I.s. observed in Aquired Immune Deficiency Syndrome (AIDS) is often (perhaps always) fatal, due to adventitious infection or tumors. A true alternative to I.s. is specific I.s., achieved by means of antigen-specific immune sera or by the induction of an immune tolerance.

Immunization: the artificial stimulation of antibody production for protection against pathogens and other antigens. There are two kinds of I., 1) active, produced by injection of live, weakened pathogens (e.g. Sabin poliomyelitis vaccine, smallpox vaccine) or killed pathogens, or purified fractions of them; and 2) passive, produced by injection of antiserum or antibodies which have been formed against the pathogen in another organism and extracted from it. While repeated active I. makes lifelong immunity possible, passive I. is effective for only a few weeks, and cannot be repeated with antiserum or antibodies from the same species of animal, because the foreign antibodies act as antigens. Passive I. therefore is used for short-term prophylaxis or therapy, to bridge the time until the body can produce its own antibodies (for example in tetanus simultaneous immunization).

Immunoassays: assays for biological molecules which derive their extreme sensitivity from the use of specific antibodies against the substance being assayed. The first step in development of I. is the purification or synthesis of the substance to be assayed. This is used to immunize an animal; serum from the immune animal may be used as the source of the antibodies, but the preferred method is now to use the animal's spleen to make hybridomas. Those hybridomas which produce monoclonal antibodies of the desired specificity can be maintained indefinitely to provide as much antibody as needed. With antibodies and purified antigen in hand, the investigator has several choices of I.

In *Radioimmunoassays* (RIA), the antibodies or antigens are made radioactive, usually by iodination (with ^{125}I), of tyrosyl residues in the protein. In *Enzyme-Linked Immunosorbent Assays* (ELISA), the antibody is linked to a convenient enzyme, e.g. peroxidase. In *homogeneous enzyme-linked immunoassays*, an enzyme is chosen which is either inhibited or (less commonly) activated when the antigen is bound by antibody. For example, an enzyme may be linked to the antigen in such a way that the antigen-enzyme is catalytically active. However, when the antigen-enzyme is bound to the appropriate antibody, it loses its catalytic activity. The difference between this type of assay and an ELISA is that the

activity of the enzyme in the ELISA is not affected by binding of the antigen to the antibody. Therefore, an ELISA, like an RIA, requires the antigen or the antibody to be adsorbed or bound to a solid support, so that unbound label can be removed by rinsing. In the homogeneous type of assay, unbound label (the enzyme) is either much more or much less active than bound label, so that the relative amounts can be determined without physical separation.

When RIA or ELISA are used to measure the titer of antibodies in serum or other fluid, the antigen is bound to some solid support, such as the plastic of the wells in a microculture tray. The amount of labelled antibody which will bind to the antigen in the well is determined and is taken as 100%. When unlabelled antibody, e.g. in a sample of serum, is added, the same total amount of antibody is bound, but the labelled antibody is diluted by the unlabelled. The percent reduction in bound labelled antibody is of course proportional to the amount of unlabelled antibody present in the sample to be assayed. Several dilutions of the unknown are mixed with a fixed amount of labelled antibody, and incubated with the antigen in separate wells of the plastic plate. The wells are then thoroughly rinsed. In RIA, the amount of labelled antibody bound to antigen is determined by placing the plastic well in a vial of scintillation fluid and counting flashes in a scintillation counter; in ELISA, by adding the substrate of the enzyme and measuring the rate of the enzymatic reaction (e.g. peroxidase-linked systems are assayed spectrophotometrically, following the addition of hydrogen peroxide and a reduced dye). A displacement curve is plotted from the several dilutions of the sample and is compared with a standard curve prepared with known concentrations of unlabelled antibody. From this the amount of antibody present in the sample may be calculated.

RIA and ELISA may also be used to determine concentrations of antigens, such as hormones, in body fluids. In this case, it is the antigen which is labelled and which competes with unlabelled antigen from the sample for binding sites on fixed antibodies. Again, the unbound labelled antigen is carefully rinsed out of the microculture well before the bound radioactivity or enzyme activity is determined.

In the homogeneous enzyme-linked assays, the procedure is similar to the above, except that there is no need to remove excess label from the incubation mixture. Other possible labels for the antigen or antibody to be determined in the assay include enzyme modulators, prosthetic groups and fluorogenic substrate which will not fluoresce or which cannot be attacked by the enzyme when it is bound to the antibody. Thus the possibilities for enzyme-linked I. are essentially inexhaustible. [T.J. Ngo & H.M. Lenhoff, eds. *Enzyme-Mediated Immunoassay* (Plenum, New York, 1985); S.B. Pal, ed. *Immunoassay Technology*, vol. **1** (1985) and **2** (1986) (Walter de Gruyter, Berlin)]

Immunochemistry: the protein chemistry of Immunoglobulins (see), including their interactions with antigen, Agglutination (see) and the Complement-binding reaction (see).

Immunoelectrophoresis: see Plasma proteins.

Immunofluorescence: a sensitive technique for

detection of antigens or antibodies in which the antibody is coupled to a highly fluorescent compound such as rhodamine or fluorescein isothiocyanate. Tissues, cells or electrophoresis gels containing the antigen are incubated with the fluorescent antibody; after thorough washing, the label indicates the presence and position of the antigen and can be seen under a fluorescence microscope. In direct I., the specific antibody is labelled. In indirect I., antibody against the specific antibody is labelled. (For example, the specific antibody might have been raised in a rabbit, and the fluorescent anti-antibodies could then be goat anti-rabbit serum.) This allows an amplification of the fluorescence: whereas only one specific antibody molecule can bind to the antigenic site, several anti-antibodies can bind to different sites on the specific antibody.

Immunoglobulins, abb. *Ig, antibodies:* specific defense proteins found in blood plasma, lymph and many body secretions of all vertebrates. The phylogenetic precursors of the Ig, the cell-bound hemagglutinins, are also found in invertebrates including annelids, crustaceans, spiders and molluscs. The most salient features of the system are its ability to respond to the presence of any foreign antigen (primary response), to respond more quickly to a previously encountered foreign antigen (secondary response), and under normal circumstances, not to respond to components of the animal's own body (self-nonself discrimination).

Lymphocytes derived from bone marrow in mammals or the Bursa of Fabricius in birds (B-cells) display Ig. on their surfaces. Upon exposure to antigen, those cells whose surface Ig. can bind the antigen are stimulated to proliferate and differentiate into plasma cells which secrete antibodies of the same specificity as the ancestor of the clone. The process of stimulation is complex and involves cooperation with other lymphocytes (T-cells) and macrophages; but B-cells can also be stimulated to proliferate by polyclonal activators such as concanavalin A (see Lectins) or bacterial lipopolysaccharide. When bound to antigen, some Ig. can fix the C1q component of the Complement system (see), thus facilitating the lysis of the foreign cell to which they are bound. Because they are multivalent, Ig. also cross-link soluble antigens and facilitate their clearance from the blood or lymph by macrophages.

Structure. The basic structure of Ig. is tetrameric, consisting of two identical light (M_r 22000 to 24000) and two identical heavy, carbohydrate-containing chains (M_r 50000 to 73000) (Fig.). There are two types of light chain, κ. and λ., each of which can be associated with any of five types of heavy chain, μ, γ, δ, α or ε. The Greek letter designating the type of heavy chain corresponds to the class of Ig. (A, D, E, G or M); these can be distinguished by electrophoresis or serologically. (See below for further discussion.)

Papain treatment of Ig. releases monovalent antigen-binding fragments, Fab (M_r 50000) and a complement-binding, Fc (M_r60000) fragment. Pepsin cleavage, on the other hand, releases a bivalent antigen-binding F(ab')$_2$ fragment and a somewhat smaller complement-binding fragment Fc'.(The terms Fc and Fc' are also applied to the corresponding region of Ig. which does not bind complement.) The F(ab')$_2$ can be dissociated by thiol reagents into monovalent

Table. Properties of human immunoglobulins (Ig.)

	IgG	IgM	IgA		IgD	IgE
			Serum	Secretions		
Sedimentation constant	6.5 ... 7S	19S	7S	11S	6.8 ... 7.9S	8.2S
M_r, of which the L-chain represent 23000	155000	940000 (Pentamer)	170000	380000 (dimer)	185000	196000
H-chain type and M_r	γ1 ... 4 50000 to 60000	μ 71000	α 64000		δ 60000 to 70000	ε 75000
Chain formula (L=κ or λ)	L$_2\gamma_2$	(L$_2\mu_2$)$_5$	L$_2\alpha_2$	(L$_2\alpha_2$)$_2$	L$_2\delta_2$	L$_2\varepsilon_2$
Carbohydrate Fraction of the serum Ig	2 ... 3% 70 ... 75%	10 ... 12% 7 ... 10%	8 ... 10% 10 ... 22%		12.7% 0.03 ... 1%	10 ... 12% 0.05%
Serum concentration mg/100 ml	1300 (800 ... 1800)	140 (60 ... 280)	210 (100–450)		3 (1 ... 40)	0.03 0.01 to 0.14)
Valence of binding	2	5(10)	1	2	?	2
Biological half-life (days)	8(IgG3) or 21	5.1	5.8		2.8	2 ... 3
Complement binding	yes	yes	no		no	no

fragments, indicating that the two H-chains in the intact molecule are held together by one or more disulfide bonds between the sites of pepsin and papain cleavage (Fig.).

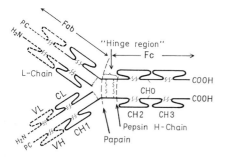

Figure. Structure of an IgG molecule. Variable (VL, VH) and constant (CL, CH) parts (domains) of the light and heavy chains, respectively. The molecule shown represents an IgG or IgA secreted Ig. Other classes have different numbers of heavy-chain domains (2 to 4), and the hinge region may not be present (IgM or IgE). The C-terminal sequence of membrane-bound Ig. is different (and slightly longer) than that of the secreted form. CHO = carbohydrate chain. VL and VH form the antigen-binding site. Papain cleavage produces two Fab and one Fc fragment; pepsin cleavage forms a (Fab')$_2$ fragment and a smaller Fc' fragment. PC = pyrrolidone carboxylic acid.

The Y shape of the Ig. molecule has been confirmed by electron microscopy and X-ray diffraction. The former shows that the molecule is flexible at the "hinge" region, as was postulated from the susceptibility of this region to proteolytic attack. Fluorescence polarization studies have shown a correlation between the flexibility of the hinge region, which varies from one class of Ig. to the next, and the ability of the Ig. to fix complement. [V. T. Oi et al., *Nature* **307** (1984) 136–140].

Historically, on the basis of their electrophoretic behavior, Ig. were divided into five classes, IgA, IgD, IgE, IgG and IgM. It was later realized that there are several genetically independent IgG "subclasses" in mice, rats and humans, and humans have subclasses of IgA as well. The classes and subclasses (or isotypes) are determined by the constant regions of the heavy chains, which are given the corresponding Greek letters (α, δ, ϵ, μ, and γ1, γ2, etc.) Each type of heavy chain can be associated with either of the two types of light chain, κ or λ, but in a given molecule, both light chains are of the same class.

Because of the extreme heterogeneity of normal Ig., it is impractical to attempt to purify them for amino acid sequencing. However, *monoclonal antibodies,* the products of a single clone of B cells, are available from several sources. Paraproteins (see), including Bence-Jones proteins (see), are found in the plasma of patients with certain types of lymphomas. In these cancers, the transformed cells retain enough of their B-cell characteristics to secrete whole or partial Igs. The rapid advances in understanding of Ig

structure and function of the past few years, however, have been made possible by *hybridomas,* which are cell lines resulting from the fusion of antibody-secreting B cells and myeloma cells maintained in tissue culture. The B-cells are taken from the pancreas of recently immunized animals, or they may be from the peripheral circulation. The hybridoma clones can be selected for production of Ig. of the desired antigenic specificity and isotype.

Comparison of the DNA and RNA from hybridomas or myelomas with that from embryonic cells has revealed that in the course of differentiation of plasma cells, genes for the constant region, a joining region (or two) and the variable region have been spliced together. Intervening sequences of DNA are lost. [M. M. Davis et al *Nature* **283** (1980) 733–739. F. R. Blattner & P. W. Tucker, "The molecular biology of immunoglobulin D" *Nature* **307** (1984) 417–422].

Immunology: the science of the biological and chemical bases of immunity, or the defense mechanisms of the human and animal organism which are activated by the invasion of antigens. I. encompasses the cellular and humoral immunity and the connections between the immune reaction and physiological and biochemical processes in the organism, in particular those which control the Immune response (see).

Immunochemistry is an essential component of I. which deals with the antigen-antibody reaction, the Immunoglobulins (see), Agglutination (see), the Complement-binding reaction (see), etc. Immunological and serological techniques are particularly important in clinical chemistry. They are also applied in genetics, molecular biology and chemotaxonomy.

Immunosensor: a Biosensor (see) which makes use of an antibody-antigen reaction to generate an electronic signal. Enzyme I. are based on competition between antigen and enzyme-linked antigen, so that sensor operation is equivalent to conventional enzyme immunoassay. See Field effect transistor.

Immunotolerance: the lack of an immune response to a) the body's own substances or antigens with which the body has had contact since before or shortly after birth (natural I.) and b) larger or smaller amounts of certain antigens (tolerogens), which the body does not recognize as foreign and therefore tolerates (acquired I.). The acquired I. can only be maintained when the tolerogens are constantly present; otherwise the immune response to this antigen arises again. Immunosuppressive measures facilitate the induction of an I. in adults.

Immunotoxins: Synthetic compounds made by coupling plant or bacterial toxins to monoclonal antibodies specific for a particular cell type. Toxins coupled to antibodies against cancer antigens should be highly specific chemotherapeutic agents; I. based on specific types of anti-lymphocyte antibodies are expected to be useful in modulating the immune response for therapeutic purposes. [E. Vitetta et al., *Science* **219** (1983) 644–650.]

IMP: abb. for inosine 5'-monophosphate (see Inosine phosphates).

cIMP: abb. for cyclic inosine 3',5'-monophosphate (see Inosine phosphates).

Inactivases: see Regulation of metabolism.

Inborn errors of metabolism: the title of a

book by Archibold Garrod, published in 1902, in which the author recognized the relationship between genes and enzymes. Many metabolic disorders are now known to be genetic in origin, and due to the absence of a protein or the synthesis of a biologically inefficient form of a protein. An I.e.m. is therefore a biochemical and genetic concept synonomous with inherited metabolic block, congenital metabolic disorder, inherited metabolic disorder, heritable disorder of metabolism, enzymopathy, and other similar terms.

The *one gene one enzyme* hypothesis (now more correctly called one cistron one polypeptide), developed later from work on auxotrophic mutants of *Neurospora* [G.W. Beadle & E.L. Tatum *Proc. Nat. Acad. Sci.* **27** (1941) 499–506] and on pigmentation mutants of insects (see Ommochromes), confirms the original concept of inborn errors. Clinically, an I.e.m. may result in the dangerous accumulation of unmetabolized material before the site of the metabolic block (e.g. in phenylketonuria) and/or failure to produce an essential metabolite (e.g. albinism). The concept now embraces nonenzymatic proteins, e.g. abnormal hemoglobins (see Hemoglobinopathy) are also the result of inborn errors. Known human and animal I.e.m. are clearly of medical interest, and a selection of these is listed in the table. Mutant microorganisms, on the other hand, have made a major contribution to the study of intermediary metabolism and molecular biology (see Auxotrophic mutants, Mutant technique). [H. Harris, *The Principles of Human Biochemical Genetics* (*Frontiers of Biology, Vol.19*) 2nd ed. (North Holland Publishing Co., 1975. Stanbury et al., eds., *The Metabolic Basis of Inherited Disease,* 5th ed. (McGraw-Hill, 1983). H. Galjaard, *Genetic Metabolic Diseases* (*Early Diagnosis and Prenatal Analysis*) (Elsevier/North Holland, 1980)]

Inborn errors of metabolism in humans.

Condition	Deficient enzyme and clinical findings
Acatalasia	*Catalase* (EC 1.11.1.6). Enzyme deficient in all tissues. In some individuals activity is less than 1% of normal with no ill effects; in others there may be ulceration of nasal and buccal mucosae, and oral gangrene.
Acid phosphatase deficiency (a lysosomal storage disease)	*Lysosomal acid phosphatase* (EC 3.1.3.2). Fatal in infancy.
Adrenal hyperplasia I	*Cholesterol monooxygenase (side-chain-cleaving)* (EC 1.14.15.6). Deficient conversion of cholesterol to pregnenolone; therefore impaired synthesis of mineralocorticoids, glucocorticoids and sex hormones (see Adrenal corticosteroids; Androgens; Estrogens).

Condition	Deficient enzyme and clinical findings
Adrenal hyperplasia II, or Nonvirilizing congenital adrenal hyperplasia	*3β-Hydroxysteroid dehydrogenase* (EC 1.1.1.51). Decreased synthesis of all three types of Adrenal corticosteroids (see), especially aldosterone, cortisol and testosterone. Severe adrenal insufficiency is manifested shortly after birth. Males show pseudohermaphroditism.
Adrenal hyperplasia III	*Steroid 21-hydroxylase* (EC 1.14.99.10). Mild form is known as simple virilizing hyperplasia. Decreased cortisol synthesis induces increased ACTH secretion, leading to overproduction of cortisol precursors and sex steroids. Excessive excretion of pregnanetriol. Over 20 other metabolites lacking 21-OH present in urine. In females the external genitalia are masculinized and internal genitalia are normal (pseudohermaphroditism). Virilization in males. Short stature in both sexes (premature fusion of bony epiphyses). Severe form is known as salt-losing form of congenital adrenal hyperplasia. It has all the features of the mild form plus severe aldosterone deficiency with resulting salt and water loss. See Adrenal corticosteroids.
Adrenal hyperplasia IV or Hypertensive congenital adrenal hyperplasia	*Steroid 11β-hydroxylase* (EC 1.14.15.4). Decreased cortisol synthesis induces increased ACTH secretion with resulting overproduction of deoxycorticosterone, which is a potent salt retaining hormone. Virilization in both sexes, and female pseudohermaphroditism. Not all patients show hypertension. See Adrenal corticosteroids.
Adrenal hyperplasia V	*Steroid 17α-hydroxylase* (EC 1.14.99.9). Decreased secretion of glucocorticoids and sex steroids. Excessive secretion of mineralocorticoids. Hyperkalemia and hypertension. Sexual infantism in females. Pseudohermaphroditism in males. See Adrenal corticosteroids.
Aldosterone deficiency	*Steroid 18-hydroxylase* (*corticosterone 18-hydroxylase*) (EC 1.14.15.5). Deficient aldosterone formation. Normal syn-

Condition	Deficient enzyme and clinical findings
	thesis of cortisol and sex hormones. Severe salt loss. See Adrenal corticosteroids.
Albinism	see Melanins
Alkaptonuria	see Phenylalanine
Drug-induced apnea	*Pseudocholinesterase* (EC 3.1.1.8). Enzyme is present in serum, but shows atypical kinetics. Condition first discovered with introduction of suxamethonium (succinyl dicholine) as muscle relaxant and electroconvulsion therapy. This drug is normally rapidly hydrolysed by pseudocholinesterase and its effects last only a few minutes. Affected subjects (1 in 2000 Europeans) develop prolonged muscular paralysis and apnea (up to 2 hours) after normal drug dose. Condition screened for by measuring inhibition of serum cholinesterase by dibucaine; percent inhibition is called dibucaine number (80% for normal enzyme, 20% for atypical enzyme, at 10^{-5}M dibucaine). Dibucaine numbers of about 62% also occur in 4% of Europeans, who possess about equal amounts of normal and atypical forms. See Acetylcholine.
Argininemia	*Arginase* (EC 3.5.3.1). Elevated blood arginine and ammonia. Neurological damage and mental retardation. Primary defect is in liver, but as in other inherited defects of urea cycle there is damage to central nervous system, probably due to toxicity of elevated ammonia. See Urea cycle.
Argininosuccinic aciduria	*Argininosuccinase* (argininosuccinate lyase) (EC 4.3.2.1). Argininosuccinate (normally present in only trace amounts) increases in serum, and is excreted in large quantities. Blood ammonia rises significantly after a protein meal. In normal subjects the enzyme does not appear to be saturated, whereas in affected subjects the decreased (less than 5% normal activity) enzyme is saturated and therefore catalyses a higher rate of arginine

Condition	Deficient enzyme and clinical findings
	synthesis. Urea production is consequently not significantly decreased, while the concentrations of intermediates preceeding argininosuccinate cleavage are altered. Some degree of mental retardation and sometimes other neurological disorders, e.g. fits. See Urea cycle.
Citrullinemia	*Argininosuccinate synthetase* (EC 6.3.4.5). Elevated blood citrulline and ammonia, and large amounts of urinary citrulline. Urea synthesis still occurs. Neurological damage and mental retardation. See Urea cycle.
Crigler-Najjar syndrome, or Constitutional nonhemolytic hyperbilirubinemia	*Glucuronyl transferase (UDP-glucuronate: bilirubin-glucuronyltransferase)* (EC 2.4.1.76). Enzyme is completely absent. Severe jaundice and brain encephalopathy. Often death in infancy. Conjugation is obligatory for excretion of bilirubin. In a milder form of the disease, some enzyme is present and treatment is possible by administration of phenobarbital, which stimulates hepatic uptake, conjugation and biliary secretion (as glucuronide) of bilirubin.
Cystathioninuria	*Cystathionase (homoserine dehydratase, cystathionine γ-lyase)* (EC 4.4.1.1). See Cysteine. High urinary excretion of cystathionine. Increased concentrations of cystathionine in serum and tissues.
Ehlers-Danlos syndrome	A group of heritable disorders. Clinical features include hyperelastic skin, hyperextensible joints, easy bruising and poor wound healing. Collagen has decreased content of hydroxylysine. Deficient enzyme may be *Lysyl oxidase* (EC 1.4.3.6), *Lysyl protocollagen hydroxylase* (EC 1.14.11.4) or *Procollagen peptidase*. See Collagen.
Formaminotransferase deficiency syndrome	*Formaminotransferase* (EC 2.1.2.5). See Histidine. Increased urinary excretion of formamino glutamic acid after oral histidine load, despite adequate serum folate. Mental

Condition	Deficient enzyme and clinical findings
	and physical retardation. Neurological abnormalities.
Fructose intolerance	*Fructose-bisphosphate aldolase* (isoenzyme B, M_r 156 000) (EC 4.1.2.13). Fructosemia, fructosuria and hypoglucosemia after intake of fructose. Intracellular accumulation of fructose 1-phosphate. Hyperuratemia. Hepatomegaly. Renal tubular dysfunction. Intraocular bleeding. Patients are symptom-free and healthy if fructose is avoided. Aldolase A (muscle and most other tissues) and aldolase C (brain and heart) are present and fully active.
Fructosuria, Essential fructosuria	*Ketohexokinase* (EC 2.7.1.3). Fructosemia and fructosuria after intake of fructose. No pathological consequences.
Galactokinase deficiency, Galactosuria, Galactose diabetes	*Galactokinase* (EC 2.7.1.6). Galactosemia and galactosuria after intake of galactose or lactose (i.e. when milk is fed). Galactitol is also produced. Severe lens cataracts from an early age. Affected individuals may otherwise be normal, but some evidence exists of neurological disturbance. Treated with galactose-free diet throughout life.
Galactose epimerase deficiency	*UDPglucose 4-epimerase* (*UDPgalactose 4-epimerase*) (EC 5.1.3.2). Enzyme activity absent (intermediate in heterozygotes) only in erythrocytes and leukocytes. Liver activity is normal. Benign. Treat by high intake of milk.
Galactosemia	*UDPglucose-hexose-1-phosphate uridylyl transferase (galactose 1-phosphate-uridylyltransferase)* (EC 2.7.1.12). Specific inability to metabolize galactose. High concentrations of galactose and galactose 1-phosphate in tissues and body fluids. Consequences generally severe. Failure to thrive. Retarded mental development. Hepatomegaly, eventually cirrhosis. Renal tubular dysfunction. Often early death, but if recognized in first days, an entirely galactose-free diet (i.e. rigorous exclusion of

Condition	Deficient enzyme and clinical findings
	milk) may permit normal growth and development.
Chronic granulomatous disease	Phagocytes fail to produce O_2^- or H_2O_2, owing to lack of cytochrome b_{245}, flavoprotein or protein kinase C, or production of defective (low affinity) NADPH oxidase. Catalase-positive bacteria are not killed. Granulomata in lymph glands. Chronic or recurrent lymphadenitis and respiratory disease. Usually death in childhood, unless maintained on antibiotics.
Histidinemia	*Histidine ammonia lyase* (*histidase*) (EC 4.3.1.3). Failure to form urocanic acid from Histidine (see). Blood histidine increased. Cerebrospinal fluid also contains high concentration of histidine. Increased urinary excretion of histidine, imidazolepyruvate, imidazolelactate and imidazoleacetate. Formiminoglutamate is not excreted after histidine load. Usually benign. Mental retardation rare.
Homocystinuria, Homocysteinemia.	*Cystathionine β-synthase* (EC 4.2.1.22). Failure to form cystathionine from homocysteine and serine (see Methionine). Elevated homocysteine and methionine in serum. Urine contains homocystine and homocysteine-cysteine disulfide. Cystathionine is virtually absent from brain tissue, where it is normally present in significant quantities. Mental retardation. Lens detachment. Skeletal abnormalities (tall stature, arachnodactyly). Arterial and venous thrombosis.
3-Methylcrotonylglycinuria, β-Hydroxyisovaleric aciduria.	*Methylcrotonyl-CoA carboxylase* (EC 6.4.1.4). See Leucine. Urinary excretion of 3-methylcrotonylglycine and 3-hydroxyisovalerate. Acidosis in early childhood, and failure to thrive. Hypotonia and muscle atrophy. Treatment is low leucine diet.
Hydroxyprolinemia	*4-Hydroxyproline dehydrogenase* (EC 1.5.99.?). Failure to convert 4-hydroxyproline to 1-proline-3-hydroxy-5-carboxylate (see 4-Hydroxyproline).

Condition	Deficient enzyme and clinical findings
	Serum 4-hydroxyproline 30–50 times normal. Urinary 4-hydroxyproline elevated. No abnormality in collagen metabolism. Severe mental retardation. Otherwise benign.
Hawkinsinuria	*4-Hydroxyphenylpyruvate dioxygenase* (EC 1.13.11.27). Enzyme appears to be defective, so that reactive intermediates are released before completion of the conversion of 4-hydroxyphenylpyruvate to homogentisate (see Phenylalanine). The condition is characterized by the urinary excretion of (2-L-cystein-S-yl-1,4-dihydroxycyclohex-5-en-1-yl)acetic acid, also known as hawkinsin, and of 4-hydroxycyclohexylacetic acid. There is mild hypertyrosinemia and acidosis. Treated with low protein diet and supplement of ascorbic acid [B. Wilcken et al., *New Engl. J. Med.* **305** (1981) 865–869].

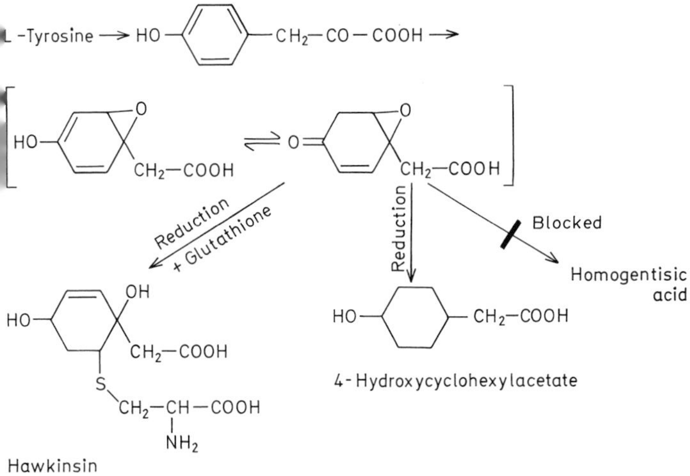

L -Tyrosine ⟶ HO—⟨ ⟩—CH₂—CO—COOH ⟶

Reduction + Glutathione
Reduction
Blocked
Homogentisic acid
4-Hydroxycyclohexylacetate
Hawkinsin

Condition	Deficient enzyme and clinical findings
Ornithine carbamoyltransferase deficiency.	(EC 2.1.3.3). Gross elevation of blood ammonia. Elevated glutamine in plasma and cerebrospinal fluid. Urea excretion is low. Urinary orotic acid increased. Uracil and uridine present in urine. The gene for the enzyme is X-linked. The condition is severe in boys (0–0.2% of normal enzyme activity in liver), who die in postnatal period (some cases of late onset have been reported). Girls have 5–10% normal liver enzyme activity. Some girls have died in later infancy or childhood, and others have survived with restricted protein intake. Abnormal EEG, mental retardation, brain atrophy and hepatomegaly.
Hyperlysinemia	*Saccharopine dehydrogenase (NADP⁺, L-lysine-forming)* or *lysine-2-oxoglutarate reductase* (EC 1.5.1.8) and *Saccharopine dehydrogenase (NAD⁺, L-glutamate-forming)* (EC 1.5.1.9). See Lysine. Elevated plasma lysine (0.2–1.5 mmol/l). Increased urinary excretion of lysine, N^ε-acetyllysine, homocitrulline and homoarginine. Usually no symptoms. Some patients mentally retarded, but this may not be a consequence of the metabolic disorder.

| Hyperammonemia type I, Carbamoylphosphate synthase deficiency | Mitochondrial *carbamoylphosphate synthase (ammonia)* (EC 2.7.2.5). Extreme hyperammonemia. Elevated glutamine in plasma and cerebrospinal fluid. Urea excretion is low. Coma and death usually in postnatal period. |
| Hyperammonemia type II, | *Ornithine carbamoyltransferase (ornithine transcarbamylase)* |

| Hyperprolinemia type I | *Proline dehydrogenase* (EC 1.5.99.8). See Proline. Elevated serum and urinary proline. Urine also contains 4-hydroxy-proline and glycine (probably due to saturation of common transport system for these amino acids in renal tubules by proline). Benign. Mental retardation in some cases. |

Condition	Deficient enzyme and clinical findings
Hyperprolinemia type II	*1-Pyrroline-5-carboxylate dehydrogenase* (EC 1.5.1.12). See Proline. Elevated serum and urinary proline. Serum pyrroline-5-carboxylate concentration 10–20 times normal. Urinary excretion of pyrroline-5-carboxylate, proline, 4-hydroxyproline and glycine increased. Mental retardation and convulsions. Low proline diet only partially effective in control.
Hypophosphatasia	*Alkaline phosphatase* (EC 3.1.3.1). Defective ossification and skeletal abnormalities. Increased urinary excretion of *O*-phosphoethanolamine and inorganic pyrophosphate, which are also abnormally high in plasma. In "pseudohypophosphatasia", the enzyme is present, but shows decreased affinity for phosphoethanolamine. Osteoblasts normally secrete extracellular vesicles containing alkaline phosphatase, which is responsible for production of inorganic phosphate by hydrolysis of pyrophosphate and organic phosphates such as *O*-phosphoethanolamine. This inorganic phosphate forms apatite crystals by reaction with calcium, leading to bone mineralization.
Isovaleric acidemia	*Isovaleryl-CoA dehydrogenase* (EC 1.3.99.10). Defective conversion of isovaleryl-CoA to β-methylcrotonyl-CoA (see Leucine). Elevated isovalerate in plasma and urine; also increased urinary excretion of isovalerylglycine, isovalerylcarnitine, and sometimes 3-hydroxyisovalerate. Ketoacidotic crises, sometimes with fatal coma. Slight mental retardation in survivors. Treated with low leucine intake and supplements of glycine and/or carnitine to increase excretion of isovaleryl conjugates. Peritoneal dialysis in crises.
Lactose intolerance	*Lactase* (intestinal) (EC 3.2.1.23). Lactase is present in infancy when it is required for digestion of lactose in the mother's milk. The enzyme

Condition	Deficient enzyme and clinical findings
	then decreases markedly with age, so that abdominal pain and diarrhea result from drinking milk, due to failure to hydrolyse lactose in the intestinal mucosa and lactose malabsorption. The condition is prevalent throughout the world, and the hereditary persistence of high intestinal lactase activity prevails only in Northern European populations (and those derived from them) and certain Arab and Hamitic races.
Pancreatic lipase deficiency	*Pancreatic lipase* (EC 3.1.1.3). Lipase of pancreatic juice reduced to about 10% of normal. Triacylglycerols in feces. Growth normal. Fat absorption about 70% of normal. Very rare condition.
Maple syrup urine disease, Leucinosis.	*Branched-chain-oxoacid dehydrogenase complex*, which is responsible for the oxidative decarboxylation of the oxo-acids derived from leucine, isoleucine and valine (see Leucine). Increased concentrations of all three branched chain amino acids and their corresponding oxo-acids in urine, plasma and cerebrospinal fluid. Serum also contains alloisoleucine (probably derived from isoleucine). Urine has characteristic odor. Marked cerebral degeneration apparent shortly after birth. Usually fatal within weeks or months of birth.
Methemoglobinemia	*NAD-methemoglobin reductase* (EC 1.6.2.?). Methemoglobin is continually formed by oxidation of hemoglobin in erythrocytes. Normally it is reduced to hemoglobin, largely by the NADH-dependent reductase (67%) and to a lesser extent by the NADPH-dependent reductase, and nonenzymatic interaction with glutathione and ascorbate. The congenital condition is very rare. Affected individuals are grayish-blue and cyanotic in appearance, due to the large proportion of circulating methemoglobin, which cannot transport oxygen. There is no

Condition	Deficient enzyme and clinical findings
	serious incapacitation. A clinically similar congenital condition has been identified, in which the hemoglobin is abnormal and cannot interact with the enzyme.
Methylmalonic aciduria, Methylmalonic acidemia.	*Methylmalonyl-CoA mutase* (EC 5.4.99.2). Failure to convert (R)-methylmalonyl-CoA into succinyl-CoA (see Leucine for degradation of branched chain amino acids). Large quantities of methylmalonic acid appear in plasma and urine. Affected children fail to thrive and show pronounced ketoacidosis. Death often occurs in early life. Hyperammonemia and intermittent hyperglycinemia are also typical. Restricted protein intake and synthetic diets are helpful, in particular low intakes of leucine, isoleucine, valine, threonine and methionine. A similar condition may also arise from a congenital deficiency of *methylmalonyl-CoA epimerase* (EC 5.1.99.1). Both conditions are unresponsive to treatment with vitamin B_{12}. Another type of methylmalonic aciduria is thought to result from an hereditary deficiency of *deoxyadenosyl transferase* (transfers 5'-deoxyadenosyl group in cobalamin synthesis), which provides the coenzyme of methylmalonyl-CoA mutase. This condition responds to injection of B_{12}. Dietary B_{12} deficiency also results in methylmalonic aciduria.
Ornithinemia, Hyperornithinemia type I	*Ornithine-oxoacid aminotransferase* (EC 2.6.1.13). Ornithine concentration increased in plasma, urine and cerebro-spinal fluid. Progressive loss of vision and usually blindness before 40th year. Treated by restriction of arginine intake, and large supplements of pyridoxine. Type II hyperornithinemia is due to an unidentified defect (possibly defective ornithine transport into mitochondria); it is characterized by hyperornithinemia, hyperammonemia, homocitrullinemia and homocitrullinuria. The

Condition	Deficient enzyme and clinical findings
	eyes are not affected, and there is lethargy, coma and mental retardation.
Orotic aciduria	*Orotidine-5'-phosphate pyrophosphorylase* (EC 2.4.2.10) and *orotidine-5'-phosphate decarboxylase* (EC 4.1.1.23). A gross deficiency of two enzymes of Pyrimidine biosynthesis (see). Abnormally high urinary orotic acid. Severe megaloblastic anemia. Marked retardation of growth and development, and slight mental retardation.
Oxalosis type I	*2-Hydroxy-3-oxoadipate synthase* (EC 4.1.3.15) (see). High urinary excretion of oxalic acid and glycolic acid. Calcium oxalate crystals form in many body tissues. Nephrocalcinosis, urolithiasis, with progressive renal insufficiency and usually death before 20 years. See Oxalic acid.
Oxalosis type II	*Glycerate dehydrogenase* (EC 1.1.1.29). high urinary excretion of oxalic acid and L-glyceric acid. Clinically similar, but milder than type I. See Oxalic acid.
Pentosuria	*NADP-specific xylitol oxidoreductase* (*L-xylose reductase*) (EC 1.1.1.10). See Glucuronate pathway. L-Xylulose is continuously excreted in large amounts in the urine. Administration of glucuronic acid increases xylulose excretion even further. The condition is benign and no treatment is needed.
Heriditary tyrosinemia type II, Tyrosinosis type II, Hypertyrosinemia type II, Richner-Hanhart syndrome.	*Cytosolic tyrosine aminotransferase* (EC 2.6.1.5). Elevated tyrosine in blood and cerebrospinal fluid. Increased urinary excretion of tyrosine, 4-hydroxyphenylpyruvate, 4-hydroxy-phenyllactate and 4-hydroxyphenyl-acetate. Slight to moderate mental retardation. Blistering and hyperkeratosis of palms and soles of feet. Photophobia. There is no hepatorenal dysfunction (c.f. Hereditary tyrosinemia type I). Controlled with diet low in phenylalanine and tyrosine. See Phenylalanine.

289

Condition	Deficient enzyme and clinical findings
Herlditary tyrosinemia type I, Tyrosinosis, Hereditary hepatorenal dysfunction.	*Fumarylacetoacetase* (EC 3.7.1.2). Failure to metabolize fumarylacetoacetate by normal route (see Phenylalanine) results in reduction of accumulated fumarylacetoacetate (or maleylacetoacetate, or both) to succinylacetoacetate, which is decarboxylated to succinylacetone. The latter inhibits 4-hydroxyphenylpyruvate: oxygen oxidoreductase (hydroxylating, decarboxylating) (EC 1.13.11.27), causing tyrosinemia (blood methionine is often also elevated), and increased urinary excretion of tyrosine, 4-hydroxy-phenylpyruvate, 4-hydroxyphenyllactate and 4-hydroxyphenylacetate. *Porphobilinogen synthase* (EC 4.2.1.24) is also inhibited, resulting in high excretion of 5-aminolevulinic acid and symptoms of acute hepatic porphyria. There is also general aminoaciduria, glucosuria, proteinuria and hypokalemia. Hypoprothrombinemia and jaundice. Death is usual in early childhood. Surviving individuals develop cirrhosis and hepatorenal dysfunction; often also malignant hepatoma, acidosis and vitamin D-resistant rickets [B. Lindblad et al. *Proc. Nat. Acad. Sci.* **74** (1977) 4641–4645].

Condition	Deficient enzyme and clinical findings
Refsum's disease, Phytanic acid storage disease, Heredopathia atactica polyneuritiformis.	*Phytanic acid α-hydroxylase,* Phytanic acid is normally formed in the body from the plant alcohol, phytol, present as an ester in chlorophyll. The presence of a branch methyl group at position 3 of phytanic acid means that the normal process of β-oxidation (see Fatty acid degradation) is blocked. Oxidation of fatty acids one carbon at a time (α-oxidation, see Fatty acid degradation) is common in plants, but also occurs to some extent in animals, especially in the brain, where it serves to initiate the degradation of phytanic acid. The resulting pristanic acid is then degraded by β-oxidation. In Refsum's disease, phytanic acid accumulates in liver and kidneys, and it may represent over 50% of total liver fatty acids. Plasma phytanic acid concentrations of 200–3100 mg/l have been reported (normal: <2 mg/l). Clinically there is peripheral neuropathy and ataxia, retinitis pigmentosa and skin and bone abnormalities. Treatment by plasma exchange and low phytol intake.
Lecithin-cholesterol acyltransferase deficiency, Norum's disease	*Phosphatidylcholine-sterol acyltransferase* (EC 2.3.1.43). The enzyme catalyses the formation of cholesterol esters by

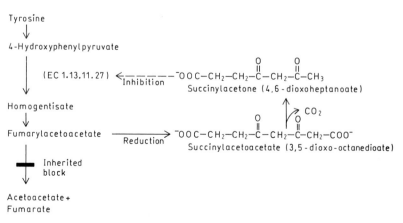

Tyrosine

4-Hydroxyphenylpyruvate

(EC 1.13.11.27) ←――――― $^-$OOC–CH$_2$–CH$_2$–C–CH$_2$–C–CH$_3$
Inhibition
Succinylacetone (4,6-dioxoheptanoate)

Homogentisate

Fumarylacetoacetate ――――→ $^-$OOC–CH$_2$–CH$_2$–C–CH$_2$–C–CH$_2$–COO$^-$
Reduction
Succinylacetoacetate (3,5-dioxo-octanedioate)

CO$_2$

Inherited block

Acetoacetate + Fumarate

...ondition	Deficient enzyme and clinical findings	Condition	Deficient enzyme and clinical findings

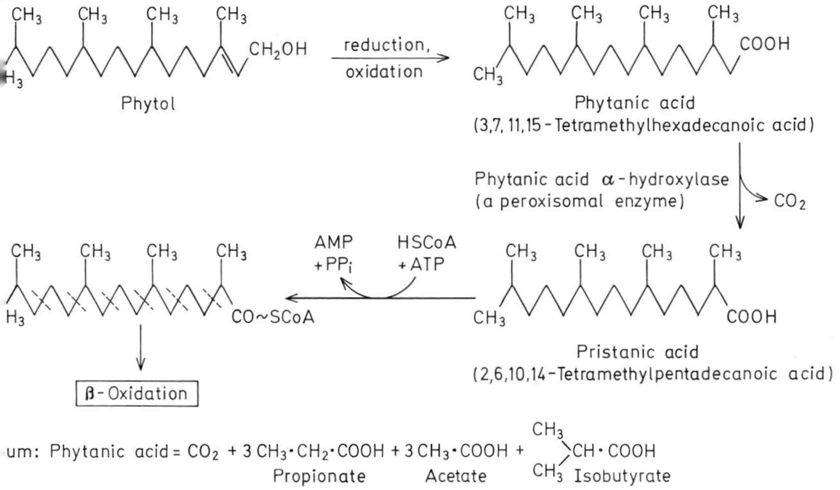

CH₃ CH₃ CH₃ CH₃

...—CH₂OH $\xrightarrow{\text{reduction, oxidation}}$

Phytol

$$CH_3 \quad CH_3 \quad CH_3 \quad CH_3$$
COOH

CH₃—

Phytanic acid
(3,7,11,15-Tetramethylhexadecanoic acid)

Phytanic acid α-hydroxylase
(a peroxisomal enzyme) $\searrow CO_2$

CH₃ CH₃ CH₃ CH₃

$\xrightarrow[\text{+ PP}_i \quad \text{+ ATP}]{\text{AMP} \quad \text{HSCoA}}$

...—CO~SCoA

↓

| β-Oxidation |

CH₃ CH₃ CH₃ CH₃

CH₃— COOH

Pristanic acid
(2,6,10,14-Tetramethylpentadecanoic acid)

...um: Phytanic acid = CO_2 + 3 CH₃·CH₂·COOH + 3 CH₃·COOH + $\begin{matrix} CH_3 \\ CH_3 \end{matrix}$ CH·COOH

Propionate   Acetate   Isobutyrate

transfer of an unsaturated fatty acid from the 2-position of lecithin to the 3-OH of cholesterol. Plasma cholesterol and triacylglycerols are increased, and lysophosphatidylcholine and cholesterol esters are decreased. The plasma is turbid or milky, and there are multiple lipoprotein abnormalities. Corneal opacities. Normochromic anemia and proteinuria due to renal damage. Therapy by enzyme replacement.

...arcosinemia	*Sarcosine dehydrogenase* (EC 1.5.99.1). Hypersarcosinemia and hypersarcosinuria. Some cases have shown mental retardation. Probably benign. See Sarcosine.
...ulfite oxidase de-...ciency, Sulfituria.	*Sulfite oxidase (sulfite dehydrogenase)* (EC 1.8.3.1). Cysteine sulfinic acid, an intermediate in the metabolism of Cysteine (see), can be transaminated to β-sulfinylpyruvate; this latter compound readily loses SO_2 in a reaction analogous to the decarboxylation of oxaloacetate (see Tricarboxylic acid cycle). The sulfite is oxidized to sulfate by sulfite oxidase (see Molybdoenzymes). Absence of the enzyme results in urinary excre-

tion of S-sulfo-L-cysteine, sulfite and thiosulfate, and a virtual absence of urinary sulfate. Progressive neurological abnormalities, mental retardation, lens dislocation. Treat with diet low in sulfur amino acids. Death in postnatal period possible.

Xanthurenic aciduria	*Kynureninase* (EC 3.7.1.3). Failure to convert 3-hydroxykynurenine to 3-hydroxyanthranilic acid, and kynurenine to anthranilic acid (see Tryptophan). Increased urinary excretion of xanthurenic acid, kynurenine and 3-hydroxykynurenine, especially after ingestion of tryptophan. Some patients have been mentally retarded, and others have been symptomless. The enzyme may not be absent, but structurally abnormal, leading to decreased affinity for its coenzyme. As evidence of this, dietary vitamin B_6 temporarily corrects the metabolic disturbance, and addition of pyridoxal phosphate to liver biopsy material increases enzyme activity to near normal.
Xanthinuria type I	*Xanthine oxidase* (EC 1.2.3.2). Failure to convert xanthine to uric acid (see Purine degradation). Xanthine therefore re-

Condition	Deficient enzyme and clinical findings
	places uric acid as end product of purine metabolism. Urinary excretion of xanthine greatly increased, and urinary uric acid abnormally low. Xanthine calculi tend to form in the renal tract.
Xanthinuria type II	Malabsorption of molybdenum. Activities of *xanthine oxidase, sulfite oxidase* and *aldehyde oxidase* are therefore deficient (see Molybdoenzymes). Urinary xanthine and uric acid are high and low, respectively (see Purine degradation). Mental retardation, seizures, cerebral atrophy, dislocation of lenses.

Lysosomal storage diseases. Absence of a specific hydrolase from the lysosomes leads to the accumulation or "storage" of its substrate. A genetic defect may result in total failure to synthesize a lysosomal hydrolase, absence of an activator protein, synthesis of a less stable enzyme, or failure to target the synthesized enzyme to the lysosomes. Lysosomal storage diseases are usually classified and named according to the nature of the accumulated material, e.g. mucopolysaccharidoses, oligosaccharidoses, mucolipidoses, sphingolipidoses, etc.

Glycogenosis type II, Pompe's disease.	See Glycogen storage diseases. There are several glycogen storage diseases, but only Pompe's disease is also a lysosomal storage disease.
Cholesterol ester storage disease.	*Acid lipase* (EC 3.1.3.2). Hepatic lipase is about 25% of normal. Cholesterol esters deposited in liver, spleen, intestinal mucosa, lymph nodes, aorta. Hepatomegaly, leading to hepatic fibrosis. Sometimes jaundice and/or splenomegaly. Relatively benign. Autosomal recessive. Wolman's disease (below) is probably the expression of a different mutant allele at the same locus [J. M. Hoeg et al., *Amer. J. Hum. Gen.* **36** (1984) 1190-1203].
Wolman's disease	*Acid lipase* (EC 3.1.3.2). Cholesterol esters and triacylglycerols deposited in adrenals, liver, spleen, bone marrow, capillaries, endothelium, ganglion cells of mesenteric plexus and mucosa of small intestine. Plasma lipids are mainly nor-

Condition	Deficient enzyme and clinical findings
	mal. Hepatosplenomegaly. Adrenal calcification and enlargement. Failure to thrive in infancy, rapid deterioration and death. Autosomal recessive. See Cholesterol ester storage disease (above).

Oligosaccharidoses. These diseases involve the degradation of asparagine-linked oligosaccharides. For the relevant reactions, see Glycoprotein degradation.

Aspartylglycosaminuria	N^4-(β-N-Acetylglucosaminyl)-L-asparaginase (EC 3.5.1.26). Impaired degradation of some glycoproteins. Severe mental retardation, motor impairment, and large amounts of urinary aspartyl-glycosamine and other glycoasparagines.

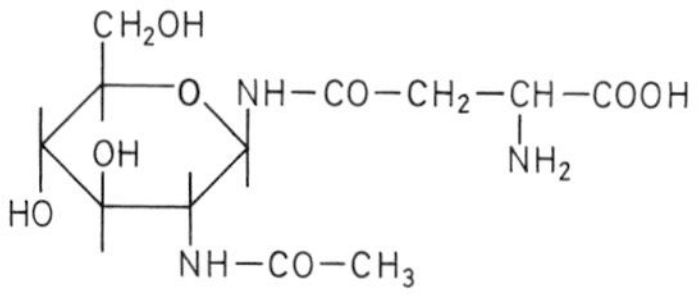

Aspartylglycosamine [2-acetamido-1-(β-aspartamido)-1,2-dideoxyglucose]

Mannosidosis	*α-Mannosidase* (EC 3.2.1.24). Accumulation of mannose-rich, glucosamine-containing oligosaccharides. Brain damage, bone abnormalities, opaque cornea and cataracts, hepatosplenomegaly. Skeletal involvement and facial features similar to Hurler's syndrome (below).
Fucosidosis	*α-L-Fucosidase* (EC 3.2.1.51). Accumulation of fucose-rich glycoproteins, sphingolipids and glycosaminoglycans. Severe progressive cerebral degeneration in infancy (rapid in type I, slow in type II). Skeletal involvement similar to Hurler's syndrome (below). Hepatosplenomegaly (type I shows cardiomegaly). Facial features different from Hurler's syndrome.
Sialidosis	*Glycoprotein sialidase (neuraminidase)* (EC 3.2.1.18). Other lysosomal activities in liver may sometimes be increased. Sialyloligosaccharides accumulate. There are several types

Condition	Deficient enzyme and clinical findings
	of differing severity. Thus mental retardation, skeletal abnormalities and Hurler-type facial features may be present in varying degrees, or absent. Hepato-splenomegaly, renal involvement and hydrops fetalis are present in some cases.
Sialic acid storage disease	Unidentified defect. Possibly a defect of lysosomal membrane transport. Free acetylneuraminic acid present in urine. Impaired motor function and ataxia. In the severe form (type II) there is progressive neurological deterioration and early death. Type II involves moderate coarsening of features, growth retardation (rickets) and hepatosplenomegaly; these effects are absent from type I, which shows some skeletal involvement (curved tibiae).

Mucolipidoses. Mucolipidosis I is also classified as an oligosaccharidosis (Sialidosis, see above). These storage diseases are due to defects in glycoprotein biosynthesis (see Glycoproteins for relevant enzymes), in contrast to the oligosaccharidoses, which are due to defects in glycoprotein degradation.

Condition	Deficient enzyme and clinical findings
Mucolipidosis II, I-cell disease.	*UDP-N-acetylglucosamine-lysosomal-enzyme precursor N-acetylglucosamine phosphotransferase* (EC 2.7.8.17). Glycosaminoglycans and glycolipids accumulate in fibroblasts, hepatocytes and Schwann cells. Clinically similar to Hurler's syndrome (below). Death in childhood.
Mucolipidosis III, Pseudo-Hurler polydystrophy.	*UDP-N-acetylglucosamine-lysosomal-enzyme precursor N-acetylglucosamine phosphotransferase* (EC 2.7.8.17). Similar to, but milder than mucolipidosis type II. Survival to adulthood.
Mucolipidosis IV	*Sialidase* (specific for gangliosides) (EC 3.2.1.?). Gangliosides, glycolipids, lipofuscin and lysobisphosphatidic acid accumulate in lysosomes of many tissues. Corneal opacities, dementia. No skeletal effects or hepatomegaly.

Mucopolysaccharidoses are lysosomal storage diseases representing a variety of genetic disorders

Condition	Deficient enzyme and clinical findings
	which affect the degradation of chondroitin sulfate, keratan sulfate, dermatan sulfate and/or heparan sulfate (see formulae below). There is no type V mucopolysaccharidosis.
Type I_H, Hurler's syndrome, Gargoylism.	*α-L-Iduronidase* (EC 3.2.1.76). Large skull, depressed nose bridge, hypertrichosis, short neck, projecting forehead, large tongue and lips, widely spaced teeth and gum hypertrophy. Thick, hairy skin, and usually clouded cornea. Coronary valves, vessels and heart muscle often affected, leading to death from heart failure before 20th year. Hepatosplenomegaly and skeletal abnormalities (dwarfism and kyphosis). Severe and progressive mental retardation. Heparan sulfate and dermatan sulfate accumulate in tissues.
Type I_S, Scheie's syndrome.	*α-L-Iduronidase* (EC 3.2.1.76). Heart involved only in some cases. No mental retardation. Skeleton less affected than in $I_{H/S}$. Facial features similar to I_H.
Type $I_{H/S}$,	*α-L-Iduronidase* (EC 3.2.1.76). Heart involved only in some cases. Little or no mental retardation. Skeleton less affected than in I_H. Facial features similar to I_H.
Type II, Hunter's syndrome	*Iduronate 2-sulfatase.* Heparan sulfate and dermatan sulfate accumulate. No clouding of cornea. Heart involvement rare. Mental retardation less than in I_H. Facial features similar to I_H. Type II_B is slightly less severe than II_A.
Type III_A, Sanfilippo A.	*Heparan N-sulfatase.* Heparan sulfate accumulates. No clouding of cornea. Heart not affected. Skeleton only slightly affected. Facial features similar to I_H, but less severe.
Type III_B, Sanfilippo B.	*α-N-Acetylglucosaminidase.* Heparan sulfate and glycosphingolipids accumulate. Facial features similar to III_A.
Type III_C, Sanfilippo C	*Glucosamine acetyltransferase.* Heparan sulfate accumulates. Facial features similar to III_A.

Condition	Deficient enzyme and clinical findings	Condition	Deficient enzyme and clinical findings
Type III$_D$, Sanfilippo D.	*N-Acetylglucosamine-6-sulfatase*. Heparan sulfate accumulates. Mild osteochondro-dystrophy, and presence of hypoplastic odontoid process. Facial features similar to III$_A$.		changes in ribs, sternum, vertebrae and bones of hands and feet. Thin enamel.
Type IV$_A$, Morquio-Brailsford syndrome (one variant)	*N-Acetylgalactosamine-6-sulfatase*. Keratan sulfate and sometimes chondroitin sulfate peptide accumulate. No mental retardation. Cornea sometimes clouded. Heart involvement rare. Marked stunting of growth. Severe and distinctive	Type IV$_B$, Morquio-Brailsford syndrome (one variant)	*β-Galactosidase* (EC 3.2.1.23). Enzyme has high K_m for keratan sulfate, and normal K_m for gangliosides. Keratan sulfate accumulates. No mental retardation. Enamel normal. Bone changes only mild. Heart involvement rare. Cornea is clouded. Facial features are normal and unaffected.

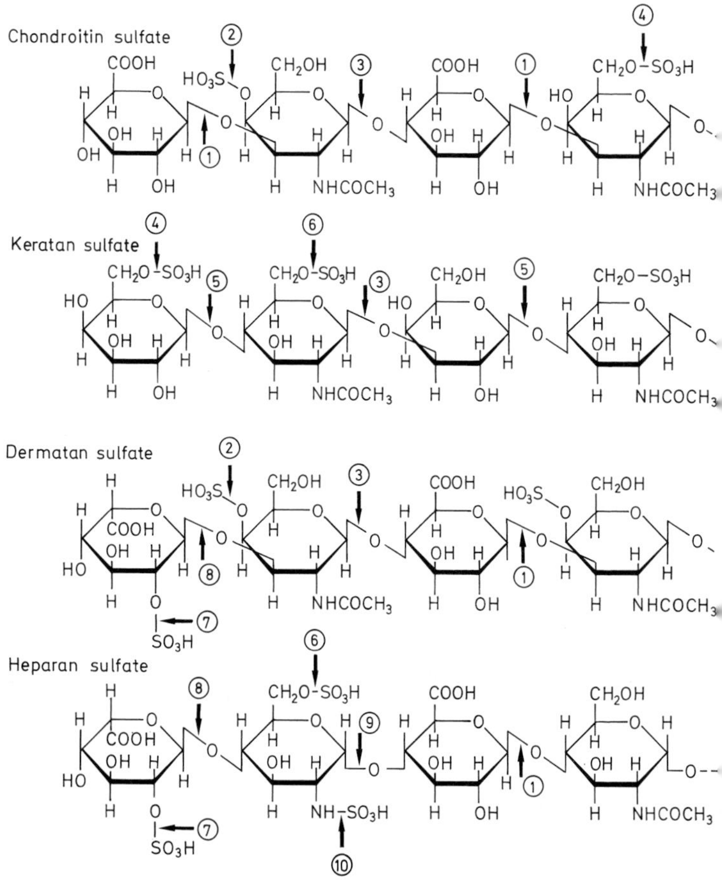

Glycosaminoglycans (mucopolysaccharides) and sites of attack by degradative enzymes
Absence or defective function of any one of these enzymes leads to accumulation of incompletely degraded mucopolysaccharide, i.e. mucopolysaccharidosis. 1. β-Glucuronidase; 2. N-Acetylgalactosamine-4-sulfatase; 3. β-N-Acetylhexosaminidase; 4. N-Acetylgalactosamine-6-sulfatase; 5. β-Galactosidase; 6. N-Acetylglucosamine-6-sulfatase; 7. Iduronate 2-sulfatase; 8. L-Iduronidase; 9. α-N-Acetylglucosaminidase; 10. Heparan N-sulfatase.

Condition	Deficient enzyme and clinical findings
Type VI$_A$, VI$_B$, VI$_C$, Maroteaux-Lamy syndrome	*N-Acetylgalactosamine-4-sulfatase.* Dermatan sulfate accumulates. No mental retardation. Heart affected in some cases. Facial features less severely affected than in I$_H$. In VI$_A$ skeletal involvement is similar to I$_H$ without vertebral deformation. VI$_B$ and VI$_C$ show moderate and mild skeletal effects, respectively.
Type VII, Sly syndrome	*β-Glucuronidase* (EC 3.2.1.31). Chondroitin 4-sulfate and chondroitin 6-sulfate accumulate. Dermatan sulfate and/or heparan sulfate sometimes also accumulate. Mental retardation not severe. Cornea sometimes clouded. Facial appearance may resemble I$_H$, or may be less severely affected or even normal. Skeletal deformities are usually very marked. Heart not affected.

Sphingolipidoses are the result of defective catabolism of glycososphingolipids (see scheme of glycososphingolipid and sphingomyelin catabolism, below). See Glycolipids.

Condition	Deficient enzyme and clinical findings
Gangliosidosis, Generalized gangliosidosis, Type I, G$_{M1}$ gangliosidosis, Neurovisceral lipidosis, Pseudo-Hurler syndrome, Maladie de Landing	*β-Galactosidase A$_1$, A$_2$ and A$_3$* (EC 3.2.1.23). Sphingolipidosis, involving accumulation of GM$_1$ ganglioside and desialo-GM$_1$ ganglioside in neurons. Keratan sulfate-related glycosaminoglycan accumulates in spleen, liver, epithelial cells of kidney glomeruli and bone marrow. Mental and motor deterioration. Hepatosplenomegaly. Invariably fatal by end of second year.
Juvenile gangliosidosis, Type II G$_{M1}$ gangliosidosis	*β-Galactosidase A$_2$ and A$_3$* (EC 3.2.1.23). Only two of the three liver β-galactosidases are absent, and sphingolipids do not accumulate in the liver. Progress of the disease is slower than that of generalized gangliosidosis, but the clinical picture otherwise similar. Death usually at 3–10 years. G$_{M1}$ deposited exclusively in central nervous system.
Adult gangliosidosis, Type III G$_{M1}$ gangliosidosis.	*β-Galactosidase* (EC 3.2.1.23). Enzyme activity is about 5% of normal. Motor deterioration. Little intellectual impairment. [A. T. Hoogeveen et al.

Condition	Deficient enzyme and clinical findings
	J. Biol. Chem. **259** (1984) 1974–1977].
G$_{M3}$ Gangliosidosis, G$_{M3}$ Sphingolipodystrophy.	*(N-Acetyneuraminyl)-galactosylglucosyl-ceramide N-acetylgalactosaminyl transferase.* Synthesis of G$_{M1}$ and G$_{M2}$ is deficient, and G$_{M3}$ accumulates in brain and liver. Clinically similar to generalized gangliosidosis (see for structures of G$_{M1}$, G$_{M2}$ and G$_{M3}$). Death in first year.
Adult G$_{M2}$ gangliosidosis	*β-N-Acetylhexosaminidase A* (EC 3.2.1.30). G$_{M2}$ accumulates intraneuronally in white and gray matter of brain; small quantities accumulate in liver and spleen. Manifested at 18–40 years (sometimes at 2–4 years followed by slow progression). Mental motor deterioration. Death in a few years after first manifestation. See generalized gangliosidosis for structure of G$_{M2}$.
Lactosyl ceramidosis, Ceramide lactoside lipidosis	*Ceramide lactoside β-galactosidase.* Accumulation of ceramide lactoside. Hepatomegaly. Splenomegaly. Progressive brain damage and neurological impairment.
Tay-Sachs disease	*β-N-Acetylhexosaminidase A* (EC 3.2.1.30). In type B of the disease, the enzyme is absent. In type AB, the enzyme is present but an activator protein is absent. In type A^{M}B, the enzyme is present but defective. G$_{M2}$ accumulates intra-neuronally in white and gray matter of the brain, with small quantities in spleen and liver. Clinical picture and prognosis similar to infantile form of Sandhoff's disease, with manifestation at 4–6 months (ganglioside accumulation begins in utero).
Fabry's disease, or Angiokeratoma corporis diffusum	*Ceramide trihexosidase (α-galactosidase)* (EC 3.2.1.22). Sphingolipidosis due to accumulation of ceramide trihexoside. Skin lesions (purple macules and papules). Corneal opacities, cataracts, retinal edema. Cardiovascular, neurological and gastrointestinal disorders. Characteristic severe burning pains in extremities. X-chromosome linked.

Condition	Deficient enzyme and clinical findings
Gaucher's disease: (a) Chronic (adult) form (at least two subtypes); (b) Infantile neuronopathic form; (c) Juvenile neuronopathic form (Norbottnian); (d) Prenatal neuronopathic form	*Glucosylceramidase* (EC 3.2.1.45). Enzyme activity about 15% of normal in (a). Enzyme totally absent in (b) and (d), and about 2.5% of normal in (c). Glucosylceramide accumulates in liver, spleen, bone marrow and leukocytes; it accumulates in the brain in (b), (c) and (d), and in the lungs in some cases of (b). Splenomegaly. Hepatomegaly. Anemia. Pancytopenia. Ostealgia and osteoporosis. Purpura. Cerebral degeneration in (b) and (c). Type (a) is manifested between 1 and 60 years, and may result in death at any time from infection or liver dysfunction. Type (b) is manifested in the first year, and is fatal by the end of the first or second year. Type (c) is manifested at 6–20 years, and is fatal in adolescence or early adulthood. Type (d) results in death in utero (ascites hydrops fetalis).
Sandhoff's disease	*β-N-Acetylhexosaminidases A and B* (EC 3.2.1.30). G_{M2} and G_{A2} accumulate intraneuronally in white and gray matter of the brain, and in spleen, and liver. Globoside accumulates in kidneys and spleen. N-Acetylglucosaminyloligosaccharides appear in tissues and urine. The infantile form is manifested between 4 and 6 months (ganglioside accumulation begins in utero), with marked hyperacusis, a cherry red spot in the macular region, progressive cerebral degeneration, and usually death before 5 years. The juvenile (late infantile) form is manifested in the first year, progresses slowly, and shows no red macular spot.
Metachromatic leukodystrophy	*Arylsulfatase A* (EC 3.1.6.8). Accumulation of galactosyl-sulfate ceramide and lactosyl-sulfate ceramide in kidneys and gall bladder, and galactosyl sulfate ceramide alone in brain white matter and peripheral nerve sheaths. Cerebral degeneration and motor disturbances. Death 5–10 years

Condition	Deficient enzyme and clinical findings
	after onset, or at 3–7 years of age (infantile form).
Krabbe's disease, Krabbe's leukodystrophy, Galactosylceramide lipidosis, Globoid cell leukodystrophy.	*Galactosylceramidase* (EC 3.2.1.46). Failure to hydrolyse galactosylceramide to galactose and ceramide. Galactosyl ceramide accumulates in axons, but not in myelin sheaths. Myelin almost totally disappears. Globoid cells infiltrate brain white matter. Rapid neurological deterioration, and always fatal in infancy.
Mucosulfatidosis (a form of metachromatic leukodystrophy)	9 distinct *sulfatases* acting on sulfatides, steroids and glycosaminoglycans; *β-galactosidase*. Accumulated metabolites include gangliosides, sulfatides, cholesteryl-sulfate and dehydroepiandrosterone sulfate in neurons and liver. Glycosaminoglycans are excreted in the urine. Moderate hepatosplenomegaly. Progressive, with motor disturbances, ataxia and convulsions. Cerebral degeneration. Onset at 1–4 years and death before 12th year.
Nieman-Pick disease	*Sphingomyelin phosphodiesterase* (EC 3.1.4.12). Sphingomyelin and cholesterol accumulate in neurons, liver, spleen, bone marrow, lymphatic tissue and lungs. Five types of the disease are recognized (types A-E), differing in severity. Type D possesses normal, and type E normal or lowered sphingomyelin phosphodiesterase activity, and the genetic defect has not been identified. Type E is benign. The other types involve hepatosplenomegaly and mental, physical and motor retardation, except type B, which has no neurological involvement. In type A, deposition of sphingomyelin begins in utero and death is usually at 2–3 years. Type C is characterized by excessive accumulation of unesterified cholesterol, accompanied by relatively less accumulation of sphingomyelin. Cholesterol accumulation appears to be due to an error of cholesterol processing and homeostasis, in which cellular uptake of cholesterol from low

Condition	Deficient enzyme and clinical findings
	density lipoproteins is excessive, and its subsequent esterification is deficient. Types C (onset during the first 6 months) and D (onset at birth) show neurological disturbances at 5–10 years, with death usually at 5–15 years. [H.S. Kruth et al. *J. Biol. Chem.* **261** (1986) 16769–16774; P.G. Pentchev et al. *J. Biol. Chem.* **261** (1986) 16775–16780].
G_{M1} Gangliosidosis with sialidosis	*β-Galactosidase* (EC 3.2.1.23) and *sialidase (neuraminidase)* (EC 3.2.1.18). Both enzymes lack a protective glycoprotein and are unstable. G_{M1} and sialyloligosaccharides accumulate in neurons, spleen, liver, bone marrow and kidneys. The infantile form (type I) is manifested in early infancy and is clinically similar to generalized gangliosidosis. The juvenile form (type II) is manifested at 3–20 years, shows progressive neurological deterioration with ataxia and moderate mental retardation, coarse facial features, bone deformities, cherry-red macular spots, and does not exhibit visceromegaly. Life expectancy is possibly normal.
Juvenile G_{M2} gangliosidosis	*β-N-Acetylhexosaminidase* (EC 3.2.1.30). Late infantile and juvenile forms are recognized. Loss of enzyme activity shows wide variations. Site of G_{M2} accumulation as in Tay-Sachs disease. Clinically similar to Tay-Sachs, but often no red macular spot, and progression of the disease is slower. Death in childhood.
Farber's disease, Disseminated lipogranulomatosis.	*Acylsphingosine deacylase* (*ceramidase*) (EC 3.5.1.23). Ceramide, glycolipids and sometimes dermatan sulfate accumulate subcutaneously, in tendons (especially at joints), and in neurons. Swollen joints and subcutaneous nodules. Growth retardation. Death usually at age 7–22 months, but some survivors.
Neuronal ceroid lipofuscinosis, Familial amaurotic	Defect unidentified. Possibly involves metabolism of dolichol-linked oligosaccharides.

Condition	Deficient enzyme and clinical findings
idiocy. Types: Santavuori (infantile) Jansky-Bielschowsky (late infantile), Batten-Spielmeyer-Vogt (juvenile), Kufs (adult).	Arachidonic acid-containing phosphoglycerides increased in brain and serum. Dolichol, dolichyl phosphate, ceroid and lipofuscin accumulate in neurons. Cerebral degeneration with optic atrophy. Mental and motor retardation with blindness. Late forms do not involve blindness.

Porphyrias. These clinical conditions are the result of genetic defects in heme biosynthesis. For the pathway of heme biosynthesis, see Porphyrins. Inborn errors have been described for seven of the eight enzymes in this pathway. Although no major genetic defect has been recorded for the first enzyme of the pathway, 5-aminolevulinate synthase (EC 2.3.1.37), low activity has been reported in a case of congenital sideroblastic anemia [G.R. Buchanan et al. *Blood* **55** (1980) 109–115]. Heme is an essential constituent of many important enzymes and hemoproteins. Absence of heme synthesis is therefore incompatible with life, and homozygotes of inherited autosomal dominant disorders of heme synthesis are not viable, unless there is residual activity of the enzyme in question. Porphyrias are classified as erythropoietic or hepatic, depending on whether the defect is located mainly in the erythroid cells or the liver.

Congenital erythropoietic porphyria, Gunther's disease	*Uroporphyrinogen III synthase* (EC 4.2.1.75). This enzyme converts hydroxymethylbilane into uroporphyrinogen III. Large amounts of uroporphyrin I and coproporphyrin I are deposited in tissues and bone marrow, and excreted in urine and feces. Hemolytic anemia and photosensitivity. Pink coloration of deciduous teeth. Early death is common. Treatment by protection from UV light, splenectomy and blood transfusion. Autosomal recessive. About 100 cases known.
Acute intermittent porphyria, Pyrroloporphyria, Swedish hepatic porphyria.	*Porphobilinogen deaminase* (EC 4.3.1.8). Enzyme activity is about half normal in liver and erythrocytes. Excessive formation of 5-aminolevulinate and porphobilinogen, resulting from decreased feedback inhibition of 5-aminolevulinate synthase. High urinary excretion of porphobilinogen and 5-aminolevulinate. Peripheral neuritis, paralysis and general demyelination. Neurosis, psychosis and abdominal colic. No photosensitivity. Mortality rate high. On-

297

Condition	Deficient enzyme and clinical findings	Condition	Deficient enzyme and clinical findings

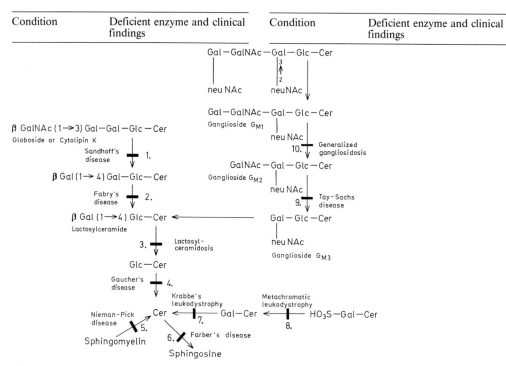

Catabolism of glycososphingolipids and sphingomyelin in humans. The sites of metabolic defects are shown by solid bars. Further details of the defects are found in the table above (Sphingolipidoses). Cer, ceramide; Gal, galactosyl; Glc, glucosyl; GalNAc, N-acetylgalactosaminidyl; neuNAc, N-acetylneuraminidyl. 1. β-N-Acetylhexosaminidase, EC 3.2.1.20; 2. Ceramide trihexosidase, EC 3.2.1.22; 3. Ceramide lactoside β-galactosidase, EC 3.2.1.?; 4. Glucosylceramidase, EC 3.2.1.45; 5. Sphingomyelin phosphodiesterase, EC 3.1.4.12; 6. Acylsphingosine deacylase, EC 3.5.1.23; 7. Galactosylceramidase, EC 3.2.1.46; 8. Arylsulfatase, EC 3.1.6.8; 9. β-N-Acetylhexosaminidase, EC 3.2.1.30; 10. β-Galactosidase, EC 3.2.1.23.

set after puberty. Treatment with intravenous hematin and high dietary carbohydrate. Many drugs are known to precipitate acute attacks, e.g. barbiturates, methyldopa, sulfonamides, sulfonylurea, pyrazinamide, griseofulvin, glutethimide, meprobamate. Other precipitating factors are ethanol, hormonal changes, starvation, mental stress and infection. Frequency is 0.015–0.1%. Autosomal dominant. In about 90% of carriers it is asymptomatic.

Variegate porphyria, Protocoproporphyria, South African hepatic porphyria, Porphyria variegata.	*Protoporphyrinogen III oxidase* (EC 1.3.3.4). Ferrochelatase activity may also be decreased. Impaired feedback inhibition of 5-aminolevulinate synthase results in excessive porphyrin production. Increased urinary excretion of porphobilinogen, 5-aminolevulinate, protopor-		

phyrin and coproporphyrin during acute attacks. Fecal protoporphyrin and coproporphyrin constantly elevated. Fecal porphyrin-peptide conjugates are increased. Mild photodermatoses; clinical picture otherwise similar to that of acute intermittent porphyria. Treatment as in acute intermittent porphyria. Rare, except in white South Africans, where frequency is 0.4%. Autosomal dominant.

		Porphyria cutanea tarda. Severe form is called hepatoerythropoietic porphyria.	*Uroporphyrinogen decarboxylase* (EC 4.1.1.37). The enzyme is unstable. Increased fecal excretion of uroporphyrin, porphyrins possessing seven carboxylic acid groups, coproporphyrin and isocoproporphyrin. Photodermatosis. Hepatic siderosis. Risk of primary hepatocellular carcinoma greatly increased. Treat-

Condition	Deficient enzyme and clinical findings
	ment by phlebotomy. Autosomal dominant. The condition may also be induced by exposure to halogenated aromatic hydrocarbons (e.g. polychlorinated biphenyls, hexachlorobenzene).
Erythropoietic protoporphyria	*Ferrochelatase* (EC 4.99.1.1). Activity greatly decreased in both bone marrow and liver. High levels of protoporphyrin in erythrocytes, normoblasts and feces. Mild photodermatosis. Erythema and pruritis. Mild edema. Usually cholelithiasis. Sometimes liver failure. Treatment by protection from UV light, and oral administration of β-carotene. Autosomal dominant.
Porphobilinogen synthase deficiency, Acute congenital porphyria.	*Porphobilinogen synthase (5-Amino-levulinate dehydrase)* (EC 4.2.1.24). Enzyme activity in erythrocytes is about 1% of normal. High urinary excretion of 5-aminolevulinate and coproporphyrin III. Porphyrin excretion in feces is normal. Neuropathic symptoms like those of acute intermittent porphyria. Autosomal recessive (?).
Hereditary coproporphyria	*Coproporphyrinogen: oxygen oxidoreductase (decarboxylating) (Coproporphyrinogen III decarboxylase)* (EC 1.3.3.3). High urinary and fecal levels of coproporphyrin III. High porphyrin content in liver. Harderoporphyrin found in feces in one variant. Resembles mild form of acute intermittent porphyria, and is often symptomless. Treatment as for acute intermittent porphyria. Autosomal dominant.

Indican: a glucoside composed of indoxyl and one molecule of glucose, m. p. 57 °C, $[\alpha]_D^{15}$-66 °C (in water). I. is found in *Indigofera* species, for example the indigo plant *(Indigofera tinctoria),* dyer's woad *(Isatis tinctoria)* and several other higher plants. I. is the precursor of natural Indigo (see).

Indicator methods: see Methods of biochemistry.

Indicaxanthin: a yellow pigment from the prickly pear *(Opuntia ficusindica)* belonging to the group of betaxanthines. I. differs from betanin by the substitution of the cyclodopa residue by L-proline. I. is the best known representative of the betaxanthines.

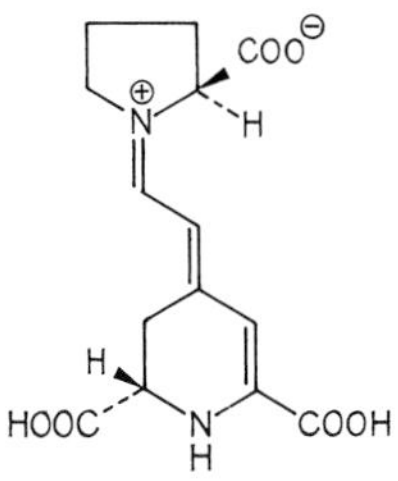

Indicaxanthin

Indigo: a dark blue vat dye, known in ancient times and particularly valued in the middle ages. I. was formerly extracted from a few *Indigofera* species, like *Indigofera tinctoria* and dyer's woad *(Isatis tinctoria).* The latter contains the colorless ester, indoxyl-5-oxofuranogluconate, while *Indigofera* contains the colorless indoxyl-β-D-glucoside (indican). When plants are damaged both of these compounds

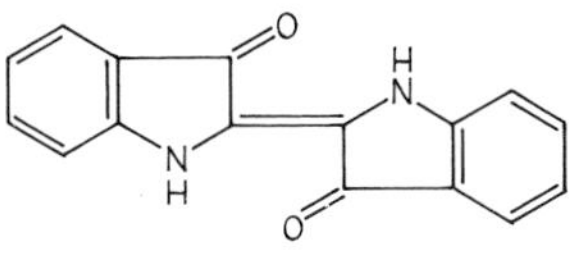

Indigo

become enzymatically hydrolysed to the corresponding sugars and indoxyl (3-hydroxyindole). Indoxyl is transformed by oxidation in the air to I. Natural I. contains up to 95% I. and the accompanying pigments indirubin and indigo brown. Today I. is produced synthetically. It is fast against light, acids and bases, and is particularly suitable for dyeing cotton and wool. I. was formerly the most important organic dye, but now halogenated derivatives and other indigoid dyes have become more important. The structure of I. was determined in 1883. [E. Epstein et al. *Nature* **216** (1967) 547–549]

Indole alkaloids: one of the largest groups of alkaloids, with more than 600 representatives. Biogenetically, almost all I. a. are derived from tryptophan; the majority also contains a monoterpene component (iridoid I. a.), which is responsible for the large variety of forms. It is advantageous to classify the I. a. according to their sources (Table), because many plant genera are characterized by particular structural forms of the alkaloids.
The cinchona alkaloids are also included in the table because they are derived from indole precursors, although their main alkaloids have a quinoline

Table. *Classification of the indole alkaloids*

Structural type	Typical representative
Carboline	Harman alkaloids
Pyrrolidinoindole	Physostigmine
Ergoline	Ergot alkaloids
Iridoid indole alkaloids	Rauwolfia, Curare (Calabash curare), Vinca Strychnos, Cinchona alkaloids

skeleton. Most of the I.a. have pronounced pharmacological properties.

Aside from the I.a., there are many natural indole compounds which are not alkaloids, e.g. the melanins, betalaines, sporidesmins and gliotoxin.

β-[Indolyl-(3)] compounds: see Auxins.

Induced fit: see Cooperativity model.

Inducer: a chemical compound which stimulates the synthesis of enzymes (see Enzyme induction). The I. may be the substrate of the enzyme or another effector, such as a hormone.

Induction: see Enzyme induction.

Infectious nucleic acids: viral nucleic acid which has entered the host cell to replicate and to have its genetic information translated by the host cell's synthetic apparatus into viral products (see Viruses; Phage development).

Informofers: protein particles which are generated when the RNA component of nucleoprotein particles (containing mRNA or its precursor) is cleaved off (see Messenger RNA).

Informosomes: according to Spirin, nucleoprotein particles containing mRNA which can be isolated from the cytoplasm of eukaryotic cells. They can be separated from ribosomes and polyribosomes by density-gradient centrifugation. I. are thought to be transport particles for mRNA. They pick up the mRNA formed in the cell nucleus, probably at the nuclear membrane, and make it available in the cytoplasm for the formation of polyribosomes. (See Messenger RNA).

INH: see Nicotinic acid.

Inhibin: a glycoprotein, M_r 31000, consisting of 2 subunits (α or A, M_r 18-20000; β or B, M_r 14-15000) cross-linked by one or more disulfide bridges. I. is secreted by the gonads, and is a specific and potent inhibitor of FSH secretion by the pituitary. It is a key hormone in the control of folliculogenesis and spermatogenesis (via its effect on secretion of follicle-stimulating hormone). It may be the missing link in the mechanism controlling differential secretion of pituitary gonadotropins, and it is of interest as a potential contraceptive. I. is present in seminal plasma and follicular fluid, and the latter has proved to be the most successful source for purification purposes.

Each subunit of I. represents the C-terminal region of a larger precursor, from which it is released by proteolytic processing. The cDNAs (derived from porcine and bovine ovarial mRNA) encoding the biosynthetic precursors of the α and β subunits have been cloned and analysed. The α and β subunits show considerable homology, suggesting their origin from distantly related genes; this is also a feature of the α and β subunits of pituitary and placental Glycoprotein hormones (see). The β subunit shows a high degree of homology with transforming growth factor β, indicating an evolutionary relationship between their genes.

Two proteins from human seminal plasma have been called α-I. (92 amino acids) and β-I. (94 amino acids); they are not of gonadal origin. α-I. is formed in the prostate, and β-I. in the seminal vesicles. They are structurally unrelated to the dimeric molecule described above that is generally accepted as I. [A.J. Mason et al., *Nature* **318** 659-663; R.G.

Forage et al., *Proc. Natl. Acad. Sci.* **83** (1986) 3091-3095].

Inhibitor peptides: low molecular weight oligopeptide-fatty acid compounds of microbial origin which irreversibly inactivate plant and animal proteases. The inhibition is stoichiometric, i.e. 1 molecule I.p. inhibits 1 molecule enzyme. The presently best known I.p. are as follows: *Leupeptin* from *Streptomyces* species, chemically acetyl-(or propionyl-)L-Leu-L-Leu-arginal, where L-leucine can also be replaced by L-isovaline or L-valine. Leupeptin inhibits cathepsin B (cathepsins are intracellular proteases), papain, trypsin, plasmin and cathepsin D, the effectiveness of the inhibition decreasing in that order. *Pepstatin*, from actinomycetes, chemically isovaleryl-L-Val-L-Val-β-hydroxy-γ-NH$_2$-ε-CH$_3$-heptanoyl-L-Ala-β-hydroxy-γ-NH$_2$-ε-CH$_3$-heptanoic acid, inhibits pepsin and cathepsin D. *Chymostatin*, structure unknown, inhibits all known chymotrypsin types, cathepsin A, B, and D and papain. *Antipain*, structure unknown, inhibits papain, trypsin and plasmin.

Inhibitor proteins: for the most part, low molecular weight, resistant proteins with compact structures and a lack of species specificity. I.p. inhibit reversibly or irreversibly either, at the molecular level, certain anabolic or catabolic metabolic processes, or, at the cellular level, growth and maturation processes of normal and malignant cells. The best known group of I.p. are the protease inhibitors, a group of protease-insensitive, disulfide-rich polypeptides widely distributed in the animal and plant kingdoms. They are especially frequent in the nutritional protein (egg white) of many eggs and in plant seeds. A large number of animal proteinase inhibitors are secretory proteins, like the trypsin inhibitor of the mammalian pancreas, blood and seminal plasma, milk, colostrum, salivary and snail mucus. The M_r of most proteinase inhibitors lies between 5000 and 25000. However, the carbohydrate-containing I.p. of human blood plasma, which belong to the group of α-globulins, have much higher M_r, for example α$_1$-antitrypsin (M_r 54000), α$_1$-antichymotrypsin (M_r 68000), inter-α-trypsin inhibitor (M_r 160000) and α-macrogloblin (M_r 820000). The proteinase inhibitors form temporary or permanent inactive enzyme-inhibitor complexes with the proteases of the digestive tract, the blood-clotting system and the blood pressure regulatory system and with cell and tissue proteases. There are also I.p. for other enzymes, including a special inhibitor of cytochrome oxidase which has been characterized. Another group of I.p., including the toxalbumins and bacterial toxins, inhibit protein synthesis. In contrast, repressor proteins, antitumor proteins and chalones block the synthesis or function of DNA or RNA, which manifests itself in the case of chalones in inhibition of mitosis. The interferons inhibit virus growth by inducing a new RNA coding for an antivirus protein. Other surface-active I.p., such as snake venom toxins, some bacterial toxins and lectins block certain receptor proteins on the synapses or the erythrocytes, producing severe disruption of the nervous system or cell agglutination. Examples of I.p. with narrowly limited effects are the apotransferrins, which inhibit microbial growth, troponin I.p., which inhibits the actomyosin-ATPase, and the peptones, which inhibit blood clotting.

Inhibitors, *antagonists:* substances which slow down or prevent chemical or biochemical reactions from occurring. A competitive inhibition is often involved, in which the inhibitor competes with the Effector (see) for the active site (receptor) according to the law of mass action. Many I. are known which specifically inhibit particular enzyme reactions. The Antibiotics (see) are a special class of I. In plants, native I. (see Growth regulators) interact in a complicated way with the phytohormones to regulate the growth and development. In addition, many synthetic I. are known which, as Retardants (see) and herbicides, are of great practical importance. Other representatives of the chemically widely varying plant I. are cinnamic acid and its derivatives, e.g. Coumarin (see) and Scopoletin (see), Heliangin (see) and phenols.

Initial rate technique: a graphic method to determine the initial reaction rate from the reaction kinetic curves. The slopes of the tangents passing through the origin $t=0$ for the reaction curves at various substrate concentrations are the initial rates at those concentrations.

Initiation: the beginning of synthesis of biopolymers; see Deoxyribonucleic acid; Ribonucleic acid; Protein biosynthesis.

Initiation codon, *start codon:* a sequence of three nucleotides in mRNA which is recognized by the anticodon of formylmethionyl-tRNA (prokaryotes) or methionyl-tRNA (eukaryotes) and thus serves as the start signal for polypeptide synthesis. The I.c. has the sequence 5'-AUG and is apparently localized in a sterically favorable position of the mRNA. See Genetic code.

Initiation factors: catalytic proteins required for the initiation of RNA synthesis (see Ribonucleic acids) and of Protein synthesis (see). At least three structurally and functionally different IF have been identified in bacterial protein synthesis: IF 1 with M_r 8000, IF 2 with M_r 75000 and IF 3 with M_r 30000. They appear to be loosely associated with the ribosomes, but can be dissociated from them with 0.5M NH$_4$Cl or KCl or isolated from the cytoplasm after homogenization. There are at least six to eight IF in eukaryotes, none of which can functionally substitute for the bacterial IF. There is thus strict class specificity in the initiation of protein synthesis.

An IF is also required for the specific initiation of RNA synthesis at the promoter. In prokaryotes this is called the sigma (σ) factor, and has a M_r of 90000.

Initiation tRNA, *starter tRNA, formylmethionyl-tRNA,* abb. *F-Met-tRNA$_F$:* a tRNA specific for methionine which differs in its primary and tertiary structure from the tRNA (Met-tRNA$_M$) which supplies the methionine incorporated into the middle of a polypeptide chain. The Met is formylated after esterification to tRNA$_F$ by a formyltransferase which transfers the formyl group from formyltetrahydrofolic acid. This enzyme is apparently absent from the cytoplasm of eukaryotes, so the Initiation tRNA for protein synthesis on 80S ribosomes is Met-tRNA$_F$. F-Met or Met is removed from most proteins during their synthesis (see Post-translational modification of proteins).

Initiator complex: see Ribonucleic acids; Protein biosynthesis.

Ino: abb. for inosine.

Inosine, abb. *Ino, hypoxanthinosine:* hypoxanthine riboside, 9β-D-ribofuranosylhypoxanthine, a β-glycosidic nucleoside of D-ribose and hypoxanthine.
$M_r = 268.23$, m.p. 218 °C (d.), $[\alpha]_D^{25} -73.6$ ($c=2.5$, 0.01 N NaOH). It occurs free, especially in meat and yeast, and is formed by dephosphorylation of inosine phosphates. It fulfills a specific function as a component of the anticodon of certain tRNA (see Rare nucleic acid bases).

R = H   Hypoxanthine

R= HOCH$_2$ Inosine

R= Inosinic acid (IMP) Inosine 5'-monophosphate

Structure of hypoxanthine, inosine and inosinic acid

Inosine phosphates: nucleotides, phosphoric acid esters of inosine. 1. *Inosine 5'-monophosphate,* abb. *IMP:* inosinic acid, hypoxanthine riboside 5'-phosphoric acid (formula, see Inosine) is made up of the purine base hypoxanthine, D-ribose and phosphoric acid. M_r 348.22, $[\alpha]_D^{25} -36.8$ ° ($c=0.87$, 0.1 M HCl). IMP is the precursor of all the other purines. It is the first intact purine compound in the course of Purine biosynthesis (see) and all other purine nucleotides arise from it. Together with guanylic acid, IMP serves as a flavoring principle and it is isolated for this purpose either from meat extract or from hydrolysates of yeast nucleic acids, or it is produced on a large scale by mutants of certain microorganisms, for example *Corynebacterium glutamicum*. 2. *Inosine 5'-triphosphate,* abb. *ITP:* M_r 508.19, can, as an energy-rich phosphate, replace ATP in certain metabolic reactions (carboxylations). It is formed by phosphorylation of IMP via inosine 5'-diphosphate, abb. IDP, M_r 428.2.

3. *Cyclic inosine 3,5'-monophosphate,* abb. *cyclo-IMP, cIMP:* a structural analog of cyclic adenosine 3',5'-monophosphate (see Adenosine phosphates) which, like the adenosine derivative, inhibits the growth of certain transplantable tumors.

Inorganic bulk elements: see Mineral nutrients.

Inosinic acid: see Inosine phosphates.

Inositol: see Cyclitols.

Inositol phosphates: phosphate esters of the cyclic alcohol *myo*-inositol. They are found in the plasma membrane as components of phospholipids, the phosphatidyl inositols. Inositol 1,4,5-trisphosphate (InsP₃) is released from phosphatidylinositol-4,5-bisphosphate (PtdInsP₂) in the plasma membrane by a variety of membrane receptors for hormones or neurotransmitters when the receptors are occupied. InsP₃ appears to be the active species responsible for the release of Ca^{2+} which follows stimulation of the cell; it is thus a second messenger (see Hormones, Calcium) for such agents as acetylcholine, vasopressin, substance P and epidermal growth factor (see Peptide hormones).

InsP₃ is soon dephosphorylated, in three steps, to *myo-inositol* (Fig.). The last of these reactions is catalysed by *myo*-inositol 1-phosphatase (EC 3.1.3.25), which is inhibited by lithium. The synthesis of *myo*-inositol de novo yields *myo*-inositol-1-P, rather than free *myo*-inositol, so that inhibition of this enzyme interrupts the cycle shown in the figure, and prevents formation of phosphatidyl inositol. It is likely that this effect of lithium is responsible for its pharmaceutical effects in manic depressive illness.

Phosphatidyl inositol is far more abundant than PtdIns4P or PtdIns4,5P, but the three species are in equilibrium, so that PtdIns4,5P removed from the membrane by cleavage is rapidly replaced. The diacylglycerol released from PtdIns4,5P by cleavage is, like InsP₃, a second messenger. Among other things, it activates the membrane C kinase, which activates a number of other proteins by phosphorylation (see Calcium). Interestingly, the diacylglycerol portion of PtdIns4,5P tends to have arachidonic acid in its 2 position; when released from the diacylglycerol by hydrolysis, the arachidonic acid may also serve as a cell activator.

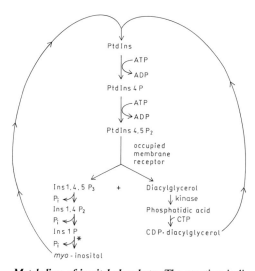

Metabolism of inositol phosphates. The reaction indicated by an asterisk is inhibited by lithium.
[M.J. Berridge *Biochem. J.* **220** (1984) 345–360; D.H. Carney et al. *Cell* **42** (1985) 479–488]

Thus the I.P., in the form of phosphatide esters, play an essential role in the activation of many types of cell. Cleavage of PtdIns4,5P releases at least three different second messengers: InsP₃, which in turn generates a surge in intracellular Ca^{2+}, diacylglycerol and arachidonic acid.

Insect attraction substances: see Pheromones.

Insect hormones: substances responsible for directing the life cycle of insects, mostly of low molecular weight. Three hormones are involved in the direction of post-embryonic development of insects 1. activation hormone (brain hormone, adenotropic factor), 2. molting hormone (see Ecdysone), and 3. juvenile hormone. Each molt is induced by the activation hormone, a polypeptide produced in the brain. Its effect is an increase in the RNA synthesis in the molting glands. This is accompanied by the secretion of a molting hormone from the prothoracic gland. The molting hormone has the structure of a steroid or its biogenetic precursor. The type of molt is decided by the juvenile hormone, a sesquiterpene from the corpora allata. High concentrations of this hormone produce a larval molt, low concentrations a pupal molt. An imaginal molt occurs when the juvenile hormone is acting alone.

Ecdysone or close structural relatives are widespread in the plant kingdom, often in high concentrations (phytoecdysones). For example, 15 kg fresh yew leaves yield 300 mg ecdysterone, an amount which could only be obtained from 6000 kg insect pupae.

Aside from the natural compounds, a large number of synthetic analogs have high activities in biological tests, especially juvenile hormone activity. It is hoped that deleterious insects can be combatted with such environmentally non-damaging biological substances. Some of these substances are already in use.

Insecticyanin: a blue biliprotein from the hemolymph and integument of the tobacco hornworm (*Manduca sexta*). With the yellow color of carotenoids, which are also present, I. is responsible for the green color of the insect larvae. I. is produced only by the larvae, but persists throughout pupation into the hemolymph of the adult female, and is then sequestered in the egg. It is a tetramer of identical subunits, each of M_r 21378, containing 189 amino acid residues of known sequence, and 2 disulfide bridges. The chromophore of I. is biliverdin IX, which is tightly bound and is only removable under denaturing conditions. High resolution X-ray crystallographic studies have been reported. [H.M. Holden et al. *J. Biol. Chem.* **261** (1986) 4212–4218]

Insertion: see Mutation.

Insulin: a polypeptide hormone, M_r 5780 (bovine), synthesized in, and secreted by the B cells of the islets of Langerhans. The first protein primary sequence ever to be elucidated was that of I. (Fig. 1). [F. Sanger et al., *Biochem. J.* **59** (1955) 509–518] I. is the only hormone that decreases the blood glucose concentration. It affects the entire intermediary metabolism, especially of the liver, adipose tissue and muscle. I. increases the permeability of cells to monosaccharides, amino acids and fatty acids, and it accelerates glycolysis, the pentose phosphate cyle, and, in the liver, glycogen synthesis. It promotes the biosynthesis of fatty acids and proteins. These indirect effects on various enzymes and metabolic processes are listed in the tables. However, the mechanism of I. action following binding of the hormone to the Insulin receptor (see) is not yet understood.

Most species possess only one type of I., but three rodents (laboratory rat, mouse, spiny mouse) and two fish (tuna, toadfish) have two distinct hormones. Accordingly, the rat possesses two I. genes, whereas only one is present in humans. Rat and human I. genes have been sequenced. [G. I. Bell et al., *Nature* 284 (1980) 26-32] The mature mRNA transcript encodes preproinsulin, which carries an amino-terminal hydrophobic, 16-residue sequence, called the signal sequence (see Signal hypothesis) (Figs. 2 and 3). As preproinsulin traverses the membrane into the lumen of the endoplasmic reticulum, the signal sequence is removed, leaving proinsulin. Proinsulin is transported to the Golgi, where proteolysis starts to remove an internal sequence known as connecting peptide (C-peptide). Proteolysis continues in storage granules; in addition to other structural homologies, C-peptide from different species contains Arg-Arg at its amino end, and Lys-Arg at its carboxyl end, representing proteolytic cleavage sites for attack by trypsin-like enzymes. Disulfide bonds are formed immediately after translation, and are present in proinsulin. For the amino acid sequence of proinsulin, see Insulin-like growth factor. Fusion of the membranes of the mature storage granules with the plasma membrane of the cell results in release (secretion) of I. Secretion of I., which occurs in response to respirable metabolites (e.g. glucose) and certain hormones (e.g. acetylcholine), is triggered by an increase in the concentration of free cytosolic Ca^{2+}, caused by an increased entry of Ca^{2+} into the cell via voltage-dependent Ca^{2+} channels, and by the mobilization of intracellular pools of Ca^{2+}. Inositol triphosphate may serve as a messenger to release Ca^{2+} from internal stores (see Inositol phosphates) and there is now considerable evidence that this event may be universal in the stimulation of various secretory cells. [S. K. Joseph et al., *J. Biol. Chem.* 259 (1984) 12952-12955; C. B. Wollheim & G. W. G. Sharp, *Physiol. Rev.* 61 (1981) 914-973]

An important physiological stimulus for I. secretion is a high concentration of blood glucose, e.g. the hyperglycemia following a meal promotes I. secretion. Other primary physiological stimuli of I. secretion are mannose, leucine, arginine, lysine, short chain fatty acids, long chain fatty acids, acetoacetate, β-hydroxybutyrate. Secondary physiological stimuli (do not promote I. release directly, but alter response to primary stimuli) include glucagon, secretin, pancreozymin, gastrin, acetylcholine, prostaglandins E_1 and E_2. I. secretion is inhibited by somatostatin, adrenalin and noradrenalin. A distinction must be drawn between biosynthesis and secretion of I. Thus, glucose concentrations higher than 2-4mM stimulate I. synthesis, while concentrations higher than 4-6mM are required for stimulation of I. secretion. I. biosynthesis is also promoted by mannose, dihydroxyacetone, glyceraldehyde, leucine, N-acetylglucosamine, α-ketoisocaproate, glucagon, methylxanthines. Adrenalin inhibits I. biosynthesis, whereas galactose has no effect. The sulfonylureas, which are used pharmacologically to stimulate I. secretion in type 2 diabetes (defective B-cell secretory response) do not stimulate I. biosynthesis.

I. is determined radioimmunologically; the normal concentration of circulating I. in human blood is 1 ng/ml.

I. is rapidly removed from the circulation and degraded. The major site of degradation (inactivation) is the liver, but most peripheral tissues contain specific insulin-degrading enzymes. Such enzymes are localized in the cells, so I. must be internalized before destruction. This occurs by receptor-mediated endocytosis, after which I. dissociates from the receptor and is degraded, while the receptor is probably recycled to the plasma membrane. A soluble I. protease is capable of attacking both the I. molecule and its separated A and B chains. In addition, the cell contains an I.-glutathione transhydrogenase, associated with intracellular membranes, which cleaves the disulfide bonds of I., producing separate A and B chains. I. induces synthesis of the degrading enzymes; the I. degrading activity of the liver decreases in starvation, and increases again on refeeding. Excessive degradation of I. may be one cause or contributing factor of diabetes mellitus (see). Two major types of this disease are recognized clinically:

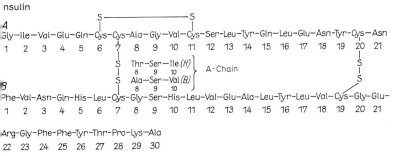

Figure 1. Primary structure of sheep insulin. Human (H) and bovine (B) insulin differ from sheep insulin in the sequence region A8 to 10; in addition, in human insulin, the C-terminal alanine of the B-chain is replaced by threonine.

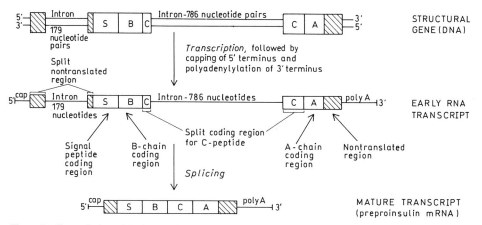

Figure 2. Transcription of the human insulin structural gene, and processing of the early RNA transcript. All processes shown here occur in the nuclei of the B-cells of the islets of Langerhans. The mature RNA transcript then passes through pores in the nuclear membrane and enters the cytoplasm, where it is translated.

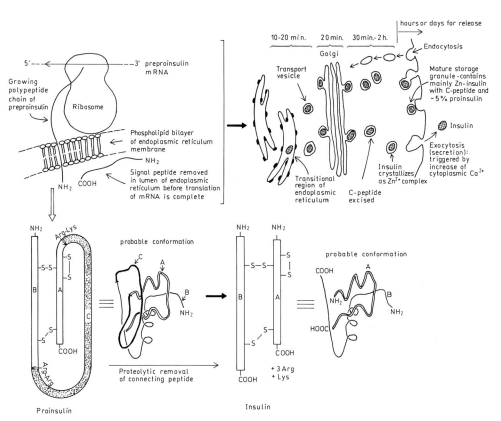

Figure 3. Post-translational modification (processing) of proinsulin in the lumen of the endoplasmic reticulum, Golgi apparatus and transport vesicles. These processes occur in the B-cells of the islets of Langerhans. Secretion of insulin occurs by exocytosis. Membranes may be recycled by endocytosis.

Type 1 or I.-dependent diabetes (IDD) is caused by a lack of functional B cells, resulting in I. insufficiency. IDD classically appears during childhood or adolescence, but can appear at any age; the patients are totally dependent on exogenous I. for survival, and are prone to ketosis. Type 2 or non-I.-dependent diabetes (NIDD) is due to a relative I. deficiency, often related to defective secretion, although synthesis of I. may be adequate. NIDD is more characteristic of middle age and older, and is also known as maturity onset diabetes; exogenous I. is not required, and control by diet alone is often possible. [W. Montague, *Diabetes and the Endocrine Pancreas – A Biochemical Approach* (Croom Helm, London, 1983); M. P. Czech, ed., *Molecular Basis of Insulin Action* (Plenum Press, New York, 1985)]

Insulin-like growth factor, *IGF, non-suppressible insulin-like activity, NSILA:* a family of polypeptides present in vertebrate blood that share considerable structural and functional similarity with insulin, but are distinguishable from it by a lack of immunological cross reactivity. Insulin has long-term growth effects, but is more remarkable for its potent short-term action, whereas IGFs show potent long-term effects. Two IGFs (IGF–I and IGF–II) are present in human serum; they possess growth-promoting activity on chick embryo fibroblasts at a concentration of 10^{-9}M.

IGF–I is also called somatomedin C. IGF–II is similar to Multiplication stimulating activity (see) of rat.

The structures of human IGF–I and IGF–II and proinsulin show obvious homologies (Fig.) The number of sequence differences between the IGFs and proinsulin suggest that duplication of the common ancestral gene occurred before the appearance of the vertebrates. The three-dimensional structures of insulin and the IGFs are probably similar, since all half-Cys and Gly residues and most of the non-polar core residues of the insulin monomer are conserved.

IGF–I receptor and Insulin receptor (see) are immunologically, structurally and functionally related. IGF–I receptor is a disulfide-linked glycoprotein heterotetramer, M_r about 400000, which can be resolved into α- and β-subunits, M_r 130000 and 98000, respectively. It binds IGF–I with high affinity, IGF–II, multiplication stimulating activity and insulin with lower affinity; it displays Protein-tyrosine kinase (see) activity. IGF–II receptor is a single chain glycoprotein, M_r about 250000, with a high affinity for IGF–II and multiplication stimulating activity, and no affinity for IGF–I or insulin. Insulin can cause a reversible desensitization of the induction of tyrosine aminotransferase by IGF–I or IGF–II, possibly by affecting a post-binding step of IGF action, which may be common to both insulin and IGF. There is evidence that Ca^{2+} has a role in the binding and subsequent cellular processing of IGF–II in pancreatic acini. [C. Thibault et al., *J. Biol. Chem.* **259** (1984) 3361–3367; K.-T. Yu &

Effects of insulin on phosphorylation and activity of various enzymes and proteins. The effects are indirect, i.e. I. binds to its receptor on the cell membrane, initiating a chain of unknown events, culminating in the phosphorylation (kinase action) or dephosphorylation (phosphatase action) of the enzyme.

Enzymes of carbohydrate metabolism	Activity	Phospho-rylation
Fructose 6-phosphate 2-kinase	↑	↓
Glycogen synthase	↑	↓
Phosphorylase	↓	↓
Phosphorylase kinase	↓	↓
Phosphoprotein phosphatase inhibitor 1	↓	↓
Pyruvate dehydrogenase	↑	↓
Pyruvate kinase	↑	↓
Enzymes of lipid metabolism		
Acetyl CoA carboxylase	↑	↑
ATP-citrate lyase	No change	↑
Diacylglycerol acyltransferase	↑	↓
Glycerol phosphate acyltransferase	↑	↑
Hydroxymethylglutaryl CoA reductase	↑	↓
Hydroxymethylglutaryl CoA reductase kinase	↓	↓
Triacylglycerol lipase	↓	↓
Other		
Insulin receptor (β-subunit)	?	↑
Ribosomal protein S6 (in 40S subunit)	?	↑
Cyclic AMP phosphodiesterase (low K_{III} type)	↑	↑
Ca-ATPase of plasma membrane	↓	↓ ?
Na/K-ATPase of plasma membrane	↑ ?	↑

↑ = phosphorylation or increase in activity
↓ = dephosphorylation or decrease in activity

Effects of insulin and other hormones on carbohydrate and lipid metabolism in muscle, adipose tissue and liver. I. stimulates amino acid uptake and protein synthesis in all three tissues.

Tissue	Process	Insulin	Adrenalin	Glucagon
Muscle	Glucose uptake	↑	↑	–
	Glycolysis	↑	↑	–
	Glycogenolysis	↓	↑	–
	Glycogen synthesis	↑	↓	–
Adipose tissue	Glucose uptake	↑	↓	(↓)
	Lipogenesis	↑	↓	(↓)
	Lipolysis	↓	↑	(↑)
Liver	Fatty acid synthesis	↑	↓	↓
	Fatty acid oxidation	↓	↑	↑
	Gluconeogenesis	↓	(–)	↑
	Glycogenolysis	↓	↑	↑
	Glycogen synthesis	↑	↓	↓
	Ketone body formation (ketogenesis)	↓	(–)	↑

↑ = increase in activity; ↓ = decrease in activity

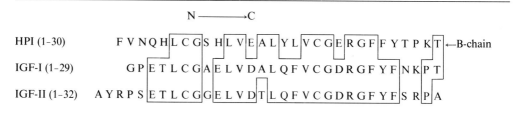

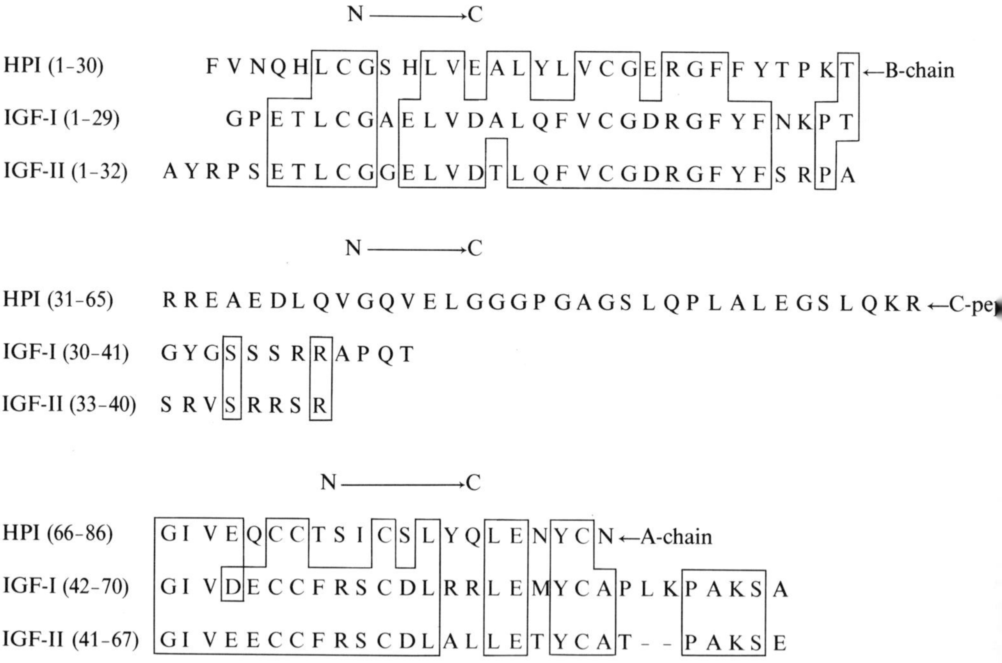

N ——————→C

HPI (1–30)   F V N Q H L C G S H L V E A L Y L V C G E R G F F Y T P K T ←B-chain

IGF-I (1–29)   G P E T L C G A E L V D A L Q F V C G D R G F Y F N K P T

IGF-II (1–32) A Y R P S E T L C G G E L V D T L Q F V C G D R G F Y F S R P A

N ——————→C

HPI (31–65)   R R E A E D L Q V G Q V E L G G G P G A G S L Q P L A L E G S L Q K R ←C-pe

IGF-I (30–41)   G Y G S S S R R A P Q T

IGF-II (33–40) S R V S R R S R

N ——————→C

HPI (66–86)   G I V E Q C C T S I C S L Y Q L E N Y C N ←A-chain

IGF-I (42–70)   G I V D E C C F R S C D L R R L E M Y C A P L K P A K S A

IGF-II (41–67) G I V E E C C F R S C D L A L L E T Y C A T - - P A K S E

Comparison of the primary structures of IGFs and human proinsulin (HPI). For the one letter code, see Amino acids.

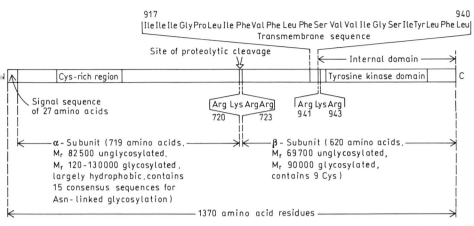

Figure 1. Structure of the precursor polypeptide of the insulin receptor. A sequence of basic amino acids (in this case $Arg_{941}Lys_{942}Arg_{943}$) at the junction of the transmembrane sequence and the cytoplasmic domain is a common feature in the structure of transmembrane proteins. It is thought that they interact with the polar groups of the phospholipids on the membrane surface.

M.P. Czech, *J. Biol. chem.* **259** (1984) 3090–3095; J.Mössner et al., *J. Biol. Chem.* **259** (1984) 12350–12356]

Insulin receptor: The short-term metabolic effects and the long-term growth-promoting activity of insulin are initiated by its binding to specific, high-affinity cell surface receptors. The I.r. is an integral membrane glycoprotein, consisting of 2 α and 2 β subunits linked by disulfide bonds. This heterotetrameric structure is derived from a single polypeptide precursor (Fig. 1), the structure of which was deduced from a cDNA clone. Removal of the signal sequence, partial glycosylation, folding of the polypeptide, and formation of the disulfide linkages (destined to link the α and β subunits) occur in the endoplasmic reticulum. Further glycosylation and proteolytic cleavage into α and β subunits occur in the Golgi apparatus, followed by transport to the plasma membrane. Analysis of the binding of ^{125}I-labelled insulin to the I.r. shows a curvilinear Scatchard plot and a binding constant of 1 nM. Mild reductive cleavage (dithiothreitol) of Triton X-100-solubilized, affinity-labelled I.r. (M_r about 440000) produces identical dimers (M_r about 220000) by cleavage of disulfide bonds between α-subunits. Complete reduction produces separate α and β subunits (M_r about 120000 and 90000, respectively) with high activity of ^{125}I-insulin attached to the α-subunit. Thus each α-subunit binds a molecule of insulin externally to the membrane surface. Only the β-subunits traverse the membrane (Fig. 2). Insulin binding causes the appearance of tyrosine kinase activity (interpreted as a large increase in the V_{max} of an existing active center) in the intracellular domain of the β-subunit. This insulin-dependent kinase catalyses phosphorylation by ATP of Tyr residues in the β-subunit itself, as well as in other peptides and proteins (see Protein-tyrosine kinase). Serine residues of the I.r. also become phosphorylated, but the serine kinase responsible is not intrinsic to the I.r. The significance of Tyr and Ser phosphorylation in the function of the I.r. is not known.

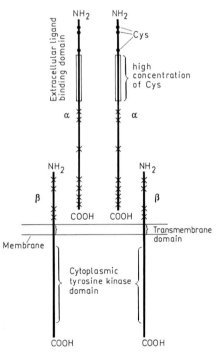

Figure 2. Proposed organization of the heterotetrameric insulin receptor complex. Single Cys residues that may be involved in disulfide linkages between subunits are represented by X.

There are many similarities between the β-subunit of the I.r., the receptors for other growth factors, and certain oncogene products. Thus tyrosine kinase activity is also present in Epidermal growth factor receptor (see), which shows considerable sequence homology with the β-subunit of the I.r., and also possesses similar domains (extracellular, transmembrane, cytoplasmic). Homologies of amino acid sequence also exist between the cytoplasmic domain of the I.r. β-subunit and products of a family of viral oncogenes, commonly known as the *src* family, which possess tyrosine kinase activity. Despite this structural similarity, no oncogene counterpart of the I.r. has yet been identified. Similarly, the receptors for platelet-derived growth factor and for insulin-like growth factor are structurally related to the I.r. β-subunit, and they possess tyrosine kinase activity, but transforming proteins derived from them by mutation have not been found. In contrast, part of the epidermal growth factor receptor is known to be a proto-oncogene product. [M.P. Czech, *Recent Progress in Hormone Research* **40** (1984) 347-377; A. Ullrich et al. *Nature* **313** (1985) 756-761]

Integrated rate equation: an equation which represents the concentration of the substrate or product as a function of time. The corresponding plots are called the progress curves, which can also be obtained by direct measurement (see Enzyme kinetics). By integrating the Michaelis-Menten equation, one obtains for example, with $(S_o-S) = P$, the integrated velocity equation $P(t) = V_m t + K_m \ln S/S_o$, where P is the product concentration, S the substrate concentration, S_o the substrate concentration at $t = 0$, and V_m and K_m are the maximal velocity and Michaelis constant, respectively.

Intercalation: a special interaction between dye molecules (e.g. proflavin) and DNA in which the dye molecule inserts itself between two neighboring base pairs of the DNA double helix. This stretching of the DNA can lead to errors in the transcription of translation of the DNA, causing either a mutation or a defective mRNA, and thus the wrong amino acid sequence of a protein.

Interconversion: in biochemistry in general, the changing of one intermediary metabolic product into another; in enzymology, in particular, the transformation of the active form of an enzyme into the inactive form, and vice versa, as a possibility for physiological regulation of metabolism.

Interconversion of enzymes: see Covalent modification of enzymes.

Interconvertible enzymes: see Covalent modification of enzymes.

Interferons: species-specific proteins which induce antiviral and antiproliferative responses in cells. I. are a major defense against viral infection and neoplasms. They are produced by cells in response to penetration by viral (or synthetic) nucleic acid, and leave the infected cell to confer resistance on other cells of the organism.

a-I.: are a group of peptides produced by macrophages. They have antiviral activity; and recent data suggests that exposure to different viruses causes the macrophages to produce different I. with different antiviral specificity. [*Nature* **305** (1983) 319-320.]

Gamma-I. (immune I.) is produced by T-cells which have been stimulated by interleukin-2 (see Lymphokines). It enhances the cytotoxic activity of T-cells, macrophages and natural killer cells, and thus has antiproliferative effects. [*Science* **221** (1983) 1362-1364.] It also increases the production of antibodies in response to antigens administered simultaneously with the α-I., possibly by enhancing the antigen-presenting function of macrophages [M.Nakamura et. al, *Nature* **307** (1984) 381-382.] I has been shown that I. have an amplification system in which I.-treated cells can transfer both antiviral and antiproliferative activity to other cells, possibly through transfer of a secondary messenger RNA. The antiviral activity elicited in target cells depends on the synthesis of antiviral proteins.

Interleukin 3: see Colony-stimulating factors.

Interleukins, Il: growth factors for blood cells and their precursors. See Lymphokines.

Intermediary metabolism: a term from early physiological chemistry, signifying all those metabolic reactions occurring between the uptake of foodstuffs and the formation of excretory products. In modern usage, I.m. is essentially identical with Primary metabolism (see).

Intermediate: any compound of Intermediary metabolism (see). In the narrow sense, an I. is a compound in a metabolic chain of reactions, excluding the starting compound and the endproduct.

Interphase: the phase between two mitoses (G_1 S and G_2 phases) in the Cell cycle (see), in which the chromosomes are completely unwound, as chromatin, in a membrane-bound nucleus.

Interphase nucleus: See Nucleus.

Interstitial cell stimulating hormone: see Luteinizing hormone.

Intervening sequence: see Intron.

Intracellular digestion: the process of digestion within the cell, in which the lysosomal system plays an important role. Macromolecular substances are taken up by pinocytosis and phagocytosis in vacuoles called phagosomes, which merge with primary lysosomes to form digestive vacuoles (secondary lysosomes). I.d. affects exogenous materials brought into the cell from outside, and endogenous cell components. One accordingly speaks of heterophagy and autophagy. Autolysis (see) is also a form of autophagy, although it occurs in dying or dead cells. In living cells, autophagy may be expressed as the I.d. of entire regions of the cytoplasm. Cytoplasmic regions bounded by membranes, which are later digested, are called cytolysomes. The fragments produced by I.d. must be removed from the digestive vacuoles and cytolysomes so that they can be further degraded. I.d. has been most thoroughly investigated in protozoa, fibroblasts in tissue culture and polymorphonuclear leucocytes.

Intrinsic facotr: see Vitamins (Vitamin B_{12}).

Intron: an intervening sequence in a eukaryotic gene. The same term is sometimes applied to the corresponding intervening sequence in the RNA transcript, but strictly speaking this should be called the *intron transcript* (IT). The term, *intervening sequence* (IVS), is also used for both the I. and IT. I. and the coding sequences (*exons*) are transcribed together; the ITs are then removed to produce functional RNA. An I. therefore makes no contribution to the final gene product of the flanking exons. I. have been found in eukaryotic mRNA, tRNA and nuclear rRNA, and mitochondrial mRNA and rRNA. Prokaryotic genes generally do not contain I., and they are less common in yeast genes than in those of "higher" eukaryotes. It might be thought

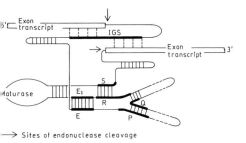

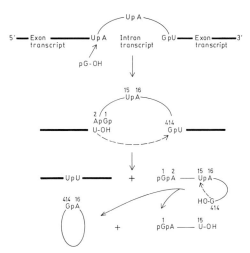

Figure 1. Secondary structure of the intron transcript (IT) of fungal mitochondrial precursor RNA (adapted from R. W. Davies et al., *Nature* **300** (1982) 719–724). Nucleotide sequence analysis of four mitochondrial introns from *Aspergillus nidulans* and five from *Saccharomyces cerevisiae* shows that all the corresponding ITs are able to form the same secondary structure. P,Q,R and S represent highly conserved regions. E and E_1 are not highly conserved, but are always complementary. This secondary structure brings the ends of the two flanking exon transcripts close to one another; precise alignment of the splicing sites is guaranteed by an internal guide sequence (IGS) in the IT. The IGS consists of two regions in tandem, which base pair with the terminal regions of the respective flanking exon transcripts. The maturase loop represents a region that translates for a mRNA maturase, a protein essential for excision; this is, however, quite exceptional and most ITs do not code for any protein. The mRNA maturase may help to stabilize the IT secondary structure against disruption by microsomes, and it may also act as a specificity factor in the splicing reaction. The diagram is not to scale; tertiary structure, not represented here, will also be present.

that the lack of I. is a more primitive condition. However, analysis of the pyruvate kinase genes of the chicken (13 kilobases, 9 or 10 I.) and yeast (1.5 kibobases, no I.) suggest that the gene ancestral to both was first assembled from smaller blocks of DNA, presumably before the divergence of pro- and eukaryotes. The subsequent loss of I. from the yeast and prokaryote genes may be seen as an adaptive response to selection for more rapid reproduction [N. Lonberg & W. Gilbert *Cell* **40** (1985) 81–90]. Genes containing I. are known as *split genes*. The process of removal of an I. from an RNA transcript and the joining of the neighboring exons is known as *splicing*. The base sequence of I. in mRNA begins with 5′GU and finishes with AG3′. These sequences act as recognition sites for *splicing enzymes* The 5′GU...AG3′ pattern is not found in tRNA precursor molecules, suggesting the existence of at least two splicing enzymes, one for mRNA, the other for tRNA. I. are often longer than exons, and I. may account for a greater proportion of a gene than exons, e.g. the gene for ovalbumin contains seven I. and 7700 base pairs, whereas the corresponding spliced mRNA contains 1859 bases.

There are possibly several different mechanisms of splicing, depending on the organism and the type of

Figure 2. Autocatalytic cleavage-ligation of Tetrahymena pre-ribosomal RNA. (Adapted from A. J. Zaug et al., *Nature* **301** (1983) 578–583.) The process is initiated by the insertion of guanosine between the 3′ end of the exon transcript (U) and the 5′ end (A) of the IT. The -UpA- phosphodiester bond is thereby cleaved, and the guanosine becomes attached as the 5′-terminus of the IT, extending the IT from 413 to 414 nucleotides. Guanosine can be replaced by GMP, GDP or GTP. For experimental purposes ^{32}P-labelled GMP is used, thereby conveniently labelling the 5′ end of the IT. The free terminus (-U-OH) of the exon transcript ligates with the 5′-terminal U of the other exon transcript (forming a new -UpU-phosphodiester bond) with concomitant release of the IT at its 3′-terminal. The 3′-terminal G (nucleotide 414) cleaves the phosphodiester bond between nucleotides 15 (U) and 16 (A), forming a cyclic RNA fragment (nucleotides 6–414) and a linear fragment (nucleotides 1–15). Thus two phosphodiester bonds are cleaved and two are formed. Cyclization probably serves to prevent this unwanted RNA fragment from taking part in the reverse reaction. The total process can be described as an autocatalytic cleavage-ligation reaction, and the RNA is known as *self-splicing RNA*. A newly coined term, *ribozyme*, has also been suggested for self-splicing RNA. Removal of ITs by splicing is one type of Post-transcriptional modification of RNA (see).

precursor RNA. A model for the splicing of fungal mitochondrial RNA has been proposed (Fig. 1). Splicing of *Tetrahymena* pre-ribosomal RNA occurs independently of ATP or protein (Fig. 2);

Inulin: high molecular weight vegetable reserve carbohydrate. m. p. 178 °C, $[\alpha]_D = -40°$. I. is a fructan consisting of 20 to 30 fructofuranose units α-glycosidically linked. Probably the reducing end of the chain terminates with glucose. I. is found as a reserve substance in the tubers and roots of many members of the *Compositae*, like dahlia and Jerusalem artichoke tubers. It is used in food for diabetics.

Inulin

Invertase, *β-D-fructofuranosidase*: a saccharide-cleaving hydrolase from yeast, fungi and higher plants. Yeast invertase is a dimeric glycoprotein (M_r 270000), *Neurospora* invertase a tetrameric glycoprotein (M_r 210000, M_r of the subunit, 51000). Aniline (fungal I.), pyridoxal (potato I.) and zinc ions (all I.) are inhibitors of I.

Invertebrate hormones, *hormones of invertebrate animals*: The best studied of these are the insect and crustacean hormones. Many of these have not been chemically characterized, but the structures and biochemical effects of two are known: Ecdysone (see) and Juvenile hormone (see).

Invert sugar: a mixture of equal parts D-glucose and D-fructose which, in contrast to dextrorotatory sucrose, is levorotatory (see Mutarotation). I.s. is generated by acid or enzymatic hydrolysis of sucrose. Because bees have the corresponding invertase, honey is 70 to 80% I.s. The sweet taste is essentially due to the fructose.

In vitro methods: see Methods of biochemistry.
In vivo methods: see Methods of biochemistry.

Ion-exchange chromatography: a liquid-solid phase chromatographic method for analytic and preparative separation and purification of mixtures of substances. I.c. is based on electrostatic binding between cations or anions of the substance to be studied or separated and the corresponding Ion exchanger (see). The components of an electrolyte solution are separated by successive adsorption and desorption through ion exchange. I.c. can be run as a front, replacement or elution process (see Chromatography). Ion exclusion and ion retardation are special techniques of I.c. Ion exclusion can be used to separate ionized from nonionized substances or substances ionized to different extents, e.g. inorganic from organic acids. Ions appear in the eluate sooner than nonions, because the latter penetrate the interstitial and the interior resin phases and thus pass through the column more slowly. Ion retardation is a delayed flow of ions over a mixed bed of cation and anion exchangers. The ions are eluted with water. I.c. can be used to separate closely related ions and compounds, e.g. inorganic ions (particularly rare earth metals), amino acids, alkaloids and optically active compounds from their racemates.

Ion exchangers: all natural and artificial substances, mostly solids, which are able to exchange bound ions for ions from the surrounding liquid medium. The structure of the solid I.e. is not significantly changed in the process. The exchange de-

pends on the properties of the ions involved; in addition, there can also be purely absorptive binding. I.e. are high molecular weight, insoluble polyelectrolytes capable of swelling. There are acidic (solid acids, macropolyacids) and basic (solid bases, macropolybases) types of widely varying chemical nature. The exchange of ions occurs stoichiometrically; accordingly, a chemical equilibrium can be established: $IG_1 + G_2 \rightleftharpoons IG_2 + G_1$. Here I is the ion exchanger, and G_1 and G_2 are the counter ions of the exchanger and the milieu. Depending on the type of ions on the exchanger, it is called a cation exchanger, which has negative functional fixed ions with positive counter-ions, or an anion exchanger, with positively charged fixed ions and negative counter-ions (anions) to equalize the charge. The anchor groups (exchange-active components) in commercially available cation exchangers are usually $C_6H_5O^-$, $-SO_3^-$, COO^-, $-PO_3^{2-}$ or AsO_3^{2-}; those of anion exchangers are usually $-N^+H_3$, $-N^+H_2R$, $-N^+HR_2$, $-N^+R_3$ (R = organic residue). Amphoteric I.e. have acidic and basic fixed ions, and metal-specific ion-exchange resins have specific fixed ions arranged in pairs so they can complex certain metal ions (specific I.e.). I.e. have large surface areas, so that their exchange capacity is usually very high, e.g. 500 m^2/g, exchange capacity 0.5 to 10 mval/g, which can lead to an ion concentration equivalent to that of a 10 N solution.

Types of I.e.: *Synthetic ion exchange resins* consist of a hydrophobic matrix capable of limited swelling. Their properties depend on their chemical type, degree of cross-linking and the number and kind of fixed ions. In the synthesis of artificial resin exchangers, the basic substance is cross-linked by bridge-forming groups, and at the same time, the anchor groups are attached. Synthetic resin exchangers consist of a carbohydrate network with many ionizable groups, either acidic or basic, which are covalently anchored to the basic substance. There are two types of resin, polycondensation and polymerization resins. Polymerization resins consisting of polystyrene cross-linked with divinylbenzene are most common. The Dowex resins and the various Wofatit resins are of this type. The chemical and physical properties of the exchange resin are changed by modification of the matrix. The polymerization resins are classified according to particle size and the degree of branching (X-number). The ion-exchange equilibrium is more quickly reached with smaller resin particles, because the speed of diffusion of the ions is inversely proportional to the third power of the particle diameter. The degree of cross-linking determines the degree of swelling and porosity of the I.e. and thus affects its molecular sieve effect and the rate of flow through the bed of resin. It also affects the exchange capacity. The degree of cross-linking is defined as the number of intermolecular bonds in the matrix, referred to 1 g I.e. The capacity of an exchanger can be expressed either as the total capacity (total number of fixed ions per gram dry exchanger, expressed in mval/g) and the useful capacity, which is the number of counter-ions which can be exchanged. It depends on several parameters. There are in some cases considerable differences between strong and weak I.e. A strongly acid resin with sulfonate groups as fixed ions is highly ionized, both in the acid and the salt form, so that ion exchange can occur over the entire pH range, while a weakly acidic, carboxyl exchanger is active only at pH 7.

2. *Cellulose ion exchangers* carry active groups near the surface of the hydrophilic network of cellulose which are introduced by substitution or impregnation. They are produced as paper or powder, and have the principal advantage that even large molecules can be exchanged which would not be able to penetrate the network of an artificial resin exchanger. They are therefore of particular importance in protein chemistry. DEAE-cellulose (diethylaminoethylcellulose) is a substituted cellulose, an anion exchanger which contains $-C_2H_4-N(C_2H_5)_2$ groups. Carboxymethylcellulose is a cation exchanger containing $-CH_2-COO^-$ groups.

3. *I.e. based on dextran gels* have active groups which are bound to the glucose residues in the hydrophobic matrix.

4. *Inorganic I.e.* are the silicic permutites, which used to be used for softening water, and zeoliths. The latter are natural crystalline aluminosilicates with a network of SiO_4 and AlO_4 tetrahedra. They are now the most common water-softening I.e., because their interstitial spaces contain not only water, but alkali and alkaline earth metal ions, which can be exchanged.

Applications. I.e. are primarily used for water softening and desalting, for simple ion-exchange processes, e.g. separation of neutral salts and salt-salt exchange, for enrichment, isolation and determination of trace elements, for separation of ions with similar properties (see Ion-exchange chromatography), for catalysis, for example in inversion of sugar. In biochemistry they are widely used for the separation and determination of charged molecules, e.g. nucleotides, proteins, amino acids (see Proteins).

Ion filtration chromatography: see Proteins.

Ionone ring: see Tetraterpenes.

Ionophore: a compound which increases the permeability of membranes to ions. I. act by delocalizing the charge of the ion and shielding it from the

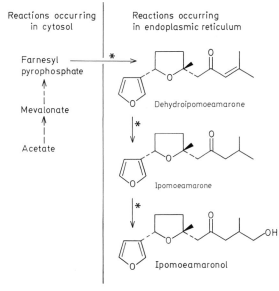

Structure of FCCP and its mode of action as an ionophore in the transport of protons across a membrane

Synthesis of ipomoeamarone and related compounds in root of sweet potato (Ipomoea batatas Lam.) infected with Ceratocystis fimbriata. The conversion of acetate into farnesyl pyrophosphate (see Terpenes) occurs in the uninfected plant, and is stimulated by infection. All the reactions marked with an asterisk are absent before infection. [P. A. Brindle & D. R. Threlfall, *Biochem. Soc. Trans.* **11** (1983) 516–522] hydrophobic region of the lipid bilayer. M_r of I. are typically in the range 500–2000. They possess both hydrophobic (confers lipid solubility) and hydrophilic (binds the ion) regions. There are two types: a) Mobile carriers, which diffuse within the membrane and catalyse the transport of up to 1000 ions per second; they show high specificity for specific ions. b) Channel-formers, which create a channel in the membrane; these are less specific, but may catalyse the transport of up to 10^7 ions per second, i.e. passage of the ion through the membrane may be limited only by the rate of diffusion.

Examples of I. are: Valinomycin (see), a mobile carrier which shows a high specificity for the transport of K^+; it can discriminate between K^+ and Na^+ in the ratio of 10000:1. Gramicidin (see) is a channel-former, with low discrimination between protons, monovalent cations and NH_4^+. Nigericin is a mobile carrier, which loses a proton when it binds a cation; it thus forms a neutral complex which diffuses through the membrane, catalysing an overall electroneutral exchange of K^+ for H^+. Uncoupling agents (see Oxidative phosphorylation) are also I., which are specific for protons, e.g. carbonyl cyanide-p-trifluoromethoxyphenylhydrazone (FCCP) contains an extensive conjugated system with delocalized π electrons; it is therefore lipid-soluble and can also form an anion (Fig.).

Ion pumps: metabolic cycles within cell membranes which can transport ions against the prevailing concentration gradient. The bioelectric membrane potential of the nerves, for example, is based on different distributions of Na^+ and K^+ ions. The high concentration of K^+ in the inside and the predominance of Na^+ on the outside of the nerve produces a normal potential of -60 mV. The energy required to maintain the ionic disequilibrium is provided by the ATP obtained from oxidative phosphorylation. When the nerve is stimulated, there is a change in the permeability of the membrane and a reversal of polarity. Na^+ ions diffuse into the cell, and K^+ ions out of it. After a few milliseconds, the original condition is restored. The Na^+ ions are pumped out of the cell. The cause of the sudden change in permeability is probably structural changes in the membrane proteins induced by released acetylcholine. Nothing is definitely known about the exact mechanism of I.p.

I.P.: abb. for isoelectric point.

IPA: 6-Δ^2-isopentenylaminopurine; 6-(3-methyl-2-butenyl)aminopurine; N^6-γ, γ-dimethylallyladenine; also known as bryokinin (see N^6-[γ,γ-Dimethylallyl]-adenosine). IPA is a Cytokinin (see).

Ipecacuanha alkaloids: iridoid Isoquinoline alkaloids (see). The most important representative of the group is Emetine (see).

Ipomoeamarone: A Phytoalexin (see) with a sesquiterpene structure. I. is formed when sweet potatoes are infected by phytopathogenic microorganisms. Synthesis, see p. 311.

Iridine: see Protamines.

Iridodial: a representative of the Iridoids (see), m. p. 90 to 92 °C, n_D^{19} 1.4782. The dialdehyde form is in equilibrium with the half-acetal form (Fig.). I. was first identified in the defensive secretion of various ants.

Iridodial

Iridoid alkaloids: see Terpene alkaloids.
Iridoid indole alkaloids: see Indole alkaloids.
Iridoid isoquinoline alkaloids: see Isoquinoline alkaloids.

Iridoids: a group of natural products characterized by a methylcyclopentanoic monoterpene skeleton. I. used to be called, incorrectly, pseudoindicans, because some of them can be transformed into intensely blue compounds in the presence of air and acids. This process has not been elucidated, but it is not chemically related to the transformation of indican glycosides into indigo. The name I. is derived from iridodial, which was discovered in the defense secretions of ants. The I. are widespread in higher plants and are often present as glycosides. In addition to the C_{10} I., e.g. verbenalin and loganin, the most important representatives, there are compounds with 9 C-atoms, e.g. aucubin, and 8 C-atoms. In a few cases, the I. have additional structural elements, such as acetate in plumieride. The Valepotriates (see) are a complete subgroup of the I. The secoiridoids, which include bitter substances like gentiopicroside, arise from the I. by cleavage of the methylcyclopentane ring. Some I. can be easily transformed into monoterpene alkaloids, and are therefore possible biogenetic precursors.

The biosynthesis of the I. starts from geranyl pyrophosphate, which is first hydroxylated and then isomerized to neryl pyrophosphate. A number of intermediate reactions (oxidations, reductions, hydroxylations or decarboxylations) lead both to the loganin type of compound (aucubin, usnedoside, actinidine) and to the gentiopicroside type, after cleavage of ring A and recyclization (gentianine, see Gentiana alkaloids).

Iridomyrmecin: a monoterpene lactone, one of the iridoids. M_r 162, m.p. 59 to 60 °C, $[\alpha]_D^{17} + 205$. I. was first isolated from a pheromone mixture from ants of the genus *Iridomyrmex*. Treatment with alkali produces the isomeric isoiridomyrmecine, m.p. 58 to 59 °C, $[\alpha]_D^{17} - 62$ °, which is a pheromone for other ant species.

Iridomyrmecin

Iristectorigenin: 7,5,3'-trihydroxy-6,5'-dimethoxyisoflavone, see Isoflavone.
Iristectorin A: iristectorigenin 7-glucoside, see Isoflavone.

Iron, *Fe:* a bioelement found in all living cells. The human body contains 4 to 5 g, of which 75% is in hemoglobin. The Fe in organisms occurs in the II and III oxidations states; in higher animals it is stored bound to protein. It is transported in the blood as a complex with transferrin (see Siderophilins), from which it is transferred enzymatically to metal-free porphyrin molecules (see Heme iron). Non-heme iron (see) is also found in a number of compounds, for example, Iron-sulfur proteins (see). The Fe metabolism of microorganisms is mediated by a group of natural products called Siderochromes (see).

Fe catalyses most redox reactions in the cell (see Cytochromes; Chlorophyll). It is involved in the reduction of ribonucleotides to deoxyribonucleotides, is a coenzyme for aconitase (EC 4.2.1.3) in the tricarboxylic acid cycle, and is a component of a number of metalloflavoproteins. It has a regulatory role in many microorganisms, for example as an inhibitor of citrate synthesis in *Aspergillus niger,* and as a promoter of antibiotic synthesis by *Streptomyces* species.

Iron metabolism: see Ferritin.
Iron porphyrin: see Heme.
Iron-sulfur proteins, abb. *Fe-S-proteins:* a group of proteins found in all organisms. They contain *iron-sulfur centers* (iron-sulfur clusters) and take part in electron transfer processes. They are involved in H_2 metabolism, nitrogen and carbon dioxide fixation, oxidative and photosynthetic phosphorylation, mitochondrial hydroxylation and nitrite and sulfite reduction. The iron in the active centers is coordinated with the sulfur atoms of cysteine residues. In addition, all Fe-S-proteins except for Rubredoxins (see) contain as many "labile" or inorganic sulfur atoms as iron atoms, and both are covalently bound in the iron-sulfur clusters. Since the iron is not bound in a porphyrin ring, this group of proteins is included in the Non-heme iron proteins (see). It can be subdivided into the *simple Fe-S-proteins* (Rubre-

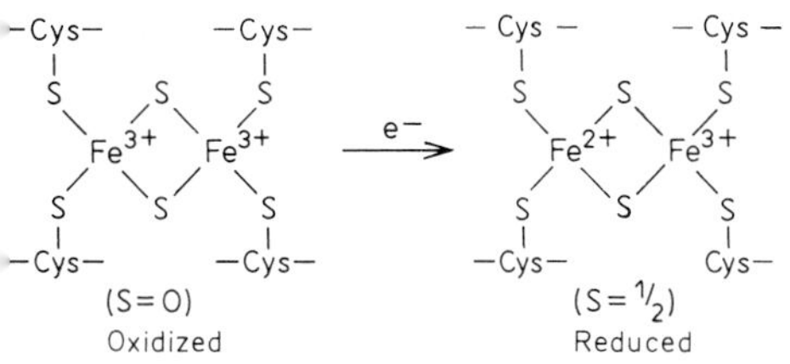

Arrangement of the iron atoms in a 2Fe-2S center

doxins, Ferredoxins [see] and others) and the *conjugated Fe-S-proteins,* like Fe-S-flavoproteins, F-Se-molybdenum proteins, Fe-S-molybdenum flavoproteins, Fe-S-heme proteins, Fe-S-heme flavoproteins, etc. A number of Fe-S-proteins cooperate with other biological electron transport system (cytochromes, flavoproteins and other oxidoreductases). The NADH dehydrogenase complex, which catalyses the following reaction in the Respiratory chain (see):
NADH + H^+ + CoQ → NAD^+ + CoQH$_2$, contains 28 Fe per molecule in seven Fe-S clusters, in addition to FMN.

Succinate dehydrogenase, which contains FAD, contains 8 Fe atoms per molecule, in two 2Fe-2S and one 4Fe-4S centers. Xanthine oxidase contains FAD, Mo and 8 Fe per molecule in two 2Fe-2S clusters.

The ubiquitous occurrence of Fe-S-proteins suggests that they arose early in the evolution of life. The active center of the Fe-S-proteins can be extracted easily by chemical means, and the iron and sulfur can be reintroduced into the apoprotein without loss of biological activity.

The apoproteins always contain at least four cysteine residues per Fe-S cluster. The type of cluster and some of its chemical and biological properties depend on the total length of the amino acid chain, the location of the cysteine residues and the amino acids located between and around the cysteines.

The iron atoms in the clusters can be replaced by ^{57}Fe, which has a nuclear spin, and studied by Mossbauer spectroscopy. The Fe atoms have a nearly tetrahedral coordination, which means they are in the "high spin" state, whether they are oxidized (Fe^{3+}) or reduced (Fe^{2+}). The Fe atoms are close to each other, so that their spins are coupled. When both are in the Fe^{3+} state, the spin coupling produces a nonmagnetic ground state (no ESR signals), but when an electron is added to the cluster it becomes paramagnetic (total spin S = ½). Due to the antiferromagnetic coupling, the g-value of this signal is lower than that of a free electron: 1.96 instead of 2.0023.

Isethionic acid: see Cysteine.

Isocitrate dehydrogenases: enzymes that catalyse the dehydrogenation of the secondary alcohol group of isocitrate, with simultaneous decarboxylation, to form 2-oxoglutarate. The NAD-specific form (E.C. 1.1.1.41) is found only in the mitochondria, where it is an enzyme of the TCA cycle. This form is activated by ADP (animal tissues) or AMP (yeast, molds), and inhibited by ATP; it catalyses the reaction only in the direction of 2-oxoglutarate and does not decarboxylate added oxalosuccinate. The NADP-specific form (E.C. 1.1.1.42) is found in both mitochondria and cytoplasm; it requires magnesium or manganese, also decarboxylates added oxalosuccinate, does not take part in the TCA cycle, and serves to produce reducing power for biosyntheses.

Isocitric acid: HOOC-CH_2-CH(COOH)-CHOH-COOH, a monohydroxy tricarboxylic acid, an isomer of citric acid, which is widely distributed in the plant kingdom and occurs in free form especially in plants of the stone-crop family *(Crassulaceae),* and in fruits. The salts of I.a., isocitrates, are important metabolically as intermediates in the Tricarboxylic acid cycle (see), where they are formed from citrate by the enzyme aconitase, then oxidized to 2-oxoglutarate. In the Glyoxylate cycle (see), isocitrate is cleaved to succinate and glyoxylate.

Isoelectric focussing: see Proteins.

Isoelectric point, abb. *I.P.* or *I.E.P.:* the pH value of a solution at which the net charge on the dissolved ampholyte is zero, i.e. the sum of the cationic charges is equal to the sum of the anionic charges. The I.P. of electrolytes, e.g. amino acids, peptides or proteins, may lie in the range from pH 1 (pepsin) to pH 11.8 (protamine), and is characteristic for each ampholyte. It depends on the ionic strength of the type of buffer used. Certain characteristic properties appear at the I.P., for example a minimal solubility and viscosity. Electrophoretic methods of separation are based on the differences in the I.P. of the individual components. The I.P. can be determined either electrophoretically at various pH values or by electrofocussing on an ampholine pH gradient.

Isoenzymes, *isozymes:* multiple forms of an enzyme with the same substrate specificity, but genetic differences in their primary structures. If there are no differences in primary structure, one speaks of Pseudoisoenzymes (see). I. often differ in their isoelectric points (charge isomers), and sometimes in their M_r (size isomers, e.g. glutaminase I. from *Pseudomonas*, glutamate dehydrogenase I. from *Chlorella*). In oligomeric I., these differences are localized in their subunits. Other differences are found in catalytic properties, for example the K_m values (see Enzyme kinetics), the pH and temperature optima and heat lability, response to effectors, immunological behavior, patterns of distribution in different organs and cell components. I. may be developmental Isoforms (see), or they may perform similar tasks within different tissues of an adult organism. They may be genetically independent products of different genes, e.g. the mitochondrial and cytosolic malate dehydrogenases, or they may arise due to differences in the control sequences in the DNA or RNA which govern transcription or RNA processing (see Post-transcriptional modification of RNA). In the case of proteins which consist of non-identical subunits, I. arise through formation of hybrid forms, e.g. heart and muscle Lactate dehydrogenase (see). I. are also due to genetic variation in the enzyme (alleles), for example human glucose-6-phosphate dehydrogenase, of which more than 50 genetic variants are known. The term I. is also used for enzymes of the same catalytic activity which can be separated by suitable methods, such as electrophoresis, although the reason for their multiplicity is not yet known. Due to their differences in charge, I. can usually be separated by electrophoresis on paper, cellulose acetate sheets, starch, agarose or polyacrylamide, by isoelectric focussing or by ion exchange chromatography. If in addition there are differences in size, the I. can be separated by zonal centrifugation in density gradients (e.g. the isocitrate dehydrogenase I.), or by gel filtration. I. with different binding afinities for inhibitors, e.g. carboanhydratase I., can be separated by affinity chromatography. Likewise, affinity methods employing monoclonal antibodies can be used to separate and identify I.

Isoflavan: a compound containing the isoflavan skeleton (Fig.), which is the most reduced of the Isoflavonoid (see) ring systems. 3 R and 3 S configurations both occur naturally. Where pterocarpans and I. are present in the same plant, both have the same configuration, thus indicating that they are biogenetically related. The first reported natural I. was the animal metabolite, equol (see Isoflavone). All other known natural I. are plant products, e.g. *(-)-duartin* [(3 S)-7,3'-dihydroxy-8,4',2'-trimethoxyisoflavan, wood of *Machaerium* spp.], *(-)-mucronulatol* [(3 S)-7,3'-dihydroxy-4',2'-dimethoxyisoflavan, wood of *Machaerium* spp.], *(+)-vestitol* [(3 S)-7,2'-di-

hydroxy-4'-methoxyisoflavan, wood of *Machaerium vestitum, Dalbergia variabilis*], *(-)-vestitol* [(3 R) configuration, *Cyclolobium claussenii, C. vecchi*], *(+)-laxifloran* [(3 R)-7,4'-dihydroxy-2',3'-dimethoxyisoflavan, *Lonchocarpus laxiflorus*], *(+)-lonchocarpan* [(3R)- 7,4'- dihydroxy- 2',3',6'- trimethoxy- isoflavan, *Lonchocarpus laxiflorus*], **Phaseolin isoflavan** [(3 R)-7,2'-dihydroxy-3',4'-dimethylchromenylisoflavan, *Phaseolus vulgaris*], *(+)-licoricidin* [5,4',2'-trihydroxy-7-methoxy-6,3-di-γ,γ-dimethylallylisoflavan, *Glycyrrhiza glabra*], and (3 S)-2'-hydroxy-7,4'-dimethoxyisoflavan (*Dalbergia ecastophyllum*).

Compared with the corresponding isoflavones and isoflavanones, I. are relatively effective fungicides [Krämer et al. (1984) *Phytochemistry* **23** 2203-2205]. General reference: *The Flavonoids*, J.B. Harborne, T.J. Mabry & H. Mabry, eds. (Chapman and Hall, 1975).

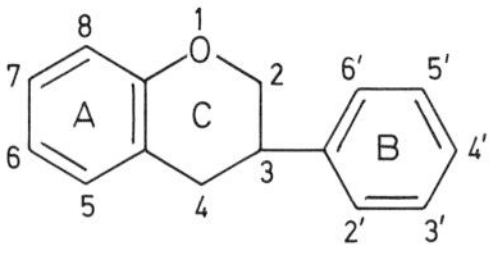

Isoflavanone ring system

Isoflavanone: an isoflavonoid with the ring system shown (Fig.). Relatively few naturally occurring I. are known. Like other Isoflavonoids (see), I. are restricted to various subfamilies of the *Leguminoseae*. Examples: *padmakastein* (5,4'-dihydroxy-7-methoxyisoflavanone, bark of *Prunus puddum*), *ferreirin* and *homoferreirin* (5,7,6'-trihydroxy-4'-methoxy- and 5,7-dihydroxy-4',6'dimethoxyisoflavanone, respectively, from heartwood of *Ferreirea spectabilis*), *sophorol* (7,6'-dihydroxy-3',4'-methylenedioxyisoflavanone, *Maakia amurensis*). [General ref.: *The Flavonoids*, J.B. Harborne, T.J. Mabry & H. Mabry, eds. (Chapman and Hall, 1975)].

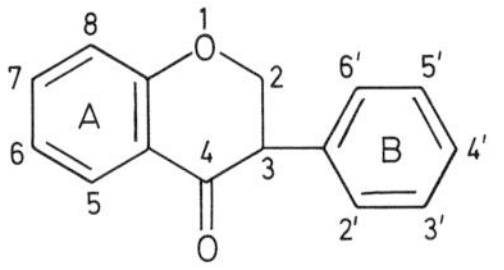

Isoflavanone ring system

Isoflav-3-ene: an Isoflavonoid (see) with the ring system shown (Fig. 1). I. have been known chemically for many years as dehydration products of isoflavonols, and the first naturally occurring I. was reported in 1974. I. are probably precursors of

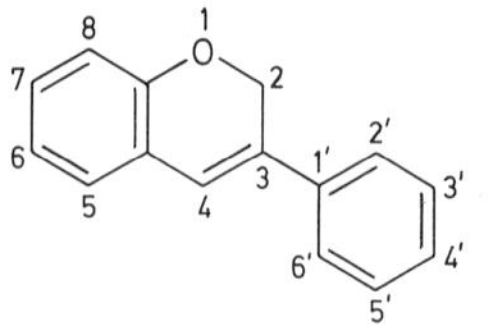

Figure 1. Isoflav-3-ene ring system

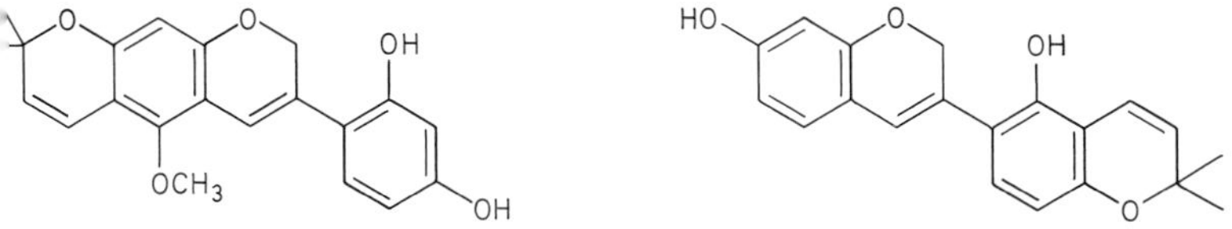

Figure 2. *Neorauflavene* (*Neorautanenia edulis*) **Glabrene** (*Glycyrrhiza glabra*)

oumestans. Examples: **haginin B** and **haginin A** 7,4′-dihydroxy-2′-methoxyisoflav-3-ene and 7,4′-dihydroxy-2′,3′-dimethoxyisoflav-3-ene, respectively, both from *Lespedeza cyrtobotrya*, **sepiol** (7,2′,3′-trihydroxy-4′-methoxyisoflav-3-ene) and **2′-methylsepiol** (both from *Gliricidia sepium*), **neorauflavene** (Fig. 2) and **glabrene** (Fig. 2). [Dewick in *The Flavonoids: Advances in Research*, J.B. Harborne & T.J. Mabry, eds. (Chapman and Hall, 1982)].

Isoflavone: one of the naturally occurring Isoflavonoids (see) based on the isoflavone ring system (Fig.). I. are biosynthesized from the corresponding chalcones by migration of the B-ring from C-2 to C-3 during formation of the oxidized chroman ring system (see Isoflavonoid).

Genistein (5,7,4′-trihydroxyisoflavone) was isolated as early as 1899 from the coloring matter of dyers broom (*Genista tinctora*), then subsequently identified in subterranean clover (*Trifolium subterraneum*), soybean, and the fruits of *Sophora japonica*. Genistein possesses estrogenic activity, and is responsible for infertility in Australian sheep grazing on pas-

tures of subterranean clover, and for the "spring flush" of milk production in dairy cows in Britain. A major metabolite of genistein in animals is equol (Fig.) [Cayen et al. (1964) *Biochim. Biophys. Acta* **86** 56–64]. Slight estrogenic activity (may increase uterine weight when administered to immature mice) is shown by **biochanin A** (5,7-dihydroxy-4′-methoxyisoflavone, *Cicer arietinum, Ferreirea spectabilis, Trifolium* spp.) and **prunetin** (5,4′-dihydroxy-7-methoxyisoflavone, *Prunus puddum, P.avium, Pterocarpus angolensis*). Definitive proof of estrogenic activity is lacking for all other I.

Further examples of I. with simple substituents are **daidzein** (4′,7-dihydroxyisoflavone, *Pueria* spp. and other legumes), **orobol** (3′,4′,5,7-tetrahydroxyisoflavone, roots of *Lathyrus montanus*), **formononetin** (7-hydroxy-4′-methoxyisoflavone, soybean, *Trifolium subterraneum, T. pratense*), **muningin** (6,4′-dihydroxy-5,7-dimethoxyisoflavone, *Pterocarpus angolensis*), **afrormosin** (7-hydroxy-6,4′-dimethoxyisoflavone, *Afromosia elata*), **tlatlancuayin** (5,2′-dimethoxy-6,7-methylenedioxyisoflavone, Mexican "tlatlancu-

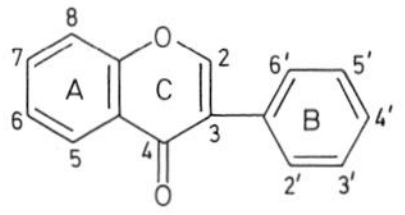

Isoflavone ring system

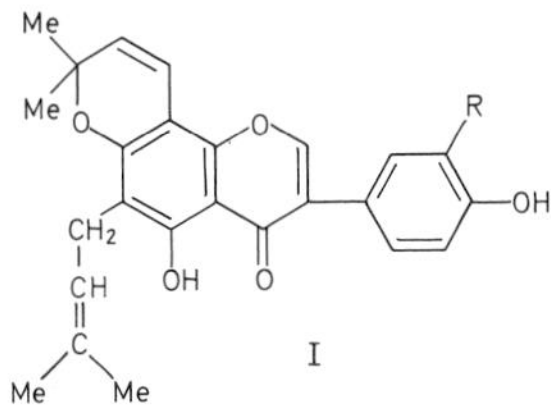

I

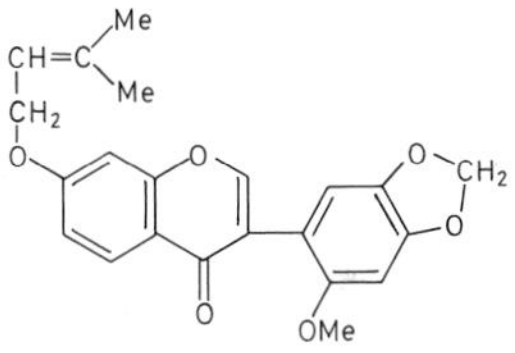

II

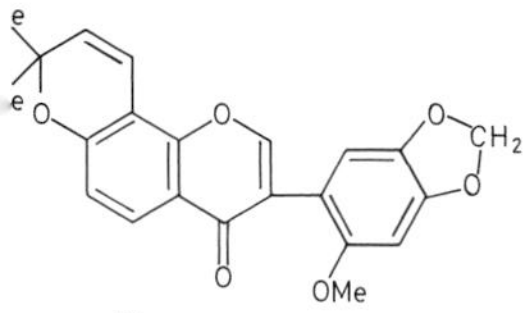

III

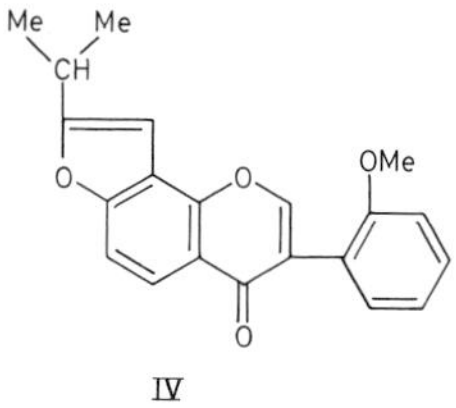

IV

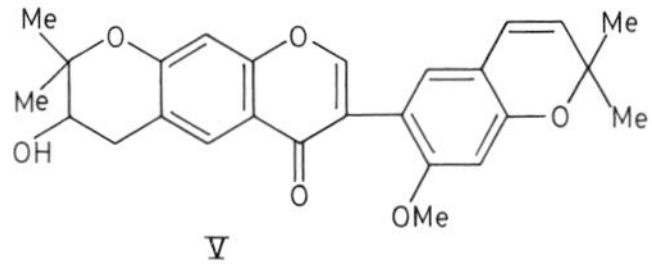

V

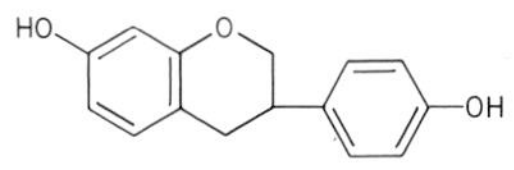

VI

Isoflavones
I: R=H, Osajin; R=OH, Pomiferin (Osaje orange tree, *Maclura pomifera*).
II: Maxima substance C (*Tephrosia maxima*)
III: Jamaicin (*Piscidia erythrina*)
IV: Munetone (*Mundulea suberosa*)
V: Mundulone (*Mundulea sericea*)
VI: Equol (metabolite of genistein in animals).

aya", *Iresine celosioides*). As shown by the foregoing list, I. are largely restricted to the *Papilionoideae*, a subfamily of the *Leguminoseae*. Notable exceptions are the I. of *Iris* spp. (*Iridaceae*): 4′,5,7-trihydroxy-3′,6-dimethoxyisoflavone (*I. germanica*), 7,5,3′-trihydroxy-6,5′-dimethoxyisoflavone (trivial name, *iristectorigenin*, *I. tectorum*, *I. spuria*), 5,7-dihydroxy-6,2′-dimethoxyisoflavone (*I. spuria*) [A. S. Shawl et al. (1984) *Phytochemistry* **23** 2405–2406]. *Orobol* and *pratensein* (5,7,3′-trihydroxy-4′-methoxyisoflavone) have been reported in *Bryum capillare*, where they exist as 7-O-glucosides, and predominantly as 7-(6″-malonyl)glucosides; this is the first report of I. in a bryophyte [S. Anhut et al. (1984) *Phytochemistry* **23** 1073–1075].

Like other flavonoids, I. exist in the plant as glycosides, the most frequent carbohydrates being glucose and rhamnose. Examples are **daidzin** (daidzein 7-glucoside), **ononin** (formononetin 7-glucoside), **genistin** (genistein 7-glucoside), **sophoricoside** (genistein 4′-glucoside), **sophorabioside** (genistein 4′-rhamnoglucoside), **prunetrin** (prunetin 7-glucoside), **iristectorin A** (iristectorigenin 7-glucoside).

The formulae of some complex I. are shown (Fig.).

Isoflavonoid: a flavonoid with the branched $C_6C_3C_6$ skeleton shown (Fig.).

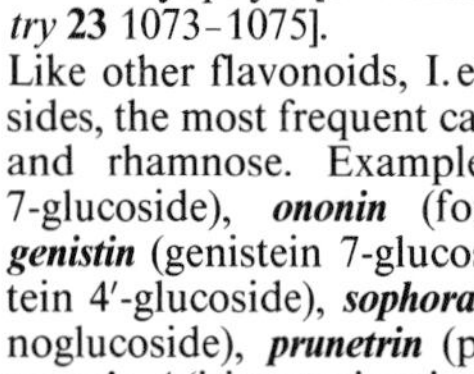

Figure 1. Skeleton of the isoflavonoids

Most I. contain the 3-phenylchroman skeleton, in which the C_3 chain is cyclized with oxygen.

I. include isoflavones, isoflavanones, isoflavans, isoflav-3-enes, rotenoids, pterocarpans, coumestans, 3-aryl-4-hydroxycoumarins, 2′-hydroxy-3-arylcoumarins, 2-arylbenzofurans, α-methyldeoxybenzoins and individual compounds like lisetin and ambanol (see separate entries). I. have a restricted botanical

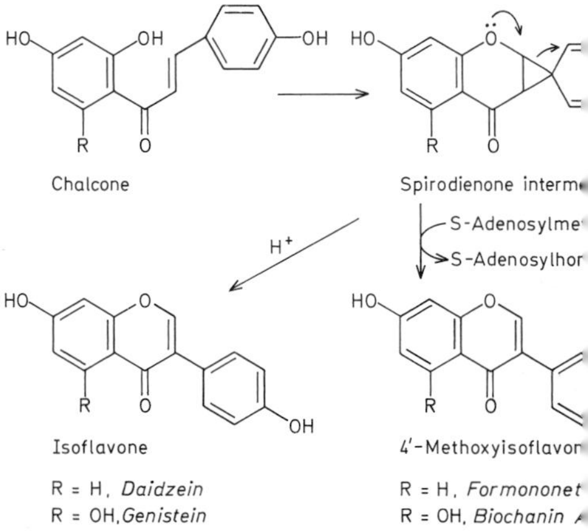

Chalcone
Spirodienone interme

—S-Adenosylme
—S-Adenosylhor

Isoflavone
R = H, *Daidzein*
R = OH, *Genistein*

4′-Methoxyisoflavon
R = H, *Formononet*
R = OH, *Biochanin A*

Figure 2. **Proposed mechanism for the cyclization an** oxidation of chalcones to isoflavones

distribution, occurring chiefly in the subfamily *Papilionoideae* of the *Leguminoseae*, and sometimes i the subfamily *Caesalpinioideae*. Their occasiona presence in other families has also been reporte (*Rosaceae, Moraceae, Amaranthaceae, Podocarp aceae, Chenopodiaceae, Cupressaceae, Iridacea Myristicaceae, Stemonaceae*). I. have also been re ported from microbial cultures (e. g. T. Hazato a *J. Antibiot. Tokyo* **32** (1979) 217–222), but in all case the culture media contained soybean meal, so th possibility remains that these I. are of plant origin.

Biosynthesis. All Flavonoids (see) are biosynthesize via chalcones. The 1,2 aryl migration which leads t the characteristic I. skeleton occurs during conver sion of the chalcone, and is accompanied by net oxi dation. In contrast, biosynthesis of all other flavo noids involves conversion of the chalcone to flavanone of the same molecular formula. An attrac tive theory for the mechanism of aryl migration in volves phenolic oxidation via a spirodienone inter mediate (Fig. 2). [A. Pelter et al. *Phytochemistry* **1** (1971) 835–850] In agreement with this propose mechanism, 4-methoxychalcones do not act as sub

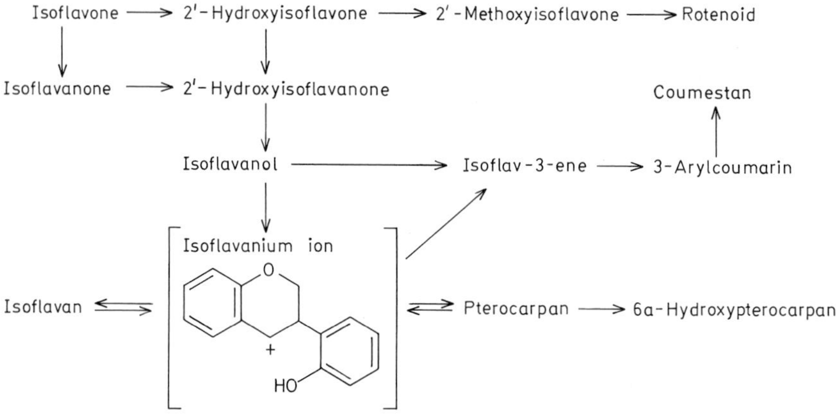

Figure 3. **Biosynthetic relationships of the main groups of isoflavonoids.** For formulae, see individual entries.

trates for aryl migration. In fact, it is possible that only two chalcones act as substrates, i. e. 2′,4′,4-trihydroxychalcone and 2′,4′,6′,4-tetrahydroxychalcone, which would give rise to daidzein and genistein, respectively. Formononetin and biochanin A can arise by methylation of daidzein and genistein, respectively, but there is strong evidence that 4′-methylation occurs mainly during conversion of the chalcone, i. e. during aryl migration. These four I. could then act as precursors of virtually all other known natural I. Although of rare occurrence, I. lacking an oxygen function at 4′ are also known. These may be formed by an analogous phenolic oxidation of 2-hydroxychalcones, leading to 2′-hydroxy (or methoxy) flavones. The B-ring substitution pattern appears to be determined in the isoflavone by hydroxylation sequences of 4′ → 2′,4′and 4′ → 4′,5′ → 2′,4′,5′. A-ring substitution at 7 or 5,7 is determined by the chalcone, but further hydroxylation at 6 is also possible in the isoflavone. The biosynthetic relationships of the main groups of I., determined chiefly by isotopic tracing (14C and 3H), are shown in Fig. 2. [P. M. Dewick in *The Flavonoids: Advances in Research* J.B. Harborne & T.J. Mabry, eds., (Chapman and Hall, 1982) 535-640; J.L. Ingham, "Naturally Occurring Isoflavonoids (1855-1981)" in *Progress in the Chem-*

istry of Organic Natural Products **43** (1983) pp. 1-266]

Isoforms: Different forms of cells or macromolecules which arise and replace each other sequentially during ontogeny. In mammals, for example, a myosin light chain present in skeletal and cardiac tissue is replaced by adult light chains which are specific for muscle type. In addition, the heavy chain of fetal myosin is replaced by a transient form of chain, neonatal heavy chain, which in turn is replaced by adult heavy chain. [A. I. Caplan, M. Y. Fizman, H. M. Eppenberger, (1983) *Science* **221** 921-927.]

28-Isofucosterol: see Avenasterol.

Isohydric principle: see Buffer, section on Buffers of body fluids.

Isoleucine, abb. *Ile:* L-α-amino-β-methylvaleric acid, $CH_3-CH_2-CH(CH_3)-CH(NH_2)-COOH$, an aliphatic, neutral amino acid found in proteins. Ile is found in relatively large amounts in hemoglobin, edestin, casein and serum proteins, and in sugar beet molasses, from which it was first isolated in 1904 by F. Ehrlich. It is an essential dietary amino acid, and is both glucoplastic (degradation via propionic acid) and ketoplastic (formation of acetate) (see Leucine). The biological synthesis of Ile starts with oxobutyric

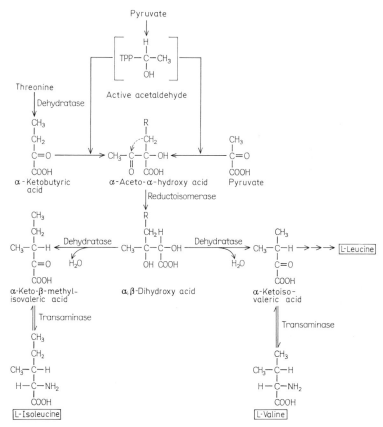

Biosynthesis of the branched-chain amino acids L-isoleucine, L-valine and L-leucine.
TPP = Thiamin pyrophosphate.

acid and pyruvate. Oxobutyrate is synthesized by deamination of L-threonine by threonine dehydratase (threonine deaminase). Ile and L-valine are synthesized by parallel pathways. The individual reaction steps (Fig.) are catalysed by the same enzyme (see Auxotrophic mutants). The biosynthesis of Ile diverges from that of the branched-chain amino acids at the level of the valine precursor oxoisovaleric acid.

Isolysergic acid: see Lysergic acid.

Isomagnolol: see Neolignans.

Isomaltose: a reducing disaccharide, composed of two molecules of D-glucopyranose linked 1,6-glycosidically. I. is formed by the enzymatic degradation of branched polysaccharides, e.g. amylopectin. I. is a stereoisomer of gentiobiose.

Isomerases: see Enzymes, Table 1.

Isonicotinic acid hydrazide: see Nicotinic acid.

Isopelletierine: see Punica alkaloids.

Isopentenylpyrophosphate: an intermediate in the biosynthesis of the Terpenes (see).

Isopentylacetate: $(CH_3)_2CH-CH_2-CH_2-O-CO-CH_3$, the most effective alarm pheromone (see Pheromones) of the honey bee. M_r 130. It is synthesized in the glandular tissue of the sting palps, and is released when the bee stings. Its odor attracts other bees.

Isopeptide bond: a covalent cross-linking bond between the ε-amino group of lysine and the side-chain carboxyl group of glutamate or aspartate, formed by condensation: $H_2N-CH(COOH)-CH_2-CH_2-COOH + H_2N-CH_2-CH_2-CH_2-CH_2-CH(NH_2)-COOH \xrightarrow{H_2O} N^\epsilon-(\gamma\text{-glutamyl})\text{-lysine}$. I.b. has been found in polymerized fibrin and native wool. It is not hydrolysed by the body's own digestive proteases, but only by the bacteria in the large intestine. Its presence in nutritional proteins therefore reduces their nutritional value.

Isoprene: see Terpenes, Hemiterpenes.

Isoprene rule: see Terpenes.

Isoprenoids: see Terpenes.

Isopycnic zone: see Density gradient centrifugation.

Isoquinoline alkaloids: a large group of alkaloids occurring widely in the plant kingdom. The heterocyclic skeleton is usually synthesized in vivo by a Mannich condensation between a phenylethylamine derivate and a carbonyl component. The resulting tetrahydroisoquinoline derivatives are converted by dehydrogenation into isoquinoline derivatives (Fig.). There are various structural types distinguished by the nature of the carbonyl group, which are typical for certain kinds of plants and are therefore named after them (Table). Compounds with more complicated ring structures may arise by sec-

ondary cyclization due to phenol oxidation, and by rearangements and ring cleavages, and these can only be identified as I.a. by their common biosynthetic pathways.

Table. Classification of the isoquinoline alkaloids

Basic structure	Typical examples
Tetrahydroisoquinoline alkaloids	Anhalonium (cactus) alkaloids
Phenylisoquinoline alkaloids	Amarylidae alkaloids
Benzylisoquinoline alkaloids	Poppy alkaloids (Papavera) Erythrina alkaloids Some curare alkaloids
Phenylethylisoquinoline alkaloids	Colchicum alkaloids
Bis-benzylisoquinoline alkaloids	Some curare alkaloids
Iridoid I.	Ipecacuanha alkaloids

Isorubijervine: a *Veratrum* alkaloid of the Jerva trum type. M_r 413.65, m.p. 237 °C, $[\alpha]_D + 6.5°$ (alcohol). I. occurs in yeasts *(Veratrum album, Veratrum eschscholtzii* and *Veratrum viride)* and differs (α-solanine) in having an additional 18-hydroxyl group. In the glycoalkaloid isorubijervosine, from *Veratrum eschscholtzii,* I. is linked to D-glucose.

Isorubijervosine: see Isorubijervine.

Isosteviol: see Stevioside.

Isotope technique, *tracer technique:* the use of radioactive and stable isotopes (more exactly, nuclides) in biological, chemical and physical research and in technology. Since atoms, groups of atoms (functional groups) or molecules are labelled by addition (in the case of elements) or chemical incorporation of indicator atoms, this technique is regarded as one of the indicator methods (see Methods of biochemistry). Both unstable radionuclides which decay with a fixed half life and emit α, β or γ radiation, and stable nuclides may be used as markers. When a stable nuclide is used, it must be added in sufficient amount so that it can be detected above its natural background level, i.e. the rarer isotope, which is to serve as a marker, must be enriched. The term tracer technique for I.t. refers to the fact that one can follow the labelled atom in a mixture with unlabelled atoms of the same element using suitable techniques for detection and measurement. The labelling guides the researcher just as a black ball among a number of white balls of the same size can always be recognized.

Radioactive or stable nuclides of all bioelements are available (Table) for use in biochemistry. There are three main advantages of I.t. for biological research 1. the specific behavior of atoms and molecules can be followed; 2. when isotope effects are excluded exact and very sensitive studies are possible; 3. the normal (physiological) behavior of a biological system is not, as a rule, affected by isotopes.

Phenylethylamine + Aldehyde → Tetrahydro-isoquinoline → Isoquinoline

Biosynthesis of isoquinolines

Radionuclides can be detected and their amounts determined by several methods (Geiger-Müller counter, thin-layer scanner, scintillation counter etc.). The practical unit of measurement is counts

Table. Selected nuclides of the bioelements

Nuclide	Symbol	Half-life	Type of radiation
Hydrogen			
Deuterium)	^2H	Stable	–
Tritium)	^3H	10.46 years	β, very soft
Carbon	^{13}C	Stable	
	^{14}C	5568 years	β, soft
Nitrogen	^{13}N	10.05 min	β
	^{15}N	Stable	
Phosphorus	^{32}P	14.3 days	β
Sulfur	^{35}S	87.1 days	β, soft

per minute (cpm), which can be converted with the help of standards and the known efficiency of the counter into the actual number of disintegrations per minute (dpm). The Curie unit (Ci) is equivalent to 3.7×10^{10} disintegrations per second. For chemical and biochemical purposes, the millicurie (mCi) (2.22×10^9 disintegrations per minute) and the microcurie (μCi) (2.22×10^6 disintegrations per minute) are more convenient units. The Curie and its multiples should now be replaced by the Bequerel (Bq), which is equivalent to one disintegration per second. The megabequerel (10^6 disintegrations per second) is more appropriate for biochemical purposes. Specific radioactivity is the number of disintegrations or impulses per time interval divided by the mass unit, e.g. cpm/mmol, cpm/g, mCi/mol, MBq/mol, etc. The specific incorporation, expressed in per cent (%), is the quotient of the specific radioactivity of the product, multiplied by 100, and the specific radioactivity of the isotopically labelled precursor.

Stable nuclides are most easily determined by mass spectroscopy. For studies of biosynthesis, or precursor-product relationships, it is often not sufficient to determine the specific incorporation, especially if it is relatively low (1% and lower), for the following reasons: an applied labelled compound can enter and be degraded by various metabolic pathways, so that the specific incorporation of the precursor into the metabolic product can only be proven by determination of the site of labelling within the product. This is achieved by chemical structural analysis, which must be developed and tested on a micro scale. If the isotope can be located at a specific position in the product, the possibility of unspecific

metabolic labelling or "smearing" of the isotope is excluded. For example, a compound labelled with ^{14}C could become metabolized to $^{14}CO_2$, which is then incorporated unspecifically by various carboxylation reactions into the reaction product.

Radioactive compounds are obtained in various ways: 1. by synthesis in the laboratory, 2. biosynthetically, using the specific synthetic capacities of organisms, 3. by radiochemical methods, e.g. tritiated compounds by Wilzbach tritiation. The second and third technique often yield unspecific, uniform labelling, but not necessarily an equal degree of labelling of all atoms. For example, in Wilzbach labelling, the C-T bonds are stable, but the O-T and N-T bonds are labile.

I.t. is applied in biochemistry as follows: 1. for the localization of metabolites, enzymes and metabolic reactions in the organism and in the cell, 2. to follow processes of uptake, transport and accumulation, 3. to determine physiological functions, e.g. thyroid activity with the aid of ^{131}I, 4. to measure the turnover of biomolecules and cell components, e.g. protein synthesis.

The combination of I.t. with chromatographic and histochemical or cytochemical techniques represents a powerful experimental technique known as autoradiography. When X-ray film is laid directly over a preparation containing radionuclides, their positions are indicated by darkening of the film. The technique has been extended to electron microscopic radioautography at the supermolecular and molecular level. Elucidation of the dark reactions of photosynthesis by M. Calvin and coworkers relied heavily on paper autoradiography, in which the radioactive products of $^{14}CO_2$ assimilation were separated by paper chromatography, and their positions determined by exposure of the chromatograms to X-ray film.

Isovaleraldehyde: $(CH_3)_2CH-CH_2-CHO$, a naturally occurring aldehyde. $\rho = 0.7977$, m.p. 92.5 °C. I. is a colorless, sharp-smelling liquid which is very reactive and polymerizes easily in the presence of acid. It occurs in many aromatic oils, especially in eucalyptus oils, and is obtained synthetically by oxidation of isoamyl alcohol.

Isovaleric acidemia: see Inborn errors of metabolism.

Isozymes: see Isoenzymes.

ITP: abb. for Inosine 5'-triphosphate.

IUB: abb. for International Union of Biochemistry.

IUPAC: abb. for International Union of Pure and Applied Chemistry.

IVS: abb. for intervening sequence. See Intron.

J

Jamaicin: see Isoflavone.

Jerveratrum alkaloids: see Veratrum alkaloids.

Jervine: a Leveratrum type of Veratrum alkaloid with a C-nor-D-homo structure. M_r 425.62, m.p. 238 °C, $[\alpha]_D$ − 147 °. J. is the main alkaloid of white and green hellebore (*Veratrum album* and *Veratrum viride*).

Juglone: see Naphthoquinones (Table).

Juniperic acid: 16-hydroxypalmitic acid, $HOCH_2-(CH_2)_{14}-COOH$, a fatty acid. M_r 272.42, m.p. 95 °C. J.a. is a typical wax acid in the waxes of many gymnosperms, e.g. juniper *(Juniperus communis)*.

Juvabione, *paper factor:* a monocyclic sesquiterpene ester from the wood of the North American balsam fir *(Abies balsamea)*. (+)-J., an oil, M_r 266,

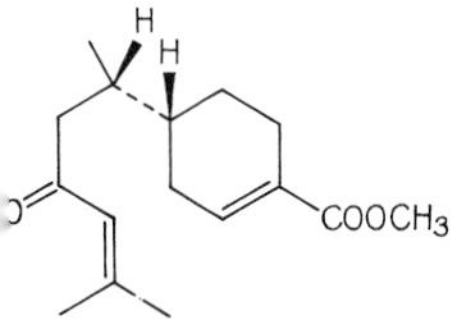

Juvabione

$[\alpha]_D^{20}$ + 79.5 ° (c=3.5, chloroform). J. was the first juvenile hormone isolated from a plant, and its structure was the first elucidated. It is specific for *Pyrrhocoris apterus* and only the dextrorotatory form

is biologically active. The search for J. was begun when it was observed that filter paper made from the wood of the balsam fir contained a factor which produced developmental anomalies in the larvae of *Pyrrhocoris*. The wood of balsam fir from other areas, contains, in addition to J., dehydro-J., M_r 264, $[\alpha]_D^{20}$ + 102.5 ° (c=3.6, chloroform). Both compounds have a steric arrangement of the side chain different from the bisabolane type.

Juvenile hormone, *larval hormone, status-quo hormone:* insect hormone responsible for the control of molting. The first J.h. to be isolated and to have its structure elucidated was obtained from the abdomens of the male silk worm moth *(Hyalophora cecropia)* in 200 μg quantities. Homologs of this compound were discovered later, and other J.h.

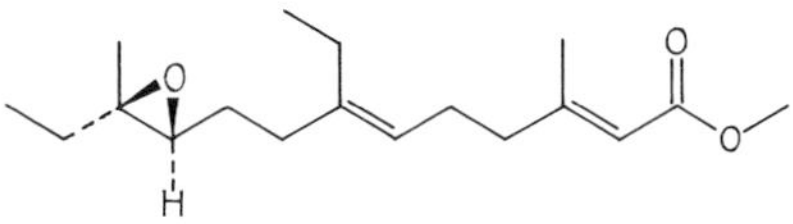

Cecropia juvenile hormone

were postulated. Juvabione (see) and farnesol derivatives are among the natural compounds with J.h. activity. Some synthetic substances are more active than the J.h., and they should be used in increased measure in the future as part of the integrated attack on insects.

Kachirachirol-B: see Neolignans.

Kaempferol: see Flavones (Table).

Kainic acid: an analog of the cyclic form of Glutamic acid (see) extracted from the seaweed *Digenea simplex*. It is an anthelminthic, and has become important as a selective neurotoxin in neurobiology. It destroys neurons, leaving axons and synapses intact. [E. G. McGeer, J. W. Olney & P. L. McGeer, *Kainic Acid as a Tool in Neurobiology* (Raven, New York, 1978)]

Kairomones: see Pheromones.

Kallidin 10: see Bradykinin.

Kallikrein, *kininogenin, kininogenase*: a proteolytic enzyme (EC 3.4.21.8) which preferentially cleaves Arg and Lys peptides. There are at least three types. Plasma K. activates the Hageman factor and kininogen (see Blood coagulation) and releases Bradykinin (see) from kininogen. Pancreatic and submandibular K. release lysyl-bradykinin (kallidin) from kininogen, as well biologically active peptides other than kinins from high molecular weight plasma proteins. Urinary K. releases kallidin from kininogen and is antagonistic to the renin-angiotensin system (see Angiotensin).

Kanamycin: an aminoglucoside antibiotic; see Streptomycin.

Kappa-casein: see Milk proteins.

Kasugamycin: an aminoglucoside antibiotic; see Streptomycin.

Kaurene: ent-kaur-16-ene, a tetracyclic diterpene which is found in the plant kingdom in both the (+) and (−) forms. (−)-K. M_r 272, m.p. 50 °C, $[\alpha]_D^{20} - 75 °$. There are various stereoisomers of K., some of them occurring naturally. The formula and significance of K. as an intermediate in the biosynthesis of the gibberellin phytohormones is given under Gibberellins.

Kava: a narcotic drink made from the roots of the kava plant *(Piper methysticum L.)* in the Pacific islands. K. has a mild pain-killing and euphoric effect (see Narcotics). The active compounds in it have not all been elucidated, but those which have been isolated are α-pyrones, including dihydromethysticin and dihydrokawain.

kDNA: abb. for kinetoplast DNA. See Catenane, Kinetoplast.

Kendall's compounds: see 11-Dehydrocorticosterone, Corticosterone, Cortisone, Cortisol.

Keracyanin: see Cyanidin.

Kerasin: see Glycolipids.

Keratan sulfate: an acidic mucopolysaccharide-composed of N-acetyl-D-glucosamine 6-sulfate and D-galactose, alternatingly β-1,3 and β-1,4 glycosidically linked. K. s. is found in the cornea of the eye, in cartilage, in the aorta and in the intervertebral discs. For formula, see Inborn errors of metabolism (Mucopolysaccharidoses).

Keratins: fibrous intracellular proteins which are components of intermediate filaments (IF) in epithelial tissues, and the proteins of hair, scales, silk, etc. At least three classes of proteins are known to form IF: acidic (Type I) keratins, neutral-basic (Type II) keratins, and vimentin, desmin and glial fibrillary acidic protein. Neurofilament proteins make up at least one other type. Around 30 different K. are known, with M_r 40,000 to 70,000. From amino acid sequence data, it can be seen that although IF subunits are diverse with respect to size and molecular properties, they all contain a central α-helical rod domain of 311–314 amino acids. The secondary structure of this domain is highly conserved; it is made up of repeated heptads (a-b-c-d-e-f-g-)$_n$ of amino acids, where the residues a and d are usually apolar. The heptads coil into an α-helix with a stripe of apolar residues spiralling around the coil; two such helices associate by apposition of the apolar stripes. Assembly of K. IF in vivo and in vitro requires equimolar amounts of Type I and Type II subunits, and it seems probable that each pair of coils has one subunit of each type. The rod domains are 45–48 nm long; values reported for the diameter range from 7 to 15 nm. It seems plausible that the end domains of the polypeptide chains protrude from the dense core formed by the α-helical rod domains. Early evidence suggested that the basic structure of the protein was a left-handed helix of three α-helices twisted around each other, but more recently this has been shown to be a four-chain structure consisting of two double helices. However, this is not necessarily the basic unit of the IF, which has yet to be elucidated.

In contrast to the α-helical structure of the α-K. discussed above, the β-K. have β-pleated sheet structure. The most prominent representative of this class is silk fibroin (M_r 365,000, 2 subunits). Here the chains run antiparallel rather than parallel, and form a zig-zag structure. The formation of hydrogen bonds between the –CH(=O)- and –NH- groups of neighboring chains stabilizes the pleated sheet structure. Together with weak hydrophobic interactions, the hydrogen bonds link pairs of polypeptides into a three-dimensional protein complex. These are additionally stabilized, in silk, by a water-soluble protein, *sericin*. The resultant fiber is very resistant and flexible, but only slightly elastic. The amino acid sequence which repeats over long stretches of the chain is, for silk fibroin, (Gly-Ser-Gly-Ala-Gly-Ala-)$_n$.

Kermesic acid: a bright red insect dye belonging to the group of anthraquinones. m.p. 250 °C (d.). It is structurally closely related to Carminic acid (see); it possesses the identical structure of a tetrahydroxylated methylanthraquinone carboxylic acid, but the *C*-glycosidic glucose on C-2 is absent. K.a. makes up 1 to 2% of kermes, the dried bodies of female scale insects *Kermococcus ilicis*. Kermes is one of the oldest known dyes and was used even in ancient times as a scarlet mordant dye (Venetian scarlet). It was supplanted in the 16th century by cochineal.

Ketoacid, oxoacid: a carboxylic acid which contains the carbonyl group –C=O in addition to the carboxyl group COOH. Depending on the position of the carbonyl with respect to the carboxyl, the acid is referred to as an α, β, or γ-ketoacid (2-, 3- or 4-oxoacid). If the two groups are adjacent, one speaks of an α-K.a. (e.g. pyruvic acid and α-ketoglutaric acid); if one CH$_2$ group is between them, it is a β-K.a. (e.g. acetoacetic acid).

α-Ketoacid dehydrogenase complex: see Multienzyme complex.

Ketoeleostearic acid: see Licanic acid.

Ketogenesis: the formation of ketone bodies. The primary product of K. is acetoacetate. It is synthesized in the liver from acetyl-coenzyme A via acetoacetyl-CoA and β-hydroxy-β-methylglutaryl-CoA. β-Hydroxybutyrate dehydrogenase catalyses the conversion of acetoacetate to β-hydroxybutyrate. The enzyme is located in the mitochondria. Acetone is formed by the spontaneous decarboxylation of acetoacetate (Fig. 1). Acetoacetate is also produced by the degradation of the ketoplastic amino acids leucine, isoleucine, phenylalanine and tyrosine.

In the normal liver, only relatively small amounts of ketone bodies are formed. Their concentration in the blood is 0.5 to 0.8 mg per 100 ml plasma. The acetoacetate produced by this physiological K. is degraded in the peripheral musculature. Coenzyme A from succinyl-CoA is transferred to the acetoacetate by acetoacetate:succinyl-CoA transferase. A direct activation of acetoacetate by coenzyme A and ATP is also known (Fig. 2). The acetoacetyl-CoA produced in either case is thioclastically cleaved into two molecules of acetyl-CoA, consuming a Co-A molecule in the process. In carbohydrate deficiency (starvation, ketonemia in the ruminants), or deficient

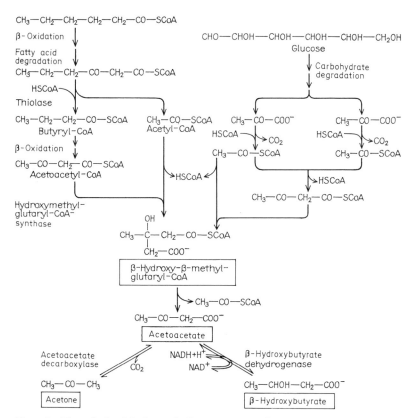

Figure 1. Biosynthesis of the ketone bodies, acetoacetate, β-hydroxybutyrate and acetone

carbohydrate utilization (diabetes mellitus), K. is greatly increased. The cause of this pathological K. is a disturbance of the equilibrium between the degradation of fatty acid to acetyl-CoA and its utilization in the tricarboxylic acid cycle. The several-fold increase in the oxidation of the fatty acids leads under these conditions to an increase in the intracellular acetyl-CoA concentration. The result is the condensation of 2 molecules of acetyl-CoA to acetoacetate via the hydroxymethylglutarate cycle (Fig. 3),

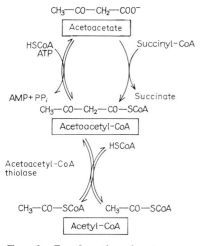

Figure 2. Transformation of acetoacetate to acetylcoenzyme A

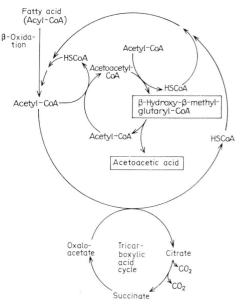

Figure 3. Hydroxymethylglutarate cycle for the regeneration of coenzyme A for the β-oxidation of fatty acids

with the release of coenzyme A. Under pathological conditions, this cycle regenerates the coenzyme A which is required for the β-oxidation of fatty acids. This series of reactions thus takes over the role of the citrate synthase in the tricarboxylic acid cycle.
In an alternative pathway for K., acetoacetyl-CoA which is a normal intermediate in the β-oxidation of the fatty acids, acts as a precursor of acetoacetate. The enzyme thiolase is used in this process. Acetoacetyl-CoA reacts with acetyl-CoA to form hydroxymethylglutaryl-CoA, which is transformed into acetoacetate.

Ketogenic amino acids: see Amino acids.

α-Ketoglutarate dehydrogenase complex: see Multienzyme complex.

α-Ketoglutaric acid, 2-oxoglutaric acid: HOOC-CO-CH$_2$-CH$_2$-COOH, a keto-dicarboxylic acid which represents an important branching point in the tricarboxylic acid cycle. m.p. 113 to 114 °C. α-K.a. is formed as its anion (α-ketoglutarate) by oxidative decarboxylation of isocitrate, by transamination of glutamate in amino acid metabolism, and by degradation of lysine via glutarate and α-hydroxyglutarate. The oxidative decarboxylation of α-ketoglutarate yields succinyl-coenzyme A. Its reductive amination leads to glutamate.

Ketone bodies: organic compounds produced by ketogenesis in the organism. K.b. are acetoacetate and the compounds formed from it, β-hydroxybutyrate and acetone. The increased production of K.b. under certain pathological conditions, e.g. diabetes mellitus, leads to acidosis, because acetoacetate and β-hydroxybutyrate are present as anions, which reduce the concentration of HCO$_3$ in the blood. Other consequences are the excretion of the K.b. by the kidneys, production of acid urine, and damage to the central nervous system.

Ketoses: polyhydroxyketones, a subgroup of monosaccharides (the other is aldoses). The characteristic feature is their non-terminal $-C=O$ group, which is given the lowest possible number in a systematic numbering. The K. can be formally derived from dihydroxyacetone. All known natural K. have the carbonyl group in the 2-position. Depending on the number of carbons in their chains, they are classified as tetruloses, pentuloses, and so forth. They are either indicated by the ending -ulose, e.g. ribulose, or have trivial names, like D-fructose.

Kidney: a paired mammalian organ with excretory and metabolic functions. Human K. are bean-shaped, weigh about 150 g. each, and lie in the posterior part of the abdominal cavity, on either side of the vertebral column. The functional unit of the K. is the nephron, consisting of a glomerular capsule and a long tubule (see Buffer, Fig.3). Each K. contains about a million nephrons. K. are responsible for maintaining the constancy of the internal milieu of the body. About 1.2 liters of blood pass through the two kidneys per minute, and this gives rise to about 120 ml of glomerular ultrafiltrate, i.e. protein-free primary urine, containing a wide range of low molecular weight nonvolatile and inorganic substances, including unmetabolized drugs. Many components of this ultrafiltrate are then passively or actively reabsorbed in the tubule: water (up to 99%), glucose, amino acids and electrolytes. The daily urinary output is about 1.5 liters. As a metabolic organ, the K. is responsible for the constancy of the acid-base balance of the body. The normal pH of the blood is 7.4, and it is important that this value is not increased (Alkalosis, see) or decreased (Acidosis, see). The K. contains amine oxidases and glutaminase, which catalyse the production of free ammonia. In acidosis, these reactions are brought into play and excess H^+ ions are consumed by formation of NH_4^+. In addition to the natural secretory processes, foreign substances, such as pharmaceuticals, can be removed from the blood by active transport. To support the large number and intensity of these active processes, the K. has a high oxygen consumption and high ATP production. K. function is regulated by the nervous system, and by three hormones: vasopressin increases water resorption in the distal part of the tubule by acting on the adenyl cyclase system; the mineralocorticoids, especially aldosterone, promote the resorption of Na; and parathormone favors the excretion of phosphate by the tubule itself.

The K. is the site of synthesis of Erythropoietin (see) and Renin (see). Renin is synthesized in the juxtaglomerular cells; it releases Angiotensin (see), which in turn stimulates the release of aldosterone by the adrenal cortex. 25-Hydroxycholecalciferol (produced in the liver from vitamin D_3 or cholecalciferol) is converted by the kidney into 1,25-dihydroxycholecalciferol, a hormone which promotes calcium uptake by the intestine and calcium mobilization in bone.

Kinases: enzymes which catalyse the transfer of a phosphate residue from ATP to another substrate, especially to the alcoholic hydroxyl groups of monosaccharides. Some important K. are hexokinase, which phosphorylates several hexoses at the C-6 position, glucokinase, which is responsible for glu-

cose 6-phosphate formation in the liver, and phosphofructokinase.

Kinesin: a soluble protein isolated from the axoplasm of squid giant axons; M_r apparently 600,000. In the presence of the non-hydrolysable ATP analog adenylyl imidodiphosphate, it binds tightly to microtubules, but it may be released from them by ATP. It apparently consists of subunits of M_r 110,000, 70,000 and 65,000. Polypeptides from bovine brain with similar binding properties and M_r 120,000 and 62,000 have also been isolated. In the presence of ATP, K. causes axoplasmic organelles to move along microtubules, but in one direction only, from the minus to the plus ends of the tubules. (In nerve axons, microtubules are usually oriented with their minus ends pointing toward the cell body and their plus ends toward the nerve terminal.) It has recently been shown that there is another soluble factor in the axoplasm which induces movement of organelles in the opposite direction; the motion induced by this factor differs from the K.-induced motion in its sensitivity to N-ethylmaleimide and vanadate. [R. D. Vale, T. S. Reese & M. P. Sheetz *Cell* **42** (1985) 39–50; R. D. Vale et al. *Cell* **43** (1985) 623–632]

Kinetic data evaluation: Data evaluation by computer is performed by a nonlinear regression analysis, using an objective procedure, in which the sums of the squares of the differences between calculated and experimental values are minimized (method of least squares). If a reaction follows Michaelis-Menten kinetics, a plot of initial velocity (v) against substrate concentration (S) gives a rectangular hyperbola of the form: $v = V_{max} S/(K_m + S)$. For a first approximation in the laboratory, this equation can be rearranged, forming the basis of several linear transformations for the determination of V_{max} (maximal velocity) and K_m (Michaelis con-

Table. **Linear transformations of the equation** $v = V_{max} S/(K_m + S)$, where v = initial velocity of reaction, V_{max} = maximal velocity when enzyme is saturated with substrate, K_m = Michaelis constant, S = starting substrate concentration. Also included is the Scatchard linear equation for ligand binding, where b = concentration of bound ligand, f = concentration of free ligand, Kd = dissociation constant, b_{max} = maximal concentration of b when ligand is saturated.

Name and No. of illustrated plot (Fig.)	Equation	Ordinate	Abscissa	Ordinate intersected at	Abscissa intersected at	Slope
Lineweaver-Burk, or double reciprocal plot (1.)	$\dfrac{1}{v} = \dfrac{K_m}{V_{max}} \times \dfrac{1}{S} + \dfrac{1}{V_{max}}$	$\dfrac{1}{v}$	$\dfrac{1}{S}$	$\dfrac{1}{V_{max}}$	$\dfrac{-1}{K_m}$	$\dfrac{K_m}{V_{max}}$
Eadie-Hofstee (2.)	$v = -K_m \dfrac{v}{S} + V_{max}$	v	$\dfrac{v}{S}$	V_{max}	$\dfrac{V_{max}}{K_m}$	$-K_m$
Hanes-Wilkinson (3.)	$\dfrac{S}{v} = \dfrac{1}{V_{max}} S + \dfrac{K_m}{V_{max}}$	$\dfrac{S}{v}$	S	$\dfrac{K_m}{V_{max}}$	$-K_m$	$\dfrac{1}{V_{max}}$
Eisenthal-Cornish Bowden, or direct linear plot (4.)	$V_{max} = v + \dfrac{v}{S} K_m$	v	$-S$	lines joining S_1-v_1, S_2-v_2, S_n-v_n; intersection at V_{max} and K_m		
Scatchard (5.)	$\dfrac{b}{f} = \dfrac{1}{K_d}(b_{max} - b)$	$\dfrac{b}{f}$	b	$\dfrac{b_{max}}{K_d}$	b_{max}	$\dfrac{1}{K_d}$

stant). All such graphical forms involve a certain degree of subjectivity.

The Lineweaver-Burk, or double reciprocal plot is widely used, but it is the least reliable of the possible linear transformations; small errors of v for small values of v result in very large errors in $1/v$; these same small errors in large values of v become almost negligible in $1/v$. The authors of this method pointed out the inherent error of high $1/v$ values, and the necessity to apply higher weightings to the low values of $1/v$, but this has often been ignored by subsequent users. It has been suggested that the Lineweaver-Burk plot should be abandoned as a method for the determination of K_m values.

A more satisfactory distribution of error is found for S/v versus S, known as the Hanes plot, or the Hanes-Wilkinson plot (Table & Fig.).

A further linear transformation is represented by v versus v/S, known as the Eadie-Hofstee plot. The error increases with v/S, but since v is a component of both coordinates, the errors vary with respect to the origin rather than the axis, i.e. all error bars converge on the origin (Table & Fig.).

A most satisfactory treatment of kinetic data is the direct linear plot of Eisenthal and Cornish Bowden (1974). Axes are drawn with $-S$ on the abscissa and v on the ordinate, but instead of making the usual hyperbolic plot of S/v (see Enzyme kinetics), corresponding points (each reading of $-S$ and its related v value) are joined by straight lines. The point of intersection of this family of lines gives the values of V_{max} and K_m. Mathematically, this plot corresponds to a rearrangement of the general equation: $V_{max} = v + vK_m/S$, in which V_{max} and K_m appear to be variables. In practice, however, only one value for V_{max} and K_m will satisfy all pairs of v and S. Owing to experimental error, the point of intersection of the lines is not always clearly defined, but the best point is easily found where most of the lines crowd closest together. The advantage lies in the fact that each observation stands alone, and it is revealed as a bad observation if its line does not conform to the majority (Table & Fig.).

The Scatchard plot is frequently used for the determination of binding constants; the other plots (1, 2, 3 and 4) are also appropriate for this purpose. Such

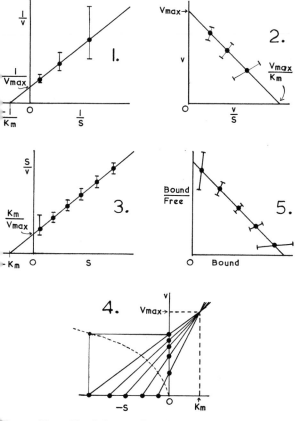

Figure. *Plots, with relative error bars, for the linear transformation of the equation* $v = V_{max} S/K_m + S$. *Also shown is the Scatchard plot (5) for the determination of ligand binding constants.* 1. Lineweaver-Burk, or double reciprocal plot; 2. Eadie-Hofstee plot; 3. Hanes-Wilkinson plot; 4. Eisenthal-Cornish Bowden, or direct linear plot.

measurements are important, e.g. for determination of the strength of association between hormones and cell membranes, between control enzymes and effector molecules, and between steroid hormones and their high affinity receptor proteins in the cells of the target organ. The Scatchard plot is a special case of the Hill equation (see), where $n_H = 1$. The concentration of one binding partner is kept constant; this will usually be a protein, perhaps an organelle or even a cell. The concentration of the smaller ligand is then varied (Table & Fig.).

Kinetic equations: a system of differential equations which describes the changes with time of the concentrations of enzyme species as a function of the reaction rate constants, the concentrations of enzyme species and the concentration variables. The K.e. for the Michaelis-Menten scheme are: $dE/dt = -k_1 S \cdot E + k_{-1} ES$, $dES/dt = k_1 S \cdot E - (k_1 + k_2) ES$ (see Michaelis-Menten equation).

Kinetin: 6-furfurylaminopurine, the model substance for cytokinins. K., in conjunction with other factors, such as auxins, induces renewed cell division in resting plant tissue. It influences the nucleic acid and protein metabolism of the plant. In addition to many other physiological effects, it prevents yellowing of isolated leaves and promotes protein synthesis at its site of application K. is obtained by hydrolysis of deoxyribonucleic acid. The 2-deoxyribose of the DNA provides the furfuryl residue of the K. It can also be made synthetically from 6-mercaptopurine and furfurylamine.

Kinetin

Kinetoplast: a structure at the base of the single flagellum of the trypanosome (a unicellular organism). The K. has a high affinity for basic dyes, and was first observed early this century by light microscopy. Electron microscopy shows that the K. is a disc-like structure in the matrix of the cell's single mitochondrion. The K. has attracted biochemical interest because it contains an unusual form of highly catenated DNA, known as kinetoplast DNA (kDNA). See Catenane.

Kinetoplast DNA: see Catenane.

King-Altman method: a method for derivation of rate equations according to simple rules. These rules come from the application of the determinant theory to the solution of inhomogeneous systems of linear equations. One first draws all possible geometric figures of the enzyme graphs which transform the various enzyme forms (enzyme species) into one another. The number of lines (edges) is 1 less than the number of enzyme forms. Circles and cycles are forbidden and are eliminated. The edges are assigned the appropriate reaction rate constants or the products of rate constants and concentration variables of the corresponding step of the reaction

(e.g. $E \xrightarrow{K_{1S}} ES$) (edge analysis). According to the King-Altman rules, the following distribution equation then holds: Enzyme form/E_t = sum of the products of the edge analysis of all pathways leading to this enzyme form, divided by Σ. Here E_t is the total enzyme concentration, Σ the sum of the numerators of all distribution equations of the enzyme graph. The rate equation is then obtained by multiplication of the product-producing enzyme form, such as EP, with the associated catalytic constant: $v = k_{cat} EP$. Here k_{cat} is the catalytic constant, EP the enzyme product complex. If there are several product-producing enzyme forms, the partial rates are added.

Kinins: see Cytokinins.

Kjeldahl technique: see Proteins.

Klenow fragment: a single polypeptide chain, M_r 76,000, produced by cleavage of *Escherichia coli* DNA polymerase I with subtilisin. It carries the $5' \rightarrow 3'$ polymerase activity and the $3' \rightarrow 5'$ exonuclease activity, but the $5' \rightarrow 3'$ exonuclease activity is lacking. K.f. is now available as a cloned protein. [H. Jacobsen et al. *Eur. J. Biochem.* **45** (1974) 623–627].

Koaburarin: (2 R)-5-hydroxy-7-*O*-β-D-glucosylflavan, see Flavan.

Kornberg enzyme: see DNA polymerase.

Koshland model: see Cooperativity model.

Krabbe's disease: see Inborn errors of metabolism.

Krebs cycle: see Tricarboxylic acid cycle.

Krebs-Henseleit cycle: see Urea cycle.

Krebs-Kornberg cycle: see Glyoxylate cycle.

Kurchi alkaloids: see Holarrhena alkaloids.

Kwashiorkor: a chronic form of malnutrition, occurring chiefly in the second year of life. The name is from the Ga language, and was introduced into modern medicine in 1933 by Cicely Williams [C. D. Williams, *Archs. Dis. Childh.* **8** (1933) 423.] The meaning of the word is "deposed child", i.e. deposed from the breast by the advent of another pregnancy. The earlier literature emphasizes the relationship between K. and various tribal custums and taboos, which result in the administration of a low protein, high-carbohydrate diet to children. In contrast, marasmus is due to a deficiency of both energy and protein. It has now been shown, however, that there are no essential differences in the dietary protein: energy ratio of marasmus and K. victims [C. Gopolan, in R. A. McChance and R. M. Widdowson, eds. *Calorie Deficiencies and Protein Deficiencies,* (Churchill Livingstone (1968) pp. 49–58]. Like all other forms of protein-energy malnutrition, K. is often precipitated and exacerbated by microbial infections and intestinal worms. It is more common in rural than urbanized areas of the developing world. K. victims show retarded growth and anemia. Plasma albumin concentrations fall below 20 or even 10 g/l (in marasmus, plasma albumin is usually about 25 g/l). The resulting decrease in the osmolarity of the blood is thought to be partly responsible for the accumulation of fluid in the body, and the watery, bloated state of the tissues; edema is a common feature of K. The hair becomes sparse, and especially in Negro children it may show patches or streaks ("flag sign") of red, blonde or grey. These hair lesions are probably due to a specific tyrosine deficiency, since the periodic administration of tyrosine to phenylketonurics causes very similar alternating bands of deeply pigmented hair. The skin shows a very characteristic dermatosis, with areas of pigmentation, depigmentation and peeling, the most severely affected areas being the lower limbs and buttocks. Muscles are always wasted, so that walking or crawling may not be possible. Fatty liver is also very characteristic of K.; an average lipid content of 390 g/kg liver has been reported (cf. 35 g/kg in marasmus); this increase is due to triacylglycerols, the level of phospholipids being relatively unaffected. In K., but not in marasmus, plasma levels of triacylglycerols and cholesterol are low. Subcutaneous fat is retained in K., in contrast to marasmus in which it is severely depleted. Changes in plasma amino acids are similar in most types of protein-energy malnutrition. A comprehensive study of plasma amino acid concentrations in K. victims has been reported [Holt et al. *Lancet* 2 (1963) 1343-1348]. In a classical study [Hansen et al. *Lancet* 2 (1956) 911-913], children suffering from K. were given various mixtures of free amino acids with or without vitamin supplementation. A mixture of all essential amino acids effected a complete cure in more than 50% of the cases; the remainder showed a partial or negative response.

Thus, K. possesses certain characteristics that are absent from other types of protein-energy malnutrition. Also, K. is common in hot, humid areas, but not in hot, dry areas. It is especially prevalent in the wet season. Furthermore, it is never found in temperate regions. These observations suggested that other factors in addition to protein deficiency may contribute to K., and that K. is associated with some metabolic peculiarity. K. is now thought to be caused by a combination of malnutrition and afla-toxin poisoning. [R.G. Hendrickse, *British Medical Journal* 285 (1982) 843-846; S.M. Lamplugh & R.G. Hendrickse, *Annals of Tropical Paediatrics* 2 (1982) 101-104.]

Aflatoxin poisoning in animals produces metabolic derangements similar to those in K., e.g. hypoalbuminemia, fatty liver, immunosuppression. Aflatoxins (see) are produced in hot, humid climates by molds growing on foods like millet, groundnuts, cassava, maize, yams, etc. It is well proven that local market foods (especially groundnuts and cooking oil derived therefrom) in hot, humid countries of the developing world contain aflatoxins. Well nourished infants can degrade and excrete these relatively small quantities of toxins, whereas protein deficiency weakens the ability of the liver to destroy these substances. It is thought that a vicious cycle develops, in which the resulting accumulation of aflatoxins causes further liver damage and a marked decrease in the protein synthesizing ability of the liver. Aflatoxin levels in the blood of Sudanese children suffering from K. are on average 10 times higher than those in properly nourished controls, whereas the urinary levels are lower in the K. victims than in controls. Autopsy liver specimens from Nigeria and South Africa show significant levels of aflatoxins in K. livers, whereas livers from marasmus or marasmic K. showed little or no aflatoxins. As a result of aflatoxin damage, K. livers cannot utilize large quantities of amino acids. The sudden and massive protein supplementation, which is sometimes given as a cure for K., may therefore be harmful. See Marasmus, Protein-energy malnutrition.

Kynureninase deficiency: see Inborn errors of metabolism.

Kynurenine: 3-anthraniloylalanine, M_r 208.2. An intermediate in Tryptophan (see) degradation.

L

Labdadienyl pyrophosphate: an intermediate in the biosynthesis of the Diterpenes (see).

Labelling site: see Isotope technique.

Laccase: see Monophenol monooxygenase.

Lac repressor protein: the first repressor substance (product of a repressor gene) to be isolated and characterized. L.r.p. is an acidic (I.P. 5.6) allosteric protein of M_r 152000. It consists of four identical subunits (M_r 38000, 347 amino acids, sequence and spatial structure known). Its primary structure has recognizable similarities to histones or to β-galactosidase. L.r.p. is transcribed from a particular gene in *Escherichia coli* which regulates the synthesis of three enzymes of lactose metabolism (the *lac* region). It inhibits the transcription of the lactose operon (see Operon) by specifically binding to the operator gene. Allolactose, which is formed from the substrate lactose, binds the L.r.p., and causes it to be reased from the operator and inactivated. See Enzyme induction.

Lac system: the region of the genome of *Escherichia coli* and other enterobacteria which controls the ability to utilize lactose and other β-galactosides. It consists of the structural genes LacZ for β-galactosidase, LacY for galactoside permease and LacA for thiogalactoside transacetylase, an operator, a promotor and a regulator gene which is responsible for the synthesis of the Lac repressor (see Repressor).

The enzymes of the L.s. are inducible, i.e. they are synthesized only when β-galactosides are present in the medium (see Enzyme induction).

The Lac operon comprises less than 0.1% of the *Escherichia coli* DNA. The gene LacZ was the first ever to be isolated in pure form, which was accomplished in 1969 by Beckwith. This section of DNA, together with the short operator region (about 27 nucleotide pairs), is about 1.4 μm long and contains about 4000 nucleotide pairs.

α-Lactalbumin: see Milk proteins.

Lactate dehydrogenase, abb. *LDH, lactic acid dehydrogenase:* an oxidoreductase and a much studied isoenzyme. LDH catalyses an NAD- or NADH-dependent side reaction of glycolysis: lactic acid ⇌ pyruvic acid. LDH is absolutely specific for L(+)-lactate; D(−)-lactate is not dehydrogenated. The highest LDH activities are found in heart muscle and liver. LDH (M_r 140000) consists of four subunits of equal size (M_r 35000). There are two different types of subunits, differing in charge, catalytic properties and organ specificity, the heart-muscle (H) and muscle (M) types. The five-fold isomerism of LDH is due to the five possible combinations of the two types of subunit in a tetramer: H_4, H_3M, H_2M_2, HM_3 and M_4 (the latter two in skeletal muscle). Because they have different I.P., the 5 forms

can be separated electrophoretically. The H_4 form has the largest negative charge. The two subunit types differ in their susceptibility to inhibition by pyruvate (H-type is strongly inhibited, M-type weakly), which serves to regulate glycolysis and the tricarboxylic acid cycle. H_4 and H_3M are, in addition, heat-labile, so that they are 100% inactivated after 5 min at 65 °C. In addition to these 5 bird and mammal LDH's, the sperm cells of these animals contain a sixth type of isoenzyme, LDH X. This LDH, which is also tetrameric, contains a third type of polypeptide chain and has a wider substrate specificity than the other LDHs. An LDH of the same size (M_r 140000), but without subunit structure or isoenzyme properties, has been demonstrated in the shrimp *(Artemia salina)*.

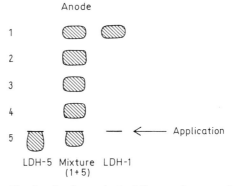

The 5 molecular variants (allomers, isoenzymes) of LDH (1 to 5, middle) which are artifically produced by random recombination of the subunits of the homogeneous tetramers LDH-5 (M-type, left) and LDH-1 (H-type, right). The diagram shows separation by starch gel electrophoresis.

LDH is used for diagnosis of heart infarction and hepatitis, since the LDH level is considerably elevated in both these diseases.

Purified LDH is used in coupled optical tests for the assay of other enzymes, e.g. pyruvate kinase, enolase, transaminases, and for the enzymatic determination of many metabolites, such as ADP, ATP, L-lactate and pyruvate.

Lactic acid: $CH_3-CHOH-COOH$, an aliphatic, optically active hydroxy acid, widely distributed in plants, especially seedlings.

DL-form: m.p. 18 °C, b.p. 119 °C at 12 Torr. L−(+)−form and D−(−)−form: m.p. 25–26 °C. The salt (lactate) of the dextrorotatory L(+)L.a. is the end product of anaerobic Glycolysis (see), and a substrate of Gluconeogenesis (see). Increased degradation of glycogen in contracting muscle can lead to an increase of blood L(+)L.a. from 5 mg% to 100 mg%. In the subsequent rest period, the majority of this L.a. is used for the synthesis of glycogen by the liver. In microbial fermentation, the DL-form of L.a. is produced.

Lactoferrin: see Milk proteins.

Lactoflavin: see Riboflavin.

β-Lactoglobulin: see Milk proteins.

Lactose, *milk sugar:* a reducing disaccharide. M_r 324.3, α-form m.p. 223 °C, $[\alpha]_D^{20}$ +89.4 ° → +55.5 ° (water), β-form m.p. 252 °C, $[\alpha]_D^{20}$ +34.9 ° → 55.3 ° (water). L. crystallizes from water as the β-form above 93 °C, and as α-lactose monohydrate below 93 °C. It consists of galactose β-1,4 glycosidically linked to glucose, both monosaccharide residues in the pyranose form. L. is not fermented by ordinary yeasts, but only by special ones, like kefir. Lactic acid bacteria convert L. to lactic acid; this is the process by which milk sours.

L. is the most important carbohydrate in the milk of all mammals. Human milk contains 6 to 8%, cow's milk 4 to 5%. L. is a component of a few oligosaccharides, and also occurs in plants, e.g. in fruits and pollen. The synthesis of L. in the milk glands of mammals is catalysed by lactose synthetase. The β-form of uridine diphosphate galactose is joined to D-glucose, and in the process the β-glycosidic galactose residue is transferred to the OH on the C-atom 4 of the glucose.

L. is cleaved by β-galactosidase or acid hydrolysis. It is prepared by evaporation of the whey from cheese production; first the lactalbumin and then L. precipitates. L. is used as a starting material in pharmaceutical preparations, and as a carbon source in the culture of microorganisms, e.g. in the production of penicillin.

β-Lactose

Lactose intolerance: see Inborn errors of metabolism.

Lactosyl ceramidosis: see Inborn errors of metabolism.

Lactotropin: see Prolactin.

Lactucerol: see Taraxasterol.

Laki-Lorand factor: see Blood coagulation (Table).

Lamp-brush chromosomes: very large chromosomes, up to 1 mm long, which occur in the meiotic prophase in newts and a few other animals. They form loops along the sides, which give them a brush-like appearance. The loops are not fixed structures. They represent unwound, individual chromomeres, consisting of DNA, protein and RNA. They appear at the sites of active RNA synthesis, thus indicating increased physiological activity in that particular chromosome section (see Gene activation).

Lanatosides: see Digitalis glycosides.

Lanolin, *wool fat, wool wax:* the fatty or more correctly waxy substance secreted by the skin of the sheep, m.p. 36–42 °C. L. constitutes up to 50% of the weight of raw wool. It is a complicated mixture of fatty acids, alcohols, fats and waxy substances. The latter are chiefly esters of steroids (cholesterol and lanosterol) and long chain aliphatic alcohol with higher fatty acids, which are δ-hydroxylated or carry a terminal isopropyl or isobutyl residue. L. is obtained from raw wool by extraction with organic solvents or soap solutions. It has the property of forming water-in-oil suspensions, and is used widely in the pharmaceutical and cosmetics industries (as Adeps Lanae) as an ointment base.

5α-Lanostane: see Lanosterol.

Lanosterol, *kryptosterol:* 5α-lanosta-8(9),24-dien-3β-ol, a tetracyclic triterpene alcohol. M_r 426.7. m.p. 140 °C, $[\alpha]_D$ +60 ° (chloroform). L. is also one of the zoosterols (see Sterols). It occurs in large amounts in the wool fat of sheep. It is derived from the hydrocarbon 5α-lanostane. It is biosynthesized from squalene, via 2,3-epoxysqualene, and is the primary product of this reaction. L. is important in the synthesis of all further tetracyclic triterpenes of the lanostane type and of the steroids.

Lanosterol

Lanthanide ion probes: The trivalent ions of the rare earth metals (lanthanides) can be monitored in biological systems, due to their magnetic and spectroscopic properties. For example, external Eu(III) and Pr(III) shift the ^{1}H-NMR resonance of the –N(CH$_3$)$_3$ head groups of lecithins in the external layer of phospholipid bilayers. This property has been used to monitor the action of local anesthetics on phospholipid bilayers, and to study the transport of lanthanide ions by ionophores. Lanthanide-induced NMR shifts have also been used to determine the conformation of 3',5'-cyclic AMP and other nucleotides in solution.

Lanthanide ions have been used notably as replacement ions for Ca(II), which itself lacks useful physical properties for the study of its behavior in biological systems. The four bound Ca(II) ions of thermolysin can be replaced by three lanthanide ions (two of these Ca(II) ions are only 3.8 Å apart and bridged by three carboxylate groups; with a small rearrangement, a single lanthanide ion is accommodated in place of the two closely paired Ca(II) ions). This exchange causes no significant change in the conformation of the polypeptide backbone, and the bound lanthanide ions serve as X-ray heavy atoms for crystallographic study of the protein. Similarly, the two bound Ca(II) ions of parvalbumin can be replaced by Eu(III) or Tb(III), which then serve as X-ray heavy atoms, or can be investigated by laser-induced luminescence. Lanthanide(III) ions have been employed as X-ray heavy atoms in many other biological macromolecules, e.g. concanavalin A, lysozyme, tRNA, and bacterial ferredoxin, but in these cases it is not established that the lanthanide ion occupies a Ca(II) binding site. Tb(III) and Eu(III) have useful luminescent properties with excited state lifetimes in the convenient range of 100–3000 μs. Sensitized luminescence of protein-bound Tb(III) has been observed in several proteins. The excitation spectrum is diagnostic of the aromatic residues responsible for sensitization (excitation maxima: 259 nm for Phe, 280 nm for Tyr, 295 nm for Trp). Gd(III) has isotropic magnetic properties and a long electronic spin-lattice relaxation time, which makes it ideal as a nuclear relaxation probe. Most of the other lanthanide ions have shorter relaxation times and fairly large magnetic anisotropies, which makes them more suitable as dipolar shift probes. Gd(III) also has a room temperature EPR spectrum, but this property has not been widely exploited. The magnetic circular dichroism of Nd(III) is quite intense, and has been used to determine Nd(III) binding to thermolysin. [W. Horrocks *Adv. Inorg. Biochem.* **4** (1982) 201–261]

Lapachol: see Napthoquinones (Table).

Larval hormones: see Juvenile hormones.

Late proteins: see Phage development.

Late RNA: see Phage development.

Latex: see Caoutchouc.

Lathyrin: see Guanidine derivatives.

Lathyrinogenic amino acids: nonproteogenic amino acids occurring in the seeds of some species of vetch *(Lathyrus)*. They include diaminobutyric acid $H_2N-(CH_2)_2-CH(NH_2)-COOH$ (neurolathrinogenic effect), β-aminopropionitrile, which occurs as the glutamyl peptide in the seeds of *Lathyrus odoratus*, and presumably the *N*-oxaloyl-α,β-diaminopropionic acid $HOOC-CH(NH_2)CH_2-NH-CO-COOH$. The disease caused in humans and animals by L.a.a. is called *lathyrism*, and takes various forms, e.g. neuro(nerve)- and osteo(bone)-lathyrism. (*N*γ-glutamyl)aminopropionitrile, for example, causes skeletal abnormalities in rats.

Lauric acid: n-dodecanoic acid, $CH_3-(CH_2)_{10}-COOH$, one of the most widespread fatty acids, a typical wax fatty acid. M_r 200.3, m.p. 44 °C, b.p.$_{100}$ 225 °C. L. is present in the seed fats of the laurel family *(Lauraceae)*, and makes up 52% of the fatty acids in palm seed oil, 48% in coconut fat, 4 to 8% in butter. It is an acid component of spermaceti.

Lawsone: see Naphthoquinones (Table).

(+)-Laxifloran: (3 R)-7,4′-dihydroxy-2′,3′-dimethoxyisoflavan, see Isoflavan.

LDH: abb. for Lactate dehydrogenase.

LDL: see Lipoproteins.

Lead, Pb: a highly toxic cumulative element in man and animals. It affects adversely nearly all steps in heme synthesis; it inhibits the mitochondrial enzyme 5-aminolevulinic acid synthase, but inhibits even more strongly 5-aminolevulinic acid dehydrase (see Porphyrins). The result of this inhibition is an increase in the blood level of 5-aminolevulinic acid, which is also detectable in the urine. Other enzymes inhibited by absorbed Pb are cytochrome P_{450} (liver), adenyl cyclase (brain and pancreas), enzymes of collagen synthesis, some ATPases, and lipoamide dehydrogenase. Clinical symptoms of Pb poisoning (plumbism) include anemia, lead-line stippling of the gums, alimentary pain, muscle weakness, and encephalopathic manifestations that occur mainly in children (convulsions, delirium, loss of memory, hallucinations). Most blood Pb is present in red cells, and one clinical characteristic of plumbism is the appearance, in the bone marrow and circulating blood, of basophilic stippled red cells (resembling reticulocytes in that they contain mitochondria; stippling caused by clumped ribosomes).

Learning: see Memory.

Lecithin: see Phospholipids.

Lecithin-cholesterol acyltransferase deficiency: see Inborn errors of metabolism.

Lectin, *phytohemagglutinin:* as defined by the Nomenclature Committee of the IUB, "sugar-binding protein or glycoprotein of nonimmune origin which agglutinates cells and/or precipitates glycoconjugates." This definition may be broadened to include similar proteins which, although they specifically bind complex saccharides, do not precipitate or agglutinate them. Such proteins can be called "monovalent L.". L. are found in almost every major taxon of flowering plants, and in some non-flowering plants as well. Vertebrate and microbial L. have also been identified. They bind to erythrocytes, leukemia cells, yeast and several types of bacteria. As the binding is saccharide-specific, L. will not agglutinate cells which do not carry the appropriate surface saccharides. It may be expected that as more kinds of surface oligosaccharide are used in screening assays, more L. will be found. Indeed they may be ubiquitous. L. may account for as much as 10% of the soluble protein in extracts from mature seeds; L. are also present in other plant tissues at lower concentrations. The L. present in vegetative tissues are often different from the seed L. in the same plant. The physiological function of plant L. is unknown. It has been suggested that they promote infection of legume root hair tips by *Rhizobium,* or that they inhibit pathogenic microorganisms. The vertebrate L. may function in development; they have a role in receptor-mediated endocytosis.

The best known plant L. are Concanavalin A (Con A) from jack beans *(Canavalia ensiformis),* which was crystallized in 1919 by Sumner, and the agglutinins from wheat germ (abb. WGA), lima beans, green beans *(Phaseolus),* castor beans *(Ricinus)* and potatoes. WGA is very well characterized, and L. with similar specificities and structure are present in rye and barley. Seeds from 90 other members of the *Triticeae* tribe of the grass family have L. which are immunochemically identical to WGA, although their specificities are not all the same. The seed L. in the grasses are found only in the embryos. Legume seeds are very rich in L., and complete amino acid sequences are known for a number of them, including Con A. There are extensive homologies among the L. from related legumes. Studies on the structure of Con A have revealed a new type of protein maturation [D.J. Bowles et al. *J. Cell Biol.* **102** (1986) 1284–1297]. Con A (consisting of 4 identical subunits, each of M_r 27500) shows maximal se-

Table. Molecular weights and subunit structures of several lectins from plant seeds

Source	M_r	Subunits No.	M_r	Remarks
Jack beans (Concanavalin)	at pH > 7 110 000	4	27 000	238 amino acids, n.c.
	at pH < 6 54 000	2	27 000	Primary structure known; not a glycoprotein but a lipoprotein
Lima beans	247 000 124 000	4 2	62 000 62 000	Glycoprotein, n.c.
	247 000 124 000	8 4	31 000 31 000	Glycoprotein, c.
Green beans (*Phaseolus*)	140 000	2α	35 000	Glycoprotein
		1β 1β	35 000 36 500	
	126 000	4	31 000	
Wheat hemag-glutinins	34 000	2	17 000	Not a glycoprotein, n.c.
Castor beans (*Ricinus*)	125 000	1	33 000	Glycoprotein, c.
		1 1	30 000 27 500	
Potato	95 000	2	46 000	Glycoprotein, n.c. (50% carbohydrate)

c. = covalent; n.c. = non-covalent binding of subunits

quence similarity with the L. of lentil, soybean and fava bean when its amino terminus is positioned near the middle of the sequences of the other L. It is now known that the primary translation product undergoes transpeptidation (Fig.).
[M. E. Etzler, *Ann. Rev. Plant Physiol.* **36** (1985) 209–234; T. C. Bøg-Hansen & E. van Driessche, eds.,

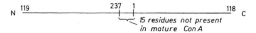

Con A precursor (Structure of Con A translation product as deduced from cloned cDNA). The numbers refer to residues in mature Con A

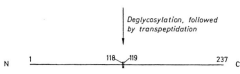

Mature Con A

Maturation of Con A precursor to form mature Con A.
Pulse-chase experiments with radioactive amino acids strongly support a mechanism of transpeptidation, rather than proteolytic cleavage followed by ligation. The only other known precedent for such a mechanism is the last step of peptidoglycan synthesis in bacteria (see Murein). Inspection of possible three-dimensional structures of Con A and its precursor shows that maturation by transpeptidation is possible without unfolding of the polypeptide chain.

Lectins: Biology, Biochemistry, Clinical Biochemistr vol. **5** (de Gruyter, Berlin, 1986); see also preceedin volumes of this series]

Leghemoglobin, Legoglobin: an autoxidizabl hemoprotein, present in the root nodules of legum nous plants. L. is structurally and functionally relat ed to hemoglobin and myoglobin. Amino acid se quence of L. shows homology with that of anima myoglobin. X-ray crystallography shows similar to pology and three-dimensional structure of L. an animal myoglobins. L. is essential for symbiotic n trogen fixation in the root nodules of leguminou plants, where it is responsible for the rapid flow o oxygen to the bacteroids (cf. role of myoglobin i transport of oxygen to respiratory enzymes of mus cle). Synthesis of the globin is under the genetic con trol of the macrosymbiont (host plant); the heme i synthesized by the microsymbiont (*Rhizobium* bac teroids). The concentration of L. shows a positiv correlation with the N_2-fixing capacity of the nodul tissue. There may be more than one L., dependin on the species of the host plant. L. from soybean ca be resolved into at least four components a–d, o which a and c are the main ones; these have molec ular weights 16 800 and 15 950; both have been crys tallized: Component c has now been further re solved into c_1 and c_2, which differ only in thei C-termini (lysine in c_1; phenylalanine in c_2). L. con tains no cysteine or methionine; and it shows no im munological cross reaction with hemoproteins from *Rhizobium*. L. is unique to the legume *Rhizobiu* symbiotic system; it is not present in free-living *Rh zobia* and uninfected leguminous plants.

Leucine

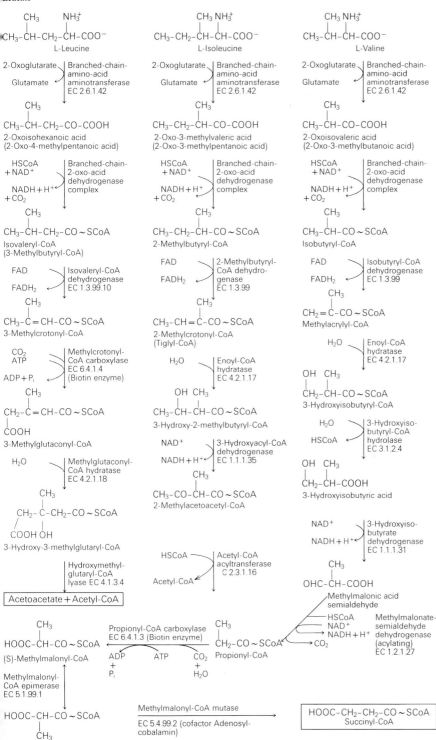

Degradation of branched-chain amino acids

L. is located between the bacteroids and the membrane envelope. (See Nitrogen fixation).

Legoglobin: see Leghemoglobin.

Legumin: an oligomeric storage protein of legume seeds, M_r about 328000, consisting of 6 pairs of subunits. Each subunit pair consists of an α-subunit (M_r about 36000) and a β-subunit (M_r about 20000) linked covalently by a disulfide bond. L. is synthesized in vitro as a single polypeptide chain, M_r about 60000, which is cleaved in vivo into α- and β-subunits. The precursor polypeptide already contains the disulfide bond, so the subunit pairs are specific. Study of *Vicia faba* L. shows that at least two different gene families (which may have arisen from a single gene) code for L. precursor polypeptide, giving rise to type A and type B subunit pairs. Type A α- and β-subunits both contain Met, which is absent from type B subunits. Since Met is an essential amino acid in animal nutrition, and is a limiting factor in the nutritional value of legume seed protein, it should be possible to improve the quality of this protein by plant breeding to increase the type A/B ratio. Further heterogeneity of amino acid composition within the subunit pairs has been detected and may result from mutation of the common ancestral gene. [C. Horstman *Phytochemistry* **22** (1983) 1861–1866]

Lettuce hypocotyl test: see Gibberellins.

L-Leucine, abb. *Leu:* L-2-amino-4-methylvaleric acid, $(CH_3)_2$-CH-CH$_2$-CH(NH$_2$)$_2$-COOH, an aliphatic, neutral amino acid found in proteins. Leu is both an essential dietary amino acid and ketogenic. It is particularly abundant in serum albumins and globulins. It is degraded to isovaleric acid by deamination and decarboxylation, then further to acetic acid via acetoacetic acid (Fig. p.335). The biosynthesis of Leu follows the scheme for branched amino acids (see L-Isoleucine) and branches off at the level of 2-oxo-3-methylvaleric acid, which undergoes condensation with an acetyl group. The subsequent reactions are analogous to those in the tricarboxylic acid cycle, and result in 2-oxo-4-methylvaleric acid, which is transaminated to Leu.

Leucinosis: see Inborn errors of metabolism.

Leucoanthocyanidins, *flavan-3,4-diols:* a class of Proanthocyanidins (see). L. are common plant constituents, especially in wood, bark, and the rind of fruits. Examples are *(+)-guibourtacacidin* (7,4'-dihydroxyflavan-3,4-diol, absolute configuration 2R:3S:4S, from *Guibourtia coleosperma* and *Acacia cultriformis;* the configuration 2R:3S:4R is also present in *A. cultriformis*), *(−)-leucofisetinidin* (7,3',4'-trihydroxyflavan-3,4-diol, abs. config. 2S:3R:4S, from *Schinopsis lorentzii*), *(+)-leucorobinetinidin* (7,3',4',5'-tetrahydroxyflavan-3,4-diol, abs.

Mollisacacidin

config. 2R:3S:4R, from *Robinia pseudoacacia*), and *(+)-mollisacacidin* (7,3',4'-trihydroxyflavan-3,4-diol abs. config. 2R:3S:4R, from heartwood of *Acacia baileyana* and sapwood of *A. mearnsii;* see Fig.). L. have been implicated in the biosynthesis of condensed tannins and condensed proanthocyanidins (see Tannins).

(−)-Leucofisetinidin: 7,3',4'-trihydroxyflavan-3,4-diol, see Leucoanthocyanidins.

Leucomycin: see Macrolide antibiotics.

Leucoplasts: colorless plastids in plant cells, generally those which are not exposed to light. The L. include the starch-storing amyloplasts, the protein-storing aleuroplasts, the fat-storing elaioplasts and the etioplasts, which are the colorless chloroplasts of sprouts and stolons (for example, potato shoots) which have been kept in the dark (etiolated). Amyloplasts are found in the non-green storage tissues, such as the endosperm of grain or the cotyledons of legumes. Reserve starch is formed and stored in them in the form of starch grains. Although the amyloplasts are colorless, they are sometimes considered to be chromoplasts.

Leucopterin: 2-amino-5,8-dihydro-4,6,7(1H)-pteridinetrione, a white pigment found in the wings of cabbage white butterflies and other butterflies. m.p. 350 °C. It is biosynthesized from guanine and two C-atoms of a pentose. It was isolated in 1926 from 200000 cabbage white butterflies by H. Wieland and C. Schöpf, and its structure was established in 1940 by R. Purrmann.

Leucopterin

(+)-Leucorobinetinidin: 7,3',4',5'-tetrahydroxyflavan-3,4-diol, see Leucoanthocyanidins.

Leukokinin: the specific leukophilic γ-globulin molecule which binds to the leukocyte membrane and acts as the precursor of Tuftsin (see).

Leukophilic γ-globulin: predominantly γG-globulin and minor amounts of γA-globulin and γM-globulin, which coat the surface of the leukocyte membrane. See Tuftsin.

Leukotrienes: lipid hormones derived from Arachidonic acid (see). Leukocytes are important sources of L. (hence the "leuko-"prefix), and all of them contain three conjugated double bonds (hence "triene"). The major L. are actually tetra-unsaturated, but the fourth double bond is not conjugated with the others. The L. are the same as the "slow-reacting substances of anaphylaxis", which mediate immune hypersensitivity. They are also known to potentiate inflammation, but the physiology of this effect is not understood. All L. cause powerful contraction of the lungs (they are hundreds to thousands of times more potent than the histamines), and they stimulate release of thromboxanes and prostaglandins. A lipoxygenase acting at C-5 of arachidonic acid produces 5-hydroperoxyeicosatetraenoic acid. Conversion of this compound to the

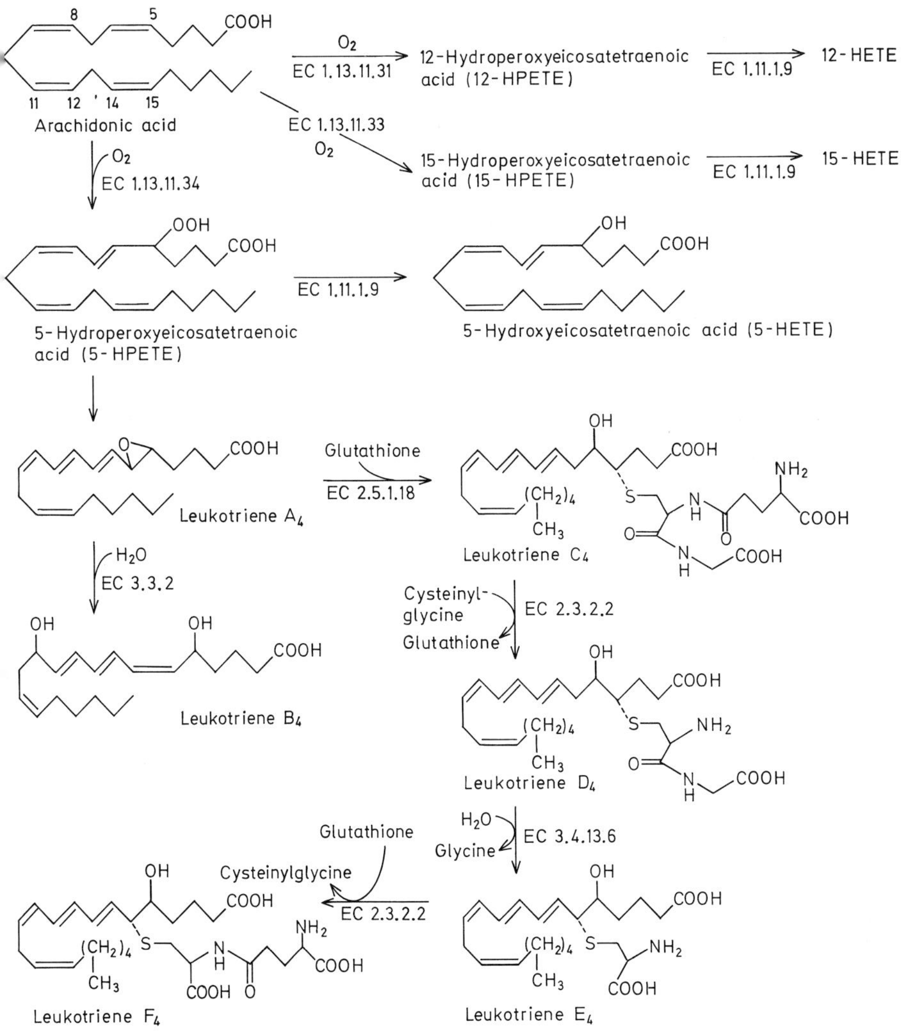

The lipoxygenase pathway of arachidonate metabolism
EC 1.11.1.9: Glutathione peroxidase
EC 1.13.11.31: Arachidonate 12-lipoxygenase
EC 1.13.11.33: Arachidonate 15-lipoxygenase

EC 1.13.11.34: Arachidonate 5-lipoxygenase
EC 2.3.2.2: γ-Glutathione transferase
EC 3.3.2. : Leukotriene-A_4 epoxide hydrolase
EC 3.4.13.6: Cysteinyl-glycine dipeptidase

5,6-epoxide produces L.A_4, which is very unstable; in buffer at pH 7.4 and at 25 °C its half-life is less than 10 sec. However, it is stabilized by alkaline conditions or by albumin. The structures of L.A_4 and other L. are given in the figure. The physiological effects of different L. are not identical. L.C_4, L.D_4 and L.E_4 are mostly myotropic, stimulating contraction of smooth muscle. L.B_4 is a chemotactic agent for macrophages and eosinophils; the former are caused to aggregate and to release superoxide and lysosomal enzymes. L.B_4 induces transformation of T lymphocytes into suppressor T cells, and may have other effects which modulate the immune response. [P. Borgeat et al., "Leukotrienes. Biosynthesis, Metabolism and Analysis", *Adv. Lipid Res.* **21**

(1985) 47–77; P. Sirois, "Pharmacology of the Leukotrienes", *ibid.* 79–101].

Leupeptin: see Inhibitor peptides.
Leurocristine: see Vincristine.
Levulose: see D-Fructose.
LH: abb. for luteinizing hormone.
Liberin: a Releasing hormone (see); a suffix (-liberin) used in the nomenclature of releasing hormones.

Licanic acid, *couepic acid, ketoeleostearic acid, oxoeleostearic acid:* 4-oxo-9,11,13-octadecatrienoic acid.
$CH_3(CH_2)_3(CH=CH)_3(CH_2)_4CO(CH_2)_2–COOH$.
Isolated from the seed fat or oil of *Licania rigida* (see Oiticica oil). The original source of the oil was

erroneously thought to be *Couepia grandiflora,* hence the name couepic acid. L.a. exists in two forms: α-L.a. is *cis*-9, *trans*-11, *trans*-13 (m.p. 74-75 °C); β-L.a. is all-*trans* (m.p. 99.5 °C). α-L.a. readily isomerizes to β-L.a. by the action of light, and in the presence of traces of sulfur or iodine. It is the only unsaturated oxoacid that has been isolated from a natural fat, and it is also present in the seed fats of other *Licania* species, e.g. *L. arborea* (Mexico), *L. crassifolia* (East Indies) and *L. venosa* (Guyana), and the seed fats of several species of *Parinarium.*

The glyceride oils of L.a. are commercially important in the manufacture of alkali and water resistant coatings in the paint industry.

Lichenin, *moss starch:* a polysaccharide serving both as storage and as structural compound. M_r 25000 to 30000, $[\alpha]_D$ +120 °. L. is composed of 150 to 200 D-glucose units linked β-1,4 with about 25% β-1,3 glucosidic linkages distributed at random in the molecule. It is found in many lichens and has antineoplastic properties.

(+)-Licoricidin: 5,4',2'-trihydroxy-7-methoxy-6,3-di-γ,γ-dimethylallylisoflavan, see Isoflavan.

Liebermann-Burchard reaction: a reaction used for colorimetric determination of sterols. The substance to be tested is dissolved in chloroform and treated with sulfuric acid and acetic anhydride. A color change from pink to blue to green indicates unsaturated sterols and is the basis for the quantitative determination of cholesterol in blood.

Ligandin: see Glutathione S-transferase.

Ligases: see Enzymes.

Light compensation point: the light intensity at which the rate of photosynthesis (CO_2 incorporation) and the rate of respiration (CO_2 production) are balanced. See CO_2-compensation point.

Lightening hormone: Pyr-Leu-Asn-Phe-Ser-Pro-Gly-Trp-NH_2, a neurohormone produced by the eye-stalk glands of crustaceans. M_r 930. The hormone is released from nerve endings in the gland in response to visual stimuli and controls the pigment granules in the hypodermal chromatophores. In this way the animal adjusts its color to match the surroundings.

Light-harvesting protein: a strongly hydrophobic, integral membrane protein, isolated from the thylakoids of many angiosperms, gymnosperms and green algae. It is chiefly associated with photosystem II, but some activity with photosystem I has also been demonstrated. Chlorophylls *a* and *b* are bound in equimolar amounts, together with lutein and β-carotene. The chlorophyll/carotenoid molar ratio is 3-7/1. M_r of protein 27000-35000, depending on the species. Complexes are also known, which contain two dissimilar subunits. Up to 6 mol Chlorophyll (see) are bound per mol protein (M_r taken as 30000). The protein serves in the transfer of light energy from chlorophyll *b* to *a,* and it is involved in (although not essential for) the stabilization or stacking of the grana. Only very small amounts of L.h.p. are found in the bundle sheath cells of C-4 plants.

Light reactions: see Photosynthesis.

Lignans: plant products (L. have also been found in primate urine; see Enterolactone, Entero-

diol) formally equivalent to two n-propylbenzene (phenylpropane) residues linked at the central carbons of their side chains:

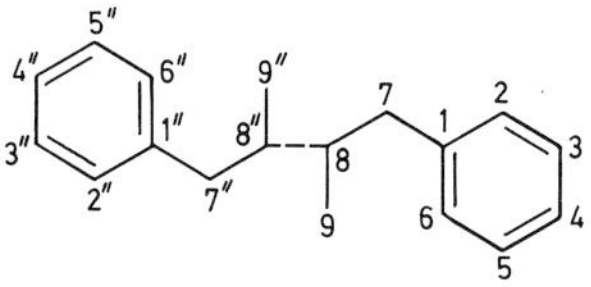

The two benzene rings are usually identically substituted, and the type of substitution is similar to that present in ring B of the C_6C_3 residues of flavonoids. It is therefore generally accepted that L. are biosynthesized by dimerization of a C_6C_3 precursor, but direct experimental evidence for this is lacking. Linear L. can be classified in four groups, depending on the type of structure between the two aromatic rings:

1. *Lignans* (derivatives of butane)

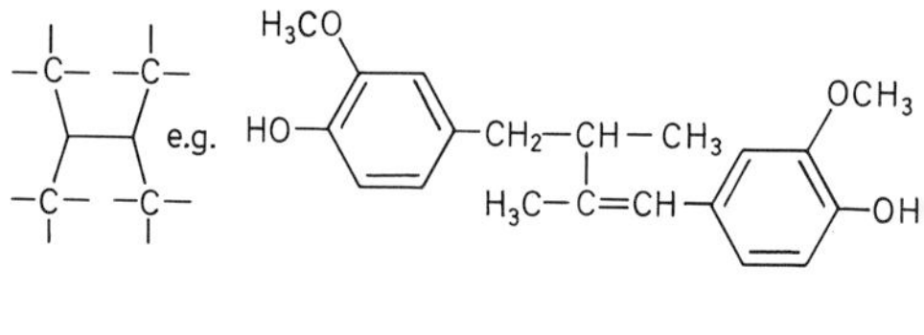

Gualaretic acid (present at about 10% in the resin of *Guaiacum officinale* L.)

2. *Lignanolides* (derivatives of butanolide)

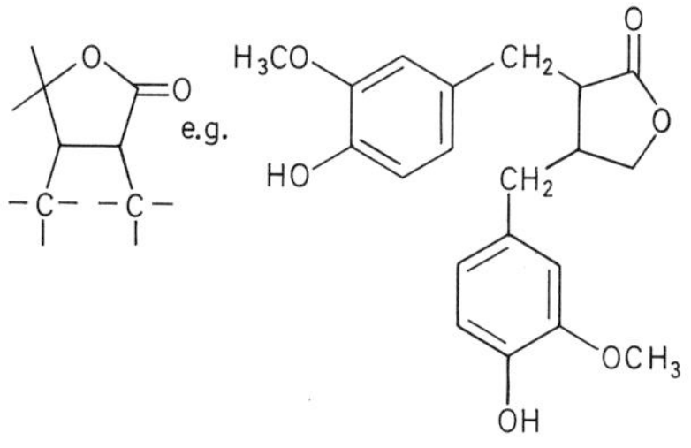

Matairesinol (from *Podocarpus spicatus,* or New Zealand "matai", and in many different woods)

3. *Monoepoxylignans* (derivatives of tetrahydrofuran)

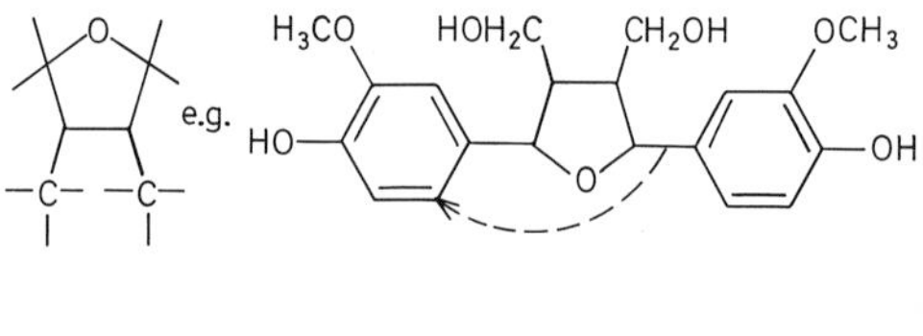

(−)-Olivil (from resin of the olive tree, *Olea europea*). The dashed arrow represents cyclization to isoolivil in presence of mineral acid (see text).

4. Bisepoxylignans (derivatives of 3,7-dioxabicyclo[3.3.0]-octane)

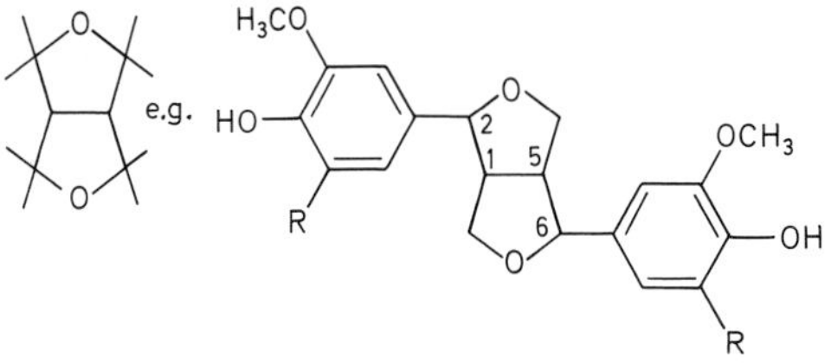

R = H: **Pinoresinol** (from exudate of *Pinus lavico* and other pines)
R = OCH₃: **Syringaresinol** (1 R,2 R,5 R,6R-form, also called *Lirioresinol A* or *Episyringaresinol*; from *Artemesia absinthum, Liriodendron* spp. and *Magnolia grandiflora;* from degradation of birch lignin). Other configurational forms occur naturally.

Further cyclization (C-7 to C-6″) produces **cyclolignans:** Cyclolignans containing a tetrahydronaphthalene ring system:

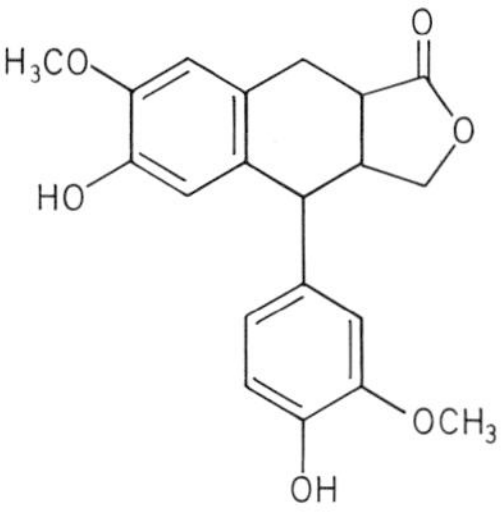

Conidendrin (from spruce; present in large amounts in waste sulfite liquor from paper making)

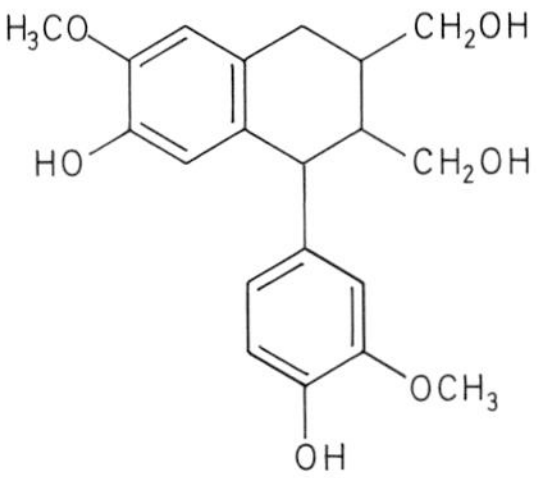

iso-Olivil (from the Australian olive, *Olea cunninghamii*)

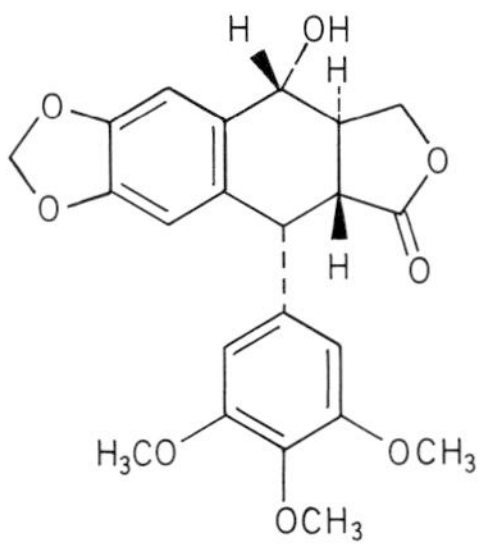

Podophyllotoxin (from the Mayapple, *Podophyllum* spp.)

Cyclolignans containing the naphthalene ring system:

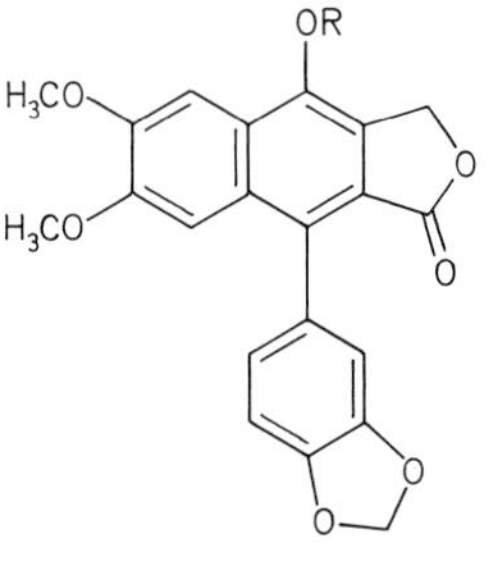

R = CH₃: **Justicidin A**
R = H: **Justicidin B** (piscicidal constituents of *Justicia hayatai*)

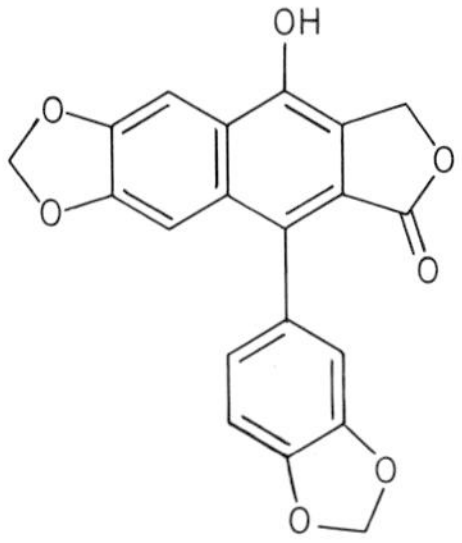

Diphyllin (from roots of *Diphylleia grayi*; also a piscicidal constituent of *Justicia hayatai*)

According to Freudenberg, L. represent stages in the synthesis of Lignin (see); enzymatic oxidation of coniferyl alcohol with a fungal laccase at pH 5 (i.e. conditions similar to those for the formation of spruce lignin in vitro) produces dehydroconiferyl alcohol and pinoresinol. Lignin formally resembles a polymer of propylbenzene (phenylpropane) units, and the bisepoxylignan structure occurs repeatedly in the lignin polymer.
Treatment of (−)-olivil with mineral acid breaks one of the benzyl ether linkages, forming a carbonium ion. The positive center then attacks the opposite guaiacol residue by electrophilic substitution, to produce iso-olivil. This type of reaction has been observed for many other L. It is probably also biologically important, because iso-olivil occurs naturally in the Australian olive.
Representatives of 55 families of vascular plants have been found to contain L. Gymnosperms feature prominently as L.-containing plants. L. occur in all parts of the plant, wood and resin being especially rich sources. L. have been isolated which possess antitumor, antimitotic, antiviral, insecticidal, piscicidal, antibacterial and fungistatic properties. Some L. inhibit mammalian cyclic AMP phosphodiesterase, and others have been shown to possess cathartic activity, cardiovascular activity, or the ability to damage DNA and inhibit nucleic acid synthesis. Podophyllotoxin and related compounds are particularly active, showing antimitotic activity and the

ability to bind to purified tubulin preparations, probably at the same site as colchicine; they also display antiviral and antitumor activity, and they inhibit DNA and RNA synthesis in mammalian cell culture. L. have been classified by the U.S. National Cancer Institute as compounds of "high interest" as potential anticancer agents, and tumor trials have been conducted on several L.

A useful source reference with literature list covering all aspects of L.: W.D. MacRae & G.H.N. Towers *Phytochemistry* **23** (1984) 1207-1220.

Lignification: see Wood.

Lignin: a polymer responsible for the thickening and strengthening of plant cell walls. The properties associated with wood are due to the incrustation of plant cell walls with L. Chemically, L. cannot be exactly defined. According to Freudenberg, L. is a highly cross linked, macromolecular, branched polymer, formed irreversibly by dehydrogenation and condensation. According to Adler and Gierer, L. is an essentially acid-resistant, polymorphic, amorphous incrustation material found in wood, consisting of methoxylated phenyl-propane units linked by ether linkages and C–C bonds. It has also been de-

scribed as a "statistical polymer of hydroxyphenylpropane units". The M_r of L. is greater than 10000, and it is insoluble in hot 70% sulfuric acid. The chemical composition of L. differs according to the plant species. Beech L. is the most extensively studied.

Biosynthesis. The primary precursors of L. are coniferyl, sinapyl and *p*-coumaryl alcohol, which are derived from 4-hydroxycinnamic acid. L. from conifers (i.e. from soft wood) is derived chiefly from coniferyl alcohol with variable but small proportions of sinapyl and *p*-coumaryl alcohol. L. from dicotyledenous angiosperms (i.e. from hard wood), particularly deciduous trees, is formed chiefly from sinapyl ($\sim$44%) and coniferyl ($\sim$48%) alcohol, with about 8% *p*-coumaryl alcohol. L. in grasses is formed from *p*-coumaryl ($\sim$30%), coniferyl ($\sim$50%) and sinapyl ($\sim$20%) alcohol. These primary L. precursors are formed from the aromatic amino acids L-phenylalanine and L-tyrosine by a series of reactions shown (Fig.). The first reaction is catalysed by L-Phenylalanine ammonia-lyase (EC 4.3.1.5) (see); this enzyme is induced by light in a process involving phytochrome, and it is of general importance in the syn-

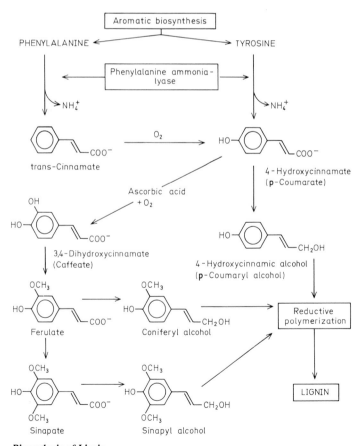

Biosynthesis of Lignin

thesis of plant phenolic compounds from phenylalanine and tyrosine.

D-Coniferin (glucoside of coniferyl alcohol), D-glucocoumaryl alcohol (glucoside of *p*-coumaryl alcohol) and D-syringin (glucoside of sinapyl alcohol) are storage forms of L. precursors. β-Glucosidases in cells between the cambium and the mature wood hydrolyse the glucosides and thereby release the alcohols for L. synthesis.

The total biosynthesis of L. represents the interplay of enzymatic phenol dehydrogenation and the nonenzymatic coupling of radicals generated by the loss of an electron from a phenolate ion; it is referred to as *reductive polymerization*.

The process can be demonstrated by the synthesis of artificial L. in a model system, in which coniferyl alcohol (80 mol.%), *p*-coumaryl alcohol (14 mol.%) and sinapyl alcohol (6 mol.%) are incubated with a phenol oxidase (e.g. fungal laccase) under strongly aerobic conditions. The product is identical with spruce L. Presumably a peroxidase or similar oxidation system is responsible for L. biosynthesis *in vivo*. When isotopes are used to study L. biosynthesis, strict criteria are applied to the identification of L., which must be clearly distinguished from other cell wall polymers. One method is to degrade the L. *in situ*, followed by the isolation of degradation products known unequivocally to be derived from L. For example, Hibbert showed that wood refluxed with ethanol containing 2% HCl gave a mixture of water-soluble aromatic ketones. These compounds are known as "Hibbert's ketones", and they have the structures: $R-CO-CO-CH_3$, $R-CH_2-CO-CH_3$, $R-CO-CH(OC_2H_5)-CH_3$, and $R-CH(OC_2H_5)-CO-CH_3$, where R is guaiacyl or syringyl. These compounds represent the intact phenylpropane structure of the original L.

Considerable quantities of L. (as L.-sulfonic acid) are formed as a byproduct in the industrial production of cellulose. The river Rhine contains 0.2-1.8 mg L. per litre, due to industrial effluent; this accounts for 20% of the total organic pollution.

Lignoceric acid: *n*-tetracosanoic acid, $CH_3-(CH_2)_{22}-COOH$, a fatty acid M_r 368.6, m.p. 84 °C. L.a. occurs as a component of glycerides (usually less than 3%) in many seed oils, such as ground nut and rape seed oil. It is a component of certain cerebrosides (e.g. kerasin), phosphatides and waxes.

Limonene: see *p*-Menthadienes.

Linalool, *coriandrol:* a doubly unsaturated monoterpene alcohol found in essential oils. Both optical isomers occur naturally. For formula, see Geraniol. L. is an oil with a scent resembling that of lily of the valley. M_r 154.25. (+)-L. b.p. $[\alpha]_D^{20}$ -19.4 °. The pure alcohol and its esters are used in perfumery.

Lineweaver-Burk plot: see Kinetic data evaluation.

Linking difference: see Linking number.

Linking number, *topological winding number, α:* a number which specifies the number of times two strands of the DNA double helix are intertwined. This is always an integer. In solution, DNA has a pitch of 10.4 base pairs per helical turn. Closed circular duplex DNA is therefore considered to be relaxed if α approximates to the number of base pairs

divided by 10.4. The totally "relaxed" value for α is represented by α^0 or β. A L.n. smaller than α^0 means that the DNA is negatively supercoiled (i.e. underwound), whereas DNA with a L.n. greater than α^0 is positively supercoiled. In both cases the DNA is under torsional strain. This strain is partitioned between twist (altered pitch of the double helix) and writhe (contortion of the double helix). Writhe is equivalent to supercoiling. The *linking difference* ($\Delta\alpha = \alpha - \alpha^0$) is an expression of strain, but is dependent on the length of the DNA. The *specific linking difference* ($\Delta\alpha/\alpha^0$), being independent of molecular length, is a more convenient measure of the strain on the DNA duplex. See also Superhelix density.

Linoleic acid: $\Delta^{9,12}$-octadecadienoic acid, $CH_3-(CH_2)_4-CH=CH-CH_2-CH=CH-(CH_2)_7-COOH$, an essential fatty acid. M_r 280.44, m.p. 5 °C, b.p.$_{14}$ 202 °C. L.a. is widely distributed in plants and animals. It occurs as a glyceride component in many fats and oils, and it is found in phosphatides. It is an essential dietary constituent for mammals.

Linolenic acid: $\Delta^{9,12,15}$-octadecatrienoic acid, $CH_3-CH_2-CH=CH-CH_2-CH=CH-CH_2-CH=CH-CH_2-(CH_2)_6-COOH$, an essential fatty acid -11.2 °C, b.p.$_{17}$ 232 °C. L.a. occurs in animals and plants, and it is especially common in plant fats and glycerophosphatides. L.a. cannot be synthesized by mammals.

Lipase deficiency: see Inborn errors of metabolism.

Lipases: a group of carboxylesterases, which preferentially hydrolyse emulsified neutral fats to fatty acids and glycerol or monoacylglycerols. Calcium ions are required for activity. Pancreatic L. also requires taurocholate. Pancreas and certain plant seeds (e.g. *Ricinus*) contain especially high activities of L. In addition, high activities are found in adipose tissue, in the stomach (especially in unweaned infants) and in the liver. M_r of pancreatic L.: 35000. Rapid removal of the third and last fatty acid residue from mixed triacylglycerols is catalysed by a specific *monoacylglycerol lipase* (EC 3.1.1.23), produced by the intestinal mucosa.

Lipidosis: see Inborn errors of metabolism.

Lipids: a heterogeneous group of biological compounds which are sparingly soluble in water, but very soluble in nonpolar solvents. They can be extracted from animal and plant tissues with a variety of organic solvents, e.g. benzene, chloroform, trichloroethene. As a class, L. are defined by their solubility; they include such chemically diverse compounds as Triacylglycerols (see), Waxes (see), and Terpenes (see; these include monoterpenes, diterpenes, carotenoids, steroids, etc.). The more complex L., such as Glycolipids (see) and Phospholipids (see) are also called *lipoids.* Triacylglycerols and waxes are known as *saponifiable L.*, whereas the terpenes are called *nonsaponifiable L.*

Lipochondria: see Dictyosomes, Golgi apparatus.

Lipofuscin, *age pigment, wear and tear pigment:* intracellular clusters of yellowish granules of universal occurrence in humans. L. is deposited continuously throughout life in nerve, heart and liver cells as well as other organs. L. is very prominent in the pyrimidal cells of the cerebral cortex, in spinal ventral horn cells and the cell bodies of the hypoglossal nuclei in elderly persons. In the neurons of the olivary nucleus, deposition of L. commences earlier and is prominent by middle age. L. appears to be derived from lysosomes, which have accumulated indigestible material. Most of these congested lysosomes, or residual bodies, are removed by exocytosis, but some persist within the cell and their constituent lipids become oxidized. The resulting mixture of proteins and partly oxidized polyunsaturated fatty acids constitutes L. It is stained intensely by Sudan black and by periodic acid-Schiff reagent.

Lipofuscinosis: see Inborn errors of metabolism.

Lipoic acid, 6-thioctic acid, (+)-5[3-(1,2-dithiolanyl)] pentanoic acid: a coenzyme of hydrogen transfer and acyl group transfer reactions. L.a. is a component of the pyruvate dehydrogenase and the 2-oxoglutarate dehydrogenase complexes (see Multienzyme complexes), which catalyse the oxidative decarboxylation of the corresponding 2-oxoacids (see also Tricarboxylic acid cycle). The natural form of L.a. is α-(+)-L.a. M_r 206.3, m.p. 47.5 °C, $[\alpha]_D^{25}$ +96.7 ° ($c = 1.88$, benzene), E_o' −0.325 V (pH 7.0, 25 °C). The asymmetric carbon has the R configuration (Fig.).

L.a. and dihydrolipoic acid form a biochemically important redox pair. Dihydrolipoic acid (reduced L.a.) is represented as $Lip(SH)_2$ and the oxidized form as $Lip(S_2)$. L.a. is frequently present in the cell as the carboxylic acid amide (lipoamide). When serving as a coenzyme, L.a. is bound covalently through an amide bond to the ε-amino group of a lysyl residue in the enzyme. When acting in the generation and transfer of acyl groups, the acyl group becomes attached to one of the sulfurs by a thiol ester linkage. In the oxidation of pyruvate by *Escherichia coli*, the intermediate has been shown to be 6-*S*-acetyl-6,8-dithioloctanoic acid (Fig.); presumably all other acyl groups are carried in the 6-S-position, since all synthesized 8-thiolacyl derivatives of dihydrolipoic acid are biologically inactive.

(3R)-1,2-dithiolane-3-pentanoic acid (the oxidized form of lipoic acid)

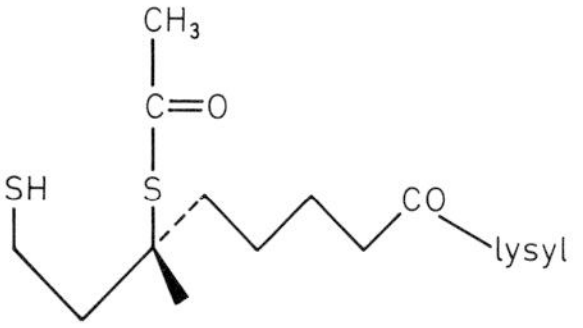

Residue of 6-S-acetyl-6,8-dithioloctanoic acid (acetyl derivative of the reduced form of lipoic acid) attached to enzyme protein by an amide linkage with the ε-amino group of a lysine residue

Lipoproteins: lipid-protein conjugates found in cellular membranes, blood plasma, cell cytoplasm and egg yolk. Blood plasma L. are responsible for the transport and distribution of lipids (hormones, dietary lipids from the intestine, fat-soluble vitamins) via the blood and lymph systems. The largest L. are the microscopically visible chylomicrons (diameter 500 nm); these are responsible for the lipemia (milky turbidity) of the blood, which follows the digestion of a fatty meal, and which disappears after about 5 hours. The other L. are always present in the blood; the apoproteins in them are produced in the liver.

In starch block electrophoresis, blood L. migrate with the β_1-, α_1- and α_2-globulin fractions. Ultracentrifugation in a high concentration of NaCl, or in a density gradient, produces a better separation and is suitable for the preparation of the various fractions. Owing to their lipid component, the L. float rather than sediment (with the exception of the very high density L. fraction, abb. VHDL), i.e. they move centripetally towards the surface of the ultracentrifuged suspension. Thus, the flotation coefficient S_f is a

Table. Classification and properties of the lipoproteins of human blood plasma. [F.T. Lindgren et al. in G.J. Nelson (ed.) *Blood Lipids and Lipoproteins: Quantitation, Composition and Metabolism* (Wiley-Interscience, New York, 1972) pp. 181-274]

Electrophoretic fraction	Ultracentrifuge fraction	Density	Flotation (S_f)	M_r ($\times 10^{-6}$)	mg/ 100 ml plasma	% protein	High content of
Chylomicrons		<0.96	10^3–10^5		0–50	1	Triacylglycerols
Pre-β	VLDL	0.960–1.006	20–400	5.0–20	150–250	7	
α_2, β_1	IDL	1.006–1.019	12–20	3.4	50–100	11	Cholesterol esters
β	LDL	1.019–1.063	0–12	2.0–2.7	315–385	21–23	
α_1	HDL	1.063–1.210		0.375	270–380	35–50	Phospholipids
α_1	VHDL$_1$	>1.210	Sediment.	0.145	?	65	Free fatty acids
Albumin	VHDL$_2$	1.210	constant 2–10S	0.280	?	97	

VLDL = Very low density lipoproeins; IDL = Intermediate density lipoproteins; LDL = Low density lipoproteins; HDL = High density lipoproteins; VHDL = Very high density lipoproteins

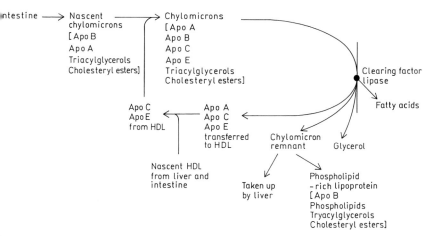

Figure 1. Synthesis and fate of chylomicrons

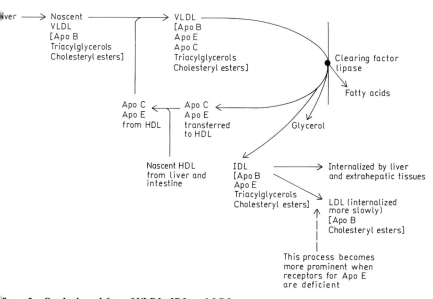

Figure 2. Synthesis and fate of VLDL, IDL and LDL

haracteristic of the L. L. contain between 1 and 2% arbohydrate.

he L. content of the blood is affected by both ge-etic and non-genetic factors, such as age, sex, diet, ormone balance, exercise and occupation. Chylo-nicrons, VLDL, IDL and LDL are involved in cho-esterol transport and deposition, and the correla-on between high blood cholesterol levels and therosclerosis has focused attention on these .. Nascent VLDL are produced in the liver, and ney contain large amounts of triacylglycerols and holesteryl esters. Nascent chylomicrons are pro-uced by the small intestine, and they also contain arge amounts of triacylglycerols and cholesteryl es-

ters. The dietary cholesterol which eventually con-trols endogenous cholesterol synthesis and LDL re-ceptor synthesis emerges from the intestine as nascent chylomicrons. The VLDL and chylomicrons transport triacylglycerols to adipose tissue and mus-cle; a lipase (lipoprotein lipase, clearing factor li-pase, EC 3.1.1.34) on the membranes of the capillary endothelial cells hydrolyses about 90% of the tri-acylglycerols. The resulting fatty acids diffuse into the tissue cells (see Acylglycerols). Some phospho-lipid is also lost, together with some apoproteins, which are transferred to HDL. The chylomicron remnants and VLDL remnants (IDL) continue in circulation (Figs. 1 and 2). HDL appears to function

343

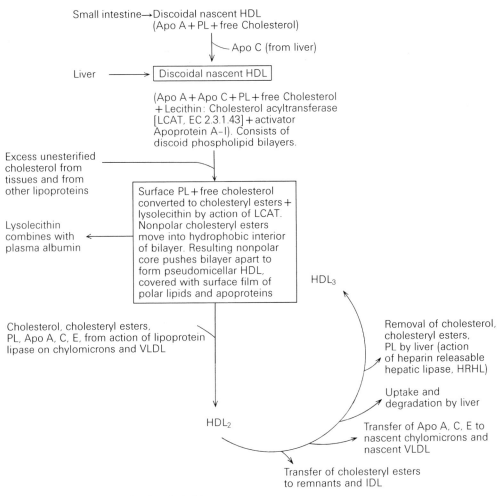

Small intestine→Discoidal nascent HDL
(Apo A + PL + free Cholesterol)

Apo C (from liver)

Liver ⟶ Discoidal nascent HDL

(Apo A + Apo C + PL + free Cholesterol
+ Lecithin: Cholesterol acyltransferase
[LCAT, EC 2.3.1.43] + activator
Apoprotein A–I). Consists of
discoid phospholipid bilayers.

Excess unesterified
cholesterol from
tissues and from
other lipoproteins

Surface PL + free cholesterol
converted to cholesteryl esters +
lysolecithin by action of LCAT.
Nonpolar cholesteryl esters
move into hydrophobic interior
of bilayer. Resulting nonpolar
core pushes bilayer apart to
form pseudomicellar HDL,
covered with surface film of
polar lipids and apoproteins

Lysolecithin
combines with
plasma albumin

HDL₃

Cholesterol, cholesteryl esters,
PL, Apo A, C, E, from action of lipoprotein
lipase on chylomicrons and VLDL

Removal of cholesterol,
cholesteryl esters,
PL by liver (action
of heparin releasable
hepatic lipase, HRHL)

Uptake and
degradation by liver

HDL₂

Transfer of Apo A, C, E to
nascent chylomicrons and
nascent VLDL

Transfer of cholesteryl esters
to remnants and IDL

Figure 3. The LDL cycle. PL, phospholipid.

largely as a transport vehicle by transferring components between the other L., and by receiving and esterifying free cholesterol from the tissues (Fig. 3). IDL contain two apoproteins, E and B-100. The former is present in several copies, and has a high affinity for cell-surface receptors called LDL receptors (a slight misnomer, as their affinity for IDL is higher). The LDL receptors congregate in clathrin-coated pits (see Clathrin), and are soon taken into the cell by endocytosis (Fig. 4), whether or not they have bound IDL or LDL. Those IDL which are not internalized by cells are degraded by loss of the apoprotein E, leaving LDL, which contain only a single molecule of apoprotein B-100 and a high content of cholesteryl esters. Apoprotein B-100 also binds the LDL receptor, so that LDL are normally cleared from the blood, albeit more slowly than IDL. Only a small proportion of chylomicron remnants is destroyed by extrahepatic tissues. Chylomicron remnants are more readily internalized than IDL by the

liver, and cholesterol derived from chylomicron remnants serves to control liver cholesterol biosynthesis and LDL receptor synthesis. The liver appears to possess specific receptors for chylomicron remnants, in addition to LDL receptors. In an animal model for familial hypercholesterolemia (Watanabe rabbit), LDL receptors are defective, but the uptake of chylomicron remnants by the liver is unaffected. The reasons for the different receptor-binding behavior of chylomicron remnants and IDL are not clear; phospholipid concentration may be important, and it may be significant that chylomicron remnants possess apoprotein B-48, which differs from B-100 of the IDL.

LDL are kept in suspension in the blood by a small amount of unesterified cholesterol, which presents its hydrophilic OH-group on the outside of the particle. This cholesterol is susceptible to loss from the particle, however, and can then settle on the inside of arteries as atherosclerotic plaque; this is thought

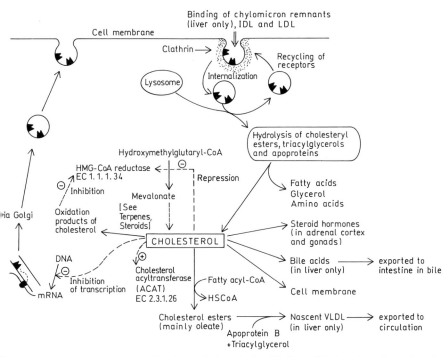

Figure 4. Control of the uptake, utilization and biosynthesis of cholesterol by hepatic and extrahepatic tissues

to be the reason for the correlation between high concentrations of LDL in the blood and atherosclerosis.

LDL receptors are present in many tissues, with the highest concentrations in adrenal glands and the largest total numbers in the liver. These tissues have especially high requirements for cholesterol, which is the precursor of the adrenal steroid hormones and of the liver bile acids. The number of LDL receptors in any given cell depends on the cell's requirement for cholesterol. In familial hyperlipidemias, the genes for the LDL receptors are defective and IDL and LDL are removed from the blood slowly or not at all. Hence a larger fraction of IDL remains in circulation long enough to be converted to LDL; high concentrations of LDL are the hallmark of this disease. Most (95%) cases of atherosclerosis are not, however, due to defective LDL receptors.

Oxidized derivatives of cholesterol, rather than free cholesterol, control cholesterol biosynthesis by inhibiting 3-hydroxy-3-methylglutaryl-CoA reductase (HMG-CoA reductase; EC 1.1.1.34) [A. A. Kandutsch et al. *Science* **201** (1978) 498–501]. Prolonged high intracellular concentrations of cholesterol also lead to decreased synthesis of HMG-CoA reductase. Endogenous cholesterol synthesis, however, cannot be totally abolished. Intracellular cholesterol also suppresses synthesis of LDL receptors (as demonstrated in fibroblast culture) (Fig. 4). High intracellular cholesterol therefore ultimately prevents uptake of LDL and IDL by the tissues, so that plasma LDL and IDL persist at abnormally high concentrations; this probably explains the connection between dietary cholesterol and atherosclerosis. LDL receptors (but not chylomicron remnant receptors) of most inhabitants of the developed world are permanently down-regulated, owing to high dietary cholesterol. On a low-cholesterol diet, the liver of a normal healthy adult synthesizes about 800 mg cholesterol per day, which replaces that lost in the feces. Free cholesterol and bile salts (derived from cholesterol) are excreted in the bile. Some of this is returned in the enterohepatic circulation, but about 250 mg/day of bile salts and about 550 mg/day of free cholesterol are lost in the feces. On a low cholesterol diet, a total of about 1300 mg of cholesterol are returned to the liver each day by the enterohepatic circulation and by HDL, which carry cholesterol from peripheral tissues to the liver. See Cholestyramine, Compactin, Mevinolin. [M. S. Brown & J. L. Goldstein, *Science* **232** (1986) 34–47]

Liposome, *phospholipid bilayer vesicle:* an aqueous compartment enclosed by a completely sealed lipid bilayer. L. are formed by ultrasonic irradiation (sonication) of phosphoglycerides (or other suitable lipids) and water. Alternatively, phospholipids are dispersed in the aqueous medium with a detergent, then the concentration of detergent is gradually decreased by dialysis until L. are formed. More simply, phospholipids are allowed to swell in an aqueous medium. Egg lecithin is often used as the phospholipid. L. are spherical or slightly elongated, and may be up to 1-2 μ in diameter. The smallest L. (about 25 nm diameter) are formed by sonication of larger ones. L. may be graded according to size by gel filtration. Rather than a simple vesicle, L. often consist of several concentric vesicles, each with a bilipid membrane, enclosing a layer of aqueous medium. If L. production is performed in the presence of salts, proteins, or other water-soluble components, these also become entrapped in the aqueous phase of the L., where they are isolated from the external environment. L. have attracted considerable interest, firstly as a model for the structure of biological membranes, and secondly as a vehicle for the encapsulation of therapeutic agents. The rationale behind the use of L. for the transport of drugs is based on the fact that L. injected into the blood stream are taken up rapidly by cells of the reticuloendothelial system. This approach is under development for the therapy of leishmaniasis (a parasitic disease of the tropics and subtropics, caused by a hemoflagellate protozoan, *Leishmania,* which invades chiefly the phagocytic cells of the reticuloendothelial system). Injection of L. containing antimonial drugs (meglumine antimonate, or sodium stibogluconate) into leishmaniasis-infected hamsters is about 1000 times more effective against the disease than injection of the drug alone.

β-Lipotropic hormone, β-LPH: a melanotropic peptide from the pituitaries of several species. The complete structure of β-melanotropin is contained within residues 41-58 of β-LPH (see Peptides, Fig. 3). Human β-MSH (see Melanotropin) is now known to be an artifact formed from β-LPH during extraction of the pituitary. [P. J. Lowry & A. P. Scott *Gen. Comp. Endocrinol.* **26** (1975) 16]

Lipotropic substances: compounds directly or indirectly involved in fat metabolism, which can prevent or correct fatty degeneration of the liver. They serve as substrates of phosphatide biosynthesis, or contribute (e. g. by methylation) to the synthesis of these substrates. Thus, choline and any substance capable of contributing methyl groups for choline synthesis (e. g. methionine) are L. s. Liver is the major site of synthesis of the plasma phosphoglycerides; when the availability of choline is restricted, the rate of phosphatidylcholine synthesis decreases, and the rate of removal of fatty acids from the liver falls below normal. If the rate of supply of fatty acids (free and esterified) to the liver remains normal, the resulting accumulation of fat gives rise to the condition of *fatty liver,* or *fatty degeneration of the liver.*

Lipotropin, *lipotropic hormone, adipokinetic hormone,* abb. **LPH:** a polypeptide hormone from the hypophysis. LPH promotes lipolysis, and it acts via the adenyl cyclase system. β-LPH contains 91 amino acid residues, M_r 9894 (porcine). γ-LPH contains 58 amino acid residues, identical in sequence to residues 1-58 of β-LPH. Corticotropin, melanotropin, ACTH and LPH all contain an identical heptapeptide in their structures. They belong to the so-called ACTH family of peptide hormones, and they are derived from a common precursor (see Peptides).

Lisetin: an Isoflavonoid (see) and the only known naturally occurring coumaronochromone (Fig.). L. was isolated from the heartwood of *Piscidia erythrina* [Falshaw et al. (1966) *Tetrahedron* Suppl. No. 7, 333-348].

Lithocholic acid: 3α-hydroxy-5β-cholanoic acid, or 3α-hydroxy-5β-cholan-24-oic acid. L. a. is a

Lisetin

monohydroxylated steroid carboxylic acid, and one of the bile acids. M_r 376.58, m. p. 185 °C, $[\alpha]_D$ +32 ° (alcohol). L. a. has been isolated from the bile of man, cow, rabbit, sheep and goat. It is normally prepared from bovine bile. L. a. is formed from chenodeoxycholate by intestinal bacteria. It is absorbed and returned to the liver for secretion. It is not readily conjugated, and it is relatively toxic to the liver. L. a. may be important in the pathogenesis of liver damage following biliary stasis.

Liver: the largest metabolic gland of vertebrates. L. and pancreas arise from the midgut in the course of embryonic development. The human L. weighs 1.5 kg and lies in the abdominal cavity in the right subphrenic space. 1 to 2 l blood per minute is supplied by the portal vein and the liver arteries, and leaves via the liver veins. The L. secretes bile, which flows out into the duodenum through a system of vessels, of which the gall bladder is a side arm. The functional unit of the L. is the hepatocyte, which is surrounded by blood vessels and bile capillaries. Metabolically, the L. is the most versatile organ in the body. Nearly all the products of digestion which are absorbed by the intestine are carried to the liver by the portal vein, there to be transformed or degraded.

Together with the muscles, the L. is the most important site of carbohydrate metabolism. The L. cell is freely permeable to glucose, in contrast to other organs whose permeability to glucose is regulated by insulin. The L. evens out the discontinuous supply of glucose provided by food by storing it as glycogen and releasing it as needed to maintain a constant blood sugar level. A number of hormones regulate these processes. Glucagon stimulates glycogenolysis, while the glucocorticoids promote gluconeogenesis, or synthesis of glucose from precursors such as lactate, which is produced in the muscles, and glucogenic amino acids. In addition, glucose is transformed into other monosaccharides in the liver, including pentoses which are produced by the pentose phosphate cycle. These sugars are needed for the synthesis of nucleic acids. The NADPH produced in the course of pentose synthesis is used in the synthesis of fatty acids. Fructose (from sucrose) and galactose (from lactose) are also metabolized in the L.

The close relationship between lipid and carbohydrate metabolism is shown by the interaction of L. and fat tissue. Fatty acids from food or the body's fat depots are degraded by β-oxidation in the L. An excess of fatty acids leads to the formation of ketone bodies. When needed, fatty acids can be synthesized from carbohydrates and transported by the blood as phosphatides to the fat tissue. Here they are converted to neutral fats. The L. is able to synthesize choleserol and to export it to other tissues, or to convert it o bile acids.

The L. has an important part in nitrogen metabolism. Amino acids absorbed from the digestive tract or produced by internal metabolism can be degraded in the L. by transamination and deamination. They can also be synthesized from α-keto acids. The L. is the site of synthesis of the blood plasma proteins (albumin, some of the globulins and fibrinogen) and the hepatogenic blood-clotting factors (prothrombin, accelerator globulin, proconvertin, Christmas factor). The nitrogen released from amino acid degradation is converted in the L. into urea, and excreted as such, via the kidneys, in the urine. The L. is also capable of synthesizing uric acid and creatine. It also degrades some of the hemoglobin to bile pigments, which are excreted via the intestines as bile.

A large number of substances, both natural (e.g. steroid hormones) and foreign (e.g. drugs) are transformed or degraded in the L. and transported to the kidneys in the blood, in the form of glucuronides or sulfates, where they are excreted in the urine. Insulin can be inactivated in the L. by reduction. Vitamin D₃, cholecalciferol, is enzymatically transformed into 25-hydroxycholecalciferol. In the kidney, this compound is further hydroxylated to 1,25-dihydroxycholecalciferol, which acts as a hormone in the regulation of calcium metabolism.

Given the number and variety of metabolic reactions occurring in the L., it follows that any L. disorder will result in a more or less severe metabolic disturbance.

Living matter: in biochemistry, the body substance of living organisms and cells. It has been determined by elemental analysis that about 40 chemical elements (see Bioelements) occur in L.m., but of these, about 10 account for more than 99% of the body substance. The dominant inorganic compound is water, the milieu in which life processes take place. Of the organic molecules of L.m. (biomolecules), the quantitatively most important are the carbohydrates, proteins and lipids, which are therefore the main classes of nutrients for humans and animals. Qualitatively, the proteins, as determinants of biological structure, function and specificity, the nucleic acids as carriers of genetic information, and the lipids as structural components of proteolipid membranes (biomembranes) are the most significant classes. Table 1 shows the average water, mineral and biomolecule contents of plant and animal bodies. The large differences in the compositions of L.m. of plants and animals reflect the basic difference in the way these organisms obtain their nutrients.

The carbon-autotrophic green plant is the primary producer of carbohydrates on the earth and uses carbohydrates extensively as storage and cell-wall substances (starch, cellulose, etc.). The nitrogen-heterotrophic animal usually consumes a rather protein-rich diet and uses proteins as supportive materials. The approximate chemical composition of the human body is shown in Table 2. Most of the mass of bacteria (see *Escherichia coli*) is made up by proteins, which may account for 70% of the dry weight. Numerically, however, the lipid molecules predominate. The number of molecules of each class of substance was calculated from the mass using Avogadro's number $(6.02 \cdot 10^{23} =$ number of molecules in one mole). The biosynthetic capacity of a bacterial cell is given under *Escherichia coli.*

Table 1. Water, mineral and organic contents (in percent of fresh weight) of animal and plant body substance. The values are rough approximations.

Class of substance	Animal (%)	Plant (%)
Water	60	75
Minerals	4.3	2.5
Carbohydrates	6.2	18
Lipids	11.7	0.5
Proteins	17.8	4.0

Table 2. Chemical composition of the human body (adapted from Rapoport). The values have been rounded off.

Class of substance	%	kg (Fresh weight)
Water	60	42.0
Minerals	4	2.8
Carbohydrates	1	0.7
Lipids	15	10.5
Proteins	19	13.3
Nucleic acids	1	0.7

LLD-factor: see Vitamins (Vitamin B₁₂).

Loading effect: see Pheromones.

Loading technique: see Methods of biochemistry.

Lobelia alkaloids: an extensive group of 2,6-disubstituted piperidine alkaloids of the genus *Lobelia,* especially from the medicinal plant, *Lobelia inflata,* cultivated in some European countries and in the USA. There are three structural types, depending on the functional groups of the substituents: *lobelidiols, lobelionols* and *lobelidiones* (Fig.). R_1 and R_2 may be $-CH_3$, $-C_2H_5$ or $-C_6H_5$, and the configuration of

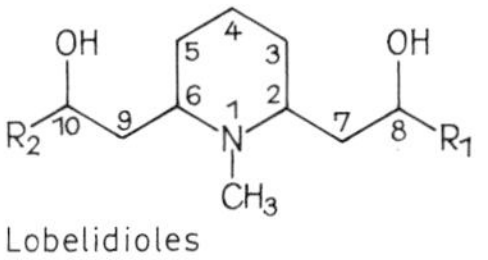

Lobelidioles

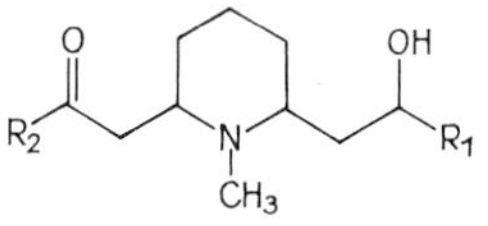

Lobelionoles

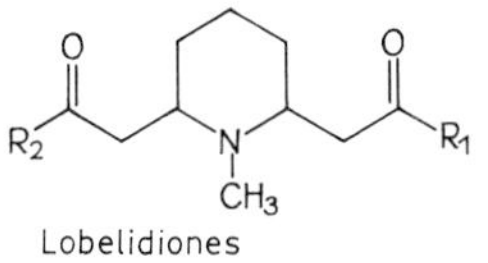

Lobelidiones

347

the side chains may be *cis* or *trans*. Compounds lacking the *N*-methyl group, and 2-monosubstituted L. a. also occur.

The chief alkaloid is Lobelin (see). In attempts to imitate the biological synthesis of L. a., lobelidiones have been synthesized under mild conditions from glutardialdehyde, methylamine and acylacetic acid. The drug is used as an antiasthmatic.

Lobelidiols: see Lobelia alkaloids.

Lobelidiones: see Lobelia alkaloids.

Lobelin: (−)-lobelin, *cis*-8,10-diphenyl-lobelionol, the main Lobelia alkaloid (see). Structurally, it is a lobelionol, in which both R_1 and R_2 are phenyl ($-C_6H_5$) groups. L. crystallizes as colorless needles, m. p. 130–131 °C, $[\alpha]_D^{15}$ − 43 ° ($c = 1$, ethanol). It is isolated from *Lobelia inflata,* and is used medicinally as a respiratory analeptic. On account of its nicotine-like properties, it is also used in the treatment of smoking addiction. Simultaneous administration of nicotine and L. has an additive effect, leading to nausea and aversion.

Lobelionols: see Lobelia alkaloids.

Loganin: an ester glucoside belonging to the iridoid group, m. p. 222–223 °C. Cleavage with emulsin produces the aglycon, loganetin. L. and the free acid, loganic acid, are found in *Strychnos* and *Menyanthus* spp. L. is a key compound in the biosynthesis of iridoids and many alkaloids. For the biosynthesis of L., see Iridoids.

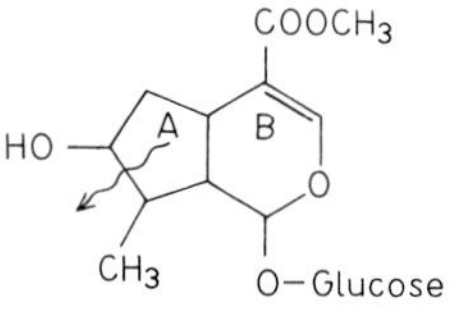

isolated from rat tissues and feces, and from the cac tus *Lophocereus schottii*. It has the structure of a tet racyclic triterpene with a 31,32-bisdemethyl-lanos tane skeleton. Its structure is therefore intermediat between lanosterol and the sterols derived from la nosterol.

Lophophorin: see Anhalonium alkaloids.

Lowry method: see Proteins.

LPH: see Lipotropin.

LRH: see Releasing hormones.

LSD: see Lysergic acid diethylamide.

LTH: see Prolactin.

Lubimin: see Phytoalexins.

Luciferase: a low molecular weight oxidoreduc tase, which catalyses the dehydrogenation of lucifer in in the presence of oxygen, ATP and magnesiur ions. During this process, 96% of the energy re leased appears as visible (mostly blue) light. Thi is the basis of Bioluminescence (see). All L. so fa purified are oligomeric, low molecular weight, thic proteins. L. of the American firefly *(Photinus pyralis* has M_r 95 000, with 2 subunits (M_r 50 000). L. o *Photobacterium* has M_r 80 000, with subunits of M 38 000 and 41 000. L. of the phosphorescent cora *(Renilla reniformis)* has M_r 34 000, with 3 subunit (M_r 12 000).

Luciferin: a collective name for the substrates o luciferases. By the action of the enzyme and in th presence of oxygen, L. gives rise to bioluminescence

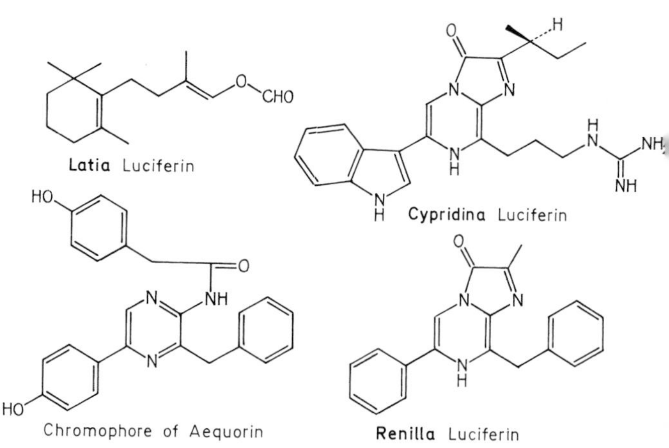

Latia Luciferin

Cypridina Luciferin

Chromophore of Aequorin

Renilla Luciferin

COOCH₃

HO

CH₃   O—Glucose

Loganin

Lohmann reaction: see Phosphagens.

Lomatiol: see Naphthaquinones.

Lophenol: 4α-methyl-5α-cholest-7-en-3β-ol, a zoo- and/or phytosterol (see Sterols). M_r 400.69, m. p. 151 °C, $[\alpha]_D$ +5 ° (chloroform). L. has been

Electron-excited states of the oxidation product o L. (thought to be peroxides) are responsible for ligh emission. The structures of five L. are known (Fig.) *Photinus* L. (for reaction mechanism and formula see Bioluminescence) is derived from fireflies of the genus *Photinus,* and it represents the only knowr naturally occurring benzthiazole derivative. Both the D- and the L-form are chemiluminescent, but only the D-compound is biologically active. It crystallizes as yellowish needles, m. p. 190 °C (dec.), $[\alpha]_D^{2}$ − 29 ° (formamide). Different species of *Photinu* emit different colors of yellow to green light, bu these differences are due to the different types of luciferase, and the L. is always the same. *Latia* L. from

he limpet, *Latia neritoides,* has an unusual sesquiterpene structure. When stimulated, the animal produces a slime, which has a powerful fluorescence
due to the interaction of L., luciferase, O_2 and a
"purple protein". In the ostracod crustacean, *Cypridina hilgendorfi,* the L. and a specific luciferase are
stored in separate glands. When stimulated, the animal secretes both components simultaneously into
the surrounding, oxygen-containing sea water,
where they produce a blue fluorescence. *Cypridina*
L. is biosynthesized from tryptophan, arginine and
isoleucine. *Renilla* L. from the coelenterate *Renilla
reniformis* (sea pansy) is an unstable compound,
which is stored as its sulfate ester. L. is released by
an L. sulfokinase. Electrical or mechanical stimulation results in bioluminescence, which spreads
over the surface of the animal in concentric waves of
green light.
Bacterial L. has not yet been characterized. The
corresponding luciferase produces luminescence in
the presence of FMNH and straight chain aldehydes
with more than 7 C-atoms.
The structural elucidation of L. was delayed on account of their low natural concentration. 30000 fireflies were required for the isolation of 15 mg L., and
40000 sea pansies yielded only 0.5 mg of the *Renilla*
L. Many L. and synthetic analogs show a spontaneous luminescence in proton-free solvents, such as
dimethyl sulfoxide, but the quantum yield is lower
than in bioluminescence.
See also Photoproteins.

Lumazine: see Pteridines.
Lumicolchicine: see Colchicum alkaloids.
Lumisterol: see Vitamins (Vitamin D).
5α-Lupane: see Lupeol.
Lupanine: see Lupin alkaloids.
Lupeol: a pentacyclic, triterpene alcohol, M_r
426.73, m.p. 215 °C, $[\alpha]_D^{20}$ +27.2 ° ($c=4\cdot8$ in
CHCl$_3$). L. has a 5α-lupane ring system. It occurs
free, esterified and as the aglycon of triterpene Saponins (see) in many plants. It has been found, e.g. in
the latex if *Ficus* spp., in the seed coats of the yellow
lupin *(Lupinus luteus)* and in the leaves of mistletoe.
It has also been detected in the cocoons of the silk-
worm, *Bombyx mori.*

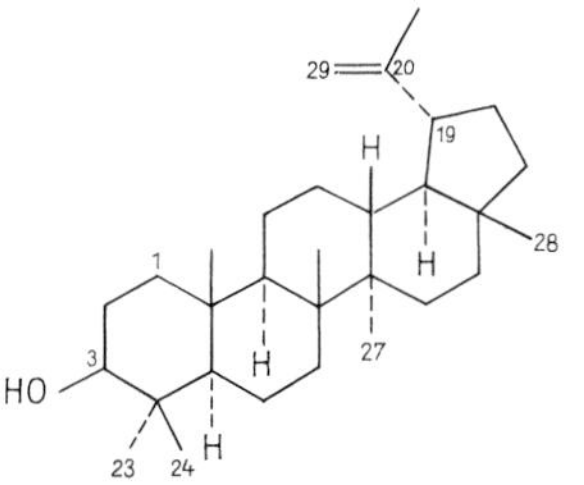

Lupeol

Lupin alkaloids: a group of quinolizidine alkaloids, containing a ring system variously known as
quinolizidine, octahydropyridocoline, norlupinane,
or 1-azabicyclo [0,4,4] decane. This ring system may
be further condensed with other N-containing ring
systems, so that tri- and tetracyclic, as well as bicyclic L. are known. Chief representatives are *lupine,
lupanine, sparteine* and *cytisine.* L. are derived biosynthetically from lysine, via its decarboxylation
product, cadaverine. Two molecules of cadaverine
give rise to lupinine, whereas sparteine is derived
from three molecules of cadaverine (Fig.). L. occur
in plants of the genus *Lupinus* (lupins), in broom
(*Sarothamnus scoparius* Koch), laburnum (*Laburnum anagyroides* Medic.) and gorse (*Ulex europaeus*
L.). Bitter lupins cannot be used directly as animal
feed, owing to the bitter and toxic L., but these can
be removed by steaming, soaking and extraction. Alternatively, the selectively bred alkaloid-poor sweet
lupins may be used.

Lupinine: see Lupin alkaloids.
Lutein, xanthophyll: (3R,3'R,6'R)-β,ε-carotene-
3,3'-diol, or 3,3'-dihydroxy-α-carotene (Fig.), a carotenoid of the xanthophyll group. M_r 568.85, m.p.
193 °C, $[\alpha]_{Cd}^{20}$ +160 (chloroform), +145 ° (ethyl
acetate). L. contains the same chromophore as α-
carotene, and it is isomeric with zeaxanthin. L. is a
yellow pigment present, together with carotene and
chlorophyll, in all green parts of plants. It is also
present in many yellow and red flowers and fruits. It
is also found in animals, e.g. in bird feathers, egg
yolk and the corpus luteum. It may be present in

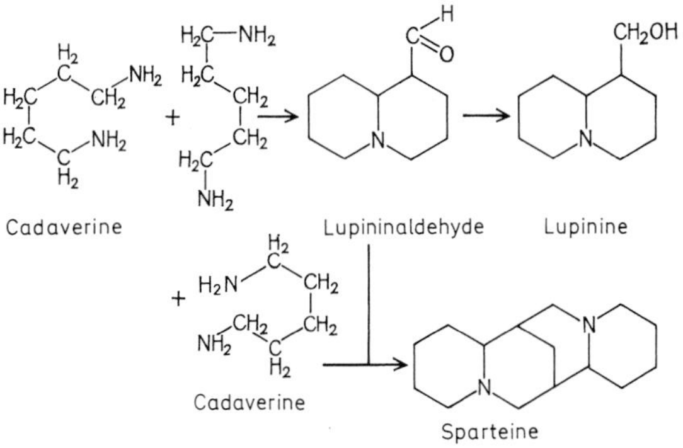

Biosynthesis of the lupin alkaloids, lupinine and sparteine

Lutein

free form, or as an ester; it has no vitamin A activity. L. was first crystallized in 1907 by R. Willstätter from stinging nettles, then later from chicken egg yolk. The structure was elucidated in 1951 by P. Karrer.

Luteinizing hormone, *LH, lutropin, gonadotropin II, prolan B, metakentrin, corpus luteum ripening hormone, interstitial cell-stimulating hormone, ICSH:* an important gonadotropin. LH is a glycoprotein (M_r 26000) containing 22% carbohydrate. It consists of a α-chain (96 amino acid residues, M_r 10791) and a β-chain (119 amino acid residues, M_r 12715) (bovine LH). The primary structures of the chains are known. The α-chain is identical with that of follicle stimulating hormone, thyrotropin and chorionic gonadotropin. The β-chain is hormone-specific. LH acts with follicle stimulating hormone to stimulate growth of the gonads and the synthesis of sex hormones.

Luteinizing hormone releasing hormone: see Releasing hormones.

Luteolin: see Flavones.

Luteolinidine: see Anthocyanins.

Luteotropic hormone: see Prolactin.

Luteotropin: see Prolactin.

Lutropin: see Luteinizing hormone.

Lyases: see Enzymes, Table 1.

Lycopene: one of the carotenoids. L. is an unsaturated, aliphatic hydrocarbon of isoprenoid origin (for formula see Carotenoids). M_r 536, m.p. 173 °C. L. is a red plant pigment, widely distributed and especially plentiful in fruits and berries. For biosynthesis of L. see Tetraterpenes. L. was crystallized from tomatoes in 1876 by A. Millardet. Its empirical formula was determined in 1910 by R. Willstätter and H. H. Escher. The final structure elucidation was performed from 1930 to 1932 by P. Karrer et al. and by R. Kuhn and C. Grundmann.

β-Lycotetraose: see α-Tomatin.

Lymphocytes: colorless, nucleated blood cells, approximately the same size as erythrocytes. L. are immunologically competent, i.e. they recognize antigens, and they play a central role in the Immune response (see). An average human possesses about 10^{12} L., which make up 1.3 kg. Only 3 g are present in the blood, the majority being found in bone marrow, spleen, lymph glands and lymph. L. are classified, according to their origin and function, into *thymus-dependent L. (T-L.)* and *thymus-independent L. (B-L.).* The T-L. include "killer cells" which form the basis of cellular immunity (e.g. T-L. are responsible for the rejection of transplants), "helper" and "suppressor" T-L. The latter interact with B-L. to promote or inhibit their response to antigens. Clonal multiplication of B-L. gives rise to plasma cells, which produce antibodies (humoral immunity). Each plasma cell produces 2000 molecules of a sin-

gle type of antibody per second. Organs rich in lymphocytes are spleen, tonsils, aggregated lymphatic nodules (Peyer's patches) of the ileum, and the vermiform appendix.

Lymphokines: proteins released from activated cells of the immune system which coordinate the immune response. *Interleukin I* is a protein of M_r 12000–16000 released from macrophages which have been stimulated by contact with an antigen. It enhances the response of T-cells which have been primed with antigen, by stimulating their production of Interleukin 2. It also stimulates the proliferation of B-cells, in conjunction with B-cell growth factor. Interleukin 1 or a very similar protein is also produced by the keratin-producing cells of the skin, corneal cells and cells lining the mouth. It has a range of effects on a variety of non-lymphatic cells, all of which contribute to inflammation. Interleukin I may be identical to endogenous pyrogen, a macrophage product which elevates body temperature by interaction with the appropriate region of the hypothalamus.

Interleukin 2 is a protein of M_r 15500 which is produced by T-cells and which induces the proliferation of T-cells, but only those which have been antigenically stimulated. *B-cell growth factor (BCGF)* is produced by activated T-cells. It stimulates proliferation of B-cells, but not their differentiation to the antibody-producing state. For this, additional L. are required. [J. L. Marx, *Science* 221 (1983) 1362–1364.]

Interleukin 3 (Il-3) regulates growth and differentiation of the pluripotent stem cells which are precursors to all types of blood cells. It appears to be identical with the following: multi-colony stimulating factor CSF, haematopoietic growth factor, burst-promoting activity, P-cell stimulating factor, mast cell growth factor, histamine-producing cell stimulating factor, Thy 1-inducing activity.

Il-3 is a glycoprotein with estimated M_r around 30000. The murine gene for Il-3 has been cloned and sequenced. It encodes a polypeptide of 166 amino acids, of which the N-terminal 32 are removed during maturation. [M. C. Fung et al. *Nature* 307 (1984) 233–237.]

Lymphotoxin: see Tumor necrosis factor.

Lys: abb. for Lysine.

Lysergic acid: a tetracyclic indole derivative, M_r 368.32. The D-form, m. p. 240 °C (d.), $[\alpha]_D^{20}$ +40 ° ($c = 0.5$, pyridine) is one of the Ergot alkaloids (see). Change of configuration at C-8 produces biologically inactive isolysergic acid. L. is a close structural relative of LSD (see Lysergic acid diethylamide).

Lysergic acid diethylamide, *LSD:* a synthetic derivative of D-lysergic acid. It is the most potent psychotomimetic substance known. Hallucinogenic derivatives of lysergic acid occur in the Mexican ritual drug, ololiuhqui (seeds of *Rivea corymbosa*);

hese are lysergic acid amide (ergine) and lysergic
cid hydroxyethylamide. LSD can be prepared semi-
ynthetically from ergot alkaloids, which also occur
n *Rivea* seeds (e.g. ergometrine). The hallucinogenic action of LSD, which is characterized by a
tate resembling schizophrenia, was discovered ac-
identally by A. Hofmann during the recrystalliza-
ion of a sample of LSD tartrate.

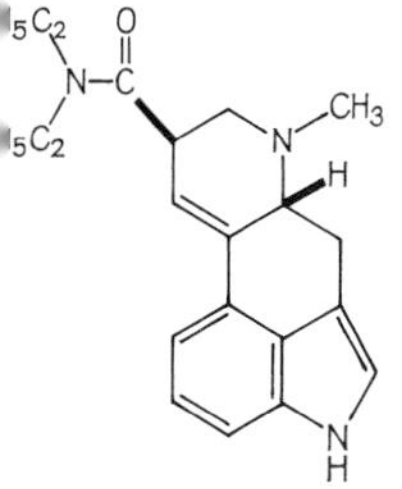

D-Lysergic acid diethylamide (LSD-25)

L-Lysine, *Lys:* 2,6-diaminocaproic acid, H_2N-
$CH_2)_4-CH(NH_2)-COOH$, a basic proteogenic, es-
ential amino acid. M_r 146.2, m.p. 224 °C (d.), $[\alpha]_D^{25}$
$+25.9$ $(c=2, 5M HCl), +13.5 (c=2, water)$. The
roteins of cereals (wheat, barley, rice) and other
egetable foodstuffs are rather poor in Lys. Children
nd young growing animals have a particularly high
equirement for Lys, since it is needed for bone for-
nation. Like threonine, Lys does not take part in re-
ersible transamination.
n rat liver, Lys degradation (Fig. 1) occurs primarily
n the mitochondria. Some features of the degrada-
ion pathway appear to be a reversal of the reactions
of Lys synthesis (Figs. 4 and 5). Saccharopine is
ormed, offering a direct route to 2-aminoadipic-
-semialdehyde, and bypassing the cyclic piperi-
leine and piperidine intermediates. The semialde-
yde is also produced by a second pathway, initi-
ted by L-amino acid oxidase. The resulting oxoacid
yclizes spontaneously to Δ'-piperideine 2-carboxyl-

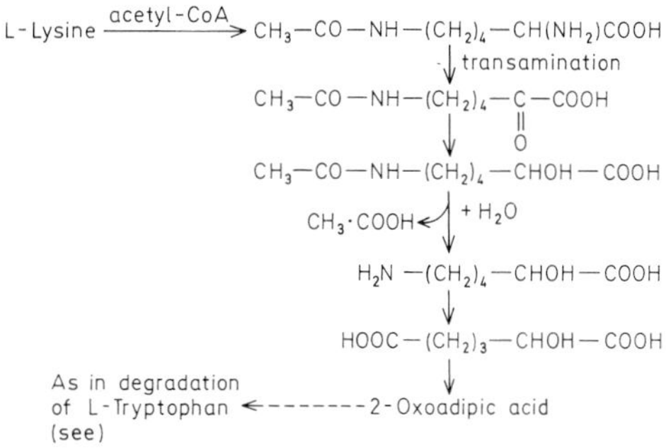

Figure 2. Degradation of Lysine by yeasts

ic acid. The semialdehyde is subsequently oxidized
to 2-aminoadipic acid, which is transaminated to
2-oxoadipic acid. Further degradation, via glutaryl-
CoA to acetyl-CoA is identical with the final stages
of L-tryptophan degradation (see L-Tryptophan).
An alternative pathway (Fig. 2) of Lys degradation
in mammals (also in yeast) involves acetylated intermediates; cyclization is thus prevented. A further
pathway exists in bacteria (Fig. 3). D-Lysine may also enter this pathway by conversion to L-lysine by a
pyridoxal phosphate-dependent racemase. Lys is
converted to 5-aminovaleramide by L-lysine oxygenase.
There are two pathways of Lys biosynthesis found
in plants and microorganisms: 1. The *aminoadipate
pathway* (Fig. 4) is found in *Neurospora crassa* and
Saccharomyces cerevisiae, and in other fungi and
yeasts. The reactions of this pathway are analogous
to those of the Tricarboxylic acid cycle (see). 2. The
diaminopimelate pathway (Fig. 5) occurs in bacteria,
cyanobacteria, green algae, higher plants and certain
fungi.
About 20000 tons of Lys per year are produced in
Japan by a specific fermentation process (Kyowa
fermentation industry). The Lys is used as a supple-

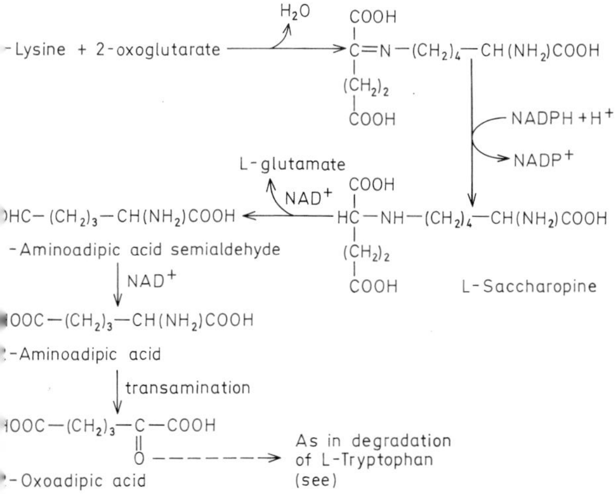

Figure 1. Degradation of lysine by animal liver

ment to foodstuffs, and to increase the biological value of low value plant dietary proteins. The growth rate of poultry and pigs is appreciably increased by the addition of 0.1–0.3% Lys to their feed. Also of technical importance is the enzymatic synthesis of Lys from DL-2-aminocaprolactam, using microbial L-aminocaprolactam hydrolase. The

remaining D-2-aminocaprolactam is racemized enzymatically, then converted entirely to L-Lys.
Lys was first isolated by Drechsel from casein in 1889.

Lysocephalins: see Phosphatides.
Lysogeny: see Phage development.
Lysolecithins: see Phosphatides.
Lysophosphatidic acid: see Inborn errors of metabolism.
Lysopine: see D-Octopine.
Lysosomal storage disease: see Inborn errors of metabolism.
Lysosomes: organelles, 0.2–2 nm diameter, found in the cytoplasm of eukaryotic cells. L. are bounded by a single lipoprotein membrane, but otherwise show no fine structure. Under the light or electron microscope, L. are markedly polymorphic. They can be characterized biochemically or histochemically, but not morphologically. L. are sites of Intracellular digestion (see), particularly of biological macromolecules, such as proteins, polynucleotides, polysaccharides, lipids, glycoproteins, glycolipids, etc. Approximately 40 different lysosomal hydrolases are responsible for this degradative activity; they all show optimal activity at acidic pH values. Marker enzyme for L. is acid phosphatase. Under anaerobic conditions, L. are destroyed, and the lysosomal enzymes are released into the cytoplasm; subsequent degradation of the cell contents by the lysosomal enzymes is known as autolysis. Autolysis is a characteristic post mortem process.
Primary L. are formed in the Golgi apparatus of the cell. Fusion of L. with phagocytosing and pinocytosing vacuoles (phagosomes) produces digestive vacuoles, known as secondary L. Excess or old cell parts, including mitochondria, may be digested by L.

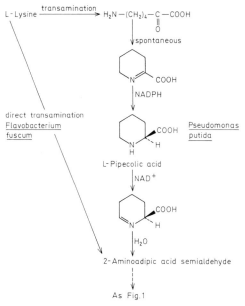

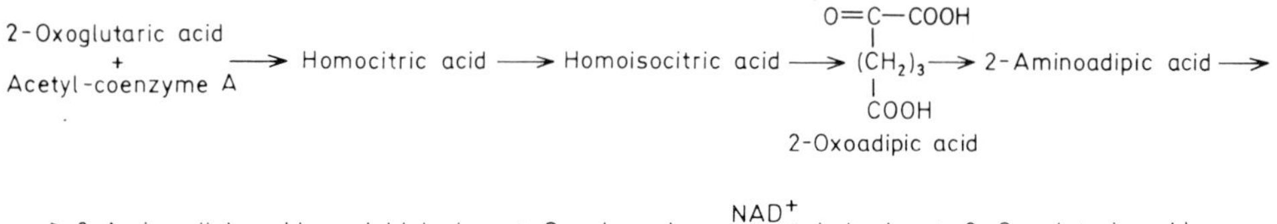

Figure 3.   Bacterial degradation of Lysine

2-Oxoglutaric acid
+
Acetyl-coenzyme A → Homocitric acid → Homoisocitric acid → 2-Oxoadipic acid → 2-Aminoadipic acid →

→ 2-Aminoadipic acid semialdehyde → Saccharopine --NAD⁺→ L-Lysine + 2-Oxoglutaric acid

Figure 4.   2-Aminoadipic acid pathway of L-lysine biosynthesis

Pyruvic acid
+
L-Aspartic acid --−H₂O→ 2,3-Dihydropicolinic acid --NADPH+H⁺→ Δ¹-Piperideine-2,6-dicarboxylic acid → N-Succinyl-2-oxo-L-6-aminopimelic acid

→ N-Succinyl-L-2,6-diaminopimelic acid → L-2,6-Diaminopimelic acid → Meso-2,6-Diaminopimelic acid --−CO₂→ L-Lysine

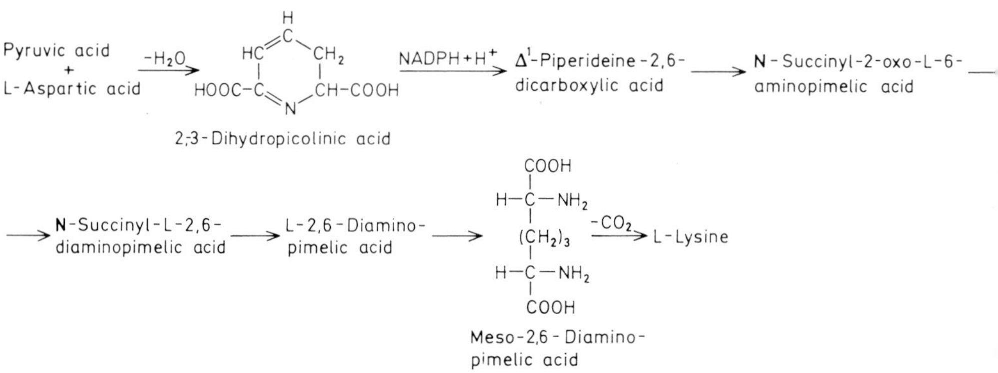

Figure 5.   Diaminopimelic acid pathway of L-lysine biosynthesis

(phagolysosomes). In ameba, L. apparently provide the digestive enzymes. L. were discovered in 1959 by De Duve (Nobel prize 1974).

Lysozyme, *endolysin, muramidase, N-acetylmuramide glycanohydrolase* (EC 3.2.1.17): a widely occurring hydrolase, found in phages, bacteria, plants, invertebrates and vertebrates. In the latter, it is found particularly in egg white, saliva, tears and mucosas. L. acts as a bacteriolytic enzyme by hydrolysing the β-1,4 linkage between *N*-acetylglucosamine and *N*-acetylmuramic acid in the proteoglycan of the bacterial cell wall. L. therefore affords protection against bacterial invasion. All animal L. consist of a single chain of 129 amino acid residues, with homologous sequence (M_r 14200-14600). There are four disulfide bridges, and the chain is folded into a known tertiary structure (42% α-helix, with hydrophobic tryptophan residues on the outer surface), with prominent hydrogen bonding between the side chains of Ser, Thr, Asn and Gln. Like other hydrolases (e.g. papain, ribonuclease), the molecule of L. has a cleft which houses the active center of the enzyme and serves for the attachment of the substrate, in this case a hexasaccharide unit of the proteoglycan molecule. There is a remarkable correspondence between the primary and tertiary structures of L. and α-lactalbumin (123 residues). It is thought that both proteins arose from a common precursor protein with lysozyme activity, an example of divergent evolution by gene duplication. On the other hand, there is no structural relationship between animal L. and bacteriophage L. The latter contain 157 residues (λ phage endolysin, M_r 17873), or 164 residues (T4- and T2-phage L., M_r 18720), and they are either endoacetylmuramidases (T4, T2), or endoacetylglucosaminidases (Streptococcal L.).

Lysylbradykinin: see Bradykinin.

M

mμ: abb. for millimicron.

Maackiain: 3-hydroxy-8,9-methylenedioxypterocarpan, see Pterocarpans.

Macdougallin: 14α-methyl-5α-cholest-8(9)-en-3β, 6α-diol, a phytosterol (see Sterols), M_r 416.69, m.p. 173 °C, $[\alpha]_D$ +72 ° (chloroform). M. was isolated from the cacti, *Peniocerius fosteriunus* and *Peniocerius macdougalli*. It has the structure of a tetracyclic triterpene with a 30,31-bisdemethyllanostane skeleton, and it represents an intermediate structure between lanosterol and the sterols derived from lanosterol.

Macroelements: see Mineral nutrients.

α₂-Macroglobulin, *α₂-antiplasmin:* an α₂-plasma protein. The first reported M_r determination by sedimentation diffusion gave a value of 820000. Later determination by sedimentation equilibrium gave values of 725000. These results, together with studies of subunit composition, indicate that the true M_r lies in the range 650000–725000. α₂-M. is a glycoprotein containing 8.2% carbohydrate. The carbohydrate moiety contains mannose, fucose, N-acetylglucosamine and sialic acid. Electron microscope studies of the protein reveal a structure resembling two beans facing each other; these two identical subunits are bound noncovalently. Each subunit consists of two peptide chains linked covalently by a disulfide bridge. α₂-M. binds tightly and inhibits a number of proteases of varying specificity and origin, e.g. trypsin, plasmin, thrombin, kallikrein and chymotrypsin. It is therefore a natural inhibitor of plasmin. Unlike other protease inhibitors, it does not block the active centers of the enzymes, so that α₂-M.-proteinase complexes are almost fully active towards low molecular weight synthetic substrates. The bound protease acts upon the α₂-M. subunit (M_r about 350000) and cleaves one protein chain near its midpoint. All the products of this cleavage are still linked covalently by disulfide bridges. Treatment of the α₂-M.-trypsin complex with urea and dithiothreitol reveals two proteins derived from α₂-M., M_r 185000 (intact subunit chain, originally linked to a similar chain by disulfide bridge), and M_r 85000 (proteolytic derivative of a 185000 M_r chain). α₂-M. selectively inhibits the growth of tumors in cell cultures and in rats. α₂-M. and ceruloplasmin are two plasma proteins with a specific binding affinity for zinc. In humans, the normal plasma concentration of α₂-M. is 220–380 mg per 100 ml.

Macrolides, *macrolide antibiotics:* a group of antibiotics from various strains of *Streptomyces*, all having the same complex macrocyclic structure. M. inhibit protein synthesis by blocking transpeptidation and the translocation on the 50S ribosomal subunit (similar to Chloramphenicol, see). Examples of M. are erythromycin (Fig.), spiramycin, oleandomycin, carbomycin, angolamycin, leucomycin, picromycin. Almost all M. are used therapeutically as broad spectrum antibiotics.

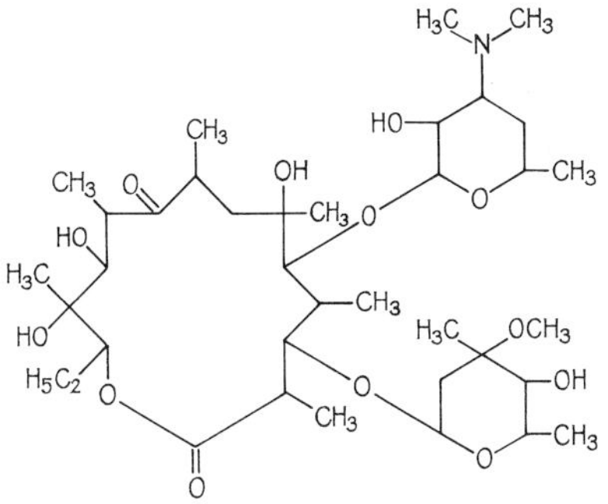

Erythromycin

Madder dyes: plant dyes from madder *(Rubia tinctorum)* and other members of the madder family *(Rubiaceae).* Important representatives are the glycosidically bound components alizarin and purpurin. The M.d. were chiefly used to make madder enamel, a colored paint with outstanding fastness to light. Now the product is made almost exclusively from synthetic alizarin.

Magnesium, *Mg:* a cation widely distributed in biological systems, with many different biological functions. Mg is the fourth most abundant cation in the vertebrate, and it has great biological significance as a constituent of the porphyrin system of chlorophyll. Mg is an essential nutrient, and it plays an important part in metabolism by acting as a cofactor of many different enzyme systems. In particular, it is involved in the reactions of energy metabolism, including the enzymatic cleavage of phosphate esters and the transfer of phosphate groups. Mg is an activator of phosphatases and a cofactor of practically all ATP-dependent phosphorylation reactions, e.g. hexokinase, phosphofructokinase and adenylate kinase; in every case an ATP-Mg complex (1:1) is formed, and this (not free ATP) is a substrate. Intracellularly, Mg is present chiefly as Mg^{2+} and $MgOH^+$. About 0.05% of the animal body is Mg, 60% in the skeleton, and 1% in extracellular fluids.

Maize factor: see Zeatin.

Major histocompatibility complex, *MHC:* several closely linked genetic loci in vertebrates which encode cell-surface glycoproteins and serum proteins known as *histocompatibility antigens*. In addition to the MHC, which have been found in all vertebrates examined for them, there are other cell surface antigens which have been little studied and are referred to as minor histocompatibility antigens (complexes).

The MHC in human beings and mice are by far the best studied. They were discovered as factors which cause rejection of organ grafts between genetically non-identical individuals, hence the adjective "histocompatibility". Their significance to the organism, however, lies in their function as moderators of the immune response. There are three classes of MHC antigens in human beings and mice which have been characterized; other MHC genes in the mouse have been mapped to the H-2 region, but little is known about them.

Class I includes the HLA-A, HLA-B, HLA-C and HLA-D gene products in humans, and the $H2K$ and $H2D$ gene products in mice. The corresponding loci in rats are Rt-1, and in chickens, B. The products of these highly polymorphic genes are components of cell-surface glycoproteins which consist of two polypeptide chains each. Only one of these polypeptides is encoded by the MHC gene; the other is a β_2-microglobulin. In the mouse, the MHC chain has M_r about 44000, while the β_2-microglobulin chain has M_r 12000. Class I antigens are found on the surfaces of all cells except sperm. They are important in the cell-mediated immune response, in which macrophages, cytotoxic T lymphocytes (CTL) and granulocytes are induced to kill body cells which bear specific foreign (viral or cancer) antigens on their surfaces. However, these specifically "programmed" killer cells attack only cells which display both the foreign antigen and the same class I antigens as the killer cells themselves. Activated CTL from one mouse cannot kill cells from a second mouse infected with the same virus unless the second mouse has the same set of class I antigens as the first. The actual function of the class I antigens in this process is still unknown.

Class II MHC antigens are found only on certain lymphoid cells. These are products of (mouse) Ia and Ir (for immune response) genes.

The Ia proteins are membrane-bound dimers of α- (M_r 55000) and β- (M_r 28000) subunits with unknown carbohydrate portions; they are responsible for activation of helper T lymphocytes. The first step of this process is ingestion of an antigen by accessory cells (macrophages), which partially digest proteins and "present" peptides to the precursors of helper T cells. (Helper T lymphocytes enhance antibody production by those B lymphocytes which recognize the same antigen as the helper cells; thus in order to be antigenic, a peptide must stimulate both T and B lymphocytes.) It has recently been discovered that these peptides must be bound to the Ia molecules. In a study of binding of various peptides to the allelic Ia molecules, it was found that the ability of a peptide to bind to a given Ia molecule was closely correlated to its antigenicity in a strain of mice which express that same Ia antigen. Peptides which do not bind to the Ia molecules of a given mouse strain are not antigenic in that strain. [S. Buus et al. Science 235 (1987) 1353-1358]

The known immune response genes influence antibody production, delayed type hypersensitivity and proliferation of T cells in vitro. Enhancing Ir genes map to the A, B, C or E subregions of the I part of the mouse H-2 (MHC) complex; suppressing Ir genes map to the J subregion of the I region.

Class III MHC genes code for serum proteins consisting of three covalently linked polypeptides, the α-, β- and γ-chains have M_r 87000, 78000 and 33000, respectively. This protein is the C4 component of the classic Complement system (see). Other complement components which are coded by MHC-linked genes are C3 in mouse, Bf in guinea pig, man and mouse, and C2 in man. [J. Klein "The Major Histocompatibility Complex of the Mouse". Science 203 (1979) 516-521; H. L. Ploegh et al. Cell 24 (1981) 287-299]

Major inorganic elements: see Mineral nutrients.

Malate: see Malic acid.

Maleic hydrazide, *MH*: 1,2-dihydro-3,6-pyridazinedione, a synthetic plant growth retardant, which is used as a herbicide against grasses. It causes a unique depression of growth, inhibition being marked but temporary. It inhibits germination of seeds, suppresses growth of roots and terminal shoots, and retards flower and bud development. It prevents the formation of suckers in tobacco and tomatoes, and prevents the sprouting of onions and potatoes. The effects are limited to green plants and there is little or no effect on any other organism. The reason for this specificity is not known. Animals are very tolerant to MH; the LD_{50} for rats is 4 g per kg, and rats are unaffected by a diet containing 1% MH throughout their lives. There is nevertheless some evidence for carcinogenic activity and MH has been listed as a suspected carcinogenic agent.

Maleic hydrazide

Malformin A: a heterodetic cyclic pentapeptide with antibiotic activity, from *Aspergillus niger*. It causes malformation of the roots of cereals. The primary structure was revised in 1974, and confirmed by total synthesis.

┌─Ile→D-Cys→D-Cys→Val→D-Leu─┐

Malformin A

Malic acid: monohydroxysuccinic acid, HOOC-CHOH-CH$_2$-COOH, a dicarboxylic acid found in many plant juices, usually in the L(+)-form. m. p. 100 °C b. p. 140 °C (d.). The malate ion is formed in the Tricarboxylic acid cycle (see) and the Glyoxylate cycle (see). Malate plays an important role in the Diurnal acid rhythm (see) of the *Crassulaceae* (stonecrop family).

Malic enzyme(s), *L-malate-NADP oxidoreductase, decarboxylating;* (EC 1.1.1.40): an important enzyme found in most organisms, which catalyses the decarboxylation of L-malate to pyruvate and CO_2, with concomitant reduction of $NADP^+$ to NADPH (or the synthesis of malate by the reverse reaction): $HOOC-CH_2-CHOH-COOH + NADP^+ \leftrightarrows CH_3-CO-COOH + CO_2 + NADPH + H^+$. M.e. has various metabolic roles: 1. Synthesis of malate by the action of M.e. may serve as an Anaplerotic reaction (see) of the TCA-cycle; 2. An important route for the total combustion of any TCA-cycle intermediate is conversion to malate, followed by decarboxylation of malate to pyruvate and CO_2 by M.e. Animal mitochondria contain two M.e., one specific for NADP, the other utilizing both NADP and NAD. Regulation is complex; when glycolysis is low, free CoA activates M.e., thereby promoting oxidation of malate, whereas rapid glycolysis increases NADH which inhibits M.e.; 3. In plants operating the C_4 pathway of photosynthesis (see Hatch-Slack-Kortschack cycle), mesophyll cells export malate to bundle sheath cells, where it is decarboxylated by M.e.; the resulting CO_2 is assimilated by the Calvin cycle while the pyruvate returns to the mesophyll cells; 4. M.e. appears to be important in a cycle for the generation of cytoplasmic NADPH: malate exported from the mitochondria is decarboxylated by cytoplasmic M.e., with the formation of NADPH. The pyruvate enters the mitochondria and is converted to oxaloacetate, followed by hydrogenation to malate by NADH. Thus, reducing power (NADH) from mitochondria is converted into reducing power (NADPH) in the cytoplasm, where it is available for NADPH-dependent biosynthetic reactions (see Hydrogen metabolism).

Malonic acid: $HOOC-CH_2-COOH$, a dicarboxylic acid, which has been found in the free form in plants, but is of only sporadic occurrence. m.p. 135.6 °C. In the cell M.a. is present as its anion (malonate); this is a known competitive inhibitor of succinate dehydrogenase in the tricarboxylic acid cycle. A metabolically important derivative of M.a. is malonyl-CoA, an intermediate of Fatty acid biosynthesis (see).

Maltase: see Disaccharidases.

Maltose, *maltobiose, malt sugar:* 4-O-α-D-glucopyranosyl-D-glucose, a reducing disaccharide, M_r 342.3. The monohydrate of M., crystallized from water or dilute ethanol, m.p. 102–103 °C, does not lose water of crystallization when dried at room temp. in vacuo over P_2O_5. $[\alpha]_D^{20}$ 112 ° → 130.4 °C ($c = 4$). M. consists of two molecules of D-glucose linked by an α-1,4-glycosidic bond; both glucose residues are present in the pyranose form. M. is a stereoisomer of cellobiose. With lactose and sucrose, it is one of the three most common naturally occurring disaccharides. It can be hydrolysed to 2 molecules of D-glucose by dilute mineral acids, or enzymatically by α-glucosidase (known previously as maltase) of yeast, malt and digestive secretions. M. is the fundamental structural unit of starch and glycogen; it also occurs free in higher plants. The monohydrate of M. is prepared technically in about 80% yield by degradation of starch with amylases (diastase). It serves as a fermentable substrate in brewing, as a component of

prepared bee food, as a substrate in microbiological growth media, and generally in food and pharmaceuticals as a nutrient and sweetner (one third as sweet as glucose).

α-Maltose

Malt sugar: see Maltose.

Malvidin: 3,5,7,4′-tetrahydroxy-3′,5′-dimethoxy-flavylium cation, the aglycon of various anthocyanins. 10 natural glycosides of M. are known, e.g. oenin (syn. primulin, ligulin), the 3-β-glucoside of M., is the pigment in the skins of black grapes and the flowers of *Primula* spp.; and malvin (3,5-di-β-glucoside of M.) from common mallow *(Malva sylvestris)* and other flowering plants.

Malvin: see Malvidin.

Mammotropin: see Prolactin.

Manganese, *Mn:* a bioelement present in all living cells, which usually contain less than 1 ppm on a dry weight basis, or less than 0.01 mM in fresh tissue. Bone contains 3.5 ppm. Bacterial spores contain high levels of Mn(II) (0.3% dry weight). Mn(II) is necessary for sporulation in *Bacillus subtilis*, and these bacteria can maintain an intracellular Mn concentration of 0.2 mM against an external concentration of 1 μM. Mn is an essential nutrient for animals and plants. Mn deficiency in animals leads to degeneration of the gonads and to skeletal abnormalities; the characteristic skeletal abnormality in chickens is called slipped tendon disease, or perosis. In plants, Mn deficiency results in chlorosis and mottling.

Many glycosyl transferases (in particular galactosyl and *N*-acetylgalactosaminyl transferases) require Mn for activity, which explains the impairment of mucopolysaccharide metabolism associated with the symptoms of Mn deficiency. Mn is required by the enzyme farnesyl pyrophosphate synthetase, which catalyses a stage in cholesterol synthesis. Mn is also required at an earlier point in cholesterol synthesis, probably at some stage in the conversion of acetate into mevalonate. Lactose synthetase requires Mn. Pyruvate carboxylase contains four atoms of Mn (i.e. one for each biotin molecule); Mn(II) is essential for the transcarboxylation, and the initial ATP-dependent carboxylation of biotin requires either Mn(II) or Mg(II). Superoxide dismutase from *Escherichia coli* (M_r 39500) contains two atoms of Mn(III). Yeast mitochondrial superoxide dismutase is a tetramer containing one atom of Mn per subunit (M_r 24000). Similar enzymes are present in chicken liver mitochondria. Presumably the Mn of superoxide dismutases alternates between the III and II states during catalysis. Manganin, a protein present in peanuts (M_r 56000), contains one atom of Mn. Avimanganin, a protein of unknown function from avian liver (M_r 89000) contains one atom of

Mn(III). Concanavalin A (M_r 190000) from jack bean contains one atom Mn(II).

Mn may also be important in the regulation of enzyme activity, e.g. nonadenylylated glutamine synthetase requires Mg(II), but the adenylylated form binds Mn(II). Also the specificity of nucleases and DNA polymerases is changed when Mg(II) is replaced by Mn(II). The physiological significance of these differences is not clear.

The reaction center of photosystem II of the chloroplast contains 2–4 Mn atoms. The effect of Mn deficiency on photosynthesis resembles poisoning by DCMU (see), i.e. evolution of O_2 by photosystem II is inhibited, while photosystem I is unaffected. It is thought that the Mn is intimately involved in the primary event of the photolysis of water, probably alternating between the III and II states.

Mannans: polysaccharides widely distributed in plants as reserve material, and in association with cellulose as hemicellulose. M. of plants consists of D-mannose predominantly in α-1,4-glycosidic linkage. Yeast cell walls contain M., consisting of a backbone of α-1,6-linked mannose with short (1–3 mannose units) branches attached by α-1,2- and α-1,3-linkages.

D-Mannitol, *mannite, manna sugar:* a hexitol related to D-mannose. M_r 182.17, HOH$_2$C–CHOH–CHOH–CHOH–CHOH–CH$_2$OH, m.p. 166 °C, $[\alpha]_D$ −2.1 ° (water). M. is found widely in plants and plant exudates, and in fungi and seaweeds. The exudate of the manna ash *(Fraxinus ornus)* contains 75% M. M. is used as a sugar substitute for diabetics.

D-Mannosamine: 2-amino-2-deoxy-D-mannose, an amino sugar related to mannose, in which the hydroxyl group on C-2 of mannose is replaced by an amino group. M_r 179.17, m.p. of the hydrochloride 180 °C, $[\alpha]_D$ −3 °. M. is a constituent of neuraminic acids, animal mucolipids and animal mucoproteins.

D-Mannose, *seminose, carubinose:* a monosaccharide hexose, M_r 180.16. α-Form: m.p. 133 °C, $[\alpha]_D^{20}$ +29.3 ° → +14.2 ° (water). β-Form: decomposes at 132 °C, $[\alpha]_D^{20}$ −17.0 ° → +14.2 ° ($c=4$). D-M. is a C-2 epimer of D-glucose. In plants, free D-M. is found only occasionally, but it is a constituent of many high molecular weight polysaccharides in algae, yeasts and higher plants. Manna (the exudate of the manna ash, *Fraxinus ornus*) contains particularly high levels of D-M. In the metabolism of D-M., the activated form of the sugar is the GDP derivative (not the UDP derivative which occurs for many other sugars). D-M. is degraded via its 6-phospho-derivative, which is converted to glucose 6-phosphate.

CH$_2$OH

D-Mannose

Mannosidosis: see Inborn errors of metabolism.

D-Mannuronic acid: a uronic acid derived from D-mannose. M_r 194.14. M. is a constituent of the polyuronide, alginic acid.

Maple syrup urine disease: see Inborn errors of metabolism.

Marasmus: a nutritional deficiency disease, mostly in children under one year, due to lack of both dietary energy and protein. The condition is usually compounded by deficiencies of minerals and vitamins, and by gastrointestinal infections. Body weight is less than 60% of the standard for the age. Dehydration is common, there is little or no subcutaneous fat, and plasma albumin levels are about 25 g/l (normal levels are above 40 g/l). The characteristic hair and skin lesions of kwashiorkor are absent. M. victims often have a history of abrupt and early weaning, followed by dilute (and often unhygenic) artificial feeds that are low in both energy and protein. Subsequent gastroenteritis may be treated by withholding food, thus precipitating and/or exacerbating the disease. A distinction is made between nutritional M. (resulting from inadequate food alone) and M. produced by infection (although inevitably against a background of poor nutrition). M. is more common in urbanized than in rural areas of the developing world. Unlike kwashiorkor, M. shows no dependence on climate, and it was once well known in industrial areas of Europe and North America. M. in children is equivalent to starvation in adults. See Kwashiorkor, Protein-energy malnutrition.

Marihuana: see Hashish.

Marker enzymes: see Enzymes.

Marker rescue: see Recombinant DNA technology.

Marker synthesis: a laboratory synthesis developed by R.E. Marker for the conversion of diosgenin into progesterone (Fig.). From 1939 to 1942, Marker undertook a series of plant collecting expeditions to find a high yield source of the starting material, diosgenin, which was already known from a species of *Dioscorea* in Japan. Two good sources were found in Mexico, i.e. *Dioscorea composita* ("barbasco") and *D. macrostachya* ("cabeza de negro"), otherwise known as "Mexican yams". By the 1950's virtually all steroid hormones had been made in the laboratory from diosgenin. Other synthetic methods are now available, but the Marker synthesis is still important. As a result of the Marker synthesis, the price of progesterone fell in the late 1940's from over 80 dollars per gram (obtained by extraction from animal ovaries) to less than 50 cents per gram.

Maroteauz-Lamy syndrome: see Inborn errors of metabolism.

Master strand: see Codogenic strand.

Mastich: see Resins.

Matrix: see Stroma.

Maturation of RNA: see Post-transcriptional modification of RNA.

Mavacurin: one of the Curare alkaloids (see).

Maximal rate: see Michaelis-Menten kinetics.

Maxima substance C: see Isoflavone.

McArdle's disease: see Glycogen storage disease.

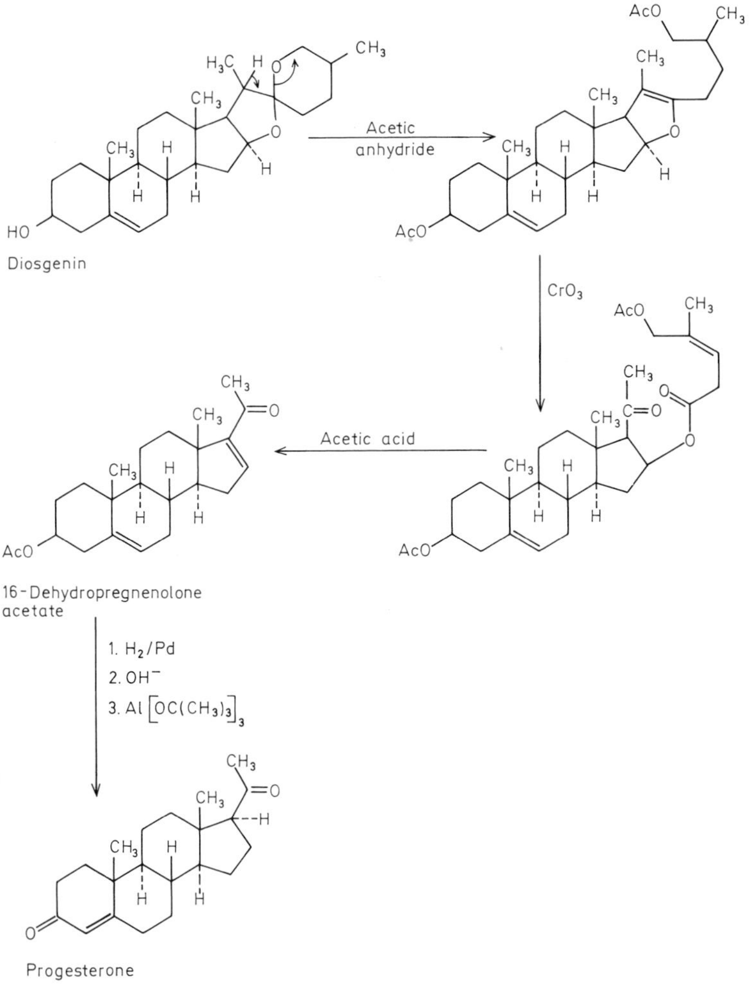

16-Dehydropregnenolone
acetate

Diosgenin

Progesterone

The Marker synthesis of progesterone from diosgenin

Mecamylamine: see Acetylcholine.

Mecocyanin: see Cyanidin.

Melanins: high molecular weight, amorphous polymers of indole quinone, empirical formula $(C_8H_3NO_2)_x$, containing 6–9% nitrogen. M. are natural pigments occurring predominantly in the animal kingdom in vertebrates and insects, and occasionally in microorganisms, fungi and higher plants. Mammalian colors are determined chiefly by two types of M.: the black or brown insoluble, nitrogenous Eumelanins (see), and the lighter colored sulfur-containing, alkali-soluble Phaeomelanins (see). A further group of low molecular weight, yellow, red and violet pigments is known as the Trichochromes (see); these are usually classified with the M., since they also serve as pigments, and are biogenetically closely related to M., i.e. they are derived from the oxidation of tyrosine (Fig.). M. are synthesized in melanosomes, which are found in cells called mel-

anocytes. M. are also synthesized in the retina of the eye. All M. are biosynthesized from L-tyrosine (I in Fig.), which is converted to indole 5,6-quinone (XI), via dopa (II), dopaquinone (III), leucodopachrome (VIII), dopachrome (IX) and 5,6-dihydroxyindole (X). XI is oxidatively polymerized to eumelanin. Alternatively, III interacts with cysteine to form cysteinyldopa (IV), which is then oxidized to cysteinylquinone (V). Cyclization of V leads to VI, which is reduced to benzothiazine (VII) by IV. Formation of V from IV requires only catalytic amounts of oxygen, and thereafter the reaction proceeds spontaneously. The dihydrobenzothiazine (VII) acts as the precursor of both the phaeomelanins and the trichochromes. In the native state, M. are often bound to proteins, forming melanoproteins with a protein content of 10–15%. In humans and other mammals, the pigments of skin, hair and eyes are almost exclusively M. The pigments of many bird feathers, the

359

skins of reptiles and fish, the exoskeletons of insects, and the colored component (sepia melanin) from the ink sac of the cuttlefish *(Sepia officinalis)* are all M. M. are sometimes distributed diffusely, or they may be present in granules. In humans, the degree of coloration of the skin, from pale through brown to almost black, depends entirely on the concentration of melanin. Moles and freckles represent areas of especially high M. concentration. Sunlight causes increased M. synthesis (sun tan). M. acts as a protective agent against excess ultraviolet irradiation of the body surface. The color change of the chameleon and other color-adaptive animals is the result of a hormonally controlled change in the distribution of M. (see Melanotropin).

Oxidative conversion of dopa into M. is catalysed by the enzyme Tyrosinase (EC 1.14.18.1) (see). Absence of tyrosinase (usually an autosomal inherited

defect in the ability to synthesize the enzyme) results in albinism; most groups of mammals occasionally produce albino individuals, which completely lack any M. pigmentation of eyes, skin, hair, feathers, etc. Albinism may also result from 1. deficient melanin polymerization, 2. failure to synthesize the protein matrix of the melanin granule, 3. lack of tyrosine, and 4. presence of tyrosinase inhibitors.

Melanotropin, *melanocyte stimulating hormone, MSH:* a peptide hormone or neuropeptide produced by opiomelanotropinergic cells of the pars intermedia of the pituitary and by neurons in the central nervous system. In species lacking the pars intermedia (e.g. chickens, porpoise, whale), α-MSH is produced by the neurohypophysis. Production of α-MSH by the pituitary is under control of MRH and MIH (see Releasing hormones) of the hypothalamus. α-MSH belongs to a family of peptide hor-

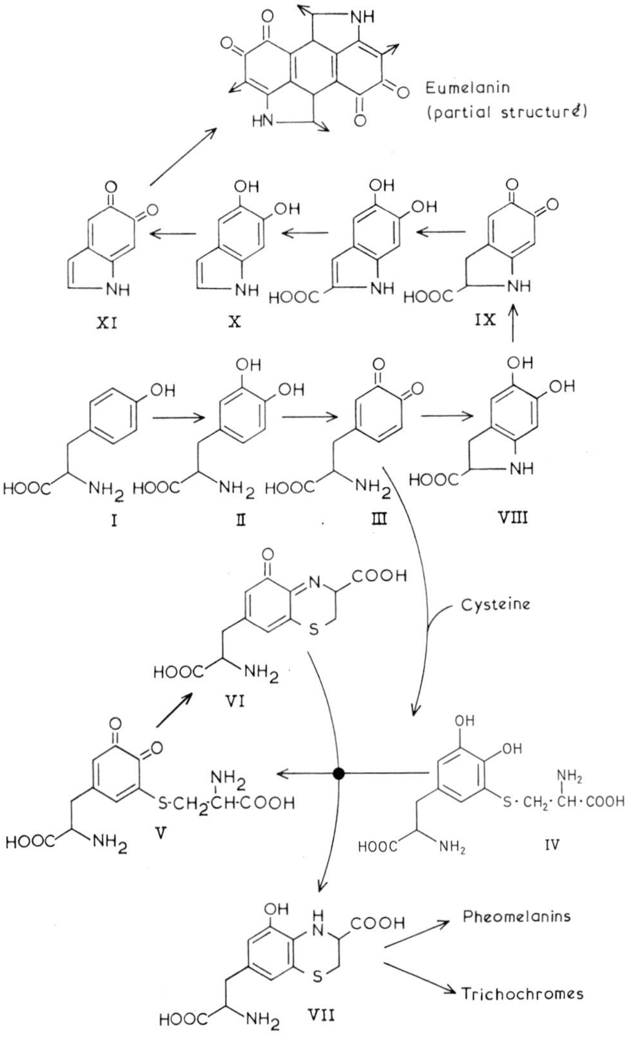

Eumelanin (partial structure)

XI   X   IX

I   II   III   VIII

Cysteine

VI

V

IV

VII   Pheomelanins

Trichochromes

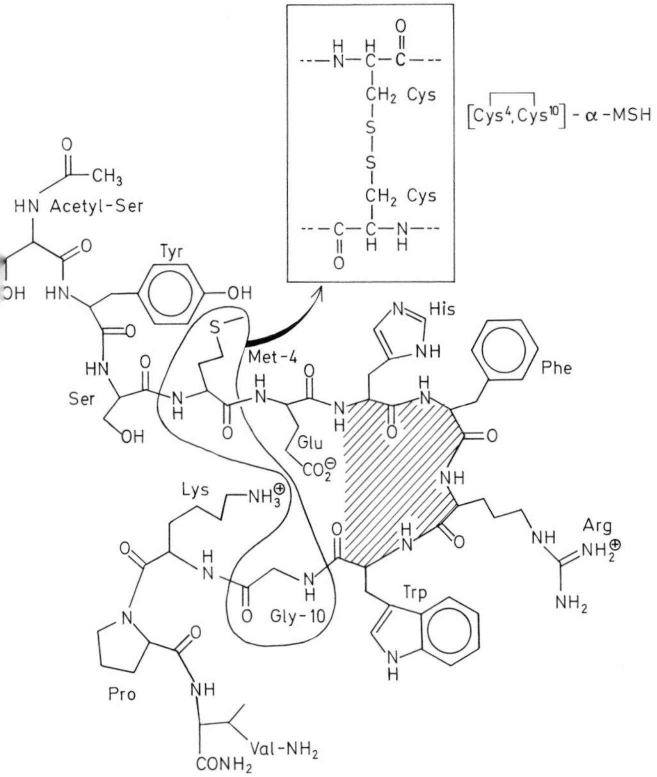

Two-dimensional representation of the reverse turn conformation of α-MSH. The shaded area represents the messenger sequence or active site. Replacement of Met-4 and Gly-10 by oxidatively coupled Cys residues results in [Cys⁴,Cys¹⁰]-α-MSH, a superpotent MSH.

mones, including ACTH, β-lipotropin, endorphins, enkephalin and others, all of which are derived from a common precursor protein, pro-opiomelanocortin (see Peptides, Fig.3). α-MSH was first recognized and named for its ability to cause dispersion of melanin in the melanophores of the skin of cold-blooded vertebrates; it also causes increased deposition of melanin in mammals, but its role in pigmentation in humans is doubtful. MSH-like activity of blood and urine increases during pregnancy in humans (the frog skin test described below was originally designed for detection of human pregnancy). This MSH-like activity may be of fetal origin (fetal pituitary is thought to produce α-MSH, which may play a role in fetal development), but its source and identity have not been established. Adult humans do not appear to possess a circulating MSH. α-MSH is, however, present extensively throughout the central nervous system of humans and other animals, where the deacetylated form is more abundant than the acetylated form. It is synthesized and secreted by an opiomelanotropinergic multineuronal transmitter system, the specific neurons of which have been located in the brain by immunological methods. Especially high concentrations of α-MSH are present in the axons and terminals of the hypophysiotropic

area of the hypothalamus, and the cell bodies containing α-MSH are localized in the acruate nucleus and in the dorsolateral region of the hypothalamus. Behavioral effects of α-MSH in mammals include arousal, increased motivation, longer attention span, memory retention and increased learning ability. α-MSH has been shown to promote somatotropin secretion, and to inhibit the secretion of prolactin and luteinizing hormone.

The same neurons produce both α-MSH and β-endorphin (which stimulates prolactin secretion). α-MSH may therefore be an endogenous antagonist of β-endorphin, thereby preventing hyperprolactinemia, but the true interplay of these two opposite effects is not clear. The classical assay for α-MSH is the in vitro frog *(Rana pipiens)* skin bioassay, which is based on the centrifugal dispersion of melanin granules in the dendritic processes of dermal melanophores, leading to skin darkening. It is capable of detecting minimal concentrations of α-MSH of 10^{-11} M, and gives dose-response curves in the range 2.5×10^{-11}–4×10^{-10}M. Lizard skin may also be used, but is less sensitive. A later method measures the increase in the conversion of [α-^{32}P]ATP into [^{32}P]cAMP by plasma membranes from mouse S-91 (Cloudman) melanoma cells in response to α-

MSH. Activation of tyrosinase or production of cAMP in whole melanoma cells have also been used as assay systems.

Mammalian α-MSH is an acetyltridecapeptidamide, identical with amino acid sequence 1–13 of ACTH (see Corticotropin): Ac-Ser-Tyr-Ser-Met-Glu-His-Phe-Arg-Trp-Gly-Lys-Pro-Val-NH$_2$. The deacetyl form is less potent and acetylation/deacetylation is probably a method of α-MSH potency control in mammals (α-MSH and β-endorphin are the only known N-acetylated peptides, although many N-acetylated proteins have been described). In salmon, only the deacetylated form has been found. Shark *(Squalus acanthias)* α-MSH is also a tridecapeptide, and differs from mammalian α-MSH in only two aspects: the N-terminal serine is never acetylated, and the C-terminus is Met (which may be amidated or possess a free carboxyl group). The major storage and release form in rat pituitary is N,O-diacetyl-α-MSH, i.e. the amino and hydroxyl groups of the N-terminal Ser are both acetylated.

Certain fragments of α-MSH have melanotropic activity with decreased potency. This has led to the concept of an active site or messenger sequence. Two such messenger sequences have been claimed for α-MSH: -Met-Glu-His-Phe-Arg-Trp-Gly- (residues 4–10 of α-MSH and ACTH) and -Lys-Pro-Val-NH$_2$ (residues 11–13 of α-MSH and ACTH), both independently capable of triggering the hormone receptor site responsible for melanin dispersion. Other authors have been unable to confirm the activity of the latter sequence, but there is general agreement that the 4–10 heptapeptide sequence represents a true messenger sequence of α-MSH.

β-MSH contains the same messenger heptapeptide sequence, but with different flanking amino acids: (bovine) Asp-Ser-Gly-Pro-Tyr-Lys-Met-Glu-His-Phe-Arg-Trp-Gly-Ser-Pro-Pro-Lys-Asp. Human β-MSH has 4 additional residues (Ala-Glu-Lys-Lys-) at the N-terminus. α- and β-MSH are derived from different regions of the same precursor, pro-opiomelanocortin. β-MSH has biological activity similar to that of α-MSH in the frog skin test; it probably occurs naturally, but is thought to have no peripheral physiological function, and to represent a byproduct of β-lipotropin degradation. β-MSH isolated from pituitaries is now known to be a fragment of β-lipotropin formed during extraction. Earlier reports of circulating human β-MSH were mistaken.

A third region of pro-opiomelanocortin contains the sequence: -Met-Gly-His-Phe-Arg-Trp-Asp-, which is closely similar to the heptapeptide messenger sequence. This suggested the existence of a third MSH. The peptide Tyr-Val-Met-Gly-His-Phe-Arg-Trp-Asp-Arg-Phe-Gly has now been found naturally and named γ-MSH, but nothing is known of its biological function, if any.

Structure-activity studies of α-MSH and its fragments strongly suggest that the active molecule possesses a reverse turn configuration in the central tetrapeptide sequence: -His-Phe-Arg-Trp (Fig,). The plausibility of this hypothesis is greatly increased by the synthesis of a conformationally restricted α-MSH analog, in which Met-4 and Gly-10 are replaced by Cys residues. Oxidative cross linkage of

these Cys residues maintains the molecule in a permanent reverse turn configuration.

The resulting [Cys^4,Cys^{10}]-α-MSH has a minimal potency in the frog skin test which is 10 000 times that of α-MSH. [V.J. Hruby et al. *Peptide and Protein Rev.* **3** (1984) 1–64; O. Khorram et al. *Proc. Nat. Acad. Sci. USA* **81** (1984) 8004–8008]

Melanotropin release inhibiting hormone: see Releasing hormones.

Melanotropin releasing hormone: see Releasing hormones.

Melanoyte stimulating hormone: see Melanotropin.

Melatonin: *N*-Acetyl-5-methoxytryptamine, M_r 232.3, a hormone of the pineal gland and retina. It inhibits the development of gonadal function in young animals and humans, and the action of gonadotropins in mature animals. The synthesis of M. is suppressed by light, acting via the eyes and nervous system, and peaks in the dark. A corresponding circadian rhythm has been observed in the activity of serotonin *N*-acetyltransferase (see below) in rodent pineal glands and the retinas of frogs and chickens. It seems likely that M. is an effector of mammalian or avian behavioral rhythms, although the molecular mechanisms are not known. [J. Redman, S. Armstrong & K. Ng, *Science* **119** (1983) 1089–1091.] Rats kept in continuous light are in continuous estrus, due to the lack of the antigonadotrophic activity of M. In amphibians, M. mediates the response of the skin pigmentation to light; it reverses the darkening effect of melanotropin by causing the melanin granules within the melanocytes to aggregate instead of disperse. M. appears also to influence the shedding of photoreceptor disks in the vertebrate retina. [J. Besharse & P.M. Iuvone, *Nature* **305** (1983) 133–135.]

M. is synthesized from serotonin by acetylation to *N*-acetylserotonin, followed by methylation (catalysed by acetylserotonin methyltransferase, (E.C.2.1.1.4) to M. In the pineal gland the biosynthesis of M. is controlled by epinephrine, which stimulates adenylate cyclase (E.C.4.6.1.1), which in turn stimulates protein kinase. The first three steps of M. synthesis (L-tryptophan → 5-hydroxytryptophan → serotonin → *N*-acetyl-5-hydroxytryptamine) are promoted in this way, but the methyltransferase is not affected. However, the rate-limiting step is the one catalysed by serotonin *N*-acetyltransferase. M. is inactivated and excreted as 6-hydroxymelatonin, or as 5-methoxyindoleacetic acid.

Melatonin

Melezitose: *O*-α-D-glucopyranosyl-(1 → 3)-*O*-β-D-fructofuranosyl-(2 → 1)-α-D-glucopyranoside, a trisaccharide consisting of a molecule of D-glucose linked to the diasaccharide turanose (an isomer of sucrose). High concentrations of M. are found in

...anna formed on the surfaces of pine trees, and in ...oney manufactured by bees which collect M. from ...is manna in times of drought.

Melibiose: 6-*O*-α-D-galactopyranosyl-D-glu-...ose, a reducing disaccharide consisting of galactose ...nd glucose linked α-1,4 glycosidically, both sugars ...eing in the pyranoid form. M. is present in plant ...ices. It represents a disaccharide grouping in the ...isaccharide raffinose.

Melissic acid: n-triacontanoic acid, CH_3-...$CH_2)_{28}$-COOH, a long-chain monocarboxylic fatty ...cid, M_r452.8. M. a. is present as an ester in beeswax ...nd montan wax.

Melissyl alcohol, *myricyl alcohol*: 1-triacontanol, ...r 1-hydroxytriacontane, CH_3-$(CH_2)_{28}CH_2OH$. M. is ...resent in plant cuticle waxes. Beeswax consists ...rincipally of the palmitate ester of M. (myricin).

Melittin: a linear, toxic (hemolytic) polypeptide ...mide, and the chief component of bee venom ...bout 50% of dried venom, and at least in 50-fold ...olar excess over other venom constituents). The ...rimary product of M. biosynthesis is prepromelit-...n, consisting of 70 amino acid residues. Removal ...f the *N*-terminal signal sequence of 21 residues (see ...ignal hypothesis) leaves promelittin. The protease ...quired for removal of the signal sequence is pre-...nt in many animal cells and is not species-specific, ...g. promelittin and not prepromelittin is the first ...etectable product when melittin mRNA is injected ...to oocytes of *Xenopus laevis*.

...he proteolytic release of M. from promelittin prob-...ly occurs outside the venom gland cells. Since ...1. disrupts phospholipid membranes, it is not sur-...rising that the promelittin inside the secretory cells ...less potent than M.

...rimary structure of M:
...ly-Ile-Gly-Ala-Val-Leu-Lys-Val-Leu-Thr-Thr-Gly-...eu-Pro-Ala-Leu-Ile-Ser-Trp-Ile-Lys-Arg-Lys-Arg-...ln-Gln-NH_2. There is a very unequal distribution ...f hydrophobic, neutral (1-20) and hydrophilic, ...ostly basic (21-26) amino acid residues. The re-...lting tenside character probably accounts for the ...harmacological and biochemical effects of M.

Melting point, T_m, t_m: the temperature, in °C, at ...hich a double-stranded nucleic acid becomes 50% ...enatured to the single-stranded form. A DNA solu-...on is heated and its absorbance at 260 nm is plot-...d against temperature. Transition from double to ...ngle-stranded DNA occurs over a relatively nar-...w temperature range, and is characterized by an ...ncrease in absorbance at 260 nm (melting of DNA ...also accompanied by a marked reduction in vis-...osity, alteration of optical rotation, and an increase ...density). T_m is taken as the temperature at the ...idpoint (half the final increase of absorbance at ...50 nm) of the S-shaped curve. The sharpness of the ...ansition indicates a cooperative alteration of struc-...re throughout the molecule (T_m is well above the ...mperature required to unstack the bases and de-...roy a single helix, but these are preserved by hy-...rogen bonding between the two helices; separation ...f the two strands occurs when the hydrogen bonds ...etween base pairs finally break). Single-stranded ...NA, on the other hand, shows only a gradual in-...rease of absorbance over a wide temperature range, ...nd has no T_m.

Under standard conditions of pH and ionic strength, T_m of DNA is proportional to the stability of the molecule. Since the base pair guanine-cyto-sine (see Base-pairing) has 3 hydrogen bonds, and adenine-thymine has only 2 hydrogen bonds, there is a linear relationship between the GC content of DNA and its T_m value: $T_m = 69$ °C $+ 0.41$ (molar % GC), i. e. the higher the degree of hydrogen bonding, the higher the temperature required to separate the strands of the double helix.

Measurement of T_m of hybrid nucleic acids (see Hybridization):

1. Cellulose nitrate discs containing known quantities of the radioactive hybrid (e. g. DNA:^{32}P-RNA) are incubated for 15 min at various elevated temperatures, cooled, treated with a nuclease specific for single-stranded nucleic acid (in this case an RNA-ase), washed and assayed for radioactivity. The melting profile is determined from the loss of radioactivity.

2. An hydroxyapatite column can be used for both large and small scale procedures. Double-stranded nucleic acids are retained by hydroxyapatite and single-stranded material is removed by washing. The column (water jacketed) is eluted at various elevated temperatures (e. g. in the range 60 °C-120 °C) and the effluent monitored (radioactivity or A_{260}) for single-stranded material.

Membrane: see Biomembrane.

Membrane enzymes: enzymes present on the surfaces or integrated into the phospholipid bilayer of the many different types of Biomembranes (see). There are, for example, the glucose 6-phosphatase of the endoplasmic reticulum, the galactosyl transferase of the Golgi apparatus, the oligomycin-sensitive ATPase of the inner mitochondrial membrane, the monoamine oxidase and rotenone-insensitive NADH-cytochrome *c*-reductase of the outer mitochondrial membrane, and the sodium-independent ATPase and the 5′-nucleotidase of the cell membrane. A very large number of reactions occur on membranes, and the M. e. play a central part in cell metabolism. M. e. are useful as markers in the isolation and identification of different types of membrane from cell homogenates.

Membranochromic pigments: plant pigments which impregnate the cell wall, e. g. the phenols and quinones that give color to heart wood.

Memory: the ability of a nervous system to store information. A functional distinction can be made between *short-term* and *long-term* M: the former lasts hours or days, while the latter is more or less permanent.

Studies on the marine snail *Aplysia* have indicated some of the molecular events associated with facilitation and depression of synaptic transmission from sensory to motor neurons. *Habituation*, a reduction in intensity and duration of a reflexive response, occurs when a weak stimulus is repeated but not coupled with either a positive or negative reinforcement. In the snail synapses, habituation is associated with a reduction in the calcium which enters the cell as a result of each action potential spike. The calcium influx determines the release of neurotransmitter from the neuron; thus a decrease in calcium influx causes fewer molecules of neurotransmitter to be released into the synaptic gap and weakens the signal transmission.

By contrast, *facilitation* of a response occurs when a weak stimulus to one region is coupled with a strong stimulus to another. (When facilitation becomes permanent, it is equivalent to classical *conditioning*). In these snails, facilitation has been traced to an ion channel for K^+ which normally assists in the return to normal membrane voltage after a voltage spike has arrived. If this channel is partially blocked, the action potential is prolonged, more Ca^{2+} can enter the neuron, and this causes more neurotransmitter to be released. Both serotonin and cyclic AMP facilitate synaptic transmission between sensory and motor neurons in *Aplysia*. Kandel and Schwartz have proposed a mechanism which would account for this: the serotonin enhances cAMP synthesis; cAMP in turn activates a protein kinase, which phosphorylates one or more proteins associated with the K^+ channel, and thus reduces the efficiency of the channel in transporting K^+ outward. These authors speculate that long-term memory might be the result of permanent facilitation of the same synapses, possibly due to the synthesis of a site-specific, highly cAMP-sensitive regulatory protein. This is consistent with the observation that learning in vertebrates is prevented by inhibitors of protein synthesis. [E. R. Kandel & J. H. Schwartz, *Science* **218** (1982) 433–443.]

Another model for M. is long-term potentiation (LTP), which occurs in the hippocampus of vertebrate brains. LTP is a stable facilitation of synaptic responses which results from very brief trains of high-frequency nervous impulses. It has been reported that injection of EGTA, a calcium chelator, into postsynaptic neurons inhibits LTP. The mechanism could be that calcium activates a proteinase which in turn increases the number of glutamate binding sites on the postsynaptic membrane in these synapses. By binding the calcium which enters the neuron, the EGTA would prevent activation of the proteinase and thus block LTP. [G. Lynch et al., *Nature* **305** (1983) 719–721.]

Menadione: vitamin K_3 and provitamin of the vitamin K group (see Vitamins).

Menaquinone-6: vitamin K_2 (see Vitamins).

p-**Menthadienes:** doubly unsaturated, monocyclic monoterpene hydrocarbons. M_r 136.24. The *p*-M. occur more commonly than the monounsaturated compounds of this structural type. There are 14 structural isomers of *p*-M., of which 9 are foun naturally in volatile oils. The commonest *p*-M. a shown in the table.

Extensive groups of oxygen-containing, monocycl monoterpenes are derived from the *p*-M.

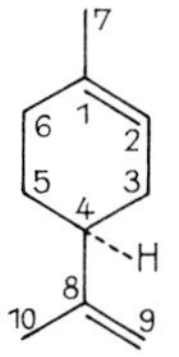

(+)-Limonene

p-**Menthane:** see Monoterpenes, Fig.

Menthol: *p*-menthan-3-ol, commercially th most important of the monocyclic, monoterpene a cohols. M_r 156.27. M. contains 3 asymmetric carbo atoms; it can exist as 4 racemates and in 8 stere isomeric forms. The naturally occurring form (−)-M., which is the chief constituent of peppermi oil. m. p. 43 °C, b. p. 216 °C, $[\alpha]_D$ −49.4 °, $\rho_{.5}^{15}$ 0.90 n_D^{20} 1.4609. M. has only limited use as a scent, b owing to its minty taste it is used in large quantitie in the preparation of flavorings, tooth paste, etc. O idation of M. produces (−)-menthone, M_r 154.2 m. p. −6 °C, b. p. 204 °C, $[\alpha]_D$ −29.6, which is als a component of peppermint oil.

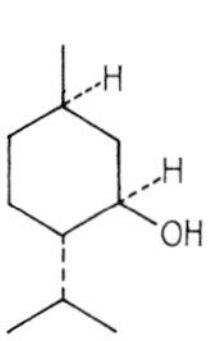

(−)-Menthol

(−)-Menthone: see Menthol.

Mercaptans: see Sulfur compounds.

Mercapto group: see Thiol group.

Mercapturic acids: see Glutathione S-transfe ase.

Meromyosin: see Muscle proteins.

Merrifield synthesis: see Peptides.

Mescaline: 3,4,5-trimethoxy-phenylethylamin the principal hallucinogenic component of the dru

Name	Δ	b. p. (°C)	$[\alpha]_D$	Occurrence
(+)-Limonene	1,8	177–178	+126.8	Orange and lemon oil
(−)-Limonene	1,8	177–178	−122.6	Peppermint oil
(±)-Limonene (Dipentene)	1,8	175–176		Pine oil
α-Terpinene	1,3	173–175		*Elettaria* spp.
β-Terpinene	1(7),3	75.5(22 mm)		*Pittosporum* spp.
γ-Terpinene	1,4	183		Widely distributed
(+)α-Phellandrene	1,5	58–59(16 mm)	+45	Eucalyptus oil
(−)α-Phellandrene	1,5	173–175	−17.7	Fennel oil
(+)β-Phellandrene	1(7),2	173	+62.5	Fennel oil
(−)β-Phellandrene	1(7),2		−74.4	Eucalyptus oil

'eyotl or Peyote, used as a ceremonial intoxicant by
Mexican Indians living near the Mexico-USA bor-
er. Peyote is the cut and dried parts of the cactus
Anhalonium lexinii. The synthetic M. derivative, 2,5-
dimethoxy-4-methylamphetamine (DOM), is a more
otent hallucagen than M.

$$H_3O-\text{benzene ring}-CH_2-CH_2-NH_2$$
$$H_3O \quad OCH_3$$

Mescaline

Meselson and Stahl experiment: the first definitive experiment showing that DNA replication is
semiconservative [M. Meselson & F. W. Stahl *Proc.
Natl. Acad. Sci.* **44** (1958) 671–682]. DNA in which
all the nitrogen atoms of the bases are ^{15}N is slightly
heavier than "normal" DNA, in which all the bases
contain only ^{14}N. The difference in buoyant density
is sufficient to permit separation of ^{14}N- and ^{15}N-la-
belled DNA by centrifugation in a density gradient
of cesium chloride. The DNA sample (<100 mg/l)
is centrifuged in a centrifugal field exceeding
100 000 × *g* in a solution of 8 M CsCl. Redistribution
of the CsCl occurs, so that the concentration increases towards the bottom of the cell, forming a
density gradient from about 1.6 to 1.8 g/ml. The
DNA, which has a buoyant density of about 1.7 g/
ml, sediments from the region of low density and
moves centripetally from the region of high density,
collecting as a narrow band where its buoyant density equals that of the matrix.

Escherichia coli was grown for several generations in
a medium containing isotopically pure $^{15}NH_4^+$ as
the sole nitrogen source. Under these conditions,
^{14}N-containing compounds of the original inoculum
are diluted out, and all the nitrogenous compounds,
including DNA, of the bacterial cells contain ^{15}N.
^{15}N-labelled DNA has a buoyant density of 1.744 g/
ml, while ^{14}N-labelled DNA has a buoyant density
of 1.704 g/ml. If replication were conservative (i.e.
the parent duplex remains intact in one daughter
cell, while the other daughter cell contains a duplex
of two newly synthesized strands), then two species
of DNA, buoyant densities 1.704 and 1.744 g/ml,
would be present after one phase of cell division in
a growth medium containing $^{14}NH_4^+$. If replication
is semiconservative (i.e. if each strand of the parent
duplex acts as a template for the synthesis of its
partner), then after one phase of cell division in the
presence of $^{14}NH_4^+$, all the daughter DNA will be
an $^{14}N:^{15}N$ hybrid of intermediate buoyant density.
After one phase of cell division in the presence of
$^{14}NH_4^+$, the band of ^{15}N-labelled DNA is replaced
by a single band of the $^{15}N:^{14}N$ hybrid (buoyant
density 1.725 g/ml) and the ^{14}N-labelled DNA
(buoyant density 1.704 g/ml) does not appear until a
second phase of cell division (Fig.). See Deoxyribonucleic acid.

Messenger RNA, *mRNA, template RNA:* RNA
which is translated into a polypeptide on the ribosome (see Protein biosynthesis). One molecule of
mRNA may carry information for the synthesis of
more than one protein, having been transcribed
without interruption from several neighboring cistrons of the DNA. This polycistronic mRNA has so
far been found only in prokaryotes. The polypeptides translated from polycistronic mRNA usually

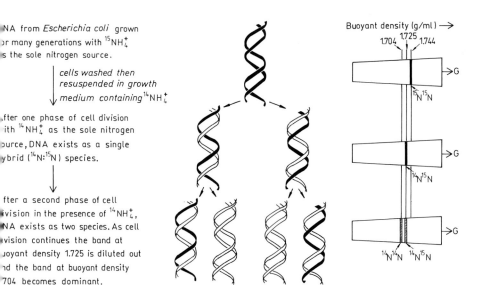

NA from *Escherichia coli* grown
or many generations with $^{15}NH_4^+$
as the sole nitrogen source.

> *cells washed then
> resuspended in growth
> medium containing* $^{14}NH_4^+$

After one phase of cell division
with $^{14}NH_4^+$ as the sole nitrogen
source, DNA exists as a single
hybrid ($^{14}N:^{15}N$) species.

After a second phase of cell
division in the presence of $^{14}NH_4^+$,
DNA exists as two species. As cell
division continues the band at
buoyant density 1.725 is diluted out
and the band at buoyant density
1.704 becomes dominant.

Buoyant density (g/ml) →
1.704   1.725   1.744

Figure. The Meselson and Stahl experiment. Bands of DNA in the cell of the analytical ultracentrifuge are detected by their absorption of UV light. → G indicates the direction of acceleration due to gravity (centrifugal
field).

have related functions, e.g. 10 enzymes in the histidine biosynthesis pathway are encoded in and translated from a polycistronic mRNA containing about 12000 nucleotides, M_r 4×10^6. Viral RNA is functionally very similar to mRNA. The entire length of viral RNA or mRNA is not translated; the start codon (always AUG) is located some distance in from the 5'-end of the molecule. For example, the start codon for β-galactosidase in the Lac mRNA of *Escherichia coli* occurs at position 39. The untranslated sequence of nucleotides at the 5'-end of prokaryotic mRNA includes nucleotides complementary to a sequence in the 16S rRNA of the ribosome, and this is believed to help bind the mRNA to the ribosome.

Functional mRNA is single-stranded. In prokaryotic cells, transcription and translation are usually coupled: mRNA becomes bound to ribosomes and translation begins before transcription is complete. The messenger is usually translated by several ribosomes at once, and thus several to many protein molecules are made from it. In prokaryotes, however, mRNA lifetimes are short, with half-times of several minutes. Eukaryotic mRNA is normally stable for hours or days.

In eukaryotic cells, the synthesis of mRNA and its subsequent translation are more complex. The first transcriptional product is a heterogeneous, very long RNA *(giant messenger-like RNA, mlRNA, heterogeneous nuclear RNA, HnRNA)*. Its M_r in animal cells is $1-15 \times 10^6$. HnRNA is synthesized in the nucleoplasm (in contrast to ribosomal mRNA, which is synthesized in the nucleolus), and it is the precursor of active, polysomal mRNA. HnRNA contains both nucleotide sequences which are ultimately translated into polypeptides (exons) and large tracts of sequences which are not translated. Those non-translated sequences which are located between exons are called "introns" (or intervening sequences); there are also repetitive sequences at the 5'-end (see Redundancy), which are not translated. The 3'-end of HnRNA and eukaryotic mRNA carries a poly(adenine) sequence (see Post-transcriptional modification of RNA), and the extreme 5'-end of most messengers carries a "cap" of 7-methylguanosine triphosphate linked 5' to 5' to the first "normal" nucleotide.

Capped messengers are more readily translated than uncapped ones in eukaryotic cell-free protein synthesis systems; removal of the caps interferes with the binding of the mRNA to the ribosome. It is not known when the cap is applied in vivo, but it may be part of the initiation of transcription, as capped mRNA fragments can be isolated from cells in which elongation of the messenger has been inhibited. [A. J. Shatkin et al. in *Messenger RNA and Ribosomes in Protein Synthesis*, C. F. Phelps & H. R. V. Arnstein, eds. (The Biochemical Society, London, Symposium 47) 1982.]

Although the 3'-poly(A) tail is regarded as a feature of eukaryotic mRNA, as distinct from prokaryotic mRNA, there are some notable exceptions, including histone mRNA, some HeLa cell mRNA and some early mRNA of the sea urchin embryo, which have no 3'-poly(A) tail.

Directly after synthesis, all HnRNA becomes bound to protein particles, which can be isolated from the nucleus. The protein components have been shown to include poly(A) polymerases and endonucleases (for processing of the HnRNA). The protein particle, without the RNA, is sometimes called an *informofer.*

HnRNA is processed in the nucleus by removal and degradation of the exons and some of the untranslated end sequences. The poly(A) sequences are largely unaffected, but some may be partially or completely degraded. The degradation is so extensive that in some cases as much as 90% of the HnRNA never enters the cytoplasm. The resulting mRNA molecules (in animal cells, M $0.05-1.5 \times 10^6$) are transported into the cytoplasm where they first appear as ribonucleoprotein particles (RNP), sometimes called *informosomes*. The mRNA then leaves the RNP and becomes associated with ribosomes to form active polysomes. Fully processed monocistronic eukaryotic mRNA still possesses some non-translated nucleotides at the 5'-end, in addition to the 3'-poly(A) tail. The storage and release of mRNA from cytoplasmic RNP may be important in the process of regulation of translation. The expression of genes is also regulated at other steps in the process (see Gene activation; Operon; Enzyme induction; Enzyme repression; Attenuation). See also Protein biosynthesis.

Mesterolone: 1α-methyl-3β-hydroxy-5α-androstan-3-one, a synthetic androgen. M. shows high activity when administered orally. Like Methyltestosterone (see), M. is used for the therapy of male gonadal insufficiency and endocrine disorders.

Mestranol: 17α-ethinyl-3-methoxyestra-1,3,5(10)-trien-17β-ol, a synthetic estrogen. M. has high biological activity when administered orally, and it is used as a component of Ovulation inhibitors (see). M. differs from Ethinylestradiol (see) by the presence of a 3-methoxy group.

Met: abb. for L-methionine.

-Met-NH₂: the abbreviation for an amidomethionyl residue at the C-terminus of a peptide or protein.

Metabolic acidosis: see Acidosis.

Metabolic alkalosis: see Alkalosis.

Metabolic block: see Mutant technique; Auxotrophic mutants.

Metabolic bypass: see Metabolic shunt.

Metabolic control, *metabolic regulation:* Metabolism is subject to control, and an analogy may be drawn with electronic and mechanical regulation processes used in technology. In a purely formal way, living systems can be regarded as cybernetic machines. Control is a fundamental principle in the organization of living organisms, and depending on the nature of the signal or method of information transfer, there are four broad types:

1. *Neural (nervous) control* The nerve impulse is an electrical signal, and the regulatory response may also be electrical (e.g. further nerve impulse to a muscle) or chemical (e.g. production of a hormone). The nervous system may be considered as a physiological broadcasting system.

2. *Hormonal (humoral) control.* Hormones (see) act as chemical signals in a regulation system that is superimposed on the more basic levels of M.c. Cycli-

Table 1. Control of enzymes in metabolism.

Control mechanism	Type of control
Chemical modification by attachment of covalently bound groups by specific enzymes.	Enzyme activity.
Physical modification by noncovalent interactions: allosteric control (feedback inhibition, precursor activation).	Enzyme activity.
Induction (derepression), repression.	Enzyme concentration controlled by enzyme synthesis.
Exposure of active center by removal of peptides.	Activation of zymogen by limited proteolysis.
Association of enzyme proteins; shift of equilibrium between de novo enzyme synthesis and enzyme degradation by group-specific proteases (inactivases).	Enzyme concentration.

Table 2. Differences between feedback inhibition and repression of enzyme synthesis.

Feedback inhibition	Repression
Inhibition of enzyme activity	Inhibition of enzyme synthesis
Allosteric interaction of enzyme and endproduct, which acts as a negative allosteric effector	Endproduct acts as corepressor and activates a repressor protein, which prevents protein synthesis at the level of transcription
Epigenetic regulation	Genetic or transcriptional regulation
The allosteric enzyme of a reaction chain (usually the first specific enzyme) is inhibited	If the relevant enzymes are controlled by an operon, coordinate regulation results in decreased synthesis of all enzymes in the reaction chain
Rapid, fine regulation	Slow, coarse regulation, in which the enzymes in question are diluted out by turnover and by cell growth and division
A reversible inhibition, depending on the endproduct concentration and the nature of the sigmoid relationship between enzyme activity and substrate concentration	Reversible by removal of the inhibitory metabolite, i.e. the repressed system can be derepressed
Can be demonstrated in vitro, i.e. with purified enzymes	Cannot be demonstrated with the purified enzymes; it depends upon protein synthesizing system of the cell. In whole cells derepression is prevented by inhibitors of transcription and translation, and by inhibitors of nucleic acid and protein metabolism

AMP acts as a second messenger for many hormones. Hormones are synthesized at specific sites (endocrine glands), then transported to the target tissue or organ. Neural and hormonal regulation represent intercellular M.c.

3. *Differential gene expression.* The signals (or triggers) of differential gene expression may be chemical (hormones) or environmental (e.g. light). Differential gene expression is responsible for the regulation, at a molecular level, of differentiation and development.

4. *Feedback and feedforward mechanisms,* in which metabolites themselves act directly as signals in the control of their own breakdown or synthesis. Feedback is negative or positive. *Negative feedback* results in inhibition of the activity or synthesis of an enzyme or several enzymes in a reaction chain by the endproduct. Inhibition of the synthesis of enzymes is called *Enzyme repression* (see). Inhibition of the activity of an enzyme by an endproduct is an allosteric effect (see Allostery; Aromatic biosynthesis); this type of feedback control is well known for amino acid biosynthesis in prokaryotic organisms, and is variously known as *endproduct inhibition, feedback inhibition* and *retroinhibition.* In positive feedback, or *feedback activation,* an endproduct activates an enzyme responsible for its production, e.g. thrombin activates factors VIII and V during blood clotting, thus contributing to the speed of the cascade system and the rapid formation of a clot. An example of *feedforward enzyme activation* is found in the activation of glycogen synthetase by glucose 6-phosphate, i.e. a metabolite activates an enzyme concerned in its utilization. *Enzyme induction* (see) is a positive feedforward process.
Mechanisms discussed under 4. are all intracellular

mechanisms of M.c., which have been studied chiefly in prokaryotic organisms.
4a. *Chemical modification of enzymes.* This occurs by forming or breaking covalent bonds. Two mechanisms have been studied in detail: phosphorylation-dephosphorylation by protein kinases and protein phosphatases (see Adenosine phosphates; Glycogen metabolism, regulation), and adenylylation-deadenylylation (see Covalent modification of enzymes).
4b. *Physical modification of enzymes* is a feature of allostery.
Both 4a and 4b result in a change in the active conformation of the enzyme protein. Chemical modification causes a change in the equilibrium: protomers $\rightleftharpoons$ oligomers, leading to the establishment or abolition of quaternery structure.

Table 3. Intracellular metabolic regulation.

Regulation system	Mechanism and consequences	Regulation system	Mechanism and consequences
Compartmentation	Differential location of enzymes in organelles, membranes, etc.	Isostery	Competitive inhibition by structural analogs
Membrane barriers	Membranes regulate the selective exchange of materials, separate cytoplasmic and noncytoplasmic compartments and create concentration and pH-gradients for osmosis and phosphorylation	Allostery	Physical modification of enzyme and therefore enzyme activity by allosteric effectors
		Chemical modification	Attachment or removal of covalently bound groups, which influence the biologically active quaternary structure of enzyme proteins
Multienzymes	Association of enzymes into multienzyme systems and complexes, i.e. metabolic compartmentation	Limited proteolysis	Exposure of active centers of peptide hormones and digestive enzymes by removal of a peptide sequence, i.e. a form of processing
Metabolic pools	Metabolic compartmentation	Regulation of enzyme concentration:	
Enzyme kinetic regulation	Competition between enzymes for common substrates and cosubstrates on the basis of their Michaelis constants; product inhibition; stoichiometric effects of metabolites on equilibria and reaction rates; changes in the type and quantity of enzyme effectors	Regulation of gene activity	Induction (derepression) and repression of enzyme synthesis by transcriptional regulation on the DNA template or the RNA polymerase
		Regulation of protein biosynthesis	Regulation at various stages of translation
Regulation of enzyme activity:		Regulation of proteolysis	Change in protein turnover by shift of balance between de novo enzyme synthesis and enzyme degradation

M.c. is also achieved by competition of enzymes for common substrates and cosubstrates, cofactor stimulation, regulation of coenzyme synthesis, product inhibition and stoichiometric inhibition by metabolites. An important principle of M.c. in eukaryotes appears to be the regulation of active enzyme concentrations by a change in the turnover of enzyme proteins. This type of control therefore depends on the regulation of Proteolysis (see), e.g. the active concentration of tryptophan synthase in yeast is controlled by group-specific proteases, called inactivases; further regulation is achieved by an inactivase inhibitor, which is also a protein. Group-specific proteases probably initiate a cascade reaction, in which they prepare proteins for further degradation by unspecific proteases. For further details of enzyme regulation by proteolysis, see Schimke, R.T. "On the Roles of Synthesis and Degradation in Regulation of Enzyme Levels in Mammalian Tissues", pp. 77–120 in *Current Topics in Cellular Regulation,* edit. Horecker, B.L. and Stadtman, E.R., Vol. 1, 1969, Academic Press; and all contributions to the Symposium: *Intracellular Protein Turnover,* edit. Schimke, R.T. and Katunuma, N., 1975, Academic Press.

Metabolic cycle: a catalytic series of reactions, in which the product of one bimolecular reaction is regenerated: A + B - - → C + A
Thus A acts catalytically and is required only in small amounts, and A can be considered as a carrier of B. The catalytic function of A and other members of the M.c. ensure economic conversion of B into C; B is the substrate of the M.c. and C is the product. If intermediates are withdrawn from the M.c., e.g. for biosynthesis, the stationary concentrations of the M.c. intermediates must be maintained by synthesis. Replenishment of depleted M.c. intermediates is called *anaplerosis.* Only one anaplerotic reaction is necessary, since the resulting intermediate is in equilibrium with all other members of the cycle. Anaplerosis may be served by a single reaction, which converts a common metabolite into an intermediate of the M.c. (e.g. pyruvate to oxaloacetate in the tricarboxylic acid cycle), or it may involve a metabolic

sequence of reactions, i.e. an *anaplerotic sequence* (e.g. the glycerate pathway which provides phospho-*enol*pyruvate for anaplerosis of the dicarboxylic acid cycle).

M.c. are *anabolic, catabolic* or *amphibolic.* The Calvin cycle (see) is an anabolic (synthetic) cycle. A truly catabolic cycle, which does not also supply intermediates for biosynthesis, probably does not exist. The Tricarboxylic acid cycle (see) is an important central metabolic pathway, serving both the terminal oxidation of substrates and the provision of intermediates for biosynthesis (e.g. biosynthesis of porphyrins and certain amino acids); it is therefore an amphibolic M.c. Similarly, the Pentose phosphate cycle (see) has a catabolic function and provides ribose phosphate for the synthesis of nucleic acids and certain coenzymes.

The first M.c. to be recognized was the Urea cycle (see), described by Krebs and Henseleit in 1932. This may be considered as an anabolic cycle since it results in the energy-dependent synthesis of urea; but with respect to its metabolic role in the degradation of protein and detoxication of ammonia, it is catabolic. However, under certain conditions, it also provides arginine for protein synthesis, and anaplerosis occurs by the synthesis of ornithine from glutamate.

Metabolic shunt, *metabolic bypass:* a metabolic pathway which bypasses some reactions and exploits others of a primary metabolic pathway, e.g. the Aminobutyrate pathway (see) and Glyoxylate cycle (see). The term is sometimes used to mean secondary metabolism (see Secondary metabolites).

Metabolism: the sum total of chemical (and physical) changes that occur in living organisms, and which are fundamental to life. Nutrients from the environment are used in two ways by living organisms; they may serve as constituents in the synthesis of components of the organism (assimilation), or they may be oxidatively degraded for purposes of energy production (dissimilation). All constituents of living organisms are subject to a process of continual breakdown and resynthesis, normally referred to as Turnover (see). Processes of breakdown are collectively called Catabolism (see), and all reactions concerned with synthesis are called anabolism. An Amphibolic pathway may serve both functions of degradation and synthesis. M. represents a cybernetic network with self-regulatory properties (see Metabolic control), containing many cyclic processes (see Metabolic cycle).

Since M. is an open system, its multistage processes continually approach a state of equilibrium that is never quite achieved, at least not while the organism is alive. M. is therefore better described as a steady state, rather than an equilibrium. Let S. be a starting substrate (e.g. a dietary component), which is converted to endproduct E via intermediates I_1, I_2, I_3-----I_n: $S \xrightarrow{1} I_1 \xrightarrow{2} I_2 \xrightarrow{3} I_3 \xrightarrow{n} I_n \rightarrow E$.

Reactions 1, 2, 3 ----- n are enzyme-catalysed conversions, leading to the endproduct, according to the principle of "organization by specificity" (Dixon). E may be accumulated or excreted. Per unit time, the quantity of S converted is proportional to the quantity of E produced (it is assumed that the quantity of available S is so large that its decrease

per unit time can be ignored). The steady state concentrations of the intermediates I_1, I_2, I_3-----I_n are constant, i.e. the rate of change of concentration is zero:

$$-\frac{ds}{dt} = +\frac{dE}{dt}; \quad \frac{dI_{1,2,3-----n}}{dt} = 0.$$

The quantity of I_2 formed per unit time is equal to the quantity of I_1 converted, and is proportional to the concentration of I_1:

$$+\frac{dI_2}{dt} = -\frac{dI_1}{dt} = k_1 I_1;$$

$$-\frac{dI_2}{dt} = k_2 I_2 \; (k = \text{rate constant})$$

At equilibrium, $k_1 I_1 = k_2 I_2$, or $\frac{I_1}{I_2} = \frac{k_2}{k_1}$.

This leads to the following conclusions: 1. Concentrations of intermediates depend on the rate constants; the larger the rate constant of a reaction, the smaller the concentration of the intermediate. 2. The rate of the conversions $S \rightarrow E$ is determined by the rate of the slowest step; Krebs called this reaction the *pacemaker* reaction. The situation is usefully illustrated by analogy with a sluice: above the sluice gate (pacemaker), the water level is high (intermediates are present in high saturating concentrations). The sluice gate determines the rate of water flow (substrate conversion). A sudden increase in the concentration of an intermediate would be self-regulated by an overflow and diversion of the intermediate into alternative metabolic channels, which occur earlier than the pacemaker reaction. After the pacemaker reaction, such an increase leads to increased synthesis of endproduct, which is stored or excreted. The endproduct may also regulate its own synthesis by actually inhibiting the synthetic process, so that the rate of production of endproduct is finely adjusted to its rate of utilization by the cell (i.e. endproduct inhibition, see Metabolic control). See also Primary metabolism; Secondary metabolism, Metabolic shunt; Intermediary metabolism; Energy metabolism, Carbon dioxide assimilation; Respiration.

Metabolite: a substance produced or consumed by metabolism. Biopolymers are not included in this definition. The precursors and degradation products of biopolymers are, however, true M. All small molecules produced or converted by enzymes during metabolism are M. According to the nomenclature of enzymology, substances (including biopolymers) that are attacked by enzymes are called substrates.

Metachromatic leukodystrophy: see Inborn errors of metabolism.

Metalloflavoproteins: see Flavin enzymes.

Metalloproteins: proteins containing complexed metals. In the metalloenzymes, the metals are functional components. In the metal-transporting M. (e.g. the blood proteins, transferrin and coeruloplasmin) and the metal-storage depot proteins (e.g. ferritin), the metal binding is reversible and the metal is a temporary component. Important iron-containing M. are the cytochromes, respiratory pigments (e.g. hemoglobin) and enzymes (e.g. catalase). Zinc is present in insulin and in carbonic anhydrase. Some glycolytic enzymes and proteases contain Manganese (see). Cu is present in certain oxidoreductases (see Copper proteins). The catalytic M. also include enzymes that require specific cations for activity, e.g. the sodium-dependent membrane ATPases. See also Molybdoenzymes.

Metallothioneins: highly conserved, sulfhydryl-rich, cytoplasmic metalloproteins, present in all vertebrates, invertebrates and fungi. Mammalian cells synthesize two isoforms, M.I and M.II. Each M. consists of 61 amino acid residues, of which 20 are Cys, and there are no disulfide bridges. Accumulation of M. is induced by Zn, Cu, Hg and especially Cd, as well as other transition metals. This induction is due to transcription of M. genes, which are regulated by metals, glucocorticoids and interferons. M. bind and detoxify several metals, and are involved in Zn and Cd resistance in mamallian cells. Gold compounds (e.g. see Auranofin) used in the treatment of rheumatoid arthritis are potent activators of M. genes in Chinese hamster ovary cells. [B.P. Monia et al. *J. Biol. Chem.* **261** (1986) 10957–10959)]

Methemoglobin: a hemoglobin in which the iron is trivalent (Fe III). M. are unable to transport oxygen.

Methemoglobinemia: see Inborn errors of metabolism.

L-Methionine, *Met:* α-amino-γ-methylmercaptobutyric acid, a sulfur-containing, essential proteogenic amino acid. M_r 149.2, m.p. 281 °C (d.), $[\alpha]_D^{25}$ +23.2 ° (c=0.5–2.0, 5M HCl), or −10.0 ° (c= 0.5–2.0, water). The nutritional value of many plant proteins is limited by their low content of Met. In the first stage of Met biosynthesis in *Escherichia coli,* cysteine and homoserine condense to form cystathionine. The latter is cleaved to give homocysteine, which is methylated to Met by the transfer of a methyl group from N^5-methyl-tetrahydrofolic acid

(Fig.). Vitamin B_{12} is a coenzyme in this methyl group transfer. The active form of Met, i.e. *S*-Adenosyl-L-methionine (see), is the methyl group donor in Transmethylation (see). Degradation of Met proceeds via L-Cysteine (see).

DL-Met is prepared on the industrial scale (about 100000 tons in 1977) by a Strecker synthesis, using β-methylmercaptopropionaldehyde prepared from acrolein and methylmercaptan. DL-Met is used to supplement poultry feed. Both D- and L-forms are effective, so that no prior separation of enantiomers is necessary.

Methods of biochemistry: usually methods for the study of metabolic processes, but in the widest sense including isolation, identification and characterization of natural substances. Methods for the study of metabolism are classified into 3 types:

1. *In vivo methods* employ whole organisms, their organs or cells, or populations of cells of microorganisms. In balance studies, substances are administered to the organism and the time course of their conversion to various products is determined by analysis of body materials or excretory products (feces, urine, expired gases) (see, e.g. Nitrogen balance). Balance studies are also performed on isolated organs, e.g. perfused liver. The load test is also a form of balance study, in which excess of a substance is administered to the organism to test the ability of an organ or organ system to deal with the substance in question. Such tests are used clinically to investigate organ function, e.g. to test kidney function by measurement of the rate of urinary excretion of injected phenol red. The most important in vivo methods are the indicator methods. Classical amongst these is the use, by Dakin and Knoop, of "non-combustible" aromatic residues on fatty acids, which led to the discovery of the β-oxidation of fatty acids (see Fatty acid degradation). This and similar chemical labelling methods have the drawback that the administered material is different from the natural material under investigation. These difficulties are now overcome by the use of isotopes; a compound in which one or more of the constituent atoms is present as a stable or radioactive isotope, is chemically and biochemically identical with the unlabelled compound (see Isotope technique). Thus isotopically labelled natural metabolites can be administered and their natural fate within the organism can be determined by isotopic tracing. A further important in vivo method is the Mutant technique (see).

A combination of isotope and mutant techniques has made a considerable contribution to the exponential growth of biochemical knowledge.

2. *In vitro methods* are performed outside the whole organism. They are essentially "test tube" methods, employing crude cell homogenates, subcellular fractions thereof, or purified enzymes. For methods of cell disruption, subcellular fractionation and enzyme purification see Proteins and Density gradient centrifugation.

The techniques of histochemistry and cytochemistry are also in vitro methods: the sites of metabolites, enzymes or metabolic reactions are identified in organ or tissue slices and in cells by characteristic chemical reactions, e.g. specific color reactions.

Methionine biosynthesis in Escherichia coli

A large proportion of in vitro studies is concerned with the activity and behavior of enzymes, coenzymes and substrates. Such studies require enriched or purified enzymes. The degree of purity necessary for the reliable determination of kinetic parameters varies according to the nature of the enzyme and the possible impurities, but all such investigations are preferably performed with a homogeneous enzyme protein. The completely pure enzyme is necessary for the further investigation of physical and chemical properties, such as photometric measurements and determination of amino acid composition. Ultimately, the crystalline enzyme is required for the study of the detailed mechanism of enzyme catalysis, and for the determination of the three dimensional structure of the enzyme and enzyme-substrate complexes.

3. *Synthetic methods* include the construction and reconstruction of e.g. multienzyme complexes and biochemical systems; and modelling and simulation, e.g. Synzymes (see) and computer simulation of glycolysis.

Methotrexate: see Aminopterin.

Methoxymellein: see Phytoalexins.

Methyl-accepting chemotaxis proteins, *MCP:* cell membrane proteins in *Escherichia coli* involved in the initiation and control of chemotactic behavior. There are at least 3 different MCP in *E. coli*, each responsive to stimuli from different types of chemoreceptor, including both attractants and repellants. Attractants elicit counterclockwise rotation of the flagella, which results in smooth swimming. Repellants elicit clockwise rotation, which changes the bacterium's direction of motion. Methylation of MCPs results in adaptation, or lessened responsiveness to stimulation. [D. Sherris & J. S. Parkinson (1981) *Proc. Natl. Acad. Sci. USA* **78**, 6051–6055].

γ-N-Methylasparagine: an amino acid residue in the β-subunit of *Anabena variabilis* allophycocyanin, formed by post-translational modification of an asparaginyl residue. [A. V. Klotz et al. *J. Biol. Chem.* **261** (1986) 15891–15894]

γ-N-Methylasparagine

Methylated xanthines: *N*-methyl derivatives of xanthine, biosynthesized by the enzymatic methylation of free xanthine (N-1,3 and 7) with *S*-adenosyl-L-methionine. Caffeine, theobromine and theophylline (Fig.) occur in certain plants and are known as purine alkaloids.
Caffeine (syn. thein, coffeine, guarine) acts as a central stimulant, and is present in tea, coffee, maté leaves, guarana paste and cola nuts. Theobromine is the principal alkaloid of cacao beans (1.5–3%) and it is also present in cola nuts and tea. Theobromine is usually prepared from cacao bean hulls, which contain 0.7–1.2%. It acts as a diuretic, smooth muscle relaxant, cardiac stimulant and vasodilator. Theo-

phylline is present in small amounts in tea. It has similar pharmacological properties to theobromine. The water solubility of M.x. can be increased by formation of molecular compounds with diethanolamine or isopropanolamine, and their solubility is also greatly increased by the presence of alkali benzoates, cinnamates, citrates or salicylates. For therapeutic purposes synthetic derivatives of M.x. are often used; these have improved water solubility compared with the natural M.x., e.g. 7-theophylline-acetic acid (1,2,3,6-tetrahydro-1,3-dimethyl-2,6-dioxopurine-7-acetic acid) and 1-theobromineacetic acid (2,3,6,7-tetrahydro-3,7-dimethyl-2,6-dioxo-1H-purine-1-acetic acid), which have the same therapeutic properties as theophylline and theobromine. The salt of 1-theobromineacetic acid with bromocholine phosphate is used as an antihypertensive. The 1-hexyl derivative of theobromine, called pentifylline, is used as a vasodilator; it has increased lipid solubility, which favors absorption.

	R_1	R_3	R_7
Xanthine	H	H	H
Theophylline	CH_3	CH_3	H
Theobromine	H	CH_3	CH_3
Caffeine	CH_3	CH_3	CH_3

Structures of methylated xanthines

O-Methylbufotenin: a toxin from the toad, *Bufo alvarius* (see Toad poisons), which has also been found in plants. In addition to its general properties as a toad poison, M. also has a psychotropic effect. The lowest fatal dose for mice is 75 mg per kg.

Methylcrotonylglycinuria: see Inborn errors of metabolism.

α-Methyldeoxybenzoins: only two naturally occurring representatives of this group are known: angolensin (Fig.) from the heartwood of *Pterocarpus* spp. and the wood of *Pericopsis* spp. (teak), and 2-O-methylangolensin (wood of *Pericopsis*). In *Pericopsis*, the M. are accompanied by the isoflavones afrormosin and biochanin A. It is therefore assumed that the M. are also of isoflavonoid origin. [W. D. Ollis et al. (1965) *Aust. J. Chem.* **18** 1787–1790. M. A. Fitzgerald et al. (1976) *J. Chem. Soc. Perkin I*, 186–191].

(−)-Angolensin

Methylglyoxal: an intermediate of carbohydrate degradation in certain organisms. In some bacteria (*Pseudomonas* spp.) glyceraldehyde 3-phosphate

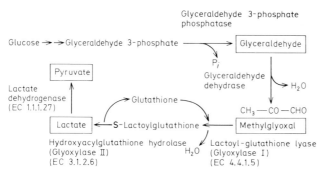

Metabolism of glyceraldehyde 3-phosphate in Pseudomonas. *Synthesis and metabolism of methylglyoxal.*

does not enter the normal glycolytic pathway; instead it is dephosphorylated to glyceraldehyde, followed by dehydration to M., which is converted to lactate (precursor of pyruvate) by a catalytic cycle involving glutathione (Fig.).

N⁶ - cis - γ - Methyl - γ - hydroxymethylallyladenosine: 6-(4-hydroxy-3-methyl-but-*cis*-2-enyl)-aminopurine, an adenine derivative, and one of the rare nucleic acid components occurring in certain tRNA. Biosynthesis occurs by modification of an adenosine residue in the nucleic acid. It is the *cis* isomer of Zeatin (see), and the free compound, like zeatin, shows cytokinin activity.

Methylmalonic acidemia: see Inborn errors of metabolism.

Methylmalonic aciduria: see Inborn errors of metabolism.

Methylnissolin: 3-hydroxy-9,10-dimethoxypterocarpan, see Pterocarpans.

Methylotrophy: the use of one-carbon compounds more reduced than CO_2 as the sole source of carbon and energy. [C. Anthony, *The Biochemistry of Methylotrophs* Academic Press, London, 1982.]

Methyltestosterone: 17α-methyltestosterone, 17α-methyl-17β-hydroxyandrost-4-ene-3-one, a synthetic androgen. M. shows high biological activity when administered orally, and it is used especially for the therapy of hypogenitalism, hormonal impotence, and peripheral circulatory disturbances. It is the 17α-derivative of Testosterone (see).

Mevaldic acid: an intermediate in Terpene (see) biosynthesis.

Mevalonic acid: an intermediate in Terpene (see) biosynthesis.

Mevinolin: 1,2,6,7,8,8a-hexahydro-β,δ-dihydroxy-2,6-dimethyl-8-(2-methyl-1-oxobutoxy)-1-naphthalene-heptanoic acid δ-lactone (Fig.), a fungal metabolite from the culture medium of *Aspergillus terreus*. The parent hydroxyacid (mevolinic acid) is a potent competitive inhibitor (K_i 0.6 nM) of 3-hydroxy-3-methylglutaryl-CoA reductase (EC 1.1.1.34). Oral administration of M. produces a 30% decrease of plasma LDL cholesterol and a moderate increase in the number of LDL receptors in human heterozygotes for familial hypercholesterolemia (see Lipoproteins). When M. is administered with Cholestyramine (see), plasma LDL cholesterol decreases by 50–60% and the increase of LDL receptors is even greater. See Compactin. [A. W. Alberts et al. *Proc. Nat. Acad. Sci.* 77 (1980) 3957–3961]

Meyerhof quotient: see Pasteur effect.

MF: abb. of maize factor (see Zeatin).

MH: abb. of Maleic hydrazide.

Michaelis constant: see Michaelis-Menten kinetics.

Michaelis-Menten kinetics: 1. The Michaelis-Menten stoichiometric model shows the relationship between free enzyme (E), substrate (S), enzyme-substrate complex (Michaelis complex, ES) and product (P):

$$E + S \underset{k_{-1}}{\overset{k_1}{\rightleftharpoons}} ES \overset{k_2}{\rightarrow} E + P$$

where k_1 and k_2 are rate constants (k_2 is also known as the catalytic constant, or k_{cat}). $(k_{-1} + k_2)/k_1$ is a kinetic constant, known as the Michalis constant and represented by K_m. If $k_2 \ll k_{-1}$, then $K_m \approx K_s = k_{-1}/k_1$ (Michaelis condition), and the equilibrium constant K_s is known as the substrate constant. If $k_{-1} \ll k_2$, then $K_m \approx k_2/k_1$ (Briggs-Haldane condition).

2. The Michaelis-Menten rate equation shows the relationship between v (rate of reaction), V_m (maximal rate when enzyme is saturated with substrate), and S (substrate concentration): $v = V_m S/(K_m + S)$. When $S \ll K_m$, the reaction is first order, and $v = (V_m/K_m)S$. When $K_m \ll S$, the reaction is zero order, and $v = V_m = $ constant. V_m and K_m are enzyme kinetic parameters. The Michaelis-Menten equation is often valid for other cases, where the derivation of the kinetic parameters from the rate constants is more complicated.

R = H, *Compactin* (see)
R = CH₃, *Mevinolin*

Mevinolinic acid

Microbiological conversions, *microbiological transformations:* Conversions of materials occurring in one or more stages, and catalysed by microorganisms. M.c. are the result of microbiological enzyme action, and often have no importance for the microbial cell. Several M.c. are important in the pharmaceutical industry. Examples are the stereospecific conversions of steroids, oxidation of sorbitol to sorbose by *Acetobacter suboxydans* (in the production of vitamin C), and the addition of acetaldehyde to benzaldehyde by *Saccharomyces cerevisiae.* The product of this last reaction is phenylacetylcarbinol, a precursor for D-ephedrine synthesis.

Microbiological industry, *fermentation industry:* a branch of industry in which materials are produced or converted by the action of micro-organisms. It includes fermentation industry (production of alcoholic beverages and organic acids), production of antibiotics, production of enzymes and biochemicals, production of baking yeasts and yeasts for foodstuff manufacture. See Industrial microbiology; Industrial biochemistry.

Microelements: see Trace elements.

β_2-Microglobulin: The smallest known plasma protein. M_r 11815; 100 amino acid residues of known primary structure [Peterson, P.A., Cunningham, B.A., Berggård, I. and Edelman, G.M. (1972) *Proc. Nat. Acad. Sci.* U.S. **69,** 1697; Cunningham, B.A., Wang, J.L., Berggård, I., and Peterson, P.A. (1973) *Biochemistry 12,* 4811]. Increased quantities of β_2-M. are found in the urine in Wilson's disease and in cadmium poisoning. There are sequence homologies between the constant region of the γGl-immunoglobulins and β_2-M., which suggest an evolutionary relationship between the two proteins. The β_2-M. gene may have evolved directly from the immunoglobulin precursor gene before its duplication.

Micronutrients: see Trace nutrients.

Microsomes: a heterogeneous fraction of submicroscopic vesicles, 20–200 nm diameter, formed during disruption of the cell by the resealing of fragments of the endoplasmic reticulum (and to some extent of the plasma membrane). Under the electron microscope, Ribosomes (see) can be seen attached to the outside of the M. Preparation of M. is by differential centrifugation of disrupted, homogenized cells; following the sedimentation of larger fragments, M. are sedimented at 100000 g.

Middle lamella: see Cell wall.

MIH: abb. for melanotropin release inhibiting hormone. See Releasing hormones.

Milk proteins: soluble proteins present in milk, consisting of caseins and whey proteins. The chief caseins are α_S, β and κ-casein. The most important whey proteins are β-lactoglobulin, α-lactalbumin and lactoferrin. In addition, milk contains several enzyme proteins, e.g. lactoperoxidase, xanthine oxidase and immunoglobulins IgG, IgA and IgM. These immunoglobulins are absorbed directly and without cleavage by the intestine of the infant, and provide it with passive immunity against those pathogens to which the mother is immune.

Millimicron, *mμ:* a unit of length, the same as a Nanometer (see).

Mineral nutrients, *major inorganic elements, inorganic bulk elements, macroelements:* inorganic nutrients required by living organisms in greater quantity than the Trace elements (see). They are absorbed as cations (Na^+, K^+, Ca^{2+} etc.) and anions (Cl^-, I^-, NO_3^-, SO_3^{2-} etc.). In some cases they may occur in the organism in large concentrations, e.g. skeletal structures such as the bones of vertebrates, or the silicic acid exoskeleton of diatoms; the epidermis of grasses, sedges, horsetails *(Equisetum)* is sometimes so completely impregnated with silicic acid that if the tissue is burned, a complete siliceous skeleton of the cells is left behind; the cell walls of the brittleworts (order *Characeae*) are incrusted with calcium carbonate. Generally, however, very high local concentrations of inorganic elements do not occur in living organisms.

The very mobile cations K^+ and Na^+ are important in membrane transport (see Ion pumps), in which K^+ is the intracellular and Na^+ the extracellular cation of the active transport mechanism. Cl^- is an intra- and extracellular anion, and it is the counterion of H^+ in the production of H^+Cl^- in the stomach. Calcium is a structural component of bone; it also occurs as calcium pectinate in the middle lamella (primordial membrane) of the plant cell wall; it acts as an oxalate trapping agent in plants, which store oxalate in the form of crystals of calcium oxalate; it is also an important ion in the process of muscle contraction. Magnesium (see) is required for the activity of many enzymes; it is involved in bone formation, and it is a component of chlorophyll.

Mineralocorticoids: see Adrenal corticosteroids.

Minimum protein (or nitrogen) requirement: the amount of complete protein required daily to compensate the nitrogen lost by excretion. Adults require 25 to 35 g complete (containing the optimal amounts of essential amino acids) protein per day. The absolute M. is the amount of nitrogen excreted on a protein-free but calorically adequate diet, about 2.4 g N = 15 g protein per day for adults. The United States Dept. of Agriculture's recommended daily allowance for adults has been revised downward in recent years from 70 g to 40 g protein per day for adults.

Minor bases: see Rare nucleic acid components.

Minor nucleic acid components: see Rare nucleic acid components.

Minus strand: see Codogenic strand.

Miraculin: a taste-modifying glycoprotein from the miraculous berry (*Synsepalum dulcificum,* family *Sapotaceae*) native to West Africa. M. does not itself taste sweet, but prior exposure of the tongue to M. causes sour substances to taste sweet; this activity may persist for several hours. Some work has been reported on amino acid composition and possible subunit structure of M. [Cagan, R.H. (1973) *Science,* **181,** 32–35].

Miraxanthins: yellow Betaxanthins (see) found in *Mirabilis jalapa.*
M.-I: betalamic acid conjugated with methionine sulfoxide.
M.-II: betalamic acid conjugated with aspartic acid.
M.-IV: betalamic acid conjugated with tyramine.
M.-VI: betalamic acid conjugated with dopamine.

Mirestrol, *miroestrol:* a potent plant estrogen from the tubers of *Pueraria mirifica* found in Thailand. The tubers of the plant are known in folk medicine for their rejuvenating and oral contraceptive properties.
[Taylor, N.E., Hodgkin, D.C. and Rollett, J.S. (1960) *J. Chem. Soc.,* 3685.
Kashemsanta, M.C.L., Savatabanduh, K. and Shaw, H.K.A. (1952) *Kew Bull.* 549].

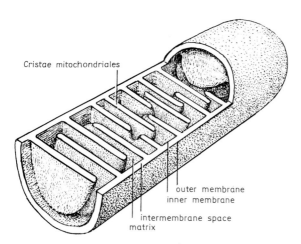

Mirestrol

Mitchell hypothesis: see Respiratory chain phosphorylation.

Mitochondria, *chondriosomes* (obsolete): organelles present in all eukaryotic cells. M. are 0.3–0.5 μm long, and they show a wide range of shapes and sizes. The average M. is rather elongated and about the same size as a cell of *Escherichia coli.* M. are on the threshold of visibility under the light microscope, and they are made more readily observable by staining with Janus green B. M. of muscle cells are known as *sarcosomes.* In the differential centrifugation of cell homogenates, M. are sedimented by centrifugation at 10000 g for 10–15 min, i.e. their sedimentation properties are intermediate between the nuclear and microsomal fractions. M. are isolated in hypertonic or isotonic solutions of nonelectrolytes, such as mannitol or sucrose.
Electron microscopy of M. shows a characteristic internal structure. There are 2 concentric membranes, each 5–7 nm thick (see Biomembranes). Between the *outer* and *inner membranes* lies the *intermembrane space* (also called *external matrix* or *outer mitochondrial space*) (Figs. 1 and 2). These 2 membranes have different submicroscopic structures, their biogenesis is different, and they are functionally distinct. The outer membrane can be removed by osmotic rupture.
The density of the outer membrane is about 1.1 g/cm^3 and it is permeable to most substances of M_r 10000 or less. It contains a high proportion of phospholipid (phospholipid to protein ratio is about 0.82 by weight); after extraction of 90% of mitochondrial phospholipid with acetone the inner membrane remains intact and its double layered structure is retained, whereas the outer membrane is destroyed. The lipid fraction contains a low concentration of cardiolipin, a high concentration of phosphoinositol and cholesterol, and no ubiquinone. The density of the inner membrane is about 1.2 g/cm^3. Neutral substances, e.g. sugars, of M_r less than 150 seem to cross the inner membrane freely, but the passage of all other substances is subject to tight control. There is a low content of phospholipids (phospholipid to protein ratio is about 0.27 by weight), containing about 20% cardiolipin. The components of the respiratory chain, including ubiquinone, are present in the inner membrane.
The inner membrane encloses the *matrix* or *stroma,* which is a contractile network of structural proteins embedded in an aqueous phase. The inner membrane also shows characteristic folds, known as *cristae,* which extend into the matrix. In some cases these structures may more closely resemble tubes, which are known as *tubuli.* The terms cristae space (matrix) and intracristae space (intermembrane space) have been used, but they are equivocal and should be avoided. The precise shape and internal structure of the M. depends on its functional state (which varies from cell to cell) and may also be influenced by the methods of isolation, fixation and staining. The number of M. per cell depends on the cell type. Thus a sperm contains 20–24 M., whereas

Cristae mitochondriales

outer membrane
inner membrane

intermembrane space
matrix

Figure 1. **Diagram of a cristae-type mitochondrion.** Each membrane consists of a bilayer.

the protozoan, *Chaos chaos*, contains 500 000. Liver parenchyma cells contain about 500 M.

Disruption of M. produces smaller fragments known as submitochondrial particles (SMP). SMP consist chiefly of fragments of inner membrane, which become resealed to form vesicles; these are sometimes referred to as "inside-out-particles", because the outer surface (i.e. exposed to the surrounding medium) corresponds to the inner surface of the membrane in the intact M. (i.e. exposed to the matrix). The method of disruption of M. (sonication, mechanical shear, detergents) and the intensity of its application determine the nature of the resulting SMP. The capacity for oxidative phosphorylation may be lost, but the particles may still actively respire (electron transport particles, ETP). On the other hand, careful and mild disruption of M. produces SMP that are still able to carry out oxidative phosphorylation.

In addition to respiration, M. are the sites of other metabolic activities, e.g. glucuronate and ascorbic acid synthesis. The intramitochondrial distribution of many enzymes is known:

The inner membrane contains all the components of the Respiratory chain (see), succinate dehydrogenase (EC 1.3.99.1), NADH dehydrogenase (EC 1.6.99.3), the enzymatic apparatus of Oxidative phosphorylation (see), ATPase (EC 3.6.1.8), 3-hydroxybutyrate dehydrogenase (EC 1.1.1.30), ferrocheletase (EC 4.99.1.1), δ-aminolevulinate synthase (EC 2.3.1.37), carnitine palmitoyltransferase (EC 2.3.1.21), fatty acid elongation system (C_{10}), $NAD(P)^+$ transhydrogenase (EC 1.6.1.1), choline dehydrogenase (EC 1.1.99.1).

The intermembrane space contains adenylate kinase (EC 2.7.4.3), nucleosidediphosphate kinase (EC 2.7.4.6) and creatine kinase (heart muscle M.) EC 2.7.3.2).

Marker enzymes for the outer membrane are flavin-containing amine oxidase (monoamine oxidase) (EC 1.4.3.4) and NADH dehydrogenase (EC 1.6.99.3) (also known as cytochrome *c*-reductase, or NADH-cytochrome b_5-reductase; this enzyme is insensitive to rotenone, amytal or antimycin A, and it is similar to the NADH-cytochrome b_5-reductase of microsomes). In addition, the outer membrane contains kynurenine 3-monooxygenase (EC 1.14.13.9), ATP-dependent fatty acyl-CoA synthetase (EC 6.2.1.3), glycerophosphate acyltransferase (EC 2.3.1.15), acylglycerophosphate acyltransferase (EC 2.3.1.51, or EC 2.3.1.52), lysolecithin acyltransferase (EC 2.3.1.23), cholinephosphotransferase (EC 2.7.8.2), phosphatidate phosphatase EC 3.1.3.4), phospholipase A_2 (EC 3.1.1.4), nucleosidediphosphate kinase (EC 2.7.4.6), and a fatty acid elongation system (C_{14}, C_{16}). It is noteworthy that the outer membrane contains many enzymes of phospholipid metabolism.

The matrix contains all the enzymes of the TCA-cycle (see), with the exception of succinate dehydrogenase. Also present are glutamate dehydrogenase EC 1.4.1.3), pyruvate carboxylase (EC 6.4.1.1), aspartate aminotransferase (EC 2.6.1.1), carbamoyl phosphate synthetase (EC 6.3.4.16) (utilizing ammonia and concerned in urea synthesis; unlike a similar enzyme in the cytosol which utilizes glutamine and

is concerned in the synthesis of pyrimidines), ornithine carbamoyltransferase (EC 2.1.3.3), GTP- and ATP-dependent fatty acyl-CoA synthetases (EC 6.2.1.10 and 6.2.1.3, respectively), and enzymes for the β-oxidation of fatty acids.

On the matrix side of the inner membrane are attached small particles (M_r about 85 000), visible under the electron microscope with negative staining (Fig. 2): these were earlier known as oxysomes, but are now more commonly called coupling factor 1 (F_1). The knob-like F_1 structure is attached to a stem, which is associated with special proteins in the membrane. This total structure is thought to constitute the ATP-synthesizing system of Oxidative phosphorylation (see).

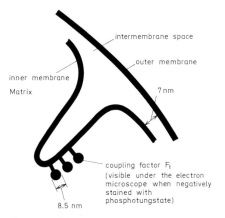

Figure 2. Diagrammatic representation of part of a mitochondrion, showing an infolding of the inner membrane to form a crista, and the arrangement of knob-like coupling factors (F_1) facing into the matrix.

Coenzymes such as NAD^+, $NADP^+$ and coenzyme A are concentrated in the aqueous phase of the matrix. The matrix also contains the protein synthesizing system of the M. Vertebrate M. contain Ribosomes (see) of size 50–55 S, whereas the M. of other eukaryotic organisms contains 70 S ribosomes. The DNA of M. is circular, histone-free, and reaches a length (in mammals) of 5 μm. It contains cistrons for ribosomal RNA and tRNA, and the genes for ATPases, for 2 of the 3 subunits of cytochrome *b*, and for 4 of the 7 subunits of cytochrome oxidase. The biosynthesis of all the other mitochondrial constituents appears to be under the genetic control of the cell nucleus. The genetic code and tRNA complement of M. are different from those of the cytoplasm (see Genetic code).

M. are formed by the fission of M., or from promitochondria. The latter are smaller, with practically no internal structure, and no enzymes of the respiratory chain or oxidative phosphorylation. Evidence is lacking for a de novo formation of M., or the biogenesis of M. from precursor components in the cell. The 2 compartments of M., one inside the other, appear to be of different evolutionary origin.

Mitomycin C: an aziridine antibiotic, M_r 334, first isolated in 1956 from *Streptomyces caespitosus*.

M. leads to an inhibition of DNA synthesis. In the natural, oxidized state, M. is inactive. In the cell it undergoes intramolecular rearrangement and reduction to form an unstable bifunctional alkylating agent. The accompanying formation of carbonium ions at positions 1 and 10 enables the compound to bind covalently to DNA and irreversibly cross link the two strands of the double helix, thus preventing the action of DNA polymerase. M. is a bacteriocide and a cytotoxin. The aziridine antibiotics include other mitomycins and porfiromycins.

Mitomycin C

Mitosis: nuclear division in the somatic cells of eukaryotic organisms, producing two identical daughter cells with the same genetic constitution. M. occurs in 4 stages, called prophase, metaphase, anaphase and telophase. In M. the daughter cells each contain the same number of chromosomes as the parent cell.

Mitosis cycle: see Cell cycle.

Mixotrophy: see Nutritional physiology of microorganisms.

mlRNA: abb. of giant messenger-like RNA. See Messenger RNA.

Moderator protein: see Calmodulin.

Modification: see Restriction endonucleases.

Modulation: 1. the change in the kinetics of an enzyme, resulting from the reversible alteration of its protein conformation by the binding of a modulating or allosteric effector (see Allostery; Covalent modification of enzymes).
2. Regulation of the rate of translation of mRNA by a modulating codon, which codes for a rare tRNA known as modulator tRNA. This is important in the regulation of the synthesis of series of enzymes from a polycistronic mRNA, where specific modulator tRNA is responsible for the continuity of translation between the individual enzymes, e.g. in the biosynthesis of the 10 enzymes of histidine biosynthesis.

Molar activity: see Enzyme kinetics.

Molar mass (symbol M): numerically the same as the Relative molecular mass (see), but with units, e.g. the molar mass of glycine is $75\,\mathrm{g\,mol^{-1}}$. It is sometimes needed instead of M_r, e.g. in the Svedberg equation: $M = R \times T \times s/D\,(1 - \bar{v}\rho)$.

Molecular genetics: a component of genetics and of Molecular biology (see). Classical genetics defined the gene, on the basis of its phenotypic expression and its mathematical distribution, as a hypothetical unit of inheritance. M.g. recognizes genes as defined segments of DNA, and shows the molecular basis for the causal relationship between the gene and its phenotypic expression.

Molecular mass: the mass of a molecule measured in Daltons (see).

Molecular weight: see Relative molecular mass.

(+)-Mollisacacidin: 7,3′,4′-trihydroxyflavan-3,4-diol, see Leucoanthocyanidins.

Molting hormone: see Ecdysone.

Molybdenum, *Mo:* an important trace element for plants and bacteria, and for the symbiotic fixation of nitrogen. Mo has not been demonstrated to be an essential nutrient in animals, but it is known to be a constituent of at least 3 animal enzymes (see Molybdoenzymes). As a constituent of various molybdoenzymes, Mo plays an important part in the nitrogen metabolism of plants and bacteria. In biological systems there is a partial antagonism between Mo and Cu.

Molybdoenzymes: At present, 6 oligomeric oxidoreductases are known, which contain Mo as an essential constituent: 1. Nitrogenase (see); 2. Nitrate reductase, EC 1.6.6.3 (see); 3. Xanthine oxidase EC 1.2.3.2 (see) from animals and bacteria; 4. Aldehyde oxidase, EC 1.2.3.1 from animal liver which catalyses the reaction $R\text{-}CHO + H_2O \to R\text{-}COOH + 2H^+ + 2e'$; 5. Sulfite oxidase, EC 1.18.3.1 from mammalian and bird liver (M_r 114000 2 subunits), which catalyses the reaction $SO_3^{2-} + H_2O \to SO_4^{2-} + 2H^+ + 2e'$; this enzyme also contains a b_5-like cytochrome and passes electrons directly to cytochrome c in the respiratory chain and 6. Formate dehydrogenase, EC 1.2.1.2, a membrane-bound protein from *Escherichia coli*, which contains one atom each of molybdenum and selenium, one heme group and nonheme iron-sulfur centers. It is NAD^+-dependent and catalyses the reaction $HCOO^- + NAD^+ \to CO_2 + NADH$.

Those molybdoenzymes that are also flavoproteins e.g. nitrate reductase, xanthine oxidase, and aldehyde oxidase, are also called molybdoflavoproteins. In all the molybdoenzymes, Mo is part of the redox system for the transfer of electrons.

Monellin: an intensely sweet tasting dimeric protein; the A-chain contains 44, the B-chain 50 amino acid residues. M. tastes 3000 times (weight basis) or 90000 times (molar basis) sweeter than sucrose. It was discovered in the fresh fruits of an African berry, *Dioscoreophyllum cumminsii* ("wild red berry") or the *Menispermaceae*. Another sweet protein is Thaumatin (see).

Monoacylglycerols: see Acylglycerols.

Monoclonal antibodies: see Immunoglobulins.

Monod model: see Cooperativity model.

Monod-Wymann-Changeux model: see Cooperativity model.

Monoglyceride: see Acylglycerols.

Monohydroxysuccinic acid: see Malic acid.

Monomer: see Subunit.

Monophenol monooxygenase, *laccase* (EC 1.14.18.1): see Oxygen metabolism; Lignin.

Monosaccharide: see Carbohydrates.

Monosomes: see Ribosomes.

Monoterpenes: aliphatic mono-, di-, or tricyclic terpenes, formed from two isoprene units ($C_{10}H_{16}$). Their occurrence is limited chiefly to plants. Most M. are volatile liquids, and they are present in volatile oils and balsams. For their separation and identification from volatile oils, M. are converted into

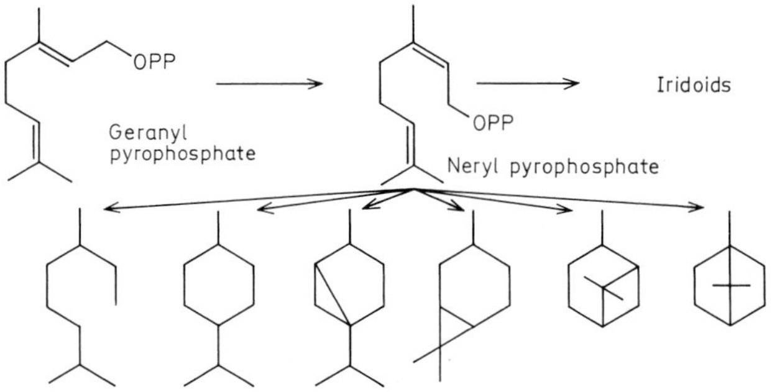

Type	3,7-Dimethyl-octane	p-Menthane	Thujane	Carane	Pinane	Bornane
Hydrocarbons	Ocimene Myrcene	Terpinene Limonene	Sabinene	Δ^3-Carene	α-Pinene β-Pinene	
Alcohols	Linalool Citronellol	Terpeneol Menthol				Borneol
Carbonyl compounds	Citral Citronellal	Menthone Carvone				Camphor

Types of monoterpene structures

nitrosochlorides by the action of nitrosyl chloride. Cantharidin (see) and the Pyrethrins (see) are solid M., which posses an unusual linkage between the two isoprene components. The many different structural types of the remaining M. arise from different types of cyclization and the introduction of substituents (Fig.). The monoterpene alkaloids are M. that also contain nitrogen; M. also participate in the synthesis of the iridoid indole alkaloids and other alkaloids (see Iridoids).
Apart from their intermediate role in the biosynthesis of other terpenes, little is known about the biological function of M. They are important volatile components of olfactorily active pheromone mixtures. In pure form or as mixtures, e.g. balsams and oil of turpentine, M. are used widely in pharmacy, food industry, perfumery and paint manufacture.
M. are biosynthesized from geranyl pyrophosphate (see Terpenes) via its isomer neryl pyrophosphate (Fig.). Aliphatic M. arise by hydrolysis of the phosphate bond, or the elimination of pyrophosphate (synthesis of 3,7-dimethyloctane type). Cyclic M. usually arise by nucleophilic substitution at C-1 of the neryl pyrophosphate, with loss of pyrophosphate. Iridoid compounds arise from the reductive cyclization of neryl pyrophosphate.

Biological significance and uses of monoterpenes.

Role	Examples
Pheromones	Citral, Iridodiol
Insecticides	Pyrethrins
Iridoids	Loganin
Alkaloids	Nuphura alkaloids
Bitter principles	Gentiopicroside
Scents	Nerol, Geraniol
Pharmaceuticals	Valtratum, Camphor

Montanic acid: n-octacosanoic acid, CH_3–$(CH_2)_{26}$–COOH, M_r 424.73, m.p. 91 °C. Esterified M. is present in various waxes, e.g. montan wax, beeswax, Chinese wax.
MOPS: see 2-(N-Morpholino)-propane sulfonic acid.
Morbus Addison: see Adrenal corticosteroids.
Morin: see Flavones (Table).
Morindone: 1,2,5-trihydroxy-6-methylanthraquinone, an orange-red anthraquinone. m.p. 285 °C (from glacial acetic acid). M. occurs in the roots, bark and wood of *Morinda* species. In some Asian countries, particularly India, it was earlier used extensively as a natural dye.
Morphactins: a group of highly potent synthetic plant growth regulators. M. are derivatives of flurenol (9'-hydroxy-flurene-[9]-carboxylic acid, R = H), chiefly by esterification of the carboxyl group and/ or substitution in the ring system, e.g. chloroflurenol (R = Cl).

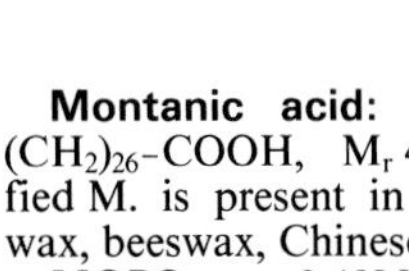

Flurenol (R = H)
Chloroflurenol (R = Cl)

Over a wide concentration range, M. inhibit and modify plant growth by acting as gibberellin antagonists. M. act synergistically with certain herbicides, e.g. 2,4-D, which enables their use as broad spectrum herbicides. Among the many physiological properies of M. are: shortening of shoot internodes, compact bushy growth habit, reversible inhibition of mitosis, effects on root growth, geotropism and phototropism, delayed induction of flowering, production of seedless fruit, and an influence on the sexual development of bisexual flowers.

Morphinane: see Benzylisoquinoline alkaloids.

Morphine: medically the most important, and quantitatively the chief member of the opium alkaloids. M. is isolated in large quantities (over 1000 tons per year) from opium, M_r 285.35. m.p. 254 °C, $[\alpha]_D^{20}$ -130.9 (methanol). Turkish and Macedonian opium contain 15-20% M. Most commercial M. is converted into codeine by methylation. M. acts as an anesthetic without decreasing consciousness, and it is the most powerful analgesic known. High doses cause death by respiratory failure. Owing to its dangerous properties as a hypnotic and narcotic, the use of M. is restricted. The diacetyl derivative of M., Heroin (see), is an even more dangerous narcotic. M. was isolated in 1806 by Sertürner as the soporific principle of poppy. It was the first alkaloid to be isolated.

For formula and biosynthesis, see Benzylisoquinoline alkaloids.

Morphogenesis: the genesis of form and shape; the ability of a molecule, group of molecules, an organ or organism to assume a certain shape. To a certain level of complexity, M. can be explained by physical-chemical laws; at a higher level, M. is regulated by chemical inducers.

The primary structure of a polypeptide determines the three dimensional structure (chain conformation) of the molecule by physical-chemical forces. The next step of M. is the formation of protein complexes (hemoglobin, multienzyme complexes) or structural elements such as collagen. Bacteriophages and ribosomes can be dissociated in vitro into their components, then reconstituted. Formation of the cell membrane, a complex of proteins and lipids, is the result of chemical affinities. If the cell membrane is fragmented by homogenization, it reseals itself into small vesicles. Thus the information for the formation of fairly complicated structures may be inherent in the separate constituents, so that a structure (e. g. ribosome) is formed from its components by a process of cooperative self assembly.

During embryonal development, the formation of organs and organ systems is induced by diffusible chemical factors, called inducers (see Embryonal inducers); such factors are responsible for, e. g. the differentiation of nerve (neural factors) or muscle and kidney tissue (mesodermal factors). Specific inducers can also be isolated from adult organisms. Growth and regeneration of nervous tissue is induced by a Nerve growth factor (see). The equilibrium between degeneration and regeneration in a fully grown organ is regulated by Chalones (see). After the operative removal of two thirds of a rat's liver, the organ regenerates to its original weight and size in three weeks. During the regeneration period, the inhibition of liver cell division is abolished by an unknown mechanism.

2-(N-Morpholino)-propane sulfonic acid, *MOPS:* M_r 209.3, used for the preparation of Buffers (see) in the pH range 6.5-7.9. pK_a 7.2 at 25 °C.

Morquio-Brailsford syndrome: see Inborn errors of metabolism.

MRH: abb. for melanotropin releasing hormone (see Releasing hormones).

mRNA: see Messenger RNA.

MSH: see Melanotropin.

Mucins: see Glycoproteins.

Mucolipidosis: see Inborn errors of metabolism.

Mucopolysaccharide degradation: see Inborn errors of metabolism.

Mucopolysaccharides: heteroglycans of animal connective tissue. Each of these acidic polysaccharides consists of an acetylated hexosamine (N-acetylglucosamine, or N-acetylgalactosamine) and an uronic acid (usually glucuronic acid, sometimes iduronic acid as in dermatan sulfate), which form a characteristic repeating disaccharide unit. The repeating structure of the polymer involves alternate 1,4 and 1,3 linkages. Many M. also contain sulfate. Examples of M. are Hyaluronic acid (see), Chondroitin sulfate (see), Dermatan sulfate (see), Keratan sulfate (see), Heparin (see), Heparitin sulfate (see). Some blood group substances and some bacterial polysaccharides are also M. In animals, M. function as supportive and protective materials and as lubricants.

M. are synthesized from the UDP-N-acetylhexosamine and UDP-D-glucuronic (or iduronic) acid. Sulfate is transferred directly from adenosine-3'-phosphate-5'-phosphosulfate after formation of the polysaccharide chain. Synthesis takes place in the endoplasmic reticulum.

Mucopolysaccharidosis: see Inborn errors of metabolism.

Mucosulfatidosis: see Inborn errors of metabolism.

(-)-Mucronulatol: (3S)-7,3'-dihydroxy-4',2'-dimethoxy-4'-methoxyisoflavan, see Isoflavan.

Multienzyme complex: an ordered association of functionally and structurally different enzymes which catalyse successive steps in a chain of metabolic reactions. The known M. c. consist of 2-7 different catalytic units associated non-covalently (M_r 160000 to a few million), with no associated lipids, nucleic acids or enzymatically inactive proteins. Some can be partly dissociated into smaller active groups of enzymes and half molecules, or even into their fundamental subunits (usually inactive) by changes of pH, temperature and ionic strength, or by treatment with neutral or anionic detergents. In many cases, dissociated M. c. have been reassociated to active forms similar to the physiological complex. M. c. represent an important form of subcellular organization (compartmentation), which is much simpler than the localization of enzymes within an organelle (e. g. tricarboxylic acid cycle). Owing to the close proximity of the active centers of the enzymes within the M. c., and their high affinity for substrates, the reactions of M. c. are rapid, but controlled. Intermediates are passed directly from one enzyme to the next, and they do not dissociate from the complex. The following M. c. have been particularly well studied biochemically, physically (hydrodynamic methods) and by electron microscopy:

2-Oxoacid dehydrogenases. Both the pyruvate and 2-oxoglutarate dehydrogenase complexes can be separated into 3 components: a dehydrogenase (or

Table. *Composition of the 2-oxoacid dehydrogenase complex of* Escherichia coli.

1. *Pyruvate dehydrogenase complex*

Enzyme	Oligomeric enzyme molecules		Subunits per enzyme molecule		Total subunits
	n	M_r	n	M_r	n
Pyruvate-DH	12	192 000	2	96 000	24
DL-Transacetylase	1	1.7×10^6	24	67 500	24
DL-DH (Flavoprotein)	6	112 000	2	56 000	12
Complex	19	4.67×10^6			60

2. *α-Ketoglutarate dehydrogenase complex*

Enzyme					
α-Ketoglutarate-DH	12	190 000	2	90 000	24
DL-Transacetylase	1	1.1×10^6	24	46 000	24
DL-DH (Flavoprotein)	6	112 000	2	60 000	12
Complex	19	2.94×10^6			60

DH Dehydrogenase, DL Dihydrolipoyl, n Number.

decarboxylase) with the dissociable cofactor thiamin pyrophosphate; a flavoprotein, dihydrolipoyl dehydrogenase; and a core enzyme containing lipoic acid which acts as a dihydrolipoyl transacetylase or transsuccinylase. The lipoic acid is bound covalently to the ε-NH_2 group of a lysine residue in the transferase. This transferase also has a special structural role; it must be present as a core for the association of the other enzymes. The two oxoacid dehydrogenases have analogous reaction mechanisms: the lipoic acid oxidizes the thiamin-bound active aldehyde, and the resulting acyl group becomes bound to the 1.5 nm long arm of the dihydrolipoic acid. Carrying the acyl group, this arm moves to the bound CoA on the transferase. The acyl-CoA is released, and the dihydrolipoic acid arm moves to the dihydrolipoyl dehydrogenase, where it becomes reoxidized. The $FADH_2$ is then reoxidized by NAD^+. For equations showing the case of 2-oxoglutarate dehydrogenase (α-ketoglutarate dehydrogenase), see Tricarboxylic acid cycle, Fig. 2.

The composition of the oxoacid dehydrogenase M.c. from *Escherichia coli* is shown in the table. The corresponding M.c. of mammalian origin are larger. For example, bovine pyruvate dehydrogenase contains a single central transacetylase molecule (M_r 8.12 × 10^6, consisting of 60 subunits), with 60 dimeric αβ) dehydrogenase molecules ($M_r\alpha = 42\,000$; $M_r\beta = 37\,000$) and 5–6 dimeric flavoprotein molecules (M_r 110 000). Pig heart 2-oxoglutarate dehydrogenase consists of a central succinyl transferase molecule with 6 molecules each of 2-oxoglutarate dehydrogenase and dihydrolipoyl dehydrogenase.

Under the electron microscope, pyruvate dehydrogenase from *E. coli* shows a polyhedral structure. *Fatty acid synthase*. Type II fatty acid synthase is a M.c. (see Fatty acid biosynthesis). 6-Methylsalicylic acid synthetase (M_r 1.3 × 10^6; 7 constituent enzymes) from *Penicillium patulinum* closely resembles fatty acid synthase.

Other examples of M.c. are anthranilate synthase and tryptophan synthase, which are involved in microbial Aromatic biosynthesis (see), and citrate lyase. The definition of M.c. may be extended to include the membrane-bound respiratory chain, the contractile protein complexes of muscle, and the ribosome (which also contains RNA), etc.

Multisubstrate enzymes: enzymes that require two or more substrates in order to catalyse a particular reaction. Accordingly, the enzyme forms a ternary (two substrates), quaternary (three substrates), etc. complex. Many enzymes are of this type, e.g. NAD-dependent dehydrogenases must bind both the substrate and NAD^+. See Cleland's short notation.

Mundulone: see Isoflavone.

Munetone: see Isoflavone.

Muningin: 6,4'-dihydroxy-5,7-dimethoxyisoflavone, see Isoflavone.

Muramic acid: N-acetyl-3-O-carboxyethyl-D-glucosamine, or O-lactyl-N-acetylglucosamine, an amino sugar derived from glucose. m.p. 151 °C, $[\alpha]_D$ 144 °. The N-acetylated form of M. is present in the mucopolysaccharide-peptide complexes (see Murein) of bacterial cell walls.

Muramidase: see Lysozyme.

Murein, *peptidoglycan:* a cross-linked polysaccharide-peptide complex of indefinite size, of the inner cell wall layer of all bacteria. It constitutes 50% of the cell wall in Gram negative, and 10% in Gram positive bacteria. M. is tough and fibrous, and it enables the cell to withstand high internal osmotic pressure. It forms an immense bag-shaped molecule, or "sacculus", which gives shape and rigidity to the bacterial cell. In Gram positive bacteria the M. layer may be up to 10 nm thick, consisting of about 20 layers of crosslinked peptidoglycan. In Gram negative bacteria, M. is probably rather more flexible and less tough than in Gram positive bacteria. M. consists of linear parallel chains of up to 20 alternating residues of β-1,4-linked residues of N-acetyl-glucosamine and N-acetylmuramic acid, extensively cross-linked by peptides.

In the M. of *Staphylococcus aureus* (Gram positive), the N-acetylmuramic acid residues carry a branched peptide of 9 amino acids. The N-terminal branch of this peptide is linked to the C-terminal branch of a similar unit in a different polysaccharide chain. The branched peptide is L-alanyl-D-isoglutaminyl-L-ly-syl-D-alanine, with an "extending arm" of pentagly-cine attached to the ε-NH$_2$ group of the lysine. The isoglutaminyl residue is so called because the α-car-boxyl group is substituted with an amino group; linkage to the lysine occurs via the γ-carboxyl group. In the M. of *Escherichia coli* (Gram negative), the lysine is replaced by *meso*diaminopimelic acid, and there is no pentaglycine "extending arm"; cross-linkage occurs between the D-alanine and the termi-nal NH$_2$ group of diaminopimelic acid. Also, a spe-cific D-alanyl carboxypeptidase removes the termi-nal D-alanine after cross-linkage has occurred through the neighboring diaminopimelic acid. Thus the scope for cross-linking is limited. Only 15-30% of the peptide side chains of *Escherichia coli* M. are involved in cross-links; the remainder are free, or (one in every 10-12 peptide side chains) they lose the D-alanine and become attached to the ε-NH$_2$ group of a lysine in a lipoprotein, via the carboxyl at the L-center of the mesodiaminopimelic acid.

Considerable variation is found in the composition of the cross-linking peptide, depending on the or-ganism: the D-isoglutaminyl residue may be re-placed by D-isoglutamyl or 3-hydroxy-D-isogluta-minyl; the alanyl residue may be replaced by L-seryl or glycyl; the lysyl residue by mesodiaminopimelyl, L,L-diaminopimelyl, ornithyl, diaminobutyryl, or homoseryl.

Biosynthesis of M. can be considered in 4 stages:
1. Synthesis of a water-soluble precursor, the UDP derivative of N-acetylmuramyl pentapeptide.

N-actylglucosamine 1-phosphate + UTP → UDP-N-acetylglucosamine + pyrophosphate.

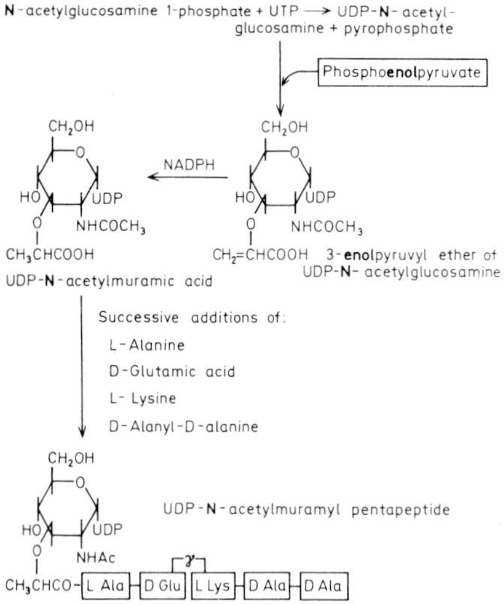

The sequence of the pentapeptide residue is a func-tion of the specificity of the enzymes concerned; there is no involvement of mRNA, ribosomes or tRNA; the addition of each amino acid and of the D-alanyl-D-alanine is accompanied by the cleavage of one ATP to ADP and phosphate.

2. Attachment of the precursor to the cell mem-brane:
The pentapeptide derivative is transferred, with loss of UMP, and becomes attached by a pyrophosphate linkage to a lipophilic carrier embedded in the mem-brane. The carrier is undecaprenyl phosphate, con-sisting of 11 isoprene units. A residue of N-acetyl glucosamine, derived from UDP-N-acetylglucos-amine, is added to the complex in 1,4-β-glycosidic linkage with the N-acetylmuramic acid. This is fol-lowed by conversion of the γ-glutamyl residue to isoglutaminyl (addition of NH$_4^+$, accompanied by cleavage of ATP to ADP and phosphate). A "space arm" of 5 glycine residues is then formed on the ε-NH$_2$ group of the lysine, by the addition of one gly-cine residue at a time from glycyl-tRNA. Ribosomes or mRNA are not involved, and in contrast to pro-tein synthesis, formation of the pentaglycine chain starts with the C-terminus and finishes with the N-terminus. The product represents the fundamental unit of M. At this stage it is still bound to the inside surface of the cell membrane by a pyrophosphate linkage to an undecaprenyl residue:

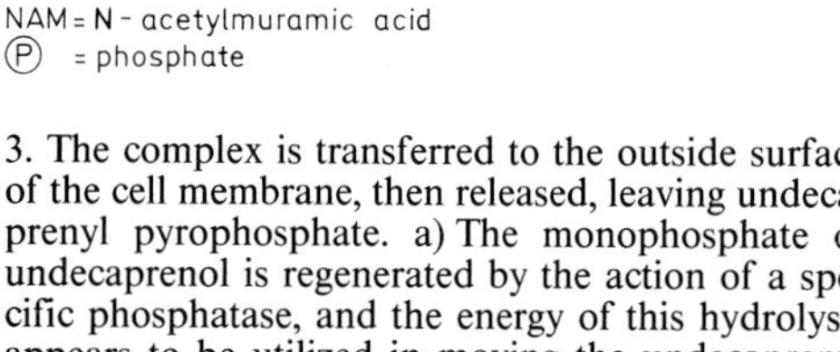

NAG—NAM—Ⓟ—Ⓟ—[CH$_2$—CH=C—CH$_2$]$_{11}$H
                                    |
                                    CH$_3$

L-Ala—D—isoGln—L-Lys—D-Ala—D-Ala—COOH
                        |
                      [Gly]$_5$
                        |
                       NH$_2$

NAG = N - acetylglucosamine
NAM = N - acetylmuramic acid
Ⓟ = phosphate

3. The complex is transferred to the outside surface of the cell membrane, then released, leaving undeca-prenyl pyrophosphate. a) The monophosphate of undecaprenol is regenerated by the action of a spe-cific phosphatase, and the energy of this hydrolysis appears to be utilized in moving the undecaprenyl monophosphate back to the inside surface of the membrane. The antibiotic activity of Bacitracin (see

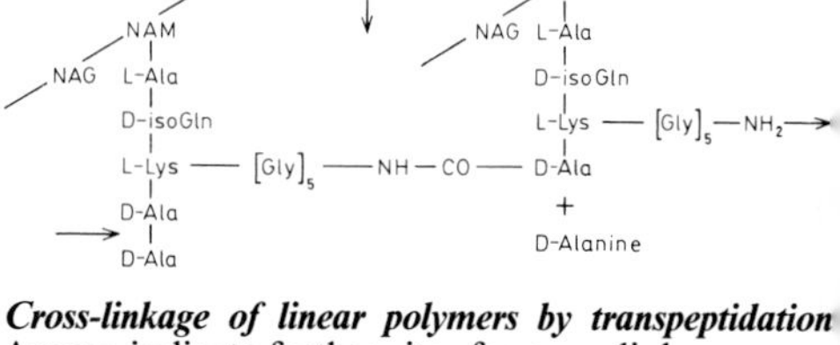

Cross-linkage of linear polymers by transpeptidation
Arrows indicate further sites for cross-linkage.

is due to its inhibition of this pyrophosphate cleavage. b) Complexes combine into linear polymers.
4. Cross-links are formed by transpeptidation, the necessary energy being derived from the cleavage of the terminal D-alanine, i.e. one peptide bond is formed at the expense of another. This stage is inhibited by penicillin.

Lysozyme (see) destroys the cell wall structure of bacteria by hydrolysing 1,4-glycosidic linkages between C-4 of N-acetylglucosamine and C-1 of N-acetylmuramic acid in M. The glycosidic bond between C-1 of N-acetylglucosamine and C-4 of N-acetylmuramic acid is not cleaved. The products of cleavage are "muropeptides", e.g.

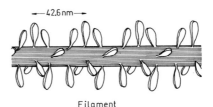

Ⓛ = site of action of Lysozyme

Muscaaurins: orange Betalains (see) present in the cap of the fly agaric. All M. are derivatives of Betalamic acid (see), which in M.I is linked to Ibotenic acid (see), and in M.II with glutamic acid. In addition to M.I to VII, the fly agaric also contains yellow muscaflavin, an isomer of betalamic acid, violet muscapurpurin and red-brown muscarubrin.

Muscarin: a biogenic amine and one of the amanita toxins. M. is a quaternary ammonium base. (+)-M. from the fly agaric *(Amanita muscaria)* is a salt of 2S,4R,5S-(4-hydroxy-5-methyltetrahydrofurfuryl)-trimethylammonium. It is concentrated in the skin of the cap of the fruiting body. (+)-M. is present in high concentrations in species of the fungal genera *Inocybe* and *Clitocybe*. The symptoms of experimental muscarin poisoning are not the same as those of fly agaric poisoning; other water-soluble compounds of fly agaric are responsible for its primary toxicity, namely Ibotenic acid (see), Muscimol (see) and Muscazone (see).

Muscarinic receptor: see Acetylcholine.

Muscazone: an α-amino acid with a heterocyclic substituent of 2(3H)-oxazolone. M_r 144, m.p. 190 °C (d.). M. occurs in the fly agaric *(Amanita muscaria)*. It is readily formed from ibotenic acid by UV-irradiation of dilute aqueous solutions, whereby the isoxazole ring of 3-hydroxyisoxazole is converted into the 2(3H)-oxazolone system of M. M. is much less toxic than Ibotenic acid (see).

Muscimol: the enol betaine of 5-aminomethyl-3-hydroxyisoxazole M_r 100, m.p. 155–156 °C (hydrate), 174–175 °C (anhydrous). M. is strongly polar, and it is probably not a native constituent of the agaric *(Amanita muscaria)*, but an artifact formed from ibotenic acid during the processing of the fungus. M. is pharmacologically very potent, and in most tests it is more active than ibotenic acid. The primary pharmacological effect is an inhibition of motor function. Experimentally, M. causes euphoric and dysphoric ill humor, and leads to a condition resembling a model psychosis, but without hallucinations.

Muscle adenylic acid: see Adenosine phosphates.

Muscle proteins: proteins present in muscle and constituting 20% of muscle tissue. The insoluble contractile proteins are organized in myofibrils which consist of organized arrays of thick and thin myofilaments. Thick filaments (about 1.5 μm long, 12–16 nm diam.) contain 200–400 molecules of the protein myosin (Fig.1).

Figure 1. Diagram of a thick filament, consisting of numerous parallel myosin molecules. During contraction, the globular heads of the myosin molecules, protruding from the side of the filament, make contact with actin molecules of the neighboring filaments.

Thin filaments (8 nm diam.) consist of polymerized Actin (see) (300–400 actin molecules per 1.0 μm length), each chain being accompanied by threadlike tropomyosin molecules and, in striated muscle, globular molecules of troponin. In smooth muscle, troponin is replaced by caldesmon. The actin and myosin are responsible for muscle contraction, while the tropomyosin and troponin or caldesmon are regulatory proteins.

In striated muscle, the myofibrils are aligned so that alternating dense and less dense bands are visible with a repeating distance of about 2.5 μm (Fig.2). The less dense *I bands* consist mainly of actin filaments which appear to be anchored at a central, darker *Z line*. The denser *A bands* consist of interdigitated myosin and actin filaments; each actin filament extends from the Z line into the A band. The center of the A band is a weakly staining line, the *M line*, which is usually visible only by electron microscopy. α-Actinin, a protein which serves as an anchor for actin in other types of cells, is found in the Z line. Among other proteins, the M lines contain Creatine kinase (see). When muscle contracts, the I bands shorten and may nearly disappear; this suggests that contraction involves sliding of the thick and thin filaments past one another.

381

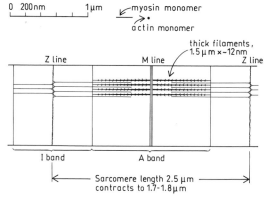

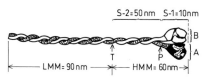

Figure 3. Diagram of a myosin molecule. T, site of attachment. P, site of attack by low concentrations of papain. LMM + S-2 = tail; S-1 = head region.

Figure 2. Interdigitation of thick and thin filaments in striated muscle. (Used with permission from D.E. Metzler, *Biochemistry. The Chemical Reactions of Living Cells*, Academic Press, 1977)

Contraction in smooth muscle appears to occur by a similar mechanism; however, here the myofibrils are not aligned so that striations are visible. Another difference between striated and smooth muscle are that actomyosin (see below) can be extracted from striated muscle at high ionic strength, but from smooth muscle at low ionic strength. [J.V. Small & A. Sobieszek in Newman & Stephens, eds. *Biochemistry of Smooth Muscle* (CRC Press, Boca Raton, 1983) pp.85–140]

Myosin accounts for one third of M.p. and two thirds of the contractile M.p. It consists of a long, slender stem (about 135 nm long, 2 nm diam.) and a globular head region (designated S-1, about 10 nm long). The stem is formed from two polypeptide chains of 2000 amino acid residues each; these are twisted together in an α-helix. The two chains are parallel with respect to their termini, and the C-termini are distal to the head end. Morphologically, there are two globular S-1 "heads", each approximately $15 \times 4 \times 3$ nm. This region contains an ATPase (part of the system producing energy for contraction) and two essential SH-groups. Treatment of extracted myosin (M_r 480000) with guanidine or dodecyl sulfate causes the molecule to dissociate into two main subunits of the tail region (each M_r 212000) and four light chains of the head region. Two of these light chains (each M_r 18000) carry essential SH-groups; the remaining two chains are designated A₁ (M_r 20700, 190 amino acids) and A₂ (M_r 16500, 148 amino acids). Myosin from molluscs contains only three light chains.

Trypsin or papain cleaves myosin into two water-soluble fragments, the meromyosins (MM). Under controlled conditions, the cleavage occurs chiefly at one point part way along the helical stem; this point is known as the "hinge" region. Light meromyosin (LMM, M_r 140000) is fibrillar and consists of two polypeptides; it represents a long C-terminal section of the original helical stem. Heavy meromyosin (HMM, M_r 340000) represents the head and part of the tail of myosin; it is globular, contains the ATPase and consists of 9 polypeptides (7 derived from

the head region). HMM has an affinity for actin and is often used to "decorate" actin filaments to reveal their orientation. Decorated actin is also used as a crystalline model in X-ray studies of the actin-troponin-myosin complex (Fig.3).

The globular S-1 heads of myosin protruding from the side of the thick filament become temporarily cross-linked with thin actin filaments to form the actomyosin complex. It is believed that the cross-bridges formed by the myosin S-1 "heads" move the actin molecules relative to the myosin molecules, but it is not known whether the myosin heads are rigidly or flexibly bound to the actin during the "power stroke" which generates tension. The difference may be visualized as follows: if the bond to actin is flexible, the myosin heads would act as rigid paddles moved by a conformational change elsewhere in the molecule; but if they are rigidly bound to the actin, the conformational changes needed to propel the actin past the myosin would include bending of the paddle-like head.

The actomyosin complex can be isolated directly from muscle, and it can also be formed in vitro from its two separate components. The complex consists of three myosin molecules and one fibrillar actin molecule (F-actin), and it has a marked ATPase activity which is activated by magnesium and calcium ions. ATP causes the complex to dissociate into actin and myosin, which then reassociate when the ATP is totally cleaved. In vitro, actomyosin fibrils can be induced to contract by addition of ATP; this phenomenon, called superprecipitation, clearly demonstrates the cooperative interaction of actin and myosin, but its significance as an in vitro model of muscle contraction is disputed.

Actin (see) constitutes 14% of total M.p. The naturally occurring thin actin filaments of striated muscle myofibrils are associated with two other M.p. which have a regulatory function, tropomyosin B and troponin. Tropomyosin B (M_r 68000) is 40 nm long and has no ATPase activity. It contains about 90% α-helix. Acting directly or indirectly via the troponin I component (see below), tropomyosin inhibits the ATPase activity of actomyosin, and thereby slows or stops contraction. In the actin filament, one tropomyosin molecule is in contact with 7 actin monomers and one troponin molecule. This complex of three components is known as α-actin.

Troponin (abb. TN; M_r 80000) is the central regulatory protein of skeletal muscle contraction in the higher vertebrates. TN is absent from molluscs, where myosin acts as the regulatory protein. TN consists of three functionally distinct components: TN-1 (M_r 24000), inhibitor of actomyosin ATPase;

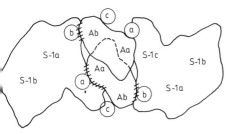

Figure 4. Positions of myosin heads (S-1), actin monomers (A) and troponin (circles) in active and relaxed states shown in relationship to a cross section of decorated actin. S-1a, S-1b and S-1c indicate the domains of the HMM. Aa and Ab are the domains of the actin monomers. The circles labelled "a" show the positions of troponin fibrils in the active position; "b" and "c" are possible positions in the relaxed state. This drawing is not a helical projection, but a schematic view of two actin monomers (the one on top is 27 Å above the other one and rotated by about 167 ° around the filament axis) and their associated S1s. The S1 on the left is therefore 27Å above the S1 on the right hand side of the drawing. In three dimensions, additional S1s would be in contact with both of the actin subunits shown. Therefore, it only appears in this two-dimensional drawing that the tropomyosin and S1 do not make equivalent contacts with the two actins. (Used with permission from E. H. Egelman *J. Musc. Res. Cell Motil.* **6** (1985) 129-151.)

TN-T (M_r 37000), which contains the binding site for tropomyosin; and TN-C (M_r 18000), which binds calcium ions. The binding of calcium by TN-C abolishes the inhibitory action of TN on the actin filaments; thus when calcium floods the sarcoplasmic reticulum in response to nervous stimulation of the muscle, ATP hydrolysis accompanied by contraction can occur.

In lower vertebrates, TN-C is absent, and the calcium-binding function is performed by structurally similar proteins, M_r 12000, known as Parvalbumins (see). The parvalbumins characteristically contain no cystine, cysteine, tyrosine or tryptophan, and therefore have no absorption maximum at 280 nm. Higher vertebrates also contain these proteins; parvalbumins have been isolated from the muscle of chicken, turkey, rabbit and human beings.

Other fibrillar elements, the paramyosins, are present in the slow contracting fibrils of smooth muscle. They are especially prominent in the "catch" muscles of molluscs, including those which hold the shells of bivalves closed without expenditure of energy by the animal. The main component of the paramyosins is tropomyosin A, which has a high α-helix content and no ATPase activity.

Other, more recently discovered M.p. are filamin, M_r 250000, which binds actin; vinculin, M_r 130000, which is part of the Z line, and titin (or connectin), M_r 250000, found in cardiac and skeletal muscle. Titin is not part of the Z line, but occupies the entire I band and the middle regions of the A bands. Its function is speculative; it may serve to attach thin filaments and possibly intermediate filaments as well. [R. M. Bagby in Newman & Stephens, eds., *Biochemistry of Smooth Muscle* (CRC Press, Boca Raton, 1983) pp. 1-84] In smooth muscle, α-actinin, vinculin and filamin anchor the thin filaments to the cell membrane [S. B. Marston & C. W. J. Smith, *J. Muscle Res. Cell Motil.* **6** (1985) 669-708; R. M. Dowben & J. W. Shay, eds. *Cell and Muscle Motility* Vol. 4 (Plenum, New York, 1983)].

Mustard oil: see Glucosinolate.

Mustard oil glucoside: see Glucosinolate.

Mustard oil thioglucoside: see Glucosinolate.

Mutagen, *mutagenic agent:* an agent which causes an increase in the rate of mutational events (see Mutation). In practice, the following M. are used: X-rays, UV-radiation at 260 nm (absorption maximum of DNA), nitrosomethylguanidine, methyl- and ethylmethane sulfonic acid, nitrite, hydroxylamine, base analogues, acridine dyes like proflavin, etc. The various physical and chemical M. have different mechanisms of action, some of which are imperfectly understood. The base analogue 5-bromouracil (abb. BU for the keto form, BU^* for the enol form) is a structural analogue of the DNA base, thymine. It is incorporated in place of thymine during DNA synthesis, resulting initially in the replacement of some A-T base pairs by A-BU pairs (see Base pairing). BU, however, shows a marked tendency to enolize; its enol form, BU^*, behaves similarly to cytosine, and base pairs with guanine. Thus A-T pairs become replaced by BU-G; effectively A is replaced by G, resulting in a change (transition) of the base (i. e. nucleotide) sequence of the DNA. Nitrite deaminates DNA bases that possess an amino group (i. e. adenine, guanine, cystosine). Thus e. g. cytosine is converted into uracil, which is not a normal DNA component. Deamination of a DNA base by nitrite always results in a false base partner, and thus leads to a different base pair in subsequent replication. Proflavin intercalates between neighboring DNA bases, so that the base sequence and therefore the instructional content of the genetic code is changed.

Mutants: organisms, which as a result of Mutation (see), show characteristic differences from the parent or wild type organism, e. g. morphological differences (changes or defects in cell wall formation), physiological differences (e. g. change in temperature sensitivity), or new nutritional requirements (see Auxotrophic mutants). Auxotrophic M. are obtained in pure culture by suitable enrichment and selection techniques, e. g. the Penicillin selection technique (see). The precise growth requirements of M. are determined by growth studies on supplemented minimal media. Further characterization includes the identification of the site of the metabolic block, and the analysis of any effects on metabolic regulation. This is achieved by the analysis of excretory products and the sizes of metabolic pools, by measurement of the activities or concentrations of individual enzymes, and by serological tests (see Immunology).

The mutation rate represents the probability of a mutation occurring in the generation of one cell; its value lies between 10^{-6} and 10^{-10}.

The mutational frequency represents the proportion

of M. in a cell population; its value lies between 10^{-4} and 10^{-11}, depending on the mutation in question.

Auxotrophic M. and M. with regulatory defects have proved very effective in the study of metabolic pathways (see Auxotrophic M., Mutant technique), the elucidation of control mechanisms (e.g. see Repressor and in the mapping of the bacterial genome.

Mutant technique: an important area of biochemical methodology, employing Mutants (see), in particular Auxotrophic mutants (see) of microorganisms. An auxotrophic mutant has a metabolic block at a certain point in the biosynthetic pathway of an essential product (e.g. an amino acid or coenzyme). Such a block is caused by the Mutation (see) of a gene, which (via transcription and translation) is responsible for the production of an enzyme:

Thus in the absence of enzyme 2, the intermediate A cannot be converted into B, and the synthesis of the essential end product is blocked. Intermediate A accumulates and is excreted, or its higher concentration may result in its conversion into other products by reactions that are normally insignificant in the nonmutant organism, e.g. phosphorylated intermediates of histidine biosynthesis are excreted by the appropriate mutants in a dephosphorylated form. In order for growth to occur, the auxotrophic mutant must be supplied with an external source of the essential product, or with any biosynthetic intermediate between (and including) B and the product. Thus two types of analysis are possible: *Accumulation analysis,* in which intermediate A, or its metabolic product(s), is isolated and identified; and *Supplementation analysis,* in which biosynthetic intermediates occurring after the metabolic block are identified by their ability to support growth (i.e. to be converted into the end product) when added to the growth medium of the auxotrophic mutant. Growth may be measured by determination of dry weight (e.g. growth of fungal mycelia), measurement of

quirement of the other. This compound was identified as L-cystathionine.

This test is performed by placing the mutants close to each other on the surface of a solidified (with Agar-agar) growth medium, containing a very low concentration of L-methionine to support suboptimal growth. After about 24 hours at 30 °, cystathionine diffuses from mutant 1 to mutant 2, which then begins to grow strongly. This phenomenon is known as cross feeding or syntrophism. With the aid of this technique, it is possible to test a large number of mutants auxotrophic for the same end product, and arrange them in the order of the metabolic blocks that they carry, e.g. if mutant 3 feeds mutant 2 (mutant 3 will also feed mutant 1), then mutant 3 is blocked later in the pathway than mutant 2, which is blocked later than mutant 1. Such an analysis can be performed with no knowledge of the nature of the intermediates involved. In the wild type organisms, biosynthetic intermediates are often present in such low concentrations that their analysis is very difficult. In contrast, they may be produced in exceptionally large quantities by auxotrophic mutants. This is probably due to the absence (or very low concentration) of the end product which in the wild type organism is responsible for the feedback control of the biosynthetic pathway. In the absence of the end product, the pathway runs out of control, at least up to the site of the metabolic block, and processes an unusually large quantity of material.

Example of supplementation analysis. The biosynthesis of L-arginine was elucidated with the aid of arginine-requiring mutants of *Neurospora crassa* From studies on perfused liver and tissue slices, the following pathways had already been proposed:
Precursors → L-Ornithine → L-Citrulline → L-Arginine
Arginine auxotrophic mutants (arg⁻) were isolated on a minimal agar medium supplemented with ornithine, citrulline and arginine. The supplementation test showed the existence of three groups:

(1) → A —|(Block) (2)→ B ------(n)------→ Product
(essential for growth)
Accumulation

turbidity (e.g. suspension cultures of bacteria or yeasts), or by total protein determination.

Example of accumulation analysis. Two groups of L-methionine-requiring (met⁻) *Escherichia coli* were isolated. Both could be supplemented with L-homocysteine, whereas only one could use L-cysteine in place of L-methionine for growth. One mutant accumulated a compound, which satisfied the growth re-

Mutant Group	Growth with Orni-thine	Citrul-line	Argi-nine	Site of Block
I	−	−	+	Cit → Arg
II	−	+	+	Orn → Cit
III	+	+	+	Precursors → Orn.

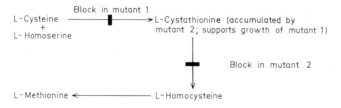

L-Cysteine + L-Homoserine —|(Block in mutant 1)→ L-Cystathionine (accumulated by mutant 2; supports growth of mutant 1)
—|(Block in mutant 2)
L-Methionine ← L-Homocysteine

The detailed and complete reaction sequence was finally elucidated by the use of further mutants (blocked at various stages between glutamate and ornithine), by the additional use of isotopic methods, and by the characterization of the individual enzymes.

The site of the metabolic block can be further ascertained by measurement of the appropriate enzyme activity. This will be absent or greatly decreased, compared with that in the wild type organism. Sometimes, serologically similar, but enzymically inactive protein is produced, i.e. the mutation has not prevented transcription and subsequent translation of a protein, but small changes (sometimes the replacement of a single amino acid) have occurred which destroy catalytic activity. Such proteins are known at CRiM proteins (Cross Reacting Material; the "i" is added for pronunciation purposes).

Elucidation of the biosynthetic pathway of the aromatic amino acids (see Aromatic biosynthesis) was greatly aided by the use of mutants of *Escherichia coli* and *Aerobacter aerogenes*. Many bacterial mutants were known to have a simultaneous requirement for phenylalanine, tyrosine, tryptophan, *p*-aminobenzoic acid and hydroxybenzoic acid, thus indicating a common biosynthetic pathway for all these compounds. This multiple requirement was satisfied by the single compound, shikimic acid, which can be isolated from the leaves of various plants (Gymnosperms, *Illiaceae,* etc.). Shikimic acid was later found to be excreted by certain auxotrophs with multiple aromatic growth requirements. Subsequently, auxotrophic mutants for each step or aromatic biosynthesis were isolated, the accumulated material was identified and shown to support the growth of mutants with earlier blocks, and the appropriate enzymes isolated and studied. The key branch point compound, chorismic acid, was identified with the aid of a triple mutant, with metabolic blocks in the conversion of chorismic acid into prephenic acid, prephenic acid into *p*-hydroxyphenylpyruvic acid, and anthranilic acid into indole 3-glycerol phosphate. In the presence of L-tryptophan (to repress the formation of enzymes converting chorismic acid into anthranilic acid), washed suspensions of this mutant (*A. aerogenes* 62 - 1) excreted chorismic acid.

The use of mutants, and in particular the supplementation test, are difficult when a single mutation of a fundamental metabolic reaction gives rise to multiple growth requirements, i.e. polyauxotrophy. This must be clearly distinguished from polyauxotrophy resulting from polygenetic mutations, i.e. single mutations in several distinct pathways.

The mutant technique has also been applied to naturally occurring mutations in animals, e.g. in the study of the degradation of phenylalanine and tyrosine (see Phenylalanine).

Mutarotation: a change in the optical rotation of an optical isomer, usually a carbohydrate, in aqueous solution. The carbohydrate molecule can exist in anomeric forms, designated α and β. These two diastereomers differ in chemical and physical behavior, such as melting point, solubility, and especially optical activity. In aqueous solution an equilibrium is gradually established between the two diastereomeric half acetal forms and the open chain form, with a consequent change in the optical rotation. Interconversion of the two diastereomers occurs via the intermediate cyclic half acetal. The attainment of equilibrium is accelerated by acids or bases. The starting value of $[\alpha]_D^{20}$ for α-D-glucose in water is $+113°$, that of β-D-glucose is $+19.7°$. After a few hours, when equilibrium between α and β forms is established, the value $[\alpha]_D^{20}$ is $+52.3°$, representing a mixture of 37% α and 63% β-glucose. A monosaccharide showing M. is often characterized by measurement of the starting and final values for rotation, e.g. $[\alpha]_D + 113° \rightarrow 52°$.

Mutation: chemical or physical changes in the genetic material of a cell or organism. A single M. represents a change in a gene, which is a defined segment of DNA (or RNA). Chemical changes in the DNA include substitution of one nucleotide for another, due to an error in copying (base pairing) or to a change such as dimerization of adjacent bases which prevents accurate copying. Physical changes include breakage and loss of part of a DNA molecule, or rearrangement of the molecule.

A point mutation affects a single nucleotide; it may consist of a) substitution of a different nucleotide, b) loss of a nucleotide (deletion) or c) insertion of an extra nucleotide. In each case the nucleotide sequence is altered; in b and c, the "reading frame", or division of the sequence into triplets to be translated into amino acid (see Genetic code; Protein biosynthesis) is shifted, with severe effects on the protein product of the gene if it is a structural gene.

The substitution of one base for another may or may not change the interpretation of a base triplet or a non-translated stretch of DNA. M. which have no phenotypic effects are called "silent M." A large majority of the DNA in a multicellular organism appears never to be translated. Some portions of this untranslated DNA regulate the expression of structural genes (see Gene expression), some apparently serve to make rearrangements such as crossing over of sister chromatids and movement of Transposons (see) possible, some may serve to promote pairing of homologous chromosomes at meiosis, and others may have no function at all or serve merely as spacers between genes (see Introns). M. in such regions are likely never to be detected.

M. which can be detected result in changes in structural or regulatory genes, and thus are reflected by absence or change in enzyme or structural proteins, metabolic reactions or overall performance of a metabolic pathway. At the morphological level, the results may be, e.g., changes in pigmentation, body structure or response to environmental changes.

M. may be random, spontaneous or induced, and in a multicellular organism, they may occur either in germ-line or somatic cells. (Naturally only the former are passed on to subsequent generations.) A random M. may be selected by environmental pressure for preservation, although it is far more likely to be deleterious. Spontaneous M. occurs because DNA replication is subject to occasional errors, and for reasons which are not understood, some stretches of DNA are more subject to such errors than others ("hot spots"). Genetic variability within a species confers the advantage of adaptability to changing environmental conditions, and thus evolution appears to have selected DNA polymerases with a low but finite tendency to err in replication. Induced M. is caused by Mutagens (see), which are defined as agents which increase the low rate of spontaneous M.

Mutational frequency: see Mutants.

Mutation rate: see Mutants.

MWC model: see Cooperativity model.

Mycobactin: a Siderochrome (see) synthesized by *Mycobacterium* spp.

Mycocerosic acids: see Fatty acid biosynthesis.

Mycorrhiza: symbiotic association between the roots of a higher plant (forest trees, orchids, etc.) and a fungus. The fungal mycelium covers the root tips, and the root hairs become reduced or disappear. The fungal hyphae take up nutrients from the soil. The hyphae may penetrate the intercellular spaces of the root (ectotrophic M.), or actually penetrate the plant cells (endotrophic M.). Endotrophic M. is necessary for the germination of orchid seeds. Ectotrophic M. are formed by many fungi with forest trees.

Mycosterols: see Sterols.

Mycotoxins: metabolic products of certain fungi and other microorganisms, which are harmful to other organisms, especially vertebrates, including man. The same M. may be produced by more than one fungal species. Out of about 100 000 described species of fungi, 50 are known to produce M.; these may damage the host directly (e.g. plant pathogenic fungi), or indirectly by causing illnesses in animals and man when M. are consumed in the diet. Ergotism is a classical example of mycotoxicosis (see Ergot alkaloids). M.-producing organisms frequently develop on improperly stored foodstuffs, leading to food poisoning. Such M. are e.g. botulinus toxins (see Toxic proteins), Aflatoxins (see) and Ochrotoxins (see). Other M. include the nephritic toxin citrinin from *Penicillium citrinum*, notatin from *Penicillium notatum* and sporidesmin from *Pithomyces chartarum* (formerly *Sporidesmin bakeri*). Other important M. producers are *Penicillium islandicum, Penicillium rubrum, Paecilomyces varioti, Fusarium sporotrichioides* and *Stachybotrys atra*. M. also include the bacterial toxins, which are subdivided into endo- and exotoxins (see Toxic proteins).

Mydriatic alkaloids: see Tropane alkaloids.

Myelin protein A1: see Encephalogenic protein.

Myoglobin: a single chain heme protein of skeletal muscle; M_r 17 200; 153 amino acid residues. Function of M. is oxygen storage and transfer (i.e. from hemoglobin to respiratory enzymes). The affinity of M. for oxygen is higher than that of hemoglobin. Muscle has a high content of M. especially the cardiac muscle of diving mammals, such as whale and seal, which contains up to 8% M., compared with 0.5% M. in the cardiac muscle of dog. High levels of M. are also found in the flight muscles of birds. M. and hemoglobin were the first globular proteins submitted to structural elucidation by X-ray diffraction analysis. M. possesses no disulfide bridges or free SH-groups. It contains 8 variously sized right-handed helical regions, joined by non-ordered or random coil regions. 121 Amino acids

are involved in helix formation, representing a total α-helix content of 77%. These 8 helices (A,B,C,D,E,F,G,H) are folded back on top of one another, and the heme is situated between helices E and F (see Proteins, Fig. 9). The molecule has the shape of a flattened ball with a pocket for the heme; the heme is almost totally buried, with just one edge exposed – that carrying the two hydrophilic propionic acid groups. The heme is held in position by a coordination complex between the central iron (II) atom and two histidine residues (on helices E and F, respectively). One of these histidines binds to the oxygen of the water molecule, which is bound to the heme. The position and the functional competence of the heme also depend upon the hydrophobic amino acids that line the inside of the heme pocket. Slight changes in the tertiary structure of M. destroy the oxygen binding function of the heme. Metmyoglobin, i.e. with Fe (III), does not bind oxygen.

The primary and even tertiary structure of M. and hemoglobins from many different species show striking similarities. M. appears to be phylogenetically the oldest known heme protein, from which hemoglobin evolved as an independent molecule 600 million years ago. M. also appears to be phylogenetically related to Leghemoglobin (see).

Myo-inositol: a cyclitol found free and combined in animals and plants. The earlier equivocal name, *mesoinositol* should be avoided. M. is biosynthesized from D-glucose, and the configuration of C-atoms 1–5 is preserved during the conversion. M. is an essential yeast growth factor. It is present in relatively large amounts in brain cells, lens, thyroid gland, muscle, lung and liver. In the biosynthesis of M., a cyclase (EC 5.5.1.4) catalyses the conversion of glucose 6-phosphate into inositol 1-phosphate, which is dephosphorylated by the action of a phosphatase (EC 3.1.3.25). Inositol oxygenase converts M. into D-glucuronate, an intermediate in the Glucuronate pathway (see) of carbohydrate degradation.

M. occurs as its mono- and diphosphate esters in phospholipids and phosphoproteins. The M. phospholipids play a role in the response of mammalian cells to external stimuli such as hormones (see Inositol phosphates). M. hexaphosphate is called phytic acid: the mixture of Ca and Mg salts of phytic acid is called phytin. Phytic acid is an important phosphate storage compound in plant tissues, e.g. in cereal grains.

Myo-inotisol

Myokinase: see Adenylate kinase.

Myosin: see Muscle proteins.

Myrcene: a triply unsaturated acyclic monoterpene hydrocarbon. M. is a pleasant smelling liquid. M_r 136.24, b.p.$_{12}$ 55–56 °C, p^{15} 0.8013, n_D^{19} 1.470. It is a component of many essential oils, and it is prepared for the perfumery industry by pyrolysis of β-

Myrcene

pinene (from oil of turpentine). M. is also used commercially for the preparation of isomeric acyclic monoterpene alcohols and their acetates.

Myricetin: see Flavones (Table).

Myristic acid: n-tetradecanoic acid, CH_3–$(CH_2)_{12}$–COOH, a fatty acid. M_r 228.4, m.p. 58 °C, b.p.$_{100}$ 250.5 °C. Glycerol esters of M.a. occur naturally in nearly all plant and animal fats, e.g. coconut oil, ground nut oil, linseed oil, rapeseed oil, milk fat and fish oils. Oil of nutmeg *(Myristicaceae)* is particularly rich in M.a. Esters of M.a. are also found in waxes.

Myrosin: see Thioglucoside glucohydrolase; Glucosinolate.

Myrosinase: see Thioglucoside glucohydrolase; Glucosinolate.

387

N

NAD: abb. of Nicotinamide adenine dinucleotide (see). NAD^+ represents the oxidized form, NADH the reduced form.

NADP: abb. of Nicotinamide adenine dinucleotide phosphate (see). $NADP^+$ represents the oxidized form, NADPH the reduced form.

Na^+K^+-ATPase: see Transport.

Nalidixic acid: an antibiotic used therapeutically against Gram negative bacteria. N.a. inhibits DNA replication in growing bacteria. It is a naphthyridine derivative, M_r 220.

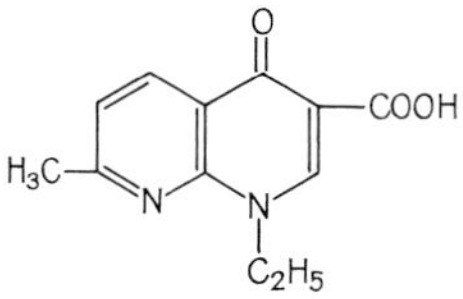

Nalidixic acid

Nanometer, *nm:* unit of length widely employed for light wavelength. It is equal to 10^{-9} m, or 10^{-7} cm.

Naphthoquinones: Naturally occurring derivatives of 1,4-naphthoquinone. N. are widely distributed, and over 120 different N. have been found in higher plants, bacteria and fungi. Examples of N. from plants are alkannin, eleutherin, juglone, lapachol, lawsone, lomatiol, plumbagin and shikonin. In the animal kingdom, N. have been found in echino-

derms, especially sea urchins (see Spinochromes). Other important N. are the K vitamins. Fungal N. are generally synthesized from acetate and malonate by the polyacetate pathway, whereas in higher plants and bacteria N. are derived from shikimic acid and a C_3 compound.

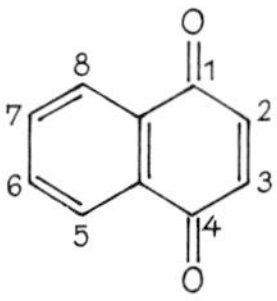

1,4-Naphthoquinone 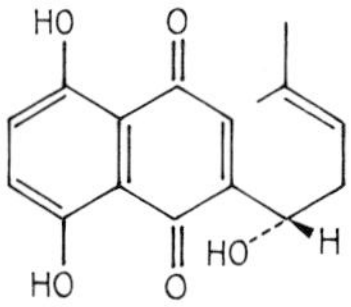   Alkannin

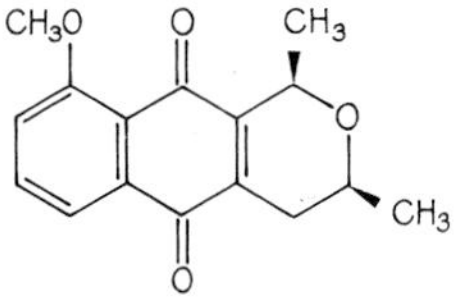

Eleutherin 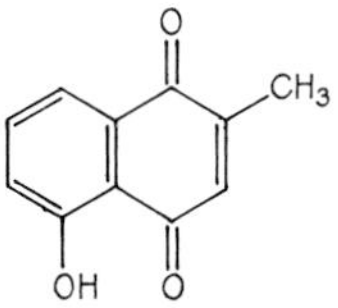   Plumbagin

Narcotics, *narcotic drugs:* substances which act predominantly on the central nervous system. Depending on the dose, N. show different phases of activity: small doses have a sedative action, some-

Naphthoquinones found in plants

Name	m.p. (°C)	Color	Occurrence
Alkannin[*]	147–149	brown-red	In the roots of *Boraginaceae*, e.g. *Alkanna tinctora*.
(9 R, 11 S)-Eleutherin	175	yellow	*Eleutherina bulbosa*.
Juglone (5-Hydroxy-1,4-N)	165	yellow to brown-red	Shells of unripe Walnuts. Leaves and roots of the Walnut tree.
Lapachol (2-Hydroxy-1,4-N, with a C_5 isoprene side-chain)	142	yellow	Members of the *Bignonaceae*, e.g. *Tecoma aralia ceaea*.
Lawsone (2-Hydroxy-1,4-N)	195	yellow to orange	Leaves of Henna *(Lawsonia inermis)*.
Lomatiol	128	yellow	The Australian plant *Lomatia ilicifolia*.
Plumbagin (5-Hydroxy-2-methyl-1,4-N)	78	yellow	Roots of *Plumbago* species.
Shikonin (an enantiomer of alkannin)	149	brown-red	*Lithospermum* species.

N = Naphthoquinone.   [*] Alkannin is the only one of these pigments still used as a dye and an indicator.

389

what higher doses are stimulatory, while even higher doses cause loss of consciousness (narcosis).

According to the 1964 WHO classification, there are 7 groups: 1. alkaloids (LSD, mescalin, opium, etc.), 2. barbiturates and other sleeping drugs, 3. alcohol, 4. cocaine, 5. hashish and marihuana, 6. hallucinogens, 7. stimulants or antidepressants (e.g. amphetamines).

N. may be used in psychiatry and psychotherapy, but misuse can lead to acute and chronic physical and mental deterioration, and to addiction. Regular use of N. leads to tolerance, i.e. an increased rate of breakdown of N. by the body, so that 10-20 times the normal dose may be necessary to achieve the desired effect.

Naringenin: 5,7,4′-trihydroxyflavanone, see Flavonoids (Fig.); Flavanone.

Naringin: naringenin 7-neohesperidoside, see Flavanone.

Natural pigments, *biochromes:* Colored organic compounds found very widely in plants and animals. Their color is due to their chemical structure, which absorbs light in the visible spectrum between 400 and 800 nm, and reflects or transmits the unabsorbed wavelengths. If all wavelengths of the visible spectrum are absorbed more or less equally, the substance appears grey to black, while the color white results from the nonabsorption and the reflection of all wavelengths in the visible spectrum. These pigment colors are distinct from Structural colors (see), which arise from light reflection and refraction effects caused by the physical structure of surfaces. Natural (and synthetic) organic pigments are unsaturated compounds containing a system of conjugated double bonds. The chromophoric groups (or chromophores) are structures such as $-CH=CH-$, $=CO$, $-N=O$, or $-N=N-$, which, when present in the conjugated system, are responsible for absorption bands in the visible region. Auxochromic groups (or auxochromes), e.g. $-NR_2$, $-NH_2$, $-OH$ do not produce color themselves, but they intensify the color effect of existing chromophores. The color may be shifted to longer or shorter wavelengths by bathochromic or hypochromic groups, respectively. Most N.p. also show some degree of fluorescence and/or phosphorescence. In some N.p. UV light may excite fluorescence at a wavelength near that of the visible color, e.g. ribitylflavin; this coincidence may increase the normal intensity of the visible color.

N.p. may be classified: 1. according to their occurrence, i.e. as plant pigments (flower, leaf, wood pigments, etc.), as animal pigments (blood, skin, hair, insect, eye, wing pigments, etc.), or as fungal, bacterial, algal or lichen pigments. Animal pigments are called zoochromes. The term, Phytochrome (see), however, has a restricted meaning and it does not refer to all plant pigments; 2. according to their chemical structure, i.e. as carotenoids, pteridines, tetrapyrroles, quinones, melanins, flavonoids, ommochromes, betalains, ergochromes indigoid pigments, etc.

Many N.p. act as attractant, warning or camouflage pigments, and are therefore important for the survival of the species. Others are protective, e.g. against UV light (melanin), or against fungal attack (some flavonoids). Some are involved in the trapping and utilization of light energy in plants, while animal pigments like hemoglobin are important for the transport of oxygen. Often, however, N.p. are end products of metabolism and have no other apparent function.

N.p. were used in antiquity as dyes. Among the oldest of these are alizarin, indigo, tyrian purple, safran, kermes, cochineal and many flavonoid-containing colored woods. In the dyeing industry they have now been largely replaced by superior synthetic compounds, but natural coloring materials are still used in the food industry.

Nearest neighbor frequency: the frequency at which a given pair of bases are immediately adjacent in the sequence of a polynucleotide chain. Since there are 4 bases, there are 16 possible pairs of nearest neighbors. Although it is now possible to determine the sequence of DNA, the parameter still gives useful information about the structure of the molecule.

Nebularine: 9-β-D-ribofuranosylpurine, a purine antibiotic (see Nucleoside antibiotics) synthesized by the mushroom *Agaricus (Clitocybe) nebularis,* and by *Streptomyces* spp. M_r 252.23, m.p. 181-182 °C, $[\alpha]_D^{25}$ -48.6 ° ($c=1$, water). N. is selectively active against mycobacteria. It has marked cytostatic properties, and in animals it is among the most poisonous of the purine derivatives.

Nebularine

Necines: see Pyrrolizidone alkaloids.

Necinic acid: Pyrrolizidone alkaloids.

Negative control: inhibition of transcription by repression, or inhibition of enzyme activity by metabolites.

Neoflavones: see Isoflavones.

Neoflavonoids: Flavonoids (see) with a 4-phenylchroman (neoflavan) skeleton, or a corresponding structure in which the C_3 chain (C-2, C-3, C-4) is not cyclized with oxygen.

Neoflavan ring system

N. include: **4-arylcoumarins**, e.g. calophyllolide from *Calophyllum inophyllum* (nuts),

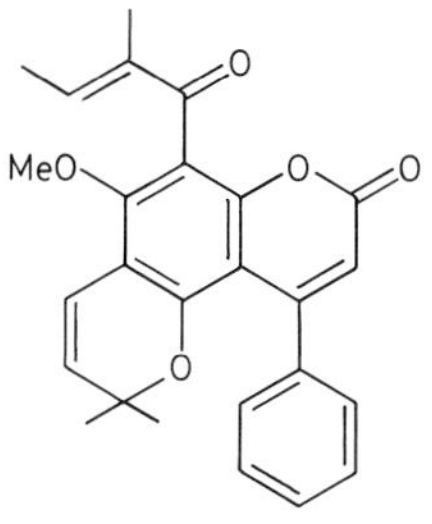

Calophyllolide

4-arylchromans, e.g. hematoxylin from *Haematoxylon campechianum* (wood),

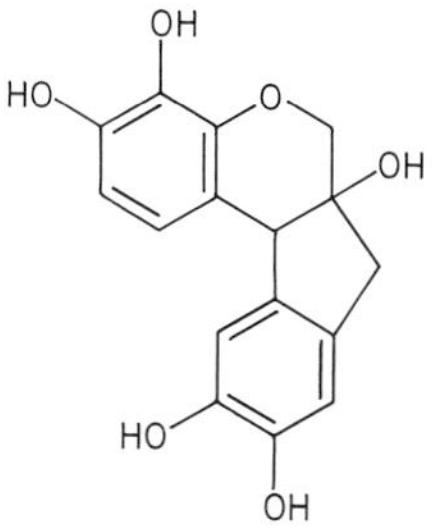

Hematoxylin

neoflavenes, e.g. 6-hydroxy-7-methoxy-3,4-dehydroneoflavan from *Dalbergia sissoo* and *D. latifolia* (bark),

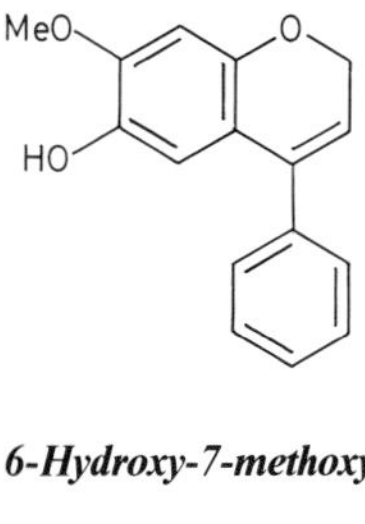

6-Hydroxy-7-methoxy-3,4-dehydroneoflavan

dalbergiquinols (3,3-diarylpropenes), e.g. latifolin from *Dalbergia latifolia* (bark, wood),

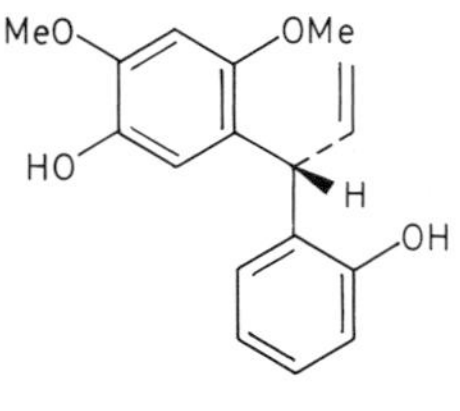

Latifolin

dalbergiones, e.g. 4-methoxydalbergione from *Dalbergia* spp. (bark, wood) (both configurations occur naturally),

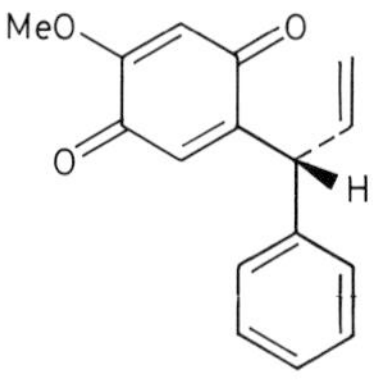

4-Methoxydalbergione

coumarinic acids, e.g. calophyllic acid from *Caesalpinia inophyllum* (nuts).

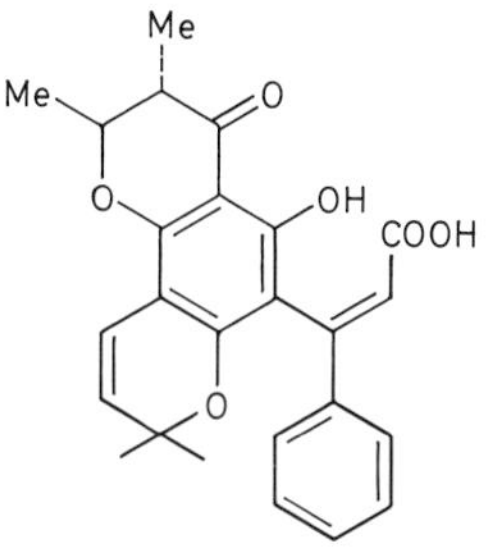

Calophyllic acid

Biosynthesis. Radioactivity from [3-^{14}C]phenylalanine is incorporated chiefly into C-4 of calophyllolide by young shoots of *Calophyllum inophyllum*, which indicates that no aryl shifts are involved, and that C-2, C-3 and C-4 of the N. are derived from C-1, C-2 and C-3, respectively, of phenylalanine. Other aspects of the biosynthesis are less certain, and are discussed in *The Flavonoids* (ed. J.B. Harborne, T.J. Mabry & H. Mabry) (Chapman and Hall, 1975).

Neolignans: bis-arylpropanoids, in which the inter-aryl linkage is other than C-8, C-8″ (see Lignans, in which the linkage is C-8, C-8″). N. are less numerous and have a more restricted phylogenic distribution than lignans. Some examples are shown (Fig.). For speculative mechanisms of N. biosynthesis, see O.R. Gottlieb *Phytochemistry* **11** (1972) 1544-1570. For comprehensive review of N., see O.R. Gottlieb, *Prog. Chem. Org. Natural Products* **35** (1978) 2-72.

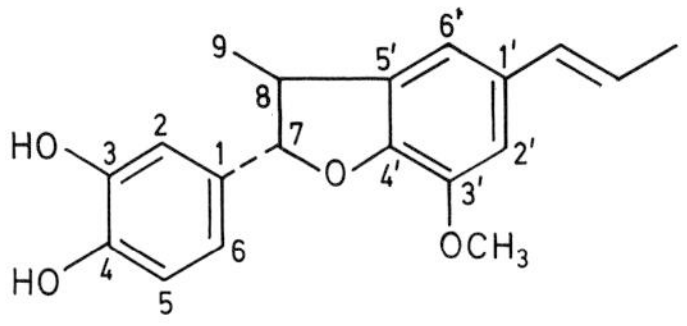

Kachirachirol-B (*Magnolia kachirachirai*) (8–5′ link-age).

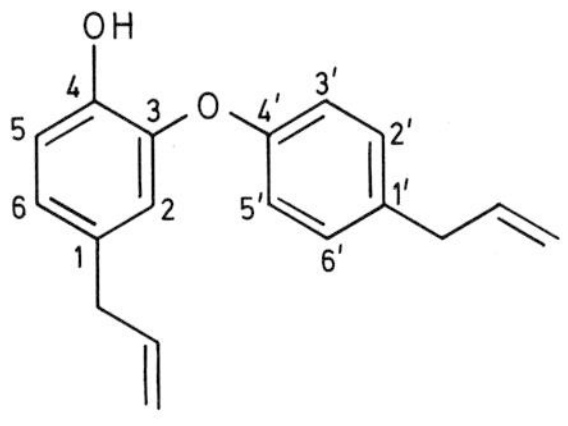

Isomagnolol (*Sassafras randaiense*) (3-O-4′ linkage)

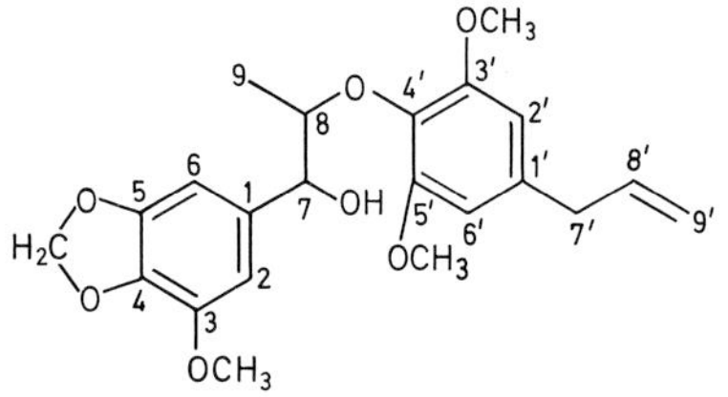

A neolignan from members of the *Myristicaceae* (8-O-4′linkage)

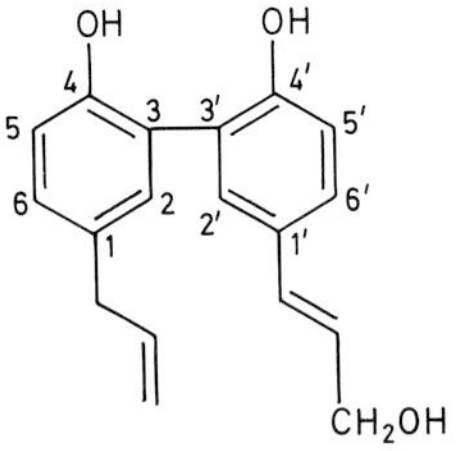

Examples of neolignans

Randainol (*Sassafras randaiense*) (3–3′ linkage) [F.S. El-Feraly *Phytochemistry* **23** (1984) 2329–2331]

Neomycin: an aminoglucoside antibiotic (see Streptomycin).

Neoplasm: see Cancer research.

Neorauflavene: see Isoflav-3-ene.

Neoretinal b: see Vitamins (Vitamin A).

Neostigmine: see Acetylcholine.

Neoxanthin: a xanthophyll, containing an al-lene group, two secondary and one tertiary hydroxyl group. The chiral centers of N. are 3 S, 5 R, 6 R, 3′ S, 5′ S, 6′ S. N. is present in the green parts of all higher plants. Together with lutein and violaxanthin, N. is one of the commonest carotenoids.

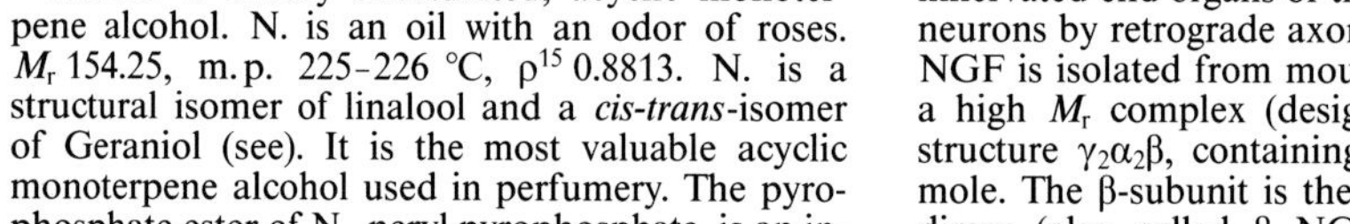

no effect after 9 days post partum in mice. A sensitive assay for NGF relies on the stimulation of the outgrowth of neurites from cultured ganglia.

NGF occurs in high concentration in adult mouse submaxillary gland, and in some snake venoms, but it is unlikely that these high local concentrations play any part in the stimulation of nerve growth. The first reported source of NGF was mouse sarcoma. It cannot be detected in innervated tissues, but denervation leads to the appearance of measurable NGF in the target tissue; when nerve regeneration is complete, NGF again becomes undetectable. NGF

Neoxanthin

Neral: the *cis*-isomer of Citral (see).

Nerol: a doubly unsaturated, acyclic monoter-pene alcohol. N. is an oil with an odor of roses. M_r 154.25, m.p. 225–226 °C, ρ^{15} 0.8813. N. is a structural isomer of linalool and a *cis-trans*-isomer of Geraniol (see). It is the most valuable acyclic monoterpene alcohol used in perfumery. The pyro-phosphate ester of N., neryl pyrophosphate, is an in-termediate in the biosynthesis of the Monoterpenes (see).

Nerve growth factor, *NGF*: a tightly (noncova-lently) associated dimer of identical polypeptides, each of M_r 13259 (Fig.), which stimulates division and differentiation of sympathetic and embryonic sensory neurons in vertebrates. The mitogenic effect of NGF is restricted to early embryonic life; it has

is therefore thought to carry information from the innervated end organs or tissues to the cell bodies of neurons by retrograde axonal transport.

NGF is isolated from mouse submaxillary gland as a high M_r complex (designated 7S), with subunit structure $\gamma_2\alpha_2\beta$, containing 1–2 g atoms Zn^{2+} per mole. The β-subunit is the hormonally active NGF dimer (also called β–NGF). The γ-subunit (also called γNGF, but with no mitogenic activity) has es-teropeptidase activity with a marked preference for Arg, and it is thought to be the processing enzyme which converts proNGF to NGF. It belongs to a family of proteases which includes the binding pro-tein for Epidermal growth factor (see). The α-sub-unit (also called αNGF, but with no mitogenic activ-ity) is structurally homologous with the γ-subunit,

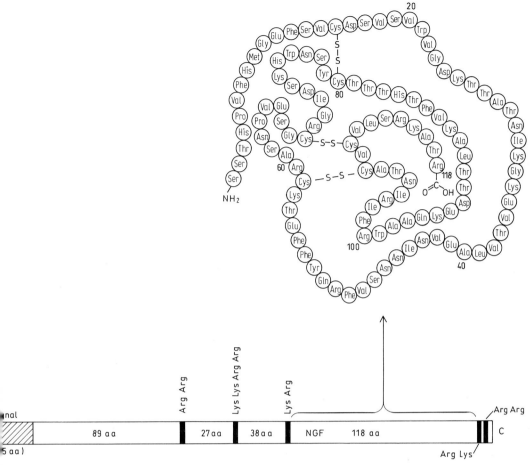

Primary structure of mouse nerve growth factor, and schematic representation of its submaxillary precursor protein (preproNGF) predicted from analysis of cDNA clones.

but it lacks proteolytic activity and appears to inhibit the activity of the γ-subunit.

The structures of human and mouse preproNGF, predicted from the analysis of cDNA clones, are highly homologous (Fig.). [R. H. Angeletti & R. A. Bradshaw, *Proc. Natl. Acad. Sci.* **68** (1971) 2417–2420; J. Scott et al., *Nature* **302** (1983) 538–540; A. Ullrich et al., *Nature* **303** (1983) 821–825; P. J. Isackson & R. A. Bradshaw, *J. Biol. Chem.* **259** (1984) 5380–5383]

Nervone: see Glycolipids.

Nervonic acid: Δ^{15}-tetracosenoic acid, CH_3-$(CH_2)_7$-CH=CH-$(CH_2)_{13}$-$COOH$, a fatty acid, M_r 336.6, m. p. 42 °C. N. a. is an acidic component of cerebrosides (see Glycolipids).

Neryl pyrophosphate: see Nerol.

Neuberg ester: old name for fructose 6-phosphate.

Neuberg fermentation: *Neuberg's first form of fermentation* is the normal "unsteered" fermentation

by yeasts, i. e. all the reactions and side reactions of the anaerobic degradation of carbohydrate, with the production of ethyl alcohol (see Alcoholic fermentation).

Neuberg's second form of fermentation (or sulfite fermentation) occurs in yeasts in the presence of sodium hydrogen sulfite. The acetaldehyde, which would normally act as a hydrogen acceptor, is converted to its bisulfite addition compound. The resulting excess NADH reduces dihydroxyacetone phosphate to glycerol 1-phosphate (or 3-phosphate), a reaction catalysed by glycerol 1-phosphate (or 3-phosphate) dehydrogenase (Baranowski enzyme) (EC 1.1.1.8). The oxido-reduction cycle (i. e. reduction of NAD and its regeneration from NADH) is thus maintained. The glycerol 1-phosphate is dephosphorylated by a phosphatase (EC 3.1.3.21), and free glycerol is produced. The process is therefore also known as a glycerol fermentation.

In the liver, free glycerol is dehydrogenated to glyceraldehyde, with the formation of NADH. The reactions of Neuberg's second form of fermentation are also important in insect flight muscle, where there is only a low activity of lactate dehydrogenase, and dihydroxyacetone phosphate serves as the normal hydrogen acceptor. In the liver, glycerol 1-phosphate is a precursor of phospholipids, and it is important for the transport of reducing power from the cytoplasm to the mitochondria (see Hydrogen metabolism). Glycerol 1-phosphate can also be produced from glycerol and ATP by the action of glycerol kinase.

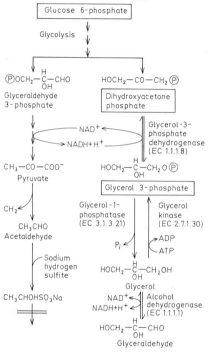

The significance of dihydroxyacetone phosphate in the 2nd. Neuberg fermentation.

Neuberg's third form of fermentation proceeds under alkaline conditions. Two molecules of glucose are converted into two molecules of glyceraldehyde and two molecules of acetaldehyde (via pyruvate by decarboxylation). The glyceraldehyde becomes reduced to glycerol, and the acetaldehyde undergoes a dismutation to produce equivalent amounts of ethanol and acetate.

Neuraminic acid: 5-amino-3,5-dideoxy-D-glycero-D-galactononulosonic acid, a C_9-compound formed from mannosamine and pyruvate, widely distributed in the animal kingdom in mucolipids, mucopolysaccharides, glycoproteins, and the oligosaccharides of milk. The *N*- and *O*-acetyl derivatives are called sialic acids.

Neuraminidase (EC 3.2.1.8): a hydrolase which cleaves *N*-acetylneuraminic acid from the nonreducing end of the heterosaccharide chain, producing glycoproteins and gangliosides. N. occurs in mixoviruses, various bacteria, blood plasma, and the lysosomes of many animal tissues. The richest source of N. is the culture filtrate of the cholera organism, *Vibrio cholerae*. M_r of the dimeric M. from *Vibrio* is 90000; N. from influenza virus has M_r 130000. Most neuraminic acid-containing proteohormones and some enzymes are inactivated by treatment with N.; on the other hand, the electrophoretic mobility of many glycoproteins, e.g. plasma proteins, may be altered after treatment with N., but their activity is unaffected. The N. of influenza virus destroys the protective mucus layer of the attacked organ.

Neuroendocrinology: the study of the interaction of nervous and hormonal systems in the regulation of metabolism, and in the adjustment of the individual organism to its environment.

Neurohormones: a group of hormones which, in conjunction with the nervous system and endocrine systems, play an important part in the regulation of somatic function, and in the adjustment of the individual organism to its environment. Examples are Releasing hormones (see), Neurohypophyseal hormones (see) and Neurotransmitters (see).

Neurohypophysial hormones: a group of hormones produced in the hypothalamus (not, as the name suggests, in the neurohypophysis; see Hypophysis). The action spectrum of these phylogenetically ancient hormones, Oxytocin (see) and Vasopressin (see), extends from an effect on the smooth muscle of the uterus, mammary gland and blood vessels, to an alteration of the permeability of the skin, urinary bladder and kidney tubules. The carrier protein of N.h. is Neurophysin (see). N.h. are small molecules (nonapeptides) and they represent an ideal model for the study of the mechanism of hormone action at a molecular biological level.

Neuronal ceroid lipofuscinosis: see Inborn errors of metabolism.

Neurophysins: small proteins associated with the pituitary hormones oxytocin and vasopressin. In bovines, oxytocin-like arginine vasopressin and N. II are synthesized from a single polypeptide precursor, as are oxytocin and N.I. Most species have more than two N., but usually there are two major N. The differences between N. occur on the ends of the polypeptides; there is a conserved middle sequence. The N. bind to the neurohormones with which they are synthesized and are secreted with them into the blood. The biological function, if any, of the N. is not known. [H. Land et al. *Nature* **302** (1983) 342–344; A. G. Robinson in D.T. Krieger & J.C. Hughes, eds., *Neuroendocrinology* (Sinauer Assoc., Sunderland, Mass., 1980)]

Neurosporene: a hydrocarbon carotenoid, containing 12 double bonds, nine of them conjugated. M_r 538, m.p. 124 °C. N. is a direct biosynthetic precursor (see Phytoene) of lycopene, and it is found in *Neurospora crassa*.

Neurotensin, NT: a tridecapeptide first discovered in and isolated from acid/acetone extracts of bovine hypothalamus.
Structure: Pyr-Leu-Tyr-Glu-Asn-Lys-Pro-Arg-Arg-Pro-Tyr-Ile-Leu.
The same peptide has been isolated and unequivocally identified from extracts of calf and human

Neurosporene

small intestine. Subsequent studies on the distribution and concentration of NT have relied on immunoassay. Antisera against the C-terminal and N-terminal regions of NT show varying degrees of cross reactivity with other naturally occurring peptides, and give different quantitative results in immunoassays. When measurements of NT are performed by immunoassay, it is therfore preferable to refer to "neurotensin-like immunoreactive material" (NTLI). The following approximate concentrations of NTLI (pmol/g fresh weight) were found in rat tissues: hypothalamus and mucosa of small intestine (50-60); thalamus, spinal cord and brainstem (12-16); posterior pituitary (30); anterior pituitary (24); cerebral cortex (2); cerebellum, esophagus, stomach and pancreas (less than 1). NT is thus a typical brain-gut peptide and a member of the Amine precursor uptake decarboxylase system (see). NTLI has been found in the gut of members of most chordate phyla, including *Uro-* and *Cephalochordata*. NT immunoreactive cells have been found in avian thymus, and NTLI has been reported in cat adrenal medulla. NT displays such a wide variety of activities in vivo that it is doubtful whether these represent true physiological functions. It seems more likely that NT acts locally, affecting only those cells near the site of release, and is then rapidly degraded (half-life of NT when injected into rats is 30 sec). It is rigidly excluded by the blood-brain barrier, so that its peripheral and cerebral effects are distinct. Peripheral injection of NT into rats has the following effects: increased vascular permeability, hypotension, cyanosis, vasodilation, peripheral blood stasis, greater than 30-fold increase in plasma histamine, increased secretion of ACTH, LH, FSH GH and prolactin, hyperglycemia, inhibition of gastric secretion of acid and pepsin, inhibition of gastric motility. Central injection causes hypothermia, enhanced phenobarbital action (NT does not itself induce sleep), and analgesia. In vitro effects include contraction of guinea pig ileum, rat fundus strip and guinea pig atrium, relaxation of rat duodenum, and release of histamine from isolated mast cells.

The amino acid sequence of NT is not present within any other known peptide or protein, but it does have certain similarities to other biologically active peptides. Thus N-terminal pyroglutamate is also present in xenopsin, thyrotropin releasing hormone, and luteinizing hormone releasing hormone, where it possibly protects against N-terminal degradation. The structures of vasopressin, LRH and NT suggest a distant evolutionary relationship for these three peptides. A similarity between the C-terminal sequence of NT (-Arg-Pro-Tyr-Ile-Leu-OH) and angiotensin I (-His-Pro-Phe-His-Leu-OH) should also be noted.

In the mucosa of the mammalian jejuno-ileum, NT is localized in endocrine-like cells called N- or NT-cells, which make contact with the intestinal lumen via a narrow apical process. It is suggested that a receptor on the cell surface responds to intraluminal lipid and signals release of NT from storage granules. Plasma NT increases after a high fat meal. Normal concentrations of plasma NTLI are 15-25 fmol/ml.

Several peptides cross-reacting with antiserum against the NT C-terminus, but not with antiserum against the N-terminus, have been isolated from chicken small intestine. Two have been analysed: Chick I: Pyr-Leu-Tyr-Glu-Asn-Lys-Pro-Arg-Arg-Pro-Tyr-Ile-Leu-OH; Chick II: H-Lys-Asn-Pro-Tyr-Ile-Leu-OH.

The amphibian peptide, Xenopsin (see) and other xenopsin-like peptides also appear to belong to the NT family.

A detailed survey of work on N. up to 1982 is given in *Annals of the New York Academy of Sciences* **400** (1982), C. B. Nemeroff & A. J. Prange Jr., eds.

Neurotoxins: toxins that act specifically on nervous tissue. They may act as antagonists of neurotransmitters, e.g. snake venom N. block acetyl choline receptors; strychnine is an antagonist of glycine (an inhibitory transmitter in the spinal cord); or they may interfere with cation transport across the nerve membrane, e.g. batrachotoxin. See Snake venoms, Bicuculline, Batrachotoxin, Curare alkaloids, Picrotoxin, Saxitoxin, Strychnine.

Neurotransmitters: small, diffusible molecules, such as acetyl choline, noradrenalin, amino acids, amino acid derivatives and peptides (e.g. substance P, enkephalin, etc.), which are released from synaptic vesicles (storage vesicles) at nerve endings, synapses and motor end plates in response to electrical stimuli. They act as chemical messengers, which transmit a signal from a nerve cell to its target organ. *Adrenergic N.,* e.g. Noradrenalin (see) and Adrenalin (see) are formed in sympathetic postganglionic synapses. *Cholinergic N.,* e.g. Acetylcholine (see) are found in pre- and postganglionic synapses of the parasympathetic nervous system. The central nervous system contains additional N., which may be excitatory, e.g. L-Glutamic acid (see), or inhibitory, e.g. γ-Aminobutyric (see) and Glycine (see), or able to show both effects, e.g. Dopamine (see) and Serotonin (see). N. are often considered to be synonymous with neurohormones, but this is debatable.

NGF: abb. for Nerve growth factor (see).

NHI-proteins: see Nonheme iron proteins.

Niacin: inclusive term for nicotinic acid and nicotinamide, which are members of the vitamin B_2 complex; see Vitamins.

Niacinamide: a member of the vitamin B_2 complex; see Vitamins.

Nickel, *Ni:* Ni occurs only in traces in living systems. In particular, it seems to be associated with

RNA. A nickel metalloprotein, named "nickeloplasmin" has been isolated from human and rabbit serum, but its function is not known. Ni protects the structure of the ribosome against heat denaturation, and it restores the sedimentation characteristics of *Escherichia coli* ribosomes that have been denatured by EDTA. Ni can activate some enzymes in vitro, e.g. deoxyribonuclease, acetyl CoA synthetase and phosphoglucomutase. Ni deficiency causes changes in the ultrastructure of the liver and alters the level of cholesterol in the liver membranes. Ni may be important in the regulation of prolactin.

Nicking-closing enzymes: eukaryotic type I Topoisomerases (see).

Nick translation: a procedure for preparing ^{32}P-labelled DNA probes for use in hybridization tests. In this context the word "translation" is not related to the translation process of protein synthesis, but to the translatory movement of a cleavage point or "nick" along a duplex DNA molecule. Nicks are introduced at widely separated sites in DNA by very limited treatment with DNAase. The resulting nicks expose free 3'OH groups. *Escherichia coli* DNA polymerase I is then used to remove concomitantly the 5'-mononucleotide terminus (i.e. 5'→3' exonuclease activity of polymerase I), and to incorporate appropriate nucleotides from ^{32}P-labelled deoxynucleoside triphosphates (α^{32}P-dNTP, where N = G, A, T or C). In the presence of all 4 radiolabelled deoxynucleoside triphosphates, label is progressively incorporated into the duplex at random, so that the

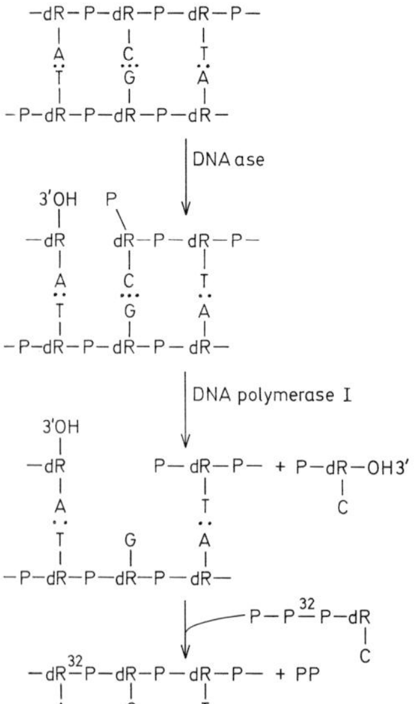

DNA effectively becomes uniformly labelled. If only one of the deoxynucleoside triphosphates is labelled, the pattern of labelling becomes nonuniform, especially with respect to homopolymer regions (Fig.) [P.W.J. Rigby et al. *J. Mol. Biol.* **113** (1977) 237–251]

Nicotiana alkaloids, *tobacco alkaloids:* a group of pyridine alkaloids occurring mainly in the tobacco plant *(Nicotiana)*. N.a. contain a pyridine ring system substituted at position 3 by pyrrolidyl or piperidyl residues. Many varieties of tobacco plant contain more than 10% N.a. Nicotine (see) is usually the chief N.a., but in *Nicotiana glauca* this is Anabasine (see) and in *Nicotiana glutinosa* the main and apparently only alkaloid is nornicotine. Other N.a. occurring in small quantities are nicotyrine, nicotelline and myosmine. All N.a. are very poisonous.

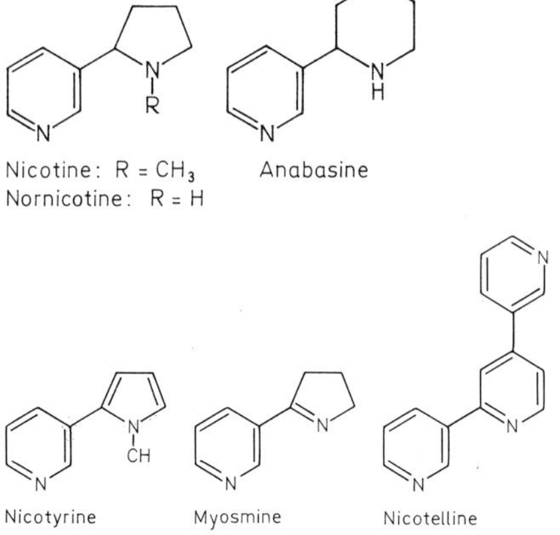

Nicotine: R = CH$_3$ Anabasine
Nornicotine: R = H

Nicotyrine   Myosmine   Nicotelline

Some Nicotiana alkaloids.

The pyridine ring system of N.a. is derived biosynthetically from nicotinic acid (or a closely related derivative; see Pyridine nucleotide cycle). The pyrolidine ring is derived from ornithine or a closely related compound in the ornithine-proline-glutamic acid family of amino acids. During the biosynthesis of nicotine, the C-3 carboxyl group of nicotinic acid is lost as CO_2, and the pyrrolidine ring is inserted in its place; this is accompanied by a labilization of the hydrogen at position C-6 of the pyridine ring of nicotinic acid. In the biosynthesis of anabasine, the precursor of the piperidine ring system is Δ^1-piperdeine, which is derived from lysine (probably by oxidation to α-oxo-ϵ-amino-caproic acid, followed by ring closure to the acid, which is then decarboxylated). The *N*-methyl group in the pyrrolidine ring is derived from methionine (see Transmethylation).

Nicotinamide: a member of the vitamin B$_2$ complex; see Vitamins.

Nicotinamide-adenine-dinucleotide, abb. *NAD, diphosphopyridine nucleotide,* abb. *DPN, cohydrogenase I, coenzyme I, cozymase:* a pyridine nucleotide coenzyme involved in many biochemical redox processes. It is the coenzyme of a large number of oxidoreductases, which are classified as pyridine nucleotide-dependent dehydrogenases. Mechanistically, it serves as the electron acceptor in the enzy-

Labelling of DNA with ^{32}P by nick translation. The diagram shows the replacement of a single nucleotide residue. In the nick translation procedure, many such replacements occur throughout the DNA molecule. dR = 2'-deoxyribose; PP = pyrophosphate.

matic removal of hydrogen atoms from specific substrates.

In the oxidized form of NAD, the pyridinium cation of nicotinamide is bound by an *N*-glycosidic linkage to C-1 of D-ribose. This nicotinamide riboside moiety is linked to adenosine via a pyrophosphate group. NAD therefore has the structure of a dinucleotide (Fig. 1). M_r of the oxidized form (NAD^+) = 663.4.

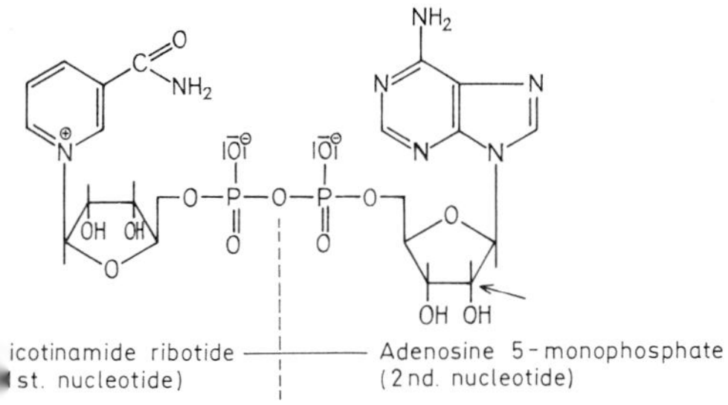

icotinamide ribotide ———— Adenosine 5-monophosphate
(1st. nucleotide)   (2nd. nucleotide)

Figure 1. Structure of nicotinamide-adenine-dinucleotide (NAD). In nicotinamide-adenine-dinucleotide phosphate (NADP), an additional phosphate group is present on C-2 of the 2nd nucleotide, as indicated by the arrow.

NAD occurs in two forms, distinguished by the configuration of the glycosidic linkage of the nicotinamide, hence α-NAD and β-NAD. Only β-NAD is enzymatically active. In both α- and β-NAD, the glydosidic linkage of the adenine has the β-configuration. $[α]_D^{23} - 34.8°$ (β-NAD); +14.3° (α-NAD).

The hydrogen reversibly carried by NAD is attached to the nicotinamide residue (Fig. 2). In view of the positive charge on the coordinated pentavalent nitrogen of the pyridinium ion, oxidized NAD is represented as NAD^+, and in accordance with the guide lines of the International Commission on Nomenclature, the reduced form is represented as NADH. When 2[H] and a pair of electrons are transferred from a substrate, the pyridinium cation loses its aromaticity and becomes reduced, and a proton is released:

Substrate-H_2 + NAD^+ ⇌ Substrate + NADH + H^+.

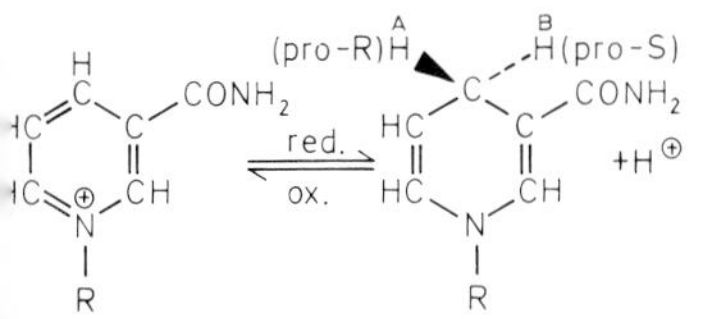

Figure 2.

Thus, of the two hydrogens transferred, one becomes covalently bound to the NAD, while the other becomes a proton in equilibrium with the protons of the aqueous medium.

Reduction of NAD introduces a prochiral center at C-4 of the pyridine ring. Transfer of the hydride ion (a proton with an electron pair) to NAD^+ is stereospecific, i.e. the newly introduced hydrogen is either *pro-R* or *pro-S* (in older nomenclature, it be-

comes attached to either the A[α] or B[β] side of the reduced nicotinamide ring). For example, ethanol, lactate and malate dehydrogenases transfer hydrogen to and from the *pro-R* position, whereas 3-phosphoglyceraldehyde and glucose dehydrogenases are *pro-S*-specific.

The oxidized and reduced forms have different spectrophotometric properties. Both show an intense absorption band in the region of 260 nm, due to the adenine; this is slightly higher in NAD^+ due to an additional contribution from the pyridine ring. NADH exhibits a broad absorption band at 340 nm, which is entirely absent from NAD^+, and is due to the quinoid structure of the reduced nicotinamide ring. Thus the reduction or oxidation of NAD is relatively easy to monitor by the change in light absorption at 340 nm. Since the absorption band is relatively wide, it may also be determined at wavelengths other than the maximum, e.g. the mercury lines 334 nm and 366 nm are suitable. This difference in absorption is exploited in the Optical test (see). $λ_{max}$ NAD^+ (pH 7.0) = 260 nm ($ε$ = 18000, i.e. 1 mole in 1 liter corresponds to an absorbance of 18000, when the light path is 1 cm).
1st $λ_{max}$ NADH (pH 10.0) = 259 nm ($ε$ = 14400).
2nd $λ_{max}$ NADH (pH 10.0) = 340 nm ($ε$ = 6200); at 334 nm, $ε$ = 6000; at 366 nm, $ε$ = 3300. In living cells, NAD is present predominantly as NAD^+, e.g. in rat liver cells $NAD^+/NADH + H^+$ = 2.6/1. A considerable proportion of this intracellular NAD is, however, not free but bound to dehydrogenases.

The nicotinamide moiety of NAD is derived biosynthetically from quinolinic acid (see Pyridine nucleotide cycle; Tryptophan). NAD is degraded by a pyrophosphatase, and by nucleosidases that cleave the glycosidic bonds to nicotinamide and adenine.

For the metabolic functions of NAD as a hydrogen carrier, see Hydrogen metabolism.

In addition to its function in hydrogen transfer, NAD^+ is the donor of the ADP-ribosyl group in the ADP-Ribosylation of proteins (see).

Nicotinamide-adenine-dinucleotide phosphate, abb. *NADP, triphosphopyridine nucleotide,* abb. *TPN, coenzyme II:* a pyridine nucleotide coenzyme, which differs from NAD by the presence of an extra phosphate residue in the 2′-position of the adenosine moiety (see Fig. 1 in Nicotinamide-adenine-dinucleotide). M_r 743.4, $E_o' = -0.317$ V (pH 7.0, 30 °C), $λ_{max}$ of the oxidized form ($NADP^+$) at pH 7.0 = 260 nm ($ε$ = 18000). 1st $λ_{max}$ of the reduced form (NADPH) at pH 10.0 = 259 nm ($ε$ = 14100). 2nd $λ_{max}$ of the reduced form = 340 nm ($ε$ = 6200). NADP is the coenzyme of dehydrogenases and hydrogenases. It is important as the agent of hydrogen transfer in reductive syntheses, e.g. in Photosynthesis (see) and in Ammonia assimilation (see). It is a fair generalization that NAD is concerned in the transfer of hydrogen to oxidative, energy-yielding processes (e.g. oxidation in the respiratory chain), whereas NADP provides reducing power in synthetic reactions. Most intracellular NADP is present in the reduced form (NADPH). In analogy with Nicotinamide-adenine-dinucleotide (see), the oxidized form is represented as $NADP^+$. Excess NADPH can be removed by transhydrogenation (see Hydrogen metabolism).

The spectrophotometric properties of $NADP^+$ and NADPH are completely analogous to those of NAD^+ and NADH. The change in extinction at 340 nm, due to oxidation and reduction, is exploited to the Optical test (see). NADP is biosynthesized by the phosphorylation of NAD in an ATP-dependent kinase reaction.
For the metabolic functions of NADP, see Hydrogen metabolism.

Nicotine: chief representative of the Nicotiana alkaloids (see). Oxidative degradation of N. produces nicotinic acid. N. is present in all parts of the tobacco plant (*Nicotiniana*), being biosynthesized in the roots of the tobacco plant, then transported to the aerial parts. Depending on the variety, the N. content of the leaves varies between 0.05 and 10%. It is also found in several plants of other families.
N. has many different physiological effects. Owing to its toxicity, it is not used therapeutically, although it is the most widely consumed addictive alkaloid in the world. Between 50 and 100 mg N. (the content of half a cigar) is fatal in man, causing respiratory paralysis. The concentration in smokers' blood plasma is 5 to 50 ng/ml. At this concentration, N. enhances the chemotactic effect of several other substances on polymorphonuclear neutrophils, which are involved in the inflammatory process. At 5 μg/ml, a concentration which may be reached in lung fluid, N. is strongly chemotactic for neutrophils. Thus N. is probably a primary etiological factor in emphysema. [N. Totti III et al., *Science* **223** (1984) 169-171.]

Nicotinic acid: pyridine 3-carboxylic acid, M_r 123.11, m. p. 234-237 °C, first obtained in 1867 as an oxidation product of the tobacco alkaloid, nicotine. It is biologically important as a member of the vitamin B_2 complex (see Vitamins). Isonicotinic hydrazide (INH), M_r 137.14, m. p. 163 °C, is an antituberculosis agent, which shows antivitamin properties against both nicotinic acid and vitamin B_6.

Nicotinic receptor: see Acetylcholine.

Nieman-Pick disease: see Inborn errors of metabolism.

NIH-shift: an anionotropy, whereby the cation intermediate produced in the hydroxylation of an unsaturated or aromatic substance by an oxygenase is stabilized. It is named after the *N*ational *I*nstitute of *H*ealth. A group (usually a hydride ion) in the position of the incoming hydroxyl group is shifted to the neighboring position.

Ninhydrin: see Amino acid reagents.

Nissolin: 3,9-dihydroxy-10-methoxypterocarpan, see Pterocarpans.

Nitrate ammonification: see Nitrate reduction.

Nitrate assimilation, *assimilatory nitrate reduction:* see Nitrate reduction.

Nitrate dissimilation, *dissimilatory nitrate reduction:* microbial reduction of nitrate to various products, which are not assimilated. See Nitrate reduction.

Nitrate reductase: enzymes which catalyse the reduction of nitrate to nitrite. All N. r. studied so far contain iron and molybdenum. In the sequence of electron transfer, molybdenum appears to be the ultimate acceptor, which then transfers electrons to nitrate; during this process the molybdenum alternates between Mo(V) and Mo(VI). Dissimilatory N. r. from *Escherichia coli* (also called respiratory N. r. (EC 1.7.99.4) is a transmembrane protein, containing Mo, inorganic sulfur and nonheme iron. Approximate M_r 220000, consisting of two subunits of M_r 150000 and 60000. Higher M_r, up to 10^6, have been reported; this reflects the problems inherent in the M_r determination of membrane proteins, which must first be solubilized, chiefly by the use of detergents. The higher M_r material probably resulted from aggregation and/or association with other proteins. During several stages of the purification procedure, the N. r. is accompanied by a small polypeptide containing cytochrome b_{556}; this is probably the natural electron donor for N. r. within the membrane. The preferred substrates for nitrate reduction in *Escherichia coli* are NADH or formate. Formate dehydrogenase is itself a molybdenum enzyme (it also contains selenium), donating electrons to another b-type cytochrome. Formate dehydrogenase and N. r. are intimately associated in the membrane, and the electron flow proceeds: Formate → [Mo,Se] → Cytochrome b → Quinone pool → Cytochrome b_{556} → Fe S → Mo → NO_3^-. This results in a transmembrane proton translocation, serving to generate ATP by chemiosmosis (see Oxidative phosphorylation). N. r. of *Escherichia coli* is induced by nitrate and repressed by oxygen; it may represent up to 15% of the total protein of the inner bacterial membrane. Similar respiratory N. r. have been found in several other bacterial genera. In the absence of an alternative nitrogen source for the organism, the nitrite produced by nitrate reduction can be reduced to ammonia (see Nitrite reductase), then assimilated (see Ammonia assimilation). Fungi, green algae and higher plants possess assimilatory N. r. (EC 1.6.6.1-3), which are high M_r (200000-500000) multimeric enzyme complexes, containing FAD, cytochrome b_{557} and molybdenum. The natural electron donor is NADH or NADPH, and most N. r. react with both. All these pyridine nucleotide-dependent N. r. also reduce cytochrome c (referred to as "diaphorase" activity) in vitro, but this property probably has no physiological significance. The sequence of electron transfer is apparently:

N. r. of *Neurospora crassa* has M_r 228000. A nitrate reductase mutant *(Neurospora crassa nit-3)* produces a subunit of N. r. containing cytochrome b_{557} and Mo (M_r 160000); this does not react with NAD(P)H, but transfers electrons to nitrate from exogenous $FADH_2$ or reduced methyl viologen. The other component is produced by *Neurospora crassa nit-1*; it is unable to reduce nitrate, but catalyses the reduction of cytochrome c by NAD(P)H.
Assimilatory N. r. from bacteria (*Azotobacter chroococcum, Ectothiorhodospira shaposhnikovii*) and Cyanobacteria (*Anabaena cylindrica, Anacystis nidulans,*

are specific for reduced ferredoxin, and they do not react with NAD(P)H.

The assimilatory N.r. have been described as soluble, cytoplasmic enzymes. They may, however, be attached to membranes or particles. Thus the N.r. of higher plants can be isolated in association with chloroplasts, or as an apparently soluble enzyme, depending on the technique of cell homogenization. Also, the nitrate reducing system of *Nostoc muscorum*, *Anabaena cylindrica* and *Anacystis nidulans* (Cyanobacteria) has been found tightly bound to photosynthetic particles, or as a soluble enzyme, depending on the technique of cell disruption. These N.r.-containing, photosynthetic particles catalyse the photoreduction of nitrate to nitrite, accompanied by the stoichiometric evolution of oxygen. In higher plants also the reducing power from the light reaction of photosynthesis may be consumed in the reduction of nitrate to nitrite (see Nitrate reduction).

A detailed study of N.r. of bacteroids from leguminous root nodules (see Nitrogen fixation; Nitrogenase) shows that the enzyme consists of two subunits: a consitutive unit containing molybdenum and nonheme iron, and an inducible unit containing iron. The latter is inducible by nitrate, whereas the constitutive subunit appears to be identical with the molybdenum-containing subunit found in xanthine oxidase (from milk, mammalian intestine and bird liver), and with fraction I of nitrogenase from *Azotobacter vinelandii* and Soybean nodule bacteroids. It has been suggested that competition between the inducible subunit of N.r. and fraction II of Nitrogenase (see) for the Mo-containing subunit is the basis of a control mechanism, whereby nitrogenase activity is manifested only when nitrate is in short supply. In the presence of nitrate, the inducible N.r. subunit is formed in such quantity that it binds all the Mo-Fe protein and renders it unavailable for the formation of nitrogenase. This may not be the only basis of control, and inhibition of nitrogenase by traces of nitrite appears to be an important factor in the suppression of nitrogenase activity by the presence of nitrate.

Nitrate reduction: In the narrow sense, N.r. is the reduction of the nitrate ion (NO_3^-) to nitrite (NO_2^-), catalysed by Nitrate reductase (see). In the broad sense, it is the reduction of nitrate to gaseous products (nitrous oxide or nitrogen), or to ammonia.

N.r. may be assimilatory or dissimilatory. Assimilatory N.r. is found in green plants, green algae and fungi, where it involves the reduction of nitrate to ammonia, and operates in conjunction with the assimilation of the ammonia. It occurs in two stages: reduction of nitrate to nitrite by Nitrate reductase (see), and the reduction of nitrite to ammonia by Nitrite reductase (see). In green plants, N.r. may occur in the roots or in the green tissue. In some species (e.g. the Field pea, *Pisum arvense*), nitrate is reduced and assimilated in the roots. The reducing power for N.r. is provided by respiration, and the xylem contains high levels of organic nitrogen relative to free nitrate. Other species (e.g. *Xanthium pennsylvanicum* [Cocklebur], *Stellaria, Trifolium*) transport high levels of nitrate in the xylem from the roots to the green tissue; in such plants the majority of reactions of ni-

trogen metabolism occur within the chloroplast, and the ATP and reducing power required can be provided directly by the light reaction of photosynthesis. Thus reduced ferredoxin in the chloroplast transfers electrons directly to nitrite reductase and to glutamate synthase, and ATP is utilized by glutamine synthetase. Nitrate reductase is remarkably sensitive to inhibition by cyanide. $10^{-9}M$ CN^- totally and reversibly inhibits nitrate reductase from *Chlorella* (c.f. $10^{-5}M$ CN^- for the inhibition of mitochondrial respiration). N.r. and assimilation in the chloroplast may be controlled by the production of HCN [L.P.Solomonson & A.M.Spehar [1977] *Nature*, **265** (1977), 373-375]. Nitrate accumulation is typical of plants growing in nitrate-rich soils, e.g. on rubbish dumps in areas of human occupation. Control of N.r. always appears to be exerted at the early stage of reduction of nitrate to nitrite, so that nitrite never accumulates. Green plants also show a light-dependent cycle of nitrate accumulation and depletion; nitrate increases in the dark and decreases during daylight. The mechanism of the light effect is probably complicated: photosynthesis supplies reducing power and ATP for N.r. and ammonia assimilation; the carbohydrate produced by photosynthesis is used for the synthesis of the carbon skeletons of amino acids; and respiration of carbohydrate also produces reducing power and ATP for N.r. and ammonia assimilation.

There are two main types of dissimilatory N.r.: 1.reduction of nitrate to ammonia, which is not assimilated (ammonification of nitrate), and 2.reduction of nitrate to gaseous compounds, such as nitric oxide, nitrous oxide and molecular nitrogen (denitrification). Ammonification and denitrification represent forms of respiration, in which nitrate replaces oxygen as the electron acceptor in an electron transport chain (see Respiratory chain). These forms of dissimilatory N.r. are therefore called nitrate respiration. Ammonification and denitrification have only one step in common, i.e. the reduction of nitrate to nitrite. In ammonification the nitrite is then reduced to ammonia, e.g. in *Clostridium perfringens, Achromobacter fischeri, Bacillus filaris, Escherichia coli*. In denitrification the nitrite is reduced to various gaseous products, according to the sequence:

$$NO_3^- \rightarrow NO_2^- \rightarrow NO \rightarrow N_2O \rightarrow N_2.$$

Many bacterial genera contain denitrifying species: *Achromobacter, Alcaligenes* (*Alcaligenes odorans* denitrifies nitrite), *Bacillus, Chromobacterium, Corynebacterium, Halobacterium, Hyphomicrobium, Moraxella, Paracoccus, Pseudomonas, Spirillum, Thiobacillus* and *Xanthomonas*. In some species of *Pseudomonas* and *Corynebacterium*, N_2O is the final denitrification product. All these bacteria are aerobes that are able to respire (denitrify) nitrate under anaerobic conditions. The only true anaerobe able to carry out denitrification is *Propionibacterium*. There is no evidence for other intermediates in the above denitrification pathway, but in the formation of nitrous oxide (N_2O) an NN bond must be formed, and there may exist transient enzyme-bound intermediates that have not yet been identified. The enzymology of denitrification from nitrite is poorly understood. It seems likely that each stage is linked to electron transport via a cytochrome system, but sites of ATP synthesis have not been unequivocally located.

Nitrate respiration: see Nitrate reduction.

Nitrite reductase: Dissimilatory N.r. (see Nitrate reduction), found in various bacteria, catalyses the reduction of nitrite, probably to nitric oxide; some authors claim that the reduction product is nitrous oxide. N.r. from *Pseudomonas aeruginosa*, *Paracoccus denitrificans* and *Alcaligines faecalis* contains both a *c*- and a *d*-type heme. The *Pseudomonas* N.r. consists of two similar polypeptides, each of M_r 63 000, and each containing two heme groups. Electrons are donated by either cytochrome c_{551} or the blue copper protein, azurin. N.r. from *Achromobacter cycloclastes* and *Pseudomonas denitrificans* are both copper proteins. The structure, mechanism of action and true intracellular location of these bacterial N.r. are poorly understood.

Assimilatory N.r. (see Nitrate reduction), found in fungi, green algae and higher plants, catalyse the reduction of nitrite to ammonia. This reduction is equivalent to the transfer of 6 electrons. There are no intermediates in the true sense, but in the presence of high concentrations of ammonium ions, the enzyme exhibits abnormal kinetics and releases traces of hydroxylamine. NADPH-dependent assimilatory N.r. (EC 1.6.6.4) from *Neurospora crassa* has M_r 290000. It is assumed to be a flavoprotein, but FAD appears to be removed during purification and must be added in vitro to achieve catalysis. The enzyme also contains Siroheme (see) as a prosthetic group, and the Siroheme component interacts with the nitrite. The sequence of electron transfer is: NADPH → FAD → Siroheme → Nitrite.

Assimilatory ferredoxin-dependent N.r. (EC 1.7.7.1) from spinach has M_r 61 000, contains one molecule of siroheme and one inorganic (Fe$_2$-S$_2$)-center. In vivo, reducing power is supplied by reduced ferredoxin from the light reaction of photosynthesis. The sequence of electron transfer is: Ferredoxin → (Fe$_2$S$_2$) → Siroheme → Nitrite.

Nitrogen: see Bioelements.

Nitrogenase (EC 1.18.2.1): the enzyme system responsible for biological nitrogen fixation. N. consists of two proteins, both of which are required for activity. One of these proteins contains iron, molybdenum and acid-labile sulfur (Mo-Fe protein,

component I; molybdoferredoxin, abb. azofermo) and the other contains iron and labile sulfur (Fe protein; component II; azoferredoxin abb. azofer.) The two components are present in the ratio I Mo-Fe protein: 2 Fe-proteins. Both protein components have been isolated and characterized from various nitrogen fixing organisms. The separated components can be reconstituted to active N. The Mo-Fe component has molecular weight in the range 200000–270000, and is tetrameric. In addition to the four subunits, the Mo-Fe protein also contains a low molecular weight component which carries the molybdenum. This same molybdenum "cofactor" may be present in other molybdenum-containing enzymes. The Fe protein has molecular weight of about 60000, and is dimeric. N. from different sources are very similar. Fe protein from one organism (e.g. *Azotobacter*) will often cross react with Fe-Mo protein from another organism (e.g. *Klebsiella*) to give a functional enzyme. On the other hand, heterologous N. components may generate catalytically inactive complexes, e.g. N. of *Clostridium* is inhibited by Mo-Fe protein of *Azotobacter*, because *Clostridium* Fe protein forms tight, inactive complex with *Azotobacter* Mo-Fe protein, even in the presence of excess *Clostridium* Mo-Fe protein.

Synthesis of N. is determined by the *nif* operon, which has been mapped in *Klebsiella* by phage P$_1$-mediated transduction. Three structural genes for N., and for the molybdenum cofactor and electron transport to N. have been identified. Another gene appears to have a regulatory product, but its precise role is still unclear. Three further genes necessary for the full expression of the *nif* operon have been mapped, but their role is unknown. The *nif* operon maps very close to the operon for histidine biosynthesis. The *nif* operon and the ability to fix nitrogen have been transferred from *Klebsiella pneumoniae* to *Escherichia coli* by R-factor mediated conjugation; this represents the first case of the transfer of nitrogen fixing ability to a non-nitrogen fixing organism. *Klebsiella nif* genes have been inserted into a plasmid (RP4, first isolated from *Pseudomonas aeruginosa*), which is transferrable to a wide variety of Gram-negative bacteria. When *Klebsiella nif* genes are transferred on RP4 to mutant *nif⁻ Azotobacter*, nitrogen fixation is restored despite the different physiology of the donor and recipient organisms.

N. is inoperative when nitrate is available as a nitro-

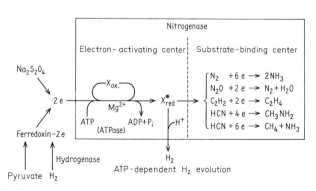

gen source, probably because N. is powerfully inhibited by traces of nitrite produced during nitrate reduction.

Transcription of the *nif* operon is promoted by active (non-adenylylated) glutamine synthetase. N. synthesis is repressed by ammonia, because an excess of ammonia initiates a control mechanism, which leads to adenylylation (inactivation) of glutamine synthetase.

N. catalyses the ATP-dependent reduction of several substrates that are of similar molecular size to, and are isoelectronic with molecular nitrogen, e.g. azide, nitrous oxide, cyanide, acetylene. Attempts to produce models for the substrate activating center of N. have resulted in the preparation of organic complexes of transition metals. These are of potential interest in chemical industry, where more efficient catalysts are sought for industrial ammonia synthesis. Such catalysts might also be exploited in the synthesis of hydrazine which is a high energy and environmentally clean fuel, but prohibitively expensive. Combustion of hydrazine produces nitrogen and water, and the heat yield is about 544 kJ/mol (130 kcal/mol). N. is regarded as having two active centers: the electron-activating, and the substrate-activating center. The electron-activating center is on the Fe protein, which has the unique property of being able to transfer electrons to the Mo-Fe protein. The ATP consumed by N. is largely required to increase the reducing potential of the electrons carried by the Fe proteins, thereby producing an extremely powerful reducing agent. Thus a ferredoxin (or flavodoxin, or rubredoxin)-dependent ATPase activates and reduces a metal-containing group X to X^*_{red}, which represents a metal hydride. The Mo-Fe protein contains the substrate-activating center. With the aid of molybdenum, the substrate is bound in such a way that it can be reduced by the electrons received from X^*_{red} on the Fe protein. The Mo-Fe protein contains 2 Mo, 28-32 Fe and about 28 acid-labile sulfurs per tetrameric protein. Up to four Fe_4S_4 centers are present. Two other centers appear to contain $MoFe_8S_6$, part of which may exist as a cubane structure similar to Fe_4S_4, but with one Fe replaced by Mo (i.e. $MoFe_3S_4$). There may also be an Fe_2S_2 center. The multinuclear Mo-containing center appears to be bound by two symmetrical polypeptide chains. Electrons are transferred singly from the ferredoxin-like Fe-protein, and stored in the Mo-Fe center; acting rather like an electrical capacitor, this center is then able to transfer several electrons in one stage (Fig.); in the case of the natural substrate, N_2, 6 electrons are transferred. Electrons may also be transferred to protons, producing molecular hydrogen. This ATP-dependent hydrogen evolution competes with ammonia synthesis. It is quite distinct from bacterial hydrogenase, which does not require ATP. In *Azotobacter* and in *Rhizobium* bacteroids, hydrogen produced by N. is taken up again by hydrogenase, then oxidized with consequent synthesis of ATP.

The activity of cell-free preparations of N. is usually measured by colorimetric assay of ammonia, following incubation with molecular nitrogen, Nitrogen fixation in general can be determined from the incorporation of $^{15}N_2$ into cell material, but the method is relatively expensive and tedious. The use of acetylene (ethyne) as an alternative substrate of N. has revolutionized studies of nitrogen fixation. Assays are performed in a closed system. The product, ethylene (ethene), is not assimilated and it can be easily assayed by gas chromatography.

Nitrogen balance: difference between the total nitrogen intake of an organism and its total nitrogen loss. Young growing animals are in positive N.b., i.e. they retain more nitrogen (as protein added during growth) than they excrete. Mature, healthy adults show a zero N.b., i.e. nitrogen intake is exactly balanced by nitrogen excretion. A negative N.b. results from a deficiency of an essential amino acid; a decrease in the concentration of any proteogenic amino acid in the body's amino acid pool impairs total protein synthesis, so that the concentration of all other free amino acids in the pool is increased; this leads to an exaggeration of degradative pathways and an increase in urea formation. The classical deletion method for the determination of the essential (indispensable) amino acid requirements of an animal involves the measurement of N.b. in adult animals receiving a complete diet, except for the omission of the amino acid under test. The daily food intake is determined, and the nitrogen content of an identical food sample is measured. Strictly, in determining the daily loss of nitrogen, dropped hair, sloughed skin and perspiration should be taken into account, but it is usually sufficient to omit these minor contributions, and use only the nitrogen content of feces and urine in the determination of N.b.

Nitrogen catabolite repression: repression by ammonia of the synthesis of various enzymes of nitrogen metabolism. See Ammonia assimilation.

Nitrogen cycle: see Nitrogen fixation.

Nitrogen excretion: see Ammonia detoxification.

Nitrogen fixation: a process in which atmospheric nitrogen is converted into ammonia. Activation of molecular nitrogen and its reduction to ammonia depend upon the catalytic activity of the enzyme Nitrogenase (see). The ammonia is then incorporated into the various nitrogenous compounds of the cell by the processes of Ammonia assimilation (see). Nitrogenase is a very unstable enzyme, especially in anaerobic organisms. N.f. is fundamentally important for the nitrogen economy of soils and waters, and it forms an essential stage in the nitrogen cycle of the biosphere. Certain free living soil microorganisms, especially of the genera *Clostridium* and *Azotobacter*, are capable of N.f. Other microorganisms fix nitrogen in symbiosis with higher plants, notably the *Leguminosae*. Many instances of N.f. by symbiotic associations between microorganisms and nonleguminous plants are also known, e.g. an Actinomycete (*Frankia* spp.) has been isolated from the nitrogen fixing root nodules of Alder *(Alnus)*. In water, especially in the ocean, the most important nitrogen fixers are the blue green bacteria (also commonly known as blue green algae, a term that is preferably avoided, because these nitrogen fixing organisms are prokaryotes, whereas the algae are eukaryotes). N.f. by blue green bacteria is of practical importance for the cultivation of rice in the tropics. The lichens (symbiotic associations between a blue-green bacterium and a fungus), and the symbiotic system *Nostoc-Gunnera* (i.e. symbiosis between a blue green bacterium and an angiosperm) are ecologically very important, since they are able to colonize habitats that have extreme climates or are poor in nutrients. The carbon and nitrogen requirement of the lichens are met by photosynthesis and N.f. Such symbiotic systems are therefore sustained largely by the atmosphere, and their nutritional demands on the remaining environment are relatively small. Lichens pioneer the exploitation of barren environments and pave the way for later colonization by plants with more exacting nutritional requirements. In poor soils, the *Nostoc-Gunnera* system can fix about 70 g atmospheric nitrogen per m^2 per year.

The ability to perform N.f. is a special characteristic of relatively few prokaryotic organisms; it has never been detected in an eukaryote. In the bacterium, *Clostridium pasteurianum,* which contributes to the nitrogen enrichment of agricultural soils, both the reducing power and the ATP required for N.f. are derived from phosphoroclastic pyruvate cleavage. In cell-free enzyme preparations pyruvate can be replaced by ATP or an ATP generating system, and a reducing agent (hydrogen or electron doner). Suitable reducing agents include sodium dithionite and potassium borohydride. Nitrogenase will also catalyse the transfer of electrons from molecular hydrogen to nitrogen in the presence of a ferredoxin-dependent hydrogenase. In most nitrogen-fixing systems, the natural electron donor is a ferredoxin; in certain cases, this is replaced by other electron transferring proteins, e.g. flavodoxin, or rubredoxin. Four molecules of ATP are required for the transfer of each pair of electrons. The stepwise reduction of nitrogen on the surface of the nitrogenase may take place via enzyme-bound intermediates, but free intermediates between ammonia (the product) and N$_2$ (substrate of N.f.) have not been observed.

So far, the most extensively studied N.f. system is the symbiotic association between members of the *Leguminosae* and *Rhizobium;* the legume often chosen for these studies is *Glycine max* (Soybean). Infection of the plant roots by virulent *Rhizobia* leads to the formation of root nodules, which have the capacity to fix nitrogen. The *Rhizobia* living free in the soil do not fix nitrogen. Under laboratory conditions, however, pure cultures of *Rhizobium* will fix nitrogen, providing a pentose (e.g. arabinose) and a dicarboxylic acid (fumarate or succinate) are present in the culture medium. During the infection process, the *Rhizobium* cells lose their rod-like shape and eventually become globular-shaped bacteroids. Reduction of nitrogen to ammonia and the assimilation of the ammonia occur in these bacteroids; carbon compounds for the respiration of the bacteroids and for the assimilation of the ammonia are provided by the plant; amino acids are exported to the host plant tissues. The natural electron donor for the symbiotic N.f. has not been identified. Leghemoglobin is necessary for N.f. by legume root nodules, but it is not required by nonlegume N.f. systems. The concentration of leghemoglobin in root nodules is an index of the nitrogen fixing capacity. The bacteroids are bathed in a solution of leghemoglobin, which is enclosed by a membrane envelope. The rate of oxygen transport through an unstirred solution of leghemoglobin is eight times higher than its rate of diffusion through water. This facilitated diffusion of oxygen to the bacteroids permits a high respiration rate, which is necessary to produce the relatively large quantities of ATP required by the nitrogenase. In contrast, oxygen interferes during the laboratory preparation of active bacteroids, due to the presence of phenols and polyphenol oxidases from the host plant tissue. These can be inactivated by adsorption onto polyvinyl pyrrolidone in the presence of ascorbic acid. Nitrogen fixing bacteroid suspensions can therefore be isolated from homogenized root nodules by using strictly anaerobic conditions, e.g. centrifugation of the homogenate under argon, or by abolishing polyphenol oxidase activity. The bacteroids can then be treated like any other bacterial source of nitrogenase. Subsequent disruption of the cells and purification of the enzyme by selective precipitation and column chromatography must be performed under strictly anaerobic conditions, because nitrogenase is irreversibly inactivated by oxygen; this is especially critical at later stages of purification, as the oxygen sensitivity of nitrogenase increases with purification. The separate protein components of nitrogenase are both inactivated by oxygen, and the Fe-protein is the more sensitive.

Nitrogen storage: see Ammonia detoxification.

nm: abb. for nanometer.

Nodule bacteria: see Rhizobia.

Noncompetitive inhibition: see Effectors.

Noncovalent bonds: various types of noncovalent bond are responsible for maintaining chain conformation and quarternary structure of proteins, and they are also important in the structure and function of nucleic acids. 1. *Hydrogen bonds* are formed between neighboring peptide bonds (separation distance 0.28 nm), between tyrosyl and carboxyl or imidazole, and between seryl and threonyl residues. Hydrogen bonding in particular is an important factor in nucleic acid structure, and in template recognition during replication, transcription and translation (see Base pairing; Ribonucleic acid; Protein biosynthesis; Deoxyribonucleic acid).

2. *Heteropolar (electrostatic) bonds* in proteins are formed between residues of opposite charge, e.g. lysyl and glutamyl.

3. *Apolar (hydrophobic) bonds* in proteins are formed between very close, uncharged groups, e.g. $-CH_3$ and $-CH_2OH$, or between more widely separated, uncharged groups, e.g. phenyl and leucyl. The effective strength of these hydrophobic bonds is increased by the entropy effect of the repulsion of the surrounding water, and they contribute to the stability of protein conformation especially at elevated temperatures.

4. *Van der Waals forces* act only at very short distances, and they represent the weak attraction between the positively charged nucleus of one atom and the negatively charged electrons of another. They are important in base stacking in the double helix of DNA (see Deoxyribonucleic acid).

Nonheme iron proteins, *NHI-proteins:* proteins containing iron that is not bound in a heme system.

In these proteins, the iron is bound by the sulfur of cyteine residues, and it is often also associated with inorganic sulfur. They are also called iron-sulfur proteins. See Ferredoxin; Rubredoxin.

Nonhistone chromatin proteins: "acidic chromatin proteins", a highly heterogeneous group of tissue-specific proteins, which are bound to certain DNA sequences. Their M_r in detergents is 30 000–70 000, i.e. markedly higher than the M_r of Histones (see). As gene derepressors, they play a part in the regulation of gene expression in mammalian cells, especially in cell proliferation.

Nonordered conformation: see Proteins.

Nonsense codon: an Amber codon (see), or Ochre codon (see).

Nopaline: see D-Octopine.

Nopalinic acid: see D-Octopine.

Noradrenalin, *norepinephrine, arterenol:* dihydroxyphenylethanolamine, a hormone and adrenergic Neurotransmitter (see) in the sympathetic nervous system. M_r 169.2. N. causes contraction of blood vessels, with the exception of coronary vessels; it therefore causes an increase in the blood pressure. It also relaxes smooth muscle, but stimulates cardiac muscle. Comparison of the effects of N. and Adrenalin (see) has led to classification of postsynaptic receptors as α or β receptors. The former are more responsive to adrenalin, and are generally excitatory, except in the intestinal smooth muscle, while the latter respond to N. and are generally inhibitory. N. is a catecholamine synthesized from L-tyrosine via dopa (see Dopamine) in the adrenal medulla and in adrenergic neurons in the nervous system. It is converted, in part, to Adrenalin (see) in these same tissues by the enzyme Noradrenalin *N*-methyltransferase (E.C.2.1.1.28). N. is deactivated by *O*-methylation; a monoamine oxidase then removes the NH_2 group to produce 3-methoxy-4-hydroxymandelic acid (vanillylmandelic acid, VMA). VMA is the major metabolite of N. in the peripheral parts of the body and in the urine. The amount of VMA in the urine is an index of parasympathetic nervous function, and is used for the diagnosis of tumors which produce N. or adrenalin. (See ascorbate shuttle.)

Noradrenalin   *Vanillylmandelic acid*

Norepinephrine: see Noradrenalin.

Norgestrel: a synthetic gestagen. N. is used in oral contraceptives, and it is the most potent of the orally active gestagens.

Norlaudanosine: see Benzylisoquinoline alkaloids.

Nornicotine: see Nicotine.

Norsteroids: see Steroids.

Northern blot: see Southern blot.

Norgestrel

Norum's disease: see Inborn errors of metabolism.

Notatin: see Mycotoxins.

Nu bodies, *nucleosomes:* subunits of Chromatin (see) containing 180 to 200 base pairs of DNA and approximately an equal weight of histone. They are released by partial digestion of chromatin by staphylococcal nuclease.

Nucleases: a group of hydrolytic enzymes, which cleave nucleic acids. Exonucleases attack the nucleic acid molecule at its terminus, whereas endonucleases are able to catalyse a hydrolytic cleavage within the polynucleotide chain. Deoxyribonucleases (DNAases) are specific for DNA, ribonucleases (RNAases) for RNA. All N. are Phosphodiesterases (see); they catalyse the hydrolysis of either the 3′ or 5′ bond of the 3′,5′-phosphodiester linkage. Ribonuclease (see) has been extensively studied.
The *N*-glycosidic bond of nucleic acids is cleaved by Nucleosidases (see).

Nucleic acid bases: constituent bases of nucleic acids. N. a. b. are fundamental to the function of nucleic acids in the storage and transfer of genetic information. They are Adenine (see), Guanine (see), Cytosine (see), Thymidine (see), Uracil (see), and others that occur less frequently (see Rare nucleic acid components). See also Nucleic acids; Genetic code; Base pairing.

Nucleic acids: polymerized nucleotides found in all cells and viruses. N. a. were first isolated in 1869 from the white blood cells of pus by Miescher, who called the material nuclein. The term N. a. was introduced in 1889 by Altman, in recognition of their acidic properties. There are two main classes of N. a., distinguished by their carbohydrate component: *Ribonucleic acid* (RNA) (see), which contains ribose, and *Deoxyribonucleic acid* (DNA) (see) which contains 2-deoxy-D-ribose. Both types have certain common structural features, but they have different biological functions. DNA stores genetic information; it is replicated during cell division, so that each daughter cell receives DNA that is identical in structure and informational content (see Genetic code). RNA is intimately involved in protein synthesis, and is primarily responsible for translating the information of the DNA into the primary structure of specific proteins (see Protein biosynthesis).
N. a. are polymers with M_r between 20 000 and approx. 10^9. They contain three structural components: the purine and pyrimidine bases, a pentose (either D-ribose in RNA or 2-deoxy-D-ribose in DNA) and phosphate esters. The five major bases are uracil (RNA only), thymine (DNA only), adenine, guanine

and cytosine (RNA and DNA); in addition, there are over 30 Rare nucleic acid components (see) which occur in various N.a. These rare components are formed by modification of existing structures within the N.a., e.g. by methylation, hydrogenation or rearrangement of normal bases.

Each mononucleotide unit is linked to its neighbor by a phosphate group, which forms an ester linkage with position 3′ of one sugar and 5′ of the neighboring sugar. This 3′,5′-linkage results in a linear chain of phosphate-linked sugar residues; a base is attached to C-1 of each sugar by an N-glycosidic linkage (Fig. 1). The linear order of the bases is statistically irregular, representing the informational code of the N.a.

The reactive -NH$_2$, -OH and -NH groups of purine and pyrimidine bases are responsible for certain properties of N.a., e.g. formation of specific hydrogen bonds between purines and pyrimidines, leading to secondary structures. Thus complementary linear chains can form a double helix (see DNA), or a linear strand can fold on itself, forming alternate linear and helical regions (RNA). Other forces involved in the spatial conformation of N.a. (see Noncovalent bonds) are homopolar cohesive forces (Van der Waals forces), hydrophobic interactions between bases and solvent, and electrostatic interactions (ionic bonds).

Sequence analysis. Since the sequence of bases in a N.a. represents the information in that molecule, determination of sequence is of utmost importance. The introduction of rapid methods for this purpose has revolutionized the life sciences; it is now a routine matter to identify new proteins in the following manner: First antibodies to the protein are raised in a suitable animal; for this purpose the protein does not have to be pure, providing the resulting antibody can be subsequently purified (e.g. removal of un-

wanted antibodies by reaction with a system which is similar to the intended experimental one but lacking the protein in question). The antibodies are used to identify an oligopeptide fragment of the desired protein, and from the sequence of this oligopeptide, a probable sequence of the RNA encoding the peptide is deduced (see Genetic code). DNA with the deduced sequence is synthesized in the laboratory and used as a Hybridization (see) probe to identify the genomic DNA which encodes the unknown protein. This DNA is then usually cloned, after which its sequence is determined by one of the methods discussed below. From the base sequence of the cloned gene, the amino acid sequence of the unknown protein can be deduced.

The two most widely used methods of sequencing DNA were developed by A. M. Maxam and W. Gilbert [see *Methods of Enzymology* **65** (1980) 499–560] and by F. Sanger, S. Nicklen & A. R. Coulson [see *Proc. Natl. Acad. Sci.* **74** (1977) 5463]. In the Maxam and Gilbert method, either single- or double-stranded DNA may be sequenced. First the DNA must be cleaved with a Restriction endonuclease (see) to obtain the desired stretch of chromosome, which must be purified by gel electrophoresis and, if it is double-stranded, separated into its two strands. The single strand of DNA to be sequenced is labelled at one end, e.g. with ^{32}P. It is then treated with a reagent which specifically modifies one or two bases in such a way that they become susceptible to attack by a second reagent, which removes the base. A third reagent is then used to break the strand at those sites where bases (but not the sugar phosphates) have been removed. For example, dimethyl sulfate methylates the 7 nitrogen of guanine, which destabilizes the ring and makes it susceptible to cleavage by a base (the second reagent). Finally, piperidine displaces the open-ring guanine and removes

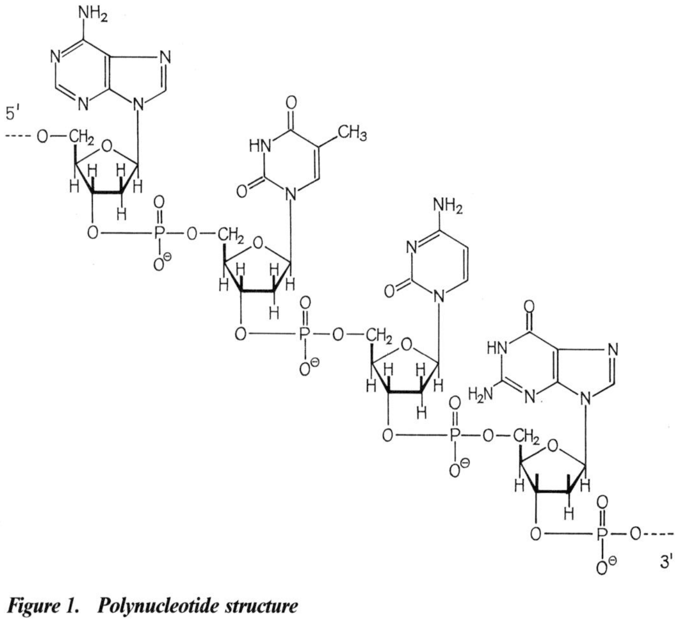

Figure 1. Polynucleotide structure

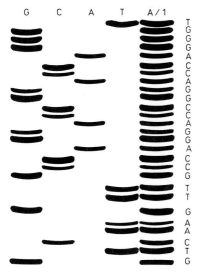

G	C	A	T	A / 1	
					T
					G
					G
					G
					A
					C
					C
					A
					G
					G
					C
					C
					A
					G
					G
					A
					C
					C
					G
					T
					T
					G
					A
					A
					C
					T
					G

Figure 2. Gel electrophoresis pattern which would be generated by sequencing of a hypothetical DNA sequence. After treatment with one of four reagents specific for the bases indicated at the top of the column, a sample of DNA is separated according to the length of the nucleotide sequence. The shortest polynucleotides are seen at the bottom of the gel. The sequence of the original sample is read off, as indicated to the right of the lane in which a mixture of the 4 treated samples is run (All).

the phosphates on its ribose, causing the strand to break. Adenine is methylated at N3 by dimethyl sulfate, but is not cleaved by base in the second step. Neither cytosine nor thymine is affected. Conditions are chosen so that the modification of bases is far from quantitative. As a result, the DNA is cleaved in many different sites, and a series of short polynucleotides is generated, each labelled at one end by the ^{32}P and ending with a base which was formerly adjacent to a guanine. In similar fashion, hydrazine treatment splits thymine or cytosine rings; hydrazine in presence of 1 M NaCl splits only cytosine; and both adenine and guanine are split by protonation (treatment with acid) followed by treatment with a base.

The method of Sanger et al. uses a primer and a DNA polymerase to make radiolabelled copies of the DNA (or RNA, if reverse transcriptase is used). One of the four nucleoside triphosphates present in the polymerization medium is deliberately mixed with an analog which, when incorporated into DNA, causes polymerization to terminate. The two classes of compound used for this purpose are 2′,3′-dideoxyribonucleotides and arabinosides. The 2′,3′-dideoxyribose moiety has no 3′-OH group with which a phosphodiester bond could form, and the arabinose has its 3′-OH group in the *trans* position, which causes it to block polymerization by prokaryotic polymerases. The ratio of chain-terminating analog to normal deoxyribonucleotide is chosen so that a mixture of polynucleotides is obtained in

which all possible fragments ending with the base in question are represented. These are separated by gel electrophoresis, and the sequence is read off, as in the method of Maxam and Gilbert. Both methods have been automated.

These sequencing methods apply to polynucleotide sequences up to 250 bases from the labeled end; thus longer stretches of DNA must be cut into appropriate fragments and these fragments then arranged in the correct order once their sequences have been determined. For this work, restriction endonucleases are indispensible. Vol. 65 of *Methods of Enzymology* is devoted to the techniques of nucleotide sequencing in entire genes and chromosomes.

Physical properties and analytical methods: The conjugated double bonds in the heterocyclic rings of the bases absorb UV light in the region of 260 nm; UV spectrophotometric analysis is therefore used for the characterization and quantitative determination of N. a. and lower M_r, related compounds, such as mono- and oligonucleotides. The intensity of UV absorption depends upon the conformation, i.e. the secondary structure of the N. a. Thus optical methods can also be used in the structural elucidation of N. a., and for detecting and monitoring structural changes. Denaturation results in an increase of absorption at 260 nm (see Hyperchromic effect). From the course of the heat denaturation curve, called the melting point curve, it is possible to assess the helical content of a N. a. and to determine the GC content of a DNA sample (see T_m value). If the heat-denatured N. a. is cooled slowly, the original structure is largely reformed (renaturation). Bacterial DNA renatures more extensively than nuclear DNA from higher organisms. Heat-denatured RNA can also be largely renatured to its original form.

Other physical methods used in the investigation of N. a. structure are Hybridization (see), electron microscopy, analytical ultracentrifugation, CsCl density gradient centrifugation, X-ray diffraction analysis, infrared spectroscopy, optical rotatory dispersion, light scattering photometry and viscosity measurements.

N. a. are purified by column chromatography (e.g. methylalbumin-silicic acid columns), by electrophoresis (e.g. polyacrylamide gels) and by density gradient centrifugation (e.g. sucrose or CsCl). A N. a. may be determined quantitatively from the UV absorption of its bases, by determination of the phosphate content, or by specific color reactions for ribose or deoxyribose (Dische reagent for DNA, Dische-Schwarz reagent for RNA). The Fuelgen reaction is used for histochemical detection of DNA.

Nucleocidin: a purine antibiotic synthesized by *Streptomyces calvus* (see Nucleoside antibiotics).

$(C_6H_{10}O_5)$

$O-SO_2NH_2$

Nucleocidin

M_r 392, $[\alpha]_D^{25} -33.3\,°$. N. is active against bacteria and fungi, and it is used therapeutically against trypanosomes. It inhibits protein synthesis.

Nucleolus (plur. *Nucleoli*): a compartment of the nucleus. The N. contains the N. organizer (see). The number of N. per nucleus varies widely. The main components of the N. are proteins (over 80% of the dry weight), RNA (over 5%) and DNA.

The N. contains the following recognizable structures: a ground material of amorphous protein, ribonucleoprotein granules, ribonucleoprotein fibrils, and the chromatin fibrils of the N. organizer.

The N. is the site of biosynthesis of ribosomal RNA (see Ribosomes). During nuclear division, the N. temporarily disappears as a visible structure.

Nucleolus organizer: a specific region on one or more eukaryotic chromosomes, where the nucleolus is formed. The DNA in this region contains genetic information for the synthesis of ribosomal RNA.

Nucleoplasm: see Nucleus.

Nucleoprotein: heteropolar complexes of nucleic acids (in particular, nuclear DNA) with basic, acid-soluble proteins (histones or protamines), and with acidic, base- or detergent-soluble non-histone-chromatin proteins. N. occur mainly in the chromatin of the cell nucleus in its quiescent state, and in the chromosomes when the nucleus is active, i. e. dividing. Many viruses consist entirely of N., but N. are absent from bacteria. N. are concerned in DNA replication, and in the control of gene function during protein biosynthesis.

The protein-RNA complexes of the ribosome are also N.

Nucleosidases: enzymes that catalyse the cleavage of the bond between the sugar residue and the base of a Nucleoside (see). The reaction is usually a phosphorolysis (not a hydrolysis), involving orthophosphate.

Nucleoside antibiotics: purine or pyrimidine nucleosides with antibiotic activity. They act as antimetabolites of natural substrates, and inhibit the growth of microorganisms by blocking the metabolism of purines, pyrimidines and proteins. Some N.a. (e.g. showdomycin) contain an analog base, others (e.g. gougerotin) contain an analog sugar, or both moieties may be modified (e.g. puromycin) (see Table).

The analog components are formed by the modification of primary metabolites. The sugars or sugar derivatives, such as cordycepose, psicose, angustose and glucuronamide are derived from D-glucose or D-ribose by various reactions, e.g. epimerization, isomerization, oxidation, reduction and decarboxylation. Methyl groups occur frequently in N.a., and are derived by transmethylation. A normal nucleoside may be modified to a N.a. without prior cleavage of the *N*-glycosidic bond, e.g. the synthesis of tubercidin. Alternatively, the free base may combine with the analog sugar, as in the synthesis of psicofuranin. Unusual amino acids occur in some N.a., e.g. amicetin contains α-methyl-D-serine. Some N.a. contain unusual bond types, such as the aza bond in 5-azacytidine, or unusual functional groups, such as the CN-group in toyocamycin. The same N.a. may be formed by systematically unrelated organisms, e.g. cordycepin from *Cordyceps* and *Aspergillus*. One organism may produce several N.a., e.g. psicofuranin and decoyinin from *Streptomyces hygroscopicus*.

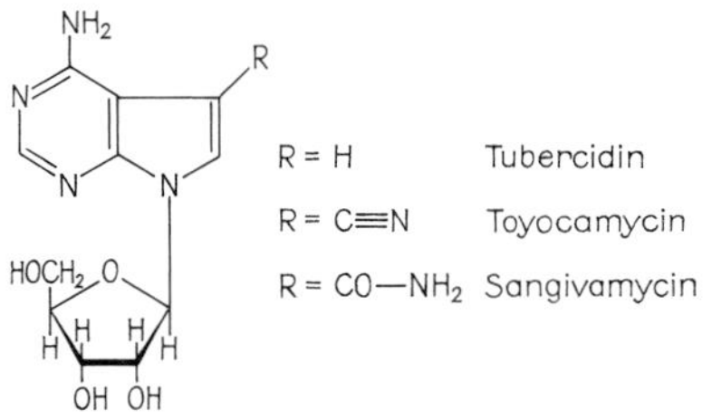

$R = H$ Tubercidin
$R = C\equiv N$ Toyocamycin
$R = CO-NH_2$ Sangivamycin

Figure 2. Structures of tubercidin, toyocamycin and sangivamycin.

Nucleoside diphosphate compounds: compounds containing a nucleoside diphosphate grouping. This grouping has an activating effect, so that the molecule has a high group transfer potential. Examples of N.d.c. are Nucleoside diphosphate sugars (see) and Cytidine diphosphate choline (see).

Nucleoside diphosphate sugars, *nucleotide sugars:* energy-rich nucleotide derivatives of mono-

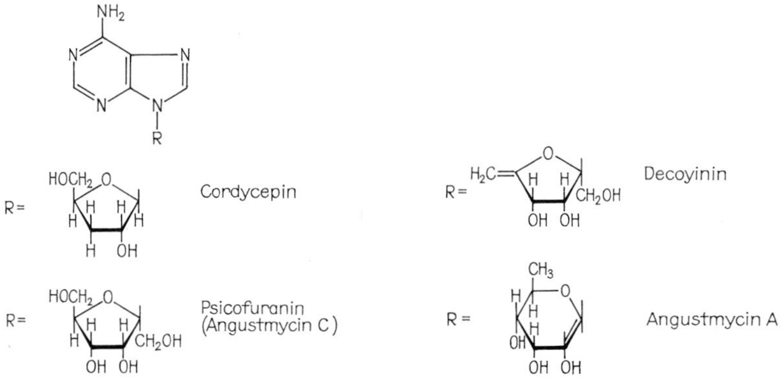

Figure 1. Structures of cordycepin, psicofuranin, decoyinin and angustmycin A.

Nucleoside antibiotics. Naturally occurring nucleoside analogs with antibiotic activity. For mode of action, see separate entry for each compound.

Antibiotic	Produced by	Base	Sugar	Antimetabolite of
3' Acetoamido-3'-deoxyadenosine	*Helminthosporium spec.*	Adenine	3-Acetamido-3-deoxyribose	Adenosine
Amicetin A	*Streptomyces fasciculatus, Streptomyces vinaceus-drappus*	Cytosine	Amicetose	Cytidine
Amicetin B (Plicacetin)	*Streptomyces plicatus*	Cytosine	Amicetose	Cytidine
3'-Amino-3'-deoxy-adenosine	*Cordyceps militaris, Helminthosporium spec.*	Adenine	3-Amino-3-deoxy-ribose	Adenosine
Angustmycin A (Decoyinin)	*Streptomyces hygroscopicus*	Adenine	L-2-Ketofucopyranose	Adenosine, Guanosine
Angustmycin C (Psicofuranin)	*Streptomyces hygroscopicus*	Adenine	Psicose (Psicofuranose)	Adenosine, Guanosine
Arabinofurano-syladenine	*Streptomyces antibioticus*	Adenine	Arabinofuranose	Adenosine
5-Azacytidine	*Streptoverticillius lakadamus*	Azacytosine	Ribose	Cytidine
Bamicetin	*Streptomyces plicatus*	Cytosine	Amicetose	Cytidine
Blasticidin S	*Streptomyces griseochromogenes*	Cytosine	4-Deoxy-4-amino-2,3-hexenuronic acid	Cytidine, Acyl-tRNA
Cordycepin	*Cordyceps militaris, Aspergillus nidulans*	Adenine	Cordycepose (3-Deoxyribose)	Adenosine
Formycin	*Nocardia interforma*	7-Aminopyra-zolopyrimidine	Ribose	Adenosine
Formycin B	*Nocardia interforma, Streptomyces lavendulae, Streptomyces roseochromogenes*	7-Hydroxypyra-zolopyrimidine	Ribose	Adenosine
Gougerotin	*Streptomyces gougeroti*	Cytosine	4-Dideoxyglucopy-ranuronamide	Cytidine, Acyl-tRNA
Nebularin	*Agricus (Clitocybe) nebularis, Streptomyces spec.*	Purine	Ribose	Adenosine
Nucleocidin	*Streptomyces calvus*	Adenine	Hexose of unknown structure	Adenosine
Puromycin	*Streptomyces albo-niger*	Dimethylamino-purine	3-Amino-3-deoxy-ribose	Acyl-tRNA
Sangivamycin	*Streptomyces spec.*	4-Amino-5-car-boxamide-7-pyr-rolopyrimidine	Ribose	Adenosine
Showdomycin	*Streptomyces showdoensis*	Maleimide	Ribose	Uridine
Toyocamycin	*Streptomyces toyocaensis, Streptomyces rimosus*	4-Amino-5-cya-no-7-pyrrolopyri-midine	Ribose	Adenosine
Tubercidin	*Streptomyces tubercidicus*	4-Amino-7-pyr-rolopyrimidine	Ribose	Adenosine

Nucleosides

Table. Nucleoside diphosphate sugars.

Activating nucleoside diphosphate group	Activated molecule	Function
Uridine diphosphate	Glucose	Involved generally in carbohydrate metabolism. Glycogen (see) synthesis. Murein (see) synthesis.
	Galactose	Galactose (see) metabolism
	Glucuronate	Glucuronate pathway (see). Glucuronate synthesis
	N-Acetylglucosamine	Metabolism of amino sugars. Chitin synthesis.
Adenosine diphosphate	Glucose	Starch (see) synthesis.
Guanosine diphosphate	Mannose	Synthesis of L-fucose, D-rhamnose and 6-deoxyhexoses.
Deoxythymidine diphosphate	Glucose	Synthesis of L-rhamnose.
Cytidine diphosphate	Ribitol Glycerol	Synthesis of Teichoic acids (see).

saccharides. The activating group is a nucleoside diphosphate. Uridine diphosphate glucose (UDP-glucose, UDPG, "active glucose") is of widespread general importance in carbohydrate metabolism. It is synthesized from glucose 1-phosphate by reaction with uridine triphosphate (Fig.). Other nucleoside diphosphate groups found in N.d.s. are listed in the table.

The activated sugar can take part in various metabolic reactions. Of particular importance is the transfer of the sugar moiety to the OH-group of another molecule, e.g. in the synthesis of oligo- and polysaccharides. (see Carbohydrate metabolism).

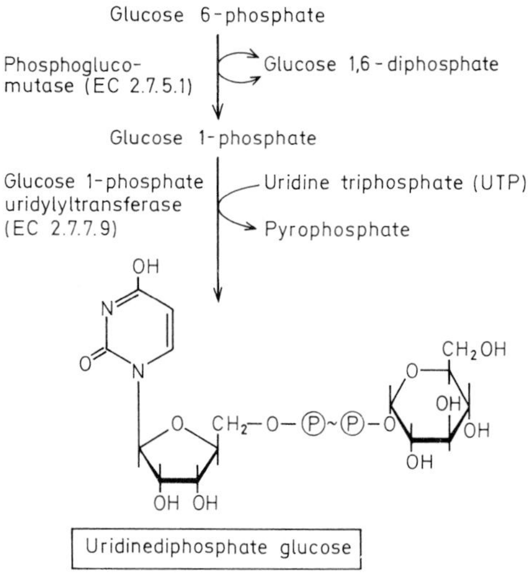

Synthesis of uridinediphosphate glucose

Nucleoside diphosphate derivatives of uronic acids are also metabolically important, e.g. uridine diphosphate glucuronic acid is formed from uridine diphosphate glucose, and it is a precursor of glucuronides (see Glucuronate pathway).

Nucleosides: N-glycosides of heterocyclic nitrogenous bases. The N-glycosides of purines and pyrimidines with pentoses are of particular biological importance. The sugar component is either D-ribose or D-2-deoxyribose, both in the furanose form (Table). C-1 of the pentose residue is linked to N-9 of the purine or N-1 of the pyrimidine by an N-glycosidic linkage (C-N bond). To distinguish between the numbering systems of the base and sugar, the numbers of the sugar atoms are characterized by a prime, i.e. C-atoms 1' to 5'. Deoxynucleosides contain D-2-deoxyribose, whereas ribonucleosides contain D-ribose. N. have trivial names derived from the component base. Pyrimidine N. end in -idine, purine N. in -osine.

The Rare nucleic acid components (see) represent nucleoside moieties in which the base or sugar is chemically modified.

N. and deoxynucleosides can be synthesized via a Salvage pathway (see). They are also produced by the hydrolysis of nucleic acids and nucleotides. Nu

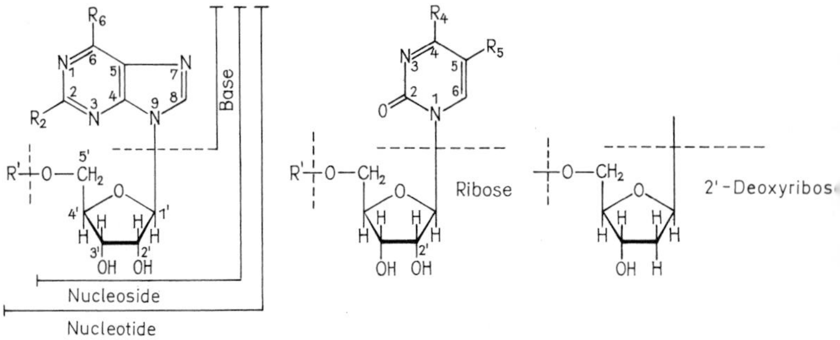

Table. **Structure of purine and pyrimidine bases, nucleosides, deoxynucleosides and nucleotides**

Purine base		$R^1 = H$ Nucleoside or Deoxynucleoside	Pyrimidine base		$R^1 = H$ Nucleoside or Deoxynucleoside
$R_2 = H$ $R_6 = NH_2$	Adenine	Adenosine Deoxyadenosine	$R_4 = OH$ $R_5 = H$	Uracil	Uridine Deoxyuridine
$R_2 = NH_2$ $R_6 = OH$	Guanine	Guanosine Deoxyguanosine	$R_4 = NH_2$ $R_5 = H$	Cytosine	Cytidine Deoxycytidine
$R_2 = H$ $R_6 = OH$	Hypoxanthine	Inosine (Deoxyinosine)	$R_4 = OH$ $R_5 = CH_3$	Thymine	Ribothymidine Thymidine*

	$R^1 = ⓟ$		$R^1 = ⓟ \sim ⓟ$		$R^1 = ⓟ \sim ⓟ \sim ⓟ$	
Nucleo- tides:	Adenosine Guanosine Inosine Uridine Cytidine Thymidine	} mono- phosphate	Adenosine Guanosine Inosine Uridine Cytidine Thymidine	} diphosphate	Adenosine Guanosine Inosine Uridine Cytidine Thymidine	} triphos- phate

* The sugar component of thymidine is 2'-deoxyribose

cleoside phosphorylases and deoxynucleoside phosphorylases catalyse the reversible, phosphate-dependent cleavage of N. and deoxyribonucleosides, forming ribose 1-phosphate or deoxyribose 1-phosphate and the free base. N. and deoxyribonucleosides can be converted into their corresponding nucleotides by the action of specific kinases.

Strictly speaking, the term nucleoside is reserved for base-sugar combinations present in nucleic acids, but the term is often applied to any base-sugar compound.

Nucleotide coenzyme: a coenzyme containing a nucleotide structure. N.c. are Pyridine nucleotide coenzymes (see), the nucleoside diphosphate moieties of Nucleoside diphosphate sugars (see) and Coenzyme A (see). The Flavin nucleotides (see) are also N.c. Strictly speaking FMN is not a nucleotide, but the term nucleotide can be generally applied to any base-sugar-phosphate group.

Nucleotides, *nucleoside phosphates:* phosphoric acid esters of Nucleosides (see). *o*-Phosphoric acid is esterified with a free OH-group of the sugar. If the sugar is D-ribose, the N. is called a ribonucleotide or ribotide. If the sugar is D-2-deoxyribose, the N. is a deoxyribonucleotide, deoxynucleotide or deoxyribotide. The phosphate may be present on position 2', 3' or 5' (in deoxyribonucleotides, only the 3'- and 5'-phosphate are possible). The 5'-nucleoside phosphates are metabolically very important; they may be mono-, di- or triphosphorylated, e.g. guanosine 5'-monophosphate, cytidine 5'-diphosphate and adenosine 5'-triphosphate.

The cyclic N. (cyclic 3', 5'-monophosphates) have important regulatory properties (see Adenosine phosphates; Guanosine phosphates; Inosine phosphates; Uridine phosphates).

Nucleoside monophosphates are synthesized de novo in the course of Purine biosynthesis (see) and Pyrimidine biosynthesis (see). They are then phosphorylated stepwise by the action of kinases, to produce nucleoside di- and triphosphates. The 2-deoxyribose moiety is formed by reduction of the ribose in ribonucleotides (see Ribonucleotide reduc-

tase). Reduction of free ribose to 2-deoxyribose does not occur in vivo.

N. and deoxynucleotides are the monomeric components of Oligonucleotides (see) and Polynucleotides (see). Enzymatic degradation of oligo- and polyribonucleotides (but not the corresponding deoxyribonucleotides) produces cyclic 2', 3'-nucleoside phosphates. N. are cleaved hydrolytically to nucleosides by the action of 5'- or 3'-nucleotidases, which function as phosphomonoesterases. N. are cleaved to free bases and phosphoribosylpyrophosphate in a pyrophosphate-dependent reaction catalysed by N. pyrophosphorylases. Certain coenzymes contain nucleotide structures (see Nucleotide coenzymes); thus NADP contains the structure of adenosine 2', 5'-diphosphate; and coenzyme A the structure of adenosine 3', 5'-diphosphate. In all living cells, N. (especially adenosine 5'-triphosphate) act as so-called high energy compounds in the storage and transfer of chemical energy.

Nucleotide sugar: see Nucleoside diphosphate sugars.

Nucleus: a large structure ($\sim 5 \, \mu m$ diam.) in eukaryotic cells, containing the bulk of the cellular DNA, and representing the chief site for the storage, replication and expression of genetic information. In the period between cell divisions, i.e. at interphase, the nucleus is densely and uniformly packed with DNA and shows few distinct structures, even under the electron microscope.

Distinguishable features are Chromatin (see), nucleoplasm, nuclear membrane and the Nucleoli (see). At nuclear division, the highly structured Chromosomes (see) are formed from the chromatin.

Chromatin and chromosomes are composed mainly of DNA, RNA and numerous proteins. Important enzyme proteins are DNA-polymerase for DNA-replication, and RNA-polymerase for transcription. Other chromosomal proteins (histones, protamines and acidic proteins) are engaged in the regulation of replication and transcription (see Chromosomes).

Other important metabolic processes occur in the

nucleoplasm, e.g. glycolysis and tricarboxylic acid cycle. NAD is synthesized only in the nucleus. The nuclear Na^+ concentration is ten times higher than that of the cytoplasm. The nucleus also contains a complete protein synthesizing system, but it is still largely unknown which nucleoproteins are synthesized in the nucleus. The nuclear membrane is important in the transfer of high and low M_r compounds between nucleus and cytoplasm.

Isolated nuclei are highly permeable to histones, protamines and other biological macromolecules, whereas ATP and Na^+ become tightly bound. The chief components of isolated and disrupted nuclei are DNA-histone complexes (nucleohistones), ribonucleic acids and poorly soluble acidic proteins (residual proteins) Nuclei also contain high concentrations of an arginase and an adenosine 5'-phosphatase of unknown function.

Nuphara alkaloids: a group of alkaloids possessing a piperidine or quinolizidine ring system, which are found in various species of water lily (*Nuphar* spp.). All N.a. contain a sesquiterpene skeleton, which is cyclized by the inclusion of other atoms (nitrogen, oxygen or sulfur). The chief N.a. are nupharidine, nupharamine and thiobinupharidine. Castoreum or castor, the secretion from the preputial follicles of the beaver, contains castoramine (M_r 247, m.p. 65–66 °C) which is related to the N.a.; it is not known whether it is synthesized by the animal, or derived from water lilies in its diet.

For the biosynthesis of N.a., see Terpene alkaloids.

Nupharamine   Nupharidine   Castoramine

Nutrient medium, *growth medium:* a medium, liquid or solid, for the cultivation of microorganisms, cells, tissues or organs. Solid media are prepared from liquid media by the addition of a gelling agent, e.g. gelatine (nowdays used only in special cases), silicic acid (when exclusion of organic compounds is necessary), or Agar-agar (see) (used widely in bacteriology, usually at a concentration of 1.5–2%). N.m. contain fairly large amounts of mineral elements, together with trace elements. Sufficient quantities of certain trace elements are often already present as impurities in the other components of the medium. The mineral constituents, or inorganic nutrients, must be correctly balanced in order to avoid competitive effects of ions, to achieve an appropriate pH value, and to establish the correct oxido-reduction status of the N.m.

The composition of a complex N.m. is more or less ill defined. This may arise when the nutritional requirements of the culture are not exactly known, and the requirements can only be satisfied by the inclusion of, e.g. yeast extract, yeast autolysate, meat extract, peptone, coconut milk, or other complicated natural mixtures. The constant aim in the use of N.m. is the definition of the minimal growth re-

quirements for the system in question; this enables the use of synthetic N.m. (i.e. made by mixing defined chemicals of known purity), or minimal N.m. (synthetic media of exactly known composition, containing only those components that are required). Some growth systems have rather complicated and exacting growth requirements, so that it is necessary to add Growth factors (see). Auxotrophic organisms also show requirements for growth factors, which are not required by the parent, wild-type prototrophic organism (see Mutant technique).

Glucose often serves as the source of carbon and energy in synthetic N.m. In the presence of glucose, the utilization of other, less efficient carbon sources is usually prevented by catabolite repression. The source of nitrogen depends on the growth system; it may be inorganic, e.g. nitrate or ammonium; or organic, e.g. urea. Table 1 shows the composition of a simple bacteriological growth medium. Tables 2 and 3 give the composition of a vitamin solution and a trace element solution used in the preparation of some N.m.

Table 1. A simple synthetic culture medium for microorganisms (after Schlegel)

Glucose	10.0 g
NH_4Cl	1.0 g
K_2HPO_4	0.5 g
$MgSO_4.7H_2O$	0.2 g
$FeSO_4.7H_2O$	0.01 g
$CaCl_2$	0.01 g
Trace element solution	1 ml
Water	1000 ml

Table 2. Vitamin solution used for the preparation of culture media for soil and water bacteria (after Schlegel). 2–3 ml of this solution are added to 1000 ml of culture medium.

Biotin	0.2 mg
Vitamin B_{12}	2.0 mg
Nicotinic acid	2.0 mg
p-Aminobenzoic acid	1.0 mg
Thiamin	1.0 mg
Pantothenic acid	0.5 mg
Pyridoxamine	5.0 mg
Distilled water	1000 ml

Table 3. The A–Z solution, or trace element solution of Hoagland.

$Al_2(SO_4)_3$	0.055 g
KI	0.028 g
KBr	0.028 g
TiO_2	0.055 g
$SnCl_2.2H_2O$	0.028 g
LiCl	0.028 g
$MnCl_2.4H_2O$	0.389 g
$B(OH)_3$	0.614 g
$ZnSO_4$	0.055 g
$CuSO_4.5H_2O$	0.055 g
$NiSO_4.7H_2O)$	0.059 g
$Co(NO_3).6H_2O$	0.055 g
Distilled water	100 ml

Nutritional physiology of microorganisms: The terms autotrophy and heterotrophy are too broad to distinguish between all the different forms of microbial nutrition. Different types of nutritional physiology are classified according to: 1. the nature of the carbon source; 2. the source and mechanism of formation of ATP; 3. the source of reducing power for the synthesis of cell constituents.

Phototrophs use light energy (see Photosynthesis), and if they obtain all their energy from light and all their carbon from CO_2, they are called *photoautotrophs*. Some phototrophs can use organic carbon sources, obtaining all or some energy from light; these are called *photoheterotrophs (i. e. they grow under mixotrophic conditions, see below)*.

Chemotrophs obtain ATP by the oxidation of inorganic or organic substrates, and assimilate CO_2 at the expense of the resulting oxidation energy (see Chemosynthesis).

Bacteria that obtain their energy from the oxidation of inorganic compounds (i. e. inorganic hydrogen donors) are called *lithotrophs* or *chemolithotrophs,* e. g. *Nitrosomonas,* which oxidizes ammonium to nitrite and nitrate; *Hydrogemonas,* which oxidizes gaseous hydrogen with oxygen; *Thiobacillus,* which oxidizes sulfide, elemental sulfur, thiosulfate or sulfite to sulfate. If all of their carbon is derived from CO_2, they are called *chemoautotrophs*. Most of these bacteria possess an electron transport chain, which is similar to that in other bacteria and mitochondria, and ATP is synthesized by "oxidative" phosphorylation during oxidation of the inorganic source of reducing power. The potentials of the reactions involved may, however, be lower than in the aerobic respiration of organic substrates, so that P/O ratios are also lower.

The use of organic sources of reducing power is known as *organotrophy*. Most microorganisms, like animals, are *chemoorganotrophs*. Cyanobacteria and purple sulfur bacteria, which carry out photosynthesis, are *photolithotrophs*.

Some autotrophs can also grow heterotrophically (facultative autotrophs), using organic energy sources. Obligate autotrophs are unable to grow heterotrophically, e. g. cyanobacteria, some species of *Thiobacillus,* and *Nitrosomonas*. On the other hand, most obligate autotrophs can assimilate organic compounds as carbon sources, but not as energy sources. This ability makes growth possible under conditions that are best described as *mixotrophic*. Under these conditions, the energy source is needed only for the generation of ATP; the organic compound provides a source of reduced carbon and, if necessary, reducing power. *Beggiatoa,* a sulfur bacterium, is unable to grow on a completely inorganic medium (reduced sulfur compounds and CO_2), and in order to utilize reduced sulfur compounds, it requires an organic carbon source such as acetate, i. e. it exhibits mixotrophy. Similarly, in the mixotrophic (photoheterotrophic) culture of green algae or euglenoids, the growth medium contains an organic energy and carbon source (e. g. glucose) and the culture is illuminated; the cells become green, and growth is supported at least in part by photosynthesis.

O

Oat coleoptile test, *Avena test:* a biotest for the quantitative determination of auxins, which is carried out as follows:
The tip of the coleoptile is cut off, and the cotyledon within it is removed by pulling on its base. An agar block with the test substance is set on one side of the coleoptile stump. The auxin diffuses into the side of the coleoptile covered by the agar and, due to the one-sided stimulation of growth, causes it to bend. The angle of the bend is a function of the auxin concentration.

Ochnaflavone: see Biflavonoids.

Ochratoxins: mycotoxins produced by *Aspergillus ochraceus* during food spoilage. In rats and mice O. cause pronounced liver damage. O. B is the dechloro derivative of O. A (Fig.). O. C is the ethyl ester of O. A.

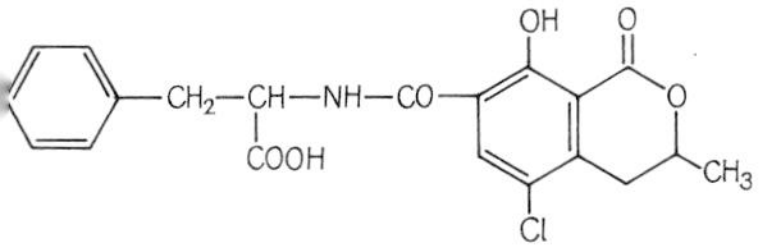

Ochratoxin A

Ochre codon: the UAA sequence in mRNA. Like the amber codon, it signals the end of protein biosynthesis. The synthesized polypeptide chain is released after the incorporation of the amino acid encoded immediately before the O. c. Ochre is probably the natural termination codon, and the one most widely employed by all living systems. See Ochre mutants.

Ochre mutants: bacterial mutants, which, as a result of a point mutation, possess a UAA codon in their mRNA (see Ochre codon). There are specific Suppressors (see) of ochre mutations.

Ocimene: a triply unsaturated monoterpene hydrocarbon. O. is an oily liquid. M_r 136.24, b.p. 176–178 °C, ρ^{15} 0.8031, n_D^{18}1.4857. It is a double bond positional isomer of Myrcene (see), and a component of many essential oils.

D-Octopine: N-α-(1-carboxyethyl)-arginine, N^2-(D-1-carboxyethyl)-L-arginine. M_r 246.3, m.p. 262–263 °C (d.), $[\alpha]_D^{24}+20.6$ ($c=1$, water). It is found in the muscles of certain invertebrates, e.g. *Octopus, Pecten maximus, Sipunculus nudus,* where it serves as a functional analog of lactic acid, i.e. the NADH produced by glycolysis is oxidized to NAD during the synthesis of D-O., and NAD can be reduced to NADH by the reversal of the same process. D-O. is biosynthesized from pyruvate and arginine by a reductive condensation, catalysed by an unspecific NADH-dependent dehydrogenase.
D-O. is also found in certain plant tumors induced by *Agrobacterium tumefaciens.* This bacterium induces tumors in dicotyledenous plants by transferring a large bacterial plasmid (called the T_i plasmid) to the eukaryotic cell. In the transformed tissue, the T_i plasmid determines the synthesis of novel amino acids, which serve as specific substrates for the bacterium. These may be D-O. and related compounds (the "octopine family"), or nopaline and nopalinic acid (the "nopaline family"), but not both.

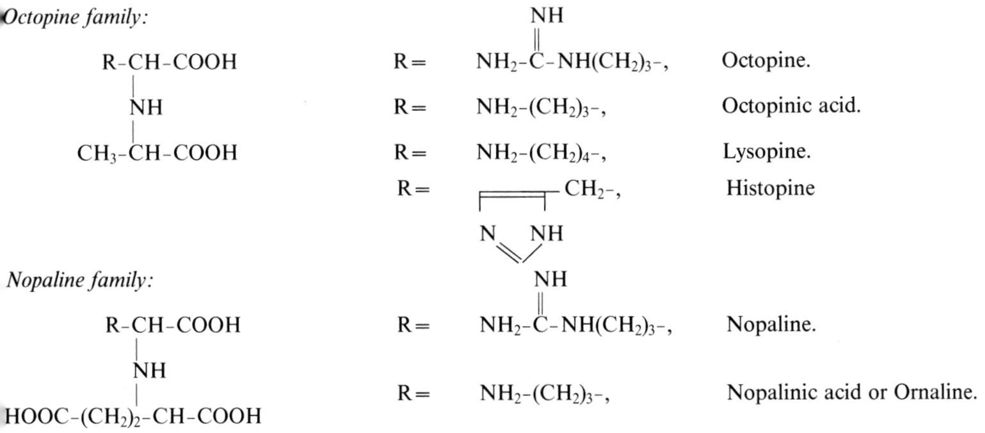

Octopine family:

R–CH–COOH   R = NH_2–$\overset{\displaystyle NH}{\overset{\|}{C}}$–NH(CH$_2$)$_3$–,   Octopine.

|
NH   R = NH_2–(CH$_2$)$_3$–,   Octopinic acid.
|
CH$_3$–CH–COOH   R = NH_2–(CH$_2$)$_4$–,   Lysopine.

            R =   ⌐———⌐–CH$_2$–,   Histopine
                   N   NH

Nopaline family:

R–CH–COOH   R = NH_2–$\overset{\displaystyle NH}{\overset{\|}{C}}$–NH(CH$_2$)$_3$–,   Nopaline.
|
NH
| R = NH_2–(CH$_2$)$_3$–,   Nopalinic acid or Ornaline.
HOOC–(CH$_2$)$_2$–CH–COOH

413

Octopinic acid: see D-Octopine.
Oenin: see Malvidin.
Oestradiol: see Estradiol.
Oestriol: see Estriol.
Oestrogen: see Estrogen.
Oestrone: see Estrone.
Oils: water-insoluble, liquid organic compounds. They are combustible, lighter than water, and soluble in ether, benzene and other organic solvents. Naturally occurring O. may be acylglycerols, e.g. the O. stored in certain seeds, or fish liver oils (see Fats); or they may be nonsaponifiable lipids, e.g. Essential oils (see).

Oiticica oil: the seed fat of *licania rigida*, a tree native to the semiarid areas of northeast Brazil. Licanic acid (see) represents 50-80% of the total esterified fatty acids of O.o. In addition, O.o. contains esterified palmitic, stearic, oleic and eleostearic acids.

Okazaki fragments: see Deoxyribonucleic acid.

Oleandomycin: a Macrolide (see) antibiotic.

5α-Oleane: see Amyrin.

Oleanolic acid: a monounsaturated, pentacyclic Triterpene (see), with a carboxylic acid group. M_r 456.71, m.p. 310 °C, $[\alpha]_D + 80$ ° (methanol). It differs structurally from β-amyrin by the presence of a carboxyl group in place of the 28-methyl group (see Amyrin). O.a. occurs free, esterified with acetic acid, or as the aglycon of triterpene saponins (see Saponins) in many plants, e.g. sugar beet, bilberry, mistletoe, cloves and cacti.

Oleic acid: Δ^9-octadecenoic acid, $CH_3-(CH_2)_7-CH=CH-(CH)_7-COOH$, the most widely distributed naturally occurring unsaturated fatty acid. M_r 282.45, m.p. 13 °C, b.p.$_{10}$ 223 °C. The double bond has the *cis* conformation. The *trans* isomer is called elaidic acid. O.a. is present in practically all the acylglycerols of depot and milk fats, and it is a component of phospholipids.

Oleoresins: see Balsams.

Oligo-1,6-glucosidase: see Dextrin 6-α-D-glucanohydrolase.

Oligonucleotides: linear sequences of up to 20 nucleotides, joined by phosphodiester bonds. Position 3' of each nucleotide unit is linked via a phosphate group to position 5' of the next unit. In the terminal units, the respective 3' and 5' positions may be free (i.e.-OH groups) or phosphorylated. O. are named according to chain length, i.e. di-, tri-, tetra-, pentanucleotides, etc. Linear sequences of more than 20 nucleotide units are called Polynucleotides (see and compare).

Oligopeptides: see Peptides.

Oligosaccharides: see Carbohydrates.

Oligosaccharidosis: see Inborn errors of metabolism.

Oligosaccharins: oligosaccharides of specific composition which act as modulators of cell behavior in plants. They are thought to control growth, development, reproduction and defense. O. are present in enzymatic and acidic digests and heat-treated preparations of plant and fungal cell wall polysaccharides. Such preparations contain small quantities of active material in the presence of large amounts of inactive substances of similar monotonous structure, so that progress in the isolation and characterization of O. has been slow. O. are specific, in contrast to other plant hormones, which are pleiotropic. Hormones, such as auxin and gibberellin, may function by activating enzymes which release O. from cell walls. For example, auxin stimulates growth of pea stems and increases 50-fold an enzyme which releases active material from cell wall xyloglycan. In turn, this material inhibits auxin-stimulated growth. Thus as auxin stimulates growth at the shoot tip, O. may be transported down the stem and inhibit the growth of lateral buds, thereby causing apical dominance.

O. have been shown to inhibit flowering, promote vegetative growth and control organ development in plant tissue culture. Effective concentrations of O. are 100-1000-fold lower than those of other plant hormones, such as auxins, cytokinins and gibberellins.

O. are also thought to be involved in elicitation (see Elicitor) of Phytoallexin (see) synthesis. In elicitation, O. may be released: 1. from the cell wall of the invading organism by host enzymes, 2. from the host cell walls by enzymes of the invading organism, and 3. from host cell walls by enzymes released from the host itself by cell damage.

An hexaglucopyranosylglucitol, released from the cell wall of *Phytophthora megasperma* f.sp. *glycinea* by acid hydrolysis, has been characterized (Fig.) and shown to be highly active (10 ng applied to 1 g of plant tissue is effective) in promoting transcription of mRNA encoding the enzymes of phytoallexin synthesis in *Glycine max*. This is the first oligosaccharide which acts as a regulatory molecule in plants to be completely characterized. It is presumed that similar O. are released enzymatically.

Other O. active in elicitation appear to be linear oligosaccharides of galacturonic acid, derived from host (e.g. *Ricinus*) cell wall by fungal enzymes. An enzyme from the plant pathogen *Erwinia carotovoris* also releases oligogalacturonides from the cell walls of *Glycine max* and other hosts. [A.G. Darvill & P. Albersheim, *Ann. Rev. Plant Physiol.* **35** (1984) 243-275; P. Albersheim & A. G. Darvill *Scientific American* **253** (1985) 58-73; D.A. Smith & S.W. Banks, *Phytochemistry* **25** (1986) 979-995]

β-D-Glcp-(1→6)-β-D-Glcp-(1→6)-β-D-Glcp-(1→6)-β-D-Glcp-(1→6)-Glucitol
```
                    3                              3
                    ↑                              ↑
                    1                              1
                β-D-Glcp                       β-D-Glcp
```

Heptaglucoside from the cell wall of *Phytophthora megasperma*. This oligosaccharide, which stimulates phytoallexin synthesis in *Glycine max*, was isolated from acid digests of the fungal cell wall [J.K. Sharp et al., *J. Biol. Chem.* **259** (1984) 11321-11336]

Ommatins: see Ommochromes.

Ommins: see Ommochromes.

Ommochromes: a class of natural pigments, which contain the phenoxazone ring system. Their colors range from yellow through red to violet. They are especially common in, but not limited to the *Arthropoda,* and were named from their occurrrence in the ommatidia of the insect eye. They are divided into two groups: the low M_r, alkali-labile, dialysable ommatins; and the high M_r, alkali-stable ommins. In the organism they are often bound to proteins as chromoprotein granules.

Crystalline dihydroxanthommatin was first prepared from the meconium (post pupal secretion) of the small tortoiseshell butterfly *(Variessa urticae).* It is found universally in insects as an eye pigment, accompanied in most orders by a greater amount of ommin. In the eyes of many Diptera *(Calliphora erythrocephala, Syrphus pyrastri, Musca domestica, Drosophila melanogaster),* xanthommatin is the only O. Rhodommatin (the *O*-glucoside of dihydroxanthommatin) and ommatin D (the sulfate ester of dihydroxanthommatin) have only been found in the *Lepidoptera;* they are present in the wings of the *Nymphalidae,* where they contribute greatly to pigmentation, but they are absent from the eyes and all other ectodermal structures. They are also present in meconium as their water soluble ammonium salts, and the first isolates were made from this source. The large amounts of xanthommatin found in meconium were probably derived from ommatin D, and it doubtful whether xanthommatin occurs in fresh secretions. Xanthommatin has also been identified in the eggs of the marine worm *Urechis caupo (Echiuridae),* and very small amounts, together with ommin, have been found in some crustaceans.

Ommin has been found in all investigated orders of the *Insecta* and *Crustaceae.* It is an especially common eye pigment in crabs, spiders, insects and cephalopods, and it is also present in the epidermis of *Cragnon, Limulus* and *Gryllus,* but not *Carcinus* and *Portunus.* Ommin is present in the eyes and skin, but not in the ink, of *Sepia officianalis (Cephalopoda).* About 75% of the pigment known as ommin contains a violet-black component called ommin A.

O. are derived biosynthetically from 3-hydroxykynurenine, an intermediate of L-Tryptophan (see) metabolism. The phenoxazine ring system is formed by the oxidative coupling of two molecules of 3-hydroxykynurenine; it is a general reaction of *o*-aminophenols that they can be oxidized, chemically or enzymatically, to phenoxazones. Further cyclization of one side chain produces the quinoline ring system that is also present in the ommatins.

Mutants of many insects are known, in which the capacity for O. synthesis is impaired. Such mutations are usually recognized from the abnormal eye color. Mutation may affect the conversion of L-tryptophan to *N*-formylkynurenine (e.g. *white eye* mutation of *Periplaneta americana), N*-formylkynurenine to kynurenine (e.g. *a* mutation of *Ephistia kühniella),* kynurenine to 3-hydroxykynurenine (e.g. *cinnabar* mutation of *Drosophila melanogaster),* 3-hydroxykynurenine to O. (e.g. *white-2* mutation of *Bombyx mori),* or the synthesis of the protein which binds the O. may be impaired (e.g. *wa* mutation of *Ephestia kühniella).*

Our knowledge of the structure and biochemistry of O. is due largely to the work of Butenandt et al.

OMP: abb. of Orotidine 5'-monophosphate.

Oncogenes: genes which are capable, under certain conditions, of inducing neoplastic transformation of cells. O. were first identified as nucleic acid sequences which are necessary for oncogenicity of certain viruses; strains of the viruses in which these genes were altered or deleted were found incapable of transforming cells in vitro or inducing tumors in vivo. Retrovirus O. (retroviruses are RNA viruses whose genomes are transcribed into DNA in the infected cells. The resulting single-stranded

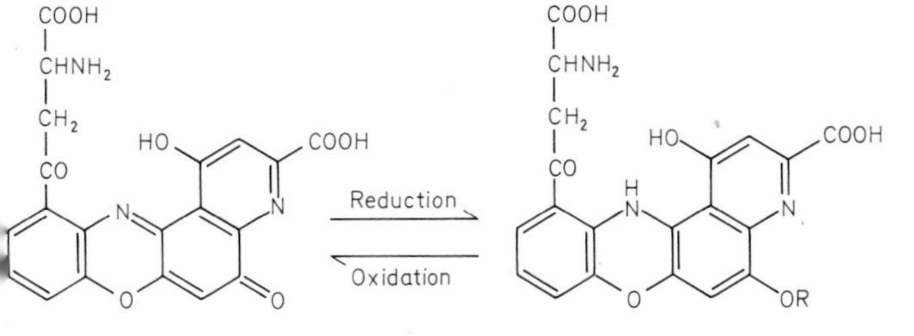

Xanthommatin (yellow-brown)

R = H, Dihydroxanthommatin (red).
R = SO₃H, Ommatin D (red).
R = 1-glucosyl, Rhodommatin (red).

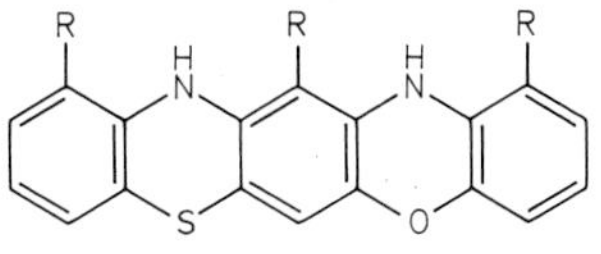

R = CO—CH₂—CH(NH₂)—COOH
Ommin A (violet-black)

DNA is replicated to form double-stranded DNA, which is then integrated into a chromosome, where it is known as a provirus) were soon found to be homologous to cellular genes which have been highly conserved in evolution. The latter are called *proto-oncogenes;* apparently a virus strain can become oncogenic by incorporating one or more proto-oncogenes into its genome. As a rule, the process does not conserve the entire proto-O. Deletions and fusion with viral genes are common, as are point mutations. DNA tumor viruses, such as polyoma, adenovirus and SV40, also contain O., and homologs have not been discovered in the vertebrate genome.

Cellular O. have also been identified in tumors by testing tumor DNA for its ability to transform cells in culture. (It is customary to refer to a viral O. as v-*onc*, and to the corresponding cellular gene as c-*onc*, e.g. v-*myc* and c-*myc*.) A c-*proto-onc* does not cause tumors; therefore, a c-*onc* must have been modified in some way through mutation or a change in the regulatory mechanism controlling its expression, or both. Activation of c-*onc* has been postulated as the mechanism by which chemical carcinogens and radiation induce cancers; another possibility is that the mutagen activates a genomic provirus which had been unexpressed.

Activation of *proto-onc* may occur by mutation, rearrangement of DNA or insertion of a segment of viral DNA into the chromosome adjacent to the c-*onc*. Either of the latter events brings the c-*onc* under the control of a foreign promoter. For example, in Burkitt's lymphoma and human chronic myeloid leukemia, chromosome exchange between non-homologous chromatids is often observed to have occurred. In some B-cell leukemias, the O. has come under the control of the immunoglobulin promoter and enhancer. It is probable that both mutation and change in the level of expression are required in most cases, as transformation in vivo usually appears to be a multistep process.

In some cases, two or more O. must be present in a virus for it to be able to transform cells in vitro (especially in the case of primary cells, i.e. cells taken directly from an animal and cultured briefly in vitro); this is usually true in vivo. Some O. are able, by themselves, to induce transformation of certain cell lines in vitro. The standard test of transforming ability for DNA is transfection of NIH 3T3 cells, which are an immortalized line of mouse cells. Some authors argue that these cells have already taken the first step towards neoplasticity, and that the ability of an O. to transform NIH 3T3 cells in vitro does not prove that it can induce a cancer in vivo in the absence of other oncogenic factors.

The normal function of most proto-oncogenes is unknown. It is believed that elucidation of the normal functions of the proto-oncogene products will suggest the mechanism(s) by which O. can induce cancers. It is thought that proto-O. are essential to cell growth and differentiation; mutations or inappropriate expression of them might thus lead to cellular immortality and/or failure to differentiate.

About 20 O. have been identified. A brief description of the more commonly studied O. is given below. In accordance with convention, the products of these genes are denoted by the letters "p", "gp" or "pp" to indicate "protein", "glycoprotein", and "phosphoprotein", respectively, followed by the approximate M_r (in kilodaltons). A superscript to the right of the symbol indicates the gene encoding the protein, e.g. pp60^{src} is a phosphoprotein of M_r 60000 encoded by the *src* gene. The products of fused genes are indicated by a hyphenated superscript, e.g. gp180$^{gag-fms}$ is a glycoprotein of M_r 180000 which is the product of a fused viral *gag* gene and the proto-oncogene *fms*. In the following list, the virus from which the v-*onc* was first isolated and the species infected by it are given in parentheses after the name of the gene.

src (Rous sarcoma virus; infects chickens). The protein product, pp60^{src}, is a Protein tyrosine kinase (see). The 19 C-terminal amino acids of pp60^{c-src} are replaced by 12 new ones in the viral gene; most of these are encoded by a sequence about 1 kb downstream from c-*src*. The 3' end of c-*src* inhibits transcription of the gene. Other O. which encode Protein tyrosine kinases are listed under that entry. pp60 is a myristylated protein, which is anchored in the plasma membrane by its myristic acid residue. Loss of myristylation by mutation also results in loss of oncogenicity.

mos (Moloney sarcoma virus; mice) shows homology to *src*, but its product appears not to be a tyrosine kinase. c-*mos* does not initiate transformation; it differs from v-*mos* at several points and its transcription is also inhibited by upstream sequences. v-*mos* requires viral LTR (long terminal repeat) sequences for transforming activity; it is present at 1–10 copies per cell, while expression of c-*mos* has not been detected.

fms (McDonough feline sarcoma virus) encodes a glycoprotein which is related or identical to the receptor for colony stimulating factor. [R.L. Mitchell et al. *Cell* **45** (1986) 497–504] It has tyrosine kinase activity in vitro, but this has not been demonstrated in vivo. The protein co-purifies with intermediate filament proteins, and its location in the cell (by double immunofluorescence) corresponds closely to that of vimentin and keratin, but not actin. c-*fms* is expressed in detectable amounts in the extra-embryonic tissues of mice, but not in the fetus or adult mouse.

v-yes (avian sarcoma virus Y73), *v-fps* (Fujinami sarcoma virus), *v-fgr* (Rasheed feline sarcoma virus), *v-abl* (Abelson murine leukemia virus) and *v-ros* (UR2 avian sarcoma virus) show a certain degree of homology with one another. All encode tyrosine kinases, and there is a cellular homolog of each. Each is capable of transforming cells (in vitro) independently. In the case of v-*fps* and the homologous v-*fes* from feline sarcoma virus, it has been shown that the tyrosine kinase activity of their products is necessary for the initiation of transformation by the viruses carrying them. The p130 product of v-*fps* is not an integral membrane protein, but appears to be linked to the membrane or cytoskeleton. Infection of chicken macrophage progenitors with retroviruses carrying v-*fps* enables them to differentiate in the absence of macrophage colony stimulating factor in the medium.

myc (avian myelocytomatosis virus). Other avian tumor viruses which contain v-*myc* cause leukemia, carcinomas and sarcomas. These have in common a 1.6 kbase sequence at the 3′ end of the gene which is shared with the c-*myc*. However, the 5′ ends of the viral and cellular genes vary. In the human genome, c-*myc* is normally on chromosome 8, near the breakpoint at which part of chromosome 8 is transferred to chromosome 14 in Burkitt's lymphoma. c-*myc* is expressed in normal avian and mammalian cells, and the gene product is located in the nucleus. Excessive expression in transgenic mice does not prevent normal development but does make the animals highly susceptible to a number of different kinds of tumor. [A. Leder et al., *Cell* **45** (1986) 485–495]

myb (avian myeloblastosis virus, which causes chicken leukemia). The viral gene encodes a protein of M_r 48 000 (located in the nucleus), while the cellular proto-oncogene encodes a p110 and is expressed in hematopoietic cells. *myb* is distantly related to *myc* and the adenovirus oncogene E1A, with homologies of 15 to 20% among the proteins.

v-mht (avian retrovirus MH2, which contains both this O. and *myc*).

erb (avian erythroblastosis (acute leukemia) virus) encodes two proteins, A and B. p75$^{erb\,A}$ is found in transformed cells; the N-terminal 1/3 of this protein is the product of the viral *gag* gene. *erb B* is a glycoprotein found in membranes; it is nearly identical to the tyrosine kinase part of the epidermal growth factor receptor molecule. However, it has no kinase activity in vitro.

v-sis (simian sarcoma virus); the protein it encodes is structurally related to the β-chain of platelet-derived growth factor. c-*cis* is located on chromosome 22 of the human genome; it is transferred to chromosome 9 in the Philadelphia translocation associated with human myelogenous leukemia.

rel (avian reticuloendotheliosis virus) c-*rel* is expressed in normal chicken cells; there are a number of mutations in v-*rel* with respect to c-*rel*.

fos (murine sarcoma virus). The N-terminal 332 amino acid residues of p$^{v\text{-}fos}$ (located in the nucleus) are identical to the corresponding part of the c-*fos* protein; however, there is a deletion in the viral gene which shifts its reading frame and causes the C-terminal 49 amino acids to be entirely different from those of the proto-oncogene. c-*fos* is expressed in the placenta and extra-embryonic membranes of fe-

tal mice. It is also expressed in neonatal mouse bones, muscles, skin and connective tissues. It is transiently expressed in resting cells which have been challenged with mitogen.

There are three forms of the *ras* gene:
Ha-*ras* (Harvey murine sarcoma virus),
Ki-*ras* (Kirsten murine sarcoma virus),
N-*ras* from some tumor cell DNA. The human genome contains cellular versions of all three genes, and the transforming genes isolated from human tumors are often derived from the *ras* family. The proteins encoded by the genes consist of 189 amino acid residues, and are called "p21^{ras}". p21$^{N\text{-}ras}$ binds GDP and phosphorylates itself; membrane proteins with these properties are often involved in the transduction of cellular signals. Injection of either c-Ha-*ras* or v-Ha-*ras* proteins into fibroblasts in vitro was found to induce membrane ruffling and pinocytosis; however, the effects of the cellular gene product were much shorter-lived than those of the oncogene product. The effect depends on calcium and apparently involves the Inositol phosphate (see) signal pathway. [D. Bar-Sagi & J. R. Feramisco, *Science* **233** (1986) 1061–1068; G. F. Vande Woude et al., eds., *Oncogenes and Viral Genes*, vol. 2 of *Cancer Cells* (Cold Spring Harbor Laboratory, 1984)]

Oncovin: see Vincristin.

One-carbon cycle: a cycle of methyl transfer and methyl oxidation involving glycine, sarcosine, dimethylglycine, betaine and choline, first proposed in 1958 from observations on the oxidation of dimethylglycine and sarcosine by rat liver mitochondria [Mackenzie & Frisell, *J. Biol. Chem.* **232** (1958) 417–427]. The existence of a one-carbon cycle is now confirmed; it involves choline oxidation in mitochondria and phosphatidylcholine synthesis in the endoplasmic reticulum (Fig.). [A. J. Wittwer & C. Wagner, *J. Biol. Chem.* **256** (1981) 4102–4108, 4109–4115]

Dehydrogenation of sarcosine and dimethylglycine is closely linked to the conversion of glycine to serine via the cycle of $N^{5,10}$-CH$_2$-THF utilization and regeneration. When isolated from rat liver, sarcosine and dimethylglycine dehydrogenases possess a tightly (but noncovalently) bound THF derivative with a chain of 5 glutamate residues. This THF(Glu)$_5$ may replace THF in the reactions shown below. In the absence of THF, oxidation of dimethylglycine and sarcosine still occurs, but the products are glycine and formaldehyde, possibly formed by hydrolysis of enzyme-bound oxidation products:

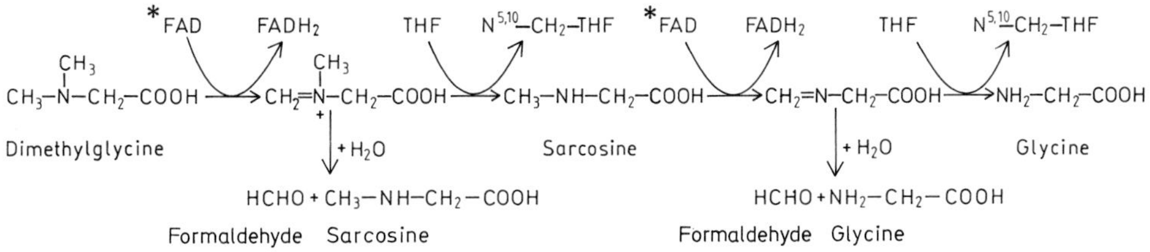

* The FAD of sarcosine and dimethylglycine dehydrogenases is covalently bound (position 8α of the isoalloxazine ring is linked to the imidazole (N3) of a histidine residue) [R. J. Cook et al., *J. Biol. Chem.* **260** (1985) 12998–13002].

One-carbon cycle

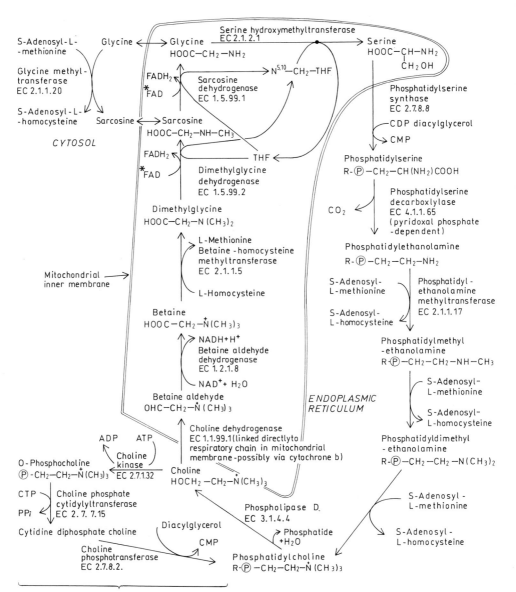

One carbon cycle in rat liver. THF, tetrahydrofolic acid (see Active one-carbon units). R, diacylglycerol.

Salvage pathway of choline utilization (reutilization of choline from phosphatidylcholine degradation, and of dietory choline)

Other products are CO_2 and formate, resulting from the breakdown of THF derivatives which fail to contribute to the conversion of glycine to serine:

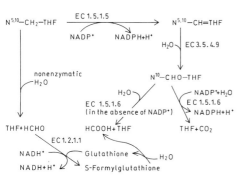

Enzymes:
EC 1.2.1.1, Formaldehyde dehydrogenase (see); EC 1.5.1.5, Methylene-THF dehydrogenase (NADP$^+$); EC 1.5.1.6, Formyl-THF dehydrogenase; EC 3.5.4.9, Methylene-THF cyclohydrolase.

Onic acids: see Aldonic acids.

Ononin: formononetin 7-glucoside, see Isoflavone.

Open system: 1. a system in dynamic equilibrium (see Steady state) with its surroundings, i.e. there is a continual exchange of material, energy and information with the environment. Application of the theory of O.s. to living systems (by Bertalanffy) involves the thermodynamics of irreversible processes. 2. In biology, plants are considered to be O.s., whereas animals are closed systems. The plant is theoretically capable of unlimited growth; certain cells remain embryonal and able to divide and differentiate, so that growth occurs from vegetative sites, such as the meristematic regions of the shoot and root tips, intercalary meristems, etc. In the animal, however, differentiation is essentially complete after the conclusion of embryonal development.

Operator: see Operon.

Operon: a group of neighboring genes, which represent a functional unit. An operon contains,
1. structural genes (S_1 to S_4 in the Fig.), which code for the primary structures of enzyme proteins catalysing successive steps in a metabolic pathway, e.g. the enzymes in the biosyntheisis of an amino acid. The primary transcription product of this group of genes is a polycistronic mRNA; therefore, during the control of transcription, all the structural genes are affected equally.
2. The promotor (P in the Fig.) is the starting point of transcription. This section of the DNA is "recognized" by RNA polymerase with the aid of the sigma factor. The affinity of the promotor for RNA polymerase (apparently determined by the structure of the promotor) is one of the factors that regulate the transcriptional frequency of the operon. When the RNA polymerase has bound to the promotor, it

must then pass through the operator region in order to reach the structural genes.

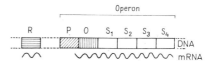

Schematic representation of an operon

3. The operator (O in the Fig.) is the control gene for the function (i.e. the transcription) of the structural genes. It is able to bind a repressor protein, which is the product of the regulator gene (R in the Fig.). R. is not a part of the operon, and it is located in a different region of the chromosome. If the specific repressor protein binds to the operator, transcription of the structural genes is blocked, i.e. the RNA polymerase cannot pass to the structural genes. If the operator is unoccupied, transcription of the structural genes can proceed. The details of these control mechanisms are described under Enzyme induction (see) and Enzyme repression (see).

The nucleotide sequence of the lactose operator was elucidated in 1973 by Gilbert and Maxam; this operator is double stranded and consists of the following base pairs:

5′ T G G A A T T G T G A G C G G A T A A C A A T T 3′
3′ A C C T T A A C A C T C G C C T A T T G T T A A 5′

The operon model was developed by Jacob and Monod, and it has only been demonstrated in prokaryotic systems. It is the basis for the explanation of Enzyme induction (see) and Enzyme repression (see). See also Attenuation.

Ophiobolanes: sesterterpene phytotoxins present in various species of *Helminthosporum* and *Cochliobolus*. The first representative isolated was ophiobolin A (ophiobolin, cochliobobin, cochliobobin A), thought to be responsible for some of the symptoms of corn blight caused by *Helminthosporum maydis*. Biosynthesis is by head-to-tail linkage of 5 isoprene units, forming all-*trans*-geranylfarnesylpyrophosphate (Fig.). It is not clear whether isomerization (to achieve the observed stereochemistry of the O.) occurs before or after cyclization of the geranylfarnesylpyrophosphate; the stereochemistry of the 7,8 double bond of ophiobolin F (Fig.) is uncertain. The bioconversions and incorporation of labels shown in the figure were studied in fungal cultures. Other ophiobolins have been isolated, differing chiefly with respect to the presence or absence of oxygen at C-14 and the degree of oxidation of C-21. [A. Stoessel in *Toxins in Plant Disease*, R. D. Durbin (ed.) (Academic Press, 1981) pp. 178–181]

Opioid peptides: a group of peptides which have been found in the brain, hypophysis, adrenal medulla, intestines, blood and urine. They act as opiates, causing, for example, a specific reduction in the contraction threshhold of the mouse vas deferens and the guinea pig ileum. The *Enkaphalins* (see) are pentapeptides first isolated from pig brain.

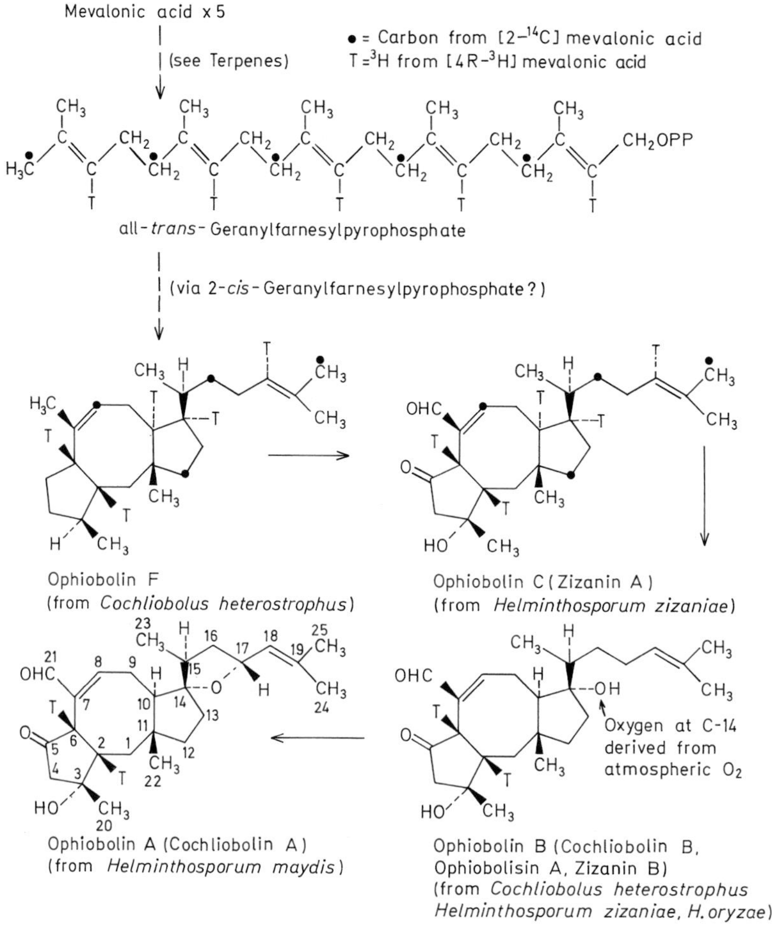

Opioid peptides

Biosynthesis of ophiobolanes

There are two of these, Met-enkephalin (Tyr-Gly-Gly-Phe-Met) and Leu-enkephalin, (Tyr-Gly-Gly-Phe-Leu). The term *"endorphin"* was coined to include all O.P., as it is a contraction of "endogenous morphine". However, it is often used to refer specifically to several larger peptides: β-endorphin, α-neo-endorphin, dynorphin, adrenorphin, etc. Each of these peptides contains the Met- or Leu-enkephalin sequence at its amino terminus, so it is thought that the pentapeptide is responsible for the specific effect of the O.P., while the remaining amino acids give the "address" at which the peptide is most effective.

To date, three precursor proteins of O.P. have been identified and sequenced. The first, *preproopiomelanocortin*, is the source of corticotropin (ACTH) and β-lipotropin (see Hormones). These hormones are lysed to smaller peptides; β-endorphin is released from β-lipotropin. The second and third precursors are preproenkephalins A and B (PPE A and PPE B). PPE A contains 4 copies of Met-enkephalin, 1 of Leu-enkephalin, and two extended Met-en-

kephalins with six and seven amino acids, respectively. It is also the source of the first O.P. amidated at the carboxy terminus, adrenorphin, and other active peptides. PPE B is the precursor of α-neo-endorphin, dynorphin, and PH-8P (a predominant O.P. in brain).

In general, the points at which the precursor proteins are hydrolysed to release O.P. are marked by pairs of basic amino acids, such as Arg-Arg, Arg-Lys, etc. In the case of adrenorphin, the C-terminal NH₂ is donated by a Gly residue immediately preceding the signal. In this case, however, and in that of PH-8P, the signal is Arg-Pro. [*Nature* **278**, 423–427 (1979); *ibid.* **295**, 202–206, 206–208 & 663–666 (1982), *ibid.* **298**, 245–249 (1982), *ibid.* **305** 721–723 (1983).] The opiate antagonist naloxon(e) relieves the O.P.-induced inhibition of contraction, suggesting a specific interaction between the O.P. and the opiate receptors. Unfortunately, the early hopes that O.P. might be used as non-addictive analgesics have not been fulfilled. The physiological function of the O.P. is not known. The enkephalins may be neurotransmitters, and the longer-chained O. P. could be neurohormones. It is of interest that large amounts of enkephalins are stored with adrenalin in the chromaffin cells of the adrenal medulla, and they are released together with adrenalin during stress. [*Science* **221** (1983) 957-960.] Opioid peptides are implicated as mediators of the suppressive effect of stress on natural killer cell cytotoxicity. [Y. Shavit et al. *Science* **223** (1984) 188-190.] Dynorphin-A-(1-8) is contained in the same neurosecretory vesicles in the hypophysis as vasopressin and neurophysin, and may therefore be involved in the antidiuretic response. [M. Whitnall et al., *Science* **222** (1983) 1137-1139] They may also be modulators of behavior, and a connection with acupuncture analgesia has been suggested. They may even be involved in the pathogenesis of schizophrenia, hallucinations, etc.

The endorphin and the opiate receptor system are also apparently involved in growth and development. Human infants and laboratory animals exposed to opiates like heroin and methadone are retarded in somatic and neurobiological growth and development. Conversely, newborn rats continuously maintained on the opiate antagonist naltrexone gained weight and passed developmental milestones earlier than litter-mate controls. [*Science* **221** (1983) 1179-1180;. *Science* **222** (1983) 1246-1248.]

Opium: the congealed or dried latex of *Papaver somniferum*. The chief active constituents of O. are the Opium alkaloids (see). O. is smoked, or preparations from it may be ingested or injected. Uncontrolled use leads to addiction and deterioration of personality. The action of O. is essentially similar to, but weaker than that of its constituent Morphine (see).

Opium alkaloids: the Papaveraceae alkaloids (see) that occur in Opium (see). There are about 40 different O. a.; the chief representatives are Morphine (see), Codeine (see) and Papaverine (see). More than 1000 tons of morphine are produced annually. Nowadays the straw of the poppy plant is also used for the isolation of O. a. This new source accounts for about on third of the annual production of morphine.

Opsin: see Vitamins (Vitamin A).

Opsonin: the name given by Wright and Douglas in 1903 to the thermostable material present in serum, which stimulates phagocytosis of bacteria. It was shown to act directly on bacteria and not on the phagocytes. It is probably identical to C3b of the complement system (see Opsonization), although other complement components may also be active to a lesser extent. In addition to its role in the opsonization of immune complexes, C3b can bind to various structures, such as foreign erythrocytes and bacterial cells, and render them more readily phagocytosed by immune adherence to phagocytes.

Opsonization: the ability of serum to render an immune complex more readily phagocytosed. O. is a property of the Complement system (see). Although immune complexes are subject to phagocytosis, interaction with complement greatly increases the rate. Component C3b (activated C3 of the complement system) has a labile binding site(s), which permits it to bind tightly to antigen-antibody complexes (op-

sonic adherence). Other stable binding sites on C3b enable this opsonized immune complex to bind to polymorphonuclear leucoytes, monocytes and macrophages (immune adherence); this binding results in enhanced phagocytosis. The phagocytosis can be abolished by metabolic inhibitors, whereas the immune and opsonic adherence (due to the physical chemical interaction of receptor and binding sites) are unaffected.

Optical density: see Absorbance.

Optical test: a method introduced in 1936 by Otto Warburg for the determination of the enzymatic activity of NAD and NADP-dependent dehydrogenases. Absorbance at 340 nm (or another suitable wavelength in that region) is measured as an index of the degree of reduction of NAD or NADP (see Nicotinamide-adenine-dinucleotide). The principle of this method is now widely used for measuring enzyme activities and the concentration of metabolites. A coupled enzyme system may be used, in which the reaction of interest does not produce or consume NAD(P)H, but is linked to one that does, e.g. the concentration of glucose can be determined from the increase in absorbance at 340 nm, when the unknown glucose concentration is incubated in the presence of excess ATP and $NADP^+$ and the enzymes hexokinase and glucose 6-phosphate dehydrogenase:

Glucose + ATP → Glucose 6-phosphate + ADP.
Glucose 6-phosphate + $NADP^+$ → 6-Phosphogluconate + NADPH + H^+.

Determinations of dehydrogenase activities by the O. t. are now conveniently performed in an automatic recording spectrophotometer, so that the change in absorbance at 340 nm is monitored continuously with time, e.g. the activity of malate dehydrogenase is measured from the rate of increase of absorbance at 340 nm due to the production of NADH in the reaction: NAD^+ + Malate → Oxaloacetate + NADH + H^+.

Orchinol: 9,10-dihydro-2,4-dimethoxy-7-phenanthrol, a phytoalexin. O. is formed by the orchid *Orchis militaris* as a defense against infection by the fungus *Rhizoctonia repens*. The structurally related hircinol (9,10-dihydro-4-methoxyphenanthrene-2,5-diol) is formed by the orchid *Himantoglossum (Loroglossum) hircinum*, against infection by Rhizoctonia. Hircinol and O. are biosynthesized via phenylalanine and *m*-hydroxyphenylpropionic acid by a so-called "m-coumaric acid pathway" (Fig.). Hircinol has also been found in the peel of *Dioscorea rotondata*, together with related compounds, the batatasins. The latter have been identified as the dormancy-inducing principle in bulbils of the yam, *Dioscorea batatas*. Batatasins, with hircinol, may also protect against fungal attack, since they inhibit the growth of *Cladosporium cladosporioides* and certain yam soft root pathogens. Several batatasins are dihydrostilbenes (e.g. batatasin III), which share a common precursor dihydrostilbene with hircinol and O. (Fig.). Batatasin I, a phenanthrene, is present in tubers of many species of *Dioscorea*, and has also been found in *Tamus communis (Dioscoreaceae)*. See Stilbenes. [K. H. Fritzmeier & H. Kindl *Eur. J. Biochem.* **133** (1983) 545-550; K. H. Fritzmeier et al. *Z. Naturforsch.* **39c** (1984) 217-221].

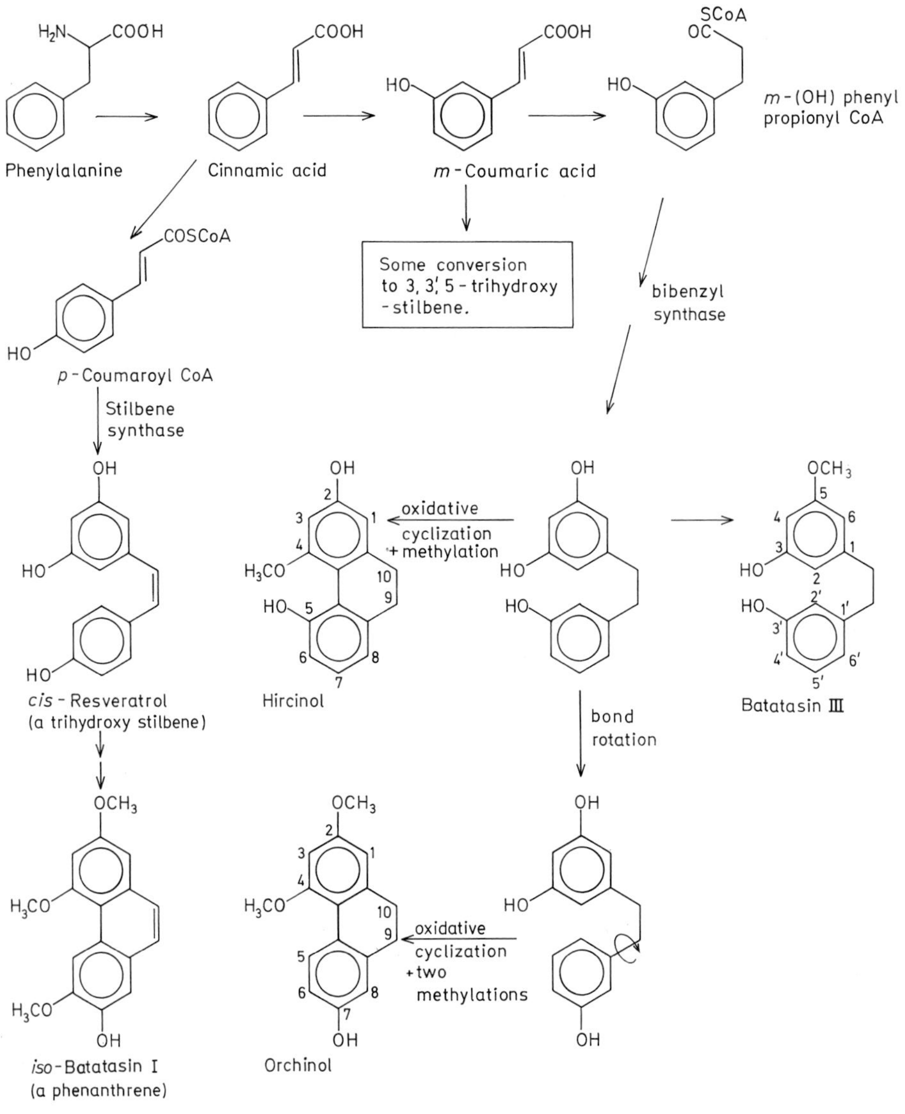

Biosynthesis of orchinol, hircinol and batatasins

ORD: abb. of Optical Rotatory Dispersion.

Ord: abb. of Orotidine.

Organotrophy: see Nutritional physiology of microorganisms.

Orn: abb. of L-Ornithine.

Ornaline: see D-Octopine.

L-Ornithine, abb. **Orn:** α,δ-diaminovaleric acid, 2,5-diaminopentanoic acid, a nonproteogenic amino acid. In mammals, Orn is an intermediate in the Urea cycle (see), and it is the precursor of the Polyamines (see). It is also an intermediate in the biosynthesis of Arginine (see) in all arginine-synthesizing organisms. In a few microorganisms, Orn can be formed from citrulline by the action of citrulline phosphorylase. Various antibiotics contain Orn, e.g. the peptide gramicidin S.

2-N-Acetyl-L-ornithine is an intermediate in the biosynthesis of arginine from glutamic acid in microorganisms, and it is an important allosteric effector of carbamoyl phosphate synthetase. 5-N-acetyl-L-ornithine, which is found in various plants, is a nonproteogenic amino acid, and a structural analog of citrulline.

Ornithine cycle: see Urea cycle.

Ornithinemia: see Inborn errors of metabolism.

Oro: abb. of Orotic acid.

Orobol: 3′,4′,5,7-tetrahydroxyisoflavone, see Isoflavone.

Orosomucoid, *α₁-seromucoid, α₁-acid glycoprotein,* abb. *$\alpha_1 AGp$:* a plasma protein of mammals and birds. O. contains about 38% carbohydrate, and is the most carbohydrate rich and water soluble of all the plasma proteins. The seromucoid fraction of blood consists of O. together with two other carbohydrate rich proteins (C1-inactivator and hemopexin). Increased blood levels of O. and of the total seromucoid fraction are associated with inflammation, pregnancy and various disease states, such as cancer, pneumonia, rheumatoid arthritis. After major surgery, increased levels of O. are produced until the wound is healed. O. also binds certain steroids, especially progesterone, but the affinity of binding is lower than that for corticosteroid binding globulin. Membranes of human platelets contain considerable quantities of tightly bound O. Human O. is a single chain glycoprotein (M_r 39000 by sedimentation equilibrium, 41600 by sedimentation diffusion; isoionic pH 3.5; O. has pronounced anion binding capacity so that the isoelectric point depends on the type and concentration of anions). The protein chain of O. contains 181 amino acid residues (primary sequence: Schmid, K. et al. *Biochemistry* (1973) *12*, 2711–2724). Five oligosaccharide units are attached to the β-carboxyl groups of asparaginyl residues at positions 15, 38, 54, 75 and 85. The native protein contains two disulfide bridges, located between cysteine residues at 5 and 147, and at 164 and 72. The amino terminus is pyroglutaminyl, and the carboxyl terminus is serine. The *C*-terminal half of O. shows great similarities of sequence with the α-chain of haptoglobin and the H-chain of immunoglobulin G. Analysis of the carbohydrate of O. [Schwarzmann, O.H.G. et al. *J. Biol. Chem.* (1978) *253*, 6983–6987] shows 14 residues of *N*-acetylneuraminic acid (sialic acid), about 34 residues of neutral hexoses (D-galactose/D-mannose in an average ratio of 1.4); 31 residues of *N*-acetylglucosamine; 2 residues of L-fucose. *N*-Acetylneuraminic acid is always located terminally via a 2-ketosidic bond; the galactose occupies a penultimate position, linked β-glycosidically to the third sugar, *N*-acetylglucosamine. O. is synthesized in the liver. Synthesis of the carbohydrate moiety is initated by transfer of an *N*-acetylglucosaminyl residue to an asparaginyl residue when the nascent polypeptide chain is still attached to the ribosome (see Post translational modification of proteins).

Orotic acid, abb. *Oro:* uracil 4-carboxylic acid. M_r 156.1, m.p. 344–347 °C (d.). O.a. is an intermediate in Pyrimidine biosynthesis (see), and it accumulates in large quantities during growth of mutants of *Neurospora crassa,* which require uridine, cytidine or uracil.

Orotic aciduria: see Inborn errors of metabolism.

Orotidine, abb. *Ord:* orotic acid-3-β-D-ribofuranoside, a β-glycosidic Nucleoside (see) of D-ribose and the pyrimidine, orotic acid. M_r 288.21, m.p.> 400 °C. Cyclohexylamine salt: m.p. 184 °C, $[\alpha]_C^{23}$ +14.3 ° (*c*=1, water). O. does not appear to be a normal intermediate of pyrimidine metabolism, but it is produced by mutants of *Neurospora crassa.*

Orotidine 5′-monophosphate, abb. *OMP:* a nucleotide of orotic acid. M_r 368.2. OMP is an intermediate in Pyrimidine biosynthesis (see). Orotidine 5′-phosphate pyrophosphorylase catalyses the synthesis of OMP from orotic acid and 5-phosphoribosyl l-pyrophosphate.

Orthonil, *PRB 8:* 2-(β-chloro-β-cyanoethyl)-6-chlorotoluene, a synthetic, stimulatory growth regulator. It is claimed that O. causes an increase in the sugar content of sugar beet.

Orthonil

Orycenin: see Glutelins.

Osajin: see Isoflavone.

Osmosis: a phenomenon associated with semipermeable membranes, especially Biomembranes (see). When two solutions are separated by a membrane which is permeable only to selected components of the solution (e.g. water), the component which can cross the membrane will flow from the side on which its partial pressure is higher to the side on which its partial pressure is lower. Cytoplasm is a concentrated solution of salts, sugars and other small molecules, most of which cannot diffuse across the plasma membrane, in water, which can cross the membrane. As a result, the cells of freshwater organisms, and of plant root hairs, experience an inflow of water until the pressure within the cell increases to the point where the partial pressure of water is equal inside and outside the cell. Fresh-water unicellular animals expel the excess water by means of contractile vacuoles; multicellular animals excrete it through their kidneys. Plant cells are surrounded by rigid Cell walls (see), which enable them to withstand the high internal pressure caused by O. without bursting. Higher land plants utilize the pressure gradient created by O. to help drive their sap up from the roots. Salt-water organisms live in an environment in which the salt concentration outside the plasma membranes is higher than inside. O. thus tends to dehydrate the cell rather than to cause it to burst, and the organism must expend metabolic energy to excrete salt in order to maintain the lower internal concentration.

Osteocalcin: A small protein (M_r 5800) which comprises 10–20% of bone protein. O. from a variety of vertebrates has γ-carboxyglutamate (Gla) residues at positions 17, 21 and 24, and a disulfide bond between cysteine residues at positions 23 and 29. In the presence of Ca^{2+}, the protein undergoes a conformational change so that all the Gla residues are on the same side of an α-helix, and 5.4 Å apart. This distance is close to the spacing between adjacent Ca atoms in hydroxyapatite crystals, which make up the mineral part of bone. Thus O. is presumed to have a role in mineralization of bone. It is synthesized in bone; small amounts are also found in blood.

Ostreasterol: see Chalinasterol.

Ouabagenin: see Strophanthins.

Ouabain: see Strophanthins.

Ouchterlony technique: see Precipitation.

Ovalbumin: see Albumins.

Ovosiston: an oral contraceptive (see Ovulation inhibitors), consisting of a mixture of the progestin, Chlormadinone acetate (see), and the estrogen, Mestranol (see).

Ovulation inhibitors: a group of steroids, which inhibit ovulation by feedback inhibition of the production of Luteinizing hormone (see) and/or Follicular stimulating hormone (see). They are used extensively as oral contraceptives (otherwise known as "the pill"). Most commercially available O.i. preparations contain a combination of a progestin and an estrogen. The progestin is the actual O.i., whereas the estrogen is included to prevent breakthrough (midcycle) bleeding. Combination O.i. consist of a progestin-estrogen combination, which is taken daily throughout the cycle. A sequential O.i. was introduced in 1965 to mimic more closely the normal rise and fall of estrogen and progestin in a woman's monthly cycle; the first 15 pills contain estrogen only, and the last 5 contain a progestin-estrogen combination. Since these were less effective than the combination products, and showed undesirable side effects, they were removed from the U.S. market in 1976. The "minipill" contains a low dose of progestin (0.075 mg norgestrel, or 0.35 mg norethindone) and no estrogen; it is free from many side effects, but users sometimes have irregular bleeding. Trials are also being conducted on the use of once-a-month and once-a-week O.i., and on injectable depot O.i. Since the natural progestin and estrogens show poor gut absorption, all the progestins and estrogens used in oral O.i. are synthetic products, which show efficient gut absorption. Synthetic progestins used as O.i. are norethindone, norgestrel, norethynodrel, ethynodiol diacetate, dimethisterone. Synthetic estrogens used in O.i. preparations are Ethinylestradiol (see), Mestranol (see), ethinylestradiol 3-cyclopentyl ether (quingestanol). The quantities used are 0.5-5 mg progestin, and 0.05-0.1 mg estrogen per tablet. There are several different commercial oral contraceptives, containing various combinations of the above synthetic progestins and estrogens. See also Chlormadinone acetate. O.i. are also used for the treatment of dysmenorrea, endometrioses, cycle-dependent migraines and sterility.

Oxalic acid: HOOC-COOH. O.a. occurs widely in plants as its calcium, magnesium and potassium salts. By forming insoluble calcium salts in the intestine, O.a. hinders the absorption of calcium. O.a. in large amounts is poisonous. Animals cannot metabolize O.a. The normal daily O.a. excretion in the human is 10-30 mg; higher levels may lead to kidney damage (formation of oxalate kidney stones).

The O.a. content of fruits and vegetables is usually less than 10 mg/100 g fresh weight; however, rhubarb leaves contain 700 mg, rhubarb stems 300 mg, celery stems up to 600 mg, and spinach 60-200 mg/100 g fresh weight.

O.a. was first prepared in 1773 from wood sorrel *(Oxalis acetosella)*. It is synthesized in plants by the oxidation of excess glyoxylate. It is also produced by the oxidative cleavage of phenylpyruvate to benzaldehyde and O.a. (a reaction in the conversion of phenylalanine to hippuric acid).

In the bacterium *Pseudomonas oxalaticus*, O.a. is metabolized (via oxalyl-CoA) by an oxidative and a

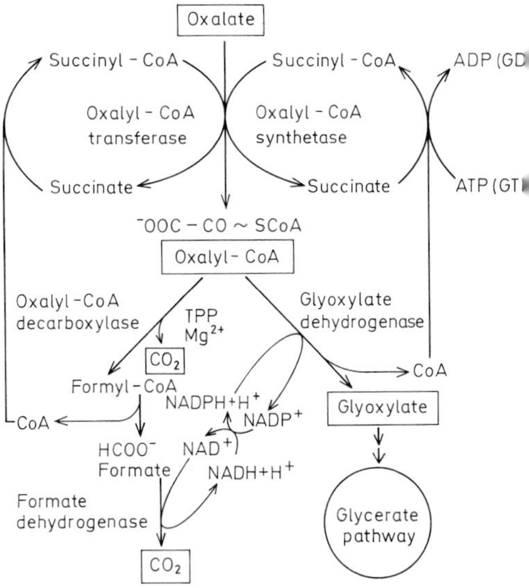

Figure 1. Oxalate metabolism in Pseudomonas oxalaticus.

reductive pathway, which are coupled and operate simultaneously (Fig.1); the balance is $2\ ^-OOC\text{-}COO^- \rightarrow 2\ CO_2 + CHO\text{-}COO^- + H_2O$.

Oxaloacetic acid: $HOOC\text{-}CO\text{-}CH_2\text{-}COOH$, an oxodicarboxylic acid present in fairly high concentrations in plants, e.g. red clover, peas. The anion (oxaloacetate) is an intermediate in the Tricarboxylic acid cycle (see), and it is the oxoacid derived from Aspartic acid (see) by Transamination (see). The enol from of O.a. may be *cis* or *trans*: hydroxymaleic acid (*cis*, m.p. 152 °C), and hydroxyfumaric acid (*trans*, m.p. 184 °C [d.]).

Oxalosis: see Inborn errors of metabolism.

Oxalosuccinic acid: a β-ketotricarboxylic acid (2-oxotricarboxylic acid):

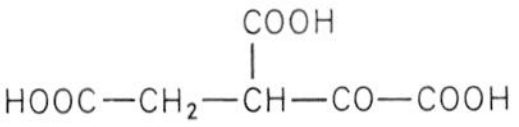

$$HOOC\text{---}CH_2\text{---}\underset{\underset{COOH}{|}}{CH}\text{---}CO\text{---}COOH$$

The anion (oxalosuccinate) is an intermediate in the isocitrate dehydrogenase reaction of the Tricarboxylic acid cycle (see). Isocitrate dehydrogenase catalyses both the oxidation of isocitrate to oxalosuccinate, and its decarboxylation to 2-oxoglutarate.

Oxidases: oxidoreductases (see Enzymes), which use molecular oxygen as an electron acceptor. See also Flavin enzymes.

Oxidation: the loss of electrons. Classically, O. was defined as combination with oxygen or removal of hydrogen. The electrons are transferred to the oxidizing agent, which becomes reduced. Therefore O. is always coupled to Reduction (see), so that any O. or reduction is part of an *oxidoreduction* process. In metabolism there are different mechanisms of enzyme catalysed O., i.e. dehydrogenation, electron transfer, introduction of oxygen, and hydroxylation (Table).

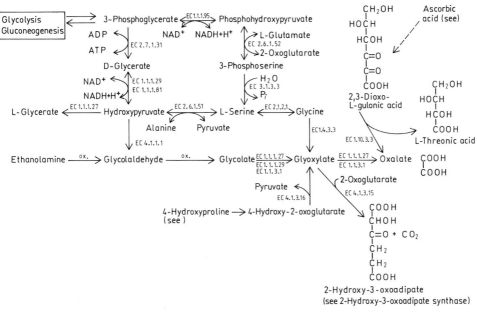

Figure 2. **Metabolic sources of oxalate.** 2-Hydroxy-3-oxoadipate is a normal excretory product in humans which diverts glyoxylate from oxalate production (see Inborn errors of metabolism, Oxalosis). EC 1.1.1.27, L-Lactate dehydrogenase; EC 1.1.1.29, Glycerate dehydrogenase; EC 1.1.1.81, Hydroxypyruvate reductase; EC 1.1.1.95, Phosphoglycerate dehydrogenase; EC 1.1.3.1, Glycolate oxidase; EC 1.4.3.3, D-Amino acid oxidase; EC 1.10.3.3, Ascorbate oxidase; EC 2.1.2.1, Glycine hydroxymethyltransferase; EC 2.6.1.51, Serine-pyruvate aminotransferase; EC 2.6.1.52, Phosphoserine-oxoglutarate aminotransferase; EC 2.7.1.31, Glycerate kinase; EC 3.1.3.3, Phosphoserine phosphatase; EC 4.1.1.1, Pyruvate decarboxylase; EC 4.1.3.15, 2-Hydroxy-3-oxoadipate synthase; EC 4.1.3.16, 4-Hydroxy-2-oxoglutarate aldolase.

The oxidizing or reducing power of a substance is indicated by its redox (reduction-oxidation) potential. The redox potential is related to the potential of the hydrogen electrode, and it is a quantitative index of electron affinity (i.e. oxidizing power) or tendency to lose electrons (i.e. reducing power). The defined standard potential in physical chemistry, or Normal potential, E_o (pH = 0, pH$_2$ = 1 atmosphere) is, however, inappropriate for biological purposes. Instead, the Normal potential E_o' at pH 7.0 is taken as the reference point. E_o' has a value of 0.42 volt on the E_o scale. The E_o' value is only valid under standard reaction conditions, i.e. all reactants at unit activity. Such conditions do not obtain in the living cell, and the ratio of concentrations of oxidized and reduced forms of the redox pair is included in the calculation. The actual redox potential E' is therefore expressed as:

$$E' = E_o' + \frac{0.06}{n} \log \frac{Ox}{Red} \quad (30\ °C).$$ The redox potential is related to free enthalpy (Gibb's potential) as follows: $\Delta G\,° = -n F E_o$, where n is the change of val-

Type of reaction	Enzymes	General reaction
Dehydrogenation	Oxidoreductases	$SH_2 \rightleftharpoons S$ D DH$_2$
Electron transfer	4 and 2 electron transferring oxidases	$SH_2 \longrightarrow S$ $\frac{1}{2}O_2$ H_2O or O_2 H_2O_2
Oxygen transfer	Dioxygenases	$S + O_2 \longrightarrow SO_2$
Hydroxylation	Hydroxylases	$S + DH_2 + O_2 \longrightarrow$ SOH + D + H$_2$O

S = substrate, D = hydrogen carrying cofactor

Oxidation reactions in metabolism

ency, and F the Faraday constant. As oxidizing potential increases E'_o becomes more positive, so that strong reducing agents are characterized by high negative E'_o values, and strong oxidizing agents by high positive values. See Table 3 under the entry for Respiratory chain; here the members of the respiratory chain and other biological redox systems are arranged in order of their redox potentials. Each component in the table is a more powerful oxidizing agent than those above it.

Oxidative metabolism: see Respiration.

Oxidative phosphorylation, *respiratory chain phosphorylation:* formation of ATP coupled with the operation of the Respiratory chain (see). Energy available from the flow of electrons from substrate to oxygen via the respiratory chain drives the synthesis of ATP from ADP and inorganic phosphate. Oxidation of one molecule of reduced nicotinamide adenine dinucleotide ($NADH + H^+$) generates three molecules of ATP, while oxidation of one molecule of reduced flavin adenine dinucleotide ($FADH_2$) yields two ATP. Complete oxidation of one molecule of glucose yields 38 ATP, 2 from glycolysis and 36 from O.p.

The mechanism of O.p., i.e. the nature of the energy transducing mechanism that converts the energy of electron flow into the chemical energy of ATP, has always been a controversial area of biochemistry. The earliest hypothesis, *chemical coupling,* proposes the existence of an energy-rich intermediate generated by electron flow and consumed in the phosphorylation of ADP:

$$A_{red} + B_{ox} + C \rightarrow A_{ox} \sim C + B_{red}$$
$$\underline{A_{ox} \sim C + ADP + P_i \rightarrow A_{ox} + ATP + C}$$
$$A_{red} + B_{ox} + ADP + P_i \rightarrow A_{ox} + B_{red} + ATP$$

A second hypothesis, *conformational coupling,* proposes that the energy is stored in a protein conformational change caused by electron transfer; return to the original conformation is linked to ATP synthesis.

Whatever the mechanism of O.p. eventually accepted, there seems little doubt that it will be based on the *chemiosmotic theory* proposed by Nobel laureate P. Mitchell. In chemiosmosis, the operation of the respiratory chain drives protons across the inner mitochondrial membrane (see Mitochondria). The resulting proton gradient then drives ATP synthesis. The F_1 "knobs" visible on the inner surface of the inner mitochondrial membrane, together with their associated hydrophobic membrane proteins, form an ion pump, which generates ATP at the expense of the proton gradient. The mechanism whereby proton transport is coupled to the operation of the respiratory chain is not clear, and the final mechanism may well invoke some version of the conformational coupling theory. The chemiosmotic theory differs from other theories of O.p. in that operation of the respiratory chains is not directly linked to ATP synthesis. Even in the absence of an operational respiratory chain, production of a proton gradient by any other method should, according to the chemiosmotic theory, promote ATP synthesis. This has been verified experimentally. Formation of a proton gradient, i.e. a decrease of pH outside the mitochondrial membrane, during operation of the respiratory chain has also been proved experimentally. The same theory also explains Photophosphorylation (see).

O.p. is normally tightly *coupled* to the flow of electrons along the respiratory chain, i.e. electrons do not flow unless ADP and inorganic phosphate are available for the synthesis of ATP. This *respiratory control* may be absent in pathological states, or destroyed artificially by *uncoupling agents.* In the brown fat tissue of some mammals, respiratory control may be partly relaxed so that electron flow without ATP production can be used to generate heat. In isolated mitochondria, lack of respiratory control indicates that the organelles are damaged. According to the chemiosmotic theory, uncoupling agents, e.g 2,4-dinitrophenol, act by rendering the inner mitochondrial membrane permeable to protons, so that the gradient collapses and is unable to drive ATP synthesis. The respiratory chain is then free of restraint, so that the respiration rate of uncoupled mitochondria is usually higher than that of optimally respiring, coupled mitochondria.

22,25-Oxidoholothurinogenin: see Holothurines.

Oxidoreductases: see Enzymes (Table 1).

Oxidoreduction: see Oxidation.

9-Oxo-*trans*-2-decenoic acid: see Queen bee substance.

2-Oxoacid dioxygenases: Oxygenases (see) which catalyse substrate hydroxylation linked to the oxidation of a 2-oxoacid (usually 2-oxoglutarate) with formation of CO_2. They all require Fe^{2+} and ascorbate, the latter being necessary to maintain the Fe^{2+} in its reduced form. The reaction can be generalized as:

$$\begin{array}{c} COOH \\ | \\ C=O \\ | \\ \underset{\text{Substrate}}{{>}C\text{-}H} + O_2 + \underset{\begin{array}{c} \\ \text{2-Oxoglutarate} \end{array}}{\overset{\begin{array}{c} \\ \end{array}}{\begin{array}{c}CH_2 \\ | \\ CH_2 \\ | \\ COOH\end{array}}} \xrightarrow{\text{Ascorbate} + Fe^{2+}} \underset{\begin{array}{c}\text{Hydroxylated} \\ \text{product}\end{array}}{{>}C\text{-}OH} + \underset{\text{Succinate}}{\begin{array}{c}CH_2 \\ | \\ CH_2 \\ | \\ COOH\end{array}} + CO_2$$

where one atom of the substrate molecular oxygen is incorporated into a carboxyl group of succinate, and the other forms the hydroxyl oxygen of the hydroxylated product.

The following 2-oxoacid dioxygenases are known:
4-Trimethylaminobutyrate, 2-oxoglutarate:oxygen oxidoreductase (3-hydroxylating), also called γ-butyobetaine, 2-oxoglutarate dioxygenase (EC 1.14.11.1). It catalyses the hydroxylation of 4-trimethylaminobutyrate to L-3-hydroxy-4-N-trimethylaminobutyrate, or Carnitine (see).

Prolyl-glycyl-peptide, 2-oxoglutarate:oxygen oxidoreductase (4-hydroxylating), also called Proline, 2-oxoglutarate dioxygenase, Protocollagen hydroxylase, and Proline hydroxylase (EC 1.14.11.2). It catalyses the hydroxylation of proline residues adjacent to glycine residues in procollagen, forming residues of 4-*trans*-hydroxyproline (see Collagen).

Thymidine, 2-oxoglutarate:oxygen oxidoreductase 2'-hydroxylating), also called Thymidine, 2-oxoglutarate dioxygenase, Thymidine 2'-hydroxylase, and Pyrimidine deoxyribonucleoside 2'-hydroxylase EC 1.14.11.3). It catalyses hydroxylation of C-2' of the deoxyribose moiety of thymidine.

Peptidyllysine, 2-oxoglutarate:oxygen 5-oxidoreductase, also called Lysine, 2-oxoglutarate dioxygenase, and Lysine hydroxylase (EC 1.14.11.4). It catalyses hydroxylation of lysine residues in procollagen, forming residues of 5-hydroxylysine (see Collagen).

Thymine, 2-oxoglutarate:oxygen oxidoreductase (7-hydroxylating), also called Thymine, 2-oxoglutarate dioxygenase, and Thymine 7-hydroxylase (EC 1.14.11.6). (The dioxygenase of EC number 1.14.11.5 has now been deleted from the EC list).

4-Hydroxyphenylpyruvate:oxygen oxidoreductase (hydroxylating, decarboxylating), also called 4-Hydroxyphenylpyruvate dioxygenase (EC 1.13.11.27). It catalyses the oxidation and decarboxylation of 4-hydroxyphenylpyruvate to 2,5-dihydroxyphenylacetate homogentisate), a reaction in the degradation of Phenylalanine (see) and tyrosine. The conversion is accompanied by migration of the side chain, and the substrate serves as its own 2-oxoacid electron donor, i.e. the enzyme may be described as an internal 2-oxoacid-dependent dioxygenase. [S. Lindstedt & M. Rundgren, *J. Biol. Chem.* **257** (1982) 11922–11931]

5-N-Trimethyl-L-lysine, 2-oxoglutarate:oxygen oxidoreductase (3-hydroxylating), also called Trimethyllysine hydroxylase (no EC no. assigned). It catalyses the hydroxylation of 6-N-trimethyl-L-lysine to 3-hydroxy-6-N-trimethyl-L-lysine, a reaction in Carnitine see) biosynthesis. [R. Stein & S. England, *Arch. Biochem. Biophys.* **217** (1982) 324–331]

Prolyl-glycyl-peptide,2-oxoglutarate:oxygen oxidoreductase (3-hydroxylating) (EC 1.14.11.?). It catalyses the hydroxylation of proline residues in procollagen to 3-*trans*-hydroxyproline (3-Hyp) residues in Collagen (see). 3-Hyp appears to occur only in the sequence -Gly-3-Hyp-4-Hyp-. [K. I. Kivirikko & R. Myllylä, *Methods in Enzymology* **82** (1982) 245–304]

Oxoeleostearic acid: see Licanic acid.

2-Oxoglutaric acid: see α-Ketoglutaric acid.

5-Oxoproline: see Pyroglutamic acid.

Oxygen: see Bioelements.

Oxygenases: enzymes that catalyse the incorporation of the oxygen of molecular oxygen into their organic substrates, i.e. the oxygen atom(s) appearing in the product is (are) derived from atmospheric O_2, and not from water. *Dioxygenases* (oxygen transferases), catalyse the introduction of both atoms of molecular oxygen. *Monooxygenases* or *hydroxylases* catalyse the introduction of one atom from molecular oxygen; the other atom becomes reduced to water. Monooxygenases therefore require a second substrate, which serves as an electron donor; for this reason they are also called *mixed function oxygenases.* They catalyse the following type of reaction: $AH + O_2 + DH_2 \rightarrow AOH + H_2O + D$, where AH is the substrate, AOH the hydroxylated substrate, DH_2 an electron donor, and D the oxidized electron donor.

Flavoprotein hydroxylases are found primarily in bacteria, e.g. 4-hydroxybenzoate hydroxylase has been crystallized from 4 different specis of *Pseudomonas.* The prosthetic group is FAD, and hydroxylation of the substrate to 3,4-dihydroxybenzoate is coupled to the oxidation of NADPH. Pteridine-dependent hydroxylases are a class of monooxygenases with a pteridine prosthetic group, e.g. L-phenylalanine-4-monooxygenase, tyrosine-3-monooxygenase of the adrenal medulla, and trytophan-5-monooxygenase of the brain.

Heme-coupled monooxygenases contain cytochrome P-450. They are present in microsomes and are responsible for many hydroxylation reactions, e.g. 11β-hydroxylation of steroids in the adrenal cortex, 2-hydroxylation of estrogens in the liver; the liver system is especially important in the hydroxylation of drugs and xenobiotics, thus rendering them water soluble, capable of conjugation and easily excretable. A cytochrome P-450 system responsible for the hydroxylation of camphor (a 5-*exo*-hydroxylase) has been purified from *Pseudomonas putida,* and named putidaredoxin; it contains FAD, an $Fe_2S_2Cys_4$ center, and a P-450 cytochrome; hydroxylation of the substrate is coupled to the oxidation of NADPH.

Certain monooxygenases contain copper (copper-containing hydroxylases), e.g. dopamine-β-hydroxylase; this enzyme is associated with the chromaffin granule of the adrenal medulla, and is responsible for the oxidation of dopamine to noradrenalin; the second substrate (electron donor) is ascorbic acid. Various phenolases are also Cu-containing monooxygenases. The mechanism of action is unclear, but it appears that a monophenol is hydroxylated to an *o*-diphenol, coupled to the oxidation of a diphenol (i.e. the second substrate) to an *o*-quinone.

Monooxygenases are responsible for the hydroxylation of proline and lysine residues in collagen. These hydroxylases contain ferrous iron and have a specific requirement for 2-oxoglutarate; the latter acts as an electron donor by becoming oxidatively decarboxylated to succinate and CO_2. 4-Hydroxyphenylpyruvate hydroxylase from liver and kidney is also an Fe-containing decarboxylating monooxygenase; hydroxylation is accompanied by migration of the side chain, while oxidative decarboxylation of the side chain serves to consume the other oxygen atom; the product is homogentisic acid (see L-Phenylalanine).

All known dioxygenases contain iron, either in heme groups or as Fe-S centers. Some also contain copper. Examples are Tryptophan-2,3-dioxygenase (see), homogentisate oxidase (see L-Phenylalanine), 3-hydroxy-anthranilate oxidase (see L-Tryptophan).

Mono and dioxygenase are employed extensively by bacteria in the degradation of aromatic compounds, and are therefore fundamentally important in the carbon cycle of the biosphere. See 2-Oxoacid dioxygenases.

Oxygen electrode: an amperometric sensor for detecting oxygen concentration changes in solution. In the polarographic O.e., a polarizing voltage of approximately $+0.6$ V, from an external source, is established between a platinum cathode and a silver/silver chloride anode bathed in saturated KCl; the current passing is proportional to the oxygen concentration in solution. In the galvanic O.e., the current is produced between a silver cathode and a lead anode bathed in saturated lead acetate; in this case the potential is generated by the electrode system and does not need to be supplied externally. In both types of O.e., the electrode is normally separated from the solution by an oxygen-permeable plastic membrane.

Oxygen metabolism: metabolic reactions involving oxygen, including 1. General oxidation of metabolites by Dehydrogenation (see), removal of electrons (oxidases, see Flavin enzymes), and addition of oxygen to the substrate (see Oxygenases). 2. Oxidation (see) of reduced coenzymes via the Respiratory chain (see).

Oxyhemoglobin: see Hemoglobin.

Oxysomes: see Mitochondria.

5'-Oxytetracycline: see Tetracyclines.

Oxytocin: a neurohypophysial, peptide hormone. M_r 1007. O. causes contraction of the smooth muscle of the uterus and of the mammary gland (milk ejection). O. is structurally related to the other neurohypophysial hormone, Vasopressin (see); these two hormones have a common ancestry and there is an overlap in their physiological activities. In lower vertebrates, four other neurohypophysial hormones have been identified, which are variants of O. and vasopressin, with different amino acid residues in positions 4 or 8, or both. [Arg[8]] Oxytocin is a hybrid analog, in which the tripeptide tail of vasopressin is attached to the ring of O.; it occurs naturally in chicken pituitary, and is considered to be the probable ancestor of all other related neurohypophysial hormones. [Ser[4], Ileu[8]] Oxytocin (also called isotocin) occurs in teleost fish; [Ser[4], Glu[8]] Oxytocin (also called glumitocin) occurs in elasmobranchs. [Ileu[8]] Oxytocin (also called mesotocin) is present in amphibians and reptiles.

$$\text{Cys—Tyr—Ile—Gln—Asn—Cys—Pro—Leu—Gly—NH}_2$$

Oxytocin

O. is synthesized in the hypothalamus in the Nucleus paraventricularis, and transported to the posterior lobe of the pituitary (hypophysis) via the Tractus paraventriculo-hypophyseus, in combination with the transport protein, Neurophysin I (see). O. is released into the blood stream in response to psychological and tactile stimulation of genitalia, or suckling stimulation of the mammary gland. O. acts on its target organ via the adenylate cyclase system. Methods for the assay of O. are chiefly biological, based on the milk ejecting activity in lactating animals, or the behavior of perfused sections or strips of mammary tissue. Radioimmunoassay is also possible. O. is inactivated and degraded in the organism by proteolysis. Over 300 analogs of O. have been synthesized, some of which have higher biological activity than the natural product. The role of O. in parturition is unclear. Although O. is probably involved in uterine contraction, it is not generally accepted that the onset of parturition is determined by the release of O. Structural elucidation and the first synthesis of O. were reported in 1953/54 by V. du Vigneaud et al. The most active analog obtained so far by amino acid substitution is [Thr[4]] Oxytocin.

P

(P): The letter P in a circle, standing alone, is sometimes used to denote inorganic phosphate (see P_i). If the circle is attached to another group by a bond (R-(P)), the encircled P may represent a phosphoryl group ($-PO_3H_2$) or a phosphate group ($-O-PO_3H_2$). If the bridge oxygen atom is shown (R-O-(P)), the encircled P must be a phosphoryl group.

~(P), ~P: a high energy or labile phosphate group. See High energy bonds; Energy-rich phosphates.

P_i: inorganic phosphate, the anion of orthophosphoric acid, PO_4^{3-}.

Pacemaker reaction: see Metabolism.

Padmakastein: 5,4'-dihydroxy-7-methoxyisoflavanone, see Isoflavanone.

Paeonidin: 3,5,7,4'-tetrahydroxy-3-methoxyflavylium cation. The 3,5-β-diglucoside (paeonin) is the chief pigment of peony.

Pahutoxin: the main toxin of the tropical box fish *Ostracion lentiginosus*.

Pahutoxin

P. is a choline derivative, highly poisonous for fish, but not for warm blooded animals.

Paleoproteins: a group of proteins from fossils, in particular the exoskeleton of molluscs (snail and mussel shells, and cuttle bone). The most studied P. is conchiolin from the hard parts of fossil mussels from the lower tertiary (25–58 million years), the jurassic (150 million years) and the silurian (440 million years). Conchiolin has been characterized chemically and by electron microscopy; it is a fibrillar scleroprotein, rich in glycine, alanine and serine.

Palindrome: a nucleic acid sequence that is identical to its complementary strand (when both are read in the same 5'-3' direction). In the region of a palindrome there is therefore a twofold rotational symmetry. Perfect palindromes, e.g. GAATTC, often occur as recognition sites for restriction enzymes. Imperfect palindromes, e.g. TACCTCTGGCGTGATA, often act as binding sites for proteins such as repressors. Interrupted palindromes, e.g. GGTTXXXXXAACC, make possible the formation of a stem with a loop (hairpin structure) as in tRNA.

Palmitic acid: *n*-hexadecanoic acid, $CH_3-(CH_2)_{14}-COOH$, a fatty acid. M_r 256.4, m.p. 63 °C, b.p.$_{100}$ 271.5 °C, b.p.$_{15}$ 215 °C. Together with stearic acid. P.a. is one of the most widely distributed natural fatty acids, and is present in practically all natural fats, e.g. 36% in palm oil, 29% in bovine carcass fat, 15% in olive oil; it is also found in phosphatides and waxes. P.a. is the raw material for the manufacture of candles, soap, wetting agents and antifoams.

Palmitoleic acid: Δ^9-hexadecanoic acid, $CH_3(CH_2)_5-CH=CH-(CH_2)_7-COOH$, an unsaturated fatty acid. M_r 254.4, m.p. 1 °C, b.p.$_{15}$ 220 °C. P.a. is present in the acylglycerols of many plant and animal fats, and in phosphatides.

Pancreas: a vertebrate organ producing a digestive secretion which enters the adjacent duodenum in response to the hormones Secretin (see) and Cholecystokinin (see). It also contains about 1 million islets of Langerhans, which have a diameter of 150 μm, a rich supply of blood and are innervated with unmyelinated nerve fibers. The islets contain various types of hormone-producing cells, the A cells which produce Glucagon (see), the B cells which make Insulin (see) and the D cells which manufacture Gastrin (see).

Pancreatic enzymes: a group of at least 12 digestive enzymes, including some of the most intensively investigated of all enzymes. Autolysis of the pancreas does not occur, because the proteolytic enzymes, trypsin, chymotrypsin A and B, elastase and carboxypeptidase A and B, and phospholipase A_2 are synthesized and stored in the pancreas as inactive zymogens. The other P.e. require effectors for optimal activity, which are present in the duodenum. Trypsin inhibitors in the pancreatic tissue and secretion afford additional protection against proteolytic destruction by active P.e. With the exception of cholesterol esterase (M_r 400000), the M_r of P.e. lie between 13700 (ribonuclease) and 50000 (α-amylase).

Pancreatic lipase deficiency: see Inborn errors of metabolism.

Pancreatin: defatted, powdered preparations of pancreas. Acetone dried pancreas powders can be stored for long periods without loss of activity. P. contains all the pancreatic enzymes in active form. It serves as a starting material for the laboratory purification of pancreatic enzymes. P. is also used in the pharmaceutical industry for the preparation of enzyme tablets, which are used in cases of secretory malfunction of the pancreas.

Pancreozymin: see Cholecystokinin.

Pantetheine 4'-phosphate, *phosphopantetheine:* the phosphate ester of *N*-(pantothenyl)-β-aminoethanethiol. A residue of P. is present in the molecule of Coenzyme A (see). P. is the prosthetic group of acyl carrier proteins in certain multienzyme complexes, e.g. fatty acid synthetase and gramicidin S synthetase, where it serves as a "swinging arm" for the attachment of activated fatty and amino acid groups.

Pantherine: see Ibotenic acid.

Pantothenic acid: a vitamin of the B_2 group (see Vitamins).

PAP: abb. for 3'-phosphoadenosine 5'-phosphate (see Sulfotransferase).

Papain (EC 3.4.22.2): a thiol enzyme from the latex and unripe fruit of *Carica papaya* (tropical melon or papaw). P. is unusually stable to high temperatures and to high concentrations of denaturing agents, e.g. 8M urea or organic solvents such as 70% ethanol or 15% dimethylsulfoxide. P. is a carbohydrate-free, basic, single chain protein (M_r 23350; 212 amino acid residues; methionine absent; IP 8.75) with 4 disulfide bridges, and catalytically important cysteine (position 25) and histidine (position 158) residues. The molecule is a rotational ellipsoid, divided by a cleft, and containing a predominance of antiparallel β-structures.

For activity, P. requires a free SH group and an operating pH of 5–5.5. SH blocking agents, like iodoacetic acid, are powerful P. inhibitors, whereas SH compounds, like mercaptoethanol, are potent activators. P. has a broad specificity, embracing endopeptidase, amidase and esterase activities. P. catalyses the cleavage of a wide variety of peptide bonds, indicating a fairly low specificity for peptide bond cleavage. Systematic studies with model peptides have, however, revealed a high degree of specificity for certain groupings. The active site of P. can be divided into 7 "subsites", each accommodating one amino acid residue of the substrate. Four of these subsites are on one side of the catalytic site, three on the other:

$$H_2N-R_1-R_2-R_3-R_4\text{—}R_5-R_6-R_7-COOH$$
$$\uparrow$$
cleavage

The site corresponding to R_3 specifically interacts with the side chain of phenylalanine (Phe). The presence of Phe in position 3 or further from the *C*-terminus increase the susceptibility of the peptide to hydrolysis, so that cleavage occurs at the peptide bond one residue removed from the Phe towards the *C*-terminus:

$$\downarrow$$
$$H_2N-Ala-Phe-Ala-Lys-Ala-CONH_2$$

For the same reason, peptides containing Phe as the second residue from the *C*-terminus are inhibitors of P., e.g. Ala-Ala-Phe-Ala, or Ala-Ala-Phe-Lys.

P. is used medically for the reatment of necrotic tissue and eczema, and in protein chemistry for cleaving proteins into large peptides.

Papaveraceae alkaloids, *Papaver alkaloids, poppy alkaloids:* a group of Benzylisoquinoline alkaloids (see), occurring especially in species of poppy *(Papaver).* They include the important Opium alkaloids (see).

Papaverine: an opium alkaloid occurring with morphine in various species of poppy. M_r 339.39, m.p. 147–148 °C. Physiologically it acts peripherally, and causes relaxation of smooth muscle; it is therefore used for the treatment of spasms in the gastrointestinal tract. P. is not addictive. For formula and biosynthesis, see Benzylisoquinoline alkaloids.

Paper autoradiography: see Isotope technique.

Paper chromatography: a chromatographic separation method, which employs a high quality filter paper (chromatography paper) as the carrier P. c. is almost pure partition chromatography on cellulose. Under certain conditions, the separation deviates from that predicted by the Nernst distribution, owing to slight adsorption and ion exchange effects. The stationary phase is a film of water or adsorbed hydration layer on the cellulose fibers of the paper. The mobile phase, an organic solvent or mixture of solvents, migrates over the stationary phase.

In a defined solvent system, each substance has a rate of migration, which is expressed as the R_f value (ratio to the solvent front):

$$R_f = \frac{\text{migration distance of substance}}{\text{migration distance of the solvent}}$$

The migration distance of the solvent is the distance from point of application of the substance to the front of the advancing solvent. Under standardized conditions (composition of solvent system; temperature; degree of vapor saturation in the chromatography tank), the R_f value is a constant.

In practice, P. c. consists of the physical separation, followed by special detection methods for the location of the individual substances. It may be performed on an analytical or preparative scale. With respect to the movement of the solvent system and position of the paper, P. c. can be ascending, descending or horizontal. Single or two dimensional procedures, and radial P. c. (circular filter paper technique) are commonly used. Additional useful techniques are multiple development (the first solvent is removed by drying and a different solvent is run in the same dimension) and flow through or run off (the solvent is allowed to run for a prolonged period by dripping from the edge of the paper).

For the separation of lipophilic substances, such as fatty acids and steroids, P. c. can be adapted to reversed phase chromatography. In this technique, the chromatography paper is made hydrophobic by impregnation with silicone or paraffin oil, or by acetylation. The phases are therefore reversed, i.e. the hydrophilic solvent system is repelled by the cellulose fibers, so that it forms the mobile phase; the more strongly hydrophobic organic solvent becomes the stationary phase. Zaffaroni systems are important for the separation of steroids: these are usually water-free, and the paper is impregnated with formamide or propylene glycol, which acts as the stationary phase.

Paper factor: see Juvabione.

PAPS: abb. for Phosphoadenosinephosphosulfate.

Paramyosin: see Muscle proteins.

Paraproteins: abnormal immunoglobulins, or normal proteins which are found in increased quantities in blood plasma in various hematological disturbances, known as paraproteinemias.

Known P. are Bence-Jones proteins, amyloid proteins, Waldenström-type IgM and cryoglobulin (a 7S globulin). Owing to their homogeneous character and relative ease of isolation on a preparative scale, P. are among the best studied immunoglobulins.

Parapyruvate: a dimer which accumulates during the storage of pyruvate.

$$^-OOC-\overset{\overset{\displaystyle OH}{|}}{\underset{\underset{\displaystyle CH_3}{|}}{C}}-CH_2-CO-COO^-$$

Parapyruvate

Parasitism: a close coexistence of two species, beneficial to one partner (the parasite) and at the expense of the other (the host). A facultative parasite (e.g. *Claviceps,* see Ergot alkaloids) can live and grow without a host, and it can be easily cultivated in vitro. An obligate parasite (e.g. rickettsias, viruses, rust and smut fungi) cannot survive separately from its host, to which it has become adapted by evolution. This adaptation occurs in the nutritional physiology of the parasite, and is often accompanied by morphological degeneration. Obligate parasites can usually be cultured in vitro, but preparation of a suitable growth medium depends upon an exact knowledge of the host-parasite relationship and the complex nutritional requirements of the parasite. Semi- or hemiparasitism is shown by some flowering plants (e.g. cow wheat, *Melampyrum pratense*), which require certain nutrients from the host, but supply others, such as carbohydrate, by their own photosynthesis. The distinction between this and other types of nutritional physiology is not always clear. Symbiosis resembles a well adapted and balanced form of parasitism. In the early stages of root nodule formation (see Nitrogen fixation) in legumes, the *Rhizobia* invade the host tissue like an attacking parasite; this is followed by a balanced period of symbiosis, after which the bacteroids are digested by the host.

Parathormone, *parathyrin*: a hormone produced by the parathyroid gland, which influences the metabolism of calcium and phosphate. P. is a single chain proteohormone with 84 amino acid residues of known primary structure [Sauer, R.T. et al. [1974] *Biochemistry, 13,* 1994-1999]. M_r 9402 (porcine). P. influences the cells that degrade bone (osteoclasts) by activation of membrane-bound adenylate cyclase and by increasing the entry of Ca^{2+} into these cells. The resulting mobilization of Ca^{2+} causes an increase in blood calcium. This is necessarily accompanied by the release of free phosphate, which is excreted via the kidneys. Thus P. favors phosphate secretion in the distal part of the kidney tubule, and inhibits phosphate resorption in the proximal tubule. P. promotes calcium absorption by the intestine. The action of P. is therefore opposite to that of Calcitonin (see). P. is degraded by the liver, and some is excreted in the urine. Absence of P. leads to a decrease of blood calcium, accompanied by neuromuscular overexcitability (tetany).

Parathyrin: see Parathormone.

Parathyroid glands, *Glandula parathyreoidea, Sandstroem's body:* endocrine organs, which develop from the endodermal lining of the third and fourth pharyngeal pouches of the embryo. P.g. are present in all vertebrates, excluding fishes. As a rule there are four P.g., but the location and number of individual P.g. may vary considerably. Parathyroid tissue is sometimes found in the mediastinum. In humans the P.g. are small, oval bodies about 5 mm long and 4 mm wide, embedded in the posterior surface of the lobes of the thyroid gland (2 in the superior, and 2 in the inferior lobes). In non-mammals, the P.g. are not embedded in the thyroid. Although the P.g. are usually physically associated with the thyroid, there is no functional relationship between the two glands. P.g. secrete parathormone, which raises the concentration of blood calcium and causes resorption of bone calcium.

Paromomycin: an aminoglucoside antibiotic; see Streptomycin.

Partition chromatography: see Chromatography.

Parvalbumins: water soluble, acidic, monomeric proteins, M_r approx. 12000. P. were originally thought to be present only in skeletal muscle of fish and amphibia, but have now been found in mammalian muscle. They have two high-affinity sites for Ca^{2+} ions ($K_{diss} = 0.1-4 \times 10^{-6}$), which also bind Mg^{2+} competitively. In relaxed muscle (Ca^{2+} approx. $10^{-8}M$), P. bind $2Mg^{2+}$ and no Ca^{2+}. In contracting muscle, intracellular Ca^{2+} increases to $10^{-6}-10^{-5}M$, and Ca^{2+} becomes bound with displacement of Mg^{2+}. Thus exchange of Ca^{2+} with Mg^{2+} seems to be important in the physiological activity of P. From rabbit, rat, chicken, carp and frog, P. are isostructural. They have 6 α-helical regions, A to F. Loops between helices C and D and between E and F are the binding sites for the 2 Ca^{2+} ions. P. have also been found in some neurons of the CNS, in Leydig cells of the testis, and in the ovary. It seems probable that P. are involved in the relaxation of fast-twitch muscle fibers in mammals. See also Muscle proteins. [C.W. Heizmann, *Experentia* **40** (1984) 910-921]

Passive hemagglutination: see Agglutination.

Passive transport: see Transport.

Pasteur effect: an inverse relationship between the rate of glucose utilization and the availability of oxygen. This was first observed in 1860 by L. Pasteur, who noted that yeasts decompose more sugar under anaerobic conditions than they do aerobically. In the anaerobic glycolysis of glucose, there is a net yield of 2 molecules of ATP per molecule of glucose, compared with 36 ATP for the complete aerobic respiration of glucose. From the ratio $36/2 = 18$, it is evident that 18 times more glucose must be consumed under anaerobic conditions than under aerobic conditions, in order to obtain an equivalent amount of energy. Thus the P.e. represents the regulation of glucose consumption to match the energy needs of the cell. Naturally, the effect is only observed in facultative cells, i.e. cells which can adjust their metabolism to either aerobic or anaerobic conditions, e.g. yeast cells, muscle cells.

In the absence of oxygen, animal cells perform anaerobic glycolysis and produce lactate; yeast cells ferment glucose to various fermentation products, e.g. glycerol and notably ethanol. When oxygen is admitted the production of lactate quickly ceases and lactate rapidly disappears; at the same time, the rate of glucose uptake is markedly decreased. The P.e. can largely be explained by the allosteric properties of phosphofructokinase (EC 2.7.1.11), the rate-limiting ("pace maker") enzyme of glycolysis. This enzyme is inhibited by ATP and activated by ADP and AMP. The ATP inhibition is overcome by

AMP, cAMP, P_i, fructose 1,6-bisphosphate and fructose 6-phosphate, but it is increased by citrate. The citrate effect is significant in yeast and in aerobic muscle, which also use fatty acids and ketone bodies as major sources of energy. The effect of cAMP may be important in adipose tissue, where it overcomes the citrate inhibition.

The P.e. is especially important in skeletal muscle; the quantity of ATP in muscle (5–7 µmoles per g fresh weight) is sufficient to support contraction for 1–5 seconds. ATP synthesized from creatine phosphate (see Phosphagens) supports further short term contraction. During this early period the muscle becomes relatively anaerobic as the oxygen stored in the myoglobins is consumed and contraction restricts the blood supply by squeezing the blood vessels. Rephosphorylation of ADP to ATP then depends upon the anaerobic glycolysis of glycogen to lactate; the activity of phosphofructokinase is markedly increased, with a corresponding increase in the overall rate of glycolysis. However, this effect cannot simply be explained by the change in level of ATP; in fact, the steady state concentration of ATP in contracting muscle is only slightly lower than in resting muscle; for example in insect flight muscle a 10% decrease in the ATP level results in a 100 – fold increase in the rate of glycolysis. Two complementary mechanisms have been proposed for this amplification: 1. Adenylate kinase (EC 2.7.4.3) catalyses the interconversion, $2ADP \rightleftharpoons ATP + AMP$ with an equilibrium constant of approximately 0.5; under the steady state conditions in the cell, the ATP concentration is much higher than those of ADP and AMP. Thus a small change in ATP concentration causes a much larger percentage change in AMP concentration. 2. Simultaneous operation of phosphofructokinase and fructose 1,6-bisphosphatase (EC 3.1.3.11) would constitute a Futile cycle (see). In resting muscle phosphofructokinase is largely inhibited by the ATP/AMP ratio. Furthermore, much of the fructose 1,6-bisphosphate that is formed is converted back to fructose 6-phosphate by fructose 1,6-bisphosphatase, i.e. the futile cycle does operate at a low rate, and the net flow rate of glycolysis is equal to the difference between the activities of phosphofructokinase and fructose 1,6-bisphosphatase. When the muscle contracts, the small decrease in ATP causes a large relative rise in AMP; the activity of phosphofructokinase increases, that of fructose 1,6-bisphosphatase decreases, and there is a marked increase in the rate of production of fructose 1,6-bisphosphate.

The first reaction of glucose utilization is catalysed by hexokinase, (EC 2.7.1.1) which is inhibited by high concentrations of glucose 6-phosphate. In cells containing a reserve carbohydrate (e.g. glycogen in muscle cells), glucose 6-phosphate is also formed via glucose 1-phosphate. Glucose 6-phosphate accumulates during the inhibition of phosphofructokinase in sufficient quantity to inhibit hexokinase, thus preventing even the first stage of glucose utilization.

A quantitative index for the effect of oxygen on glycolysis and fermentation is given by the Meyerhof oxidation quotient:

which is usually about 2. The consumption of one molecule of oxygen generally inhibits the production of two molecules of lactate or ethanol.

The P.e. is not observed, or only slightly, in malignant tumors, intestinal mucosa, red blood cells (mammalian) and retina (mammalian and especially avian). These tissues obtain their energy by the anaerobic degradation of glucose, even in the presence of oxygen.

Uncoupling agents (e.g. dinitrophenol) abolish the P.e., so that the full of anaerobic carbohydrate utilization is maintained in the presence of oxygen.

Pathocidin: see 8-Azaguanine.

Pathological proteins: see Paraproteins.

PCA: abb. for pyrrolidone carboxylic acid. Other abb. are now preferred (see Pyroglutamic acid).

P-cell stimulating factor: see Colony stimulating factors.

P700 Chlorophyll a-protein: a strongly hydrophobic, integral membrane protein isolated from the thylakoids of many angiosperms, gymnosperms and green algae. It is a component of photosystem I. The ratio chlorophyll a/P700 is in the range 40–45/1, and preparations are also associated with cytochromes f and b_6. The complex contains 14 mol chlorophyll per mol protein (M_r 110000). Thus only one in three of the complex molecules contains P700, and the total pigment system must contain at least 3 molecules of the protein. Photosynthesis-deficient mutants are known, which lack P700 Chlorophyll a-protein, e.g. mutants of *Scendesmus* and *Antirrhinum*.

PCM: abb. of Protein-calorie malnutrition.

Pectin esterase: see Enzymes. Table 2.

Pectin lyase: see Enzyme, Table 2.

Pectins: high M_r polyuronides, consisting of α-1,4 glycosidically linked D-galacturonic acid residues. Some of the carboxyl groups are present as their methyl esters. The free acids are called pectinic acids. Varying extents of esterification and polymerization give rise to a wide variety of P. They are distributed widely in plants, and are found in association with cellulose; they are structurally important as cement and support substances, particularly in the middle lamellae and primary walls of plant cells. P. are especially plentiful in fleshy fruits, roots, leaves and green stems. They are prepared from sliced sugar beet and from apple and citrus residues (following juice extraction) by gentle extraction with hydrochloric, lactic or citric acids.

Due to the presence of hydrophilic groups, P. have a high capacity for binding water. They are therefore powerful gelling agents, a property which is widely exploited in the food industry (P. are used as setting

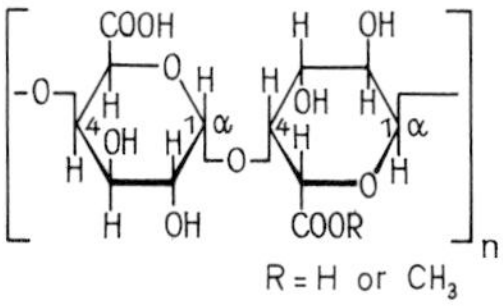

$R = H$ or CH_3

Pectin

$$\frac{(\text{Rate of anaerobic fermentation}) - (\text{Rate of aerobic respiration})}{\text{Rate of } O_2 \text{ uptake}}$$

agents in jams, etc.), and in the preparation of pharmaceuticals and cosmetics.

Pelargonidin: 3,5,7,4'-tetrahydroxyflavylium cation, an aglycon of many Anthocyanins (see). M.p. > 350 °C. P. is widely distributed in higher plants as its various glycosides, which are responsible for the rose, orange-red and scarlet colors of petals and fruits. In addition to some acylated derivatives, the structures of about 25 natural glycosides are known, e.g. callistephin (3-β-glucoside) of the red aster; fragarin (3-β-galactoside) of the strawberry; and pelargonin (3,5-di-β-glucoside) of various *Pelargonium* species and garden dahlias.

Pelargonin: see Pelargonidin.

Pellagra: see Vitamins (Nicotinic acid).

Pellotine: see Anhalonium alkaloids.

PEM: abb. for Protein-energy malnutrition.

Pempidine: see Acetylcholine.

Penicillinamidase: see Enzymes, Table 2.

Penicillinases (EC 3.5.2.6): enzymes which catalyse the hydrolysis of the β-lactam ring of penicillin; the resulting penicilloic acids have no antibiotic activity. The production of P. is adaptive, and P.-producing (i.e. penicillin-resistant) strains can be isolated by treating bacterial populations with penicillin. P. are single chain enzymes. The primary structures of soluble P. from *Staphylococcus aureus* (257 amino acid residues, M_r 28800) and *Bacillus licheniformis* (265 amino acid residues, M_r 29500) are known [Ambler, R.P. and Meadway, R.J. *Nature*, Lond., *222*, (1969) 24; Meadway, R.J. *Biochem. J.* 115, (1969) 12P–13P]. *Bacillus licheniformis* possesses a membrane-bound P. (M_r about 33000), which is a precursor of the soluble exopenicillinase [Yamamoto, S. and Lampen, J.O. *J. biol. Chem.* 251, (1976) 4095–4101].

Penicillins: sulfur-containing antibiotics produced by fungi of the genera *Penicillium, Aspergillus, Trichophyton* and *Epidermophyton*. All P. contain a condensed β-lactam-thiazolidine ring system, whereas the acyl group (R) is variable.

Name	R
Benzylpenicillin (penicillin G)	$C_6H_5-CH_2-CO-$
Pentenylpenicillin (penicillin F)	$CH_3CH_2-CH=CH-CH_2-CO-$
n-Heptylpenicillin (penicillin K)	$CH_3(CH_2)_6-CO-$
Penicillin N	$HOOC-CH(NH_2)(CH_2)_3-CO-$
6-Aminopenicillanic acid ("6 APS")	$H-$
Penicillin X	$HO-C_6H_4-CH_2-CO-$
Penicillin V	$C_6H_5O-CH_2-CO-$
Ampicillin	$C_6H_5-CH(NH_2)-CO-$

The growth inhibitory properties of P. were first observed in 1928 in a *Staphylococcus* culture by Alexander Fleming and its use against bacterial infections in humans was first tested in 1941. Against sensitive bacteria, penicillin G is active at a dilution of 1:50000000; one mg of penicillin G corresponds to 1670 international units. The P. are, however, unstable to acids, and they are readily hydrolysed to inactive penicilloic acids by the action of Penicillinases (see). Commercially, P. are prepared by large scale fermentation in steel vats (up to 100 m³), using high yielding strains. These commercial strains produce about 20000 times more P. than the original isolates, and the yield is further increased by the addition of precursors, e.g. phenylacetic acid for penicillin G.

6-Aminopenicillanic acid (6 APS) is an important precursor for the organic synthesis of new P. The compound itself has no antibiotic activity; it is isolated as a fermentation product from cultures of *Penicillium chrysogenum*, or prepared by the enzymic hydrolysis of benzylpenicillin. Thousands of new P. have been prepared by the acylation of 6 APS, but only a few of these are therapeutically useful, e.g. Penicillin V is relatively stable to acid and is not hydrolysed in the stomach, so that it may be administered in tablet form; Ampicillin (the aminophenylacetyl derivative of 6 APS), has a wider spectrum of activity than most other P., including activity against various Gram negative bacteria (*Typhus, Escherichia coli*, etc.).

Penicillin V serves as the starting material for the semisynthesis of a group of antibiotics related to cephalosporin C, which are active against penicillin-resistant strains.

P. are biosynthesized from the amino acids α-aminoadipic acid, cysteine and valine, which become linked by peptide bonds. The residue of α-aminoadipic acid is usually subsequently lost and replaced by a different acyl residue, but it is retained in the molecule of penicillin N.

P. act by inhibiting cross linkage of the muropeptide in the Murein (see) layer of the cell wall. The cell wall is thus weakened and cannot withstand the high internal pressure of the bacterial cell (about 30 atmospheres). P. are the most widely and intensively used antibiotics; they are well tolerated by the animal organism and have a relatively broad spectrum of activity.

Penicillin selection technique: an aid to the isolation of chosen auxotrophic mutations in a bacterial population. It is based on the fact that penicillin kills growing bacterial cells, but does not affect nongrowing cells.

In a medium containing only growth requirements for the wild type organism, the wild type will grow, whereas nutritional auxotrophs will fail to grow (at a suitable concentration of cells over a suitable period of time, to avoid cross feeding of auxotrophs by wild cells). Addition of penicillin then kills the wild cells. After washing away the penicillin, the cells are cultured on a minimal medium supplemented by the auxotrophic requirement to favor growth of the auxotrophs (e.g. histidine in the isolation of histidine auxotrophs). The alternating cycles of penicillin treatment and growth in the presence of the auxotrophic requirement may be repeated several times, resulting in a considerable enrichment of auxotrophs. Cultures are finally plated on solid media, and individual colonies tested for their auxotrophic requirements.

Pentamethonium: see Acetylcholine.

Pentamethylenediamine: see Cadaverine.

Pentitols: C_5-sugar alcohols. Naturally occurring P. are D- and L-arabitol, ribitol and xylitol.

Pentolinium: see Acetylcholine.

Pentosans: polysaccharides consisting of pentose residues, e.g. arabans and xylans. P. are widely distributed in the plant kingdom, and they are important as cell wall and storage materials.

Pentose metabolism: see Pentose phosphate cycle; Phosphoketolase pathway.

Pentose phosphate carboxylase: see Ribulose 1,5-diphosphate carboxylase.

Pentose phosphate cycle, *hexose monophosphate shunt, Warburg-Dickens-Horecker pathway, phosphogluconate pathway:* an oxidative pathway of carbohydrate metabolism, in which glucose 6-phosphate (derived from glucose by phosphorylation; see Kinases) is totally degraded to carbon dioxide, accompanied by the reduction of $NADP^+$ to $NADPH + H^+$. Overall equation: $C_6H_{12}O_6 + 7H_2O + 12NADP^+ + ATP \rightarrow 6CO_2 + 12NADPH + 12H^+ + ADP + P_i$. The importance of P. lies in the production of reduced NADPH which is required for biosynthesis (e.g. of fatty acids), and in the production of pentoses required for the synthesis of nucleosides, nucleotides and nucleic acids. By the action of a transhydrogenase system, or NADPH cytochrome c reductase, the NADPH can be reoxid-

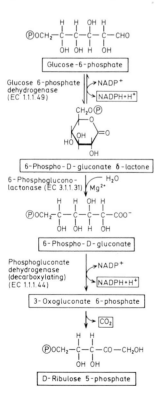

Figure 1. Oxidative phase of the pentose phosphate cycle.

ized to produce energy (36 molecules ATP per molecule glucose), but operation of P. for the sole purpose of energy production in unusual.

P. consists of an oxidative phase (resulting in the production of NADPH, CO_2 and D-ribulose 5-phosphate) and a nonoxidative phase (a fairly complex series of sugar interconversions). The experimental evidence for the oxidative phase (Fig. 1) belongs to classical biochemistry, and the reactions leading from glucose 6-phosphate to D-ribulose 5-phosphate are well proven. Glucose 6-phosphate dehydrogenase (EC 1.1.1.49) ("Zwischenferment" in the older literature) was discovered in 1931 by Otto Warburg. In 1934, Warburg and Christian discovered $NADP^+$ as a result of their investigation of the oxidation of glucose 6-phosphate by erythrocytes. Further contributions to the study of the conversion of glucose 6-phosphate into D-ribulose 5-phosphate were made by Dickens, Horecker, Racker and Kornberg.

In the scheme shown in Fig. 2, tentatively proposed by Horecker in 1954, the nonoxidative phase results in the conversion of 3 molecules of D-ribulose 5-phosphate into 2 molecules of fructose 6-phosphate and one molecule of glyceraldehyde 3-phosphate. Evidence for this nonoxidative phase was obtained from buffered extracts of acetone-dried rat liver, pea root and pea leaf. The pathway became generally accepted and it has been published widely in biochemical texts. However, it has long been known that the results of certain isotopic labelling studies are inconsistent with this scheme.

In the light of more recent evidence [Williams, J.F., Clark, M.G. & Blackmore, P.F. *Biochem. J. 176,* (1978) 241-256; Williams, J.F., Blackmore P.F. & Clark, M.G. *Biochem. J. 176* (1978) 257-282], it is necessary to revise the accepted scheme of sugar interconversions in the nonoxidative phase of P. The new scheme (Figs. 3 and 4) is consistent with all the isotopic labelling data. Evidence for the revised pathway was obtained from buffered extracts of acetone-dried rat liver. This later work resulted in the discovery of five additional intermediates of pentose phosphate metabolism: D-manno-heptulose 7-phosphate, D-altro-heptulose 1,7-bisphosphate, D-glycero-D-ido-octulose 1,8-bisphosphate, D-glycero-D-altro-octulose 1,8-bisphosphate, and D-arabinose. A new enzyme, arabinose phosphate 2-epimerase was also tentatively identified, which catalyses the reaction D-arabinose 5-phosphate ⇌ D-ribose 5-phosphate.

The new scheme requires high aldolase activity and has no transaldolase, whereas the old scheme has no aldolase and dependes on transaldolase. D-arabinose 5-phosphate inhibits transaldolase, but if the transaldolase is present in sufficient quantity, some activity probably remains in the presence of D-arabinose 5-phosphate.

Clearly it is incorrect to consider the nonoxidative phase of the P. as a fixed, albeit fairly complex, mechanism for the conversion of 3 molecules of pentose phosphate into a triose phosphate and 2 molecules of hexose phosphate. There exists a network of possible reactions, which can change in emphasis, depending on the tissue, and possibly on the physiological state of the tissue. Thus, in rat epididy-

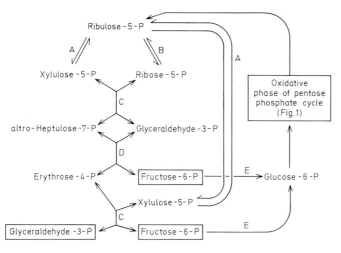

Sum of reactions in nonoxidative phase:

3 Ribulose –5–P ⟶ Glyceraldehyde –3–P + 2 Fructose –6–P

Figure 2. Scheme of the nonoxidative phase of the pentose phosphate cycle proposed in 1954 and widely accepted up to 1978.

A: Ribulosephosphate 3-epimerase (xylulosephos-
phate 3-epimerase) (EC 5.1.3.1)
B: Ribosephosphate isomerase (EC 5.3.1.6)

C: Transketolase (EC 2.2.1.1)
D: Transaldolase (EC 2.2.1.2)
E: Glucosephosphate isomerase (EC 5.3.1.9)
P = phosphate

Sum of reactions:

3 Ribulose – 5 – P ⟶ Dihydroxyacetone – P + Fructose – 6 – P + Glucose – 6 – P

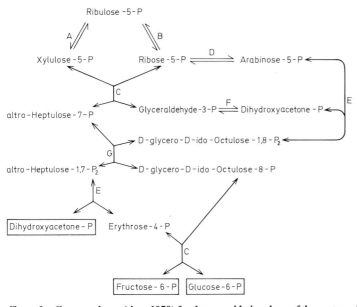

Figure 3. Current scheme (since 1978) for the nonoxidative phase of the pentose phosphate cycle (see also Fig. 4).

A: Ribulose phosphate 3-epimerase (xylulose phos-
phate 3-epimerase) (EC 5.1.3.1)
B: Ribose phosphate isomerase (EC 5.3.1.6)
C: Transketolase (EC 2.2.1.1)
D: Arabinose phosphate 2-epimerase

E: Fructose-bisphosphate aldolase (EC 4.1.2.13)
F: Triose phosphate isomerase (EC 5.3.1.1)
G: A phosphotransferase
P = phosphate

Pentose phosphate cycle

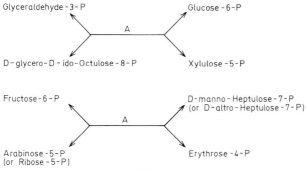

Ribose -5-P + Dihydroxyacetone-P $\overset{B}{\rightleftharpoons}$ D-glycero-D-altro-Octulose -1,8-P$_2$

Figure 4. Additional reactions that contribute to the nonoxidative phase of the pentose phosphate cycle shown in Fig. 3.
A: Transketolase (EC 2.2.1.1)
B: Fructose-bisphosphate aldolase (EC 4.1.2.13)
P = phosphate

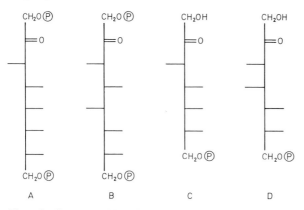

Figure 5. Structures of 7- and 8-carbon sugars involved in the pentose phosphate cycle.
A: D-glycero-D-altro-octulose 1,8-bisphosphate.
B: D-glycero-D-ido-octulose 1,8-bisphosphate.
C: altro-heptulose 7-phosphate (or sedoheptulose 7-phosphate).
D: manno-heptulose 7-phosphate.

mal fat pad, isotopic labelling suggests that the older scheme of Horecker operates for the metabolism of pentose phosphates. The significance of the new scheme has not yet been assessed in plants; it would provide an explanation of the Gibbs effect (see). Total acceptance of the new scheme, however, must await independent confirmation. Other workers [Wood, T. and Gascon, A. *Archives of Biochemistry and Biophysics, 203* (1980) 727–733] have reported their failure to demonstrate the interconversion of D-arabinose 5-phosphate and D-ribose 5-phosphate, or the role of D-arabinose 5-phosphate as an acceptor for transketolase in baker's yeast, *Candida utilis,* or rat liver.

The rate at which C-1 of glucose, as compared with C-6, is converted to CO$_2$ is used as an index for the quantitative importance of different pathways of carbohydrate metabolism. In Glycolysis (see), C-1 and C-6 become equilibrated at the level of the triose phosphates. Subsequent oxidation and decar-

boxylation in the tricarboxylic acid cycle therefore results in equal rates of production of $^{14}CO_2$ from [1-^{14}C] glucose and [6-^{14}C] glucose.

i.e. $\dfrac{^{14}CO_2 \text{ from } [1-^{14}C] \text{ glucose}}{^{14}CO_2 \text{ from } [6-^{14}C] \text{ glucose}} = 1.$

In the oxidative phase of the pentose phosphate cycle, C-1 of glucose is removed totally by the decarboxylation of 3-oxogluconate 6-phosphate, so that a ratio greater than 1 is evidence for the functioning of the pentose phosphate cycle. In the Glucuronate cycle (see), however, decarboxylation of 3-oxo-L-gluconate represents the removal of C-6 of glucose; thus ratios less than 1 indicate that the glucuronate pathway is operating.

In mammals the pentose phosphate cycle is especially important in the cornea, lens, liver and lactating mammary gland. It is also a common pathway in invertebrates, bacteria and plants.

The pentose phosphate cycle is completed when glucose 6-phosphate (some of which is derived from fructose 6-phosphate by the action of glucose phosphate isomerase) is converted into ribulose 5-phosphate by the oxidative phase (Fig. 1) of the cycle.

Pentoses: aldoses containing five carbon atoms. P. are an important group of monosaccharides (see Carbohydrates). Naturally occurring P. include D- and L-arabinose, L-lyxose, D-xylose, D-ribose and 2-deoxy-D-ribose, and the ketopentoses (pentuloses) D-xylulose and D-ribulose. P. occur chiefly in the furanose form. They are not fermented by the usual yeasts. By distillation with dilute acids. P. are converted into furfural, a reaction which serves for the detection of P. and their differentation from hexoses.

Pentosuria: see Inborn errors of metabolism.

Pepsin (EC 3.4.23.1): a protease in the stomach of all vertebrates with the exception of stomachless fish (e.g. carp). Purified P. shows maximal activity at pH 1–2, but in the stomach the optimal pH is 2–4. Above pH 6, P. is inactivated by denaturation. P. preferentially catalyses the hydrolysis of peptide bonds between two hydrophobic amino acids (Phe-Leu, Phe-Phe, Phe-Tyr). With the exception of protamines, keratin, mucin, ovomucoid and other carbohydrate-rich proteins, most proteins are attacked by P. The products of P. action are peptone, i.e. mixtures of peptides in the M_r range 300–3000. P. is a highly acidic (IP 1), single chain phosphoprotein (327 amino acid residues of known primary sequence, M_r 34500), which is released from its zymogen (pepsinogen, M_r 42500) by autocatalysis in the presence of hydrochloric acid. The largest of the cleavage peptides (M_r 3000) produced in this activation process is a P. inhibitor. Further peptic degradation of this inhibitor to inactive products is therefore necessary for the full realization of pepsin activity. At least 4 different active pseudo-isoenzymes of porcine and human P. are known. These are P.A. (EC 3.4.23.1) (the classical alkali-labile P.), B (EC 3.4.23.2) (the alkali-stable gelatinase, M_r 36000, 332 amino acid residues), C (EC 3.4.23.3) (the alkali-stable gastricsin, M_r 31500, 298 amino acid residues) and D. Probably B and D are autolysis products of the zymogen of the predominant A-form, since no pepsinogens of B and D have been detected. Pepsinogen C, the zymogen of P.C., has M_r 41000.

Pepstatin: see Inhibitor peptides.

Peptidases: see Proteases.

Peptide antibiotics: oligopeptides with antibiotic activity. They are usually cyclic. In addition to L-amino acids, they often contain nonproteogenic D-amino acids and unusual amino acids, and other components like branched hydroxy-fatty acids. The cyclic structures are closed by amide or (in depsipeptide antibiotics) ester bonds. Owing to their ring structures and high content of nonproteogenic constituents, P.a. are resistant to proteolysis. Most P.a. are relatively toxic, and only a few are used clinically (notably Penicillins, see), particularly for treatment of Gram negative infections (*Proteus, Pseudomonas*). Most known P.a. are of bacterial origin; a few are produced by *Streptomycetes* or lower fungi. Their antibiotic action is exerted in different

areas of metabolism, e.g. cell wall biosynthesis, nucleic acid and protein biosynthesis, energy metabolism, nutrient uptake. The biosynthesis of P.a. does not involve ribosomes or mRNA (see Gramicidins). The following important P.a. are listed separately: Penicillins, Gramicidins, Bacitracins, Tyrocidins, Polymyxins, Actinomycins.

Peptide bond: the most important type of covalent bond between amino acid residues in peptides and proteins. Formally, a P.b. is an amide bond between the carboxyl group of one amino acid and the amino group of another. Shortening of the C–N bond by mesomerism of the P.b. has been confirmed by X-ray crystallographic analysis. Since free rotation around the bond axis is not possible, two planar conformations of the P.b. are possible, i.e. *cis* and *trans* (Fig.). In native peptides and proteins, the *trans* P.b. predominate, but P.b. to the nitrogen atom of proline are always *cis*. The relative rigidity of the P.b. strongly affects secondary structure in Proteins (see). Cyclic dipeptides (2,5-dioxopiperazines) and polyproline contain only *cis* P.b.

Peptide hormones: a group of hormones with the chemical structure and physical properties of peptides. The smallest P.h., thyreotropin releasing hormone, is a tripeptide. Most P.h. are oligopeptides with 5 to 30 residues; see Oxytocin, Vasopressin, Releasing hormones, Opioid peptides. Although P.h. are considered separately from Proteohormones (see), there is a range of hormonally active compounds extending from P.h. to protein hormones. There are, however, differences as well as similarities: some P.h. are synthesized from amino acids by specific synthetases (e.g. releasing hormones), and many are formed by proteolysis of protein precursors (e.g. Angiotensin, see). P.h. act chiefly via the adenylate cyclase system (see Adenosine phosphates) and they are degraded by proteolytic enzymes.

Sequence analysis of P.h. reveals that many are capable of adopting stable amphiphilic secondary structures at membrane/aqueous interfaces. That such structures are important to the action of the P.h. is indicated by studies with synthetic analogs with similar or dissimilar amphiphilicity but otherwise different primary structures from the natural P.h. [E.T. Kaiser & F.J. Kezdy, *Science 223* (1984) 249–255]

Peptides: organic compounds consisting of two or more amino acids joined covalently by peptide bonds (see Peptide bond). The number of amino acid residues in a peptide is indicated by a Latin prefix, e.g. dipeptide, tripeptide. *Oligopeptides* contain

10 or fewer residues; more than 10 residues constitute a *polypeptide*. P. with M_r 10,000 (about 100 amino acid residues) lie on the borderline between polypeptides and proteins. P. dialyse through natural membranes, whereas proteins do not.

In the names of P., an amino acid contributing its carboxyl group to the peptide bond is given the ending *-yl*, and only the amino acid retaining its free carboxyl group retains its usual name, e.g. glycylhistidyl-alanine. Using the accepted abbreviations for Amino acids (see), the name of this tetrapeptide becomes Gly-His-Lys-Ala. It is customary to write the formula horizontally, always starting with the *N-terminal amino acid* (the one with the free α-amino group) on the left and finishing with the *C-terminal amino acid* on the right. The use of this abbreviated notation presupposes that amino acids with extra functional groups (e.g. lysine or glutamic acid) are linked by α-peptide bonds. Fig. 1 shows the methods of representing the α- and γ-peptide bonds in the tripeptide glutathione.

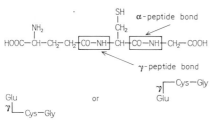

Figure 1. Glutathione (reduced)

The complete amino acid sequence of a P. is shown by linking the 3-letter amino acid symbols by hyphens. If only part of a sequence is shown, the terminal symbols must also carry a second hyphen, e.g. -Ala-Ile-Val-Lys-. If part of a sequence is unknown, the symbols are enclosed in parentheses and separated by commas, e.g. Ser-Phe-Gly-(Tyr, Asn, Val)-Pro-Ala. Using the one-letter notation (see Amino acids), this peptide would be represented as SFG(Y,B,V)PA, where the absence of punctuation between letters indicates a known sequence.

P. which contain only amino acid residues are called *homomeric P.*, whereas P. with non-amino acid constituents are known as *heteromeric* P. Each of these classes is further subdivided into *homodetic P.*, which contain only peptide bonds, and *heterodetic P.*, which contain other bonds, e.g. ester, thioether or disulfide. Heterodetic P. possessing one or more ester bonds as hetero link are called *Depsipeptides* (see). The classification of depsipeptides does not fit in entirely with the homomeric/heteromeric system; it has arisen from the fact that the synthesis of these peptides presents rather special problems. There are two types of homomeric depsipeptides: *O-peptides* contain ester-linked hydroxyamino acids, and they may be linear or cyclic (i.e. lactones); in *S-peptides* or *thiodepsipeptides*, the mercapto group of a cysteine residue is acylated with an amino acid. Heteromeric depsipeptides contain hydroxy acids as the hetero component, and they are also known as *peptolides*. P. with non-amino acid components cova-

lently bound to terminal amino or carboxyl groups, or to side chains are called *peptoids*. Important representatives of this group are glyco-, nucleo-, lipo-, phospho and chromopeptides.

In *linear P.*, it is not usually considered necessary to denote the direction of the peptide bond (from -CO to NH-), but in *cyclic P.*, the arrow should always be used.

Synthetic analogs of naturally occurring P. are named according to the following IUPAC-IUB rules:

If one or more amino acid residues are replaced, the full name of the new amino acid(s) and its(their) sequence position(s) are written in square brackets before the trivial name of the peptide, e.g. [7-alanine]oxytocin, [3-leucine, 7-alanine]oxytocin, or in abbreviated notation, [Leu³, Ala⁷]oxytocin.

If the peptide chain is extended at the terminus, the additional amino acid is represented in the normal way, e.g. valyl-bradykinin (abb. Val-bradykinin) has valine added to the *N-terminus* of bradykinin, while bradykinyl-valine (abb. bradykinyl-Val) has valine added to the *C-terminus*.

When an extra amino acid is introduced into the chain, this is indicated by the prefix *endo*, e.g. endo-6a-glycine-bradykinin (abb. endo-Gly⁶ᵃ-bradykinin) has a glycyl residue inserted between residues 6 and 7 of bradykinin.

Omission of amino acids is shown by the prefix *de*, e.g. de-3-proline-bradykinin (abb. de-Pro³-bradykinin).

P. representing partial sequences of larger P. are represented by placing the sequence numbers of the first and last amino acids in brackets after the trivial name, followed by the Greek root for the number of residues in the partial sequence, e.g. bradykinin-(3-9)-heptapeptide.

P. occur widely and have many different biological functions. Many are hormones, e.g. Corticotropin, Melanotropin, Oxytocin, Vasopressin, Releasing hormones, Insulin, Glucagon, Gastrin, Secretin, Angiotensin, Bradykinin, Endorphins, Opioid peptides (see entries for each of these). Many microorganisms produce P., often with antibiotic activity (see Peptide antibiotics). Some P. are very toxic, e.g. Phallotoxins (see), Amatoxins (see) and Melittin (see). Very simple P. are found in muscle, e.g. carnosine (β-Ala-His), anserine (β-Ala-MeHis). A wide variety of biologically active P. have been isolated from plants, e.g. γ-glutamyl-β-cyanoalanine (a neurotoxin from *Vicia sativa*.) Some P. inhibit enzymes, e.g. pancreatic Trypsin inhibitor (see).

Some P. are synthesized by reaction of an acyl phosphate (e.g. glutamate + ATP → γ-glutamyl phosphate + ADP) with the amino group of the next amino acid (γ-glutamyl phosphate + cysteine → glutathione see Glutathione). In a second type of P. synthesis, activated aminoacyl groups (amino acid + ATP → aminoacyl-AMP + PP$_i$) are transferred to -SH groups to form intermediate thioesters. See Gramicidins for details. Thirdly, some P. are formed by the controlled hydrolysis of proteins (see Protein biosynthesis, Post translational modification of proteins.) Thus peptide hormones (e.g. see Insulin, Bradykinin) are formed by enzymatic degradation of protein precursors. Stepwise degradation of the orig-

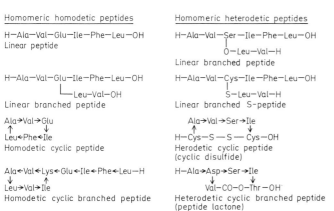

Homomeric homodetic peptides

H—Ala—Val—Glu—Ile—Phe—Leu—OH
Linear peptide

H—Ala—Val—Glu—Ile—Phe—Leu—OH
     └—Leu—Val—OH
Linear branched peptide

Ala→Val→Glu
↑   ↓
Leu←Phe←Ile
Homodetic cyclic peptide

Ala←Val←Lys←Glu←Ile←Phe←Leu—H
↓   ↑
Leu→Val→Ile
Homodetic cyclic branched peptide

Homomeric heterodetic peptides

H—Ala—Val—Ser—Ile—Phe—Leu—OH
            |
            O—Leu—Val—H
Linear branched peptide

H—Ala—Val—Cys—Ile—Phe—Leu—OH
            |
            S—Leu—Val—H
Linear branched S-peptide

Ala→Val→Ser→Ile
↑   ↓
H—Cys—S — S — Cys—OH
Herodetic cyclic peptide
(cyclic disulfide)

H—Ala→Asp→Ser→Ile
        ↓   ↓
        Val—CO—O—Thr—OH
Heterodetic cyclic branched peptide
(peptide lactone)

Figure 2

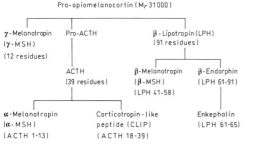

Pro-opiomelanocortin (M_r 31 000)

γ-Melanotropin Pro-ACTH   β - Lipotropin (LPH)
(γ-MSH)   (91 residues)
(12 residues)

        ACTH   β-Melanotropin   β-Endorphin
        (39 residues) (β-MSH)   (LPH 61-91)
                       (LPH 41-58)

α-Melanotropin   Corticotropin-like   Enkephalin
(α-MSH)   peptide (CLIP)   (LPH 61-65)
(ACTH 1-13)   (ACTH 18-39)

Figure 3. Derivation of the ACTH family of peptides from a single precursor M_r 31 000.

inal protein precursor may produce a family of hormones, in which one active hormone acts as the precursor of another, e.g. Releasing hormone (see), Pro-Leu-Gly-CONH$_2$, is produced from oxytocin by a membrane-bound exopeptidase. The so-called ACTH family is derived from a single precursor molecule (Fig. 3).

The half-life of most peptide hormones is less than 30 minutes; rapid inactivation results from the concerted action of various endo- and exopeptidases.

The primary structure of P. is elucidated by the standard methods of sequence analysis (see Proteins). The conformation of P. is an important determinant of their biological activity. It is stabilized by peptide and disulfide bonds. In P. which contain unusual bonds and/or constituents, other factors also stabilize the conformation. Although X-ray crystallography (see Proteins) was used in the elucidation of the three-dimensional structures of insulin and gramicidin S, the conformational analysis of P. is now by preference carried out in solution, using the spectroscopic methods ORD, CD, IR, NMR and ESR.

With respect to their general properties, P. are intermediate between amino acids and proteins (the following generalizations do not always apply to cyclic peptides). The acid-base behavior and solubility of linear P. depend on the amino acid sequence; the ampholite properties depend on the number and distribution of basic and acidic groups, and solubili-

ty is influenced by the relative numbers of hydrophobic and hydrophilic side chains. Compared with amino acids, the acidic character of P. is more pronounced. As a rule, the amount of P. can be determined by alkalimetric titration in 50% alcohol. Like amino acids, P. produce a blue to blue-violet color with ninhydrin, which can be used for their detection in chromatography and electrophoresis. P. can be differentiated from amino acids by the biuret test. P. are hydrolysed to their constituent amino acids by proteases, acids or bases, and their quantitative amino acid composition can be determined by conventional methods (see Proteins).

Chemical synthesis of P. serves a number of purposes: 1. Confirmation of the primary structure of natural P. and proteins. Some of these, e.g. releasing hormones, exist in such low concentrations that structural elucidation must be performed in conjunction with trial syntheses. 2. Investigation of the relationship between structure and biological activity. 3. Modification of the activity and specificity of naturally occurring, biologically active P. 4. Preparation of model P. for conformational analyses, e.g. polyamino acids. 5. Preparation of test P. for immunological studies.

The basic principle of P. synthesis is the acylation of the amino group of one amino acid with the carboxyl function of a second amino acid. This must be designed to give the correct sequence of the two amino acids, and to ensure that there is no racemization at the optically active center of either amino acid. For the synthesis of a defined P., it is therefore necessary temporarily to block (protect) all the functional groups which would take part in undesired side reactions, including the carboxyl, amino and sulfhydryl groups of side chains. Some protective groups are listed in the table. In the formation of a peptide bond, the carboxyl C atom of one amino acid undergoes nucleophilic attack by the amino N atom of the other amino acid. That is, the unbonded electrons on the N atom are attracted to the electrophilic (slightly positive) C nucleus. The process is facilitated by the presence of a substituent (X) on the C which draws electrons away from the C nucleus; in this case the reaction can take place under mild con-

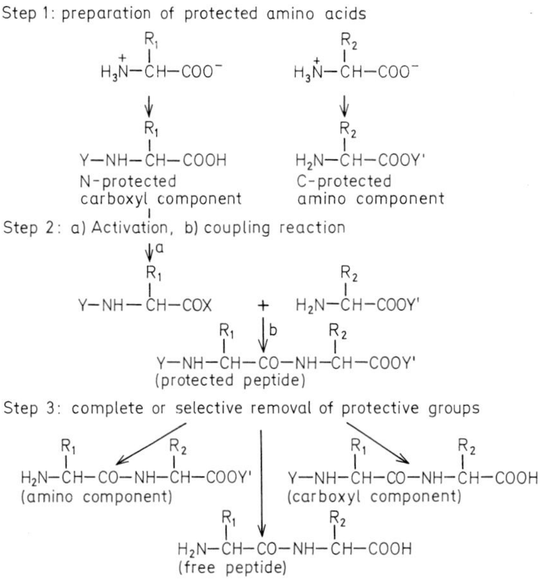

Peptides

Step 1: preparation of protected amino acids

$$H_3\overset{+}{N}-\overset{\overset{\displaystyle R_1}{|}}{C}H-COO^- \qquad H_3\overset{+}{N}-\overset{\overset{\displaystyle R_2}{|}}{C}H-COO^-$$

$$Y-NH-\overset{\overset{\displaystyle R_1}{|}}{C}H-COOH \qquad H_2N-\overset{\overset{\displaystyle R_2}{|}}{C}H-COOY'$$

N-protected
carboxyl component

C-protected
amino component

Step 2: a) Activation, b) coupling reaction

$$Y-NH-\overset{\overset{\displaystyle R_1}{|}}{C}H-COX \quad + \quad H_2N-\overset{\overset{\displaystyle R_2}{|}}{C}H-COOY'$$

$$Y-NH-\overset{\overset{\displaystyle R_1}{|}}{C}H-CO-NH-\overset{\overset{\displaystyle R_2}{|}}{C}H-COOY'$$
(protected peptide)

Step 3: complete or selective removal of protective groups

$$H_2N-\overset{\overset{\displaystyle R_1}{|}}{C}H-CO-NH-\overset{\overset{\displaystyle R_2}{|}}{C}H-COOY' \qquad Y-NH-\overset{\overset{\displaystyle R_1}{|}}{C}H-CO-NH-\overset{\overset{\displaystyle R_2}{|}}{C}H-COOH$$
(amino component) (carboxyl component)

$$H_2N-\overset{\overset{\displaystyle R_1}{|}}{C}H-CO-NH-\overset{\overset{\displaystyle R_2}{|}}{C}H-COOH$$
(free peptide)

Figure 4.

Azide method

$$R-\overset{\overset{\displaystyle O}{||}}{C}-NH-NH_2 + HNO_2 \longrightarrow R-\overset{\overset{\displaystyle O}{||}}{C}-N_3 \xrightarrow{+H_2N-R'} R-\overset{\overset{\displaystyle O}{||}}{C}-NH-R' + HN_3$$

hydrazide   azide   peptide

Mixed anhydride method

$$R-\overset{\overset{\displaystyle O}{||}}{C}-OH + H_5C_2-O-\overset{\overset{\displaystyle O}{||}}{C}-Cl \xrightarrow{-HCl} \text{mixed anhydride} \xrightarrow{+H_2N-R'} R-\overset{\overset{\displaystyle O}{||}}{C}-NH-R' + CO_2 + H_5C_2-OH$$

ethyl chloroformate   mixed anhydride   peptide

Dicyclohexylcarbodiimide (DCC) method

$$R-\overset{\overset{\displaystyle O}{||}}{C}-OH + DCC \longrightarrow \text{O-acyl-lactim} \xrightarrow{H_2N-R'} R-\overset{\overset{\displaystyle O}{||}}{C}-NH-R' + \text{dicyclohexylurea}$$

DCC   O-acyl-lactim   peptide   dicyclo-
hexylurea

p-Nitrophenylester method

$$R-\overset{\overset{\displaystyle O}{||}}{C}-OH + HO-\!\!\!\bigcirc\!\!\!-NO_2 \xrightarrow{+DCC} R-\overset{\overset{\displaystyle O}{||}}{C}-O-\!\!\!\bigcirc\!\!\!-NO_2 \xrightarrow{+H_2N-R'} R-\overset{\overset{\displaystyle O}{||}}{C}-NH-R' + HO-\!\!\!\bigcirc\!\!\!-NO_2$$

p-nitrophenylester   peptide

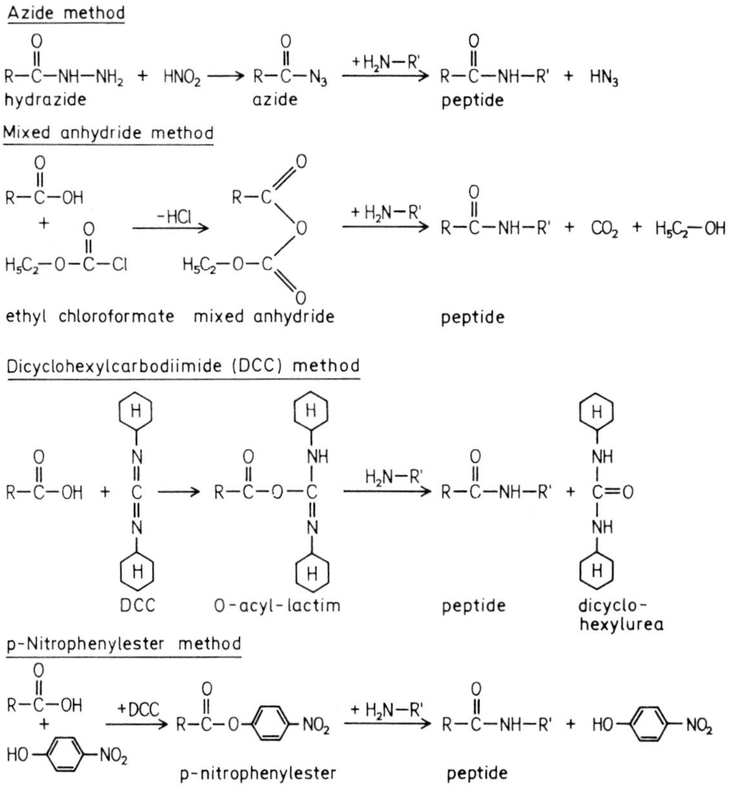

Figure 5. Some methods of peptide synthesis.

440

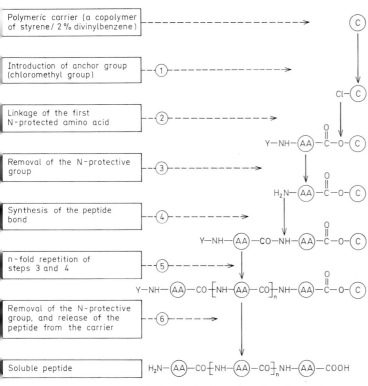

Figure 6. Principle of the Merrifield synthesis. Ⓒ carrier; Ⓐ amino acid

ditions. The last step of P. synthesis consists of the removal of protective groups, either from all the reactive groups or from only one of them, if the protected peptide is to serve as an intermediate in the synthesis of a higher P. (Fig. 4). Fig. 5 shows some of the available methods for activation. For the synthesis of higher P., two different strategies are usually applied. In order to avoid racemization, C-protected P. are extended one residue at a time by reaction with N-protected amino acids. At a later stage, fragment condensation may be used; N-protected P. are condensed with C-protected P. This has the advantage that the product of the condensation reaction is markedly different from the starting materials, whereas stepwise synthesis results in a product which is rather similar to the precursor P. Purification and characterization of the product is therefore much easier after fragment condensation.

By exploiting a number of different protective groups and methods of activation, it is possible to attain maximal blocking of side chain functions and a minimum of undesirable reactions. This synthetic strategy is limited by problems of solubility, especially for P. with more than about 30 amino acid residues. However, its success has been shown by the total synthesis of insulin, glucagon, corticotropin and other P.

The difficulties of conventional P. synthesis are theoretically overcome by synthesis on carrier polymers (Merrifield synthesis), introduced in 1962. In princi-

ple, an amino acid is bound via its carboxyl group to an insoluble, easily filtered, polymeric carrier, then the P. chain is built up stepwise from the C-terminal end (Fig. 6). Thus the laborious purification steps of the conventional P. synthesis are reduced to simple filtration and washing. With complete reaction at each stage of the synthesis, cleavage of the final P. from the carrier should give a pure product. The technical operation of the method is therefore very simple. In addition, the total synthesis can be automated, and machines for this purpose are commercially available. In practice, however, 100% yields are not realized, and the final product is accompanied by incomplete and wrong P., making the final purification especially difficult. Assuming a conversion rate of 99% at each coupling stage, polypeptides with 100 to 200 amino acid residues can still be synthesized, albeit in low yield. At the present state of development, despite many variations in the technique and attempts at optimization, an unequivocal synthesis is only only possible for P. or fragments containing up to 10 to 20 amino acids. On the other hand, the Merrifield synthesis has been used for the total synthesis of ribonuclease A with high enzymatic activity and the same specificity as the native enzyme; as has an analog of human growth hormone with 188 amino acids. The success of these syntheses was, however, assessed on the basis of biological activity, and it was not possible to apply classical criteria of protein or P. purity.

Table. *Some protective groups in general use.*

	IUPAC abb.	Cleavage reactions
Amino-protective groups		
Benzyloxycarbonyl	Z	HBr/glacial acetic acid; H_2/Pd; Na/liquid NH_3
tert-Butyloxycarbonyl	Boc	M HCl/glacial acetic acid; trifluoroacetic acid
Trifluoroacetyl	Tfa	M piperidine; NaOH/acetone
2-(*p*-Biphenylyl)-iso-propyloxy-carbonyl	Bpoc	5% trifluoroacetic acid/CH_2Cl_2
Carboxyl-protective groups		
Alkyl esters	-OMe -OEt	NaOH/acetone
Benzyl ester	-OBzl	H_2/Pd; Na/liquid NH_3; NaOH
tert-Butyl ester	-OBu^t	trifluoroacetic acid
Side chain protective groups		
S-Benzyl	Bzl-	Na/liquid NH_3
S-Acetamidomethyl	Acm-	Hg^{2+} (pH 4)
O-tert-Butyl	Bu^t	trifluoroacetic acid; HCl/diethylether
O-Benzyl	Bzl	NH_3

Peptidoglycan: see Murein.

Peptidyltransferase (EC 2.3.2.12): an integral enzymatic activity of the large ribosomal subunit. P.t. catalyses peptide bond formation between the NH_2 group of the amino acid of an aminoacyl-tRNA and the COOH group of the terminal amino acid of the peptidyl-tRNA. The reaction appears to be sterically favored by the relative positions of the two reaction partners at the acceptor and donor sites of the ribosome. It is thought that P.t. is located on the surface of the 50S subunit, where it makes contact with the 30S subunit.

Peptolides: see Depsipeptides.

Peptones: mixtures of polypeptides produced by the partial degradation of protein. P. are prepared by the acid or enzymatic (pepsin, trypsin) hydrolysis of dietary proteins, like albumins, caseins and muscle proteins. Autolysed and trypsin-treated yeast contains vitamins and other growth factors in addition to P. Owing to their low M_r of 600–3000, P. are not coagulable and cannot be salt-precipitated. P. inhibit blood coagulation and act as lymphagogues (i.e. they increase the flow of lymph). P. are used in the formulation of many microbiological growth media.

Perezone: see Benzoquinones.

Permeability factor: see Vitamins (Vitamin P).

Permeases: a term applied to carrier proteins involved in the transport of materials across membranes. P. is a misnomer and should be avoided.

Peroxidase (EC 1.11.1.7): an oligomeric oxidase which utilizes hydrogen peroxide as an oxidant in the dehydrogenation of various substrates. The H_2O_2 is reduced to water:

$$AH_2 + H_2O_2 \rightarrow A + 2H_2O$$

In the cell, P. is often accompanied by catalase and localized in the peroxisomes. The prosthetic group of plant P. is ferriprotoporphyrin (a red hemin); animal P. contain an unidentified green hemin. The most widely studied P. is the crystallizable horse radish P., which can be separated into apoenzyme and hemin under mild conditions by treatment with acetone-hydrochloric acid. The nonenzymatic, heat-stable P. activity of hemoproteins is called pseudo-peroxidase, e.g. as shown by hemoglobin and its degradation products. The P. reaction is diagnostically important, because the P. activity of myelotic granulocytes is greatly increased, whereas lymphocytes, reticular cells, carcinoma, sarcoma and myeloma cells have no P. activity. The P. of thyroid gland (M_r 200000) consists of 3 subunits (M_r 67000).

Pestalotin: a metabolic product from the culture filtrate of the fungus *Pestalotia cryptomeriaecola Sawada*. M_r 214.25, m.p. 88–88.5 °C. P. was isolated in 1971 by Japanese workers. It has the properties of a synergist of gibberellins.

Pestalotin

Petite mutants: spontaneous mutants, chiefly yeasts, with chemical or physical defects in the respiratory chain. P.m. grow very slowly and form small colonies on nutrient agar ("petite" colonies). The same phenotype can be produced by a chromosomal mutation (segregational petite), or a mutation in the mitochondrial DNA (vegetative or neutral petite). In the latter case, mitochondrial structure is considerably altered, largely due to changes in the amino acid composition of the structural proteins of the inner mitochondrial membrane. Since these structural proteins are important for the correct arrangement and conformation of the respiratory chain enzymes, the effect of petite mutation on the respiratory chain is probably secondary.

Petroselinic acid: *cis*-Δ^6-octadecenoic acid, CH_3-$(CH_2)_{10}$-CH=CH-$(CH_2)_4$-COOH, an unsaturated fatty acid. M_r 282.5, m.p. 33 °C, b.p.$_{18}$ 238 °C. P. is an acylglycerol component in the seed fats of many aromatic plants. e.g. parsley, celery, caraway.

Petunin: a blue anthocyanin present in the flowers of the blue garden petunia. m.p. 178 °C (3,5,7,4′,5′-pentahydroxy-3′-methoxyflavylium cation). The structures of more than 10 other naturally occurring glycosides of petunidin have been determined.

PGE$_1$: see Prostaglandins.

PGF$_{1\alpha}$: see Prostaglandins.

PGQ: abb. for pyrroloquinoline quinone.

pH-activity profilie: see Enzyme kinetics.

Phaeophorbide: see Chlorophyll.

Phaeophytin: see Chlorophyll.

Phaeoplasts: photosynthetic organelles in the brown algae and in diatoms. P. are brown due to the presence of carotenoids like fucoxanthin and β-carotene.

Phage, *bacteriophage* [Greek, phagein = to devour]: viruses that attack bacteria. As a generalization, they all consist of a protein coat (the capsid or capsomer) surrounding the genetic material, which is DNA or RNA. The nucleic acid is usually double stranded, but ΦX 174 (phi-chi, but usually called phi-ex) contains circular, single stranded DNA.

Length and M_r of circular DNA from some coliphages

Phage	Length of DNA (μm)	$M_r \times 10^6$
ΦX 174	1.77	1.7
λ	17	32
T2	56	130
T3	11.6	23
T4	50	125
T7	12.5	25

P. have various shapes and sizes, e.g. ΦX 174 and Qβ (both coliphages) are spheroidal particles, and apical knob-like structures are also discernible on ΦX 174; coliphage fd is elongated, best described as a flexuous strand; PM2 of *Pseudomonas* is spheroidal with apical knobs and enclosed by an additional envelope. The most complicated P. structures are found among certain well studied coliphages (i.e. P. that attack *Escherichia coli*); these have been given arbitrary names based on letters and numbers, such as T_1, T_2, P_1, λ (lambda), etc. (T_2, T_4 and T_6 are called the T-even P.).

The structure of coliphage T_2 (Fig.) may be taken as an example of structure of T-even P. Under the elec-

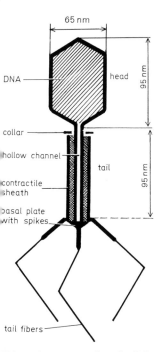

Schematic representation of coliphage T_2.

tron microscope, the head and tail parts can be distinguished, together with thread-like appendages. The head contains DNA plus putrescine and spermidine, which neutralize about 30% of the acidic groups of the DNA. The tail (a hollow cylinder of contractile protein) serves as an injection tube for the DNA. The collar and the base plate of the tail have a hexagonal symmetry, which is related to the symmetry of the head (a deformed icosahedron). The base plate contains a special lysozyme and 6 molecules of the coenzyme, 7,8-dihydropteroyl-hexaglutamate. The fibers, which serve to attach the P. to the bacterial surface (a process called adsorption), are appendages of the base plate. Adsorption represents a specific interaction of the tail fibers with molecular structures on the bacterial surface; this process closely resembles an immunochemical reaction, and infection cannot occur unless specific receptors are present on the bacterial surface. There is therefore a high degree of specificity between a P. and its host bacterium. With attachment of the tail fibers, the base plate is brought to the cell surface. Lysozyme in the base plate catalyses a local dissolution of the bacterial cell wall, the protein of the tail contracts, and the DNA is injected into the bacterial cell.

T_1, T_3, T_5, T_7 and λ have similar structures to those of the T-even phages, but the tail is less rod-like (i.e. it is more flexuous), and it does not contract during the infection process.

Phage development: entry of phage DNA (or RNA) into the host cell initiates a series of processes leading to new phage progeny. P.d. can be divided into three phases: 1. synthesis of early phage RNA and early phage proteins, and termination of the synthesis of all host nucleic acids and proteins; 2. synthesis of late RNA and late proteins; 3. morphogenesis of new phages. This complicated interaction of host cell and phage is regulated largely on the level of gene expression.

Development of virulent T-phages (e.g. T7): The protein coat of the phage remains outside the bacterial cell, and only phage DNA is injected. Upon entry of the phage DNA, synthesis of host DNA, RNA and protein immediately stops. Phage nucleic acids and proteins are then synthesized, using phage DNA and the transcription und translation apparatus of the host cell. The progress of P.d. can be represented as follows (all genes are transcribed from the 5′ end of the phage genome):

Synthesis of early RNA is catalysed by the host RNA polymerase, which recognizes the operator at the 5′ end of the phage genome. Owing to the relatively unspecific binding of the host RNA polymerase, the synthesis of early RNA is relatively slow, representing 75% of the total time of P.d. for the transcription of only 20% of the phage genome. The products of early RNA, the early proteins, are responsible for the termination of host nucleic acid synthesis. The first protein is an RNA polymerase specific for the phage, possibly accompanied by a specific sigma factor. The phage RNA polymerase permits the transcription of further phage genes for DNAase and RNAase (for the degradation of host nucleic acids), and DNA polymerase and DNA ligase (for the replication of phage DNA). Late RNA, transcribed from the second half of the phage genome, codes for late protein, e.g. lysozyme and proteins of the head and tail. All these phage components are then assembled into new, viable phage progeny, a process which may require some enzymatic control, but which also involves a high degree of cooperative self assembly. Finally, the lysozyme lyses the bacterial cell wall; depending on the particular strains of phage and host, and on the culture conditions, 100–200 viable phage particles are released per infected bacterial cell.

both leftwards and rightwards of the promotor during the early stages of induction; several genes have overlapping sequences; and there are various positive and negative controls.

Phagosomes: see Intracellular digestion.

Phallotoxins: heterodetic, cyclic heptapeptides, present in *Amanita phalloides*. Together with the Amatoxins (see), P. are the main toxic components of this fungus. The chief P. are phalloidin, phalloin and phallacidin (Fig.). Phallacidin is similar to phalloidin, but the D-threonine-alanine grouping is re-

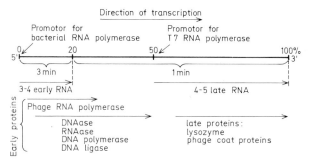

Simplified scheme for the transcription of the T7 phage genome.

Development of the temperate coliphage λ: Infection can lead to immediate production of new phage progeny and lysis of the bacterial cell, as described for T7 (see above); this is called a nonlysogenic reaction. Usually, however, the λ phage initiates a lysogenic reaction, i.e. P.d. is repressed and the phage DNA becomes integrated into the bacterial chromosome. In lysogeny, the host DNA and the integrated phage DNA are replicated synchronously, and the phage DNA is present in the DNA of all subsequent daughter cells. This integrated phage DNA is called a prophage. A regulator gene in the prophage controls the synthesis of a specific repressor, which prevents the independent replication of phage DNA. P.d. can be induced by various chemical or physical stimuli (e.g. temperature rise, UV irradiation, osmotic shock, chemical mutagens). Under the influence of an appropriate stimulus, synthesis of the repressor ceases and certain genes of the prophage become derepressed. Transcription of these derepressed genes requires a sigma factor from the host cell, which "recognizes" the promotor of the phage (prophage) DNA. The first products of these derepressed genes are an "antirepressor" protein, then an "antitermination" factor. The latter exerts a Positive control (see) by enabling the transcription (dependent on the bacterial RNA polymerase) of two further λ genes. The product of one of these genes modifies the host DNA polymerase, so that it becomes specific for the replication of λ DNA. The product of the other gene is a phage-specific sigma factor, which enables the RNA polymerase to transcribe further phage genes to polycistronic late RNA. The latter is translated into late proteins (head and tail proteins, lysozyme, etc.) by the translation apparatus of the host. The scheme described above is simplified: in reality several operons of the phage genome share one promotor; genes are transcribed

placed by valyl-D-erythro-β-hydroxyaspartic acid. P. are not as toxic as the amatoxins, but they act more quickly. The structural requirements for toxicity are the cycloheptapeptide structure, and the characteristic thioether bridge between the indole residue of the tryptophan and the mercapto group of the cysteine. LD_{50}-values in white mice: phalloin 1.35 mg, phalloidin 1.85 mg, phallacidin 2.5 mg per kg body weight.

Phaseolin, *phaseollin:* 3-hydroxy-9,10-dimethylchromenylpterocarpan (Fig.) (see Pterocarpans). P. is a phytoalexin (see) produced by various bean species in response to infection by phytopathogenic microorganisms, e.g. *Phytophthora, Monilinia,* and to other forms of stress like treatment of the plant with heavy metal salts. P. production can also be induced by a water-soluble polypeptide (monilicolin A, M_r 8000, 65 amino acids) isolated from *Monilinia fructicola.* The polypeptide is not fungi- or phytotoxic, and it induces P. at concentrations as low as 2.5×10^{-9}M; it appears to be a specific inducer of P.

Table. Structures of naturally occurring phenazines

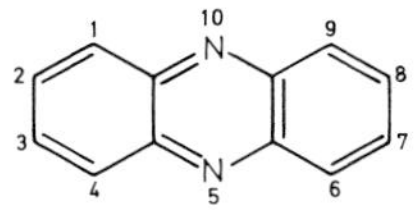

Phenazine numbering (positions 1,2,3,4 on one ring; 5 = N; 6,7,8,9 on other ring; 10 = N):

Name	Positions of substituents (C8 always carries hydrogen in known phenazines)									Source and comments
	1	2	3	4	5	6	7	9	10	
Aeruginosin A	COOH				CH₃		NH₂			Pseudomonas aeruginosa (red strain)
Aeruginosin B	COOH		SO₃⁻		CH₃		NH₂			Pseudomonas aeruginosa (red strain)
Iodinin	OH				→O	OH			→O	First identified from Chromobacterium iodinum (renamed Brevibacterium iodinum), later from many other genera. Purple crystals with coppery sheen from CHCl₃, m.p. 236°C (d.).
Pyocyanine	OH				CH₃					Pseudomonas aeruginosa
Chlororaphin	CONH₂									Pseudomonas chlororaphis. 1:1 mixture of dihydro- and dehydro-forms. Green crystals from acetone, m.p. 228-235°C. Sublimes at 210°C in absence of oxygen. Rapidly forms oxychlororaphin in air.
Griseolutein A	COOH					CH₂OC=O / HOCH₂ / CH₂OH		OCH₃		Broad spectrum antibiotics from Streptomyces griseoluteus. Relatively low toxicity against mammals. Especially active against Staphylococcus aureus.
Griseolutein B	COOH					CH₂OH		OCH₃		
Griseoluteic acid	COOH				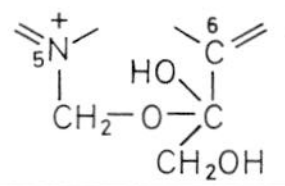 ⁺N=S CH₂–O–C / HO–C / CH₂OH			OCH₃		Streptomyces griseoluteus
Lomofungin	COOCH₃			OH		CHO	OH	OH		Streptomyces lomondensis
Myxin	OH				→O	OCH₃			→O	Sorangium. Red needles from acetone, m.p. 120-130°C; also given as 149°C (d.).
2(OH)Phenazine 1-carboxylate	COOH	OH								Many different species of Pseudomonas. Colors ranging yellow to orange.
Phenazine 1-carboxylate	COOH									
2(OH)Phenazine		OH								
Phenazine 1,6-dicarboxylate	COOH					COOH				

and does not, for example, induce pisatin production in pea plants. Induction of P. by any means is always accompanied by an increase in the measurable catalytic activity of phenylalanine ammonia lyase.

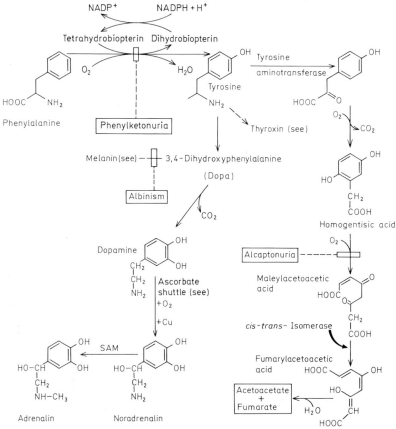

Phaseolin.

Phaseolin isoflavan: (3 R)-7,2'-dihydroxy-3',4'-dimethylchromenylisoflavan, see Isoflavan.

Phe: abb. for L-phenylalanine.

Phellandrene: see *p*-Menthadienes.

Phenazines: compounds based on the phenazine ring system (Table, p. 445). All known naturally occurring P. are produced only by bacteria, which excrete them into the growth medium. Both six-membered carbon rings of P. are derived biosynthetically from the shikimate pathway of aromatic bio-synthesis, via chorismic acid (not from anthranilate, as reported earlier). The earliest identified biosynthetic intermediate after chorismate is phenazine 1,6-dicarboxylate, which has been isolated from *Pseudomonas phenazinium* and from nonpigmented mutants of other organisms. Phenazine 1,6-dicarboxylate is thought to be the key branch point compound at the beginning of at least two pathways accounting for the biosynthesis of a wide variety of different P. [Byng, G.S. *J. Gen. Microbiology,* **97** (1976) 57–62. Byng, G.S. and Turner, J.M. *Biochem. J., 164* (1977) 139–145].

Phenosulfate esters: see Sulfate esters.

Phenotype: see Mutation.

L-Phenylalanine, abb. *Phe:* L-α-amino-β-phenylpropionic acid, an aromatic proteogenic amino acid. M_r 165.2. Phe is essential in the animal diet, and it is both glucogenic and ketogenic. The first stage in the catabolism of Phe is hydroxylation to L-tyrosine, which is the precursor of Melanin (see), the neurotransmitter Dopamine (see), the hormones Adrenalin, Noradrenalin and Thyroxin (see entries on each), and other compounds. This first step is catalysed by Phe hydroxylase (a monooxygenase), EC 1.14.16.1. Excess L-tyrosine is broken down to fumarate and acetoacetate (Fig. 1).

Classical phenylketonuria is an hereditary defect in

Figure 1. Metabolism of Phenylalanine

the synthesis of Phe hydroxylase (the enzyme may be absent or inactive), which affects about 1 infant in 10,000. These individuals are unable to convert Phe into tyrosine, and the major route of Phe metabolism is thus blocked. Phenylpyruvate and phenylacetic acid are excreted in the urine. The condition is accompanied by defective pigmentation, and if untreated, by severe mental retardation. (Hence the other name, phenylpyruvic oligophrenia, also known as Fölling's syndrome.) The urine of newborn infants is now routinely tested (the Guthrie test) for the presence of phenylketones; the condition can be compensated by a diet low in phenylalanine, and the typical mental retardation is thereby avoided. Other types of phenylketonuria are due to defective reduction or synthesis of dihydrobiopterin (see Inborn errors of metabolism).
A further hereditary defect results in the absence of homogentisate oxidase (a dioxygenase), EC 1.13.11.5. This condition is known as alcaptonuria. It is harmless, but the urine of affected individuals turns black on standing, due to the autoxidation of the excreted homogentisic acid.
In plants and bacteria, Phe and L-tyrosine are synthesized by the shikimate pathway of Aromatic biosynthesis (see). Chorismic acid, the branch compound of aromatic biosynthesis, is converted into prephenic acid, which is the precursor of both L-tyrosine and Phe (Fig. 2). Pretyrosine dehydrogenase, pretyrosine dehydratase, prephenate dehydrogenase and prephenate dehydratase may serve as taxonomic markers for microorganisms. Thus, in Cyanobacteria, some Coryneform bacteria, *Hansenula henrici*, *Halobacterium halobium*, *Trichococcus*, *Sulfolobus*

acidocaldarius, *Methanosarcina barkeri*, *Methanococcus voltae*, *Micrococcus luteus*, *Leptothrix* spp., *Sphaerotilus* spp., *Euglena gracilis* and *Zea mais*, tyrosine is synthesized exclusively via pretyrosine. In Mung bean and *Pseudomonas* spp., however, tyrosine is synthesized by two pathways, i.e. via pretyrosine and via 4-hydroxyphenylpyruvate. In some organisms, pretyrosine may be converted into Phe by pretyrosine dehydratase, but this is the exception, and the majority of Eubacteria, Cyanobacteria, Archaebacteria and eukaryotes (yeasts, green plants) synthesize Phe exclusively via phenylpyruvate [F. Lingens & E. Keller *Die Naturwissenschaften* **70** (1983) 115–118]. The detailed enzymology of tyrosine synthesis in *Pseudomonas aeruginosa* has been reported [Patel et al. *J. Biol. Chem.* **252** (1977) 5839–5846].
L-Phenylalanine ammonia-lyase, *PAL* (EC 4.3.1.5): a fungal and plant enzyme catalysing the conversion of L-phenylalanine into *trans*-cinnamic acid and ammonia by nonoxidative deamination (Fig.), an early key reaction in Flavonoid (see) and Lignin (see) biosynthesis. From most plant sources, PAL is a tetrameric protein, M_r about 330000, consisting of 4 identical subunits of M_r 83000. Exceptions include PAL from mustard cotyledons (4 subunits, each of M_r 55000) and PAL from wheat leaves (2 subunits of M_r 75000 plus 2 of M_r 85000). PAL possesses two functionally active sites per tetramer, and, like histidine ammonia-lyase, the active sites contain a dehydroalanine residue (not part of an orthodox peptide chain, but present in a Schiff's base linkage at the active site). It is thought that the amino group of Phe adds to the β-position of the dihy-

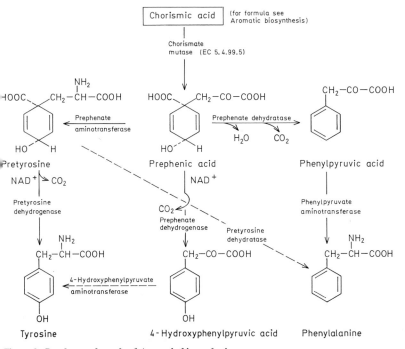

Figure 2. Prephenate branch of Aromatic biosynthesis.

droalanyl double bond, resulting in the formation of enzyme-ammonia and enzyme-cinnamate intermediates. This would account for the correct stereochemical elimination of the pro-3S hydrogen of Phe, leading to *trans* elimination of ammonia, which has been proven by using Phe stereospecifically labelled with isotopic hydrogen at C-3. PAL from many sources shows negative rate cooperativity (Hill coefficient < 1.0) with respect to Phe. Since PAL is competitively inhibited by *trans*-cinnamic acid, the enzyme in vivo is probably more sensitive to cinnamic acid concentration than to the pool size of Phe. Cinnamic acid therefore appears to be important in regulating the flux through the phenylpropanoid pathway. PAL from monocots and microorganisms also catalyses deamination of Tyr to 4-coumaric acid and ammonia. PAL increases rapidly when biosynthesis of flavonoid Phytoalexins (see) is initiated; it therefore represents an ideal system for investigation of de novo gene regulation in plants.
[R. A. Dixon et al. *Advances in Enzymology* (ed. Alton Meister) **55** (1983) 1-136].

appropriate P. Generally the quantity of P. produced by one animal is less than 1 μg. The quantity of Bombykol (see) (the sexual attractant of the silk worm moth) produced by one female moth is sufficient to attract all male members of the species in existence, if they were within range. The activity of a sexual attractant is expressed in attractant units (AU). One unit is the concentration of material (μg/ml) in light petroleum (b. p. 30–50 °C), so that when a glass rod is dipped in the solution and dried it will attract 50% of experimental males from a distance of 1 m, and stimulate them to vigorous wing buzzing. One unit is in the range of 10^{-6} μg and lower.

Scientifically, P. are useful for the investigation of biological information transfer, receptor theories and structure-activity relationships. Some P. are being used increasingly in the integrated control of insect pests; their potency serves the interest of science and efficiency, while they have no harmful effect on the environment.

P. must be distinguished from allelochemicals,

L-Phenylalanine   *trans*-Cinnamic acid

Phenylketonuria: see L-Phenylanine.
Phenylpyruvic oligophrenia: see L-Phenylalanine.
Pheophorbide: see Chlorophyll.
Pheophytin: see Chlorophyll.
Pheromones: predominantly low M_r substances produced by animals, especially insects (insect attractants), and secreted outside the body for purposes of communication (chemical biocommunication) with members of the same species. The main structural types are lower terpenes and higher unbranched fatty acids, and there are also ali- and heterocyclic representatives.

Depending on their mode of perception, P. are classified as oral or olfactory. An animal may show a direct reaction to P., which ceases when the P. disappears (i. e. a releasing effect), or the response may constitute a long term physiological change (i. e. a priming effect). In most cases, a mixture of P. is necessary, together with appropriate biotic und abiotic environmental factors.

For individual members of a species, the most important P. are the Sexual attractants (see). For social insects, e. g. honey bees, ants and termites, aggregation P., alarm P. and trail P. are also important. The activity of P. can be studied by electrophysiological measurement of the nerve impulse of isolated receptor organs (e. g. electroantennograms, abb. EAG); known quantities of P. (10^{-6}–10^{-2} μg) are blown over the isolated antennae in a stream of air, and the resulting cell potential is fed to a recorder with the aid of microelectrodes. The olfactory cells of the silk worm moth or the cockroach produce a measurable nerve impulse in response to one molecule of the

which serve as signal substances between individuals of different species. These are subdivided into allomones (of benefit to the producer, e. g. warning secretions), and kairomones (e. g. flower scents). A strict subclassification is, however, difficult, especially for substances with multiple functions.

Phlein: a high M_r reserve carbohydrate in plants. P. is a straight chain polymer of fructofuranose units joined by 2,6-glycosidic linkages. There is probably a D-glucose unit at the reducing end of the chain.

Phleomycins: see Bleomycins.

Phlorhizin, *phloridzin:* a dihydrochalcone found in the root bark of pears, apples and other members of the *Rosaceae*. P. specifically blocks resorption of glucose by kidney tubules, thus inducing glucosuria. It therefore finds use in experimental physiology. Its activity may be due to inhibition of mutarotase.

Phlorhizin

Phlorin: 5,22-dihydroporphyrin.
Phomine: see Cytochalasins.
Phorbol esters: esters of phorbol, $C_{20}H_{28}O_6$. Some are tumor-promoting compounds originally discovered in oil expressed from the seeds of *Croton tiglium*, but not all P. e. are tumorigenic. [Hecker & Schmidt *Fortschr. Chem. Org. Naturst.* **31** (1974) 377]

There is a direct correlation between tumorigenicity and ability to stimulate an amiloride-sensitive Na^+/H^+ exchange across the plasma membranes of cultured human leukemia cells. [J.M. Besterman & P.Cuatrecasas *J. Cell. Biol.* **99** (1984) 340–343]. A similar exchange appears to be associated with initiation of proliferation or differentiation in cultured cells.

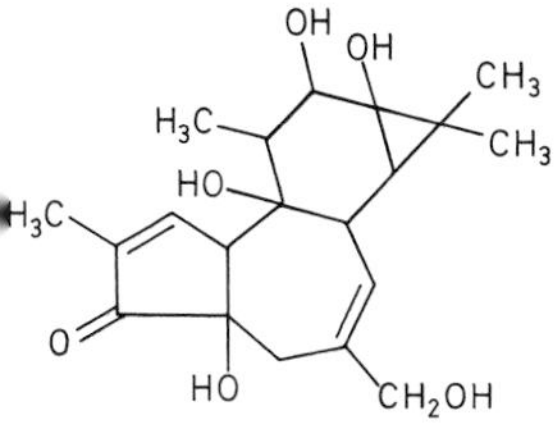

Phorbol

Phosphagens: energy rich guanidinium or amidine phosphates, which function as storage depots for high energy phosphate in muscle. Excess energy rich phosphate (i.e. ATP) is transferred to P., from which ATP can be regenerated when required. In invertebrates, arginine phosphate (phosphoarginine: M_r 254.2) is the commonest P. The phosphate group is attached to the guanidine nitrogen of the amino acid:

$$ATP + \text{Guanidine derivative} \rightleftharpoons$$
$$ADP + \text{Phosphagen} (^-_H^N \sim \textcircled{P})$$

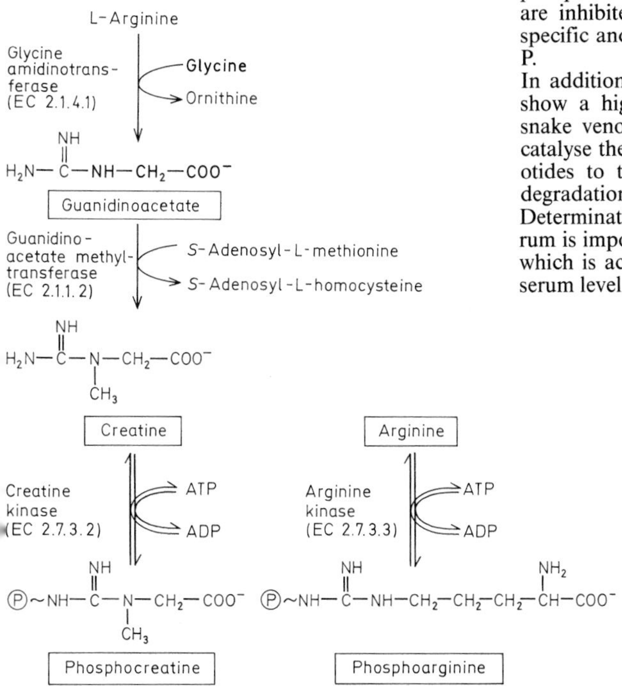

Synthesis of phosphocreatine (creatine-5-phosphate) and phosphoarginine (arginine phosphate).

The reaction is catalysed by arginine kinase (EC 2.7.3.3) (Fig.). There are several other invertebrate P., particularly in worms, e.g. lombricine phosphate and taurocyamine phosphate. The P. of vertebrate muscle is creatine phosphate (phosphocreatine: M_r 211.1), which is formed from creatine and ATP by the action of creatine kinase (EC 2.7.3.2) (Lohmann reaction, Fig.): reversal of the reaction regenerates ATP. The system, creatine phosphate + creatine kinase + ADP, is often used in vitro for the continual generation of ATP, if the enzyme under investigation is inhibited by substrate levels of ATP.

Phosphatases, *phosphoric monoester hydrolases* (EC sub-sub-group 3.1.3): esterases that catalyse the hydrolysis of monophosphate esters. P. are widely distributed in living organisms, and they are mostly dimeric proteins with a catalytically important serine residue in the active center. They are classified on the basis of their pH optima as acid P. (EC 3.1.3.2) e.g. the single chain P. of liver (M_r 16000) and erythrocytes (M_r 10000), and prostate P. (M_r 102000; 2 chains, each of M_r 50000), or alkaline P. (EC 3.1.3.1), e.g. intestinal mucosa P. (M_r 140000; 2 chains, each of M_r 69000), placenta and bone P. (M_r 120000; 2 chains, each of M_r 60000) and the P. of *Escherichia coli* (M_r 85000; 2 chains, each of M_r 43000). The alkaline P. contain one or two essential zinc atoms per subunit, and also require Mg(II) ions for full activity; these cations have no effect on acid P. In contrast to other esterases, P. are only slightly inhibited by diisopropylfluorophosphate. On the other hand, the alkaline P. are strongly inhibited by ethylenediaminetetraacetic acid (EDTA), inorganic phosphate and L-phenylalanine, while the acid P. are inhibited by fluoride. D(+)-Tartaric acid is a specific and highly potent inhibitor of prostate acid P.

In addition to the relatively unspecific P., some P. show a high specificity, e.g. 5'-nucleotidase from snake venom and 3'-nucleotidase from rye, which catalyse the hydrolysis of 5'-nucleotides or 3'-nucleotides to their respective nucleosides (see Purine degradation; Pyrimidine degradation).

Determination of the acid tartrate-sensitive P. in serum is important in the diagnosis of prostate cancer, which is accompanied by a marked increase in the serum level of this enzyme. Measurement of alkaline

P. in serum is used in the diagnosis of bone disease, especially bone tumors, and diseases of the liver (hepatitis) and the gall bladder (e.g. obstructive jaundice), all of which result in a several fold increase in serum P. The isoenzymes of both acid and alkaline P. differ on the basis of their neuraminic acid contents. The most widely used substrate for P. assay is p-nitrophenyl phosphate; the released p-nitrophenol can be determined directly by photometry (λ_{max}405 nm).

Phosphatidases: see Phospholipases.

Phosphatide degradation: the hydrolytic cleavage of phosphatides (see Phospholipases).

Phosphatides: see Phospholipids.

Phosphatidic acid: see Phospholipids.

Phosphatidylcholine: see Phospholipids.

Phosphatidylcholine-sterol acyltransferase deficiency: see Inborn errors of metabolism.

Phosphatidylethanolamine: see Phospholipids.

Phosphatidyl inositol: see Inositol phosphates.

Phosphatidylserine: see Phospholipids.

Phosphoadenosinephosphosulfate, *3'-phosphoadenosine-5'-phosphosulfate,* *adenosine-3'-phosphate-5'-phosphosulfate,* abb. **PAPS, active sulfate:** the product of sulfate activation. PAPS is a key intermediate in the reduction of sulfate and the formation of sulfate esters by sulfokinases. It is produced in a two stage reaction:

ATP + SO_4^{2-} → APS + PP_i (ATP sulfurylase)
(EC 2.7.7.4)
APS + ATP → PAPS + ADP (APS kinase)
(EC 2.7.1.25)
Sum: $\overline{2ATP + SO_4^{2-} → PAPS + ADP + PP_i}$

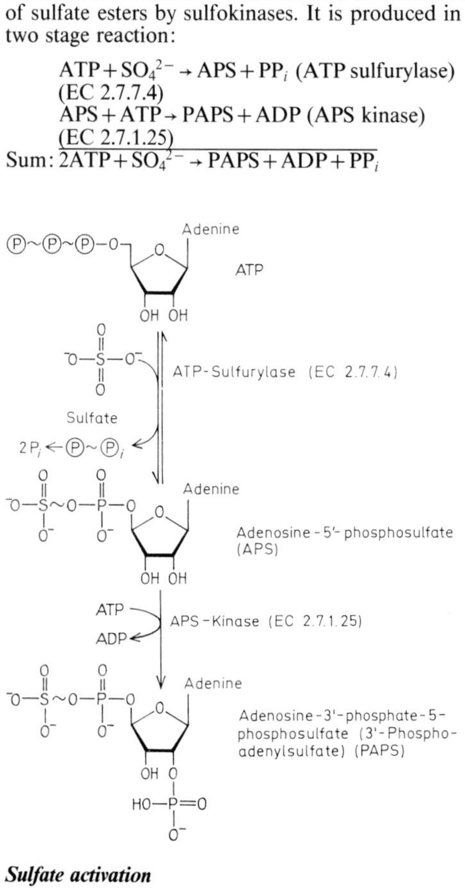

Sulfate activation

Synthesis of PAPS (Fig.) is known as Sulfate activation (see). In the first stage, the terminal pyrophosphate of ATP is replaced by sulfate, forming adenosine phosphosulfate (APS); this reaction is catalysed by sulfate adenylyltransferase (ATP sulfurylase, EC 2.7.7.4). In the second stage, the 3' position of the adenosine residue of APS is phosphorylated by ATP (catalysed by adenylsulfate kinase, EC 2.7.1.25) Formation of the anhydride bond between adenylic acid and the sulfate anion is a strongly endergonic reaction. The energy balance of the overall process becomes weakly negative only after the hydrolysis of the pyrophosphate and the consumption of two molecules of ATP. As mixed acid anhydrides of sulfate and adenylic acid, PAPS and APS have high group transfer potentials.

Phosphoarginine: see Phosphagens.

Phosphocreatine: see Phosphagens.

Phosphodiester: a phosphate ester in which two hydroxyl groups of the phosphoric acid are esterified with organic residues:

$$RO-PO_2H-OR'.$$

For example, R and R' may be nucleosides. All polynucleotides and nucleic acids are P., in which the 3' and 5' positions of neighboring pentose units are linked by esterification with a phosphate residue.

Phosphodiesterases: enzymes that catalyse the hydrolytic cleavage of phosphodiesters, e.g. endonucleases, Ribonuclease (see) and Deoxyribonuclease (see), and the less specific exonucleases. The latter degrade both DNA and RNA stepwise in the 3' → 5' direction, producing 5'-mononucleotides (snake venom P., EC 3.1.4.1), or in the 5' → 3' direction, producing 3'-mononucleotides (spleen P., EC 3.1.16.1). P. have been used for the sequence determination of nucleic acids (especially RNA). 3':5'-Cyclic-nucleotide P. (EC 3.1.4.17) catalyses the hydrolysis of cyclic AMP (see Adenosine phosphates).

Phospho*enol*pyruvate: see Pyruvate.

Phospho*enol*pyruvate carboxylase: see Photosynthetic carboxylation.

Phosphofructokinase, *6-phosphofructokinase* (EC 2.7.1.11): an oligomeric phosphotransferase, and a key control enzyme of glycolysis. P. is induced by insulin, and its activity is subject to allosteric control; it is activated by AMP, fructose 6-phosphate, fructose 1,6-bisphosphate, magnesium, potassium and ammonium ions, and inhibited by ATP and citrate (see Pasteur effect). P. catalyses the phosphorylation of fructose 6-phosphate by ATP in the presence of magnesium ions to form fructose 1,6-bisphosphate. The reaction is irreversible, and the conversion of fructose 1,6-bisphosphate to fructose 6-phosphate is catalysed by the fluoride-sensitive enzyme, fructose bisphosphatase, EC 3.1.3.11 (fructose diphosphatase). P. of yeast, muscle, liver and erythrocytes have been isolated and studied. Yeast P. consists of six subunits, 3α and 3β, which are immunologically unrelated. Subunit M_r is 130000 in the native protein, and 96000 in the protein modified by partial degradation by yeast proteases. The hexameric enzyme has M_r 755000 (native), or 570000 (proteolytic modification). In 6 M guanidinium chloride, these subunits dissociate into two equal sized chains of M_r 63000 (native) or 59000 (modified).

The degree of association of muscle P. depends on its concentration. At 7 mg protein/ml it exists as an active hexameric aggregate (M_r 2×10^6); at 0.5 mg protein/ml as an active monomer (M_r 340000). In 6 M guanidinium chloride and 0.1 M mercaptoethanol, the latter dissociates into its 4 inactive chains (M_r 80000). A similar behavior has been described for the octameric erythrocyte P. (M_r 500000). Chicken liver P. (M_r of the smallest active form 400000) is cold sensitive; it dissociates reversibly at 0 °C into four inactive protomers (M_r 100000), each consisting of two identical chains. *Clostridium* species contain a P. of M_r 144000 (4 subunits, M_r 35000).

Phosphogluconate pathway: see Pentose phosphate cycle.

Phosphoglyceric acids: monophosphate esters of Glyceric acid (see).

Phosphoglycerides: see Phospholipids.

Phosphohexoketolase pathway: see Phosphoketolase pathway.

Phosphoketolase pathway: a pathway of carbohydrate degradation found in various microorganisms, especially *Lactobacillus,* in which a ketopentose phosphate undergoes phosphorolytic cleavage to triose phosphate and acetyl phosphate (Fig.). The key enzyme is phosphoketolase (EC 4.1.2.9), which catalyses the TPP-dependent, irreversible cleavage of D-xylulose 5-phosphate to D-glyceraldehyde 3-phosphate and acetyl phosphate. The balance of pentose metabolism in *Lactobacillus* species is Pentose ($C_5H_{10}O_5$) + 2ADP + 2P$_i$ → Acetate ($C_2H_4O_2$) + L-Lactate ($C_3H_6O_3$) + 2ATP. There is a net yield of 2ATP per molecule of pentose; one molecule of ATP is required for the phosphorylation of the pentose, which is ultimately converted to D-xylulose 5-phosphate (Fig.); 2ATP are derived from the Glycolysis (see) of glyceraldehyde 3-phosphate, and one from the acetyl kinase reaction with acetylphosphate.

In *Acetobacter xylinum,* carbohydrate is degraded by a phosphohexoketolase pathway; a specific phosphohexoketolase catalyses the phosphorolytic cleavage of fructose 6-phosphate to D-erythrose 4-phosphate and acetylphosphate.

Phosphokinases: see Kinases.

Phospholipases, *phosphatidases:* a collective name for the carboxylic acid esterases, P. A$_1$, A$_2$ and B, and the phosphodiesterases, P. C and D, which are specific for lecithins. P. A$_1$ (EC 3.1.1.32) catalyses the release of the fatty acid at C-1 of the glycerol, producing a lysophosphatide (a 2-acylglycerophosphocholine) which hemolyses erythrocytes. P. A$_2$ (EC 3.1.1.4) removes the unsaturated fatty acid at C2. P. B (EC 3.1.1.5) also removes the unsaturated fatty acid, but only from lysophosphatides. P. C (EC 3.1.4.3) releases the base in its phosphorylated

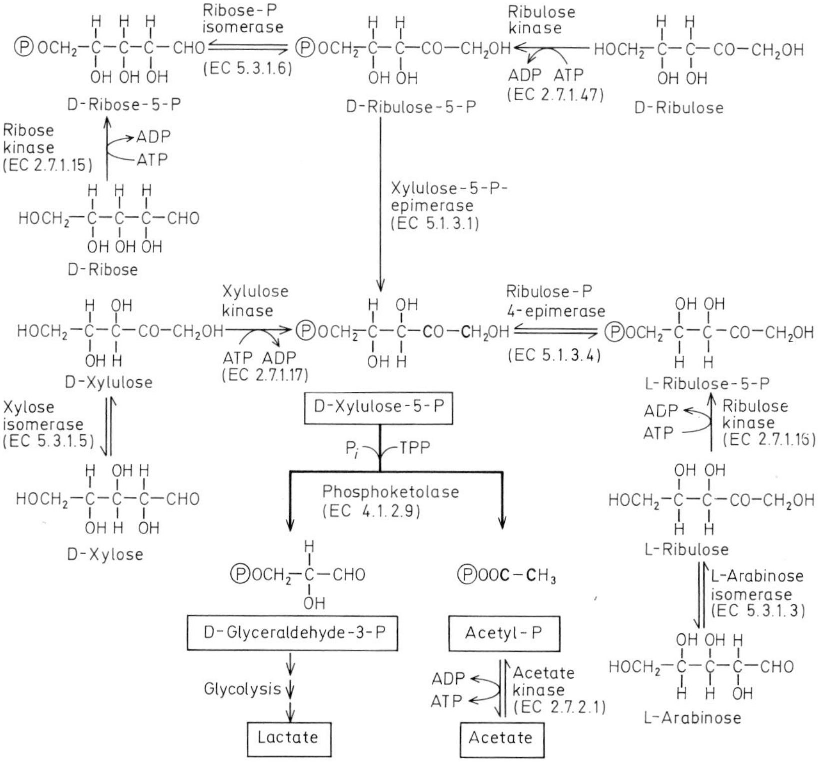

The position of phosphoketolase in pentose metabolism. P phosphate, TPP thiamin pyrophosphate.

form. P. D (EC 3.1.4.4) cleaves on the other side of the phosphoryl group and releases the nonphosphorylated base (Fig.). P. activity is particularly high in liver and pancreas (P. A_1), in bee and snake venom (P. A, and P. A_2), in bacteria (P. C) and plants (P. D). Due to their compact structures (6–15 disulfide bridges), P. A_2 and P. C have unusual heat stability (5 min at 98 °C) and are insensitive to diethyl ether, chloroform and 8 M urea. P. A_2 from porcine pancreas is synthesized as a zymogen (130 amino acid residues of known sequence, M_r 14660), which is converted to active P. A_2 (123 amino acid residues) by tryptic removal of the *N*-terminal heptapeptide, Pyr-Glu-Gly-Ile-Ser-Ser-Arg-. P. A_2 from snake venom *(Crotalus atrox)* (M_r 14500) and bee venom (M_r 14550, 129 amino acid residues) are about the same size as pancreas P. A_2. Both P. A_2 isoenzymes from the venom of *Crotalus adamanteus* (266 amino acid residues, M_r 29865) are twice the size of other P. A_2. Bacterial P. C have been isolated from *Clostridium welchii* and *Bacillus cereus*.

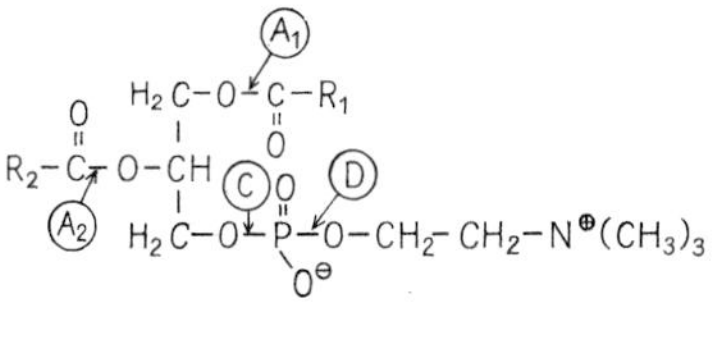

Sites of attack by lecithin-cleaving enzymes. R_1 saturated fatty acyl residue. R_2 mono- or multi-unsaturated fatty acyl residue.

Phospholipid: any lipid containing phosphoric acid as a mono- or diester. P. are the basic constituents of Biomembranes (see) and are especially abundant in the brain and myelin sheaths of the nerves. P. are derived either from glycerol (the generic term "glycerophospholipid" is recommended by the IUPAC-IUB; "Phosphoglyceride" is often used) or sphingosine (sphingophospholipids).
(1) Glycerophospholipids (phosphoglycerides, phosphatides) are derivatives of *phosphatidic acid* (Fig. 1), a derivative of glycerol phosphate in which both the remaining hydroxyl groups are esterified to fatty acids. The phosphate group may be additionally ester-

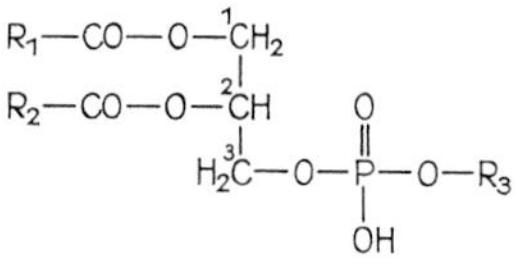

Figure 1. Phosphoglycerides.

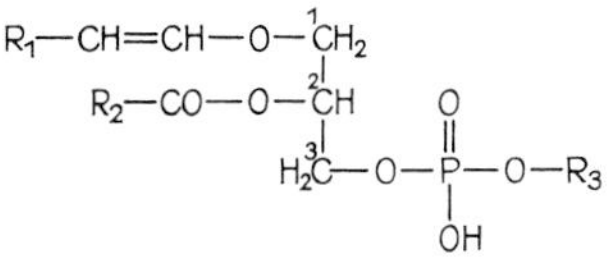

R_1 = Alkyl

R_2 = Alkenyl

$R_3 = CH_2CH_2-NH_2$ or $CH_2CH_2-\overset{\oplus}{N}(CH_3)_3$

Figure 2. Plasmalogens.

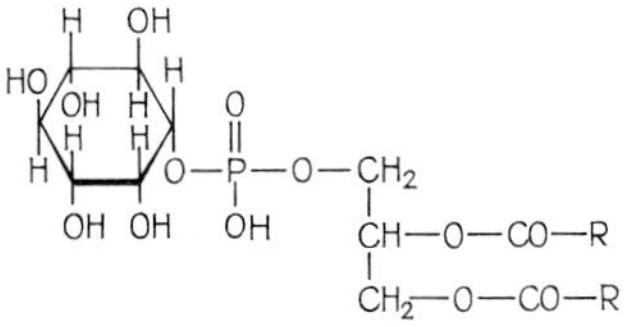

Figure 3. Inositol phosphatides.

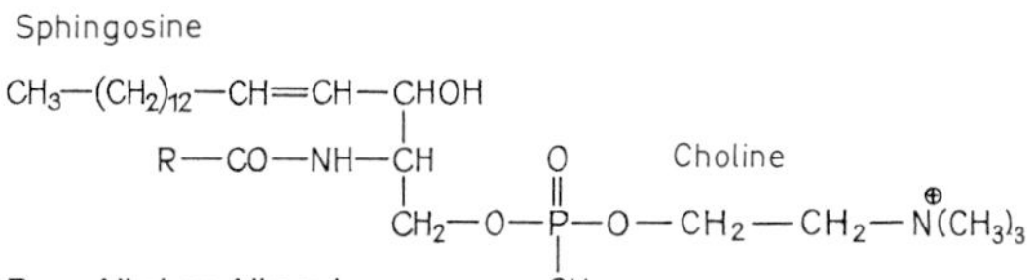

Sphingosine

$CH_3-(CH_2)_{12}-CH=CH-CHOH$

R = Alkyl or Alkenyl

Figure 4. Sphingosine phosphatides.

ified to one of the following alcohols: choline, ethanolamine, serine, glycerol or inositol. The resulting compounds are called phosphatidylcholine *(lecithin)*, phosphatidylethanolamine, phosphatidylserine (these two groups are called *cephalins* in older literature), phosphatidylinositol and phosphatidylglycerol. *Lysophosphatidic acids (lysolecithins* and *lysocephalins)* contain an unesterified hydroxyl group. They are produced from phosphatides by the enzymes phospholipase A_2 (EC 3.1.1.4) and A_1 (EC 3.1.1.32). *Plasmologens* Fig. 2) are glycerophospholipids in which the glycerol bears a 1-alkenyl ether group. *Plasmenic acids* are compounds in which glycerol 3-phosphate has a 1-alkenyl ether group and a fatty acid esterified to the second hydroxyl. In analogy to the derivatives of phosphatidic acid, these compounds are called plasmenylethanolamine, plasmenylserine, etc. In *diphosphatidylglycerol* (Cardiolipin, see), two phosphatidic acid moieties are esterified to a single molecule.
(2) *Sphingomyelins* (sphingophospholipids, ceramide 1-phosphorylcholine) are derivatives of the long, unsaturated amino alcohol *sphingosine* rather than glycerol. In sphingomyelins, the amino group of the sphingosine forms an amide linkage to a fatty acid, and the primary hydroxyl is esterified to phosphorylcholine (Fig. 4).
All phospholipids are thus composed of a strongly hydrophobic portion, the fatty acids and the sphingosine chain, and a highly hydrophilic portion, which in many cases (lecithins and cephalins) is zwitterionic. This accounts for their tendency to form bilayers in aqueous media and for their excellent emulsifiying properties.

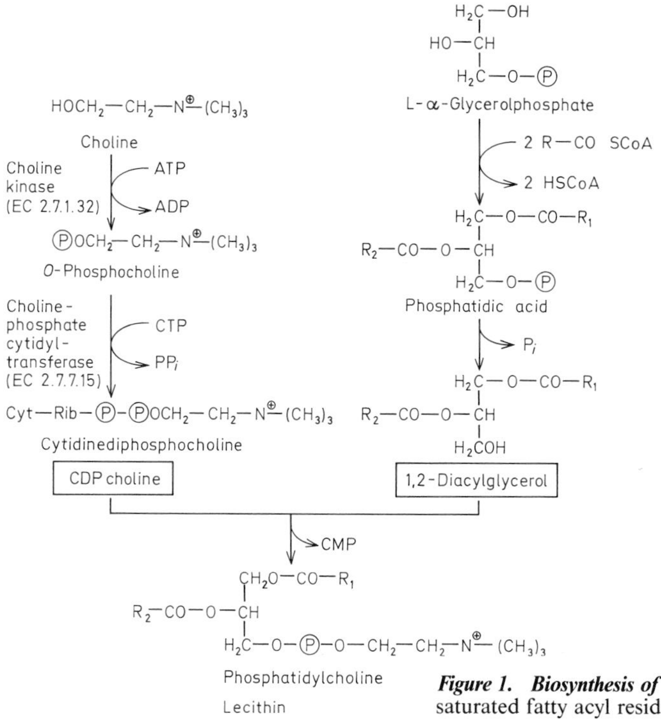

Figure 1. Biosynthesis of lecithin. Cyt cytosine, Rib ribose, R₁ saturated fatty acyl residue, R₂ unsaturated fatty acyl residue.

Phospholipid biosynthesis: of particular importance is the biosynthesis of lecithins and cephalins, which are glycerophosphatides. The phosphatidic acid and 1,2-diacylglycerol precursors of glycerophosphatides are synthesized as described under Fat biosynthesis (see).

In the biosynthesis of lecithin (Fig. 1), a saturated fatty acid (e.g. palmitic or stearic) is introduced at the α'-position of glycerol phosphate, and an unsaturated fatty acid (e.g. oleic or linolenic) is introduced at the β-position. A phosphorylcholine unit is then transferred to the diacylglycerol from cytidine diphosphate choline (CDP-choline) by the action of a phosphorylcholine transferase (EC 2.7.7.15). The resulting CMP is rephosphorylated to CTP by ATP, and the CTP serves to activate another molecule of phosphorylcholine.

The biosynthesis of cephalins is analogous to that of lecithins. The biosynthesis of phosphatidylserine proceeds from phosphatidylethanolamine.

In the biosynthesis of sphingomyelins, sphingosine is formed from palmitaldehyde (hexadecanal) and serine, via dihydrosphingosine (Fig. 2). Sphingomyelin is formed from N-acylsphingosine (ceramide) by condensation with CDP-choline and release of CMP. Sphingosine is also a precursor for the synthesis of cerebrosides, sulfatides and gangliosides.

Phosphon D: 2,4-dichlorobenzyltri-n-butylphosphonium chloride, a synthetic plant growth retardant. P. inhibits, e.g. the growth of chrysanthemum stems and induces flowering.

Phosphopantetheine: see Pantetheine 4'-phosphate.

Phosphoproteins: conjugated proteins, containing phosphate esterified with the hydroxyl

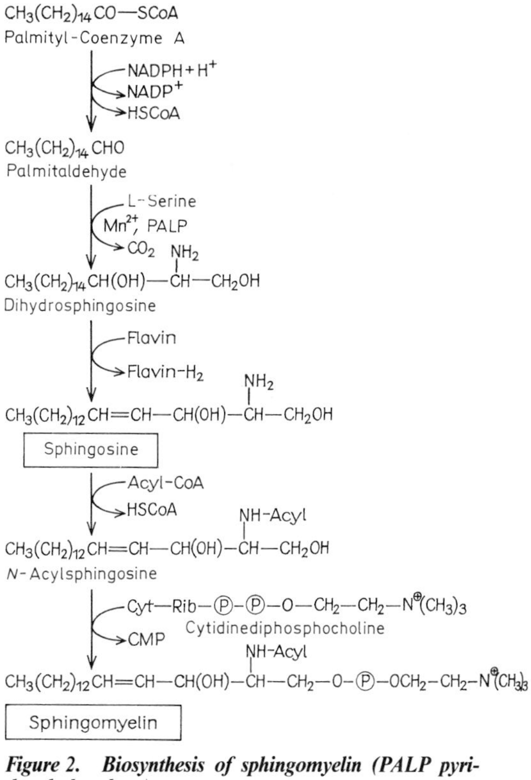

Figure 2. Biosynthesis of sphingomyelin (PALP pyridoxal phosphate).

453

groups of serine or (less often) threonine residues. Well known P. are casein (see Milk proteins) and ovalbumin (see Albumins). The latter can contain one or two phosphate groups; this microheterogeneity (see Heterogeneity) is due to variations in the production and activity of the phosphoprotein phosphokinase in the hen oviduct. Other P. are phosphovitin and vitellin of egg yolk, and pepsin of gastric juice.

Phosphopyruvate kinase: see Pyruvate kinase.

5-Phosphoribose I-diphosphate: see 5-Phosphoribosyl l-pyrophosphate.

5-Phosphoribosylamine, abb. *PRA:* an intermediate in Purine biosynthesis (see).

5-Phosphoribosyl l-pyrophosphate, *5-phosphoribose 1-diphosphate,* abb. *PRPP:* an energy-rich sugar phosphate formed by the transfer of a pyrophosphoryl residue from ATP to ribose 5-phosphate. M_r 390.1. PRPP is concerned in various biosynthetic reactions, e.g. biosynthesis of purines, pyrimidines and histidine.

Phosphoroclastic fission of pyruvate: a special mechanism for the cleavage of pyruvate found only in saccharolytic Clostridia. It is responsible for the synthesis of ATP during nitrogen fixation. The first stage is the synthesis of acetyl phosphate:

$$CH_3-\overset{\overset{\displaystyle O}{\|}}{C}-COOH + P_i \longrightarrow$$
Pyruvate

$$CH_3-\overset{\overset{\displaystyle O}{\|}}{C}-O-PO_3H_2 + CO_2 + H_2$$
Acetyl phosphate

followed by synthesis of ATP from acetyl phosphate catalysed by acetokinase:
Acetyl phosphate + ADP $\rightleftharpoons$ ATP + Acetate. The first stage requires several enzymes (in the form of a multienzyme system analogous to pyruvate dehydrogenase, together with phosphotransacetylase and hydrogenase) and cofactors (thiamin pyrophosphate, ferredoxin and coenzyme A).

Phosphorus: see Bioelements.

Phosphotransacetylase: see Acetyl phosphate.

Phosphotransferase system: see Active transport.

Phosphovitin: an egg yolk protein containing 10% phosphate (M_r 180000). Serine constitutes 50% of the total amino acid content, and all the serine residues are phosphorylated. P. is synthesized in the liver of the laying hen and transported in the blood to the developing egg.

Photocitral: a cyclization product of Citral (see).

Photoheterotrophism: see Nutritonal physiology of microorganisms.

Photolysis of water: cleavage of water by the light reaction of Photosynthesis (see). P. is a property of photosystem II. It is not a simple photodissociation of water, but is the physiological counterpart of the Hill reaction (see). Electrons are withdrawn from water or OH^- ions, then transported to $NADP^+$ via an electron transport chain; this results in the production of molecular oxygen with the formation of $NADPH + H^+$. Since Hill reagents oxidize water by withdrawal of electrons, $NADP^+$ is the natural Hill reagent.

Photophosphorylation: the synthesis of ATP in Photosynthesis (see). The mechanism of P. is similar to that of Oxidative phosphorylation (see) by the respiratory chain; in both cases cytochromes are involved in electron transport. A distinction is drawn between cyclic and noncyclic P. Both forms are found in green algae and higher plants. Cyclic P. involves cyclic electron transport. Under the influence of light, electrons emitted from chlorophyll *a* return to chlorophyll *a* via an electron transport chain, thereby giving rise to ATP synthesis. Thus the positive holes left in the chlorophyll structure by the loss of electrons are refilled, and the electron excitation energy is transduced to the chemical energy of ATP. Cyclic P. involves cytochrome *f,* and the only product is ATP. In noncyclic P., ATP synthesis is linked to the transport of electrons from water (see Photolysis of water) to $NADP^+$, thus producing both ATP and a reducing agent (NADPH). The production of molecular oxygen, the byproduct of water photolysis, is characteristic of photosynthesis in green plants and algae. Noncyclic P. can be considered as a Hill reaction (see) coupled to the synthesis of ATP. Whereas cyclic P. requires only photosystem I, noncyclic P. depends upon the joint operation of both photosystems, which are connected in series; electrons are transported from OH^- ions of water to $NADP^+$ in an open-chain (noncyclic) system. The two kinds of P. are functionally and structurally separate in the chloroplast, but they are closely interrelated. They can be separated experimentally, e.g. by the inhibitors *o*-phenanthroline or dichlorophenyldimethylurea (DCMU), which block photosystem II. There is considerable uncertainty regarding the ATP yield (P:2e ratio) of both cyclic and noncyclic P. Most experimentalists obtain values of approximately 1.3 (molecules ATP synthesized per 2 electrons transported to $NADP^+$) for noncyclic P., and the yield from cyclic P. is probably lower.

Photoproteins: proteins responsible for luminescence in many light-emitting coelenterates. Light emission by P. does not involve a luciferin-luciferase system (see Luciferin), and the reaction proceeds in the absence of oxygen. Aequorin, the P. of the jelly fish *Aequorea,* contains a substituted 2-aminopyrazine as chromophore; light production (λ_{max} 469 nm) is activated specifically by Ca^{2+}. A similar Ca^{2+}-activated P., obelin, has been isolated from *Obelia geniculata* (λ_{max} of emitted light 475 nm). Both of these proteins have been employed as sensitive probes for measuring intracellular Ca^{2+} concentrations, with a sensitivity of at least 10 nM Ca^{2+}. [Campbell, A. K. and Simpson, J. S. A., Chemi- and bioluminescence as an analytical tool in biology, in *Techniques in Metabolic Research,* B 213, pp. 1–56, Elsevier (1979)].

Photoreactivation: repair of biological systems damaged by UV-irradiation, in a process promoted by light of a different wavelength. Pyrimidine dimers, which result from UV-irradiation, can be monomerized by the action of UV light of shorter wavelength or light of longer wavelength, which promotes the action of repair enzymes. The enzyme binds only to UV-damaged DNA and converts pyrimidine dimers to monomers when irradiated with light of an appropriate wavelength.

Photorespiration: light enhanced respiration in photosynthetic organisms. Illumination of C3-plants markedly increases the rate of oxygen utilization; this increase in respiration can be as high as 50% of the net photosynthetic rate. P. thus results in a loss of yield in the photosynthesis of C3-plants. In C4-plants, P. is either absent or extremely low. P. is largely due to the oxygenase activity of Ribulose-bisphosphate carboxylase (see), which oxidatively cleaves ribulose 1,5-bisphosphate into phosphoglycolate and 3-phosphoglycerate. Glycolate (derived from the phosphoglycolate) leaves the chloroplasts and enters the peroxisomes, where it is oxidized (by a flavoprotein oxidase) to glyoxylate. Hydrogen peroxide from the action of the flavoprotein oxidase may oxidize some of the glyoxylate to formate and CO_2, but the majority is destroyed by peroxidases and catalase. Most of the glyoxylate is transaminated to glycine, which enters the mitochondria. Glycine may be decarboxylated and/or converted into serine, some of which may reenter the peroxisomes and become oxidized to hydroxypyruvate and D-glycerate. Thus various reactions occur that result in loss of carbon as CO_2. The process depends on light because light is required for the operation of the Calvin cycle, which supplies the ribulose 1,5-bisphosphate. Some of these reactions are illustrated in the diagram of the glycolate cycle shown under Glycine (see), but here the phosphoglycolate is shown as coming from active glycolaldehyde. See also CO_2-compensation point; Light compensation point.

Photosynthesis: 1. Any light-dependent synthesis. 2. The reductive synthesis of carbohydrate in green plants and Photosynthetic bacteria (see). P. was formerly defined as the assimilation of carbon dioxide, but it is now recognized as primarily a process of energy transduction, in which light energy is converted into the chemical energy of oxidizible organic carbon compounds. P. in green plants (but not bacterial P.) can be represented by:

$$6CO_2 + 6H_2O \xrightarrow[\text{Chlorophyll}]{h\nu} C_6H_{12}O_6 + 6O_2.$$ In principle, this general equation is the reverse of Respiration (see). The reaction is catalysed by chlorophyll a, which is structurally bound in the Thylakoids (see). In photosynthetic bacteria, chlorophyll a is replaced by bacteriochlorophyll a (see Bacterial photosynthesis). Other Photosynthetic pigments (see) serve as auxilliary pigments for light absorption and energy transfer. P. is the most important process for the production of organic material in the biosphere. All nonphotosynthetic organisms are directly or indirectly dependent on P. of phototrophic organisms. By comparison, Chemosynthesis (see) plays a quantitatively insignificant role in the carbon cycle of the biosphere.

P. is an energy-dependent (endergonic) process, in which light energy is converted into the chemical energy of ATP by Photophosphorylation (see). In addition, P. in green plants (not in photosynthetic bacteria) produces reduced NADP, a reducing agent

used by the cell in reductive biosyntheses. Synthesis of ATP, NADPH and H^+ occurs in noncyclic photophosphorylation, which is linked to the photolysis of water. ATP and NADPH, known as the primary products of P., have a transitory existence (i.e. they are rapidly utilized) and they do not accumulate. They are the earliest relatively stable products of the light reaction of P., in which light energy, via electron excitation energy, is transduced to the chemical energy of an energy-rich compound (ATP) and a reducing agent (NADPH). Strictly speaking, the light reaction consists of purely photochemical events, and the chemical reactions of photophosphorylation in which light plays no direct part should be described as dark reactions. By convention, however, the photochemical events in structurally bound chlorophyll, together with the reduction of $NADP^+$ and the synthesis of ATP (i.e. the production of assimilatory power, a term coined by Arnon) constitute the light reaction of P. It is characteristic of the light reaction processes (in contrast to the purely chemical processes of the dark reaction) that they are not temperature dependent. The dark reaction of P. includes the reactions responsible for the reductive synthesis of carbohydrate, e.g. synthesis of starch or sucrose, and it includes the initial reactions of CO_2 fixation in C4-plants. Thus the dark reaction includes all reactions from the initial binding of carbon dioxide to the formation of carbohydrate reserves, e.g. starch. For a description of these processes, see Calvin cycle; Hatch-Slack-Kortschack cycle; Starch.

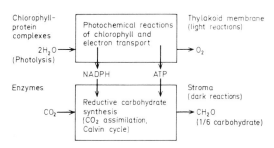

Figure 1. Relationship between light and dark reactions of photosynthesis.

In green plants, P. is localized in the Chloroplasts (see). All the processes of the reaction occur in the membrane system of the thylakoids, whereas the dark reaction proceeds in the stroma of the chloroplasts, which (with the exception of the thylakoid carboxydismutase) contains the enzymes of carbon dioxide assimilation (Fig. 1). The stroma also contains the Protein biosynthesis (see) system of the chloroplast.

Photosynthetic bacteria possess only one photosystem, known as photosystem I, whereas green plants possess photosystem I and II (see Photosystems). Cyclic phosphorylation requires only photosystem I. In noncyclic phosphorylation, both systems are coupled together, and their interrelationship is often shown diagrammatically by the so-called "zig-zag" energy diagram (Fig. 2). Chlorophylls a_I and a_{II} absorb light and emit electrons. The electrons emitted by chlorophyll a_I are received by ferredoxin and transferred to NADP$^+$. Electrons emitted by chlorophyll a_{II} are transported to chlorophyll a_I via an electron transport chain containing plastoquinone, plastocyanin and cytochrome f; these electrons enter the positive holes (electron vacancies) of chlorophyll a_I. Operation of the electron transport chain is coupled to the synthesis of ATP. The positive holes of chlorophyll a_{II} are filled by electrons from the photolysis of hydroxyl ions. The joint operation of the two photosystems leads to the synthesis of ATP and NADPH. Electrons from chlorophyll a_I may also take part in cyclic electron transport via a cytochrome chain and back to chlorophyll a_I, thereby producing ATP, but not NADPH (see Photophosphorylation). The natural electron acceptor of photosystem I has not been identified; it is referred to as ferredoxin reducing substance (FRS). The nature of electron transfer between photosystem II and plastoquinone is also unclear; the primary electron acceptor of photosystem II (which quenches the fluorescence of chlorophyll, and is therefore called substance Q) is unidentifed.

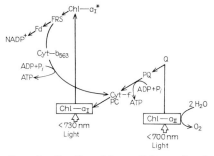

Figure 2. Coupling of the two light reactions in noncyclic photophosphorylation. Chl chlorophyll, Cyt cytochrome, Fd ferredoxin, FRS ferredoxin reducing substance, Q quenching substance.

Some herbicides act as artificial electron donors or acceptors which short circuit the electron transport chain of P., e.g. the bipyridylium herbicides, paraquat and diquat, compete with ferredoxin for electrons from the ferredoxin reducing substance (FRS); the reduced bipyridylium system is then spontaneously oxidized, forming hydrogen peroxide which damages the plant tissues. Other herbicides block electron flow from water in photosystem II, e.g. monuron (3-(4-chlorophenyl)1,1-dimethylurea, or CMU) and diuron (dichlorophenyldimethylurea, or DCMU).

Nomenclature: Photosystems I and II are responsible for light reactions 1 and 2, or photoevents 1 and 2. On the other hand it is customary to refer to the sum of the light photoevents, including ATP synthesis and NADP reduction, as *the* light reaction of P. Similarly, one may speak of the dark reactions or the overall dark reaction of P. Photosystem I is so called because it arose first in evolution (i.e. in photosynthetic bacteria), whereas photosystem II and the ac-

companying photolysis of water were later evolutionary developments.

Photosynthetic bacteria: phototrophic bacteria, e.g. green sulfur bacteria *(Chlorobacteriaceae),* purple sulfur bacteria *(Thiorhodaceae)* and nonsulfur purple bacteria *(Athiorhodaceae).* The *Chlorobacteriaceae* include *Chlorochromatium consortium* and species of *Chlorobium.* The *Thiorhodaceae* are represented by *Chromatium okenii* and *Thiospirillum jenense.* *Athiorhodaceae* include three genera: *Rhodopseudomonas,* *Rhodospirillum* and *Rhodomicrobium.* All P.b. are deeply colored, due to the presence of photosynthetic pigments.

In place of chlorophyll *a,* P.s. contain bacteriochlorophyll *a* (2-devinyl-2-acetyl-3,4-dihydrochlorophyll *a*). Green sulfur bacteria also contain bacteriochlorophylls *c* and *d* (formerly called *Chlorobium* chlorophylls), which absorb between 700 and 760 nm. In addition to bacteriochlorophylls, P.b. contain Carotenoids (see), which serve as auxilliary photosynthetic pigments. Due to their different pigmentation, the various types of P.b. are able to exploit different areas of the spectrum for photosynthesis.

P.b. neither produce nor consume molecular oxygen. They do not perform the photolysis of water, and they can live anaerobically, most of them being strict anaerobes. They contain one photosystem, analogous to the photosystem I of higher green plants, algae and blue-green bacteria. Green and purple sulfur bacteria perform the photolysis of hydrogen sulfide:

$$2H_2S + CO_2 \xrightarrow{\text{light}} [CH_2O]_n + H_2O + 2S$$
$$\text{Carbohydrate}$$

Sufficient energy for this process is achieved in one photoevent (in contrast to the photolysis of water, which requires two photoevents coupled in series; see Photosynthesis). Reducing power is trapped by ferredoxin, but some of the energized electrons are recycled (see Photophosphorylation) for the synthesis of ATP. Nonsulfur purple bacteria use various substances as donors of hydrogen and electrons, e.g. isopropanol:

$$2CH_3CHOHCH_3 + CO_2 \xrightarrow{\text{light}} [CH_2O]_n +$$
$$\text{Isopropanol} \qquad\qquad \text{carbohydrate}$$
$$2CH_3COCH_3 + H_2O$$
$$\text{Acetone}$$

Photosynthetic carboxylation: the enzymatic fixation of carbon dioxide in photosynthesis. In C-3 plants, the photosynthetic carboxylation enzyme is ribulose bisphosphate carboxylase (EC 4.1.1.39). In C-4 plants it is phospho*enol*pyruvate carboxylase (EC 4.1.1.31). P.c. is the first step of carbon dioxide assimilation in photosynthesis, and one of the dark reactions.

Photosynthetic cycle: see Calvin cycle.

Photosynthetic experimental organisms and systems: for technical reasons, certain systems are preferred for the investigation of photosynthesis, e.g. green algae such as *Chlorella,* photosynthetic bacteria and isolated chloroplasts. Green algae and euglenoids (e.g. *Euglena gracilis*) can be cultured under defined conditions in an illuminated chemostat (see Fermentation techniques), and they can also be grown in Synchronous culture (see). It is relatively easy to obtain chlorophyll-deficient mutants of these organisms, which must grow heterotrophically (see Mutant technique). The production of plastids can be prevented by culture in the dark; the admission of light induces the formation of the photosynthetic apparatus. Such systems are therefore especially suited to the study of the regulation of autotrophism and heterotrophism.

Photosynthetic pigments: pigments that take part in the trapping and utilization of light in Photosynthesis (see). Seed plants *(Spermatophyta),* ferns *(Pteridophyta),* mosses *(Bryophyta),* green algae *(Chlorophyta,* e.g. *Chlorella),* euglenoids *(Euglenophyta,* e.g. *Euglena)* and brittleworts *(Characeae)* contain both chlorophylls *a* and *b,* and carotenoids, but no biliproteins. The latter are found in red algae (see Rhodoplasts) und Blue-green bacteria (see). Certain algae lack chlorophyll *b (Chrysophyta, Pyrrophyta* and *Cryptophyta).* Table 1 lists the P.p. of various organisms, and table 2 shows the thylakoidal (see Thylakoids) carotenoids of the red beech *(Fagus silvatica)* and the green alga, *Chlorella pyrenoidosa.*

All P.p. are either hydrophobic, or they posses a strongly hydrophobic grouping, e.g. the phytol residue of chlorophyl. A simple model, in which P.p. are associated with the lipid layer of the thylakoid membrane, is however, unsatisfactory. It is necessary to propose a certain degree of ordered structure for P.p., and this is not possible if P.p. are subject to the random mobility of the lipid membrane components, as demanded by the fluid-mosaic model for membrane structure. The binding of a P.p. molecule to a protein would also be an unsatisfactory model, because the various P.p. would then be too widely separated for the efficient transfer of photons or resonance energy. A more feasible model would involve the binding of several P.p. molecules to one protein, and there is much evidence for a system of this kind; e.g. several chlorophyll-binding proteins have been isolated from thylakoid membranes, in particular P700-Chlorophyll *a*-protein (see) and Light-harvesting protein (see).

Photosystems, *pigment systems:* structural-functional units of the light reaction of Photosynthesis (see). The quantum efficiency of photosynthesis in chloroplasts falls sharply at wavelengths longer than 680 nm, although chlorophyll still absorbs light from 680 to 700 nm. This phenomenon is known as the "red drop". However, the quantum efficiency of light above 680 nm is increased by the simultaneous presence of shorter wavelength light. This Emerson effect (see) led to the proposal that photosynthesis depends upon the interaction of two light reactions (i.e. two photosystems), both driven by light less than 680 nm, but only one by light of longer wavelengths.

As a working hypothesis for the distribution of

Table 1. Photosynthetic pigments and their occurrence in the plant kingdom.

Organism	Chlorophylls					Biliproteins				Carotenoids	
	a	*b*	*c*	*d*	*e*	B*a*	B*c/d*	Per	Pcy	Carotenes	Xanthophylls
Higher plants*	+	+								+	+
Green algae	+	+								+	+
Brown algae	+		+							+	+
Diatoms	+		+							+	+
Red algae	+		+	+				+	+	+	(+)
Blue-green bacteria	+							+	+	+	+
Green sulfur bacteria						+	+			+	(+)
Purple sulfur bacteria						+				+	+

Footnotes to Table 1: –
* seed plants, ferns and mosses.
B*a* = bacteriochlorophyll *a,* B*c/d* = bacteriochlorophylls *c* and *d.*
Per = phycoerythrin, Pcy = phycocyanin, (+) = a trace.

Table 2. Percentage composition of thylakoidal carotenoids (from Wiessner).

	Red beech (Fagus sylvatica) %	Chlorella pyrenoidosa %
α-Carotene		4
β-Carotene + Lycopene	34	15
Sum of carotenes	34	19
Lutein	45	50
Violaxanthin	14	10
Neoxanthin	7	12
Sum of xanthophylls	66	72

Ratio xanthophylls/carotenes 1.95 *(Fagus),* 4.3 *(Chlorella).*

chlorphylls in the two photosystems (photosystems I and II, also called pigment systems I and II), it is proposed that the pigments are stacked in order of their maximum absorption wavelenths, arranged to trap light rather in the way that a rain gauge is designed to catch water. Thus in photosystem I, chlorophyll *b* (λ_{max} 650 nm) absorbs light; by resonance transfer, quanta are then passed to chlorophyll *a* (λ_{max} 670 nm), from there to chlorophyll *a* (λ_{max} 680 nm), then to chlorophyll *a* (λ_{max} 695 nm); finally all the collected energy is concentrated into and trapped by the relatively few molecules of P700. Light is also directly absorbed and transferred by each chlorophyll in the system. There are relatively few molecules of P700 (see Pigment 700), which is the ulitmate acceptor or energy trap. Similarly photosystem II contains chlorophyll b_{650}, chlorophyll a_{670}, chlorophyll a_{680} and a final energy trap known as P690. Photosystem II is responsible for the photolysis of water. It has been shown that 4 quanta are required for the production of one oxygen molecule by PS II. Removal of an electron from PS II by photoexcitation creates a cation, which in turn withdraws an electron from water via a water-cleaving protein. The reaction center of this protein contains two manganese ions, which are only 0.27 nm apart; they presumably undergo reversible oxidoreduction: $Mn^{3+} \rightleftharpoons Mn^{4+}$.

The two photosystems are coupled in series (see Photosynthesis).

Phototropism: see Nutritional physiology of microorganisms.

Phrenosine: see Glycolipids.

Phycobilins: see Phycobilisome.

Phycobiliproteins: see Phycobilisome.

Phycobilisome: a light-gathering structure in blue-green bacteria and red algae. P. are directly attached to photosynthetic membranes, but are not part of them. They consist of three types of chromophore, all of which are *phycobiliproteins* (or biliproteins or phycobilins): phycoerythrin (PE), phycocyanin (PC) and allophycocyanin (APC). A model of the P. structure is given in the figure. The rods of PE and PC are thought to be held together by noncolored proteins with M_r around 30000. The P. are somewhat larger than ribosomes, and may form a regular 2-dimensional array adjacent to the photosynthetic membrane.

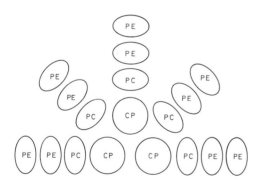

Model of phycobilisome structure
CP = core protein, probably allophycocyanin
PC = phycocyanin
PE = phycoerythrin

Light energy is passed efficiently from PE to PC to APC to photosystem I or II. The amounts of PE and PC in the cell may vary, depending on the light conditions. The absorbance maxima of PE are from 498–568 nm; of PC, around 625 nm; and of APC, 618–673 nm. The relative amounts of PE and PC present are determined by the spectral characteristics of the available light; in redder light, more PC is present. The total phycobiliproteins may amount to 60% of the soluble protein in the cell.

Phycobiliproteins. Like the phytochromes in higher plants, phycobiliproteins (PBC) have open tetrapyrrole chromophores. These may be biosynthesized from protoporphyrin IX or heme, but the pathway has not been elucidated. Each PBC consists of α- and β-chains. The M_r of these polypeptides varies among species, but the α-chain (M_r 12–20000) is usually smaller than the β-chain (M_r 15–22000). The smallest stable unit is a trimer, (α β)$_3$, but the basic unit in vivo is the hexamer. This consists of two trimers, each of which is disk-shaped. [A. N. Gantt, *Ann. Rev. Plant Physiol.* **32** (1981) 327–347]

Phylloquinone: see Vitamins; Vitamin K.

Physalaemin: Pyr-Ala-Asp-Pro-Asn-Lys-Phe-Tyr-Gly-Leu-Met-NH₂, an amphibian tachykinin from the skin of the South American frog, *Physalaemus fuscumaculatus.* [V. Erspamer et al., *Experentia* **20** (1964) 489–490] P. has also been identified as a tumor-related peptide in extracts of human lung small cell carcinoma [L. H. Lazarus et al. *Science* **219** (1983) 79–81]. It affects blood vessels and extravascular smooth muscle, stimulates exocrine secretion, and has a distinct neurotropic effect. The C-terminal hexapeptide from C. elicits full biological response in guinea pig ileum, while the presence of the five N-terminal amino acids is necessary for full expression of sialogogic activity. Conformational studies have been reported. [J.-L. Bernier et al. *Eur. J. Biochem.* **142** (1984) 371–377]

Physostigmine, *eserine:* an indole alkaloid from calabar beans, the ripe seeds of *Physostigma venenosum* (a woody vine indigenous to the west coast of Africa). P. contains a pyrrolidinoindole ring structure and a urethane group (Fig.), and exists in a stable form, m.p. 105–106 °C, and an unstable form, m.p. 87 °C. M_r 275.35 $[a]_D$ −82 ° (chloroform). It occurs with its *N*-oxide, geneserine, m.p. 129 °C, $[a]_D$ −175 ° (acetone).

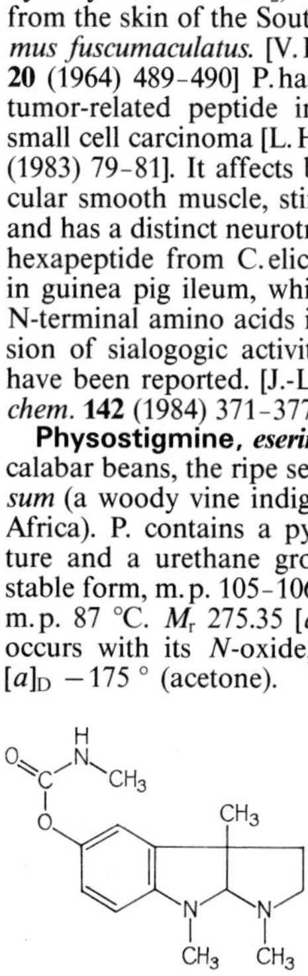

Physostigmine

P. is used in opthalmic practice in the same way as pilocarpine, for pupil contraction and for the reduction of intraocular pressure. It is an inhibitor of acetylcholinesterase, a property shared by certain other basic urethanes, such as neostigmine (prostigmine); these urethanes presumably acylate, and therefore block, the enzyme. P. and neostigmine have been used in surgery to counteract the action of curare, and both have been used for the relief of Myasthenia gravis, a disease characterized by muscular weakness associated with a rapid breakdown of acetylcholine. These have now been largely replaced by other synthetic compounds.

The fatal human dose of P. is about 10 mg. It was first isolated in 1864, when forty-six children in Liverpool were poisoned by eating calabar beans hrown on a rubbish heap from a West African car-go ship.

Phytanic acid: see Inborn errors of metabolism Refsum's disease).

Phytanic acid storage disease: see Inborn er-ors of metabolism (Refsum's disease).

Phytic acid: see Myo-inositol.

Phytin: see Myo-inositol.

Phytoalexins, *phytoncides, stress compounds:* sub-tances with antibiotic activity produced by plants in esponse to injury or stress, e.g. infection with fungi,)acteria or viruses, mechanical wounding, UV irra-liation, cold and treatment with phytotoxic chemi-als (e.g. heavy metals) or Elicitors (see). They func-ion as growth inhibitors of pathogens, chiefly fungi. They have also been defined as novel post-infection-il metabolites produced by plants in response to ungal infection, and which, because of their anti-ungal activity, protect the plant from attack by fun-;i. Under certain conditions P. may also be antibac-erial. P. do not include antifungal compounds al-eady present in the uninfected plant, e.g. protocate-:huic acid and catechol (onion bulbs) or chlorogenic icid (many plants). The original concept of P. in-:luded compounds formed from immediate precur-ors, e.g. the antifungal dihydroxymethoxybenzoxa-

zine released from its glucoside, which is present in cells of wheat and maize. The modern definition of P. excludes such compounds, and is restricted to compounds synthesized by a remote pathway, which is first activated by infection or stress (see e.g. Fig. 2). Since there are no immediate precursors, the appearance of P. is delayed until the biosynthetic pathway has been activated.

Most known P. have been discovered in the subfami-ly *Papilionoideae* of the *Leguminoseae*, followed by the *Solanaceae*, while many examples are known from other families. P. do not belong structurally or biosynthetically to any one class of compounds. Many are isoflavonoids and several of these are pterocarpans, e.g. pisatin (see Pterocarpans), Phase-olin (see), Glyceollins (see). There are also terpenes (see Fig 2; see Ipomoeamarone), acetylenic com-pounds (see Safynol; Wyerone acid), α-hydroxydi-hydrochalcones, stilbenes and polyketides (Fig. 1).

In orchids P. are formed only when the roots and shoot of the germinating orchid seed are colonized by fungi. The fungi penetrate the plant, but are pre-vented from spreading by P. (see Orchinol). In most plants producing P., infection or stress cause the for-mation of a number of different P., i.e. there is a multicomponent response. For example, medicarpin

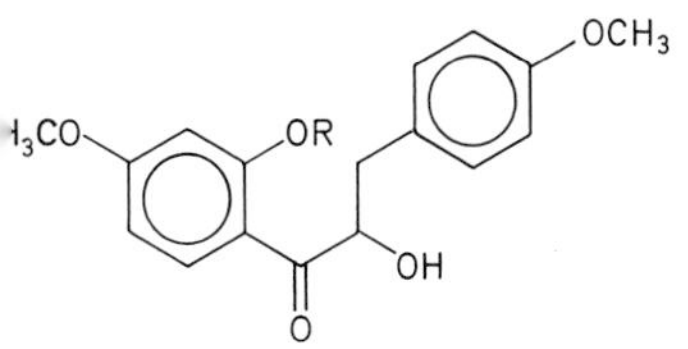

R = H: *Odoratol,*
R = CH₃: *Methylodoratol,* two α-hydroxydihydro-:halcones from pods and cotyledons of *Lathyrus* *)doratus* infected with *Phytophthora megasperma* A. Fuchs et al. *Phytochemistry 23* (1984) 2199–2201]

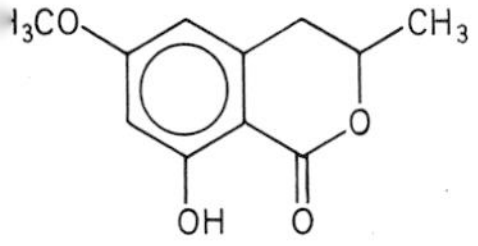

5-*Methoxy-mellein,* a polyketide from carrot infected Jith *Ceratocystis fimbriata* [J. A. Bailey & J. W. Mansfield (eds.) *Phytoalexins* (Blackie & Son Ltd., 982)]

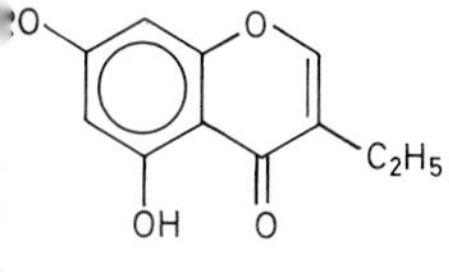

R = H: *Lathodoratin*
R = CH₃: *Methyl-lathodoratin,* two chromones from Lathyrus odoratus and *L. hirsutus* infected with *Bo-rytis cinerea* or *Helminthosporum carbonum* [D.J. Robeson et al. *Phytochemistry 19* (1980) 2171–2173]

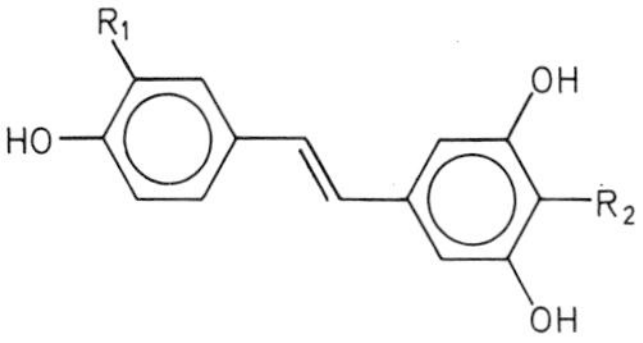

$R_1 = OH; R_2 =$ CH₃ >CH–CH=CH–, CH₃

4-(3-methylbut-1-enyl)-3,5,3′,4′-tetrahydroxystil-bene

$R_1 = H; R_2 =$ CH₃ >CH–CH=CH–, CH₃

4-(3-methylbut-1-enyl)-3,5,4′-trihydroxystilbene

$R_1 = H; R_2 =$ CH₃ >C=CH–CH₂–, CH₃

4-(3-methylbut-2-enyl)-3,5,4′- -trihydroxystilbene (or 4-isopentenyl resveratrol)

Stilbenes from *Arachis hypogaea* seeds, sliced and allowed to become infected with natural microflora. The compounds are active against *Cladosporium cu-cumerinum*. [G. Aguamah et al. *Phytochemistry 20* (1981) 1381–1384]

Figure 1. Examples of phytoalexins. Other P. are de-scribed in Fig. 2, and cross referenced from the text as individual entries.

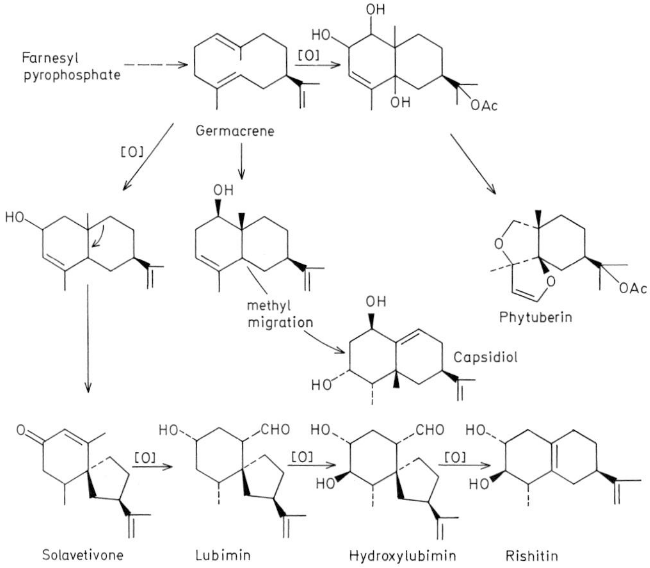

Figure 2. Biosynthesis of some sesquiterpene phytoalexins. Phytuberin, solavetivone, lubimin, hydroxylubimin and rishitin are produced by *Phytophthora*-infected potatoes (*Solanum tuberosum*) and by potato tuber discs treated with Elicitor (see). Capsidiol is a P.from *Capsicum*. [P.A. Brindle & D.R. Threlfall *Biochem. Society Transactions* **11** (1983) 516-522]

(a pterocarpan) and a variety of compounds related to wyerone acid are all produced by *Vicia faba*, while rishitin, lubimin and phytuberin are produced by potato.

[J.L. Ingham "Phytoalexins and Other Natural Products as Factors in Plant Disease Resistance" *The Botanical Review* (New York Botanical Garden), *38* (1972) 343-424. R.A Dixon et al. *Advances in Enzymology* (ed. Alton Meister) *55* (1983) 1-136. J.A. Bailey & J.W. Mansfield (eds.) *Phytoalexins*, Blackie & Son Ltd. 1983. P.A. Brindle & D.R. Threlfall *Biochemical Society Transactions* **11** (1983) 516-522. G.A. Cooper-Driver, T.Swain & E.E. Conn (eds.) *Chemically Mediated Interactions between Plants and Other Organisms* (Recent Advances in Phytochemistry *19* (Plenum Press, 1985). D.A. Smith & S.W. Banks, *Phytochemistry* 25 (1986) 979-995; M.S. Kemp & R.S. Burden, *Phytochemistry* 25 (1986) 1261-1269]

Phytochelatin: a plant peptide produced in response to heavy metals, e.g. cadmium, copper, mercury, lead and zinc. The structure is (γ-Glu-Cys)$_n$-Gly ($n=3$-7). Like Metallothionein (see), P.form metal-thiolate bonds and thus sequester toxic metal ions. They are probably derived from glutathione rather than RNA-directed protein synthesis.

Phytochemistry: the chemistry of natural products from plants (see Natural product chemistry). It is part of plant biochemistry, and it is concerned chiefly with secondary metabolites.

Phytochrome: A ubiquitous plant pigment which mediates the light sensitivity of many growth and developmental processes. P.exists in two forms, one which has a light-absorption maximum around 660 nm (P_r), and one which absorbs maximally

around 730 nm (P_{fr}). The pigment consists of a tetrapyrrole chromophore (Fig.) in an apoprotein of 120000-127000 M_r. The chromophore is related to c-phycocyanin, which is found in blue-green bacteria.

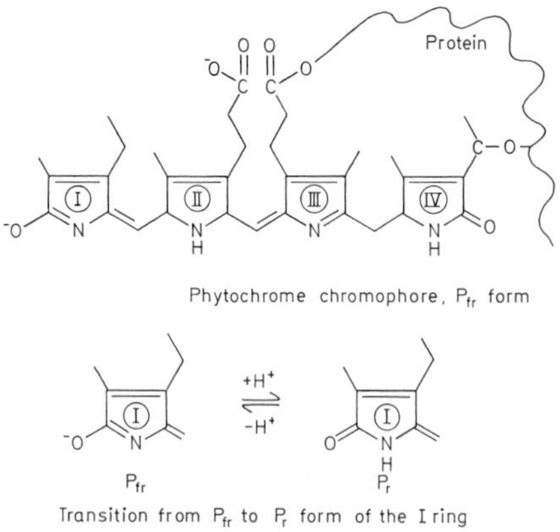

Phytochrome chromophore, P_{fr} form

Transition from P_{fr} to P_r form of the I ring

Phytochrome chromophore

P. is synthesized as P_r, which is thought to be the inactive form. Exposure to red light of 660 nm wavelength converts P_r to P_{fr}, the active, and also more labile form. Exposure of P_{fr} to far-red light (730 nm) converts it back to P_r. Dark-grown tissues accumulate P_r, but because of the faster rate of destruction or inactivation of P_{fr}, the total P. content of light-grown tissues is only 1-3% that of dark-grown tissues.

These conversions correspond to the physiological effects of red and far-red light. In lettuce seeds, for example, red light promotes germination, while far-red light inhibits it. Similarly, plants which flower when the days are short (short-day plants) can be prevented from flowering if their long dark period at night is interrupted by a single five-minute interval of irradiation with red light. This effect can be reversed, however, by an immediately subsequent irradiation with far-red light. It is thought that the plants accumulate P. at night, and when the nights are long enough to accumulate a threshold amount of it, the plant is induced to flower. The reverse situation exists in long-day plants; here it can be assumed that high P. levels inhibit flowering.

P_{fr} promotes or inhibits a number of different processes in different tissues. The mechanisms involved are unknown. One possibility is that P_{fr} binds to membranes and influences their permeability to ions. As in animal neurosecretory cells, the influx of ions could affect synthesis and/or secretion of plant hormones. [Kendrick & Frankland, *Phytochrome and Plant Growth* The Institute of Biology's Studies in Biology no. 68 (Edward Arnold Ltd., London, 1976)]

Phytoecdysone: see Ecdysone.

Phytoene: an aliphatic, colorless hydrocarbon carotenoid, M_r 544. P. is a polyisoprenoid. It contains six branch methyl groups, two terminal isopropylidene groups and nine double bonds, three of them conjugated. Only the Δ^{15} double bond has *cis* configuration. Biosynthetically, P. is derived from two molecules of geranylgeranyl pyrophosphate, and it serves as a C_{40}-starter molecule in the biosynthesis of other carotenoids: phytofluene, carotene, neurosporene and lycopene are formed by the stepwise dehydrogenation of P. It is found widely in plants, and is especially plentiful in tomatoes and carrot oil. It was isolated in 1946 from tomatoes by J. W. Porter and F. P. Zscheile, and its structure was elucidated in 1956 by W. J. Rabourn and F. W. Quackenbush. For structure and biosynthesis, see Tetraterpenes.

Phytofluene: an aliphatic, polyisoprenoid hydrocarbon carotenoid., M_r 548. P. contains ten double bonds (five of them conjugated), six branch methyl groups and two terminal isopropylidene groups. As in phytoene, the central double bond between C-15 and C-16 has *cis* configuration. P. is found widely in plants, e.g. tomatoes and carrots, and it is an intermediate in lycopene biosynthesis (see Tetraterpenes). Structural elucidation by L. Zechmeister in 1954.

Phytohemagglutinins: see Lectins.

Phytohormones, *plant hormones:* The classic P. are endogenous regulators of plant growth and development. They are analogous to animal hormones in that the sites of their synthesis may be remote from the sites of their action, but in contrast to animal hormones, P. have multiple activities and low action specificities. The newly discovered Oligosaccharins (see) are much more specific and are effective at concentrations similar to those at which animal hormones are effective (10^{-8} or 10^{-9} M), but they may not be transported from the site of their synthesis to the site of their action.

The classic P. are Auxins (see), Gibberellins (see), Antheridiogen (see), Cytokinins (see), Abscisic acid (see), Flowering hormone (see) and Fruit ripening hormone (see). These are reviewed by T. C. Moore, *Biochemistry and Physiology of Plant Hormones* (Springer, New York, Heidelberg, Berlin, 1979).

Phytokinins: see Cytokinins.

Phytol: see Chlorophyll.

Phytoncides: see Phytoalexins.

Phytosterols: see Sterols.

Phytotoxin: a compound produced by a fungal or bacterial plant parasite, which is toxic to the plant. The term should not be confused with Plant toxin (see). Compared with the large number of plant parasites with demonstrable phytotoxic activity, only relatively few phytotoxins have been identified. See separate entries: *Alternaria alternata* toxins, Coronatine, Fusicoccin, Gibberellins, Ophiobolanes, Stemphylotoxins, Tabtoxins.

Phytuberin: see Phytoalexins.

Picromycin: see Macrolides.

Picrotoxin, *cocculin:* a molecular compound of one molecule picrotoxin and one molecule picrotin. P. is a neurotoxin, which occurs in the seeds of *Anamirta coculus,* and is also found in *Tinomiscium philippinense.* As a specific antagonist of 4-Aminobutyric acid (see), it acts as a central and respiratory stimulant, and is an antidote to barbiturates. It is extremely toxic to fish. Picrotoxinin is the active component of the molecular compound; picrotin is physiologically inactive.

Picrotoxinin, R: $H_3C-\overset{|}{C}=CH_2$

Picrotin, R: $H_3C-\overset{|}{\underset{|}{C}}-OH$
CH_3

Piezoelectric sensors: Quartz crystals may be induced to resonate under electrical control. In this state they are sensitive to the mass of absorbing material. Piezoelectric sensors make use of this effect for the detection of very small amounts of inorganic or biological material, e.g. by use of surface immobilized enzymes or antibodies.

Pigment 700, P_{700}: a Chlorophyll (see) with an absorption maximum at 700 nm. P. 700 is a component of photosystem I and seves as an energy sink or trapping center in the light reaction of this system. It is important in the primary energy transformation of Photosynthesis (see): the highly negative redox potential of P. 700 (following its excitation by light) results in the transfer of electrons to the primary acceptor, which then transfer electrons to NADP via ferredoxin. The transiently oxidized P. 700 (due to emission of electrons) is reduced again by electrons from cytochrome f. The reaction center of photosystem I consists of about 500 molecules of closely packed P. 700 in a quasicrystalline state, resembling the crystal of a semiconductor. Owing to the high degree of structural order, the electron excitation energy can be transferred by resonance.

Pigment systems: see Photosystems.

PIH: abb. for prolactin release inhibiting hormone (see Releasing hormones).

Pilocarpine: an imidazole alkaloid, and the chief alkaloid from the leaves of Brazilian *Pilocarpus* species. M_r 208.26, m.p. 34 °C, b.p.$_5$ 260 °C, $[\alpha]_D^{20}$ – 100.5 ° (chloroform). P. is used therapeutically as a diaphoretic, i.e. to induce sweating, and especially in nephritis to relieve the kidneys and remove toxic metabolites. It is also used in opthalmology as an antagonist of atropine, and for regulating the intraocular pressure in glaucoma.

Pilocarpine

Pimaradiene type: see Diterpenes (Fig.).
Pinane type: see Monoterpenes (Fig.).
Pineal gland, *pineal body, epiphysis, Corpus pineale:* a small, cone-shaped, unpaired organ situated between the cerebral hemispheres on the roof of the third ventricle of the mammalian brain. Phylogenetically, the P.g. is a vestigial parietal eye, the light sensitive organ of reptiles. It produces the hormone, Melatonin (see).

Pinecembrin: 5,7-dihydroxyflavanone, see Flavanone.

Pinestrebin: 5-hydroxy-7-methoxyflavanone, see Flavanone.

Ping-Pong mechanism: see Cleland notation.
Pinitol: see Cyclitols.
L-Pipecolic acid: piperidine-2-carboxylic acid, a nonproteogenic amino acid. It is formed from L-lysine, either by α-deamination followed by cyclization and reduction, or as a normal intermediate in the degradation of lysine to α-aminoadipic acid. The 4- and 5-hydroxy derivatives of L-P.a. are found especially in mimosas and palms.

L-Pipecolic acid   *L-Baikiain*

L-Baikiain (1,2,3,6-tetrahydropyridine-α-carboxylic acid), a rare nonproteogenic amino acid first isolated from the wood of *Baikiaea plurijuga*, is structurally related to L-P.a.

Piper alkaloids: a group of alkaloids occurring in various species of *Piper*, especially black pepper *(Piper nigrum)*. Structurally, P.a. consist of an aromatic carboxylic acid with an unsaturated side chain

(e.g. piperic acid, sinapic acid) in amide linkage with a basic component, usually piperidine. The chief representative is Piperine (see).

Piperidine alkaloids: a group of alkaloids containing the piperidine ring system. Simple P.a. are the alkyl substituted piperidines which occur sporadically. The other P.a. are classified according to their origin, e.g. Conium alkaloids (see), Punica alkaloids (see), Sedum alkaloids (see) and Lobelia alkaloids (see). These various groups are structurally different and have different mechanisms of biosynthesis. Other P.a. are found in water lilies, and are biosynthesized from mevalonic acid (see Nuphar alkaloids). A dehydropiperidine structure is present in the Areca alkaloids (see) and the Betalains (see).

Piperine: piperic acid piperidide, a Piper alkaloid (see), the chief alkaloid of black pepper *(Piper nigrum)* and responsible for its sharp taste. M_r 285.35, m.p. 16 °C. Both double bonds are *trans* (Fig.). The *cis-cis* isomer, earlier called chavicine, does not occur naturally.

Piperine

Pisatin: 3-methoxy-6a-hydroxy-8,9-methylenedioxypterocarpan, see Pterocarpans.

Pituitary gland: see Hypophysis.
pK: see Buffers.
PL: abb. for placentalactogen.
Placentagonadotropin: see Choriogonadotropin.
Placentalactogen: see Choriomammotropin.
Plant hormones: see Phytohormones.
Plant mucilages: high M_r complex, colloidal polysaccharides, which form gels and have adhesive properties. They are widely distributed in the plant kingdom, being found as secondary membrane thickening and as intercellular and intracellular material. They occur in root, bark, cortex, leaves, stalks, flowers, endosperm and seed coat. Some bulbs contain special mucilage cells. Some P.m. may function as food reserves. On account of their high affinity for water, certain P.m. may be used as water reservoirs (i.e. as antidessicants) by plants that live under very dry conditions; mucilagenous seed coatings may have a similar function. Together with the structurally related plant gums, P.m. form an ideal material for sealing damaged tissue; owing to their often heterogeneous carbohydrate composition, P.m. are relatively resistant to microbial attack.

P.m. can be roughly classified into three groups: 1. Neutral, containing one or more types of sugar, but no uronic acids, e.g. a linear polymer of 1,4-linked mannopyranose isolated from the tubers of certain orchids.

2. Acidic, resembling the plant gums, but usually containing D-galacturonic acid as the acidic residue, e.g. a P.m. from the bark of *Ulmus fulva* (slippery elm), which contains D-galactopyranose, D-galacturonic acid, L-rhamnopyranose and 3-methyl-D-ga-

lactose. In this type of P.m., the ratio of uronic acid residues to neutral sugar residues is usually about 1:3.

3. P.m. present in algae (notably seaweeds), often of highly complicated structure and very high M_r, and often containing esterified sulfate. This group is well exemplified by the agars from seaweeds. The agars range from a neutral species, agarose (see Agar-Agar), to the highly acidic, sulfated Carrageenans (see). The primary sequence of the entire agar family is based on a repeating disaccharide unit of galactose derivatives. Other seaweed P.m. are Alginic acid (see), laminarin (β-1,3 linked glucose residues with some β-1, 6 branch points; isolated from *Laminaria*) and fucoidin (contains sulfated fucose residues; isolated from *Laminaria* and *Fucus* spp.).

Plant pigments: see Natural pigments.

Plant toxin: a compound produced by a plant, which adversely affects man or animals, e.g. ricin from Castor bean, amanitin from the Death cap fungus. See also Toxic proteins.

Plaque: a transparent area in a lawn of bacteria on the surface of a solidified growth medium. P. is caused by lysis of bacteria in that area by bacteriophage. Under controlled conditions, each P. represents a center of infection initated by one infective bacteriophage particle. The number of P. produced after evenly spreading a known volume of phage suspension over the surface of the bacterial culture is used as a simple assay of the number of infective phage particles.

The term is also used in a general sense for an area of lysed cells in a lawn. In immunology, a lawn of red blood cells in used to detect immunoglobulin-producing cells in a P. assay. When complement is added to the system, those erythrocytes that have bound to the immunoglobulin molecules are lysed, leaving a transparent P. Since most antigens can be artificially bound to erythrocyte membranes, the technique can be used for practically any antibodies capable of fixing complement.

Plasma albumin: see Albumins.

Plasma factors: see Blood coagulation.

Plasmakinins: physiological, highly active oligopeptides, with hormone-like properties. P. act upon the smooth muscle of blood vessels, gastrointestinal tract, uterus and bronchi. Important representatives are Bradykinin (see), kallidin and methionyllysyl-bradykinin.

Plasmalemma: the cell membrane. See Biomembrane.

Plasmalogens: see Phospholipids.

Plasma proteins: a complex of predominantly conjugated proteins present in the blood plasma of vertebrates. The number of P.p. is estimated to be more than 100. In mammalian plasma the concentration of P.p. is 6–8%. Serum proteins lack Fibrinogen (see) and Prothrombin (see), but are otherwise essentially the same as P.p. Approximately 60 P.p. have been isolated and characterized. Of these, only albumin, prealbumin, retinol binding protein and a few trace proteins (e.g. lysozyme) are free from carbohydrate. The remaining P.p. are glycoproteins e.g. Orosomucoid (see), Hemopexin (see), Haptoglobin (see), C 1-inactivator, Immunoglobulins (see); some may also contain lipids (see Lipoproteins). P.p. help

to regulate the pH-value and osmotic pressure of the blood; they transport ions, hormones, lipids, vitamins, metabolic products, etc. P.p. are also responsible for blood coagulation, for defense against foreign proteins or microorganisms (see Immunoglobulins), and for certain enzyme reactions. With the exception of immunoglobulins, the P.p. are synthesized in the liver. The 5 main groups of P.p., i.e. albumin, α_1, α_2-, β- and γ-globulin (in order of increasing migration rate from the anode), can be separated by electrophoresis, either in the Tiselius carrier-free buffer system, or on a carrier such as filter paper or cellulose acetate. A better separation is achieved by exploiting the additional effect of electroosmosis (in agarose or starch gel), or sieving (polyacrylamide). The best single stage separation of P.p. into over 30 fractions is achieved by immunoprecipitation. Two dimensional immunoelectrophoresis after Laurell gives an even higher resolution: in this method the gel for the second electrophoretic direction contains the antibody mixture. See Fig. illustrating various, separation methods. The smallest known P.p. is β_2-Microglobulin (see).

P.p. are usually isolated on a preparative scale without loss of biological activity, by precipitation with ammonium sulfate, ethanol or rivanol. To an increasing extent, however, these methods are being replaced, even on an industrial scale, by efficient modern column methods, such as gel filtration, ion exchange, affinity and immunoadsorption chromatography, and by ultrafiltration.

Plasmid: extrachromosomal DNA in the bacterial cell. P. carry genetic information and replicate independently of the bacterial chromosome. A P. which can become reversibly integrated into the host DNA is called an episome. P. are circular molecules of duplex, supercoiled DNA, ranging in size from M_r about 1.5×10^6 to 1.5×10^8. P. DNA differs physically and chemically from DNA of the bacterial chromosome, e.g. in its GC-content. P. DNA represents 1 to 3% of total cell DNA, but in exceptional cases it may reach 20%. P. genes are responsible for various nonessential properties of bacteria. Resistance to drugs or antibiotics is determined by P. called drug resistance factors or R-factors (see). Conjugation in bacteria is determined by P. called bacterial sex factors, fertility factors or F-agents. Bacterial virulence also depends upon P. Colicinogenic factors are P. in *Escherichia coli,* which determine the synthesis of colicins (see Toxic proteins).

Plasmin, *fibrinolysin* (EC 3.4.21.7): a trypsin-like enzyme, containing two polypeptide chains. Its inactive precursor or zymogen (plasminogen, profibrinolysin) occurs in blood at a concentration of 50–100 mg/100 ml serum. P. is responsible for fibrinolysis, i.e. dissolution of blood clots by the proteolytic degradation of fibrin to soluble peptides. In addition, the products of fibrinolysis inhibit the conversion of fibrinogen into fibrin, thus forming a delicately balanced mechanism for the prevention of clotting. The natural inhibitor of P. is α_2-Macroglobulin (see).

Plasminogen (a single chain β-globulin, M_2 81 000, containing 8–10% carbohydrate) is converted into P. (i.e. "activated") by hydrolysis of a specific arginyl-

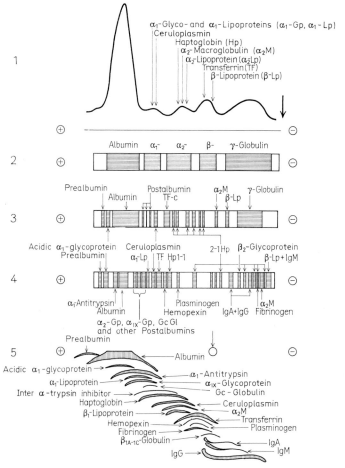

Separation of human plasma proteins by five electrophoretic methods at pH 8.
1. Carrier free (Tiselius) electrophoresis. 2. Paper electrophoresis, 3. Starch gel electrophoresis, 4. Polyacrylamide gel electrophoresis (pH 9), 5. Immunoelectrophoresis (in practice the precipitation bands against polyvalent antihuman serum overlap one another, but have been drawn widely separated for legibility). Arrows in 1. and 5. indicate the site of application of the plasma sample.

valine bond. The resulting P. consists of two chains held together by a single disulfide bond. The proteolytic activity of P. depends upon the light chain of 233 amino acid residues; this contains the catalytically important serine and histidine residues that are characteristic of the serine proteases, and it also contains sequences homologous with those of trypsin. P. further resembles trypsin in that it splits arginyl linkages, and it is also inhibited by soybean and pancreatic trypsin inhibitors. Plasminogen activation is catalysed by specific tissue activators, which are released under various conditions, e.g. emotional stress, during exercise, following the injection of adrenalin, and it is released from blood vessel walls on vascular injury. The best known activator is the kidney enzyme, urokinase (EC 3.4.21.31) (M_r 53000). Streptokinase (M_r 47000, 416 amino acid residues, cysteine-free), produced by β-hemolytic streptococci, is of medical interest; although it is not

generally considered to be a proteolytic enzyme, it is one of the most potent exogenous activators of human (not animal) plasminogen; it forms a complex with plasminogen, which then releases active P.

Plasminogen: see Plasmin.

Plasmochromic pigments: plant pigments contained in plastids, e.g. Chlorophyll (see) and Carotenoids (see).

Plasmon: the total extrachromosomal hereditary complement of a eukaryotic cell. See Chondrome; Plastome; Cytoplasmon.

Plast: see Cell, 2.

Plastids: organelles in the cells of eukaryotic plants. P. contain DNA, and are self replicating. Division of P. and replication of P. DNA are not synchronized with nuclear division and replication of nuclear DNA. P. are usually ellipsoid, 1–10 μm long. They can, however, be larger and have various shapes, e.g. in green algae. Chloroplasts (see) are P.

Chromoplasts (see) contain various pigments responsible for plant (especially flower) coloration. P. devoid of pigments are called Leucoplasts (see); these act as storage sites for starch in the root and shoot. P. are derived biogenetically from Proplastids (see)

Plastome, *plastidom:* the total genetic information contained in the DNA of the plastids of a eukaryotic plant cell.

Plastoquinone: a polyisoprenoid quinone. P. is structurally similar to ubiquinone, but P. contains a methyl group (not a methoxy group) in the aromatic ring. P. can be isolated from chloroplasts; it acts as a reversible redox component in photosynthetic electron transport (see Photosynthesis).

Platelet activating factor, *PAF:* a phospholipid released by IgE-sensitized leucocytes in the presence of antigen, and possibly endogenous to platelets as well. It is present during anaphylactic shock, and appears to mediate inflammation and allergic responses. In vitro, PAF causes platelets to change shape, aggregate and release the contents of their granules at concentration of 10^{-11} to 10^{-10} M. PAF is structurally similar to an antihypertensive polar renomedullary lipid (APRL) from kidneys, which reduces the blood pressure of artificially hypertensive animals.

Platelet-derived growth factor, *PDGF:* an hydrophobic, heat-stable (100 °C), cationic (P.I. 9.8–10.2) polypeptide (2 homologous chains linked by disulfide bonds), M_r 32000 contained in the α-granules of platelets and released into serum during blood clotting. PDGF is a potent mitogen, which stimulates growth of fibroblasts, glial cells, monocytes, neutrophils and smooth muscle cells; it represents the major growth factor of human serum. PDGF is the product of a proto-oncogene, c-*sis*. The oncogene counterpart is v-*sis*, an insertion sequence carried in the genome of simian sarcoma virus (SSV). The v-*sis* sequences are essential for the transforming activity of the virus, and they give rise to a M_r 28000 oncogene product known as p28^{sis}. PDGF and p28^{sis} compete for the same receptor, possess striking sequence homology (sequences in PDGF were determined on the polypeptide, whereas the sequence of p28^{sis} was predicted from analysis of the SSV genome), and are immunologically cross reactive.

Binding of PDGF to its specific, saturable cell surface receptor stimulates Protein-Tyrosine kinase (see) activity in the receptor molecule, leading to phosphorylation of the receptor itself (autophosphorylation) as well as other cell proteins. PDGF receptor has M_r about 170000; it has been isolated by affinity cross linking to ^{125}I-labelled PDGF (using disuccinimidyl tartrate or disuccinimidyl suberate), followed by precipitation with a monoclonal phosphotyrosine antibody from detergent (Triton)-treated cells after stimulation with PDGF. [M.D. Waterfield et al., *Nature 304* (1983) 35–39; R.F. Doolittle et al., *Science 221* (1983) 275–277; J.Y.J. Wang & L.T. Williams, *J. Biol. Chem.* **259** (1984) 10645–10648; A.R. Frackelton, *J. Biol. Chem.* **259** (1984) 7909–7915]

Plicacetin: see Amicetin.

Plumbagin: see Naphthoquinones.

Plumierid: an Iridoid (see).

Pluripoietin: see Colony stimulating factors.

PMS: see Pregnant mare serum gonadotrophin.

Poisons, *toxins:* compounds which damage an organism in relatively small amounts, and which may kill it at high enough doses. *Plant poisons* are found in all parts of certain plants; they are not produced by specialized cells. They tend to affect the heart and circulation. Many types of compound are represented - proteins, alkaloids, steroids, amines, unusual amino acids, etc. Since plants synthesize about 100 times as many natural products as animals, it is not surprising that the number of plant poisons is very large. Of those whose structures are known, the most important include the proteins ricin and the viscotoxins, nearly all alkaloids, especially the Aconitum, Colchicum, Conium, Curare, Strychnos and Tropan alkaloids, the cardiac glycosides, the unusual amino acids like djencolic acid and lathryin, and mushroom poisons. The toxins of lower fungi and bacteria (mycotoxins and antibiotics) are particularly diverse in structure. Not enough is known about the biological significance of plant poisons. They may serve as a protection against insects and grazing animals, however. Plant poisons tend to be less toxic than animal poisons.

Animal poisons are produced by the animals under normal conditions. They are structurally diverse, including peptides and proteins, acids, amines, betaines, quinones, alkaloids, terpenes and steroids. Some are the same as vertebrate hormones: the defensive secretion of the water beetle *Dytiscus marginalis* contains 0.4 mg deoxycorticosterone. To obtain this amount from bovine adrenal glands, one would need 1000 animals!

Most animal poisons are mixtures. They occur in almost every group of animals, excepting perhaps some molluscs, tunicates and birds. Passive poisons are contained in the tissues or organs of the animal and are only effective if the flesh is eaten. Actively poisonous animals, however, usually produce the poison in special glands and have various means of delivering it: secretion from skin glands (see Amphibian toxins), ejection from the mouth or abdominal region by muscle pressure on the poison glands, or injection through teeth (see Snake venoms), sting (see Bee toxin) or hairs (caterpillars). Some snakes and all skunks can spray their poison in a chosen direction. The bombadier beetle's spray is ejected by a chemical explosion which occurs when the animal adds catalase to a mixture of quinones and 23% hydrogen peroxide. The heat of reaction brings the mixture almost to the boiling point, and the pressure of the released oxygen causes it to spray out. Animal poisons have various effects, but the most common are heart, blood and nerve poisons. Some, like the snake toxins, are used therapeutically, or to produce antisera.

The toxicity of natural poisons, expressed by their LD$_{50}$ (see Dose), is much higher than that of conventional inorganic poisons (see Table).

Modern toxicology is concerned with the relationships between the structure and effects of poisons and drugs. The site and mode of action, and detoxifying reactions are of particular interest. Many poisons inhibit vital enzymes. Cyanide ions, for exam-

465

Table. Comparison of the toxicities of various poisons.

Substance	Occurrence or Significance	LD_{50} (μg/kg mouse)
Inorganic poisons		
Potassium cyanide KCN		3000
Arsenic As_2O_3		10000
Mercuric chloride $HgCl_2$		23000
Plant poisons		
Strychnine	Strychnos alkaloid	500
C-Toxiferin I	Curare alkaloid	25
Colchicine	Colchicum alkaloid	3000
Nicotine	Tobacco alkaloid	16000
Coniine	Conium alkaloid	60000
k-Strophantine	Cardiac glycoside	15000
Digitonin	Cardiac glycoside	150000
Fungal poisons		
α-Amanitin	*Amanita phalloides*	200
Ergotoxin	Ergot alkaloid	100000
Bacterial toxins		
Botulin toxin	*Clostridium botulinum*	0.00003
Tetanus toxin	*Clostridium tetani*	0.0001
Diphteria toxin	*Corynebacterium diphtheriae*	0.3
Animal poisons		
Batrachotoxin	Arrow-poison frog	2
Tetrodotoxin	Puffer fish	8
Cobra toxin	Cobra	0.3
Crotalus toxin	Rattlesnake	0.2
Bufotoxin	Toad poison	400
Melittin	Bee venom	4000
Samandarine	Salamander alkaloid	1500

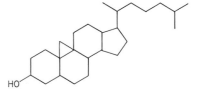

Pollinastanol

ple, bind to the iron of cytochrome oxidase and block electron transport to oxygen, while carbon monoxide blocks the oxygen-transport function of hemoglobin by binding more tightly to the oxygen-binding sites than oxygen does. Other poisons interfere with the basic metabolism of the cell and affect its structure. Snake venom enzymes, for example, remove specific fatty acid components from phospholipids. The lysolecithins and lysocephalins formed in this way emulsify the lipids of the erythrocyte membranes and cause the cells to lyse. Melittin, the main component of bee venom, has the same effect. The amanitins and phalloidins irreversibly destroy the endoplasmic reticulum of liver cells. The organism responds to toxins by attempting to detoxify them. Foreign proteins are attacked by specific antibodies, and most poisons are modified structurally in the liver. The inactivation occurs in the simplest cases by conjugation to form glycosides which are excreted. An animal may become tolerant of a P. if the detoxifying enzyme system is induced, so that the P. is quickly inactivated.

Pollinastanol: 4,4-demethylcycloartenol, a plant constituent related structurally to the sterols and to cycloartenol. Particularly good sources of P. are pollen from members of the *Compositae*, the fern *Polypodium vulgare*, and roots of sarsaparilla (*Smi-*

lax medica). In these plants, P. is an important intermediate in the biosynthesis of cholesterol.

Poly A: abb. for Polyadenylic acid (see).

Polyacrylamide gel electrophoresis: see Proteins.

Polyadenylic acid, abb. *Poly A:* a Homopolymer (see) consisting entirely of residues of adenylic acid. Poly A sequences of varying lengths are found at the 3′ end of many eukaryotic mRNA molecules (see Post-transcriptional modification of RNA, Poly A polymerase). The length of the poly A sequence depends on the source of the RNA and the physiological state of the cell, and is inversely related to the half-life of the mRNA.

The function of poly A sequences is not clear. They may be required for transport of mRNA from nucleus to cytoplasm or for binding to the ribosome. However, mRNA is only attacked by hydrolytic enzymes after the loss of all poly A; thus the metabolic half-life of a mRNA molecule may be regulated by the length of its poly A sequence. The degradation of a poly A sequence may be controlled by proteins which are actually bound to it.

The histone mRNA of higher eukaryotes does not have a poly A tail. A short sequence of oligo A has been found in the 3′-region of bacteriophage T7 RNA, and a short oligo A sequence has also been detected at the 3′-end of RNA from *Escherichia coli*. The mRNAs of the slime mold *Dictyostelium* and of HeLa cells contain two poly A segments separated by a short sequence of other nucleotides.

The presence of the poly A sequence has been exploited for the isolation and purification of eukaryotic mRNA, by using affinity columns of oligo-dT-cellulose.

Poly ADP-ribose: see ADP-ribosylation of proteins.

Polyamines: a group of aliphatic, straight-chain amines derived biosynthetically from amino acids. They include spermine, spermidine, cadaverine and putrescine. The latter two are diamines produced by decarboxylation of lysine or ornithine, respectively. In plants, putrescine can also be synthesized by decarboxylation of arginine to agmatine, followed by hydrolytic removal of urea. Animals do not have arginine decarboxylase (EC 4.1.1.19), but in plants it is often more abundant than ornithine decarboxylase (EC 4.1.1.17). Putrescine is converted to spermidine, and spermidine to spermine, by the addition of an aminopropyl group. This group is provided by decarboxylated *S*-adenosyl methionine (Fig.)

P. are required for cell division, and probably for differentiation. Spermine apparently stabilizes the DNA which is tightly packed in the heads of sperm cells. Mutant Chinese hamster ovary cells which are auxotrophic for putrescine have been developed.

When these are starved for P., they lack 90% of their actin filament bundles and have essentially no microtubules. It may be inferred that P. are essential for the stability of these structures. P. stimulate growth of higher plants and appear in some cases, e.g. tomatoes, to be required for formation of fruit. Putrescine accumulates in plants subjected to stress, especially deficiency of K^+ and Mg^{2+}, NH_4^+ feeding, acidification and high salinity.

In organisms which lack arginine decarboxylase, putrescine can be synthesized only by decarboxylation of ornithine. Many inhibitors of ornithine decarboxylase are known, and these have a number of medical applications against tumors, viruses, trypanosomes (e.g. *Trypanosoma brucei*, which causes African sleeping sickness) and plasmodia (e.g. *Plasmodium falciparum*, which is the agent of malaria). [*Polyamines in Biology and Medicine*, D.R. Morris & L.J. Marton, eds. (Dekker, New York, 1981); A.E. Pegg & P.P. McCann *Am. J. Physiol.* 243 (1982) C212-221; T.A. Smith, *Ann. Rev. Plant Physiol.* 36 (1985) 117-143]

$$H_2N-CH_2-CH_2-CH_2-CH-COOH \qquad \text{Ornithine}$$
$$| \atop NH_2$$

Ornithine decarboxylase

$$H_2N-CH_2-CH_2-CH_2-CH_2-NH_2 \qquad \text{Putrescine}$$

$$H_2N-CH_2-CH_2-CH_2-\overset{CH_3}{\overset{|}{\underset{\oplus}{S}}}-Ado$$

$$H_3C-S-Ado$$

$$H_2N-(CH_2)_3-NH-(CH_2)_4-NH_2 \qquad \text{Spermidine}$$

$$H_2N-CH_2-CH_2-CH_2-\overset{CH_3}{\overset{|}{\underset{\oplus}{S}}}-Ado$$

$$H_3C-S-Ado$$

$$H_2N-(CH_2)_3-NH-(CH_2)_4-NH-(CH_2)_3-NH_2$$
$$\text{Spermine}$$

Figure. Biosynthesis of polyamines.

Polyamino acids: naturally occurring or synthetic polymers, consisting of identical amino acids linked by peptide bonds. P.a. in the capsular substance of anthrax bacteria contain residues of D-glutamic acid linked by γ-peptide bonds. Poly-γ-D-glutamic acid precipitates antibodies against anthrax, a property not shared by poly-γ-L-glutamine acid. P.a. of M_r 10^3-10^6 are easily prepared by the polymerization of aminocarboxylic acid anhydrides. P. are structurally similar to Sequence polymers (see).

Poly A polymerase: an enzyme which specifically catalyses the synthesis of poly A sequences, a process which does not require DNA. P. from calf thymus has M_r about 150000. It is localized in the nucleus, and is responsible for the covalent attachment of poly A sequences to RNA (see Polyadenylic acid).

Polyisoprenes: see Polyterpenes.

Polyisoprenoids: see Polyterpenes.

Polyketides, *acetogenins:* natural products containing several recurring two-carbon units, formally equivalent to the condensation products of several molecules of acetate. Biosynthesis of P. occurs on a multienzyme complex. The first stage involves interaction of a starter molecule with a peripheral SH-group of the enzyme complex. A common starter molecule is acetyl-CoA (e.g. synthesis of anthraquinones and griseofulvin), i.e. HSCoA is lost and the acetyl group becomes attached via a thiol ester linkage to the peripheral acyl carrier protein (HSp). Other starter molecules are propionyl-CoA, cinnamoyl-CoA (synthesis of flavonoids and stilbenes), malonic acid amide-CoA (synthesis of tetracyclines), the CoA derivatives of the oxoacids of leucine, isoleucine and valine, nicotinyl-CoA, etc. Following the introduction of the starter group, a malonyl residue (from malonyl-CoA) is attached to the SH-group of a pantetheine residue, which forms the prosthetic group of a protein situated centrally in the multienzyme complex (HS_c). The starter group (e.g. acetyl) is then transferred with simultaneous decarboxylation of the malonyl residue, forming an acetoacetyl group on HS_c. This acetoacetyl group is then transferred to HSp, and another malonyl residue is introduced to HS_c. There is thus an extremely close analogy with the inital stages of fatty acid synthesis, but the two reduction steps of fatty acid synthesis do not occur in P. synthesis. The synthesis of P. proceeds by the stepwise introduction of malonyl groups. The recurring two carbon units evident in the structure of all P. are therefore derived biosynthetically by the incorporation and decarboxylation of malonyl-CoA. The β-polyketone intermediate does not accumulate; it cyclizes by ester or aldol condensation, forming the ring systems found in various P., and carbonyl groups are usually present in the resonance-stabilized enol form. Further modifications include introduction of extra oxygen by hydroxylation, methylation of hydroxyl groups to form methoxy groups, or direct methylation of carbon atoms (methyl donor is S-adenosylmethionine), reduction, substitution with polyisoprenoid residues, and chlorination of an aromatic ring.

Examples of P. are Tetracyclines (see), Griseofulvin (see), Macrolide antibotics (see), Cycloheximide (see), and various fungal products such as orsellinic acid, 6-methylsalicylic acid and cyclopaldic acid.

For a comprehensive review, see "The Biosynthesis of Acetate-Derived Phenols (Polyketides)", by Packter, N.M., pp. 535-570, in *The Biochemistry of Plants*, Vol. 4, 1980 (Edit. Stumpf, P.K.), Academic Press.

Polymerases: a collective term for enzymes that catalyse the formation of macromolecules from simple components, e.g. see separate entries for each of the following: DNA-polymerase, RNA-polymerase, RNA-synthetase, RNA-dependent DNA-polymerase, Polynucleotide phosphorylase, Poly A-polymerase.

Polymerization: see Ribonucleic acid.

Polymorphism: a genetically determined Heterogeneity (see) of proteins, especially enzymes, P. occurs when the frequency of a genetic variant in a population is greater than 1%. Frequencies of this order develop by positive selection or by the effect of incidental genetic drift on rare mutations that have a heterozygotic advantage. The resulting allelomorphs (gene pairs) of a protein differ from each other by substitution or deletion of an amino acid at one or more sites in the peptide chain. P. is shown by, e.g. the β-chain of hemoglobin, haptoglobin, transferrin, adenosine deaminase, and glucose 6-phosphate and 6-phosphogluconate dehydrogenases. P. is found especially in enzymes not involved in primary metabolic processes, and in enzymes with broad in vitro substrate specificity

Polymyxins: heteromeric, homodetic, cyclic, branched peptides, produced by *Bacillus polymyxa*, possessing antibiotic activity against Gram-negative bacteria. P. are used to treat infections of the intestinal and urinary tracts, sepsis and endocarditis. Various P., designated A, B$_1$, B$_2$, C, D, E and M are known. Polymyxin B$_1$ is a cyclic decapeptide, consisting of a ring of 7 amino acids with a tripeptide side chain linked via the γ-amino group of a diaminobutyric acid residue. The *N*-terminus (D-α, γ-diaminobutyric acid) is acylated by (+)-6-methyloctanoic acid (isopelargonic acid), abb. MOA.

Polymyxin B$_1$

Polymyxin B$_1$

Polynucleotide ligase, *DNA-ligase:* an enzyme that joins two DNA fragments by catalysing the formation of an internucleotide ester bond between phosphate and deoxyribose (see Single strand breakage). P. are active in the repair of damaged DNA, and in the linkage of Okazaki fragments during DNA replication. They have been used in vitro, in the cell-free synthesis of genes (DNA) (see Gene synthesis).

Polynucleotide-methyltransferase, *polynucleotide methylases:* specific enzymes that catalyse the methylation of purine and pyrimidine bases, or sugars in the intact polynucleotide chain by transfer of methyl groups from S-adenosyl-L-methionine. Methylated bases and/or sugars (see Rare nucleic acid components) are found in tRNA, mRNA, rRNA and DNA. 5-Methylcytosine and 6-methyladenine are particularly common. tRNA (see) is notable for its high content of modified components, many of them methylated. The 5'-terminal cap structures of eukaryotic and viral mRNA are methylated

after transcription, and *N*^{6}-methyladenosine is typically found internally between the 5' and 3' ends of the mRNA molecule (see mRNA). Methylation of rRNA is essential for ribosome maturation. Methylation of DNA serves in the recognition of specific regulatory proteins, and it protects DNA from the action of restriction endonucleases produced by its own cell. 1-1.5% of mammalian DNA and 5-6% of plant DNA is methylated. The degree of methylation is often organ-specific. Phage DNA is also methylated. DNA from T2, T4, T7 and P1 contains 6-methyladenine, whereas that from T3 and T5 contains methylthymine; the latter, however, arises by the methylation of deoxy-TPP during DNA replication, and not by the action of P.m. on the intact DNA.

Polynucleotide phosphorylase: an enzyme that catalyses the synthesis in vitro of polyribonucleotides. 5'-Nucleoside diphosphates are added to oligonucleotide starter molecules (primers), with the release of phosphate. The resulting sequence depends upon the availability of components for the reaction. Thus it is possible to synthesize homopolymers (poly A, poly U, etc.) or Copolymers (see) (e.g. poly AU). The function of P.p. in the cell is not clear. Since it also acts as an exonuclease and catalyses the reverse reaction, it is possible that the enzyme functions in the synthesis and remobilization of storage polynucleotides.

Polynucleotides: linear sequences of at least 20 nucleotides, in which the 3'-position of each monomeric unit is linked to the 5' of the neighboring unit via a phosphate group. The valency angles between phosphate and sugar residues are such that the sugar-phosphate backbone forms a helix, with the bases projecting sideways. The sugar may be D-ribose (in ribopolynucleotides) or 2-deoxy-D-ribose (in deoxyribopolynucleotides). The Nucleic acids (see) are specific high M_r P. Various types of P. structure are found e.g. single-stranded, double-stranded, or internally folded (alternating regions of single and double strands). The base sequence of P. is denoted by the base letters, A, G, C, U, T. The letter p is used before the base letter for 5'-substitution, and after the base letter for 3'-substitution, e.g. pGpApU represents 5'-phosphate → 3'-hydroxyl.

Polynucleotide thioltransferases: thiolases that catalyse the specific thiolation of purine and pyrimidine bases in the synthesis of Rare nucleic acid components (see).

Polypeptides: see Peptides.

Polyporenic acid, *polyporenic acid A:* a tetracyclic triterpene carboxylic acid. M_r 486.74, m.p. 200 °C $[α]_D + 74$ °. P.a. is structurally related to 5α-lanostane (see Lanosterol). It occurs in the fungus *Polyporus* spp. growing on birch trees.

Polyporinic acid: see Benzoquinones.

Polyprenols, *polyprenyl alcohols:* acyclic, polyisoprenoid alcohols. P. occur free or esterified with higher fatty acids in microorganisms, plants and animals. The natural source and the number of isoprene units is usually indicated in the names of P., e.g. betulaprenol-8 is formed from 8 isoprene units and is found in *Betula verrucosa*.

P. also differ from each other in the order of double bonds and by the fact that some are partially hy-

Table. Polyprenols.

Name (Origin)	No. of iso-prene units	Conformation of double bonds
Solanesol (Tobacco)	9	all *trans*
Castaprenols (Chestnut)	11–13	3 *trans*, remainder *cis*
Ficaprenols (Ornamental rubber tree)	10–12	3 *trans*, remainder *cis*
Betulaprenols (Silver birch)	6–13	3–4 *trans*, remainder *cis*
Undecaprenol (Bacteria)	11	2 *trans*, 9 *cis*
Dolichols (Mammals, Microorganisms)	14–24	3–4 *trans*, remainder *cis*

drogenated. Dolichol-24 has the highest M_r of known P.

Undecaprenol (bactoprenol) from *Salmonella* contains eleven isoprene units, and two *trans* and nine *cis* double bonds. In the form of undecaprenyl phosphate, it acts as a carrier of carbohydrate residues in the biosynthesis of bacterial antigenic polysaccharides; synthesis of Murein (see) also depends upon undecaprenyl phosphate. In eukaryotes the Dolichol phosphates (see) function in the transfer of carbohydrate residues in the synthesis of glycoproteins and glycolipids. Probably the long lipid chains of these P. serve to anchor them in membranes, while the phosphate group acts as a carrier by protruding into the cytoplasm. It is not known whether all P. function as carbohydrate carriers. The structural relationship between solanesol and plastoquinone-9 and ubiquinone-9 and the joint occurrence of these compounds suggest a precursor role for P. Biosynthesis of P. proceeds from mevalonic acid and the conformation of all double bonds is predetermined in early precursors.

Polyribosomes, *polysomes, ergosomes* (obsolete): the structural unit of Protein biosynthesis (see), consisting of several to many Ribosomes (see) attached along the length of a strand of mRNA. P. occur free, or attached to the endoplasmic reticulum. The individual ribosomes are separated by 60–90 nucleotide units of the mRNA, i.e. the distance between them is 20–30 nm. Each ribosome covers about 35 nucleotide residues. Length of the P. is approximately proportional to the size of the synthesized polypeptide. P. are formed by association of mRNA with ribosomal subunits (produced at the termination of translation) under the influence of initiation factor IF 3, or by direct association of mRNA with newly formed ribosomes.

Polysaccharides: see Carbohydrates.

Polysaccharide sulfate esters, *polysaccharide sulfates:* sulfate esters of polysaccharides, synthesized by the action of sulfokinases. There are two possible mechanisms of biosynthesis: 1. Sulfation of the polysaccharide chain by phosphoadenosinephosphosulfate (PAPS); 2. Polymerization of sugar sulfates, which occur as sulfated sugar nucleotides.

P.s.e. are found especially in algae, where they act as cell wall components. They act as supportive material in the vegetative forms of the *Thallophyta*, and protect them from desiccation; this is particularly important in terrestial forms and in those that grow in intertidal zones. P.s.e. of marine algae represent a wide range of different structures. The in vitro synthesis of P.s.e. from marine algae has not yet been achieved. Many Plant mucilages (see) are P.s.e., e.g. Agar-agar (see) and Carrageenans (see). Other examples are Chondroitin sulfates (see) and Heparin (see).

Polysaccharide sulfates: see Polysaccharide sulfate esters.

Polysomes: see Polyribosomes.

Polytenic chromosomes: see Giant chromosomes.

Polyterpenes, *polyisoprenes, polyisoprenoids:* acyclic, unsaturated, terpene hydrocarbons or alcohols, of the general formula $(C_5H_8)_n$, and consisting of a large number of isoprene units. They are classified as shown in the table.

Table. Classification of polyterpenes.

Compound	No. of isoprene units	Double bonds
Polyprenols	6–24	*cis/trans*
Gutta, Balata	about 100	*trans*
Natural caoutchouc	10 000	*cis*

All P. are unbranched molecules, and the double bonds may be *cis* or *trans* (Fig.). For the biosynthesis of P. see Terpenes. With the exception of certain Polyprenols (see), P. are found only in plants, in latex and latex cells.

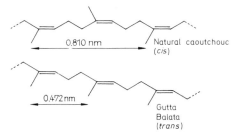

Polyterpenes, showing conformation of double bonds.

Poly U: abb. for polyuridylic acid.

Polyuridylic acid, abb. *poly U:* a Homopolymer (see) of uridylic acid, containing an indeterminate number of nucleotide units. In a cell-free, ribosomal protein biosynthesis system, poly U acts as synthetic mRNA, and its translation product is polyphenylalanine. It is often used to determine the synthetic capacity of such systems. Poly U played an important part in early work on the deciphering of the genetic code.

Polyuronides: macromolecular compounds found in plants, consisting of units of uronic acids in the pyranose form (see Carbohydrates). The chief components are D-glucuronic acid, D-galacturonic acid and D-mannuronic acid. P. contain free carboxyl groups and are consequently more strongly hydrated than polysaccharides formed from neutral monosaccharides. Examples are pectins, alginic acid and plant mucilages.

Pomiferin: see Isoflavone.

Pompe's disease: see Glycogen storage disease.

Ponasterone A: 2β,3β, 14α, 20α(R),22β(R)-pentahydroxy-5β-cholest-7-en-6-one, a phytoecdysone (see Ecdysone). M_r 464.65, m.p. 260 °C, $[\alpha]_D +90°$ (methanol). P.A. was isolated from plants, e.g. the fern *Podocarpus nakaii* Hay, and it has molting hormone activity in insects. The closely related ponasterone B and ponasterone C were also isolated from *Podocarpus;* they differ from P.A. in that P.B. has the opposite configurations at C2 and C3, while P.C. possesses an additional hydroxyl group at C24.

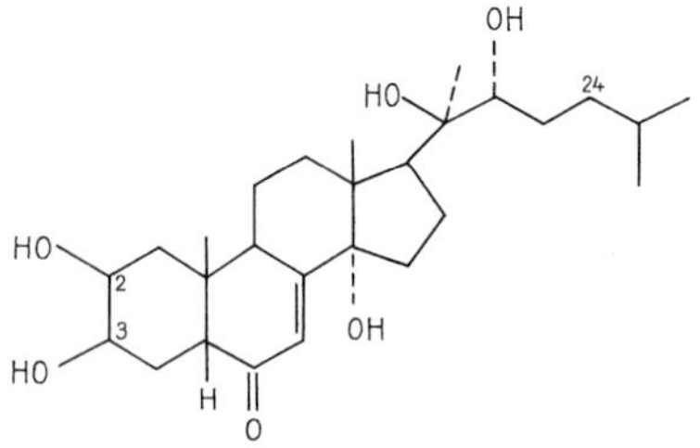

Ponasterone A

Poppy alkaloids: see Papaveraceae alkaloids.

P/O quotient: see Respiratory chain.

P/O ratio: see Respiratory chain.

Porifersterol: (24S)-5α-stigmasta-5,22-dien-3β-ol, a marine zoosterol (see Sterols). M_r 412.7, m.p. 156 °C, $[\alpha]_D -49°$ (chloroform). P. is a characteristic sterol of sponges, and has been isolated from, e.g. *Haliclona variabilis, Cliona celata* and *Spongia lacustris*. It differs from Stigmasterol (see) by an altered configuration at C24.

Porphin: the parent tetrapyrrole of the Porphyrins (see). Porphin is a term from the original Fischer nomenclature. It is synonymous with porphyrin. See also Chlorin.

Porphobilinogen: see Porphyrins.

Porphodimethene: 5,10,15,22-tetrahydroporphyrin.

Porphomethene: 5,15-dihydroporphyrin.

Porphyria: see Inborn errors of metabolism.

Porphyrinogen: 5,10,15,20,22,24-hexahydroporphyrin. See Porphyrins.

Porphyrins: cyclic tetrapyrroles, which can be considered as derivatives of the parent tetrapyrrole, porphin. The eight β-hydrogens of the parent porphin are completely or partly substituted by side chains, e.g. alkyl, hydroxyalkyl, vinyl, carbonyl or carboxylic acid groups. The different P. are classified on the basis of these side chains, e.g. protoporphyrin, coproporphyrin, etioporphyrin, mesoporphyrin, uroporphyrin. In all naturally occurring P., the two side chain substituents of each pyrrole ring are dissimilar; therefore when two types of side chain substituents are present (e.g. etioporphyrin, coproporphyrin, uroporphyrin) there are only 4 possible isomers (designated I to IV). With three different side chains (e.g. protoporphyrin) the number of possible isomers increases to 15. Protoporphyrin occurs naturally only as the IX isomer. Since it can be considered as a derivative of coproporphyrin III, protoporphyrin IX is also called protoporphyrin III. Of the many possible P. isomers, only types I and III occur naturally. Type I P. have no known useful function and are excreted. Protoporphyrin IX is particularly abundant in nature, occurring as the corresponding heme in hemoglobin, myoglobin and most cytochromes. Many metal ions are complexed by P., forming metalloporphyrins. The chelate complex of protoporphyrin IX with Fe(II) is called protoheme or Heme (see). With Fe(III), the complex is called hemin or hematin. Chlorophylls (see) are magnesium complexes of various P. In cobalamin (see under Vitamins), cobalt is complexed by a corrin ring structure, which is structurally and biosynthetically

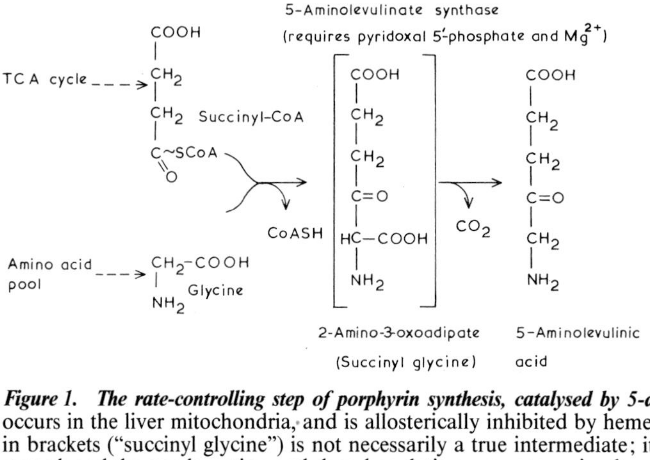

Figure 1. The rate-controlling step of porphyrin synthesis, catalysed by 5-aminolevulinate synthase. This reaction occurs in the liver mitochondria, and is allosterically inhibited by heme and hemin. The intermediate shown in brackets ("succinyl glycine") is not necessarily a true intermediate; it has not been unequivocally demonstrated, and the condensation and decarboxylation may occur simultaneously.

related to the P. The Bile pigments (see) are derived metabolically from P. Porphyrinogens (which are intermediates in P. biosynthesis) represent P. in which both pyrrolenine nitrogens and all the bridge methene carbons are hydrogenated.

The biological metalloporphyrins share a common pathway of biosynthesis to the stage of protoporphyrin IX (Figs. 1, 2 and 3). The parent porphyrin structure is derived biosynthetically from succinyl-CoA and glycine. Two molecules of the intermediate 5-aminolevulinic acid condense to form the substituted monopyrrole, porphobilinogen. The enzyme catalysing this reaction, 5-aminolevulinate dehydrase (porphobilinogen synthase, EC 4.2.1.24), is very sensitive to lead; in lead poisoning there is a significant increase in the urinary excretion of 5-aminolevulinic acid, and the activity of 5-aminolevulinate synthase in the blood is decreased. Four

molecules of phorphobilinogen react via their aminomethyl side chains, with loss of ammonia, to form the tetrapyrrylmethane intermediate, hydroxymethylbilane. During metabolism of hydroxymethylbilane to uroporphyrinogen III, ring IV undergoes rearrangement (compare positions of ring IV propionate and acetate side chains in hydroxymethylbilane and uroporphyrinogen III, Fig. 2). In contrast, no rearrangement of ring IV occurs in the non-enzymatic conversion of hydroxymethylbilane to uroporphyrinogen I. The resulting uroporphyrinogen III is decarboxylated to coproporphyrinogen III. Further decarboxylation and dehydrogenation lead to the key substance of P. synthesis, protoporphyrinogen IX.

There is a lack of uniformity in the nomenclature of P. All types of nomenclature are current, and appear to depend upon individual editorial policy. Naming

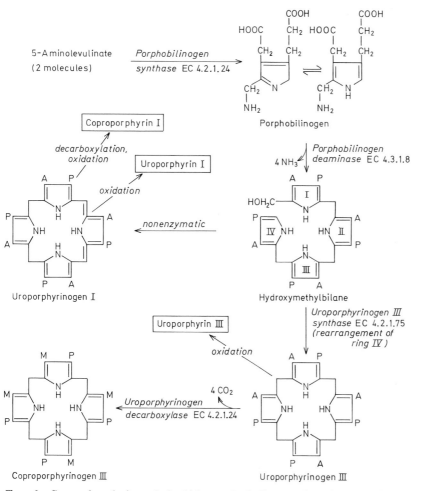

Figure 2. Stages of porphyrin synthesis which occur in the liver cytoplasm. Coproporphyrin I is excreted in constant but small amounts in the urine (40–190 mg/day) and feces (300–1100 mg/day); its rate of excretion increases during hemolytic disorders. The further conversion of coproporphyrinogen III occurs in the mitochondria.

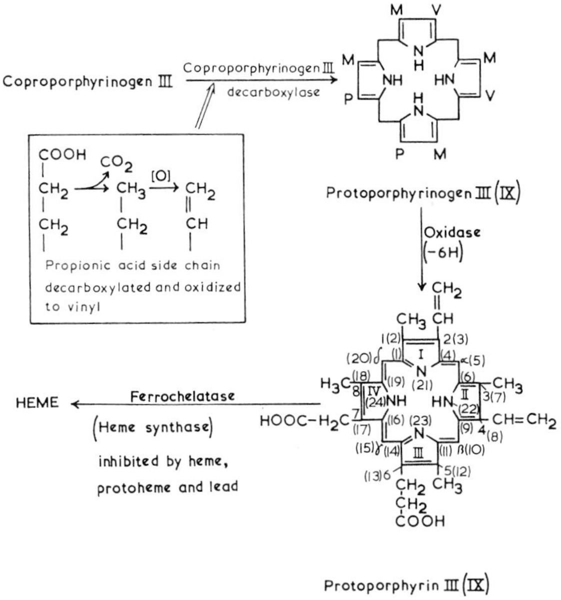

Figure 3. Final stages in the synthesis of protoporphyrin III (IX) in liver mitochondria, and its conversion into heme. The bracketed numbers and the Greek letters correspond to the Fischer system. The unbracketed numbers represent the 1–24 porphyrin numbering system.

Table. *Porphyrins*

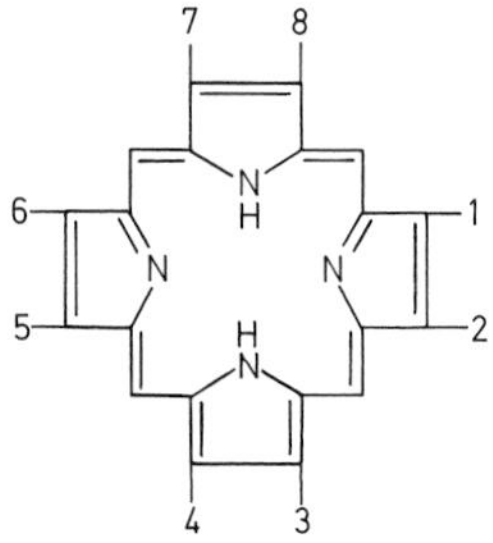

	Substituent No.							
	1	2	3	4	5	6	7	8
Coproporphyrin I	M	P	M	P	M	P	M	P
Coproporphyrin III	M	P	M	P	M	P	P	M
Deuteroporphyrin I	M	H	M	H	M	P	M	P
Deuteroporphyrin III	M	H	M	H	M	P	P	M
Etioporphyrin I	M	E	M	E	M	E	M	E
Etioporphyrin III	M	E	M	E	M	E	E	M
Hematoporphyrin I	M	HE	M	HE	M	P	M	P
Hematoporphyrin III	M	HE	M	HE	M	P	P	M
Mesoporphyrin I	M	E	M	E	M	P	M	P
Mesoporphyrin III	M	E	M	E	M	P	P	M
Protoporphyrin I	M	V	M	V	M	P	M	P
Protoporphyrin III	M	V	M	V	M	P	P	M
Uroporphyrin I	A	P	A	P	A	P	A	P
Uroporphyrin III	A	P	A	P	A	P	P	A
Isocoproporphyrin	M	E	M	P	P	P	P	M

M = methyl, $-CH_3$; P = propionic acid, $-CH_2CH_2-COOH$; H = hydrogen; HE = hydroxyethyl, $-CH_2CH_2OH$; V = vinyl, $-CH=CH_2$; E = ethyl, $-CH_2CH_3$; A = acetic acid, $-CH_2-COOH$. The corresponding porphyrinogens (see Fig. 2 for porphyrinogen ring system) have identical patterns of substitution.

and numbering systems were proposed by the Commission of Nomenclature of Biological Chemistry in 1960, and in the recommendations of the IUPAC/ IUB Commission for Biochemical Nomenclature and the IUPAC Commission on Nomenclature in Organic Chemistry (*Pure and Applied Chemistry 51* (1979) 2251–2304). The original system of numbering and trivial names introduced by Fischer is, however, still widely favored. [D. Dolphin, ed., *The Porphyrins, Vol. 1. Structure and Synthesis, Part A* (Academic Press, 1978)]

Positive control: activation of transcription, or the activity of enzymes by activators or metabolites.

Post-transcriptional modification of RNA, *maturation of RNA, processing of RNA:* Most eukaryotic RNA is modified after transcription. The genes for transfer RNAs are about 100 nucleotides long, whereas the mature molecules contain only about 80, the remainder having been removed during P. Eukaryotic ribosomal RNA (rRNA) consists of two species, with Sedimentation coefficients (see) of 18 and 28 S. It is transcribed, however, as a 45 S molecule. Nearly half of this is excised and degraded, leaving the two smaller molecules of mature rRNA. Prokaryotic rRNA is less severely pruned, losing only about 10% of its nucleotides during P.

The mRNA of bacteria is probably not processed, as translation begins before the mRNA has been completely transcribed. In eukaryotes, however, P. of the precursors of mRNA (heteronuclear RNA, hnRNA) involves at least three operations: capping, addition of a "tail" of polyadenosine, and excision of Introns (see). The "cap" is a 7-methyl-guanosine residue linked via 3 (rather than the usual 1) phosphate groups to the 5′-OH group of the 5′-terminal ribose of the RNA. This 5′-terminal ribose is also methylated at its 2′-OH group. The cap is added by nuclear enzymes very early in transcription. Shortly after transcription is complete, another enzyme locates a signal sequence which includes AAUAAA, cuts the RNA about 20 base pairs beyond the signal, and adds a length of about 200 base pairs of polyadenosine. This process does not occur on all mRNA precursors, but mRNA lacking the tail (i.e. histone mRNA) has a very short lifetime in the cytoplasm. Finally, the non-coding sequences (introns) are removed from the RNA. Both the introns and the sequences "downstream" from the polyadenylation site are degraded. In some cases, only about 10% of the original hnRNA appears in the cytoplasm as mRNA. [C. Montell et al., *Nature 305* (1983) 600–605.]

The catalysts for at least some of the P. are Small nuclear riboproteins ("snurps") (see). [*Nature 307* (1984) 412–413.]

Post-translational modification of proteins, *processing:* In a general sense, any difference between a functional protein and the linear polypeptide sequence encoded between the the initation and termination codons of its structural gene can be regarded as a P. t. m. Thus, the folding of the polypeptide chain, stabilized by weak, noncovalent interactions, and the association of the subunits of an oligomeric protein both represent a modification that occurs after translation. It is more usual, however, to restrict the meaning of P. t. m. to modifica-

tions that involve the making or breaking of covalent bonds. Examples of P. t. m. by Proteolysis (see) are: 1. The first amino acid introduced during initiation of Protein biosynthesis (see) is *N*-formylmethionine in prokaryotic systems and methionine in eukaryotic systems. Few final proteins contain the formyl or even the methionine residue. For example, no completed protein isolated from *Escherichia coli* contains an *N*-terminal *N*-formylmethionine and only 30% have *N*-terminal methionine.

2. Removal of the signal sequence responsible for the transport of secretory proteins through the membrane of the rough endoplasmic reticulum (see Signal hypothesis).

3. Activation of proenzymes (zymogens) (see Pepsin; Trypsin; Chymotrypsin).

4. Activation of the Complement (see) system.

5. Proteolysis in the cascade system of Blood coagulation (see) (see Prothrombin).

6. Activation of proteo- and peptide-hormones (see Insulin).

7. Proteolysis of larger protein precursors, e. g. proteolysis of procollagen.

Another type of P. t. m. is the attachment of prosthetic groups, e. g. the insertion of heme into hemoproteins, and the attachment of carbohydrates to glycoproteins.

A further important form of P. t. m. is the modification of amino acid residues. Only 20 amino acids are encoded during translation, yet well over 100 different amino acid residues are known from various proteins. Examples are: hydroxylation of proline and lysine residues to form the hydroxyproline and hydroxylysine residues in Collagen (see); phosphorylation of serine to phosphoserine residues (see Phosphoproteins); adenylylation of tyrosine residues (see Metabolic control); carboxylation of glutamate to γ-carboxyglutamate residues (see γ-Glutamylcarboxylase); amide groups are sometimes found on *C*-terminal residues, e. g. glycinamide; the ε-amino group of lysine residues may be modified by methylation, acetylation or phosphorylation, and by the formation of peptide linkages with, e. g. biotin or lipoic acid; lysine residues are the precursors of desmosine (see Elastin). Clearly, the difference between the modification of an amino acid residue and the attachment of a prosthetic group is sometimes purely semantic.

Most amino acid modifications occur after release of the polypeptide from the ribosome, and a large number of processing enzymes (in particular those responsible for glycosylation, disulfide bridge formation and iodination) are found in the endoplasmic reticulum and Golgi apparatus. Limited proteolysis occurs in the extracellular space and in secretory granules. Cross linking reactions (e. g. cross linking of lysine and hydroxylysine residues in collagen; formation of isopeptide cross links between carboxyl groups of glutamate or aspartate and the ε-amino group of lysine; formation of ether cross links involving iodotyrosine) are extracellular.

Modification sometimes occurs at an earlier stage. Formylmethionyl-tRNA is formed by the formylation of methionyl-tRNA before it is incorporated as the *N*-terminal amino acid in prokaryotic translation. It appears that Pyroglutamic acid (see) and possibly certain acetylated amino acids are also formed at the aminoacyl-tRNA stage. Some modifications occur in the nascent polypeptide chain, i.e. while the polypeptide is still attached to the ribosome. Hydroxylation of lysine and proline occur in the nascent chain of the collagen precursor protein. N-terminal formylmethionine and methionine are removed when the nascent chain is about 50 amino acid residues long, as demonstrated in the synthesis of bacteriophage f2 coat protein in *Escherichia coli*, and in the synthesis of rabbit hemoglobin. Some disulfide bridge formation, some N-terminal acylation and some glycosylation also occur in the nascent polypeptide chain. Many different kinds of P.t.m. are exemplified in the synthesis of Collagen (see).

PP-factor: pellagra preventative factor: biologically equivalent to nicotinic acid and nicotinamide. See Vitamins.

PP$_i$: inorganic pyrophosphate; the anion of pyrophosphoric acid, $P_2O_7^{4-}$.

℗℗: inorganic pyrophosphate (see PP$_i$).

PRA: abb. for phosphoribosylamine.

Pratensein: 5,7,3'-trihydroxy-4'-methoxyisoflavone, see Isoflavone.

PRB 8: see Orthonil.

Precalciferol: see Vitamins (Vitamin D).

Precipitation: formation of an insoluble, inactive antigen-antibody complex from a soluble antigen and a bivalent specific antibody (the precipitin). Monovalent or incomplete antibodies do not precipitate the soluble antigens; the resulting complexes cannot form cross-linkages and therefore remain soluble. In the quantitative determination of the precipitation reaction (due to Heidelberger), increasing quantities of antigen are added to a constant amount of antibody. At the Equivalence point (see) the supernatant above the precipitate contains neither antigen nor antibody. In most studies, only the antigen titre is determined, i.e. the dilution of antigen at which P. is still just visible. For this purpose, the *Ouchterlony technique* is used, in which both reaction partners are allowed to diffuse into one another in an agar gel, forming precipitation lines in their area of contact; this is an economical method which requires only very small quantities of antigen and antibody.

Precipitation curve, *Heidelberger curve:* a plot of the quantity of precipitate formed during the titration of an antibody with an antigen, or vice versa. The antibody must be at least bivalent. The P.c. reaches a maximum at the Equivalence point (see), then decreases as the concentration of one component exceeds its optimum (Fig.)

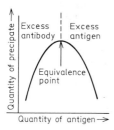

Precipitation curve. An increasing quantity of antigen is added to a constant amount of antibody. The resulting precipitate is removed by centrifugation, and antibody and antigen are determined in the supernantant.

Precursor: a starting compound in the biosynthesis of a Metabolite (see).

Prednisolone: 1,2-dehydrocortisol, 11β,17α,21-trihydroxy-pregna-1,4-diene,3,20-dione, a synthetic steroid prepared by chemical or microbiological dehydrogenation of cortisol. It has powerful antiinflammatory and antiallergic activity, very similar to that of prednisone (1,2-dehydrocortisone, the corresponding dehydro derivative of cortisone); both compounds are used in the treatment of disorders such as asthma, arthritis and eczema.

Prednisolone

Prednisone: see Prednisolone.

Pregnancy hormones: see Progesterone.

Pregnane: see Steroids.

5β-Pregnane-3α-diol, *pregnanediol:* a biological degradation product of progesterone. M_r 320.52, m.p. 224°C, $[\alpha]_D + 27°$. It occurs as the glucuronide, particularly in pregnancy urine.

5β-Pregane-3α, 20α-diol-glucuronide

Pregnant mare serum gonadotropin, abb. *PMS:* a hormone produced by the uterine endometrium. M_r 28000. PMS is poorly excreted by the kidneys, and it accumulates in the blood. Its action is similar to that of Follicle stimulating hormone (see).

Pregnenolone, *3β-hydroxy-pregn-5-en-20-one:* an intermediate in the biosynthesis of progesterone from cholesterol, and in the biosynthesis of androgens in the adrenal cortex. See Steroids, Androgens.

Prephenic acid: see L-Phenylalanine.

Pre-proprotein: a precursor in the biosynthesis of a secretory protein (see Signal hypothesis), which possesses an N-terminal signal peptide sequence in addition to the proprotein sequence, e.g. pre-proinsulin, pre-protrypsin, pre-proparathormone, pre-promellitin, etc.

Preprotein: the biosynthetic precursor of a secretory protein containing a very short-lived signal peptide sequence (see Signal hypothesis), e.g. preovomucoid and prelysozyme.

Pretyrosine, *arogenic acid*: see L-Phenylalanine.

PRH: abb. for Prolactin Releasing Hormone (see Releasing hormones).

Pribnow box: a sequence of seven nucleotide pairs, which is the same or very similar in all promoters. It is located five to seven nucleotide pairs from the initiation point of RNA transcription.

The structure of the P.b. is:

$$5'\text{-TATPuATG-}3'$$
$$3'\text{-ATAPyTAC-}5'$$

where Pu is a purine and Py a pyrimidine.

Primary metabolism: those metabolic processes (see Metablism) that are basically similar in all living cells, and are necessary for maintenance and survival. P.m. includes the fundamental processes of growth (synthesis of biopolymers and their constituents, synthesis of macromolecular superstructures of cells and organelles, energy production and transformation, and the Turnover (see) of body and cell constituents.

Primary structure: see Proteins.

Primer: 1. a small polymer that is required as a starter for the synthesis of a larger biopolymer. In nucleic acid synthesis, for example, an oligonucleotide serves as a primer, and it is extended by the enzyme-catalysed addition of further nucleotide units from nucleoside triphosphates. The P. molecule is therefore incorporated into the final product. A P. or starter must not be confused with a Template (see), which is not incorporated into the product.
2. a Pheromone (see) that causes a long term physiological change.

Primetin: see Flavones (Table).

Primin: see Benzoquinones.

Primobolane: 1-methyl-17β-acetoxy-5α-androst-1-en-3-one, a synthetic Anabolic steroid (see). It is a less potent (one tenth) androgen, but a more potent (five times) anabolic agent than testosterone propionate (see Testosterone), and it is used therapeutically.

Pristanic acid: see Inborn errors of metabolism (Refsum's disease).

Pro: abb. for L-proline.

Proaccelerin: see Blood coagulation (Table).

Proanthocyanidins: colorless substances from plants which are converted into anthocyanidins by heating with acid. This conversion is chemical and does not imply a biogenetic relationship. P.are subdivided into 1. monomeric flavan-3,4-diols, also known as Leucoanthocyanidins (see), and 2. dimers and higher oligomers of flavan-3-ols, known as condensed P.(also classified as condensed vegetable tannins; see Tannins). The four major, naturally occurring dimeric flavan-3-ols are all configurational isomers (Fig.1), representing dimers of catechin and/or epicatechin (see Catechins). These compounds are also known as procyanidins, because cyanidin is the product of their treatment with acid. Early proposals for procyanidin biosynthesis implicated a symmetrical flav-3-en-3-ol intermediate. [E.Haslam et al. *J. Chem. Soc. (Perkin I)* (1977) 1637-1643] According to later work, however, the biosynthesis is catalysed by multienzyme complexes, a flav-3-en-3-ol is not involved, and the 2,3-*cis* stereochemistry of (−)-epicatechin is achieved by the action of an epimerase on the precursor (+)-dihydroquercetin (Fig.3). [H.A. Stafford *Phytochemistry 22* (1983) 2643-2646.]

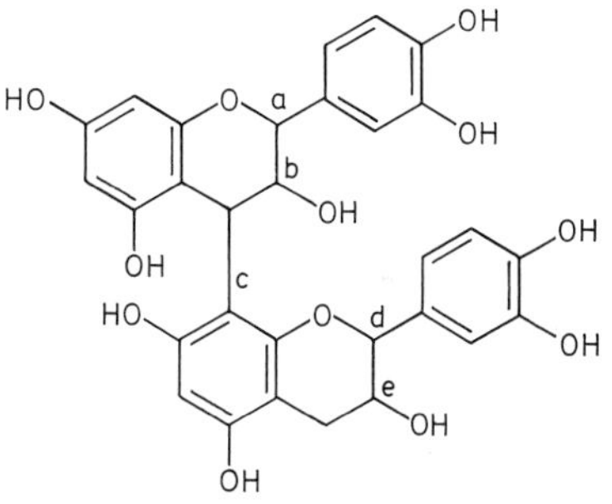

Name and source	Bond configuration				
	a	b	c	d	e
Epicatechin-(4β→8)-epicatechin (Apple, hawthorn, cocoa bean, *Cotoneaster*, quince, cherry, horse chestnut)	R ---	R ---	β ◄	R ---	R ---
Epicatechin-(4β→8)-catechin (Grape, cranberry, sorghum)	R ---	R ---	β ◄	R ---	S ◄
Catechin-(4α→8)-epicatechin (Raspberry, blackberry)	R ---	S ◄	α ---	R ---	R ---
Catechin-(4α→8)-catechin (Willow and poplar catkins, strawberry, hops, rose hips)	R ---	S ◄	α ---	R ---	S ◄

Figure 1. Configurations and sources of naturally occurring procyanidins

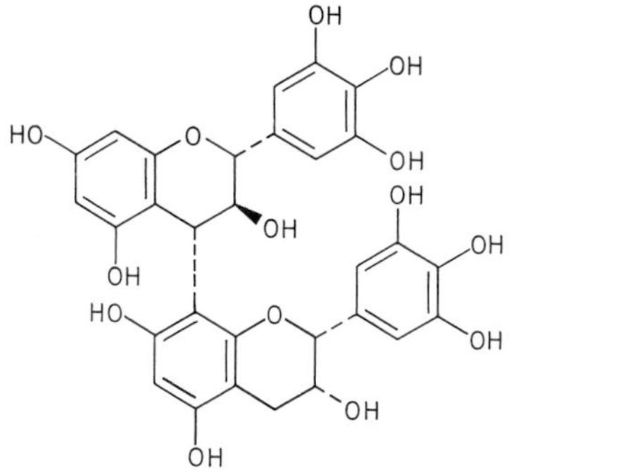

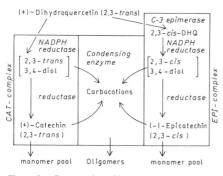

Gallocatechin-(4α→8)-epigallocatechin (a prodelphinidin from *Ribes sanguineum*)

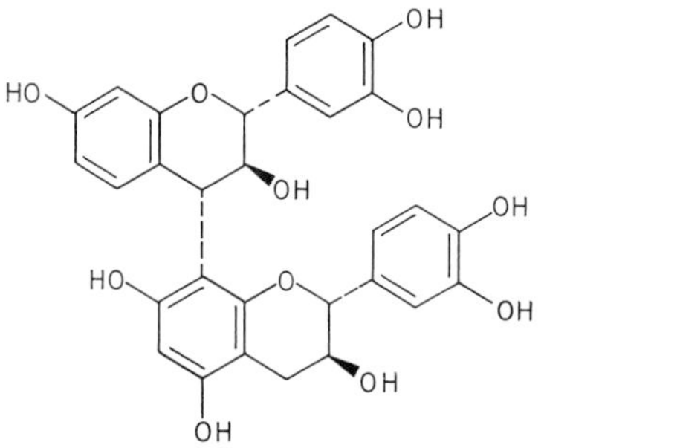

Fisetinidol-(4α→8)-catechin (a profisetinidin from *Acacia baileyana*)

Figure 2.

Less widely distributed are the prodelphinidins, which yield delphinidin with acid, e. g. a prodelphinidin from *Ribes sanguineum* (Fig. 2). A profisetinidin (fisetinidol-(4α→8)-catechin) has also been reported from the heartwood of *Acacia baileyana* [L. Y. Foo *Phytochemistry 23* (1984).] (Fig. 2).

A new, convenient system of nomenclature (used in Figs. 1 and 2) has been proposed for P. [L. J. Porter et

Figure 3. Suggested multienzyme complexes in procyanidin biosynthesis. DHQ = dihydroquercetin. After H. A. Stafford, *Phytochemistry 22* (1983) 2643–2646.

al. *J. Chem. Soc. (Perkin I)* (1982) 1209–1216] In it the names of common catechins with the most frequently encountered 2R configuration (see Catechins) are retained for the individual units of the dimer. Where 2S configuration occurs, it is denoted by the prefix *enantio*. The configuration at C-4 is denoted by α or β.

Proazulenes, azulenogens, hydroazulenes: a group of natural cyclic sesquiterpenes, which can be termally dehydrogenated or dehydrated to Azulenes (see). P. are chiefly compounds of the guaiane type, e. g. guaiol.

Proazulene type: see Sesquiterpenes (Fig.).

Process control: a term used in industrial biochemistry, particularly with reference to control of production of microbial fermentation products, e. g. by control of nutrient supply, air supply, etc., and choice of continuous or batch culture. See Fermentation techniques.

Processing: modification of protein molecules after translation (see Post-translational modification of proteins), or modification of RNA after transcription (see Post-transcriptional modification of RNA).

Prochirality: A molecule (or atom) is prochiral if it becomes chiral (asymmetrical or dissymmetrical) by the replacement of one point ligand by a new point ligand. A prochiral carbon atom possesses two identical (a, a) and two different (b, c) substituents. It has a one-fold alternating axis, and a mirror plane of symmetry, i.e. the plane through C, b and c cuts the molecule into mirror image halves. A classical example is the prochirality of citric acid, which is discussed in greater detail under Tricarboxylic acid cycle (see). NADH and NADPH are also prochiral, even though these molecules have perfect bilateral symmetry. Enzymes distinguish between the two sides of NAD(H) and NADP(H), so that hydrogen is added to or removed from the 4R or 4S position of C4 of the nicotinamide ring.

The R and S system [Cahn, R. S., Ingold, C. and Prelog, V. (1966). Specification of molecular chirality. *Angew. Chem. Int. Ed. Engl. 5,* 385–415] is used to specify configuration at prochiral centers. Each substituent is given a certain priority (based on atomic number). The model of the molecule is viewed from the side furthest from the lowest priority group. If the sequence of groups in decreasing priority is clockwise, the configuration is R (rectus: right handed); if the sequence is anticlockwise the configuration is S (sinister: left handed). Naming of pairs of hydrogen atoms at prochiral centers is particularly important. Clearly any atom or group replacing a hydrogen will have a higher atomic number than the hydrogen it replaces. If the replacement results in S-chirality, then the replaced hydrogen is designated H_S; similarly, the other hydrogen must be H_R [Goodwin, T. W. "Prochirality in Biochemistry", in *Essays in Biochemistry* Vol. 9 (1973) pp. 103–160. Published for the Biochemical Society by Academic Press.

Bentley, R. *Molecular Asymmetry in Biology,* 2 Vols. Academic Press (1970)].

Proconvertin: see Blood coagulation.

Prodigiosin: a red pigment and secondary metabolite of the bacterium, *Serratia marcescens*. In the biosynthesis of P., L-proline enters intact, and conributes a greater number of carbon atoms than any other amino acid. All the carbon atoms of ring A and the associated carbon atom in ring B are derived directly from L-proline. The biosynthetic origin of the remainder of the molecule is less well understood. Carbon atoms 3 and 2 of alanine contribute the methyl group and its associated ring carbon atom in ring C.

Prodigiosin

Progesterone, *corpus luteum hormone, luteohormone:* pregn-4-ene-3,20-dione; Δ^4-pregnane-3,20-dione. M_r 314.47, m.p. 128 °C (ethanol/water), 121 °C (pentane/hexane), $[\alpha]_D + 192$ ° (ethanol). P. is structurally related to the parent hydrocarbon, pregnane (see Steroids). It is the natural progestin, and an antagonist of the Estrogens (see). It promotes proliferation of the uterine mucosa. Having thus prepared the uterus, P. promotes implantation and further development of the fertilized ovum in the uterine mucosa (secretion phase). During pregnancy, it prevents further maturation of follicles, and stimulates development of the lactatory function of the mammary gland. In mammals, including humans, P. is produced by the corpus luteum; during pregnancy the production of P. by the corpus luteum is markedly increased, and some is also produced by the placenta. Absence of or failure to produce P. causes abortion.

P. is determined biologically by the Clauberg test. Young female rabbits (600–800 g) receive estrogens subcutaneously daily for 8 days, which results in proliferative changes in the uterine mucosa. The P. under test is then injected subcutaneously in oil. One rabbit unit is the smallest quantity of P. required to change the endometrium to the secretory phase (detected histologically).

P. is degraded in the liver and kidneys; the chief products are hydroxylated pregnanes, e.g. 5β-pregnane-3α, 20α-diol.

P. is biosynthesized from cholesterol via pregnenolone. By conversion to 17α-hydroxyprogesterone, P. acts as a precursor of androgens and adrenocortical hormones.

P. was first isolated in 1934 from corpus luteum by Butenandt et al. It was later shown to present in the plant *Holarrhena floribunda*. It is used in veterinary medicine to correct irregularities of estrus and to control habitual abortion.

Progesterone

Progress curve: in enzyme kinetics, a plot of the concentration of a reactant, or several reactants (e.g. substrates, products, enzyme-substrate complexes) or an enzymatic reaction as a function of the time for which the reaction has progressed. A P.c. may be derived theoretically, preferably with the aid of a computer (e.g. simulated from the Michaelis-Menten scheme, making certain assumptions about the values of rate constants), or obtained experimentally. The experimental determination may be discontinuous (analysis of discrete samples at different time intervals) or continuous (continuous monitoring of parameters in a single sample, e.g. by spectrophotometric or polarographic methods). Detailed mathematical treatment and design of computer programs for theoretical P.c. have been reported [W. W. Cleland *Biochim. Biophys. Acta, 67* (1963) 104; *67* (1963) 173; *67* (1963) 188. *Nature, 198* (1963) 463). None of these publications contains a graphical representation of theoretical P.c. The originals of such plots first appeared in Cleland's lecture notes, and have been reproduced, e.g. in all editions of Biological Chemistry by Mahler and Cordes, published by Harper and Row.

Prokaryote, *prokaryotic organism:* see Cell.

Prolactin, *lactotropin, lactogenic hormone, mammotropin, luteomammotropic hormone, luteotropic hormone,* abb. *LTH, luteotropin:* a gonadotropin. Phylogenetically, P. is one of the oldest adenohypophysial hormones. It acts primarily on the mammary gland by promoting lactation in the postpartal phase, and in rodents it also acts on the ovary. Its activity in males is not clear. Bovine P. is a single chain polypeptide (198 amino acid residues, M_r 22 500, three disulfide bridges), of known primary structure.

P. synthesis in the α-cells (eosinophils) of the anterior pituitary (adenohypophysis) is under the control of the hypothalamus hormones PRH and PIH (see Releasing hormones). It is produced in increasing amounts during pregnancy and during suckling; it can be detected in the blood (ng/ml range) by radioimmunological assay. P. promotes metabolism and growth, and it influences osmoregulation, pigment metabolism, parental behavior and it suppresses reestablishment of the menstrual cycle post partum. P. acts on its target cells via the adenylate cyclase system.

Prolactin release inhibiting hormone: see Releasing hormones.

Prolactin releasing hormone: see Releasing hormones.

Prolamines: a group of simple (unconjugated) proteins, soluble in 50–90% ethanol. They occur in cereals, and contain up to 15% proline and 30–45% glutamic acid, but they have only low contents of essential amino acids. The chief representatives are gliadin (wheat and rye), zein (maize; contains no tryptophan or lysine) and hordein (barley: contains no lysine). Oats and rice do not contain P.

L-Proline, abb. *Pro:* pyrrolidine-2-carboxylic acid, a proteogenic amino acid. Pro is very soluble in water, but is also soluble in ethanol, so it can be separated from other amino acids by ethanol extraction. Being an imino acid, it forms a yellow color with

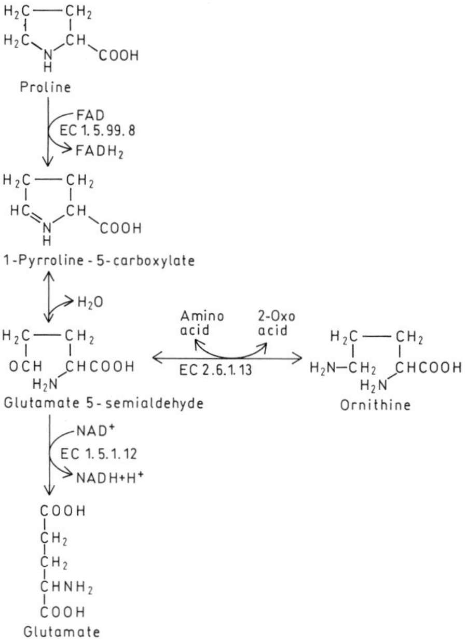

Proline

FAD
EC 1.5.99.8
FADH$_2$

1-Pyrroline-5-carboxylate

H$_2$O

Glutamate 5-semialdehyde

Amino acid 2-Oxo acid
EC 2.6.1.13

Ornithine

NAD$^+$
EC 1.5.1.12
NADH+H$^+$

COOH
CH$_2$
CH$_2$
CHNH$_2$
COOH
Glutamate

4-Hydroxyproline

FAD
EC 1.5.99
FADH$_2$

1-Pyrroline-3-hydroxy-5-carboxylate

H$_2$O

4-Hydroxyglutamate 5-semialdehyde

NAD$^+$
EC 1.5.1.12
NADH+H$^+$

2-Oxo acid   Amino acid
EC 2.6.1.23

EC 4.1.3.16

Glyoxylate

4-Hydroxyglutamate   4-Hydroxy-2-oxoglutarate

Pyruvate

Degradation of ornithine and proline.
EC 1.5.1.12, 1-pyrroline-5-carboxylate dehydrogenase.
EC 1.5.99.8, proline dehydrogenase
EC 2.6.1.13, ornithine-oxoacid aminotransferase.

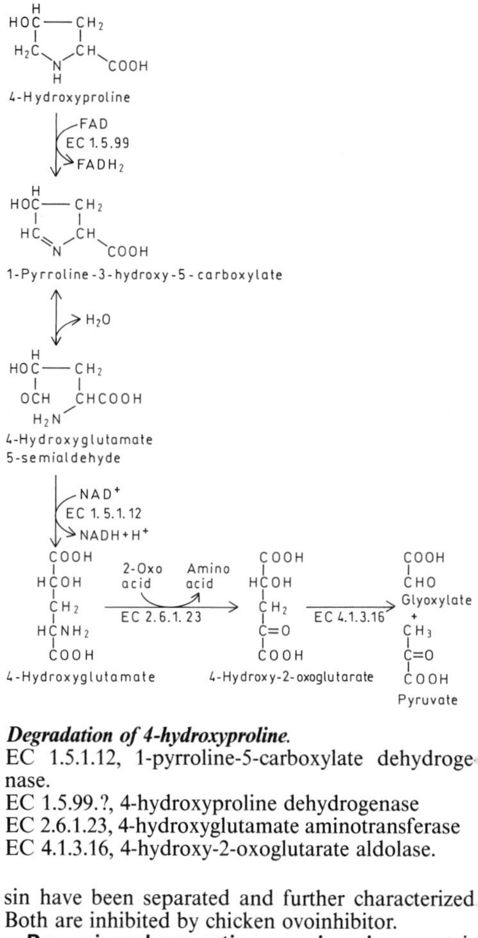

ninhydrin, rather than the purple color which is characteristic of α-amino acids.
Together with hydroxyproline (Hyp), Pro is an essential component of collagen, gliadin and zein. Collagen yields 15% Pro on hydrolysis.
Pro is glucogenic. It does not participate in α-helix formation, and therefore has special importance in protein tertiary structure (see Proteins). Pro is biosynthesized chiefly from L-glutamate via glutamic-γ-semialdehyde. Some may also be formed from exogenous ornithine via pyrroline carboxylic acid. It is a nonessential amino acid in mammals, but is essential for the growth of chickens. Azetidine-2-carboxylic acid is a Pro antagonist.
Proline derivatives: trans-4-hydroxy-L-proline (Hyp) is an important component of animal supportive and connective tissues. Free all-*cis*-4-hydroxy-L-proline occurs in *Santalum album* and other plants. Small quantities of 3-hydroxy-L-proline are also present in collagen. 4-Hydroxy-L-proline is formed mainly by hydroxylation of ribosome-bound peptidylprolyl-RNA, a reaction requiring ascorbic acid and catalysed by proline hydroxylase (see Oxygenases). Free 4-hydroxy-L-proline is formed by cyclization of γ-hydroxy-L-glutamate. Other derivatives of Pro are 4-methylproline (found in certain antibiotics) and 4-hydroxymethylproline (apple skin).

Promitochondria: see Mitochondria.

Promotor: see Operon.

Pronase: a mixture of at least 4 proteolytic enzymes from *Streptomyces griseus*. Two peptide esterase components resembling chymotrypsin and tryp-

Degradation of 4-hydroxyproline.
EC 1.5.1.12, 1-pyrroline-5-carboxylate dehydrogenase.
EC 1.5.99.?, 4-hydroxyproline dehydrogenase
EC 2.6.1.23, 4-hydroxyglutamate aminotransferase
EC 4.1.3.16, 4-hydroxy-2-oxoglutarate aldolase.

sin have been separated and further characterized. Both are inhibited by chicken ovoinhibitor.

Pro-opiomelanocortin, *pro-opiocortin:* a protein (M_r 31000) from the pars distalis and pars intermedia of the pituitary. P. contains the sequences of ACTH and β-LPH (see Peptides, Fig. 3). In the pars distalis, ACTH is released from P. to regulate adrenocortical function. In the pars intermedia, P. is cleaved to unprocessed α-MSH, which then undergoes amidation and acetylation to α-MSH (see Melanotropin).

Prophage: see Phage development.

Propionic acid: CH$_3$-CH$_2$-COOH, a simple fatty acid. M.p. -22 °C, b.p. 140.9 °C. P.a. occurs as its salts (propionates) and esters in many plants. It is especially important in the metabolism of propionic bacteria, which perform a propionic acid fermentation. *Propionibacterium shermanii* synthesises P.a from pyruvate (Fig.). Propionylcoenzyme A (see) is an important metabolic derivative.

Propionyl-coenzyme A, *propionyl-CoA:* activated propionic acid, formed by attachment of propionic acid to coenzyme A by a thioester linkage. Propionyl-CoA is important in fatty acid biosynthesis and in fatty acid degradation; it is formed during the β-oxidation of fatty acids with odd numbers of carbon atoms, or with branched chains.

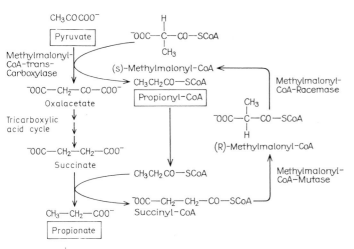

Biosynthesis of propionate from pyruvate by Propionibacterium shermanii.

Proplastids: rounded (0.2–1 µm diam.), colorless and largely structureless organelles in the meristematic tissues of higher plants, or in unicellular algae cultured in the dark. P. are the biogenetic precursors of Plastids (see). They can be converted into chloroplasts by illumination. During this transformation, the contents of nucleic acids, ribosomes and enzymes of transcription and translation (already present at low levels in P.) are increased, whereas thylakoid membranes and the enzymes of photosynthesis are synthesized de novo. Like plastids, P. cannot be formed de novo, and they reproduce by division.

Proproteins: inactive protein precursors, which are activated by the removal (a highly specific reaction) of a peptide sequence. Examples are various Secretory enzymes (see) (procarboxypeptidase, proelastase, prothrombin, etc.) hormone precursors (proinsulin, proparathormone, etc.), peptide toxins (promellitin, etc.) and Zymogens (see).

Prostacyclin, *PGI₂*: see Prostaglandins.

Prostaglandins: a group of biologically active, unsaturated C-20 fatty acids derived primarily from Arachidonic acid (see). The PG are related structurally and metabolically to the Leukotrienes (see) and Thromboxanes (see), which are treated separately. All PG can be considered as formal derivatives of prostanoic acid (which does not occur naturally, however). The conventional abbreviation is PG, with an additional letter and number. The E series are β-hydroxyketones, the F series are 1,3-diols; the A type are α,β-unsaturated ketones. All PG have a double bond at C-13 and an OH group at C-15. The series number indicates the number of double bonds, which depends on the fatty acid precursor. Series 1 are biosynthesized from 8,11,14-eicosatrienoic acid, series 2 from arachidonic acid, and series 3 from 5,8,11,14,17-eicosapentaenoic acid. The series 2 compounds are most abundant. Some PG have little or no biological activity and are presumably metabolic products of the active species. The active species are very unstable, and it is often difficult to determine whether a given compound is active in its

own right or is merely a precursor of a more active PG. However, it appears that the most active species are PGE_2, $PGF_{2\alpha}$, PGD_2, PGG_2, PGH_2, PGI_2 and the thromboxanes.

The biosynthesis of the PG is catalysed by a multienzyme complex (Fig.), prostaglandin synthase. The endoperoxides PGG_2 and PGH_2 are very active, but are also rapidly metabolized. PGH_2 is the precursor for both prostacyclin (PGI_2) and the thromboxanes. Very generally, the PG prevent aggregation of platelets and clotting (at high concentrations) while the thromboxanes promote aggregation and clotting. Since PGI_2 and thromboxanes are formed from a limited amount of common precursor, the branch point is an important site for homeostatic control. In mammals, there are receptors for PGI_2 on platelets, vascular muscle and other cells; these receptors are coupled to the adenylate cyclase. The endothelial cells of mammalian blood vessels produce PGI_2 in response to injury or irritation. In birds, however, the thrombocytes have receptors for PGE_2, and it is PGE_2 rather than PGI_2 which is produced by their endothelial cells [J. M. Ritter et al. *Lancet* 1983-I, 317]. In mammals, PGI_2 inhibits aggregation of platelets and causes elevation of cAMP in them. At concentrations high enough to activate the adenylate cyclase, PGI_2, PGE_1 or PGD_2 will inhibit clotting, but recently it has been found that at concentrations which are too low to activate the cyclase, these PG activate factor X and thus initiate blood clotting [A. K. Dutta-Roy, T. K. Ray & A. K. Sinha, *Science* **231** (1986) 385–388]. The effects of different PG in various tissues are not identical; for example, PGI_2 does not cause diarrhea, but PGE_2 does. PGE_1 and PGI_2 stimulate adenylate cyclase in platelets, while thromboxane A_2 and PGH_2 inhibit the stimulation. Most of the PG cause smooth muscle in blood vessels, the gastrointestinal tract and the uterus to contract, but PGH_2 is generally vasodilatory, as is PGI_2. The relation of the PG to inflammation and pain is not clear, but it is very likely that the antiinflammatory and analgesic effects of aspirin are due to its inhibition of PG synthesis. Aspirin inhibits the synthe-

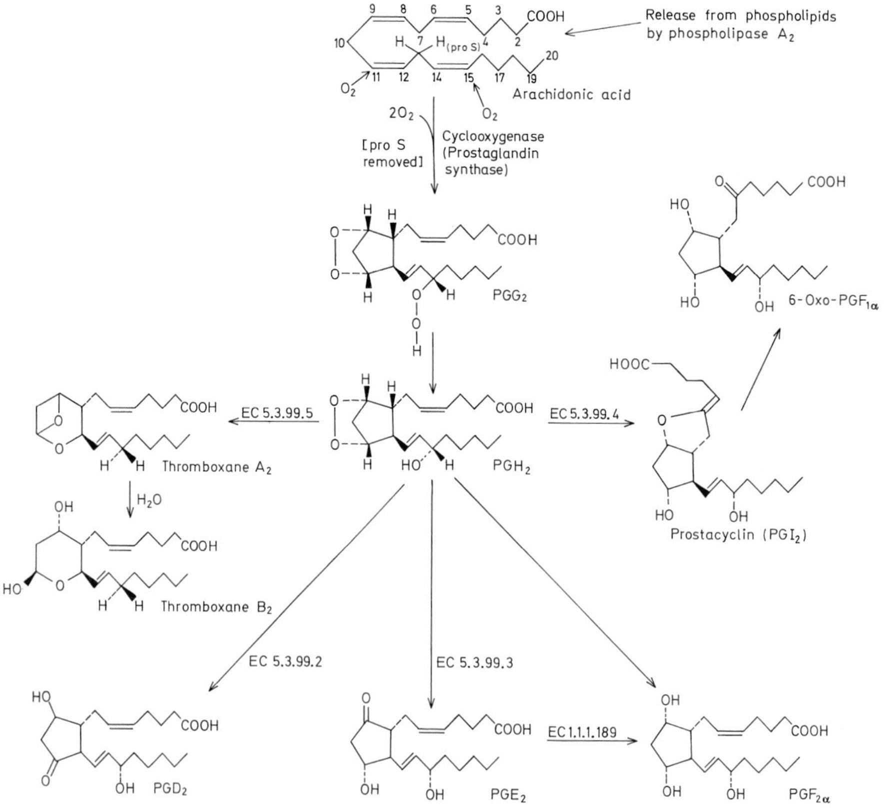

The cyclooxygenase pathway of arachidonate metabolism.
EC 1.1.1.189: Prostaglandin-E_2 9-oxoreductase
EC 5.3.99.2: Prostaglandin-H_2D-isomerase
EC 5.3.99.3: Prostaglandin-H_2E-isomerase

EC 5.3.99.4: Prostacyclin synthase
EC 5.3.99.5: Thromboxane synthase

sis of thromboxanes at lower concentrations than are required to inhibit PG synthesis; thus low concentrations of aspirin enhance bleeding by shifting the metabolism of PGH_2 toward PGI_2 and away from thromboxanes. [S. Moncada & J. R. Vane, "Pharmacology and endogenous roles of Prostaglandin Endoperoxides, Thromboxane A_2 and Prostacyclin" *Pharmacological Rev.* **30** (1979) 293–331; P. B. Curtis-Prior, *Prostaglandins, an Introduction to Their Biochemistry, Physiology and Pharmacology* (North-Holland, 1976).]

PG exhibit a wide variety of pharmacological properties. Of particular importance are: bronchoplasmolytic activity (treatment of acute asthma); control of gastric secretion (possible ulcer therapy); antagonism of the diuretic action of angiotensin (treatment of essential hypertension and cardiovascular disorders); initiation of ovulation, e. g. in cows, pigs and sheep; relief of pain during parturition; and finally, very small amounts of PG cause abortion.

Prostanoic acid: see Prostaglandins.

Prosthetic group: nonproteogenic, low M_r groups present in conjugated proteins. In an enzyme, a P. g. is a catalytically active group attached to the enzyme protein (apoenzyme). In the wide

sense, P. g. are therefore Coenzymes (see), but many authors make a clear distinction, i. e. P. g. are covalently bound (e. g. heme, biotin, phosphopantetheine), whereas coenzymes are freely dissociable and can be removed by dialysis (e. g. pyridoxal phosphate). The flavins (FAD and FMN) of flavoproteins and flavoenzymes are usually regarded as P. g.

Protamines: a group of strongly basic, low M_r, simple (unconjugated) proteins associated with DNA in the cell nucleus. They replace the somatic Histones (see) in sperm, at least during spermogenesis. P. have been prepared from fish and bird sperm; they contain 80–85% arginine, the remaining amino acids being alanine, glycine, proline, serine and valine (or isoleucine). M_r of P. is between 4050 (clupein, 30 amino acid residues) and 4420 (iridin, 33 amino acid residues). The sequence of salmin A-I is Pro-Arg$_4$-Ser$_3$-Arg-Pro-Val-Arg$_5$-Pro-Arg-Val-Ser-Arg$_6$-Gly$_2$-Arg$_4$. Protamine-like proteins, containing 40–80 amino acid residues, have been isolated from invertebrate sperm.

Protease inhibitors: see Snake venom.

Proteases: all enzymes that catalyse the exergonic hydrolysis of peptide bonds in proteins and peptides (Fig.). P. are divided into two groups, depending on their site of attack on the polypeptide chain:

$$-\underset{\underset{H}{|}}{N}-\underset{\underset{H}{|}}{\overset{R_1}{\overset{|}{C}}}-\underset{\overset{||}{O}}{C}\rule{1cm}{0.4pt}\underset{\underset{H}{|}}{N}-\underset{\underset{H}{|}}{\overset{R_2}{\overset{|}{C}}}-\underset{\overset{||}{O}}{C}-$$

$$+\,H_2O$$

$$\downarrow$$

$$-\underset{\underset{H}{|}}{N}-\underset{\underset{H}{|}}{\overset{R_1}{\overset{|}{C}}}-\underset{\overset{||}{O}}{C}-OH\;+\;H_2N-\underset{\underset{H}{|}}{\overset{R_2}{\overset{|}{C}}}-\underset{\overset{||}{O}}{C}-$$

1. *Endopeptidases (proteinases)* catalyse the hydrolysis of bonds within the peptide chain, forming variously sized cleavage peptides. They can be further subdivided into acidic, neutral and basic endopeptidases. Neutral and basic types can each be divided into Serine proteases (see) and thiol proteinases (see Thiol enzymes). Examples of animal proteinases are Pepsin (see), Rennet enzyme (see), Trypsin (see), Elastase (see), Thrombin (see), Plasmin (see) and Renin (see). For examples of plant and bacterial proteinases see Papain, Subtilisin, Bromelain. There are also yeast and fungal proteases.
2. *Exopeptidases* catalyse the hydrolytic removal of only terminal amino acids from the polypeptide chain. They can therefore be classified into *N*-terminal exopeptidases (aminopeptidases, e.g. leucine aminopeptidase) and *C*-terminal exopeptidases (e.g. carboxypeptidases A, B, Y etc.). Tri- and dipeptidases are also classified as exopeptidases.
Owing to their central importance in protein degradation and turnover, P. are widely distributed in the organism. Especially high concentrations are found in the digestive tract and in lysosomes, where they catalyse total degradation of dietary and cellular proteins, respectively, to amino acids. Most known P. are unconjugated metalloenzymes, containing or requiring a metal (e.g. zinc) as an activator or stabilizer. Self digestion (autolysis) is avoided by synthesis as inactive precursors (see Zymogens), by the presence of specific protease inhibitors, or by storage in special cell organelles called lysosomes.
Substrate specificity of P. may be high, e.g. rennet enzyme, carboxypeptidase B, enterokinase, etc., or low, e.g. pronase, pepsin and intracellular proteinases. Often P. are specific for certain amino acid residues, e.g. trypsin catalyses hydrolysis of arginyl and lysyl bonds. Pancreatic P. are among the most extensively studied of all enzymes.
Proteid: an obsolete name for a conjugated protein (see Proteins).
Protein: a naturally occurring polymer of high M_r, consisting predominantly of amino acids linked by peptide bonds. P. account for more than 50% of the organic constituents of protoplasm. By weight, they are the major component of the dry material of living organisms, and they are among the most important functional components of the living cell. All enzymes are P., and enzymes catalyse the many biochemical reactions that constitute the metabolism of the living cell; these reactions are controlled by modification of the activity and/or quantity (i.e. rate of synthesis) of appropriate enzymes. Some of the

regulators and transmitters in these control processes are also P., e.g. the proteohormones (see Hormones), repressor and activator molecules (see Gene activation; Operon), and the membrane P. which determine intracellular concentrations of many enzyme substrates and products (see Active transport). Structural P. contribute to the mechanical structure of organs and tissues (e.g. elastin, collagen), or they may constitute the bulk of a natural structure (e.g. the silk of insect cocoons and spiders' webs is the protein fibroin; feathers, hair, nails and hoofs are mainly α-keratin). Contractile P. (e.g. actin and myosin of muscle, and dynein of cilia and flagella) are responsible for active movement in living organisms. Examples of storage P. are ovalbumin (from egg white), casein (from milk), gliadin (wheat seeds) and zein (maize seeds). Examples of transport P. are hemoglobin (transport of oxygen in vertebrate blood) and serum albumin (transport of fatty acids and many other substances, including hormones, drugs, etc.). P. are also involved in biochemical defense, e.g. antibodies, complement, interferons and various P. of the blood-clotting system. Certain bacterial and plant toxins are P. (see Toxic proteins). Snake venoms (see) contain various enzymes and toxic P. P. exposed on the surfaces of cells permit recognition of one cell type by another, and thus have a role in morphogenesis and the recognition of foreign tissue as such (as in graft rejection.)
The behavior and properties of P. are determined by their structures. Theoretically, there is no limit to polypeptide chain length, and all permutations and combinations of the 20 constituent amino acids are possible (see Genetic code). Further possibilities for structural variation are offered by Post-translational modification of proteins (see), attachment of nonprotein prosthetic groups, and different levels of quaternary structure, so that the possible diversity of structure (and therefore function) is almost limitless.
P. in solution are neither rigid nor motionless. The bonds between the carbon atoms in the protein backbone and amino acid side chains allow considerable rotational and torsional flexibility, and the thermal motions of the individual atoms therefore produce writhing motions of the chain. These motions are almost certainly important to the enzymatic activity of the P.; and mutations which affect the flexibility of the chain could be expected to affect the kinetics of the enzyme's catalysis as well. (See Allostery.) [R. H. Pain, Nature 305 (1983) 581–582]
Classification: A logical and systematic classification of the many thousand (about 40000 in the human) P. from all living species would only be possible on the basis of their primary and tertiary structures. A simple division into enzyme P. and nonenzyme P. is meaningful and practical. A further useful classification is: simple (pure, or unconjugated) P. and conjugated P. The simple P. can be further subdivided, according to their solubility properties and molecular shape, into water-soluble globular and water-insoluble fibrous P. (Table). Alternatively, P. may be classified according to their origin, e.g. viral, bacterial, plant and animal P. Different P. within one organism may be classified as blood, milk, cerebrospinal, secretory, muscle or structural P., etc. Subcellular lo-

calization is also used for classification, e.g. ribosomal, microsomal, mitochondrial, lysosomal, nuclear, cytosol and membrane P. For over 30 years electrophoresis has been an especially valuable tool for the differentiation of the P. of serum and cerebrospinal fluid in clinical biochemistry and diagnosis. The different electrophoretic migration of the P. of these fluids permits their classification into prealbumin, albumin, α_1-, $\alpha_2\beta_1$-, β_2-, and γ-globulins.

Table. Classification of proteins

I. Simple or unconjugated proteins

1. Globular proteins: soluble in water or dilute salt solutions; rotational ellipsoids with frictional coefficients $f/f = 1.1$–1.5. Representatives are Albumins (see), Globulins (see), Protamines (see), Histones (see), Prolamines (see) and Glutelins (see).

2. Fibrous proteins: insoluble in water and salt solutions; greatly extended molecular structures with very high degree of molecular asymmetry; structural proteins, very stable to acids, alkalis and proteolytic enzymes.

II. Conjugated proteins: In addition to amino acids they contain a nonprotein moiety, called the prosthetic group, which is usually essential. It is attached by covalent, heteropolar or co-ordinate linkage. Prosthetic groups include carbohydrates, lipids, nucleic acids, metals, chromogens, heme groups and phosphate residues. The P. containing the above are called glyco-, lipo-, nucleo-, metallo-, chromo-, hemo- and phospho-P., respectively.

Physical chemical properties:

1. *Ampholyte properties.* The ampholyte character of P. is determined by the presence of free acidic and basic groups in the P. molecule. Depending on the pH of the solvent, P. may have acidic or basic properties. The excess positive or negative charge results in an increase in the hydration and solubility of P.

The degree of hydration depends upon the size of the net charge, irrespective of whether the charge is negative or positive. On the other hand, the direction of migration in electrophoresis does depend upon the sign of the charge. At its isoelectric point (IP), a P. has no net charge (i.e. it exists as a zwitterion), so that its solubility and degree of hydration are minimal. The important buffering action of P. is due to their ampholyte character.

2. *Solubility.* This is determined by amino acid composition, the distribution of polar and nonpolar amino acids on the surface of the molecule, and the surrounding milieu (pH, ionic strength, temperature). Hydrophilic P. are surrounded by a layer of water (hydration layer) and they are able to occlude hydrophobic substances and prevent them from separating from aqueous solution. This protective colloid function is responsible for the stability of many body fluids. Addition of polar solvents such as alcohol or acetone, or high concentrations of neutral salts, results in loss of the hydration layer and causes precipitation (salting out) of P. Many P., however, require a low salt concentration to prevent their precipitation. This salting-in effect is due to the suppression of the ordered (association) or random (aggregation) assembly of P. molecules by accumulation of electrolyte ions on the protein surface. The solubility diagram is a sensitive criterion of protein purity: for a homogeneous protein, the plot of the quantity of added versus dissolved protein is linear up to a sharply defined saturation point.

3. *Denaturation.* Loss of native structure and biological activity of P. by destruction of tertiary and quarternary structure, with the formation of random coil or metastable forms, is known as denaturation. In denaturation, noncovalent bonds are broken, in particular hydrogen bonds. Disulfide bridges are only broken in the presence of reducing agents (reductive denaturation). (As a prelude to amino acid sequence

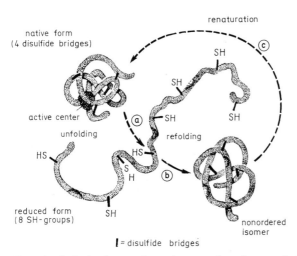

Figure 1. Reductive denaturation and renaturation of pancreatic ribonuclease

determination, disulfide bridges are often cleaved oxidatively with performic acid, producing two residues of cysteic acid). P. can be denatured with 6-8M urea, 6M guanidinium chloride, 1% sodium dodecylsulfate and other detergents, extreme acidic or alkaline pH-values, heat and by UV- or X-irradiation. Denaturation is often reversible (renaturation), even after treatment with reducing agents. Fig. 1 is a diagrammatic representation of a) the reductive denaturation of ribonuclease in 8 M urea with thioethanol (2-mercaptoethanol), b) its reoxidation in 8 M urea to one of the 105 possible disulfide bridge isomers, and c) its renaturation to the enzymically active form in urea-free medium, by disulfide exchange in the presence of traces of thioethanol. Formation of completely disordered structures results in irreversible denaturation, e.g. heat denaturation of ovalbumin (Fig. 2.), where the conformation of the polypeptide chain is irreversibly destroyed.

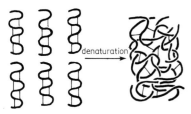

Figure 2. Irreversible heat denaturation of ovalbumin. Weak intramolecular forces (not disulfide bridges) are broken, and the unfolded polypeptide chains become tangled to form a gel, as in meringue or cooked egg white.

Determination of M_r: Several physical chemical methods are available for the determination of the M_r of P. They can be classified into two main types, i.e. kinetic and equilibrium methods. Kinetic methods depend on the measurement of particle transport; they include techniques for the measurement of rates of diffusion, rates of sedimentation in the ultracentrifuge, viscosity (e.g. in the Ubbelohde viscometer), rates of migration in electrophoresis, and rates of migration in gel chromatography on polyacrylamide or dextran gels of known porosity (molecular sieve, or gel chromatography). Equilibrium methods are applied to P. solutions that are in thermodynamic equilibrium. Such methods include the measurement of osmotic pressure in a membrane osmometer, measurement of light scattering (the Tyndall effect, which increases markedly with increasing particle size, is measured in a light scattering photometer), and measurement of small angle, X-ray diffraction (which depends upon the mass radius of the protein) with a special camera. Electron microscopy represents a further specialized method for the determination of molecular shape and subunit composition; it is particularly suited to the determination of molecular shape and subunit composition of multienzyme complexes and P. with quarternary structure; the preparation is usually negatively stained with osmium tetroxide or other heavy metals, and the resolution can be as high as 1.5 nm. If the amino acid composition and the number of

tryptic peptides, or the primary structure of a P. are known, the absolute M_r can be calculated. Currently, the most preferred methods for M_r determination of native and dissociated P. are the various ultracentrifuge methods, gel filtration and polyacrylamide gel electrophoresis.

1. *Ultracentrifuge methods*; M_r can be determined in the ultracentrifuge by measurement of the rate of sedimentation, or of the sedimentation equilibrium.
In the rate methods, very high centrifugal forces are used (e.g. rotor speeds of 40-50000 rpm, representing forces on the order of 100000 g). Under these conditions the colloidal protein molecules have a higher density than the solvent, their rate of diffusion is high in comparison to the rate of back diffusion, and they are completely sedimented. The rate of sedimentation is directly proportional to M_r, i.e. each protein migrates to the bottom of the ultracentrifuge cell at a different rate, depending on its size. The resulting changes in protein concentration along the length of the cell during ultracentrifugation are monitored by optical methods, e.g. with schlieren optics, or Rayleigh interference optics, and by UV absorption. The mass unit for sedimentation per unit time is the Svedberg (S). The S-values of P. lie in the range 1.2S (insulin) to 185 S (tobacco mosaic virus). P. mixtures become crudely separated into their major components. For example, serum is separated into the most rapidly sedimenting macroglobulins (M_r around 10^6), globulins (M_r about 160000) and albumin M_r 67500) or prealbumin (M_r 61000) (Fig. 3). It should be borne in mind that at the start of ultracentrifugation, the ultracentrifuge cell contains a homogeneous solution of P. Thus the peaks shown in Fig. 3 are not peaks of protein concentration; they represent boundaries of P. concentration, which are formed as P. become concentrated towards the bottom of the ultracentrifuge cell at different rates; the schlieren optical system, employing cylindrical lenses, shows the rate of change of concentration at each boundary. M_r is calculated from the Svedberg equation, $M_r = RTS/D(1 - v\rho)$, where S is the sedimentation constant corrected for viscosity and density of water at 20 °C, R is the gas constant, T is the absolute temperature (K), D is the diffusion constant, v is the partial specific volume of the protein (= reciprocal of the density of the molecule; for most P. it is between 0.71 and 0.74) and ρ is the density of the solvent system. D. is determined

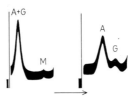

Figure 3. Sedimentation pattern of normal human serum in the ultracentrifuge, measured with a schlieren optical system. Left hand picture taken 51 min after start of centrifugation; right hand picture after 125 min. Centrifugation at 59800 rpm. A = albumin, 4.5S; G = globulin, 7S; M = macroglobulin, 19S.

separately by measurement of peak (boundary!) broadening caused by diffusion during low speed centrifugation.

In equilibrium methods, the P. solution is subjected to a centrifugal force of only 10000-15000g. Under these conditions, sedimentation is slow and the effects of back diffusion are no longer negligible. After a fairly long period (hours to days) a stationary state is achieved, in which the net flow of protein molecules is zero. From the resulting concentration gradient between the meniscus and the bottom of the cell, the M_r can be calculated without knowing the diffusion coefficient. The time required for the low speed method can be shortened by using the approach-to-equilibrium method of Archibold (1947), in which the concentration gradient is measured at the top of the cell when the P. zone is only partly pulled away from the meniscus. The centrifugation time can be further shortened to 2-4 hours by using the meniscus depletion method of Yphantis (1964) under high speed conditions (high speed equilibrium method): only small volumes of P. solution are placed in the ultracentrifuge cell (0.1 ml of a 5% P. solution, forming a liquid column of 3 mm). The sedimentation equilibrium can then be determined at high rotor speeds (up to 40000 rpm) when the mensicus becomes depleted of P. Sedimentation equilibrium methods are not suitable for P. mixtures, because overlapping concentration gradients would be formed. The Yphantis method is particularly suitable for the investigation of the dissociation and association of oligomeric P.

In density gradient centrifugation (Martin and Ames, 1961), P. migrate in a sucrose gradient of increasing density, and each P. comes to rest in a discrete zone where particle density matches that of the solvent. Since the distance moved by a P. in the gradient is inversely proportional to its M_r, an approximate M_r of an unknown P. can be determined by comparison with known standards. The same method can be used to characterize subcellular fractions, and these can be isolated in preparative quantities by applying the principle to large rotors (zonal centrifugation).

2. Gel filtration, or molecular sieve chromatography. Gels of chosen pore size are prepared from beads or granules of cross-linked dextrans, agarose, or polyacrylamide. Analyses are performed on columns, or by thin-layer chromatography. The gel particles contain pores of defined size. The solvent, salts and other low M_r material enter these pores and permeate the gel particles; but macromolecules above a certain size are excluded and they can only migrate in the solvent between the gel particles. The method is cheap, only simple apparatus is required, gels may be reused many times, and the separation times are short. The thin-layer version of the method is especially economical in the use of gel material and P. The parameter for M_r determination is the ratio of the elution volume, V_e, of the P. and the exclusion volume, V_o (the void, or dead volume, i.e. the volume of the interstitial liquid between the gel particles, which is the same as the elution volume of a substance that is completely excluded from the gel). In thin-layer methods, the distance moved on the gel plate is compared with that of standards. A calibration curve is prepared with the aid of several standard P.: V_e versus log M_r, or V_e/V_o versus log M_r. Depending on the type of gel, the method measures approximate M_r in the range 500 to several million, with an accuracy of 5-10%. Gel chromatography is also useful for the M_r determination of P. subunits: the gel consists of a cross-linked dextran or agarose, and it is equilibrated with the dissociating medium, which contains 0.1-1% dodecylsulfate, or 4-6M guanidinium chloride. The analytical M_r range is then limited to 5000 to 70000.

3. Polyacrylamide gel electrophoresis., abb. PAGE. A gel with the desired pore size is prepared by the simultaneous polymerization and cross-linking of acrylamide with bis-acrylamide. Electrophoresis is performed in tubes, or on slabs of gel, with a gel length of 7-10 cm. In the methods of Fergusun (1964) and of Hedrick and Smith (1968), the rate of migration of P. is measured in gels of different concentrations (i. e. different pore size). Migration distance is plotted against gel concentration (%) for a series of standard proteins. A linear calibration is obtained by plotting the logarithms of the M_r. Alternatively, electrophoresis is performed in the presence of sodium dodecylsulfate (SDS). The method was introduced by Shapiro (1967) and is commonly known as SDS-PAGE. The anionic detergent molecules form a layer around the polypeptide chain with their negative charges exposed to the surrounding medium. Globular P. are unfolded (it may be necessary to add 2-mercaptoethanol to cleave disulfide bridges), forming structures that approximate to rigid rods. The SDS molecule binds to the polypeptides with a constant weight ratio, corresponding to about one SDS molecule per three peptide bonds, so that the charge per unit length is constant (charges on the native protein are smothered and can be ignored). Thus in any given electrical field, the acceleration is the same for every SDS-coated protein molecule. Separation depends solely on the size of the molecule and the sieving effect of the gel. Log M_r versus relative distance of migration is linear, and the system can be calibrated with P. of known M_r. Since most oligomeric P. are dissociated by SDS, the method is often used for the M_r determinion of subunits. It is the most rapid method (2-4 hours), and the most economical in use of test material (10-50 µg P. per analysis), with an accuracy of 5-10%.

Single-chain P. have M_r between 10000 and 100000; oligomeric (multichain) P. have M_r between 50000 and several million.

Determination of molecular symmetry: Molecular symmetry of P. can be determined from measurements of viscosity, streaming birefringence, rates of sedimentation and diffusion, or directly by electron microscopy. For a known M_r, the frictional coefficient can be calculated from ultracentrifugal measurements, e. g. from the sedimentation:

$$f = \frac{M_r(1 - v\rho)}{S}$$

where v is the partial specific volume, ρ is the density and S is the sedimentation constant. The axial ratio a/b of a P. can be derived from the frictional ratio f/f_o, where f_o is the f of a spherical molecule. The value of a/b for most globular P. is between 2 and 20, and greater than 20 for fibrous P., e.g. the axial ratio of fibrinogen is 30.

Purification and isolation: Biological material usually contains a great variety of P., together with carbohydrates, lipids and nucleic acids. Purification is therefore necessary as a prelude to the detailed characterization of any P. The insoluble, fibrous structural P. and the soluble P. that occur in high concentrations (e.g. hemoglobin from erythrocytes, casein from milk, trypsin from pancreatic juice) can be isolated relatively easily. For most globular P., however, separation from impurities requires multi-step purification procedures. Artifacts caused by the action of proteases are avoided by using protease inhibitors (e.g. diisopropylfluorophosphate) and/or by working at 4 °C so that proteases are relatively inactive. With the exception of P. in serum, cerebrospinal fluid, milk, egg white and various secretions, P. are localized inside cells and tissues. The first stage of P. purification therefore consists of extraction from biological material, following disruption of cell walls and/or membranes by mechanical methods (high speed rotary cutting blade; abrasion between closely fitting glass, or glass and teflon surfaces, e.g. Potter-Elvehjem homogenizer), by physical methods (ultrasound; alternate freezing and thawing; grinding of frozen tissue with fine Al_2O_3 granules; shaking with glass beads; change of pH; osmotic shock; treatment with detergent; forcing a frozen paste of cells between closely fitting steel surfaces, e.g. Hugh's press for disrupting bacteria; forcing a cell suspension at high pressure through a fine orifice, e.g. French press), or by enzymic methods (treatment with proteases such as trypsin, papain, lipases, neuraminidase, hyaluronidase, or lysozyme). The resulting cell homogenate can be extracted with various reagents, such as salt solutions, glycerol, dilute acids, detergents and a wide variety of buffers. Cellular debris is removed by coarse filtration or low speed centrifugation. The second step is usually a preliminary separation by ammonium sulfate fractionation. If necessary, nucleic acids are removed by precipitation with protamine sulfate, or they can be hydrolysed to low M_r products by the addition of ribonuclease and deoxyribonuclease. Subsequent purification steps include chromatography (gel filtration, ion exchange, adsorption), and possibly preparative electrophoresis with or without carrier, or isoelectric focussing. Some rapid and high performance methods are affinity chromatography (1968) and ion filtration chromatography (1972), the protective crude separation by ultrafiltration membranes (1964), preparative electrophoresis (1959), isoelectric focusing (1962), ammonium sulfate gradient stabilization chromatography (1972), gradient sieve chromatography (1973) and fractional precipitation with polyethylene glycol (1973).

Affinity chromatography is a column technique with wide application. It exploits the reversible and bi-ospecific interaction between the functional groups of the protein under investigation and a ligand (either a small molecule or an antibody, receptor or other specific protein) which is covalently bound to an insoluble, inert carrier. If the ligand is too close to the matrix, interaction with the protein may be sterically hindered. In order to obtain a high binding capacity, small ligands are usually attached to the matrix via a hydrocarbon chain (flexible arm, or spacer group) of variable length. For maximal stability of the protein-ligand complex, a modified substrate or specific inhibitor (usually competitive) is used as the ligand; for the purification of antibodies, the appropriate antigens are used as ligands. Alternatively, antibodies specific for the desired P. may be immobilized on the matrix. Owing to the high selectivity of the protein-ligand interaction, it is often possible to isolate a P. from a mixture in a single step of affinity chromatography; all other substances, including P., pass unretarded through the column. The desired P. is released (i.e. the protein-ligand complex is dissociated) by changing the pH or ionic strength, or by adding a competing ligand to the elution solvent. Columns can be regenerated and used repeatedly. One of the many examples of affinity chromatography is the purification of NAD-dependent dehydrogenases, e.g. glyceraldehyde 3-phosphate dehydrogenase, using NAD as the affinity ligand covalently bound to Sepharose (matrix) via a spacer arm of 6-aminohexanoic acid. After washing a crude enzyme mixture through the column, dehydrogenases are eluted with buffers containing NAD (see Chlorotriazine dyes). In addition to the purification of a large number of enzymes, the method has also been used for the purification of antibodies and antigens, hormones, macromolecular inhibitors, transport and receptor P. (e.g. the hitherto intractable insulin receptor was purified 250 000-fold), and even cell populations.

Ion filtration chromatography is a time-saving combination of ion exchange and gel chromatography without a salt gradient. Separation of P. is performed on diethylaminoethyl-(DEAE)-dextran (cross-linked) at a constant pH within 2 to 3 hours.
Ultrafiltration is a rapid and protective technique, by which a P. solution is simultaneously concentrated and freed from small molecular contaminants. It is usually operated with excess, rather than reduced pressure. By using the appropriate membrane filter (pore sizes range from 15 μm upwards), P. mixtures can also be crudely fractionated according to M_r. Biologically active P. can be prepared on an industrial scale by using tubular membranes with large surface areas and a daily throughput capacity of several thousand liters.

High performance liquid chromatography (HPLC) has now largely replaced other methods (e.g. thin layer chromatography) for separating and identifying PTH-amino acids from sequencing studies (see below). HPLC of whole proteins on hydroxyapatite is possible, but recovery is often very low. Silane-coated silica is used in reversed phase HPLC systems for whole proteins (silica alone is unsuitable, but its low compressibility makes it ideal as a packing for

HPLC columns). Low compressibility organic polymers suitable for HPLC of proteins are under development.

With new carrier systems and refrigerated apparatus with voltages in the thousands, electrophoretic methods can now be used for the preparative separation of P. Carrier-free systems (*continuous electrophoresis*), or zone electrophoresis with a carrier (*starch block*, or *polyacrylamide gel electrophoresis*) are used. Continuous electrophoresis is usually performed on paper; P. move vertically with a downward flow of buffer, and are at the same time subjected to horizontal separation by a potential difference across the paper. In starch block and continuous electrophoresis, P. are mostly negatively charged (buffer pH 7-9), and they migrate according to their net charge, whereas migration in polyacrylamide gel electrophoresis includes the additional effect of molecuar size (molecular sieve effect). After zone electrophoresis, the starch block or gel is sectioned, and more or less pure proteins are eluted from the individual slices. In *isoelectric focussing*, P. are separated electrophoretically in a pH-gradient stabilized by a sucrose gradient, so that separation depends upon their different IPs. During isoelectric focussing, each P. migrates to a position corresponding to its own IP. The current is switched off and the column is carefully emptied to collect each band of P. directly and in preparative quantity.

Discontinuous electrophoresis (abb. disc electrophoresis; the term disc also conveniently describes the sharp, disc-shaped bands of P. that are formed in the tubular column version of this method) exploits a discontinuity of gel pore size and pH. It has a very high resolving power in comparison with other methods of P. separation. Nowadays any claim for P. homogeneity should be supported by evidence from disc electrophoresis. The principle was first applied to separations in tubular columns of polyacrylamide gel. Later it was made even more incisive by the introduction of a slab gel technique, in which several P. samples can be run alongside each other on the same gel. As described here, the conditions were originally designed for the investigation of serum P. A short length of large pore gel buffered at pH 6.7 (known as the stacking gel, containing about 3% acrylamide) is layered on top of a small pore gel, buffered at pH 8.9 (known as the running gel, containing 7.5% acrylamide). Both gels contain a highly mobile ion (the leading ion), which is usually chloride. The buffer (pH 8.3) in the anode and cathode compartments contains glycine (the trailing ion), which has a net negative charge, but is less mobile than chloride. The P. are (is) incorporated into a gel mixture identical to that of the stacking gel, and layered on top of the latter. Conditions of pH are chosen so that the P. have a mobility betwen the leading ion and the trailing ion. As the glycine enters the stacking gel, the P. become sandwiched between the two ions and concentrated into very tight bands, one stacked on top of the other in order of decreasing mobility, with the last one followed by glycine. In this state they enter the running gel and are thus subjected to a discontinuity of both pore size and pH. The mobility of the fastest protein now falls,

glycine overtakes all the P. and runs directly behind the chloride, and each thin starting zone of P. is in a linear voltage gradient. Subsequent migration depends upon both the charge and the size of each P. molecule.

The criterion of P. purity is homogeneity. A so-called pure P. preparation (i.e. only one P. is present) may contain inorganic salts, other small organic molecules (e.g. substrates, coenzymes) and water. Even a crystallized P. will contain much water and possibly other small molecules and ions. A special question of P. homogeneity arises in the study of quaternary structure, when it is necessary to determine whether subunits are identical or nonidentical. Homogeneity can be monitored by analytical disc electrophoresis (one P. band), analytical ultracentrifugation (a single symmetrical boundary), solubility diagram and IP. The amino acid composition of a homogeneous P. shows all the constituent amino acids in quantities of more or less the same order, and in sensible ratios; the presence of a contaminating P. of different amino acid composition is therefore easily recognized, especially if the amino acid composition of the pure P. is already known. Detection of more than one *N*-terminal amino acid (by Edman degradation, by dansylation, or by 2,4-dinitrophenylation) is indicative of contaminating P., since a single P. chain contains only one *N*-terminus. Similarly, heterogeneity is indicated by the presence of more than one *C*-terminal amino acid (determined by treatment with carboxypeptidase). If a P. is known to contain x lysine and arginine residues (i.e. trypsin-sensitive sites), the theoretical maximum number of tryptic peptides is $x + 1$. Inhomogeneity is indicated if this number is exceeded. The ability of a P. to crystallize shows that it is fairly pure. For enzymes, further evidence of purity is given by pH and temperature optima, substrate specificity, kinetic parameters and specific catalytic activity.

Detection and measurement: P. can be detected qualitatively by a variety of denaturation methods: precipitation with trichloroacetic acid, perchloric acid, picric acid and heavy metal salts (Cu, Fe, Zn Pb), or by warming at their IP. There are also color reactions for certain amino acid residues (aromatic, phenolic, SH and guanido groups) and for the peptide bond. The biuret color reaction, which is specific for peptide bonds, has been combined with the Folin reagent, which is specific for tyrosyl and tryptophanyl residues, to produce the sensitive Lowry method for the colorimetric assay of P. (lower detection limit 5-10 µg P. per ml); The blue color complex has an absorption maximum at 750 nm and is stable for several hours. The method is usually standardized against human or bovine serum albumin. The oldest quantitative procedure for P. is the Kjeldahl method, in which the sample is digested with sulfuric acid to convert the P. nitrogen into ammonium sulfate; after alkalization of the digest, the ammonia is steam-distilled into 0.1 or 0.01M acid, then determined by back titration. The quickest and most suitable method for routine measurements (e.g. in a flow-through spectrophotometer cell) is the measurement of UV absorption at 280nm, which depends chiefly on the presence of tyrosyl residues in the P. Using the measured absorption coefficient, $A_{280,1\ cm}^{1\%}$, of a purified P. (e.g. the value for trypsin is 15.4), the absorption at 280 nm can be used to determine unknown concentrations of the same P.

Composition and structure: The sequence of amino acids in a P. is genetically determined. The correct order or sequence of all the amino acid residues is known as the *primary structure* (Fig. 4).

From 1 mg P., the quantity of each amino acid (corresponding to the area under each peak) can be determined with an accuracy of a few percent. Analysis is complete in about 20 hours. On-line computation of results, with direct readout of amino acid composition, is being used increasingly often. The most common method of hydrolysis consists of heating the P. in 6 M HCl at 105–110 °C for 20–70 hours in a sealed tube. A drawback of acid hydroly-

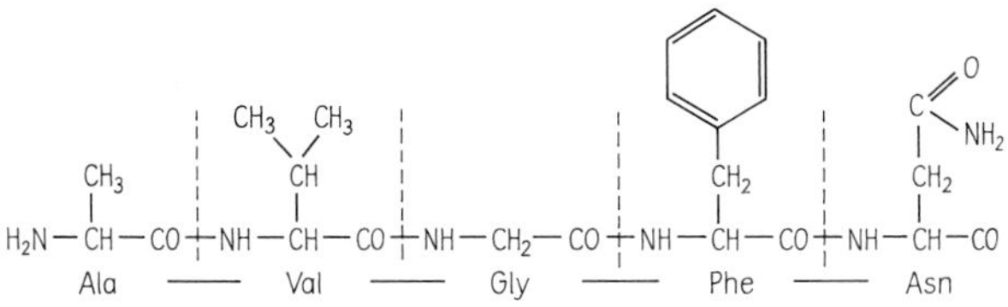

Figure 4. Primary structure of the N-terminal end of a protein chain

The first primary P. structure to be determined was that of insulin in 1953. By the end of 1980, over 700 complete primary P. structures had been reported. Determination of primary structure is usually preceeded by the determination of amino acid composition, i. e. total hydrolysis of the P., followed by quantitative determination of all the amino acids in the hydrolysate. Several methods are available for the separation and quantitative determination of amino acids in a mixture, e.g. paper chromatography, high voltage electrophoresis and gas-liquid chromatography of volatile amino acid derivatives. However, the separation of all the P. amino acids and several other substances, using a 100 cm column of sulfonated polystyrene (Dowex 50-X8) and buffers of progressively increasing pH from 3.4 to 11 (Stein and Moore, 1951) represented a major advance in P. chemistry. The automated version of this method (together with other improvements, such as micropowder in place of bead resin, use of 8% cross-linked Amberlite IR-120, and use of two columns, one for neutral plus acidic, and one for basic amino acids) revolutionized the study of P. structure and composition. The machine automatically collects column effluent, adds ninhydrin, heats for color development, records color intensity, and plots the intensity on a graph (Fig. 5) (see Amino acid reagents).

sis is that some amino acids are altered: tryptophan is totally destroyed, asparagine and glutamine are hydrolysed to aspartic and glutamic acid, respectively, and cysteine is oxidized to cystine. There is also some loss of serine, threonine, cistine and tyrosine; many of these destructive processes are encouraged by the presence of carbohydrate and may be at least partly avoided by performing the hydrolysis under nitrogen and in the presence of stannous chloride. Hydrolyses are performed for varying periods and the values for labile amino acids are extrapolated to zero time. Alkaline hydrolysis (5M NaOH or Ba(OH)$_2$ at 110 °C for 20 hours) is especially destructive and causes racemization, but it is particularly suitable for the determination of tryptophan alone. Most of these problems can now be avoided by hydrolysing the P. with 3 M *p*-toluene-sulfonic acid plus 2% thioglocollic acid in a sealed tube. Unfortunately, the residual *p*-toluene-sulfonic acid is nonvolatile and must be loaded onto the analytical column with the amino acids; it emerges early and tends to interfere with the separation of the acidic amino acids. Tryptophan may also be determined from the UV absorption of the P., after making allowance for the tyrosine content. Some P. can be hydrolysed for analytical purposes by a mixture of proteolytic enzymes, such as papain or subtilisin (both fairly nonspecific) in conjunction with leucine

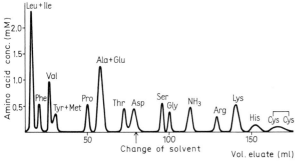

Figure 5. Column chromatographic separation of an acid hydrolysate of serum albumin in the automatic amino acid analyser (after Moore and Stein).

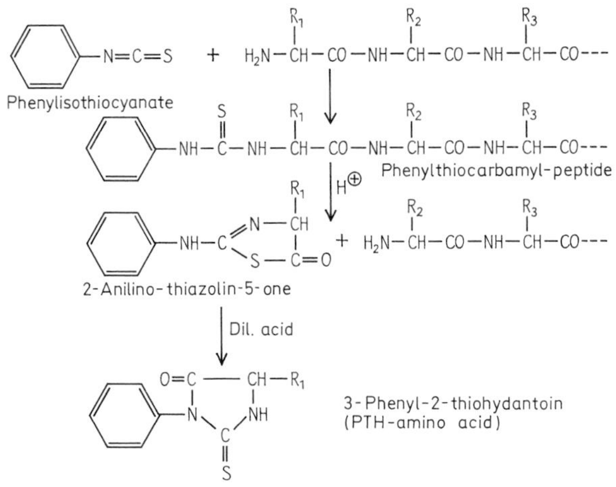

Figure 6. Edman degradation of a peptide

aminopeptidase and proline iminopeptidase, but such hydrolyses are often incomplete.

The next stage in primary structure determination is the reductive (2-thioethanol) or oxidative (performic acid) cleavage of disulfide bridges, followed by enzymatic (e.g. with trypsin) and/or chemical (e.g. with cyanogen bromide) cleavage of the polypeptide chain. Peptides from this cleavage are separated by ion-exchange chromatography or by the finger print technique (a two-dimensional separation on paper or thin-layer plates, in which electrophoresis is used in the first dimension and partition chromatography in the second). Finally, the amino acid composition and the *N*- and *C*-terminal amino acids of each cleavage peptide are determined. In the Edman degradation, the peptide is reacted with phenylisothiocyanate at pH 8–9 at 40 °C; acid treatment removes the *N*-terminal amino acid (R_1) as a substituted phenylthiohydantoin (PTH-amino acid), which is identified chromatographically (Fig.6); the next amino acid (R_2) becomes the *N*-terminus of the new peptide. The method has been automated, and with the aid of a specially built sequenator, it is possible to determine the sequence of amino acids in a peptide, starting from the *N*-terminus. Sequences of more than 100 amino acids can be determined in this way, using only 1 nmole of protein. Each operational cycle (i.e. derivatization and separation of each residue) requires 20–30 min. In the original liquid phase method, reagents washed over a thin film of reactants adsorbed on the inside surface of a spinning glass cup. In the later gas-phase sequencers, the protein is dried on a glass fiber disc, and some reagents (triethylamine, trifluoroacetic acid) are passed as vapors over the adsorbed reactants, thus avoiding build-up of reagent impurities and permitting higher sensitivity. In both methods, the protein is adsorbed noncovalently to a glass support, and wash-out is always a danger (especially with less hydrophilic membrane proteins). A new generation of sequenators is under development, in which the protein is attached covalently to an activated glass. Identification of the *N*-terminus may also be performed with Sanger's reagent or dansyl chloride. Once the amino acid sequence of each peptide is known, their order in the original P. must be established. Peptides at the *N*- and *C*-termini can be identified by comparison with the termini of the original P. The positions of interior peptides are established by cleavage of the P. by more than one method; e.g. proteolytic enzymes of different specificities can be used. Comparison of the amino acid sequences of the two (or more) sets of peptides permits alignment because the termini of one set are overlapped by peptides of the second set. Positions of disulfide bridges are determined after the linear sequence has been established. Tryptic digestion is performed with the disulfide bridges intact, and cystine-containing peptides are isolated. Their disulfide bridges are then cleaved; the resulting two peptides are identified by comparison with the tryptic peptides of the linear sequence. During this procedure, any free SH-groups are blocked with iodoacetate while the disulfide bridges are still intact.

An early method of sequence determination employs partial hydrolysis of the P. with acid or nonspecific endopeptidases. The resulting hydrolysate contains a large number of relatively small peptides, representing many overlapping regions of the sequence. By analysis of a sufficient number of peptides, and comparison of their amino acid sequences, it is possible (but very laborious) to construct the linear sequence of the original P. The classical determination of the primary structure of the A chain of Insulin (see) was performed in this way.

Comparative studies on the primary structures of homologous P. from different species (e.g. hemoglobin from vertebrates; see Homologous proteins) or analogous P. (e.g. subtilisin from *Bacillus subtilis* and mammalian trypsin) have made a valuable biochemical contribution to questions of divergent and convergent evolution. However, for an explanation of P. function and behavior, especially the mechanism of enzyme action, the primary structure alone is insufficient and a knowledge of secondary and tertiary structure is needed.

Secondary structure refers to the way in which the polypeptide chain is folded. Folding is due to the formation of hydrogen bonds between peptide bonds in close juxtaposition to one another (separated by 0.28 nm). These bonds involve the carbonyl oxygen of one peptide bond and the amide hydrogen of the other. In helical structures, hydrogen bonds exist within the peptide chain, whereas pleated sheets or β-structures result from hydrogen bonding between two lengths of polypeptide chain (which may or may not be part of the same chain). The most common helical structure is the α-helix, with 3.6 amino acid residues per revolution (Fig. 7).

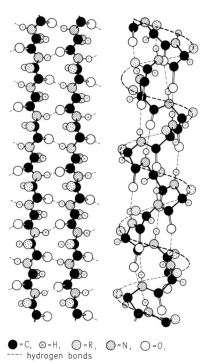

● = C, ☉ = H, ◉ = R, ◨ = N, ○ = O,
---- hydrogen bonds

Figure 7. Secondary structure of proteins. Left, β-pleated sheet structure of a fibrous protein; right, α-helix structure of a polypeptide chain.

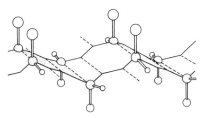

Figure 8. β-Pleated sheet structure; antiparallel arrangement of two polypeptide chains.

Collagen (see) has a specialized structure containing interchain, hydrogen-bonded, left-handed helices. Otherwise, all known P. helices are right-handed, but the possibility remains that left-handed helices might be found in P. which have not yet been analyzed by X-ray diffraction, e.g. membrane P. In the pleated sheet structure, the polypeptide chain is more or less stretched out, and neighboring lengths of chain can be parallel or antiparallel (with respect to their *N*- and *C*-termini), e.g. the pleated sheet structure of silk fibroin (Fig. 8) consists of antiparallel polypeptide chains.

Most P. also contain a large number of amino acid residues which cannot be assigned to any regular structure. Only certain peptide conformations are possible; others would bring neighboring unbonded atoms too close together (see Ramachandran plot). The fully denatured state of a polypeptide is referred to as a random coil; this completely disordered structure is also found in helix-coil transitions and in synthetic polyamino acids.

The percentage of α-helical structure in a P. can be estimated by measurements of optical rotatory dispersion (ORD) and circular dichroism (CD), because the helices are themselves optically active. The slow rate of exchange of P. hydrogen with deuterium oxide or tritium oxide depends inversely on the amount of intramolecular hydrogen bonding; estimates of α-helix based on this method often disagree with the amounts obtained by optical methods. This could be because there are other forms of hydrogen bonding (e.g. in β-sheets or loops) than α-helix in the molecule, because of the existence of left-handed helices (which have optical activity of opposite sign from right-handed helices), or because hydrogen exchange is retarded by factors other than hydrogen bonds, e.g. areas so surrounded by hydrophobic bonds that hydrogen exchange is difficult. X-ray crystallography, which is used for the determination of tertiary structure, also provides detailed information about secondary structure. Hemoglobin and myoglobin have about 75% helical content as determined by X-ray crystallography and ORD, but these two methods show poor agreement on the α-helical content of other P., such as cytochrome c and carbonic anhydrase. This probably signifies differences in the secondary structures of dissolved and crystalline P.

Tertiary structure defines the overall molecular shape and gives detailed information about the spatial arrangement of reactive amino acid residues, e.g. at the active sites of enzymes, or at the antigen-binding site of antibodies. With a knowledge of tertiary structure it is possible to visualize the three-dimensional shape of a P. to a high resolution (0.2 nm or better) and to observe the changes in molecular architecture that accompany the formation of an enzyme-substrate or enzyme-inhibitor complex. Tertiary structure is determined from the X-ray diffraction of isomorphic crystalline heavy metal derivatives of the P. The resulting diffraction diagram, and the electron density map constructed from it, provide information on the type and position of amino acid residues; at higher resolution (0.15 nm) it is even possible to determine the distance between the

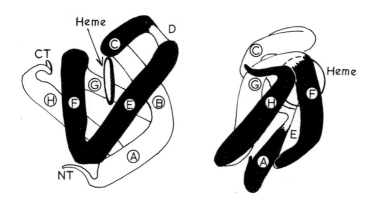

Front   Side

Figure 9. Chain conformation (tertiary structure) of myoglobin. The molecule consists of eight stretches of α-helix, lettered A to H. NT is the *N*-terminus, containing two nonhelical amino acids. CT is the *C*-terminus, containing four nonhelical amino acids. The number of amino acid residues in each helical stretch is A(16), B(16), C(7), D(7), E(20), F(10), G(19), H(26). The nonhelical regions between the helical stretches are AB (1 amino acid residue), CD(8), EF(8), FG(4), GH(5).

atoms in the P. molecule. Despite the very specialized and demanding nature of this technique, the spatial structures of numerous P. have been determined (Fig. 9). The occurrence and stabilization of the three-dimensional P. structure is the result of several forces: hydrogen bonds between tyrosyl and carboxyl or imidazole groups, and between seryl and threonyl residues; disulfide bridges, which have a primary function in the stabilization of conformation; Van der Waals forces; noncovalent mutual attraction between the uncharged ($-CH_3$, $-CH_2OH$) or hydrophobic (phenyl, leucyl) residues separated by about 0.3 nm; electrostatic attraction between polar side groups (e.g. $COO^-...^+NH_3$), which are also involved in the solvation of the molecule; interaction of P. and aqueous solvent, which favors the formation of hydrophobic bonds (interatomic distance 0.31–0.41 nm) in the nonpolar interior of the molecule, and thus makes an important contribution to natural P. conformation. Secondary and tertiary structures are referred to jointly as chain conformation. In fact, it is sometimes difficult to make a clear distinction between secondary and tertiary structure. According to nuclear magnetic resonance (NMR) studies, the chain conformation of a protein is changeable within certain limits, so that the conformation determined by X-ray crystallography represents one of several possible states "frozen" by crystallization.

Aggregation or association of two or more identical or different polypeptide chains by noncovalent interaction leads to stable oligomeric (or multimeric) P. These ordered associations represent *quaternary structure*, and the individual polypeptide chains are called subunits (Fig. 10). In rare cases, quaternary structure may also be maintained by disulfide bridges. P. with quaternary structure are of widespread occurrence. Most of the known multimeric P. contain either 2 or 4 similar sized subunits. Far less

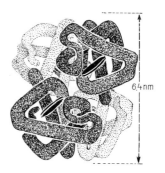

Figure 10. Quaternary structure of hemoglobin, showing spatial arrangement of the α- and β-chains. Black discs represent heme groups.

common are P. with uneven numbers of subunits, with different sized subunits, with subunits capable of independent enzymatic acitvity, or with both regulatory and catalytic subunits. Possession of quaternary structure appears to confer a flexibility of shape and activity, which is necessary for the physiological role of the P. Monomers derived from multimeric enzymes are usually inactive. The subunit composition of oligomeric P. can be determined by dissociation of the aggregate and investigation of the separate subunits by ultracentrifugation, polyacrylamide disc electrophoresis, gel filtration, ion exchange chromatography or viscometry. Alternatively, the structure of the intact molecular aggregate can be studied by electron microscopy, or by low angle X-ray or neutron diffraction. Aggregates are dissociated by treatment with sodium dodecyl sulfate (1%), 8 M urea or 6 M guanidium chloride, by changes of pH, temperature or protein concentration, or by chemical modification (succinylation, maleylation, removal or attachment of cofactors).

Protein A: a protein produced by most strains of *Staphylococcus aureus.* It is mainly covalently linked to cell wall peptidoglycan, and a small proportion of the free protein is secreted into the growth medium. Certain methicillin-resistant strains of *S. aureus* produce only free P.A. One of the richest sources of P.A is the cell wall of *S. aureus* strain Cowan I, which contains 6.7% by weight of the protein. P.A is usually isolated by digestion of bacteria with lysostaphin, followed by ion exchange, gel filtration, and finally affinity chromatography on IgG-Sepharose. P.A is a single polypeptide, M_r 42000, with little or no carbohydrate, C-terminal lysine, and a blocked N-terminus. Tryptophan, cysteine and cystine are absent. There are 4 highly homologous domains (each containing a tyrosine residue, which serves as an iodination site for labelling purposes), and a C-terminal domain which is bound to the cell wall.

P.A induces synthesis of polyclonal antibodies in B lymphocytes of human and mouse origin, and it is probably a T-cell-regulated polyclonal activator of human B cells. It is chemotactic, blocks heat-labile and heat-stable opsonins, activates or inhibits complement fixation (depending on the dose) and has hypertensive activity when injected in some animals and humans.

P.A binds IgG of all mammals. It does not bind avian IgG, and ruminant IgG is bound only weakly. P.A and IgM of some species are also bound. This property is exploited in the isolation and purification of IgG subclasses, which are eluted separately from P.A-Sepharose with a stepwise pH gradient. I-labeled P.A is especially useful as a revealing agent for bound antibodies (see, e.g. Recombinant DNA technology, subsection on Screening). [J.W. Goding *J. Immunol. Methods* 20 (1978) 241–253; A. Surolia et al. *Trends Biochem. Sci.* 7 (1982) 74–78.]

Protein biosynthesis: a cyclic, energy-requiring, multistage process, in which free amino acids are polymerized in a genetically determined sequence to form polypeptides. P.b. represents the translation of genetic information carried by mRNA. It requires the presence of mRNA, Ribosomes (see), tRNA, amino acids and a number of enzymes and protein factors, some of which are integral components of the ribosome. In addition, certain cations are required, and ATP and GTP are necessary as sources of energy. The mechanism and the individual steps of P.b. were elucidated mainly by careful preparation of the functional components, by the use of synthetic polynucleotides, and in particular by the use of specific inhibitors in cell-free systems. The process is similar in prokaryotes and eukaryotes, although it involves more factors in the latter. The differences will be mentioned in the text where appropriate. A standardized nomenclature exists for the "factors" involved in initiation (IF-1, IF-2, etc.), elongation (EF-1, etc.) and termination (release factor, RF). To distinguish them from the prokaryotic factors, eukaryotic factors are designated "eIF-1", etc.

There are four phases of protein biosynthesis.

Activation of amino acids. Each amino acid is esterified with a specific tRNA (see Aminoacyl-tRNA synthetase). This phase does not require the presence of polysomes. The next phases occur successively on each ribosome of the polysome.

Initiation must start with a small ribosomal subunit; 80S or 70S ribosomes are inactive and their spontaneous dissociation into subunits under physiological conditions is slow. In *Escherichia coli,* the factor IF-1 promotes the dissociation of the ribosomes, but this has not been demonstrated in other bacteria or eukaryotes. The protein IF-3 (or eIF-3) prevents the association of the subunits, and thus makes the small one available for initiation. In eukaryotes, the actual first step is the binding of initiator tRNA (Met-tRNA$_f$) along with eIF-2 and GTP to the 40S subunit. In prokaryotes, this may be the first step, or it may come after the binding of mRNA. The assembled complex of 30S subunit, IF-1, 2 and 3, GTP and mRNA with the fMet-tRNA$_f$ is called the 30S-initiation complex. The corresponding initiation complex in eukaryotes involves another factor, eIF-4, and ATP, which are required for binding the mRNA. How much ATP is required is not known. The 40S initiation complex is thus 40S:eIF-3:Met-tRNA$_f$:eIF-2:GTP:eIF-4:mRNA. (eIF-4 consists of several proteins; see the Table.)

The initiation codon for both pro- and eukaryotes is AUG, but the mechanism by which the ribosome locates it differs. Prokaryotes seem to recognize sequences in the messengers to the left (toward the 5' end) of the AUG codon; these sequences are complementary to part of the 16S rRNA molecule. The 70S ribosomes can start either at the first AUG in the messenger or at an internal AUG, and can even translate circular messengers. Eukaryotic ribosomes, however, must start with the AUG nearest the 5'-end of the messenger. The m7G5'pppX "cap" (see Messenger RNA) appears to guide the system, but some uncapped messengers can also be translated. The role of the cap-binding protein (see Table) in this context is not clear. It is thought that the ribosome must "walk" down the messenger until it comes to the first AUG. It is possible that ATP provides the energy for the "walk".

In prokaryotes and eukaryotes, the initiation complex is prepared for the addition of the large ribosomal subunit by the release of initiation factor 3. In bacteria, the 50S subunit appears simply to replace IF-3, with IF-1 and IF-2 leaving the complex afterward. In eukaryotes, another factor, eIF-5, catalyses the departure of the previous initiation factors and the joining of the 60S subunit. (The order of these reactions is not known.) In both cases, the release of initiation factor 2 involves the hydrolysis of the GTP bound to it. Also, the Met-tRNA$_f$ is bound to the P site of the large ribosomal subunit.

Elongation. The ribosome can accomodate two tRNA molecules at once. One of these carries the Met-tRNA or the peptide-tRNA complex, and is thus called the P site; the other accepts the incoming aminoacyl-tRNA and is therefore called the A site. What binds to the A site is actually a complex of GTP, elongation factor TU, and aminoacyl-tRNA. The tRNA, of course, must be aligned with the next codon on the messenger which is to be read; the elongation factor is presumably responsible in some

Table. *Prokaryotic and eukaryotic protein synthesis factors.*

Factor	M_r	Function
IF-1		Equilibration of 70S $\rightleftharpoons$ 50S + 30S units. Stabilization of initiation complex
IF-2		Binding of fMet-tRNA$_f$ to 30S subunit. Process may or may not require mRNA
IF-3		Prevents association of 30S and 50S subunits
eIF-1	15000	Stabilization of initiation complex
eIF-2	α subunit, 32000–38000 β subunit, 47000–52000 γ subunit, 50000–54000	Binding of Met-tRNA$_f$ to 40S subunit; process requires GTP and occurs before mRNA is bound
eIF-2A	50000–96000	Binding of Met-tRNA$_f$ to 40S subunit; process requires mRNA but not GTP
eIF-3	500000–750000 (complex of 7–11 polypeptides)	Prevents association of ribosomal subunits, stabilizes initiation complex.
eIF-4A	48000–53000	Binding of mRNA to 40S initiation complex
eIF-4B	80000–82000	Binding of mRNA to 40S initiation complex
eIF-4C	19000	Stabilization of initiation complex
eIF-4D	17000	Stabilization of initiation complex
eIF-5	125000–160000	Release of eIF-2 and eIF-3 from initiation complex; binding of the 60S subunit to the 40S complex.
Cap recognition protein	24000	Not clear; binds to mRNA cap.
EF-TU	43000	GTP-EF-TU binds aminoacyl-tRNA to ribosomal A site
EF-TS	35000	Displaces GDP from EF-TU-GDP which has been released from the ribosome; EF-TU:EF-TS complex reacts with GTP to regenerate EF-TU-GTP
EF-G	80000	Involved in translocation of peptidyl-tRNA from the A to the P site; GTPase
eEF-TU (EF-1α)	53000	GTP-eEF-TU binds aminoacyl-tRNA to ribosomal A site
eEF-TS (EF-1β)	30000	Displaces GDP from eEF-TU-GDP which has been released from the ribosome; eEF-TU:eEF-TS complex reacts with GTP to regenerate eEF-TU-GTP
eEF-G (EF-2)		Involved in translocation of peptidyl-tRNA from the A to the P site
RF-1	47000	Recognizes UAA and UAG termination codon; releases peptide from ribosome-bound tRNA
RF-2	35000–48000	Recognizes UAA and UGA termination codons; releases peptide from ribosome-bound tRNA
RF-3	46000	Stimulates RF-1 and RF-2 activities
eRF	56000–105000	Recognizes all three termination codons; has ribosome-dependent GTPase activity

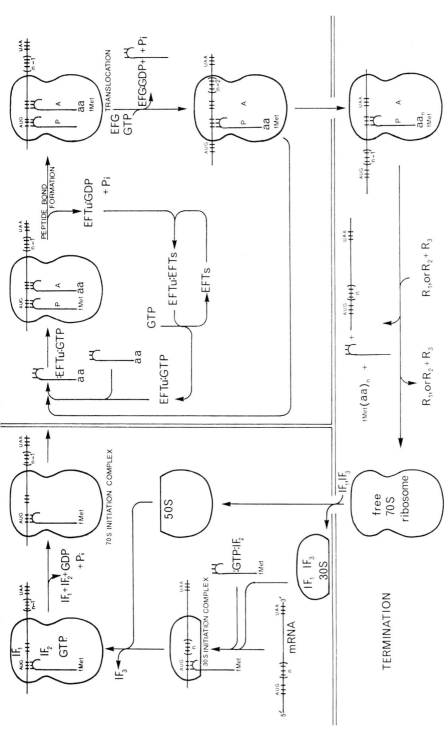

Diagrammatic representation of translation on prokaryotic ribosomes. The elongation cycle starts by interaction of the 70S initiation complex with fMet-tRNA: EFT$_U$:GTP. In all subsequent rounds of the cycle, fMet-tRNA: EFT$_U$: GTP interacts with the mRNA: ribosome complex carrying the growing polypeptide chain. Termination occurs when n amino acids have been incorporated, where n represents the number of codons between the initiation codon AUG and the termination codon (in this example UAA).

way for guiding it to precisely the right nucleotide triplet. The GTP is then hydrolysed to GDP, and the EF-TU:GDP complex leaves the ribosome. The GDP is released from the factor when the latter forms a complex with elongation factor TS: EF-TU:EF-TS. Although the affinity of EF-TU for GDP is higher than its affinity for GTP, the complex with EF-TS has a higher affinity for GTP, which replaces the EF-TS. The EF-TU-GTP is then ready to pick up another aminoacyl tRNA and to recycle.

Meanwhile, on the ribosome, a reaction is catalysed between the carboxyl of the P-site occupant and the (free) amino group of the A-site occupant. The peptidyl transferase activity which catalyses this is intrinsic to the ribosome.

The final step of elongation is the motion of the ribosome relative to the mRNA, which is accompanied by the translocation of the peptidyl-tRNA from the A to the P site. Elongation factor G is involved in this step, although translocation can occur, slowly, in the absence of this factor. A complex of EF-G and GTP binds to the ribosome, and GTP is hydrolysed in the course of the reaction. The deacylated tRNA is also released from the P site at this time.

It is interesting that IF-2 and EF-TU have analogous functions: IF-2 recognizes only Met-tRNA$_f$, while EF-TU recognizes all other aminoacyl-tRNAs; but each serves to introduce an aminoacyl-tRNA to a specific site on the ribosome. What is more, the two proteins also have a stretch of homology about 100 amino acids in length. It must also be pointed out that the hydrolysis of the GTP bound to EF-TU or IF-2 does not provide the energy, per se, for the binding of the aminoacyl-tRNA to the ribosome. Instead, the GTP hydrolysis provides for the release of the tightly bound EF-TU or IF-2 from the ribosome. The energy for peptide bond formation is invested at the stage of loading the amino acids onto the tRNAs, when two phosphate bonds of ATP are hydrolysed in the course of forming the ester bond between the amino acid and the $2'$ or $3'OH$ of the tRNA. All the other ATPs and GTPs hydrolysed in the course of P.b. apparently serve to increase the accuracy of the process.

Termination. The end of a polypeptide synthesis is signaled by a Termination codon (see) at the A site. Three prokaryotic release factors are known: RF-1 is specific for termination codons UAA and UAG, while RF-2 is specific for UAA and UGA. RF-3 stimulates RF-1 and RF-2, but does not itself recognize the termination codons. RF-3 also has GTPase activity; it is likely that it accelerates termination at the expense of GTP. Only one eukaryotic release factor is known; it has GTPase activity. It is possible that eIF-3 interacts with the ribosome as it leaves the mRNA, causing the two subunits to separate. This is suggested by the observation that eukaryotic cells may have a large pool of 80S ribosomes which are not active in protein synthesis, and do not exchange with the pool of actively cycling ribosomal subunits. The preceeding account describes P.b. for one ribosome, but it should be pointed out that the functional system is the Polysome (see). At any one time, several ribosomes are positioned along the mRNA; those nearest the $3'$-end carry the longest newly synthesized polypeptide chains, whereas those at the $5'$-end have translated fewer codons and therefore carry a relatively shorter length of peptide. Thus initiation, elongation (at various stages) and termination proceed simultaneously on the same length of mRNA.

The tertiary structure of the final protein begins to appear during P.b., before completion of the polypeptide. In many cases, the protein is subjected to further reactions, which convert it into its biologically active form, e.g. by covalent attachment of certain groups, or by removal of certain sequences of amino acids (see Post-translational modification of proteins). The initial translation product of secretory proteins contains a metabolically short-lived *N*-terminal peptide sequence, which functions in the attachment of the ribosome to the membrane of the rough endoplasmic reticulum, and in the transfer of these proteins across the membrane into the tubules of the endoplasmic reticulum (see Signal hypothesis). [*Ribosomes, Structure, Function and Genetics* Steenbock Symposium on Ribosomes, 9 th. (University of Wisconsin Press, Madison) 1979. *DNA Makes RNA Makes Protein* T. Hunt, S. Prentis & J. Tooze eds. (Elsevier Biomedical Press, Amsterdam) 1983.]

Proteinase inhibitors: see Inhibitor proteins.

Proteinases: a group of Proteases (see).

Protein-calorie malnutrition: see Protein-energy malnutrition.

Protein deficiency: see Protein-energy malnutrition, Kwashiorkor, Marasmus.

Protein degradation: see Proteolysis.

Protein-energy malnutrition, *PEM, protein-calorie malnutrition, PCM:* a spectrum of nutritional deficiency states, occurring characteristically in children under five years, although no age is immune. Marasmus and kwashiorkor are regarded as the two extremes of this spectrum. For many years, kwashiorkor was attributed to protein deficiency with adequate energy intake, in contrast to marasmus, which was attributed to deficiencies of both energy and protein. It is now known that there is no essential difference in the dietary protein:energy ratio of marasmus and kwashiorkor victims, and the characteristic clinical presentation of kwashiorkor is thought to be due to aflatoxin poisoning (see Kwashiorkor for further details). Three intermediate forms of PEM have been defined: marasmic kwashiorkor, nutritional dwarfism and "underweight child". PEM is often complicated by vitamin and mineral deficiencies, and by infection. Plasma albumin concentrations below 35 g/l are characteristic of protein deficiency and of PEM. Gamma globulin, on the other hand, may be raised in response to the presence of infection. There is always a decrease in metabolic rate, but this is roughly in proportion to the decrease in cell mass. Plasma concentrations of branched chain amino acids and tyrosine are lower than normal, whereas the concentrations of some nonessential amino acids may be increased; similar abnormal patterns of plasma amino acids appear after only 4 days on a protein-free diet. Therapeutic administration of protein causes an increase in the concentrations of plasma amino acids, sometimes accompanied by overflow aminoaciduria.

Loss by diarrhoea is probably largely responsible for

the low plasma potassium often observed in PEM; values less than 2.5 mmol/l have been recorded. Plasma magnesium may be low for the same reason. Plasma sodium is usually normal. Total body water increases to 65-80% of body weight (60% is normal).

Endocrine hypofunction is not a feature of PEM. Increased levels of certain hormones are sometimes observed, e.g. increased growth hormone has been reported in kwashiorkor; plasma cortisol and other adrenocorticosteroids may be raised in PEM. Thyroid function is usually normal, but fasting concentrations of plasma insulin may be low. For a review of the effect of malnutrition on circulating hormones, see R.D.G. Milner *Mem. Soc. Endocr. 18* (1970) 191.

In PEM there appear to be no adaptive changes toward a more economic utilization of energy. Decreased protein intake results in increased levels of amino acid activating enzymes and decreased rates of urea synthesis, but normal amino acid and protein metabolism is rapidly restored when adequate dietary protein becomes available (except in kwashiorkor, where aflatoxin poisoning causes liver dam-

ters (see) and other tumor promoters. It is thought that activation of P. by these esters mimics the effects of growth or proliferation signals, but is not subject to the same controls as the latter.

Proteinoids: heteropolyamino acids; artificially prepared polypeptides ($M_r > 1000$) formed in 20-40% yield by heating a mixture of several amino acids for 16 h at 170 °C (thermal condensation). P. show many similarities with globular proteins, e.g. relative quantities of individual amino acids, solubility, spectral properties, denaturation, degradation by proteases, catalytic activity (e.g. esterase, ATPase, decarboxylase activities) and hormone action (MSH activity). They can be regarded as models for the first information-carrying molecules. In water, they become organized into microsystems with a typical ultrastructure (microspheres), which share a number of properties with living cells, i.e. they possess a bilayer membrane, which exhibits a certain degree of semipermeability; they can also multiply in the absence of nucleic acids, by budding.

Protein-tyrosine kinase, *PTK:* an enzyme that catalyses the ATP-dependent phosphorylation of tyrosyl residues in proteins:

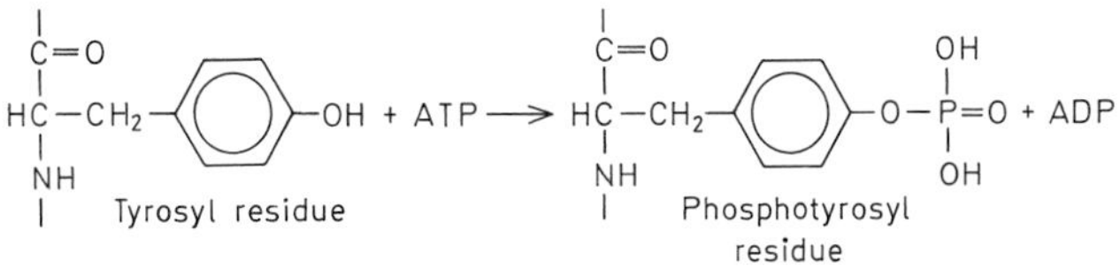

age). Rapid recovery therefore proceeds immediately on receipt of a balanced diet; during recovery, energy utilization may be nearly 40% higher than normal until the correct body weight for the age is achieved. See Kwashiorkor, Marasmus.

Protein kinase C: a membrane protein apparently involved in several types of cellular signal transduction. It phosphorylates serine and threonine residues and depends on calcium and phospholipid for activity. At physiological calcium concentrations, it is activated by diacylglycerol, a product of the action of phospholipase C on inositol phospholipids (see Inositol phosphates).

P. purified from bovine brain has M_r 81000; the primary structure of the molecule has been deduced from cloned cDNA [P.J. Parker et al., *Science 233* (1986) 853-859]. Hybridization of cDNA to libraries of human, bovine and rat genomes revealed that there are several distinct forms of the enzyme, which in the human genome are located on separate chromosomes. It is postulated that the existence of a family of similar P. may explain the diversity of effects of the enzyme and also account for differentiation of its functions in different tissues. [L. Coussens et al., *Science 233* (1986) 859-866.]

Binding of a number of growth factors, hormones and extrinsic agents to their receptors causes an increase in cellular calcium and cleavage of inositol phospholipids, which thus serve as second messengers. P. is one of the targets of these messengers, presumably regulating the activity of the proteins it phosphorylates and thus affecting the cellular activities. P. is evidently the site of binding of Phorbol es-

PTK activity is characteristic of certain membrane proteins, and about one third of known oncogenes and their proto-oncogene counterparts code for PTKs (see Table, p.496). The phosphotyrosine content of cell proteins may increase up to 10-fold in response to transformation by viral PTK-containing oncogenes, with the exception of v-*erb*, v-*fms* and v-*ros*. Among the substrates which become phosphorylated are: the glycolytic enzymes, enolase, phosphoglycerate mutase and lactate dehydrogenase, three cytoskeletal proteins (vinculin of adhesion plaques; p36, which is part of a Ca^{2+}-sensitive complex in the submembranous cortical cytoskeleton, and perhaps in ribonucleoprotein particles; p81 which, like p36, is submembranous, but present in microvillar cores), a 50000 M_r protein associated with pp60^{v-src}, and some membrane glycoproteins. PTKs become autophosphorylated by acting as substrates to their own PTK activity; in some cases autophosphorylation appears to be necessary to sustain PTK activity towards other substrates. In the case of membrane receptor PTKs (see Insulin receptor), binding of the specific ligand on the outer surface of the cell promotes the appearance of PTK activity in the cytoplasmic domain. See Oncogene. [T. Hunter & J.A. Cooper, *Ann. Rev. Biochem. 54* (1985) 897-930; J.A. Cooper, *BioEssays 4* (1986) 9-15.]

Proteoglycans: high M_r compounds of carbohydrate and protein, found in animal structural tissues, e.g. the ground substance of cartilage and bone. The ground substance and gel fluids of these tissues owe their viscosity, elasticity and resistance to infective organisms to the presence of P. Each P.

Table. Some normal cellular proteins and oncogene products which possess protein-tyrosine kinase activity.

Normal protein	Normal gene	Oncogene product	Oncogene	Oncogene carrier
Insulin receptor	Insulin receptor gene	–	–	–
Insulin-like growth factor (IGF-l)	IGF-l gene	–	–	–
Platelet derived growth factor	c-*sis*	p28^{sis}	v-*sis*	Simian sarcoma virus
Epidermal growth factor receptor	c-*erb*-B	gp68/74$^{v\text{-}erb\text{-}B}$	v-*erb*-B	Avian erythroblastosis virus
pp60$^{c\text{-}src}$	c-*src*	pp60$^{v\text{-}src}$	v-*src*	Rous sarcoma virus
p98$^{c\text{-}fps}$	c-*fps*	Pl40$^{gag\text{-}fps}$	v-*fps*	Fujinami avian sarcoma virus
p92$^{c\text{-}fes}$	c-*fes*	P85$^{gag\text{-}fes}$	v-*fes*	Many feline sarcoma viruses
p-150$^{c\text{-}abl}$	c-*abl*	P120$^{gag\text{-}abl}$	v-*abl*	Abelson murine leukemia virus
–	c-*ros*	P68$^{gag\text{-}ros}$	v-*ros*	Avian sarcoma virus UR2
–	c-*yes*	P90$^{gag\text{-}yes}$	v-*yes*	Y73 Avian sarcoma virus
–	c-*fgr*	P70$^{gag\text{-}fgr}$	v-*fgr*	Gardner-Rasheed feline sarcoma virus
Colony stimulating factor (CSF-1) receptor	c-*fms*	gp140$^{v\text{-}fms}$	v-*fms*	McDonough feline sarcoma virus

contains 40–80 acidic mucopolysaccaride chains (glucosaminoglycans) bound to protein via *O*-glycosidic linkages to serine or threonine. In contrast to the Glycoproteins (see), the prosthetic group of P. has M_r 20000–30000, consisting of many (100–1000) unbranched, regularly repeating disaccharide units. The disaccharides are composed of *N*-acetylhexosamine (which may or may not be sulfated) linked to a uronic acid or to galactose. In the chondroitin sulfates, the linkage region between polysaccharide and protein contains xylose linked *O*-glycosidically to serine; this is followed by two galactose residues and one glucorunic acid, to which the first repeating disaccharide is attached. In corneal keratan sulfate, the polysaccharide chain is linked to the protein via a glucosamine residue, which is joined to an asparagine side chain by a glycosylamine linkage. In cartilage keratan sulfate, most of the carbohydrate-protein linkages are *O*-glycosidic bonds between *N*-acetylglucosamine and the hydroxyl of serine or threonine.

The chondroitin sulfates, together with collagen, form the major component of cartilage. Mammalian skin contains proteodermatan sulfate, and the intestinal mucosa contains protein-bound heparin. Heterogeneity of P. is due to differences in polypeptide chain length, and to the number and distribution of the attached polysaccharide chains. Microheterogeneity also exists, due to small differences in the chain lengths of the polysaccharide chains, and the distribution of sulfate residues.

P. can be extracted from cartilage under mild conditions with 4 M guanidinium chloride. The resulting P. subunit has $S_w = 16S$ (see Sedimentation coefficient), and M_r 1.6×10^6. In the tissue, P. exist as giant molecular aggregates ($S_w = 70S$ und 600S), which are formed by noncovalent association of P. subunits with a glycoprotein.

Proteohormones, *protein hormones:* proteins (often glycoproteins) with hormonal function. Like other proteins, they are synthesized by the translation of appropriate mRNA, and degraded by proteolysis. M_r of P. are between 5000 and 25000 for the monomers and correspondingly higher for the dimers and polymers. See Choriogonadotropin; Follicle stimulating hormone; Luteinizing hormone; Thyreotropin. Although a close relationship exists between P. and Peptide hormones (see), a distinction is made between these two.

Proteolysis, *protein degradation:* hydrolysis of proteins by the action of proteolytic enzymes, or nonenzymatically by acids (e. g. 6 M HCl at 110 °C for 24 h or longer) or alkalis. Ultimate products of P. are amino acids. Dietary proteins are hydrolysed to L-amino acids by P. in the intestine. After absorption, these amino acids are used in the synthesis of new proteins specific to the organism; these, in turn, are eventually hydrolysed to amino acids by P. as part of the continual process of synthesis and degradation of cellular constituents (see Turnover). A distinction is made between extracellular P. (e. g. digestion and blood coagulation) and intracellular P. The latter occurs at neutral and acidic pH values, and the relevant endopeptidases (called cathepsins) are localized chiefly in the lysosomes. The cathepsins from protease-rich organs, like liver, spleen and kidney, can be separated into cathepsins A, B, C, D, E and L. Cathepsins A to E have pH optima in the range 2.5–6, and (with the exception of D and E) they also hydrolyse synthetic, low M_r substrates. Other cathepsins are active only at neutral pH, and they attack only proteins. M_r of cathepsins are between 25000 (cathepsin B) and 100000 (cathepsin E). Cathepsins B, B₂, C and some neutral cathepsins are SH-enzymes. In the intact cell P. is controlled and it occurs in the lysosomes (autophagy). In damaged cells, the same cathepsins are released from the ruptured lysosomes and are responsible for autolysis, i. e. the uncontrolled total degradation of the cell.

Table. *Formation of biologically active proteins from inactive precursors by limited proteolysis.*
AA, amino acids; Chy, chymotrypsin; CP, carboxypeptidase

Inactive protein	M_r	No. of AA	Active fragment	M_r	No. of AA
Pepsinogen	42500	362	Pepsin	34500	327
Prorennin	36200	321	Rennin	30700	272
Trypsinogen	24000	229	Trypsin	23400	223
Chy-ogen	25666	245	Chy A	25170	241
Pro-CP A	90000	850	CP A	34300	307
Prothrombin	72000	560	Thrombin	39000	309
Proinsulin	9100	84	Insulin	6000	51
Fibrinogen	340000	3400	Fibrin monomer	327000	~3270

In limited P. only certain peptide bonds of a protein are hydrolysed; this results in the production of biologically active (e.g. enzymes or hormones) or inactive (e.g. para-κ-casein) proteins or peptides. Limited P. occurs in digestion, blood coagulation and milk clotting; it is responsible for the activation of zymogens and for the release of certain peptide hormones, e.g. insulin, angiotensin, vasopressin, oxytocin and various kinins (see Table).

Prothrombin, *factor II:* the precursor of Thrombin (see) in the Blood coagulation (see) system. The protein is a single polypeptide chain, M_r 72000 (582 amino acid residues), with a carbohydrate content of 14.7% (bovine) or 11.8% (human). Synthesis of P. occurs in vertebrate liver and requires vitamin K for the synthesis of γ-carboxyglutamate (Gla) residues in the N-terminal region of the chain. During blood coagulation, P. is converted to thrombin by factor X_a (EC 3.4.21.6). This reaction is accelerated by a factor of 20000 by active factor V and acidic phospholipid; the latter is normally sequestered on the inside of cell membranes, and its presence in plasma would thus signal injury. However, the reaction is accelerated by a factor of 100000 in the presence of platelets and factor Va.

P. binds to the phospholipids by its 10 Gla residues; the binding is mediated by Ca^{2+}. The region (called F1) containing these residues is removed by thrombin in vitro; however, in vivo, this cleavage is suppressed. The first cleavage by factor X_a removes the segment called F1·2, with a total M_r of 32000. In addition to the Gla region, this segment contains the F2 region, which mediates binding to factor V_a. Finally, the Arg-Ile bond between peptides A and B is cleaved to produce mature thrombin. [L. M. Jackson

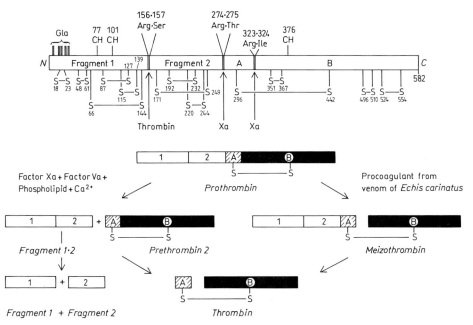

Structure of prothrombin and its conversion into thrombin. The pathway on the left operates during normal blood coagulation in response to injury, whereas the right-hand pathway is promoted by the procoagulant of snake venom. Fragments 1 and 2, formed from fragment 1·2 by the action of thrombin, have no known physiological function. Gla = 4-carboxyglutamate residues.

& Y. Nemerson, *Ann. Rev. Biochem.* **49** (1980) 765–811; J. Rosing et al. *J. Biol. Chem.* **261** (1986) 4224–4228; S. Magnusson et al. in E. Reich, D. B. Rifkin and E. Shaw, eds., *Proteases and Biological Control* (Cold Spring Harbor Conferences on Cell Proliferation) (Cold Spring Harbor Laboratory, 1975) pp. 123–149.]

Prothromboplastin: see Blood coagulation.

Protoalkaloids: see Biogenic amines.

Protocyanin: see Cyanidin.

Protoheme, *heme, ferroheme, ferroprotoporphyrin, protoheme IX:* [7,12-diethenyl-3,8,13,17-tetramethyl-21H, 23H-porphine-2,18-dipropanoate (2-)-N^{21}, N^{22}, N^{23}, N^{24}]-iron; or 1,3,5,8-tetramethyl-2,4-divinylporphine-6,7-dipropionic acid ferrous complex. $C_{34}H_{32}FeN_4O_4$. M_r 616.48. Protoheme crystallizes as fine brown needles with a violet sheen; $\varepsilon_{572} = 5.5 \times 10^3$; in phosphate buffer pH 7.0, absorption maxima occur at 575 nm and about 550 nm. It is the prosthetic group of a number of hemoproteins, e.g. hemoglobins, erythrocruorins, myoglobins, some peroxidases, catalase and cytochromes *b*. The four coordinate bonds of the iron lie in the plane of the nearly planar porphyrin ring structure, while the two unoccupied sites of the iron are perpendicular to it.

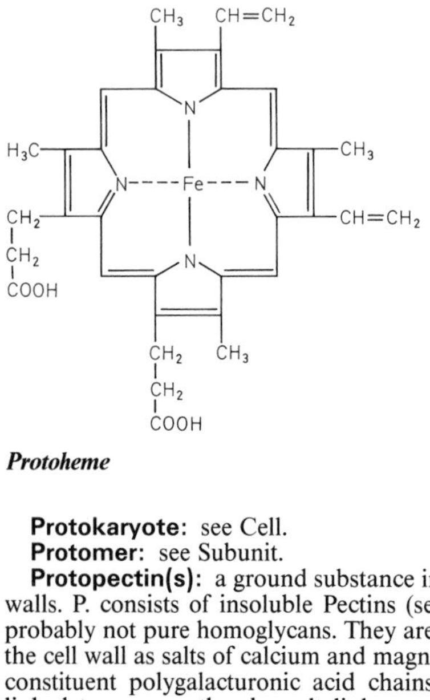

Protoheme

Protokaryote: see Cell.

Protomer: see Subunit.

Protopectin(s): a ground substance in plant cell walls. P. consists of insoluble Pectins (see) and are probably not pure homoglycans. They are present in the cell wall as salts of calcium and magnesium. The constituent polygalacturonic acid chains of P. are linked to one another by salt linkages, phosphate bonds and esterification with arabinose.

Protoplast: see Cell, 2.

Prototrophism: the property of being able to grow at the expense of usual or common nutrients (see Nutrient medium), with no special requirement for Growth factors (see). P. is shown by prototrophic organisms.

Provitamin D$_2$: see Ergosterol.

Provitamin D$_3$: see 7-Dehydrocholesterol.

Provitamins: inactive precursors of Vitamins (see). P. are mostly of vegetable origin, and are converted into active vitamins after absorption from the diet.

PRPP: abb. for 5-phosphoribosyl 1-pyrophosphate.

Prunetin: 5,4'-dihydroxy-7-methoxyisoflavone, see Isoflavone.

Prunetrin: prunetin 7-glucoside, see Isoflavone.

Prunin: naringenin 7-glucoside, see Flavanone.

Pseudoalkaloids: a group of alkaloids earlier assigned to other groups (e.g. some were grouped with the terpenes) with which they show a close structural relationship. At the time, their nitrogen content seemed incidental.

Pseudobaptigen: see Pterocarpans.

Pseudogibberellin A$_1$: see Gibberellins.

Pseudohermaphroditism: see Inborn errors of metabolism (Adrenal hyperplasia).

Pseudoindicans: an old name for Iridoids (see).

Pseudo-isoenzymes: multiple forms of an enzyme, which catalyse the same reaction. They have similar properties to isoenzymes, but do not have genetically determined differences of primary structure. Their multiplicity is the result of enzymatic or nonenzymatic modification of one original primary sequence, either in vivo or in vitro (i. e. during isolation). Examples of P. formed in vivo are the different chymotrypsins, trypsins, pepsins and carboxypeptidases, each group being derived from a single zymogen. Also in this category are the different degrees of aggregation shown by oligomeric enzymes that consist of identical subunits, e.g. glutamate dehydrogenase. P. formation in vitro is responsible for the occurrence of the 4 α-amylases, the 2 yeast phosphofructokinases, as many as 13 heart muscle lipoyl dehydrogenases, and the numerous forms of phosphoglucose isomerase.

Pseudopelletierine, *ψ-pelletierine, pseudopunicine, 9-methyl-3-granatanone:* 9-methyl-9-azabicyclo-[3,3,1] nonan-3-one, the most important representative of the Punica alkaloids, present in the root bark of *Punica granatum*. M_r 153.21, m.p. 54 °C, b.p. 246 °C. Its structure is based on the meso form of granatane (9-azabicyclo [3,3,1] nonane). For biosynthesis, see Punica alkaloids.

Pseudotropine: see Tropane alkaloids.

Pseudouridine, *5-β-D-ribofuranosyluracil, 5-ribosyluracil, ψ, ψ rd:* a structural analog of uridine containing a C-C bond between C-5 of uracil and C-1 of ribose. M_r 244.2, m.p. 223–224 °C. ψ is a Rare nucleic acid component (see) found in tRNA. Despite earlier conflicting evidence, it is now clear that ψ is formed by rearrangement of uridine after assembly of the tRNA chain, i.e. by post-transcrip-

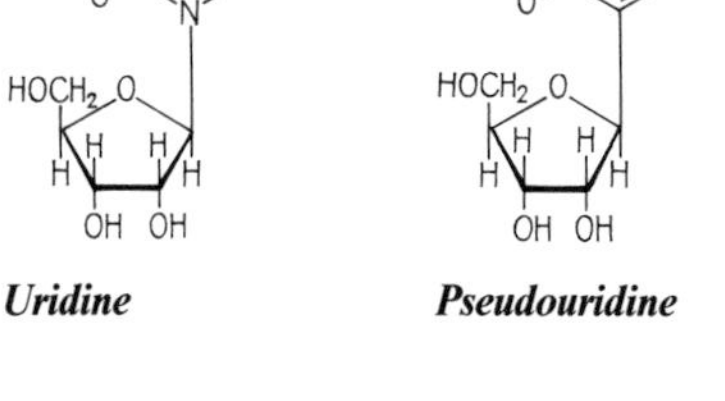

Uridine   *Pseudouridine*

tional modification of the tRNA. An enzyme that can modify specific uridine residues in the anticodon region of many species of tRNA has been purified from *Salmonella typhimurium* and *Escherichia coli.*

Psicofuranin: see Angustmycin.

Psi (ψ) factor: a protein responsible for the specific initiation of the RNA polymerase reaction at the promotor sites of the genes for rRNA in bacteria.

Psilocin: see Psilocybin.

Psilocybin: 4-phosphoryloxy-*N, N*-dimethyltryptamine, m.p. 220-228 °C. P. and the related compound psilocin (4-hydroxy-*N,N*-dimethyl-tryptamine, m.p. 173-176 °C), are jointly responsible for the psychotropic action of the fruiting body of the Mexican hallucinogenic fungus Teonanacatl *(Psilocybe mexicana).* P. is the first naturally occurring phosphorylated indole derivative to be isolated. It can be hydrolysed to psilocin. Both compounds are slightly poisonous. Administered by mouth or by intramuscular injection, they cause hallucinations similar to those caused by LSD; the latter is, however, about 100 times more potent.

Psilocybin : R = H_2PO_3
Psilocin : R = H

Psychodelic drugs: see Hallucinogens.
Psychodysleptic drugs: see Hallucinogens.
Psycholitic drugs: see Hallucinogens.
Psychotomimetic drugs: see Hallucinogens.

Psychotropic agents: chemical compounds that influence the human psyche. P.a. are used in psychiatry. They include Narcotics (see) and Hallucinogens (see), and are almost exclusively of vegetable origin. About 50 different modes of psychotropic activity are found among about 20000 natural plant products. In addition, there are semisynthetic (chemically modified natural products) and synthetic P.a., some of which are used in medicine. Some P.a. are also used as narcotics. All P.a. eventually produce a personality change. Use and possession of P.a. are therefore subject to legal controls, designed to prevent misuse. P.a. have scientific as well as medical uses; psychotropic natural products serve as model substances in pharmacy, pharmacology and toxicology, and P.a. in general are used in the biochemical investigation of the function of the central nervous system.

Pteridines: a group of compounds containing the pteridine ring system (Fig.). The majority of naturally occurring P. are chemically related to pterine (Fig.). A smaller number are derived from lumazine (Fig.). Both folic acid (see Tetrahydrofolate; Vitamins: B₂ complex) and Tetrahydrobiopterin (see) are P. and serve as hydrogen transfer cofactors. Folic acid is a vitamin for mammals, but they are able to synthesize tetrahydrobiopterin; in spite of their chemical similarity, these compounds are synthesized by different pathways.

Because of their role as enzyme cofactors, P. are ubiquitous. The cofactors are metabolized and ex-

Pteridine

Pterine

Lumazine

creted or deposited as pigments, e.g. as Xanthopterin (see), Leucopterin (see), sepiapterin, etc. in the wings of insects. (The compounds were discovered by G. Hopkins in 1890, in butterfly wings. The name is derived from "pteron", the Greek for "wing".) Mammals excrete bio-, xantho-, neopterin and others in their urine. The details of biosynthesis have not been completely worked out (1986), as there are some highly unstable reaction intermediates (see Tetrahydrobiopterin, Folic acid).

Elevated excretion of neopterin is correlated with certain malignant diseases, viral infection and graft rejection, apparently because neopterin is secreted by macrophages in the course of T-lymphocyte activation. [I. Ziegler, in *Biochemical and Clinical Aspects of the Pteridines*, vol. *4* (1985) 347-361] Activation of lymphocytes in vitro is stimulated by sepiapterin, dihydro- and tetrahydropterins, but repressed by xantho- and isoxanthopterins. These P. are thus lymphokines.

Pterins: see Pteridines.

Pterocarpans, *coumaranochromans:* isoflavonoids with the ring system shown (Fig.)

Pterocarpan ring system (systematic name: 6a, 11a-dihydro-6-H-benzofuro[3,2-c][1]benzopyran; the systematic Ring Index numbering is shown)

Examples: ***pterocarpin*** (3-methoxy-8,9-methylenedioxypterocarpan, *Sophora japonica*), ***homopterocarpin*** (3,9-dimethoxypterocarpan), ***ficifolinol*** (3,9-dihydroxy-2,8-di-γ,γ-dimethylallylpterocarpan, *Neorautanenia ficifolia*). Some P. are Phytoalexins (see), e.g. ***Phaseolin*** (see), ***Glyceollins*** (see), ***pisatin*** (3-methoxy-6a-hydroxy-8,9-methylenedioxypterocarpan, *Pisum sativum*), ***maackiain*** (3-hydroxy-8,9-methylenedioxypterocarpan, *Lathyrus* spp.), ***variabilin*** (3,9-dimethoxy-6a-hydroxypterocarpan, *Lathyrus* spp.), ***medicarpin*** (3-hydroxy-9-methoxypterocarpan, *Lathyrus* spp.), ***nissolin*** (3,9-dihydroxy-10-methoxypterocarpan, *Lathyrus nissolia*), ***methylnissolin*** (3-hydroxy-9,10-dimethoxypterocarpan, *Lathyrus nissolia*) [D.J.

Robeson & J.L. Ingham *Phytochemistry* **18** (1979) 1715–1717]. Many P., however, have been isolated from healthy, unstressed plants, especially from the heartwood of tropical genera of the *Leguminoseae*. The relative configuration of the 6 a,11 a chiral center is *cis*. The configuration of (–)-P. is 6 aR, 11 aR, and that of (+)-P. is 6 aS, 11 aS. [K.G.R. Pachler & W.G.E. Underwood *Tetrahedron* **23** (1967) 1817–1826.]
Like other isoflavonoids, P. are largely restricted to the *Leguminoseae*. Structural relationships governing the phytoalexin (i.e. fungicidal) properties of P. have been studied [H.D. Van Etten *Phytochemistry* **15** (1976) 655–659]. 6 a,11 a-dehydro-P. (which have a fundamentally different shape from that of P.) are also active fungicides. A common 3-dimensional molecular shape, therefore, does not seem to be necessary.
Biosynthesis. Investigation of the biosynthesis of phytoalexin P. is greatly helped by the fact that their synthesis can be induced by treatment of the plant with fungi, Elicitor (see), UV light or heavy metal salts. Isotopically labelled precursor can be added at the time of maximal synthesis, and a high incorporation of isotope is observed. It has been shown, for example, that 2′,4′,4-trihydroxychalcone and formononetin are excellent precursors of medicarpin in lucerne seedlings. In the same system, medicarpin and the isoflavan vestitol are interconvertible, presumably via an isoflavanium ion (see Isoflavonoids). [1,2-^{13}C$_2$]Acetate has been administered to CuCl$_2$-treated *Pisum sativum*, and the ^{13}C-^{13}C coupling in the resulting pisatin analysed by ^{13}C-NMR. This shows that C-atoms 1 and 1 a, 2 and 3, and 4 and 4 a are incorporated as intact C$_2$ units, i.e. the carbons are not randomized by rotation of the acetate-derived ring, and the absence of oxygen on C-1 is due to loss of oxygen before cyclization. In contrast, other flavonoids (e.g. apigenin, kaempferol) are synthesized with free rotation of the acetate-derived ring (see Chalcone synthase). The biosynthetic sequence shown (Fig.) is strongly supported by the observation that all the represented compounds act as efficient precursors of pisatin in *Pisum sativum*. [P.M. Dewick in J.B. Harborne & T.J. Mabry, eds., *The Flavonoids: Advances in Research* (Chapman and Hall, 1982) pp.535–640]

Pterocarpin: 3-methoxy-8, 9-methylenedioxy-pterocarpan; see Pterocarpans.
Pteroylglutamic acid: see Vitamins (Folic acid).
Puff: see Giant chromosomes.
Punctuation codon: see Termination codon.
Punica alkaloids: a group of piperidine alkaloids, originally isolated from the bark of the pomegranate tree (*Punica granatum* L., official drug *Cortex granati*), and subsequently isolated from other plant families. Decoctions of the drug, or the isolated alkaloids, have some use as vermifuges. Biosynthesis of the predominant P.a., pseudopelletierine and isopelletierine, are shown in the Figure. *N*-Methylisopelletierine is also an important member of the group. P.a. are higher homologs of the tropane or pyrrolidine alkaloids, which have an analogous mode of biogenesis.

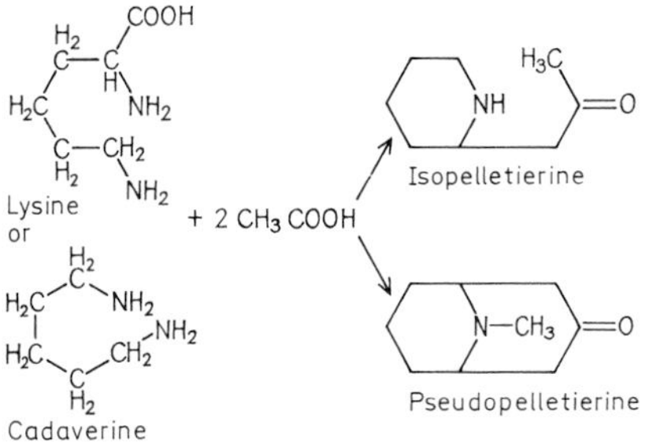

Biosynthesis of Punica alkaloids, isopelletierine and pseudopelletierine.

Purine: a heterocyclic compound with a condensed pyrimidine-imidazole ring system. M_r 120.1, m.p. 217 °C. P. was prepared from uric acid by Emil Fischer in 1884. The free compound is not known to occur naturally, but the otherwise unsubstituted P. ring system is found in combination with ribose in the nucleoside antibiotic, Nebularine (see).
Various substituted and oxidized purine derivatives occur naturally, and are of considerable biological importance. The purine derivatives, Adenine (see) and Guanine (see), are present in DNA and RNA, and they are commonly referred to as purine bases.

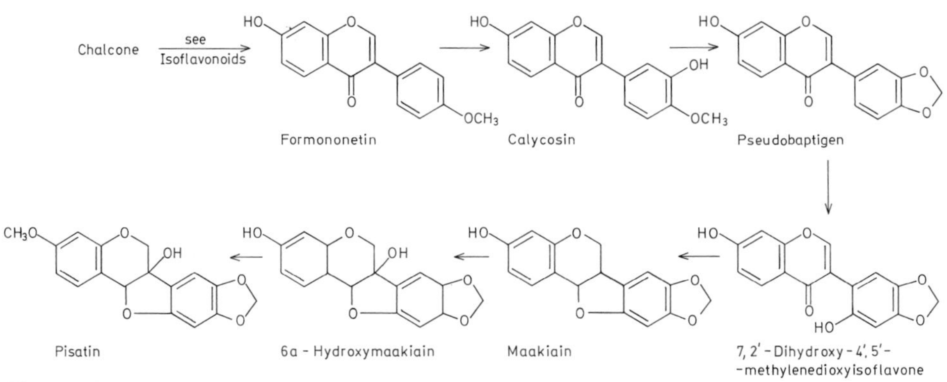

Biosynthesis of pisatin in Pisum sativum

Certain Rare nucleic acid components (see) are formed by modification of the P. bases in the polynucleotide chain. P. analogs, like 8-Azaguanine (see), can also be incorporated in place of natural P. bases during nucleic acid biosynthesis. P. bases are found in some low M_r Nucleotide coenzymes (see), and they are also components of other biologically active compounds, e.g. Nucleoside antibiotics (see), alkaloids (see Methylated xanthines), vitamins (see Vitamin B_{12}) and Cytokinins (see). The purine nucleotides, ATP (see Adenosine phoshates) and GTP (see Guanosine phosphates), are key compounds in biological energy metabolism. In some animals (birds, reptiles, insects), the chief route of nitrogen excretion is via purine synthesis; common excretory products are the purine oxidation product, Uric acid (see), and its further oxidation product, Allantoin (see). Spiders excrete guanine. See also Purine deg-

radation; Purine biosynthesis. Purines are weak bases with specific light absorption in the UV between 230 and 280 nm. They display lactam-lactim and/or enamine-ketimine-tautomerism.

Purine

Purine antibiotics: purine derivatives with antibiotic activity. They occur as nucleosides (see Nucleoside antibiotics), polypeptides (see Viomycin), or free bases (see 8-Azaguanidine).

Purine bases: see Purine; Adenine; Guanine.

Purine biosynthesis, *de novo purine biosynthesis:* a common pathway for the biosynthesis of the pu-

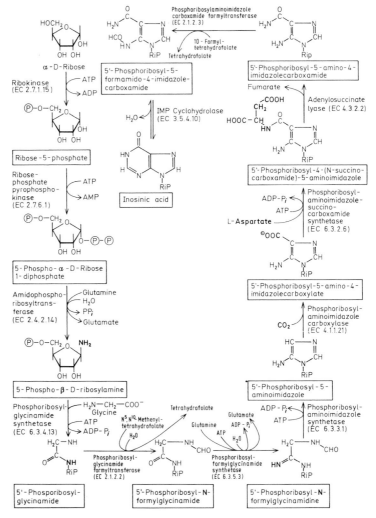

Figure 1. Formation of the purine ring system in the biosynthesis of inosinic acid.

501

rine ring system found at all levels of evolutionary development. α-D-Ribose 5-phosphate is pyrophosphorylated to 5-phosphoribosyl 1-pyrophosphate. The pyrophosphate group is then replaced by an

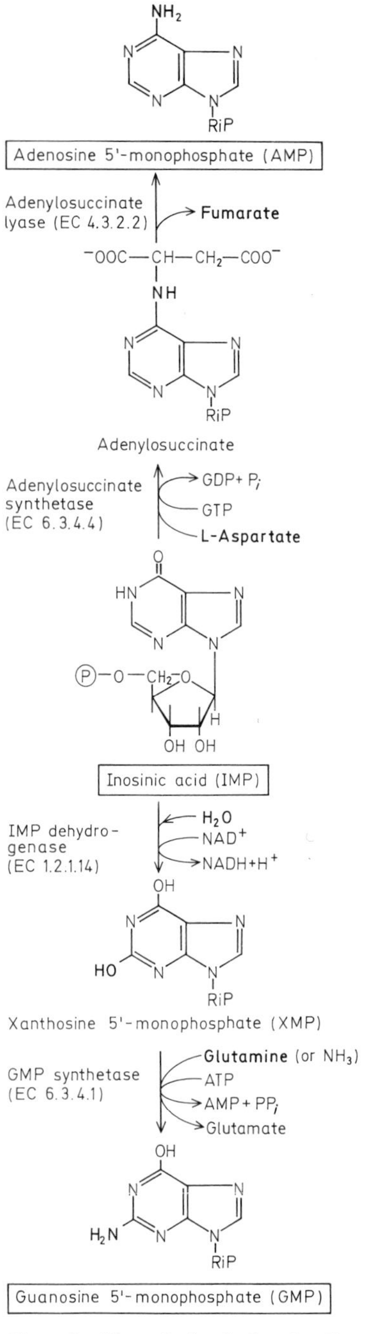

amino group, which is transferred from the amide group of L-glutamine. The nitrogen of this amino group is destined to become N-9 of the purine ring system. The ribose phosphate remains attached throughout the successive enzyme catalysed steps which eventually lead to the complete purine ring system. Thus the purines are synthesized as their nucleoside monophosphates. The first product with a complete purine ring system is inosinic acid, which serves as the precursor of the other purine nucleotides. All the stages of P.b. are shown in Fig. 1, and the subsequent interconversions leading to the synthesis of AMP and GMP are shown in Fig. 2.

The nucleoside monophosphates are converted to the triphosphates (the direct precursors of RNA) by two kinase reactions. These kinases have a low specificity, and they catalyse the phosphorylation of nucleotides of adenine, guanine and the pyrimidines (Fig. 3). An alternative route for the synthesis of purine nucleotides is the Salvage pathway (see).

P.b. is regulated by both end products, AMP and GMP, which jointly inhibit phosphoribosylpyrophosphate amidotransferase (EC 2.4.2.14). GMP also inhibits IMP-dehydrogenase (1.2.1.14); AMP inhibits adenylsuccinate synthetase (EC 6.3.4.4). Further control is exerted by the requirement for GTP in AMP synthesis (Fig. 2). ATP inhibits GMP reductase, which converts GMP into IMP in one step.

The foundation work on P.b. was performed by Buchanan and Greenberg with pigeon and chicken liver extracts. This led to the formulation of the scheme of P.b. by Greenberg in 1953. The postulated intermediates were isolated from mutant microorganisms.

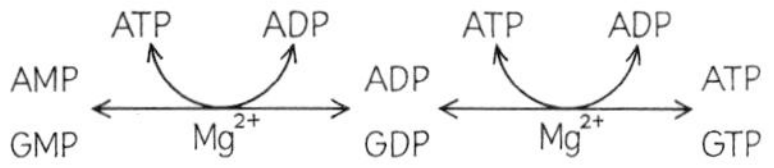

Figure 3.

Purine catabolism: see Purine degradation.
Purine cycle: see Glycine-Allantoin cycle.
Purine degradation, *purine catabolism:* a series of reactions in which purines are degraded by cleavage of the purine ring. P.d. is usually aerobic, but anaerobic P.d. occurs in certain microorganisms.

Aerobic P.d. The amino groups of adenine and guanine are removed hydrolytically by specific deaminases, which attack the free bases, the nucleosides or the nucleotides (Fig.). Uric acid is then produced by the action of xanthine oxidase (EC 1.2.3.2), which is the key enzyme of aerobic P.d. In humans and apes, the uric acid is excreted largely unchanged. In most reptiles and mammals, it is oxidized to allantoin by uricase (EC 1.7.3.3) (uricolysis).

In other organisms, including most fish and amphibians, allantoin is converted into allantoic acid, which is further degraded in two stages to yield 1 molecule of glyoxylic acid and 2 molecules of urea.

The inherited metabolic disease, xanthinuria, is caused by the absence of xanthine oxidase; xanthine and hypoxanthine are excreted instead of uric acid. Gout is caused by an increase in the rate of purine biosynthesis. The resulting increase in the concentration of blood uric acid leads to the deposition of crystalline uric acid in the joints.

Figure 2. *Biosynthesis of adenosine 5′-monophosphate and guanosine 5′-monophosphate from inosinic acid.*

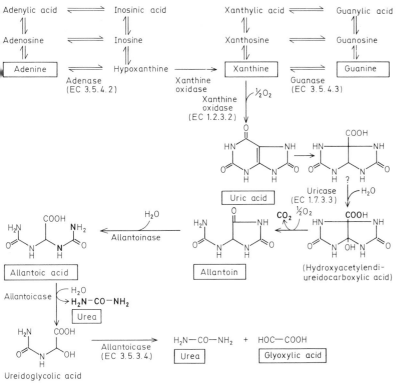

Aerobic purine degradation.

Anaerobic P.d., anaerobic xanthine degradation. In certain microorganisms, e.g. *Micrococcus* and *Clostridium,* the substrate of nonoxidative P.d. is xanthine. Hydrolysis between C6 and N1 of the 6-membered ring of xanthine produces ureidoimidazolylcarboxylic acid. Further hydrolytic removal of ammonia and CO_2 produces aminoimidazolecarboxylic acid. This is decarboxylated to aminoimidazole. The ring of aminoimidazole is opened by the simultaneous loss of ammonia and addition of 2 molecules of water, to produce formiminoglycine. The latter is hydrolysed to glycine, ammonia and formate.

Purine interconversion: see Purine biosynthesis.

Puromycin: a nucleoside antibiotic from *Streptomyces alboniger.* M_r 472. P. inhibits protein biosynthesis on 70S and 80S ribosomes. It is a structural analog of the 3'-terminal end of aminoacyl-tRNA (Fig.). It replaces the latter during the elongation phase of protein biosynthesis, and a peptide bond is formed between the free amino group of P. and the COOH group of the C-terminal amino acid of the preceding peptidyl-tRNA (see Fragment reaction). Further elongation of the polypeptide is prevented, and the polypeptide fragment with the attached P. becomes separated from the ribosome. Other aminoacylnucleoside antibiotics, e.g. gougerotin and blasticidin S, have a similar action mechanism.

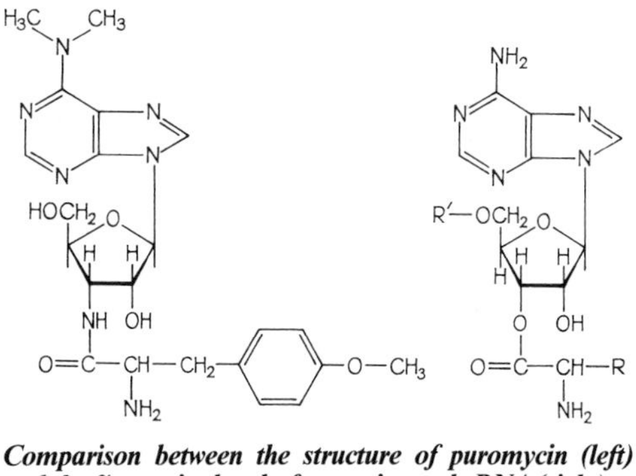

Comparison between the structure of puromycin (left) and the 3'-terminal end of an aminoacyl-tRNA (right).

Purpurin: 1,2,4-trihydroxyanthraquinone, a red anthraquinone pigment, m.p. 263 °C. The glycoside of P. occurs in madder root *(Rubia tinctorum)* (accompanied by alizarin), and in other members of the *Rubiaceae.* P. is formed from its glycoside during storage, and there is no appreciable quantity of P. in the fresh root. It is used as a reagent for the detection of boron, for the histological detection of insoluble calcium salts, and as a nuclear stain. It forms colored lakes with various metal salts, and is used as a fast dye in cotton printing.

503

Putrescine: tetramethylenediamine, an ubiquitous Polyamine (see). It is formed by decarboxylation of ornithine and, in some organisms, by decarboxylation of arginine to agmatine followed by cleavage of urea. It is the precursor of spermine and spermidine in ordinary metabolism, and is essential for cell division. It accumulates during bacterial degradation of arginine, and increased protein decomposition (e.g. in cholera) leads to the appearance of P. in urine and feces.

Pyr.: abb. for Pyroglutamic acid (see for other recommended abbreviations).

Pyranose: see Carbohydrates.

Pyrethrins: diterpene insecticides present in the flowers of *Chrysanthemum cinerariaefolium* (syn. *Pyrethrum cinerariaefolium*). The dried flowers, known as "pyrethrum", also have insecticidal activity and serve as starting material for the preparation of P.P. are esters of chrysanthemic acid (giving series I compounds) or pyrethric acid (giving series II compounds) with the alcohol pyrethrolone (giving pyrethrins), cinerolone (giving cinerins) or jasmololone (giving jasmolins) (Fig. and Table).

Pyridine alkaloids: a group of alkaloids containing the pyridine ring system, which occur in various unrelated plants, and as metabolic products of microorganisms. Important examples are Nicotiana alkaloids (see), Areca alkaloids (see), Gentiana alkaloids (see), and Valeriana alkaloids (see). They are biosynthesized either from nicotinic acid, or as products of terpene synthesis.

Pyridine nucleotide coenzymes: Nicotinamide-dinucleotide (see), and Nicotinamide-adenine-dinucleotide phosphate (see).

Pyridine nucleotide cycle: a salvage pathway in which nicotinamide produced by degradation of NAD$^+$ is reutilized to synthesize more NAD$^+$. The P.n.c. probably operates in all organisms, whether or not they are capable of synthesizing the pyridine ring system, and irrespective of the pathway of synthesis (from L-tryptophan in animals, *Neurospora* and *Xanthomonas pruni*; from aspartate and dihydroxyacetone phosphate in plants and most bacteria).

Pyridine nucleotide transhydrogenase: see Hydrogen metabolism.

Pyridoxal phosphate, *PalP*: the coenzyme form of Vitamin B$_6$ (see). M_r of PalP, 247.1.
PalP is stable in aqueous solution when kept refrigerated and protected from light. It is particularly sensitive to photodecomposition in the solid state and in alkaline solution. PalP plays an important

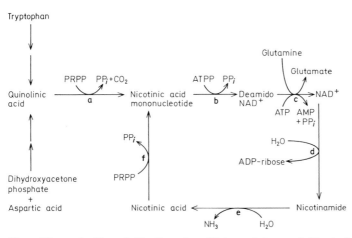

Structures of the naturally occuring pyrethrins.

R$_1$	R$_2$	Name
-CH$_3$	-CH$_3$	Cinerin I
-CH$_3$	-C$_2$H$_5$	Jasmolin I
-CH$_3$	-CH=CH$_2$	Pyrethrin I
-CO$_2$CH$_3$	-CH$_3$	Cinerin II
-CO$_2$CH$_3$	-C$_2$H$_5$	Jasmolin II
-CO$_2$CH$_3$	-CH=CH$_2$	Pyrethrin II

The pyridine nucleotide cycle. For formulae of intermediates, see L-Tryptophan; Nicotinamide adenine dinucleotide.

a Nicotinate-nucleotide pyrophosphorylase (carboxylating) (EC 2.4.2.19) (also called quinolinate transphosphoribosylase)
b Nicotinate mononucleotide adenylyltransferase (EC 2.7.7.18).
c NAD$^+$ synthetase (glutamine hydrolysing) (EC 6.3.5.1).

d NAD$^+$ nucleosidase (EC 3.2.2.5).
e Nicotinamidase (EC 3.5.1.19).
f Nicotinate phosphoribosyltransferase (EC 2.4.2.11).
PRPP = 5-phosphoribosyl 1-pyrophosphate, or 5-phospho-α-D-ribose 1-diphosphate.

central role in amino acid metabolism, acting as the coenzyme in many different metabolic conversions of amino acids. It is formed from pyridoxal by a kinase reaction (see Vitamin B_6 for formulae): pyridoxal + ATP $\xrightarrow{Mg^{2+}}$ pyridoxal 5-phosphate + ADP. With amines and amino acids, PalP forms Schiff's bases (azomethines). The substrate of a pyridoxal phosphate enzyme is the Schiff's base of the amino acid with PalP; the action specificity, i.e. transamination, decarboxylation, racemization etc. is determined by the apoenzyme. The initial reaction of PalP with an enzyme involves the formation of a Schiff's base between the ε-amino group of a lysine residue in the active center; an exchange process then results in the formation of a Schiff's base between the substrate amino acid and PalP, with release of the free ε-amino group of lysine (see Transamination). In the Schiff's base, the electrophilic form of the positive pyridine nitrogen favors the formation of a mesomeric structure, but this is only possible by removal of a substituent from the α-C atom as a cation, e.g. + CH_2OH in the L-serine hydroxymethyl transferase reaction, H^+ in racemization and transamination, or CO_2 in amino acid decarboxylation. In another group of reactions (cata-

lysed by L-serine dehydratase, L-cysteine desulfhydrase, L-tryptophan synthase, and tryptophanase) labilization of the α-hydrogen is followed by removal of the β-substituent (Fig.).

Pyridoxamine phosphate: see Transamination.
Pyridoxine: see Vitamin B_6.
Pyridoxol: see Vitamin B_6.
Pyrimidine: 1,3-diazine, a heterocyclic compound, consisting of a six-membered ring with 2 nitrogen atoms (Fig. 1). M_r 80.1, m.p. 20–22 °C, b.p. 124 °C. The P. ring system is present in many natural compounds, e.g. antibiotics (nucleoside antibiotics), pterins, purines and vitamins; it is especially important in the pyrimidine bases, Cytosine (see), Uracil (see) and Thymine (see), which are constitu-

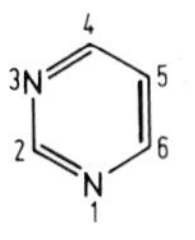

Figure 1. Numbering system of pyrimidine. An older system is based on that of the pyrimidine ring of Purine (see).

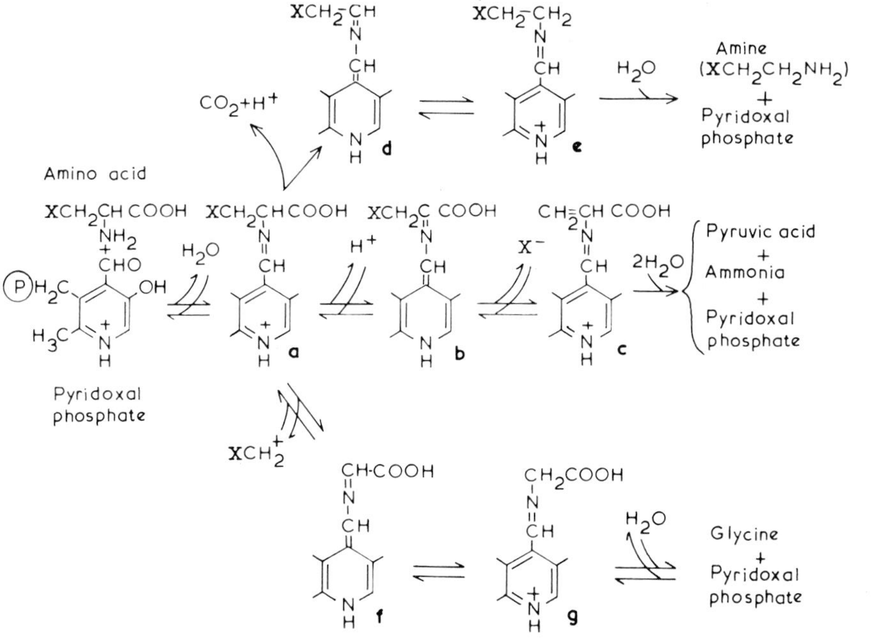

Coenzyme role of pyridoxal phosphate (PalP) in various reactions of amino acid metabolism.
In all reactions, the first stage is formation of Schiff's base *a* by condensation of PalP and the amino acid. Schiff's bases *a* and *b* represent part of *transamination,* but for the complete mechanism, see Transamination. *Racemization:* $a \to b$, followed by $b \to a \to$ amino acid + PalP, with addition of the proton in the opposite configuration. *Amino acid decarboxylation:* $a \to d \to c \to$ amine + PalP. *Serine hydroxymethyltransferase* (EC 2.1.2.1): X = OH; L-Serine + PalP $\to a \to f \to g \to$ glycine + PalP. Reversal of these reactions leads to L-serine synthesis from glycine. The hydroxymethyl group is carried by tetrahydrofolic acid. *Cysteine desulfhydrase* (EC 4.4.1.1): X = SH; Cysteine + PalP $\to a \to b \to c \to$ pyruvic acid + ammonia + PalP. *Serine dehydratase* (EC 4.2.1.13): X = OH; L-Serine + PalP $\to a \to b \to c \to$ pyruvic acid + ammonia + PalP. *Tryptophanase* (EC 4.1.99.1): X = indole; L-Tryptophan + PalP $\to a \to b \to c \to$ pyruvic acid + ammonia + PalP. *Tryptophan synthase* (EC 4.2.1.20): 1st stage: X = OH; L-serine + PalP $\to a \to b \to c$. 2nd. stage: X = indole; $c \to b \to a \to$ L-tryptophan + PalP.

ents of nucleic acids. Pyrimidine itself does not occur naturally. Pyrimidine analogs (see) can also become incorporated into nucleic acids.

The pyrimidine bases of nucleic acids possess an amino or hydroxyl group at position 6, and always an oxygen function at position 2. This gives rise to tautomeric structures, in which hydrogen is bound to oxygen or to ring nitrogen (shown for uracil in Fig. 2).

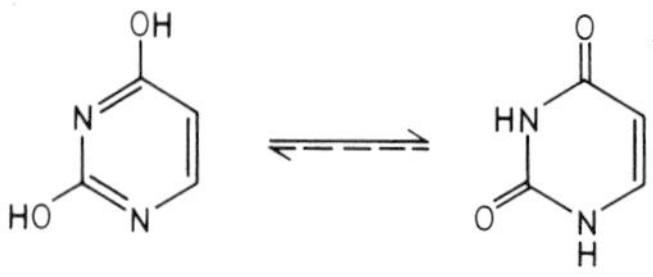

Lactim   Lactam

Figure 2. Tautomers of uracil.

Pyrimidine analogs, *antipyrimidines:* pyrimidines and pyrimidine nucleosides structurally related to, but different from the natural compounds. They therefore act as antimetabolites and selectively inhibit certain biochemical pathways, especially nucleic acid synthesis. Most P.a. are modified bases or their nucleosides, but there are also pyrimidine nucleoside analogs with modified sugar components. The most common chemical modifications are the introduction of substituents (e.g. halogens on C5 of uracil and cytosine), replacement of an OH with an SH group (e.g. 2-thiouracil), and replacement of a ring carbon with nitrogen (5-azauracil). Arabinonucleosides (see) (steric inversion of the OH at C2 of the ribose) and Xylosylnucleosides (see) (inversion at C3) of natural pyrimidine bases are also active P.a. Incorporation of 5-bromouracil (an analog of thymine) into DNA was first reported in 1952. The 5-fluorouracil compounds were developed in 1957; 5-fluorouracil is incorporated into RNA in place of uracil.

[*Antimetabolites of Nucleic Acid Metabolism.* The Biochemical Basis of their Action with Special Reference to their Application in Cancer Therapy. by Peter Langen. 1975. Gordon and Breach (London, New York, Paris)]

Pyrimidine antibiotics: structurally modified pyrimidine derivatives with antibiotic activity. They occur as nucleosides (see Nucleoside antibiotics), polypeptides (e.g. albomycin and grisein), or free bases (e.g. bacimethrin). The P.a., Toxoflavin (see) and Fervenulin (see), are biosynthesizied from purines.

Pyrimidine bases: see Pyrimidine.

Pyrimidine biosynthesis, *de novo pyrimidine biosynthesis:* total synthesis of the pyrimidine ring of uracil, thymine, cytosine and their derivatives from carbamoyl phosphate and aspartate in all living cells. (The pyrimidine ring of thiamin [vitamin B₁] has a different biosynthetic origin; see below).

Biosynthesis of uridine and cytidine nucleotides. This is shown in Fig. 1. The first pyrimidine nucleotide to appear de novo in this pathway is uridine 5'-monophosphate (UMP; uridylic acid). This is phosphorylated to uridine 5'-triphosphate (UTP). By donation of an amino group from ammonia or glutamine (in

animal tissues), UTP is converted into cytidine 5'-triphosphate (CTP).

Biosynthesis of thymine nucleotides. This is shown in Fig. 2. Since thymine is a constituent of DNA, the corresponding nucleotides contain 2-deoxyribose. Thymidylic acid (TMP) is therefore more correctly dTMP (deoxythymidine 5'-monophosphate). The reaction sequence is: CMP→CDP→dCDP→ dCMP→dUMP→TMP (dTMP)→TDP (dTDP)→ TTP (dTTP). Methylation of dUMP to TMP is catalysed by thymidylate synthase (EC 2.1.1.45). The cofactor, N^5,N^{10}-methylenetetrahydrofolic acid, serves to transfer the active C1 unit to C5 of dUMP, and it also functions as a reducing agent in the formation of the methyl group from the active C1 unit.

Biosynthesis of ribothymidylic acid. The unit occurs as a minor component (see Rare nucleic acid components) in many species of tRNA. It is formed by the methylation (from *S*-adenosyl-L-methionine) of C5 of uracil in the existing nucleic acid molecule.

Biosynthesis of 5-hydroxymethyldeoxycytidylic acid. This component of the DNA of T-even phages is biosynthesized from deoxycytidine 5'-monophosphate (Fig. 3).

Pyrimidine nucleotides may also be produced by the Salvage pathway (see).

Regulation of P.b. In *Escherichia coli,* carbamoyl phosphate synthetase is activated by the purine nucleotides, IMP and XMP, and it is inhibited by the pyrimidine nucleotides, UMP and UDP.

The key control point is the synthesis of *N*-carbamoylaspartic acid, catalysed by aspartate carbamoyltransferase (aspartate transcarbamylase, EC 2.1.3.2). In *Escherichia coli* and *Aerobacter aerogens* this enzyme is inhibited by CTP, and the inhibition is prevented by ATP. In *Pseudomonas fluorescens,* the enzyme is inhibited by UTP, whereas in higher plants the regulatory inhibitor is UMP.

Aspartate carbamoyltransferase from *Escherichia coli* is one of the most thoroughly studied allosteric enzymes. It has M_r 310000, and can be dissociated into two identical *catalytic* subunits (each of M_r 100000, and containing three polypeptide chains, called C-chains, of M_r 34000), and three identical *regulatory* subunits. Each regulatory subunit contains two R-chains, M_r 17000; each R-chain binds one molecule of CTP. The catalytic subunits are active in the absence of the regulatory subunits, but regulation by CTP only occurs in the complete oligomeric enzyme. The control mechanism is explained by the Cooperativity model (see) of allosteric enzymes.

In some cases, uracil may also repress the synthesis of aspartate carbamoyltransferase and dihydroorotate oxidase (EC 1.3.3.1).

Biosynthesis of the pyrimidine ring of thiamin (vitamin B₁) from aminoimidazoleribonucleotide. The 2-methyl-4-amino-5-hydroxymethyl-pyrimidine ring present in thiamin is synthesized from aminoimidazoleribonucleotide, which is an intermediate in purine biosynthesis (Fig. 4).

Pyrimidine degradation, *pyrimidine catabolism:* reductive or (in special cases) oxidative reactions leading to the cleavage of the heterocyclic ring of natural pyrimidines.

1. *Reductive P.d.* (Fig.). To a certain extent, this process represents a reversal of Pyrimidine biosyn-

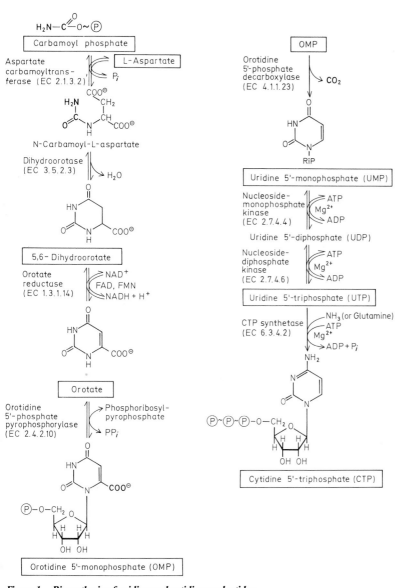

Figure 1. Biosynthesis of uridine and cytidine nucleotides.

thesis (see). The pyrimidine ring is partially hydrogenated, and the resulting dihydro-compound is
cleaved hydrolytically. Cytosine is converted to uracil by deamination, and uracil is degraded to β-alanine. Thymine is degraded to β-aminoisobutyrate.
These endproducts are transaminated and metabolized to common metabolic intermediates (Fig.).
2. *Oxidative P.d.* In *Corynebacterium* and *Mycobacterium.* Uracil is oxidized to barbituric acid, which is
cleaved hydrolytically to urea and malonic acid.
Thymine is oxidized to 5-methylbarbituric acid, followed by hydrolysis to urea and methylmalonic acid.

Pyrimidine dimers: see Dimers.

Pyroglutamic acid, *pyrrolidone carboxylic acid,
5-oxoproline, pyrrolid-2-one-5-carboxylic acid,* abb.
Pyr, PCA, < Glu, ⌐*G:* an internal cyclic lactam of
glutamic acid, representing a condensation of the α-
amino with the γ-carboxyl group. Pyroglutamic acid
occurs as the *N*-terminal residue of some proteins. It
was first reported as an *N*-terminus in the heavy
chain of rabbit IgG. [Wilkinson et al., *Biochem J.
100* (1966) 303–308.] The abbreviation PCA was
used in this early work, but the IUPAC–IUB Commission on Biochemical Nomenclature (Recommendations, 1971) recommend the use of < Glu or ⌐G.

507

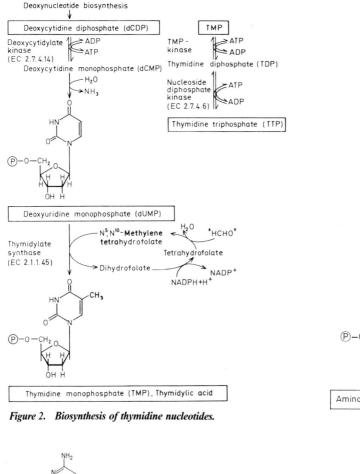

Figure 2. Biosynthesis of thymidine nucleotides.

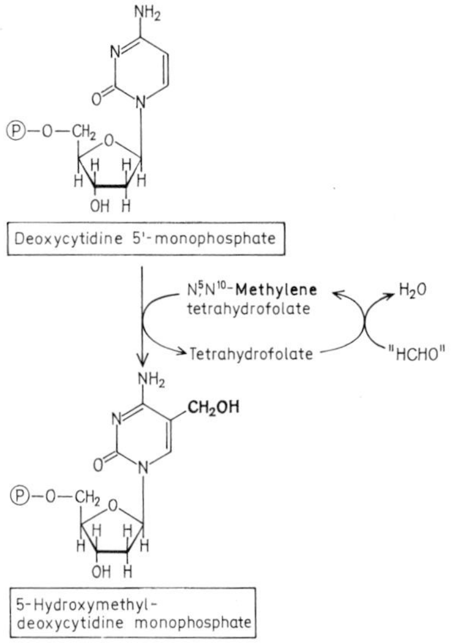

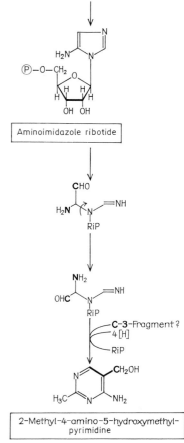

2-Methyl-4-amino-5-hydroxymethyl-
pyrimidine

*Figure 4. Synthesis of 2-methyl-4-amino-5-hydroxyme-
thyl-pyrimidine from aminoimidazole ribotide (RiP=
ribose 5-phosphate).*

*Figure 3. Biosynthesis of 5-hydroxymethyl-deoxycyti-
dylic acid.*

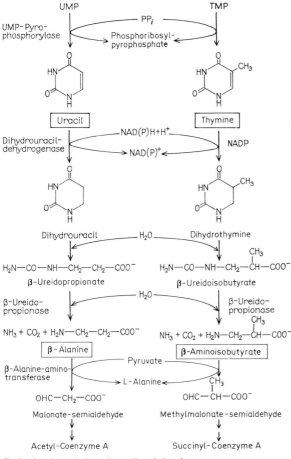

Reductive degradation of uracil and thymine.

Pyr is frequently used as a nonstandard abbreviation.

Pyrroles, *pyrrole derivatives:* compounds containing the pyrrole ring. They are subdivided into mono, di, tri and tetrapyrroles. The tetrapyrroles may be noncyclic or cyclic. Bile pigments (see) and the chromophores of Biliproteins (see) are linear, while Porphyrins (see) and Corrinoids (see) are cyclic tetrapyrroles.

Pyrrolidine-2-carboxylic acid: see L-Proline.

Pyrrolidine alkaloids: a group of Alkaloids (see) with simple structures. P.a. are either derivatives of proline (e.g. stachydrin and its diastereoisomer, betonicin), or they are derived from a *N*-methyl-2-alkylpyrrolidine (e.g. hygrin and cuskhydrin). The latter occur together with the tropane

alkaloids, with which they share the same biogenetic precursors, ornithine and acetate.

Pyrrolid-2-one-5-carboxylic acid: see Pyroglutamic acid.

Pyrrolizidine alkaloids, *Senecio alkaloids:* a group of ester alkaloids, in which amino alcohols (necines) are esterified with necic acids. The necines are derivatives of the pyrrolizidine ring system (also known as 1-azabicyclo [0,3,3] octane) (see Alkaloids, Table) and they possess one or two alcoholic hydroxyls, e.g. retronecine (Fig.).

The necic acids (esterified with the hydroxyls of the necines) are branched aliphatic, mono or dibasic acids, containing 5–10 carbon atoms, e.g. angelic, monocrotalic, senecic and tiglic acids. The various combinations of necines and necic acids, together

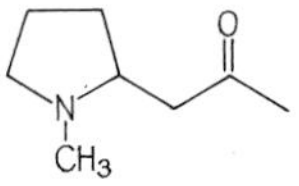

Hygrin

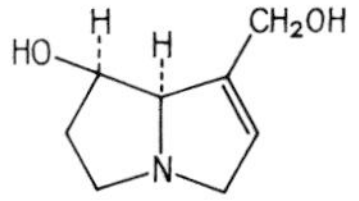

Retronecine

with amine oxides and isomers gives rise to a very large number of P.a. Free necines also occur naturally. P.a. occur chiefly in species of *Senecio*, the largest genus of the *Compositae*.They are hepatotoxic, and can cause liver cirrhosis in grazing animals.

Pyrrolo quinoline quinone, *PQQ:* 2,7,9-tricarboxy-1*H*-pyrrolo [2,3-*f*] quinoline-4,5-dione, the cofactor of the enzyme methanol dehydrogenase (EC 1.1.99.8) from *Hyphomicrobium X* and *Methylophilius methylotrophus,* and of glucose dehydrogenase (EC 1.1.99.17) from *Acinebacter calcoaceticus.* [Duine, J.A., Frank, J., and Verwiel, P.E.J. (1980) *Eur. J. Biochem.* **108,** 187–192].

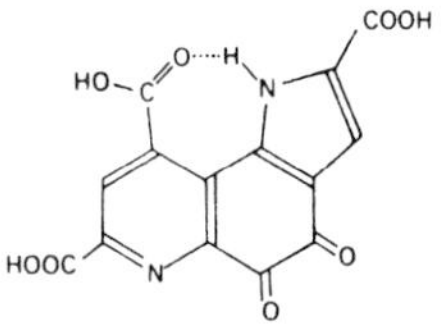

Pyrrolo quinoline quinone

Pyruvate: the anion of pyruvic acid. P. is an important metabolic intermediate in aerobic and anaerobic metabolism (Fig.).

P.synthesis: 1. P. is synthesized from phospho*enol*pyruvate in Glycolysis (see). Phospho*enol*pyruvate is an enol ester and an energy-rich compound with a free energy of hydrolysis of 50.24 kJ (12 kcal) per mol. During catalysis by pyruvate kinase, this free energy is exploited for the transfer of the phosphate group to ADP, resulting in the synthesis of ATP and P.

2. P. is produced in the metabolism of certain amino acids, in particular transamination of alanine, dehydration of serine, and desulfhydration of cysteine.

P.metabolism: 1. P. is reduced to lactate in anaerobic glycolysis.

2. It is converted to ethanol in anaerobic Alcoholic fermentation (see).

3. By the action of the pyruvate dehydrogenase complex, under aerobic conditions, P. is oxidatively decarboxylated to acetyl coenzyme A. The latter is an important metabolite in various other biosynthetic and biodegradative processes. Equation for oxidative decarboxylation of P.:

$$CH_3COCOO^- + HSCoA + NAD^+ \rightarrow$$
(Pyruvate)   (Coenzyme A)
$$CH_3CO\text{-}SCoA + CO_2 + NADH + H^+$$
(Acetyl coenzyme A)

Complete oxidation of one molecule of P. via the TCA-cycle results in the production of 15 molecules of ATP (14 from the respiratory chain + 1 from the substrate level phosphorylation in the conversion of succinyl-CoA to succinate).

4. P. is carboxylated to oxaloacetate (see Carboxylation), representing the first stage in Gluconeogenesis (see).

5. During Nitrogen fixation (see) in *Clostridium,* P. undergoes phosphoroclastic fission into acetyl phosphate and CO_2.

Pyruvate carboxylase (EC 6.4.1.1): a biotin-dependent ligase, in animals and plants, which catalyses the addition of CO_2 to pyruvate:

Pyruvate + CO_2 + ATP + $H_2O \xrightarrow{Mn^{2+}}$ Oxaloacetate + ADP + P_r The enzyme is practically inactive in the absence of its positive allosteric effector, acetyl-CoA. This reaction is an important early stage of Gluconeogenesis (see), and is an example of CO_2 fixation in the animal organism. For the mode of attachment of the coenzyme, biotin, and the mechanism of CO_2 transfer, see Biotin enzymes. The active form of P.c. is a tetramer, M_r 600000 (yeast), 650000 (chicken liver), or 520000 (porcine liver), which is in equilibrium with the corresponding dimers and monomers. The dimers and monomers (M_r 130000, 3 chains, each of M_r 47000) of the porcine enzyme are also enzymatically active. Avian P.c. is cold sensitive, and reversibly dissociates into inactive monomers (M_r 160000) at 0 °C. Like all biotin enzymes,

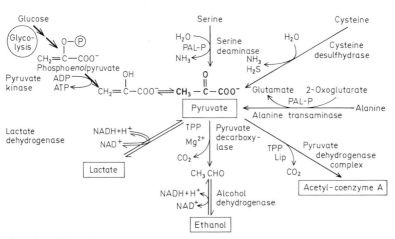

Central position of pyruvate in metabolism. TPP Thiamin pyrophosphate, PAL-P pyridoxal phosphate, Lip lipoamide.

P. c. is inactivated by avidin, due the binding of the coenzyme as an avidin (biotin)$_4$-complex. On the basis of their structural homologies, P. c. and acetyl-coenzyme-A-carboxylase are thought to have evolved from a common enzyme.

Pyruvate decarboxylase, *carboxylase* (EC 4.1.1.1): a thiamin pyrophosphate (TPP)-dependent lyase, absent from animals, and present in high activity in yeast and wheat seedlings. P. d. is a specific enzyme of alcoholic fermentation, which catalyses the cleavage of pyruvate (via active acetaldehyde) into acetaldehyde and CO_2. The cofactors are Thiamin pyrophosphate (see) and magnesium ions. In the plant cell, P. d. competes with the pyruvate dehydrogenase complex for pyruvate. M_r of P. d. from yeast and *Escherichia coli* is 190 000 (two indentical subunits, M_r 95 000).

Pyruvate dehydrogenase: a Multienzyme complex (see) which is responsible for the formation of acetyl-CoA from pyruvate, one of the central metabolic reactions (see Pyruvate, Acetyl-coenzyme A). It is subject to three types of control: 1. The enzyme complex is inhibited by acetyl-CoA and NADH; the transacetylase is inhibited by acetyl-CoA, and NADH inhibits the dihydrolipoyl dehydrogenase. These inhibitions are reversed by CoA and NAD$^+$, respectively. 2. Enzyme activity is influenced by the energy state of the cell; the complex is inhibited by GTP and activated by AMP. 3. The complex is inhibited when a specific serine residue in the pyruvate decarboxylase is phosphorylated by ATP. This phosphorylation is inhibited by pyruvate and ADP. The complex is reactivated by removal of the phosphoryl group by a specific phosphatase.

Pyruvate kinase, *phosphopyruvate kinase* (EC 2.7.1.40): a widely distributed, metal-ion dependent phosphotransferase, present in yeast, muscle, liver, erythrocytes and other organs and cells. It catalyses the last reaction of glycolysis: Phospho*enol*pyruvate (PEP) + ADP; ⟶ Pyruvate + ATP (substrate level phosphorylation). Each subunit of P. k. forms an intermediate, cyclic, ternary metal bridge complex

$$P. k. - Mn - ADP$$
$$\diagdown PEP \diagup$$

in which the PEP and ADP are bound to the enzyme via a manganese (II) ion. Tetrameric P. k. from muscle and erythrocytes (M_r 230 000) shows Michaelis-Menton type kinetics (plot of initial velocity against substrate concentration is a rectangular hyperbola), whereas the yeast enzyme (M_r 190 000, 4 or 8 subunits) is an allosteric enzyme, showing sigmoid kinetics. The nucleotide base sequence of the P. k. gene from the chicken and a discussion of the evolutionary implications of the gene structure has been published by N. Lonberg & W. Gilbert [*Cell 40* (1985) 81-90].

Pyruvate phosphate dikinase: see Hatch-Slack-Kortschak cycle.

Pyruvic acid: $CH_3-CO-COOH$, the simplest and most important α-ketoacid (2-oxacid), m. p. 11.8 °C, b. p. 165 °C (d.). For the role of P. a. in metabolism, see Pyruvate.

Pythocholic acid: 3α,7α,16α-trihydroxy-5β-cholan-24-oic acid, a bile acid possessing three hydroxyl groups. M_r 408.58, m. p. 187 °C. P. a. is a characteristic component of the bile of many snakes, and has been isolated from the bile of the tiger snake, python and boa-constrictor, among others.

Q

Q: abb. for coenzyme Q. See Ubiquinone.

Quantasome: the smallest structural unit of photosynthesis; small elementary units of the thylakoid, measuring $18 \times 15 \times 10$ nm, M_r 2 million, containing 230 chlorophyll molecules, cytochromes, copper and iron. Q. are obtained by ultrasonic disintegration of isolated chloroplasts, and they can be visualized in the electron microscope. They may also be observed as granular units in the chloroplast lamella. The functional status of Q. is not clearly defined; they may be involved in both electron transport (see Photosynthesis) and photosynthetic ATP synthesis (see Photophosphorylation), and therefore analogous to the electron transport particles of the respiratory chain.

Quantum efficiency, *quantum yield:* see Quantum requirement.

Quantum requirement: the number of light quanta required for the formation of one molecule of O_2 in Photosynthesis (see). Two quanta are required per electron. The theoretical value of Q.r. is eight, since the production of one molecule of O_2 proceeds according to the following equation, with transfer of four electrons from water (see Photolysis of water) to NADP$^+$.

$2 H_2O \rightarrow O_2 + 4H^+ + 4e^-$.

The experimentally determined value of Q.r. lies between 8 and 10 for leaves, and between 10 and 14 for isolated chloroplasts. It is influenced by the physiological state of the experimental system. The inverse of Q.r. is the quantum yield (or quantum efficiency). The quantum yield therefore represents the number of molecules transformed (i.e. CO_2 molecules liberated) per quantum of light absorbed.

Quantum yield, *quantum efficiency:* see Quantum requirement.

Quaternary structure: see Proteins.

Queen substance: originally a term for the entire mandibular gland secretion of the queen bee, which contains about 30 substances. It is now the trivial name for 9-oxo-*trans*-2-decenoic acid. This compound, together with 9-hydroxy-*trans*-2-decenoic acid, is very important as a pheromone for the maintenance of the division of labor within the hive. In the course of caring for the young, the worker bees lick the pheromone mixture off the queen. This causes their ovaries to shrink, and they are inhibited from building queen cells. Larvae in queen cells are not fed honey, but royal jelly, a mixture of pollen and secretions. Royal jelly does not contain Q. It is recommended as a health product, but its effectiveness is disputed.

Quebrachitol: see Cyclitols.

Quercetin: see Flavones (Table).

Quercitrin: see Flavones (Table).

Quinazoline alkaloids: a group of about 30 alkaloids, which occur in higher plants (in families which are taxonomically very distant from one another), animals and bacteria. They are derived biosynthetically from anthranilic acid. The simplest representative of the Q. is Glomerine (see) which is a very rare animal Q. Of the plant Q., Febrifugine (see) has some significance. In the wide sense, tetrodotoxin from the puffer fish can be included among the Q., as it is a zwitterionic polyhydroxy-2-iminoperhydroquinazoline.

Quinidine: see Quinine.

Quinine: the most important of the cinchona alkaloids. M_r 324.21, m.p. 57 °C (trihydrate), 174 to 175 °C for dehydrated crystals. $[\alpha]_D^{17} - 284.5$ (0.05 M H_2SO_4). In Q., a quinoline ring system is connected via a secondary hydroxyl on C4 to a quinuclidine structure (see Cinchona alkaloids, Fig.). Q. occurs in nature in association with its stereoisomers quinidine, the C-9 epimer, m.p. 172.5 °C, $[\alpha]_D^{15} = +334.1$ (0.05 M H_2SO_4), and epiquinine and epiquinidine, the C 8′ epimers. Q. forms bitter-tasting salts and has many physiological effects. It is used therapeutically as a drug against malaria and bacterial influenza. By reducing the rate of tissue respiration, it has an antipyretic effect.
It also acts as an analgesic and inhibits heart excitation, although quinidine surpasses it in this last respect. Q. is toxic, leading to deafness and blindness, and in quantities of 10 g it is fatal.

Quinoline alkaloids: a group of alkaloids based on the quinoline skeleton. They are found both in microorganisms (see Viridicatine) and in higher plants. The most important therapeutically are the Cinchona alkaloids (see). The starting material for the biosynthesis of some Q. is anthranilic acid (see Viridicatine); for others it is tryptophan (see Cinchona alkaloids).

Quinolizidine alkaloids: a group of alkaloids based on the quinolizidine (norlupinane) skeleton. The most important Q. are the Lupin alkaloids (see), which are synthesized from lysine via cadaverine. The Nuphara alkaloids (see), in contrast, also possess a Q. ring system, but are synthesized by the terpene pathway.

Quinones: aromatic dioxo compounds derived from benzene or multiple-ring hydrocarbons such as naphthalene, anthracene, etc. They are classified as Benzoquinones (see), Naphthoquinones (see) An-

R = O   Queen substance

R = H, OH   9 - Hydroxy - *trans* - 2 - decenoic acid

thraquinones (see), etc. on the basis of the ring system. The $C=O$ groups are generally ortho or para, and form a conjugated system with at least two $C=C$ double bonds; hence the compounds are colored, yellow, orange or red. This type of chromophore is found in many natural and synthetic pigments.

The Q. are a large and varied group of natural products found in all major groups of organisms. Those with long isoprenoid side chains, such as plastoquinone, ubiquinone and phytoquinone are involved in the basic life processes of photosynthesis and respiration. Q. are biosynthesized from acetate/malonate via shikimic acid. A few Q. are used as laxatives and worming agents, and others are used as pigments in cosmetics, histology and aquarell paints.

R

Radioimmunoassay: see Immunoassays.

Raffinose, *melitose:* a nonreducing trisaccharide. M.p. 120 °C, $[\alpha]_D^{20} + 123$ °C (water). R.contains units of D-galactose, D-glucose and D-fructose. The galactose and glucose are linked by an α-1,6-glycosidic bond, and the fructose is linked to the glucose by an α, β-1,2-glycosidic bond. R.is easily fermented by yeasts. Yeast enzymes hydrolyse R.into D-fructose and melibiose; emulsin hydrolyses R.into sucrose and D-galactose. In plant metabolism, R.may function in place of sucrose as a transport carbohydrate. R.is found widely in many higher plants, where it is the second (the first is sucrose) most commonly occurring free sugar. Sugar beet, molasses and many seeds, e.g. cotton seeds are especially rich in R.

Ramachandran plots, *conformational maps:* plots of rotation about the αC-carbonyl-C bond (ψ) in a peptide linkage against rotation about the αC-amino-N bond (Φ). A general R.p. is constructed with the aid of models and computers. Using the accepted atomic radii of C, N, O and H, possible combinations of the two angles (i.e. regions of no steric hindrance) are indicated by blocked out areas on the graph or map. With only a small decrease in contact distance, these permissible regions become larger and a new permissible region appears. The permissible conformations include antiparallel β-pleated sheets, parallel β-pleated sheets, polyproline helix, collagen supercoil, right and left handed α-helices, right handed ω-helix, 3_{10} threefold helix, and the π-helix (4.4 residues per turn). An R.p. for a given protein can be constructed from Ψ and Φ values determined experimentally by X-ray diffraction and model building.
[Ramachandran, G.N. in *Aspects of Protein Structure,* Academic Press (1963) p.39]

Randainol: see Neolignans.

Random coil: see Proteins.

Rapanone: see Benzoquinones.

Raphanatin: 7-glucosylzeatin. R. is formed from the cytokinin, zeatin, and it has no cytokinin activity. It is a storage form of zeatin, present in radish seedlings. Glucosylation at position 7 of the purine ring probably serves to protect zeatin from enzymatic degradation.

Rapoport-Luerbing shuttle: part of the Embden-Meyerhof pathway of glycolysis, constituting a self-regulating system which maintains the concentrations of 2,3-diphosphoglycerate and ATP at the expense of each other in the erythrocyte. A decrease in pH increases the activity of 2,3-diphosphoglycerate phosphatase, so that the concentration of 2,3-diphosphoglycerate decreases. The activity of diphosphoglycerate mutase therefore increases due to relief of the inhibition by 2,3-diphosphoglycerate and ex-

tra provision of its cofactor, 3-phosphoglycerate. The net result is an increased flux through 2,3-diphosphoglycerate in the conversion of 1,3-diphosphoglycerate to 3-phosphoglycerate, and less conversion of ADP to ATP by phosphoglycerate kinase (Fig.). Hypoxia increases the 2,3-diphosphoglycerate concentration by increasing pH, and at the same time an increased amount of the ester is bound to deoxyhemoglobin, thereby decreasing the pool of the free ester. See Hemoglobin; Glycolysis. [E.Gerlach & J.Duhm, *Scand. J.Clin. Lab. Invest.* **29** (1972) Suppl.126 5.4a-5.4h].

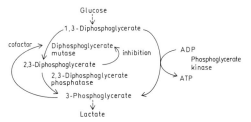

Rapoport-Luerbing shuttle of the erythrocyte

Rare nucleic acid components, *unusual nucleic acid components, minor nucleic acid components:* nucleic acid components of relatively infrequent occurrence, formed by the enzyme-catalysed modification of either the base or sugar of the usual nucleic acid constituents, i.e. modification of adenine, guanine, cytosine, uracil, thymine or ribose. With the exception of 5-hydroxymethyldeoxycytidylic acid (see Pyrimidine biosynthesis), all R.n.c. are formed by modification of residues in the intact polynucleotide chain of the nucleic acid. Modified nucleic acid bases are also called *minor* bases. Enzymic derivatization of free purines and pyrimidines does not occur, except in the formation of purine alkaloids in plants (see Methylxanthines).
The chief types of modification are acetylation with acetyl-CoA (formation of N^4-acetylcytidine and 5-acetyluridine), glucosylation with UDPG (glucosylation of 5-hydroxymethyl-cytidine to 5-glucosylhydroxymethylcytosine), isoprenylation with γ, γ-dimethylallylpyrophosphate (conversion of adenosine to N^6-isopentenyladenosine), reduction (uridine to 5,6-dihydrouridine), thiolation with cysteine (formation of 2-thiouridine), cleavage of a C-C bond (conversion of uridine to pseudouridine), and methylation, which is particularly common. During methylation, methyl groups are transferred from S-adenosyl-L-methionine (SAM) to C, N, or O-atoms on bases (e.g. 5-methyluridine) or sugar moieties (e.g. 2-O-methyluridine). Other important R.n.c. are inosine and ribothymidine. About 40 R.n.c. are known. The

515

enzyme-catalysed modifications are species-specific. Different types of nucleic acids differ significantly with respect to their contents of R.n.c. Transfer RNA contains an especially large number of R.n.c.; they occur in specific positions, largely in the single-stranded regions, e.g. within and directly adjoining the anticodon loop. DNA and rRNA contain methylated nucleotides. Prokaryotic and phage mRNA appear to contain no methylated residues, i.e. if present their concentration is lower than the detection limit (less than 1 in 3500 residues). Eukaryotic and viral mRNA contain unique 5-terminal cap structures with methylated residues, and some internal 6-methyladenylic acid residues (see mRNA).
In the methylation of DNA, specific methylases catalyse the transfer of methyl groups from SAM to the 6-amino groups of adenine residues and C5 of cytosine. A specific pattern of methylation serves to protect DNA from the cell's own Restriction endonucleases (see); these enzymes destroy the DNA of invading viruses. The DNA of viruses that are able to replicate within a particular host are protected from the host endonucleases by methylation of bases at the endonuclease-sensitive sites.
Carcinogenic alkylating agents alkylate chiefly guanine residues on the N 7 atom. Some of the R.n.c. in tRNA are active cytokinins in the free state, e.g. N^6-(γ, γ-dimethylallyl)-adenosine.

Rate equation: in Enzyme kinetics (see), an equation expressing the rate of a reaction in terms of rate constants and the concentrations of enzyme species, substrate and product. When it is assumed that steady state conditions obtain, the Michaelis-Menten equation (see) is a suitable approximation. R. are represented graphically (see Enzyme graph); they may be derived by the King-Altman method (see).

Rauwolfia alkaloids: a group of about 50 structurally related indole alkaloids from the roots and rhizomes of various species of *Rauwolfia, Aspidosperma* and *Corynanthe.* All R.a. contain a β-carbolene skeleton; they are classified into 3 types: 1. yohimbine (corynanthine), 2. ajmaline, 3. serpentine. The large number of R.a. is due to the existence of stereoisomers. Thus *Rauwolfia* contains seven stereoisomers of yohimbine.
Some R.a. have valuable pharmacological properties. They may act centrally (e.g. see Reserpine) or peripherally (e.g. see Yohimbine; Ajmaline). In addition to the pure alkaloids or their synthetic analogs, extracts of the drug Radix Rauwolfiae and combination preparations are also used. The drug has been known since early times in Indian folk medicine, and its systematic investigation began in 1930.

Yohimbine   *Ajmaline*

Reagin: see Immunoglobulins.
Receptor: see Hormones.
Receptor proteins: mostly membrane-bound, but also soluble proteins with high specific affinity for hormones, antibodies, enzymes and other biologically active compounds. Binding of proteohormones to membrane-associated R.p. represents the first stage in the expression of hormonal activity; this is followed by activation of the membrane-bound adenylate cyclase, which catalyses formation of the second messenger, cyclic AMP. Detailed studies have been reported on the membrane R.p. for acetyl choline in the electric organ of the electric eel, and in skeletal muscle; on the insulin receptor of fat cells (adipocytes); on the adenylate cyclase receptor of adrenergic neurotransmitters (chemical transmitters at vegetative nerve endings); and on the R.p. for *colicin M* (M_r 85000) in the membrane of *Escherichia coli.* The R.p. for steroid hormones are soluble and present in the nucleus. R.p. for estrogens, cortisol and progesterone exist as interconvertible dimers (8 S) or monomers (4 S), depending on the salt concentration. Interaction with the specific steroid apparently changes the conformation of the R.p., so that it interacts with certain regions of DNA and influences the transcription of specific genes, thus resulting in the expression of hormonal activity. R.p. exist in extremely small quantities, but their isolation is possible by the use of affinity chromatography.

Recombinant DNA technology, *genetic engineering:* the isolation and study of single genes, and the reintroduction of these genes into cells of the same or different species. *Gene cloning* is essential to R.D.t.
If a single gene is cloned in sufficient quantity, it can be sequenced (see Deoxyribonucleic acid) and the amino acid sequence of the gene product can be predicted from the nucleotide sequence of the DNA. For example, the amino acid sequences of the insulin receptor, nicotinic and muscarinic cholinergic acetylcholine receptors, epidermal growth factor precursor and many other proteins have been predicted in this way.
Factors controlling gene expression can be studied by gene cloning. For example, eukaryotic genes for tyrosine aminotransferase (TAT), together with flanking nucleotides, including initiation and promotion elements, can be incorporated into a plasmid adjacent to, and upstream of, the bacterial gene for chloramphenicol acetyltransferase (CAT). After initiation, transcription proceeds through the structural TAT gene and continues through the adjacent CAT gene. Thus, factors affecting TAT transcription exert a similar influence on transcription of the CAT gene, and they can be studied by measurement of the CAT catalytic activity of the bacterial host. With the aid of this system, a glucocorticoid control region of the TAT gene has been identified and shown to consist of two glucocorticoid receptor binding sites.
The ultimate goal of the genetic engineer is the expression of purified genes in commercially exploitable, rapidly growing species. Human insulin, ovalbumin, fibroblast interferon, somatostatin, human growth hormone and other eukaryotic proteins can

all be produced by expression of their cloned genes in *Escherichia coli.* Stages in the isolation and cloning of a gene are outlined in Fig. 1, and the strategy for cloning human growth hormone is shown in Fig. 2.

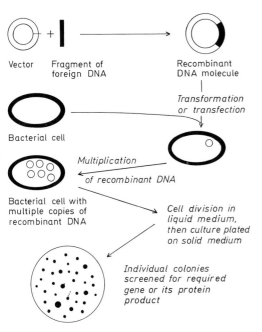

Figure 1. Essential stages in gene cloning using a bacterial plasmid vector.

R.D.t, together with hybridization and blotting techniques, can be used to screen for a variety of genetic defects (see Hemoglobinopathy, Inborn errors of metabolism). Fetal cells, obtained by amniocentesis of the amniotic fluid during the 16th week of pregnancy, are cultured and their DNA is analysed. Mutation often destroys a restriction site or creates a new one, so that the pattern of nucleotides obtained after restriction analysis may change, thereby revealing the existence of the mutation. Thus, the gene for normal hemoglobin contains GAG for the glutamate residue at position 6 of the β-chain. This forms part of a restriction site for Dde I (which attacks $\frac{CTNAG}{GANTC}$) and for Mst II (which attacks $\frac{CCTNAGG}{GGANTCC}$). The mutation responsible for sickle cell anemia (see Sickle cell hemoglobin) eliminates this restriction site by converting GAG to GTG, resulting in valine at position 6 of the β-chain. The gene for sickle cell anemia can therefore be detected early in fetal development by digesting embryo-derived DNA with Dde I or Mst II and performing a Southern blot (see), using cloned normal β-globin DNA as the hybridization probe.

DNA vectors. These may be Plasmids (see), viruses (see Phage, Phage development) or Cosmids (see).

Plasmids used in gene cloning usually carry one or more genes which confer characteristics on their host cells, enabling them to be distinguished from cells not carrying the plasmid (i.e. selection markers). For example, antibiotic resistance of a bacterium is often due to the presence of a plasmid carrying the genes for the resistance. All plasmids carry at least one replication initiation site, so that they can multiply independently of the host chromosome. Most plasmids are bacterial. Some strains of *Saccharomyces cerevisiae* contain a 2 μm circular plasmid, which has facilitated the construction of cloning vectors for this industrially important organism. Plasmids have not been found in other eukaryotes. Cassette vectors are the most advanced generation of vectors, which carry all the signals needed for gene expression (promoter, terminator, ribosome binding site) in the form of a cassette. The foreign gene is inserted into a unique restriction site in the cluster of expression signals. To construct a cassette vector, an entire *Escherichia coli* gene with its expression signals is inserted into a vector. The reading frame is then removed, leaving the expression signals intact. Initial difficulties in the construction of cassette vectors are now overcome by separate oligonucleotide synthesis of promoter, terminator and ribosome binding signals, which are ligated to form a cassette, then inserted into a plasmid. Imperfect cassette vectors (i. e. vectors retaining some of the original *Escherichia coli* reading frame) are still widely used and have made a significant contribution to studies on the production of recombinant proteins.

Sources of DNA for cloning. cDNA clone bank. To produce a cDNA clone bank, mRNA is isolated from an organ, tissue or organism, then used as a template to synthesize cDNA. In multicellular eukaryotes, any one type of specialized cell does not express its total genome. Relatively few proteins are produced, and the corresponding mRNA is therefore present in relatively high proportion. For example, mRNA from pancreas contains a high proportion of mRNA for preproinsulin, legume root nodules contain a high level of leghemoglobin mRNA, and the mRNA for silk fibroin is predominant in the silk-synthesizing glands of the silk worm. An additional advantage of starting with mRNA is that it represents the end product of processing, in which intron transcripts (see Intron) may have been removed. The cDNA nucleotide sequence therefore corresponds to the amino acid sequence of the gene product. Eukaryotic recombinant genes containing introns may not be processed after being genetically engineered into a prokaryotic environment. Since most eukaryotic mRNA carries a 3'-polyA tail (see Messenger RNA), it can be isolated by affinity chromatography on oligo-dT cellulose.

mRNA used for cDNA synthesis can be enriched with respect to the message of interest by sucrose density gradient centrifugation, or by high performance liquid chromatography. Each mRNA fraction is tested in an in vitro translation system (reticulocyte lysate or wheat germ systems are used for eukaryotic mRNA), using radioactive amino acid precursors. Usually the translation product is analysed by precipitation with specific antibody, fol-

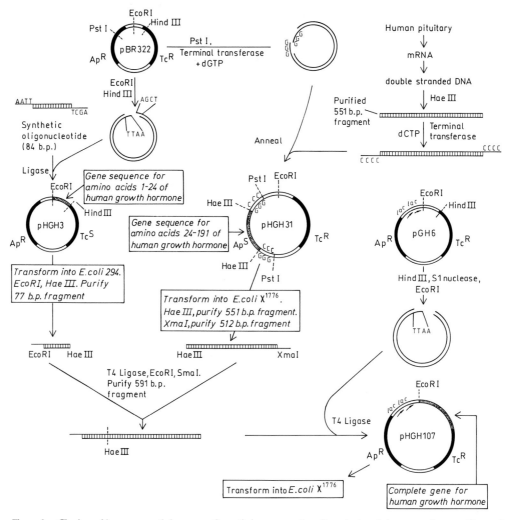

Figure 2. Cloning of human growth hormone. Growth hormones (see Somatotropin) are species-specific, and human cadavers were previously the only source of human growth hormone for medical use, e.g. correction of dwarfism. One liter of a culture of *Escherichia coli* χ^{1776} carrying pHGH107 plasmids produces about 2.5 mg of human growth hormone; previously, about 200 human cadavers would have been needed to prepare this quantity. The protein is not secreted, and it must be isolated and purified from the harvested cell paste. Rigorous purification is necessary to avoid the presence of antigenic bacterial proteins, if the product is to be injected into humans.

Hae III restriction sites are present in the 3′ noncoding region of the gene, and in the sequence coding for amino acids 23 and 24. Hae III restriction therefore cleaves the structural gene into two fragments, which are cloned separately and later ligated. The smaller cleavage fragment was discarded and replaced by a chemically synthesized oligonucleotide containing an ATG initiation codon. In ligation, restriction enzymes are also used to cleave unwanted dimerization products, e.g. in the formation of the 591 b.p. fragment, EcoI cleaves dimers formed from the EcoI sites, and SmaI cleaves XmaI dimers. *Escherichia coli* strain χ^{1776} is a "safe" host, specially constructed to meet safety requirements of genetic engineering; it cannot colonize or survive in the intestinal tract of warm-blooded animals, and it can only survive under specially controlled laboratory conditions. Ap^R and Tc^R represent the genes for resistance to ampicillin and tetracyclin, respectively. If resistance is lost by inactivation of the gene, the same region is denoted by Ap^S or Tc^S (S = sensitive). b.p. = base pair. [D.V. Goeddel et al. *Nature* **281** (1979) 544–548].

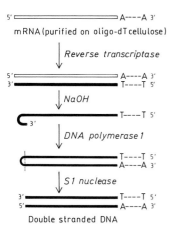

Figure 3. In vitro synthesis of single-stranded and double-stranded cDNA. The small hairpin structure at the 3' end of the initially synthesized single strand primes the synthesis of the second DNA strand.

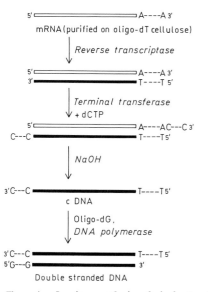

Figure 4. In vitro synthesis of single-stranded and double-stranded cDNA. Synthesis of the second strand of DNA is primed by oligo-dG, which associates with a 3' oligo-dC tail added at an earlier stage by the action of terminal transferase and dCTP.

lowed by sodium dodecyl sulfate gel electrophoresis and subsequent fluorography.

The first stage of cDNA synthesis is performed with reverse transcriptase (see RNA-dependent DNA polymerase), resulting in a mRNA-cDNA hybrid (fig. 3). Such hybrids have been successfully inserted into bacterial plasmids, but usually the mRNA template is removed by alkaline hydrolysis, and the single strand of cDNA is converted to the double

stranded form. A region of self complementarity at the 3' end of the cDNA results in the formation of a small hairpin structure. Further action of reverse transcriptase or DNA polymerase I in the presence of deoxyribonucleotide triphosphates results in continued 3'→5' elongation, forming a double stranded structure with a closed end. Cleavage of the hairpin loop with S1 nuclease produces double stranded cDNA suitable for cloning. DNA polymerase I also possesses nuclease activity, which may partially destroy the synthesized DNA. It is therefore usual to use the Klenow fragment (see) of DNA polymerase I, or a different polymerase, such as bacteriophage T4 DNA polymerase.

Cleavage of the hairpin loop with S1 nuclease necessarily removes part of the DNA sequence corresponding to the 5' end of the mRNA. An improved method for full length duplex cDNA synthesis overcomes this problem. The first cDNA strand is tailed with oligo-dC so that second strand synthesis can be primed with oligo-dG (Fig. 4).

Genomic library or gene bank. A genomic library is prepared from total cellular DNA, which is fragmented by cleavage with restriction endonucleases or by mechanical shear. Long, thin DNA molecules are easily broken by shearing forces. Intense irradiation with ultrasound produces DNA fragments of about 300 nucleotide pairs, while treatment in a high speed blender (1500 rev/min for 30 min) produces fragments of about 8 kb pairs. To establish a genomic library, the total complement of DNA fragments is cloned in a suitable vector, usually a phage or cosmid. This approach is sometimes referred to as a *shot gun method*, since it is nonselective and relies on the statistical probability that the required gene is contained within at least one of the DNA fragments. The number of clones needed to ensure that a genomic library contains all the genes of the cellular genome can be calculated from the formula, $N = \frac{\ln(1-P)}{\ln(1-a/b)}$, where N is the number of clones required, P is the probability that any gene is present (e.g. this value can be set variously at 95%, 80%, etc.), *a* is the average size of the DNA fragment inserted into the vector, and *b* is the size of the total genome.

If bacteriophage DNA is used as a cloning vector, it is first purified then treated with a restriction endonuclease. The fragmented cellular genome and phage DNA restriction fragments are mixed, annealed and ligated, producing a population of recombinant DNA phage molecules, in which all the restriction fragments of the cellular genome are randomly distributed. This hybrid phage DNA is transfected into a host bacterium (most studies have been performed with bacteriophage λ and *Escherichia coli*), eventually producing a phage population carrying the genomic library. Bacteriophages may act as *insertion vectors* or *replacement vectors*. Insertion vectors are opened at a single restriction site, so that insertion of the foreign DNA increases the size of the vector. Replacement vectors are cleaved at two restriction sites and the intervening sequence is replaced by the fragment of foreign DNA. The insertion vector, phage λNM607, can carry up to 9 kb of new DNA, whereas the replacement vector, phage λEMBL4, can carry up to 23 kb fragments. This compares with a maximal insert of 5 kb or less for most plasmid vectors. Cosmids, however, possess a much higher capacity, with a maximal limit of about 52 kb and a working limit of about 40 kb. They can therefore carry DNA fragments that are too large to be handled by plasmid or viral vectors, and they are useful for the establishment of genomic libraries of eukaryotic organisms which have very large ge-

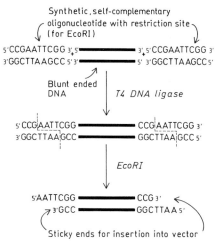

Figure 7. Generation of sticky ends on double-stranded, blunt-ended DNA by the use of linkers. Linkers are self-complementary, synthetic oligonucleotides containing appropriate restriction sites.

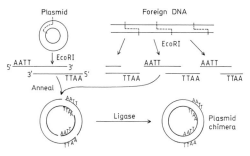

Figure 5. In vitro generation of a chimeric plasmid containing foreign DNA. The required sticky ends are created by cleaving both the plasmid and the foreign DNA with the same restriction endonuclease (in this example, EcoRI).

nomes. Cellular genome fragments may be screened at an earlier stage, thereby avoiding the need to screen all the clones of a genomic library (see below: Blotting and hybridization techniques).

Cleavage and rejoining of DNA. These processes are fundamental to R.D.t. and gene cloning. With the aid of restriction endonucleases of different specificities, the DNA vectors and cellular genomes can be cleaved at chosen sites. Large DNA molecules may also be cleaved by mechanical shear. Most restriction enzymes generate cohesive or "sticky" ends of 1 to 4 nucleotides, so that when vector and donor DNA are cleaved by the same enzyme, the DNA termini can be annealed, then covalently linked by DNA ligase (DNA ligase seals single strand nicks between adjacent nucleotides in a duplex chain) (Fig. 5). Such simple rejoining of the two DNAs is not always possible, because 1. the restriction enzyme may leave a "blunt" end, 2. it may be necessary to use different restriction enzymes, so that the potentially sticky ends are noncomplementary, 3. DNA fragments produced by mechanical shear, enzymatic cDNA synthesis or chemical DNA synthesis do not have sticky ends. However, sticky ends can be generated by addition of nucleotide "tails" to the 3′ end of DNA chains (a process known as *tailing*), by the action of terminal transferase from calf thymus (Fig. 6). Alternatively, synthetic DNA *linkers* may be ligated to the vector and/or donor DNA. The linker is a short DNA fragment containing the recognition sequence for one or more restriction enzymes, which is not present in the DNA receiving the linker. After ligation of the linker to blunt ended DNA, it is cleaved with an appropriate restriction enzyme to generate a sticky end (Fig. 7). Clearly, the two sticky ends produced by restriction are potentially capable of reuniting with each other, rather than participating in the intended hybridization. This tendency is overcome partly be performing hybridization at a

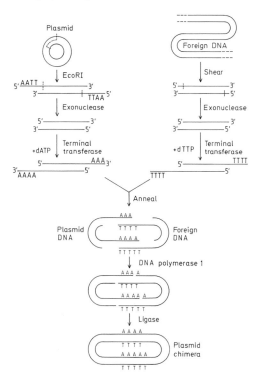

Figure 6. In vitro generation of a chimeric plasmid containing foreign DNA. Sticky ends are created by the action of terminal transferase and an appropriate deoxyribonucleotide triphosphate.

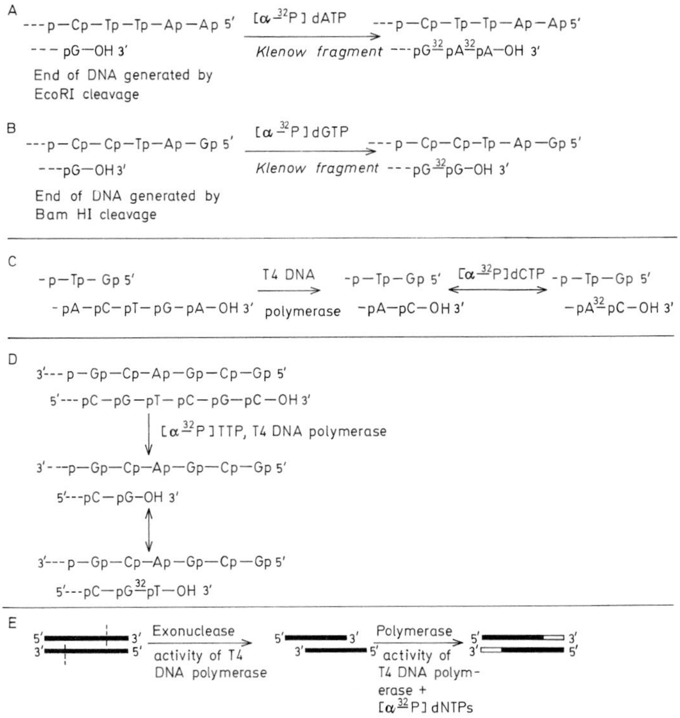

A
```
---p—Cp—Tp—Tp—Ap—Ap 5'        [α-³²P] dATP        ---p—Cp—Tp—Tp—Ap—Ap 5'
--- pG—OH 3'          ───────────────────>    ---pG³²pA³²pA—OH 3'
                        Klenow fragment
End of DNA generated by
EcoRI cleavage
```

B
```
---p—Cp—Cp—Tp—Ap—Gp 5'        [α-³²P]dGTP        ---p—Cp—Cp—Tp—Ap—Gp 5'
---pG—OH 3'          ───────────────────>    ---pG³²pG—OH 3'
                        Klenow fragment
End of DNA generated by
Bam HI cleavage
```

C
```
-p—Tp—Gp 5'           T4 DNA          -p—Tp—Gp 5'   [α-³²P]dCTP   -p—Tp—Gp 5'
-pA—pC—pT—pG—pA—OH 3'  ───────>   -pA—pC—OH 3'  ───────>   -pA³²pC—OH 3'
                      polymerase
```

D
```
3'---p—Gp—Cp—Ap—Gp—Cp—Gp 5'
5'---pC—pG—pT—pC—pG—pC—OH 3'

        │ [α-³²P]TTP, T4 DNA polymerase
        ▼

3'---p—Gp—Cp—Ap—Gp—Cp—Gp 5'
5'---pC—pG—OH 3'

        ▲
        │
        ▼

3'---p—Gp—Cp—Ap—Gp—Cp—Gp 5'
5'---pC—pG³²pT—OH 3'
```

E
```
5'━━━━━3'   Exonuclease    5'━━━━━3'   Polymerase   5'━━━━━3'
3'━━━━━5'   ───────>      3'  ━━━━5'   ───────>    3'━━━━━5'
          activity of T4              activity of
          DNA polymerase              T4 DNA polym-
                                      erase +
                                      [α-³²P] dNTPs
```

Figure 8. Methods for the 3' end labelling of double stranded DNA. The reactions are similar to those used for filling recessed ends of double-stranded DNA after restriction cleavage, except that only one [α-³²P]deoxyribonucleotide triphosphate is used.
A and B: Klenow fragment (see) possesses only polymerase activity, and it efficiently catalyses addition of the appropriate nucleotide residue at the recessed 3' end of DNA.
C: T4 DNA polymerase also possesses exonuclease activity. The protruding 3' arm is degraded one nucleotide at a time, until a residue is encountered corresponding to the nucleotide substrate present in the incubation. The polymerase activity of the enzyme is then manifested, and an exchange reaction occurs.
D: The reactions shown in C. may also be applied to blunt-ended DNA.
E: The combined exonuclease and polymerase activities of T4 DNA polymerase are exploited in the O'Farrell method for the general labeling of double-stranded DNA. dNTP = deoxyribonucleotide triphosphate.

relatively high DNA concentration to increase the frequency of intermolecular reactions. Also, treatment of linearized vector DNA with alkaline phosphatase removes 5'-terminal phosphate and prevents recircularization and dimer formation by plasmids; circularization of the vector can then occur only by insertion of non-phosphatase-treated DNA, which brings with it one 5'-terminal phosphate to each junction. The resulting nick is repaired by the host after transformation.

Introduction of recombinant DNA into a host cell. Plasmids enter the bacterial cell by Transformation (see). It is usually necessary to make the bacterial cell receptive to transformation, i.e. the cells must be competent. For *Escherichia coli*, this is usually achieved by treatment with a solution of 50 mM CaCl₂. This causes the DNA to bind more effectively to the cell surface. Movement of DNA into the competent cells is then stimulated by raising the temperature briefly (2 min) to 42 °C, a process known as heat shock. Phage and cosmid DNA enter

the bacterial cell in a process equivalent to transformation, but usually referred to as *transfection*.

Probes. ³²P-labelled mRNA, cDNA or synthetic oligonucleotides are used as probes in Southern blotting analysis. Double-stranded DNA can be labelled with ³²P by Nick translation (see), by 3' end labeling (Fig. 8), by the random priming process (Fig. 9), or by 5' end labelling (Fig. 10). Synthetic oligonucleotide probes need contain only about 15–20 nucleotides (about 6 gene codons). Short sequences (about 6 amino acids) of the gene product are selected. If the total sequence is not known, sequences are chosen from those of the peptides produced by proteolysis. The corresponding oligodeoxyribonucleotide is synthesized chemically (see DNA synthesis). Owing to the degeneracy of the genetic code, there are several possible oligonucleotide structures for one amino acid sequence. Oligopeptide sequences containing tryptophan or methionine are therefore favored, because these two amino acids have single codons, thereby decreasing the possible number of

521

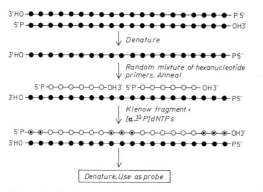

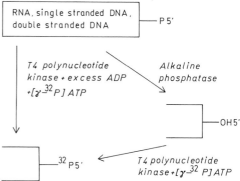

Figure 9. Random priming for the in vitro labelling of DNA probes. Single stranded DNA is incubated with a random mixture of hexanucleotides. Some of these will be complementary to regions of the DNA, and they act as primers for the synthesis of a complete complementary strand, using the Klenow fragment of DNA polymerase I and [α-³²P]deoxyribonucleotide triphosphates (dNTPs). ● = Nucleotide residues of the original DNA. ○ = Nucleotide residues of hexanucleotide primers. ⊙ = ³²P-labelled nucleotide residues. Random priming gives a relatively low concentration of probe, but the specific activity is high. It is therefore suitable for preparing probes for single copy gene detection in a Southern blot. By using 200 µCi of [α-³²P]dNTPs at 6000 Ci/mol, probes with specific activities greater than 5×10^9 dpm/µg can be prepared. To achieve the equivalent probe specific activity by Nick translation (see), it would be necessary to use 1 mCi labelled dNTP per µg DNA. [A.P. Feinberg & B.Vogelstein *Anal. Biochem. 132* (1983) 6–13; A.P. Feinberg & B.Vogelstein, Addendum *Anal. Biochem. 137* (1984) 266–267]

Figure 10. 5′ End labeling of DNA and RNA. Labeling is most efficient with single-stranded molecules and at the 5′ ends of double-stranded DNA with recessed 3′ ends. Blunt ended DNA and DNA with recessed 5′ ends react more slowly. After removal of the 5′ phosphate with alkaline phosphatase, the 5′OH-group is rephosphorylated by the action of polynucleotide kinase and ATP. Alternatively, in the presence of excess ADP, the reverse reaction is encouraged, effectively leading to exchange between the 5′ phosphate and the γ-phosphate of ATP.

probe structures. In practice, all of the possible oligonucleotide sequences are not synthesized individually. During each polymerization step where the sequence is degenerate, a mixed addition reaction is performed, resulting in a mixed probe, which is radiolabelled and used directly. If the gene structure is known, however, single oligonucleotide probe sequences can be selected and synthesized.

Blotting and hybridization techniques. DNA fragments from restriction digests are separated according to their size by agarose gel electrophoresis. The electrophoretogram is screened for the presence of specific nucleotide sequences by performing a Southern blot (see), followed by hybridization with ³²P-labelled specific RNA or DNA probes, then autoradiography. In this way, the total fragment mixture from a cellular genome can be screened for those fragments carrying a particular gene; this dispenses with the need later to screen a total genomic library (Fig. 11). A screening procedure for detection of DNA sequences in transformed bacterial colonies by hybridization in situ with radioactive probe RNA was developed by Grunstein and Hogness in 1975. A single colony among thousands can be detected by this technique. Colonies to be screened are replica plat-

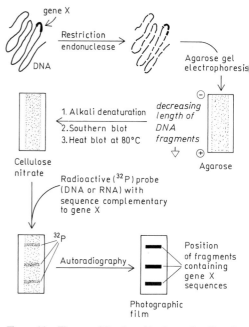

Figure 11. The use of Southern blotting and radioactive probes for screening fragmented DNA.

ed onto a nitrocellulose filter disc on the surface of an agar culture plate, and allowed to grow. The nitrocellulose filter is then treated with alkali to lyse the bacterial colonies and denature their DNAs. Protein is removed with a proteinase, leaving denatured DNA bound to the nitrocellulose. Single-stranded, but not double-stranded, DNA binds very strongly to nitrocellulose. The DNA is then fixed firmly to the nitrocelluose by heating at 80 °C. To identify the original sites of colonies carrying the required transformed gene, the nitrocellulose disc is incubated with ^{32}P-labelled, specific RNA. Only the specified DNA hybridizes with the RNA, and this is located by autoradiography. Comparison of the original culture plate with the autoradiogram permits identification of any colony carrying the gene of interest (Fig. 12).

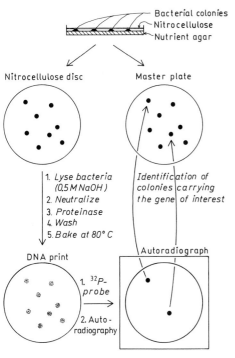

Figure 12. Screening procedure of Grunstein and Hogness.

Gene expression. Synthesis of a functional protein depends on transcription of the gene, translation of the mRNA, often processing of the mRNA, and often post-translational processing of the initial translation product (see Post-translational modification of proteins). Transcription of a cloned insert requires the presence of a promoter which is recognized by the host RNA polymerase. Translation requires a ribosome binding site on the mRNA. If the mRNA is translated in *Escherichia coli*, the ribosome binding site includes the translation start codon (AUG or GUG) and a ribonucleotide sequence complementary to the bases at the 3′ end of the 16S ribosomal RNA. These are known as S-D sequences (named after Shine and Dalgarno, who first postulated these sequences in 1975), and they are present in almost all *Escherichia coli* mRNAs. S-D sequences vary in length from 3 to 9 nucleotides, and they are situated

3 to 12 bases before the translation start codon. Although a foreign protein product may be synthesized from its own N-terminus under the control of an *Escherichia coli* promoter (e.g. human growth hormone), usually a fusion hybrid or chimeric protein is produced. For example, the eukaryotic gene for β-endorphin was attached to the bacterial gene for β-galactosidase and expressed in *Escherichia coli.* The resulting protein was therefore a hybrid of β-galactosidase and β-endorphin, from which the β-endorphin (a polypeptide of 31 amino acids) could be cleaved proteolytically: 1. Lysine residues were protected by reaction with citraconic anhydride; 2. citraconylated endorphin was released by treatment of the hybrid with trypsin, which attacks an arginyl bond immediately preceding the β-sequence (β-endorphin contains no arginyl residues); 3. native endorphin was prepared by removal of citraconyl groups at pH 3.0. Several genes have been cloned as partners of the β-galactosidase gene, resulting in their initial expression as fusion proteins, e.g. somatostatin and human insulin. Several strategies can be employed for final cleavage of the chimeric protein. For example, if the gene is entirely synthetic (see DNA synthesis), an additional gene codon for N-terminal methionine can be included, thereby creating a CNBr-sensitive cleavage site in the chimeric protein. A particular problem was encountered in the expression of the human insulin gene. Proinsulin contains a 35-residue C-peptide, which must be removed after translation (see Insulin). A gene was constructed encoding an analogue of proinsulin, in which the C-peptide sequence is replaced by 6 amino acids (Arg-Arg-Gly-Ser-Lys-Arg), which is easily removed by proteolysis in vitro.

Oocytes are able to transcribe exogenous DNA, and *Xenopus* oocytes in particular have been used experimentally as a surrogate genetic system for the expression of cloned DNA, e.g. genes for ovalbumin, sea urchin histones, and various proteins of phage SV40 have been expressed in this way. The oocyte is a relatively large cell with a large nucleus, so that microinjection of DNA into the nucleus is technically easy, e.g. 20–40 nanoliters of DNA solution can be injected via a fine glass capillary.

Mouse embryos have also been used as expression systems for foreign DNA. Microinjection of SV40 DNA into preimplantation blastocysts, followed by implantation into the uteri of foster mothers, resulted in progeny which were mosaics, i.e. only a proportion of cells in each tissue contained integrated SV40 DNA; but the subsequent generation produced genetically defined substrains, showing that SV40 DNA had become integrated into germ line cells. To insure integration into a host chromosome at an early stage of development, and to avoid mosaicism, the DNA can be injected into the male pronucleus of the newly fertilized egg. Between 3% and 40% of animals from these embryos are transgenic mice, i.e. their DNA contains the integrated foreign gene. Mouse metallothionein (MMT) gene has been fused with other genes and reintroduced into mouse embryos. Induction of the MMT gene with heavy metals (e.g. zinc or cadmium) results in expression of the hybrid partner as well. Thus, transgenic mice carrying fused (hybrid) genes for [MMT-*Herpes sim-*

plex thymidine kinase], [MMT-rat growth hormone] and [MMT-human growth hormone] have been reared.

Selection. In direct selection methods, only the desired recombinants grow after plating transformants on an agar growth medium. For example, if a gene specifying antibiotic resistance is cloned in a sensitive host, then only transformants carrying the resistance gene will survive and form colonies on growth medium containing the antibiotic. In a direct selection technique known as *marker rescue*, auxotrophic mutants are used as hosts for the vector carrying a biosynthetic gene, e. g. if the A gene for tryptophan synthase is cloned in a *trp*A⁻ auxotroph, only those cells containing a plasmid-borne copy of the A gene are able to form colonies on growth media lacking tryptophan. Selection for the desired gene can be performed by hybridization, either on the initial digest of the cellular genome, or on the plated colonies of the final transformed population (see Blotting and hybridization techniques, above). Enrichment of the desired mRNA prior to cDNA synthesis is also a form of screening.

If the cloned gene is expressed, screening can be performed by exploiting some property of the synthesized protein. For example, in the immunochemical method of Broome and Gilbert, the transformed cells are first plated in the conventional way on nutrient agar, and a replica plate is prepared. The colonies on one of the plates are lysed (by exposure to chloroform vapor, by spraying with an aerosol of virulent phage, or by thermo-induction of an inducible prophage already present in the host) to release antigen from the positive colonies. The surface of the plate is then placed against a sheet of polyvinyl coated with antibody to the gene product. The washed polyvinyl sheet, carrying bound gene product, is exposed to ¹²⁵I-labelled IgG, which reacts with the bound antigen at different sites from those involved in the initial binding. Positive areas are detected by autoradiography. Peptide acetyltransferase has been cloned in phage λ from a cDNA library derived from rat brain mRNA, and selected by a novel procedure known as *enzyme immunodetection assay* (EIDA) [J. H. Eberwine et al. *Proc. Natl. Acad. Sci. 84* (1987) 1449-1453]. The t-butoxycarbonyl (Boc) derivative of [Leu⁵]enkephalin hydrazide was reacted with paper carrying diazotized aminophenylthioether groups. After removal of the Boc groups with acid, the resulting substrate paper possessed covalently bound [Leu⁵]enkephalin with a free amino terminus. Bacterial plates with patterns of phage plaques (derived from λ phage carrying the cDNA library) were placed in contact with the substrate paper. The covalently bound enkephalin became acetylated at the site of any plaque expressing peptide acetyltransferase activity. After incubation of the paper with antisera specific for the acetylated form of enkephalin, the acetylated areas retained bound IgG. Finally the IgG was revealed by incubation with ¹²⁵I-labelled Protein A (see) and autoradiography.

[T. Shine et al. *Nature 285* (1980) 456-461; R. Wetzel et al. *Gene 16* (1981) 63-71; H. Yoshioka et al. *J. Biol. Chem. 262* (1987) 1706-1711; H. M. Jantzen et al. *Cell 49* (1987) 29-38; R. W. Old & S. B. Primrose,

Principles of Gene Manipulation 3rd edtn. (Blackwell, 1985); M. Grunstein & D. S. Hogness, *Proc. Natl. Acad. Sci. 72* (1975) 3961-3965. S. Broome & W. Gilbert *Proc. Natl. Acad. Sci. 75* (1978) 2746-2749; T. Moniatis, E. F. Fritsch & J. Sambrook, eds., *Molecular Cloning* (Cold Spring Harbor Laboratory, 1982).]

Redoxases: see Ferredoxin.

Redoxin: an electron-transferring protein. R. that contain iron, bound as a functional group to S, N or O-ligands of the protein, are known as Ferredoxins (see). An R. containing no metal is Thioredoxin (see). See also Rubredoxin.

Redox potential: see Oxidation.

Reductases: see Flavin enzymes.

Reduction: the addition of electrons. R. is the converse of Oxidation (see). In biochemical systems R. may involve electron transfer only (e. g. R. of cytochromes and ferredoxin), but the majority of biochemical R. involve the addition of hydrogen (hydrogenation). The pyridine nucleotide coenzymes, NAD⁺ and NADP⁺, play an important part in R., and hydrogen transfer in reductive biosynthesis is usually mediated by NADPH. In green plants, the reducing power from the light reaction of photosynthesis appears as NADPH, which is then utilized in various reductive biosyntheses, notably the conversion of CO_2 to carbohydrate; in photosynthetic bacteria, the reductant formed in photosynthesis is NADH.

R. must be accompanied by a corresponding oxidation, so that enzymes catalysing these reactions are generally known as oxidoreductases (see Enzymes, Table 1). See also Dehydrogenation.

Redundancy: 1. The occurrence (frequent in eukaryotes, only in isolated instances in prokaryotes) of linearly arranged, largely identical, repeated sequences of DNA. Between 50 and 10^7 such repeated base sequences have been demonstrated, depending on the source of the DNA. The degree of R. in a sample of DNA can be determined from the rate of renaturation following heat denaturation: the rate of reannealing increases with the degree of R. Repetitive DNA may constitute 20-80% of the total DNA. Satellite DNA has especially high R., whereas genes coding for proteins normally have low R.; an exception is found in the high R. of the histone genes of the sea urchin. The nontranscribed part of giant messenger-like RNA also contains repetitive sequences, but their function is unknown. Genes for tRNA, 5S-RNA and rRNA (but not mRNA) show R., the R. value for rRNA lying between 100 and 7500.

In prokaryotes, gene R. has only been detected in isolated cases, and the values are low. In *Bacillus subtilis* only 0.4% of the genome is redundant (e. g. 9-10 genes for rRNA and 40 genes for tRNA).

2. *Terminal redundancy:* the existence of identical genetic information at each end of a viral chromosome. In λ-phages this constitutes up to 20 nucleotide pairs, whereas up to 6000 nucleotide pairs may be terminally redundant in T-even phages.

Refsum's disease: see Inborn errors of metabolism.

Regulator gene: see Operon.

Regulon: a group of structural genes, whose

gene products (enzymes) are involved in the same reaction pathway, and which are regulated together. The individual genes are in different regions of the chromosome, i.e. they do not lie on neighboring sequences of DNA, as in an operon. In *Escherichia coli*, the 8 genes for arginine synthesis constitute a R. In *Neurospora crassa*, the genes for histidine biosynthesis are found on four different chromosomes, but they are still subject to a concerted regulation, and therefore constitute a R.

Reichstein's substances F, H, M, Q and S: see Cortisone, Corticosterone, Cortisol, Cortexone, Cortexolone.

Relative molecular mass (symbol M_r): the mass of a molecule relative to the Dalton (see). M_r has no units. It is equivalent to "molecular weight", a term used widely in physical and biological sciences. The word "weight" is, however, dimensionally incorrect, so that "molecular weight" should be avoided and replaced by relative molecular mass M_r).

Relaxation protein: a type I eukaryotic topoisomerase (see Topoisomerases) isolated from the nuclei of LA9 mouse and HeLa cells, and characterized by its ability to remove superhelical turns from closed, circular DNA. [H-P. Vosberg et al. *Eur. J. Biochem.* 55 (1975) 79–93.]

Relaxin: a female sex hormone, formed in mammals during pregnancy. R. is a heterodetic, cyclic polypeptide, M_r 12000; the A-chain contains 22, the B-chain 26 amino acid residues. It is probably produced by the ovary and placenta and/or uterus under the influence of progesterone. It causes relaxation of collagenous connective tissue (sensitized by estrogens) in the symphysis and ileosacral joints, thus enabling enlargement of the pelvic girdle during pregnancy and at birth.

Release factors, *termination factors:* catalytically active proteins necessary for the termination step of RNA synthesis and protein biosynthesis. For further details, see Ribonucleic acid; Protein biosynthesis (subheading Termination).

Release inhibiting hormone: see Releasing hormones.

Releaser: see Pheromones.

Releasing factors: see Releasing hormones.

Releasing hormones, *releasing factors, liberins, statins:* a group of peptide neurohormones, synthesized in various distinct nuclei of the hypothalamus. They are released into the capillaries of the portal vessels in the median eminence of the hypothalamus, then carried to the anterior pituitary (adenohypophysis) where they regulate the production and secretion of tropic hormones (e.g. thyreotropin, somatotropin, gonadotropins). R.h. are produced in nerve cells, and their synthesis represents the conversion of an electrical stimulus into hormonal signals. In the median eminence of the hypothalamus, R.h. are stored in nanogram quantities. In the anterior pituitary, R.h. act via the adenylate cyclase system, and are then degraded by proteolysis.

Thyrotropin releasing hormone, thyrotropic hormone releasing factor (TRF), thyroliberin:
Pyr-His-Pro-NH₂ (Pyroglutamyl-L-histidyl-L-prolinamide), M_r 262. The identical hormone is found in all hitherto investigated species. Secretion of TRF is promoted by neurotransmitters, e.g. noradrenalin, and inhibited by serotonin. TRF stimulates the anterior pituitary to synthesize and secrete thyrotropin, which in turn stimulates the thyroid gland to secrete thyroxin and triiodothyronin. The latter two hormones exert a negative feedback on the secretion of TRF and thyrotropin. TRF also stimulates the secretion of prolactin and acts as a neurotransmitter in the central nervous system.

Luteinizing hormone releasing factor (LRH), luliberin:
Pyr-His-Trp-Ser-Tyr-Gly-Leu-Arg-Pro-Gly-NH₂, a decapeptide, M_r 1182, which stimulates the synthesis and secretion of luteinizing hormone (LH) and follicle stimulating hormone (FSH) by the anterior pituitary. The names, gonadotropin releasing factor (GRF) and gonadoliberin, have therefore been suggested for LRH. Synthesis of LRH is regulated by the central nervous system and is subject to environmental influences (time of year, light, olfactory stimuli, sexual stimulation). LH and FSH stimulate the production of sex hormones by the gonads. LRH acts as a neurotransmitter in the central nervous system; it can be measured by radioimmunological techniques.

Follicle stimulating hormone releasing factor (FRH), folliliberin: this is chemically identical with LRH.

Corticotropin releasing factor (CRF), corticotropin releasing hormone: a polypeptide of 41 amino acids [J. Spiess et al., Proc. Nat. Acad. Sci. U.S.A. 78, 6517–6521 (1981)] which stimulates both β-endorphin and corticotropin secretion in the anterior pituitary and in some areas of the central nervous system. Corticotropin stimulates synthesis and secretion of glucocorticoids by the adrenals; the glucocorticoids inhibit the secretion of CRF or the response of the anterior pituitary to the releasing hormone (negative feedback). β-Endorphin, in turn, inhibits the secretion of LRH in neurons which control sexual behavior.

Three of the seven anterior pituitary hormones do not act on endocrine glands, but influence tissue metabolism directly. There is no hormone feedback mechanism for the control of these three hormones, and control is exerted by the action of release-inhibiting hormones.

Somatotropin releasing factor (SRF), somatotropin releasing hormone, somatoliberin: The identity of the native hormone is unknown. Val-His-Leu-Ser-Ala-Glu-Gln-Lys-Glu-Ala, a decapeptide, M_r 1112 (porcine), has been found to stimulate synthesis and secretion of somatotropin and is therefore called SRF.

Somatotropin release inhibiting hormone (SIH), somatostatin: a polypeptide, M_r 1638, which inhibits the secretion of somatotropin, thyrotropin, insulin, glucagon, gastrin and cholecystokinin. SIH has a broad activity spectrum; it acts outside the hypothalamus as a neurotransmitter in the central nervous system, and as a hormone in the intestinal tract; it has also been found in the thyroid and pancreas.

Ala-Gly-Cys-Lys-Asn-Phe-Phe-Trp-Lys-Thr-Phe-Thr-Ser-Cys

Prolactin releasing factor (PRF), prolactin releasing hormone, prolactoliberin: thought to be identical with thyrotropin releasing factor. The structure of *Prolactin release inhibiting hormone (PIH) (prolactostatin)* is unknown. PRF promotes, whereas PIH inhibits the synthesis and secretion of prolaction by the anterior pituitary.

Melanotropin releasing factor (MRF), melanotropin releasing hormone, melanoliberin: corresponds structurally to the open hexapeptide ring of Oxytocin (see), without the tripeptide side chain, i. e. Cys-Tyr-Ile-Gln-Asn-Cys.

Melanotropin release inhibiting hormone (MIH), melanostatin: structurally equivalent to the peptide side chain of oxytocin, Pro-Leu-Gly-NH_2 (bovine, rat), or to Pro-His-Phe-Arg-Gly-NH_2 (bovine). MRF stimulates, whereas MIH inhibits synthesis and secretion of melanotropin by the anterior pituitary, leading respectively to darkening or lightening of amphibian skin.

Renaturation: conversion of a denatured protein or nucleic acid into its native configuration. See Nucleic acids; Proteins.

Renin (EC 3.4.99.19): an endopeptidase, M_r 43000, formed in the juxtaglomerula cells (cells next to the glomerula) of the kidney. In the plasma, it releases angiotensin I (a decapeptide) from the *N*-terminal sequence of angiotensinogen (an α_2-plasma globulin). Another enzyme (M_r 210000) in lung (3 subunits, each M_r 70000), kidney (8 subunits, each M_r 25000) and plasma removes histidine and leucine from the *C*-terminus of angiotensin I, releasing the highly hypertensive octapeptide, angiotensin II.

Rennet enzyme, *rennin, chymosin:* a pepsin-like proteinase (see Proteases) which is formed as inactive prorennin in rennet bags, and probably in the stomach of all nursing mammals. It is converted to active rennin by pepsin or autocatalytically. In the process its M_r is reduced from 36200 (prorennin) to 30700 (rennin), by cleavage of an activation peptide (M_r 5500, 49 amino acids) from the *N*-terminal end R.is the milk-coagulating enzyme of young mammals and requires calcium ions for its effect (pH optimum, 4.8). It specifically attacks a single peptide bond in κ-casein (-*His*-Leu-*Ser*-Phe-Met-Ala-), and splits κ-casein into insoluble para-κ-casein and a *C*-terminal glycopeptide (M_r 8000). The *Ser* residue next to the Phe-Met bond to be hydrolysed serves as a binding site for R., while the *His* 3 residues from the bond serves as a proton acceptor and donor. This process also occurs in the absence of proteases when the protein is heated at slightly acid pH. In both cases, the protective colloid function of κ -casein in the milk is destroyed.

Repair enzymes: enzymes that catalyse stages in the repair of DNA. They include exo- and endonucleases, DNA-polymerases and ligases. See Deoxyribonucleic acid (Repair of DNA).

Repeatability: In accordance with the International Organization for Standardization (Precision of test methods – Determination of repeatability and reproducibility. Draft International Standard ISO/Dis 5725, October 1977), *repeatability* refers to measurements in one laboratory over a short time peroid, whereas *reproducibility* refers to measurements over longer periods of time and/or in different laboratories.

Replication: DNA replication. See Deoxyribonucleic acid.

Replication site: the site of DNA replication in vivo. In bacteria, the R.s. is on the cell membrane The circular DNA is bound to the cell membrane at its initiation region, together with DNA-polymerase and initiation proteins. See Deoxyribonucleic acid.

Replicator: see Replicon.

Replicon: a term proposed by Jacob and Brenner for the replication unit. In prokaryotes or viruses, it is the complete circular or linear DNA (or RNA in RNA viruses), representing the total chromosome of the organism. If a cell contains more than one R., they are replicated independently. R.are subject to positive control, i. e. they contain a gene for the production of an initiator molecule which initiates replication by interacting with a replicator (the initiation site on the DNA).

Bacterial chromosomes, episomes or plasmids, viruses, DNA molecules of chloroplasts and mitochondria are R. Eukaryotic chromosomes consist of a large number of R., i.e. one DNA molecule has several replication sites. See Deoxyribonucleic acid.

Repression: see Enzyme repression.

Repressor: an allosteric protein, which regulates the transcription of structural genes, and is encoded by a regulator gene. R.binds to the operator region of DNA, thereby preventing synthesis of the mRNA for the structural genes of an operon or regulon (see Enzyme repression). The *Lac repressor*, which regulates transcription of the Lac operon in *Escherichia coli* (see Enzyme induction), is a tetrameric protein (M_r 152000) consisting of 4 identical subunits (each of M_r 38000, containing 347 amino acid residues). It was the first R.to be purified and structurally elucidated (Bayreuther et al. 1973).

In an inducible enzyme system, the R.is inactive in the presence of the effector (inducer); binding to the inducer apparently changes the conformation of the R., so that it no longer binds to the operator (see Enzyme induction). Synthesis of mRNA can therefore proceed only when the inducer is present. In enzyme repression, the situation is reversed; R.is activated by a corepressor (the endproduct of a biosynthetic pathway, e.g. an amino acid) so that it can bind to the operator. In this case, synthesis of mRNA proceeds only in the absence of corepressor (see Enzyme repression). See also Derepression.

Reproducibility: This is not the same as Repeatability (see).

Reserpine: therapeutically the most important Rauwolfia alkaloid. M_r 608.9, m.p. 265 °C $[\alpha]_D^{25}$-123 ° (chloroform). Hydrolysis of R.with alcoholic KOH produces reserpinic acid (structurally similar to yohimbine), trihydroxybenzoic acid and methanol. R.is found widely in the genus *Rauwolfia* and is responsible for the sedative properties of these plants. After a latent period, R. causes long lasting sedation, with decrease of blood pressure and decreased pulse rate. It is used as a powerful neuroleptic drug in psychiatry.

Reserve cellulose, *storage cellulose:* see Cell wall.

Resilin: a Structural protein (see) from the exoskeleton of arthropods, especially insects. R. has a high glycine content, and no cystine. R. is located betwee the chitin lamellae, and endows the arthropod exoskeleton with a certain elasticity. A notable component of R. is trityrosine (Fig.), which is formed in the protein after translation by cross linking the tyrosine side chains of one or more polypeptide chains. This results in an irregular three dimensional lattice, which is responsible for the rubberlike properties of R.

Trityrosine residue in resilin

Resin acids, *resinic acids:* hydroaromatic diterpenes, the acid components of resins. Colophony (rosin) consists of up to 90% R. a. The most important representatives are abietic acid, neoabietic acid, dextro-pimaric acid and neopimaric acid. Salts and esters of the R. are called *resinates.* The alkali salts are also called *resin soaps.*

Resinates: salts and esters of resin acids.

Resinoids: see Essential oils.

Resinols: see Resins.

Resinotannins: see Resins.

Resins: largely amorphous, solid or half-solid, transparent, odorless and tasteless organic substances, usually of vegetable origin. Tree R. are classified according to age into fossil R., such as amber, recent fossil R. (several years to centuries old), e.g. copal R., and recent R., which occur mostly as balsams fresh from injured trees. Caoutchouc (see) is included with the R. Herbaceous plants produce R., e.g. mastic, but not in any considerable quantity. Mixtures of R. with mucin are called gum R. Solutions of R. are referred to as balsams. The most important animal R. is shellac, produced by the female East Asian scale insect *(Tachardia lacca).*

R. are supercooled melts, soluble in nonpolar solvents. Like essential oils, they are complex mixtures in which terpenes and aromatics predominate. The structures of resin components are not well clarified, but they can be classified according to their chemical properties into: *Resinotannols:* aromatic phenylpropane resin alcohols allied to the tannins, occurring partly free, but more generally combined with aromatic acids or with umbelliferone in the form of esters (tannol R.); *Resinols:* crystalline, colorless, resin alcohols, e.g. terpene alcohols, occurring partly free, partly as esters; *Resin acids:* partly crystalline and mostly free; they combine with alkalis to form soaps, and they form crystalline salts or esters called resinates (see Resin acids); *Resenes:* indifferent substances that are neither esters nor acids; *Resines:* esters, e.g. coniferyl benzoate.

Crude R. and refined resin components are widely used in the production of paints, varnishes, textile conditioners, cosmetics and pharmaceuticals.

Resin soaps: see Resin acids.

Respiration: *oxidative metabolism:* a process by which cells derive energy in the form of ATP from the controlled reaction of hydrogen with oxygen to form water. The hydrogen is derived from the degradation of organic substrates by oxidases or dehydrogenases, which in turn release it to a system of redox catalysts (see Respiratory chain) located in the inner mitochondrial membrane, or in the cell membrane of bacteria. The electron is stripped from the hydrogen atom and passed along the respiratory chain; three of the transfers from one redox catalyst to the next power the formation of ATP from ADP and P_i (see Oxidative phosphorylation). The last redox catalyst passes the electrons to a terminal acceptor, which is usually oxygen, but may be nitrate (nitrate respiration). The hydrogen ions produced at the beginning of the chain then join the O^{2-} to form H_2O.

Respiratory acidosis: see Acidosis.

Respiratory alkalosis: see Alkalosis.

Respiratory chain, *electron transport chain:* a series of redox catalysts which transport electrons from respiratory substrates to oxygen. The energy of this electron flow may be used in the synthesis of ATP. The coupling of ATP synthesis with electron transport in the R.c. is known as Oxidative phosphorylation (see). The R.c. complex is located in the inner mitochondrial membrane of eukaryotes, and in the cell membrane of prokaryotes. It is functionally closely associated with the enzymes of the Tricarboxylic acid cycle (see) which supply reducing equivalents, mostly in the form of NADH, with some $FADH_2$. Depending on the tissue and its metabolic state, other pathways, e.g. Fatty acid degradation (see) may be more important than the tricarboxylic acid cycle in providing NADH or $FADH_2$ for the R.c.

The overall reaction is $NADH + H^+ + \frac{1}{2}O_2 \rightarrow H_2O + NAD$, $\Delta G^\circ = -221.7\,kJ/mol = -53\,kcal/mol$. The standard electrochemical potentials of individual steps of the R.c. are given in Table 1. In three of the steps (Fig.), the potential difference is large enough to supply the energy required for the phosphorylation of ADP; these are recognized as the three sites of Oxidative phosphorylation (see). The first site is the transfer of electrons from NADH to ubiquinone. Since electrons from $FADH_2$ enter the chain at the level of ubiquinone, they promote the formation of only two molecules of ATP per electron pair. The amount of ATP generated by electrons from any given substrate can be expressed as the P/O ratio, i.e. the number of molecules of ATP generated per atom of oxygen consumed. For dehydrogenations involving NAD as the coenzyme, the P/O ratio is 3. For substrates oxidized by flavin nucleotide enzymes, the P/O ratio is 2. Electrons flow singly along the chain of cytochromes, yet each phosphorylation requires that 2 electrons pass the site, and 4 electrons are needed to reduce one molecule of oxygen. The mechanisms for the accommodation of these requirements are not clear. Since the R.c. is a membrane-bound system, its components can only be purified after disruption of the membrane (see Electron transport particle; Mitochondria). Using sonic oscillation, E. Racker produced submitochondrial particles capable of oxidative phosphorylation coupled to electron transport. Treatment with urea removes a component F_1 from the particles and prevents further phosphorylation, but electron transport, i.e. respiration, continues. Since isolated F_1 structures have ATPase activity, it seems likely that they form part of the ATP synthesizing apparatus in situ. Treatment with detergents has been used to isolate four enzyme complexes from mitochondrial membranes, designated I, II, III and IV. The components and activities of these complexes are shown in Table 2. The best studied components of the R.c. are ubiquinone and cytochrome c, which are soluble and easily removed, and thus accessible. They are relatively small molecules that appear to serve as carriers transporting electrons between immobilized components.

Because the mitochondrial membrane is impermeable to ATP/ADP and NAD/NADH, two transport systems are necessary for the function of the R.c. One is the ATP/ADP carrier, which effects facilitated exchange diffusion, and the other is a metabolic shuttle (see Hydrogen metabolism) which allows hydrogen generated in the cytoplasm to enter the mitochondrion.

Some organisms use hydrogen acceptors whose redox potentials are higher than that of NADH (e.g. sulfide, thiosulfate, nitrite). The thermodynamic work required to reduce these substrates is derived from participation of ATP in Reversed electron transport (see) along a cytochrome chain. In general, such organisms are obligate anaerobes, and their electron transport chains lack cytochrome oxidase.

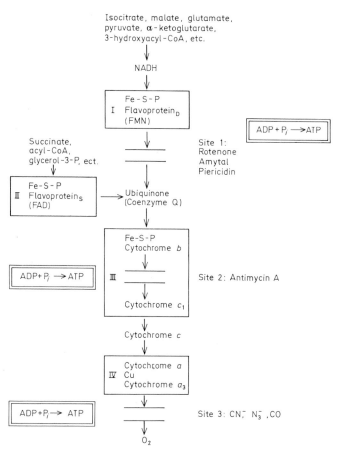

The respiratory chain. Boxes indicate the compositions of complexes I to IV, Electron flow is shown by arrows. Sites of action of some inhibitors are labelled 1, 2 and 3 and indicated by horizontal bars. Sites 2 and 3 are also coupled with ATP synthesis, i.e. according to the chemiosmotic theory, transfer of electrons from cytochrome b to cytochrome c_1, and from cytochrome a_3 to oxygen provides enough energy for the creation of a proton gradient for the synthesis of one molecule of ATP. The first site of ATP synthesis may not be identical with inhibition site 1 as shown here, but it is known to exist on the ubiquinone side (rather than the substrate side) of complex I.

Table 1. E_o *values for some biological redox pairs*

Redox system	E'_o in volts
H^+/H_2	-0.420
$Ferredoxin_{ox}/Ferredoxin_{red}$	-0.420
$NAD^+/NADH+H^+$	-0.320
$Glutathione_{ox}/Glutathione_{red}$	-0.230
$FMN/FMNH_2$	-0.122
Cytochrome b (Fe^{3+}/Fe^{2+})	$+0.075$
Fumarate/succinate	0.031
Ubiquinone/ubihydroquinone	$+0.100$
Cytochrome c (Fe^{3+}/Fe^{2+})	$+0.254$
Cytochrome a (Fe^{3+}/Fe^{2+})	$+0.290$
Fe^{3+}/Fe^{2+}	$+0.770$
$1/2\ O_2/O^{2-}$	$+0.810$

Respiratory chain phosphorylation: see Oxidative phosphorylation.

Respiratory control: see Respiratory chain; Oxidative phosphorylation.

Respiratory inhibitor: a compound which interferes in some way with the respiratory chain. There are three types, 1. Uncouplers (see) which prevent the synthesis of ATP without stopping the flow of electrons, 2. inhibitors of oxidative phosphorylation in the narrower sense, and 3. inhibitors of electron transport along the respiratory chain. The second type of A. inhibits both electron transport and the generation of ATP. The prototype of this group is oligomycin, which interferes with the use of energy-rich intermediates for ATP synthesis. Atractyloside, a plant poison, and the antibiotic bongkrekic acid inhibit the carrier which mediates the exchange of ADP and ATP across the mitochondrial membrane. Oxidative phosphorylation is then stopped for lack of ADP in the mitochondria. Another form of inhibition is exercised in the presence of certain monovalent cations by ionophores like valinomycin, nigericin, nonactin and gramicidin, etc. which divert the energy of respiration to mitochondrial ion transport. The points of attack of the 3rd type of inhibitor are shown in the Fig. to Respiratory chain (see). The order of the electron carriers in the respiratory chain was deduced by using this type of inhibitor. The last

Some palindromic sequences of double-stranded DNA recognized by specific restriction endonucleases (↓ indicates site of cleavage) and/or modification enzymes (*indicates methylated base).

Enzyme	Target palindromic sequence
Eco RI (from *Escherichia coli* strain R)	3'-CTTAAG-5' 5'-GAATTC-3'
Eco RI' (*E. coli* strain R)	3'-PyPyTAPuPu-5' 5'-PuPuATPyPy-3'
Eco RII (*E. coli* strain R)	3'-GGACC-5' 5'-CCTGG-3'
Eco (PI)	3'-TCTAGA-5' 5'-AGATCT-3'
Hin DII *(Haemophilus influenzae)*	3'-CAPuPyTG-5' 5'-GTPyPuAC-3'
Hpa I *(Haemophilus parainfluenzae)*	3'-CAATTG-5' 5'-GTTAAC-3'
Hae III *(Haemophilus aegypticus)* and Bsu x5 *(Bacillus subtilis)*	3'-CCGG-5' 5'-GGCC-3'
Hpa II *(Haemophilus parainfluenzae)*	3'-GGCC-5' 5'-CCGG-3'

member of the chain in aerobic organisms, cytochrome oxidase, is inhibited by cyanides, azides and carbon monoxide. The toxicity of the latter in mammals, however, is due to its affinity to hemoglobin, not to its inhibition of the respiratory chain.

Respiratory poison: see Respiratory inhibitor.

Restriction endonucleases (EC 3.1.23.1 to EC 3.1.23.45): enzymes present in a wide variety of prokaryotic organisms, where they serve to cleave foreign DNA molecules (e.g. phage DNA). R.e. recognize specific palindromic sequences (see Palindrome) in double-stranded DNA. Many R.e. are known, all with different specificities (Table). These represent a powerful set of tools, that are used for the analysis of chromosome structure. Host DNA is

Table 2. *Components of the mitochondrial respiratory chain.*

Complex	Component	Functional group	Activity
	Pyridine nucleotide-dependent dehydrogenases	NAD^+ or-$NADP^+$	
I	Flavoprotein$_D$	FMN Non-heme-iron	NADH: ubiquinone oxidoreductase
II	Flavoprotein$_s$	FAD Non-heme iron	Succinate: ubiquinone oxidoreductase
–	Ubiquinone (Coenzyme Q)	Reversibly reducible quinone structure	
III	Iron-sulfur proteins Cytochrome b (b_K and b_T) Cytochrome c_1	Iron sulfur centers Heme (noncovalent) Heme (covalently bound)	Ubiquinone: cytochrome c oxidoreductase
–	Cytochrome c	Heme (covalently bound)	
IV	Cytochrome a Cytochrome a_3	Heme a Copper protein Heme a	Cytochrome oxidase

protected from the activity of the host R. e. by methylation (by S-adenosyl-L-methionine). This methylation process is known as *modification* (Table). The phenomenon whereby foreign DNA introduced into a prokaryotic cell becomes ineffective owing to the action of R. e. is known as *restriction.*

Two types of *restriction-modification system* have been found in bacteria. In type I systems, the methylase and R. e. are both associated with a complex containing three different polypeptide chains: an α-chain with R. e. activity, a ß-chain with methylase activity and a γ-chain with the recognition site for the DNA sequence. Type I systems require S-adenosyl-L-methionine and ATP for both R. e. and methylase activities, they are less specific, and cleavage sites may be random and far removed (1000 base pairs) from the 5′ side of the recognition site. In type II systems, methylases and R. e. are separate, S-adenosyl-L-methionine donates the methyl group but does not take part in DNA cleavage. Type II systems do not require ATP, and they are also highly specific.

Restriction fragment length polymorphisms: variations in the lengths of nucleic acid fragments produced by endonuclease digestion, resulting from polymorphism in the genome. After digestion of DNA with an appropriate restriction endonuclease, the electrophoretic pattern of DNA fragments is analysed by the Southern blot procedure. R. f. l. p. have been used for assessing the structural heterogeneity of genes. They are also used in linkage studies as markers for the detection of disease, as tracers for the identification of alleles in pedigree analysis, and for determining the clonal origin of tumors. In linkage studies, the polymorphism may not exist within the gene in question, but is closely linked to it, e. g. the normal human globin gene and the sickle cell globin gene are linked to different *Hpa* I sites (*Hpa* I is a restriction endonuclease from *Haemophilus parainfluenzae*, with the target sequence GTT↓AAC). Thus *Hpa* I digestion of DNA containing the sickle gene produces a 13 kilobase fragment that is absent from digests of normal DNA, and the digest of normal DNA contains a 7.6 kilobase fragment that is absent from digests of DNA containing the sickle gene (Fig.). [D. Botstein et al., *Am. J. Hum. Genet. 32* (1980) 314–331; B. Vogelstein et al., *Science 227* (1985) 642–645]

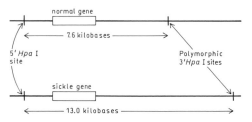

Polymorphic 3′Hpa I sites linked to normal and sickle β-globin genes

Restriction-modification system: see Restriction endonucleases.
Reticuline: see Benzylisoquinoline alkaloids.
Retinal: see Vitamins (Vitamin A).

Retinoic acid: see Vitamins (Vitamin A).
Retinol: see Vitamins (Vitamin A).
Retronecine: see Pyrrolizidine alkaloids.
Reversed electron transport: reversal of Oxidative phosphorylation (see) in which NAD^+ is reduced by an ATP-dependent reverse transport of electrons. R. e. t. occurs in organisms that oxidize hydrogen donors whose redox potential (see Oxidation) is more positive than that of the pyridine nucleotide coenzymes, and it operates in the oxidation of substrates not specific for NAD (see Respiratory chain), e. g. Succinate $+ NAD^+ \xrightarrow{ATP}$ Fumarate $+ NADH + H^+$. The redox system succinate/fumarate (E_o 0.00 V) is 320 mV more positive than the redox system $NAD^+/NADH + H^+$ (E_o −0.32 V); electrons are passed from succinate to flavoprotein in the respiratory chain, then via NADH-dehydrogenase to NAD^+. R. e. t. has been shown in nitrate bacteria *(Nitrobacter)*, insect flight muscle mitochondria and kidney mitochondria under anaerobic conditions. It is a feature of bacterial Photosynthesis (see).

Reversed phase chromatography: see Paper chromatography.
Reverse transcriptase: see RNA-dependent DNA-polymerase.
R_f-value: see Paper chromatography.
L-Rhamnose: 6-deoxy-L-mannose, a deoxyhexose. M_r 164.16, m. p. 122 °C, $[\alpha]_D^{20} + 38° \rightarrow +8.9°$. L.-R. is a component of many glycosides, e. g. anthocyanins, and of plant mucilages. It is biosynthesized from glucose.

Rhein: 1,7-dihydroxy-3-carboxyanthraquinone, a yellow anthraquinone, m. p. 321 °C, present free or as a glycoside in the roots of many higher plants, particularly rhubarb. R. has a purgative action, and various R.-containing drugs *(Radix Rhei, Folia Sennae)* are used therapeutically.

Rhesus factor, abb. *Rh-factor:* several closely related, blood group-specific erythrocyte antigens, present in 85% of Europeans (Rh-positive). The antigen is absent from the remaining 15%, who are Rh-negative. The natural antibody does not normally occur in humans, but is formed in rabbits or guinea pigs after immunization with rhesus monkey erythrocytes. If Rh-negative individuals come into contact with Rh-antigen, e. g. by blood transfusion, or from an Rh-positive fetus (the antigen crosses the placenta), Rh-antibodies are formed. Repeated transfusion may then lead to hemolysis, or an Rh-positive fetus may suffer hemolytic damage (erythroblastosis of the newborn).

Rh-factor: see Rhesus factor.
Rhizobia: bacteria of the genus *Rhizobium,* which can live free in the soil, or enter into a symbiotic relationship with leguminous plants. As leguminous symbionts, R. are responsible for the formation of root nodules and the fixation of atmospheric nitrogen (see Nitrogen fixation). In the nodule of the host plant, R. are present as bacteroids, which differ morphologically (ill defined globular shape) and biochemically (high rate of nitrogen fixation, and absence of ribosomes) from the free living form. Under defined laboratory conditions, i. e. in the presence of a dicarboxylic acid (e. g. succinate or fumurate) and a pentose (ribose, xylulose, arabinose),

pure cultures of free living R. show low rates of nitrogen fixation, but the free form probably does not fix nitrogen in the soil. Host specificity is fairly strict, and species are named accordingly, e.g. *Rhizobium leguminosarum, R. meliloti, R. trifolii, R. lupini, R. japonicum*, etc.

Rho (ρ) factor: see Ribonucleic acid.

Rhodoplasts: photosynthetic organelles of the red algae *(Rhodophyta)*. R. are red or red-violet, due to the presence of important light-trapping photosynthetic pigments, called Biliproteins (see). R. are responsible for the characteristic color of these marine algae.

Rhodopsin: see Vitamins (Vitamin A).

Rhodoxanthin: 3,3'-dioxo-β-carotene, a xanthophyll, M_r 562, m.p. 219 °C, found as a pigment in brown-red ("copper") leaves, in the needles of various conifers (e.g. yew) and in bird feathers.

RIA: abb. for Radioimmunoassay. See Immunoassays.

Ribitol: an optically inactive, C_5-sugar alcohol, biosynthesized by reduction of ribose. M.p. 102 °C. R. is a component of the flavin molecule, e.g. riboflavin (see Flavin-mononucleotide).

Riboflavin: see Vitamins (Vitamin B_2).

Riboflavin-adenosine-diphosphate: see Flavin-adenine-dinucleotide.

Riboflavin 5'-phosphate: see Flavin-mononucleotide.

Ribonuclease (EC. 3.1.27.5): a pancreatic phosphodiesterase specific for RNA. R. catalyses hydrolysis of the phosphate ester bond between pyrimidine nucleoside 3-phosphate residues and the 5-hydroxyl group of the neighboring ribose residues. The cleavage products are 3'-ribonucleoside monophosphates and oligonucleotides possessing a terminal pyrimidine nucleoside 3'-phosphate. Although R. is present in most vertebrates, high activities are only found in the pancreas of ruminants, certain rodents and some herbivorous marsupials. In ruminants, R. is required for digestion of the nucleic acids present in the rumen microflora, which account for a large proportion of the dietary nitrogen and phosphorus. Bovine pancreas contains four types of R. (A, B, C and D). *R. A*, which contains no carbohydrate, is the predominant form. The other R. are glycoproteins, which differ only in their sugar composition. Variations in the carbohydrate moiety gives rise to microheterogeneity and to the existence of pseudo-isoenzymes.

Primary structure (single chain, basic protein, 124 amino acid residues, M_r 13700), secondary structure (4 disulfide groups, about 15% α-helix, about 75% β-structures), tertiary structure (active center containing His_{12}, His_{119}, Lys_7 and Lys_{41}, located in a cleft, which divides the molecule into two halves, Fig.), and the mechanism of action of R.A. are known. The pH optimum is 7.0–7.5, and the isoelectric point is 7.8. It is inhibited by penicillin, basic dyes (e.g. acridine), EDTA and divalent cations (Cu^{2+}, Zn^{2+}). R.A. is relatively heat stable (heating to 85 °C causes no loss of activity). In the presence of 2-mercaptoethanol in 8 M urea, it is denatured to a random coil, which can be reoxidized under mild conditions to reform the fully active R. (reversible denaturation). Limited proteolysis with subtilisin

forms a peptide containing residues 1–20 (S-peptide): the remaining protein, known as S-protein or *R.S* (residues 21–124), has full enzymatic activity, providing it is still associated with S-peptide. Removal of S-peptide, by treatment with urea or acid, causes a reversible loss of activity. Only residues 2 and 14 of the S-peptide are necessary for maintaining the normal configuration and function of R. Even Val_{124} is unnecessary for R. activity. The only known dimeric R. was isolated from bovine seminal fluid; it consists of two identical polypeptide chains (each of M_r 14500 and containing 124 amino acid residues identical in sequence with pancreatic R.) joined by two disulfide bridges.

R. have also been isolated from fungi and bacteria. *R. T1* (EC 3.1.27.3) and *R. T2* (EC 3.1.27.1) are present as impurities in *Aspergillus* Taka-diastase. RT.1 (primary structure known, 104 amino acid residues, M_r 110902, IP 2.9, pH optimum 7.4) catalyses hydrolysis of internucleotide bonds between 3'-guanylic acid and 5'-hydroxyl groups. R.T2 catalyses hydrolysis between 3'-adenylate and 5'-hydroxyl groups of neighboring nucleotides. R.T1 and R.T2 were used in the sequence elucidation of tRNA. The bacterial nuclease of *Staphylococcus aureus* (primary and tertiary structures known, 149 amino acid residues, M_r 16800) is an extracellular, unspecific phosphodiesterase, which degrades both DNA and RNA to 3'-phosphomononucleotides and dinucleotides.

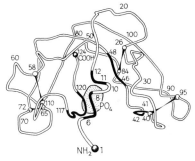

Schematic representation of the tertiary structure of the ribonuclease molecule. The continuous band represents the conformation of the α-C atoms of the peptide chain. Numbers show the localization of amino acid residues. Blacked-in parts of the peptide chain represent catalytically important regions. PO_4 = position of the phosphate residue between His_{119} and His_{12}.

R. is used therapeutically to inhibit tumor growth, for the treatment of certain chronic leukemias, and as an analgesic. In biochemistry, it is used in the investigation of nucleic acid structure, and for the removal of RNA from DNA or proteins. As a model protein, R. has made a considerable contribution to the understanding of protein chain conformation.

Ribonucleic acid, abb. *RNA*, obsolete *pentose nucleic acid:* a biopolymer of ribonucleotide units, present in all living cells and some viruses.

Structure. The mononucleotides of RNA consist of ribose phosphorylated at C3, and linked by an N-glycosidic bond to one of four bases: adenine, guanine, cytosine or uracil. Many other bases (chiefly methylated bases) also occur, but are less common (see Rare nucleic acid components). The mononucleotides form a linear chain via 3′,5′-phosphodiester bonds (see Nucleic acids). The base composition of RNA shows wide variations, but these do not lend themselves to mathematical interpretation and prediction, as for DNA. Sequence analysis of RNA has become a standard technique. (In many cases, the amino acid sequences of proteins are predicted from the sequence of the corresponding mRNA [or DNA] because it is much easier to clone the nucleic acid than to isolate the protein.)

RNA does not form a double-stranded α-helix. The single chains show partial folding into α-helical regions, probably by hydrogen bonding between complementary bases. These α-helical regions are separated by regions of single-stranded, unordered RNA.

Types of RNA and their occurrence. There are three main types of RNA, classified on the basis of their function: Messenger RNA (mRNA) (see), ribosomal RNA (rRNA) (see Ribosomes) and Transfer RNA (tRNA) (see); they also have different secondary and tertiary structures. Viral RNA is structurally and functionally very similar to mRNA.

In eukaryotic cells, RNA is found in the nucleus, cytoplasm and in the cytoplasmic organelles (ribosomes, mitochondria, chloroplasts). The nucleus is the chief site of RNA synthesis. rRNA is synthesized by the nucleoli, whereas high M_r, polydisperse RNA (precursor of cytoplasmic mRNA) is transcribed on the DNA of the chromatin. Low M_r RNA is also synthesized in the nucleus, consisting partly of tRNA and partly of RNA which has a regulatory function in gene activation.

In addition to tRNA, the cytoplasm also contains rRNA (in the ribosomes) and mRNA (in polysomes). Other nucleoprotein particles have also been demonstrated, which represent transport forms of mRNA (see Informosomes). All cytoplasmic RNA is synthesized in the nucleus. Mitochondria and plastids also contain mRNA, tRNA and rRNA, which are transcribed from the DNA of the organelle.

The less structured bacterial cell also contains tRNA (cytoplasm), rRNA (ribosomes) and mRNA (polysomes).

Function. In all cells, RNA functions in the transfer of genetic information from DNA to the site of protein biosynthesis (mRNA) and in the translation of this information during protein biosynthesis (mRNA, rRNA and tRNA). In addition, the RNA in *Escherichia coli* and *Bacillus subtilis* ribonuclease P has been shown to be responsible for the catalytic activity of the riboprotein. [C.Guerrier-Takada & S.Altman, *Science 223* (1984) 285–286.] The RNA in Small nuclear ribonuclear proteins (see) also appears to be catalytically active; other examples are likely to be discovered.

Biosynthesis. 1. *DNA-dependent RNA synthesis (transcription).* With the exception of viral RNA, all types of RNA are synthesized on a template of DNA. According to the principle of Base pairing (see), the base sequence of the DNA determines the synthesis of a complementary base sequence in the RNA (Fig.) The growing RNA chain is released from the template, so that the process can start again, even before synthesis of the previous molecule is complete. Nucleoside triphosphates, positioned next to each other by complementary pairing of their bases along the transcribing strand of DNA, become linked to form a polynucleotide by the removal of pyrophosphate from each triphosphate group. This reaction is catalysed by RNA polymerase (EC 2.7.7.6):

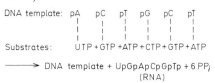

Subsequent hydrolysis of the pyrophosphate helps to shift the equilibrium of the reaction in favor of RNA synthesis. The process is well understood in bacteria, and can be considered in three stages:

a) *Initiation.* RNA becomes bound to a specific site on the DNA, known as the promotor (see Operon), which contains high concentrations of cytosine and thymidine. Sigma factor, a subunit of RNA polymerase (see), is required for the start of synthesis. Sigma factor also ensures that only the codogenic strand of the DNA is transcribed (strand selection). A short section of the double-stranded DNA comes apart and the first nucleotide (always ATP or GTP) becomes bound. An initiation complex, consisting of DNA, RNA polymerase and the first nucleotide, is formed.

b) *Polymerization.* RNA polymerase moves along the DNA. As it progresses, it opens the double helix and links the ribonucleoside monophosphate units in the 5′-3′ direction in the order dictated by the base of the codogenic DNA strand. As bases in the synthesized RNA strand leave their complementary sites on the DNA, the DNA base pairs reform and the double helix closes again.

c) *Termination.* The end of RNA synthesis and complete release of RNA from the DNA template are signalled by termination codons in the DNA, which are recognized by a termination factor, known as rho-(ρ)-factor.

There are several specific inhibitors of RNA synthesis, e.g. actinomycin, rifamycin, α-amanitin. Often, especially in eukaryotes, the primary transcription products are not identical with the functional, mature RNA molecules. The primary transcription products are usually high M_r and subject to successive stages of degradation, known as maturation or processing (see Post-transcriptional modification of RNA).

Transcription of each gene is regulated according to the requirements of the cell (see Gene activation; Operon; Enzyme induction; Enzyme repression; Messenger RNA.)

2) *RNA-dependent RNA synthesis.* Multiplication of RNA viruses depends upon an RNA-dependent reaction, in which viral RNA acts as a template for the synthesis of new RNA. The process is catalysed by RNA synthetase (see).

3) *Starter-dependent RNA synthetase.* In vitro, polynucleotides can be synthesized by the polymeriza-

tion of 5'-nucleoside diphosphates with the release of phosphate. The process requires the presence of an oligonucleotide, which serves as a starter (primer), and is catalysed by polynucleotide phosphorylase. The base sequence of the resulting polynucleotide is not encoded by a template, and it depends upon the relative concentrations of the different nucleoside diphosphates in the reaction mixture.

RNA degradation. RNA is continually degraded in the cell. It is cleaved by various ribonucleases, polynucleotide phosphorylases and phosphodiesterases. In strong acid, RNA is hydrolysed completely to bases, phosphate and ribose; alkaline hydrolysis produces 2'- and 3'-nucleoside monophosphates.

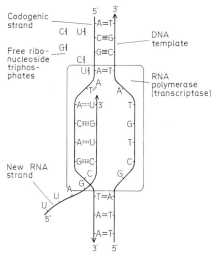

Schematic representation of transcription on a DNA template. A = Adenine, C = cytosine, G = guanine, U = uracil, T = thymine.

Ribonucleotide reductase (EC 1.17.4.1 or 1.17.4.2): an enzyme system that catalyses reduction of ribonucleotides to 2-deoxyribonucleotides. This is a stage in the biosynthesis of DNA precursors, and is the only metabolic route for the reduction of ri-

bose to deoxyribose. R. r. is subject to a complicated control mechanism, in which an excess of one deoxyribonucleotide inhibits reduction of all other ribonucleotides. This means that DNA synthesis can be inhibited by an excess of any one deoxyribonucleotide or deoxyribonucleoside (subsequently phosphorylated in the cell), and the enzyme has been widely investigated as a possible target for anticancer drugs. R. r. from *Escherichia coli* and mammals catalyse reduction of nucleoside diphosphates; in some bacteria, *Lactobacillus* and *Euglena,* the enzyme requires vitamin B_{12} and reduction occurs at the nucleoside triphosphate level. The oxygen at C2 is reduced and removed as water; the immediate reducing agent is Thioredoxin (see), which is itself reduced by NADPH and thioredoxin reductase (a flavoprotein). During reduction of the ribose moiety, C2-hydrogen exchanges with protons from water without loss of configuration.

Ribonucleotides: see Nucleotides.

D-ribose: a monosaccharide pentose. M_r 150.13, m. p. 87 °C, $[\alpha]_D^{20}$ −23.7 °C. It is not fermented by yeasts, and the dry solid occurs as the pyranose form. At 35 °C in aqueous solution it forms an equilibrium mixture: 6% α-furanose, 18% β-furanose, 20% α-pyranose and 56% β-pyranose. It is a component of RNA, some coenzymes, vitamin B_{12}, ribose phosphates and various glycosides. D-R. is prepared by acid hydrolysis of yeast nucleic acids, or by the chemical conversion of arabinose.

β-D-Ribose

Ribose phosphates: phosphorylated derivatives of ribose. Ribose 1-phosphate, and ribose 5-phosphoribosyl 1-pyrophosphate are metabolically important. Ribose is phosphorylated in position 5 by the action of ribokinase (EC 2.7.1.15) and ATP. Ribose 5-phosphate is also produced in the Pentose phosphate cycle (see), and in the Calvin cycle (see) of photosynthesis. Phosphoribomutase catalyses the interconversion of ribose 5-phosphate and ribose 1-phosphate, and the cosubstrate of this reaction is

Synthesis of ribose phosphates

ribose 1,5-bisphosphate. 5-Phosphoribosyl 1-pyrophosphate donates a ribose 5-phosphate moiety in the de novo biosynthesis of purine and pyrimidine nucleotides (see Purine biosynthesis; Pyrimidine biosynthesis), in the Salvage pathway (see) of purine and pyrimidine utilization, in the biosynthesis of L-Histidine (see) and L-Tryptophan (see) and in the conversion of nicotinic acid into nicotinic acid ribotide (see Pyridine nucleotide cycle). Ribose 1-phosphate can also take part in nucleotide synthesis (see Salvage pathway).

Ribosomal proteins: integral proteins of ribosomes. Prokaryotic ribosomes contain 35-40% protein, and eukayotic ribosomes contain 48-52% protein. The most extensively studied R.p. are those of *Escherichia coli*. The 50S-subunit contains 34 different L-proteins (L=large subunit), and the 30S-subunit contains 21 different S-proteins (S=small subunit) (Fig.). All 55 R.p. are immunologically distinct, except for L7 and L121. Each protein has been isolated and characterized with respect to M_r, amino acid composition, pK-value, stoichiometry within the ribosome, and specific interaction with rRNA. With the exception of protein S1 (M_r 65000), the M_r are all in the range 9000-28000. The pK-values are basic (pK > 9), except for S6 (pK 4.9), L7 (pK 4.8) and L12 (pK 4.9). The strong basic character of R.p. is due to their high contents of lysine and arginine. L7 and L12 have been sequenced; both contain 120 amino acid residues, and differ only by an *N*-terminal acetyl group in L7. The chemical and physical properties of L7 and L12 closely resemble those of contractile proteins, such as myosin and flagellin. 50S-Ribosomal subunits lacking L7 and L12 have no GTPase activity, but EFG-dependent GTPase is restored by the addition of either protein. When antibodies to L7 and L12 are added to a reconstitution mixture (see Ribosomes), the resulting particles lack GTPase activity. Antibodies to L7 and L12 also prevent formation of a complex between EFG, GTP, 50S-subunit and Fusidic acid (an inhibitor of translocation; see). It is therefore suggested that L7 and L12 are involved an GTP hydrolysis at translocation by acting as the binding site for EFT_U, EFT_S, and EFG on the 50S-ribosomal subunit (see Protein biosynthesis); they may even serve as contractile protein in the physical movement of the ribosome along the mRNA during translocation. The use of specific antibodies is a powerful tool in the study of the function and location of R.p. within the ribosome. All the S-proteins and most of the L-proteins are accessible to antibody, which indicates that these proteins are at least partly exposed on the surface of the ribosome. They are either bound covalently to rRNA (primary binding proteins), or they interact strongly with it. Each protein is probably dispersed within a matrix of RNA, so that there is very little, if any surface interaction between the different R.p. (see ribosome, Fig.2). This model is supported by studies with cross linking reagents; when the intersite distance of the reagent is 5 Å there is practically no cross linkage, relatively little cross linkage when the distance is 9 Å, and much more with an intersite distance of 12 Å. In order to study the function of ribosomal proteins, self assembly has been performed in the absence of the protein in question, or

in the presence of its mutant form. Spatial relationships have also been investigated by attempting reconstitution with cross linked proteins. Studies on the effect of the order of addition of each protein during reconstitution have been particularly informative. The grouping of proteins has also been studied by mild ribonuclease digestion of the intact 30S-subunit. This produces ribonucleoprotein fragments (Brimacombe fragments) of varying size, representing subsets of the original ribosomal subunit. For example, one such fragment contains S7, S9, S10, S13, S14 and S19, while another contains S4, S5, S6, S8, S11, S15, S16, S17, S18 and S20, representing two clusters on opposite sides of the 30S-subunit. Particles reconstituted in the absence of S4, S7, S8, S9, S16 or S17 are nonfunctional and show a grossly altered sedimentation rate; particles reconstituted in the absence of S3, S5, S10, S11, S14 or S19 may show small alterations in sedimentation behaviour, but have greatly impaired function; S1, S2, S6, S12, S13, S18, S20 and S21 are not required for assembly, but they are necessary or stimulatory for function.

S4, S8, S15 and S20 are bound particularly tightly to the 16S rRNA. As far as it is possible to attribute function to the various R.p., the following relationships have been shown: S1 (mRNA binding); S2, S3, S10, S14, S19, S21 (fMet-tRNA binding); S3, S4, S5, S11, S12 (codon recognition); S1, S2, S3, S10, S14, S19, S20, S21 (function of A and P sites); S9, S11, S18 (binding of aminoacyl-tRNA); S2, S5, S9, S11 (close proximity to GTPase). Functional ribosomes can be reconstituted from rRNA and R.p. from different organisms, e.g. 16S rRNA from *Bacillus stearothermophilus* and S-proteins from *Escherichia coli* form active hybrid 30S-subunits.

Proteins involved in binding the 50S- to the 30S-subunit are S20 (actually binds to 50S), S5 and S9 (both restore ability of 30S to combine with 50S, and antibodies to S9 prevent combination of 30S with 50S), S16 (becomes cross linked to 50S by cross linking reagents), S12 (binds to 23S rRNA), S11 (antibodies to S11 prevent combination of 30S with 50S, and S11 binds to 23S rRNA).

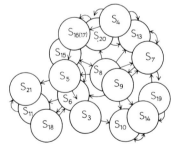

Three dimensional arrangement of the 21 S-proteins of the 30S subunit of the Escherichia coli ribosome. Arrows indicate the nature and intensity of the interactions between individual subunits. 16S RNA provides the framework for the assembly process. For further details concerning the order and nature of the interactions, see Nomura, M. (1973) *Science, 179,* 869.

The replicase of virus Qβ (an RNA-phage) consists of four subunits: one is encoded by the viral genome, while the other three are the host proteins EFT_U, EFT_S and the ribosomal protein S1.
The Fig. shows a three dimensional model for the arrangement of R.p. in the 30S-subunit. There is insufficient evidence for the construction of a three dimensional model including all 34 L-proteins. Many antibiotics act by combining with R.p., e.g. streptomycin interacts with S12, and bacterial mutants resistant to streptomycin have been shown to have an altered S12. Erythromycin interacts with L22, spiramycin with L4.

Ribosomal RNA: see Ribosomes.

Ribosomes, *monosomes:* the sites of Protein biosynthesis (see) in the cell. R. resemble giant multienzyme complexes. They are spherical to ellipsoid, highly hydrated cell organelles, 15 to 30 nm in diameter, normally present in the cytoplasm as Polysomes (see). They were first described in 1953 by Palade (Nobel Prize 1974).

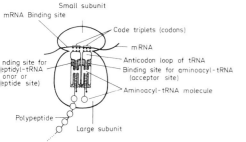

Figure 1. Schematic representation of a ribosome.

The number of R. in a cell is directly correlated with its capacity for protein synthesis. There are two main types of R., depending on size and origin (Table): 80S-R. from the cytoplasm of eukaryotic cells; and 70S-R. from prokaryotic cells, plastids and some mitochondria. The mitochondria of vertebrates contain 55S-R.
The subunit composition of R. depends essentially on the ionic concentration of the suspension medium, especially the Mg^{2+} concentration. If this is less than 0.001 M, the R. dissociate into two morphologically and functionally dissimilar subunits (Fig. 1). Thus 70S-R. consist of a large 50S and a small 30S subunit, whereas 80S-R. consist of 60S and 40S subunits. In the complete absence of Mg^{2+} (or in the presence of about 1 M monovalent cations, such as Li, Cs or K), the subunits dissociate into still smaller discrete ribonucleoprotein particles, known as core particles; at the same time, certain proteins (called split proteins) are removed. Core and split proteins are inactive in protein biosynthesis, but they can reassociate into functionally competent ribosomal subunits.

Composition and fine structure of R.: The three-dimensional structure of R. has been studied by electron microscopy (whole R., subunits, and subunits cross-linked by antibodies against individual ribosomal proteins); by low angle diffraction of X-rays and neutrons; and by Fourier reconstruction of elec-

Table. *Comparison of some properties of 70S- and 80S-robosomes*

	70S (Escherichia coli)	80S (mammals)
M_r ($\times 10^6$)	2.7	4.0
S-values of subunits	50 + 30	60 + 40
% RNA	65	50
S-values of high M_r rRNA	23 + 16	28 + 18
M_r of high M_r rRNA ($\times 10^6$)	1.1 + 0.56	1.7 + 0.7
GC content of high M_r rRNA (%)	54 + 54	67 + 59
Number of ribosomal proteins	34 + 21	about 70
Initiation of protein biosynthesis by	formyl-Met-tRNA_F	Met-tRNA_Met
Inhibition of protein biosynthesis:		
by chloramphenicol	+	−
by cycloheximide	−	+

tron micrographs of crystallized subunits. Repeated confirmation of the main features of R. structure by different authors has led to a "consensus" structure, in particular for the R. of *Escherichia coli* (Fig. 2). Proposed shapes of ribosomal subunits depicted in the first edition of the *Concise Encyclopedia of Biochemistry* (H.-G. Wittmann, *European J. Biochem.* 61 (1976) 1–13) are now transcended by the consensus structure. Both subunits are asymmetric. The large (50S) subunit consists of a central protuberance (head) and two side arms or protrusions (the ridge and the L7/L12 stalk) inclined at about 50 ° on either side of the head. The small (30S) subunit possesses a cleft or indentation which divides the structure into unequal parts, resulting in regions termed platform, head and base. In whole R. the platform and much of the base of the 30S subunit is in contact with the 50S subunit.
Using antibodies against purified ribosomal proteins, subunits can be cross-linked by bivalent antibody attachment. Examination of electron micrographs of the resulting subunit dimers permits identification of the site of attachment of the antibody molecule, a site which presumably represents an exposed part of the antigenic protein. In this way, several ribosomal proteins have been mapped on subunit surfaces (Fig. 2).
R. structure has been highly conserved during evolution, and it is similar for the most distantly related organisms. Nevertheless, electron microscopy reveals small but distinct differences in the shapes of R. from eubacteria, archebacteria, eocytes and eukaryotes (cytoplasmic R.), which have been used to interpret phylogenic relationships between these groups. [J. A. Lake, *Ann. Rev. Biochem.* 54 (1985) 507–530.]
R. contain only RNA and proteins, 65% RNA and 35% protein in the case of 70S and 50% each in the case of 80S R. Ribosomal RNA (rRNA) contains helical regions (about 70% of the total rRNA) which alternate with nonhelical regions. The latter are

linked to ribosomal proteins by specific ionic inter-actions and hydrogen bonding between nucleotides and amino acids (Fig. 3).

Total reconstitution of each ribosomal subunit from its separate RNA and protein components was first achieved by Nomura et al. between 1969 and 1972. Reassociation is spontaneous, and proceeds by a process of cooperative self assembly, i.e. all the in-formation needed for the correct assembly of a ribo-some is contained in the structure of its components. Reconstitution of the 30 S subunit proceeds at 40 °C

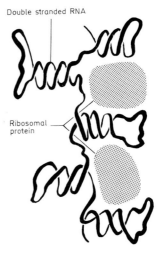

Double stranded RNA

Ribosomal protein

Figure 3. Schematic representation of the arrangemen *of ribosomal RNA and ribosomal protein within th ribosome.*

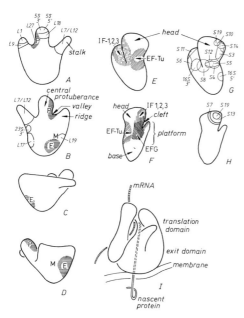

Figure 2. Consensus structure of the Escherichia coli 70S ribosome and its subunits. A, B, C, D are different orientations of the large (50S) subunit. *E, F, G, H* are two orientations of the small (30S) subunit. On the large subunit, E, M and P represent the nascent protein exit site, the membrane binding site, and the peptidyl transferase site, respectively. 23 S3′ indicates the position of the 3′ terminus of 23 S rRNA. On the small subunit, IF-1,2,3 represents the probable loca-tion of initiation factors 1, 2 and 3. EF-Tu represents the binding site of the EF–Tu:GTP:aminoacyl-tRNA complex (see Protein biosynthesis). EF–G represents the binding site of elongation factor G (see Protein biosynthesis) near the interface area with the large subunit. 16 S3′ and 16 S5′ indicate the positions of the 3′ and 5′ termini of 16 S rRNA. Numbers preceded by S and L represent ribosomal proteins of the small and large subunits, respective-ly, which have been mapped by electron microscop-ic visualization of subunit-antibody complexes.

I is a diagrammatic representation of the whole rib-osome, showing the probable location of messenger RNA and newly synthesized polypeptide, and the position and orientation of the ribosome with re-spect to the membrane of the endoplasmic reticulum during synthesis of secreted proteins.

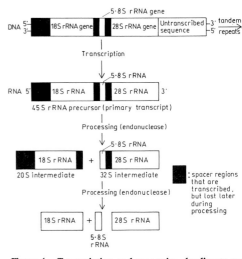

Figure 4. Transcription and processing, leading to ma ture mammalian 5.8S, 18S and 28S rRNA.

at high KCl concentration, and takes about 10 mi Reconstitution of the 50S subunit is slower and re quires higher temperatures. The reconstitution of eu karyotic ribosomal subunits has not been reported.

The various rRNAs can be separated by chromato graphy or electrophoresis. Prokaryotic rRNA con sists of 3 fractions: M_r 0.56×10^6 ($\approx$ 16S) from th 30S subunit; M_r 1.1×10^6 ($\approx$23S) and M_r 5000 ($\approx$5S) from the 50S subunit. Eukaryotic rRNA con sists of 4 fractions: M_r 0.7×10^6 ($\approx$ 18S) from th small subunit, and from the large subunit, 5S, 5.8 and a fraction whose size depends on its origin, i. M_r 1.3×10^6 ($\approx$25S) to 1.75×10^6 ($\approx$29S). 5.8 rRNA has no size counterpart in prokaryotes, and is generally thought to be characteristic of eukary tes. Its sequence corresponds to that of the first 150 approx.) nucleotides of prokaryotic 23 S rRNA. The parasitic eukaryote *Vairimorpha necatrix* (Microsporidia), however, has no 5.8S rRNA. Microsporidia ribosomes are also prokaryotic in size (70 S; subunits 30 S and 50 S, containing 16 S- and 23 S-like RNAs). 23 S-like *V. necatrix* rRNA, 25 S rRNA of *Saccharomyces cerevisiae*, and *Escherichia coli* 23 S RNA all have a certain degree of structural and sequence homology. [C. R. Vossbrinck & C. R. Woese, *Nature 320* (1986) 287–288.]

Currently determined rRNA sequences are published regularly by V. A. Erdmann et al. in supplements to the journal *Nucleic Acids Research*. These rRNA sequences are part of the Berlin Databank, which is online accessible worldwide, and is continually updated as soon as a new RNA sequence is available.

Biogenesis and processing of rRNA: 28 S, 18 S and 5.8S eukaryotic rRNA are transcribed as a single large 45 S RNA bound to protein in the Nucleolus (see). The 45 S RNA is rapidly degraded by specific endonucleases, in several stages, to produce mature 28 S, 18 S and 5.8S rRNA (Fig. 4). Thus after processing, only about 45% of the transcribed 45 S precursor RNA appears in R. The other cleavage products are apparently destroyed.

In prokaryotic cells, 23 S and 16 S rRNA are formed in tandem; the precursor molecule is only 10% longer than the two mature rRNAs, and processing consists of the removal of 200 to 250 nucleotides from the 5' ends of the 23 S and 16 S precursors. Prokaryotic and eukaryotic 5 S rRNA is always transcribed separately from the other rRNA. Methylated bases are characteristic components of all rRNA.

tive, and a Rare nucleic acid component (see) found in tRNA (as the mononucleotide unit of ribothymidylic acid, more correctly thymidylic acid). See Pyrimidine biosynthesis; Thymidine phosphates.

Ribothymidylic acid: see Ribothymidine; Pyrimidine biosynthesis.

Ribotides: see Nucleotides.

Ribozyme: see Intron.

D-Ribulose: a monosaccharide pentulose. M_r 150.13, $[\alpha]_D^{21} + 16$ °. The 5-phosphate and 1,5-diphosphate of D-R. are important intermediates of carbohydrate metabolism. Ribulose 1,5-bisphosphate is the CO_2 acceptor in the dark reaction of photosynthesis (see Calvin cycle), and ribulose 5-phosphate is an intermediate in the Pentose phosphate cycle (see).

Ribulosebisphosphate carboxylase, carboxydismutase, pentose phosphate carboxylase (EC 4.1.1.39): the photosynthetic carboxylation enzyme of green plants and purple and green bacteria. It is structurally bound in the thylakoids, but is easily solubilized during isolation. The enzyme consists of two types of subunit; the larger subunit is encoded in the nuclear genome of the plant, the smaller subunit in the plastid DNA. A high proportion of total leaf protein is accounted for by R. c.; in spinach it makes up 16% of leaf protein. The spinach enzyme has M_r 560000, consisting of eight pairs of nonidentical subunits and eight active sites. The K_m of the isolated enzyme for total CO_2 $(CO_2 + HCO_3^-)$ is surprisingly high (11–30 mM), but it is lower (0.6 mM) in the chloroplast, probably due to allosteric activation by fructose 6-phosphate and ribose 5-phosphate within the chloroplast. The catalysed reaction is:

Ribulose
1,5-bisphosphate

3-Phosphoglycerate

At every stage of maturation, the various RNA fractions are associated with proteins (see Ribosomal proteins), most of which are basic. Some of these proteins may be endonucleases involved in the maturation process, but the nature and function of these proteins is as yet poorly understood. During the maturation process, proteins are lost and exchanged, the true ribosomal proteins appearing at the end of maturation. It is unclear to what extent the ribosomal proteins are synthesized within the nucleus, or transported into the nucleus after synthesis in the cytoplasm. It is certain, however, that ribosomal subunits are first assembled in the nucleolus, then exported into the cytoplasm.

Ribosylthymine: see Ribothymidine.

Ribothymidine, more correctly *thymidine, ribosylthymine:* 5-methyluridine, a pyrimidine deriva-

In the presence of high concentrations of O_2, R. c. acts as an oxygenase, and catalyses the oxygenation and cleavage of ribulose 1,5-bisphosphate into phosphoglycolate and 3-phosphoglycerate. Subsequent metabolism of glycolate (derived from phosphoglycolate) contributes to the phenomenon of Photorespiration (see).

Richner-Hanhart syndrome: see Inborn errors of metabolism.

Ricin: a toxalbumin phytotoxin from *Ricinus* seeds. M_r 66000, 493 amino acid residues. R. inhibits protein biosynthesis (causes dissociation of polysomes) and has antitumor properties. It consists of an A-chain (M_r 32000) and a B-chain (M_r 34000) joined by disulfide bridges. After reductive separation by 2-mercaptoethanol, both chains show increased inhibitor activity but markedly decreased

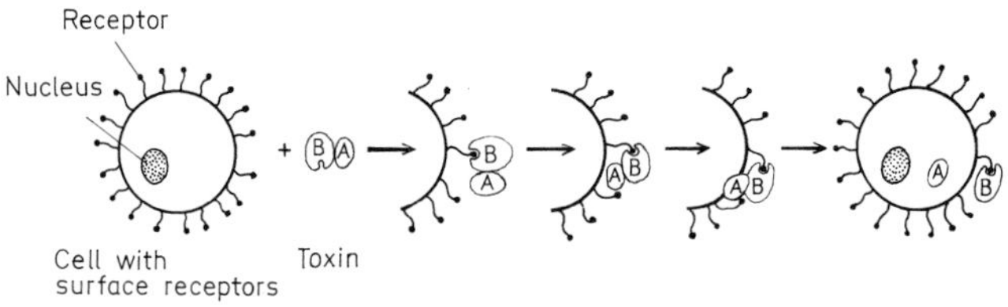

Ricinin

Hypothetical scheme for the interaction of a toxin (ricin or abrin) and a eukaryotic cell. A A-chain, B B-chain.

toxicity. Toxic activity is carried by the A-chain (effectomer), and the B-chain (haptomer) binds the toxin to the cell surface. Similar action and structure are possessed by **abrin** (M_r 65000; A-chain 30000, B-chain 35000), a toxalbumin from the red seeds of *Abrus precatorius*. Abrin is used in opthalmology.
Entry of R. or abrin into the cell occurs in two stages (Fig.). In the first stage, the toxin becomes bound by its B-chain to the terminal galactose of the receptor on the cell surface. In the second stage, a disulfide cleaving system of the cell releases the A-chain, which enters the cell by endocytosis. The A-chain binds to the 60S ribosomal subunit, which is then incapable of reacting with elongation factor EF2, and protein synthesis stops. Other toxic proteins are thought to act by an analogous two-stage mechanism.
Ricinin: a poisonous pyridine alkaloid from seeds of *Ricinus communis*. M_r 164.17, m.p. 201°C, b.p.$_{20}$ 170–180°C. R. is an exceptional alkaloid, in that it occurs in only one type of plant and is not accompanied by other alkaloids. R. is biosynthesized from nicotinic acid. Biosynthetic precursors of nicotinic acid are aspartic acid and a 3-carbon compound (probably hydroxyacetone phosphate). Administration of radioactive precursors to *Ricinus*, shows a high incorporation of ^{14}C from aspartate and glycerol into R.

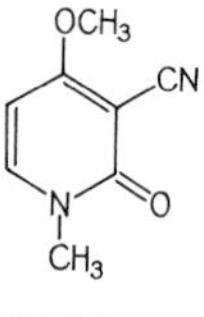

Ricinin

Ricinoleic acid: 12-hydroxyoleic acid, CH_3 $-(CH_2)_5-CHOH-CH_2-CH=CH-(CH_2)_7-COOH$, a fatty acid, M_r 298.45, m.p. (α-form) 7.7°C, (β-form) 16.0°C, b.p.$_{15}$ 250°C. R.a. is present in the acylglycerols of castor oil (*Ricinus* oil) where it accounts for 80–85% of the total fatty acids. It is also present in maize oil, wheat oil and various other vegetable oils.
Rickets: vitamin D deficiency disease. See Vitamins.
Rieske protein: a very electropositive iron-sulfur protein, containing a 2Fe/S center, and accompanying the *b* cytochromes and cytochrome c_1 in the respiratory chain complex III. The R.p. is in rapid equilibrium with cytochrome c_1.
Rifamycins: a group of antibiotics produced by *Streptomyces mediterranei*. They contain a naphthalene ring system bridged between positions 2 and 5

by an aliphatic chain. Rifamycin SV and rifampici inhibit DNA-dependent RNA synthesis in prokary otes, chloroplasts and mitochondria, but not in th nuclei of eukaryotes. Inhibition is due to the forma tion of a stable complex between RNA-polymeras and R.; binding of the enzyme to DNA still occur but incorporation of the first purine nucleotide int RNA is prevented. Thus R. specifically inhibit ir itiation of RNA synthesis, but not chain elongation Some R. also inhibit eukaryotic and viral RNA-po lymerases.

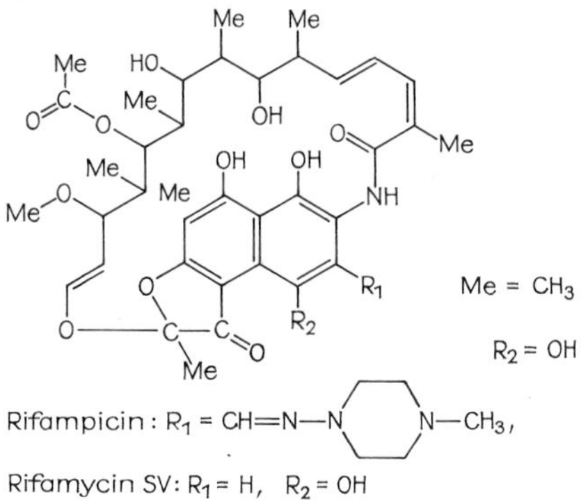

Me = CH_3

R_2 = OH

Rifampicin: R_1 = CH=N—N   N—CH_3,

Rifamycin SV: R_1 = H, R_2 = OH

Rishitin: see Phytoalexins.
RNA: abb. for Ribonucleic acid (see).
RNA-dependent DNA-polymerase, *revers transcriptase*: an enzyme present in retroviruse: some of which cause cancer, e.g. avian myeloma vi rus and various leukemia viruses. These are RNA vi ruses, and the enzyme catalyses the synthesis of th provirus DNA, using the viral RNA as a template The resulting DNA is then incorporated into the ge nome of the infected cell. Study of the process o RNA-dependent DNA synthesis, involving severa virus-specific enzymes, may therefore contribute t the understanding of malignant transformation o cells by RNA viruses and the problem of cancer i general. Occurrence of the enzyme in cells or viruse can be used for the diagnosis of oncogenic viruses.
DNA synthesis is analogous to transcription or re plication, i.e. RNA acts as a template for the forma tion of a base-complementary molecule of single stranded DNA. The latter is replicated by the actio of a DNA-dependent DNA-polymerase, thus form ing double-stranded DNA, which can be furthe replicated (this is the provirus). Host DNA i cleaved by a viral endonuclease, and the proviru DNA is inserted by the action of a ligase.
RNA-dependent RNA-polymerase: see RN/ Synthetase.

RNA-dependent RNA synthesis: see RNA Synthetase.

RNA polymerase, *DNA-dependent RNA polymerase, nucleoside triphosphate:RNA nucleotidyltransferase, transcriptase* (EC 2.7.7.6): an enzyme which catalyses the synthesis of RNA on a DNA template. The base sequence of the resulting RNA is complementary to that of the DNA template (see Ribonucleic acid).
There are at least four types of RNA-P. from eukaryotic cells: RNA-P. I is found in the nucleolus and preferentially catalyses the synthesis of rRNA. RNA-P. II is present in the nucleoplasm, is mainly responsible for the synthesis of mRNA, and is specifically inhibited by α-amanitin. RNA-P. III transcribes tRNA and other classes of small RNA. The fourth RNA-P. is smaller than the other three and transcribes RNA from mitochondrial DNA; however, it is encoded by a nuclear gene. The RNA-P. found in chloroplasts and mitochondria show similarities with prokaryotic RNA-P., e.g. inhibition by rifamycins.
There is extensive sequence homology between the largest subunits of RNA-P. II and III from yeast; these subunits also display great homology with the largest subunit (β') of prokaryotic RNA-P. [L. A. Allison et al. (1985) *Cell* **42** 599–610]. The largest subunit of *Drosophila* RNA-P. II has been shown to be involved with chain elongation and may, like the homologous β' subunit of *Escherichia coli*, interact with promoters of genes. [J. Biggs, L. L. Searles & A. L. Greenleaf (1985) *Cell* **42** 611–621.]
The structure of bacterial RNA-P. is known; the native enzyme has M_r 500000 and is a complex of five subunits: $\alpha_2\beta\beta'\sigma$. M_r of the subunits: α_2 40000, β 155000, β' 165000, σ 90000. The enzyme has different functional parts: the initiation site binds nucleoside triphosphate, the polymerization site catalyses the formation of internucleotide linkages. The β'-subunit is responsible for binding the enzyme to the template DNA. The σ factor contains a binding site for specific nucleotide sequences in the codogenic DNA strand (promoter region); it therefore acts as the initiation factor for the synthesis of RNA exclusively on the codogenic strand. Termination of RNA synthesis, signalled by specific regions of the DNA, depends on the rho (ρ) factor, a protein which is not considered to be part of the enzyme.
Other extra protein factors appear to be necessary for the specific function of bacterial RNA-P., e.g. psi (ψ) factor for promoter initiation in ribosomal RNA synthesis, or kappa (κ) factor as a further termination factor.
Infection of cells with DNA phage may lead to the production of a new, phage-specific RNA-P. (e.g. T7 phage; RNA-P. is a single-chain protein, M_r 107000), or the host RNA-P. may be modified with the addition of new protein subunits encoded by the phage (e.g. T4 and λ phage); in each case the resulting RNA-P. is specific for transcription of phage DNA.

RNA-Replicase: see RNA-Synthetase.

RNA-Synthetase, *RNA-replicase, RNA-dependent RNA-polymerase:* an enzyme that appears in bacterial, animal and plant cells, following infection with RNA viruses. Using a template of single-

stranded (viral) RNA, the enzyme catalyses the synthesis of a complementary strand of RNA. The substrates are 5'-nucleoside triphosphates, which become linked by a phosphodiester bond between positions 3' and 5' of neighboring ribose units, with the release of pyrophosphate. The newly synthesized RNA is hydrogen-bonded by base-pairing to the template RNA, thus forming double-stranded RNA (replicative form). This consists of a minus strand (which serves for the synthesis of a new plus strand), and a plus strand (the original viral RNA). At least some of the subunits of the RNA-S. are encoded by the viral genome (see Phage development).

Robinin: see Flavones (Table).

Robinson ester: see Glucose 6-phosphate.

Robustaflavone: see Biflavonoids.

Rocellic acid: (S)-2-dodecyl-3-methylbutanedioic acid; (s)-2-dodecyl-3-methylsuccinic acid, a branched chain dicarboxylic acid found in lichens. M.p. 132 °C, $[\alpha]_D + 17.4$ ° (ethanol). R.a. is biosynthesized from itaconic acid.

Rotenoids: chromanochromanones present in plants (especially roots) of the *Papilionaceae*, a subfamily of the *Leguminoseae*, notably in the related genera, *Derris*, *Lonchocarpus*, *Tephrosia* and *Mundulea*. R. are powerful insecticides, and preparations of R.-containing plants are used as garden sprays and dusts (derris dust). R. are toxic to humans and animals only in large doses, and they are absorbed more efficiently by the lungs than the alimentary canal. R. are also active piscicides. R.-containing plants are among those used by native fisherfolk for paralysing river fish (*Derris* in Asia, *Lonchocarpus* in South America, *Tephrosia* on all four continents).
The most important R. is rotenone (Fig.), which gives its name to the group.

R = H, **Rotenone** (*Derris* spp., *Tephrosia* spp., *Lonchocarpus* spp., *Piscidia erythrina*, *Neorautanenia ficifolia*, *Pachyrrhizus erosus*).
R = OH, **Sumatrol** (*Derris malaccensis*, *Piscidia erythrina*).
Biosynthesis. Tracing studies (^{14}C) with *Derris elliptica* plants and germinating seeds of *Amorpha fructicosa* have shown that ring A of rotenone is derived from the aromatic ring of phenylalanine, and that C-1 and C-2 of the phenylalanine side chain provide C-12 and C-12a of rotenone, respectively; this indicates that an aryl migration occurs from C-3 to C-2 of phenylalanine. Structural comparison suggests that R. are derived from isoflavonoids by addition of the methylene carbon C-6, which closes ring B. C-6

of rotenone is derived from the methyl group of methionine. [L. Crombie, *J. Chem. Soc.* (C) Org. (1963) 3029–3032.]

They are always accompanied by isoflavones.

All R. so far examined have positive Cotton effects, and therefore form one stereochemical series with respect to the 6a and 12a chiral centers (6aS, 12aS). [J. Claisse et al., *J. Chem. Soc.* (1964) 6023–6036.]

Rotenone is used experimentally as an inhibitor of mitochondrial respiration. It has little effect on the mitochondrial oxidation of succinate, but it powerfully inhibits all oxidations which operate via NADH dehydrogenase. The site of inhibition by rotenone has been located on the oxygen side of the nonheme iron of NADH dehydrogenase (see Respiratory chain). Piericidin and amytal appear to act at or very close to the same site. [W. W. Wainio, *The Mammalian Mitochondrial Respiratory Chain* (Acadamic Press, 1970); T. P. Singer & M. Gutman, *Adv. Enzymol.* **34** (1971) 79–153; J. B. Harborne, T. J. Mabry & H. Mabry, eds. *The Flavonoids* (Chapman & Hall, 1975)]

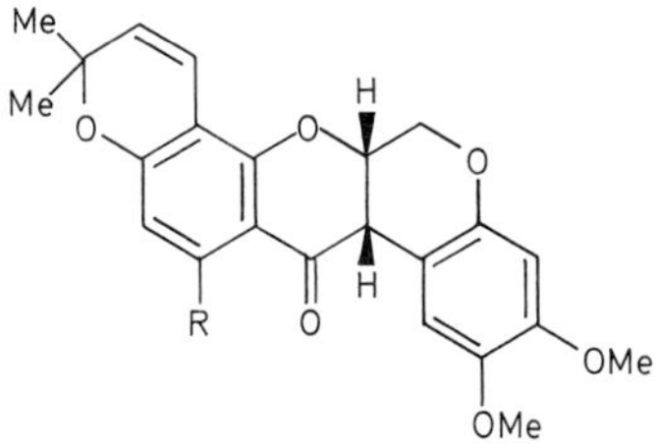

Rubixanthin

atrum type. M_r 413.65, m.p. 242 °C, $[\alpha]_D + 19$ ° (ethanol). It occurs in hellebore *(Veratrum album, V. nigrum* and *V viride)* and differs structurally from solanidine (see α-Solanine) by the presence of a 12α-hydroxyl group.

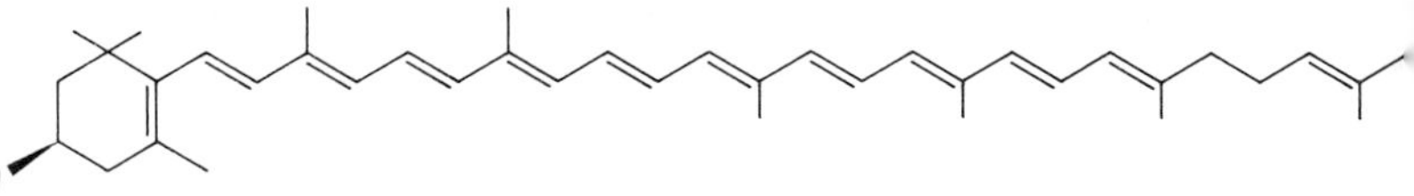

Rubixanthin: 3(R)-β,ψ-carotene-3-01; 3(R)-hydroxy-γ-carotene, a xanthophyll. M_r 552, m.p. 160 °C. R. is a copper-red pigment of various rose species and some other higher plants. The 5'-*cis*-isomer, gazaniaxanthin, is the pigment of various *Gazania* species.

Rubredoxin: a redoxin, functionally similar to Ferredoxin (see). M_r 6000. R. was isolated from *Clostridium pasteurianum;* synthesis of R. appears to be promoted by a relative deficiency of iron. It contains less iron than ferredoxin, usually containing one Fe atom/molecule of protein. Under acid conditions it is more stable than ferredoxin, and it has a more positive redox potential (E_o −0.057 V); thus

R = H, Deguelin *(Tephrosia* spp., *Derris* spp., *Lonchocarpus nicosa)*
R = OH, α-Toxicarol *(Tephrosia* spp., *Derris* spp.)

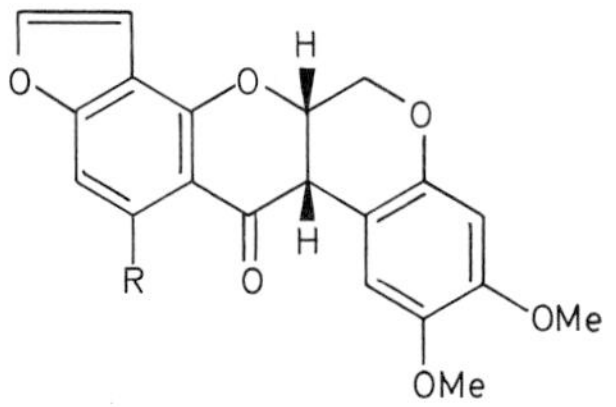

R = H, Elliptone *(Derris elliptica)*
R = OH, Malaccol *(Derris malaccensis)*

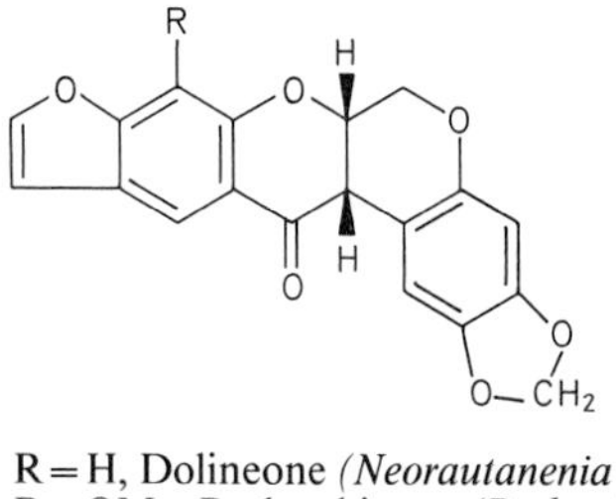

R = H, Dolineone *(Neorautanenia pseudopachyrhiza)*
R = OMe, Pachyrrhizone *(Pachyrrhizus erosus)*

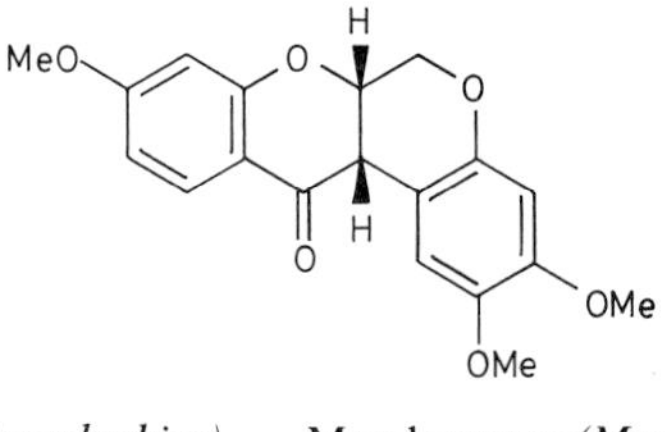

Munduserone *(Mundelea sericea)*

Some typical rotenoids

rRNA: ribosomal RNA (see Ribosomes).

Rubber: 1) vulcanized caoutchouc. 2) the water-soluble components of rubber resin. (See Caoutchouc).

Rubijervine: 12α-hydroxysolanidine; solanid-5-ene-3β, 12α-diol, a Veratrum alkaloid of the jerver-

when R. replaces ferredoxin in a ferredoxin-dependent reaction the reaction rate is decreased. The iron is bound by coordination with 4 cysteinyl residues; other possible ligands are tyrosine and lysine. Redoxins similar to R. have been isolated from *Peptostreptococcus elsdenii* and other bacteria. R. from *Mi-*

crococcus aerogenes contains 53 amino acid residues of known sequence.

Rubrosterone: 2β,3β,4α-trihydroxy-5β-androst-

Rubrosterone

7-ene-6,17-dione, a plant steroid. M_r 334.42, m.p. 245 °C, $[\alpha]_D + 119$ ° (methanol). R. was isolated, together with ecdysterone, from the roots of *Amarantha obtusifolia* and *A. rubrofusca,* and is considered to be a biogenetic degradation product of ecdysone.

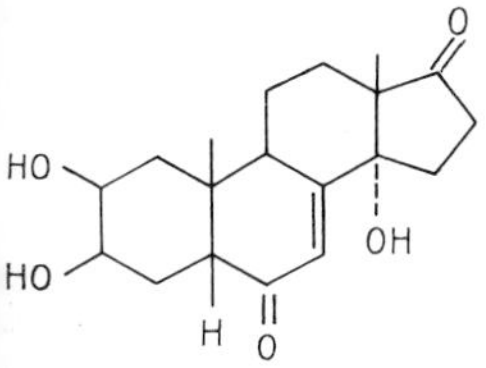

Rutaceae alkaloids: alkaloids from common rue (*Ruta graveolens* L.) and other members of the *Rutaceae.* They include quinoline, furanoquinoline, pyranoquinoline and acridine compounds, and are biosynthesized from anthranilic acid. The furanoquinolines have spasmolytic activity, and the drugs are sometimes used therapeutically.

Rutin: see Flavones (Table).

S

S, *Svedberg unit:* see Sedimentation coefficient.

Sabininic acid: 12-hydroxylauric acid, $HOCH_2-(CH_2)_{10}-COOH$, a fatty acid, M_r 216.31, m.p. 84 035, present as a typical wax acid in the wax of many pines.

Safynol: *trans,trans*-3,11-tridecadiene-5,7,9-tri-ine-1,2-diol, a Phytoalexin (see). S. and Δ^3-dehydro-safynol are formed by safflower (*Carthamus tinctorius*) following infection by *Phytophthora*. ED_{50} is 12 µg/ml for S. and 1.7 µg/ml for the dehydro derivative.

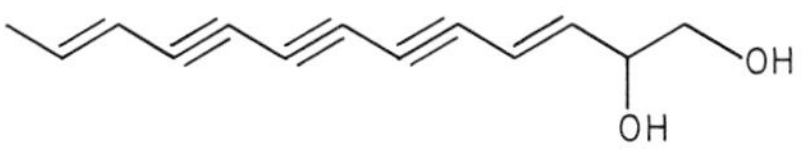

Safynol

Salamander alkaloids: toxic steroid alkaloids secreted by the skin glands of salamanders, e.g. *Salamandra maculosa* (European fire salamander). S.a. are modified steroids, in which the A-ring is extended to a seven membered ring by a nitrogen between C2 and C3 (A-azahomosteroids). They excite the central nervous system and cause paralysis. The chief representative, **samandarine** (1α, 4α-epoxy-3-aza-A-homoandrostane-16β-ol) (Fig.) also causes hemolysis (lethal dose for mice 1.5 mg/kg). *Samandarone* possesses a keto group in place of the hydroxyl group at position 16. *Samandaridine* (contains a lactone bridge (-CH₂-CO-O-) between C16 and C17, and also possesses local anesthetic properties.

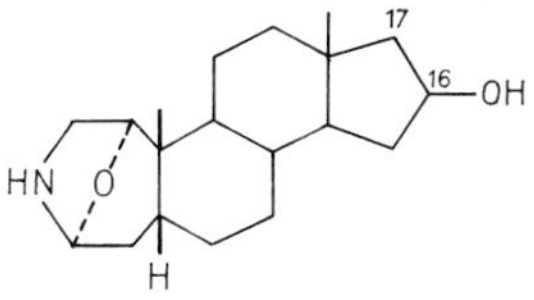

Samandarin

Salamander toxins: toxins secreted by the skin glands of *Salamandra maculosa* (European fire salamander) and *Salamandra atra* (alpine salamander). S.t. include the Salamander alkaloids (see), biogenic amines (tryptamine, 5-hydroxytryptamine), and high M_r substances that cause skin irritation and hemolysis.

Salmine: see Protamines.

Salsola alkaloids: a group of simple isoquinoline alkaloids, occurring in *Salsola* spp. Chief representative is salsoline (1-methyl-6-hydroxy-7-methoxy-1,2,3,4-tetrahydroisoquinoline), which occurs in the D-form, m.p. 215-216 °C, $[α]_D+40°$ (Water) and in the DL-form.

Salsoline: see Salsola alkaloids.

Salting in: see Proteins.

Salting out: see Proteins.

Salvage pathway: utilization of preformed purine and pyrimidine bases for nucleotide synthesis. In addition to de novo synthesis, the S.p. represents an alternative pathway for formation of purine and pyrimidine nucleotides. In mutant microorganisms lacking de novo purine and pyrimidine synthesis,

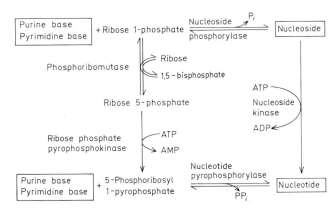

Synthesis of nucleosides and nucleotides from preformed bases.

the S.p. is the only route for nucleotide synthesis, following administration of exogenous purine and pyrimidine bases.

In liver, specific pyrophosphorylases catalyse the synthesis of nucleotides from free bases and 5-phosphoribosyl 1-pyrophosphate; alternatively, nucleosides may be formed first from bases and ribose 1-phosphate by the action of a nucleoside phosphorylase (Fig.). Deoxynucleosides can be formed from bases and deoxyribose 1-phosphate by the action of deoxynucleoside phosphorylase.

SAM: abb. for S-adenosyl-L-methionine.

Samandaridine: see Salamander alkaloids.

Samandarine: see Salamander alkaloids.

Samandarone: see Salamander alkaloids.

sAMP: abb. adenylosuccinate.

Sandhoff's disease: see Inborn errors of metabolism.

Sanfilippo syndrome: see Inborn errors of metabolism.

Sangivamycin: 4-amino-5-carboxamide-(D-ribofuranosyl)-pyrrolo-(2,3-d)-pyrimidine, a deazaadenine-type antibiotic from *Streptomyces spec.* (see Nucleoside antibotics). The antibiotic activity of S. is similar to that of Toyocamycin (see).

Sapogenins: see Saponins.

Saponins: a large and widely distributed group of plant substances, named from their ability to form strongly foaming, soap-like solutions with water. They are all glycosides, and are classified according to the nature of the aglycon (also called a genin), i.e. steroid, triterpene and steroid-alkaloid saponins; the latter are also known as glycoalkaloids. Aglycons of the steroid saponins are also called sapogenins; these are spirostane-type, C_{27}-steroids. They all posses a 3β-hydroxyl group (spirostanols), which forms a glycosidic linkage with the sugar, e.g. D-glucose, D-galactose or L-arabinose. The triterpene saponins are less well studied; their aglycons are chiefly tetra and pentacyclic triterpenes.

S. are powerful surfactants, cause hemolysis and are potent plasma toxins and fish poisons, but they have no toxic effects when ingested by humans, because they are not absorbed. Many S. have antibiotic activity, mainly against lower fungi. With steroids, e.g. cholesterol, S. form poorly soluble 1:1 molecular compounds, which can be used for the analytical separation of S. or steroids.

S. are biosynthesized by cyclization of 2,3-epoxysqualene (see Squalene); in the case of the steroid S., cholesterol is a subsequent intermediate in the biosynthesis of the spirostane skeleton. The stage at which the sugar becomes attached and the mechanism of this attachment are not known.

S. have been found in over 100 plant families; they occur, e.g. in soapwort, rape, soybean, foxglove, sycamore. Important representatives are Digitonin (see) and Dioscin (see). S. are used as detergents and foaming agents, and have been used since antiquity as fish poisons.

Saprophytism: a heterotrophic mode of nutrition, in which dead, organic material serves as the substrate. Many bacteria and most fungi are saprophytic; it is nevertheless possible to grow some fungi and to achieve fruiting body formation on entirely synthetic media.

Many carbohydrates can serve as carbon sources for saprophytes, and in some cases, alcohols, fats, organic acids or hydrocarbons are used. Proteins are also used as a carbon source.

Biological decay and decomposition are due to S., which is therefore important in the cycling of elements in the biosphere. Such processes may be primarily concerned with the incorporation of nutrients for growth, or with energy production, e.g. fermentation. For utilization for growth of the saprophyte, all nutrients must ultimately be converted to primary metabolites, like glucose, or intermediates of glycolysis or the TCA-cycle.

There are many different nitrogen sources. Some molds, yeasts and bacteria are able to utilize inorganic nitrogen, e.g. nitrate or ammonium salts. Certain soil bacteria can even assimilate atmospheric nitrogen. Other saprophytes require organic nitrogen in the form of amino acids, peptones or proteins. Many intermediate stages exist between S. and Parasitism (see).

Sarcine: an obsolete name for hypoxanthine.

Sarcosine, abb. *Sar, N-methylglycine:* $CH_3-NH-CH_2-COOH$, an intermediate in the metabolism of choline in liver and kidney mitochondria (see One-carbon cycle). Sar has also been isolated from starfish and sea urchins, where it appears to be a major metabolite.

Sarcosine dehydrogenase (EC 1.5.99.1): a mitochondrial flavoprotein (the flavin is covalently bound) catalysing conversion of sarcosine to glycine and a one-carbon unit at the oxidation level of formaldehyde (or bound $-CH_2-$). The metabolic fate of the one-carbon unit depends on the availability of tetrahydrofolate (THF). In the absence of THF, the products are glycine and formaldehyde. See One-carbon cycle.

Sarcosinemia: see Inborn errors of metabolism.

Sarcosomes: see Mitochondria.

Sargasterol: (20S)-stigmasta-5,24(28)-diene-3β-ol, a phytosterol (see Sterols), M_r 412.7, m.p. 132 °C, $[\alpha]_D$ −48 ° (chloroform), in brown algae, e.g. *Sargassum ringolianum.* It differs from Fucosterol (see) by the opposite configuration at C20.

Satellite DNA: DNA fractions that can be separated from the main DNA by CsCl-gradient centrifugation. S.DNA has been demonstrated in the nuclei of many eukaryotes, accounting for about 1% of nuclear DNA in man, and about 10% in the mouse. It consists of a linear double helix, and differs markedly from the rest of the nuclear DNA with respect to base composition and density. It contains more 5-methylcytosine than nuclear DNA. The DNA of the nucleolus is S.DNA, containing the cistrons for ribosomal RNA. With the aid of cytological hybridization experiments, it has also been demonstrated in various parts of the chromosomes, where its function is unknown. S.DNA often shows a high level of redundancy: Mouse S.DNA contains about 10^6 copies of each 150-300 nucleotide pairs; in guinea pig S.DNA, there are about 10^7 copies of the 6 base sequence, GGGAAT. This high level of redundancy results in a rapid rate of renaturation, following heat denaturation.

Extrachromosomal DNA is also considered to be S.DNA.

Saxitoxin: a neuromuscular blocking agent, which prevents nerve transmission by blocking sodium pores in postsynaptic membranes. S. is produced by dinoflagellates of the genus *Gonyaulax,* found in "red tides". S. accumulates in shellfish that ingest the dinoflagellates, hence cases of poisoning from eating the Californian sea mussel *(Mytilus californianus),* the Alaskan butterclam *(Saxidomus giganteus)* and the scallop.

Saxitoxin

Schardinger enzyme: see Xanthine oxidase.
Scheie's syndrome: see Inborn errors of metabolism.
Schemochromes: see Structural colors.
Schiff's bases: see Pyridoxal phosphate.
Schottenol: 5α-stigma-7-ene-3β-ol, a widely distributed phytosterol (see Sterols), M_r 414.72, m.p. 151 °C, isolated, e.g. from the cactus *Lophocerues schottii.* S. is an essential dietary constituent for the insect *Drosophila pachea,* which lives on this plant.
Scillabiose: see Scillaren A.
Scillaren A, *glucoproscillaridin A, transvaalin:* a bufadienolide cardiac glyoside, M_r 692.78. m.p. 184-186 °C (prisms), 208-211 °C (leaflets), $[\alpha]_D^{23}$ −71.9 ° (c = 1.011, methanol). The aglycon, *scillarenin* (Fig.) has M_r 384.52, m.p. 232-238 °C, $[\alpha]_D^{20}$ −16.8 ° (c = 0.357, methanol), +17.9 (c = 0.39, chloroform). The carbohydrate residue is the disaccharide, scillabiose (6-deoxy-4-O-β-D-glucopyranosyl-L-mannose; 4-O-β-D-glucopyranosyl-L-rhamnose), attached glycosidically at C3 of the aglycon. S.A. is the chief active component of *Scilla maritima* (squill), used since antiquity as a diuretic, cardiac stimulant and mouse poison.

Scillarenin

Scillarenin: see Scillaren A.
Scleroproteins: see Structural proteins.
Scopolamine: α-(hydroxymethyl)benzeneacetic acid 9-methyl-3-oxa-9-azatricyclo [3.3.1.0^{2,4}] non-7-yl ester; 6β,7β-epoxy-3α-tropanyl S-(-)-tropate, a tropane alkaloid, M_r 303.36, from *Solanaceae* spp., especially *Datura metel* L. and *Scopola carniolica* Jacq. L-Scopolamine (Hyoscine) is a viscous liquid, $[\alpha]_D^{20}$ −28 ° (water), −18 ° (ethanol), soluble in water at 15 °C, forming crystalline monohydrate, m.p. 59 °C. DL-Scopolamine (atroscine), forms an effluorescent hydrate, m.p. 55–57 °C (also reported as 82–83 °C). S. has similar pharmacological activity to that of hyoscamine, but has comparatively less activity on the peripheral nervous system. It is a highly toxic, anticholinergic agent. The hydrobromide has been used to sedate mental patients, as a preanesthetic, and to control motion sickness. For formula and biosynthesis, see Tropane alkaloids.
Scopoletin: see Coumarin.
Scorpamines: see Scorpion venoms.
Scorpion venoms: secretions of the scorpion stinging apparatus. Active principles of S.v. are the neurotoxic scorpamines. They are similar to cobra toxins (see Snake venoms) with respect to M_r (M_r 6800-7200; 4 disulfide bridges; 63-64 amino acid residues of known sequence), amino acid composition (high contents of basic and aromatic amino acids) and activity (both peripheral and central nervous system). The toxin from the North African scorpion, *Androctonus australis,* is one of the most potent known nerve poisons.
Scotophobin: Ser-Asp-Asn-Gln-Gln-Gly-Lys-Ser-Ala-Gln-Gln-Gly-Gly-Tyr-NH₂, a pentadecapeptide isolated from the brains of rats trained to avoid the dark. It induces dark avoidance in untrained animals. Isolation, biological activity and proof of structure of S. have been the subject of controversy: Ungar, G., Desiderio, D.M. & Parr, W. "Isolation, Identification and Synthesis of a Specific-Behaviour-Inducing Brain Peptide", *Nature, 238* (1972), 198–202; Stewart, W.M. "Comments on the Chemistry of Scotophobin", *Nature, 238* (1972), 202–210.
Scurvy: Vitamin C deficiency disease. See Vitamins (Ascorbic acid).
Scyllitol: see Cyclitols.
Scymnol: see Bile alcohols.
SD 8339: see Cytokinins.
S-D sequence: see Recombinant DNA technology.
Second messenger: see Hormones.
Secondary metabolism: see Secondary metabolites.
Secondary metabolites: Substances such as pigments, alkaloids, antibiotics, terpenes, etc., which occur only in certain organisms, organs, tissues or cells, and are the products of *secondary metabolism.* They are thus distinct from *primary metabolites* (products of *primary* or *general metabolism*), which are concerned in the energy metabolism, growth and structure of all, or at least very large groups of, organisms, e.g. glycolysis and TCA cycle intermediates, amino acids and their biosynthetic precursors, proteins, purines and pyrimidine bases, nucleosides, nucleotides, nucleic acids, sugars, polysaccharides, fatty acids, triacylglycerols, etc.
Many S.m. have no apparent biological function; some, however, have been exploited during the course of evolution (Table 2), and have even become fundamentally important to the life of their producing organisms (e.g. plant and animal hormones). Others have ecological importance, by acting as attractants (scents, colors), antifeedants and toxic defense or attack substances, like salamander

Secondary metabolites

Table 1. The relationship of secondary metabolism to total metabolism.

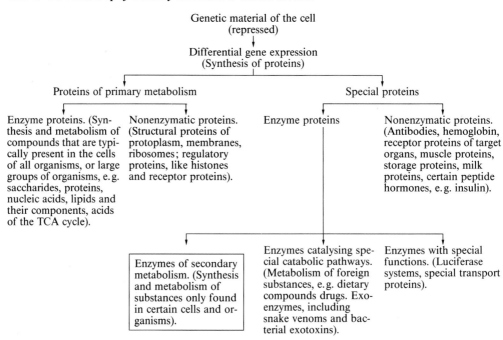

Genetic material of the cell
(repressed)

Differential gene expression
(Synthesis of proteins)

Proteins of primary metabolism

Special proteins

Enzyme proteins. (Synthesis and metabolism of compounds that are typically present in the cells of all organisms, or large groups of organisms, e.g. saccharides, proteins, nucleic acids, lipids and their components, acids of the TCA cycle).

Nonenzymatic proteins. (Structural proteins of protoplasm, membranes, ribosomes; regulatory proteins, like histones and receptor proteins).

Enzyme proteins

Nonenzymatic proteins. (Antibodies, hemoglobin, receptor proteins of target organs, muscle proteins, storage proteins, milk proteins, certain peptide hormones, e.g. insulin).

Enzymes of secondary metabolism. (Synthesis and metabolism of substances only found in certain cells and organisms).

Enzymes catalysing special catabolic pathways. (Metabolism of foreign substances, e.g. dietary compounds drugs. Exoenzymes, including snake venoms and bacterial exotoxins).

Enzymes with special functions. (Luciferase systems, special transport proteins).

alkaloids, cardiac glycoside toad poisons, and physiologically active compounds produced by insects, e.g. HCN, formic acid, *p*-cresol, *p*-benzoquinone. Despite the great chemical diversity of S.m., they are assembled from relatively few precursors, e.g. acetate, shikimate, isopentenyl pyrophosphate, etc. (Table 3), which usually occupy a key branchpoint position in primary or general metabolism. The S.m. of animals are not always synthesized de novo by the organism, but may be derived from the diet.

The milkweed *(Asclepias curassavica)* produces several cardiac glycosides (e.g. calotropin) within its tissues as a defense against insect feeding. These substances are bitter and toxic, but certain insects, notably the caterpillar of the monarch butterfly, have become adapted to them and feed upon the plant. The glycosides are sequestered during feeding and subsequently passed on to the tissues of the adult butterfly. If a bird (commonly a blue jay) eats the butterfly, it is caused to vomit by the bitter cardiac glycosides; thereafter it avoids feeding on this butterfly, which it recognizes from its typical wing pattern. Not only the cardiac glycosides, but the butterfly wing pigments are S.m.; moreover, other butterflies, which contain no cardiac glycosides, have achieved protection against bird attack by evolving a wing pattern and coloration similar to that of the monarch butterfly. Thus S.m. may be intimately involved in complicated ecological interactions.

In multicellular organisms, S.m. are produced by specific organs, tissues or cells, which contain the appropriate enzymes (see, e.g. Salamander alkaloids); they are usually formed only during certain periods of development or differentiation of the producing organism or cell. In microorganisms, S.m. (e.g. antibiotics, acetate-derived compounds) are usually synthesized at the end of the exponential growth phase (trophophase) or at the beginning of the stationary phase (idiophase), and their formation is usually repressed during rapid growth. This is probably due to catabolite represssion of enzyme synthesis during rapid growth on readily assimilated carbon sources. The risk of "metabolic suicide" is thus diminished, since growth and cell division are complete before toxic secondary products (e.g. antibiotics) are produced.

Table 2. Biological functions of secondary natural products.

Effectors within the synthesizing cell, i.e. intracellular messengers.

Effectors of other cells of the same organisms, i.e. intercellular messengers (plant and animal hormones, neuroendocrine transmitters).

Effectors of other organisms (blood pigments, flower scents, pheromones, antibiotics, insecticides, phytoalexins, toxins, antifeeding agents, sexual attractants).

Factors for the exploitation of specific ecological situations (chelating agents, e.g. siderochromes).

Storage forms of waste products from primary metabolism.

Table 3. Relationships of primary and secondary metabolites.

Primary metabolites	Secondary products
Sugars	Unusual sugars (amino, deoxy and methyl sugars, and sugars with branched chains). Reduction products (sugar alcohols, cyclitols, streptidine). Oxidation products (uronic acids, aldonic acids, sugar dicarboxylic acids).
Acetate/Malonate	Fatty acid derivatives (n-alkanes, acetylene derivatives). Polyketides (anthracene derivatives, tetracyclins, griseofulvin, phenolcarboxylic acids from fungi and lichens, pyridine derivatives).
Isopentenyl pyrophosphate	Hemiterpenes (isoprene). Monoterpenes (iridoid components of volatile oils). Sesquiterpenes (bitter principles, components of volatile oils). Diterpenes (components of resins, gibberellins, phytol). Triterpenes (squalene, sterols, etc.). Tetraterpenes (carotenoids, xanthophylls). Polyterpenes (caoutchouc, gutta percha).
Propionate	Methyl fatty acids. Macrolide antibiotics.
Acids of the TCA and glyoxylate cycles	Alkyl citric acids.
Shikimate pathway of aromatic synthesis	Naphthoquinones, anthraquinones, quinoline and quinazoline alkaloids, phenazines.
Amino acids	Amines, methylated amino acids, betaines, cyanogenic glycosides, mustard oils, alkaloids, glycine conjugates, glutamine and ornithine, S-alkylcysteine derivatives, dioxopiperazines, peptides (penicillins), hydroxamic acids.
Phenylpropane amino acids	Cinnamic acid, coumarins, lignin, lignans, flavan derivatives, stilbenes, phenolcarboxylic acids, phenols, components of volatile oils.
Porphyrins	Bile pigments.
Purine	Methylated purines, purine antibiotics, pteridines, benzopteridines, pyrrolopyrimidines.

The majority of known S.m. are synthesized by plants. More than 5000 plant alkaloids have been identified, compared with about 50 animal alkaloids. This difference may be related to excretory metabolism. Animals are able to remove from their bodies the endproducts and byproducts of metabolism, whereas plants employ "metabolic excretion", i.e. products are accumulated in vacuoles, cell walls and (lipophilic substances) in special excretory cells or spaces (volatile oil cells, resin ducts, etc.). Sites of synthesis and accumulation of S.m. are therefore often different. In animals, S.m. are usually stored in special organs (e.g. the salamander alkaloids and toad poisons are stored in skin glands). They may also be present in body fluids (e.g. cantharidine in insects is present in the lymph), or in hair and skin (e.g. melanins).

"Metabolic excretion" is only one of several theories put forward to explain the synthesis of S.m., especially those with no known function, like plant alkaloids and acetate-derived compounds (see Polyketides) from microorganisms. Any explanation must embrace the fact that so-called "degenerate" mutants can be isolated which do not produce the S.m., yet grow and divide as usual: nonalkaloid-producing strains of alkaloid plants are known (e.g. nicotine-less tobacco plants), and pigment, antibiotic and polyketide production by microorganisms shows great strain variation with no apparent effect on viability. More recent explanations, especially relevant to microorganisms, place emphasis on the production of S.m. rather than S.m. themselves, i.e. it is thought that the operation of secondary metabolism is advantageous by keeping metabolism running at a low rate, rather than closing it down completely after growth ceases. The unbalanced growth hypothesis put forward by Bu'lock (see Woodruff, H.B. "The Physiology of Antibiotic Production: The Role of the Producing Organisms", in Biochemical Studies of Antimicrobial Drugs, *16th Symposium of the Society For General Microbiology,* 1966, pp. 22–46. Cambridge University Presss) suggests that mechanisms controlling primary metabolism are not adequate to prevent overproduction of some compounds when balanced growth ceases. Since these compounds may be toxic to the cell, secondary metabolism diverts synthesis to the production of harmless products which are excreted. According to this theory, secondary metabolism should increase long-term viability; there is some evidence that *Pseudomonas aeruginosa* loses viability when grown under conditions that prevent secondary metabolism.

Secondary structure: see Proteins.

Secosteroids: see Steroids.

Secretin: a polypeptide hormone, M_r 3050, containing 27 amino acid residues. S. shows considerable sequence homology with Glucagon (see), Vasoactive intestinal peptide (see) (VIP) and Gastric inhibitory peptide (see) (GIP), and it is thought that these four hormones evolved from a common ancestral protein by a process of gene multiplication.

Production of S. by the duodenal mucosa is stimulated by the acidic pH of the chyle, large peptides ("secretogogues") from the incomplete hydrolysis of dietary protein, fat and alcohol, i.e. it is generally

			5					10				
Secretin:	His-	Ser-	Asp-	Gly-	Thr-	Phe-	Thr-	Ser-	Glu-	Leu-	Ser-	Arg- Leu- Arg-
Glucagon:	His-	Ser-	Gln-	Gly-	Thr-	Phe-	Thr-	Ser-	Asp-	Tyr-	Ser-	Lys- Tyr- Leu-
VIP:	His-	Ser-	Asp-	Ala-	Val-	Phe-	Thr-	Asp-	Asn-	Tyr-	Thr-	Arg- Leu- Arg-
GIP:	Tyr-	Ala-	Glu-	Gly-	Thr-	Phe-	Ile-	Ser-	Asp-	Tyr-	Ser-	Ile- Ala- Met-

	15				29					25			
Asp-	Ser	Ala-	Arg-	Leu-	Gln-	Arg-	Leu-	Leu-	Gln-	Gly-	Leu-	Val-	NH$_2$
Asp-	Ser-	Arg-	Arg-	Ala-	Gln-	Asp-	Phe-	Val-	Gln-	Trp-	Leu-	Met-	Asp- Thr
Lys-	Gln-	Met-	Ala-	Val-	Lys-	Lys-	Tyr-	Leu-	Asn-	Ser-	Ile-	Leu-	Asn- NH$_2$
Asp-	Lys-	Ile-	Arg-	Gln-	Gln-	Asp-	Phe-	Val-	Asn-	Trp-	Leu-	Leu-	Ala- Gln- Gln

Amino acid sequences of secretin, glucagon, VIP and GIP. Amide groups are present at the C-termini of secretin and VIP. The sequence of GIP is shown incomplete; it continues: -Lys-Gly-Lys-Lys-Ser-Asp-Trp-Lys-His-Asn-Ile-Thr-Gln (total 43 residues).

stimulated by food with the exception of carbohydrate. S. is secreted into the blood. It stimulates formation and secretion of NaHCO$_3$-rich pancreatic juice and NaHCO$_3$-rich bile, and inhibits HCl production by the stomach.
["Structure of Porcine Secretin", Mutt, V., Jorpes, J.E. & Magnusson, S. *Eur. J. Biochem. 15*, (1970) 513–519.
"Biological Activities of Synthetic Peptides Corresponding to Fragments of and to the Entire Sequence of the Vasoactive Intestinal Peptide", Bodansky, M., Klausner, Y.S. & Said, S.I. *Proc. Nat. Acad. Sci USA. 70*, (1973) 382–384].

Secretory enzymes: see Secretory proteins.

Secretory proteins: proteins synthesized intracellularly, often in specialized secretory organs (e.g. digestive glands), then secreted. S.p. that are also enzymes are called *secretory enzymes*. In cells actively engaged in synthesizing S.p., the rough endoplasmic reticulum (RER) is highly developed. As a generalization, it can be stated that proteins retained by the cell are synthesized on polysomes that are not attached to membranes, whereas S.p. are synthesized on polysomes bound to the endoplasmic reticulum. Our concept of S.p. synthesis and secretion are due largely to the work of Palade (see Palade, G. "Intracellular aspects of the process of protein synthesis", *Science, 189* (1975) 347–358); his studies on the synthesis and secretion of digestive enzymes by the guinea pig pancreatic exocrine cell provided a conceptual model for this process in all secretory cells. During synthesis on the RER, S.p. pass throught the reticular membrane into the lumen; they then pass to the Golgi apparatus, where they become condensed into secretory granules, which leave the cell by exocytosis. In other cells (e.g. hepatocytes), the smooth endoplasmic reticulum (SER) may function in the transfer of S.p. from the lumen of the RER to the Golgi apparatus. Synthesis and secretion into the lumen of the RER depend upon the formation and ultimate removal of a signal peptide sequence (see Signal hypothesis). Other modifications, such as glycosylation (see Post-translational modification of proteins) are also associated with the intracellular production of S.p. (many S.p. are glycoproteins). Most proteolytic S.p. are synthesized as inactive precursors (e.g. trypsinogen), thus avoiding self digestion (autolysis) by the producing cell. S.p. are secreted into the blood (e.g. serum albumin, serum cholinesterase and blood coagulation enzymes are synthesized in the liver), or into ducts from their producing glands (e.g. salivary and pancreatic amylases).

Sedamine: see Sedum alkaloids.

Sedimentation coefficient, *sedimentation constant:* a measure of the rate of sedimentation, used in the determination of M_r of macromolecules by ultracentrifugation. The sedimentation coefficient *(s)* is equal to the rate of sedimentation of a macromolecule per unit centrifugal field; specifically,

$$s = \frac{dx/dt}{\omega^2 \cdot FEx},$$ where s is the sedimentation coefficient, ω is the angular velocity of the centrifuge rotor (radians/sec), x is the distance from the center of rotation, dx/dt is the velocity of sedimentation. The sedimentation coefficient has the dimensions of time per unit force, and usually lies between 1×10^{-13} and 200×10^{-13}. The factor 1×10^{-13} is called the Svedberg unit (S), i.e. $1S = 10^{-13}$s. Sedimentation is monitored in the centrifuge cell by schlieren optics (Philpot-Svensson method), or by UV-absorption. Ideally, sedimentation constants are determined at a number of different macromolecule concentrations, and the *s*-values are extrapolated to zero concentration, where the activity coefficient becomes unity. In addition, sedimentation coefficients are corrected to a standard state with respect to solvent viscosity, which is taken as that of water at 20 °C; this gives the *standard sedimentation coefficient*, or S $°_{20w}$-value. S $°_{20w}$ values for most proteins and nucleic acids lie between 4 and 40 S (Svedberg units), for ribosomes and ribosomal subunits between 30 and 80S, and for polysomes above 100S.

D-Sedoheptulose, *D-altro-2-heptulose:* a monosaccharide from *Sedum* (stonecrop). M_r 210.19, m.p. (monohydrate) 102 °C. The 7-phosphate is an intermediate of carbohydrate metabolism (see, e.g.

CH$_2$OH
|
CO
|
HO—C—H
|
H—C—OH
|
H—C—OH
|
H—C—OH
|
CH$_2$OH

D-Sedoheptulose

Pentose phosphate cycle); aldolase reactions of sedo-heptulose 7-phosphate give rise to D-erythrose 4-phosphate, which is a precursor of Aromatic biosynthesis (see).

Sedum alkaloids: a group of piperidine alkaloids from *Sedum* spp. They are 2- or 2,6-substituted piperidine derivatives, similar in structure and biosynthesis to the Punica and Lobelia alkaloids. Chief representative is **sedamine**: N-methyl-2-(β-hydroxy-β-phenylethyl)-piperidine, M_r 219; the L-form has m.p. 61-62 °C, $[\alpha]_D$ −82 ° (methanol).

Seed germination test: see Gibberellins.

Selenium, *Se*: an element toxic in large quantities, but an essential micronutrient for mammals, birds, many bacteria, probably fish and other animals. A requirement by higher plants is uncertain. Se is an essential component of the enzyme glutathione peroxidase, which is important in the protection of red cell membranes and other tissues from damage by peroxides:

2 GSH + H_2O_2 (or R-OOH) → GSSG + 2H_2O (or H_2O + R-OH)

Normal sheep muscle contains a low M_r selenoprotein of unknown function, which is absent from the muscle of Se-deficient sheep suffering from dystrophic white muscle disease.

Residues of the selenoamino acid, selenocysteine, are present in bacterial formate dehydrogenase *(Escherichia coli)*, and glycine reductase *(Clostridium)*.

[Stadtman, T.C. (1979) *Advances in Enzymology, 48,* 1-28]

Selenoamino acids: amino acids containing selenium (Se) in place of sulfur (S), e.g. Se-methylselenocysteine. They are formed in plants growing on Se-rich soils. See Selenium.

Self-splicing RNA: see Intron.

Semidehydroascorbate: a free radical formed from ascorbate when the latter serves as a reducing agent. The free radical has a highly unstable electronic configuration, and it is able to act as an oxidant in the NADH system. It is also re-reduced to ascorbate by cytochrome b_{561} (see Ascorbate shuttle). Formation of the free radical can be demonstrated by EPR sepctroscopy [I. Yamazaki & L. H. Piette, *Biochim. Biophys. Acta 50* (1961) 62-69] and by a scavenger method involving one electron transfer to oxidized cytochrome *c* [I. Yamazaki, J. Biol. Chem. *237* (1962) 224-229]. In the absence of an efficient trap, two free radicals dismute to form one molecule of ascorbate and one of dehydroascorbate.

Senecio alkaloids: see Pyrrolizidine alkaloids.

Sephadex: a trade name for a series of polydextrans used in gel filtration chromatography. See Dextrans, Proteins.

Sequence: see Proteins.

Sequence polymers: synthetic amino acid polymers, consisting of multiple repeats of one short sequence. In contrast to Polyamino acids (see), S.p. contain more than one type of amino acid residue. They are prepared by the self condensation of activated peptides (di to hepta). *p*-Nitrophenyl esters (see Peptides) and other activated esters can be used; ring closure is avoided by using high concentrations of reactants in polar solvents in the presence of organic bases. The average M_r of the resulting

S.p. are considerably lower than those of the synthetic polyamino acids. They are used to investigate molecular aspects of antigenicity and antibody specificity, as model enzymes, and as models in the study of protein structure.

Sequential mechanism: see Cleland short notation.

Sequential therapy: see Ovulation inhibitors.

Ser: abb. for L-serine.

L-Serine, abb. **Ser:** L-α-amino-β-hydroxypropionic acid, HO-CH_2-CH(NH_2)-COOH, a proteogenic, glucogenic amino acid. M_r 105.1, m.p. 223-228 °C, $[\alpha]_D^{25}$ -7.5 ($c = 2$, water), +15.1 ($c = 2$, 5M HCl). Ser is a major component of silk fibroin. In phosphoproteins, phosphate is esterified chiefly with the hydroxyl groups of Ser residues. During acid hydrolysis of proteins, a large proportion of Ser is destroyed. It is converted quantitatively into formaldehyde by periodate oxidation. Metabolically, Ser and glycine are interconvertible by the action of tetrahydrofolate-5,10-hydroxymethyltransferase; in this reaction, the β-C atom of Ser is transferred as an active hydroxymethyl group, which is a very important metabolic source of one carbon units. L-Serine dehydratase (a pyridoxal phosphate enzyme, EC 4.2.1.13) catalyses the conversion of Ser into pyruvate and ammonia. Ser is synthesized from glycine, or from 3-phosphoglycerate (a glycolytic intermediate). In liver and in *Escherichia coli*, 3-phosphoglycerate is dehydrogenated (NAD$^+$-dependent) to phosphohydroxypyruvate, which is transaminated to 3-phosphoserine. The latter is dephosphorylated by a specific phosphatase. In plants 3-phosphoglycerate (a photosynthetic product) is dephosphorylated to glyceric acid, which is dehydrogenated to hydroxypyruvate; the latter transaminates with L-alanine, forming Ser and pyruvate. Transsulfuration of Ser with L-homocysteine produces L-cysteine.

Serine cephalins: see Phospholipids.

Serine hydrolases: hydrolases which have a catalytically active serine residue in their active center, e.g. trypsin, chymotrypsin A, B and C, thrombin and B-type carboxylic acid esterases. See Serine proteases.

Serine hydroxymethyltransferase: see Active one-carbon units.

Serine phosphoglycerides: see Phospholipids.

Serine proteases: a group of well studied animal and bacterial endopeptidases (see Proteases), which have a similar action mechanism, and a catalytically active serine residue in their active centers (serine residue 195 in chymotrypsin). In all S.p. catalysis involves formation of an ester between the hydroxyl group of the catalytically active serine and the carboxyl group of the cleaved peptide bond (acyl-enzyme intermediate); this is hydrolysed in the deacylation stage of the reaction, restoring the free hydroxyl group of the serine and releasing the cleavage peptide. The active serine residue is selectively and irreversibly acylated by organic phosphate esters, like diisopropylfluorophosphate (DFP), or phenylmethanesulfonylfluoride (PMSF), which therefore inhibit S.p. Trypsin, chymotrypsin A, B and C, pancreatic elastase, invertebrate trypsins and chymotrypsins, thrombin, plasmin and kallikrein of the blood, the elastase-like α-lytic protease from

Myxobacter and the trypsin-like S. p. from *Streptomyces griseus* are all thought to be homologous proteins, derived from a common ancestral protein by gene multiplication. Subtilisin (see) from *Bacillus subtilis* is a S. p.; it resembles chymotrypsin in the hydrogen bonding of the charge-relay system, but is otherwise structurally dissimilar; it is thus an example of convergent evolution of a catalytic center in two different groups of proteins.

Serine sulfhydrase: see Sulfate assimilation.

Serotonin: 5-hydroxytryptamine, M_r 176.2, a plant and animal hormone. It is produced by hydroxylation of L-tryptophan to 5-hydroxytryptophan, followed by decarboxylation. The synthesis occurs in the central nervous system, lung, spleen and argentaffine "light" cells of the intestinal mucosa. S. is stored in thrombocytes and mast cells of the blood. It acts as a Neurotransmitter (see), stimulates peristalsis of the intestine, and causes a dose-dependent constriction of smooth muscle. It stimulates the release from arterial endothelium of a dilator substance which counteracts its primary constricting effect. [T. M. Cocks & J. A. Angus, *Nature 305* (1983) 627–630.] S. is a precursor of the hormone Melatonin (see). It is inactivated and degraded by monoamine oxidases and aldehyde oxidases to 5-hydroxy-indoleacetic acid.

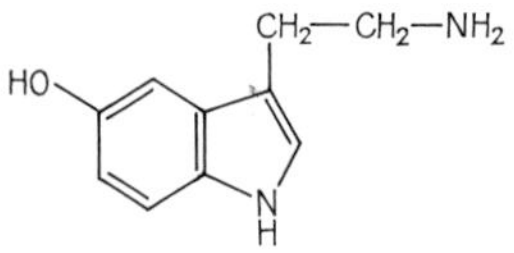

Serotonin

Serratomolide: a cyclic depsipeptide produced by *Serratia marcescens*. Chemically, it is the cyclic dimer of serrataminic acid (D-β-hydroxydecanoyl-L-serine).

Serum albumin: see Albumins.

Serum proteins: see Plasma proteins.

Sesquiterpenes: aliphatic, mono, di or tricyclic terpenes, formed from three isoprene units ($C_{15}H_{24}$). About 100 structural types are known, and about 1000 natural representatives, forming the largest class of terpenes. Most are found in the volatile oils of plants. Little is known of the physiological significance of S.; some compounds have an ecological role (Table). Some are isolated technically for use in perfumery.

Table. Some sesquiterpenes and their functions.

Function	Sesquiterpenes
Juvenile hormones	Juvabione, farnesyl derivatives
Phytohormones	Abscisic acid
Plant sex hormones	Sirenin
Pheromones	Farnesol
Antibiotics	Trichothecin
Proazulenes	Guaiol
Alkaloids	Nupharidine
Scents	Santalols, cedrenes
Bitter principles	Cnicin
Phytoalexins	Ipomeamarone

S. are biosynthesized from farnesylpyrophosphate (see Terpenes). Acyclic S., e.g. farnesol, are formed by hydrolytic removal of the pyrophosphate group. The various types of cyclic S. are formed by elimina-

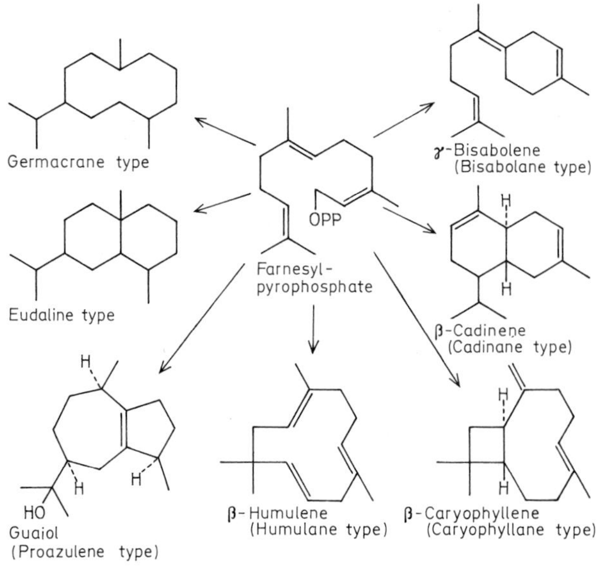

Some important sesquiterpene structural types

tion of the pyrophosphate residue to form an unstable cation, which stabilizes by loss of a proton.

Sesterterpenes: terpenes formed from five isoprene units ($C_{25}H_{40}$). They have a tricyclic skeleton, and have been isolated from insect secretions and lower fungi. See Ophiobolanes.

Sexual attractants: natural products involved in sexual interaction.

The S. a. of insects (see Pheromones) are particularly numerous and well studied examples of *animal S. a.* They are usually produced by the sexually mature female in order to attract and predispose the male to copulation.

Plant S. a. are called **gamones**, or **plant sex hormones**. They occur when at least one of the gametes involved in fertilization has a free existence, i.e. in many algae, lower fungi, mosses and ferns. The gamones that have been investigated, e.g. sirenin and ectocarpene, are produced by the female gametes and act as chemotactic agents to attract the male gametes. In contrast to the pheromones, gamones only influence interaction of the gametes (i.e. they act at the cellular level) and do not affect the behavior of the whole organism.

SF: abb. for Sulfation factor (see Somatomedin).

Shellac: see Resins.

Shemin cycle: see Succinate-glycine cycle.

SH-enzymes: see Thiol enzymes.

Shikimic acid: see Aromatic biosynthesis.

Shikonin: see Naphthoquinones (Table).

Showdomycin: 2-β-D-ribofuranosylmaleinimide, a *C*-substituted Nucleoside antibiotic (see) from *Streptomyces showdoensis,* structurally related to uridine and pseudouridine. M.p. 153 °C, $[\alpha]_D^{23} + 50$ ° ($c = 1$, water). It selectively inhibits enzymes of uridine and orotic acid metabolism; the maleinimide moiety reacts with sulfhydryl groups of the affected enzymes. S. is especially active against *Streptococcus haemolyticus.*

Showdomycin

Sialic acids: see Neuraminic acid.

Sialic acid storage disease: see Inborn errors of metabolism.

Sickle cell hemoglobin, abb. *HbS:* one of the most frequently occurring abnormal hemoglobins, especially in negroids. As a result of a point mutation, the glutamic acid residue at position 6 in the β-chain (normal hemoglobin) is replaced by valine. The α-chain is normal ($\alpha_2\beta_2^{6Glu \rightarrow Val}$). There are no marked differences in the conformations of HbS and normal hemoglobin. DeoxyHbS undergoes self association and forms a liquid crystalline phase, which distorts the erythrocytes into a sickle shape. In this phase, Deoxy-HbS monomers are in eqilibrium with polymers. The polymers consist of 6–8 heli-

cal deoxy-HbS strands lying side by side to form a tubular shape of diameter 140–148 A. Deformation of the erythrocytes into sickle shapes ("sickling") leads to their aggregation and to a decrease in blood circulation. Clinical symptoms are anemia and acute ischemia, tissue infarction and chronic failure of organ function. In heterozygotic carriers, the condition known as sickle cell trait may be without serious clinical effects; it may even go unrecognized, but sickling occurs when the individual is subjected to abnormally low oxygen tension. Negroid trainee airline or airforce pilots must therefore be screened for heterozygotic sickling. Homozygotes die from extensive hemolytic anemia.

Sideramines: see Siderochromes.

Siderochromes: iron-containing, red-brown, water soluble secondary metabolites produced by microorganisms. S. include a series of antibiotics, the *sideromycins* (albomycin, ferrimycin, danomycin, etc.) and a class of compounds with growth factor properties for certain microorganisms, the *sideramines* (ferrichrome, coprogen, ferrioxamine, ferrichrysin, ferrirubin, etc.). The sideramines competitively inhibit the antibiotic activity of the sideromycins.

S. are specific ligands for iron. Their synthesis and secretion by microorganisms is increased under conditions of iron defiency. S. of the *catechol* type are produced by anaerobic microorganisms, whereas the *hydroxamate* type are produced by aerobic microorganism. They typically contain a central iron(III)-trihydroxamate complex (Fig.), and specifically bind Fe^{3+}.

As metal chelating agents, sideramines fulfill two functions: they transport iron into the microbial cell and/or make the chelated iron available for heme synthesis. In analogy with animal transferrin, the iron of microbial ferrichrome is transferred enzymatically into the porphyrin molecule during heme synthesis.

Other microbial chelating agents, e.g. mycobactin, aspergillic acid and schizokinen, are sometimes classified with the sideramines.

Sideromycins: see Siderochromes.

Siderophillins: nonheme, iron-binding, single chain animal glycoproteins, M_r about 77000, carbohydrate content about 6%. On the basis of their occurrence, they are classified as transferrin (vertebrate blood), lactoferrin (mammalian milk and other body secretions) and conalbumin or ovotransferrin (avian blood and avian egg white). S. differ in their physical, chemical and immunological properties, but each possesses two binding sites for iron (III).

The iron is bound less firmly than in Ferritin (see). Transferrin (a β-globulin) is the best studied S.; 15 genetic variants are known, the most common being transferrin A, B and C. Its main function is the transport of absorbed dietary iron (III) to iron depots (liver and spleen), or to the reticulocytes and their precursors in the bone marrow. Transferrin becomes bound to surface receptors of the reticulocytes, enters the cell and releases its iron, then is returned to the blood as iron-free apotransferrin. Of the total 7–15 g transferrin in the body, only one third is complexed with iron (III). By virtue of their ability to chelate iron, all S. inhibit bacterial growth; this property is important in avian eggs and in milk.

SIF: abb. for Somatotropin release inhibiting factor. See Releasing hormones.

Sigma (σ) factor: see Ribonucleic acid; Initiation factors.

Signal hypothesis: A mechanism proposed by Blobel for the segregation of secretory proteins during their biosynthesis. It has since been found applicable to most proteins which are either inserted into membranes or transported across them, including organelle proteins which are synthesized in the cytosol. Such proteins are transcribed with a signal or leader peptide of 15 to 30 amino acid residues. About 70 amino acid residues must be incorporated into a peptide before it can protrude from the ribosome; at this point the translation of secretory proteins is arrested until the polypeptide chain binds a soluble signal recognition particle (SRP). The complex of SRP and nascent peptide subsequently binds to a "docking protein" in the appropriate membrane: endoplasmic reticulum, mitochondrial, (bacterial) plasma membrane, etc. The SRP is released, while the translation of the polypeptide resumes. The polypeptide is extruded through or into the membrane. In many cases, the signal peptide is cleaved on the other side of the membrane (the intralumenal space, in the case of the endoplasmic reticulum) by a specific signal peptidase. However, this is not always the case; some membrane-spanning proteins retain their signal peptide sequences. The chains of some membrane-spanning proteins cross the membrane several times; these are supposed to contain internal signal sequences which promote spontaneous insertion into the lipid bilayer. It is also possible for membrane proteins to be translated completely before insertion into the membrane. [W.T. Wickner & H.F. Lodish *Science* **230** (1985) 400–407. R. Gilmore & G. Blobel *Cell* **42** (1985) 497–505]

Proteins with cleavable signal peptides are called *preproteins* prior to cleavage of the signal. Since many secretory proteins are synthesized as inactive *proproteins*, the corresponding precursors are called *preproproteins*.

SIH: abb. for Somatotropin release inhibiting hormone. See Releasing hormones.

Silicon, *Si*: an essential trace element in human nutrition [Carlisle, E.M. *Science,* **178,** (1972) 619–612; Carlisle, E.M. *Fed. Proc. Fed. Amer. Soc. Exp. Biol.,* **32,** (1973) 930]. Si is a cross-linking agent in connective tissue. It is thought that Si is bound via oxygen to the C-skeleton of mucopolysaccharides, thus linking parts of the same polysaccharide, or linking acidic mucopolysaccharides to proteins. Si may also serve a matrix or catalytic role in bone mineralization. High levels of Si (as SiO_2) are present in plants and diatoms (see Mineral elements).

Silk fibroin: see Keratins.

Silybin: a flavanolignan from *Silybum marianum* Gaert. (milk thistle). S. protects animals against poisoning by phalloidin (see Phallotoxins). Crystallographic studies of S. indicate that the spacing and alignment of the aromatic rings A and B (Fig.) are almost identical with those of the Phe residues 9 and 10 in Antamanide (see), which also protects against phalloidin poisoning. It is also suggested that the correct arrangement of these two aromatic rings is essential for attachment to a target receptor on the liver cell membrane, thus preventing entry of phalloidin into the cell. [H. L. Lotter, *Zeitschrift für Naturforschung* **39c** (1984) 535–542]

Silybin

Sinalbin: see Glucosinolate.

Sinapine, *sinapic acid choline ester,* **2{[3-(4-hydroxy-3,5-dimethoxyphenyl)-1-oxo-2-propenyl]oxy}-N,N,N-trimethylethanammonium hydroxide:** an alkaloidal base occurring widely in the *Cruciferae*, first isolated in 1825 from black mustard seeds. It serves as the cation of sinalbin (see Glucosinolate). The component alcohol, choline, is a common plant metabolite involved in phospholipid synthesis and transmethylation; sinapic acid is also a widely distributed plant constituent implicated in Lignin (see) biosynthesis. The combination of these two compounds to form S., however, appears to be a peculiarity of the *Cruciferae*. S. behaves as a storage compound; during germination of mustard seeds, S. is hydrolysed to choline and sinapic acid, which are then further metabolized. Two weeks after germination S. is no longer detectable. The esterase responsible for S. hydrolysis has been purified 20-fold from white mustard seedlings. [A. Tzagoloff, *Plant Physiology* **38** (1963) 207–213.]

Sinapine

Single-strand break: a break in a double-stranded DNA molecule which involves only one of the two strands, so that the molecule remains together. One S.s.b. is required for the initiation of unwinding the double helix during replication. S.s.b.

are caused by endonucleases, physical conditions or chemicals. Such breaks are repaired by the polydeoxyribonucleotide synthetases (EC 6.5.1.1 and 6.5.1.2).

Single-substrate enzymes: enzymes which catalyse reactions involving only one substrate. They are usually isomerases or hydrolytic enzymes; in the latter case, the water involved in the reaction is regarded as a constant, and there is frequently no special enzyme-water complex formed.

Sinigrin: see Glucosinolate.

Sirenin: the first plant sexual attractant or gamone to be structurally elucidated. S. is a sesquiterpene, M_r 236. It occurs naturally in the L-form, $[\alpha]_D^{23}$ $-45°$ ($c = 1.0$, chloroform), but the DL-form is also biologically active. S. is produced by the female gametes of the fungus *Allomyces*, which lives in damp soils. Gametes swim from the mycelium, and S. acts as a chemotactic stimulus to attract the male to the female gametes. It is active at a concentration of 10^{-10} M.

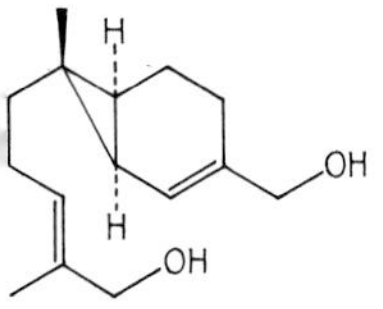

L(-)-Sirenin

Siroheme: the heme prosthetic group found in sulfite reductase of *Escherichia coli* and nitrite reductase of green plants.

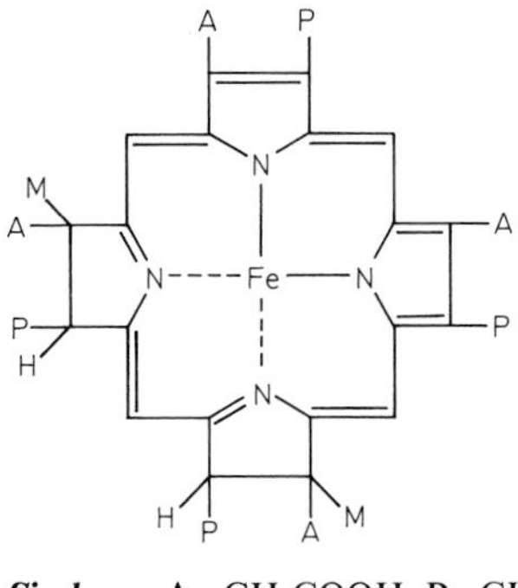

Siroheme. A -CH$_2$COOH, P -CH$_2$CH$_2$COOH, M -CH$_3$.

Site of labelling: see Isotope technique.

Sitosterols: a group of phytosterols (see Sterols) structurally related to the parent hydrocarbon stigmastane (see Steroids). They are widely distributed in higher plants, and usually occur as mixtures that are difficult to resolve; sometimes they occur as glycosides. Chief representative is β-sitosterol (stigmast-5-ene-3β-ol, Fig.), M_r 414.7, m.p. 140 °C, $[\alpha]_D$ $-37°$ (chloroform), which occurs, e.g. in cotton seed, germinating wheat and rye, sugar cane wax, potatoes, tobacco and pine bark.

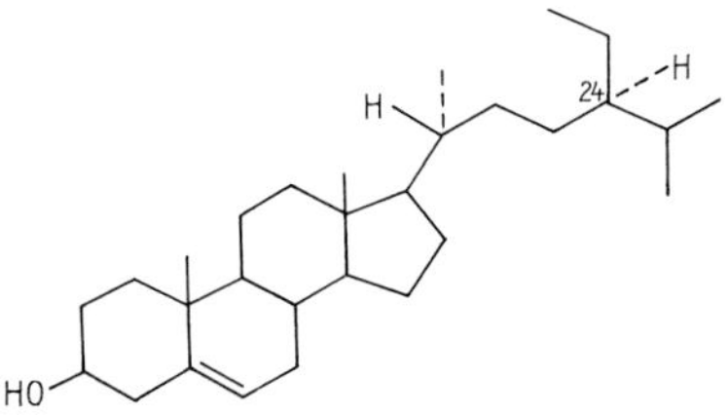

β-Sitosterol

Skimmiol: see Taraxerol.

(+)-Skyrin: see Emodin.

Slow reacting substance A: abb. for slow reacting substance of anaphylaxis. It is produced by sensitized cells as part of the immune response to antigens. See Leukotrienes.

Slow reacting substances: substances that cause smooth muscle to contract slowly in vitro. See Leukotrienes.

Sly syndrome: see Inborn errors of metabolism.

Small nuclear ribonucleoproteins: although not all S.n.p. belong to the U class, the latter are the best characterized. See U snRNP.

Snake venoms: a mixture of toxins produced in the venom glands (parotid gland, or salivary gland of the upper jaw) of venomous snakes (asps or hooded snakes, e.g. the cobra; sea snakes; vipers, e.g. puff adder, rattlesnake). They consist of highly toxic, antigenic polypeptides and proteins (which cause paralysis and death of the prey), and enzymes (which facilitate the spread of the toxins, and initiate digestion of undivided swallowed prey). The enzymes include hyaluronidase (promotes spread of toxins), ATPase and acetylcholine esterase (paralysis), phospholipases (hemolysis), proteinases and L-amino acid oxidases (tissue necrosis and blood clotting).

Important toxins are cobramine A and B from cobra toxin; and crotactine and crotamine from crotoxin, the toxin of the North American rattlesnake.

The toxic proteins are classified according to their mode of action; cardiotoxins, neurotoxins and protease inhibitors (with inhibitory activity towards chymotrypsin and trypsin). *Cardiotoxins* (heart muscle poisons) cause an irreversible depolarization of the cell membranes of heart muscle and nerve cells. *Neurotoxins* (nerve poisons) show curare-like activity; they prevent neuromuscular transmission by blocking the receptors for the transmitters at the synapses of autonomic nerve endings and at the motor end plate of skeletal muscle. *Protease inhibitors* inhibit acetylcholine esterase and similar enzymes involved in nerve transmission.

By the end of 1973, the primary structures of about 5 cardiotoxins, 2 protease inhibitors and about 35 neurotoxins were known. The cardiotoxins and protease inhibitors consist of 60 amino acid residues (M_r 6900), whereas the best studied neurotoxins (cobra) can be subdivided into long chain (71-74 amino acids; M_r 8000) and short chain (60-62 amino acids, M_r 7000) toxins. Despite their structural and pathophysiological differences, the cardio- and neurotoxins of the asps show sequence homologies. Neurotoxins so far isolated from the vipers have high M_r (e.g. rattlesnake crotoxin, M_r 30000); they may even show subunit structure, e.g. the most potent S.v., taipoxin, consists of two nonidentical polypeptide chains. Owing to the presence of disulfide bridges (4 at M_r 7000, 5 at M_r 8000 and 7 at M_r 13500), the neurotoxins are extremely stable. They suffer no loss of activity when treated with 8 M urea for 24 hours 25 °C, or for 30 min at 100 °C. But they are rapidly inactivated by strong alkali, probably by disulfide exchange or desulfuration. In addition to the disulfide groups, a tryptophan and a glutamate residue are also necessary for the neurotoxic activity of cobra toxins.

Cobra toxins are similar to Scorpion venoms (see).

The yearly death rate from snake bites is estimated at 30000-40000, about 50 being registered in Europe, the remainder chiefly in Asia.

Many animals (hedgehog, ichneumon) have a natural immunity against S.v. Large quantities of S.v. are used for the preparation of specific antisera (by the periodic active immunization of horses) for immunization against snake bite. Some S.v. are used therapeutically for the treatment of neuralgic pain, rheumatic diseases and epilepsy.

Snakes are farmed for the production of S.v. The animals are allowed to bite a membrane and inject the venom into a glass container, or the venom is expressed by pressure on the venom glands. This produces between 10 (vipers) and a few hundred mg (asiatic snakes) S.v. from each snake.

snRNA: see U snRNP.

snRNP: see U snRNP.

Snurp: see U snRNP.

α-Solamargine: see α-Solasonine.

α-Solamarine: a Solanum glycoalkaloid, first isolated from woody nightshade *(Solanum dulcanamara)*. The aglycon is tomatidenol (22S: 25S)-spirosol-5-ene-3β-ol, M_r 413.67, m.p. 239 °C, $[\alpha]_D$ −37.8 (chloroform), and the carbohydrate residue is a trisaccharide of one molecule of D-glucose and two molecules of L-rhamnose. Tomatidenol differs from tomatidine (see α-Tomatine) by the presence of a double bond at position 5; it can be used as a starting material for the laboratory synthesis of steroid hormones. β-S. inhibits growth of mouse Sarcoma 180.

Solanesol: see Polyprenols.

Solanidane: see Solanum alkaloids.

Solanidine: see α-Solanine.

α-Solanine, *solanine:* a Solanum alkaloid, and the chief toxic alkaloid of the potato *(Solanum tuberosum)*, also present in other *Solanum* spp. It is a glycoalkaloid containing the aglycon solanidine (solanid-5-ene-3β-ol), M_r 397.65, m.p. 218 °C, $[\alpha]_D$

D - Glucose
|
D - Galactose — O-
|
L - Rhamnose

α-Solanine

−27 ° (chloroform), and the trisaccharide β-solatriose (Fig.). Potato tubers contain no more than 0.01% α-S., which is harmless in this quantity. Potato shoots contain 0.5% α-S. and therefore cannot be used as food.

Solanum alkaloids: steroid alkaloids that occur in plants of the nightshade family *(Solanaceae),* belonging to the genera *Solanum, Lycopersicon, Cyphomandra* and *Cestrum.* S.a. are structurally related to the parent hydrocarbon, cholestane (see Steroids). In addition, they contain nitrogen, representing structural derivatives of either *spirosolane* (contains a secondary amino group), or *solanidane* (contains a tertiary amino group) (Fig.). In the plant S.a. occur mostly as glycoalkaloids (e.g. see α-Solanine; α-Tomatine); the free aglycons can be released by acid hydrolysis. These glycoalkaloids are also Saponins (see). They form insoluble 1:1 addition products with cholesterol. The aglycons of S.a. are biosynthesized from cholesterol. Certain S.a. serve as starting material for the laboratory synthesis of steroid hormones.

5α-Solanidane

Solasodine: see α-Solasonine.

α-Solasonine, *solasonine:* a Solanum steroid alkaloid occurring in many *Solanum* spp., e.g. *S. sodomeum, S. aviculare, S. laciniatum* and *S. nigrum.* It is a glycoalkaloid containing the aglycon, *solasodine* [(22R:25R)-spirosol-5-ene-3β-ol], M_r 413.67, m.p.201 °C, $[\alpha]_D$ −107 ° (chloroform); and a branched trisaccharide.

Other solasodine glycosides are *β-solasonine* (with L-rhamnose and D-glucose), and *α-solamargine,* the first glycoside isolated from *Solanum marginatum* (with two molecules of L-rhamnose and one molecule D-glycose). Solasodine is used as starting material for the laboratory synthesis of steroid hormones.

β-Solasonine: see α-Solasonine.

β-Solatriose: see α-Solanine.

Solavetivone: see Phytoalexins.

Soluble RNA: see Transfer RNA.

Solvent accessible area, *solvent accessible surface:* that part of the surface of a macromolecule which is accessible to a solvent molecule, i.e. that part of the van der Waals surface which can be in contact with a probe sphere representing a solvent molecule. [*Science* 221 (1983) 709-713]

Somatomedin: a collective term for a group of peptides. S.C. is the same as insulin-like growth factor (IGF) I, and S.A. is a mixture consisting largely of IFG II and some IGF I. (see Insulin-like growth factor). S.A. causes an increase in sulfur incorporation into cartilage and is therefore called *sulfation factor.* [E.M. Spencer, ed., *Insulin-Like Growth Factors, Somatomedins* (de Gruyter, Berlin, 1983)]

Somatostatin: see Releasing hormones.
Somatotropic hormone: see Somatotropin.
Somatotropin, *somatotropic hormone,* abb. *STH, growth hormone,* abb. *GH:* a fundamentally important hormone, which in conjunction with other hormones (insulin, thyroxin, etc.) controls growth, differentiation and the continual renewal of body substances. GH is a single chain polypeptide containing 190 amino acid residues and two disulfide bridges (for primary sequence, see Fig.). It is synthesized in the anterior pituitary in response to SRF and SIH (see Releasing hormones), which are produced in the hypothalamus. The production of these latter hormones is in turn affected by low blood sugar levels, increased levels of blood amino acids and by a variety of neural influences (stress situations). The growth stimulating activity of STH is especially apparent in pituitary dwarfism. STH acts directly on cellular metabolism, without the intermediate participation of other glandular hormones; as a prerequisite of growth stimulation, it promotes nucleic acid and protein biosynthesis. Treatment with STH results in increased growth of muscle, skeleton and organs, and an increased functional capacity of kidneys and liver).

bic acid; it is used in the food industry as a preservative and as a softening agent in sweets. Since it is well tolerated, it is used as a sweetner in diabetic diets.
By heating in the presence of acid catalysts, D-S. undergoes an intramolecular loss of water with the formation of internal ethers. One such ether, 1,4-sorbitan, is converted commercially into dispersants and emulsifying agents by partial esterification with fatty acids and reaction with ethylene oxide.

Sorbitol   *1,4-Sorbitan*

L-Sorbose: a monosaccharide hexulose, M_r 180.16, m. p. 162 °C, $[\alpha]_D^{20} - 42.9$ °, present in certain plant juices, e.g. rowan berries, and biosynthesized

Complete amino acid sequence of somatotropin

NH$_2$–Phe–Pro–Thr–Ile–Pro–Leu–Ser–Arg–Leu–Phe–Asp–Asn–Ala–Met–Leu–Arg–Ala–His–Arg–Leu–
　　　　　　　　5　　　　　　　　　　10　　　　　　　　　　15　　　　　　　　　　20

His–Gln–Leu–Ala–Phe–Asp–Thr–Tyr–Gln–Glu–Phe–Glu–Glu–Ala–Tyr–Ile–Pro–Lys–Glu–Gln–
　　　　　　　25　　　　　　　　　　30　　　　　　　　　　35　　　　　　　　　　40

Lys–Tyr–Ser–Phe–Leu–Gln–Asn–Pro–Gln–Thr–Ser–Leu–Cys–Phe–Ser–Glu–Ser–Ile–Pro–Thr–
　　　　　　45　　　　　　　　　　50　　　　　　　　　　55　　　　　　　　　　60

Pro–Ser–Asn–Arg–Glu–Glu–Thr–Gln–Lys–Ser–Asn–Leu–Gln–Leu–Leu–Arg–Ile–Ser–Leu–Leu–
　　　　　　65　　　　　　　　　　70　　　　　　　　　　75　　　　　　　　　　80

Leu–Ile–Gln–Ser–Trp–Leu–Glu–Pro–Val–Gln–Phe–Leu–Arg–Ser–Val–Phe–Ala–Asn–Ser–Leu–
　　　　　　85　　　　　　　　　　90　　　　　　　　　　95　　　　　　　　　　100

Val–Tyr–Gly–Ala–Ser–Asn–Ser–Asp–Val–Tyr–Asp–Leu–Leu–Lys–Asp–Leu–Glu–Glu–Gly–Ile–
　　　　　　105　　　　　　　　　110　　　　　　　　　115　　　　　　　　　120

Gln–Thr–Leu–Met–Gly–Arg–Leu–Glu–Asp–Gly–Ser–Pro–Arg–Thr–Gly–Gln–Ile–Phe–Lys–Gln–
　　　　　　125　　　　　　　　　130　　　　　　　　　135　　　　　　　　　140

Thr–Tyr–Ser–Lys–Phe–Asp–Thr–Asn–Ser–His–Asn–Asp–Asp–Ala–Leu–Leu–Lys–Asn–Tyr–Gly–
　　　　　　145　　　　　　　　　150　　　　　　　　　155　　　　　　　　　160

Leu–Leu–Tyr–Cys–Phe–Arg–Lys–Asp–Met–Asp–Lys–Val–Glu–Thr–Phe–Leu–Arg–Ile–Val–Gln–
　　　　　　165　　　　　　　　　170　　　　　　　　　175　　　　　　　　　180

Cys–Arg–Ser–Val–Glu–Gly–Ser–Cys–Gly–Phe–COOH
　　　　　　185　　　　　　　　　190

Somatotropin release inhibiting hormone: see Releasing hormones.
Somatotropin releasing factor: see Releasing hormones.
Somatotropin releasing hormone: see Releasing hormones.
Sophorabioside: genistein 4′-rhamnoglucoside, see Isoflavone.
Sophoricoside: genistein 4′-glucoside, see Isoflavone.
Sophorol: 7,6′-dihydroxy-3′,4′-methylenedioxyisoflavanone, see Isoflavanone.
D-Sorbitol: a C_6-sugar alcohol found widely in plants. M_r 182.17, m. p. 97 °C, $[\alpha]_D^{20} - 2$ ° (water). It can be prepared by catalytic or electrochemical reduction of the configurationally related D-glucose, D-fructose or L-sorbose. Technically, it is prepared by catalytic hydrogenation of D-glucose. D-S. is the starting material for the technical synthesis of ascor-

from D-sorbitol. L-S. is an intermediate in the commercial synthesis of ascorbic acid.
SOS response: a collective response to DNA damage observed in bacteria. It includes the appearance of new, error-prone DNA repair activity, so that DNA repair is associated with mutagenesis. It also includes massive synthesis of recA protein, lytic development of certain prophages in lysogenic strains, and inhibition of cell division leading to filamentous growth in nonlysogenic strains. The total collective SOS response can be completely abolished by mutation at either the *recA* or *lexA* loci. Alternatively, mutation at these loci may result in temperature-sensitive (42 °C) gratuitous expression of the SOS response (*tif* at the *recA* locus, and *tsl* at the *lexA* locus).
Incubation of nonlysogenic *tif* or *tsl* strains at 42 °C results in a total SOS response, including the formation of long, multinucleate filaments. Filamentous

growth is abolished by *sfi*A mutations, but all other components of the SOS response are unaffected, thereby defining a factor involved in the inhibition of cell division.

The hypothesis that *rec*A and *lex*A jointly control inducible functions related to DNA damage was originally known as the **SOS hypothesis** (SOS for the international distress signal).

Chemical or physical DNA damage can be monitored by the **SOS function**, which is measured by the **SOS chromotest**. This test therefore serves as a screen for potential mutagens in prokaryotes, which are inferred to be potential carcinogens in eukaryotes.

The SOS chromotest uses a strain of *Escherichia coli* in which the structural gene for β-galactosidase (see) is linked to the promoter of the *sfi*A operon [by operon fusion, using phage Mud(ap,*lac*)]. In the proper orientation, the *lac* genes of the prophage are expressed from the adjacent bacterial promoter. DNA damage induces a high level of expression of the *sfi* gene (i.e. part of the SOS response) leading to synthesis of β-galactosidase, which can be monitored colorimetrically using the chromogenic substrate, *o*-nitrophenyl-β-D-galactoside. [E. M. Witkin, *Bacteriological Reviews 40* (1976) 869–907; O. Huisman & R. D'Ari, *Nature 290* (1981) 797–799]

Southern blot: a method for transferring fragments of DNA from electrophoresis gels to cellulose nitrate membranes, first described by E. M. Southern. [*J. Mol. Biol. 98* (1975) 503–517]. Similar procedures for replicating electrophoretograms of RNA or proteins have been capriciously named Northern or Western blots, respectively. As yet, there is no Eastern blot.

The gels used for electrophoresis (agarose, polyacrylamide) are often unsuitable for further study of the resolved substances, whereas the replica (or blot) can be submitted to a variety of analytical procedures. For example, extremely small quantities of DNA can be located by hybridization with radioactive RNA; and proteins can be located and identified with antibodies. In the original method of Southern, the nitrocellulose membrane is layered on top of the gel (DNA is first denatured by immersion of the gel in buffered ethidium bromide), followed by moist filter paper and then by several layers of dry filter paper. DNA fragments move from the gel by capillarity and are trapped in the nitrocellulose. For the replication of protein separations, zones may be transferred to nitrocellulose, or to diazobenzyloxymethyl (DBM) paper or diazophenylthioether (DPT) paper. Transfers can be performed more rapidly by electrophoresing the zones from the gel into the receiving layer; the replica is then called an electroblot or "blitz" blot.

Soybean trypsin inhibitor, abb. *STI:* the best known plant trypsin inhibitor. With bovine trypsin, at pH 8.3, it forms a stoichiometric, enzymatically inactive, stable complex with an association constant of 5×10^9 per mol STI. It also inhibits other vertebrate and invertebrate trypsins and plasmin. Chymotrypsin is inhibited to a small extent, and other endopeptidases not at all. STI is a single polypeptide chain, M_r 21100, 181 amino acid residues (of known sequence) and two disulfide bridges. The molecule is compact and has a low α-helix content

due to the presence of proline residues. It is consequently resistant to proteases and to denaturation. The reactive center contains a specific peptide bond, $Arg_{63}-Ile_{64}$, which is hydrolysed when the trypsin-STI complex is formed. The ensuing interaction with the active site of trypsin, which also involves Arg_{65} of STI, results in firm binding of STI and inactivation of the trypsin.

Soybean also contains the Bowman-Birk inhibitor, which consists of 78 amino acid residues (M_r 8000) and is much smaller than STI. It is a "double headed" inhibitor, i.e. in addition to inhibiting trypsin, it can simultaneously inhibit a second protease, e.g. chymotrypsin.

Since the nutritional value of protein-rich soybean products may be reduced by the presence of STI and other inhibitors, these are inactivated by heating the milled beans.

Specific activity: see Enzyme kinetics.

Specific incorporation rate: see Isotope technique.

Specificity constant, *physiological effectivity:* a measure of the turnover of a substrate. S. c. is the ratio of the catalytic constant and the Michaelis constant: k_{cat}/K_m. It is equal to the rate constant of a reaction for the rate equation:
$v = k_{cat}E_o S/(K_m + S)$, where E_o is the total enzyme concentration, and S is the substrate concentration, when $S \ll K_m$.

Specific linking difference: see Linking number.

Specific radioactivity: see Isotope technique.

Spectinomycin: see Streptomycin.

Spectrin: a protein which makes up about 75% of the "skeleton" of the erythrocyte membrane; there are about 10^5 copies of each subunit per cell. S. is a heterodimer or tetramer of two polypeptides, the α-subunit having an M_r of 260,000 and the β-subunit, 225,000. (These were originally called "band 1" and "band 2", respectively, from their electrophoretic behavior.) Under normal conditions, S. appears to exist as a mixture of dimers and tetramers, linked into a loose network by actin. This flexible structure lies just inside the lipid bilayer of the membrane (see Biological membranes). The dimers consist of one α and one β chain aligned parallel to one another to form a rod about 100 nm long. These associate end-to-end to form the tetrameric rod, which is about 200 nm long. Both subunits consist of repeated 106-amino-acid homologous units; and the α repeat unit is also homologous to the β repeat unit. S. is non-covalently bound to Ankyrin (see), by a site on the β chain. The β chain also has four sites for phosphorylation close to its *C*-terminus. The significance of these is unknown. The physical state of the S. network is sensitive to the intracellular Ca^{2+} levels and the ATP charge of the cell; when the Ca^{2+} level is high or ATP is depleted, it is more highly polymerized, and the erythrocyte loses its typical shape. Thus S. is probably responsible for maintaining the concave-disc shape of the erythrocyte, and for allowing the cell the extreme flexibility it needs to pass through capillaries. A number of hemolytic anemias are related to abnormalities in the S. structure which make the cells more fragile.

Non-erythroid forms of spectrin have also been reported. [V. Bennett, "The Membrane Skeleton of Human Erythrocytes and Its Implications for More Complex Cells" (1985) *Ann. Rev. Biochem.* **54** 273–304]

Spermaceti: a solid animal wax, m.p. 45-50 °C, obtained from the head of the sperm whale, *Physeter macrocephalus.* An oily, liquid, crude sperm oil is secreted in a special large cylindrical organ in the upper region of the huge jaw and above the right nostril of the whale. After capture, this cavity is emptied of its oil, which, on cooling, deposits crystalline S. This is separated by pressure and purified by remelting and washing with dilute NaOH to remove the last traces of oil. S. consists chiefly of cetyl palmitate, accompanied by smaller quantities of cetyl laurate and myristate. It is used in the pharmaceutical and cosmetic industries as a basis for creams.

Spermidine: see Polyamines.

Spermine: see Polyamines.

Spheroidine: see Tetrodotoxin.

Sphingolipidosis: see Inborn errors of metabolism.

Sphingomyelins: see Phospholipids.

Sphingosine: 2-amino-4-octadecene-1,3-diol, M_r 299.48, m.p. 67 °C, a long-chain amino alcohol which is a component of sphingomyelins (see Phospholipids) and Glycolipids (see). It is not found in free form in plants or animals.

Spider toxins: toxic substances produced in the venom glands of many spiders. They serve to paralyse and kill prey, and are dangerous to humans only in rare cases, e.g. the toxin of the South European *Latrodectus tredecimguttatus,* or the American black widow *(Latrodectus mactans).* The active principles of S.t. are proteinaceous and related to those of snake and scorpion venoms. They contain hyaluronidase and proteolytic activity, but phospholipases and hemolytic or blood clotting activities are absent.

Spinasterols: a group of very similar phytosterols (see Sterols), found in higher plants. Chief representative is *α-spinasterol* (5α-stigmasta-7, 22-diene-3β-ol; Fig.), M_r 412.7, m.p. 172 °C, isolated e.g. from spinach, senega root and lucerne. α-, β-, γ- and δ-S. differ in the position of the side chain double bond.

α-Spinasterol

Spinochromes: derivatives of 1,4-naphthoquinone (see Naphthoquinones), responsible for the red or orange color of sea urchin shells. Over 20 different hydroxylated S. are known. In the native state, they are present as calcium and magnesium salts. They differ from the echinochromes, which occur in the eggs, perivisceral fluid and internal organs

of the sea urchin. Echinochrome A is a red pigment in the eggs and skeleton of the sea urchin; it is a pentahydroxy-1,4-naphthoquinone with an ethyl group at C6, m.p. 223 °C.

Spinulosine: see Benzoquinones.

Spiramycin: a Macrolide (see) antibiotic.

Spirographis porphyrin: see Chlorocruoroporphyrin.

Spirosolane: see Solanum alkaloids.

Spirostane: the oxygen-containing parent structure of the steroid saponins (see Saponins). The S. system is formally derived from the parent hydrocarbon cholestane (see Steroids). The name S. embraces the configuration of all asymmetric centers, with the exception of positions 5 and 25.

(25 S)-5α-Spirostane

Spirostanols: see Saponins.

Splicing: see Intron.

Splicing enzymes: see Intron.

Split gene: see Intron.

Split proteins: see Ribosomes.

Spongonucleosides: see Arabinosides.

Spongosine: 9-β-D-ribofuranosyl-2-methoxy-adenine, a nucleoside with a modified base, isolated from sponges.

Spongosterol: (24R)-5α-ergost-22-ene-3β-ol, a marine zoosterol (see Sterols), M_r 400.66, m.p. 153 °C, $[\alpha]_D$ +10 °C (chloroform), occurring as a typical sterol of sponges *(Spongia)* and isolated, e.g. from *Suberitis domuncula* and *Suberitis compacta.*

Spongothymidine: see Arabinosides.

Spongouridine: see Arabinosides.

Sporidesmolides: cyclic depsipeptides from the fungus *Pithomyces chartarum.* Sporidesmolide I is cyclo-(-Hyv-D-Val-D-Leu-Hyv-Val-MeLeu-); sporidesmolide II contains D-allo-isoleucine in place of D-valine, while sporidesmolide III contains L-leucine in place of L-N-methylleucine. (Hyv represents a residue of α-hydroxyisovaleric acid).

Sporopollenin: the material of the outermost cell wall layer (exine) of pollen grains and spores of pteridophytes, and also present in small amounts in fungal zygospore walls (e.g. zygospore wall of *Mucor mucedo* contains 1% S.). S. is extremely resistant to physical, chemical and biological degradation. Pollen grains are therefore well preserved in geological strata, and have proved archeologically useful as quantitative and qualitative markers of previous plant life and agriculture. P. is an intimate mixture or complex of 10-15% cellulose, 10% of an ill-defined xylan fraction, 10-15% of a lignin–like material, and 55-65% of a lipid. It is now generally accepted that the lipid material is formed by the oxidative polymerization of carotenoids; a virtually identical material can be synthesized in the laboratory by the catalytic oxidation of β-carotene.

["Sporopollenin", by Shaw, G., pp. 31–58 in *Phytochemical Phylogeny* (Edit. Harborne, J.B.) 1970, Academic Press]

S-protein: a cleavage product of ribonuclease, representing amino acid residues 21–124 of the ribonuclease primary sequence. It is produced, together with S-peptide (positions 1–20), by the action of subtilisin on ribonuclease.

Squalene: biochemically the most important aliphatic triterpene. M_r 408, b.p.$_{10}$ 262–264 °C, ρ 0.8584. For formula see Triterpenes. S. was first isolated from fish liver oils, and later found in many plant oils. It is the intermediate in the biosynthesis of all cyclic triterpenes. Cyclization of S. is catalysed by a mixed function oxygenase, and proceeds via 2,3-epoxysqualene (squalene epoxide).

SRH: abb. for somatotropin releasing hormone. See Releasing hormones.

sRNA: abb. for the obsolete name, soluble RNA, now known as Transfer RNA (see) or tRNA.

SRS: abb. for slow reacting substance. See Leukotrienes.

Stachydrine: see Pyrrolidine alkaloids.

Stachyose: a nonreducing tetrasaccharide (see Carbohydrates) found in plants. M.p. 170 °C, $[\alpha]_D^{20} + 149$ °C. The four sugar residues are linked in the order D-galactose-D-galactose-D-glucose-D-fructose.

Staphylococcal protein A.: see Protein A.

Starch: a high M_r polysaccharide, formula $(C_6H_{10}O_5)_n$; the chief storage carbohydrate in most higher plants, consisting of about 80% water-insoluble Amylopectin (see) and 20% water-soluble Amylose (see). In plant metabolism, S. first appears as an assimilation product in the chloroplasts. It is then degraded, the products of degradation are translocated, and S. is resynthesized as storage S. (S. grains or granules) in storage organs, e.g. roots, tubers or pith. S. grains are classified as compound, simple, centric or acentric, depending on their characteristic shapes and stratifications. On the basis of these characteristic forms, flour from different sources

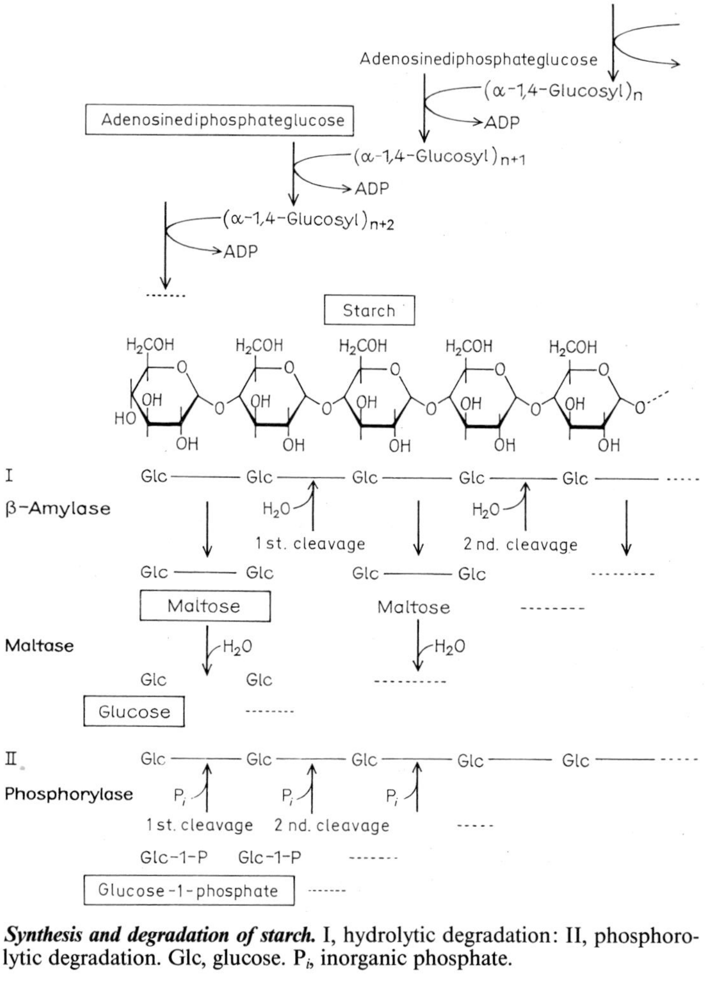

Synthesis and degradation of starch. I, hydrolytic degradation: II, phosphorolytic degradation. Glc, glucose. P$_i$, inorganic phosphate.

(e.g. maize, rice, wheat, rye) can be identified microscopically.

S. is biosynthesized from adenosine-diphosphate-glucose (Fig.). During animal digestion, S. is hydrolysed by amylases. α-Amylase (EC 3.2.1.1) hydrolyses $\alpha(1 \to 4)$ linkages at random, forming a mixture of glucose and maltose. Maltose is hydrolysed to glucose by α-D-glucosidase (maltase, EC 3.2.1.20). Amylose is thus completely degraded to glucose. α-Amylase cannot hydrolyse $\alpha(1 \to 6)$ linkages at the branch points of amylopectin; the product of α-amylase action on amylopectin is therefore a large, highly branched core or limit dextrin, representing 40% of the original amylopectin. The small intestine also contains an oligo-$\alpha(1 \to 6)$-glucosidase, which hydrolyses the $1 \to 6$ linkages, and completes the total degradation of amylopectin. β-Amylases (EC 3.2.1.2) are found especially in germinating seeds (e.g. malt); they act on the nonreducing end of the polysaccharide chain, removing successive maltose units. β-Amylase is also unable to attack $\alpha(1 \to 6)$ branch points. In plant cells, storage S. is remobilized by phosphorolysis to glucose 1-phosphate.

S. is very important in human nutrition, supplying most of the dietary carbohydrate requirement (humans require about 500 g carbohydrate per day). Potatoes, cereals and bananas are particularly rich in S. Metabolism of 1 g S. supplies 16.75 kJ (4 kcal). S. is prepared commercially from plant sources, in particular potatoes, wheat rice and maize, and it has many uses in the food industry and in technology.

Start codon: see Initiation codon; Protein biosynthesis.

Starter: see Primer.

Starter tRNA: see Initiation tRNA.

Statins: see Releasing hormones.

Status-quo hormones: see Juvenile hormones.

Steady state: a chain of chemical reactions is in a steady state when the concentration of all intermediates remains constant, despite a net flow of material through the system, i.e. the concentration of intermediates remains constant, while a product is formed at the expense of a substrate. The term is used in two ways:

1. *Kinetic S. s.* is part of the kinetic response of an enzyme-catalysed reaction during transition from the initial state to the equilibrium state (steady state region) (Fig.). It is characterized by a slow change in the concentration of intermediates (see Enzyme species), while the concentration variables alter with the rate of the overall reaction.

In the S.s. region, $(\mathrm{d}ES/\mathrm{d}t)/(\mathrm{d}S/\mathrm{d}t) \ll 1$, and in the kinetic equation $\mathrm{d}ES/\mathrm{d}t$ can be approximated to zero.

2. In the *induced S. s. (true S. s.)* the concentrations of all reactants remain constant, since the rates of formation and removal are equal. For the simple case of a chain of metabolic reactions:

$$\overset{v}{\to} X_1 \overset{k_1}{\longrightarrow} X_2 \overset{k_2}{\longrightarrow} - - - \overset{k_{n-1}}{\longrightarrow} X_n \overset{v}{\to},$$

it can be stated that $\mathrm{d}X_i/\mathrm{d}t = 0$ and $k_i X_i = v$, where $i = 1, - - - -, n-1$.

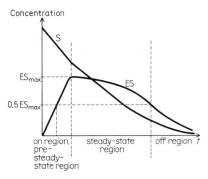

Concentration

Time course of the concentrations of substrate and enzyme-substrate complex during an enzymatic reaction. S. substrate concentration; ES, concentration of enzyme-substrate complex.

Stearic acid: *n*-octadecanoic acid, $CH_3-(CH_2)_{16}-COOH$, a fatty acid, M_r 284.5, m.p. 71.5 °C, b.p.$_{15}$ 232 °C. Together with palmitic acid, S.a. is one of the most plentiful and most widely distributed fatty acids, occurring in practically all animal and plant oils and fats, e.g. 34% in cocoa butter, 30% in mutton fat, 18% in beef fat, 5-15% in milk fat. It is used in the manufacture of candles, soaps, detergents, antifoams, pharmaceuticals and cosmetics. Commercial stearin is a mixture of S.a. and palmitic acid.

Stemphylotoxins: phytotoxins (Fig.) isolated from culture filtrates of *Stemphylium botryosum* Wallr. f. sp. *lycopersici*, the causative agent of leaf spot and foliage blight disease of tomato. The symptoms of these diseases are caused by external application of S. Toxicity of S. is measured by their inhibition of incorporation of ^{14}C-amino acids into protein in exponentially growing tomato cell suspensions. S. also inhibit rootlet elongation of tomato, and they inhibit growth of *Spirodella oligorrhiza* (duckweed). S. I is about 100 times more toxic than S. II. Conversion of I into II occurs readily by acid or base catalysis. S. show a high affinity for ferric, but not for ferrous iron (apparent stability constants for the Fe^{3+} complexes: 1.7×10^{24} (I) and 1.6×10^{24} (II); cf. $10^{30}-10^{34}$ for hydroxamate Siderophores (see)). S. may therefore function as siderophores, by sequestering ferric iron from the host plant for use by the fungus. Biosynthesis of S. is regulated by iron, the optimal concentration for S. production being 2 mg/l (Fe^{2+} or Fe^{3+}).

Stemphylotoxin I   *Stemphylotoxin II*

Sterane: an earlier name for gonane. See Steroids.

Stercobilin: see Bile pigments.

Steroid alkaloids: nitrogen-containing steroids present in plants. S. a. are especially common in the plant families, *Solanaceae* (nightshades), *Liliaceae* (lilies), *Apocyanaceae* (periwinkles) and *Buxaceae* (boxwoods), where they often occur as glycoalkaloids, or esterified as ester alkaloids. They are classified, according to their origin, into Solanum alkaloids (see), Veratrum alkaloids (see), Funtumia alkaloids (see) Holarrhena alkaloids (see) and Buxus alkaloids (see), but each group may contain different structural types. The Salamander alkaloids (see) are of animal origin and represent a special group. S. a. are formal derivatives of either cholestane or pregnane, and they are further subdivided on the basis of structure, e. g. Solanum alkaloids comprise spirosolanes and solanidanes. Some S. a., e. g. certain Veratrum alkaloids, are used pharmacologically.

Steroid alkaloid saponins: see Saponins.

Steroid hormones: groups of steroids which function as hormones. They comprise Adrenal corticosteroids (see), sex hormones (see Gonadal hormones), Ecdysone (see) and related molting hormones, and the plant sex hormone Antheridiol (see). 1,25-Dihydroxycholecalciferol$_3$ (an active metabolite of the seco-steroid vitamin D$_3$) is also considered to be a S.

It was earlier thought that the S. bind to specific receptors in the cytosol of target tissues, followed by entry of the receptor-steroid complex into the nucleus. This widely held hypothesis appears to be no longer tenable. The receptor is bound in the nuclear matrix. Cytosolic receptors represent a reproducible artifact, due to release of the nuclear receptor by homogenization and dilution of the cell contents with homogenizing buffer.

Sex steroid-binding globulin (SBG) may play an active role in the entry of S. into target cells. Thus, ^{125}I-labelled SBG (purified from retroplacental blood serum) specifically binds to plasma membranes of human decidual endometrium only when it is carrying estradiol. [J. Gorski et al. *Mol. Cell. Endocrinol.* **36** (1984) 11–15; P. J. Sheridan et al. *Nature 282* (1979) 579–582; H. R. Walters et al. *J. Biol. Chem.* **255** (1980) 6799–6905; K. Liimpaphayon et al. *J. Obstetr. Gynecol.* III, No. 8 (1971) 1064–1068; N. I. Zhuk et al. *Biochemistry* (USSR) **50** (No. 7, part 1) (1986) 936–939]

Steroids: a large group of terpenoids, including many biologically important compounds, e. g. Sterols (see), Steroid hormones (see), Bile acids (see), Cardiac glycosides (see), Steroid alkaloids (see) and steroid saponins (see Saponins). Synthetic S., e. g. Ovulation inhibitors (see) and Anabolic steroids (see), and structurally modified steroid hormones are pharmacologically important. Structurally, S. are derivatives of the hydrocarbon cyclopentanoperhydrophenanthrene, which contains 6 asymmetrical centers (marked with asterisks in Fig. 1). There are thus $2^6 = 64$ theoretically possible stereoisomers, but relatively few of these are known among the S. The nomenclature of the tetracyclic ring structure and the numbering of the carbon atoms are shown in Fig. 2. There are a number of fundamental type-hydrocarbons, defined by the substituents at C-atoms 10, 13 and 17 (Table). The simplest member is gonane (earlier called sterane) in

which rings B/C and C/D are *trans* to one another, as in the majority of natural S. The configuration at C-5 depends on the *cis-trans* relationship of rings A and B; thus 5α-gonane is A/B *trans*, and 5β-gonane is A/B *cis*. With substitution at C-13, the nomenclature becomes 5α- or 5β-estrane. Introduction of a further methyl group at C-10 gives rise to 5α-androstane (earlier called testane) or 5β-androstane (earlier called etiocholane), etc. Both of the angular methyl groups are present in most natural S. Sterically, they are above the fixed plane of the S. ring system (Fig. 3); all such atoms or groups above the plane of the molecule are designated by the prefix β. In drawn or printed formulas, the bond linking a β-methyl group to the ring carbon is shown as a solid line. Bonds to α-groups or atoms (i. e. below the plane of the ring) are shown as dotted lines. Wavy lines are used for bonds of unknown configuration, and the Greek letter Xi (ζ) is used instead of α or β. Structural differences among S. arise from the type, number, position and configuration of the substituents, as well as the number and position of double bonds. Substituents containing oxygen are denoted by an appropriate suffix, e. g. -ol and -dione, in the name of the type structure. Double bonds are denoted by the suffix -ene, e. g. cholest-5-ene. Ring contraction is shown by the prefix nor- (norsteroids); ring expansion by the prefix homo- (homosteroids). Ring cleavage is indicated by the prefix seco- (secosteroids). More than 20000 S. are known, and about 2% of these have some medical significance.

S. are biosynthesized ultimately from acetyl-CoA. The early stages of this biosynthesis, leading to the open chain triterpene intermediate, squalene, are discussed under Terpenes (see). Squalene is oxidized to 2,3-epoxysqualene (Fig. 4). 2,3-Epoxysqualene is then cyclized to lanosterol in nonphotosynthetic organisms, and to cycloartenol in photosynthetic organisms (Fig. 4). Cycloartenol is the biosynthetic precursor of the phytosterols (Fig. 6), whereas lanosterol is the biosynthetic precursor of mycosterols and zoosterols (Fig. 5). Details of reactions involved in the biosynthesis of sterols from cycloartenol and

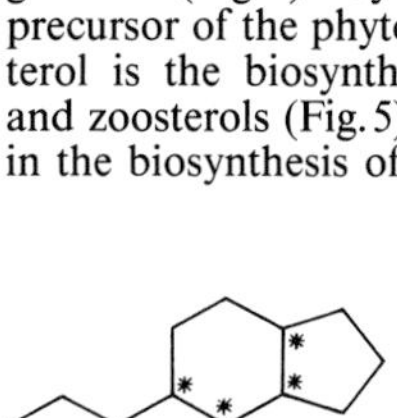

Figure 1. Cyclopentanoperhydrophenanthrene.

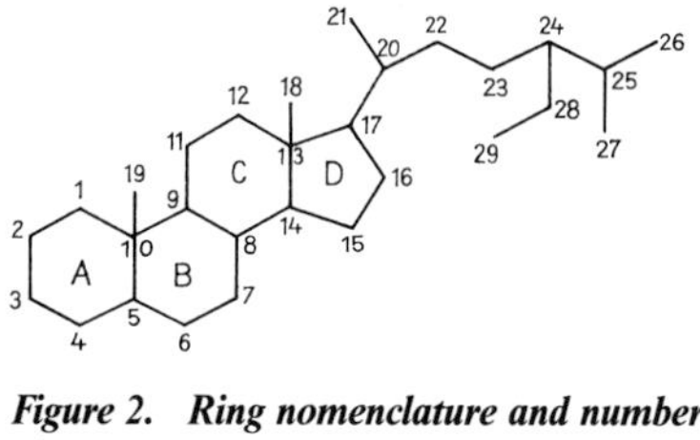

Figure 2. Ring nomenclature and numbering of carbon atoms.

Table. Parent hydrocarbon structures of the steroids.

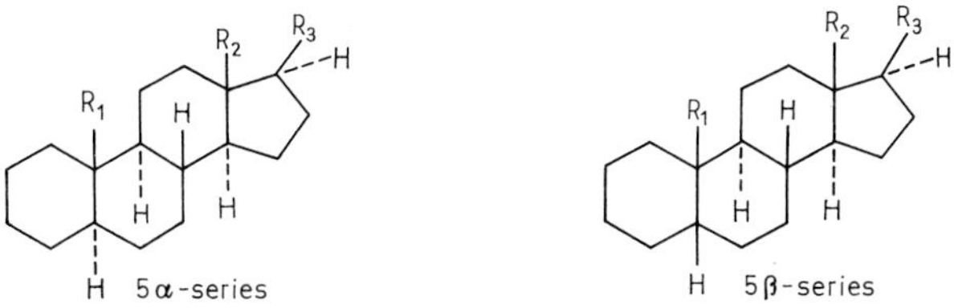

Name	R_1	R_2	R_3
5α-Gonane (formerly Sterane) 5β- Gonane	H	H	H
5α-Estrane 5β-Estrane	H	CH₃	H
5α-Androstane (formerly Testane) 5β-Androstane (formerly Etiocholane)	CH₃	CH₃	H
5α-Pregnane (formerly Allopregnane) 5β-Pregnane	CH₃	CH₃	C₂H₅
5α-Cholane (formerly Allocholane) 5β-Cholane	CH₃	CH₃	CH(CH₃)CH₂CH₂CH₃
5α-Cholestane 5β-Cholestane (formerly Coprostane)	CH₃	CH₃	CH(CH₃)CH₂CH₂CH₂CH(CH₃)₂
5α-Ergostane 5β-Ergostane	CH₃	CH₃	CH(CH₃)CH₂CH₂CH(CH₃)CH(CH₃)₂
5α-Stigmastane 5β-Stigmastane	CH₃	CH₃	CH(CH₃)CH₂CH₂CH(C₂H₅)CH(CH₃)₂

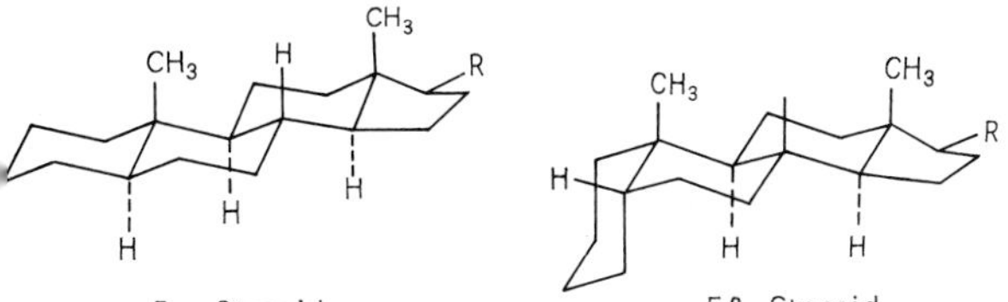

Figure 3. Conformations of the steroid molecule.

lanosterol are shown in Figs. 7–11. Further biological conversion of cholesterol to a variety of important animal steroids commences with side chain cleavage to pregnenolone (Fig. 12). [E. I. Mercer, *Pestic. Sci.* 15 (1984) 133–155]

In ergosterol biosynthesis, the $\Delta^8 \rightarrow \Delta^7$ transformation involves loss of the 7α-hydrogen (derived biosynthetically from the 2-*pro-R* hydrogen of mevalonic acid), whereas in animals, higher plants and chlorophyte algae, the 7β-hydrogen (2-*pro-S* of mevalonic acid) is lost.

Insertion of the C-5,6 double bond ($\Delta^7 \rightarrow \Delta^{5,7}$) involves removal of the 5α (4-*pro-R* in mevalonic acid) and 6α (5-*pro-S* in mevalonic acid) hydrogens. In rat liver and yeast microsomal preparations, the reaction requires aerobic conditions and NADH or NADPH. This suggests a mechanism with hydroxylation at C-5 or C-6, followed by dehydration, or a fatty acid-type desaturation. There is evidence for the former mechanism in yeast, whereas the Δ^5-desaturase of rat liver appears to be an oxygenase-type enzyme with participation of cytochrome b₅.

Sterol Δ^{22}-dehydrogenase introduces a *trans* double bond. By using ³H-labelled mevalonate as a substrate for ergosterol synthesis in fungi, it has been shown that the 22-*pro-S* and 23-*pro-S* hydrogens are eliminated during desaturation. Higher plants also eliminate 22-*pro-S* and 23-*pro-S* hydrogens in sterol biosynthesis. The mechanism is assumed to be the same in chlorophyte algae, but confirmation is difficult, because these organisms are impermeable to mevalonic acid.

Steroids

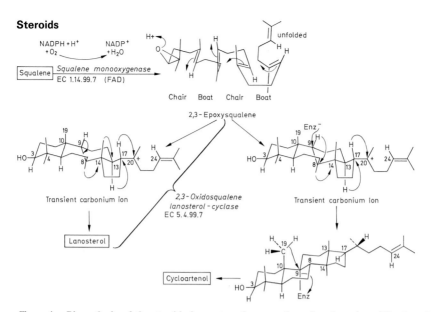

Figure 4. Biosynthesis of the steroid ring system from squalene. Squalene is oxidized to 2,3-epoxysqualene which interacts with a proton and undergoes forward cyclization to form a transient carbonium ion. Backward rearrangement occurs by a series of four Wagner-Meerwein shifts: *1.* 17β-H is transferred to C-20 (H at C-20 takes up *R* configuration); *2.* 13α-H is transferred to C-17; *3.* 14β-methyl group is transferred to C-13; *4.* 8α-methyl group is transferred to C-14. In the synthesis of lanosterol, the resulting C-8 carbonium ion is stabilized by loss of 9β-H as a proton, and formation of a double bond between C-8 and C-9. In the synthesis of cycloartenol, the C-8 carbonium ion is not stabilized by loss of 9β-H. Instead, a Wagner-Meerwein shift occurs so that the 9β-H becomes 8β-H, and the resulting C-9 carbonium ion adds to a nucleophilic group (probably provided by the cyclizing enzyme). *Trans* elimination of H⁺ from the C-19 methyl group, and of the nucleophile leads to formation of the C-9,10,19 cyclopropane ring of cycloartenol.

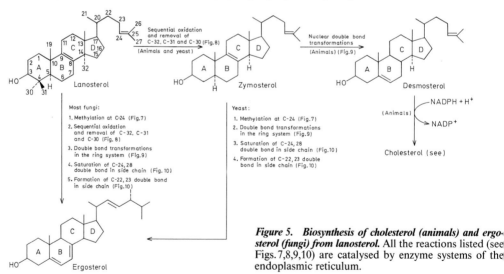

Figure 5. Biosynthesis of cholesterol (animals) and ergosterol (fungi) from lanosterol. All the reactions listed (see Figs.7,8,9,10) are catalysed by enzyme systems of the endoplasmic reticulum.

Figure 7. Methylation at C-24 of the C-24,25-unsaturated side chain of various sterols. R represents the ring system of the appropriate sterol. Δ²⁴-Sterol methyltransferase has been purified from the microsomal fraction of yeast grown aerobically at the expense of glucose. When ethanol serves as the carbon source, the enzyme is found in the mitochondria. If mitochondrial development is repressed by anaerobiosis in the presence of 10% glucose, the enzyme is present in the promitochondrial fraction.
The substrate must possess a C-24,25 double bond. Evidence for the mechanism depicted above was obtained in several fungi, using [*methyl*-²H₃]methionine and [2-¹⁴C,(4R)-4-³H₁]mevalonic acid as sterol precursors. The

562

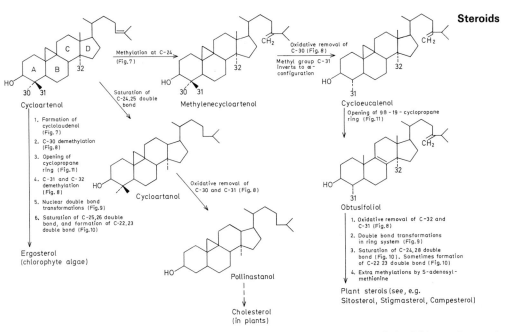

Figure 6. Some biological conversions of cycloartenol. In green plants all sterols are derived from cycloartenol. Cholesterol, the predominant animal sterol, is also widespread in plants, where it may even occur in large quantities (e.g. in Compositae pollen). In plants, cholesterol is the biological precursor of phytoecdysones, cardenolides and steroid hormones, which may play a part in defense mechanisms against herbivorous animals.

methylation process in those chlorophyte algae which synthesize ergosterol differs from that in fungi. Both cycloartenol and cyclolaudenol are present in *Chlorella*, and studies with [*methyl-*[14]C]-*S*-adenosylmethionine and [*methyl-*[2]H₃]methionine support the mechanism shown above. Confirmatory studies with labelled mevalonic acid are not possible, because green algae are impermeable to mevalonic acid.

In the biosynthesis of plant sterols, the side chain may be subject to further methylations, all dependent on *S*-adenosylmethionine. Branch ethyl groups are formed by the additional methylation of branch methyl groups. See, e.g. Stigmasterol, Sitosterol, Campesterol.

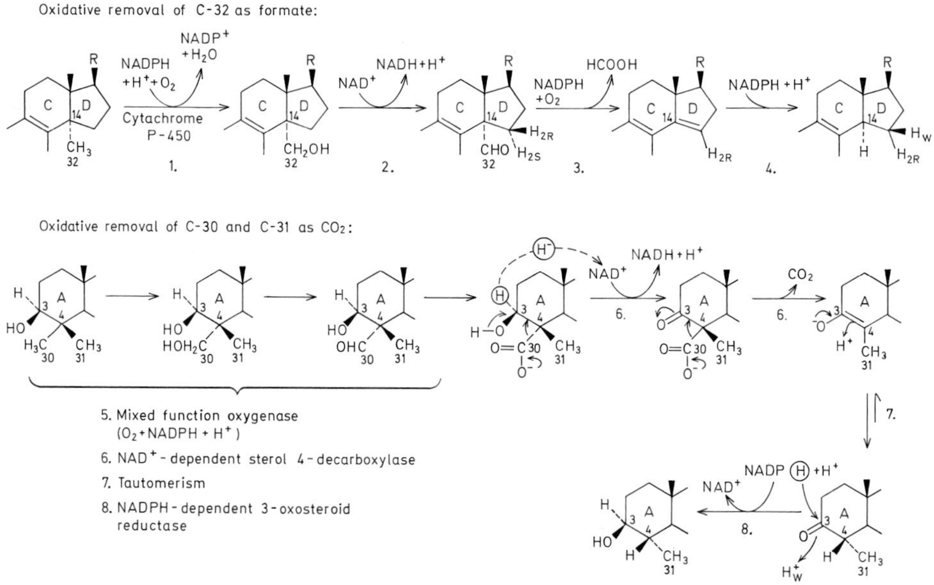

Oxidative removal of C-32 as formate:

Oxidative removal of C-30 and C-31 as CO_2:

5. Mixed function oxygenase
   $(O_2 + NADPH + H^+)$
6. NAD^+-dependent sterol 4-decarboxylase
7. Tautomerism
8. NADPH-dependent 3-oxosteroid
   reductase

Figure 8. Sequential oxidation and removal of C-32, C-31 and C-30 of sterols. In fungi and animals, C-32 (C-14 methyl group) is removed first. The outline process of C-32 removal is clear, and the reactions depicted above together with indicated stereochemistry are not in dispute. Some detailed mechanisms, however, remain to be elucidated. The next carbon to be removed is C-30 in the 4α position. The remaining methyl group (C-31, originally in position 4β) then undergoes an inversion of configuration and takes up the 4α position. The decarboxylation reaction is analogous to the decarboxylation of a β-keto (3-oxo) acid. The cycle of reactions is then repeated, and C-31 is lost as CO_2.

In photosynthetic organisms, the 4α-methyl group is removed first, producing a 4,14-dimethylsterol (4-demethylcyclolaudenol in chlorophyte algae, and cycloeucalenol in tracheophytes). This is followed by opening of the 9β,19-cyclopropane ring to give the corresponding Δ^8-sterols (Fig. 11). The remaining two methyl groups are then removed by the mechanisms shown above.

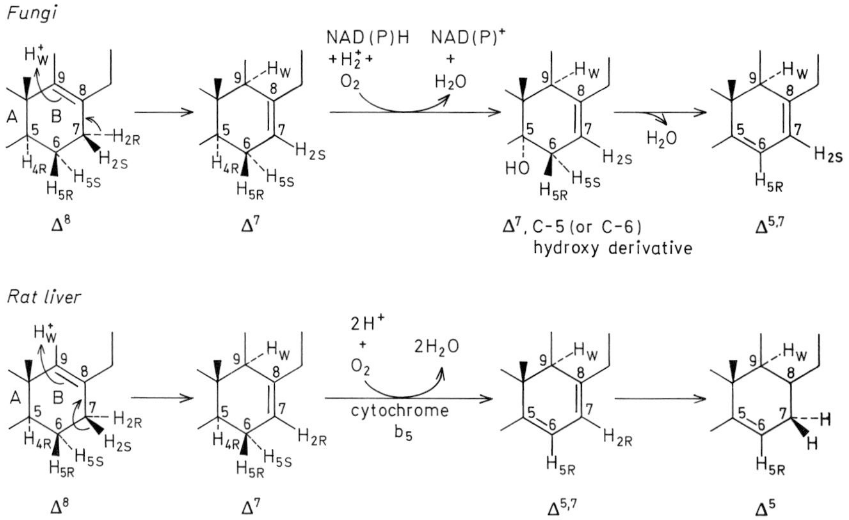

Fungi

Rat liver

Figure 9. Double bond transformations in the sterol ring system. The sequence of nuclear double bond transformations is well established in animal sterol biosynthesis, and the same sequence is assumed to occur in green plants. The sequence is: $\Delta^8 \rightarrow \Delta^7 \rightarrow \Delta^{5,7} \rightarrow \Delta^5$. The last stage does not occur in ergosterol biosynthesis, which retains the $\Delta^{5,7}$ double bonds.

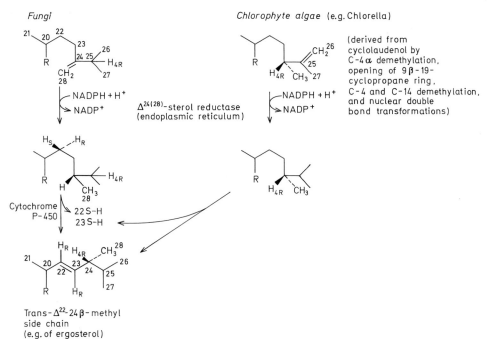

Fungi

Chlorophyte algae (e.g. Chlorella)

(derived from cyclolaudenol by C-4α demethylation, opening of 9β-19-cyclopropane ring, C-4 and C-14 demethylation, and nuclear double bond transformations)

$\Delta^{24(28)}$-sterol reductase (endoplasmic reticulum)

Cytochrome P-450

22S–H
23S–H

Trans-Δ^{22}-24β-methyl side chain (e.g. of ergosterol)

Figure 10. *Saturation of the C-24,28 double bond (and the C-25,26 double bond), and formation of a C-22,23 double bond in the sterol side chain.* The yeast NADPH-dependent $\Delta^{24(28)}$-sterol reductase is a cytochrome P-450 system of the endoplasmic reticulum. Saturation of the C-25,26 double bond in chlorophyte algae is assumed to be mechanistically similar, but it carries no implications for chirality.

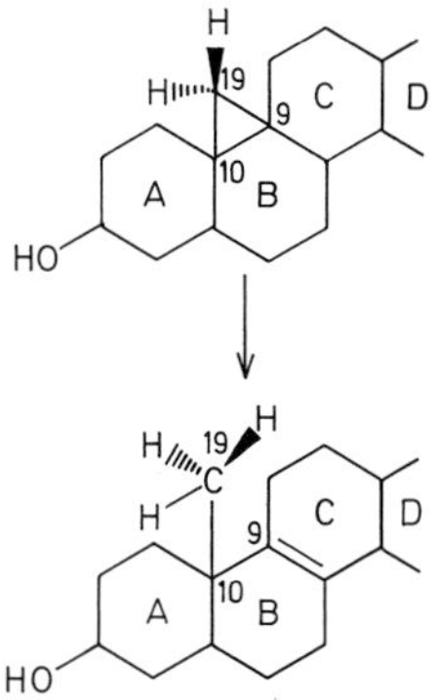

Figure 11. *Opening of the 9β-19-cyclopropane ring during sterol biosynthesis in photosynthetic organisms.*

565

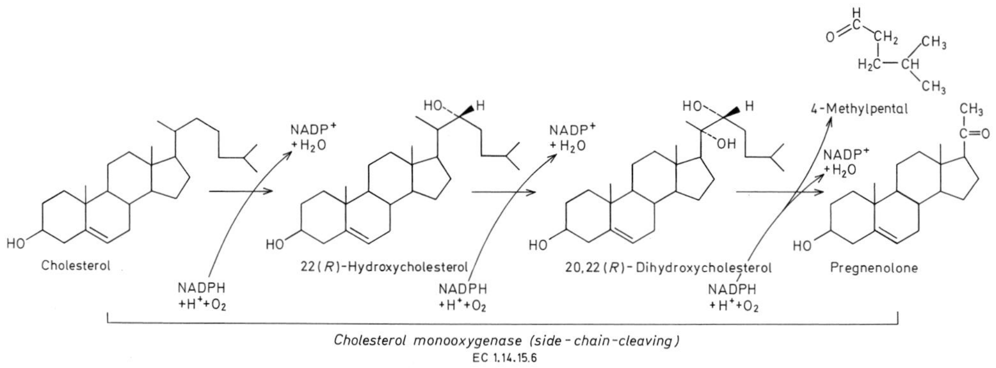

Figure 12. Conversion of cholesterol to pregnenolone by oxidative side chain cleavage. This is an essential first stage in the biosynthesis of progesterone, androgens, estrogens and adrenal cortical steroids from cholesterol.

Steroid saponins: see Saponins.

Sterols: a group of naturally occurring steroids, possessing a 3β-hydroxyl group and a 17β-aliphatic side chain. S. are structural derivatives of the parent hydrocarbons, cholestane, ergostane and stigmastane. They occur in animal and plant cells as free S., and as glycosides and esters. According to their origin, they are classified as zoosterols (animals), phytosterols (plants), mycosterols (fungi) and marine S. (marine animals and plants). More recently, S. have also been found in bacteria. Some important S. are Cholesterol (see), Stigmasterol (see) and Ergosterol (see). S. are essential dietary components for insects. With digitonin, S. form insoluble addition compounds, and inhibit the hemolytic action of saponins. For biosynthesis, see Steroids. S. are isolated from the nonsaponifiable fraction of neutral fat. They are separated and identified by adsorption chromatography on silicic acid (column and thin layer) and by gas chromatography. Mass spectrometry is widely used for identification of S. With strong acid, S. give typical color reactions, which are used for their quantitative determination, e.g. Liebermann-Burchard reaction (see).

Steviol: the aglycon of Stevioside (see).

Stevioside: a tetracyclic diterpene, M_r 804, m.p. 196-198 °C, $[\alpha]_D$ −39.3 ° ($c=5.7$, alcohol), from the bush, *Stevia rebaudiana*, which is native to Paraguay. S. is both a glycoside and a Glucose ester (see). Enzymatic hydrolysis releases the aglycon, steviol ($R_1 = R_2 = H$) M_r 318, m.p. 215 °C $[\alpha]_D$ −94.7 ° (ethanol). Acid hydrolysis causes a Wagner-Meer-wein rearrangement of rings C and D. to produce isosteviol, m.p. 234 °C, $[\alpha]_D$ −78 ° (ethanol). S. is 300 times sweeter than sucrose and would be an ideal sweetening agent; commercial exploitation is difficult on account of its low concentration (about 6 g S. per kg dried leaves). It possesses a gibberellin-like growth stimulating activity.

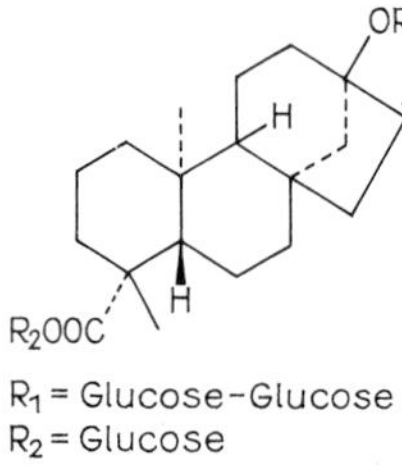

STH: abb. for somatotropic hormone.

STI: abb. for Soybean trypsin inhibitor (see).

Stigmastane: see Steroids.

Stigmasterol: 5α-stigmasta-5,22-diene-3β-ol, M_r 412.7, m.p. 170 °C, $[\alpha]_D$ −49 ° (Chloroform), a widely distributed phytosterol (see Sterols), first isolated from calabar beans, and later from many other sources, e.g. soybeans, carrots, coconut oil and sugar cane wax. It is used as a starting material in the technical synthesis of steroid hormones.

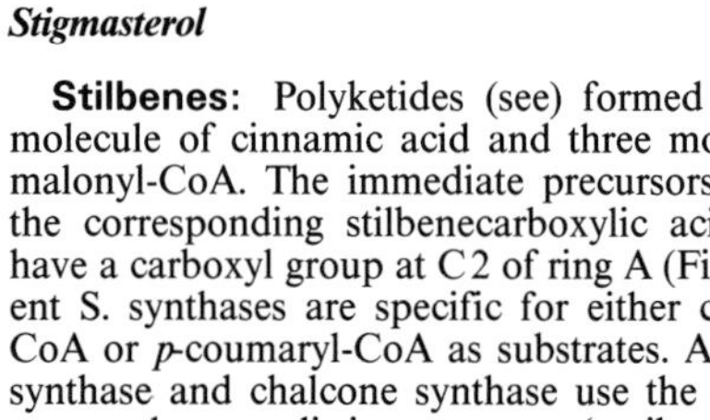

Stigmasterol

Stilbenes: Polyketides (see) formed from one molecule of cinnamic acid and three molecules of malonyl-CoA. The immediate precursors of S. are the corresponding stilbenecarboxylic acids, which have a carboxyl group at C2 of ring A (Fig.). Different S. synthases are specific for either cinnamoyl-CoA or p-coumaryl-CoA as substrates. Although S. synthase and chalcone synthase use the same substrates, they are distinct enzymes (antibodies to one do not cross-react with the other). S. synthase has been purified from cell suspension cultures of *Arachis hypogaea* (peanut). It is a dimer of M_r 90000 (monomer M_r 45000), which converts 1 mol p-coumaroyl-CoA and 3 mol malonyl-CoA into 3,4′,5-trihydroxystilbene (resveretrol) (see Orchinol). [A. Schoeppner & H. Kindl *J. Biol. Chem.* **259** (1984) 6806-6811.]

R_1 = Glucose-Glucose
R_2 = Glucose

Stevioside

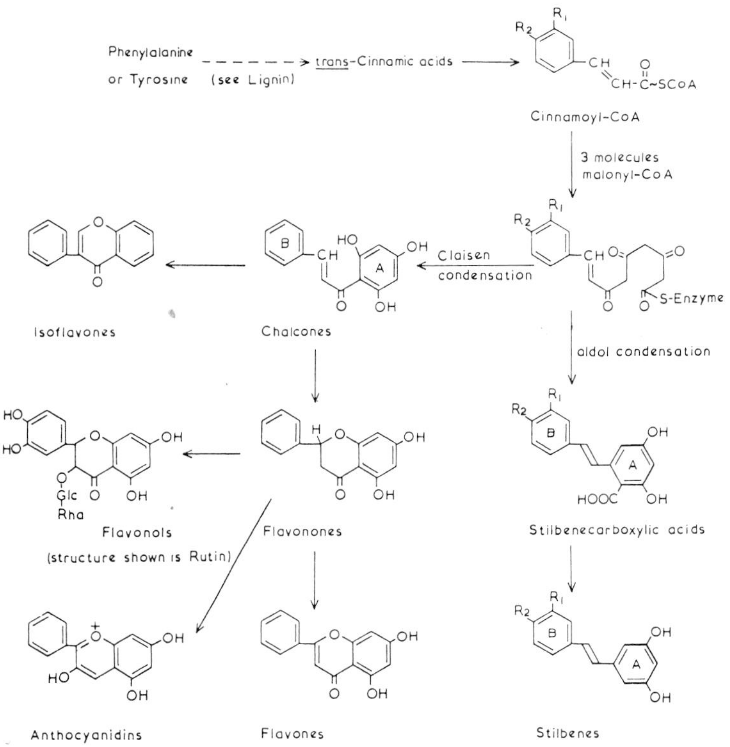

Biosynthesis of stilbenes and related compounds

Cinnamic acid precursor of ring B	R₁	R₂	Stilbene
Cinnamic acid	H	H	Pinosylvin
p-Coumaric acid	H	OH	Resveratrol
Caffeic acid	OH	OH	Piceatannol
Isoferulic acid	OH	OCH₃	Rhapontigenin

Stilbestrol, *stilboestrol, trans-stilbestrol:* a synthetic, nonsteroid Estrogen (see), used in estrogen therapy.

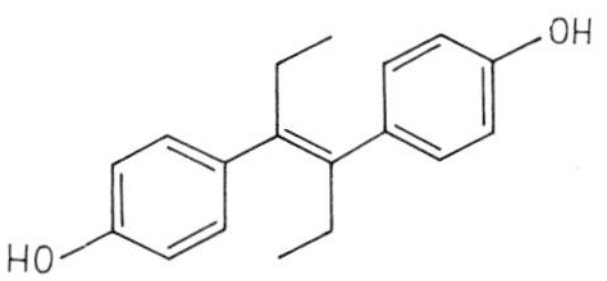

Stilbestrol

Stimulant amines: see Antidepressants.

Stoichiometric model, *stoichiometric reaction scheme:* in enzyme kinetics, a chemical reaction equation in which molecules involved in the reaction, including the enzyme, are represented as letters. These also indicate the molar concentrations. The numbers in front of the letters are the stoichiometric coefficients, which indicate how many moles or molecules reactant are involved in the corresponding reaction step (the coefficient is usually omitted). The simplest example of a S.M. in enzyme kinetics is: $E + S \rightleftharpoons ES \rightarrow E + P$. More complicated reactions are more conveniently expressed by enzyme graphs.

Stop codon: see Termination codon.

Strand polarity, *antiparallel conformation:* the polarity of nucleotide chains, with reference to the sequence of 3',5'-phosphodiester bonds. Polynucleotide chains have a 3'-end (the terminal sugar residue is linked to the preceding residue via its 5'-hydroxyl, and the 3'-hydroxyl is free or phosphorylated) and a 5'-end (5'-hydroxyl is free or phosphorylated). In DNA and other double-stranded nucleic acids, the two strands always lie antiparallel to one another. During replication and transcription, the newly synthesized strand is always antiparallel to its template. The polarity of a nucleotide strand is indicated by $3' \rightarrow 5'$ or $5' \rightarrow 3'$.

Strand selection: the ability of DNA-dependent nucleic acid polymerases to choose the codogenic strand of the double-stranded DNA.

Streptogenin peptides: natural products (e.g. liver extracts, peptones, partial hydrolysates of proteins) or synthetic peptides, which stimulate the growth of microorganisms, especially lactic acid bacteria. The unknown growth stimulant is called *streptogenin*. The growth stimulation of *Lactobacillus casei* by 1 mg standard liver extract is equivalent to one streptogenin unit. In synthetic peptides streptogenin activity depends on the presence of cysteine, and high activity is guaranteed by a neighboring or N-terminal leucine residue.

Streptokinase: see Plasmin.

Streptomycin: an antibiotic from *Streptomyces griseus*. M_r 581.6. S. is an aminoglucoside, in which streptidine is linked glycosidically to the disaccharide, streptobiosamine (Fig.). S. inhibits protein biosynthesis on 70S ribosomes. It becomes bound to the 23S core protein of the 30S ribosomal subunit. This protein appears to be responsible for the binding of mRNA, which is prevented by S.

Similar modes of action are shown by the aminoglucoside antibiotics, kanamycin and neomycin (Fig.) and by paromomycin, kasugamycin, spectinomycin and gentamycin. All are used therapeutically. S. was one of the first antibiotics used against tuberculosis. However, it has been shown to have toxic side effects, e.g. damage to the auditory nerves *(Nervus acusticus)*. Total synthesis of S. was reported in 1974.

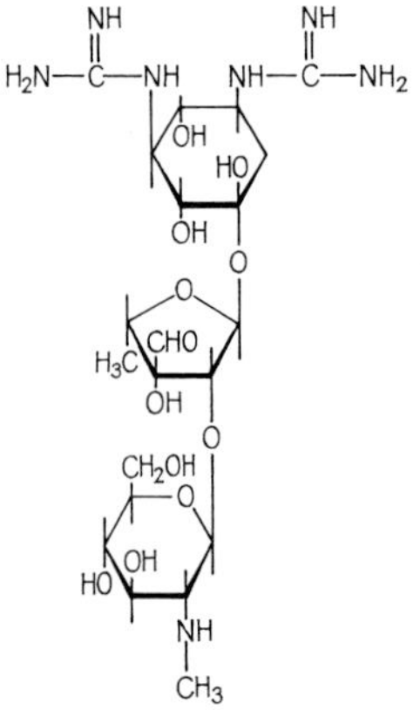

g-Strophanthin contains the aglycon, ouabagenin (g-strophanthidin), M_r 438.52, m.p. 255 °C, $[\alpha]_D + 11°$, linked glycosidically at position 3 with L-rhamnose. Ouabain is a specific inhibitor of the membrane-bound $(Na^+ + K^+)$-ATPase, which is responsible for maintaining high intracellular concentrations of K^+ and low intracellular concentrations of Na^+. Ouabain therefore inhibits sodium ion transport out of the cell and potassium ion transport into the cell, and simultaneously inhibits entry of glucose and amino acids. The aglycon of k-strophanthin is strophanthidin, M_r 401.51, m.p. 136 and 235 °C, $[\alpha]_D + 41°$, which is linked to one molecule each of D-glucose and D-cymarose. Strophanthidin is the aglycon of other cardiac glycosides, e.g. convallatoxin of *Convallaria majalis* (lily of the valley); it differs structurally from digitoxin (see Digitalis glycosides) by the presence of an 18-aldehyde group. S. are highly potent toxins, which have been used in Africa since antiquity as arrow poisons.

Streptomycin   *Kanamycin*   *Neomycin C*

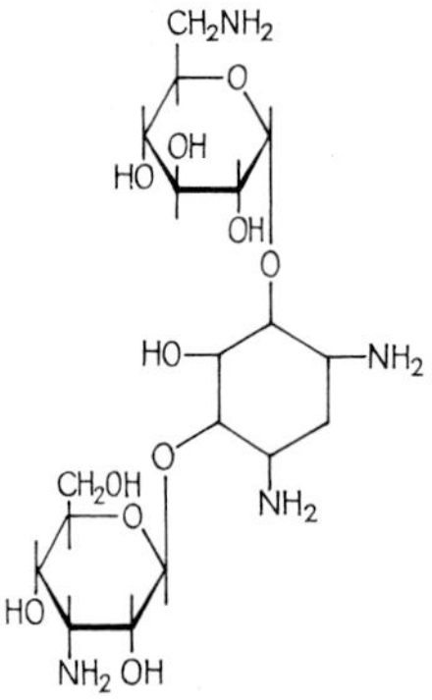

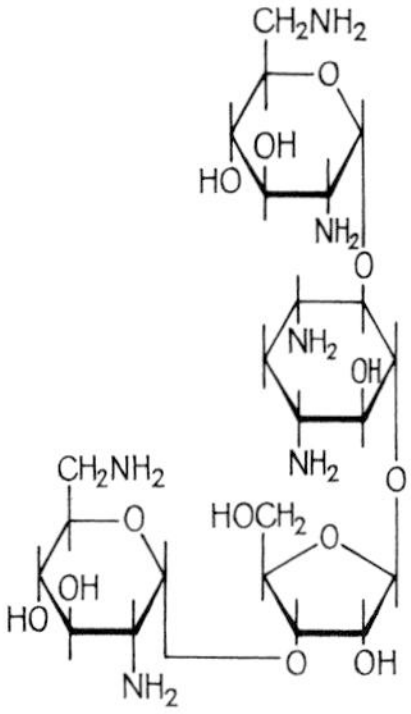

L-Streptose: 5-deoxy-3-formyl-L-lyxose, a monosaccharide, M_r 162.14, with a branched carbon chain. In combination with 2-deoxy-2-methylamino-L-glucose, L-S. forms streptobiosamine, the disaccharide component of Streptomycin (see). L-S. is biosynthesized by the rearrangement of an unbranched hexose.

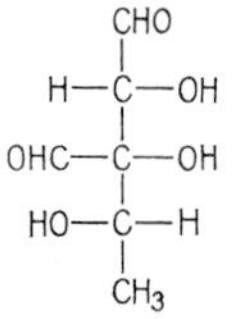

They specifically alter cell permeability. In medicine they are therapeutically very important as cardiac stimulants with more rapid activity than the Digitalis glycosides.

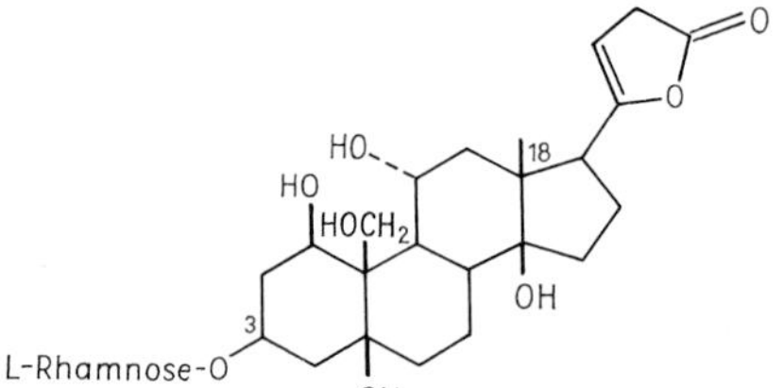

L-Streptose

Ouabain

Stroma, *matrix:* colorless, homogeneous (by light microscopy) ground substance of cell organelles, like Chloroplasts (see) and Mitochondria (see).

Strophanthidin: see Strophanthins.

Strophanthins, *strophanthosides:* cardenolide cardiac glycosides from *Strophanthus* spp., e.g. g-strophanthin (ouabain) from *Strophanthus gratus,* and k-strophanthin from *Strophanthus kombe.*

Strophanthosides: see Strophanthins.

Structural colors, *schemochromes:* colors created by optical effects, due to the physical nature of surfaces, e.g. interference, refraction and diffraction on very thin layers. All spectral colors can be produced in this way, including white (total reflection) and black (total absorption). A change of color with the viewing angle gives rise to iridescence. S. c. occur frequently in Nature, e.g. the color effects of pearls and the inner layers of sea shells, which result from light interference at thin calcium carbonate layers. Bird feather and butterfly wing colors are due to chemical pigments with a contribution from S.c.

Structural genes: see Operon.

Structural proteins, *skeletal proteins, scleroproteins, fibrous proteins:* a group of simple animal proteins with structure and support functions. They are generally insoluble in water and salt solutions: The best known S.p. are the cystine-rich Keratins (see). Others are Collagen (see), Elastin (see), Crystallins (see), silk fibroin, chondrin, spongin, etc. They are subdivided on the basis of chain conformation (see Proteins) into: 1. S.p. with α-helical structure, e.g. α-keratins; 2. S.p. with β-pleated sheet structure, e.g. β-keratins, silk fibroin; 3. S.p. with triple helical structure, e.g. collagen. Conformation is related to amino acid composition. The amino acid composition of S.p. with β-pleated sheet structure shows 90% of the simple amino acids, glycine, alanine and serine, whereas α-helical S.p. contain a much higher percentage of bulky side chains; β-keratin also contains a large number of cystine residues. Collagen characteristically contains a high concentration of the nonhelix forming amino acids, proline and hydroxyproline. Owing to their unusual amino acid composition and fibrous structure, and their resulting insolubility in water, S.p. are not attacked by most proteolytic enzymes. Poor digestibility and absence of essential amino acids make them unsuitable as dietary proteins.

Strychnine: a Strychnos alkaloid. M_r 334.42, m.p. 286–288 °C, b.p.$_5$ 270 °C, $[\alpha]_D^{20}$ –139 ° (chloroform). S.is the chief alkaloid in the seeds of tropical *Strychnos* spp., from which it is obtained commercially; the total synthesis by Woodward is uneconomic and only of scientific interest. It has many therapeutic properties, but its high toxicity (see Neurotoxins) precludes its extensive use in medicine; 5 mg in children and 30–100 mg in adults are sufficient to cause death from muscle rigor. It is used as a rodent poison (mixed with red dyed wheat grains). For formula and biosynthesis see Strychnos alkaloids. S.causes a blockade of central inhibition by selectively antagonizing the effect of glycine. (Glycine is an inhibitory transmitter of 4-aminobutyrate receptors in the central nervous system of vertebrates, particularly in the spinal cord). S.is thought to interact with receptor sites on postsynaptic membranes, thus preventing access of the inhibitory transmitter, glycine.

Strychnos alkaloids: a group of indole alkaloids from the tropical plant genus *Strychnos*. The highly toxic chief alkaloids, Strychnine (see) and Brucine (see), contain a heptacyclic ring structure (Fig.). They are accompanied by Yohimbine (see) -type alkaloids. S.a. are biosynthesized from tryptophan and a terpenoid C_{10}-unit.

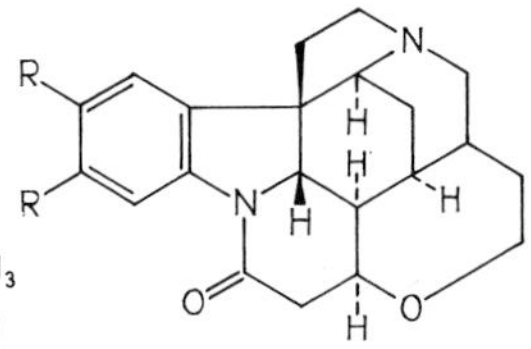

Brucine R=OCH$_3$

Strychnine R=H

SU: abb. for Subunit.

Suberic acid: octanedioic acid, HOOC–(CH$_2$)$_6$–COOH, a higher saturated dicarboxylic acid. m.p. 142 °C, b.p. 300 °C (d.) S.a. is formed by the oxidation of cork or ricinus oil with nitric acid, along with its homologs azelaic and sebacic acids. S.a. is obtained in higher yields from cyclooctene (technical synthesis).

Submersed culture: see Fermentation techniques.

Submitochondrial particle: see Electron transport particle; Mitochondria.

Substance P, *SP:* an undecapeptide: Arg-Pro-Lys-Pro-Gln-Gln-Phe-Phe-Gly-Leu-Met-NH$_2$, present in brain and intestine, and a member of the Amine precursor uptake decarboxylase system (see). Studies were first performed on dried, powdered acid-alcohol extracts of equine brain and intestine; the active principle was called substance P (P for powder), a name that is now widely accepted. The highest concentrations of substance P-like immunoreactivity (SPLI, measured by radioimmunoassay) in the brain are found in the substantia nigra, habenula, hypothalamus, spinal cord sensory ganglia, dorsal roots and substantia gelatinosa. It is now established that SP acts as a neurotransmitter, probably throughout the central nervous system, and certainly in the sensory neurons of the spinal cord. SPLI has also been found in one type of enterochromaffin cell (EC$_1$) present in the gastrointestinal

Table. *Amino acid composition of some structural proteins.*
Numbers represent amino acid residues per 1000 total residues.

Amino acid residues	Skin collagen round worms	calf	Cuticlin round worms	Elastin *Lig. nuchae*	Fibroin *B. mori*	Keratin wool	Resilin locust
Glycine	274	330	150	360	445	85	376
Gly + Ala	346	441	317	597	739	146	487
Pro + Hypro	312	221	301	124	3	85	79
Hypro	16	98	0	12	0	0	0
Lys + Hylys + Arg + His	82	88	36	12	9	104	45
Asp + Glu	136	117	146	39	23	173	152
Cys/2	27	0	24	0	0	106	0
Trp	0	0	0	0	2	12	0

tract and bile duct. SP stimulates contraction of isolated rabbit jejunum and causes transient hypotension when injected intravenously into anesthetized rabbits. It shows close structural similarities to Physolaemin (see) and Eledoison (see). [*Substance P in the Nervous System, Ciba Foundation Symposium 91*, R. Porter and M. O'Connor, eds. (Pitman, 1982)]

Substrate binding center: see Active center.

Substrate binding site: see Active center.

Substrate constant: see Michaelis-Menten kinetics.

Substrate level phosphorylation: adenosine triphosphate (ATP) synthesis not involving photosynthetic phosphorylation or oxidative phosphorylation in the respiratory chain. S.l.p. occurs in Glycolysis (see) and in the oxidative decarboxylation of 2-oxoglutarate in the Tricarboxylic acid cycle (see). In some bacteria, S.l.p. can also occur from carbamoyl phosphate, formed in the degradation of L-citrulline or allantoin. The fine mechanism of S.l.p. is best studied for the glycolytic enzymes, glyceraldehyde 3-phosphate dehydrogenase and 3-phosphoglycerate kinase. The enzyme-substrate complex is oxidized to an energy-rich acylthioester by NAD^+; the acyl group is transferred to phosphoric acid to form 1,3-diphosphoglycerate, which contains a low energy alcohol-phosphate ester and a high energy acyl (carboxyl)phosphate. By the action of 3-phosphoglycerate kinase, the energy-rich phosphate on Cl is transferred to ADP, forming ATP and 3-phosphoglycerate (for formulae, see Fig. 2, Glycolysis). A thioester is also formed in the oxidative decarboxylation of 2-oxoglutarate in the TCA-cycle, i.e. succinyl-CoA. In this case the first product of S.l.p. is GTP, which can give rise to ATP by transphosphorylation: $GTP + ADP \rightarrow ATP + GDP$. Synthesis of ATP from phospho*enol*pyruvate and ADP in glycolysis is also S.l.p. There is no unifying general mechanism of S.l.p.

Substrate mediated transhydrogenation: see Hydrogen metabolism.

Substrate specificity: the ability of an enzyme to recognize and specifically bind its substrate. S.s. is a function of protein structure. *High* or *strict S.s.* indicates the ability to bind and convert only one or a limited number of substrates. An enzyme with *broad S.s.* converts a wide range of related substrates.

Subtilisin (EC 3.4.21.4): an extracellular, single chain, alkaline serine protease from *Bacillus subtilis* and related species. S. are known from four different species of *Bacillus: S. Carlsberg* (274 amino acid residues, M_r 27277), *S. BPN′* (275 amino acid residues, M_r 27537), *S. Novo* (identical with S.BPN′) and *S. amylosacchariticus* (275 amino acid residues, M_r 27671). The observed sequence differences between different S. represent conservative substitutions and are limited to the surface amino acids. Like the pancreatic proteinases, S. has catalytic Ser 221, His 64 and Asn 32 residues, but it is structurally very different from the other serine proteases, e.g. the active center of S. is -Thr-Ser-Met-, whereas that of the pancreatic enzymes is -Asp-Ser-Gly-; pancreatic enzymes contain 4-6 disulfide bridges, whereas S. contains none; S. contains 31% α-helical structure and 3 spatially separated domains, where-

as the pancreatic enzymes have 10-20% α-helical structure and a high content of β-structures; in both types, the active center is a substrate cleft. S. also have a broader substrate specificity than the pancreatic enzymes. This is a notable example of the convergent evolution of catalytic activity in two structurally completely different classes of proteins. S. is used in the structural elucidation of small peptides; and on account of its stability to anionic detergents, it is used in biological washing powders.

Subunit: 1. In protein chemistry the smallest protein or polypeptide unit that can be separated from an oligomeric protein without breaking covalent bonds. S. may be identical or nonidentical. In allosteric enzymes, e.g. aspartate transcarbamylase, nonidentical S. can be further classified into regulatory and catalytic S. Sometimes, the smallest active combination of S. is called a *monomer*, and the smallest identical subgrouping of S. in an oligomer is called a protomer (e.g. the αβ-protomer of hemoglobin), producing the sequence: subunit-monomer-protomer-oligomer (quaternary protein). According to this definition polypeptides linked by disulfide bridges are not S., e.g. in insulin, chymotrypsin, fibrinogen and immunoglobulins, and such proteins are not considered to have quaternary structure (see Proteins). There are several ways of dissociating quaternary proteins. Usually the protein is treated with 1% sodium dodecylsulfate (SDS), 6 M guanidine hydrochloride, or 8 M urea. Other treatments for dissociation of quaternary proteins into individual S. include increase or decrease of ionic strength, pH, protein concentration or temperature, removal or addition of a cofactor, or chemical modification (reaction with succinic or maleic anhydride). Demonstration and M_r determination of S. are performed by polyacrylamide gel electrophoresis, gel filtration and various other physical chemical methods (see Proteins) in the presence of the appropriate denaturing agent.

2. The term is also applied to the two subunits of Ribosomes (see).

Succedaneaflavanone: see Biflavonoids.

Succinate dehydrogenase (EC 1.3.99.1): an oligomeric flavoenzyme of the TCA-cycle, which catalyses the oxidation of succinate to fumarate. M_r (bovine heart) 100000. S.d. consists of two nonidentical, iron-containing subunits (M_r 70000 and 30000). Coenzyme of S.d. is FAD, which receives hydrogen directly (without the participation of NAD or NADP) and transfers it to ubiquinone or cytochrome *b* of the respiratory chain, or (in vitro) to redox dyes, e.g. methylene blue. S.d. occurs only in mitochondria (and prokaryotic cell membranes), where it is firmly bound in the membrane in association with the respiratory chain. It has a regulatory role in the TCA cycle, and can be isolated from the succinoxidase complex.

Succinate-glycine cycle, *glycine-succinate cycle*, *Shemin cycle*: a bypass of the TCA-cycle of particular importance in the metabolism of red blood cells. It converts succinyl-CoA and glycine into 5-aminolevulinate, which is the biosynthetic precursor of the Porphyrins (see). Alternatively, 5-aminolevulinate is deaminated to 2-oxoglutarate semialdehyde,

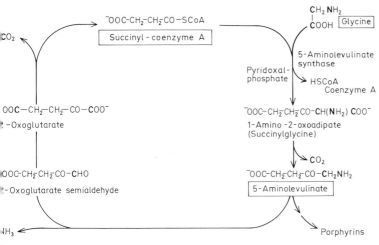

Synthesis of 5-aminolevulinate via the succinate-glycine cycle

and the cycle is completed by the formation of succinyl-CoA via 2-oxoglutarate. One turn of the cycle converts glycine into 2 molecules CO_2 and 1 molecule NH_3. 2-Oxoglutarate semialdehyde can also be converted into succinate and a C1-unit.

Succinic acid, *ethane dicarboxylic acid:* $HOOC-CH_2-CH_2-COOH$, m.p. 185 °C, b.p. 235 °C (d.). Succinate is an important intermediary metabolite generated both in the Tricarboxylic acid cycle (see) and in the Glyoxylate cycle (see). S.a. can be used via the reactions of the tricarboxylic acid cycle for the synthesis of amino acids and carbohydrates. Succinyl-coenzyme A is the starting point for the synthesis of porphyrins from intermediates of the Succinate-glycine cycle (see). Succinyl-coenzyme A is also formed by carboxylation of propionyl-coenzyme A in the degradation of valine and isoleucine, and from 2-methylmalonyl-coenzyme A in the degradation of fatty acids with an odd number of carbon atoms. The conversion of succinyl-coenzyme A to succinate in the tricarboxylic acid cycle generates one molecule of GTP, which is equivalent to 1 ATP.

Succinic acid 2,2-dimethylamide, *Alar 85, B9:* $HOOC-(CH_2)_2-CO-N(CH_3)_2$, a synthetic growth regulator. It is used to stimulate blossoming in apples and to accelerate the development of fruit color and fruit loosening in cherries.

Succinic acid 2,2-dimethylhydrazine, *dimethazide, B-995:* $HOOC-(CH_2)_2-CO-NH-N(CH_3)_2$, a synthetic growth regulator. It acts as a gibberellin antagonist, and retards growth especially actively in dicotyledonous plants. It increases the number of blossoms on fruit trees and prevents the loss of immature fruit.

Succinoylcholine, *suxamethonium:* see Acetylcholine.

Succinylacetoacetate: see Inborn errors of metabolism (Hereditary tyrosinemia type I).

Succinylacetone: see Inborn errors of metabolism (Hereditary tyrosinemia type I).

N-Succinyladenylate: adenylosuccinate, see Purine biosynthesis.

Succinylcholine: see Acetylcholine.

Succinyl-coenzyme A: see Succinic acid.

Sucrase, *invertase:* see Disaccharidases.

Sucrose, *cane sugar, beet sugar:* α-D-glucopyranosyl-β-D-fructofuranoside, a nonreducing disaccharide (see Carbohydrates). M_r 342.31, decomposes 160–186 °C with charring, $[\alpha]_D^{20} + 66.5$ °, biosynthesized from fructose 6-phosphate and UDP-glucose (Fig.). S. is a trehalose-type disaccharide and therefore does not give typical sugar reactions, i.e. does not form an osazone or oxime, or show mutarotation. Hydrolysis with dilute acid or enzymes, e.g. α-glucosidase (maltase, EC 3.2.1.20), or β-D-fructofuranosidase (invertase, EC 3.2.1.26) cleaves S. into equal parts of D-glucose and D-fructose (see Invert sugar). As a weak acid ($K \approx 10^{-13}$), S. (and saccharides derived from S.) forms salts (saccharates) with alkali and alkaline earth metal hydroxides. S. is widely distributed in plants, where it represents a transport form of soluble carbohydrate. It is an important metabolic product in all chlorophyll-containing plants, and cannot be synthesized by animals.

S. is obtained commercially from two economically important plants: the expressed juice of sugar cane *(Saccharum officinarum)*, which grows in the tropics, contains 14–21% S.; the expressed juice of sugar beet *(Beta vulgaris)*, grown in temperate climates, contains 12–20% S. The chemical synthesis of S. was reported in 1953, but is economically unimportant.

S. is used as a food and sweetener. It is fermentable, but in high concentrations it inhibits growth of microorganisms, and is therefore used as a preservative. It is used as the substrate in the industrial fermentative production of ethanol, butanol, glycerol, citric and levulinic acids, and as a preservative, sweetener and demulcent in pharmaceutical preparations. Esters, ethers and other chemical modification products of S. are used in the preparation of detergents, soaps and plastics.

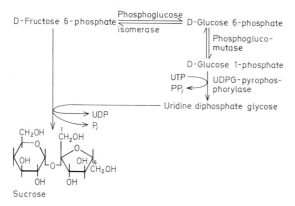

Biosynthesis of sucrose

Sucrose density gradient centrifugation: see Density gradient centrifugation.

Sugar: in the narrow sense a general name for commercially available Sucrose (see). In the wider sense a term used for Carbohydrates (see), in particular mono- and oligosaccharides.

Sugar alcohols: polyhydric alcohols, which occur widely as metabolic reduction products of monosaccharides. They are named by replacing the -ose ending of the parent monosaccharide by -itol. According to the number of carbon atoms they are classified as pentitols, hexitols, etc. S. a. show only low optical activity, are not fermented by yeasts, and do not react with Fehling's solution or phenylhydrazine. They are biosynthesized by reduction of the corresponding monosaccharide with NADH or NADPH. Important natural representatives are glycerol, erythritol, ribitol, xylitol, D-sorbitol, D-mannitol, dulcitol.

Sugar anhydrides: internal acetals formed by the intramolecular removal of water from a sugar molecule. They are chemically similar to glycosides, and can be hydrolysed to the corresponding sugars by water or dilute acids.

Sugar esters: esters of mono- or oligosaccharides with organic or inorganic acids. The phosphate esters, e.g. glucose 6-phosphate, are fundamentally important in intermediary carbohydrate metabolism. Chondroitin and mucoitin sulfates, and heparin are typical animal sulfate esters.

Sulfamates: see Sulfur compounds.

Sulfatases: see Esterases.

Sulfate activation: formation of the active sulfates, Phosphoadenosinephosphosulfate (see) (PAPS) and adenosine 5'-phosphosulfate (APS). PAPS is the substrate of sulfotransferases in the synthesis of sulfate esters. APS is the substrate of sulfate assimilation and sulfate respiration.

Sulfate assimilation, *assimilatory sulfate reduction:* the reductive assimilation of oxidized inorganic sulfur (see Sulfate reduction), leading to the biosynthesis of L-cysteine. S. a. is a property of plants and bacteria, whereas protozoa and animals can only perform the first step of S. a., i.e. sulfate activation (see Phosphoadenosinephosphosulfate). The sulfate ion SO_4^{2-} is reduced to the oxidoreduction level of the sulfide ion, S^{2-}, or the mercapto group of L-cys-

teine. According to one view, Sulfate reduction (see) proceeds via free intermediates to hydrogen sulfide, H_2S, which is incorporated into the SH-group of cysteine by the action of serine sulfhydrase (cysteine synthase, EC 4.2.1.22):

$$H_2S + HOCH_2CH(NH_2)COOH \xrightarrow{\text{PalP}}$$
L-serine
$$HSCH_2CH(NH_2)COOH + H_2O$$
L-cysteine

Serine sulfhydrase is a pyridoxal phosphate enzyme. According to a different view of this process, the free intermediates of S. a. are byproducts, and S. a. proceeds via enzyme- and carrier-bound intermediates (Fig. 1). Sulfate is converted to adenosine 5'-phosphosulfate (APS), which is phosphorylated to PAPS. The latter takes part in the synthesis of sulfate esters, or it can be converted back to APS by a specific 3'-nucleotidase. The substrate of S. a. is not PAPS, but APS (according to work on *Chlorella*, yeasts and spinach chloroplasts). An APS-transferase transfers sulfate from APS to a hypothetical carrier, Car-S$^-$, which is possibly identical with thioredoxin, although only one thiol group is apparently required. This carrier is possibly a coenzyme of the APS-transferase. The product of the transferase reaction is probably Car-S-SO$_3^-$; this is reduced to Car-S-S$^-$ by a ferredoxin-dependent reductase. Car-S$^-$ (or the coenzyme of APS-transferase) is regenerated by reaction of Car-S-S$^-$ with O-acetyl-L-serine, with the formation of L-cysteine. Free intermediates could arise from this reaction pathway by reductive cleavage of bound intermediates. Free sulfide might arise from reductive dismutation of Car-S-S$^-$ or some similar enzyme-bound intermediate. Exogenous sulfite might enter the system by combining with Car-S$^-$ to form Car-S-SO$_3^-$, which would agree with results from defective mutants of *Salmonella, Escherichia coli, Chlorella,* yeast, *Aspergillus* and *Neurospora*. It is, however, possible that two alternative pathways of S. a. exist: via free intermediates (see Sulfate reduction), and a pathway via enzyme- and carrier-bound intermediates (see Fig. 1), in which organic or bound thiosulfate, -S-SO$_3^-$, plays a central role.

S. a. is regulated by enzyme induction and repression, and by feedback inhibition (Fig. 2). In *Escheri-*

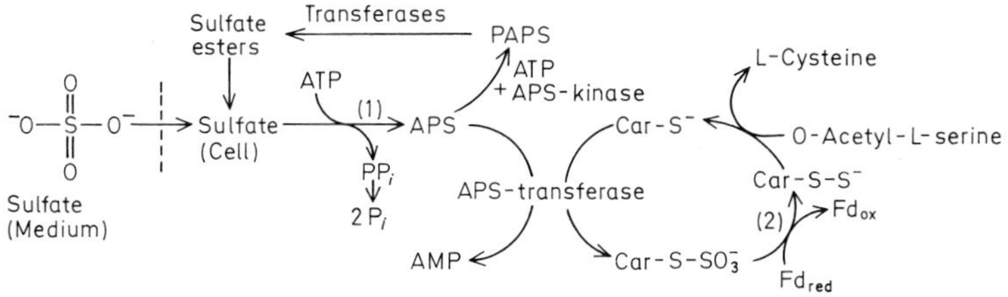

Figure 1. ***Sulfate assimilation via enzyme- and carrier-bound intermediates.*** (1) ATP-sulfurylase, (2) APS-reductase.

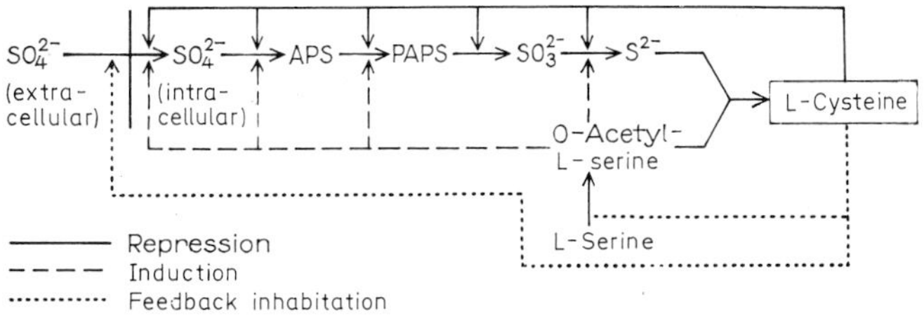

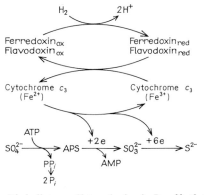

Figure 2. ***Sulfate assimilation in*** *Escherichia coli* ***via free intermediates, with control mechanisms.***

chia coli, L-cysteine represses all steps of S.a. (assuming S.a. to proceed via free intermediates), starting with the uptake of sulfate by the cell. *O*-Acetyl-L-serine induces uptake of sulfate, synthesis of APS and PAPS and synthesis of sulfite reductase, but not of sulfate reductase. Acetylation of L-serine to *O*-acetyl-L-serine is feedback inhibited by L-cysteine.

Sulfate esters: products formed by the transfer of a sulfuryl group to the oxygen function of organic compounds. Donor for the sulfuryl transfer is Phosphoadenosinephosphosulfate (see) (PAPS). Naturally occurring S.e. are Polysaccharide sulfate esters (see), and sulfate esters of phenols, steroids, choline, cerebrosides and flavonoids. Synthesis of phenol sulfate esters in mammalian liver was the first biological sulfate ester synthesis to be recognized, and the first to be performed in vitro. It is an important reaction for the detoxication of phenols in animals and man. Transfer of the sulfuryl group from PAPS to the acceptor is catalysed by phenolsulfotransferase. Ascorbic acid 2-sulfate may also function as a donor of sulfuryl groups.

Sulfate reduction: reduction of the sulfate ion SO_4^{2-} to the sulfide ion S^{2-}, in which the hexavalent, positive sulfur of the sulfate is converted to the divalent, negative form. S.r. must be preceded by Sulfate activation (see). The substrate of enzymatic S.r. is therefore either adenosine phosphosulfate (APS) or phosphoadenosinephosphosulfate (PAPS).

The enzymology of S.r. has been studied in particular in enzyme preparations from baker's yeast *(Saccharomyces cerevisiae)* and the anaerobic bacterium, *Desulfovibrio desulfuricans*. The former organism performs assimilatory S.r. (see Sulfate assimilation), the latter dissimilatory S.r. (see Sulfate respiration).

Sulfate respiration, ***dissimilatory sulfate reduction:*** a form of respiration in which the sulfate ion replaces oxygen as the terminal electron acceptor (see Sulfate reduction). The sulfate ion must first be activated (see Sulfate activation). S.r. is an anaerobic process in which sulfate is reduced to hydrogen sulfide, which is excreted. Ecologically, S.r. contributes to desulfurication, and is important for the sulfur cycle of the biosphere.

S.r. is limited to the members of two bacterial genera: *Desulfovibrio* (Fig.) and *Desulfotomaculum.* The process is described by the equation: $8[H]+SO_4^{2-} \rightarrow H_2S+2H_2+2OH^-$. Hydrogen donors are alcohols, organic acids (e.g. pyruvate), or molecular hydrogen (many strains have a constitutive hydrogenase). With pyruvate, the process can be represented as: $4\,\text{pyruvate}+H_2SO_4 \rightarrow 4\,\text{acetate}+4CO_2+H_2S$. Since acetic acid appears as a fermentation product, this S.r. can be considered as an anaerobic acetic acid fermentation.

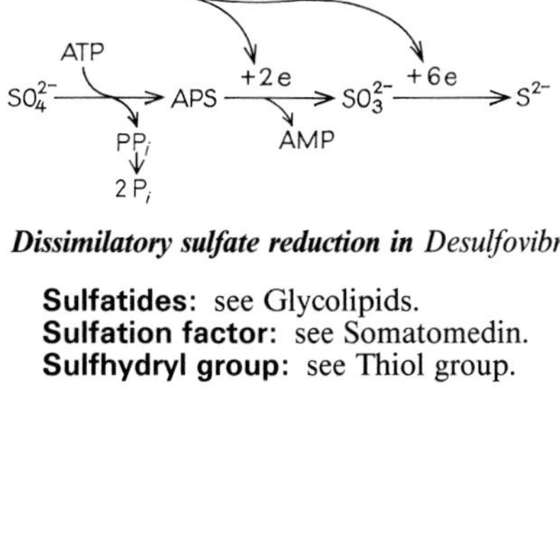

Dissimilatory sulfate reduction in *Desulfovibrio.*

Sulfatides: see Glycolipids.
Sulfation factor: see Somatomedin.
Sulfhydryl group: see Thiol group.

Sulfhydryl reagents, *SH-reagents:* substances that react with thiol groups (syn. sulfhydryl or SH-groups) of proteins. In vivo they cause metabolic changes or alterations of function and they are generally toxic. In vitro they are used to detect, titrate and characterize SH-groups. If the catalytic activity of an enzyme is inhibited by SH-reagents, the enzyme is eligible for designation as an SH-enzyme or Thiol enzyme (see). The affected SH-groups may be in the active center of the enzyme, or they may be elsewhere, so that inhibition of activity results from a change in the conformation or solubility of the enzyme.

Three types of cell constituents react with SH-reagents: 1. low M_r thiols, such as lipoic acid, coenzyme A, glutathione and cysteine; 2. non-enzyme proteins, such as actomyosin, membrane proteins and structural proteins; 3. enzymes. The toxic action of SH-reagents in vivo is due chiefly to their effect on SH-enzymes.

The most potent SH-reagents are the mercurials, which react with SH-groups according to the equation: Protein-SH + $^+$Hg-X → Protein-SHg-X + H$^+$ (see Fig. 1).

4-Chloromercuribenzoate is the most commonly used mercurial for the spectrophotometric titration of SH-groups; reaction with SH-groups at pH 7 results in an increase in absorbance at 250 nm. 4-Chloromercuriphenylsulfonic acid has the advantage of greater water solubility, but its poor lipid solubility makes it unsuitable for general biological purposes. Phenylmercury chloride and ethylmercury chloride show good lipid solubility and are therefore able to penetrate into the interior of subcellular particles and into cells. The various colored mercurials, which possess azo- or nitro-groups, are useful for the histochemical detection of SH-groups (e.g. f and g, Fig.1), and some are also used for the spectrophotometric titration of SH-groups (e.g. h, Fig.1). Formation of mercuric derivatives has been used for the purification and crystallization of enzymes, e.g. crystalline Hg complexes of enolase from yeast, lactate dehydrogenase from Jensen sarcoma and rat muscle, and papain have been prepared. These complexes are enzymatically inactive, but Hg^{2+} can be removed and activity restored by dialysis against cyanide.

Compared with the mercurials and heavy metals, other SH-reagents are much less potent. On the other hand, they are often very useful for their selective activity against certain enzymes (e.g. the classical inhibition of glyceraldehyde 3-phosphate dehydrogenase, and therefore glycolysis, by iodoacetate), and as reagents in protein chemistry (e.g. ^{14}C-labelled iodoacetate is used to label cysteine residues in proteins: I^{14}CH$_2$-^{14}COOH + Protein-SH → Protein-S-^{14}CH$_2$-^{14}COOH + HI). Two reactive iodine groups are present in 2,2′-dicarboxy-4,4′-diiodoaminoazobenzene, so that this reagent is able to cross-link SH-groups.

Certain disulfides are also active SH-reagents, reacting with SH-groups by a disulfide-sulfhydryl interchange: X-SS-X + Protein-SH → Protein-SS-X + XSH. A well known compound of this type is 5,5′-dithiobis(2-nitrobenzoate), or Ellman's reagent. The substituted maleimides are used for both labelling and inhibition studies (Fig. 4).

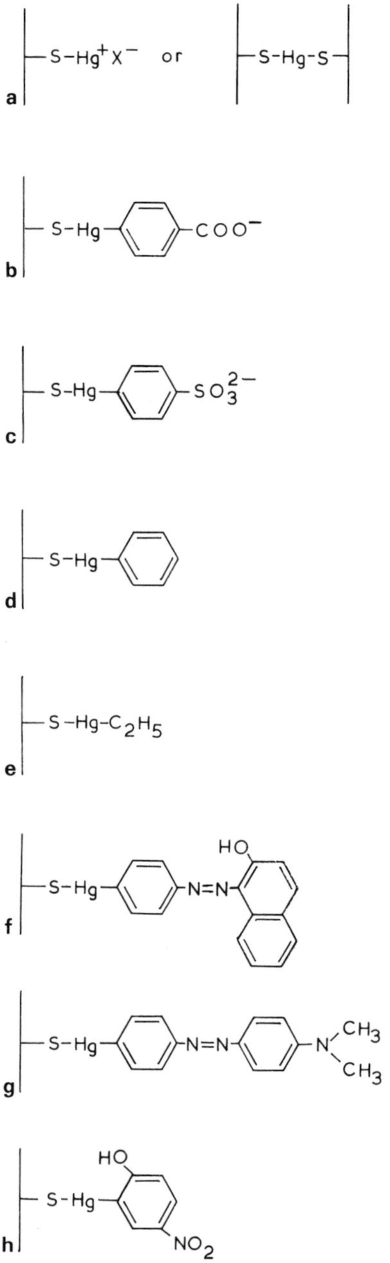

Figure 1. Reaction products of some mercurials with the SH-group of a protein. The reagents are: mercuric salts (giving product **a**), 4-chloromercuribenzoate (**b**), 4-chloromercuriphenylsulfonic acid (**c**), phenylmercury chloride (**d**), ethylmercury chloride (**e**), 4-chloromercuriphenylazo-β-naphthol (**f**), 4-chloromercuri-4′-dimethylaminobenzene (**g**) and 2-chloromercuri-4-nitrophenol (**h**). These reagents are usually employed as their chlorides (as in the above names) or as their acetates.

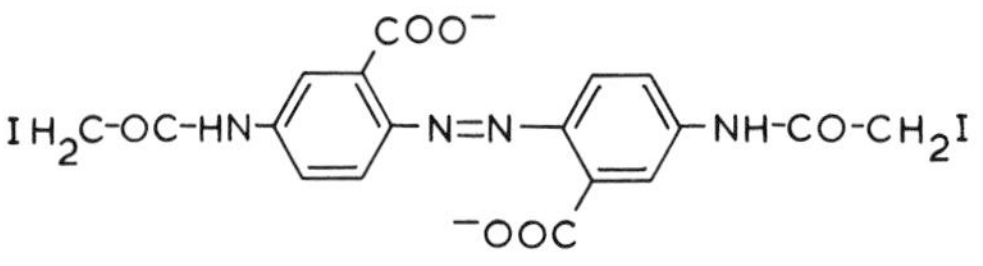

Figure 2.   2,2'-Dicarboxy-4,4'diiodoaminoazobenzene

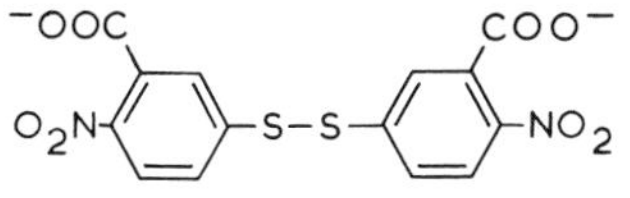

Figure 3.   Ellman's reagent

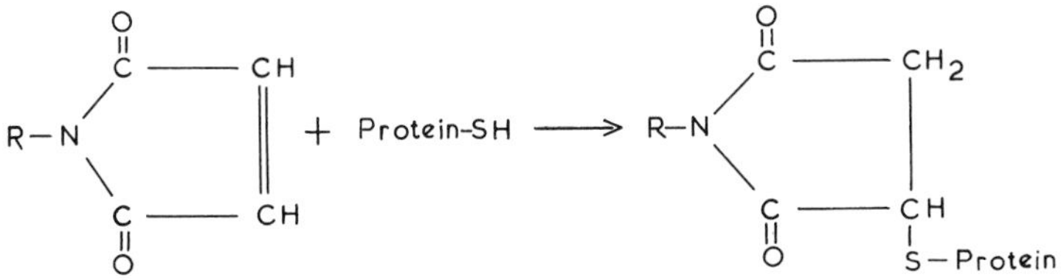

xyl groups) produces sulfate esters. Sulfated nitrogen functions are found in mustard oil glycosides and arylsulfamates; PAPS has been shown to be the sulfuryl donor in the formation of sulfamates. S.t. are very important in the synthesis of polysaccharide sulfate esters, and in the detoxication of phenols by sulfate ester formation in animals and man. The other product of the sulfuryl transfer reaction is 3'-phosphoadenosine 5'-phosphate (PAP):

PAPS + acceptor → PAP + product.

Detailed studies on purified S.t. are lacking. More recent studies indicate that ascorbic acid 2-sulfate may also act as a biological donor of sulfuryl groups.

Figure 4.   Reaction of a substituted maleimide with the SH-group of a protein.

The most widely used of these is *N*-ethylmaleimide ($R = C_2H_5$ in Fig. 4), which can be very selective for certain SH-groups of an enzyme. It is used as a general SH-reagent for the detection of SH-functions in purified proteins and in biological processes such as hormone release, metabolite uptake, membrane transport, etc.

Alloxan, well known for its ability to specifically damage the insulin secreting β-cells in the pancreatic islets, is an SH-reagent. It shows high specificity, being quite inactive towards many known thiol enzymes.

Arsenite and the arsenicals are among the earliest known SH-reagents, but the exact mechanism of their reaction with SH-groups is still unclear. The reaction often involves formation of a crosslink between two SH-groups, and may therefore depend upon the correct juxtaposition of groups within the protein.

Sulfhydryl shuttle: see Auranofin.

Sulfides: see Sulfur compounds.

β-Sulfinylpyruvic acid: see Cysteine.

Sulfite dehydrogenase deficiency: see Inborn errors of metabolism.

Sulfite fermentation: see Neuberg fermentation.

Sulfite oxidase: see Molybdoenzymes.

Sulfite oxidase deficiency: see Inborn errors of metabolism.

Sulfituria: see Inborn errors of metabolism.

Sulfokinase: see Sulfotransferase.

Sulfonic acids: see Sulfur compounds.

Sulfonolipids: see Capnine.

Sulfotransferase, *sulfokinase:* an enzyme that catalyses the transfer of sulfuryl groups from phosphoadenosinephosphosulfate (PAPS) to oxygen and nitrogen functions of suitable acceptors. Transfer to an oxygen function (alcoholic and phenolic hydro-

[Kresse, K. & Newfield, E.F. (1972) *J. Biol. Chem.* **247**, 2164-2170]

Sulfoxides: see Sulfur compounds.

Sulfur: see Bioelements.

Sulfur compounds: compounds containing reduced or oxidized sulfur, seldom both (Table). Biochemically important S.c. are the sulfur amino acids (see L-Cysteine; L-Methionine), biotin (thiophane ring) and thiamin (thiazole ring) (see Vitamins), sulfatides (complex lipids of the nervous system; see Glycolipids), and thiol peptides (see Glutathione; Vasopressin; Oxytocin; Insulin). S.c. also include Penicillin (see) and the sulfonamides, which are important synthetic therapeutic agents. The mustard oil glycosides contain both oxidized and reduced sulfur. Sulfate esters (see) are excreted by animals.

Sulfur cycle: see Sulfur metabolism.

Sulfuretin: 6,3',4'-trihydroxyaurone. See Aurones.

Sulfur metabolism: the metabolism of sulfur containing compounds in living organisms. Sulfur in various forms (see Sulfur compounds) is needed by all living organisms for the synthesis of biomolecules. Many microorganisms can utilize inorganic sulfur sources, like sulfide, sulfite, sulfate, thiosulfate, and in some cases even elemental sulfur. Plants assimilate inorganic sulfate (see Sulfate assimilation), and to a limited extent they can also incorporate atmospheric sulfur dioxide into cysteine and methionine. Sulfur dioxide cannot, however, satisfy the total sulfur requirement of plants – higher concentrations cause damage, as frequently observed in woodlands in industrial areas. Sulfur dioxide is oxidized in the leaf to sulfate, which can be transported and assimilated. Sulfate is taken up from the soil, reduced (see Sulfate reduction), then assimilated. The chief source of sulfate in the soil for the nutrition of most plants is gypsum ($CaSO_4 \cdot 2H_2O$) and anhydrite

Table. Naturally occurring sulfur compounds.

Compound	General structure	Examples
Thiols (mercaptans)	RSH	L-Cysteine; coenzyme A
Sulfides or thioethers	RSR_1	L-Methionine
Sulfoxides	$RSOR_1$	Allicin; formation from alliin by alliinase when onions *(Allium cepa)* are grated
Methylsulfonium compounds	$(CH_3)_2S^+R$	*S*-Adenosyl-L-methionine; dimethyl-β-propiothetin
Sulfate esters	$R-O-\overset{\overset{O}{\|\|}}{\underset{\underset{O}{\|\|}}{S}}-O^-$	Phenol sulfates; polysaccharide sulfates
Sulfamates	$R=N-O-\overset{\overset{O}{\|\|}}{\underset{\underset{O}{\|\|}}{S}}-O^-$	Aryl sulfamates; mustard oil glycosides
Sulfonic acids	$R-\overset{\overset{O}{\|}}{\underset{\underset{O}{\|}}{C}}-\overset{\overset{O}{\|\|}}{\underset{\underset{O}{\|\|}}{S}}-O^-$	Glucose 6-sulfonate; cysteic acid; taurine and methyltaurine

($CaSO_4$). Sulfides, like FeS and FeS_2 (pyrites) can become oxidized to sulfur by purely chemical processes in the soil; the sulfur is then oxidized to sulfate by sulfur bacteria. Colorless sulfur bacteria (e.g. *Beggiatoa, Thiobacillus, Thiothrix*) oxidize reduced forms of sulfur, and thus play an important part in the circulation of sulfur in the biosphere (i.e. the sulfur cycle, see Fig.). Reduction of sulfate is specific to microorganisms (see Sulfate respiration) and plants (see Sulfate assimilation), i.e. these organisms perform inorganic S.m. In contrast, animals are dependent upon a dietary source of organically bound sulfur, chiefly in the form of the proteogenic amino acids, L-cysteine and L-methionine. Cysteine can also be synthesized from dietary methionine by animals. In plants and animals the same mechanism is used for the synthesis of methionine from cysteine (see L-Methionine). Some coenzymes also contain sulfur (see Sulfur compounds).

Sulfate respiration and sulfate assimilation involve the initial formation of active sulfate (see Phosphoadenosinephosphosulfate), a process which also occurs in animals (see Sulfotransferase). Oxidized organic sulfur compounds arise chiefly from the degradation of L-cysteine or secondary sulfur-containing metabolites.

Supercoil: see Superhelix, Topoisomerases.

Superhelix: (see also Collagen). The DNA double helix (see Deoxyribonucleic acid) is itself coiled helically to form a superhelix. This is referred to as the tertiary structure of DNA (secondary structure is represented by the double helix, primary structure by the linear sequence of nucleotides). There is an even higher level of organization, in

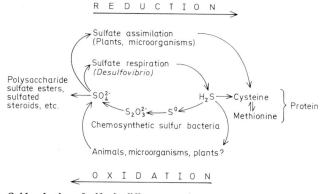

Oxidoreduction of sulfur in different organisms

which the superhelix is further coiled to form a super-superhelix. Supercoiling of DNA enables this large molecule to be packed into a relatively small space, and is essential for the formation of the DNA-histone complexes of chromatin. (Chromosomal DNA is compacted about 8000-fold in length, while DNA in a nucleosome is folded to give a 6.8-fold contraction of length, with formation of a unit fiber of length $100 \text{\AA}$.) In addition, the superhelix appears to serve in the control of gene expression.

The superhelix is left-handed, whereas the double helix is right-handed. This can be compared to the Möbius strip, in which a strip of paper is twisted before joining the ends, thus producing a higher order of twist, or a kink, in the completed circle. In conformity with this model, unwinding of the superhelix also causes unwinding of the double helix. Supercoiled, circular DNA denatures more easily than nonsupercoiled DNA, i.e. supercoiled DNA is under strain. Enzymes called Topoisomerases (see) are responsible for converting nonsupercoiled DNA into a superhelix and for regulating the degree of supercoiling.

Studies in vitro show that replication and transcription do not occur unless the DNA is supercoiled. It appears that the uncoiling of the superhelix promotes separation of the strands of the double helix, thus providing access for DNA- or RNA-polymerases. The strain of the DNA supercoil can be released by "nicking" with gamma rays. Such "relaxed" supercoiled DNA shows a decreased ability to promote RNA transcription, or to bind RNA polymerase. It is therefore suggested that gene expression is controlled by the degree of supercoiling of DNA; transcribed genes are supercoiled to the correct degree, so that an RNA-polymerase can cause local unwinding, whereas untranscribed genes are coiled to the wrong degree.

Binding of one molecule of ethidium bromide unwinds the helix by 26 °; the quantity of bound ethidium can be determined fluorimetrically, thus forming a basis for the determination of the degree of superhelicity. Supercoiling can also be measured by velocity sedimentation or equilibrium sedimentation in the presence of intercalating dyes, and by agarose gel electrophoresis (supercoiled DNA migrates more rapidly than relaxed DNA). For other information on superhelices and for references, see Topoisomerases, Linking number.

Superhelix density, σ: according to W. Bauer & J. Vinograd [*J. Mol Biol. 33* (1968) 141–172], this is the number of superhelical turns per 10 base pairs of the DNA circular duplex. It is therefore a measure of the strain on the DNA due to supercoiling. Negative S. d. ranging from 0.04 to 0.09 have been reported.

According to A. Nordheim et al., [*Cell 31* (1982) 309–318], superhelical density (denoted by these authors also as σ) is equivalent to specific linking difference (see Linking number).

Superoxide dismutase, *SOD, hyperoxide dismutase, superoxide: superoxide oxidoreductase,* (E.C. 1.15.1.1): When it was found that the copper and zinc-containing proteins hemocuprein (bovine erythrocytes), hepatocuprein (bovine liver) and cere-

brocuprein (human brain) have a common identity, they were given the single name, cytocuprein [R. J. Carrico & H. F. Deutsch, *J. Biol. Chem. 245* (1970) 723–727.)] Cytocuprein is now known to be identical with SOD.

SOD catalyses the dismutation (disproportionation) of superoxide: $O_2^{\cdot -} + O_2^{\cdot -} + 2H^+ \rightarrow O_2 + H_2O_2$. There are two main types of SOD: 1. cyanide-sensitive, Cu- and Zn-containing, eukaryotic enzymes, M_r 31000–33000 (2 subunits, M_r 16000); and 2. cyanide-insensitive, Fe or Mn-containing, prokaryotic enzymes, M_r about 40000 (2 subunits, M_r 20000). SOD from liver mitochondria contains Mn and has M_r 80000 (4 subunits, M_r 20000). It has been shown that the designations "prokaryotic" and "eukaryotic" are not applicable in every case: Mn-type SOD has been demonstrated in eukaryotes, while cyanide-sensitive SOD is absent from many algae. In contrast to the widely held view in the early 1970s, there is also a lack of correlation between the presence of SOD and aerobicity. Many anaerobes possess SOD, while some aerobes do not (however, these organisms do have high levels of catalase).

There has been debate over the biological significance of SOD, due to the realization that the superoxide radical is not highly reactive in aqueous solution. Some authors have suggested that SOD is primarily a metal storage protein. However, recent studies on the structure and mechanism of Cu-Zn-SOD have shown that its structure has been highly conserved through evolution. [J. A. Tainer et al., *Nature 306* (1983) 284–286; E. D. Getzoff et al., *ibid* 286–290] Others have demonstrated that SOD specifically catalyses the dismutation of superoxide, and that this radical, if its concentration is elevated, may react with hydrogen peroxide to form a much more reactive species, possibly the hydroxyl radical. [B. Halliwell in R. Lontie, ed. *Copper Proteins and Copper Enzymes, Vol. 2* (CRC Press, Boca Raton, 1984) pp. 63–102].

Supplementation test: see Mutant technique.

Suppressor: with reference to Amber mutants (see) and Ochre mutants (see), a *S. mutation* is a secondary mutation in a tRNA cistron, which restores the ability of the tRNA to recognize the nonsense codon. The resulting new tRNA is called *suppressor tRNA.* The nonsense codon then becomes a sense codon and codes for a specific proteogenic amino acid during translation.

In a general sense, a S. mutation is any mutation that partly or completely restores a genetic function, lost by another mutation. By definition, a S. mutation must be at a different site from that affected by the first mutation. It may be in the same gene (intragenic S. mutation) or a different gene (intergenic S. mutation). A *S. gene* reverses the effect of a mutation in another gene.

Suprarenin: see Adrenalin.

Svedberg unit: see Sedimentation coefficient.

Swivelase: a type I topoisomerase. See Topoisomerases.

Symbiosis: close spatial coexistence of different species of advantage to all partners. It is therefore also called *mutualistic S.* The distinction between S., Commensalism (see) and Parasitism (see) is not always clear. Biochemically relevant symbiotic systems are root nodule bacteria - leguminous plants (see Nitrogen fixation); alga-fungus in lichens; rumen flora - ruminants; fungus - higher plant (see Mycorrhiza). Partners showing large differences in size are called macrosymbionts and microsymbionts.

Symport: see Transport.

Synchronous culture: a synchronized cell population, in which all cells divide and pass through subsequent phases of the cell cycle at the same time. Synchronization can be achieved in various ways, e.g. by nutrient limitation, light stimulation, temperature change, treatment with antimetabolites of nucleic acid metabolism. In S.c. the cell count increases stepwise. Synchrony is usually lost after a few synchronous divisions, i.e. the cell count reverts to a continual increase. S.c. techniques have been applied to various bacteria, *Chlorella, Euglena gracilis,* etc. Light synchronization is important in the study of endogenous biological rhythms.

Syndein: see Ankyrin.

Synergists: substances or factors that increase the biological activity of another substance or factor, but are inactive alone in the same quantity.

Synthases: see Enzymes (Table).

Synthetases: see Enzymes (Table).

Synzymes, *enzyme analogs:* synthetic macromolecules with enzymatic activity. S. may be prepared by polymerization of amino acids or their derivatives, or by the attachment of catalytic groups to nonprotein materials. Examples of amino acid-derived S. are: a glutamic acid-phenylalanine copolymer (Glu:Phe=9:1) with one third of the activity of natural lysozyme; a copper(II)-poly-ε-carbobenzoxy-L-lysine (M_r 440000) with specific alcohol dehydrogenase activity. For non-amino acid S., methylenimidazole (nucleophilic, catalytically active) and dodecyl (apolar substrate binding site) side chains are attached to a polymethylenimine carrier. This globular macromolecule has a high esterase activity for phenylsulfate esters, which is 100 times greater than that of natural aryl sulfatase. The cyclodextrins are S. with phosphatase activity. Low M_r peptides with enzymatic activity are also S. The decapeptide, Glu-Phe-Ala-Ala-Glu-Glu-Ala-Ala-Ser-Phe shows glycosidase activity for dextran and chitin.

T: abb. for thymine.

T₃: see Thyroxin.

T₄: see Thyroxin.

T_m, t_m: the temperature at the mid-point of a temperature-dependent transition. T_m is commonly used to denote the Melting point (see) of DNA.

Tabtoxins: chlorosis-inducing dipeptides produced by several species of phytopathogenic *Pseudomonas*; e.g. "wild fire", a highly infectious and destructive leafspot disease of tobacco, is caused by *Pseudomonas tabaci*, which produces T. T. consists of threonine linked to tabtoxinine β-lactam [2-amino-4-(3-hydroxy-2-oxo-azocyclobutan-3-yl)butanoic acid]. In [2-serine]T., the threonine is replaced by serine (Fig.). Tabtoxinine β-lactam is the actual phytotoxin; it is released from T. by plant proteases, and acts as an inhibitor of glutamine synthetase. [P. A. Taylor et al. *Biochim. Biophys. Acta* **286** (1972) 107–117; T. F. Uchytil & R. D. Durbin *Experentia* **36** (1980) 301–302]

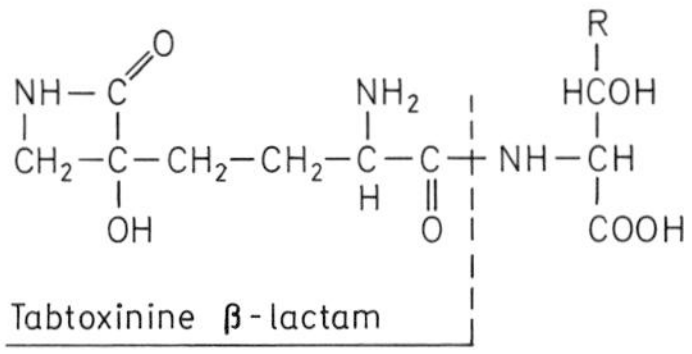

Tabtoxinine β-lactam

Tabtoxins

R = CH₃: tabtoxin

R = H: [2-serine]tabtoxin

Tachysterol: see Vitamins (Vitamin D).

Taipoxin: see Snake venoms.

Taiwaniaflavone: see Biflavonoids.

Taka amylase: a bacterial α-amylase (EC 3.2.1.1) isolated and crystallized from *Aspergillus oryzae* taka diastase preparations. T-a. (M_r 50000) is a calcium-containing, single-chain protein with N-terminal alanine and C-terminal serine. Like the tetrameric *Bacillus subtilis* α-amylase (M_r 96000), T-a. is resistant to sodium dodecylsulfate but is reversibly denatured by 6 M guanidine or 8 M urea. T-a. must not be confused with Adenosine deaminase (see), which is also present in taka diastase.

Tannic acid: Chinese gallotannin. See Tannins.

Tannins: originally, substances of vegetable origin capable of converting animal skin into leather. A more modern and appropriate definition is naturally occurring compounds of M_r 500–3000, containing a sufficient number of phenolic hydroxyl groups (about 2 groups per M_r 100) to form cross links between macromolecules, such as proteins, cellulose and/or pectin.

By cross-linking proteins, T. can inhibit the activities of plant enzymes and organelles. Polyvinylpyrrolidone is therefore often added to adsorb T. during the isolation of plant enzymes.

The chief reactive centers responsible for cross-linking and complex formation by T. are ortho-dihydroxy phenolic groups; isolated phenolic hydroxyl groups do not seem to make a significant contribution to these reactions.

There are two classes of vegetable T.:

1. Hydrolysable T., which can be hydrolysed to glucose (or another polyhydric alcohol) and gallic acid (gallotannins) or ellagic acid (ellagitannins). The simplest known gallotannin is 1-O-galloyl-β-D-glucopyranose from *Rheum officiale* (Chinese rhubarb). In contrast, Chinese gallotannin ("tannic acid") (widely distributed in *Hamamalidaceae*, *Paeonaceae*, *Aceraceae* and *Anacardiaceae*, and sporadically in the *Ericaceae*) may contain up to 8 galloyl groups (Fig.).

Ellagitannins are derivatives of hexahydroxydiphenic acid, which becomes lactonized to ellagic acid during hydrolysis. The simplest known ellagitannin is corilagin from *Caesalpina coriaria* and other plants (Fig.). Other phenolic components are sometimes found in place of gallic or hexahydroxydiphenic acid in T., e.g. chebulic acid (in myrobalans T.) and brevifelin carboxylic acid (in algarobilla T.) (Fig.).

2. Condensed T. are polymers in which the monomeric unit is a phenolic flavan, usually a flavan-3-ol, and in which the flavan units are linked by 4:8 (C-C) bonds. Many higher oligomers and polymers of Proanthocyanidins (see) are therefore condensed vegetable T. An example is the precyanidin polymer from the seed coat of sorghum (Fig.). Synthesis of condensed T. by biomimetic condensation reactions has been reported. The condensation sequence is initiated by flavan-3-ols acting as nucleophiles, and flavan-3,4-diols as potential 4-carbenium ions [J. J. Betha et al. *Phytochemistry* **21** (1982) 1289–1294].

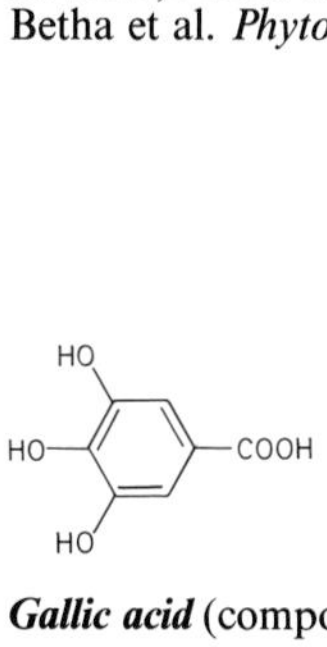

Gallic acid (component of gallotannins)

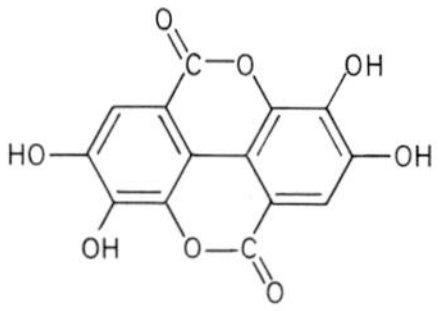

Tannins

Ellagic acid (derived from hexahydroxydiphenic acid during hydrolysis of ellagitannins)

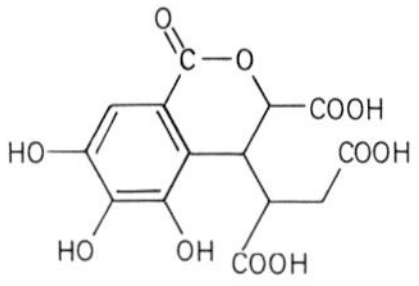

Chebulic acid (component of some hydrolysable tannins)

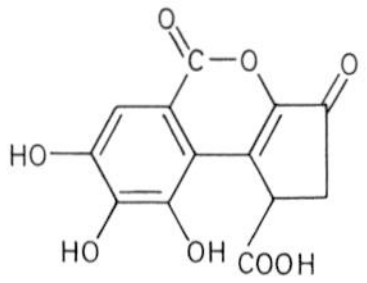

Brevifolin carboxylic acid (component of some hydrolysable tannins)

hexahydroxydiphenic   glucose   gallic acid
acid residue   residue   residue

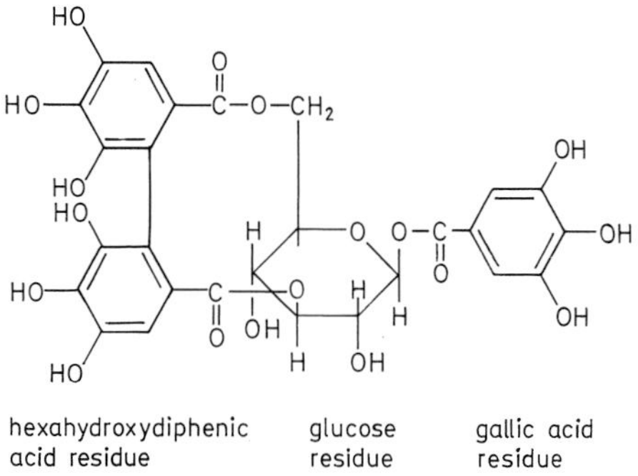

Corilagin (an elligatannin from *Caesalpinia coriaria*)

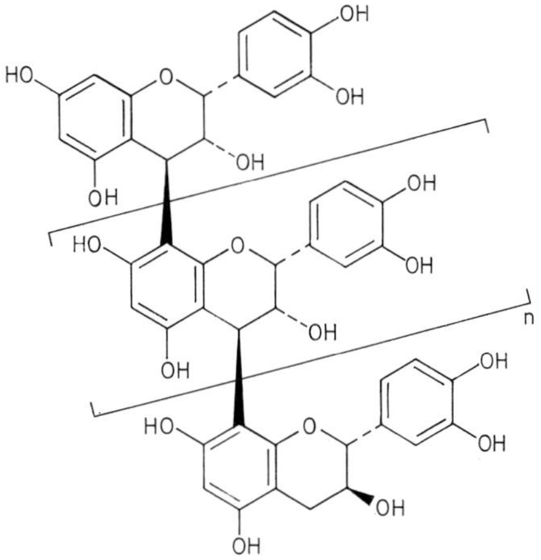

Condensed tannin (or procyanidin polymer) from seed coat of sorghum. $n=4-5$. M_r 1700–2000

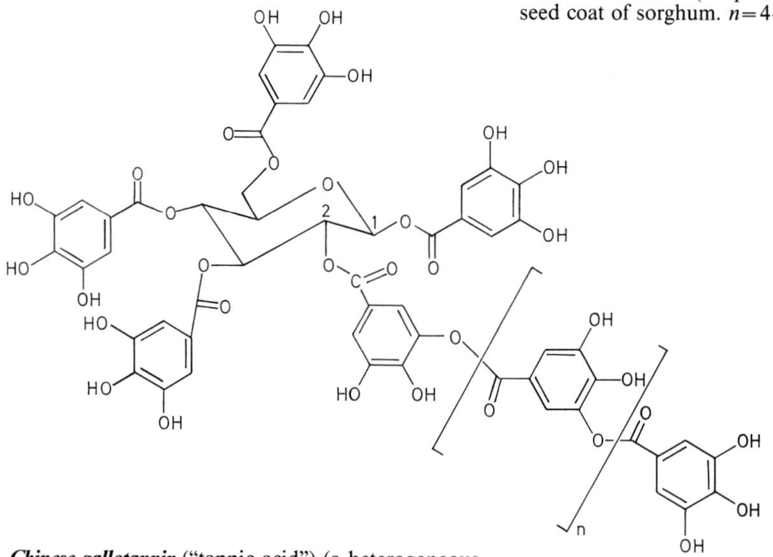

Chinese gallotannin ("tannic acid") (a heterogeneous tannin, in which $n=0$, 1 or 2, and the C-1 galloyl group may be absent)

5α-Taraxastane: see Taraxasterol.

Taraxasterol, α-lactucerol, α-anthesterol: a simple, unsaturated, pentacyclic triterpene alcohol, M_r 426.73, m.p. 227 °C $[\alpha]_D + 50$ °, structurally a derivative of the parent hydrocarbon, 5α-taraxastane. It occurs as the acetate in the latex of the dandelion *(Taraxacum officinale)* and other members of the *Compositae*.

5α-Taraxerane: see Taraxerol.

Taraxerol, alnulin, skimmiol: a simple, unsaturated, pentacyclic triterpene alcohol, M_r 426.73, m.p. 285 °C, $[\alpha]_D + 3$ °, structurally a derivative of the parent hydrocarbon, 5α-taraxerane. It occurs in many members of the *Compositae*, e.g. dandelion *(Taraxaum officinale)*, and in alder *(Alnus)* bark.

Tarichatoxin: the main toxin of North American salamanders *(Taricha torosa, T. rivularis)*. It is identical with Tetrodotoxin (see).

Tartronate-semialdehyde synthase (EC 4.1.1.47), *glyoxylate carboligase:* a plant enzyme which converts two molecules of glyoxylate to tartronate semialdehyde and CO_2. Hydroxymethyl thiamin pyrophosphate is a reactive intermediate in the reaction. The enzyme plays a role in the biosynthesis of carbohydrates from C_2 compounds.

Taurine: see L-Cysteine.

Taurocholic acid: see Bile acids.

Tay-Sachs disease: see Inborn errors of metabolism.

TDP: see Thymidine phosphates.

Tectoquinone: 2-methylanthraquinone, a yellow anthraquinone, m.p. 179 °C, present in teak wood. It is one of the few nonhydroxylated, naturally occurring anthraquinones. Owing to its content of T., teak is largely resistant to termites and fungi.

Teichmann's crystals, chlorhemin crystals: rhombic crystals formed by heating hemoglobin with sodium chloride and glacial acetic acid. T.c. are used for the microscopic detection of blood.

Teichoic acids: polymers present in the cell walls of Gram-positive bacteria. They consist of chains of glycerol or ribitol residues joined by phosphate groups; in addition sugars are linked to the glycerol or ribitol, and some of the hydroxyl groups are esterified with residues of D-alanine. For example, T.a. from *Staphylococcus aureus H* consists of eight ribitol units joined $1 \rightarrow 5$ by phosphodiester linkages; the sugar, *N*-acetylglucosamine, is attached to position 4 of the ribitol chiefly by β-linkages with some α-linkages. Glycerol T.a. are found more widely than ribitol T.a. In a few cases little or no sugar is present so that alkaline hydrolysis gives mainly alanine, glycerol and its phosphates; but the presence of other sugars is more usual, e.g. *N*-acetylglucosaminyl (T.a. from periplasmic space of *Staphylococcus aureus H*), glucosyl (T.a. from periplasmic space of *Lactobacillus arabinosus*), α-*N*-acetylgalactosaminyl (T.a. from wall of *Staphylococcus lactis*). The glycerol units are joined $1 \rightarrow 3$ by phosphodiester linkages, and the D-alanine or sugar residues are carried on C 2 of the glycerol.
Biosynthesis of ribitol T.a. is by progressive transfer of D-ribitol 5-phosphate units from CDP-ribitol to position 1 of the previous unit; it is not known at which stage the sugar and alanine residues are added. Similarly, glycerol T.a. are biosynthesized from CDP-glycerol.

Template: a macromolecule that determines the structure of another macromolecule. In the synthesis of DNA on a template of DNA (DNA replication), or RNA on a template of DNA (transcription), or DNA on a template of RNA (reverse transcription), the relationship between the template and the newly synthesized molecule depends on base complementarity (see Base pairing). In protein translation, 3 nucleotides in the mRNA determine the nature of the amino acid at each position in the polypeptide (see Genetic code).

Template RNA: see Messenger RNA.

Terminal: adjective for the chain end component of a biopolymer, e.g. *N*- and *C*-terminal amino acids (see Peptides).

Terminal oxidase: the terminal enzyme of the respiratory chain. In most organisms it is Cytochrome oxidase (see), but in various plant systems other T.o. are present or have been proposed. In aerobic nitrate respiration, the T.o. is nitrate reductase.

Termination: the final phase in the biosynthesis of biopolymers. See Biopolymers; Protein biosynthesis; Ribonucleic acid.

Termination codon, stop codon, punctuation codon: a sequence of three nucleotides in mRNA, which signals the end of polypeptide synthesis and release of the polypeptide in the process of Protein biosynthesis (see). 5'-UAA (see Ochre codon), 5'-UAG (see Amber codon) and UGA are T.c.

Termination factors: see Release factors.

Terminus: the chain of a biopolymer, e.g. *N*- or *C*-terminus of a protein, meaning the *N*- or *C*-terminal amino acid.

Terpene alkaloids, isoprenoid alkaloids: alkaloids containing a terpene structure, with 10–30 carbon atoms. They are conveniently classified according to the genera in which they occur (Table).

Table. Classification of terpene alkaloids.

Terpene type	Name
Monoterpene	Gentiana alkaloids Valeriana alkaloids
Sesquiterpene	Nuphara alkaloids Dendrobium alkaloids
Diterpene	Aconitum alkaloids Erythrophleum alkaloids
Triterpene (Steroid)	Solanum alkaloids Veratrum alkaloids Funtumia alkaloids Holarrhena alkaloids Buxus alkaloids Salamander alkaloids

T.a. are biosynthesized from mevalonic acid, but the origin of the nitrogen is not known. They therefore differ from the iridoid, isoquinoline and indole alkaloids, in which a monoterpene is linked to an amino acid.

Terpenes, terpenoids, isoprenes, isoprenoids: an extensive group of natural products whose structures are composed of isoprene units. The number of carbon atoms is usually a multiple of 5. T. are biosynthesized from the active 5-carbon unit, isopentenyl pyrophosphate. Inspection of the formulae of terpenes shows that they can be built up from a C_5 unit, which Wallach suggested might be isoprene. Ruzicka (1921) put forward the *isoprene rule,* in which the hydrocarbon skeleton of many open chain and cyclic terpenes is constructed from isoprene units arranged head-to-tail. The rule has proved useful in the assignment of structure, although there are exceptions, e.g. Squalene (see) contains a tail-to-tail arrangement. Free isoprene has not been demonstrated as a naturally occurring substance, but biosynthetic studies with the biochemically equivalent isopentenylpyrophosphate ("active isoprene") have confirmed the validity of the isoprene rule.

Originally only compounds with 10 carbon atoms (monoterpenes) were considered as T., and the oxygen-containing T. were classified as camphors. According to the mechanism of biosynthesis, however, all compounds derived from "active isoprene" are now classified as T. or isoprenes, including steroids, carotenoids, etc.

Structure. T. are subdivided according to the number of C_5 units in their structure, i.e. Hemiterpenes (see), Monoterpenes (see), Sesquiterpenes (see), Diterpenes (see), Sesterterpenes (see), Triterpenes (see) and Polyterpenes (see) (Table). Each of these groups contains different structural types, resulting from unusual linkages of C_5 precursors (e.g. head-to-head), cyclizations, various convolutions of large open chain precursors prior to cyclization, introduction of functional groups (alcohols, aldehydes, ketones, carboxylic acids, lactones), formation of epoxides, introduction of heteroatoms (e.g. terpene alkaloids contain nitrogen), cleavage of cyclic compounds, rearrangements, etc. Some T., e.g. steroids, lack certain carbon atoms, whereas others, e.g. juvenile hormones, contain additional carbon atoms. Isoprene components or small T. can also become linked to other structures, e.g. in lysergic acid, mycelianamide, humulone, indole alkaloids, chlorophyll, vitamins E and K and ubiquinone.

Owing to the regular occurrence of branch methyl groups, and the need to indicate the steric configuration of hydrogen atoms on alicyclic rings, methyl groups are indicated by a line and exocyclic methylene groups as a double line. For the configuration of substituents, the conventions previously used only for steroids are now used for T., i.e. β-oriented groups project above the plane of the paper and their attachment bond is shown as a wedge or a solid line, whereas α-groups lie below the plane of the

Table. Classification of terpenes.

Group	No. of C_5-units	Examples
Hemiterpenes	1	"Active" isoprene
Monoterpenes	2	Citral, iridoids, camphor
Sesquiterpenes	3	Abscisic acid, proazulenes
Diterpenes	4	Gibberellins, resin acids
Sesterterpenes	5	Cochliobolin
Triterpenes	6	Steroids, sterols, ecdysone
Tetraterpenes	8	Carotenoids
Polyterpenes	up to 10000	Caoutchouc, gutta percha, polyprenols

paper and are attached by a dashed bond. Similarly, symbols for carbon atoms in a chain are omitted, and adjoining double bonds are separated by a point (Fig. 1).

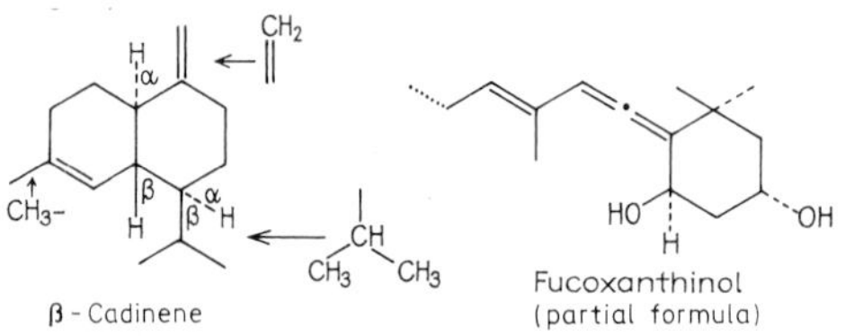

Occurrence and Function. Structures of over 5000 naturally occurring T. are known. They occur in all living forms, but biological functions are known for only a few. Carotenoids are important as accessory photosynthetic pigments, and several groups of T. act as hormones in plants, insects and higher animals. Increasingly large numbers of pheromones are being isolated and shown to be T. Some T. are also important in medicine and as raw materials in the industrial preparation of foods, perfumes, varnishes and rubber.

Biosynthesis. Precursors of T. are synthesized from acetyl-CoA according to the reactions shown in Fig. 2. Key intermediates are mevalonate and isopentenylpyrophosphate ("active isoprene"). Isopentenyl-pyrophosphate is in equilibrium with dimethylallylpyrophosphate, which serves as a starter for polycondensation reactions. Thus dimethylallylpyrophosphate and isopentenylpyrophosphate form geranylpyrophosphate, the precursor of the monoterpenes. Condensation with two more molecules of isopentenylpyrophosphate produces first farnesylpyrophosphate (precursor of the sesquiterpenes) then geranylgeranylpyrophosphate (precursor of the diterpenes). Two molecules of farnesylpyrophosphate undergo tail-to-tail condensation to form the Triterpenes (see), while two molecules of geranylgeranylpyrophosphate react tail-to-tail to form the Tetraterpenes (see). Polyterpenes (see) are formed by multiple head-to-tail condensation of isopentenylpyrophosphate units.

Terpenoids: see Terpenes.

Terpinene: see *p*-Menthadienes.

Tertiary structure: see Proteins.

Testane: earlier name for 5α-androstane. See Steroids.

Testosterone: 17β-hydroxy-androst-4-en-3-one, an important androgen synthesized in the interstitial cells of the testicle. It stimulates growth of the prostate and seminal vesicles, and promotes sperm maturation and development of male secondary sexual characteristics. Apart from mammalian testes, T. also occurs in blood and urine. It was first isolated in

β - Cadinene

Fucoxanthinol (partial formula)

Figure 1. Terpenes.

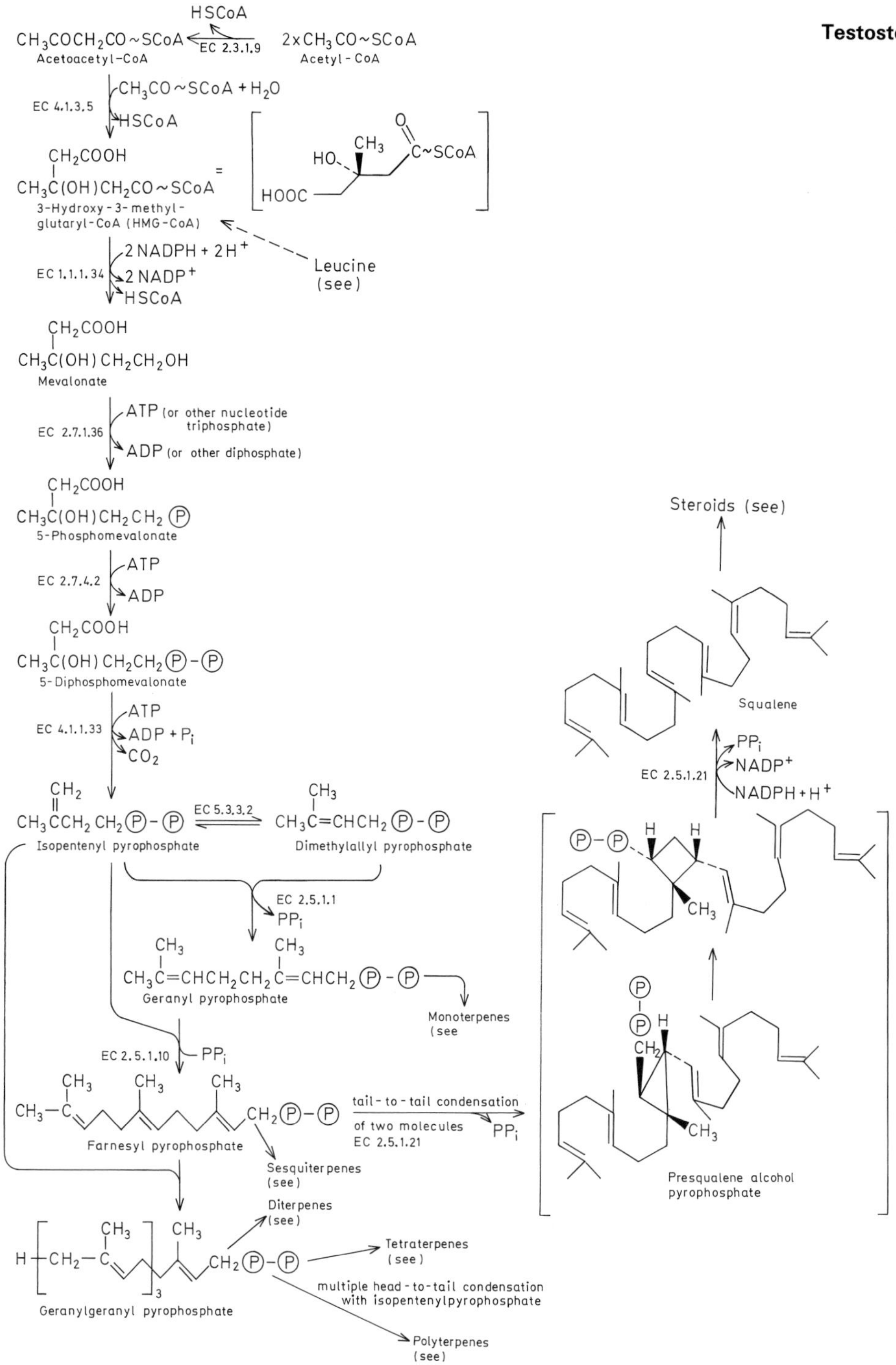

HSCoA

$CH_3COCH_2CO\sim SCoA \xleftarrow{EC\ 2.3.1.9} 2\times CH_3CO\sim SCoA$
Acetoacetyl-CoA   Acetyl-CoA

$CH_3CO\sim SCoA + H_2O$

EC 4.1.3.5   HSCoA

CH_2COOH
$CH_3C(OH)CH_2CO\sim SCoA$
3-Hydroxy-3-methyl-glutaryl-CoA (HMG-CoA)

$\left[\begin{array}{c} HO. \quad CH_3 \quad \overset{O}{\overset{\|}{C}}\sim SCoA \\ HOOC \end{array} \right]$

2 NADPH + 2H$^+$
EC 1.1.1.34   2 NADP$^+$
HSCoA

Leucine (see)

CH_2COOH
$CH_3C(OH)CH_2CH_2OH$
Mevalonate

ATP (or other nucleotide triphosphate)
EC 2.7.1.36   ADP (or other diphosphate)

CH_2COOH
$CH_3C(OH)CH_2CH_2$ Ⓟ
5-Phosphomevalonate

ATP
EC 2.7.4.2   ADP

CH_2COOH
$CH_3C(OH)CH_2CH_2$ Ⓟ–Ⓟ
5-Diphosphomevalonate

ATP
EC 4.1.1.33   ADP + P$_i$ CO$_2$

CH_2
$CH_3\overset{\|}{C}CH_2CH_2$ Ⓟ–Ⓟ $\xrightleftharpoons{EC\ 5.3.3.2}$ $CH_3\overset{CH_3}{\overset{|}{C}}=CHCH_2$ Ⓟ–Ⓟ
Isopentenyl pyrophosphate   Dimethylallyl pyrophosphate

EC 2.5.1.1
PP$_i$

$CH_3\overset{CH_3}{\overset{|}{C}}=CHCH_2CH_2\overset{CH_3}{\overset{|}{C}}=CHCH_2$ Ⓟ–Ⓟ
Geranyl pyrophosphate

Monoterpenes (see

EC 2.5.1.10   PP$_i$

$CH_3-\overset{CH_3}{\overset{|}{C}}\ \ \ \overset{CH_3}{}\ \ \ \overset{CH_3}{}\ CH_2$ Ⓟ–Ⓟ
Farnesyl pyrophosphate

tail-to-tail condensation of two molecules   PP$_i$
EC 2.5.1.21

Sesquiterpenes (see)
Diterpenes (see)

$H\left[CH_2-\overset{CH_3}{\overset{|}{C}}\ \ \overset{CH_3}{}\right]_3\ CH_2$ Ⓟ–Ⓟ
Geranylgeranyl pyrophosphate

Tetraterpenes (see)

multiple head-to-tail condensation with isopentenylpyrophosphate

Polyterpenes (see)

Steroids (see)

Squalene

EC 2.5.1.21
PP$_i$
NADP$^+$
NADPH + H$^+$

Ⓟ–Ⓟ

CH$_3$

Presqualene alcohol pyrophosphate

Figure 2. Biosynthesis of terpenes from acetyl-CoA.
EC 1.1.1.34, hydroxymethylglutaryl-CoA reductase (NADPH); EC 2.3.1.9, acetyl-CoA acetyltransferase; EC 2.5.1.1, dimethylallyltransferase; EC 2.5.1.10, geranyltransferase; EC 2.5.1.21, farnesyltransferase (also catalyses reduction of presqualene alcohol pyrophosphate to squalene); EC 2.7.1.36, mevalonate kinase; EC 2.7.4.2, phosphomevalonate kinase; EC 4.1.1.33, pyrophosphomevalonate decarboxylase; EC 4.1.3.5, hydroxymethylglutaryl-CoA synthase; EC 5.3.3.2, isopentenyldiphosphate Δ-isomerase.

1935 from bovine testes. Esters of T., e.g. T. propionate, are used in the treatment of male sex hormone deficiency, endocrine disorders in gynecology and in geriatrics. For structure and biosynthesis, see Androgens.

19-nor-Testosterone: a synthetic anabolic steroid. 19-nor-T. differs from Testosterone (see) by the absence of the C 19 methyl group. It has a higher anabolic but lower androgenic activity than testosterone.

Tetanus toxin: see Toxic proteins.

Tetracyclins: a group of antibiotics from various *Streptomyces* spp. T. contain four linearly fused six-membered rings; individual T. differ according to the nature of substituents (Fig. and Table).

T. inhibit protein biosynthesis by preventing the binding of aminoacyl-tRNA to ribosomes. Next to the penicillins, T. were one of the most widely used antibiotics, particularly in the treatment of bronchitis, pneumonia, bile duct and urinary infections, plague and cholera. They are also widely employed as additives in animal feedstuffs. On account of side reactions and increasing resistance of bacteria to T., their use is declining.

Table. The structures of some tetracyclins.

Name	R_1	R_2	R_3	R_4	R_5
Chlortetracyclin (aureomycin)	H	H	OH	CH_3	Cl
Oxytetracyclin (terramycin)	H	OH	OH	CH_3	H
Tetracyclin	H	H	OH	CH_3	H
Methacyclin (rondomycin)	H	OH	$CH_2=$		H
Doxycyclin (vibramycin)	H	OH	H	CH_3	H

Tetraethylammonium: see Acetylcholine.

Tetrahydrobiopterin, BH_4: a hydrogen transfer cofactor of a number of aromatic amino acid hydroxylases, including phenylalanine, tyrosine and tryptophan hydroxylases. These are necessary for synthesis of the neurotransmitters dopa, 5-hydroxytryptophan, dopamine, adrenalin, noradrenalin and serotonin; thus BH_4 is necessary for neurological function. Brain tissues from adults with Down's syndrome, senile dementia of the Alzheimer type and severe endogenous depression were found to have very low capacity to synthesize BH_4. [J.A. Blair et al., in *Biochemical and Clinical Aspects of the Pteridines* vol. **3** (de Gruyter, Berlin, 1984)]

Δ^1-Tetrahydrocannabinol, abb. *THC:* the psychotropic principle of hashish. The $(-)$-form is 10–15 times more potent than the $(+)$-form. In humans, THC is degraded and excreted relatively slowly. For forensic purposes, 10^{-7} mol THC can be detected in 0.02 ml body fluid of hashish smokers, using immunochemical methods.

Δ^1-Tetrahydrocannabinol

Tetrahydrofolic acid, abb. *THF, folate-H_4,* coenzyme F: 5,6,7,8-tetrahydropteroylglutamic acid, the coenzyme responsible for the binding, activation and transfer of all active one carbon units, with the exception of carbon dioxide (the F in coenzyme F stands for formylation). M_r 445.4, (after chemical preparation involving acetic acid, it retains two molecules CH_3–COOH and has M_r 565.4), λmax 298 nm in 0.01 M phosphate, pH 7.0, $\varepsilon_{298} \leqq 28000$. In the solid state, THF is slowly oxidized by air and must therefore be stored under vacuum or in an inert atmosphere. In solution, THF is oxidized rapidly to *dihydrofolic acid,* abb. *DHF, folate-H_2* (7,8-dihydropteroylglutamic acid). DHF is also produced as a byproduct in the enzymatic synthesis of thymidylic acid by thymidylate synthase (see Pyrimidine biosynthesis). THF/DHF forms a redox system of E_o -0.19 V. Stability of THF is strongly influenced by pH, being highest at pH 7.4. Ascorbic acid (34 mM) and, to a lesser extent mercaptoethanol (10 mM) have a stabilizing effect on THF solutions. Autoxidation of THF or its metabolic conversion into DHF can be following spectrophotometrically by the shift of absorption from 298 nm to 282 nm.

THF is formed from folic acid (for structure, see Vitamins, folic acid) by enzymatic reduction. For details of biological function of THF, see Active one carbon units.

Tetrahydrofolic acid conjugates contain three to seven residues of glutamic acid. They function in methionine synthesis in some microorganisms.

Tetraterpenes: terpenes comprising eight isoprene units ($C_{40}H_{64}$). Naturally occurring T. are almost all carotenoids, and the group contains no polycyclic compounds.

Biosynthesis. Tail-to-tail condensation of two molecules of geranylgeranylpyrophosphate (see Terpenes) gives phytoene, which undergoes stepwise dehydrogenation to produce the all-*trans* configuration of the true carotenoids (Fig. 1). The ionone rings of cyclic carotenoids arise by addition of a proton at C_3, and formation of a bond between C2 and C7. This is followed by removal of hydrogen from C7 to form a β-ionone ring, or from C5 to form an α-ionone ring (Fig. 2).

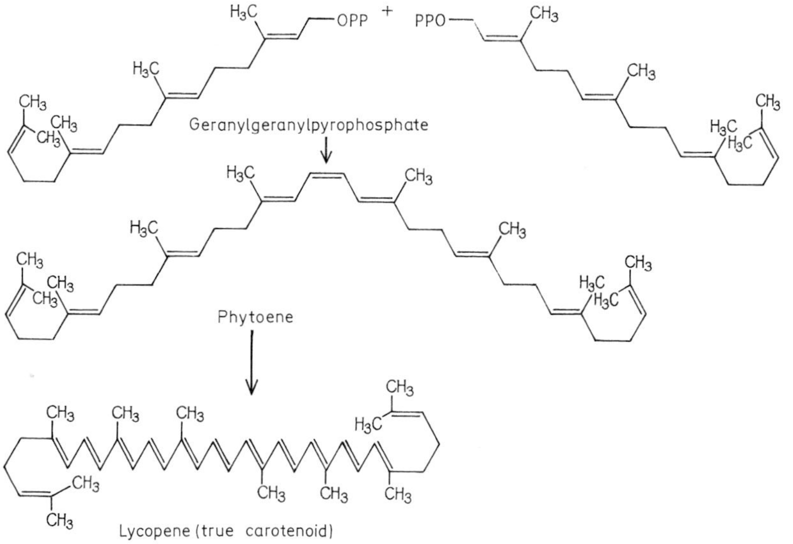

Figure 1. Synthesis of carotenoids from two molecules of geranylgeranylpyrophosphate.

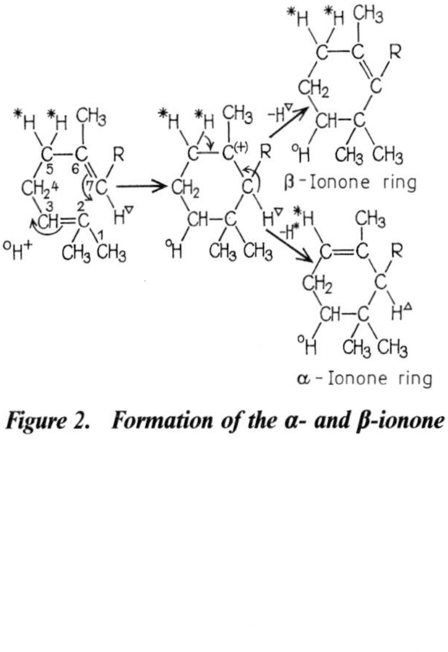

Figure 2. Formation of the α- and β-ionone ring.

Tetrodontoxin: see Tetrodotoxin.

Tetrodotoxin, *tetrodontoxin; spheroidine, taricha-toxin, fugu poison:* octahydro-12-(hydroxymethyl)-2-imino-5,9 : 7,10a-dimethano-10a H-[1,3] dioxocino [6,7-d] pyrimidine-4,7,10,11,12-pentol, a guanidine derivative that exists in 2 tautomeric forms (Fig.). T. is an extremely potent toxin from the ovaries, liver and skin (but not present in the blood) of many species of *Tetrodontidae,* especially the globe fish, *Spheroides rubripes.* M_r 319.28, darkens above 220 °C, $[\alpha]_D^{25}$ − 8.64 ° (c=8.55, dil. acetic acid), LD_{50} i.p. in mice 10 µg/kg. T. acts on the membranes of nerve fibers, and is an antagonist of Batrachotoxin (see).

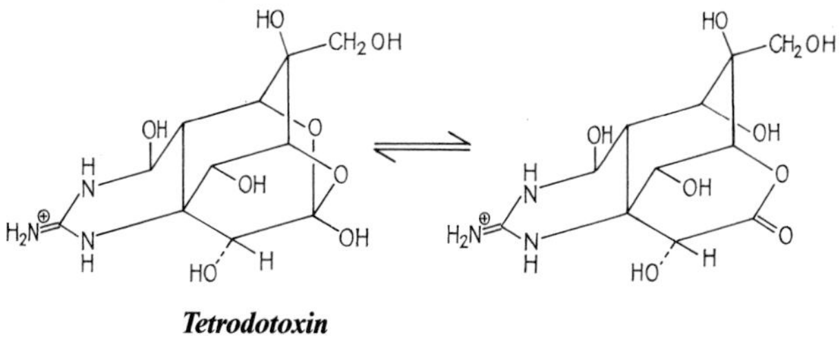

Tetrodotoxin

Tetrose: a monosaccharide containing four carbon atoms, e.g. threose, erythrulose. T. occur as intermediates in carbohydrate metabolism, usually as their phosphates.

Thalassemias: a group of genetically determined disorders of hemoglobin synthesis, characterized by partial or total absence of the synthesis of one or another of the hemoglobin chains. The resulting imbalance in globin chain synthesis may lead to the precipitation of globins produced in excess. Such precipitated proteins appear as inclusions, which are responsible for defective maturation and decreased survival of erythroid cells. At both the genetic and molecular level, T. represent a very heterogeneous family of blood disorders. They are broadly classified according to the globin chain which is inefficiently synthesized, i.e. α, β, $\delta\beta$, δ and $\gamma\delta\beta$ T. Some (e.g. the $\delta\beta$ T.) are characterized by abnormally high levels of fetal hemoglobin ($\alpha_2\gamma_2$) and have much in common with the group of conditions known as *hereditary persistence of fetal hemoglobin* (HPFH). See Hemoglobin. [D.J. Weatherall & J.B. Clegg, *The Thalassemia Syndrome* 3rd edn. (Blackwell Scientific Publications, Oxford, 1981)]

Thaumatin: a sweet tasting, strongly basic, histidine and carbohydrate-free, single chain protein, M_r 21000, (270 amino acid residues, 8 disulfide bridges), IP 11.5. T. is 750–1600 times (weight basis) or 30000–100000 times (molar basis) sweeter tasting than sucrose. It occurs in the fruits of *Thaumatococcus daniellii* (a monocotyledon of the arrowroot or *Maranta* family). T. shows considerable sequence homology with the B-chain of another sweet protein, Monellin (see). These two proteins are immunologically related, and it is thought that a tripeptide sequence (-Glu-Tyr-Gly-) near the surface of each molecule is a common antigenic determinant.
[Iyengar, R.B., Smits, P., Van der Oudera, F., Vander Wel, H., Van Brouwershaven, J., Ravestein, P., Richters, G. and Van Wassenaar, P.D. *Eur. J. Biochem.* **96** (1979) 193–204]

THC: abb. for Δ^1-tetrahydrocannabinol.

Thd: abb. for Ribothymidine.

Thebaine: see Benzylisoquinoline alkaloids.

Theobromine: see Methylated xanthines.

Therapeutic index: see Dose.

Therapeutic range: see Dose.

Thermolysin (EC 3.4.24.4): a heat-stable, zinc and calcium-containing neutral protease, M_r 37500, from *Bacillus thermoproteolyticus,* with a substrate specificity similar to that of Subtilisin (see). After one hour at 80 °C, T. still has 50% original activity. This high heat stability of T. is attributed to the large number of hydrophobic regions and the presence of four bound calcium ions, which serve in place of disulfide bridges (T. contains no disulfide bridges) to maintain the compact shape of the molecule. T. is neither a thiol nor a serine enzyme.

Thermostable enzymes, *heat stable enzymes:* a small number of enzymes, mostly hydrolases, whose temperature optima lie between 60 and 80 °C. T. e. usually have a compact structure, stabilized by many disulfide bonds and/or extensive hydrophobic regions, and a low α-helix content. Examples are particularly common amongst the bacterial enzymes, e.g. thermolysin and certain amylases. T.e. are used in the preparation of biological washing powders, and in the food industry.

Thetins: sulfonium compounds, e.g. dimethylthetin, which can be used as a methylating agent (see Transmethylation). T. also occur naturally, e.g. dimethyl-β-propiothetin is present in algae and higher plants. Dimethylsulfide is a decomposition product of T.

$(H_3C)_2\overset{\oplus}{S}-CH_2-COO^{\ominus}$

Dimethylthetin

$(H_3C)_2\overset{\oplus}{S}-CH_2-CH_2-COO^{\ominus}$

Dimethyl-β-propiothetin

THF: see Tetrahydrofolic acid.

Thiamin: see Vitamins (Vitamin B$_1$).

Thiamin pyrophosphate, abb. *TPP, aneurin pyrophosphate,* abb. *APP, cocarboxylase:* the pyro phosphoric ester of thiamin (aneurin, vitamin B$_1$, Fig. 1), and the prosthetic group (or coenzyme) of various thiamin pyrophosphate enzymes, e.g. Pyruvate decarboxylase (see), pyruvate dehydrogenase and 2-oxoglutarate dehydrogenase (see Multienzyme complex), Transketolase (see), glyoxylate carboligase and oxalyl-CoA decarboxylase (see Oxalic acid). The free cation of TPP has M_r 425.3, while the chloride (M_r 460.8) crystallizes from ethanol with one molecule of water (M_r 478.8). M.p. of TPP 240–244 °C (d.), λ_{max} 245 and 261 nm (in phosphate buffer, pH 5.0); 231.5 and 266 nm (pH 8.0), λ_{min} 248 nm.

In its role as a coenzyme, TPP reacts with the substrates of TPP-enzymes to form *active aldehydes:* 1. Active acetaldehyde (hydroxyethylthiamin pyrophosphate, abb. HETPP); 2. Active glycolaldehyde [2-(1,2-dihydroxyethyl)-thiamin pyrophosphate, abb. DETPP]; 3. Active formaldehyde (2-hydroxymethylthiamin pyrophosphate, abb. HMTPP). Active pyruvate (pyruvylthiamin pyrophosphate) and active glyoxylate [2-(hydroxycarboxymethyl)-thiamin pyrophosphate] are postulated as intermediates in the formation of active aldehydes. The hydrogen atom at position 2 of the thiazolium ring of TPP (between the sulfur and the nitrogen) (Fig. 2) has a high pK_a of about 12.6. It is thought that the thiazolium dipolar ion (or ylid) (i.e. C 2 forms a carbanion stabilized by the positive charge on the nitrogen) is the key intermediate in the coenzyme function of TPP. Addition of the ($\delta+$) C-atom of a substrate carbonyl group to the C2 carbanion produces an active intermediate. Electrons can then flow from the attached substrate into the ring system of the TPP and bond cleavage occurs between the attached substrate carbon atom (the original carbonyl carbon) and a neighboring carbon atom. All reactions catalysed by TPP-enzymes conform to this mechanism.

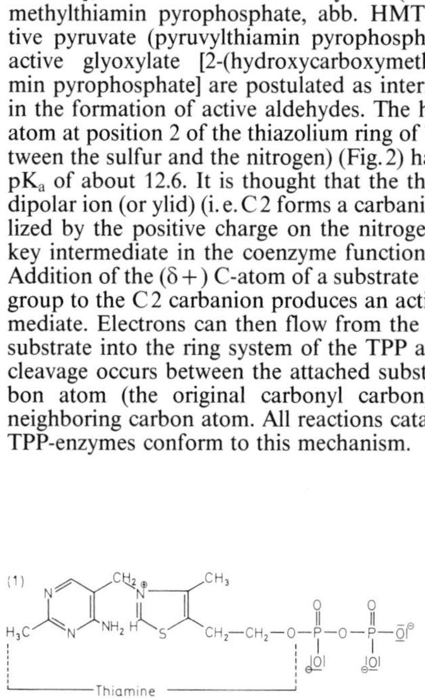

Figure 1. Thiamin pyrophosphate.

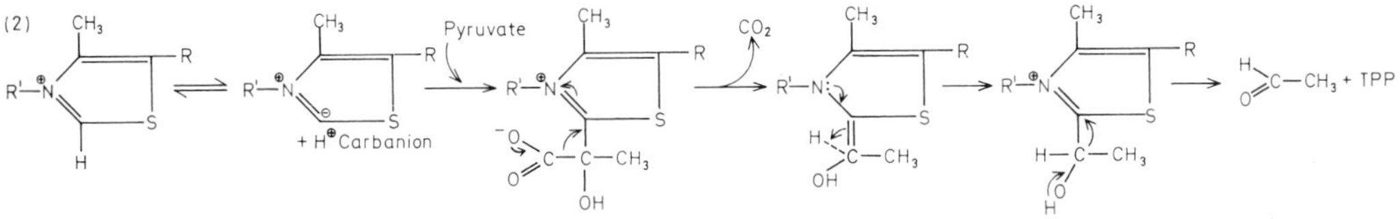

(2)

Figure 2. Mechanism of thiamin pyrophosphate (TPP) catalysis for the decarboxylation of pyruvate.

Thin-layer chromatography, *TLC*: a form of chromatography (see) in which the solid carrier material is spread in a thin layer on a glass or plastic plate. The advantages of the method are the short distances required for good separation and the correspondingly short development times, high sensitivity, separation of very small amounts of substances, and, if inorganic carrier materials are used, the possibility of using caustic detection reagents. Prespread thin-layer plates are available commercially, as is a spreading device with which one can spread any desired thickness of carrier from 0 to 2 mm. With appropriate equipment, the plates can be used for ascending, descending, horizontal or multiple chromatography.
TLC can be used preparatively as an "open column" (see Column chromatography). Depending on the size of the plate, up to 100 g of material can be separated by preparative TLC while analytical TLC can be used for amounts between 10 ng and 10 mg.
TLC was originally developed for the separation of lipophilic substances on inorganic carrier materials. However, if cellulose powder is used as carrier, hydrophilic substances, including amino acids, nucleotides and carbohydrates, can also be separated. Lipophilic substances can be separated on aluminum oxide, silica gel, acetylated cellulose and polyamide; hydrophilic substances on cellulose, cellulose ion exchangers, diatomaceous earth and polyamides. The cellulose ion exchangers used in TLC have shorter cellulose fibers than those used for column chromatography. *Polyamide TLC* makes use of hydrophilic or hydrophobic polyamides for separation of a wide range of substances. It depends on the reversible formation of hydrogen bonds between the substances and the amide groups of the carrier. The eluant (water, methanol, formamide, etc.) displaces these hydrogen bonds, forming its own with the carrier. The separation thus depends on the differences in the strengths of the hydrogen bonds formed by the substances to be separated.

Thiobinupharidine: see Nuphara alkaloids.

Thioctic acid: see Lipoic acid.

Thioester, *acylmercaptan*: a compound of the general formula RS~CO-R_1. The thioester (acylmercaptan) bond is energy-rich. All fatty acyl coenzyme A derivatives (activated fatty acids, e.g. acetyl-CoA) are T. During substrate phosphorylation on glyceraldehyde 3-phosphate dehydrogenase a thiol group of the enzyme forms an energy-rich intermediate T.

Thioethers: see Sulfur compounds.

Thioglucoside glucohydrolase, *β-thioglucosidase, myrosin, myrosinase, sinigrinase, sinigrase* (EC 3.2.3.1): a plant enzyme responsible for the conversion of glucosinolates into isothiocyanates (see Glucosinolate).

Thiol enzyme, *SH-enzyme*: an enzyme whose activity depends upon the presence of a certain number of free thiol groups. T.e. are found amongst the hydrolases, oxidoreductases and transferases. Known T.e. are bromelain, papain, urease, various flavoenzymes, pyridine nucleotide enzymes, pyridoxal phosphate enzymes and thiolproteinases. T.e. are typically inhibited by Sulfhydryl reagents (see).

Thiolesterases: see Esterases.

Thiol group, *sulfhydryl group, mercapto group*: -SH, the functional group of thiols (mercaptans), RSH, where R is the remainder of the molecule. T.g. may be structurally important as in Thiol enzymes (see), or functionally important as in Coenzyme A (see), Pantetheine-4′-phosphate (see), Lipoic acid (see), Thioredoxin (see), etc. The functional form of lipoic and thioredoxin is a dithiol.

Thiols: see Sulfur compounds.

2-Thiomethyl-N^6-isopentenyladenosine: 2-methylmercapto-6-isopentenyladenosine, an adenosine derivative and one of the rare nucleic acid components found in tRNA from wheat. It is an active cytokinin. The hydroxylated derivative, *2-methylmercapto-6-(4-hydroxy-3-methyl-cis-2-enylamino)-purine* has also been found in some species of tRNA.

Thioredoxin: a heat-stable, acidic, metal-free redoxin, M_r 12000. T. is a component of deoxyribose synthase, in which T. and thioredoxin reductase form a hydrogen transfer system linked to the reduction of ribose or ribonucleoside phosphates by $NADPH + H^+$. T. is a single polypeptide chain of 109 amino acid residues with *N*-terminal serine. The functional group of the oxidized form consists of a disulfide bridge between two cysteine residues, which are separated by 10 amino acid residues. Ribonucleoside triphosphate reductase from *Lactobacillus leishmanii* (Figure 1) also requires a cobalamine

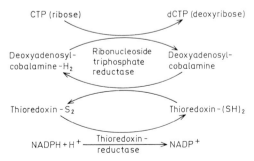

Figure 1. Ribonucleoside triphosphate reductase and thioredoxin reductase in Lactobacillus leishmanii.

coenzyme (cobamide) as a hydrogen carrier. The co-balamine coenzyme (DBC = oxidized, DBCH$_2$ = re-duced form) mediates an intramolecular hydrogen transfer: hydrogen from T. is transferred to the DBC-coenzyme in a ternary complex of T./reduc-tase/DBC; hydrogen is then transferred from the complex to the ribonucleoside triphosphate, e. g. cyt-idine triphosphate (CTP) (Figure 2). In its mecha-nism of action, T. resembles Lipoic acid (see).

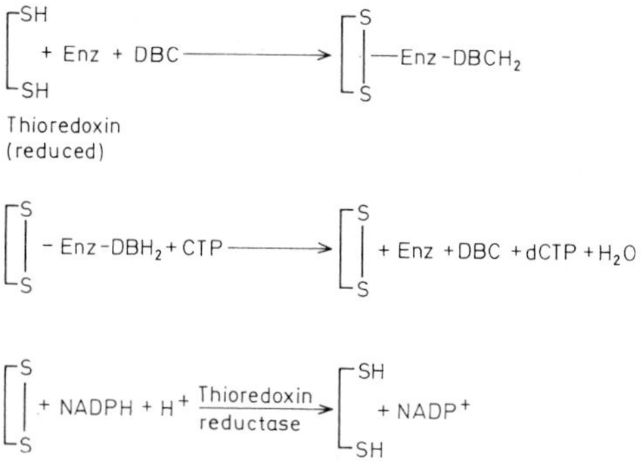

Figure 2. Hypothetical role of cobalamine coenzyme (DBC) in the action of ribonucleoside triphosphate re-ductase.

Thr: abb. for L-threonine.

L-Threonine, abb, *Thr:* L-threo-α-amino-β-hy-droxybutyric acid, H$_3$C-CH(OH)-CH(NH$_2$)-COOH, a proteogenic, essential amino acid with two asymmetric C-atoms. M_r 119.1, m.p. 253 °C (d.), $[\alpha]_D^{25}$ -28.5 (c = 1-2, water). A useful reaction for the determination of L-T. is oxidation with per-iodate to acetaldehyde, glyoxylate and ammonia. Enzymatic hydrolysis of peptide bonds involving L-T. appears to be particularly difficult, which may be relevant to the nutritional physiology of this amino acid. The principal degradative reaction of L-T. in most organisms is conversion to 2-oxobutyrate and ammonia by the pyridoxal phosphate-dependent en-zyme, L-threonine dehydratase (EC 4.2.1.16). This degradative enzyme (also called threonine deami-nase) is distinct from biosynthetic threonine dehy-dratase needed for the production of 2-oxobutyrate in the biosynthesis of isoleucine; in *Escherichia coli,* the latter enzyme is allosterically inhibited by isoleu-cine. L-Threonine acetaldehyde-lyase (threonine al-dolase; EC 4.1.2.5) is a pyridoxal phosphate enzyme which converts L-T. into glycine and acetaldehyde. It is present in various organisms, including mam-mals, and appears to be a purely degradative en-zyme. Oxidation of L-T. to 2-amino-3-oxobutyrate, followed by decarboxylation, produces aminoace-tone, a urinary constituent. In microorganisms, ami-noacetone is converted to R-1-amino-2-propanol, an intermediate in the biosynthesis of vitamin B$_{12}$. Aminoacetone may also be oxidatively deaminated to methylglyoxal, which can be attacked by glyoxa-lase to form D-lactate.

In plants and microorganisms, L-T. is biosynthe-sized from phosphohomoserine by a γ-elimination of phosphate followed by β-replacement with an OH-group. This total reaction is catalysed by the pyridoxal phosphate enzyme, threonine synthase (EC 4.2.99.2). The phosphohomoserine is derived from aspartate via aspartyl phosphate, aspartate semialdehyde and homoserine.

Thrombin: a blood coagulation enzyme, respon-sible for the conversion of fibrinogen to fibrin. T. is a glycoprotein (5% carbohydrate), M_r 39000, pro-duced by activation of Prothrombin (see). T. is a typ-ical Serine protease (see) with catalytically impor-tant residues His$_{58}$, Asp$_{102}$ and Ser$_{195}$ in the B-chain, and considerable sequence homology with trypsin, chymotrypsin and elastase. Autolysis of T., leading to a decrease of M_r from 39000 to 26000 at the ex-pense of the B-chain (M_r 33000 to 19500) causes no loss of activity.

Thrombospondin: a glycoprotein, M_r 430000, consisting of 3 disulfide-linked chains (each of M_r 142000). T. is present in very low levels in normal plasma, and it is a major constituent of platelet α-granules. T. is named from the observation that it is released from platelets in response to thrombin treatment. After release, some T. becomes associated with platelet surfaces, where it is at least partly re-sponsible for the lectin-like activity of platelets. Antibodies to T. inhibit thrombin-induced platelet aggregation. Under the electron microscope, puri-fied T. appears as filaments, 7 × 70 nm. It contains 1.9% neutral sugars, 1.4% amino sugars, 0.7% sialic acid, and no hexuronic acid. T. has many features in common with other "adhesive" proteins: Fibrinogen (see), Fibronectin (see), von Willebrand factor (see). [J. W. Lawler et al., *J. Biol. Chem.* **253** (1978) 8609–8616; R. Wolff et al., *J. Biol. Chem.* **261** (1986) 6840–6846]

Thrombosthenin: see Muscle proteins.

Thromboxanes: derivatives of the Prostaglan-dins (see) which induce aggregation of platelets, for-mation of clots and smooth muscle contraction. TXA$_2$ is more active than TXB$_2$, but it decays so rapidly (fig.) that experimental analysis is difficult. TXB$_2$ does not cause an increase in platelet cAMP levels (in contrast to the prostaglandins), and TXA$_2$

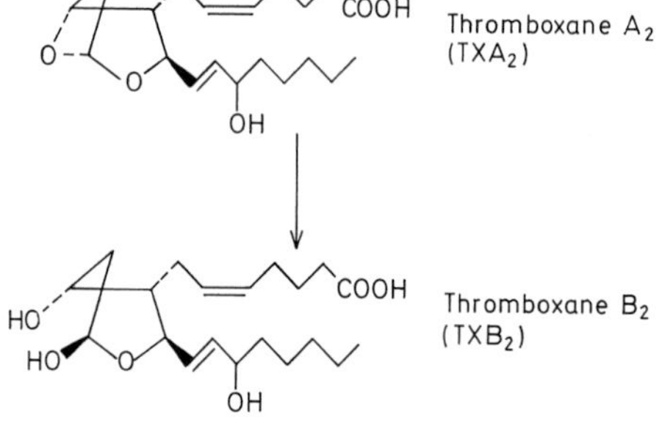

Prostaglandin H$_2$ (PGH$_2$)

Thromboxane A$_2$ (TXA$_2$)

Thromboxane B$_2$ (TXB$_2$)

inhibits such an increase in the presence of prostaglandins.

The immediate precursor of the TX is prostaglandin H_2, which is ultimately derived from Arachidonic acid (see). TXA_2 is rapidly converted to TXB_2; however, it is stabilized somewhat by albumin. The formation of TXA_2 is inhibited by low doses of aspirin; this accounts for the anti-coagulant effect of aspirin. [S. Moncada & J.R. Vane *Pharmacol. Rev.* **30** (1979) 293-331]

Thujane: see Monoterpenes (Figure).

Thy: abb. for thymine.

Thylakoids: internal membrane structures of the chloroplast. Under the electron microscope, T. appear as disc-shaped, flattened vesicles, about 600 nm diameter. These are arranged in stacks, which are the grana observable under the light microscope. In addition to these granal T., there are also stromal T., which pass singly through the stroma of the chloroplast, joining together various stacks of granal T. The functional unit of T. is thought to be the Quantasome (see). The T. membrane is about 9 nm thick, enclosing a thin internal space or *loculus;* it contains approximately equal quantities of protein and lipid, and is notable for its high content of galactosyl diacylglycerol, digalactosyl acylglycerol and sulfolipid. There is probably a greater proportion of lipid molecules on the inside of the membrane and more protein on the outside, but distinct layering into protein and lipid seems to be absent. The inside lipid layer contains the chlorophylls and carotenoids, and the chlorophyll is present largely, if not entirely, in the form of protein complexes. The protein subunits of the outer layer have a diameter of 4 nm.

Thymidine, more correctly *deoxythymidine,* Abb. *dThd, thymine deoxyriboside:* a Nucleoside (see) of thymine and D-2-deoxyribose. M_r 242.33, m. p. 185-186 °C $[\alpha]_D^{16} + 32.8$ ($c = 1.04$, 1MNaOH). dThd should not be confused with Ribothymidine, which is also (and more correctly) called thymidine. For metabolic importance, see Thymidine phosphates.

phosphate, abb. *TDP* (more correctly *deoxythymidine 5'-diphosphate,* abb. *dTDP*), M_r 402.2, which serves as the activating group in certain Nucleoside diphosphate sugars (see); and to *thymidine 5'-triphosphate,* abb. *TTP* (more correctly *deoxythymidine 5'-triphosphate,* abb. *dTTP*), M_r 482.18, a substrate of DNA synthesis.

Thymidylic acid: see Ribothymidine; Thymidine phosphates.

Thymin: see Thymopoetin.

Thymine, abb. *T* or *Thy:* 2,6-dihydroxy-5-methylpyrimidine, 5-methyluracil, a pyrimidine base present in DNA. M_r 126.1, m.p. 321-326 °C (d.). T. was first isolated in 1893 from thymonucleic acid. It is formed by the degradation of thymidine 5'-monophosphate; the methyl group of T. does not arise from methionine, but from an active one carbon unit (see Pyrimidine biosynthesis).

Thymine deoxyriboside: see Thymidine.

Thymine dimer: see Dimers.

Thymonucleic acid, *thymus nucleic acid:* nucleic acid from the thymus gland; effectively an obsolete term for DNA.

Thymopoietin, *thymin, thymosin:* a family of largely identical monomeric polypeptide hormones from the thymus, all consisting of 49 amino acid residues (Fig.). T. induces differentiation of prothymocytes to thymocytes. It impairs neuromuscular transmission, and binds with high affinity to the acetylcholine binding region of the nicotinic acetylcholine receptor of *Torpedo californica.* It has been implicated in the pathogenesis of myasthenia gravis. The synthetic pentapeptide thymopentin, Arg-Lys-Asp-Val-Tyr, corresponds to residues 32-36 of T.; it possesses all the biological properties of T. and represents the active site of the latter. Splenin, a variant of T. III (Fig.), has Glu in place of Asp at position 34, and has no effect on neuromuscular transmission. [G. Goldstein et al., *Science* **204** (1979) 1309-1310; T. Audhya et al., *Biochemistry* **20** (1981) 6195-6200; K. Venkatasubramanian et al., *Proc. Natl. Acad. Sci.* **83** (1986) 3171-3174]

10
NH$_2$-Residue 1-Residue 2-Phe-Leu-Glu-Asp-Pro-Ser-Val-Leu-Thr-Lys-Glu

30   20
Glu-Gly-Ala-Pro-Leu-Thr-Val-Asn-Asn-Ala-Val-Leu-Glu-Ser-Lys-Leu-Lys

40
Gln-Arg-Lys-Residue 34-Val-Tyr-Val-Glu-Leu-Tyr-Leu-Gln-Residue 43-Leu

49
HOOC-Arg-Lys-Leu-Ala-Thr

Thymidine phosphates, more correctly *deoxythymidine phosphates:* Nucleotides (see) of thymine; phosphate esters of deoxythymidine. Although T. p. contain deoxyribose, the prefix *deoxy* is usually omitted, because the corresponding ribose derivatives hardly ever occur naturally. *Thymidine 5'-monophosphate* abb. *TMP, thymidylic acid* (more correctly *deoxythymidine 5'-monophosphate,* abb. *dTMP, deoxythymidylic acid*): a component of DNA, and an intermediate in the synthesis of TPP (see Pyrimidine biosynthesis). M_r 322.2, m. p. 225-230 °C. Stepwise phosphorylation of TMP leads to *thymidine 5'-di-*

Residue Protein	1	2	34	43
TPI (thymus)	Gly	Gln	Asp	His
TPII (thymus)	Pro	Glu	Asp	Ser
TPIII (spleen)	Pro	Glu	Asp	His
Splenin (spleen)	Pro	Glu	Glu	His

Structures of thymopoietins and splenin.

Thymosin: see Thymopoetin.

Thyrocalcitonin: see Calcitonin.

Thyroid gland: Glandula thyreoidea, a well vasculated gland at the front of the neck. It is paired in amphibians and birds, and unpaired in elasmobranch fish and mammals, weighing 20-60 g in the human. The T.g. synthesizes, stores (in the thyroid follicles) and secretes Thyroxin (see) and triiodothyronin, under the influence of the anterior pituitary hormone thyrotropin. It also synthesizes Calcitonin (see) in the parafollicular C-cells.

Thyroid stimulating hormone: see Thyrotropin.

Thyrotropin, *thyroid stimulating hormone,* abb. **TSH:** a glycoprotein hormone, M_r 25 000 (bovine), containing 23% carbohydrate. Primary structure of some TSH molecules is known. It consists of an α- and a β-chain, and the α-chain is identical to that of Luteinizing hormone (see). Synthesis occurs in the basophilic cells of the anterior pituitary. Both synthesis and secretion are stimulated by thyrotropin releasing hormone (see Releasing hormones) from the hypothalamus, and inhibited by thyroxin. TSH generally stimulates the thyroid gland; blood circulation of the thyroid gland is increased, uptake of iodine is promoted, the rate of synthesis of thyroglobin, triiodothyronine and thyroxin is increased, and the secretion of thyroid hormones is stimulated. Inactivation occurs in the kidney. Blood concentrations of TSH are in the order of ng/ml, and can be measured radioimmunologically.

Thyrotropin releasing hormone: see Releasing hormones.

Thyroxin, *3,5,3',5'-tetraiodothyronine,* abb. T_4: a hormone produced by the thyroid gland and absolutely essential for growth and development. M_r 776.9. T_4 and the second thyroid hormone, 3,5,3'-triiodothyronine (abb. T_3, M_r 651.0) are synthesized from L-tyrosine residues in thyroglobulin, a dimeric glycoprotein (M_r 670000) that constitutes the bulk of the thyroid follicle. Tyrosine residues in thyroglobulin become iodinated, so that the protein contains several mono- and diiodotyrosine residues. The nature of the subsequent coupling reaction is uncertain, but it is equivalent to the transfer of iodinated rings from some iodotyrosine residues to form ether linkages by reaction with the hydroxyl functions of other iodinated tyrosine residues. T_4 and T_3 are released by proteolysis of thyroglobulin. Synthesis and release of T_4 and T_3 from thyroid epithelial cells, together with a parallel increase in the uptake of iodine by the thyroid gland, are stimulated by the hormone thyrotropin from the anterior pituitary. Both hormones are carried in the blood to all body cells, partly in the free form and partly bound to prealbumin and glycoprotein, T_4 being more tightly bound than T_3.

Metabolic action of T_3 and T_4: increased oxygen uptake by mitochondria, and increased heat production (calorigenic effect); in physiological concentrations both hormones increase synthesis of RNA and protein; in higher doses they act catabolically, causing negative nitrogen balance and mobilization of fat depots. Independently of their calorigenic effect, they increase the rate of cell differentiation and metamorphosis, e.g. development of tadpoles into frogs. Biological half life of T_4 is 7-12 days (based on the sustained activity of a single T_4 dose). Degradation consists of removal of iodine (reused by the thyroid gland), deamination and coupling with glucuronic acid or sulfate in the liver, followed by urinary excretion.

Hyperthyroidism is caused by overactivity of the thyroid gland leading to an excess of T_4 and T_3. Hypothyroidism results from decreased hormone production; this may be caused by iodine deficiency, administration of goitrogens, defective enzymes in hormone synthesis, autoimmune thyroiditis (antibodies are formed against the body's own thyroid tissue), etc. Prolonged hypothyroidism may result in dwarfism, mental deficiency, goiter and myxedema.

Thyroxin

Tin, Sn: a metal occurring in many tissues and dietary components. The redox potential of $Sn^{2+} \rightleftharpoons Sn^{4+}$ is 0.13 volt, near to the redox potential of the flavin enzymes, suggesting a possible biological role. It is still not certain that Sn is biologically essential, and its presence in tissue may be environmental contamination. It has been reported that Sn is essential for the growth of rats.

Tingitanin: see Guanidine derivatives.

Tissue hormones: hormones produced in specialized, single cells scattered through a tissue rather than clumped in a gland (see Hormones). They fall into three groups: 1. Secretin (see), Gastrin (see), and Cholecystokinin (see) from the gastrointestinal tract; 2. Angiotensin (see) and Bradykinin (see), which occur as inactive precursors in the blood; and 3. Biogenic amines, such as Histamine (see), Serotonin (see), Tyramine (see) and Melatonin (see). This last group is an exception to the rule that hormones act at sites removed from the cells which produce them; they affect the immediately surrounding tissue.

Tlatlancuayin: 5,2'-dimethoxy-6,7-methylenedioxyisoflavone, see Isoflavone.

TMP: see Thymidine phosphates.

TN: abb. for troponin; see Muscle proteins.

TNF: abb. for Tumor necrosis factor.

Toad toxins: poisons found in the secretions of the skin glands of toads *(Bufonidae)* (Figure). These are classified as 1. bufadienolides (bufogenins) with digitalis-like effects on the heart (see Cardiac glycosides), e.g. bufotoxin. They strengthen and slow the heartbeat. The bufadienolides are present in toad blood at a dilution of 1:5000 to 1:20000, and they are necessary for normal heart activity. 2. Alkaline toxins which are alkaloids derived from tryptamine or indole, e.g. bufotenine, dehydrobufotenine and O-methylbufotenine. The alkaline toxins of some species of toads also contain adrenalin and similar substances. Bufotenines increase the blood pressure and have a paralysing effect on the motor centers of the brain and spinal column.

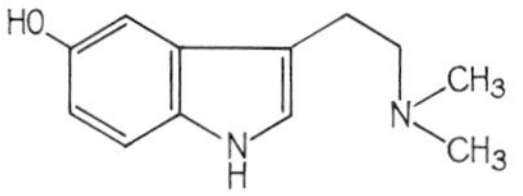

Bufotenine

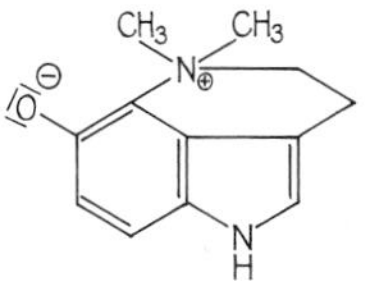

Dehydrobufotenine

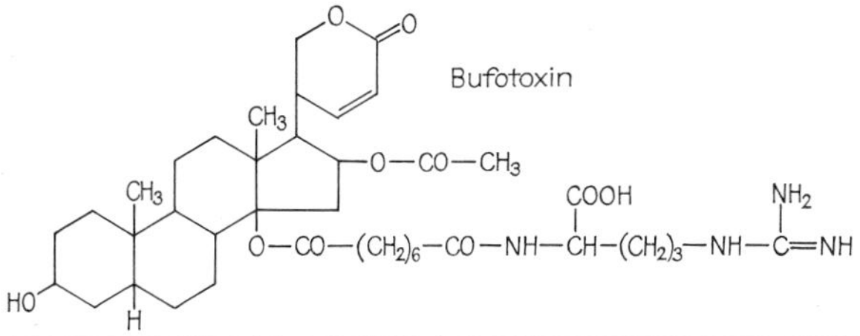

Bufotoxin

T.t. have an anesthetic effect which is several times as potent as that of cocaine.

Tobacco alkaloids: see Nicotiana alkaloids.

Tobacco mosaic virus: see Virus coat proteins.

Tocopherol: see Vitamins (Vitamin E).

Tocoquinone: see Vitamins (Vitamin E).

Tolerogens: see Immunotolerance.

Tomatidenol: see β-Solamarine.

Tomatidine: see α-Tomatine.

α-Tomatine, *tomatine:* a Solanum alkaloid, and the chief alkaloid of the tomato *(Lycopersicon esculentum),* also occurring in other *Lycopersicon* and *Solanum* spp. T. is a glycoalkaloid of the aglycon *tomatidine* [(22S:25S)-5α-spirosolane-3β-ol, M_r 415.7, m.p. 210 °C, $[\alpha]_D + 6.5$ ° (chloroform)] and the tetrasaccharide β-*lycotetraose*. T. imparts a bitter taste and protects the tomato plant from attack by the Colorado potato beetle. It also has antibiotic activity against the causative agents of tomato wilt and other pathogenic fungi.

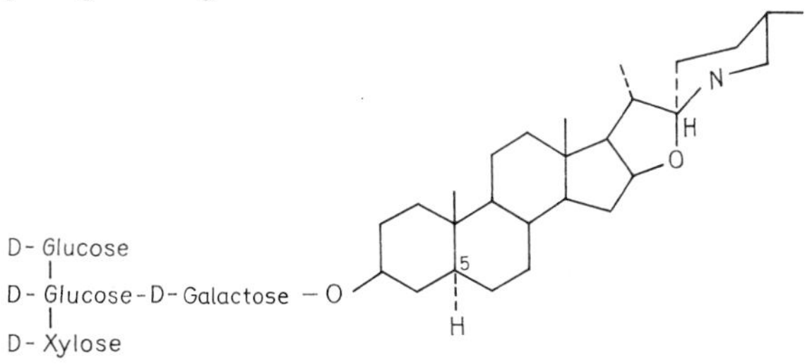

integration, transposition and renaturation of single-stranded DNA circles. In addition, topoisomerization may also be important in the chemomechanical activity of DNA in such processes as the filling of phage heads, transfer of DNA in bacterial conjugation, and mitosis. The role of T. in eukaryotes is poorly understood. T. also catalyse the formation and resolution of knotted and catenated circular duplex DNA.

Type I T. transiently break one strand of the DNA double helix, so that the Linking number (see) changes in steps of one. ATP is not required. Reactions catalysed by type I T. can be explained by the rotation of a transiently broken strand around its unbroken, complementary neighbor (Fig. 1); alternatively the reaction can be represented by the passing of the unbroken strand through the transient break. Prokaryotic type I T. become covalently bound to

D-Glucose
|
D-Glucose-D-Galactose — O
|
D-Xylose

Tomatine

Tonoplast: see Vacuole.

Topoisomerases: enzymes which interconvert topological isomers of circular duplex DNA by altering the degree of supercoiling (superhelicity; see Superhelix). In prokaryotes, an appropriate degree of supercoiling is thought to be important in the control of replication, transcription, recombination,

the broken 5'-end by a phosphotyrosine bond, thus conserving the energy of the cleaved phosphodiester bond, and permitting the two ends of the broken strand to be resealed after topoisomerization. Negatively, but not positively supercoiled DNA is relaxed by prokaryotic type I T., and Mg^{2+} is required for activity.

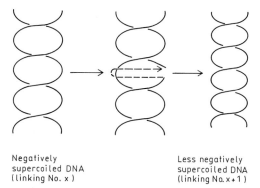

Negatively supercoiled DNA (linking No. x)	Less negatively supercoiled DNA (linking No. x + 1)

Figure 1. A model for the catalytic action of type I topoisomerase. A single strand of the DNA double helix becomes transiently broken, and it rotates around the unbroken strand before resealing.

Eukaryotic type I T. relax both positively and negatively supercoiled DNA; during catalysis, the enzyme attaches at the 3′-end of the break via a phosphotyrosine linkage. Rat liver type I T. catalyses the formation in vivo of chromatin-like material from relaxed circular DNA and core histones. This suggests a role in vivo for eukaryotic type I T. in the formation of chromatin, which correlates well with the fact that the eukaryotic enzymes are found almost entirely in the chromatin fraction. It is proposed that supercoiling or unwinding of DNA with-

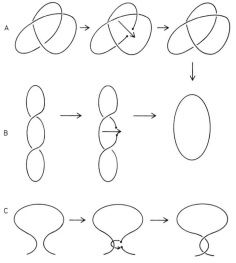

Figure 2. Reactions catalysed by type II topoisomerases. In each case the single line represents duplex DNA. Transient breakage occurs in both strands of the double helix.
A: resolution of a knotted circular duplex.
B: relaxation of supercoiling of a circular duplex.
C: one loop of a circular duplex, showing generation of supercoiling.

in the nucleosome causes increased torsional strain on the rest of the DNA molecule, which in turn is relaxed by the action of type I T. This theory must still be reconciled with the fact that chromatin may be assembled in vivo from newly replicated, discontinuous DNA (see Chromatin).

The following enzymes described in the literature are now known to be type I T.: *Escherichia coli* ω-protein (identical with Eco DNA T., *E. coli* swivelase, and *E. coli* type I T.), untwisting enzymes, nicking-closing enzymes, Relaxation protein (see), DNA-Relaxing enzyme (see).

Type II T. promote a double-strand breakage, through which another section of unbroken double helix passes before the break is resealed (Fig. 2.) The linking number is therefore changed in steps of two. Type II T. can also catenate and decatenate closed circles of DNA, as well as relaxing supercoils.

DNA gyrase is a type II T., which differs from all other T. in being able to increase the torsional strain on the DNA circular duplex, i.e. it can convert relaxed circular DNA into a superhelix. This involves an increase in free energy, which is supplied by ATP. In the absence of ATP, supercoiled DNA is relaxed by DNA gyrase. Supercoiling by gyrase is always negative, i.e. the same as intracellular DNA. Other reactions catalysed by gyrase are the DNA-dependent hydrolysis of ATP, and the knotting, unknotting, catenation and decatenation of circular duplex DNA.

Gyrases are inhibited by a group of antibiotics comprising novobiocin, coumermycin and clorobiocin, and by a different group represented by nalidixic acid and oxolinic acid. Mutants of *Escherichia coli* resistant to one or the other of these groups of antibiotics have been isolated, and the corresponding sites of mutation have been mapped. Since resistant mutants contain antibiotic-resistant DNA gyrase, and the drug resistance of both the bacterium and its gyrase are cotransduced by phage P1, the site of attack of these antibiotics appears to be the gyrase enzyme itself. For this reason, the designation of the locus for nalidixic and oxolinic acid resistance (48 min on the *E. coli* map) has been changed from *nal A* to *gyr A*. Similarly, the locus for resistance to coumermycin, novobiocin and clorobiocin (82 min) has been renamed *gyr B* (formerly *cou*). Significantly, all these antibiotics are inhibitors of DNA replication, which is supporting evidence for the participation of DNA gyrase in this process. Gyrase is an equimolar complex of two proteins (A and B) and probably exists in solution as the A_2B_2 tetramer. Each protein is the site of attack of one family of antibiotics. DNA gyrase activity has not been found in any eukaryotic organism.

Other type II T. catalyse an ATP-dependent relaxation of supercoiled DNA, and are described in the literature as ATP-dependent DNA-relaxing enzymes. The first representative of this group to be described was isolated from *E. coli* infected with T4 phage [L. F. Liu, C-C. Liu and B. M. Alberts, *Nature* **281** (1979) 456–461]. The purified enzyme has protein components of M_r 63 000 (product of phage gene 39) and 52 000 (phage gene 52). In vivo, it appears to be responsible for fork initiation (but not fork movement) in DNA replication. In vitro, it relaxes both negatively and positively supercoiled DNA. When large amounts of T4 DNA T. are incubated with circular duplex DNA in the absence of ATP, knotted DNA molecules are produced. These knotted circles are restored to a simple circular form by incubation with catalytic amounts of enzyme in the presence of ATP. Similar type II T. have now been isolated from many different eukaryotic sources (e.g. *Drosophila* embryo, *Xenopus laevis* germinal vesicles, rat liver mitochondria, calf thymus, HeLa cell nuclei, yeast).

An earlier nomenclature distinguished between type I and type II T., according to whether they relaxed positive and negative, or only negative supercoiling; this is not widely used and should be discontinued. Since all topological alterations of DNA involve transient breakage and rejoining of DNA strands, any enzyme that causes breakage of DNA may also show T. activity. For example, T. activity has been demonstrated for ΦX174 cistron A protein (which breaks the replicative form of DNA and becomes attached to it at a specific site) and for the *int* protein of phage λ (which catalyses an intermolecular strand transfer during integrative recombination). Conversely, enzymes first recognized for their T. activity may later be found to have other physiological roles involving transient DNA cleavage. [C. W. Wang and L. F. Liu, in *Molecular Genetics*, part III *Chromosome structure*, J. H. Taylor, ed. (Academic Press, 1979) pp. 65–88. M. Gellert *Ann. Rev. Biochem.* **50** (1981) 879–910. K. Geider & H. Hoffmann-Berling, *Ann. Rev. Biochem.* **50** (1981) 233–260]

Topological winding number: see Linking number.

Toxalbumins: see Toxic proteins.

Toxic proteins: mostly low M_r, single chain, nonenzymic proteins, produced especially by snakes and invertebrate animals, but also by some plants (phytotoxins) and virulent strains of bacteria. With the exception of bacterial enterotoxins, and *Botulinus* toxins, T. p. show practically no oral activity, and are only toxic when injected, i. e. when the digestive tract is bypassed.

Known *plant toxins* are 1. the homologous *viscotoxins* (M_r 4840, 46 amino acid residues of known sequence, 3 disulfide bridges) from leaves and branches of the European mistletoe, which have hypotensive activity and cause a slowing of the heart beat; 2. the *toxalbumins* Ricin (see) and abrin, which inhibit protein biosynthesis.

The best studied of the *bacterial toxins* are the thermolabile *exotoxins* of Gram positive bacteria, which are secreted into the surrounding medium: 1. Five *enterotoxins* are secreted by *Staphylococcus aureus* in the gastrointestinal tract, causing diarrhea and vomiting. Enterotoxin B is of known primary structure (M_r 28370, 293 amino acid residues, one disulfide bridge). 2. The *Diphtheria toxin* from *Corynebacterium diphtheriae* is an acidic, single chain protein (M_r 62000), of high toxicity (1 μg/kg body weight is fatal). It inactivates peptidyl transferase II in eukaryotic cells, by promoting the attachment of ADP-ribose to the enzyme. 3. *Tetanus toxin* (M_r 150000) from *Clostridium tetani* exists in two forms, filtrate and cell toxin. The former has two subunits (M_r 95000 and 55000), and the latter has one. In the

mouse, 0.01 ng/kg is fatal. Tetanus toxin partially inhibits the binding of botulinum Type A neurotoxin to nerve endings. 4. The five highly toxic *Botulinus toxins* from *Clostridium botulinum* are SH proteins and require the presence of one free SH-group for neurotoxic activity. They are resistant to proteolytic digestive enzymes, but are destroyed by boiling. 0.03 ng/kg is fatal in the mouse. Type A (M_r 140000) binds specifically to the membranes of peripheral nerve terminals, where it irreversibly inhibits the release of acetylcholine. [J. O. Dolly et al., *Nature* **307** (1984) 457–460.]

The relatively heat-stable *endotoxins* are released by autolysis of the bacteria. *Cholera toxins* (M_r 84000–102000) are endotoxins released from the Gram-negative *Vibrio cholerae* in the intestine. They are composed of two functionally different subunits, L and H. The L subunit has a high affinity for gangliosides of the membranes of nerve cells, adipocytes, erythrocytes, etc., while the H subunit is responsible for toxicity. The *colicins* (M_r 60000) are endotoxins produced by intestinal bacteria. Their toxicity is due to inhibition of cell division and inhibition of DNA and RNA degradation (colicin E$_2$), or to inhibition of protein biosynthesis by inactivation of the 30S ribosomal subunit (colicin E$_3$).

Toxoflavin: 3,8-dimethyl-2,4-dihydroxypyrimido(5,4-e)-*as*-triazine, an antibiotic from *Pseudomonas coccovenans*, with high antibacterial activity, but no activity against fungi. M. p. 171 °C (d.). In the biosynthesis of T., C8 of a purine precursor is removed, and the *as*-triazine ring is formed by introduction of the aminomethyl group of glycine (Figure). Both methyl groups are introduced by transmethylation. Xanthotricin from *Streptomyces albus* is identical with T. It interferes in the transport of electrons in the cytochrome system.

Biosynthesis of toxoflavin by Pseudomonas coccovenans.

Toyocamycin: 4-amino-5-cyano-7-(D-ribofuranosyl)-pyrrolo-(2,3-d)-pyrimidine, 6-amino-7-cyano-9-β-D-ribofuranosyl-7-deazapurine, a 7-deazaadenine antibiotic from *Streptomyces toyocaensis* and *S. rimosus*. M. p. 243 °C. Biosynthesis is analogous to that of Tubericidin (see), i. e. the carbon atoms of the pyrrole ring are derived from 5-phosphoribosyl 1-pyrophosphate. T. is particularly active against *Candida albicans*, *Saccharomyces cerevisiae* and *Mycobacterium tuberculosis*.

TPN: abb. for triphosphopyridine nucleotide. See Nicotinamide-adenine-dinucleotide phosphate.

TPP: abb. for Thiamin pyrophosphate (see).

Trace elements, *microelements:* elements required in very small quantities by living organisms. They act catalytically, or are components of catalytic systems. A clear distinction between T.e. and other mineral nutrients is not always possible, e.g. in the case of iron. A further classification into T.e. and ultratrace elements is sometimes used.

Deficiency of T.e. can lead to characteristic deficiency symptoms or diseases, thus indicating the essential nature of these nutritional factors. For example, iodine is a component of the thyroid hormones and essential for thyroid function. Iodine deficiency is responsible for endemic goiter, and certain types of cretinism; it can be avoided by addition of iodides to the drinking water. Other T.e. are chromium, copper, fluoride, magnesium, manganese, nickel,

the synthesis and degradation of fatty acids, synthesis of conjugated bile acids via cholic acid-CoA compounds, and other reactions such as acetylation of amino acids and amines.

Transaldolase (EC 2.2.1.2): see Transaldolation.

Transaldolation: a reaction of carbohydrate metabolism, in which a C_3-unit (equivalent to a dihydroxyacetone unit) is transferred from a ketose to an aldose. T. is catalysed by transaldolase (EC 2.2.1.2). The C_3-unit does not exist in the free state, but remains bound to the ε-group of a lysine residue in the enzyme (Fig.). Only fructose 6-phosphate and sedoheptulose 7-phosphate are cleaved by transaldolase. Acceptors for the C_3-unit are the aldose phosphates, D-glyceraldehyde 3-phosphate, D-erythrose 4-phosphate and more rarely ribose 5-phosphate. There is no coenzyme and the mechanism of reaction is similar to that of aldolase (EC 4.1.2.13).

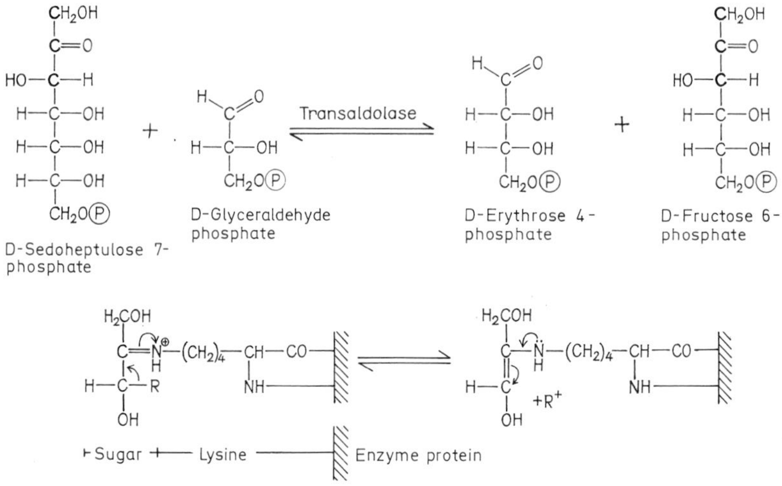

Transaldolase reaction (above), and binding of the ketose to the ε-amino group of a lysine residue of the enzyme (below).

vanadium, silicon, tin, selenium, zinc (see individual entries).

Trace element solution: see Nutrient medium (Table 3).

Trace nutrients, *micronutrients:* a general term for any essential dietary component required in small quantities, like Trace elements (see) and Vitamins (see). Deficiency of T.n. leads to deficiency symptoms, e.g. vitamin deficiency diseases. T.n. act catalytically or are precursors of catalytically active substances in the organism. Essential amino acids therefore have an equivocal status in this classification. Flavoring principles are definitely not T.n.

Tracer technique: see Isotope technique.

Transacylases: see Transacylation.

Transacylation: reversible transfer of acyl groups (R–CO–) from a donor to an acceptor, e.g. transfer of the acyl residue CH_3–CO– by acetyl-CoA to an acceptor Y:
CH_3–CO~S–CoA + Y → CH_3–CO–Y + CoA   T. is catalysed by transacylases, which are important in

Transamidase: see Transamidation.

Transamidation: transfer of the amide nitrogen of Glutamine (see) as an NH_2-group. T. is catalysed by transamidases. All glutamine transamidases so far investigated have a catalytically important thiol group in their active centers and are inhibited by the glutamine analogs, azaserine, 6-diazo-5-oxonorleucine (DON) and L-2-amino-4-oxo-2-chloropentanoic acid ("chloroketone"), e.g. anthranilate synthase (EC 4.1.3.27), carbamoyl phosphate synthetase (EC 6.3.5.5), transglutaminase (EC 2.3.2.13), 5'-phosphoribosyl-N-formylglycinamide synthetase (EC 6.3.5.3), glutamate synthase (EC 1.4.1.13).

Transamidinases, *amidinotransferases:* enzymes catalysing Transamidination (see). T. catalyse transfer of the amidine group of arginine in the synthesis of creatine and other Phosphagens (see). T. from *Streptomyces griseus* and *S. baikiniensis* catalyses amidine transfer in the biosynthesis of streptidine. T. are also involved in the synthesis of certain Guanidine derivatives (see). Transfer of the intact amidine group from L-arginine has been proved by double labelling with ^{14}C and ^{15}N. T. also has hydrolytic activity and is therefore a potential Arginase (see).

Transamidination: reversible enzymatic transfer of the amidine group,

NH
‖
-C-NH$_2$, between guanidines. T. is a group transfer reaction of nitrogen metabolism, which occurs in two stages and involves an intermediate enzyme-amidine complex:

$$R-NH-\overset{\overset{\text{NH}}{\|}}{C}-NH_2 + Enzyme-SH \rightleftharpoons R-NH_2 + Enzyme-S-\overset{\overset{\text{NH}}{\|}}{C}-NH_2$$

$$Enzyme-S-\overset{\overset{\text{NH}}{\|}}{C}-NH_2 + R-NH_2 \rightleftharpoons R-NH-\overset{\overset{\text{NH}}{\|}}{C}-NH_2 + Enzyme-SH$$

In the absence of a suitable acceptor, the enzyme-amidine complex is stable; on standing in aqueous solution or on heating it releases urea. T. is catalysed by Transamidinases (see). Formamidine disulfide, a SH-blocking agent, is a powerful inhibitor of T. The most important amidine donor in T. is L-arginine; biosynthesis of L-arginine is equivalent to the de novo synthesis of the amidine group. T. is important in the biosynthesis of Phosphagens (see).

Transaminases, *aminotransferases* (EC sub-subgroup 2.6.1): enzymes catalysing transamination, i.e. the reversible transfer of the amino group of a specific amino acid to a specific oxoacid, forming a new amino acid and a new oxoacid. Coenzyme of T.

is pyridoxal 5′-phosphate, which becomes bound to the apoenzyme by condensation of its carboxyl group with the ε-amino group of a lysine residue, forming a Schiff's base or internal aldimine. During transamination, however, the coenzyme reacts with the incoming amino acid, which displaces the lysine and forms an external aldimine or *primary Schiff's base* (Figure). Formation of a chelate ring by a bridge proton between the amino nitrogen and the phenolic oxygen of the coenzyme helps to maintain the conjugated system of the Schiff's base in a planar conformation. Rearrangement, with loss of the α-hydrogen as a proton, produces a quinonoid ketimine *(transitional Schiff's base),* containing a conjugated system extending from the carboxyl group to the ring nitrogen. At this stage, the catalytic lysine residue, or another basic group is thought to act as an electron sink. Further rearrangement produces a nonquinonoid ketimine *(secondary Schiff's base),* which is hydrolysed to the new oxoacid and pyridoxamine 5′-phosphate. This represents one half of the transamination process. Another oxoacid condenses with the pyridoxamine 5′-phosphate, and the sequence of reactions is reversed to form a new amino acid, thus completing the amino group transfer.

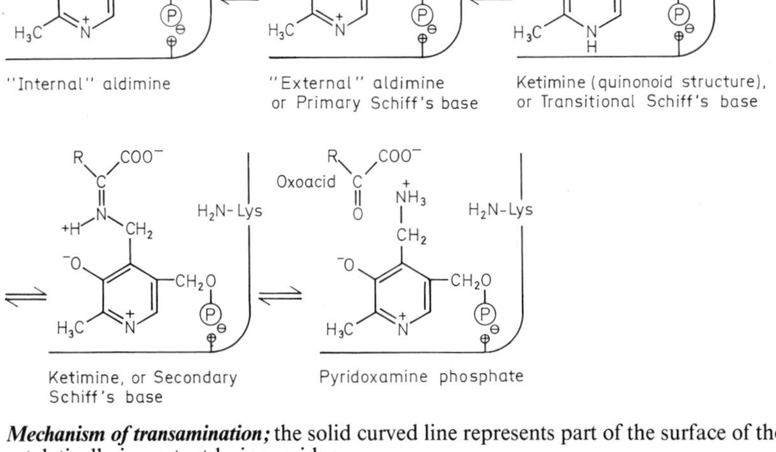

Mechanism of transamination; the solid curved line represents part of the surface of the apoenzyme, showing a catalytically important lysine residue.

Interconversion of the tautomeric Schiff's bases is the rate limiting step of transamination.

Practically all amino acids and oxoacids can take part in transamination. The specificity of most transaminases, however, demands that one of the reaction partners should be an acidic amino acid (i.e. glutamate or aspartate) or its corresponding oxoacid. It should be noted that transamination is freely reversible, anergonic process, i.e. no high energy compound (e.g. ATP) is produced, or required, and the direction of transamination depends entirely upon a mass action effect of its substrates. Thus, in the liver, when amino acids are in excess, transamination converts them to oxoacids (which enter carbohydrate metabolism) and the amino groups appear in glutamate and aspartate (and are subsequently incorporated into urea, see Urea cycle). In plants and bacteria, most pathways of amino acid synthesis involve elaboration of the oxoacid, which is finally transaminated (usually with glutamate) to the required amino acid. At no stage in transamination is free ammonia produced; the amino nitrogen is always covalently bound in an amino acid or in the pyridoxamine phosphate coenzyme.

Animal tissues, especially liver and heart muscle, contain very high activities of glutamate-oxaloacetate T. (GOT) (preferred name, aspartate aminotransferase, EC 2.6.1.1) and glutamate pyruvate T. (GPT) (preferred name, alanine aminotransferase, EC 2.6.1.2). GPT occurs in the liver as a cytosolic enzyme and shows only very low activity in heart muscle, whereas GOT is higher in heart than liver. GOT is about equally distributed between cytosol and mitochondria in both organs, M_r 90000 (2 identical subunits, M_r 45000). The primary sequence of pig heart cytoplasmic GOT is known (each subunit contains 412 amino acid residues). The serum activity of T. is very low, but increases markedly in certain illnesses associated with tissue damage. The value and ratio of GOT and GPT activities are used in the early diagnosis of, and for following the progress of treatment of, different liver diseases (greatly increased in acute liver inflammation, moderately increased in chronic cases, and hardly increased in obstructive jaundice) and heart muscle infarction. Both T. are determined by coupled optical tests. *GPT:* a serum sample is added to a buffered mixture of L-alanine, 2-oxoglutarate, lactate dehydrogenase (excess) and NADH. As pyruvate is formed it is reduced to lactate by the action of lactate dehydrogenase and NADH. The rate of formation of NAD^+, measured from the rate of decrease of absorption at 340 nm (or other suitable wavelength, e.g. 366 nm), is a measure of GPT activity. *GOT:* the reaction mixture contains L-aspartate, 2-oxoglutarate, malate dehydrogenase (excess) and NADH; the resulting oxaloacetate acts as the substrate of malate dehydrogenase, and the procedure is completely analogous to that described for GPT.

Transamination: reversible transfer of amino groups, between two amino acids and their respective keto acids, catalysed by transaminases. T. is fundamentally important in the amino acid metabolism of all living organisms. For detailed mechanism, see Transaminases.

Transcarbamylation: transfer of the carbamyl group of Carbamoyl phosphate (see).

Transcarboxylation: see Biotin enzymes.

Transcortin: see Cortisol.

Transcriptase: see RNA-polymerase.

Transcription: the DNA-dependent synthesis of RNA. See Ribonucleic acid.

Transdeamination: conversion of the amino group of an amino acid to ammonia by the combined action of a transaminase (TA) and L-glutamate dehydrogenase (GDH) (EC 1.4.1.2; 1.4.1.3):

Amino acid + 2-oxoglutarate $\overset{TA}{\rightleftharpoons}$ 2-oxoacid + Glutamate;

Glutamate + NAD^+ + H_2O $\overset{GDH}{\longrightarrow}$ 2-oxoglutarate + $NADH + H^+ + NH_4^+$.

T. is an important process in ureotelic organisms, where it accounts for most of the ammonium entering the Urea cycle (see) via carbamoyl phosphate (the other nitrogen atom incorporated into urea is derived directly from aspartate, formed by transamination of amino acids with oxaloacetate). Alternatively, the ammonia may be assimilated as the amido nitrogen of L-glutamine, then used in a variety of other processes (see Ammonia assimilation; Transamidination), depending on the biosynthetic capabilities of the organism in question.

Transduction: transfer of DNA from one bacterial cell to another by bacteriophage. There are two types. In *generalized T.* the phage infects the bacterial cell (the donor) and enters a nonlysogenic cycle leading to lysis of the cell and release of phage progeny (see Phage development). Most of the phage progeny are normal, but during the process of phage assembly within the infected bacterial cell, pieces of degraded bacterial DNA occasionally become falsely packaged into phage heads; in the case of *Escherichia coli* and phage P_1 this falsely packaged DNA cannot be larger than 3% of host genome, and it represents an entirely random sample of fragmented host DNA. The new phage population is then used to infect a second bacterial culture (recipient); the majority of phage particles, being normal, kill the bacterial cells that they infect. Under correctly chosen conditions (one phage particle per bacterial cell), most of the cells in the recipient culture are killed; the remaining viable cells are those that received falsely packed bacterial DNA instead of phage DNA during the infection process. This DNA is integrated into the recipient genome by genetic recombination, and is expressed by the recipient cell. In *specialized T.* the phage (e.g. λ) becomes integrated at a specific site in the DNA of the recipient, under lysogenic conditions. The lytic cycle is then initiated (by temperature change, UV-light, etc.), and the phage DNA of some phage progeny carries small fragments of bacterial DNA from the specific integration site. Following infection of the recipient under lysogenic conditions, phage DNA and any attached bacterial DNA become integrated into the recipient DNA. The transfered fragment of bacterial DNA does not represent a random sample, and can only be derived from donor DNA in the region of the specific integration site. Thus a λ-phage is known that integrates in the region of the histidine utilization genes *(hut)* of *Salmonella*. Similarly, studies on the organization of the tryptophan synthase operon were aided by transduction with a λ-phage that specifically integrates into *Escherichia coli* DNA in the region of the genes for tryptophan synthesis.

Transferases: see Enzymes, Table 1.

Transfer factors: see Elongation factors.

Transferrin: see Siderophilins.

Transfer-RNA, *tRNA, soluble RNA, sRNA, acceptor RNA, transport RNA:* The smallest known functional RNA, present in all living cells and essential for Protein biosynthesis (see). Different tRNAs contain between 70 and 85 nucleotide residues; average M_r is 25000. There is at least one specific tRNA per cell for each of the 20 protein amino acids. There may be between 50 and 70 tRNA species within one cell; this multiplicity is the result of organelle specificity, and the fact that there may be two or more different but specific tRNAs for one amino acid. The source and specificity of a tRNA species is indicated by a code, e.g. tRNA$^{Val}_{yeast}$ is the valine-specific tRNA from yeast. [^{14}C-Val]tRNA$^{Val}_{yeast}$ represents the named tRNA esterified with ^{14}C-labelled valine.

Function. tRNA is esterified with its specific amino acid by the action of Aminoacyl-tRNA synthetase (see). The resulting aminoacyl-tRNA becomes bound to the acceptor site of the 50S-subunit of a ribosome, where antiparallel base pairing occurs between the anticodon of the tRNA and the complementary codon of the associated mRNA. The specificity of this base pairing ensures that the amino acid is incorporated into the correct position in the growing polypeptide chain. During translation the deacylated tRNA is released from the ribosome and becomes available for recharging with its amino acid.

Structure. The primary structures of more than 50 different tRNAs from various organisms are known. The first tRNA structure was determined by Holley in 1965 for tRNA$^{Ala}_{yeast}$; indeed, this was the first reported primary sequence of any nucleic acid. Sequence determination is facilitated by the occurrence of unusual nucleic acid components, which act as markers in the identification of oligonucleotide fragments produced by nuclease degradation of the tRNA. Maximal base pairing of the primary structure gives rise to a secondary structure, known as the clover leaf model (Figure 1), which contains three main loops and four stems: 1. The anticodon loop contains the Anticodon (see), a sequence of 3 nucleotides specific for the relevant amino acid. Uracil is always found next to the 5′-end of the anticodon, and a purine or purine derivative is always

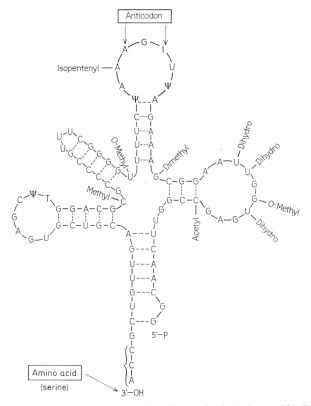

Figure 1. Clover leaf model of a tRNA molecule (serine-specific tRNA from yeast) A, adenine; C, cytosine; G, guanine; I, inosine; U, uracil; T, thymine; Ψ, pseudouracil.

next to the 3'-end. In tRNA from plants, this purine derivative is often N⁶-(γ,γ-dimethylallyl)-adenine, also known as triacanthine. 2. The dihydrouracil (di-HU) loop always contains dihydrouracil. 3. The thymine-pseudouracil (TΨC) loop, characterized by the sequence 5'-GTΨC-3' appears to be involved in the binding of aminoacyl-tRNA to the 50S ribosome. A fourth loop may be present, and it may even have an associated stem with base pairing, but it is not an invariable feature of every tRNA.

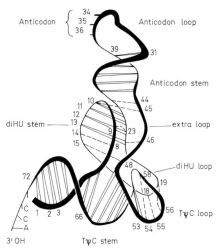

Figure 2. Three dimensional structure of yeast tRNA^Phe as determined by X-ray crystallography. Double solid lines represent hydrogen bonds between bases in double helical stems. Dotted lines represent hydrogen bonding between bases outside the helices, i.e. 8.....14, 9.....23, 10.....45, 15.....48, 18....55, 19.....56, 22.....46, 26.....44, and 54.....58.

The 3'-terminus of all tRNAs is 3'-ACC. The 3'-terminal adenine is always separated from the first nucleoside in the TΨC loop (ribothymidine or uridine) by 21 nucleosides. The 5'-terminus is always phosphorylated. The aminoacyl-tRNA species that binds to the ribosome, and is active in protein biosynthesis, carries the amino acid esterified at position 3' of the 3'-terminal adenosine. The free aminoacyl-tRNA, however, represents a tautomeric mixture of 2'- and 3'-aminoacyl-tRNA. Moreover, the initial product of aminoacyl-tRNA biosynthesis is either the 2'- or the 3'-aminoacyl derivative, depending on the specificity of the Aminoacyl-tRNA synthetase (see). The highly specific recognition of a tRNA by the corresponding aminoacyl-tRNA synthetase (amino acid activating enzyme) apparently depends upon the three-dimensional structure of each tRNA. X-ray analysis of crystalline tRNA confirms the existence of four regions of base pairing and three loops, and shows how these are arranged in a three-dimensional structure (Figure 2). The TΨC and di-HU loops and stems are folded back, so that the molecule has a compact shape, 9.0 nm long and 22.5 nm wide. The anticodon loop and the 3'-ACC

end are still widely separated as in the clover leaf model. In addition to hydrogen bonding between bases in the helical stems, the structure is stabilized by an extensive network of hydrogen bonds involving specific interactions between bases and the ribose phosphate backbone, and by some hydrogen bonding between base pairs outside the helices.

Synthesis and processing. A precursor molecule of tRNA is synthesized by transcription from DNA, then processed to the functional tRNA. For example, the precursor of tRNA^Tyr_coli consists of 126 nucleotides and is 41 nucleotides longer than the mature, active molecule. The extra sequence is removed by a specific endonuclease. Certain nucleotide residues also undergo further post-transcriptional modifications, resulting in several Rare nucleic acid components (see), which are partly responsible for the specific spatial structure of each tRNA.

Transformation: conceptually the simplest form of genetic transfer. "Naked" DNA from a donor cell enters a recipient cell and is incorporated into the recipient DNA by genetic recombination. There is no other carrier substance or structure involved; small fragments of the donor DNA simply penetrate the membrane (and wall if it is present) of the recipient cell. T. was first described in 1944 by Avery (USA) for the transformation of R (rough)-type nonpathogenic *Pneumococci* into S (smooth)-type pathogenic *Pneumococci* by treatment with killed S-type cells. The "transforming principle" was eventually shown to be DNA; this work gave the first proof that the genetic material of the cell is DNA. Frequency of T. may be low (<1%) owing to rapid degradation of donor DNA before genetic recombination can occur.

Transglycosidation: transfer of a glycosidically bound sugar residue to another molecule with a suitable recipient OH-group. T. is catalysed by transglycosidases, e.g. galactosidase catalyses transfer of a galactose residue from lactose to the C6-hydroxyl group of glucose, or to a further lactose molecule to form the trisaccharide, 6-galactosidolactose. T. is sometimes involved in the synthesis of oligo- and polysaccharides.

Transhydrogenase: see Hydrogen metabolism.

Transhydrogenation: see Hydrogen metabolism.

Transit peptides: amino-terminal extensions of precursors to chloroplast proteins which are encoded by nuclear DNA and synthesized in the cytoplasm. The T.p. are removed by Post-translational modification (see) before the protein takes on its mature configuration inside the chloroplast. The overall amino acid sequences of the T.p. are not conserved between species, but the positions of proline and the charged amino acids are highly conserved. The T.p. appear to mediate the transport of the chloroplast protein precursors into the organelles. [G. Van den Broeck et al. *Nature* **313** (1985) 358–363.]

Transition: replacement of a purine by a different purine, or a pyrimidine by a different pyrimidine in the polynucleotide chain of DNA. T. results in a gene mutation. It may occur spontaneously, or it may be promoted experimentally with mutagens.

Transketolase (EC 2.2.1.1): an enzyme that catalyses transketolation, an important process of carbohydrate metabolism, especially in the Pentose phosphate cycle (see) and Calvin cycle (see). T. has been found in a wide variety of cells and tissues, including mammalian liver, green plants and many bacterial species. The enzyme contains divalent metal cations and the coenzyme, thiamin pyrophosphate. Transketolation involves transfer of a C_2-unit (often called active glycolaldehyde or a ketol moiety) from a ketose to C1 of an aldose. Only ketoses with the L-configuration at C3 and preferably a

trans configuration on the next carbon (i.e. C1, 2, 3 and preferably 4 as in fructose) can serve as donors of the C_2-unit. The acceptor is always an aldose. Transketolation is reversible; a list of known reactions is shown in the Table. Details of the reaction in which xylulose 5-phosphate serves as the donor of the C_2-unit, and ribose 5-phosphate as the acceptor are shown in the Figure: The C_2-unit becomes bound to the thiamin pyrophosphate as 2-(α,β-dihydroxyethyl)-thiamin pyrophosphate, and the remainder of the molecule is released as glyceraldehyde

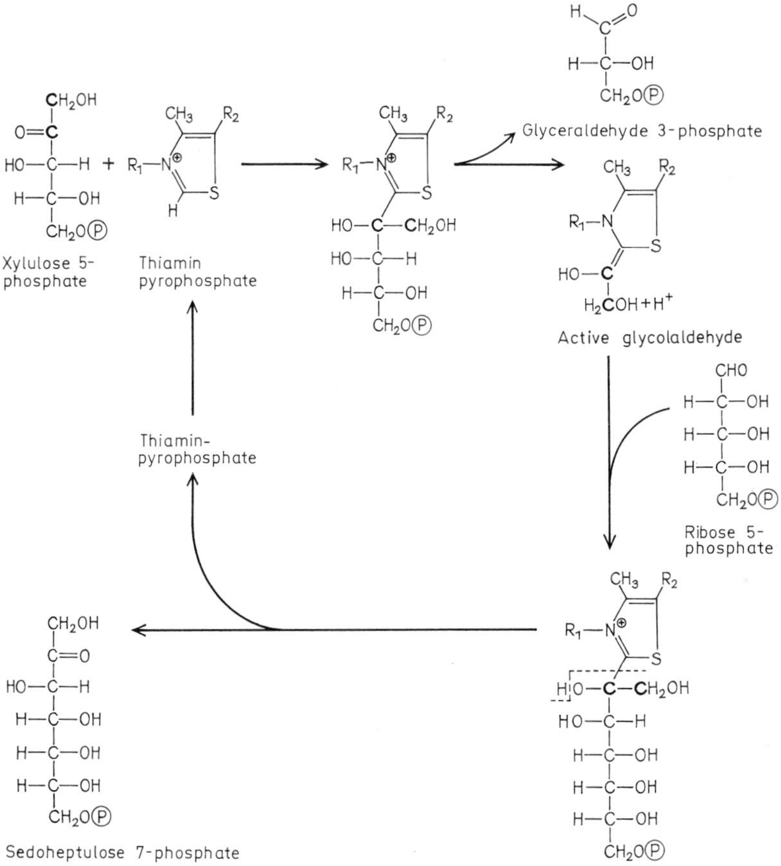

Mechanism of action of transketolase (only the thiazole group of thiamin pyrophosphate is shown).

Table. Transketolase reactions (P = phosphate)

Donor (R-CHOH-CO-CH$_2$OH)	+ Acceptor (R$_1$- CHO)	⇌ Donor (R$_1$- CHOH-CO-CH$_2$OH)	+ Acceptor (R-CHO)
L-Erythrulose	+ Glycolaldehyde	⇌ L-Erythrulose	+ Glycolaldehyde
D-Xylulose 5-P	+ D-Glyceraldehyde 3-P	⇌ D-Xylulose 5-P	+ D-Glyceraldehyde 3-P
D-Fructose 6-P	+ D-Erythrose 4-P	⇌ D-Fructose 6-P	+ D-Erythrose 4-P
D-Sedoheptulose 7-P	+ D-Ribose 5-P	⇌ D-Sedoheptulose 7-P	+ D-Xylulose 5-P
D-Fructose 6-P	+ D-Glyceraldehyde 3-P	⇌ D-Xylulose 5-P	+ D-Erythrose 4-P
D-Sedoheptulose 7-P	+ D-Glyceraldehyde 3-P	⇌ D-Xylulose 5-P	+ D-Ribose 5-P

3-phosphate; transfer of the C_2-unit to ribose 5-phosphate produces sedoheptulose 7-phosphate. If the ribose 5-phosphate were replaced by erythrose 4-phosphate, the products would be glyceraldehyde 3-phosphate and fructose 6-phosphate.

Transketolation: see Transketolase.

Translation: in the wider sense equivalent to Protein biosynthesis (see). In the narrower sense, T. is the decoding process whereby each codon (see Genetic code) in mRNA is translated into one of 20 amino acids during protein synthesis on polysomes.

Translocation: see Transport.

Transmethylases: see Transmethylation.

Transmethylation: transfer of a methyl group ($-CH_3$) from a physiological methyl donor to C-, O- and N-atoms of biomolecules. T. to oxygen produces the methoxy group ($-OCH_3$). The most important methyl donor is S-Adenosyl-L-methionine (see). Thus the methyl groups in a wide variety of methylated natural products originate from the methyl group of methionine. The thioether group of methionine itself does not participate directly in T.; it must first be activated to a sulfonium group by S-adenosylation, i.e. by synthesis of S-adenosyl-L-methionine. Betaines (see) and Thetins (see) have limited physiological significance as methyl donors. Methylated nucleic acid components are formed by methylation of the polynucleotide chains of nucleic acids by S-adenosyl-L-methionine, catalysed by specific transmethylases (methyltransferases). Exceptions are thymine and 5-hydroxymethylcytosine, where the methyl (or hydroxymethyl) group is derived from $N^{5,10}$-methylenetetrahydrofolic acid (see Active one-carbon units).

Transphosphatases: see Kinases.

Transplantation antigens, *histocompatibility antigens*: antigens on the surface of nucleated cells, particularly leukocytes and to a lesser extent thrombocytes. A lack of correspondence between T.a. of donor and recipient leads to transplant rejection. To ensure optimal compatibility between donor and recipient, the T.a. of both are typed by the histocompatibility test.

Transport: passage of ions and certain molecules through biological membranes. Most polar molecules do not pass freely across biomembranes. Exchange of essential metabolites between a cell and its surroundings, or between cytoplasm and organelles, therefore depends on T. mechanisms within the membranes. All T. mechanisms catalysed by biomembranes have three characteristic properties: saturation, substrate specificity and specific inhibition. T. may be active or passive.

Passive T. can only operate in the presence of an appropriate concentration gradient, which acts as the driving force in transporting the material through pores in the membrane. The commonest form of passive T. involves a carrier mechanism. According to this model, the process is capable of on-off regulation, and the carrier is a specific protein which attaches to the substance to be transported on one side of the membrane, takes it across the membrane, releases it on the other side, then returns to the starting position. Depending on the concentration gradient, passive T. is reversible and can occur in either direction. Important passive T. systems in animal tissues are, e.g. the glucose carrier in human erythrocyte membranes, and the ATP-ADP carrier of the mitochondrial membrane, which normally transports one molecule of ADP into the mitochondrial matrix, and one molecule of ATP (formed by oxidative phosphorylation) from the matrix to the cytoplasm.

In *active T.* material is transported against a concentration gradient, and the process is linked to the cleavage of ATP. In recognition of the involvement of ATP, active T. systems ("pumps") are also called ATPases. In the model for active T., the substance in question becomes attached to a complementary binding site on the protein carrier, then transported to the other side of the membrane by diffusion, rotation, or a change of conformation. The free energy of ATP cleavage is required for release of the substrate from the carrier; this probably involves a conformational change in the carrier protein, which alters the binding affinity for the substrate. E.g. in animal tissues: the Na^+ and K^+ pumps, and the active transport mechanisms for glucose and other sugars, and for amino acids. The so-called Na^+ K^+-ATPase also appears to be important for the T. of glucose and amino acids. Furthermore, ATPases play an important role in nerve transmission, muscle control and sensory perception. The Na^+ and K^+ transporting ATPase system (Na^+K^+ pump) transports Na^+ ions out of the cell against the electrochemical gradient (intracellular Na^+ concentration < 10 mM; extracellular conc. about 150 mM) and K^+ ions from the surrounding milieu into the cell (extracellular K^+ concentration < 4 mM; intracellular conc. 120–160 mM). The high internal K^+ concentration is biochemically important, being essential for protein biosynthesis and for the maximal activity of various enzymes, etc. The Na^+/K^+ gradient across the cell membrane is necessary for the excitation response of muscle cells and for signal transmission by nerve cells. A two-stage process is postulated for the transport mechanism:

$$ATP + Na^+ \text{internal} + ⒠ \xrightleftharpoons{Mg^{2+}} Na\text{-}Ⓔ \sim P + ADP$$
$$Na\text{-}Ⓔ \sim P \rightleftharpoons Na\text{-}\triangle \sim P$$
$$K^+ \text{external} + H_2O + Na\text{-}\triangle \sim P \rightleftharpoons Ⓔ + P_i$$
$$+ Na^+ \text{external} + K^+ \text{internal}$$

Two different conformations (O and △) are postulated for the ATPase. The intermediate phosphorylated form of the enzyme contains an acyl phosphate group bound to an aspartic acid residue at the active center of the ATPase. Na^+ K^+-ATPase has M_r 250000–300000 and consists of two different subunits. Kidney and brain cells use about 70% of their synthesized ATP for the exchange transport of Na^+ and K^+ catalysed by Na^+K^+-ATPase.

A mechanism in which glucose transport is coupled with a Na^+ gradient is postulated for the active transport of glucose, against a concentration gradient, from the small intestine in the blood stream, and from the glomerular filtrate through the epithelial cell layer of the kidney tubuli into the blood. Na^+ ions are pumped out of the cell by Na^+K^+-ATPase, thus forming a higher Na^+ concentration outside than inside cell. Glucose and Na^+ are then transported into the cell by a passive carrier that has binding sites for both glucose and Na^+ (cotransport or symport).

Active amino acid T. has certain similarities with the active T. of glucose, especially from the intestine into the blood. More than five different T. systems are known. In some cells amino acid T. also appears to be coupled with a Na^+ gradient.

The ATP-dependent, intracellular T. of Ca^{2+} from the sarcoplasm into the sarcoplasmic reticulum is an essential process in the initiation of muscle relaxation. This process depends on the activity of a Ca^{2+}-ATPase system (Ca^{2+} pump). Mitochondria of animal cells are also able to accumulate Ca^{2+} against a steep gradient.

Specific proteins in the periplasmic space of bacteria can bind sugars, amino acids and inorganic ions, and therefore play an important part as carriers associated with T. systems. Especially aerobic bacteria possess true active T. systems.

Group translocation is a special type of membrane T. A membrane-bound enzyme catalyses a reaction between substrates on opposite sides of the membrane, and the product accumulates on one side of the membrane. E.g. various bacteria take up glucose by phosphorylating it to glucose 6-phosphate. Group translocation is also involved in the T. of amino acids. In contrast to active T., group translocation involves modification of the transported material.

Transport antibiotics: see Facilitated diffusion.

Transport RNA: see Transfer RNA.

Transposable elements: mobile genetic elements found in several types of eukaryotic organism. [H. Eibel & P. Philippsen, "Preferential integration of yeast transposable element Ty into a promoter region" *Nature* **307** (1984) 386–388].

Transsulfuration: exchange of sulfur between L-homocysteine and L-cysteine, with L-cystathionine as an intermediate (Figure). Strictly speaking, T. is not a group transfer reaction, because the sulfur bond formed in the synthesis of cystathionine is different from that broken in the formation of L-cysteine. T. operates in the biosynthesis of L-cysteine from L-methionine, and in the biosynthesis of L-methionine. The methionine precursor, L-homocysteine, is formed by T. as follows:

L-Homoserine + Succinyl-CoA → CoA + O-Succinyl-L-homoserine

O-Succinyl-L-homoserine + L-Cysteine → L-Cystathionine + Succinate

L-Cystathionine + H_2O → L-Homocysteine + Pyruvate + NH_3

In some organisms L-homoserine is initially acylated to O-acetyl-L-homoserine (by acetyl-CoA), and in plants it is converted to oxalyl-L-homoserine by reaction with oxalyl-CoA.

Transvaalin: see Scillaren A.

Transversion: replacement of a pyrimidine by a purine, or a purine by a pyrimidine in the polynucleotide chain of DNA. T. results in a gene mutation. It may occur spontaneously, or it may be promoted experimentally with mutagens.

Trehalose: a nonreducing disaccharide, consisting of two glucopyranoside residues. M_r 342.30. There are three forms of T., depending on the nature of the glycosidic linkage: α, α-T. (m. p. 204 °C, $[\alpha]_D^{20} + 197$ °); α, β-T. or neotrehalose ($[\alpha]_D^{20} + 95$ °); β, β-T. or isotrehalose (m. p. 135 °C, $[\alpha]_D^{20} - 42$ °). T. is present in algae, bacteria, numerous lower and higher fungi, and occurs sporadically in nonphotosynthetic tissues of higher plants. It is the "blood sugar" of insects. T. is cleaved by many fungal enzymes, and it is fermented by certain yeasts.

α, α-Trehalose

TRF: abb. for thyrotropin releasing factor (see Releasing hormones).

TRH: abb. for thyrotropin releasing hormone (see Releasing hormones).

Triacylglycerol: see Acylglycerols, Fats.

Triamcinolone: 16α-hydroxy-9α-fluoroprednisolone; 9α-fluoro-11β,16α,17α,21-tetrahydroxypregna-1,4-diene-3,20-dione, a synthetic steroid prepared from cortisol. The antiinflammatory activity of T. is 50 times greater than that of cortisone acetate (see Cortisone), and it does not cause undesirable salt retention. It is used in the treatment of arthritis, allergies etc.

Tricarboxylic acid cycle, *TCA-cycle, citric acid cycle, Krebs cycle:* a fundamentally important cycle of reactions in the terminal oxidation of proteins, fats and carbohydrates (Figure 1). CO_2 is formed by the decarboxylation of oxo-acids (oxalosuccinate, not shown in Figure 1, is an intermediate between isocitrate and 2-oxoglutarate; see Isocitrate dehydrogenase). Together with the Respiratory chain (see), operation of the TCA-cycle leads to the synthesis of the energy-rich compound ATP. In addition to energy production, the TCA-cycle provides intermediates for biosynthesis. Several important groups of substances are derived from intermediates of the TCA-cycle, and various other metabolic cycles are linked with the TCA-cycle. In eukaryotes, the TCA-cycle operates in the mitochondria, where it is structurally and functionally integrated with the respiratory chain, and with the degradation of fatty acids. In prokaryotes, the enzymes of the TCA-cycle are localized in the cell cytoplasm.

The TCA-cycle is responsible for the oxidation and cleavage of the acetyl group of acetyl-CoA, with the

Formation of cysteine from homocysteine (derived from methionine)

Tricarboxylic acid cycle

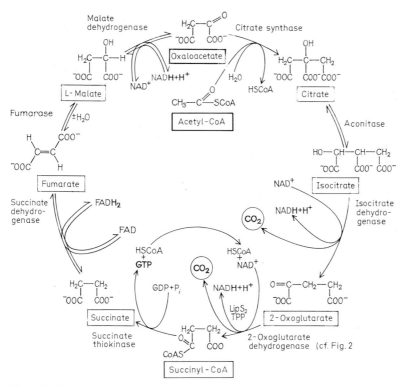

Figure 1. *The tricarboxylic acid cycle (TCA-cycle).*

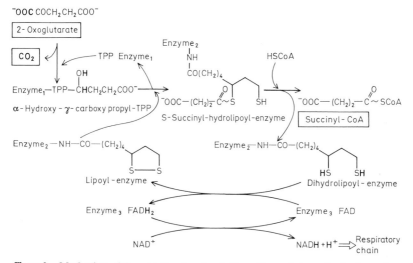

Figure 2. *Mechanism of the oxidative decarboxylation of 2-oxoglutarate by the 2-oxoglutarate dehydrogenase complex (EC 1.2.4.2).*
Enzyme$_1$ = 2-oxoglutarate decarboxylase
Enzyme$_2$ = lipoyl-reductase-transacetylase (lipoyl reductase + transsuccinylase)
Enzyme$_3$ = dihydrolipoyl dehydrogenase
TPP = thiamin pyrophosphate
HSCoA = coenzyme A

Table 1. *The reactions of the tricarboxylic acid cycle.*

Reaction number	Equation	Name of enzyme (see also under separate entries)	Inhibitors	$\Delta G^{\circ\prime}$ in kJ/mol (kcal/mol)
1	Acetyl-CoA + oxaloacetate + $H_2O \rightarrow$ citrate + HSCoA + H^+	Citrate *(si)*-synthase (EC 4.1.3.7)	none	-38.04 (-9.08)
2a	Citrate $\xrightarrow{Fe^{2+}, GSH}$ isocitrate	Aconitate hydratase (EC 4.2.1.3)	Fluorocitrate*, trans-Aconitate*	$+6.66$ ($+1.59$)
2b	Citrate $\xrightarrow{Fe^{2+}, GSH}$ cis-aconitate	Aconitase		$+8.54$ ($+2.04$)
2c	cis-Aconitate $\xrightarrow{Fe^{2+}, GSH}$ isocitrate	Aconitase		-1.89 (-0.45)
3	Isocitrate + NAD^+ $\xrightarrow{Mg^{2+} (Mn^{2+}), ADP}$ 2-oxoglutarate + NADH + $H^+ + CO_2$	Isocitrate dehydrogenase (EC 1.1.1.41)	ATP	-7.12 (-1.70)
4	2-Oxoglutarate + HSCoA + NAD $\xrightarrow{Mg^{2+}, TPP, LipS_2}$ succinyl-CoA + CO_2 + NADH + H^+	2-Oxoglutarate dehydrogenase complex (EC 1.2.4.2)	Arsenite, Parapyruvate*	-36.95 (-8.82)
5	Succinyl-CoA + GDP + P_i $\xrightarrow{Mg^{2+}}$ succinate + GTP + HSCoA	Succinyl-CoA synthetase (EC 6.2.1.4)	Hydroxylamine	-8.85 (-2.12)
6	Succinate + FAD $\xrightarrow{Fe^{2+}}$ fumarate + $FADH_2$	Succinate dehydrogenase (EC 1.3.99.1)	Malonate*, Oxaloacetate*	~ 0
7	Fumarate + $H_2O \rightarrow$ L-malate	Fumarate hydratase (EC 4.2.1.2)	meso-Tartrate*	-3.68 (-0.88)
8	L-Malate + $NAD^+ \rightarrow$ oxaloacetate + NADH + H^+	Malate dehydrogenase (EC 1.1.1.37)	Oxaloacetate*, Fluoromalate*	$+28.02$ ($+6.69$)

Sum of equations 1–8, i.e. balance of the TCA-cycle without the respiratory chain: $\quad -60.00$

Acetyl-CoA + $3NAD^+$ + FAD + GDP + P_i + $2H_2O$ $\xrightarrow{TCA\text{-}cycle}$ $2CO_2$ + HSCoA + 3NADH + H^+ + $FADH_2$ + GTP $\qquad (-14.32)$

Abbr.: HSCoA = Coenzyme A; GSH = Glutathione; AM(D)(T)P = Adenosine mono(di) (tri)phosphate; TPP = Thiamin pyrophosphate; $LipS_2$ = Lipoic acid amide; GD(T)P = Guanosine di(tri)phosphate; P_i = Inorganic phosphate; $FAD(H_2)$ = Enzyme-bound oxidized (reduced) Flavin-adenine-dinucleotide; NAD^+ (H) = Oxidized (reduced) Nicotinamide-adenine-dinucleotide. Compounds with* are competitive inhibitors.

formation of two molecules of CO_2. During one operation of the TCA-cycle, there are four dehydrogenations (each equivalent to 2 hydrogen atoms) in which the hydrogen is transferred to NAD^+ or FAD. The reduced coenzymes are then reoxidized by the respiratory chain, which catalyses the oxidation of the hydrogen to water.

Oxidation in the TCA-cycle is achieved by the addition of water and subsequent removal of hydrogen; there is no direct involvement of oxygen: $CH_3CO \sim SCoA + 3H_2O \rightarrow 2CO_2 + 8[H] + HSCoA$. The initiating reaction of the TCA-cycle is the condensation of acetyl-CoA with oxaloacetate, catalysed by citrate synthase. One molecule of water is consumed, and the products are citrate and coenzyme A. Citrate is then converted into oxaloacetate by 7 consecutive, enzyme-catalysed steps (Table 1). Reactions 3 and 4 involve decarboxylation. Figure 2 shows the mechanism of action of the 2-oxoglutarate dehydrogenase complex.

Energy balance of the TCA-cycle. A total of about 900 kJ (215 kcal) free chemical energy is available from the oxidation of acetyl-CoA by the TCA-cycle, with the involvement of the respiratory chain. Operation of the respiratory chain alone accounts for -810 kJ (-193.4 kcal) of this free energy, i.e. three operations of the chain from NADH (NADH + 0.5 O_2 + $H^+ \rightarrow NAD^+ + H_2O$, $\Delta G^{\circ\prime} = -219.4$ kJ [-52.4 kcal]), and one operation of the chain from $FADH_2$ ($FADH_2$ + 0.5 $O_2 \rightarrow$ FAD + H_2O, $\Delta G^{\circ\prime} = -151.6$ kJ [36.2 kcal]). A portion of this energy is used in the synthesis of 12 molecules of ATP, representing about 40% of the total free energy: Reactions 3, 4 and 8 give 3×3 ATP by the oxidation of $3 \times$ NADH in the respiratory chain; Reaction 6 gives 2 ATP by $FADH_2$ oxidation in the respiratory

Tricarboxylic acid cycle

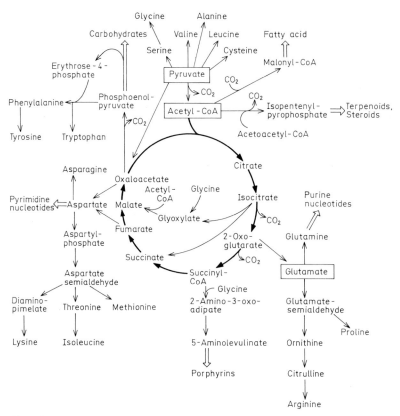

Figure 3. The biosynthetic functions of the tricarboxylic acid cycle.

chain; and the GTP produced in reaction 5 (see substrate level phosphorylation) is energetically equivalent to ATP, i.e. $GTP + ADP \rightleftharpoons GDP + ATP$ has an equilibrium constant of 1.0. The overall equation, including the respiratory chain and oxidative phosphorylation is: $CH_3 \cdot CO \sim SCoA + GDP + 11ADP + 12P_i + 2O_2 \rightarrow 2CO_2 + 13H_2O + GTP + 11ATP + HSCoA$.

The TCA-cycle is linked to Gluconeogenesis (see) by the conversion of oxaloacetate to phospho*enol*pyruvate. It is also the source of intermediates in the synthesis of many amino acids, especially aspartic and glutamic acid and other amino acids derived from them. Succinyl-CoA is a precursor of the porphyrins, e.g. heme, chlorophyll, vitamin B_{12}. The function of the TCA-cycle may be modified by integration with other pathways, e.g. γ-Aminobutyrate pathway (see), Glyoxylate cycle (see), and the Succinate-glycine cycle (see). Carboxylation of pyruvate (see Pyruvate carboxylase; see Carboxylation) serves as one stage in gluconeogenesis from pyruvate; it is also an anaplerotic reaction (see Metabolic cycle) of the TCA-cycle, i.e. it maintains the level of oxaloacetate, which would otherwise be depleted by removal of intermediates of the TCA-cycle for biosynthesis. In animals the net synthesis of carbohydrate from acetyl-CoA (and therefore from fatty acids) is

not possible. In plants and bacteria, however, the presence of the glyoxylate cycle permits the incorporation of a second acetyl group from acetyl-CoA, resulting in the net synthesis of TCA-cycle intermediates (and therefore of carbohydrate) from two-carbon units. This is important in the utilization of oil reserves in seeds for the synthesis of carbohydrate (e.g. cellulose of cell walls) during germination, and in the growth of bacteria at the expense of simple carbon sources, such as acetate.

Reactions analogous to those of the TCA-cycle are found in the biosynthesis of leucine and lysine (Figure 4). Similarly, the sequence of reactions represented by the dehydrogenation of succinate by a flavoenzyme, followed by hydration of fumarate to malate, then dehydrogenation of malate by a NAD-linked dehydrogenase, finds a counterpart in the initial stages of fatty acid degradation (see).

Regulation of the TCA-cycle. ADP/ATP ratios and $NAD/NADH + H^+$ ratios have a regulatory influence on the TCA-cycle, particularly via the regulation of the allosteric enzyme isocitrate dehydrogenase. The enzyme is activated by ADP, and inhibited by ATP and NADH (Table 2). Other sites of regulation are the syntheses of acetyl-CoA, oxaloacetate and citrate. Oxaloacetate is the catalytic intermediate of the TCA-cycle for the oxidation of

Table 2. Sites of regulation of the TCA-cycle.

Reaction number	Name of enzyme	Location	Require-ment for	Release of	Activated by	Inhibited by	Remarks
1	Citrate synthase	Mito-chondria	Acetyl-CoA, Oxaloace-tate	Citrate HSCoA		Long-chain acyl-CoA	Control point for uti-lization of acetyl-CoA
3a	NAD-de-pendent isocitrate dehy-drogenase	Mito-chondria	NAD^+	NADH, CO_2	ADP	ATP, NADH	For high rates of TCA-cycle see reac-tion 3b.
3b	NADP-dependent isocitrate dehydro-genase	Cyto-plasm and Mito-chondria	$NADP^+$	NADPH, CO_2	Oxalo-acetate?		The extramitochon-drial enzyme is im-portant in the produc-tion of NADPH
9	Glutamate dehydro-genase	Mito-chondria	NADPH or NADH, NH_3	$NADP^+$ or NAD^+	ADP	GDP^+ NADH	
10	Pyruvate carboxy-lase	Cyto-plasm	ATP, CO_2	ADP	Acetyl-CoA		Control of carbohydrate metabolism
11	Acetyl-CoA carboxy-lase	Cyto-plasm	ATP, CO_2	ADP	Citrate	Long-chain acyl-CoA	Control of fat synthesis
12	Citrate lyase	Cyto-plasm	Citrate	Acetyl-CoA, Oxaloace-tate			Important for the ex-tramitochondrial syn-thesis of Acetyl-CoA
13	Isocitrate lyase	Cyto-plasm	Isocitrate	Glyoxy-late, Dicar-boxylic ac-ids		Phospho-*enol* pyruvate	Only found in bacte-ria and plants

acetyl-CoA to $CO_2 + H_2O$; it also inhibits succinate dehydrogenase and malate dehydrogenase. Since the TCA-cycle only works in conjunction with the respiratory chain, its activity is also regulated by the supply of oxygen. When an aerobically respiring cell is made anaerobic, the TCA-cycle comes to a halt with the accumulation of the reduced coenzymes NADH and FADH₂.

Asymmetrical metabolism of citrate. Although the ci-trate molecule has perfect bilateral symmetry, it is degraded asymmetrically (see prochirality). Accord-ing to the stereochemical numbering proposed by Hirschmann, C-1 is at the end of the chain occu-pying the *pro-S* position. Citrate synthesized in the TCA-cycle from oxaloacetate and [1-¹⁴C]acetyl-CoA contains ¹⁴C in position 1, and is referred to as *sn*-[1-¹⁴C]citrate. Aconitase catalyses the removal of the OH-group from C-3 and the H$_R$ proton from C-4. After rehydration of the *cis*-aconitate (see Aconitate hydratase; see Aconitic acid), the resulting isocitrate carries its OH-group on the carbon originating from C-4 of citrate. The CO_2 from the conversion of isoci-trate into 2-oxoglutarate is therefore derived from

the original oxaloacetate and not from the acetyl group of the acetyl-CoA. Subsequent decarboxyla-tion of succinyl-CoA removes yet another carbon of the original oxaloacetate. Thus, the new "catalytic" molecule of oxaloacetate, formed after one round of the TCA-cycle, contains only two carbons of the original oxaloacetate plus both carbon atoms of the original acetyl group. Since fumarate and succinate are metabolized symmetrically, the original acetyl C-1 atom becomes equally distributed between C-1 and C-4, and the original acetyl C-2 between C-2 and C-3 of the new oxaloacetate. In the next round of the cycle, all of the original acetyl C-1 is removed as CO_2. The C-2 of the original acetyl group be-comes distributed amongst all four carbons of the new oxaloacetate, so that it theoretically can never be entirely removed by decarboxylation in the TCA-cycle: it does not contribute to CO_2 until the third round of the cycle. Patterns of labelling from the in-corporation of ¹⁴C-labelled acetyl-CoA are therefore complex, but the fundamental experimental obser-vation is that some ¹⁴C from labelled acetyl-CoA is retained by intermediates of the TCA-cycle.

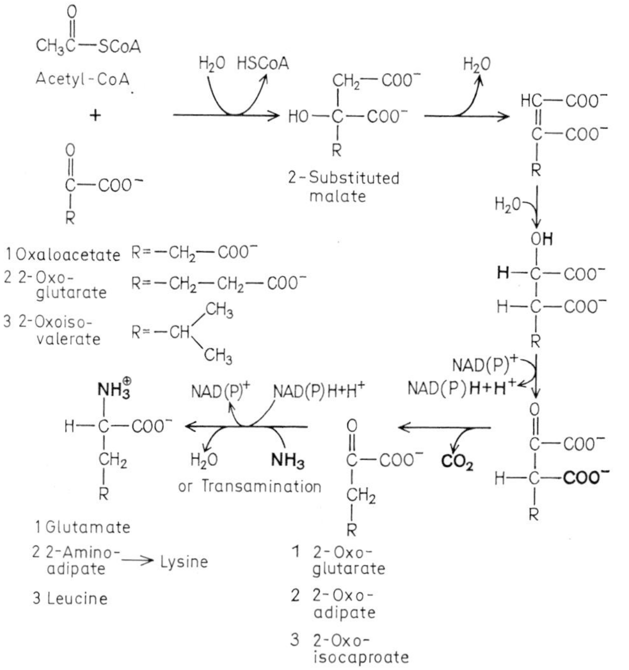

Tricetinidin

Figure 4. Some metabolic reaction sequences, which are analogous to part of the TCA-cycle.

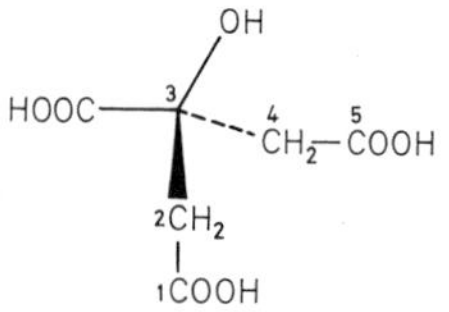

Figure 5. Formula of citrate with prochiral numbering of the carbon atoms.

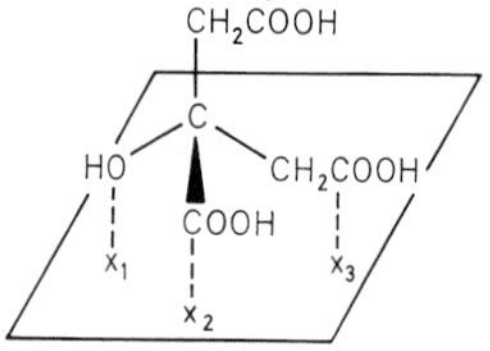

Figure 6. Diagrammatic representation of the three point attachment theory of Ogston, where X represents a side of attachment of citrate to the active center of aconitase.

Historical. The TCA-cycle was discovered almost simultaneously in 1937 by Krebs, and by Martius and Knoop. Green coined the term "cyclophorase" for the total multienzyme complex of the TCA-cycle and the associated respiratory chain. Early studies with tissue suspensions showed that ^{14}C from C-1 of acetate was found only in the 4-carboxyl carbon of 2-oxoglutarate, whereas the symmetrical metabolism of citrate would demand the presence of label in both the 2- and 4-carboxyl groups. It was therefore suggested that citrate could not be an intermediate in the TCA-cycle. In 1948, however, Ogston proposed his three-point attachment theory, in which the active sites of citrate synthase and aconitase are asymmetrical: three functional groups of citrate must become attached to three complementary, asymmetrically arranged binding sites. With the recognition of prochirality, it is now no longer necessary to propose that the site of asymmetry lies within the enzyme. Ogston's theory is, however, not disproved, and the precise nature of citrate binding to aconitase will not be known until the active center of the enzyme has been mapped.

Tricetinidin: 3-deoxydelphinidin, see Anthocyanins.

Trichochromes: yellow-orange and violet natural pigments containing a substituted $\Delta^{2,2'}$-bis (1,4-benzothiazine) ring system. Color is due to the conjugated chromophore system, $-S-C=C-C=N-$. Biosynthetically, T. are related to the Melanins (see). Together with the phaeomelanins, T. are responsible for the red and auburn colors of human hair and bird feathers. T. B and C are yellow-orange; T. E and F are violet.

606

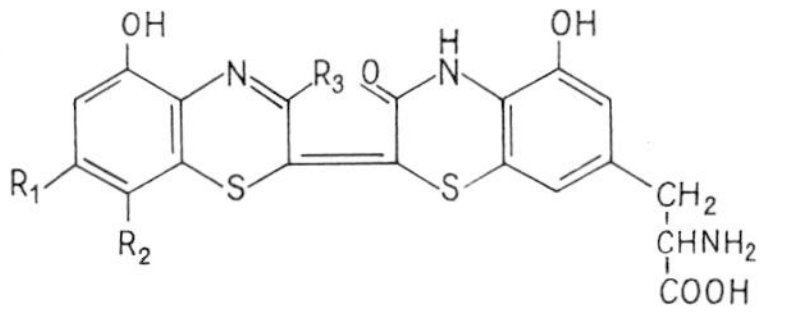

Trichochrome **B**: $R_1=H$, $R_2=CH_2-CHNH_2-COOH$; $R_3=COOH$

Trichochrome **C**: $R_1=CH_2-CHNH_2-COOH$, $R_2=H$, $R_3=H$

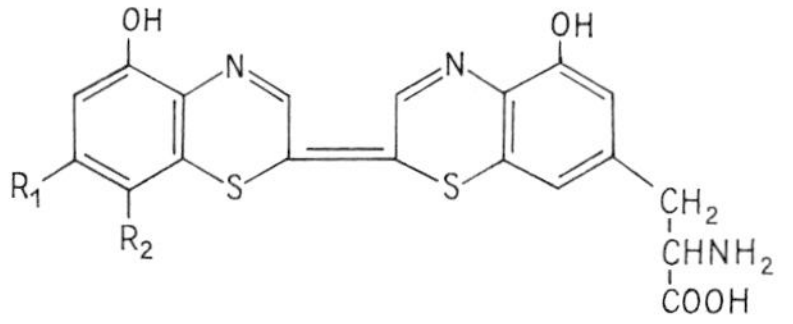

Trichochrome **E**: $R_1=H$, $R_2=CH_2-CHNH_2-COOH$

Trichochrome **F**: $R_1=CH_2-CHNH_2-COOH$, $R_2=H$

Tricholomic acid: *erythro*-dihydroibotenic acid, a compound isolated from the Basidomycete (mushroom) *Tricholoma muscarium*. It can also be prepared by reduction of ibotenic acid. It possesses flavor promoting activity similar to, but much more active than, that of sodium glutamate. It also has a synergistic effect on the flavor improving property of inosinic acid and guanosine 5′-phosphate.

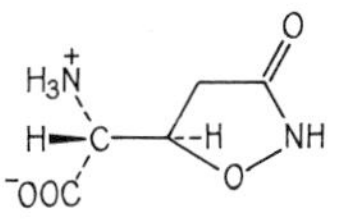

Tricholomic acid

Triethylcholine: see Acetylcholine.

Trifluoromethanesulfonic acid, *triflic acid*, *TFMS*: CF_3SO_3H, M_r 150.7, b.p. 162 °C. TFMS is one of the strongest acids known, and has proved effective for the simultaneous deprotection and resin cleavage at the conclusion of the solid-phase synthesis of peptides, e.g. tuftsin, enkephalin, bovine pancreatic RNase and chicken neurotensin. [H. Yajima et al. *Chem. Pharm. Bull.* **29** (1981) 2587; H. Yajima & N. Fujii *Biopolymers* **20** (1981) 1958; Y. Kiso et al. *Chem. Pharm. Bull.* **27** (1979) 1472; M.K. Chaudhuri & V.A. Najjar *Anal. Biochem.* **95** (1979) 305]

Triglyceride: see Acylglycerols, Fats.

Trigonellin: 1-methylnicotinic acid, a metabolite of nicotinic acid or nicotinamide found in many plants. It is both a hormone and a storage form of nicotinic acid. It is apparently not a niacin metabolite in animals, although it is found in the urine of coffee drinkers. Green coffee beans contain relatively large (> 500 mg/kg) amounts of T.; roasting the beans converts T. to nicotinic acid. Coffee is a significant dietary source of niacin (see Vitamins) in South and Central America.

Tri(hydroxymethyl)methylamine, *TRIS*: $H_2N-C(CH_2OH)_3$, M_r 121, a widely used buffer substance, suitable for the pH range 7-9. The required pH is usually obtained by adding HCl to TRIS dissolved in water. TRIS buffers have a high temperature/pH gradient, e.g. 0.05 M TRIS, pH 7.05 (adjusted with HCl) at 37 °C has pH 7.20 at 23 °C. The pH of a TRIS buffer should therefore be established at the intended working temperature. See Buffers.

3,5,3′-Triiodothryonin: see Thyroxin.

Trimethylglycine: see Betaines.

Triose phosphates: D-glyceraldehyde 3-phosphate ($PO_3H_2-OCH_2-CHOH-CHO$) and dihydroxyacetone phosphate ($PO_3H_2-OCH_2-CO-CH_2-OH$), important intermediates in Glycolysis (see) and Alcoholic fermentation (see). The two T.p. are interconvertible via the ene-diol form, by the action of T.p. isomerase; the equilibrium mixture contains 96% ketotriose phosphate and 4% aldotriose phosphate. The T.p. hold a key position in carbohydrate metabolism, being intermediates of gluconeogenesis and photosynthetic CO_2 fixation.

Trioses: glyceraldehyde and dihydroxyacetone phosphate. They contain 3 C-atoms and are the simplest monosaccharides. Their phosphates are important metabolic intermediates (see Triose phosphates).

Triphosphomonoesterases: see Esterases.

Triphosphopyridine nucleotide: see Nicotinamide adenine dinucleotide phosphate.

Triple helix: see Collagen.

Triplet code: see Genetic code.

TRIS: see Tri(hydroxymethyl)methylamine.

Trisaccharides: see Carbohydrates.

Trisporic acids, *fungal sex hormones*: structurally similar C_{18}-terpene carboxylic acids from heterothallic fungi of the *Mucorales* type, e.g. *Blakeslea trispora* or *Mucor mucedo*. T.a. are only produced when (+)-strains are mixed with (−)-strains; a prohormone from the (−)-strain is transformed into T.a. by the (+)-strain. The T.a. then induce formation of zygospores in the (−)-cells. *Trisporic acid C* is the most important member of the group, M_r 306; 20 µg are sufficient to induce zygospore formation. Unlike other diterpenes, T.a. are biosynthesized by the cleavage of β-carotene.

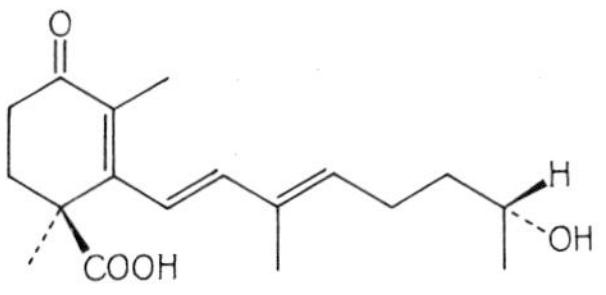

Trisporic acid C

Triterpenes: an extensive group of terpenes biosynthesized from six isoprene units. Apart from squalene, which is acyclic, most of this group are tetra or pentacyclic hydroaromatic compounds based on the parent hydrocarbon, sterane, i.e. they are steroids. Included with T. are those terpenoid natural products with fewer than 30 C-atoms, which are biosynthesized via a C_{30} intermediate, but with the subsequent loss of one or more C atoms. Addition of extra C atoms and incorporation of heteroatoms are also possible, e.g. the steroid alkaloids. Many

Table. Triterpenes.

Type of compound	No. of carbon atoms	Examples
Sex hormones		
Estrogens	18	Estradiol
Androgens	19	Androsterone
Progestins	21	Progesterone
Adrenal corticosteroids	21	Corticosterone
Cardiac glycosides	21, 23, 24	Digitoxigenin
Steroid alkaloids	21, 27	Tomatidine
Bile acids	24, 27	Cholic acid
Sapogenins	27	Digitogenin
Vitamin D	27, 28	Vitamin D_2
Molting hormones	27	β-Ecdysone
Sterols		
Mycosterols	27, 28	Ergosterol
Zoosterols	27, 28, 30	Cholesterol
Phytosterols	29, 30	β-Sitosterol
Cucurbitanes	30	Cucurbitacin D

T. have high biological activity, in particular the steroid hormones.

Biosynthesis. Tail-to-tail condensation of two molecules of farnesylpyrophosphate produces squalene. This, and subsequent reactions leading to steroids, are discussed under Terpenes (see) and Steroids (see). In the biosynthesis of pentacyclic T., the 3-hydroxysqualene cation cyclizes to a prosterol cation, which has not yet been identified in the free state. A Wagner-Meerwein rearrangement and ring closure leads to the expansion of ring D (compounds I and II, Fig.). A further Wagner-Meerwein rearrangement results in expansion of ring E in III, to form IV, which is the precursor of a large number of pentacyclic T. (germanicol, friedelin, multiflorenol, taraxol, etc.) T. isolated between 1977 and 1981 have been reviewed by M. C. Das & S. B. Mahato *Phytochemistry* **22** (1983) 1071–1095.

Triterpene saponins: see Saponins.

Trityrosine: see Resilin.

tRNA: abb. for transfer-RNA.

tRNA methylases: see Polynucleotide methyltransferases.

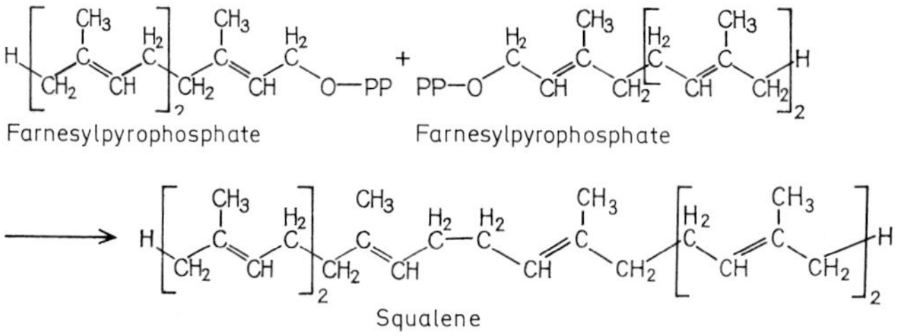

Figure 1. Formation of squalene from two molecules of farnesylpyrophosphate.

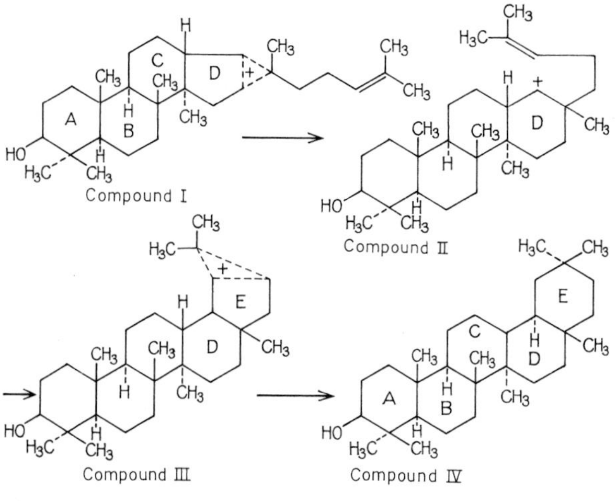

Figure 2. Formation of pentacyclic triterpenes.

Tropane alkaloids: a group of Alkaloids (see). T. a. are esters of various amino alcohols based on a substituted tropane (*N*-methyl-8-azabicyclo-3,2,1-octane) ring system. This system is not optically active, because the two rings can only be linked in a *cis* configuration, thus resulting in a *meso* form. Introduction of an hydroxyl group at position 3 produces the geometric isomers, tropine (tropane-3α-ol; the OH-function is *trans* to the CH₃N-group, Figure 1) and pseudotropine (ψ-tropine; tropane-3β-ol; Figure 1). The wide variety of T. a. is the result of substitution of the base moiety, esterification with different acids and the occurrence of isomers and demethylated derivatives. Tropane-3β-ol-2-carboxylic acid (ecgonine, Figure 1) is important as the base component of the coca alkaloids. The acid component is usually an aromatic carboxylic acid (Table).

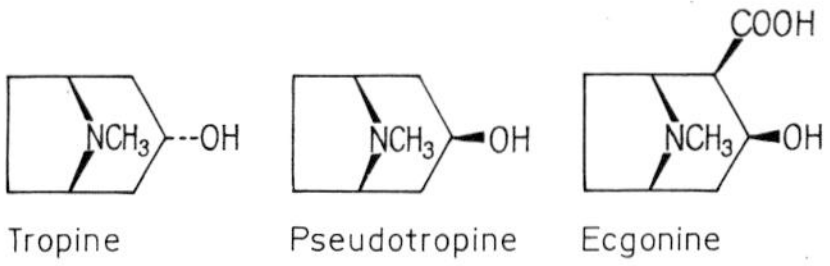

Table. Structure of tropane alkaloids.

Alkaloid	Amino alcohol	Acid
Atropine	Tropine	DL-Tropic acid
L-Hyoscyamine	Tropine	L-Tropic acid
Littorine	Tropine	Phenyllactic acid
Scopolamine	Scopine	Tropic acid
Noratropine	Nortropine	Tropic acid
Tropacocaine	Pseudotropine	Benzoic acid
Cocaine	Ecgonine methyl ester	Benzoic acid

Tropine   Pseudotropine   Ecgonine

Figure 1

According to their plant of origin T. a. can be classified as Belladonna or Datura and Coca alkaloids. The **Belladonna** or **Datura** type occur in *Atropa belladonna,* and other members of the *Solanaceae,* e.g. thorn apple (*Datura* spp.) and henbane (*Hyoscyamus* spp.), and their amino alcohol component is tropine. The **Coca alkaloids** (e.g. cocaine and tropacocaine) are esters of ecgonine and pseudotropine, which, together with other T. a., are found in the coca shrub, *Erythroxylum coca,* cultivated in Peru and Bolivia, also in Java, and to a limited extent in Sri-Lanka. In all cases the T. a. are accompanied by pyrrolidine bases, e. g. hygrine. T. a. for therapeutic purposes are derived almost entirely from plant sources. The tropane ring is derived biosynthetically from compounds of the glutamic acid/proline/ornithine group, via an *N*-methylpyrroline cation which condenses with acetoacetic acid to form ketones (of the tropinone type). Following reduction of the ketone function, the resulting amino alcohol becomes esterified with an appropriate aromatic acid (ecgonine is esterified with benzoic acid, with additional methylation of the 2-carboxyl function to form cocaine; tropine is esterified with tropic acid to form hyoscyamine, which can be further epoxidized to scopolamine). See Figure 2.

Plants containing T. a. have been used in folk medicine since ancient times. The T. a. are powerful poisons. Atropine poisoning is characterized by distension of the pupil of the eye, dryness of the mouth, delirium and double vision. Higher doses result in general paralysis, owing to the action of the drug on involuntary muscle, and finally death. Atropine is used widely in ophthalmology for causing dilation of the pupil of the eye (mydriatic activity). The bel-

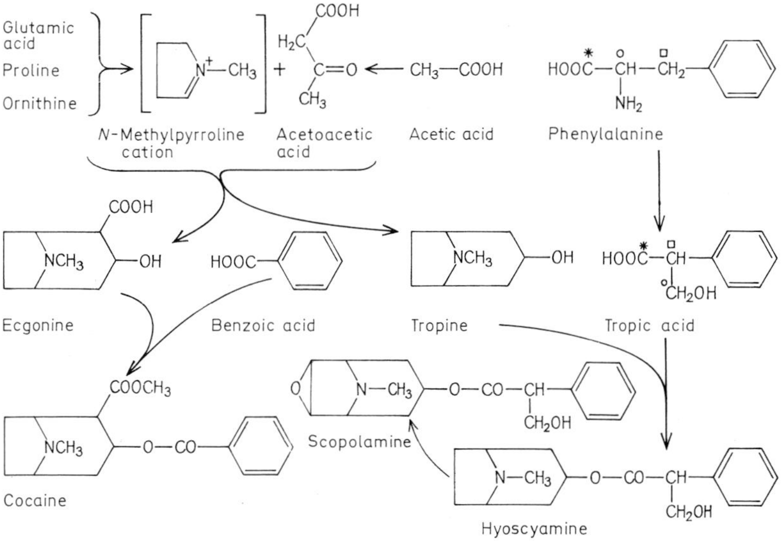

Figure 2. Biosynthesis of tropane alkaloids.

609

ladonna alkaloids were therefore previously called *mydriatic alkaloids*. Cocaine was used widely as a local anesthetic, but has now been largely replaced by other synthetic drugs.

Tropan-3-one: see Tropinone.

Tropic acid: see Tropane alkaloids.

Tropine: see Tropane alkaloids.

Tropinone, *tropan-3-one:* a possible biosynthetic precursor of the Tropane alkaloids (see), present in *Solanaceae* spp. M_r 139.19, m. p. 42 °C, b. p. 224–225 °C. Robinson's laboratory synthesis of T. (Figure) from succindialdehyde, methylamine and the calcium salt of acetone-dicarboxylic acid in aqueous solution at ordinary temperatures (yield 42%, 2 molecules of CO_2 are readily lost on subsequent treatment with acid), i.e. under mild or apparently physiological conditions, gave impetus to modern studies on alkaloid biosynthesis.

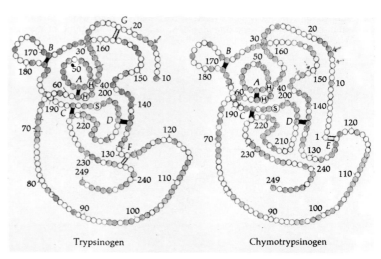

| Succin-dialdehyde | Methylamine | Acetone-dicarboxylic acid | | Tropinone |

Tropocollagen: see Collagen.

Tropomyosin: a protein associated with Actin (see), both in muscle (see Muscle proteins) and in the cytoskeleton of other cell types. There are two very similar forms in striated muscle, α-T. and β-T.

Both have 284 amino acid residues (M_r 33000) per subunit. The molecule is a two-chain coiled-coil α-helix with the two subunits twined around one another. α-α, α-β and β-β dimers have been observed; different proportions of the two types appear in different types of muscle and may reflect specialization. T. dimers polymerize head-to-tail to form a fiber which lies in the groove of an F-actin helix. Each dimer spans 7 (muscle) or 6 actin subunits. In vertebrate striated muscle, T. binds tightly to troponins, proteins which regulate contraction. In other tissues, T. is associated with some but not all microfilaments (see Cytoskeleton), where it probably stabilizes F. actin. Non-muscle cells synthesize a number of different T. from a family of genes; the differences between the isoforms are not great but probably have regulatory function. Horse platelet T., for example, has only 247 amino acid residues, binds only slightly to troponins, and probably spans only 6 actin units instead of 7. [M. R. Payne & S. E. Rudnick in *Cell and Muscle Motility*, **6** (J. W. Shay, ed.) (Plenum Press, New York & London, 1985) pp. 141–184]

Troponin: see Muscle proteins.

Trp: abb. for L-tryptophan.

Trypsin (EC 3.4.21.4): a serine protease present as a zymogen (trypsinogen) in the pancreas of all vertebrates. Trypsinogen is released via the pancreatic duct into the duodenum. Conversion of trypsinogen into T. is initiated in the small intestine by enterokinase, and accelerated autocatalytically by traces of T. Activation consists of the removal of the *N*-terminal acidic hexapeptide, Val(Asp)$_4$Lys (IP 9.3), from trypsinogen. Calcium ions accelerate this

Comparison of the structures of bovine trypsinogen and chymotrypsinogen A. Solid arrows show the hydrolysis sites that result in the activation of trypsinogen to trypsin and chymotrypsinogen to chymotrypsin. Broken arrows are additional hydrolysis sites to form α-chymotrypsin. Shaded circles represent amino acids that are identical or similar in the two proteins. Disulfide bridges are lettered A to G. H represents the histidine residues, and S the serine residue of the active site that reacts with diisopropylfluorophosphate. Deletions are shown by lines between the circles. To aid comparison, the residue numbers of both structures are based on those of chymotrypsinogen. (From Hartley, B. S., Brown, J. R., Kauffman, D. L. and Smillie, L. B. (1965) *Nature, 207,* 1157) process and make it more specific for cleavage of the Lys_6-Ile_7 bond, thus avoiding formation of inactive byproducts through other less selective cleavages. The resulting strongly basic β-T. (IP 10.8) is cleaved by limited autolysis of the Lys_{131}-Ser_{132} bond to give the 2-chain structure of α-T.; further cleavage of Lys_{176}-Asp_{177} produces a 3-chain active enzyme called pseudo T.; the individual chains are held together by 6 disulfide bridges. The amino group of the N-terminal Ile of T. (originally Ile_7 of trypsinogen) forms an ion pair with Asp_{182} which is important in the conformation of T. The ion pair, Lys_{95} and Asn_{223}, protects the C-terminal residue from attack by carboxypeptidase. Bovine T. has M_r 23300 and 223 amino acid residues; 101 of these residues (41%), including 4 of the 6 disulfide bridges, are identical with corresponding sequence positions of Chymotrypsin (see) (Figure). The conservative regions include the sequences around the two catalytically important residues, His_{46} and Ser_{138}. There is also considerable correspondence between the chain conformation (secondary structure) of T. and chymotrypsin and elastase (low α-helix content, extended β-structures, similar design of the substrate cleft). T. is stable at pH 2-4 and +4 °C for several weeks, but at pH 9 and 30 °C it undergoes progressive autolysis of its intact Lys- and Arg-bonds, and becomes totally inactive within 24 h. Calcium ions delay this process but do not prevent it. T. activity is not inhibited by 6 M urea. Of all the digestive endopeptidases, T. has the most pronounced substrate specificity, catalysing hydrolysis of only Lys- and Arg-bonds; this is due to the presence of an ionic binding site (Asp_{189}) in the active center. In addition to endopeptidase activity, T. has a high activity towards N-acylated arginyl or lysyl peptide esters and N-acylated arginyl- or lysylarylamides. These synthetic substrates are used for the assay of T. and in the study of T. kinetics. T. is irreversible inhibited by many naturally occurring T. inhibitors (e.g. trasylol and soybean T. inhibitor), and by synthetic inhibitors like diisopropylfluorophosphate (forms an ester with the hydroxyl group of the reactive Ser_{183}) or tosyl-L-lysylchloromethylketone (alkylates N3 of the reactive His_{46}).

Enzymes similar to T. have been found in many invertebrates, like crabs and insects, e.g. cocoonase produced by silk moths to attack their proteinaceous cocoon and aid escape.

N. B. All sequence positions quoted above are for trypsinogen.

Tryptamine: β-indolyl-(3)-ethylamine, a biogenic amine, M_r 160.2, produced by decarboxylation of tryptophan. T. stimulates the contraction of smooth muscle of blood vessels, uterus and central nervous system. It is found in both plants and animals, and as a bacterial degradation product of tryptophan.

Tryptamine

L-Tryptophan, abb. *Trp, α-amino-β-indolepropionic acid:* an aromatic, essential amino acid. M_r 204.2, m. p. 281-282 °C. $[\alpha]_D^{25}$ −33.7 (c=1-2, water), or +2.8 (c=1-2, 1M HCl). Trp is nutritionally very important, although it is present in relatively small amounts in proteins. Acid hydrolysis of proteins completely destroys Trp. p-Dimethylaminobenzaldehyde or xanthydrol gives a violet coloration with Trp, which is used in its determination.

Trp is the precursor of several physiologically important metabolites. Cleavage of the pyrrole ring of Trp by oxidation (see L-Tryptophan 2,3-dioxygenase) to produce N-formylkynurenine is the first reaction in a major metabolic pathway of Trp in animal liver (Figure 1). The open chain intermediate 2-amino-3-carboxymuconic acid-semialdehyde (IX, Figure 1) serves as the starting point of two different pathways. Spontaneous cyclization of IX produces quinolinic acid, which is the precursor of the nicotinamide moiety of NAD. Enzymatic conversion of IX to 2-aminomuconic acid (X, Figure 1) initiates the pathway for the complete degradation to CO_2 and H_2O via acetyl-CoA and the TCA-cycle. The extent to which the spontaneous cyclization of IX can be exploited for the synthesis of NAD depends upon its relative rates of enzymatic formation from VI and conversion into X. In rats, for example, nicotinic acid is not essential in the diet (i.e. it is not a vitamin) because the total requirement for NAD and NADP can be satisfied by synthesis from Trp. On the other hand, in the cat, practically no synthesis of NAD occurs from Trp, and there is a total dependence on a dietary source of nicotinic acid. In the human, part of the NAD is synthesized from Trp, but the rate of cyclization of IX is insufficient for the synthesis of the total NAD requirement, and the remainder must be synthesized from dietary nicotinic acid or nicotinamide (see Vitamins).

Trp degradation is markedly affected by vitamin B_6 deficiency. Kynureninase, which catalyses the cleavage of 3-hydroxykynurenine into 3-hydroxyanthranilic acid and L-alanine, is a Pyridoxal phosphate (see) enzyme. The activity of liver kynureninase (EC 3.7.1.3) decreases rapidly in vitamin B_6 deficiency, compared to the decrease shown by other pyridoxal phosphate enzymes. For example, after 14 weeks of experimental B_6 deficiency in rats, the liver kynureninase is reduced to about 16% of its original value, whereas kynurenine transaminase (EC 2.6.1.7) still shows 60% activity. Consequently, compounds derived from the Trp degradation pathway prior to the hydrolysis of 3-hydroxykynurenine are excreted in greater quantity (e.g. xanthurenic acid and kynurenic acid), whereas the urinary level of nicotinic acid derivatives is decreased. Exaggeration of this picture by feeding extra Trp forms the basis of the *tryptophan load test* for vitamin B_6 deficiency: 2 g of Trp are given in a single dose suspended in a beverage before breakfast. The extra Trp induces the synthesis of L-Tryptophan 2,3-dioxygenase (EC 1.13.11.11), so that there is a greatly increased flow of Trp metabolites into the degradation pathway. A 24-hour urine sample is analysed for Trp metabolites. Xanthurenic acid is normally measured for clinical purposes, because it is easily assayed photometrically as a green ferric-xanthurenic acid complex, or by an improved

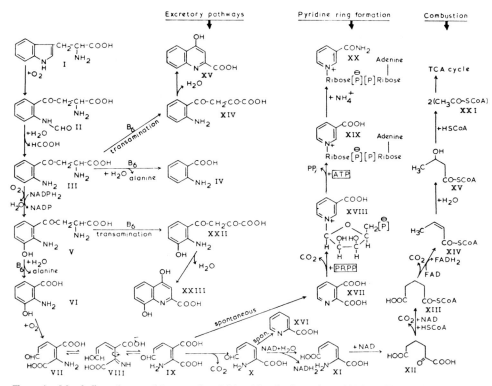

Figure 1. Metabolic pathways of L-tryptophan initiated by the formation of N-formylkynurenine in animal liver. PRPP = 5-phosphoribosyl 1-pyrophosphate. *I* L-tryptophan, *II* N-formylkynurenine, *III* kynurenine, *IV* anthranilic acid, *V* 3-hydroxykynurenine, *VI* 3-hydroxyanthranilic acid, *VII VIII IX* 2-amino-3-carboxymuconic acid-semialdehyde, *X* 2-aminomuconic acid-semialdehyde, *XI* 2-aminomuconic acid, *XII* 2-oxoadipic acid, *XIII* glutaryl-CoA, *XIV* crotonyl-CoA, *XV* 3-hydroxybutyryl-CoA, *XVI* picolinic acid, *XVII* quinolinic acid, *XVIII* nicotinic acid mononucleotide, *XIX* deamido-NAD, *XX* NAD, *XXI* acetyl-CoA, *XXII* o-amino-m-hydroxybenzoylpyruvic acid, *XXIII* xanthurenic acid, *XXIV* o-aminobenzoyl-pyruvic acid, *XXV* kynurenic acid.

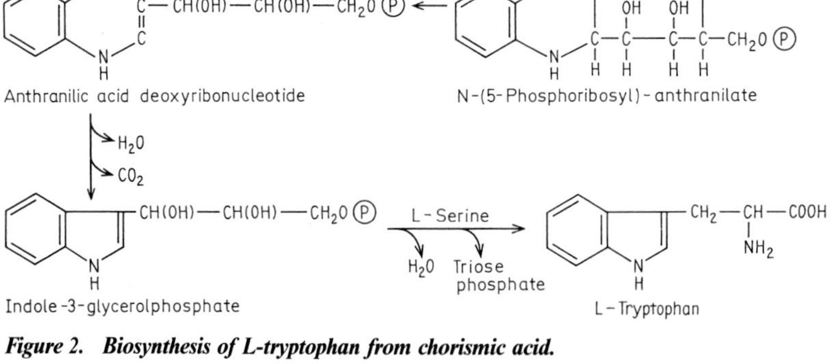

Figure 2. Biosynthesis of L-tryptophan from chorismic acid.

fluorimetric technique. In patients receiving deoxy-pyridoxine (see Vitamin B_6), the tryptophan load test reveals a 20-fold increase in the excretion of kynure-nine, hydroxykynurenine and xanthurenic acid, and a 3-fold increase in the excretion of kynurenic acid, compared to normal controls. At the same time, the excretion of N-methyl-2-pyridone-5-carboxamide (a metabolite of nicotinic acid) is decreased.

The following metabolites of Trp are listed sepa-rately: Actinomycins, Auxin, Indican, Indigo, In-dole alkaloids, Kynurenine, Melatonin, Ommo-chromes, Phallotoxins, Serotonin, Toad toxins, Tryptamine, Violacein.

Trp is synthesized in bacteria and plants via shiki-mate and chorismate (Figure 2) (see Aromatic bio-synthesis; see L-Tryptophan synthase).

Tryptophan 2,3-oxygenase, *tryptophan oxyge-nase, tryptophan pyrrolase, L-tryptophan: oxygen 2,3-oxidoreductase (decyclizing),* (EC 1.13.11.11): an enzyme catalysing the oxidation of L-tryptophan to N-formylkynurenine, the first stage in the total deg-radation of L-Tryptophan (see) in animals and mi-croorganisms; this is also the first stage in the con-version of L-tryptophan into the nicotinamide moie-ty of NAD and NADP in molds and in some mammals, and in the conversion of L-tryptophan in-to Ommochromes (see) in insects.

Studies with $^{18}O_2$ and $H_2^{18}O$ show that both atoms of oxygen appearing in the product originate from molecular oxygen and not from water. Inhibition of the reaction by superoxide dismutase suggests that the substrate oxygen is activated to the superoxide ion prior to participation in the oxygenase reaction. In *Pseudomonas,* synthesis of tryptophan oxygenase is apparently induced by L-tryptophan, but the true inducing agent is kynurenine, formed from the L-tryptophan by low constituent levels of the enzyme. Normal adult mammalian liver always contains fair-ly high activity of the enzyme, but higher levels of synthesis (due to increased synthesis of mRNA) are induced by glucocorticoids. Administration of L-tryptophan also leads to an increase in the level of the enzyme, but this appears to be due to a de-creased rate of breakdown of the enzyme protein. Tryptophan oxygenase is absent from the liver of normal rats up to the tenth post-natal day, but it can be induced at any time during this ten day period by the administration of L-tryptophan or glucocorti-coids.

The rat liver enzyme has M_r 167 000 (4 subunits of M_r 43 000, comprising 2 types: $\alpha_2\beta_2$). The enzyme from *Pseudomonas* has M_r 122 000 (4 subunits of M_r 31 000; structure of oligomer not known). Both bac-terial and mammalian enzymes contain protopor-phyrin IX, which is essential for catalytic activity. Two moles of heme are present per mole of enzyme; in vitro, the enzyme is inactive unless this heme is reduced by a reducing agent such as H_2O_2, ascorbic acid, or superoxide anion. The presence of 2 g at-oms of Cu per mole of enzyme has been claimed, but after some dispute it has now been proved that Cu is not essential in the enzymatic mechanism [Ishimura, Y. et al. *J. Biol. Chem.* **255** (1980) 3835–3837]. The enzymatic mechanism is ordered bi-uni; L-tryptophan binds to the active (reduced) enzyme, then with molecular oxygen, forming a ter-

nary complex of enzyme-substrate-oxygen. This ter-nary complex can be detected during the reaction by its spectral properties (maxima at 418, 545 and 580 nm) which are similar to those of the oxygen-ated ferroheme forms of hemoglobin, myoglobin or peroxidase.

Hartnup's disease, a hereditary defect associated with mental retardation, is due to a deficiency of tryptophan oxygenase. The $v^+ \rightarrow v$ mutation in *Dro-sophila melanogaster,* manifested as defective om-mochrome synthesis, is due to a deficiency of trypto-phan oxygenase; it can be reversed by the injection of kynurenine.

L-Tryptophan synthase, *tryptophan desmolase, L-serine hydro-lyase (adding indoleglycerol-phosphate),* (EC 4.2.1.20): the enzyme catalysing the synthesis of L-tryptophan from L-serine and indole 3-glycerol phosphate. T.s. from *Escherichia coli* (M_r 149 000) has $\alpha_2 \beta_2$ subunit composition. The enzyme sepa-rates easily into monomeric subunit α (also called protein B) (M_r 29 000) and dimeric subunit β_2 (also called protein B) (M_r of dimer 90 000) when eluted from DEAE cellulose with a sodium chloride gra-dient. The separated subunits catalyse partial reac-tions of L-tryptophan synthesis:

Indole 3-glycerol phosphate $\rightharpoonup$ indole + 3-phospho-glyceraldehyde

Indole + L-serine $\overset{b}{\rightharpoonup}$ L-tryptophan + H_2O.

In the presence of the reconstituted $\alpha_2 \beta_2$ complex, the rates of these partial reactions are 30 to 100 times greater than with the individual subunits. The sum of the two partial reactions:

Indole 3-glycerol phosphate + L-serine $\rightarrow$ L-trypto-phan + 3-phosphoglyceraldehyde + H_2O, is cata-lysed by the $\alpha_2 \beta_2$ complex, and free indole is not de-tectable as an intermediate in this overall reaction. T.s. from *Escherichia coli* is therefore a simple mul-tienzyme complex. Pyridoxal phosphate is essential for enzymatic activity; each β subunit binds one molecule of the coenzyme. The β_2 dimer is cleaved to the monomer by urea concentrations greater than 4M.

The primary sequence of the 268 amino acid resi-dues of protein A is known [Guest, J.R. et al., *J. Bi-ol. Chem.* **242** (1967) 5442–5446]. Studies on the ge-netic analysis of this sequence are now classical. In fact the genetics and structure of T.s. from *Escheri-chia coli* are probably more thoroughly investigated than those of any other multienzyme complex. Out-standing among these studies was the comparison of the linear positions of mutational sites in the cistron for protein A (determined by phage-mediated three point genetic crosses between mutant strains) with the linear position of the sites of amino acid replace-ment in the corresponding CRiM proteins. This pro-vided the first proof of the colinearity of gene and polypeptide structure [Yanofsky, C. et al. *Proc. Nat. Acad. Sci.,* **51** (1964) 266–2272].

Elucidation of the primary sequence, and the identi-fication of active sites of protein B are in progress [Higgins W., Miles, E.W. & Fairwell, T. *J. Biol. Chem.* **255** (1980) 512–517].

In *Neurospora crassa,* only one T.s. protein is pro-duced (M_r 150 000, consisting of 2 identical mono-mers of M_r 75 000), but genetic and biochemical analysis show that the T.s. gene in *Neurospora* is subdivided into two regions homologous with the A and B regions of *Escherichia coli* [Matchett, W.H. & DeMoss, J.A. *J. Biol. Chem., 250* (1975) 2941-2946]. Studies on T.s. from other microorganisms indicate that T.s. from *Escherichia coli* is typical of prokaryotes, while the *Neurospora* enzyme serves as a model of the eukaryotic type. Among the prokaryotic T.s. there is some degree of cross reactivity between α and β subunits from different organisms. The literature on T.s. up to 1972 has been reviewed by Yanofsky, C. and Crawford, I.P. in *The Enzymes,* pp. 1-31, vol VII, 3rd. edtn. (1972), edit. Boyer, P.D., Academic Press.

The review by E.W.Miles (pp. 127-186 in *Advances in Enzymology* (1979) vol *49,* edit. Alton Meister) deals with studies on the catalytic mechanism and structure of the enzyme up to 1979, but not with the genetic analysis.

Tryptophol: see Auxins.

TSH: abb. for thyroid stimulating hormone, or thyrotropin.

TTP: see Thymidine phosphates.

Tubercidin: 6-amino-9-β-D-ribofuranosyl-7-deazapurine, a purine antibiotic (see Nucleoside antibiotics) from *Streptomyces tubercidicus,* and one of the group of 7-deaza-adenine-nucleoside analogs. M_r 266.25, m. p. 247-248 °C (d.), $[\alpha]_D^{17} - 67$ ° $(c = 1,$ 50% acetic acid). The N7 of adenine is replaced by a methylene group. T. is biosynthesized from adenosine (Figure): the C-atoms of the pyrrole ring are derived from a ribose moiety, which is introduced from 5-phosphoribosyl 1-pyrophosphate. As an antimetabolic of adenosine, T. interferes with purine metabolism. T. can also be converted into nicotinamide-deaza-adenine-dinucleotide, which inhibits glycolysis. T. is particularly active against *Mycobacterium tuberculosis* and *Candida albicans.*

Tubocurarine: see Acetylcholine.

Tubulins: dimeric proteins of two closely related subunits (subunit M_r 60000). T. are the major components of microtubules, accompanied by smaller quantities of higher M_r proteins. Microtubules are composed of a series of parallel filaments, formed by end-to-end aggregation of T. molecules. Each T. molecule strongly binds one molecule of GTP and loosely binds a second molecule. There is thus an analogy with actin, but the two proteins are structurally dissimilar. The alkaloid, Colchicine (see) binds strongly to T. and causes disassembly of labile microtubules, such as the mitotic spindle. Colchicine therefore has the effect of blocking dividing eukaryotic cells in metaphase, and results in the production of highly polyploidic daughter cells.

Tuftsin: a naturally occurring phagocyte activating tetrapeptide: Thr-Lys-Pro-Arg. T.stimulates phagocytosis and pinocytosis, and promotes motility and migration of phagocytes. It causes a chemotactic response in phagocytes and generates chemiluminescence in the absence of a target particle, presumably due to formation of H_2O_2, superoxide and hydroxyl radicals. It increases the killing ability of phagocytes, and enhances the rate of clearance of bacteria from the blood by phagocytosis. T.also stimulates the immunogenic function of phagocytes, presumably by enhanced antigenic processing and signal generation to antibody-forming lymphocytes. It also stimulates macrophages and granulocytes, so that they become highly tumoricidal in vitro and vivo. All these properties are probably due to membrane activation, which is sensitive to nanomolar concentrations of T.

T.is excised from the H-chain of a precursor γ-globulin molecule, known as leukokinin, which binds to the phagocyte membrane. Release of T.occurs in

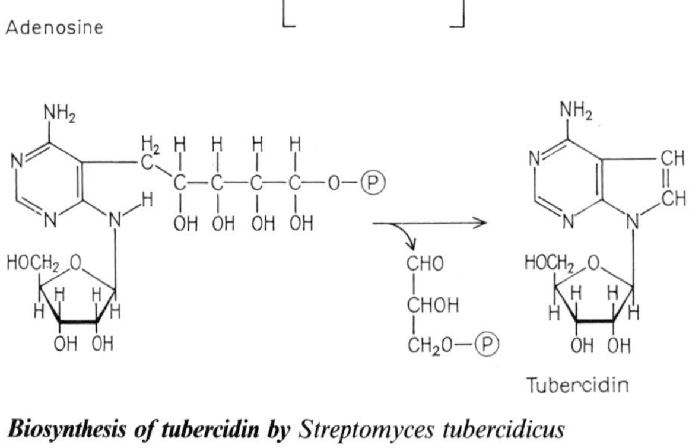

Biosynthesis of tubercidin by *Streptomyces tubercidicus*

two stages: the C-terminus is released by endocarboxypeptidase activity in the spleen, followed by release of the N-terminus by the action of leukokinase, an enzyme present on the outer surface of the phagocyte membrane.

Patients genetically deficient in T. (human tuftsin deficiency syndrome) have a high frequency of very severe infections. These patients produce a mutant peptide (as yet unidentified), which is a strong inhibitor of natural T. After splenectomy, leukokinase is still produced and coats the leukocytes, but it is inactive in T. formation. Activity reappears and reaches normal values several months later, indicating that the necessary enzyme activity arises elsewhere in the body. [V. A. Najjar, *Adv. Enzymology* **41** (1974) 129–178].

Tulipanin: see Delphinidin.

Tumor antigens: carcinoembryonic antigens, which serve as an aid to the early recognition of liver carcinoma and teratoblastomas (tumors of reproductive cells, especially in the testes and ovaries). T. a. are embryonal plasma proteins produced in the placenta during pregnancy and in some organs of the embryo, but no longer detectable shortly after birth. They can be formed again later in life in response to malignant tumors. Three T. a. are important in clinical tumor diagnosis: α-fetoprotein (M_r 65 000), embryonogenic colon antigen (ECA), and the Regan isoenzyme of the placenta. Appearance of α-fetoprotein in the serum is a definite indication of liver carcinoma or teratoblastoma. Benign liver disorders cause no increase in tumor antigen. Determination of these T. a. permits detection in the early phase of 60–80% of liver carcinoma cases, and 20–25% (80–90% in young people) of teratoblastoma cases; 2 ng α-fetoprotein per ml serum can be detected by a radioimmunological test. The Regan isoenzyme of placental alkaline phosphatase increases particularly in cases of malignant tumors of the female genital tract.

Tumor necrosis factor, *TNF*:

TNF-α, cachectin: a protein (human, M_r 17 000, two Cys in disulfide linkage, no carbohydrate) originally found in the serum of animals sensitized with *Bacillus* Calmette-Guerin or with *Corynebacterium parvum*, and challenged with bacterial endotoxin. TNF-α produces hemorrhagic necrosis of some tumors in experimental animals, and cytolytic or cytostatic effects on tumor cells in culture. The major cellular source of TNF-α is the macrophage, and it is thought to be a mediator of the cytotoxic activity of macrophages against tumor cells in culture. It has been implicated in the induction of shock and cachexia resulting from invasive stimuli, and it is a potent pyrogen, causing fever by direct action on the hypothalamic thermoregulatory centers and by induction of the synthesis of interleukin-1.

TNF-β, lymphotoxin: a glycoprotein (human, one aspartate-linked glycosylation site, M_r of the unglycosylated protein 18 664 containing 171 amino acid residues) produced by mitogen-stimulated lymphocytes, which causes cytostasis of some tumor cell lines and cytolysis of other transformed cells.

Both TNF-α and TNF-β have been purified, sequenced and cloned by recombinant DNA technology in *Escherichia coli*. According to studies with human cervical carcinoma (cell line ME-180), TNF-α and TNF-β bind to the same receptor (2000 receptor sites per cell). Preincubation of cells with γ-interferon increases the number of receptor sites two- to three-fold, which accords with the observation that interferons act synergistically with TNF in tumor necrosis in vitro. Studies with TNF-α show that binding is followed by receptor-mediated internalization of TNF by endocytosis. The amino acid sequences of TNF-α and TNF-β show about 50% homology, but no significant homology with other lymphokines, e.g. γ-interferon, interleukin-2 or interleukin-3. The TNF-α gene resides on chromosome 6 in humans, and is closely linked to the gene specifying TNF-β; presumably both genes arose from a common ancestor by tandem duplication. [B. Bharat et al. *Nature* **318** (1985) 665–667; M. Tsujimoto et al. *Proc. Natl. Acad. Sci. USA* **82** (1985) 7626–7630; B. Beutlet & A. Cerami, *Nature* **320** (1986) 584–588]

Tunicamycin: a mixture of homologous, nucleoside antibiotics, produced by *Streptomyces lysosuperificus*, and active against viruses, Gram-positive bacteria, yeast and fungi. The structure of T. consists of one residue each of uracil, a C_{11}-aminodeoxy-dialdose (tunicamine), N-acetylglucosamine and a fatty acid (Figure). T. differ from one another in the chain

Tunicamycin

length of the fatty acid component. The major fatty acid components are *trans* α,β-unsaturated *iso*-acids. The names T. A, B, C, D have been proposed for those T. whose fatty acid component contains 15, 16, 17 and 18 C-atoms, respectively (i.e. $n = 9$, 10, 11, 12 in the formula shown in Figure) [Takatsuki et al., *Agric. Biol. Chem.* **41** (1977) 2307–2309]. High pressure liquid chromatography of T. shows two major and eight minor components, all active in the inhibition of protein glycosylation [Mahoney, W. C. and Duksin, D. *J. Biol. Chem.* **254** (1979) 6573–6576]. Four of these components presumably correspond to T. A, B, C and D of the Japanese authors. Commercially available T. is always a mixture of all possible homologous components. It is used as the unresolved mixture for biochemical studies on the inhibition of protein glycosylation. T. is reported to be a selective inhibitor of glycoprotein synthesis, and it has been shown that the transfer of *N*-acetylglucosamine to dolichylphosphate is sensitive to T. in microsomal fractions of calf liver and chicken embryo, and in yeast. The initial reaction: UDP-Glc*N*Ac + Dol-P → Dol-PP-Glc*N*Ac + UMP, is sensitive to T., but subsequent extension of the carbohydrate chain by addition of residues to Dol-PP-Glc*N*Ac is not sensitive to T. (Glc*N*Ac = *N*-acetylglucosamine, Dol = dolichyl, P = phosphate). ["Uses of Tunicamycin", a colloquium held at the 586th. meeting of the British Biochemical Society, pub. in *Biochemical Society Transactions*, *8* (1980) 163–171]

Tunichrome: a green chromogen in the blood cells of tunicates, e.g. *Ascidia nigra, Ciona intestinalis, Molgula manhattensis.*

Turbidostat: see Fermentation techniques.

Turgorins, *leaf movement factors:* plant hormones which bring about movement (nastic responses to light, temperature, physical shock, etc.) by causing changes of turgor. T. are active at concentrations of $10^{-5} - 10^{-7}$M or lower. The archetypal plant showing movement is the "sensitive" plant, (*Mimosa pudica* L.), which serves as a test system for T. If a cut mimosa shoot is placed in a solution of T., the hormones are transported to the pulvini, which contain T. receptors. The pulvini then undergo marked turgor changes, each pair of pinnules collapses, and the whole leaf assumes the typical irritated posture. Using this system, T. have been isolated from several plants which show nastic responses (Table). Most known T. are glycosides of 3,4-di, or 3,4,5-trihydroxybenzoic acid with an acidic monosaccharide (glucose 6'-sulfate, glucose 3',6'-bisulfate or glucuronic acid) (Table). The leaf movement factor from *Glycine max* (Fig.) is an exception, being an ester of 4-coumaric acid and galactaric acid. [H. Schildknecht, *Angew. Chem. Int. Ed. Engl.* **22** (1983) 695–710; *Endeavour* **8** (1984) 113–117.]

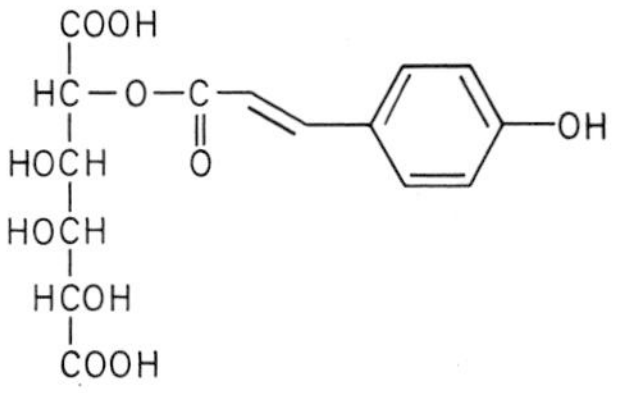

Turgorin from Glycine max

Turnover: the balance of synthesis and degradation of biomolecules in living organisms. All cell components are subject to continual degradation and resynthesis, i.e. they are subject to T. See Steady state.

Turpentine: see Balsams.

Turpentine oil: a volatile oil obtained by steam distillation of turpentine from various *Pinus* spp. T.o. is a colorless liquid, most of which boils between 155 and 162 °C. ρ 0.865–0.870, n_D^{25} 1.465–1.480. On exposure to air, T.o. alters rapidly and finally resinifies. Its composition varies with origin, chief components being bicyclic monoterpenes of the carane and pinene type. Lower grades are obtained by steam distillation of the wood, roots, stumps, sawdust, etc. (wood turpentine), and also as a byproduct in the manufacture of sulfite cellulose (sulfite turpentine). Pinene is isolated on a large scale from T.o. for the synthesis of camphor. T.o. is used widely as a cleaning agent and solvent in the manufacture of shoe polish, varnish and paints. It is also used occasionally in pharmacy under the name Oleum terebinthinae.

Tyr: abb. for L-Tyrosine.

Tyramine: β-hydroxyphenylethylamine, a biogenic amine, M_r 137.2, found in plants (ergot, broom and other legumes) and animals (blood, urine, bile, liver), and as a bacterial degradation product of L-tyrosine. T. is produced by the decarboxylation of L-tyrosine.

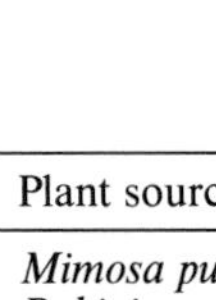

Tyramine

Table. Structures and sources of turgorins.
(P)LMF = (periodic) leaf movement factor

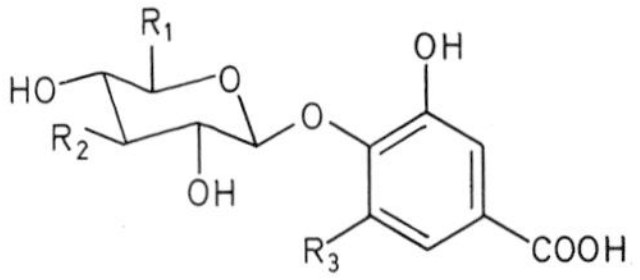

Turgorin	R_1	R_2	R_3	Plant source
PLMF 1	CH₂OSO₃H	OH	OH	*Mimosa pudica, Acacia karoo, Oxalis stricta, Robinia pseudacacia*
PLMF 2	CH₂OSO₃H	OSO₃H	OH	*Acacia karoo*
S-PLMF 2	CH₂OSO₃H	OH	H	*Oxalis stricta*
M-LMF 5	COOH	OH	OH	*Aimosa pudica*

Tyrian purple: 6,6'-dibromoindigotin, a red-violet pigment containing two brominated and oxidized indole ring systems. T.p. is present in marine molluscs of the genera *Murex* and *Nucella,* and a few related whelks. The isolation and structural elucidation of T.p. from the hypobranchial body of the purple snail *Murex brandaris* were performed by Friedlander between 1909 and 1911. In antiquity and later in the middle ages, T.p. was one of the most expensive dyes.

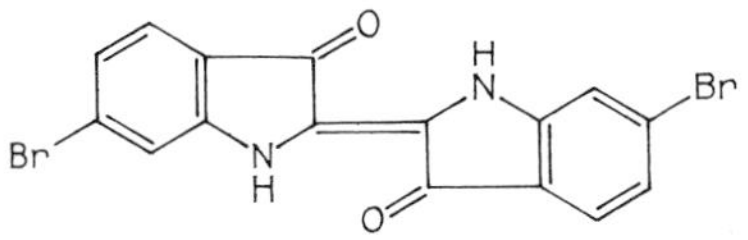

Tyrian purple

Tyrocidins: homodetic, homomeric, cyclic peptide antibiotics active against Gram positive bacteria. Like the Gramicidins (see), T. are produced by *Bacillus brevis.* The structure of tyrocidin A, confirmed by total synthesis, is cyclo-(-Val-Orn-Leu-D-Phe-Pro-Phe-D-Phe-Asn-Gln-Tyr-). In tyrocidin B, Phe_6 is replaced by Trp. In tyrocidin C, Phe_6 is replaced by Trp, $D-Phe_7$ by D-Trp, and Asn_8 by Asp. Tyrocidin E differs from tyrocidin A by replacement of Asn_8 by Asp, and Tyr_{10} by Phe.

L-Tyrosine, abb. *Tyr:* L-α-amino-β-(p-hydroxyphenyl)-propionic acid, an aromatic, ketogenic, proteogenic amino acid. M_r 181.2, m. p. 342–344 °C (d.), $[\alpha]_D^{25}-10.0$ ($c=2$ in 5M HCl). L-T. is nonessential in humans, since it can be synthesized by hydroxylation of L-phenylalanine. Millon's color reaction for proteins is specific for L-T. residues. The widely used Lowry protein assay also depends largely on the specific reaction of the phenolic residues of L-T. For the biosynthesis of L-T. see Aromatic biosynthesis. For all other metabolism of L-T. see L-Phenylalanine. L-T. is used therapeutically in disturbances of thyroid function.

Tyrosine kinase: see Protein-tyrosine kinase.

Tyrosinemia: see Inborn errors of metabolism.

Tyrosinosis: see Inborn errors of metabolism.

U

U: abb. for uracil.

UALase: abb. for urea amidolyase.

Ubiquinone, *coenzyme Q*, abb. *Q:* a low M_r electron transport component of the respiratory chain. Structurally, it is a 2,3-dimethoxy-5-methylbenzoquinone carrying an isoprenoid side chain. There are various types designated according to the number of carbon atoms, or the number of isoprene units in the side chain: U-30 (=U-6), U-35 (=U-7), U-40 (=U-8), U-45 (=U-9) and U-50 (=U-10), where the first figure refers to the number of carbon atoms, and the figure in parentheses to the number of isoprene units. Other abbreviations are, e.g. coenzyme Q_{10}, UQ-50, UQ_{10}, Q-10 and CoQ_{10} (Fig. 1). U. are slowly destroyed by oxygen, UV light and sunlight, and they are rapidly oxidized in alkaline solution, except in the presence of pyrogallol to remove oxygen.

U./dihydroubiquinone is a redox system in the respiratory chain. Reversible reduction of the benzoquinone to hydroquinone is a stepwise reaction: the semiquinone (hydroquinone radical) is first formed by electron transfer; further transfer of a single electron forms the hydroquinone anion or phenolate, which takes up two protons to form the hydroquinone (Fig. 2). The reverse reaction, dehydrogenation of the hydroquinone to the quinone, is initiated by dissociation of the hydroquinone to the hydroquinone anion by release of two protons; oxidation then proceeds in two stages by removal of electrons. Other dehydrogenations are analogous, e.g. dehydrogenation of ethanol to acetaldehyde via an intermediate alcoholate. Thus in the enzymatic dehydrogenation of ethanol by alcohol dehydrogenase (see Alcoholic fermentation) one proton (H^+) and an electron pair (2e) are transferred together as a hydride ion, and the second proton equilibrates with the protons of the aqueous medium (see Nicotinamide adenine dinucleotide).

U. is so named from its ubiquitous occurrence.

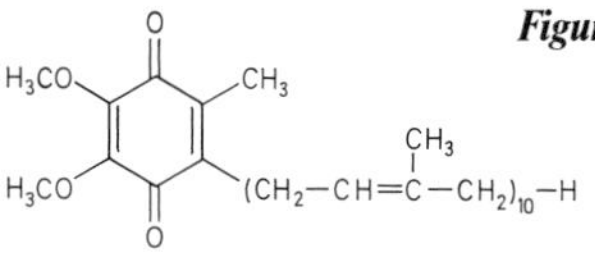

Ubiquitin, *ATP-dependent proteolysis factor 1, APF-1:* A small polypeptide, M_r 8500, first isolated during the purification of polypeptide hormones from thymus. U. was then found by radioimmunoassay in vertebrates, invertebrates, plants and yeasts. Earlier reports that it induces differentiation of thymocytes and stimulates adenylate cyclase have not been confirmed; the name, ubiquitous immunopoietic polypeptide (UBIP) is therefore inappropriate. The primary structure of U. is almost identical in insects, trout, cattle and humans. [J. G. Gavilanes et al., *J. Biol. Chem.* **257** (1982) 10267–10270]

N-terminal ubiquitination appears to be necessary and sufficient for U.-dependent degradation of proteolytic substrates. Acetylation of *N*-termini in vivo blocks U.-dependent proteolysis (measured in vitro). *N*-terminal acetylation in vivo may therefore protect proteins from degradation by preventing ubiquitination. U.-dependent proteolysis has an absolute requirement for ATP. The ATP is required for synthesis of the U.-protein complex and for the subsequent degradation of the proteolytic substrate (Fig. 1), but the nature of this proteolysis is unclear. The reticulocyte cell-free system has been used for many studies on U.-dependent proteolysis. The physiological function of the U. pathway in reticulocytes may be to remove globin molecules containing biosynthetic errors, and it may be important in removing the proteins of mitochondria and ribosomes and various other proteins that are lost during erythrocyte maturation. Studies have been reported on an ATP-dependent protease from rabbit reticulocyte lysate, which degrades U.-lysozyme conjugates, but not free lysozyme even in the presence of U. [R. Hough & M. Rechsteiner, *J. Biol. Chem.* **261** (1986) 2391–2399]

Chromosomal protein A24, now called uH2A (ubiquitin-H2A semihistone) is the most prominent of a family of branched proteins in which the *C*-terminal glycine (76) of U. is linked by an isopeptide bond to the $\varepsilon-NH_2$ of lysine 119 of histone 2A. Ubiquitination of many intracellular proteins at the ε-amino groups of their Lys residues to yield branched U.-protein conjugates may represent a physiological role for U., distinct from its established function in protein degradation. The U. of uH2A and uH2B is in rapid equilibrium with free U. in interphase cells. As mitotic chromosome condensation reaches completion, the levels of uH2A and uH2B in chromosomes show a marked decrease, probably due to enzymatic deubiquitination; then normal levels are rapidly restored during postmitotic chromosome decondensation.

Figure 1. Coenzyme Q_{10}.

Hydroquinone Hydroquinone Hydroquinone Benzoquinone
            anion   radical
         (phenolate) (semiquinone)

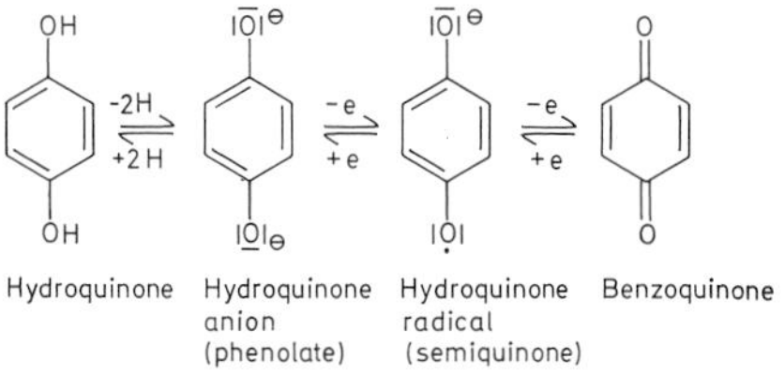

Figure 2. Oxidation of hydroquinone.

DNA sequences for U. have been cloned from a variety of eukaryotes. U. is generated by processing of a poly-U. precursor protein. U.-coding elements are typically organized in spacerless head-to-tail arrays, the number of coding repeats varying according to the organism, e.g. 6 in yeast, 9 in human (Fig. 2). [A. Hershko *Cell* **34** (1983) 11–12; D. Finley & A. Varshavski *Trends Biochem. Sci.* **10** (1985) 343–347] High endothelial venule (HEV) homing receptor (also known as gp90^{MEL-14}, a glycoprotein, M_r 90000, specifically recognized by a rat monoclonal antibody, MEL-14, against murine lymphocytes) on the cell surface of murine lymphocytes has been identified as a branched structure with two amino termini, and consisting of a core polypeptide modified by glycosylation and ubiquitination. cDNA clones for U. possess no code for a signal sequence, so it is unlikely that U. can target the membrane on its own; probably ubiquitination of the prospective membrane receptor protein occurs while it is being synthesized, i.e. while it is still accessible to the U. in the cytoplasm. Ubiquitination is not an exclusive feature of the HEV-homing receptor; monoclonal antibodies to U. reveal other ubiquitinated proteins on the lymphocyte surface. HEV-homing receptor binds specifically to the lining of postcapillary high endothelial venules, thereby enabling lymphocytes to enter lymphoid organs, a process necessary for the immune response. The U. moiety of the HEV-homing receptor may be critical for this cell surface recognition. [M. Siegelman et al. *Science* **231** (1986) 823–829; T. St. John et al. *Science* **231** (1986) 845–850; M. Gallatin et al. *Cell* **44** (1986) 673–680]

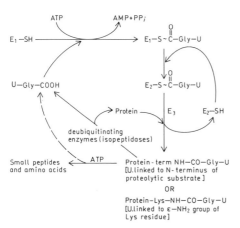

Figure 1. Role of ubiquitin in the degradation of proteins and branched proteins E$_1$–SH (U.-activating enzyme), E$_2$–SH (a transferase which transfers U. to the conjugation site) and E$_3$ (a ligase which catalyses amide bond formation) have all been purified and their properties studied [A. Hershko, *J. Biol. Chem.* **258** (1983) 8206–8214]. U–Gly–COOH represents U. with its *C*-terminal glycine. Two deubiquitinating enzymes have been characterized [C. M. Pickart & I. A. Rose *J. Biol. Chem.* **261** (1986) 10210–10217; S.-I. Matsui et al. *Proc. Nat. Acad. Sci. USA* **79** (1982) 1535–1539]

```
1    76  1    76  1    76  1    76  1    76  1    76
Met – Gly-Met – Gly-Met – Gly-Met – Gly-Met – Gly-Met – Gly-Asn
      ↑        ↑        ↑        ↑        ↑        ↑
```

Figure 2. Structural organization of poly-ubiquitin precursor protein in Saccharomyces cerevisiae, deduced from the nucleotide sequence of the ubiquitin gene. Arrows indicate sites of proteolytic cleavage in the maturation process. The *C*-terminus (here Asn) varies according to the organism, e.g. Val in human, Tyr in chicken.

UDP: abb. for uridine 5′-diphosphate.

UDPG: see Nucleoside diphosphate sugars.

UDP-glucose: see Nucleoside diphosphate sugars.

Ultracentrifugation: see Methods of biochemistry; Proteins.

Ultrafiltration: see Proteins.

UMP: abb. for uridine 5′-monophosphate.

cUMP: abb. for cyclic uridine monophosphate. See Uridine phosphates.

Uncompetitive inhibition: see Effectors.

Uncoupler: a chemical compound which prevents oxidative phosphorylation of ADP to ATP in the respiratory chain without affecting electron transport. It thus uncouples respiration and ATP formation; respiration continues or is even increased, but yields no energy. Some important U. are 2,4-dinitrophenol (abb. DNP), dicumarol, carbonyl cyanophenylhydrazone, salicyl-anilide, etc. These compounds affect only oxidative phosphorylation, but not substrate phosphorylation in glycolysis. An U. probably destroys or discharges an energy-rich state or intermediate generated by the electron transport. It is thought that uncoupling of oxidative phosphorylation may be involved in heat production in brown fat tissue. See Oxidative phosphorylation; Ionophore.

Unit membrane: see Biomembranes.

Untwisting enzymes: eukaryotic type I Topoisomerases (see).

Unusual nucleic acid components: see Rare nucleic acid components.

Ura: abb. for uracil.

Diketo pH < 8.5　　　Keto-enol pH 13　　　Enol pH > 13

pH-Dependent tautomeric forms of uracil

Uracil, abb. *U* or *Ura:* 2,4-dihydroxypyrimidine, a 1,3-diazine, which occurs as a pyrimidine base in all ribonucleic acids. M_r 112.09, m.p. 335 °C (d.). U. is formed by degradation of U. nucleotides and nucleosides (see Pyrimidine biosynthesis), and it is the starting point for reductive and oxidative Pyrimidine degradation (see).

Urate oxidase: see Uricase.

Urates: salts of Uric acid (see).

Urd: abb. for uridine.

Urea, *carbamide:* H$_2$N-CO-NH$_2$, the diamide of carbonic acid. M_r 60.01, m.p. 132.7 °C. Solutions of U. in water develop high concentrations of reactive cyanate ions on standing. These can be removed by acidification. U. is the product of ammonia detoxification in the ureotelic animals. It is produced by several metabolic pathways: 1.via the urea cycle, 2.via oxidative purine degradation and enzymatic hydrolysis of allantoic acid and glyoxylurea (ureidoglycollate), 3.by hydrolysis of L-arginine and other guanidine derivatives by arginase and other enzymes of the EC sub-sub class 3.5.3 (see L-Arginine, Guanidine derivatives), 4.via various other metabolic pathways of limited distribution or importance, e.g. by the rare oxidative pyrimidine degradation. U.is hydrolysed by urease and urea amidolyase. Other mechanisms of urea hydrolysis are of doubtful significance and have not been enzymologically proven.

U. is accumulated by many higher fungi *(Basidiomycetes),* particularly by *Agaricus* species and Gasteromycetes like the puff-balls *(Lycoperdon)* and the bovists *(Bovista).* In the cultivated champignon, for example, U.is a true nitrogen excretion product, and in bovists and puff-balls it serves for the storage and translocation of nitrogen for the formation of protein and chitin in the spores (see Ammonia detoxification). U.can be used in osmoregulation by the marine cartilagenous fish, in which it is accumulated in the tissue fluids and blood in relatively high concentrations. U.is also stored in the body fluids of lungfish *(Dipnoi)* during estivation, when the animals enclose themselves in a cocoon-like structure to survive the dry period. U.is formed in this case via the urea cycle and by the degradation of purines (see Glycine-Allantoin cycle).

U.is used in high concentrations as a denaturing agent for proteins. The toxic effect of U.on the skin is not understood. U.has long been used as a nitrogen fertilizer, but a urease-catalysed, sudden release of ammonia can occur in the soil, leading to nitrogen loss and poisoning. U.is therefore applied in the form of U.-aldehyde condensation compounds, which are "slow-release" sources of nitrogen for plants. The difficulty can also be overcome, in principle, by a suitable application form (granulates) and by the use of urease inhibitors (see Urease). There are similar difficulties in the use of U.as a nonprotein source of nitrogen for cattle feed, and for this reason, suitable compounds, from which the urea is released slowly, must be used. In the ruminant stomach, U.is degraded by the symbiotic ruminant microorganisms (see Symbiosis).

Urea amidolyase, *ATP: urea amido-lyase,* abb. *UALase, urea carboxylase (hydrolysing)* (EC 6.3.4.6): a urea splitting enzyme present in some yeasts (*Saccharomyces, Candida* etc.) and green algae (*Chlorella,* etc.), where it replaces urease. It is a biotin enzyme and is inhibited by avidin. The catalysed reaction is an ATP-dependent cleavage of urea to CO$_2$ and NH$_3$:

$$NH_2-CO-NH_2 + HCO_3^- + ATP \rightarrow 2HCO_3^- + 2NH_4^+ + ADP + P_i$$

UALase is a Multienzyme complex (see) and consists of at least two enzyme proteins: 1. *urea carboxylase,* which catalyses carboxylation of urea to the anionic form of *N*-carboxyurea or allophanic acid, NH$_2$-CO-NH-CO$_2^+$, a reaction requiring biotin and ATP; 2. *allophanate hydrolase,* an amidase which cleaves the allophanate into two hydrogen carbonate ions and two ammonium ions. The regenerated hydrogen carbonate acts catalytically, being continually reused as more urea is converted. The carboxylase component appears to be constitutive, whereas the allophanate hydrolase is inducible. Allophanate is an inducer of the enzymes of Purine degradation.

Urea carboxylase: see Urea amidolyase.

Urea cycle, *arginine-urea cycle, ornithine cycle, Krebs-Henseleit cycle:* a metabolic cycle present in mammals and other ureotelic animals (e.g. adult amphibians), which results in the synthesis of urea from carbon dioxide, ammonia and the α-amino nitrogen of L-aspartic acid (Fig.). The process is energy-dependent; synthesis of one molecule of urea or L-arginine requires 3 molecules of ATP, and involves the expenditure of 4 high energy bonds (2 molecules of ATP are cleaved to ADP and inorganic phosphate, and one is cleaved to AMP and pyrophosphate, the latter being further hydrolysed to inorganic phosphate). The U.c. is catalytic, and is based on the recycling of the catalytic molecule, L-ornithine. The primary function of the U.c.is to convert waste nitrogen into nontoxic, soluble urea, which can be excreted. The cycle may also be completed, not by hydrolysis of L-arginine to urea, but by transfer of the amidine group to glycine, to form L-ornithine and guanidinoacetic acid (the precursor of creatine; see L-Arginine; Phosphagens). A further function is the synthesis of the proteogenic amino acid, L-arginine. The U.c.has, in fact, evolved from an original pathway for L-arginine synthesis. In animals, the U.c.is more or less primed by synthesis of L-ornithine from L-glutamate, and to some extent by synthesis of L-ornithine from the products of degradation of L-proline. Dietary L-arginine may serve to supplement the cycle, or the synthesis of L-ornithine and its conversion to L-arginine in the U.c. may supplement the dietary requirement for L-arginine; the balance of these two processes depends on the animal species in question, its physiological state and its diet, e.g. many young, growing animals have a dietary requirement for L-arginine, whereas the adult appears able to synthesize its total requirement.

L-Arginine synthesis itself represents the primary synthesis of the amidine group; other naturally occurring guanidino compounds, chiefly Phosphagens (see), are synthesized by transfer of the amidine group from L-arginine to the appropriate amino receptor.

The chief site of the U.c. is the liver. Conversion of L-ornithine to L-citrulline and the synthesis of carbamoyl phosphate occur in the mitochondrial matrix, and all the other reactions of the U.c.occur in the cytoplasm. Kidney cytoplasm contains the enzymes for the conversion of L-citrulline to L-ornithine, but kidney mitochondria lack the necessary enzymes for converting L-ornithine to L-citrulline, and for synthesizing carbamoyl phosphate. Some L-citrulline is transported from the liver to the kidneys, where it is converted to L-ornithine and urea.

Fumarate produced by the action of argininosuccinate lyase can enter the mitochondria and be converted in the TCA cycle to oxaloacetate; transamination of the latter to aspartate represents the channelling of waste nitrogen into the amino group of aspartate, which then transfers this nitrogen to the U.c. by the action of argininosuccinate synthetase. This series of reactions links the U.c. with the TCA cycle. One reaction of this extra cycle (malate + NAD^+ → oxloacetate + NADH + H^+) represents a source of 3 molecules of ATP by Oxidative phosphorylation (see). The chief source of ammonia consumed in the synthesis of carbamoyl phosphate is the oxidative deamination of L-glutamate by L-glutamate dehydrogenase: L-glutamate + $NAD^+ + H_2O$ → 2-oxoglutarate + NADH + H^+ + NH_3; here again, the oxidation of NADH provides 3 molecules of ATP. The energy requirement of the U.c. is therefore amply covered by the energy production of associated processes. See also Ammonia assimilation.

Urease (EC 3.5.1.5): an enzyme of high catalytic activity that catalyses the hydrolysis of urea to CO_2 and NH_3:

$$O=C{\overset{\textstyle NH_2}{\underset{\textstyle NH_2}{}}} + 2H_2O \rightleftharpoons H_2CO_3 + 2NH_3.$$

U. is found especially in plant seeds and microorganisms, as well as invertebrates (crabs, mussels), and shows a high degree of substrate specificity; apart from urea, it only attacks urea derivatives like hydroxy- and dihydroxyurea, which also act as non-competitive inhibitors of U. The U. of soybean was the first enzyme to be crystallized (Sumner, 1926). *Properties of soybean U:* pH-optimum 7.0; IP 5.0; M_r 489 000. It consists of two enzymatically active half molecules (M_r 240 000), which separate at pH 3.5. At pH 9.0 in the presence of 0.1% sodium dodecyl sulfate and 45% thioglycerol, they dissociate further into quarter molecules (M_r 120 000). At neutral pH in 0.1% sodium dodecyl sulfate, U. dissociates into its 8 subunits (M_r 60 000), which comprise two covalently bound chains (each chain of M_r 30 000). U. is remarkably resistant to denaturation by its own

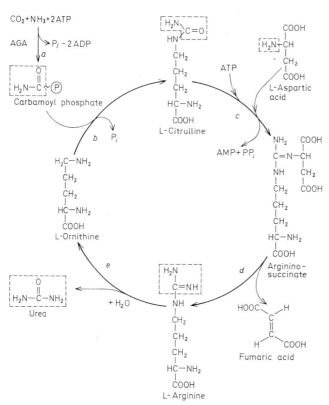

Urea cycle

a Carbamoyl-phosphate synthetase (ammonia), EC 6.3.4.16.
b Ornithine carbamoyltransferase, EC 2.1.3.3.
c Argininosuccinate synthetase, EC 6.3.4.5.

d Argininosuccinate lyase, EC 4.3.2.1.
e Arginase, EC 3.5.3.1.
AGA = *N*-Acetylglutamic acid, a stimulatory allosteric effector of carbamoyl phosphate synthetase.

substrate, urea, which at 8 M concentration can be used for the denaturation of numerous proteins. In 8-9 M urea, U.dissociates into M_r 60000-subunits, which still possess urease activity. Furthermore, U.-antiurease complexes are still catalytically active. Bacterial U. is a smaller molecule than soybean U.

Ureide plants: plant families that accumulate allantoin and/or allantoic acid, and use these compounds as nitrogen reserves. U.p.are members of the *Aceraceae, Boraginaceae, Hippocastanaceae* and *Platanaceae.*

Ureotelic organisms: see Ammonia detoxification.

Uric acid: 2,6,8-trihydroxypurine, a purine derivative which is the most important end product of nitrogen metabolism. M_r 168.1, m.p. 400 °C. U.a.was discovered in urine in 1776 by Scheele. Its salts are called urates. It is excreted in the urine in certain animal families, particularly birds and reptiles. It has been isolated from bird excrements (guano). Humans and the great apes usually excrete U.a. unchanged. In the adult human, U.a.contains 1 to 3% of the total nitrogen in the urine. Elimination of nitrogen in the form of uric acid (uricotelic) is one method of ammonia detoxification.

U.a.is generated from xanthine by the enzyme xanthine oxidase in aerobic purine catabolism. The amino nitrogen from the degradation of amino acids can also be transferred to U.a. The enzyme uricase converts U.a. to allantoin (uricolysis; see Purine degradation).

Lactim form   Lactam form

Tautomeric forms of uric acid

Uricase, *urate oxidase* (EC 1.7.3.3) a copper-containing aerobic oxidase, which, in the presence of oxygen, catalyses the oxidation of poorly soluble uric acid or urates to soluble allantoin, with the formation of hydrogen peroxide. U.occurs in all vertebrates and in all invertebrates with the exception of insects (only flies and related groups possess U.). The chief site of uric acid oxidation is the liver, where U. is stored in special uricase-rich microbodies called uricosomes. U. is used for the clinical diagnosis of increased uric acid levels, particularly in gout. Porcine U. has M_r 125000 (4 subunits, M_r 32000), IP 6.3, pH-optimum 9-9.5. Analogs of uric acid are powerful inhibitors of U.

Uricolysis: oxidation and decarboxylation of uric acid to allantoin, catalysed by uricase as part of aerobic Purine degradation (see).

Uricotelic organisms: see Ammonia detoxification.

Uridine, abb. *Urd:* 3-β-D-ribofuranosyluracil, a β-glycosidic Nucleoside (see) of D-ribose and the pyrimidine base, uracil. M_r 244.20, m.p. 165-167 °C, $[\alpha]_D^{16} + 9.6$ ° ($c = 2.0$, water). Uridine phosphates (see) are metabolically important.

Uridine diphosphate glucose: see Nucleoside diphosphate sugars.

Uridine phosphates: Nucleotides (see) of uracil; phosphate esters of uridine. *Uridine 5'-monophosphate,* abb. *UMP, uridylic acid,* M_r 324.2, m.p. 198.5 °C is produced de novo in Pyrimidine biosynthesis (see), or by degradation of nucleic acids. UMP is the starting point for the synthesis of other pyrimidine nucleotides. *Uridine 5'-diphosphate,* abb. *UDP,* M_r 404.2, serves as the activating group of many Nucleoside diphosphate sugars (see) involved in transglycosidation. *Uridine 5'-triphosphate,* abb. *UTP,* M_r 482.2, is a structural analog of ATP, and is required for the synthesis of uridine diphosphate sugars (see Nucleoside diphosphate sugars).

Cyclic uridine 3,5'-monophosphate, abb. *cyclo-UMP, cUMP:* M_r 306.2, is a cyclic nucleotide, and like cyclic adenosine 3,5'-monophosphate (see Adenosine phosphates), it is involved in metabolic regulation. cUMP inhibits growth of some transplantable tumors. A specific cUMP-degrading enzyme is present in heart muscle.

Uridylic acid: see Uridine phosphates.

L-Urobilin: see Bile pigments.

Urocanic acid: see L-Histidine.

Urokinase: see Plasmin.

Uronic acids: aldehyde carboxylic acids formed by oxidation of the terminal primary alcohol group of aldoses. U.a. are named by adding the ending "-uronic acid" to the stem of the parent monosaccharide, e.g. D-glucuronic acid, D-galacturonic acid, D-mannuronic acid. U.a. tend to form lactones, usually γ-lactones. They give the usual reactions for sugars (see Carbohydrates), and are widely distributed as components of glycosides, polyuronides, polysaccharides and mucopolysaccharides.

5α-Ursane: see Amyrin.

Ursodeoxycholic acid: 3α,7β-dihydroxy-5β-cholan-24-oic acid, a dihydroxylated steroid carboxylic acid, belonging to the bile acids. M_r 392.58, m.p. 203 °C, $[\alpha]_D + 57$ ° (alcohol). U.is a characteristic component of bear bile, and is also present in human bile.

Ursolic acid: a simple unsaturated pentacylic triterpene carboxylic acid, M_r 456.71, m.p. 292 °C, $[\alpha]_D + 72$ ° (chloroform). U.a. is a structural derivative of α-Amyrin in which the 28-methyl group is replaced by a carboxyl group (see Amyrin). It occurs widely in plants, as the free acid, esterified, or as the aglycon of triterpene saponins (see Saponins), e.g. in the wax layer of apples, pears and cherries, in the skin of bilberries and cranberries, and in the leaves of many members of the *Rosaceae, Oleaceae* and *Labiatae.*

U-snRNP, *snurp:* one of a group of small nuclear ribonuclear proteins (snRNP). Rapid progress of elucidation began with the discovery, by Lerner and Steitz [*Proc. Natl. Acad. Sci. USA* **76** (1979) 5495-5499] that U1, U2, U5, U4/U6 are precipitated by human systemic lupus erythematosus (autoimmune) sera of the Sm type. The RNA in U-snRNP is rich in uracil; hence the designation U. By 1978, 6 types of RNA (snRNA) associated with U-snRNP had been identified and designated U1 through U6 snRNA. At this writing, an additional 4 U-snRNA have been identified: U7-U10. All these species except U6 are characterized by a trimethylguanosine "cap" at the 5' end, and all (except U3) are precipitated by Sm antibodies. The classical U1, U2, U4/U6-snRNP are found in the nucleoplasm; U3-snRNP are found in the nucleolus. U4 and U6 are found in a single snRNP. There are 7 to 8 proteins associated with the particles; the contributions of protein and RNA to the function are not well characterized, but it is very likely that it is the RNA which is catalytically active, and the protein which provides structural stability. The U7-RNA from sea urchins has 56 or 57 nucleotides and the corresponding U7-snRNP has a M_r of 200-250000. The *Drosophila* U7-snRNP has M_r 140000.

Those U-snRNPs for which a biological function is known participate in the processing of RNA. U1-snRNP is required for cleavage of pre-mRNA at the 5'-splice site [A. Krämer et al. *Cell* **38** (1984) 299-307]. Antiserum specific for U2-snRNP can precipitate the RNA in the complex which carries out splicing in a cell-free system; hence U2-snRNP is probably also involved in removal of introns from pre-mRNA [P.J. Grabowski et al. *Cell* **42** (1985) 345-353]. U4 and U6 RNA are associated in a single RNP. Sm antisera prevent polyadenylation of pre-mRNA in vitro, but specific anti-U1 and anti-U2 antibodies do not; hence it is likely but not proven that U4/U6 snRNPs are involved in polyadenylation of mRNA. Sea urchin U7 snRNA is required for 3' processing of histone pre-mRNA; it contains a sequence complementary to CAAGAAAGA, which is conserved in histone genes [M.L. Birnstiel et al. *Cell* **41** (1985) 349-359]. In contrast to the other U-snRNPs, which are found in the nucleoplasm, U3-snRNP is found in the nucleolus, where it is associated with 5.8 S, 18 S and 28 S RNA. It may therefore have a role in the processing of ribosomal RNA. [I.W. Mattaj, *Trends Biochem. Sci.* **9** (1984) 435-437; O. Georgiev & M.L. Birnstiel *EMBO J.* **4** (1985) 481-489]

Uteroferrins: a class of purple acid phosphatases from mammalian sources, including porcine uterus, bovine and rat spleen, and hairy cell leukemia cells of human spleen. They are glycoproteins, M_r 35-40000, containing 2 Fe atoms per mol of enzyme. In the fully oxidized ($2Fe^{3+}$) form, they are enzymatically inactive, whereas the one-electron reduced form (Fe^{2+}, Fe^{3+}) catalyses hydrolysis of phosphate esters. Earlier controversy concerning the nature of phosphate binding is now resolved: The binuclear iron cluster binds inorganic phosphate strongly in the oxidized state; reduced U. binds phosphate weakly, accompanied by a red shift in the tyrosinate-Fe^{3+} charge transfer band and loss of characteristic EPR signals of U. The physiological significance of U. is unknown. [J.W. Pyrz et al. *J. Biol. Chem.* **261** (1986) 11015-11020]

UTP: abb. for uridine 5'-triphosphate.

Utter reaction: see Gluconeogenesis.

V

Vaccenic acid: Δ^{11}-octadecanoic acid, $CH_3-(CH_2)_5-CH=CH-(CH_2)_9-COOH$, an unsaturated fatty acid, M_r 282.5, m.p. 44 °C *(trans)*, 14.5 °C *(cis)*. The *trans* form occurs in the glycerides of animal fats, e.g. beef, mutton and butter fat, and in vegetable oils. The *cis* form is hemolytic, occurring in plasma and various animal tissues, and in *Lactobacillus*. V.a. is the principal unsaturated acid in *Escherichia coli*.

Vacuoles: structures within plant cells composed of a three-layered membrane *(tonoplast)* enclosing the *cell sap*. In early stages of development, several small V. may be present, and they occupy a relatively small proportion of the cell volume. Mature, differentiated cells usually contain one large central V., and the cytoplasm is a relatively thin layer pressed firmly between the plant cell wall and the V. The cell sap within the V. is an aqueous solution of numerous substances in true or colloidal solution; in addition to sugars and salts these include inner secretions, so that the V. is considered as an excretory organ. Alkaloids accumulate in V. and are neutralized by salt formation with inorganic and organic anions. V. of acid and ammonium plants contain accumulated ammonium salts of organic acids. The V. is important in the maintenance of turgor (inner pressure) of the plant cell, by acting as an osmotic system (see Osmosis). The tonoplast is semipermeable; due to the high osmotic pressure of the cell sap, the V. expands and compresses the cytoplasm as a thin layer tightly against the inner surface of the cell wall; further expansion is prevented physically by the cell wall. The resulting turgor is important for the mechanical strength of herbaceous plants.

Val: abb. for L-valine.

Valepotriates: iridoids from *Valeriana* and *Kentranthus* spp. *Valeriana officinalis* contains up to 5% V., which are responsible for the sedative properties of this drug. Hydroxyl groups on the iridoid structure are esterified with isovaleric acid. Hydrolysis of the esters with HCl causes decomposition of the unstable alcohol moiety with production of a blue color. The most important representative is Valtratum (see).

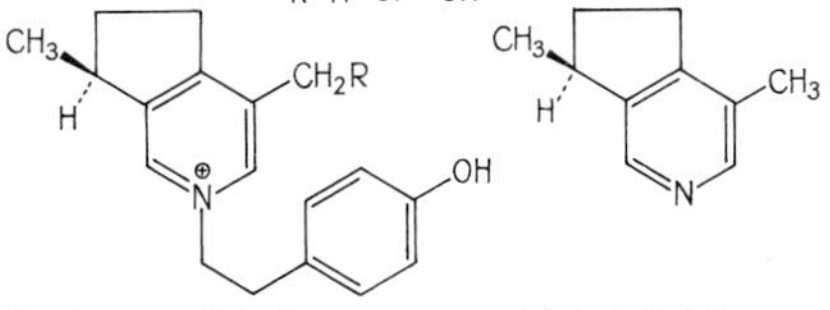

Quaternary Valerian alkaloids

L(+)-Actinidine

Valerian alkaloids: terpene alkaloids containing a pyridine ring (therefore also considered as pyridine alkaloids) from valerian *(Valeriana officinalis)*. The quaternary V.a.(Fig.) are responsible for the excitory action of valerian on cats.

L-Valine, abb. *Val:* L-α-aminoisovaleric acid, $(CH_3)_2CH-CH(NH_2)-COOH$, an aliphatic, neutral, essential, glucogenic, proteogenic amino acid. For biosynthesis, see L-Isoleucine. For degradation, see Leucine. The intact molecule of Val is incorporated in the biosynthesis of Penicillin.

Valinomycin: cyclo-(-D-Val-Lac-Val-D-Hyv-)$_3$, an antibiotic, cyclic depsipeptide, especially active against *Mycobacterium tuberculosis*. In addition to valine, it contains the heterocomponents, L-lactic acid (Lac) and D-α-hydroxyisovaleric acid (D-Hyv). It is an Ionophore (see), which selectively transports potassium ions across membranes.

Valtratum, *valepotriatum*, *valtrate*, *valepotriate*: the chief member of the Valepotriates (see) from valerian root, M_r 422; an oil unstable to acid, alkali

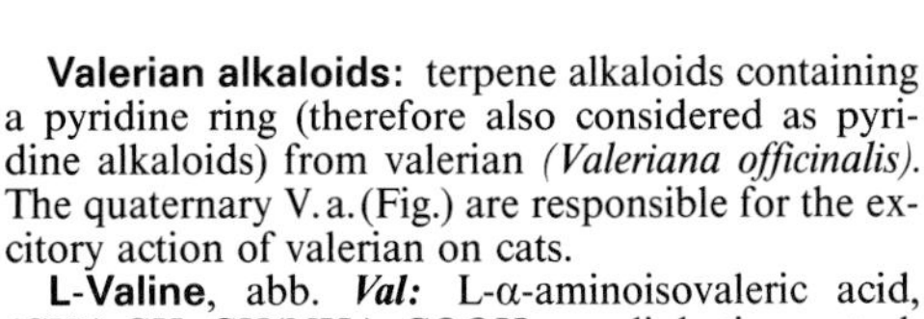

Valtratum   Baldrianal

R = —CO—CH$_2$—CH(CH$_3$)$_2$

and heat, n_D^{20} 1.4906, $[\alpha]_D^{20}$ + 172.2 ° (methanol). Acid hydrolysis produces isovaleric acid and 4-acetoxymethyl-7-formylcyclopenta[c]pyran (baldrianal), the latter being formed by dehydration and rearrangement of the unstable constituent iridoid alcohol (Fig.). Baldrinal has m.p. 112-113 °C, M_r 218. Structure of V.was revised in 1968 ["Die Konstitution der Valepotriate" P.W. Thies, *Tetrahedron, 24* (1968) 313-347] *Valtratum* is accepted as the non-proprietary name by the World Health Organization.

Vanadium, V: a trace element required for normal growth by animals; most studies have been performed on rats and chicks. A reliable estimate of human V requirement is lacking; most dietary items contain less than 100 ng V/g. V.is taken up as V^{5+} and reduced to V^{3+} in the cell. V.-deficiency causes an increase in plasma cholesterol and triglycerides. V.stimulates the oxidation of phospholipids and decreases cholesterol synthesis by inhibiting squalene synthase (a liver microsomal enzyme system); it also stimulates acetoacetyl-CoA deacylase in liver mitochondria. It has also been implicated in bone metabolism or formation. V.-deficiency leads to abnormal bone growth, and injected radioactive V. shows a high incorporation into areas of active mineralization in dentine and bone.

V. is required for optimal growth of some green algae, and it inhibits growth of *Mycobacterium tuberculosis.* Nitrogen fixation (see) by *Azotobacter* is increased by V. Certain ascidians show a remarkable ability to concentrate V. from the surrounding sea water (see Heavy metals).

Variabilin: 3,9-dimethoxy-6a-hydroxypterocarpan, see Pterocarpans.

Vasoactive intestinal peptide, *VIP:* an octacosapeptide from porcine small intestine, which causes vasodilation, lowers arterial blood pressure, increases cardiac output, enhances myocardial activity, increases glycogenolysis and relaxes the smooth muscle of trachea, stomach and gall bladder. For structure, see Secretin.

Vasopressin, *antidiuretic hormone,* abb. *V., antidiuretin, pitressin:* a neurohypophysial peptide hormone. V. has a direct antidiuretic action on the kidneys. It also causes vasoconstriction of the peripheral vessels, with slowing of the heart beat and increase of blood pressure. The amino acids in positions 3, 4 and 8 of the nonapeptide are variable. The phylogenetic precursor of V. and of Oxytocin (see) is [8-arginine]vasotocin. V. occurs as [8-arginine]V., M_r 1084 and [8-lysine]V., M_r 1056. [8-Arginine]V. is physiologically and pharmacologically one of the most active substances; 2 ng are sufficient to cause pronounced antidiuresis in the human. V. is synthesized in the supraoptic nucleus of the hypothalamus, then transported in granules, in combination with neurophysin II, down nerve fibers (supraoptico-hypophysis) to the posterior pituitary (neurohypophysis), where it is stored. Release of V. depends on the degree of hydration of the organism, and is stimulated by thirst and lack of water. It activates the adenyl cyclase system in the distal tubule of the kidney, resulting in increased resorption of water and increased excretion of Na^+. Determination of V. is performed biologically by measurement of blood pressure and/or diuresis, following injection into test animals. A radioimmunological assay is also used. [E.G. Beardwell, *J. Clin. Endocrin. Metab.33* (1971) 254–260] [8-Arginine]V. also influences learning and memory, brain development, cardiovascular control, thermoregulation, development of tolerance to and dependence on opiates and ethanol, and drug-seeking behavior. Proteolysis of [8-arginine]V. to the heptapeptide pGlu-Asn-Cys(Cys)-Pro-Arg-Gly-NH$_2$ or to the corresponding hexapeptide lacking the terminal Gly-NH$_2$ removes the pressor activity of V., but the derivative peptides are even more active than the parent in facilitating memory consolidation in rats. [J.P.H. Burbach et al., *Science 221* (1983) 1310–1312]

Veracevine: see Germine.

Veratramine: a Veratrum alkaloid (see) of the jerveratrum type, with C-nor-D-homo structure, M_r (anhydrous) 409, m.p. (monohydrate) 209.5–210.5 °C $[\alpha]_D^{19}$ (anhydrous) −70 ° (methanol), found in hellebores *(Veratrum album, V. eschscholtzii, V. viride).* *V. viride* also contains *veratrosine,* a glycoalkaloid in which the 3-β-hydroxyl group of veratramine is linked glycosidically to D-glucose.

Veratrosine: see Veratramine.

Veratrum alkaloids: a group of steroid alkaloids found in the *Solanaceae,* and in the genera *Veratrum* (hellebores) and *Fritillaria* (fritillary). V.a. are structural derivatives of the parent hydrocarbon cholestane (see Steroids); in some members ring C is contracted and ring D is expanded (C-nor-D-homo type). V.a. are subdivided into *jerveratrum alkaloids,* e.g. Jervine (see), Rubijervine (see), Isorubijervine (see), Veratramine (see), and *ceveratrum alkaloids,* e.g. Germine (see). Members of the first group contain 2–3 oxygen atoms and occur free or linked glycosidically to a molecule of D-glucose (glycoalkaloids). Ceveratrum alkaloids contain 7–8 oxygen atoms, and occur chiefly as esters; the commonest ester acids are acetic, angelic and veratric. V.a. have a positive ionotropic action on the heart, and cause a decrease of blood pressure by reflex inhibition of the vasomotor centers. They were used for treatment of hypertension, but have been replaced by Rauwolfia alkaloids and other drugs. V.a. have also been used as insecticides. They are biosynthesized from cholesterol.

Verbenaline: see Iridoids.

Vernaline: see Flowering hormone.

Vernine: obsolete name for guanosine.

Versene: see Ethylenediaminetetraacetic acid.

Vertebrate hormones: hormones of vertebrate animals. On the basis of studies on the phylogenetic relationships of certain proteins, proteohormones and peptide hormones, the separate classification of V.h. and Invertebrate hormones (see) appears to be justified. Chemically, V.h. are an heterogeneous group, which can be subdivided into Steroid hormones (see), hormones derived from amino acids (see Thyroxin, Adrenalin, Melatonin), Peptide hormones (see), Proteohormones (see), and hormones derived from fatty acids (see Prostaglandins). There is, however, no fundamental difference between V.h. and invertebrate hormones, with respect to types of chemical structure, or biochemical mode of action.

Vestitol: 7,2′-dihydroxy-4′-methoxyisoflavan, see Isoflavan.

VHDL: see Lipoproteins.

Vimentin: a fibrous protein which makes up intermediate filaments in cells of mesenchymal origin. Subunit M_r, 53 000. The structure is similar to that of the α-keratins. See Keratins; Cytoskeleton (intermediate filaments).

Vinblastine, *vincaleucoblastine:* a dimeric indole-indoline alkaloid, m.p. 211–216 °C (d.), $[\alpha]_D$ +42 ° (chloroform). Structurally, V. is equivalent to a combination of the alkaloids vindoline and catharidine (see Vinca alkaloids). Very low concentrations of V., accompanied by vindoline and catharidine are present in *Vinca rosea.* V. is one of the most effective naturally occurring antitumor agents, and is used primarily in the treatment of Hodgkin's disease.

Vinca alkaloids, *Catharanthus alkaloids:* a group of about 60 iridoid indole alkaloids from *Vinca (Catharanthus)* spp. Structurally, they are tetra- or pentacyclic indole derivatives with an iridoid component, e.g. vindoline, m.p. 174–176 °C, $[\alpha]_D$ −18 ° (chloroform), and vincamine, m.p. 232–233 °C, $[\alpha]_D$ +41 ° (pyridine). These are accompanied in the

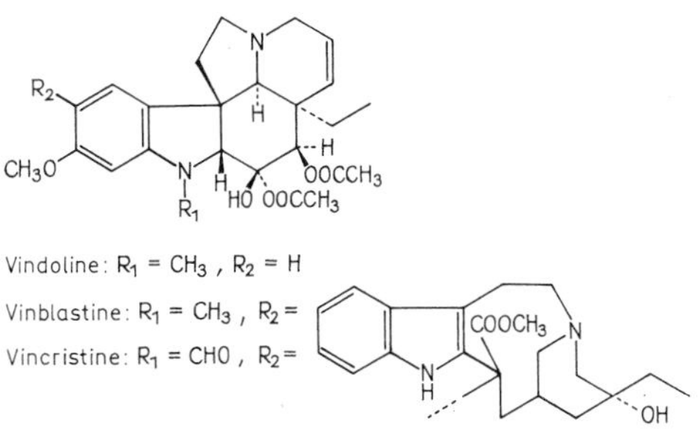

Vindoline: $R_1 = CH_3$, $R_2 = H$

Vinblastine: $R_1 = CH_3$, $R_2 =$

Vincristine: $R_1 = CHO$, $R_2 =$

leaves by small quantities (about 0.005%) of two dimeric V.a., i.e. Vinblastine (see) and Vincristine (see) (Fig.). Tryptophan and mevalonic acid are biosynthetic precursors of V.a. Vincamine has hypertensive activity and is used pharmaceutically, partic-

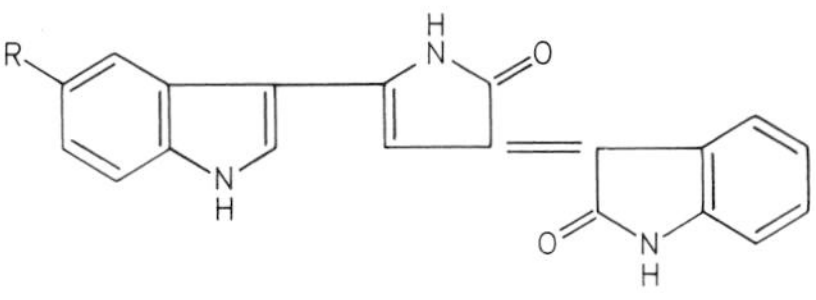

R = OH, Violacein
R = H, Deoxyviolacein

Violaxanthin: 3(S),3'(S)-dihyroxy-β-carotene-(5R,6S,5'R, 6'S)-5,6,5', 6'-diepoxide, a xanthophyll, M_r 600.85, m. p. 208 °C, $[\alpha]_{Cd}^{20}$ +35 ° ($c=0.08$, chloroform). V. is one of the most important plant carotenoids, present as an orange or brown-yellow pigment in all green leaves, and especially plentiful in flowers and fruits of *Viola tricolor, Taraxacum, Tagetes, Tulipa, Citrus, Cytisus*, etc.

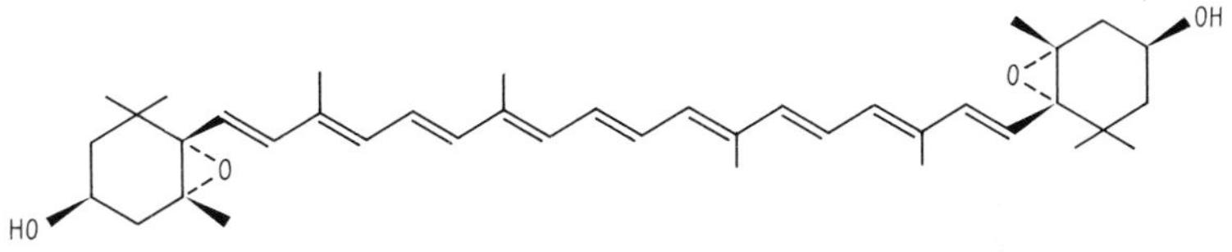

Violaxanthin

ularly in Hungary. The dimeric V.a. show good oncolytic properties and are used in the treatment of carcinomas.

Vincaleucoblastin: see Vinblastin.

Vincamine: see Vinca alkaloids.

Vincristine, *leurocristine:* a dimeric indole alkaloid closely related to vinblastine, from *Vinca rosea* (see Vinca alkaloids). It is used mainly for the treatment of acute leukemia in children, and against various other neoplasmic growths.

Vindoline: see Vinca alkaloids.

Violacein: the major purple pigment of *Chromobacterium violaceum,* accompanied by smaller amounts of deoxyviolacein. Every C-atom of V. is derived biosynthetically from tryptophan. DL-tryptophan labelled with ^{14}C in the ring or side chain is incorporated with very little dilution into V. by nonproliferating cells of *C.violaceum.* [DeMoss, R.D. and Evans, N.R. (1960) *J. Bact., 79,* 729]

Violanin: see Delphinidin.

Viomycin, *celiomycin, florimycin, tuberactinomycin B:* a polypeptide antibiotic from various *Streptomyces* spp., including *S. floridae, S.puniceus* and *S.vinaceus,* containing a 7-deazaadenine ring. M_r 685.71, m.p. 280 °C (d.) (sulfate), $[\alpha]_D^{25}$ −32 ° ($c=1$, water). Degradation products of V. include the guanidino compound, viomycidin. V. inhibits both nucleic acid and protein biosynthesis. It is active chiefly against Gram negative bacteria, and is used therapeutically against *Mycobacterium tuberculosis.* Structure: Noda et al. *J. antibiot. 25* (1972) 427.

Viridicatine: a quinoline alkaloid from moulds of the genus *Penicillium.* V. is biosynthesized from anthranilic acid, phenylalanine and the methyl group of methionine (Fig.). Biosynthesis is catalysed by the enzyme cyclopenase.

Viridicatol: see Viridicatine.

Viridogrisein: see Etamycin.

Virus coat proteins, *capsids:* proteins with the largest known M_r values (up to 40×10^6). They com-

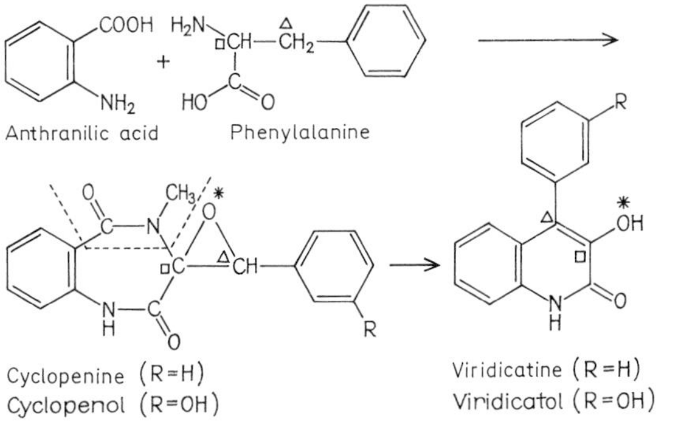

Anthranilic acid   Phenylalanine

Cyclopenine (R=H)
Cyclopenol (R=OH)

Viridicatine (R=H)
Viridicatol (R=OH)

Biosynthesis of the quinoline alkaloids, viridicatine and viridicatol

prise many, usually identical, subunits, called *capsomeres* (M_r 13000-60000). V.c.p. of tobacco mosaic virus consists of 2130 capsomers of M_r 17500. V.c.p. lie on the exterior of the virus particle, enclosing the DNA or RNA. Primary structures of capsomeres from several strains to tobacco mosaic virus (M_r 17400-17600; 157-158 amino acids) and of turnip yellow mosaic virus (M_r 20000, 188 amino acids) are known. M_r of bacteriophage capsids lie in the range M_r 5168 (49 amino acids) to M_r 14034 (131 amino acids).

Viruses: infectious particles composed of nucleic acid and protein. V. are various shapes (cubic, spherical or helical) and sizes (30 nm − 1 μm). M_r of V. lie in the range $<1-40 \times 10^6$, and V. therefore pass through bacterial filters. They can be collected by ultracentrifugation and visualized under the electron microscope. Between 60 and 70 °C, they become heat-denatured; at − 70 °C they remain infective for years. V. can be crystallized; the first crystallized V. was tobacco mosaic V. (Stanley, 1935) (Fig.).

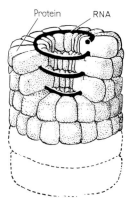

Protein   RNA

Schematic representation of part of tobacco mosaic virus, showing spirally arranged RNA and protein subunits

V. are parasites of animals, plants and microorganisms. They possess no metabolism of their own. Their nucleic acid encodes the necessary genetic information for self-reproduction, but they are dependent upon the transcription and translation apparatus of the host cell. Synthesis and assembly of new V. progeny must therefore occur in the host cell. V. nucleic acid may be DNA or RNA. Many V. are the causative agents of disease, e.g. the V. of smallpox, chicken pox, certain skin diseases, and some cancers (oncogenic V.) are all DNA-V., whereas rabies, infective hepatitis, influenza, measles, poliomyelitis, meningitis, leukemia, foot and mouth disease, avian myeloma, and various plant diseases are caused by RNA-V. Many V. are transmitted by insects.

The V. of bacteria are called Phages (see); these provide the best understood examples of V. development.

α-Viscol: see Amyrin.

Viscotoxins: see Toxic proteins.

Visual purple: see Visual process.

Visual process: the process by which light induces a nerve impulse in a photoreceptive cell. Light is absorbed by a visual pigment, a chromoprotein consisting of an apoprotein (opsin) and the chromophore, 11-*cis*-retinal (neoretinal b). The aldehyde group of 11-*cis*-retinal forms a Schiff's base (see) with the δ-amino group of a specific lysine residue of the opsin. In vertebrate retinas, there are two classes of photoreceptors, rods and cones. The rods, which are responsible for black-and-white vision at low light intensities, have an outer segment in which there is a stack of flattened membrane discs. These and the plasma membrane contain rhodopsin (visual purple). Color vision in the cones is mediated by closely related opsins; three in the case of normal human eyes. These have absorption maxima at 430 nm, 540 nm or 575 nm, respectively. All three pigments (iodopsins) contain 11-*cis*-retinal, but their opsins (cone-type opsins) are different. Other vertebrates have different numbers of opsins with somewhat different sensitivities. Each rod or cone contains only one type of opsin. The amino acid sequences of the human color pigments have been deduced by molecular genetics methods. [J. Nathans et al. *Science* **232** (1986) 193-202; *ibid.* 203-210]

In the primary event of visual excitation, light isomerizes the 11-*cis*-retinal of the opsin to all-*trans*-retinal. Since the all-*trans*-retinal does not fit the binding site for 11-*cis*-retinal, the opsin molecule becomes unstable and undergoes a series of conformational changes, followed by hydrolysis of the Schiff's base linkage between all-*trans*-retinal and opsin (Fig. 1).

Further events in the process are known with some confidence only for rod cells, in which the rhodopsin is bleached to metarhodopsin I. Metarhodopsin I appears about 10^{-5} sec after the illumination of rhodopsin, and metarhodopsin II at 10^{-3} sec. Red rhodopsin is thus bleached, the final mixture of opsin and all-*trans*-retinal being yellow in color (formerly called "visual yellow"). Bleached rhodopsin is able to diffuse freely within the membrane, where it interacts with a "G-protein" (*transducin*). Transducin consists of three subunits, one of which (α) carries a GDP molecule. (The subunits of transducin have the following molecular weights: α, 39000; β, 36000; γ, about 10000 [B.K.-K. Fung et al. *Proc. Natl. Acad. Sci. U.S.A.* **78** (1981) 152-156].) Activated rhodopsin binds to the transducin, causing the α-subunit to exchange the GDP for a GTP. The entire rhodopsin-transducin-GTP complex now dissociates. The rhodopsin is still catalytically active and interacts with more molecules of transducin (about 100 in 0.5 sec). The α-transducin subunit and the GTP remain associated and bind to a molecule of a cGMP-phosphodiesterase (PDE), forming the "G(α)-GTP-PDE complex". The resulting hydrolysis of cyclic GMP is observed within milliseconds of the onset of illumination.

In the rod cells, there are membrane channels which allow a constant inflow of Na^+ ions. Light absorption by rhodopsin interrupts this flow, thus inducing a voltage impulse in the cell membrane. Both Ca^{2+} and cGMP are involved in this process. There are two hypotheses regarding the mechanism: (1) Ca^{2+}

stored within the discs is released by light to block the Na$^+$ channels; and (2) cGMP keeps the channels open. The light-induced hydrolysis of cGMP causes the channels to close. It is known experimentally that cGMP decreases the ability of Ca^{2+} or Co^{2+} to block the Na$^+$ channels, and increases their ability to permeate the membrane. It is also known that several outer rod proteins are phosphorylated in the dark by a cyclic-nucleotide-dependent kinase, and that this phosphorylation is reversed in the light. It has also been found that 1 photon causes 100–300 Na$^+$ channels to close.

A combined hypothesis has recently been proposed by E.A. Schwartz: (1) The concentration of cGMP in the cell is controlled by feedback mechanisms. (2) The Na$^+$ channels, which also admit a small amount of Ca^{2+}, have binding sites for Ca^{2+} on the cytoplasmic side of the membrane. When bound to these sites, Ca^{2+} causes the channels to close. (3) The affinity of the Ca^{2+} sites for Ca^{2+} is modulated by cGMP; a drop in the cGMP concentration causes the affinity of the sites for Ca^{2+} to rise. Thus hydrolysis of cGMP would indirectly cause the Na$^+$ channels to close. (4) The increase in extracellular Ca^{2+} which is observed during photoexcitation would be explained if the Ca^{2+} influx is prevented by closure of Na$^+$ channels, while Ca^{2+} eflux through other channels is not decreased.

pigment porphyropsin in place of rhodopsin. The function of porphyropsin is analogous to that of rhodopsin, and it contains 3-dehydroretinal in place of retinal. Cyanopsin (dehydroretinal+cone-type opsin) is found in the retinas of most fresh water fish.

Vitamins [Latin *vita*+amine]: substances present in the animal diet in only small quantities, and indispensable for the growth and maintenance of the organism. A dietary requirement is implicit in the definition of V. Most V. are essential for the metabolism of all living organisms (some fat-soluble V. may have metabolic roles unique to animals), and they are synthesized by plants and microorganisms. The dietary requirement in the animal results from the evolutionary loss of this biosynthetic ability. The biosynthetic abilities, and thus the dietary requirements of different species, vary. Ascorbic acid, for example, is a V.(V.C) only for the primates and a few other animals (e.g. the guinea pig); most animals can synthesize it, and for them it is therefore not a V. Some V. can be synthesized from provitamins obtained from the diet. Some of the V. requirement of humans and higher animals is supplied by the intestinal flora, e.g. most of the V. K required by humans is provided in this way.

The role of V. is largely catalytic; most V. serve as coenzymes and prosthetic groups of enzymes. For most of these, the nature of the biocatalytic function

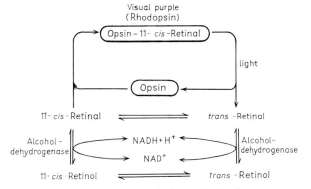

Visual purple
(Rhodopsin)

Figure 1. Synthesis of visual purple during the visual process.

In the recovery of the visual apparatus, the G(α)-GTP-PDE complex is inactivated by spontaneous cleavage of the GTP. Activated rhodopsin is the substrate for a kinase which may de-activate it. The kinase is inhibited by cGMP, so that the amount of activated rhodopsin may be subject to feedback regulation. The *trans*-retinal may be isomerized directly to 11-*cis*-retinal, which recombines with opsin to form rhodopsin; or it may first be reduced to retinol by NADH-dependent alcohol dehydrogenase, followed by isomerization of the retinol and reoxidation to 11-*cis*-retinal (Fig. 1). Some of the retinal is continually lost from the cells of the retina, so that the continuation of the visual cycle depends upon a continual replacement from the blood.

Land and marine animals possess rhodopsin. Certain fresh water fish and some amphibians have the

has been elucidated (Table). V. D, however, acts as a regulator of bone metabolism, and is thus similar to hormones. As a component of the visual pigments, V. A acts as a prosthetic group; however, it is not known whether it is associated with catalytic proteins in its other functions. Nicotinamide and riboflavin are constituents of the hydrogen-transferring coenzymes (see Respiratory chain). Biotin, folic acid, pantothenic acid, pyridoxine, cobalamin and thiamin are (or are precursors of) coenzymes of group-transfer reactions. The low daily requirement (Table) for V. reflects their catalytic and/or regulatory roles. Thus V. are nutritionally quite different from fat, carbohydrate or protein, which are required in the diet in considerable quantities as substrates of tissue synthesis and energy metabolism.

The biological activity of a pure V. can be expressed in International Units abb. IU. 0.3 µg V. A (retinol),

8 µg thiamin hydrochloride, 0.18 µg biotin, 50 µg L-ascorbic acid, 0.025 µg ergocalciferol or 1 mg DL-α-tocopherol acetate each corresponds to 1 IU. The system of IU has been retained, even though the structures of all V. are known, because in most cases a family of closely related compounds all have V. activity, but the biological activities of different members of the family can vary considerably.

The lack or deficiency of V. as a result of unbalanced nutrition leads to characteristic metabolic disturbances. Complete absence of a V. leads to avitaminosis, with typical clinical symptoms. Relative deficiency of a V. causes hypovitaminosis. Such conditions are reversible by administration of the appropriate V. Excessive intake of a V., e.g. V. A and V. D, can lead to hypervitaminosis.

Formerly, V. were named after the diseases they cured, e.g. antiscorbutic V., antirachitic V., antiberiberi factor. Not all V., however, have such a pronounced specificity, and the clinical pictures of many avitaminoses and hypovitaminoses are complex and variable. A nomenclature based on letters of the alphabet was developed simultaneously; the designations A, B, C, D, and E were applied in the historical order of discovery. Subscripts were applied as more refined chemical analysis revealed that the originally isolated substances were in fact complex mixtures; this was especially pronounced among the B vitamins. Partly because of the confusion over the "B complex", trivial names which give an indication of the chemical structure of the V., e.g. pyridoxine or pyridoxol for V. B$_6$, are now preferred. The V. represent an heterogeneous group of substances which are classified into two main subgroups: fat-soluble and water-soluble (Table), depending on whether they can be extracted from foodstuffs with organic solvents or water.

Ascorbic acid (Vitamin C; antiscorbutic V.): a water-soluble V. with a wide natural distribution, especially in fresh vegetables and fruit. V. C is the γ-lactone of 2-oxo-L-gulonic acid, derived from carbohydrate metabolism. In most mammals it is synthesized from D-glucuronate by reactions of the Glucuronate pathway (see) (Fig. 1). Higher primates, including human beings, and guinea pigs cannot synthesize V. C because the enzyme that catalyses the conversion of L-gulono-γ-lactone to 2-oxo-L-gulono-γ-lactone is absent. For these animals it is therefore a true V. and must be supplied in the diet.

V. C is a powerful reducing agent on account of its ene-diol grouping. Ascorbic acid oxidase (a copper enzyme) catalyses the removal of hydrogen from ascorbic acid ($C_6H_8O_6$) to produce dehydroascorbic acid ($C_6H_6O_6$). V. C is involved in several metabolic

Fat-soluble vitamins	First described	Recommended daily intake (mg)	Biochemical action	Clinical activity
Calciferol (V. D)	1922	0.01–0.025	calcium and phosphate metabolism	antirachitic V.
Phylloquinone (V. K) (menaquinone	1935	1	cofactor for γ-carboxylation of Glu residues in coagulation proteins	antihemorrhagic V.
Retinol (V. A)	1913	2.7	visual process	epithelial protection V. antixerophthalmic V.
Tocopherol (V. E)	1922	5	antioxidant; otherwise unknown	antisterility V.
Water-soluble vitamins				
Ascorbic acid (V. C)	1925	75	reducing agent for some monooxygenases	antiscorbutic V.
Biotin (V. H)	1935	0.25	coenzyme of various carboxylation reactions	the "skin" V.
Cobalamin (V. B$_{12}$)	1948	0.003	coenzyme of various metabolic reactions	antianemic V., extrinsic factor
Folic acid	1941	1–2	one carbon unit transfer	therapy of certain anemias
Niacin	1937	18	respiration, hydrogen transfer	pellagra preventative
Pantothenic acid	1933	3–5	transfer of acyl groups	chick antidermatitis factor; anti-gray-hair factor
Pyridoxine (V. B$_6$)	1936	2	amino acid metabolism, transfer of amino groups	weakness, nervous disorders, depression
Riboflavin (V. B$_2$)	1932	1.7	respiration, hydrogen transfer	antidermatitis V.
Thiamin (V. B$_1$)	1926	1.2	carbohydrate metabolism, transfer of active aldehyde	antineuritic V.

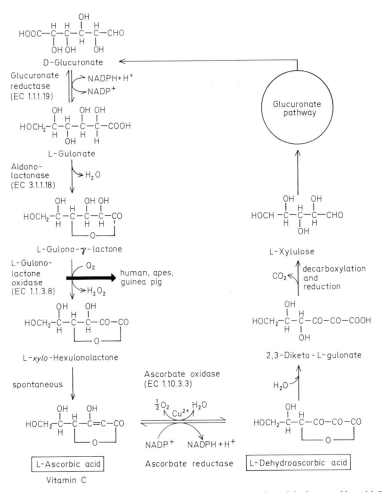

Figure 1. *Biosynthesis of ascorbic acid and its conversion into dehydroascorbic acid.* See Oxalic acid (Fig. 2).

hydroxylation reactions, e.g. hydroxylation of the amino acid proline in collagen. (See 2-Oxoacid dioxygenases; Ascorbate shuttle.)

V.C is particularly important for the maintenance of the inner walls of blood vessels. Deficiency results in scurvy, a long-known avitaminosis, characterized by rupture of blood capillaries, hemorrhage of the skin and mucosas, inflammation of the gums, loosening of the teeth and painful swellings of the joints. Resistance of the organism to infectious diseases is also reduced. The recommended daily intake is 75 mg, which is considerably higher than for other V. (Table).

The commercial synthesis of V.C starts from glucose, which is converted to the sugar alcohol D-sorbitol by hydrogenation. The bacterium *Acetobacter ruboxydans* then performs the specific oxidation of D-sorbitol into L-sorbose. A carboxyl group is introduced at C-1 of the diacetone derivative of L-sorbose, and the resulting diacetone-2-oxo-L-gulonic acid is heated with HCl to give ascorbic acid.

Biotin (Vitamin H; bios II; coenzyme R): a sulfur-containing water-soluble V. Chemically it is a cyclic urea derivative: 2'-oxo-3,4-imidazoline-2-tetrahydro-thiophene-*n*-valeric acid, (Fig. 2).

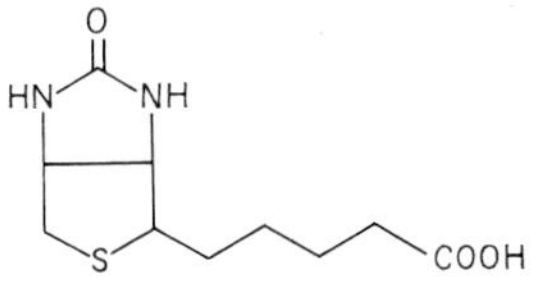

Figure 2. *Biotin (Vitamin H)*

Biotin has been isolated from liver extracts and egg yolks, and it is a yeast growth factor. There are 8 stereoisomers; the most important biological isomer is D-biotin. Biotin is biosynthesized from cysteine, pimelic acid and carbamoyl phosphate. It acts as the prosthetic group of carboxylation enzymes (see Biotin enzymes): the carbonyl group of biotin forms an

631

amide bond with the ε-amino group of a lysine residue in the enzyme protein. The corresponding compound of biotin and lysine is called biocytin. In animals, biotin deficiency causes skin disorders (seborrhea) (hence V."H", for "Haut" [German]=skin), and loss of hair. Excessive intake of raw eggs causes avitaminosis in humans; egg whites contain the protein avidin, which specifically binds biotin and prevents its absorption.

The binding of biotin to avidin is so tight that the biotin-avidin complex is utilized to visualize specific proteins, lipids and carbohydrates. The avidin molecule is linked to an enzyme which converts a colorless, soluble substrate to a colored, insoluble product. The biotin is linked to a soluble probe, such as an antibody, the substrate of a particular receptor molecule, or a nucleic acid. The probe interacts with the corresponding structure within the cell, and the avidin-enzyme complex then produces a deposit of colored enzyme product at the site where the probe-biotin complex is immobilized. [*Virology*, 126 (1983) 32; *Proc. Natl. Acad. Sci. U.S.A.* **80** (1983) 4045]

Calciferol (Vitamin D; antirachitic vitamin): a group of fat-soluble V. chemically related to the steroids. They are produced from provitamins, $\Delta^{5,7}$-unsaturated sterols, by UV-irradiation. If an individual receives adequate exposure to sunlight, dietary V. D is unnecessary. Calciferol is thus a V. only for people who, due to confinement indoors or highly pigmented skin at higher latitudes, are unable to synthesize a sufficient amount. In the conversion of sterols to calciferol, ring B of the steroid is opened between C-9 and C-10, forming precalciferol, which acts as a precursor of the V. D group (Fig.3). Important members are V. D_2 (ergocalciferol), derived from ergosterol, and V. D_3 (cholecalciferol), derived from 7-de-

hydrocholesterol. Lumisterol and tachysterol are byproducts of the synthesis. V. D_3 is converted enzymatically in the liver and kidneys to 25-hydroxycholecalciferol, followed by hydroxylation to the highly active 1α,25-dihydroxycholecalciferol (1α,25-dihydroxy-V. D_3); this compound represents the active form of the vitamin in the human, and its behavior resembles that of a hormone, rather than a biocatalyst. A receptor for this compound has been found on the surfaces of myeloid leukemia cells, which it induces to differentiate into macrophages or granulocytes. The receptor has also been found on monocytes from normal humans, but is absent from non-dividing B and T lymphocytes. This suggests an association between the presence of the receptor and the mitotic activity of the lymphocytes. [D.M. Provvedini et al., *Science* **221** (1983) 1183–1185]

V. D_1 is a molecular compound of lumisterol and ergocalciferol.

V. D_4 is 22-dihydroergocalciferol, produced from 22-dihydroergosterol by UV-irradiation. V. D_3 is present in cod liver oil in particularly large quantities, and it is also formed in human skin from 7-dehydrocholesterol by the action of sunlight.

The V. D complex is also present in e.g. herrings, egg yolk, butter, cheese, milk, pig liver and edible fungi.

V. D is important in calcium metabolism. It is required for calcium absorption and the mineralization of bone. The V. D deficiency disease, known as rickets, is characterized by a softening and malformation of the bones. It results from a poor absorption of calcium, coupled with deficient incorporation of calcium into bone tissue. Rickets can be cured by exposure to sunlight and by the adminis-

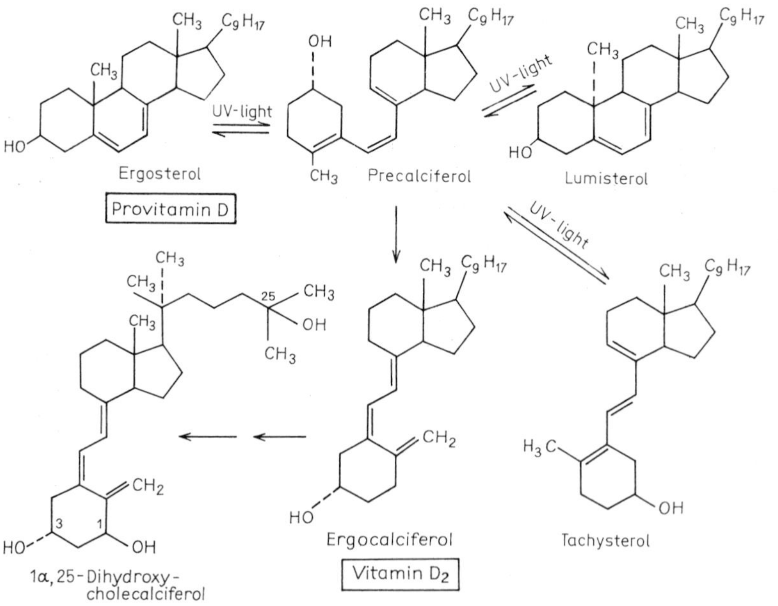

Figure 3. Biosynthesis of vitamin D_2.

tration of synthetic V.D. Because V.D is stored in the liver, overdosage can lead to hypervitaminosis, with disturbances of calcium and phosphate metabolism and withdrawal of calcium from the bones. Exposure to sunlight never leads to hypervitaminosis D.

Cobalamin (Vitamin B_{12}; extrinsic factor; animal protein factor): a group of water-soluble corrinoids required in very small amounts. The corrinoid structure consists of a complex corrin ring system, with a centrally bound trivalent cobalt atom, a base-nucleotide moiety, and a monovalent group (called the cobalt ligand) bound to the cobalt (Fig. 4). In biolog-

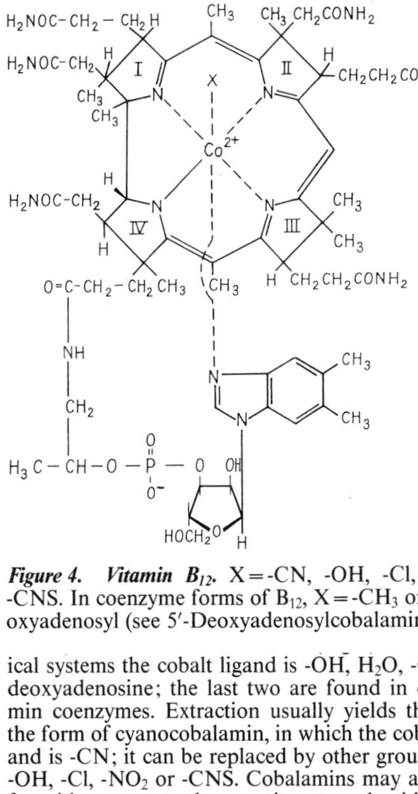

Figure 4. Vitamin B_{12}. X = -CN, -OH, -Cl, -NO$_2$, -CNS. In coenzyme forms of B_{12}, X = -CH$_3$ or 5'-deoxyadenosyl (see 5'-Deoxyadenosylcobalamin).

ical systems the cobalt ligand is -OH⁻, H$_2$O, -CH$_3$ or deoxyadenosine; the last two are found in cobalamin coenzymes. Extraction usually yields the V.in the form of cyanocobalamin, in which the cobalt ligand is -CN; it can be replaced by other groups, e. g. -OH, -Cl, -NO$_2$ or -CNS. Cobalamins may also differ with respect to the constituent nucleotide base, but in most types of V.B_{12} it is 5,6-dimethylbenzimidazole. The biosynthesis of this base is closely related to that of riboflavin. The 4 pyrrole rings of the corrin ring system are synthesized via 5-aminolevulinic acid (see Porphyrins). The 6 methyl groups of the corrin ring system are derived from methionine. V.B_{12} was the first natural product containing cobalt to be discovered. 5'-Deoxyadenosylcobalamin (see) is the coenzyme of various isomerization reactions, and of ribonucleotide triphosphate reductase (see Thioredoxin). Methylcobalamin is the coenzyme of homocysteine transmethylase (see Methionine), and the methylation of tRNA. The coenzyme of methane production by methanogenic bacteria (*Methanobacillus, Methanosarcina*) is a methylcobalamin, containing the base 5-hydroxybenzimidazole. V.B_{12} oc-

curs predominantly in animal tissues and animal products, such as egg yolk and milk. It is synthesized principally by bacteria, and green plants contain little or none. Deficiency symptoms are sometimes observed in strict vegetarians, most often in breast-fed infants whose mothers consume no animal products. The body reserves of cobalamin are usually so large that an adult can survive for many years on them in the absence of dietary intake. Plants may nevertheless contain a variety of undiscovered corrinoids, which for various reasons (e. g. they may contain different bases) do not support the growth of V.B_{12}-dependent bacteria or animals. V.B_{12} is identical with the LLD-factor, required for the growth of the bacterium *Lactobacillus lactis*. Cyanocobalamin is also known as antipernicious anemia factor. Pernicious anemia is characterized by a severely reduced production of red blood cells, deficient gastric secretion and disturbances of the nervous system. It is not usually caused by a dietary deficiency of V.B_{12}, but by poor absorption. Cure is effected by injection of small amounts (3–6 µg) of V.B_{12}. Excretion of methylmalonic acid is used for the diagnosis of V.B_{12} deficiency.

Absorption of V.B_{12} requires an intrinsic factor, which is normally present in the gastric mucosa. Intrinsic factor is a neuraminic acid-containing glycoprotein, M_r 60 000. It forms a pepsin-resistant complex with V.B_{12}, which is absorbed in the lower part of the intestinal tract. Absence of this factor causes pernicious anemia.

Various bacteria, e. g. *Propionibacterium shermanii,,* and species of *Streptomyces*, such as *Streptomyces olivaceus*, synthesize V.B_{12}, but do not secrete it into the growth medium. Preparation of V.B_{12} by industrial fermentation (various antibiotic-producing strains of *Streptomyces* are used) therefore necessitates extraction of the bacterial cell mass.

Folic acid (pteroylglutamic acid; the old name vitamin B_c is now obsolete): a pteridine derivative, especially plentiful in liver, yeast and green plants. The chemical structure of folic acid contains 3 moieties: 2-amino-4-hydroxypteridine, *p*-aminobenzoic acid and one or more glutamic acid residues, linked by peptide bonds via their γ-carboxyl groups. (Fig. 5). Folic acid is a growth factor for some bacteria. The biochemically active form of folic acid is Tetrahydrofolic acid (see), which is a coenzyme in the metabolism of Active one-carbon units (see). In human beings, folic acid avitaminosis is more often caused by faulty uptake and/or utilization, than by dietary deficiency. It usually results in abnormalities of the blood, e. g. megaloblastic anemia, thrombocytopenia. Antimetabolites of folic acid are aminopterin and methopterin, which are used therapeutically in the treatment of leukemia. The sulfonamides are antimetabolites of *p*-aminobenzoic acid, and therefore act as inhibitors of bacterial folic acid synthesis.

Niacin (*nicotinic acid* and *nicotinamide*), (pellagra preventative factor, Vitamin PP): these simple pyridine derivatives (Fig. 6) are widely distributed in nature, and are especially plentiful in liver, fish, yeast and germinating cereal grains. Nicotinamide and nicotinic acid are nutritionally equivalent, since both can be assimilated for the purposes of NAD(P) synthe-

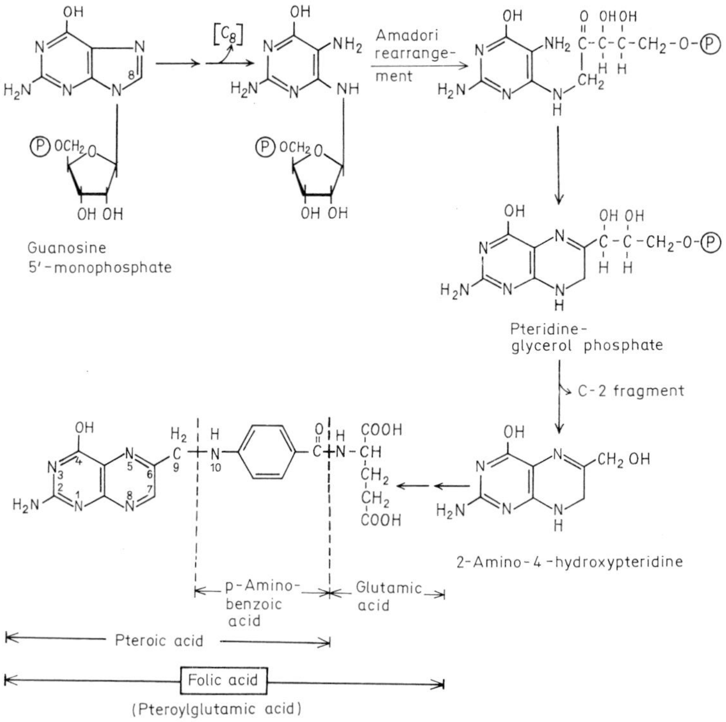

Figure 5. *Biosynthesis of folic acid from guanosine monophosphate.*

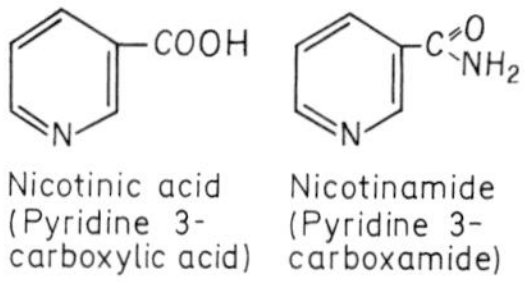

Nicotinic acid
(Pyridine 3-
carboxylic acid)

Nicotinamide
(Pyridine 3-
carboxamide)

Figure 6.

sis. For therapeutic purposes, however, nicotinamide is preferred because large doses of nicotinic acid may have undesirable side effects.

In many mammals and in fungi, the nicotinamide moiety of NAD(P) can be derived from L-Tryptophan (see). The extent to which the dietary nicotinamide requirement of animals can be spared by dietary tryptophan varies according to the species. Thus, if the definition of a V. implies a dietary requirement, nicotinamide is not a V. for the rat, which can satisfy its total requirement by the degradation of tryptophan. Synthesis of the nicotinamide moiety of NAD(P) in bacteria and plants occurs by a different pathway, in which aspartic acid and dihydroxyacetone phosphate act as precursors. In mutants (e.g. of *Escherichia coli*) which have lost the ability to synthesize the pyridine ring of NAD(P), nicotinamide (or nicotinic acid) is an essential growth factor.

Under certain nutritional conditions, e.g. when maize forms the bulk of the human diet, the deficiency of niacin leads to pellagra. (Treatment of maize with lime, as practiced by Central Americans,

releases niacin precursors. Pellagra is therefore not common among them, in spite of their monotonous diet, but was common among poor Europeans who did not treat their maize in this way.) This deficiency disease affects the skin (brown coloration), the digestive system (diarrhea) and the nervous system (dementia). Pellagra can be cured by feeding tryptophan; or nicotinamide may be administered therapeutically. The only other animal in which a characteristic nicotinic acid deficiency syndrome has been described is the dog. In other species, the deficiency state is not easily differentiated from other nutritional deficiencies. In the dog, the condition is known as canine black tongue; the tongue becomes bluish black, and lesions, accompanied by degeneration of the nerves, occur on the inner surfaces of the gums, lips and mouth. There are also histological changes in the spinal cord and other regions of the central nervous system.

Pantothenic acid: consists of 2,4-dihydroxy-3,3-dimethylbutyric acid (pantoic acid) linked to β-alanine by an amide bond. It is widely distributed in animals and plants. Pantoic acid is biosynthesized from valine (Fig. 7). Humans are unable to perform the condensation reaction between pantoic acid and β-alanine. Only the D(+) form of pantothenic acid is biologically active. It is required as a precursor of Pantetheine (see) for the synthesis of Coenzyme A (see).

Non-experimental human deficiency states have not been observed. Pantothenic acid is presumably pre-

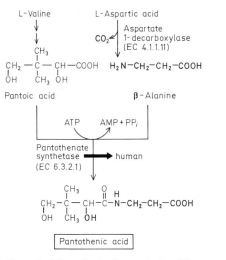

Figure 7. Biosynthesis of pantothenic acid.

sent in sufficient quantity in all known diets. Experimental human deficiency results in burning sensations, muscle weakness, abdominal disorders, vasomotor instability and depression. Pantothenic acid deficiency in chicks causes pellagra-like skin changes (hence the name chick antidermatitis factor), growth failure and degeneration of the spinal cord. In rats, pantothenic acid deficiency results typically in adrenal necrosis and hemorrhage; there are also widespread general pathological changes and the hair turns gray (hence the term anti-gray-hair factor). In the dog, pantothenic acid deficiency results in hypoglycemia, respiratory and cardiac distress, convulsions and sudden coma. Pathological changes include fatty liver, kidney degeneration and mottled thymus.

Phylloquinone (Vitamin K; antihemorrhagic vitamin; coagulation vitamin): a group of fat-soluble, naphthoquinone compounds with varying sizes of isoprenoid side chain (Fig. 8). Mammals can synthesize the side chain, but not the naphthoquinone moiety. V. K$_1$ is especially plentiful in green plants. V. K$_2$ (farnoquinone; menaquinone-6; 2-methyl-1,4-naphthoquinone) is found chiefly in bacteria. V. K$_3$ (menadione; 2-methyl-1,4-naphthoquinone) is actually a provitamin.

In many bacteria V. K is a component of the respiratory chain, in place of ubiquinone. V. K deficiency causes the deficient production of blood coagulation factors, in particular prothrombin, leading to abnormally long clotting times and a marked tendency to hemorrhage. V. K serves as a cofactor in the carboxylation of glutamic acid residues during post translational modification of prothrombin and other blood coagulation proteins (see Post-translational modification of proteins; see 4-Glutamyl carboxylase). Avitaminosis is rare in human children and adults, because sufficient V. K is provided by the gut flora. However, it does not cross the placenta well, so that neonates are at risk for avitaminosis. Fatal hemorrhage is sometimes observed in breast-fed infants whose mothers' milk does not contain adequate amounts of the vitamin. V. K$_3$ preparations are used in the treatment of hemorrhages and liver diseases.

2-Methylnaphthohydroquinone and phytol, or isophytol are the starting compounds for the technical synthesis of V. K$_1$.

Warfarin is a V. K antagonist used as a rodenticide; it causes death by hemorrhage after the animals have fed on it repeatedly. It is also used clinically as an anticoagulant; patients taking it may need supplemental V. K. Dicoumarol is an important antagonist of V. K. It is present in moldy clover hay, and thereby responsible for hemorrhage in cattle.

Pyridoxine (Vitamin B$_6$; pyridoxol; adermine): several naturally occurring compounds have V. B$_6$ activity in animal nutrition: pyridoxol (2-methyl-3-hydroxy-4,5-di[hydroxymethyl]-pyridine) is the chief form of V. B$_6$ in vegetables, whereas pyridoxal, pyridoxal phosphate, pyridoxamine and pyridoxamine phosphate are present in animal tissues (Fig. 9). V. B$_6$ is water-soluble and occurs in liver, kidney, yeast, vegetables and cereals. The biocatalytically active compounds are pyridoxal phosphate and pyridoxamine

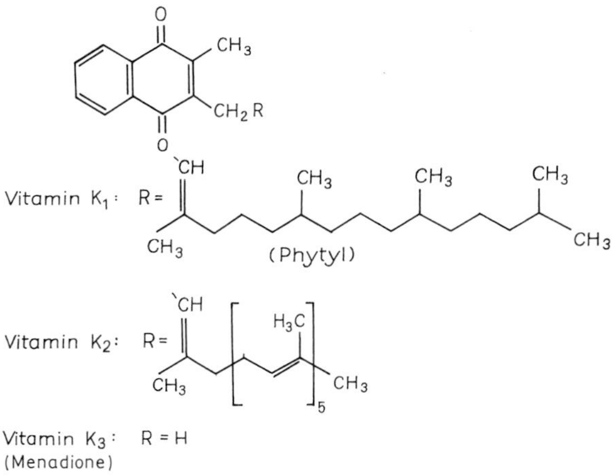

Figure 8. Vitamin K active compounds.

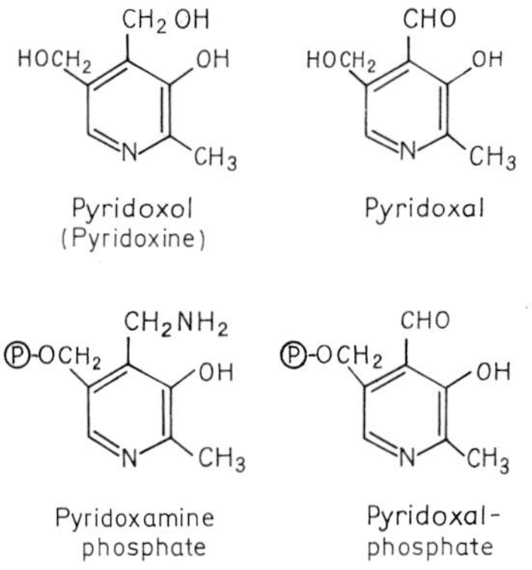

Figure 9. Vitamin B₆-active compounds.

phosphate (see Transamination; Pyridoxal phosphate), but all forms of V.B₆ are interconvertible within the cell.

In humans, V.B₆ deficiency is not a very specific condition, and symptoms are easily confused with those of most deficiencies of the V.B group; occasionally there may be specific nervous disorders and anemia. Experimental V.B₆ deficiency in animals results in V.B₆ pellagra, characterized by loss of hair, edema and red scaly skin. V.B₆ deficiency inhibits the degradation of L-tryptophan, and the excretion of xanthurenic acid is used as an index of V.B₆ deficiency (see L-Tryptophan); it may also result in a lowered conversion of glutamate into γ-aminobutyrate in the brain, since decreased levels of γ-aminobutyrate have been found in the brains of animals deficient in V.B₆.

Carbonyl reagents, e.g. semicarbazide, hydroxylamine, hydrazine, condense with pyridoxal phosphate to form relatively stable derivatives. A large part of the toxic activity of such compounds therefore probably represents a severe form of B₆-avitaminosis. Isonicotinyl hydrazide (isoniazid) inhibits pyridoxal kinase, the enzyme that catalyses the phosphorylation of pyridoxal (see Pyridoxal phosphate). It is a most effective drug against tuberculosis because the pyridoxal kinase of *Mycobacteria* has low activity and is easily blocked. The drug also forms a hydrazone with pyridoxal, so that its long-term use may produce symptoms of V.B₆ deficiency. L-Cycloserine is toxic and inhibits many pyridoxal phosphate enzymes. Penicillamine causes a decrease in the activity of brain glutamate decarboxylase and causes convulsions, presumably by forming a stable derivative with pyridoxal phosphate. The classical competitive inhibitor of V.B₆ is 4-deoxypyridoxine, which leads to convulsions and other symptoms of V.B₆ deficiency. Another competitive inhibitor is the structural analogue toxopyrimidine, which produces running fits when fed to rats or mice.

Retinol (Vitamin A, axerophthol, xerophthol) (obsolete names: epithelial protection vitamin; growth vitamin): a fat soluble V. with polyisoprenoid structure. The alcohol retinol is also known as V.A₁; 3-dehydroretinol, V.A₂, is characterized by an additional double bond between C-3 and C-4 in the ring. V.A is essential to the Visual process (see), and also

for growth, skeletal development, normal reproductive function and maintenance of differentiation in tissues.

V.A occurs predominantly in animal products, such as milk, butter, egg yolk, cod liver oil and the body fat of many animals. All the Carotenes (see), which are abundant in green plants and fruits, have provitamin A activity (see Fig. 10). Conversion of carotenes to V.A occurs in the small intestine, but other organs, such as muscle, lungs and serum also function to a limited extent in this capacity. Invertebrates cannot perform this conversion. β-Carotene is oxidatively cleaved by the intestinal mucosa into two molecules of retinal (α- and γ-carotenes yield only 1 molecule of V.A) which are then reduced to all-*trans*-retinol and esterified with a fatty acid, chiefly palmitate. This V.A palmitate is transported in the lymph to the liver for storage. Free retinol is released by hydrolysis and transported from the liver by a retinol-binding plasma protein. Prealbumin also becomes bound to the retinol-protein complex and is thought to exert a protective function. Retinol is removed from the plasma by the cells of the retina, where it is oxidized to all-*trans*-retinal.

Retinol is oxidized to retinaldehyde by alcohol dehydrogenase; retinaldehyde is oxidized to retinoic acid by aldehyde oxidase. Although retinoic acid plays no part in vision, it can stimulate growth in animals with V.A deficiency, promote cell differentiation and suppress carcinogenesis. It is thought that the retinoic acid is converted to a more active compound, which has yet to be discovered.

An early symptom of V.A deficiency in humans is night blindness, caused by the deficient regeneration of rhodopsin. Later the deficiency leads to hyperkeratosis of the epithelia of the eye (xerophthalmia), skin follicles, respiratory tract and digestive tract. The respiratory epithelia are then highly susceptible to infection. V.A deficiency is a world health problem the magnitude of which is second only to the magnitude of undernourishment in general. In Asia alone, about 10 million children annually are stricken by xerophthalmia; most do not survive to adulthood because of their susceptibility to infections. In young animals, V.A deficiency arrests growth; in adults, it causes resorption of fetuses, stillbirths and birth defects. In males, the germinal epithelia degenerate and the testes atrophy.

V.A can be stored in the organism for several months, chiefly in the liver in the form of its palmityl ester.

V.A was formerly prepared from saponified fish oils, but since the 1950's practically all commercial V.A has been derived chemically from citral.

Riboflavin (Vitamin B₂; lactoflavin; 6,7-dimethyl-9-(D-1'-ribityl)-isoalloxazine: a yellow flavin derivative, which occurs chiefly in a bound form in flavin nucleotides or flavoproteins in yeasts, animal products and legume seeds. Milk contains free riboflavin. It is required as a precursor of Flavin mononucleotide (see) and Flavin-adenine-dinucleotide (see), which are coenzymes of Flavin enzymes (see).

In rats, experimental riboflavin deficiency causes growth failure and dermatitis around the nostrils and the eyes. In humans, riboflavin deficiency (ariboflavinosis) is characterized by lip lesions, a sebor-

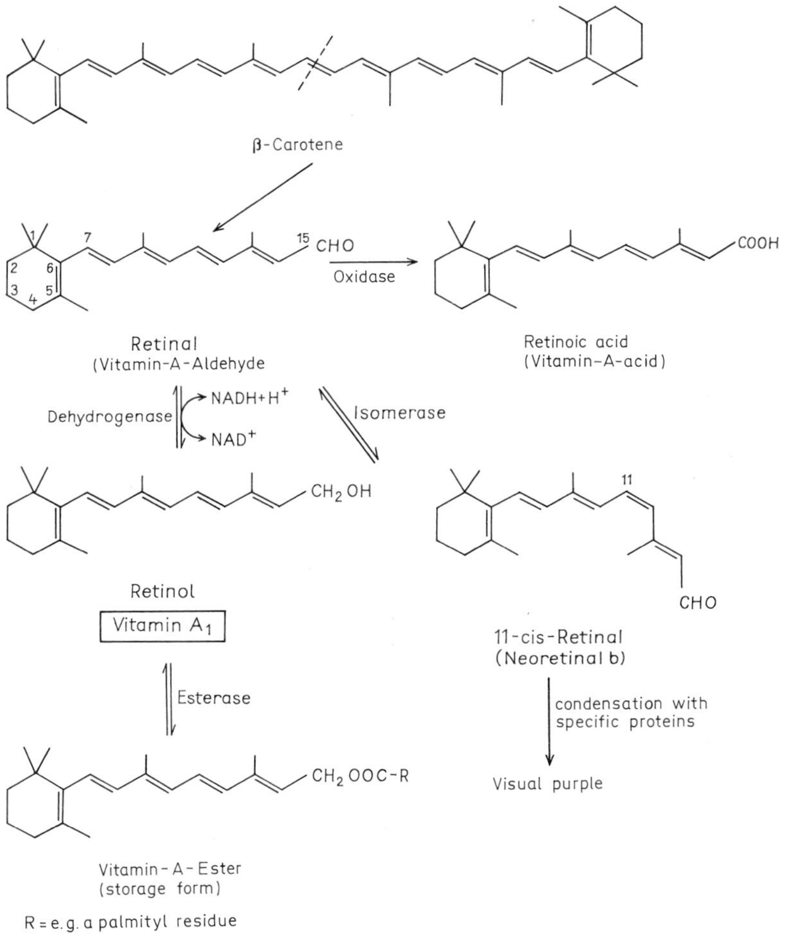

Figure 10. Synthesis of vitamin A-active compounds from β-carotene.

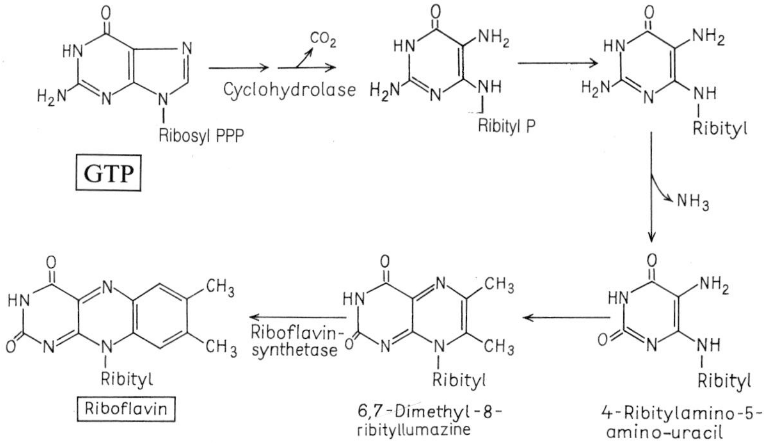

Figure 11. Biosynthesis of riboflavin from guanine, ribitol and diacetyl.

637

rheic dermatitis about the nose, ears and eyelids, and loss of hair. Angular stomatitis, glossitis, cheilosis, and ocular changes such as photobia, indistinct vision and corneal vascularization have also been reported as typical of human ariboflavinosis. Most human diets contain adequate riboflavin.

The biosynthetic precursors of riboflavin are a purine (probably guanine), ribitol (a sugar alcohol) and diacetyl (a diketone) (Fig.11).

Riboflavin is produced commercially either by a total chemical synthesis starting from 3,4-dimethylaniline and ribose, or by microbial fermentation with the fungi *Eremothecium ashbyii* or *Ashbya gossypii,* which produce up to 5 mg free riboflavin per ml of culture medium. Riboflavin is used in veterinary and medical practice, for the supplementation of animal feeds, and as a natural coloring agent in food products.

Thiamin (Vitamin B₁, aneurin, antiberiberi factor, antineuritic V.): a water-soluble V.of widespread occurrence in natural materials, especially in yeasts and germinating cereal grains. V.B₁ contains a pyrimidine and a thiazole ring. Both ring systems are synthesized separately as phosphorylated derivatives, which then become linked via a quaternary nitrogen. The pyrimidine ring is synthesized from aminoimidazole-ribotide (an intermediate of purine synthesis) and a C-3 unit. Alanine and methionine are precursors of the thiazole ring (Fig.12). The pyrophosphorylated form of V.B₁, Thiamin pyrophosphate (see), is a coenzyme of Decarboxylases (see), Transketolases (see) and 2-oxoacid-dehydrogenases (see Multienzyme complex).

Deficiency of V.B₁ results in disturbances of carbohydrate metabolism, accompanied by an increase in the concentration of blood oxoacids (mostly pyruvate), which reflects the role of thiamin pyrophosphate as a coenzyme of pyruvate dehydrogenase. The typical deficiency disease, beriberi, results from a diet exclusively of polished rice. It is characterized by disturbances of the central and peripheral nervous system (polyneuritis) and of cardiac function. Other neurological disorders, such as secondary symptoms of alcoholism (alcohol polyneuritis) also respond favorably to V.B₁. High concentrations of V.B₁ are used in the treatment of diabetic acidosis, cardiac infarction and other disorders of cardiac function. Large quantities of V.B₁ are produced chemically from the precursors, 2-methyl-4-amino-5-aminomethylpyridine, carbon disulfide and 3-chloro-5-hydroxypentanone-(2). Antimetabolites of V.B₁ are pyrithiamin, hydroxythiamin, thiamin propyldisulfide, and the antibiotic bacimethrin.

Tocopherol (Vitamin E; antisterility factor): a group of fat-soluble V.containing a chromane ring with a polyisoprenoid side chain. Eight compounds with V.E activity are known: α-, β-, γ-tocopherol etc., which differ in the number and positions of the methyl groups in the aromatic ring. Biologically, the most important member is α-tocopherol (Fig.13).

Since tocopherol is easily oxidized to a quinone, tocoquinone, V.E acts as a naturally occurring antioxidant. It prevents the spontaneous oxidation of highly unsaturated materials, e.g. unsaturated fatty acids.

V.E occurs in wheat seedlings, and has been isolated from wheat seedling oil. It is also present in lettuce, celery, cabbage, maize, palm oil, ground nuts, soybeans, castor oil and butter. In animal experiments, V.E deficiency results in the death of the embryo in pregnant females. In the male, there is atrophy of the gonads and muscle dystrophy. Neither

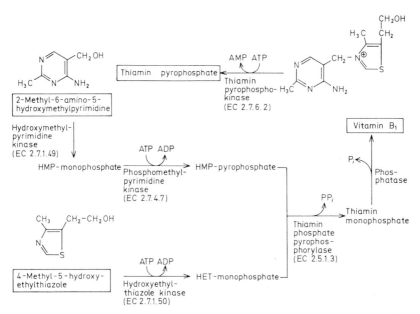

Figure 12. Biosynthesis of vitamin B₁ and its coenzyme form thiamin pyrophosphate. HMP = 2-methyl-6-amino-5-hydroxy-methylpyrimidine, HET = 4-methyl-5-hydroxyethylthiazole.

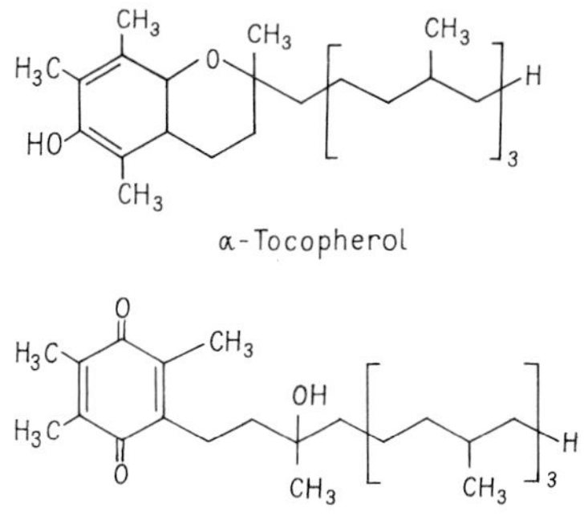

α-Tocopherol

Tocoquinone

Figure 13. Vitamin E-active compounds.

deficiency states nor V. E-hypervitaminosis have been described in humans.

Starting materials for the commercial synthesis of V. E are trimethylhydroquinone and isophytol (from acetone).

Vitamin B₂ complex: a group of water-soluble V. consisting of folic acid, niacin, pantothenic acid and riboflavin. The term V. B_2 is now reserved for riboflavin alone.

Vitamin B_T: see Carnitine.

Vitamin F: the essential fatty acids, which cannot be synthesized in the body. These include the unsaturated fatty acids, in particular linoleic, linolenic and arachidonic acid. They occur in high concentrations in vegetable oils. Unsaturated fatty acids are also the precursors of the Prostaglandins (see), but these are not counted as V. Deficiency of essential fatty acids in rats leads to loss of hair, disturbances of water balance, and reproductive failure. In the human, unequivocal deficiency symptoms have not been described.

Vitamin G.: an obsolete term for riboflavin, a component of the V. B₂ complex.

Vitamin P (permeability factor): a term applied formerly to a group of plant Flavones (see), e.g. hesperidin, eriodictin, and particularly quercetin. These flavones decrease the permeability of blood vessels, and for this reason have been used pharmaceutically. There is no evidence that they are essential in the diet, and they are no longer counted as vitamins.

Vitamin PP (PP-factor): an old name for nicotinamide, a component of the V. B₂ complex.

Historical: By the beginning of this century, many illnesses which had been known since antiquity were recognized as dietary deficiency states. The founder of the vitamin concept was Lumin (1853-1937). Historically, V. B₁ played a large part in the discovery of V. In 1896, Ch. Eijkmann proposed that beriberi was a "deficiency disease". During subsequent decades, the importance of "accessory food factors" for normal growth and development was gradually recognized. In 1912, C. Funk proposed the name "vitamin" for these accessory factors, reflecting the fact that the first of these factors to be studied (V. B₁) contains nitrogen. In 1933, O. Warburg and R. Kuhn showed that certain V. are integral components of certain enzymes.

The demonstration that riboflavin is an integral component of the old respiratory "yellow enzyme" introduced a new era of vitamin research with the emphasis on molecular biology.

Most V. were purified between 1920 and 1950. The last one was V. B₁₂ in 1948, whose structure was elucidated by A. R. Todd in 1955. Chemical syntheses are known for all V.

Clearly, animal and human nutritional physiology are central to the V. concept. Microorganisms, however, have also played an important part in V. research. Bacteria, yeasts and fungi with specific growth requirements are still widely used in the microbiological assay of V., e.g. the yeast *Kloekera apiculata* has a growth requirement for pyridoxine; the bacterium *Lactobacillus* requires many different V. for growth, and is chiefly used for the assay of nicotinic acid. Such organisms may be isolated from natural sources, or they may be selected in the laboratory, following mutagenesis (see Mutant technique). Mutant strains of *Escherichia coli* with specific requirements for each of the water-soluble B-vitamins have been isolated. Conversely, some V. are produced commercially by industrial fermentation with microorganisms. [W. Friedrich, *Vitamins* (Walter de Gruyter, Berlin · New York, 1988)]

Vitamin solution: see Nutrient medium (Table 2).

Vitellin: a lipophosphoprotein present in egg yolk together with Phosphovitin (see). It is present in a higher concentration than phosphovitin, but in contrast to this phosphorus-rich protein, V. contains only 1% phosphate. In yolk, and in neutral salt solutions, V. exists as a dimer (M_r 380000; 16-22% lipid). The monomer (M_r 190000) consists of two dissimilar chains (L₁, M_r 31000; L₂, M_r 130000); only L₁ contains phosphate.

Volutin: see Blue-green bacteria.

Von Gierke's disease: see Glycogen storage disease.

von Willebrand factor: a series of self-aggregated structures, all derived from a common glycoprotein subunit, M_r 225000, synthesized in endothelial cells and megakaryocytes. In plasma the smallest aggregates are dimers (M_r 450000), while the largest aggregates may have M_r 20 million. Aggregates are held together by disulfide bonds. vWf is also present in platelet α-granules, and it is secreted from stimulated platelets.

The primary human and bovine vWf translation product in endothelial cells is an intracellular precursor, M_r 240-260000, which is cleaved to the M_r 225000 subunit immediately before secretion. vWf binds strongly (but noncovalently) to circulating factor VIIIc (antihemophilic factor; see Blood coagulation), thereby stabilizing the latter. It mediates interaction of platelets with damaged epithelial surfaces, and interaction of vWf with platelets is essential for normal primary hemostasis, as shown by prolonged bleeding times in patients with von Willebrand disease or Bernard-Soulier platelet defect. There is evidence for the participation of vWf in platelet aggregation, but a preaggregation step promoted by ADP, thrombin or collagen may be necessary.

vWf has many features in common with other "adhesive" proteins: Fibrinogen (see), Fibronectin (see), Thrombospondin (see). [D.C. Lynch, *J. Biol. Chem.* **258** (1983) 12757–12760; S.E. Senogles & G.L. Nelsestuen, *J. Biol. Chem.* **258** (1983) 12327–12333; J.L. Miller et al., *J. Clin. Invest.* **72** (1983) 1532–1542]

Vulgaxanthin: a yellow betaxanthin from sugar beet *(Beta vulgaris)*. In *V.I,* a betalamic acid moiety is linked to glutamic acid; in *V.II,* the betalamic acid is linked to glutamine.

W

Warburg-Dickens-Horecker pathway: see Pentose phosphate cycle.

Warburg's Atmungsferment: cytochrome oxidase.

Warburg's respiratory enzyme: cytochrome oxidase.

Water: H_2O, quantitatively the most important inorganic constituent of living cells. M.p. 0 °C, b.p. 100 °C. A normal living cell contains about 80% W. Plants can contain up to 95%, jelly fish 98%, higher animals 60–75% (Tables 1 and 2). All life processes depend upon W.

There are several theories regarding the structure of W. The dipole properties of W. (Fig. 1) result in an interaction of W. molecules; hydrogen bonding gives rise to molecular aggregates, such as the tetrahydrol structure (Fig. 2). Most cluster hypotheses postulate networks of four-fold linked W. molecules separated by areas of monomeric molecules (Fig. 3). The average half life of such a cluster is only 10^{-11}s. The dipole character of the water molecule determines the physical and chemical properties of W., and is essential to its biological function. Dipolar W. molecules interact with macromolecules, especially proteins and nucleic acids, to form a hydration layer. The activity of enzymes in cytoplasm is very dependent on the degree of hydration. W. dissolves organic and inorganic substances, and is responsible for their extra and intracellular transport. Organic compounds are classified as hydrophilic (dissolve in W., e.g. amino acids, proteins, nucleic acids, carbohydrates) and hydrophobic (insoluble or poorly soluble in W., e.g. fats, lipids).

W. is a reactant in enzyme (hydrolase)-catalysed hydrolytic cleavage of macromolecules (proteins, carbohydrates, fats), representing the first stage in the biological degradation of these substances. W. is formed metabolically by the operation of the respiratory chain (Table 3) (respiratory W.), and is the substrate of photosynthesis.

In animals, W. is important in the regulation of body temperature. Heat is removed by evaporation of W. from body and respiratory tract surfaces.

Heavy water (D_2O) contains deuterium (^2_1H or D) in place of ^1_1H. D_2O causes a decrease in metabolic activity and leads to cytological and morphological changes, sometimes resulting in the death of the organism. This isotope effect can be used to study the role of W. in biological systems. W. is formed metabolically from the reaction of oxygen with the hydrogen of substrates (see Respiratory chain), and can be produced from gaseous hydrogen and oxygen by *Hydrogenomonas* (see Hydrogen metabolism). It is also formed by dihydrases from substrates, e.g. dehydration of malic acid by the action of malate dehydrase, forming fumaric acid.

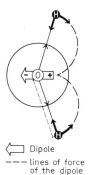

Dipole
lines of force of the dipole

Figure 1. Dipole of water.

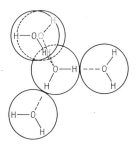

Figure 2. Tetrahydrol structure of water.

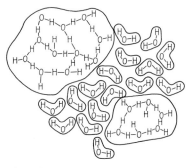

Figure 3. Cluster structure of liquid water (after Némethy and Scheraga).

Table 1. Water content (%) of some human organs and tissues.

Total	60
	(representing 42 kg)
Skin	58
Skeleton	28
Muscle	70
Adipose tissue	23
Liver	71
Brain	75

Table 2. Distribution of water in human organs and tissues (total = 100%).

Muscle	50.8
Skeleton	12.5
Skin	6.6
Blood	4.7
Stomach, intestines	3.2
Liver	2.8
Brain	2.7
Lungs	2.4
Adipose tissue	2.3
Kidneys	0.6
Spleen	0.4
Remainder	11.0

Table 3. Water formed (g) in the degradation of foodstuffs (per 100 g)

Protein	41.3
Carbohydrate	55.9
Fats	107.0
(Alcohol	117.4)

Watson-Crick model: see Deoxyribonucleic acid.

Wax acids: $CH_3-(CH_2)_n-COOH$, long chain, even numbered monocarboxylic acids, which occur esterified in waxes, e.g. lauric acid (C_{12}), myristic acid (C_{14}), palmitic acid (C_{16}), carnaubic acid (C_{24}), cerotic acid (C_{26}), montanic acid (C_{28}), melissic acid (C_{30}) and other higher fatty acids.

Wax alcohols: long chain, monohydroxy, aliphatic alcohols of the general formula $CH_3-(CH_2)_n-CH_2OH$, which occur esterified in waxes, e.g. cetyl alcohol (C_{16}), carnaubyl alcohol (C_{24}), ceryl alcohol (C_{26}) and myricyl alcohol (C_{30}).

Waxes: esters of long chain, even numbered fatty acids (see Wax acids) and monohydric, straight chain, aliphatic alcohols (see Wax alcohols), or sterols. W. are secreted by animals and plants, usually as ester mixtures that are difficult to resolve, often accompanied by free fatty acids, or high M_r unbranched hydrocarbons. The wax alcohols and wax acids usually have a similar number of carbon atoms, e.g. cetyl palmitate, ceryl cerotate. W. are very hydrophobic. They form a waterproof layer on the aerial parts of plants, e.g. leaves and fruits, which prevents water loss, excessive wetting of surfaces and attack by microorganisms. In animals, W. are found on skin and feathers, and are used by bees to build the honeycomb. Animal waxes are spermaceti, shellac, beeswax and wool wax (lanolin). Carnauba wax is obtained from the leaves of the palm tree, *Copernica cerifera.* W. are solid at room temperature, becoming softer at higher temperatures and eventually liquid. On the basis of their closely similar physical properties, certain mineral substances are also classified as W., e.g. solid hydrocarbons (paraffin wax), montan wax (extracted from lignite), ozocerite (a fossil resin), etc. These products and natural W. are used widely in industry.

Western blot: see Southern blot.

WGA: abb. for wheat germ agglutinin. See Lectins.

Whey proteins: see Milk proteins.

Wieland-Gumlich aldehyde: see Curare alkaloids.

Withaferol A: see Withanolides.

Withanolides: a group of C_{28} plant steroids based on the parent hydrocarbon, ergostane (see Steroids), and containing a characteristic withanolide ring system with a δ-lactone side chain. The approximately 50 known representatives have all been isolated from members of the *Solanaceae*, belonging to the genera *Withania, Dunalia, Datura* and *Nicandra.* The earliest and most important W. is **Withaferol A** (22R)-4β-27dihydroxy-1-oxo-5β,6β-epoxywitha-2,24-dienolide (Fig.) from *Withania somnifera* and *Acnistus arborescens;* this compound is bacteriostatic and inhibits the growth of certain tumors. Other representatives are **27-deoxywithaferol A, 27-deoxy-14α-hydroxywithaferol A** and **withanolide D** (OH-group on C20 instead of C27).

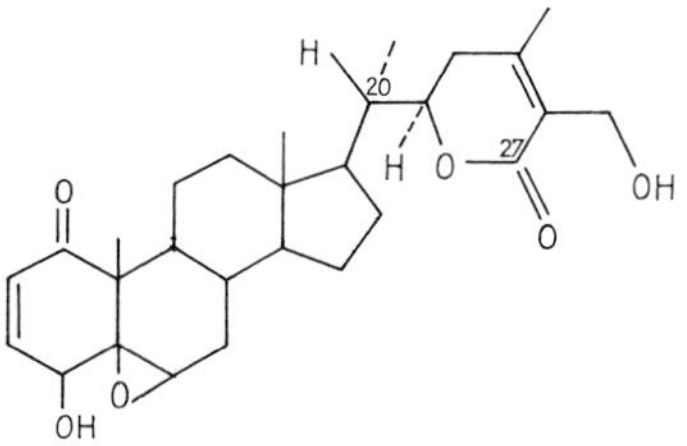

Withaferol A

Wobble base: see Wobble hypothesis.

Wobble hypothesis: a hypothesis proposed by Crick (1966) to explain the degeneracy of the Genetic code (see) with respect to codon-anticodon base pairing. The codon for practically every amino acid

Table. Comparison of the tRNA-anticodon with the corresponding codon for some amino acids.

Amino acid	Anticodon (3' → 5')	Codon – wobble base (5' → 3')
Ala	CGI	GC – U, C, A
Ser	AGI	UC – U, C, A
Phe	AAG^{Me}	UU – U, C
Val	CAI	GU – U, C, A
Tyr	AψG	UA – U, C
Met	UAC	AU – G

is specified by the first two bases, which show strict complementary base pairing with the third and second bases of the anticodon in the corresponding tRNA. The 3'-codon base is less strictly specified, and is called the *wobble base*. Hydrogen bonding between this base and the 5'-anticodon base shows a relative lack of base pairing specificity; thus the purine, inosine, is often found in position 5' of the anticodon (Table). The wobble base is often a pyrimidine, which like inosine forms only two hydrogen bonds.

Wolman's disease: see Inborn errors of metabolism.

Wood: a mixed polymer with lignin as a structural component. The deposition of lignin in the cellulose matrix of the cell wall is called lignification. It is in principle comparable to an irreversible swelling. Cellulose and lignin are physically and chemically bound together in W.

Wood sugar: see D-Xylose.

Wood-Werkmann reaction: see Carboxylation.

Wool fat: see Lanolin.

Wool wax: see Lanolin.

Wyerone acid: a Phytoalexin (see). W. a. and its methyl ester wyerone are produced by the broad bean (*Vicia faba*) in response to *Botrytis* infection. ED_{50} against various *Botrytis* spp. 9-45 µg/ml. *Vicia*

also produces wyerone epoxide. [Hargreaves et al. *Phytochemistry* **15** (1976) 1119-1121]

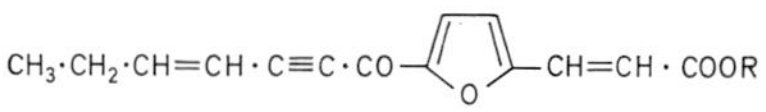

$CH_3 \cdot CH_2 \cdot CH=CH \cdot C \equiv C \cdot CO$ —〔O〕— $CH=CH \cdot COOR$

R=H : Wyerone acid
R=CH₃ : Wyerone R

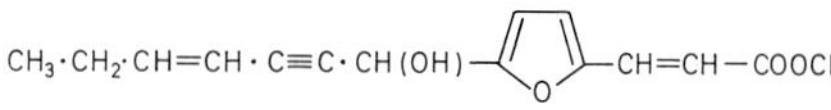

$CH_3 \cdot CH_2 \cdot CH=CH \cdot C \equiv C \cdot CH(OH)$ —〔O〕— $CH=CH-COOCH_3$

Wyerol

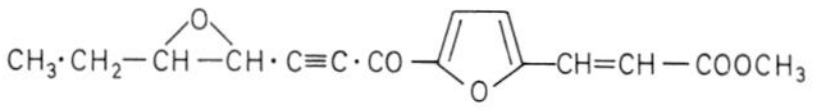

$CH_3 \cdot CH_2 - CH - CH \cdot C \equiv C \cdot CO$ —〔O〕— $CH=CH-COOCH_3$

Wyerone epoxide

$CH_3 \cdot CH_2 \cdot CH_2 \cdot CH_2 \cdot C \equiv C \cdot CO$ —〔O〕— $CH=CH-COOR$

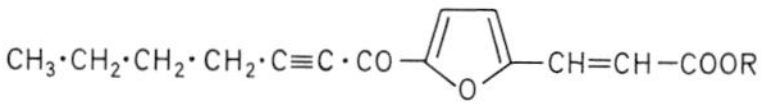

R=H : Dihydrowyerone acid
R=CH₃ : Dihydrowyerone

Wyerone acid and related acetylenic compounds produced by Vicia faba in response to infection. The proportions of the various compounds depends on the infected tissue, species of infecting fungus, and time after initiation of synthesis.

X

Xan: abb. for xanthine.

Xanthine, abb. *Xan:* 2,6-dihydroxypurine, a purine and the starting point for Purine degradation (see). M_r 152.1, m.p. > 400 °C (d.). Xan was discovered in 1817 in renal stones. It is found free, together with other purines. Some derivatives are physiologically important, especially xanthosine phosphates and the Methylated xanthines (see).

Xanthine dehydrogenase: see Xanthine oxidase.

Xanthine oxidase, *xanthine dehydrogenase, Schardinger enzyme:* an enzyme of aerobic purine degradation, which catalyses the oxidation of hypoxanthine to xanthine, and xanthine to uric acid:

Hypoxanthine $+ H_2O + O_2 \rightarrow$ Xanthine $+ H_2O_2$
Xanthine $+ H_2O + O_2 \rightarrow$ Uric acid $+ H_2O_2$

It is a dimeric enzyme, M_r 275000, pH-optimum 4.7, IP 5.35, containing 2 FAD, 2 Mo and 8 Fe (data for the enzyme from milk). The substrate specificity is low; it catalyses the oxidation of other purines (e.g. adenine), aliphatic and aromatic aldehydes, pyrimidines, pteridines and other heterocyclic compounds. In animal tissues (e.g. calf liver) X.o. is in the Golgi apparatus; it is also a secretory enzyme present in milk, where its activity can be used to differentiate between fresh and heated or pasteurized milk. X.o. and xanthine dehydrogenase are sometimes considered as separate enzymes, X.o. being the original Schardinger enzyme from milk (EC 1.2.3.2), and xanthine dehydrogenase an enzyme from chicken liver (EC 1.2.1.37). Together with aldehyde oxidase (EC 1.2.1.3), these three enzymes have very similar composition and presumably mechanism of action, although there are differences in substrate specificity. The molybdenum appears to be involved in the initial hydroxylating attack on the substrate. Electrons are then transferred to the FAD, and from the FAD to the iron which is present as a nonheme Fe-S center. Molecular oxygen is reduced by the nonheme iron center. Superoxide is produced in addition to hydrogen peroxide; these are decomposed by superoxide dismutase and catalase, respectively.

Xanthinuria: see Inborn errors of metabolism; Purine degradation.

Xanthochymusside: see Biflavonoids.

Xanthocillin: 1,4-di-(4-hydroxyphenyl)-2,3-diisonitrilobutadiene(1,3), a bacteriostatic antibiotic used against local infections due to Gram positive and Gram negative organisms.

Xanthophylls: a group of Carotenoids (see). See also Lutein.

Xanthoprotein reaction: a qualitative test for protein, using concentrated nitric acid. The resulting yellow color is due to the nitration of aromatic amino acid residues.

Xanthopterin: 2-amino-4,6-dioxotetrahydro-

pteridine, first isolated as the yellow pteridine wing pigment of the brimstone butterfly and related species. M.p. > 400 °C. X. is also the yellow pigment of bees, wasps and hornets. It is biosynthesized from guanine and two carbon atoms of a pentose.

Xanthopterin

Xanthorrhone: see Flavan.

Xanthosine, abb. *Xao:* 9-β-D-ribofuranosylxanthine, a β-glycosidic nucleotide of D-ribose and the purine base, xanthine, M_r 284.23, carbonizes at > 300 °C, $[\alpha]_D^{30} - 51.2$ ° ($c = 1$, 0.1 M NaOH). Xao is formed by the deamination of guanosine. Xanthosine 5'-monophosphate is metabolically important (see Xanthosine phosphates).

Xanthosine phosphates: phosphate esters of xanthosine, or xanthine nucleotides. *Xanthosine 5'-monophosphate,* abb. *XMP, xanthidylic acid, xanthylic acid,* M_r 364.22 is an intermediate of Purine biosynthesis (see).

Xanthothricin: see Toxoflavin.

Xanthoxin: an endogenous growth regulator, occurring widely in higher plants, possessing inhibitory properties similar to those of abscisic acid. It occurs in both the *cis, trans,* and in the biologically less active *trans, trans* form. Plant tissues can probably convert *cis, trans*-X. into $(R) - (+)$-abscisic acid.

*cis, trans-**Xanthoxin***

Xanthurenic aciduria: see Inborn errors of metabolism.

Xanthylic acid: see Xanthosine phosphates.

Xao: abb. for xanthosine.

Xenopsin, *XP:* an octapeptide from the skin of the African frog, *Xenopus laevis.* The structure of XP: Pyr-Gly-Lys-Arg-Pro-Trp-Ile-Leu-OH, is very similar to that of Neurotensin (see). Antiserum raised to the C-terminal region of neurotensin cross reacts with XP. Neurotensin and XP both characteristically increase the hematocrit and induce cyanosis in anesthetized rats. SP therefore appears to be an amphibian counterpart of neurotensin. Other peptides immunologically similar to neurotensin and SP have been demonstrated in skin, brain and intestine of *Xenopus laevis, Rana catesbeiana* and *Bufo marinus.* Isolation and structure: K. Araki et al. *Chem. Pharm. Bull. (Tokyo)* **21** (1973) 2801; physiology: R. Carraway et al. *Endocrinology* **110** (1982) 1094–1101.

Xerophthol: vitamin A, see Vitamins.

XMP: abb. for xanthosine 5′-monophosphate.

Xylans: high M_r polysaccharides of xylose. The xylose residues are in the pyranose form (see Carbohydrates), and the linkages are usually β-1,4-glycosidic. Next to cellulose, X. are the commonest of all plant substances, occurring as the main constituent of hemicelluloses.

Xylitol: $CH_2OH-(CHOH)_3-CH_2OH$, an optically inactive C_5-sugar alcohol, related to xylose. M.p. 61.5 °C. X. is a byproduct of wood saccharification, and can also be prepared by the catalytic hydrogenation of xylose. It is fully utilized by the human organism, and can therefore be used as a sugar substitute in diabetic diets.

D-Xylose, *wood sugar:* a monosaccharide pentose, M_r 150.13, m.p. 153 °C, $[\alpha]_D^{20}$ +94 ° → +19 °, not fermented by yeast. Reduction of X. gives xylitol, mild oxidation gives xylonic acid. It is produced by acid hydrolysis of Xylans (see). An important dietary component for herbivores, especially ruminants.

Xylosyl nucleosides: Nucleosides (see) in which the sugar component is xylose. They may act as analogs of purine or pyrimidine ribosides (see Pyrimidine analogs). Adenine xyloside is phosphorylated in the cell, and inhibits the growth of various transplantable animal tumors.

Xylulose: a monosaccharide pentulose, occurring naturally in both the D- and L-form. The 5-phosphate of D-X. is an important intermediate in the Pentose phosphate cycle (see) and serves as a C_2-donor in Transketolation (see). L-X. is an intermediate of the Glucuronate pathway (see), and it is normally metabolized via xylitol and D-X. In pentosuria, an inherited metabolic disease, L-X. is excreted in the urine.

Y

Yeast adenylic acid: see Adenosine phosphates.

Yellow enzymes: see Flavin enzymes.

Yield coefficient: see Cultivation of microorganisms.

Yohimbine: a Rauwolfia alkaloid (see). M_r 354.45, m.p. 235–236 °C, $[\alpha]_D$ +106 ° (pyridine). Y. has five chiral centers and therefore many stereoisomers; seven of these occur naturally, the most important being *Corynanthine,* m.p. 225–226 °C (d.), $[\alpha]_D$ −82 ° (pyridine), obtained chiefly from the bark of the tropical tree, *Corynanthe yohimbe.* Y. is vasodilatory and has been used in the treatment of arteriosclerosis, in veterinary medicine and by African natives as an aphrodisiac.

Z

Zaffaroni system: see Paper chromatography.

Zeatin: 6-(4-hydroxy-3-methyl-but-*trans*-2-enyl)-aminopurine, a naturally occurring Cytokinin (see for formula). Z. occurs free in many plants, especially in immature maize kernels, and is identical with the previously described *maize factor* (abb. *MF*). Its derivatives, Dihydrozeatin (see), zeatin riboside and zeatin ribotide are also cytokinins. The *cis* compound (see N^6-*cis*-γ-Methyl-γ-hydroxymethylallyl-adenosine) is a rare nucleic component in certain species of RNA.

Zeaxanthin: (3R,3'R)-β,β-carotene-3,3'-diol; 3,3'-dihydroxy-β-carotene, a xanthophyll, M_r 568.85, m.p. 206 °C (d.), isomeric with lutein; one of the commonest plant pigments (yellow-orange), especially plentiful in maize and fruits of the sea buckthorn; also found in algae and bacteria. It occurs free and esterified as the dipalmitate, and shows no vitamin A activity. The 5,6-monoepoxide of Z., antheroxanthin, is also a common plant pigment.

Humans contain 2-4 g Zn, the majority being intracellular. Blood contains only 7-8 µg Zn/ml, of which 85% is in the erythrocytes, where it is required for the activity of carbonic anhydrase.

Zizanin B: see Sesterterpenes.

Zoochromes: see Natural pigments.

Zoosterols: see Sterols.

Zwischenferment: Glucose 6-phosphate dehydrogenase. See Pentose phosphate cycle.

Zymase: an old name for a mixture of 11 enzymes of glycolysis isolated from yeast after mechanical disruption of the cell wall. It catalyses alcoholic fermentation.

Zymogens: inactive precursors of enzymes, usually proteolytic enzymes. Z. are converted into active enzymes by limited proteolysis. Best known examples include the Z. of digestive enzymes (pepsinogen, trypsinogen, chymotrypsinogens A, B and C, proelastase, and procarboxypeptidases A and B), and of blood coagulation enzymes (prothrombin

Zeaxanthin

Zein: see Prolamines.

Zinc, Zn: an essential bioelement for the growth and development of plants, animals and microorganisms. Zn has a high affinity for nitrogen and sulfur ligands, and occurs in the cell in association with many different compounds, e.g. proteins (insulin), amino acids, nucleic acids.
Zn is a tightly bound component of zinc-metalloenzymes, and it stimulates in vitro the activity of zinc-metal-enzyme complexes. There are about 20 zinc-metalloenzymes, e.g. dehydrogenases, phosphatases, carboxypeptidases, carbonic anhydrase. Zn also activates the enzymatic synthesis of tryptophan.

and plasminogen). Z. can be stored at their sites of synthesis without danger of causing self digestion of cells or tissues. Certain proteohormones are also produced as inactive precursors, e.g. proinsulin. Local autolysis at the synthesis sites of proteolytic hormones is also prevented by the presence of specific enzyme inhibitors, or by the storage of enzymes in particles. For further details of Z. activation, see individual entries for each Z. named above.

Zymosterol, *5-α-cholesta-8(9),24-dien-3β-ol:* a mycosterol (see Sterols) present in yeast. Z. is an intermediate in the biosynthesis of cholesterol from lanosterol. See Steroids.